BIOLOGY

BIOLOGY
SECOND EDITION

Donald D. Ritchie
BARNARD COLLEGE OF COLUMBIA UNIVERSITY

Robert Carola

ADDISON-WESLEY PUBLISHING COMPANY

Reading, Massachusetts
Menlo Park, California
London
Amsterdam
Don Mills, Ontario
Sydney

This book is in the Addison-Wesley Series in Life Science

Sponsoring Editor: Nancy J. Kralowetz
Production Editor: Laura R. Skinger
Copy Editor: Nancy Shapiro
Text and Cover Designer: Catherine L. Dorin
Illustrators: Oxford Illustrators
Phil Carver and Friends
Donald D. Ritchie
Photo Researchers: Diann Korta
Mary Dyer
Cover Photographer: Charles Steinhacker
Production Manager: Herbert Nolan

The text of this book was composed in Palatino by York Graphic Services, Inc.

Library of Congress Cataloging in Publication Data

Ritchie, Donald D.
Biology.
Bibliography: p.
Includes index.
1. Biology. I. Carola, Robert. II. Title.
QH308.2.R57 1983 574 82-11318
ISBN 0-201-06356-5

Reprinted with corrections, June 1983

ISBN 0-201-06356-5
BCDEFGHIJ-DO-89876543

This book is dedicated to Ann Carola

A great man of science . . . knows everything about everything, except why a hen's egg don't turn into a crocodile, and two or three other things.

Charles Kingsley
The Water Babies

Preface

COMPLEX PROCESSES HAVE BEEN EXPLAINED STEP-BY-STEP

This second edition of *Biology* has been written for introductory biology courses. Our most important considerations were to give the student the information needed to have a comprehensive view of modern biology, while at the same time making the material *understandable*. If it took two paragraphs instead of one to explain a concept properly, we used two. If four pictures provided a better explanation than one, we used four. Whenever it was possible to take a difficult process and break it down into its sequential steps, we did it, and we placed a picture at each step. (See "How Are Proteins Synthesized?" starting on page 116, "Mitosis" on page 131, "Meiosis" on page 137, "The Mechanism of Nerve Action" on page 398, and "The Mechanics of Heart Action" on page 488, for example; there are several other similar sections.)

"TO SUM UP" SECTIONS HAVE BEEN ADDED

Whenever a section contained a particularly complex idea, we added an itemized summary, "To Sum Up," at the end of it. We tried in every way possible to anticipate the students' questions and to answer them simply and clearly.

TECHNICAL TERMINOLOGY HAS BEEN REDUCED

Many so-called technical terms have been omitted, and the overall word count of the book has been decreased substantially. More pictures have been added, and they have been made larger.

CHAPTERS HAVE BEEN REWRITTEN

Many chapters have been redone completely. The ecology chapters, for instance, have been rewritten after discussion with biology teachers all over the country, and the pictures, especially the photographs, were carefully chosen to complement the text and improve the discussion. Large, full-color photographs help improve the ecology chapters. The Chemistry of Life, Cellular Respiration, The Genetic Code and Protein Synthesis, and other chapters have been modified to make them more accessible to the student. The final part, "Evolution, Ecology, and Animal Behavior" has been reorganized to integrate the chapters within.

KEY POINTS HAVE BEEN OUTLINED

Each chapter begins with an itemized list entitled "Some Key Points." Written without using any technical terms, these sections are meant to introduce the student to the important points of the chapter.

CHAPTER SUMMARIES HAVE BEEN RETAINED

At the same time, itemized, detailed Chapter Summaries have been placed at the end of each chapter. These summaries were especially well received in the first edition.

CHAPTER-END QUESTIONS HAVE BEEN ADDED

As many as 15 questions have been added at the end of each chapter. These questions, entitled Ask Yourself, have been written to help the student review and understand the text. For the convenience of the instructor, answers to all questions appear in the Instructor's Manual. An exception to the general format occurs in the two chapters dealing specifically with genetics, where Genetics Problems and Answers of a more quantitative nature appear. Additional questions and answers dealing with genetics are found in the Instructor's Manual.

RECOMMENDED READINGS HAVE BEEN UPDATED

An updated list of Recommended Reading, again chosen for the student who would like to do some follow-up reading of relatively nontechnical material, has been placed in the students' Study Guide and in the Instructor's Manual.

THE INDEX HAS BEEN EXPANDED AND THE GLOSSARY HAS BEEN UPDATED

The Index in the second edition contains about twice as many entries as the previous one. Its approximately 6000 entries make it the most comprehensive and useful index available for a book of this type. Also, the already comprehensive Glossary has been updated and made more complete.

COLOR PORTFOLIOS EXPAND BASIC INFORMATION

Like the first edition, the second edition contains several full-color portfolios that expand on themes introduced within the text. These color portfolios (such as "How Animals Move," following page 430, and "What Animals See," following page 414) were one of the favorite features of both teachers and students. We are gratified that the portfolios have proved to be entertaining, but perhaps even more important, they are also instructive and supportive.

BOXED ESSAYS HAVE BEEN UPDATED

Users of the first edition told us that they enjoyed the boxed essays. The reviewers of the present edition urged us to include the essays in the new edition. So we did, updating, adding, shortening, and generally trying to make them even more useful and interesting.

CAPTIONS ARE SHORTER AND LABELS HAVE BEEN ADDED

The length of the captions has been reduced, so that the student receives the necessary information quickly and clearly. To supplement this change, many labels have been added to photographs to increase their usefulness. On the other hand, if a label on a drawing seemed unnecessary or overly technical, it was deleted.

MANY PICTURES HAVE BEEN ADDED

Finally, the number of pictures has been increased markedly. Many new scanning electron micrographs are included, full-color photographs are used when they enhance the discussion, and several simplified drawings have been inserted in place of (or in addition to) single, comprehensive ones. In general, the pictures are larger, clearer, and more instructive. Photographs have been grouped into compact picture essays when several photographs did the job better than one (see Chapter 26, The Mechanisms of Evolution, for example, especially pages 526, 527, and 537). All in all, we tried to be consistently aware of the power of well-chosen pictures, and did our best to use them in the students' best interests.

Preparing a revision can be terribly boring, but we felt only positive feelings about this revision. Every time we discovered an opportunity to improve the book we were exhilarated by the actual process of finding a better way. We look forward to a third edition, and we ask you to help us make it even better than this one. If you have any suggestions, write to us in care of Addison-Wesley.

ACKNOWLEDGMENTS

We were fortunate to have the assistance of many reviewers, all of whom spent a great deal of time reading the manuscript and scrutinizing the pictures in the course of making many intelligent and insightful suggestions. We would like to thank them all:

Teresa Audesirk
University of Missouri

Sara Bennett
Chesterfield, New Hampshire

Richard Boohar
University of Nebraska

William Bowen
University of Arkansas

Robert C. Evans
Rutgers State University

Kathleen Fields
Cambridge, Massachusetts

Paul Hertz
Barnard College

Craig Himes
Bloomsburg State College

Joseph Hindman
Washington State University

Joyce Maxwell
California State University–Northridge

Charles R. Noback
College of Physicians and Surgeons
Columbia University

Richard St. John
Widener College

John Schmitt
Ohio State University

Genevieve Tvrdik
Slippery Rock State College

Charles K. Wagner
Clemson University

We would also like to thank the staff of Addison-Wesley, especially our editors, James Funston and Nancy Kralowetz.

New York
December 1982

D.D.R.
R.C.

Abridged Contents

Contents

PART THREE
THE DIVERSITY OF LIFE

9
The Living Kingdoms: The Lower Groups and Plants 155

10
The Living Kingdoms: The Animals 186

PART FOUR
GENETICS AND ANIMAL DEVELOPMENT

11
Mendelian Genetics 233

12
Genetics and Chromosomes 245

Prologue

A DEFINITION OF BIOLOGY COMES EASILY: IT IS the science of life. That is short and seems simple, but it gives no definition of either science or life. Neither science nor life can be defined simply. We sometimes think we know intuitively whether a thing is alive or not, but intuition can be notoriously misleading. The most obvious way to find if an unknown object is alive is to poke it and see if it moves, that is, to find if it is responsive. But many things that we would never think of as living are responsive. An automobile responds if we move the proper controls, and a violin responds if a bow is drawn across the strings. Indeed, musicians commonly refer to instruments as "responsive." And movement is not restricted to animate beings. Clouds move and waves move and the heavenly bodies shift position, so that primitive peoples have ascribed a kind of life to them; but in our society we regard clouds and waves and heavenly bodies as nonliving. When it comes to defining life, the best that biologists can do is to list a number of features that are common to many living things, and then add that only by combinations of those features can we decide whether or not a thing is alive.

What Is Life?

We can make a reasonable start at defining life by saying that it is the total of all the actions and attributes of the world's organisms. Such a definition involves circular reasoning, as organisms are living things, and our definition can be reduced to the unsatisfactory minimum of "life is what living things do." Nevertheless, we can add some specific details.

Living things move. The movement of plants may be imperceptible, but it is there. We have no doubts about the mobility of most animals, especially such graceful ones as these female impalas, which can leap eight feet high and 25 feet forward. However prejudiced we may be, the impalas are no more alive than these trees in Vermont.

Responsiveness to a changing environment is one of the striking features of a living being. This bush baby peers at the photographer from its tree hole in an African forest.

The most obvious characteristic of animals is **movement;** but movement alone is not a dependable criterion for life, as many lifeless things (stars, clouds, machines) move, and many living ones scarcely move at all (moss, some parasitic animals), or move so slowly that the movement is hardly perceptible (most plants). **Responsiveness** is the ability of things to react to change, and is sometimes known as irritability; but not all living things are responsive, or at least not immediately so (winter buds of trees), and machines can be made exquisitely responsive (calculators). **Growth,** specifically the peculiar internal growth of organisms, is a more dependable standard, even though not all living organisms are growing all the time (frozen seeds), and some inanimate things grow, even though their growth is by external rather than internal addition (crystals, volcanoes).

Definite form is so characteristic of most organisms that it is by their form alone that we recognize them. A seagull silhouetted against the sky is easily identifiable from several hundred feet away, because of its overall form. Size is less dependable, but limits to size are readily apparent. Upper limits are imposed by the strength of bones and muscles in animals and by the strength of wood in plants. Still we are accustomed to ranges of sizes, and are surprised at a guinea pig as large as a small calf. There is such a thing: a South American capybara. We would be astonished at a kitten-sized elephant.

Metabolism includes a continuous turnover of materials, along with continuous input and output

A quartz crystal is more precisely formed than any living cell, and is just as clearly not alive. These crystals of rose quartz grew, but not in the way living organisms do. New material was merely added to the outside of each crystal until it reached its present size.

(a)

All the parent organisms with their offspring shown here exemplify a special ability of living things: reproduction. (a) Minutes after the birth of this Burchell's zebra, it must be able to stand, feed, and run away from possible predators. (b) Although we may not think of sea anemones as having babies, here we see a beadlet anemone with its young. (c) Crocodiles are commonly regarded as vicious, but this Nile crocodile mother carefully carries her newly hatched young in her mouth. The young may continue to be transported this way for months, alternating between male and female parents. (d) Female greater flamingoes feed their chicks with a nutritional fluid secreted from their digestive tracts.

(b)

(c)

(d)

of energy, and applies to all the chemical events that go on in living things. Metabolism, however, is not always easy to detect, and most of the activities that make up metabolism can be duplicated in the laboratory, using nonliving substances. **Reproduction** is an almost certain indication of a living condition. Not every living organism reproduces, but something that does reproduce is almost certainly alive. As some wag has said, you need not fear that some morning you will find your desk crawling with little typewriters. But it is theoretically possible—indeed, it would be realistically possible if anyone wished to spend the money on such a project—to construct a machine that would make a machine like itself, and so on. Such a self-reproducing machine would not be alive any more than a bottle-making machine is.

Trying to define life was never easy, but was made even more difficult when viruses were discovered. They do not move about on their own, they can hardly be called responsive, they have no metabolism, and reproduce only by making some living organism build copies of themselves. They do have definite form and size, and are composed of the same kinds of chemical compounds that are typical of plants and animals. Viruses can be thought of as borderline organisms with more nonliving than living features. Consideration of viruses brings out a principle that we shall meet again and again in biology: Neat categories with sharp boundaries are rare, and possibly do not exist except in human imagination. Nevertheless, we can recognize that a cat slithering down an alley and a buttercup blooming in a meadow are definitely alive, and we can set ourselves to studying their structures and activities and relations to the world without letting the borderline cases divert us from the task of learning what we can about life.

(a)

× 150,000

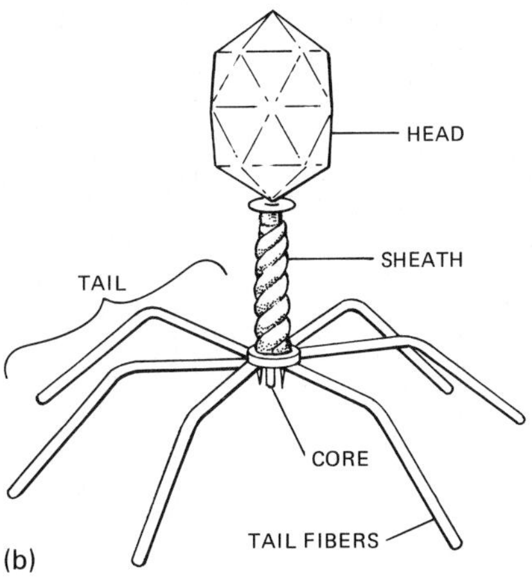

(b)

A virus particle floating in the air shows none of the usual characteristics of a living thing, but if it makes contact with a suitable host cell it can attach itself and inject its nuclear material (DNA) into the cell (movement?), and the host cell may then make copies of the virus (reproduction?). But is the virus alive? *(a) The head of the T_4 virus contains its DNA, which is forced through the long tail into the host like liquid through a syringe. (b) A model of the T_4 virus.*

What Is Science?

Next we try to define science, and immediately run into difficulties. As in trying to define life, however, we do not let difficulties, some vagueness, or differences of opinion blind us to the main ideas. Science has been defined as common sense carried to extremes. It has also been defined as a way of thinking by which reasonable people can arrive at agreement. Whether science is or is not a strictly experimental exercise is open to differences in opinion. Geology and astronomy are largely observational rather than experimental, yet they are accepted as sciences.

Sciences are supposed to be objective to the extent that what is found out about a subject must be accepted whether we like it or not. If you object to the force of gravity, you jump off a high building at your peril. This does not mean, though, that science progresses without hunches, luck, imagination, fads, intrigue, faith, or human emotion. So does language, and so does art, but science differs from both in one important way: It recognizes and corrects its own mistakes. For example, it was once thought that blood was the important material that passed hereditary features from generation to generation. Observation and experiment proved that idea wrong, and the result is that now we can come closer to achieving two of the main objectives of science: to *predict* and to *control* events. We can now tell within reasonable limits what will be the outcome of many matings, and can produce at will an impressive number of desired organisms. Art does not work that way, but relies on subjective evaluation and the indefinable "taste" of a period of time.

What Principles Unify Biology?

The idea of **evolution** affects all biological thinking. If we start with the most obviously related and most clearly similar creatures, identical twins, we find that they are alike in sex, color, structure, and even in such minute details as those of blood chemistry. As less close relationships are considered, more and more differences appear. A Swedish citizen is not exactly like one from the western Australian desert, though both are human beings and share many common features. They are not as similar as twins, but they are more like each other than they are like horses. Yet in their bony and muscular structure and reproductive methods they are more like horses than they are like starfish. Carrying this thinking further, humans, horses, and starfish in their developmental patterns are more similar to one another than any of them is like a butterfly. An important fact comes to light, however, when we inspect organisms that seem to be enormously different. If we look carefully at such various organisms as birds, mushrooms, mosses, flowers, and fishing worms, we discover many points of great similarity, especially in details of growth and chemical make-up.

All these demonstrable facts lead to the recognition of a number of basic features common to *all* living things. If we accept the idea that similarity indicates relationship, we conclude that all living beings, with their many common features, are related by descent from an ancient common ancestor. That is evolution.

Recognizing evolution as a phenomenon gives meaning to many biological facts that would otherwise be nothing more than isolated scraps of information. When we learned, for example, how yeast handles its energy problems, we had knowledge

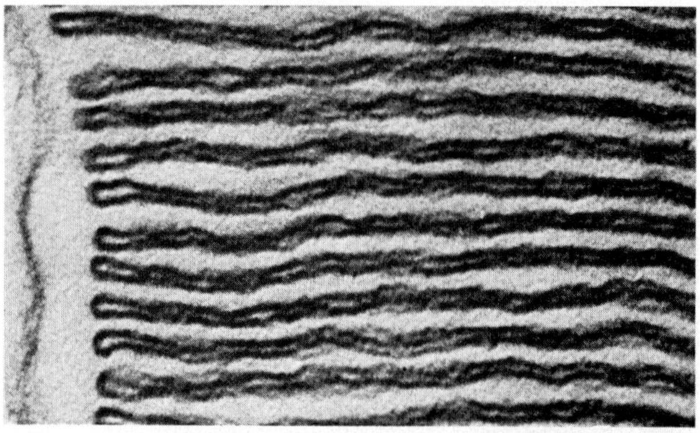

(a) × 35,000

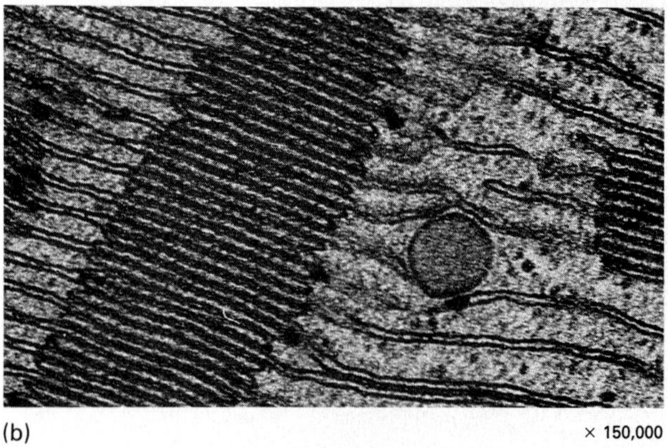

(b) × 150,000

The light-trapping apparatus of (a) a human eye, shown in an electron micrograph, and (b) grana of a chloroplast in a green leaf. Such startling similarities of structure related to function indicate a fundamental unity of life.

that was interesting to bakers and brewers but not necessarily to medical doctors or to the rest of humanity. But if we think of the relatedness of all living things, and look at more complex organisms—most of them harder to study than yeast—we begin to find that what we learned from yeast is applicable to such widely separated things as cacti and people. Hundreds of equally convincing examples could be cited.

When we think of evolution as the major unifying principle in biology, then uncounted bits of knowledge make sense. Snippets of information may be interesting in themselves, as words listed in a dictionary may be, but their interest is immensely greater when they can be seen to fit into a general scheme. A dictionary may list all the necessary words, but a dictionary is a long way from a *Hamlet* or an *Old Testament*. Looking at biological structures and processes as results of evolutionary change, and regarding living things as products of such change, we find that biological information can be integrated and made easier to understand. Without some such all-encompassing idea, the study of biology would be little more than picking up a confusing collection of technical fragments, about as useful and attractive as the shards of a shattered vase.

Evolution of plants and animals has not gone on in an empty space. The environment, literally "the surroundings," has been the shaping force to which the actions of organisms responded. Thus we find a second unifying idea that helps integrate biological thinking. That is the recognition of **ecology** as a major part of biological study. In its broadest sense, ecology is concerned with the reactions of all living things to all aspects of their environment through all time and habitable space.

Plants and animals and the world they live in are constantly reacting to one another. Elk in Wyoming migrate each spring from winter ranges like this one to higher ground where new plant growth is abundant for food.

(a)

(b)

(c)

Every living thing needs energy to survive. Green plants can convert light energy from the sun into chemical energy, but animals get their primary energy from the food they eat. (a) A corn snake suffocates a white-footed mouse before swallowing it. When the snake feels the mouse's heart stop beating, it will swallow the mouse head first. (b) An African elephant practically stands on tiptoe to reach part of its enormous daily vegetarian diet. (c) Lions receive so much usable energy from meat that they may not have to eat more than three or four times a week.

The inner stability of living organisms, homeostasis, is usually taken for granted. If we are healthy, we stay the same even though the external conditions change. These young king penguins can survive a frigid winter in Antarctica by warming their bodies with precise internal regulatory mechanisms.

The things any organism does are constantly being affected by other organisms, by warmth or cold, wetness or dryness, light or dark, the pull of gravity, and the presence or absence of countless substances. During the life of any individual, all those ecological factors are acting, frequently changing. They determine whether the individual will thrive, barely survive, or die, whether it will or will not reproduce, and if it does, what will happen to its offspring. As generation after generation lives and reproduces, the endless variations of which organisms are capable are subjected to the forces of the environment. The end result has been the evolution of the many life forms we now see on earth. Evolution and ecology are thus inseparable. The study of each helps the understanding of the other, and together they strengthen the concept of the unity of life.

One practical aspect of life is its requirement for **energy.** Few facts can be called universally applicable to living things, but one is indisputable: Everything that every living creature does requires a power supply. Most animals obtain their energy from the food they eat. Most plants are powered by sunlight. A few bacteria and some deep-sea dwellers use chemical energy. But not one escapes the rule that every life process must have its energy source, or life ceases. Indeed, most poisons work either directly or indirectly by blocking an organism's energy flow.

Another concept that strengthens the idea of a unity in all living things is that of **homeostasis,** from the Greek for "standing the same." That means the phenomenon of retaining a degree of stability in spite of changing external forces. The earth and everything on it changes; the "everlasting hills" and "perpetual snow" are everlasting and perpetual only when thought of by us upstarts, the human species. Day follows night, seasons come and go, rainfall patterns are altered, entire climates change, rivers and coastlines shift, new species of organisms arise and old ones become extinct. Yet life continues, and as each individual goes through its hours, days, and years, it makes endless adjustments, warming or cooling itself, moving from one place to another, bending, eating, voiding, breathing, or otherwise meeting the demands of a changing environment. Constant internal and external rearrangements are necessary to ensure homeostasis. A living thing is like a ballet dancer standing on tiptoe, perhaps seeming motionless, but really keeping balance by means of small, delicate, controlled changes. Organisms must maintain continuous homeostatic control, or they die.

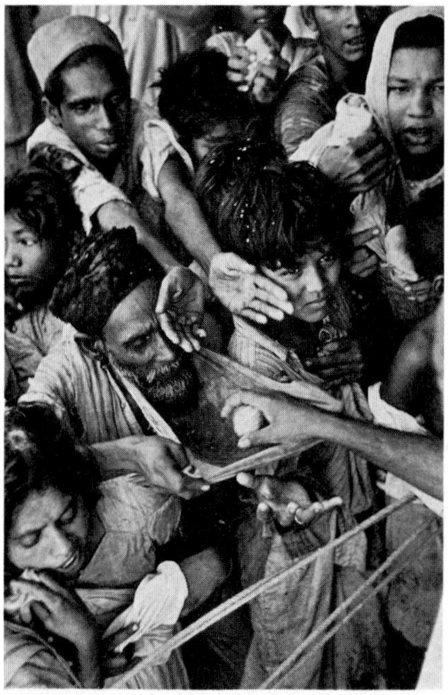

Tommorrow's biologists may have to face such problems as too little food, too many people, human-made pollution that seems to be everywhere, and an ever-increasing population of elderly people who are ultimately dependent upon society.

What Does Biology Offer the Nontechnical Citizen?

People, being themselves alive, are constrained by the forces that regulate all living things. Reach out and touch some selected spot. How did you know where to reach? What made your hand move? How did you know when you made the correct contact? Where did you get the strength to lift the weight of your arm? Biologists have found answers—or at least partial answers—to these and hundreds of other questions about the actions of plants and animals. There are practical personal reasons for knowing biological principles.

Biology is becoming increasingly political. Today's students will be tomorrow's makers of social and political decisions, and to make sensible decisions they should have the kind of background that biology provides. This does not mean so much an accumulation of factual information as an improved ability to use facts in a rational way, and to be able to separate verifiable information from conjecture. Many unresolved questions are already being asked. How should the world's food be distributed?

Is the human population to be arbitrarily limited? Are the current advances in genetic engineering going to be constructive or destructive? Are wildernesses and wild creatures worth their cost to the human species? How should we balance the evils of water and air pollution against immediate human desires? To what extent is animal experimentation justified? What is to be done for old people who are beginning to make a larger proportion of the population? These are a few of the problems that need attention. They all have biological bases, but they will have to be dealt with by social-minded laypeople rather than by laboratory-bound scientists.

Finally there is human curiosity. Some people are driven to learn as much as they can for the same reason that mountain climbers attack a difficult peak: because it is there. If you are not naturally drawn to the unknown, this will not make sense to you; but if you are one who loves to know just for the sake of knowing, you will find in biology some of the most intellectually satisfying material that human ingenuity has brought to light.

As mountains evoke a compulsive response from some people to climb them, so does the challenge of biology attract some people to explore the living world.

PART ONE
THE MECHANICS OF LIFE

Scanning electron micrograph of sodium chloride crystals (table salt). × 40

1
The Chemistry of Life

SOME KEY POINTS

1. An atom's structure determines how the atom may form chemical bonds with other atoms.
2. All biological actions involve changes in electron positions and consequent gains or losses of energy.
3. Covalent bonds are formed when atoms share electrons. Ionic bonds are formed when an atom gains or loses electrons, thereby obtaining an electrical charge that attracts other atoms. Hydrogen bonds are formed through a weak interatomic attraction.
4. In a chemical reaction, bonds between atoms are broken or formed, and different combinations of atoms result.
5. An ion is a particle with an electrical charge. An acid is a substance that, when dissolved in water, releases hydrogen ions. A base, or alkali, releases hydroxyl ions when dissolved in water. The pH scale measures acidity and alkalinity.
6. The many different compounds formed by organisms can be placed into a small number of categories, mainly water, the carbohydrates, the lipids and related compounds, amino acids and proteins, and the nucleic acids.
7. Water makes up the bulk of living things, and all organisms need water to keep their cells functioning properly.
8. An enzyme is a protein with a specific three-dimensional molecular shape that permits it to facilitate a chemical reaction without permanently entering into the reaction itself.

AN INDIVIDUAL'S AWARENESS OF THE WORLD comes from observation. We see the leaves turn colors in the autumn and fall to the ground. We watch a baby grow and develop. If these observations are followed by careful descriptions of the phenomena, we have the beginning of science. But such descriptions do not explain the phenomena. A good description is a basis for asking a question: What causes leaves to turn color? Why do leaves fall? What causes a baby to grow and change? These questions spring quite naturally out of observations and descriptions. To answer them, biologists are increasingly drawn to chemical explanations.

Much of the excitement of biology in the twentieth century is the result of discoveries of chemistry and the application of that knowledge to living systems. We want you to participate in and understand this excitement. In this chapter, we try to show the importance of chemical events that lie behind the familiar phenomena that characterize the living world.

BIOLOGICALLY USEFUL CHEMICAL ELEMENTS

Everything in the universe is made up of **matter,** anything that occupies space. Matter is composed of **elements,** substances that cannot be broken down by ordinary chemical means into simpler particles. An **atom** is the smallest unit of an element that retains the chemical characteristics of that element.

Of about 100 known elements, about three dozen are used by living organisms. Of these three dozen, only four are used in quantity: *oxygen,* makes up about 65 percent of living matter; *carbon*, about 18 percent; *hydrogen*, about 10 percent; and *nitrogen*, about 3 percent. The remaining 4 percent is accounted for by all the other elements incorporated in living material.

THE STRUCTURE OF ATOMS

Each element is composed of a specific type of atom that is different from the atoms of every other element. Despite these differences, however, all atoms have basic structures that are similar enough so that we can study atoms in general, instead of studying the atoms of each element separately.

Atoms consist mostly of space, even the atoms of the hardest surface you can think of. The word "atom" was proposed at a time when atoms were thought to be the smallest particles of matter. (The word is from the Greek, and can be roughly translated as "uncuttable.") Since the beginning of the twentieth century, however, we have known that atoms are made up of a number of smaller components called *subatomic particles*.

ISOTOPES AS AIDS TO BIOLOGISTS

The atomic weight of carbon might seem to be exactly 12 if a carbon atom has six protons and six neutrons. Indeed, most carbon atoms on earth do have six of each, for a total of twelve nuclear particles, and the symbol for the commonest carbon atom is C^{12}. But not *all* carbon atoms have six neutrons. A few, with eight neutrons, have a weight of 14 (written C^{14}), and those heavier atoms are averaged in when atomic weights of elements are calculated and published. Since most carbon atoms are C^{12}, the accepted atomic weight of carbon is close to 12, that is, 12.01115.

Those carbon atoms that have six protons but *not six neutrons* are **isotopes.** Isotopes of any element are atoms with the same number of protons and electrons, but different numbers of neutrons. Although the commonest carbon atom has six protons and six neutrons in its nucleus, other isotopes of carbon have seven or eight neutrons, giving them atomic weights of 13 and 14 instead of the usual 12. Since all forms of carbon have six electrons and six protons, they all have the *atomic number* 6.

Some isotopes are naturally *radioactive;* others are made so by laboratory procedures. Radioactive isotopes are unstable because their nuclei are undergoing changes called *decay. Radioisotopes* emit radiation that can be detected. For example, radiocarbon (C^{14}) emits fast-moving electrons called *beta particles* from the carbon nucleus. Radiation can be detected and localized by several physical or photographic means, and thus an investigator can know where a compound containing a radioactive isotope has gone. Since radiocarbon (C^{14}) and ordinary carbon (C^{12}) act the same chemically, an experimental compound containing C^{14} can be fed or injected into an organism, and the path of the compound can be followed by "watching" where the radioactiv-

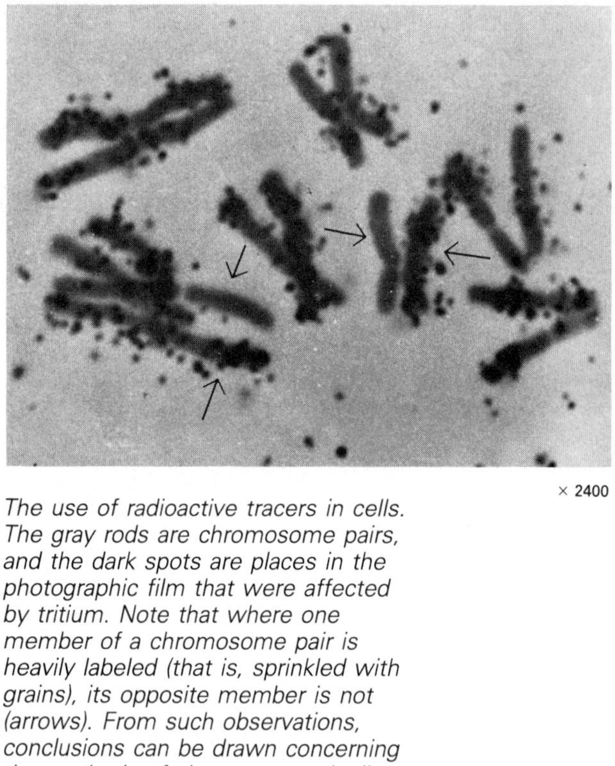

× 2400

The use of radioactive tracers in cells. The gray rods are chromosome pairs, and the dark spots are places in the photographic film that were affected by tritium. Note that where one member of a chromosome pair is heavily labeled (that is, sprinkled with grains), its opposite member is not (arrows). From such observations, conclusions can be drawn concerning the methods of chromosome duplication (Chapter 8).

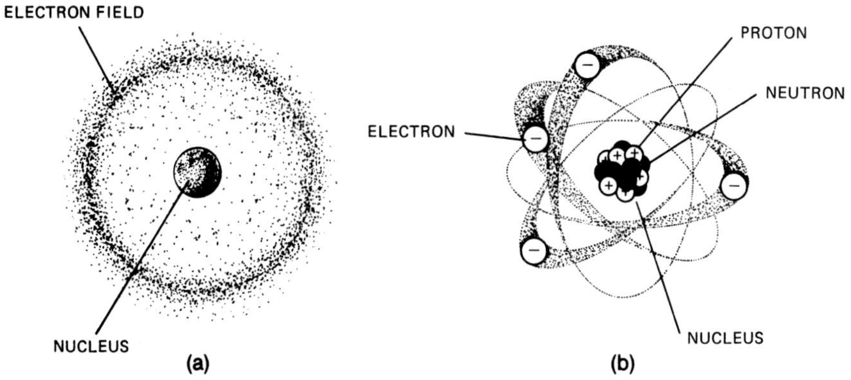

Figure 1.1
The structure of an atom. (a) Electrons orbiting around the nucleus form an electron field, indicating that no one electron can be located at a particular place. (b) Another version of an atom, showing the negatively charged electrons in orbital paths around the nucleus, which is composed of electrically neutral neutrons and positively charged protons.

ity goes. Common radioactive isotopes used in biological experiments are sulfur (S^{35}), phosphorus (P^{32}), carbon (C^{14}), and tritium (H^3).

Isotopes of oxygen and carbon were essential in discovering the details of photosynthesis. Heavy oxygen was used to trace the fate of water, H_2O^{18}. Radiocarbon was used to study the buildup of carbohydrate from carbon dioxide $C^{14}O_2$ (Chapter 4). Isotopes of nitrogen helped explain the replication of DNA (Chapter 6).

The presence of radioisotopes can also be used to determine the age of samples. Cosmic radiation from space generates a certain amount of C^{14}, which gets into the air as radioactive carbon dioxide, $C^{14}O_2$. The radioactive carbon dioxide is incorporated into a plant during photosynthesis (Chapter 4). When a plant dies, it ceases to accumulate C^{14}, but the radioactive carbon continues to emit beta particles as the C^{14} decays. The rate at which C^{14} decays is such that after 5730 years, half of the original radioactivity is lost. Such a time span is called the *half-life* of the isotope. After another 5730 years, half the remaining radioactivity is lost. The amount of radiocarbon detectable in samples of unknown age can be determined, and the age of the sample can be estimated. Radiocarbon dating is a common method of determining the ages of biological material, especially wood. It is useful up to about 10,000 years and has been reasonably well correlated with dates obtainable from historically dated artifacts, but the accuracy of the method lessens with older samples. Other elements can be used at other age ranges.

The best-known subatomic particles are the relatively heavy, central **nucleus,** which contains electrically charged* (positive) **protons,** and uncharged **neutrons.** (The hydrogen atom is the only one without any neutrons in the nucleus.) Orbiting around the nucleus are negatively charged subatomic particles called **electrons.** Modern atomic theory describes these whirling electrons as existing in an electron "cloud" or "field" (Figure 1.1a), which means that an electron cannot be predictably located at any one place at any one time.

Atoms are frequently compared to small solar systems, with the nucleus like the sun at the center, and the electrons orbiting like the planets (Figure 1.1b). The comparison between an atom and the solar system is imperfect, however, because unlike planets, electrons can shift orbits. We will discuss this idea further in a later section, Energy-Level Shells.

Atomic Number and Atomic Weight

The chemical and physical properties of an atom are determined by the number of protons and neutrons in its nucleus, and by the number and arrangement of electrons surrounding the nucleus. Every element has an **atomic number** that tells the number of protons in the nucleus of its atoms. An element also has an **atomic weight,** which is the weight of an atom of that element compared to that of a carbon atom, which has an assigned weight of about 12 (six protons and six neutrons). Thus, an atom of hydrogen, which is about $\frac{1}{12}$ as massive as an atom of

* Protons exert attractive or repulsive forces, depending on what they are attracting or repelling. Protons repel other protons, but attract electrons. A proton is therefore said to be *charged,* and that charge is called *positive* (+). An electron is said to carry a *negative* charge (−). Positive charges repel other positive charges, and attract negative ones.

TABLE 1.1
The Atomic Structure of Some Common Elements

ELEMENT (AND SYMBOL)	ATOMIC NUMBER	NUMBER OF PROTONS	NUMBER OF NEUTRONS	NUMBER OF ELECTRONS	ATOMIC WEIGHT
Hydrogen (H)	1	1	0	1	1.008
Carbon (C)	6	6	6	6	12.01
Nitrogen (N)	7	7	7	7	14.01
Oxygen (O)	8	8	8	8	15.99
Sodium (Na)	11	11	12	11	22.99
Phosphorus (P)	15	15	16	15	30.97
Sulfur (S)	16	16	16	16	32.06
Chlorine (Cl)	17	17	18	17	35.45
Potassium (K)	19	19	20	19	39.10
Calcium (Ca)	20	20	20	20	40.08

carbon, has an atomic weight of about 1, and an atom of calcium, which is about 3.3 times as massive as carbon, has an atomic weight of about 40.

Note that the atomic weights listed in Table 1.1 are not expressed in whole numbers. That is because atomic weight is actually computed by averaging the weights of the different forms in which an element may exist in nature. (See the essay on p. 16.) For instance, *most* carbon atoms have six protons and six neutrons, but not *all* carbon atoms have six neutrons. Since most carbon atoms have six protons and six neutrons, the accepted atomic weight of carbon is 12, even though its actual atomic weight is slightly higher.

Energy-Level Shells

The number of electrons and protons in any given atom determines its chemical activity, but the way the electrons are distributed is also important. The electrons orbiting an atomic nucleus are not distributed randomly. Rather, they are distributed as though they were in layers, known as **energy-level shells** (Figure 1.2). The word "energy" is important; a critical point is that when an electron moves toward the nucleus from a more-distant level, energy is *released,* whereas raising an electron to a shell more distant from the nucleus *requires* energy. *All biological actions involve changes in electron positions and consequent gains or losses of energy.*

There are no more than two electrons in the shell nearest the nucleus, as many as eight in the

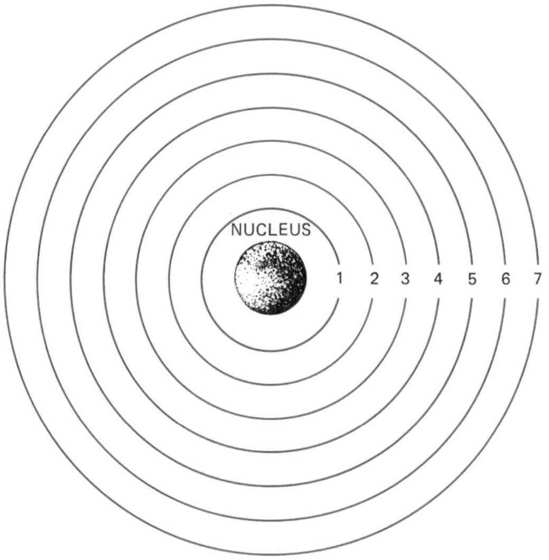

Figure 1.2
Diagram of the possible orbits of an electron around the nucleus of an atom. In its resting state, the electron remains at the first level, but if sufficient energy is pumped into the atom, the electron may be raised to higher energy levels, that is, pushed into orbits at greater distances from the nucleus. It requires ever-greater amounts of energy to move the electron to more distant orbits, known as "shells." Just as energy is required to raise an electron to a more distant orbit, energy is emitted if an electron falls to a lower level.

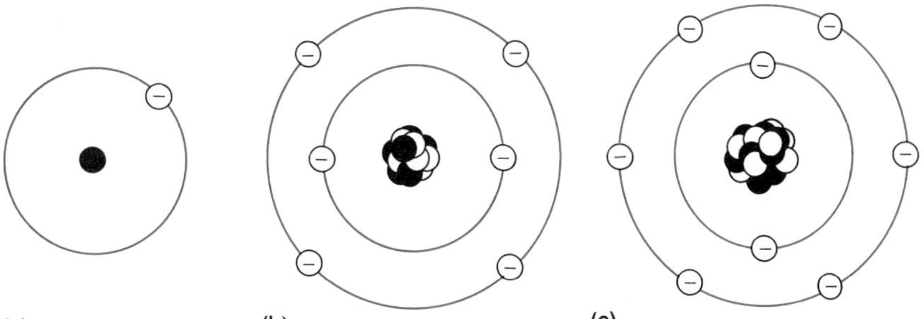

Figure 1.3
Diagrams of the structure of atoms. (a) A hydrogen atom, with one proton in its nucleus and a single electron in the first orbit. (b) A carbon atom, with six protons (black balls) and six neutrons (white balls) in its nucleus, two electrons in its first orbit, and four electrons in the second orbit. (c) An oxygen atom, with eight protons and eight neutrons in its nucleus, two electrons in its first orbit, and six electrons in its second orbit. The diagrams are drawn without regard to actual proportions because it is impossible to give a realistic picture of an atom on a piece of paper. If the picture of an oxygen atom, for example, were drawn with a nucleus one centimeter in diameter, the first orbit would be about 150 meters away, the second would be about 600 meters away, and the electron would be too small to print. An atom, even of the heaviest element, is mostly "space."

second, and larger numbers in still more distant shells (Figure 1.3). As many as seven shells have been identified in the largest atoms.

Atoms seek the most stable form. An atom is most stable, and will not combine with other atoms, when its outermost energy-level shell is full. For instance, if an atom has seven electrons in its outermost (second) shell, it becomes more stable by the capture of one more electron. In picking up an additional electron, the atom becomes negatively charged, since it now has one more negatively charged electron than it has positively charged protons. The combining of atoms with other atoms is the result of the tendency to attain the most stable number of electrons in the outer shell.

The need to understand the action of electrons and electron shells can be shown by some reactions of the element carbon. Carbon is an especially important element because it is a part of all biological molecules and participates in all biological reactions. Having four electrons in its outermost shell, carbon must *gain* four electrons to fill its outer shell to capacity (Figure 1.3b). A carbon atom may gain four electrons by sharing electrons with other atoms. This sharing process results in a bond between the atoms. By sharing electrons and forming bonds, carbon atoms combine to form intricate chains and ring structures. These structures appear complicated, but they are determined by the relatively simple "combining capacities" of the combining atoms, determined in turn by the distribution of electrons.

The gaining, losing, and sharing of electrons in their outer shells give atoms their capacity to combine with other atoms. Combinations of atoms can result in large molecules consisting of thousands of atoms. It is important to realize that large molecules have specific three-dimensional shapes because the combining atoms have specific relationships to each other. Large molecules made up of the same number of atoms combined in the same way will have identical three-dimensional shapes. A molecule that contains different kinds of atoms or the same kinds combined in different ways will have a different three-dimensional shape. Shapes of molecules are important in practically every cellular activity.

TO SUM UP ENERGY-LEVEL SHELLS

1. Electrons orbit the nucleus of an atom as though they were distributed in layers; these layers are known as *energy-level shells.*
2. The shell nearest the nucleus contains no more than two electrons, with as many as eight in the second, and larger numbers in still more distant shells.
3. Raising an electron to a more-distant shell uses up energy; dropping an electron to a lower shell releases energy.
4. An atom is most stable, and will not combine with other atoms, when its outermost energy-level shell is full. Because atoms with their outer shells fully occupied are in their most stable form, atoms tend to lose, gain, or share electrons in order to complete their outer shells.

THE UNION OF ATOMS: CHEMICAL BONDS

Substances in this world rarely exist as elements. Instead, some atoms join with others to make larger particles. Such a particle is a **molecule,** which is the smallest unit into which a substance can be divided and still retain the chemical nature of that substance. A few rare gases are one-atom molecules, but most molecules have two or more atoms; for example, oxygen (O_2) has 2 atoms, methyl alcohol has 6, and a simple sugar has 24, while others have various numbers up to the tens of millions (*macromolecules; macro* means "big" in Greek). When a molecule has more than one kind of atom, it is a **compound,** as is water (H_2O).

When atoms interact chemically to form molecules or compounds, the atoms are held together by electrically attractive forces called *bonds,* specifically **chemical bonds.** We will discuss three of the types of bonding important in living systems. Certain kinds of atoms form certain kinds of bonds when they combine with other atoms. These bonds have unique characteristics that give unique properties to the compounds so formed.

Covalent Bonds

Most of the bonds that stabilize biological compounds are strong **covalent bonds** that owe their strength to the *sharing of electrons* between atoms. When we appreciate the strength of human muscles and the toughness of our skin, we are indirectly appreciating the strength of covalent bonds in molecules making up the muscles and skin. In molecules of water, starch, or fat, the bonds between the atoms are also covalent, meaning that an oxygen atom, for example, has in its outer electron shell not only its own electrons but some of those from an adjacent atom as well (Figure 1.4). The number of electrons shared is always a multiple of two.

Ionic Bonds

Another kind of bonding occurs when an atom does not have as many electrons in its outer shell as it might and has access to an extra electron from a different atom. That different atom may contribute an electron to the first atom. In that case, the first atom has one negative electron more than it has positive protons, and the atom acquires a negative charge.

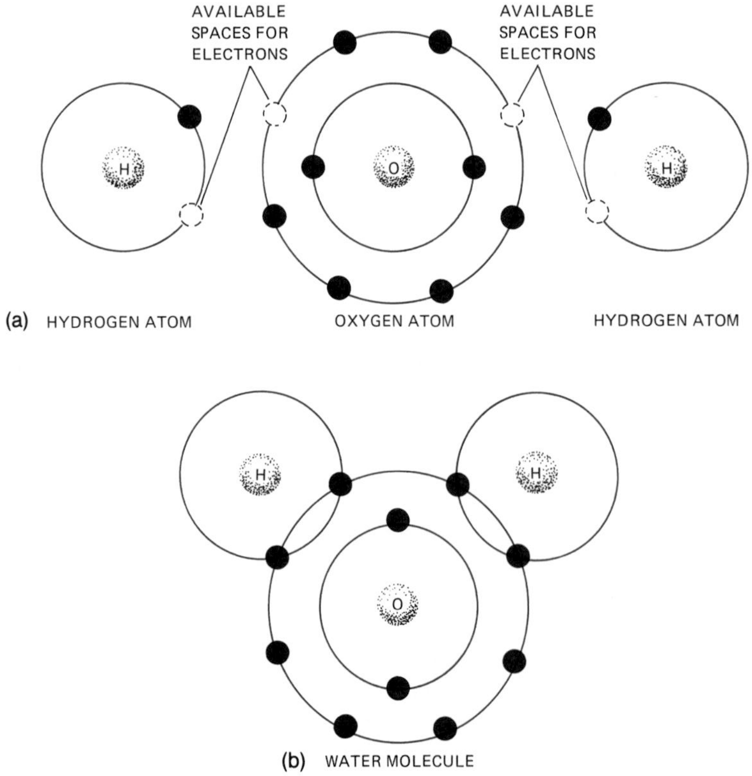

Figure 1.4
The formation of *covalent bonds* between an atom of oxygen and two atoms of hydrogen, resulting in a molecule of water. (a) The hydrogen and oxygen atoms are shown before they combine. The oxygen atom has two spaces available for electrons in its outer shell, and the hydrogen atoms each have one available space in their shell. (b) Each hydrogen atom, initially with one electron in its outer shell, shares its electron with the outer shell of oxygen, and at the same time, two electrons from the outer shell of oxygen are shared with the two hydrogen atoms.

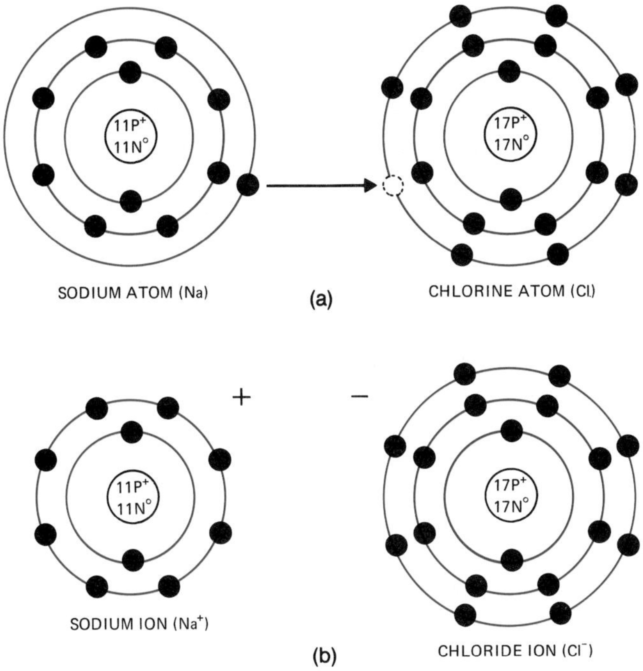

Figure 1.5
The formation of an *ionic bond* between an atom of sodium and an atom of chlorine. (a) An "extra" electron from the outer shell of the sodium atom (black dot with arrow) is transferred to the "incomplete" outer shell of the chlorine atom; the dotted circle indicates an available space for an electron. (b) After losing its "extra" electron, the sodium atom acquires a positive charge, and is now a sodium *ion,* represented by Na^{+}. The chlorine atom, having received an electron to stabilize its outer shell, acquires a negative charge, and becomes a chloride *ion,* (Cl^{-}). Both atoms, neutral at the start, were not attracted to each other, but having developed opposite charges, they are now mutually attractive. Note that the outer orbit of the sodium *atom* in (a) is no longer represented in the diagram of the sodium *ion* in (b).

The atom that contributes an electron has one electron less than the number of its protons, and thus has a net positive charge. A positively charged atom is attracted to a negatively charged atom, and the result is a rather weak bond between two atoms of opposite charge. This is an **ionic bond** (Figure 1.5). (A particle with an electrical charge is called an *ion.*)

Hydrogen Bonds

Still a third type of bond, a **hydrogen bond,** is formed when a hydrogen atom, essentially a single proton, acts as a bridge between two electrically negative atoms (Figure 1.6). In certain covalently bonded substances such as water, the nonhydrogen (oxygen) atoms tend to attract the shared electrons more than the hydrogen atoms do, and thus acquire a slight negative charge. The hydrogen acquires a slight positive charge. As a result, the hydrogen atom, while still maintaining its covalent bond with oxygen, becomes attracted to negatively charged atoms in the vicinity. When this happens, a weak hydrogen bond is formed. As shown in Figure 1.6, the hydrogen atom in one water molecule frequently forms a hydrogen bond with the oxygen atom in another water molecule until many water molecules are bonded together.

Hydrogen bonds are weaker than covalent or ionic bonds, but in many molecules found in living

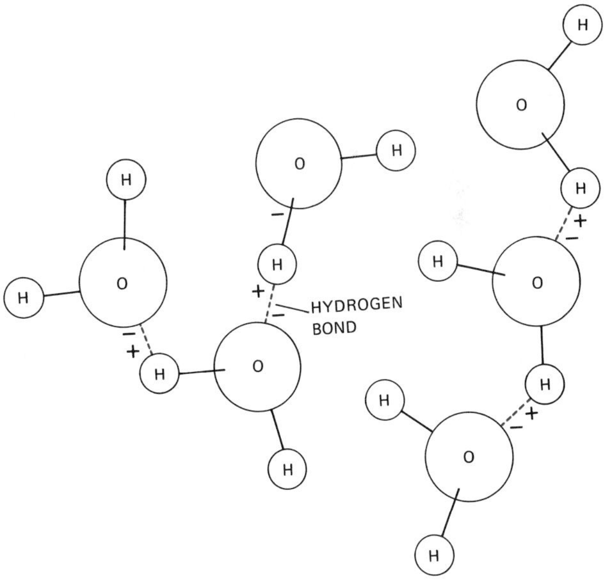

Figure 1.6
Hydrogen bonding between water molecules, shown as dashed lines in color joining oxygen (large circles) of one molecule with hydrogen (small circles) of another. Weak cross-bonding is thus formed within a mass of water.

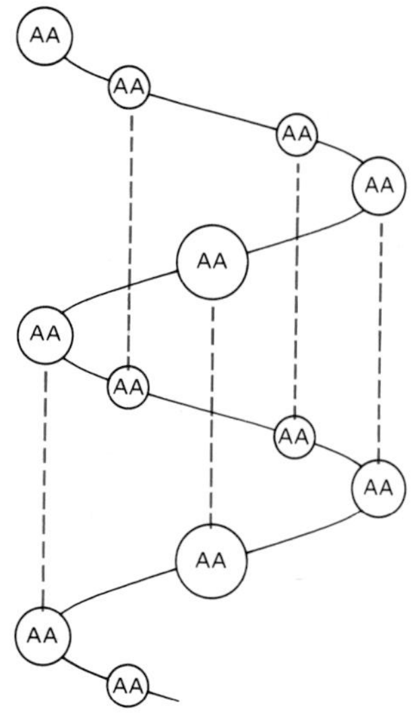

Figure 1.7
Hydrogen bonding between atoms within a molecule. The circles represent amino acid subunits (AA) of a protein molecule, with the hydrogen bonds holding the subunits in place in a spiral called a *helix.* The dashed lines in color indicate the hydrogen bonds.

systems there are enormous numbers of hydrogen bonds that aid in maintaining a stable three dimensional shape. Hydrogen bonds are like sewing threads: A single stitch is not strong, but a great many stitches can hold a seam securely.

Later in this book, we will discuss structural protein, DNA, enzymes, and hemoglobin molecules. All these very large molecules are held in proper shape by hydrogen bonds (Figure 1.7). Maintaining the shape of such molecules is necessary for them to perform their biological roles.

TO SUM UP BONDING

1. Chemical bonds *hold atoms together.* The main types of bonds are covalent, ionic, and hydrogen bonds.
2. Some atoms gain stability by *sharing* electrons, or forming covalent bonds. If an atom has one electron in its first shell, it can fill the shell with an electron that is shared with another atom. Covalent bonding always occurs with *pairs* of electrons.
3. Ionic bonds are formed when an electrically charged (either negative or positive) atom or group becomes *attracted* to an oppositely charged atom or group.
4. A hydrogen bond occurs when a covalently bonded hydrogen atom with a positive charge acts as an electrically charged *bridge* between two negatively charged atoms.

CHEMICAL REACTIONS

As we just saw, atoms combine by forming bonds. The act of combining two or more atoms is a chemical change called a chemical reaction. Chemical reactions also occur when bonds are broken, so we can say that in a **chemical reaction,** bonds between atoms are broken or formed, and different combinations of atoms or molecules result.

A simple chemical reaction is one in which two elements *combine* to make a new compound. Hydrogen and oxygen combine to make water, or in the symbolism of chemistry,

$$2H_2 + O_2 \rightarrow 2H_2O.$$

This expression merely tells us that hydrogen reacts with oxygen to produce water. Specifically, two molecules of hydrogen (made up of two atoms each) react with one molecule of oxygen (made up of two atoms) to produce two molecules of the compound water. Most chemical reactions are *reversible,* that is, water can also be broken down to hydrogen and oxygen.

Some Important Chemical Reactions

Some chemical reactions are especially important in biological systems, and we will briefly discuss three of them here: (1) oxidation-reduction, (2) hydrolysis, and (3) condensation.

Oxidation-reduction **Oxidation-reduction reactions** occur when an atom or molecule loses electrons (oxidation) or gains electrons (reduction) (see p. 44). Oxidation and reduction always occur together, because when one atom or molecule gains electrons, another has lost them. The relevance of oxidation-reduction reactions in living cells is that when electrons are transferred from one compound to another, energy is transferred as well, and in biology, energy transfer is of prime importance. Every organism needs a steady source of energy, and the destruction of any organism's oxidation-reduction power means immediate death.

Hydrolysis Another kind of chemical reaction of special interest to biologists is **hydrolysis** (Greek, "water loosening"). In a *hydrolytic reaction,* a molecule of water is added to the reacting molecule, and the bonds of the reacting molecule are broken. Sucrose (cane sugar) can be hydrolyzed to two simple sugars, glucose and fructose (which have the same numbers and kinds of atoms, but in different arrangements; see p. 26):

$$\underset{\text{sucrose}}{C_{12}H_{22}O_{11}} + \underset{\text{water}}{H_2O} \rightarrow \underset{\text{glucose}}{C_6H_{12}O_6} + \underset{\text{fructose}}{C_6H_{12}O_6}.$$

Later, when we consider one of the most important molecules of living cells, adenosine triphosphate (ATP), we will be dealing with hydrolysis whenever ATP is broken down to provide energy for cellular work (p. 44). Most digestive reactions are hydrolytic, producing smaller, simpler, more soluble molecules from large, complex, insoluble ones. Proteins in the food you eat, for instance, are useless to your body until they are broken down into amino acids by hydrolysis.

Condensation **Condensation reactions** are essentially hydrolytic reactions in reverse. Small, simple molecules are united in a condensation reaction into larger, more complex ones, and one or more molecules of water are eliminated. See Figure 1.25, which shows how two amino acids are condensed to form a protein (p. 35).

IONS, ACIDS, AND BASES

As we saw above, an **ion** is a particle with an electrical charge. It may be a single atom or a group of atoms; it may have an extra electron, making it a *negative* ion, or it may lack an electron, making it *positive.* Ions function in many biological activities, including the balance of salts in blood, the transmission of nerve impulses, the uptake of minerals in the roots of plants, and the contraction of muscles.

One special instance of ion formation is found in acids and bases. In common experience, an acid is something that tastes sour; it can turn blue litmus paper red or keep red cabbage juice red. At an atomic level, an **acid** is a substance that, when dissolved in water, releases positively charged hydrogen ions (H^+). **Bases,** usually called *alkalis,* feel slimy and can turn red litmus paper and red cabbage juice blue. They release negatively charged hydroxyl ions (OH^-) when dissolved in water.

The proportional amount of ionic hydrogen or hydroxyl released by a compound determines its strength. A strong acid, such as hydrochloric acid, is completely ionized in water, and consists of hydrogen ions (H^+) and chloride ions (Cl^-). Acetic acid, the familiar acid of vinegar, is a weak acid, as only about one percent of acetic acid molecules are ionized in water, forming hydrogen ions and the rest of the molecule (CH_3COO^-) (Figure 1.8). The separation of a molecule into ions is *dissociation,* and the union of ions into a molecule is *association.* Every acid and base has its own degree of association and dissociation, depending on its molecular nature and the immediate environment.

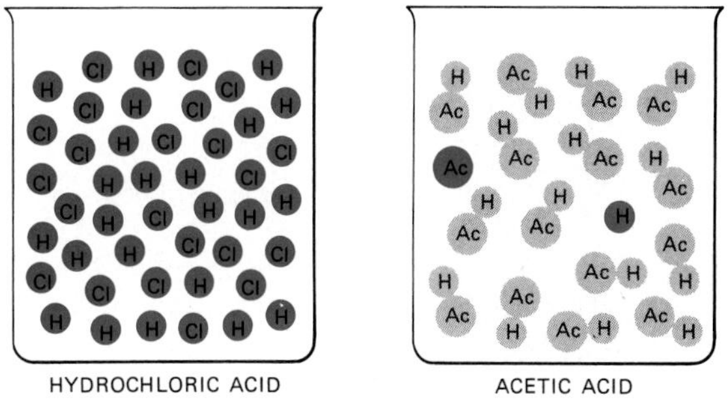

Figure 1.8
Diagram of the dissociation of two compounds: hydrochloric acid and acetic acid. Undissociated molecules are gray; those dissociated into ions are in color. All the particles in the hydrochloric acid vessel are shown as ionized into H^+ or Cl^- because hydrochloric acid is practically 100 percent dissociated in water solution. In the acetic acid vessel, only one ion pair is dissociated into H^+ and Ac^-, and even that is proportionately too many, because only about 1 percent of the acetic acid molecules are dissociated. The difference in ionization is the difference in acidity between the two compounds; that is, the greater the ionization, the greater the acidity.

Measurement and Effects of Acidity

In everyday life when we think about strong acids, we think about their dangerous properties. Strong acids have a capacity to inflict severe burns because they are chemically very reactive. This reactivity is primarily due to the hydrogen ion. Recall that hydrogen is merely a nucleus consisting of one proton orbited by one electron. In the absence of the electron, the hydrogen ion is a "naked proton." It is extremely small and can move very rapidly and alter the associations of other atoms in other molecules.

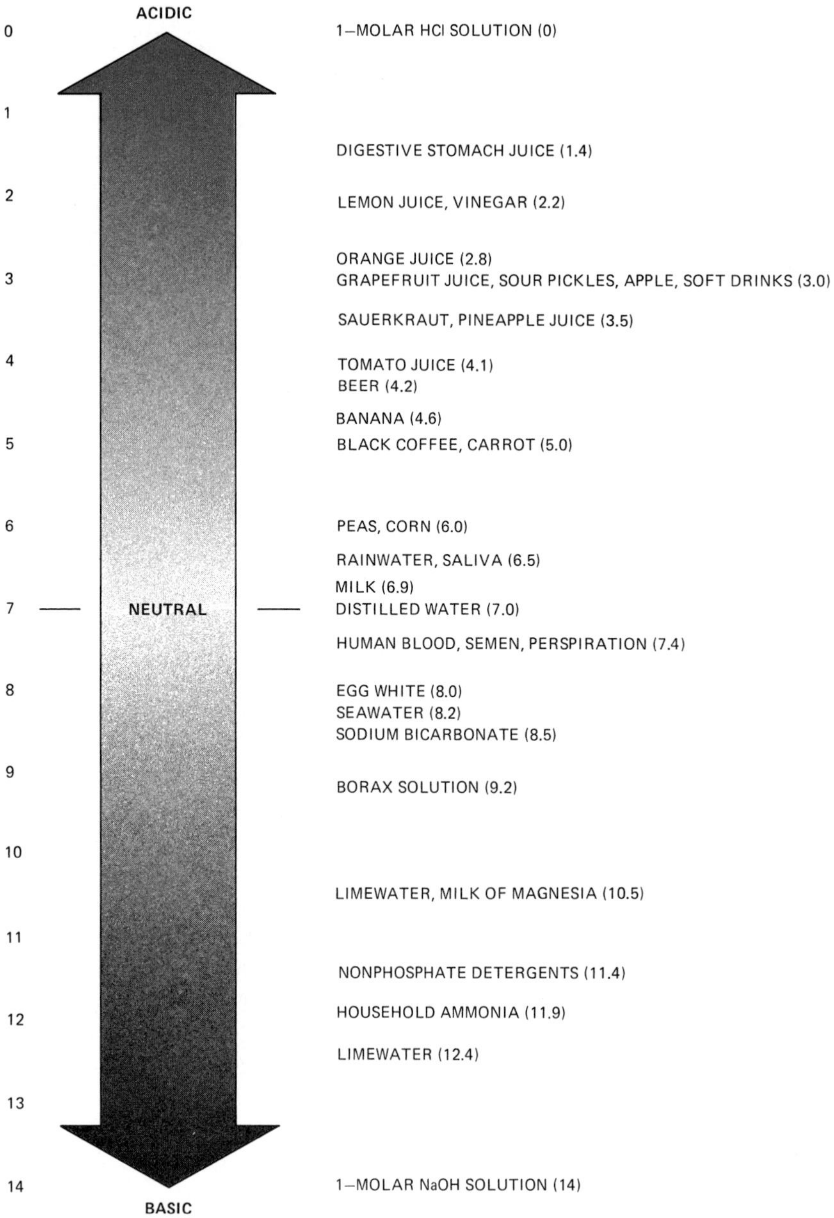

Figure 1.9
The pH scale. Note that pure distilled water is neutral, rather than being either acidic or basic; this occurs because distilled water has equal numbers of hydrogen ions and hydroxyl ions. Each number on the scale represents a tenfold change in acidity. Black coffee at pH 5 contains 10 times the acidity of corn at pH 6.

Figure 1.10
The action of a buffer. (a) When 5 ml of the base sodium hydroxide is added to 20 ml of an acid solution at pH 1, there is a marked change in the acidity (hydrogen ion concentration). The pH value increases by one unit. (b) When 5 ml of the base sodium hydroxide is added to 20 ml of an acid solution that contains a buffer, there is a much smaller change in the acidity. The pH value increases by only 0.5 unit.

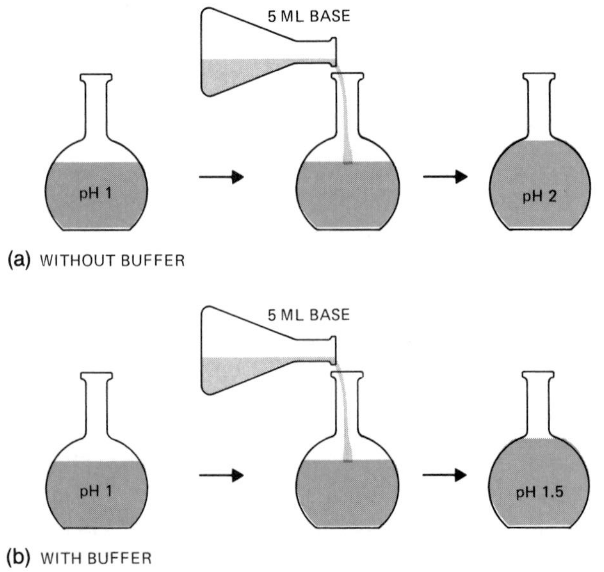

Hydrogen ions are important in the processes of cell life. We must be alert to their importance when discussing biochemical aspects of energy conversion within the cell. And we can see why it is important to keep the number of hydrogen ions "under control."

Acidity is measured as the concentration of hydrogen ions or hydroxyl ions in a solution. A numerical scale, from 0 to 14, indicates acidity or alkalinity, with 7 the neutral point, where the number of hydrogen ions equals the number of hydroxyl ions (this state occurs in distilled water). This is the **pH scale** (p for potential, H for hydrogen) (Figure 1.9). The lower the number is below 7, the more acidic is the solution, and the higher the number is above 7, the more basic (alkaline) is the solution. The numbers in the pH scale are logarithmic, each whole number representing a tenfold change in acidity, with a solution at pH 4 containing 10 times the H^+ concentration of one at pH 5.

At first reading, ions and the pH scale may seem remote from ordinary living, but in fact, almost every biological activity depends upon careful regulation of the ionic concentration of cellular fluids. If you do not believe that, try blowing as hard as you can for one full minute. Unless you are accustomed to heavy breathing, you will get dizzy, maybe find it impossible to take another breath, or even pass out. (That is all right, you will come to quickly.) All this is because if you breathe abnormally fast you will remove more than the customary amount of carbon dioxide from your blood, thereby reducing the carbonic acid content of your blood and slightly raising the pH level. The effect of that small shift will affect your central nervous system immediately. Within seconds, however, the normal pH of your blood will be reestablished, and you will be as you were before the heavy breathing. Blood in humans is close to pH 7.4; if it is not, you are in trouble.

One of the ways that pH is held reasonably steady is by means of buffers. A *buffer* is a combination of compounds that resists changes in pH, even if a strong acid or base is added (Figure 1.10).

SOME IMPORTANT BIOLOGICAL MOLECULES

Despite the enormous number of different compounds formed by organisms, there is a relatively small number of categories into which the compounds can be placed. The main types of compounds are water, the carbohydrates, the lipids and related compounds, amino acids and proteins, and the hereditary material, the nucleic acids. Except for water, these substances make up the *organic compounds* of living things. **Organic compounds** always contain carbon. They are typically large and complex, are usually held together by covalent bonds, and are the principal materials that form living organisms. Although water is not an organic compound, it is so important that it must be considered along with the organic compounds.

Water

Although water is familiar and common and seems simple chemically, it deserves special attention. Water makes up the bulk of living things. Most animals consist of 80 to 90 percent water. All organisms either live in water or obtain enough water to keep cellular machinery functioning.

The cellular functions of water depend on its unique properties, which grow out of the bonding of two hydrogen atoms to one oxygen atom (Figure 1.11). Oxygen is covalently bonded to hydrogen, but these bonds are distorted because the heavier oxygen atom tends to draw electrons to itself and away from the hydrogen atoms. As a result, the oxygen bears a slight negative charge, and the two hydrogens bear slight positive charges. Because the two ends of the water molecule bear different charges, it is sometimes said to be a *polar molecule.* As a polar molecule, water can form hydrogen bonds with other water molecules and with a variety of other kinds of compounds.

Nearly all large molecules in living systems (carbohydrates, nucleic acids, proteins) have sites

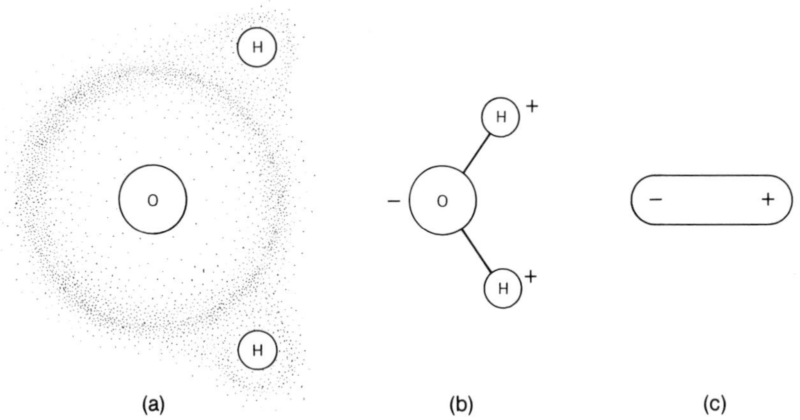

Figure 1.11
Three ways to represent a water molecule. (a) The oxygen and hydrogen atoms are shown as having distinct nuclei with hazy clouds of electrons surrounding them. Since there is at any moment a greater likelihood that there will be more electrons at the "oxygen end" of the molecule than at the "hydrogen end," the oxygen end is negative with respect to the hydrogen end. This makes the molecule *polar.* (b) The atoms are shown as skeletons connected by lines, with the charges indicated by plus and minus signs. (c) The whole molecule is drawn as a simple stick with one plus end and one minus end.

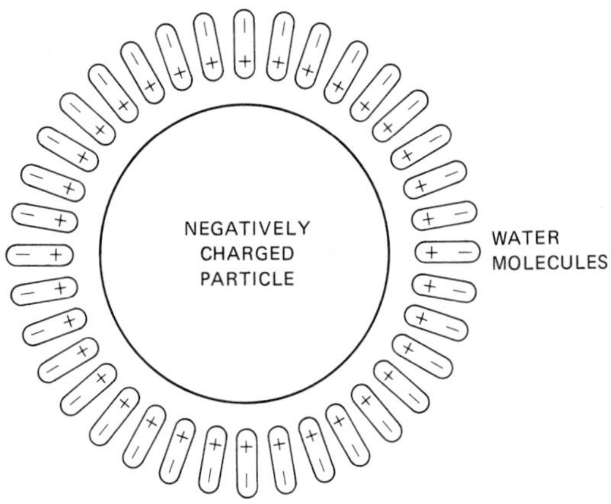

Figure 1.12
A "water shell" surrounding a small charged particle with an electrically negative-charged surface. The plus ends of the water molecules are attracted to the negative surface and form an orderly film over the surface of the particle. Such a particle, shown here as if it were sliced open, would in reality be a sphere, and the water film would cover the surface like a piece of velvet over a ball. Such water shells help keep small particles in suspension in cells.

that are slightly charged. Water will be attracted to the charged sites and will thereby contribute to keeping such large molecules suspended in a water environment. The polar water molecule will of course be attracted to ions. Many surfaces within cells are charged, and water molecules will be attracted to those surfaces (Figure 1.12). The neatly packed water molecules add stability to the surfaces of cell membranes and proteins. One reason why cells are damaged by freezing and heating is that such extremes of temperature disrupt the water layers surrounding cytoplasmic particles.

Because of the structure and polarity of the water molecule, the physical properties of water are also special. One important property unique to water is that in solid form it is less dense than it is as a liquid at 4°C. Consequently, ice will float in water. Among chemical molecules this is a unique property. If water acted like other substances, ice would sink to the bottom of rivers and oceans. This property of water was probably a requirement for the origin and evolution of life on earth.

Another important physical property of water is its high *heat of vaporization.* It takes a great deal of heat to turn liquid water into water vapor (the gas). When a person perspires, evaporation from the surface results in a loss of heat. Because of the high heat of vaporization, evaporation of only a little water has a large cooling effect. Such evaporation of water has a cooling effect that can be very important in maintaining the inner stability of the organism.

Water enters into the many chemical reactions of the cells directly. For example, the hydrolysis of a protein yields amino acids. Water also plays an indirect role in providing the medium for the chemical interaction of all other biochemical molecules. We begin to appreciate this function when we remember that about 80 percent of most living cells is water.

The polarity, size, and shape of the water molecule make it particularly effective in suspending small molecules such as ions, sugars, and amino

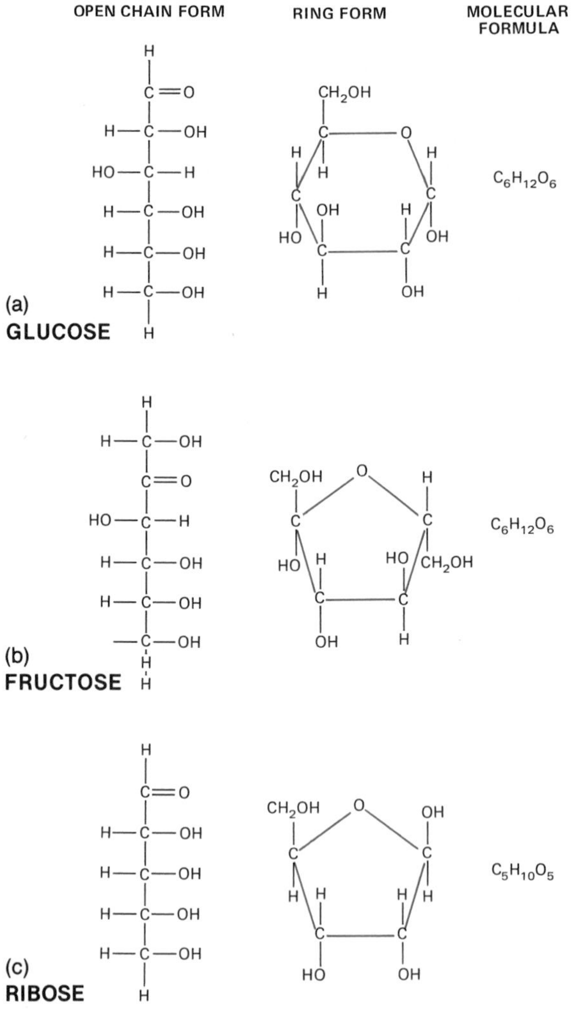

Figure 1.13
Structures of glucose, fructose, and ribose. Note that although glucose and fructose both have the same molecular formula, $C_6H_{12}O_6$, they have different molecular *structures* or arrangements of atoms.

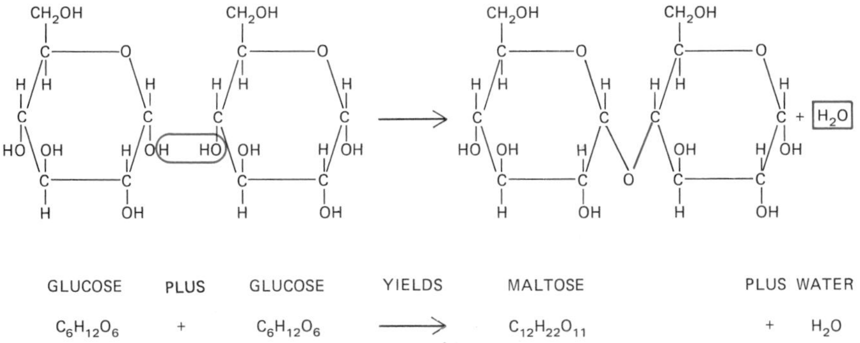

Figure 1.14
A dehydration synthesis or condensation reaction. When a molecule of water (circled in color) is removed during the reaction, a bond (color line) is formed between the two monosaccharide molecules, producing a single molecule of the disaccharide maltose and a molecule of water (in color box).

acids. When suspended in water, these molecules can still move, and when they collide with an enzyme or another molecule, they may undergo a chemical reaction, one that would not occur in the absence of water. For example, an absolutely dry digestive enzyme mixed with an absolutely dry food would be ineffective. This is one reason why dehydration, the removal of water, contributes to the preservation of foods and other materials.

Later in this text we will talk about ions and other molecules "in solution." By this term we refer to the capacity of water molecules to surround ions and other molecules completely and hold them in suspension but *still available for chemical reaction.*

Carbohydrates

Carbohydrates are an important source of energy. They are carbon compounds in which the ratio of hydrogen to oxygen is usually 2 to 1; the general formula for a carbohydrate is CH_2O.

Carbohydrates are usually classified according to the complexity of their molecules. The simplest carbohydrates are called **monosaccharides** (Gr. *monos,* single; *sakkharon,* sugar). Monosaccharides are so simple that they cannot be decomposed by adding water to them. Three common monosaccharides are glucose, fructose, and ribose* (Figure 1.13).

Monosaccharides are the building blocks of more complex sugars. When two monosaccharides are chemically united, they form a **disaccharide** (*di,* two). Disaccharides all have the molecular formula $C_{12}H_{22}O_{11}$. (Many compounds have the same molecular formulas, but different molecular *structures,* that is, different arrangements of atoms; see Figure 1.13.) Disaccharides are formed when a molecule of water is removed from two monosaccharide molecules. Such a chemical reaction is called *dehydration synthesis,* or *condensation.* In Figure 1.14 we see how two monosaccharide glucose molecules combine to form the disaccharide maltose (malt sugar).

When carbohydrate molecules are made up of more than two simple sugars they are called **polysaccharides** (*poly,* many). Polysaccharides are formed by the same type of dehydration synthesis that forms disaccharides, and can be broken down into their many simple-sugar components by the process of *hydrolysis,* which requires the addition of water to break the chemical bonds formed during condensation. Polysaccharides, such as starch and glycogen, are bonded together in long chains that may be branched or unbranched (Figure 1.15). Poly-

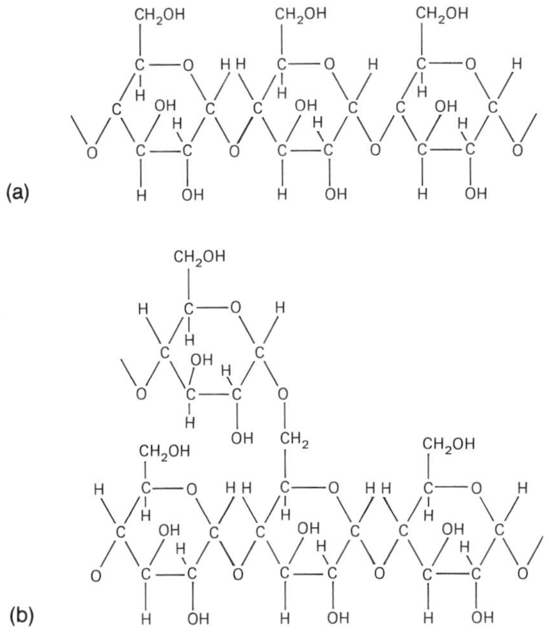

Figure 1.15
Polysaccharide chains. (a) A portion of a straight chain of amylose, which contains 250 to 500 linked molecules of glucose, and (b) a portion of a branched chain of amylopectin, which is a highly branched glucose polymer consisting of about 1000 glucose molecules.

* The suffix "-ose" usually stands for sugar. Prefixes such as *tri-* (three), *pent-* (five), and *hex-* (six) indicate the number of carbon atoms in the molecule. Pentoses, for example, are monosaccharides containing five carbons.

THE ROLE OF ENZYMES

Louis Pasteur, the great nineteenth-century French scientist, nearly always guessed right, but he did miss once. He believed that fermentation is a strictly biological activity that can proceed only in the presence of living cells. Proof that living cells are not essential to fermentation came in 1898, when the German chemist Eduard Büchner demonstrated fermentation in a liquid from which all cells had been removed. The fermenting substance came from yeast, and the active material was called an **enzyme,** a word that literally means "in yeast."

During the next three-quarters of a century, knowledge of enzymes increased. We now know that they are proteins of a very large molecular weight. There are an estimated 10,000 different enzymes in human cells. All the molecular changes that occur in these cells are controlled by enzymes.

How do enzymes function? An enzyme is a protein that *speeds up* a chemical reaction without permanently entering into the reaction itself. Any substance that can function in this way is called a *catalyst.* Therefore,

$$X + Y \longrightarrow Z \text{ slow,}$$

$$X + Y \xrightarrow{\text{enzyme}} Z \text{ fast.}$$

The enzyme is unchanged by the reaction it controls, and it can be used over and over again to activate the same reaction. The function of each enzyme is specific, that is, an enzyme able to assist one chemical reaction has no effectiveness in another reaction or with other reactants.

It is the shape—the molecular structure—of an enzyme that permits it to join both reactants together, atom to atom, and then withdraw after the chemical union has been completed. Reactant X will fit into one site on an enzyme and Reactant Y will fit into the adjoining site. After X and Y have been locked together by way of the enzyme "mold," the new Z molecule will separate from the enzyme to go about its new business. With-

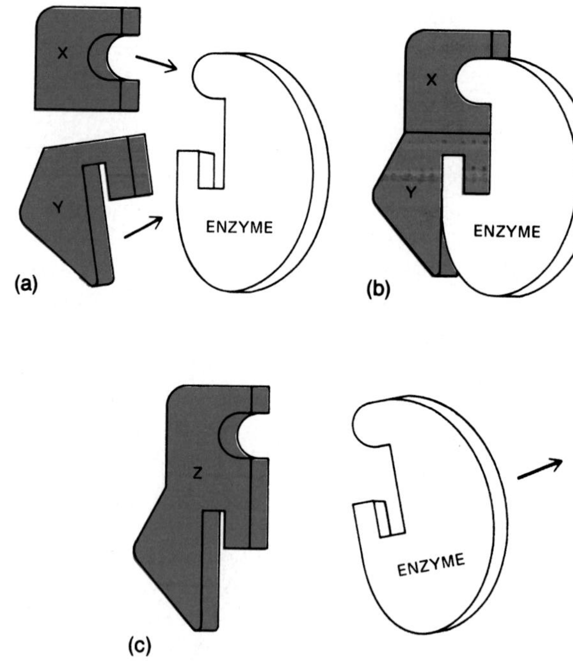

(a) Reactant X fits into one portion of the enzyme, and Reactant Y fits into the adjoining portion. (b) Reactants X and Y are locked into position by the enzyme mold. X and Y are now side-by-side and are chemically bonded to each other. (c) The new molecule Z is formed, and the enzyme is freed to perform the same catalytic action elsewhere.

out the help of enzymes the same reaction would take place, but only at a uselessly slow pace.

It is believed that *enzyme inhibitors,* such as poisons, may prevent normal enzyme action by fitting themselves into one or both "slots" and clogging the space normally occupied by the two reactants.

If a cell is to function, it must have an adequate supply of enzymes, it must be supplied with some necessary additional substances, and it must be provided with a suitable cellular environment.

What is "an adequate supply" of enzymes? Here "adequate" means first the correct enzyme for a given job. Being proteins, different enzymes are constructed out of different amino acids. The amino acid composition and primary structure of an enzyme give it a specific three-dimensional shape that fits it for contact with one special molecule or a group of similar molecules. The *specificity* of enzyme action is a result of the shape of the enzyme molecule, which allows it to come into close contact with the molecule or molecules, that is, the *substrates,* on which it works. Because enzymes are so large, there is usually one particular place, or *active site,* on an enzyme that plays the catalytic role. An enzyme succeeds by making such intimate contact with the reactant molecule that it can alter chemical bonds. Sometimes this means breaking one molecule into two, as when sucrose is converted into two simple sugars. It may also mean bringing two compounds together, as when two amino acids are combined.

What "additional substances" are necessary for enzyme function? Many enzymes cannot catalyze a reaction unless they are accompanied by extra molecules. These substances may be metals, like magnesium or zinc, or they may be organic compounds. Such a helping compound is a *coenzyme,* which can be experimentally removed from its enzyme. When it is removed, the enzyme is incapacitated. Many vitamins are in fact coenzymes.

What is a "suitable" cellular environment? Several requirements must be met. First, the temperature is critical. If it is too cold, the general reaction rate will be too slow; chemical reactions generally slow down with lowered temperature. If the temperature rises too high, the hydrogen bonds and water shells of the enzyme (a protein) will be disrupted. If such a disruption occurs, the protein is *denatured,* and its enzymatic ability is permanently destroyed. A temperature of 55°C is enough to denature many enzymes. Second, the acidity of the cellular environment affects the enzyme reaction rate. Too low or too high a pH will inhibit enzyme action. Each enzyme has an *optimal pH* at which it works fastest, and consequently pH regulation in a cell is essential to its survival. Third, the concentration of substrate must be sufficient. One might compare the enzyme-substrate relation to bricklayers building a wall. Three bricklayers (the enzymes) can build a wall out of bricks (the substrate) three times as fast as one bricklayer can, provided there are plenty of bricks. But having more bricklayers is not going to speed up the construction if each bricklayer has to spend most of the time scouting around for bricks.

An additional feature of enzymes is their *reversibility,* or ability to drive a reaction in either of two directions. Under proper circumstances, the same enzyme is capable of breaking a compound into its components or recombining them to make the original compound:

$$X + Y \xrightarrow{\text{enzyme}} Z,$$

and

$$Z \xrightarrow{\text{enzyme}} X + Y.$$

Because cellular activity depends on enzyme action, one can say that all life does too.

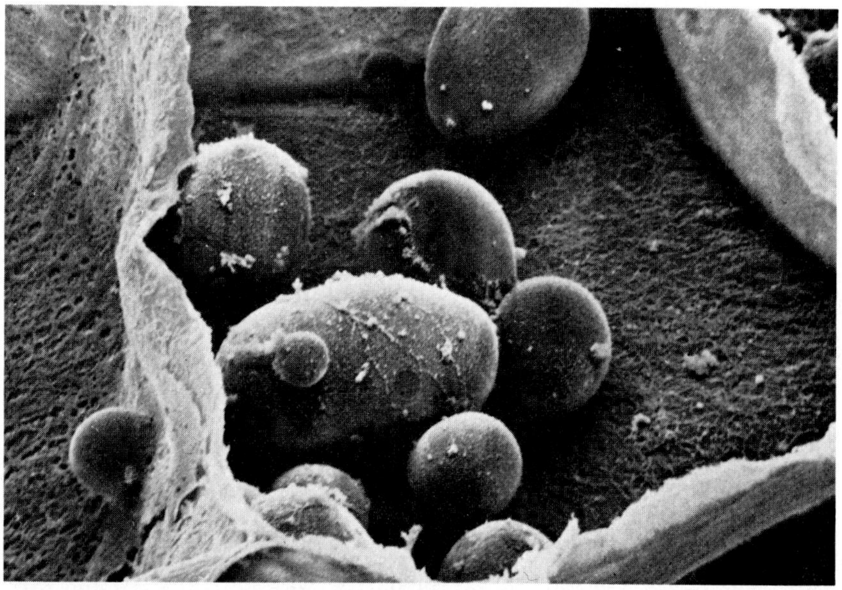

(a) × 10,000

Figure 1.16
Three important polysaccharides. (a) Starch granules from a potato are shown in this scanning electron micrograph. Plants store glucose in the form of starch. (b) Cellulose fibers in the wall of a simple green plant (an alga) are magnified 20,000 times in this electron micrograph. Cellulose gives strength, flexibility, and elasticity to plant walls. (c) Glycogen is the stored form of sugar in human beings and many other animals. It is converted to usable glucose when energy is needed. This electron micrograph of a liver cell shows glycogen as dark spots.

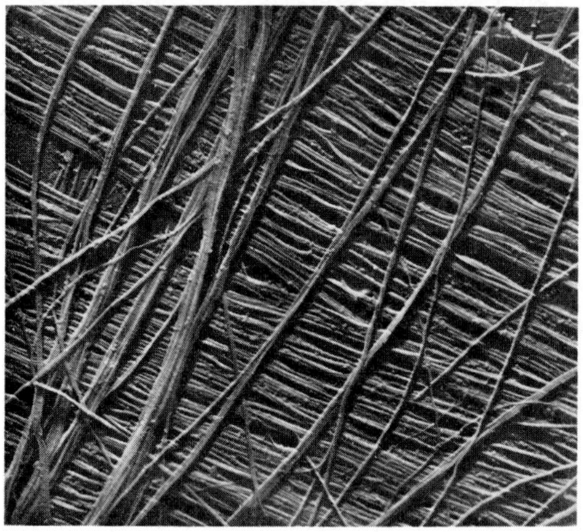

(b) × 20,000

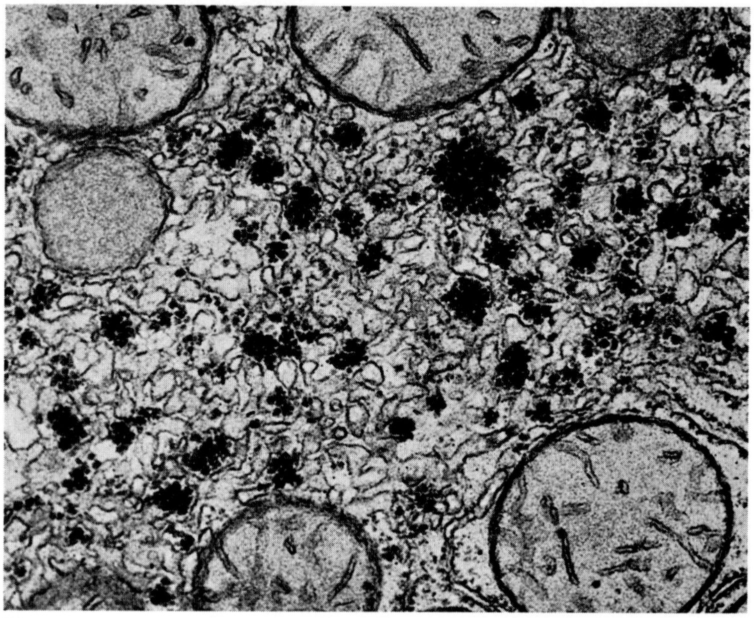

(c) × 12,000

saccharide chains (also called *polymers*) may contain hundreds or thousands of simple carbohydrate molecules. The general formula for such a polysaccharide as starch is $(C_6H_{10}O_5)_n$, where n equals the number of glucose units in the molecule.

In the living world there are three very important polysaccharides, all of which are made up of glucose: (1) starch (the storage form of glucose in plants), (2) cellulose (the most abundant material in plant cell walls), and (3) glycogen (the primary form of glucose storage in animal cells) (Figure 1.16). Although all of these polysaccharides are made up of units of glucose, they have very different chemical and physical properties because the glucose units in the three polysaccharides are linked to one another in different ways. Both starch and glycogen can be used as sources of food by animal cells, but cellulose is not usually digestible by animals. Only a very few organisms (snails, wood-rotting fungi, and some bacteria) are able to digest cellulose to produce glucose molecules that can be used for food. Human beings and most other animals would starve on a

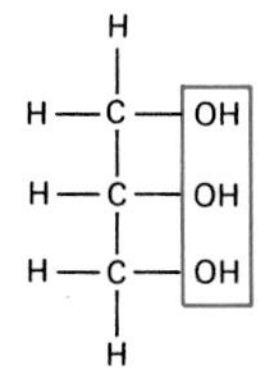

(a) GLYCEROL, $C_3H_8O_3$

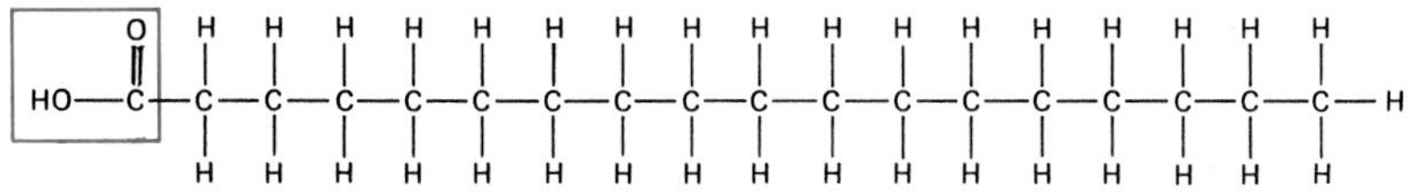

(b) FATTY ACID (STEARIC ACID, $C_{17}H_{35}COOH$)

Figure 1.17
Glycerol and fatty acid structures. (a) A glycerol molecule with its three hydroxyl (OH) groups outlined in color. (b) A fatty acid, stearic acid, showing its carboxyl group (COOH) outlined in color. Glycerol and fatty acids are the building blocks of fats.

diet of cellulose in spite of its glucose content. There is no human enzyme that can hydrolyze cellulose to produce glucose units.

Lipids

Lipids are a diverse group of biological compounds. One cannot readily characterize them except to say that they will not dissolve in water. They serve as concentrated food storage materials, as structural components of cells, and as regulatory chemicals. Familiar lipids are fats, oils, and waxes; less commonly known but equally important are the phospholipids and steroids.

Fats **Fats** (usually known as *neutral fats* because they are not electrically charged) are mainly reserve foods. Each fat molecule contains the building blocks *glycerol* (also called *glycerine*), and usually three molecules of *fatty acids* (Figure 1.17). The glycerol molecule contains three hydroxyl (OH) groups, and fatty acids contain a carboxyl (COOH) group, shown in Figure 1.17 as

$$HO-\overset{\overset{\displaystyle O}{\|}}{C}-$$

It is the hydroxyl and carboxyl groups that make the formation of a fat possible (Figure 1.18).

Oils are fats that are liquid at room temperature. *Waxes* are fats compounded with some "backbone" other than glycerol; lanolin, used in cosmetics, is an example of a wax, and waxes also occur as protec-

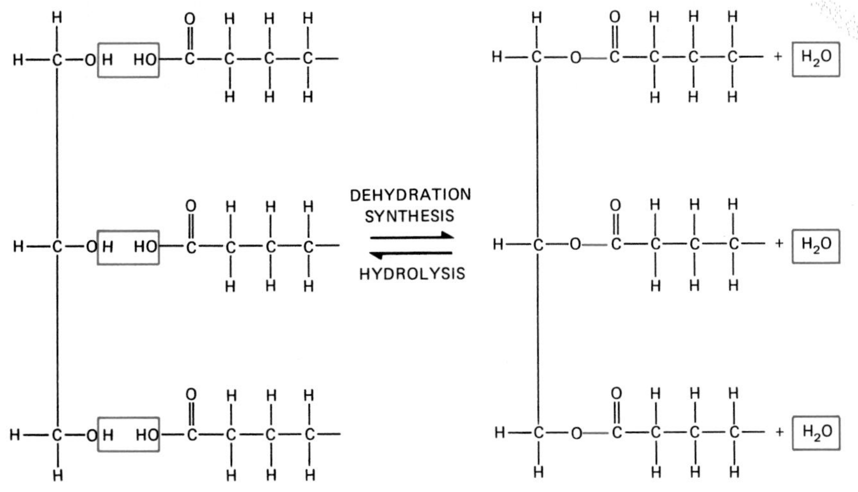

GLYCEROL + FATTY ACIDS = FAT + WATER

Figure 1.18
The dehydration synthesis of a fat. Three water molecules (outlined in color on the left) are removed from glycerol and three fatty acids. The bonding of the fatty acids to glycerol yields a fat and three molecules of water (each outlined in color on the right).

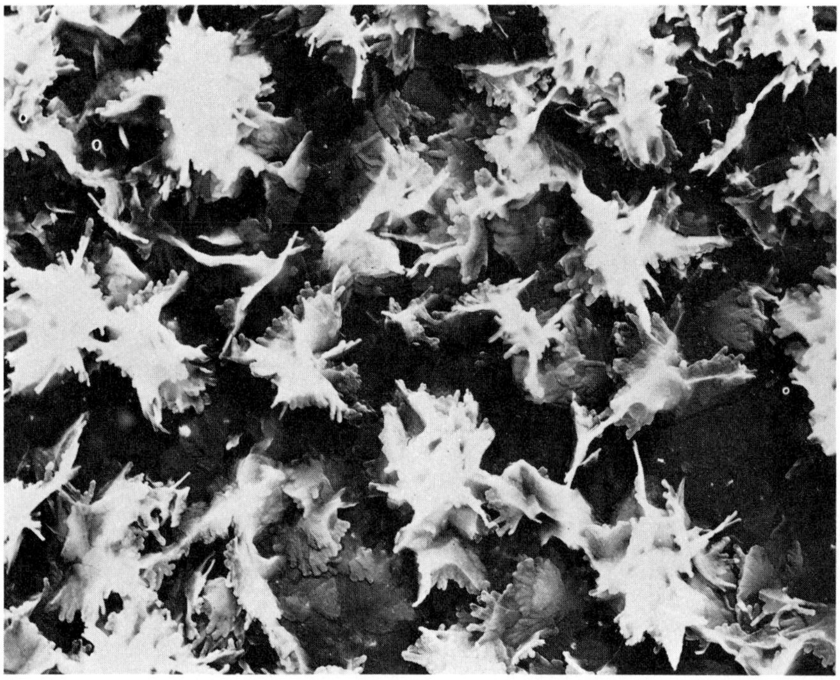

× 14,000

Figure 1.19
Waxes are lipids that usually have a protective function in plants and animals. This scanning electron micrograph shows wax on the surface of a eucalyptus leaf. The waxy deposits, which are produced by the leaves of land plants, prevent excessive water evaporation.

tive coatings on leaves, fur, feathers, and other surfaces (Figure 1.19).

Because there is an apparent relation between dietary fats and human circulatory diseases, emphasis is sometimes put on the kinds of fats eaten, with unsaturated fats being preferred by some dietitians. A *saturated fat* is one whose carbon atoms all have at least two hydrogen atoms, with no double bonds between adjacent carbons. (In single bonds, two electrons are shared; in double bonds, four electrons are shared. A double bond is indicated by a double line between the symbols for the two atoms.)

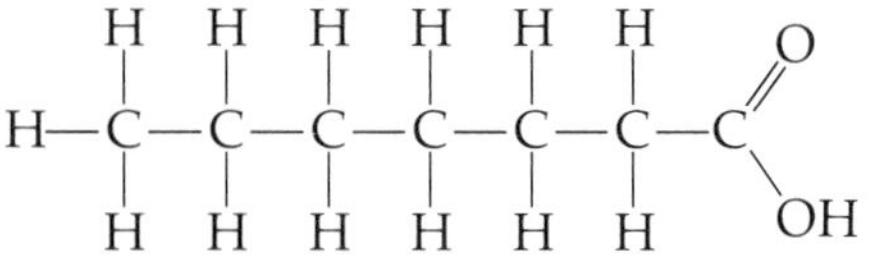

An *unsaturated fat* has double bonds between some adjacent carbons instead of having hydrogen atoms attached. Unsaturated fats are liquid at room temperature, and are considered oils.

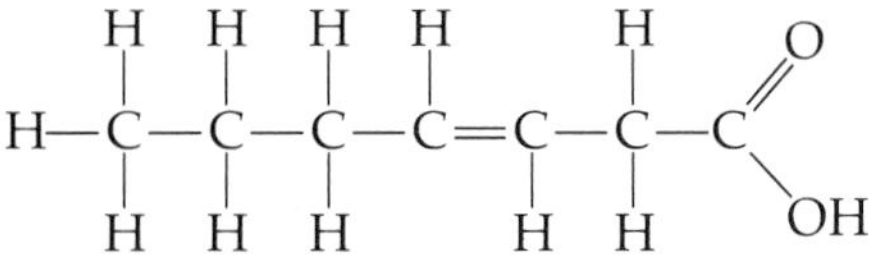

The greater the number of double bonds, the more unsaturated or oily the fats will be. For this reason, unsaturated vegetable oils are referred to as *poly*-unsaturates. Animal fats typically contain more saturated fatty acids than vegetable fats do. *Essential fatty acids* are those that must be obtained from food because the human body cannot synthesize them.

Phospholipids **Phospholipids** are a group of fatty compounds that contain fatty acids, glycerol, and a negatively charged phosphate (Figure 1.20). The long carbon and hydrogen chains of the fatty acids have no attraction for water, but the charged phosphate group has a strong attraction for polar water molecules. If a single phospholipid molecule is placed in water, water will tend to be attracted to the charged phosphate group and avoid the hydrocarbon chains. If many phospholipid molecules are placed in water, their long hydrocarbon chains will tend to associate together and expose the phosphate groups to the surrounding water.

Cell membranes are built up of phospholipids and proteins. However, it is the phospholipids that give membranes their unique sheetlike structure and many of their functional properties. In cell membranes, phospholipids are arranged so that the charged phosphate groups are on the outside and the fat portions are in contact with one another in the interior of the membrane (Figure 1.21). Later we will discuss further the structure and many roles of cell membranes (p. 53). But it is important to remember that a membrane's shape and many of its

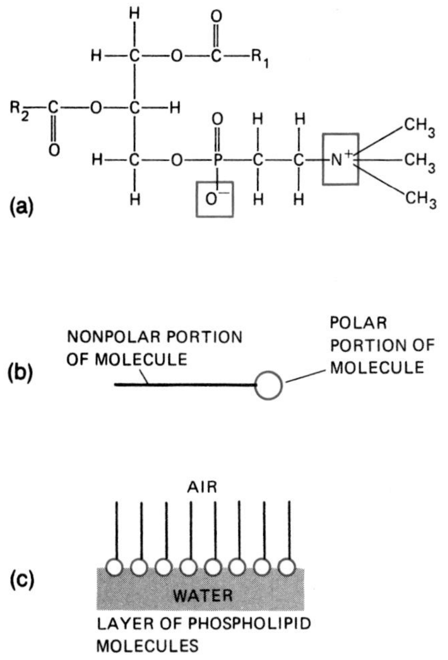

Figure 1.20
Phospholipids. (a) Structural formula of a phospholipid. The usual glycerol backbone of three carbon atoms is present, and two of them bear fatty acids, represented by R_1 and R_2 in the figure. The third carbon atom has attached to it a group of atoms (of the compound *choline*) containing nitrogen and phosphorus. As shown in the colored rectangles, two of the atoms may be charged, giving this portion of the entire molecule the quality of polarity. The bulk of the phospholipid molecule is nonpolar. (b) One commonly used model of a phospholipid molecule. (c) Diagram of the formation of a *monomolecular layer* of a phospholipid on a water surface. The polar portion of the phospholipid is in the water (because water, too, is polar), and the nonpolar portion of the phospholipid, which is not attracted to water, sticks up into the air. Such films are similar to those formed in cells.

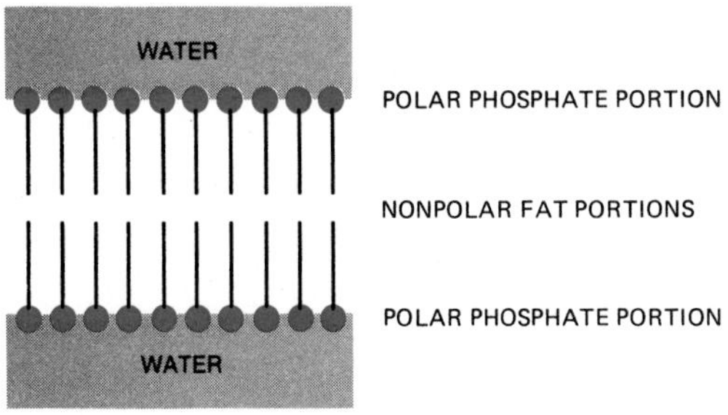

Figure 1.21
One end of a phospholipid molecule is polar and soluble in water, while the other end is nonpolar and insoluble. This tendency to be soluble at one end and insoluble at the other is important in cell membranes, which are built of phospholipids and proteins.

properties are due to the shape, size, and properties of the phospholipid molecules out of which it is made.

Steroids **Steroids** are lipids that do not contain fatty acids. They are complex molecules composed of four interlocking carbon rings. Three of the carbon rings are cyclohexane (six-sided) rings, and the fourth is a cyclopentane (five-sided) ring (Figure 1.22). Each particular steroid has its own side-group combination of carbon and hydrogen.

Steroids are important to our bodies in many ways. The male and female sex hormones, and the adrenal gland hormones are steroids; Vitamin D, which helps to regulate calcium metabolism, is a steroid; cholesterol, which is important in membranes and in brain and nervous tissue, is a steroid; and bile acids, which aid in digestion and the absorption of fats, are also steroids.

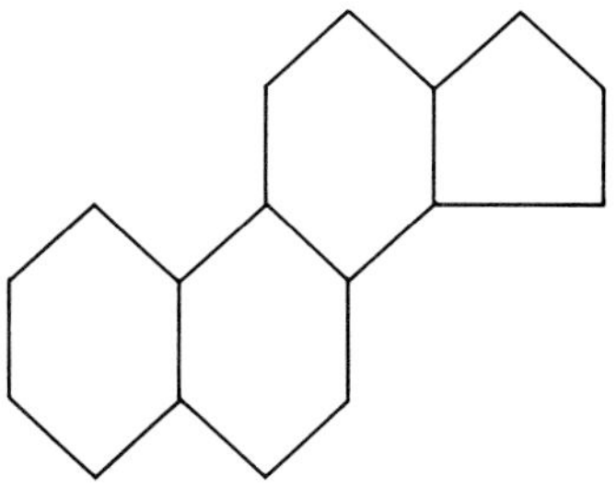

Figure 1.22
Steroids are formed with four interlocking carbon rings. Note that one of the rings is five-sided, while the other three are six-sided.

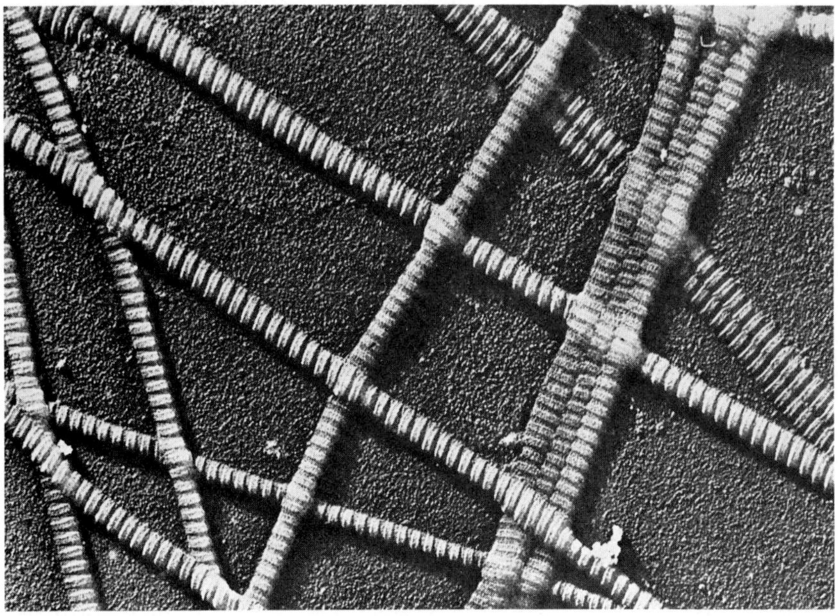

× 35,000

Figure 1.23
Collagen (KAHL-uh-jehn) is a complex structural protein that is often arranged in interlacing bundles of tough fibers to provide strength and flexibility to skin, ligaments, bones, and other such connective tissues that undergo great physical stress. It is thought to be the most abundant animal protein.

Proteins

Proteins are extremely important to our modern understanding of how living things are structured and how they work. Protein molecules are large compared with such molecules as sucrose (Figure 1.23). A protein molecule may be a million times heavier than a sucrose molecule. For this reason the proteins are often referred to as *macromolecules* (*macro* = large).

Amino acids A protein is made up of units called **amino acids** that link together, one to another, forming a chain. Such a chain of amino acids is a *polypeptide.* Twenty different amino acids are used in the cellular synthesis of proteins. Actually, chemists can make more than 20 amino acids, but only 20 are important in living systems. All amino acids have certain chemical characteristics in common: a carboxyl group, —COOH, which gives them their acidic nature, and a nitrogen-containing amino group, $—NH_2$. Figure 1.24 shows the general structure of an amino acid, where R could be any one of 20 different chemical groups.

The 20 amino acids in living systems differ in the particular chemical nature of the R (for *radical* or *root*) group. In the amino acid glycine, R is a single hydrogen atom. In other amino acids the R group is much larger, consisting of a chain or a ring of carbon atoms to which other acidic, basic, or polar (charged) groups may be attached. Because of their different chemical composition and structure these groups have very different chemical properties. It is the character of the R group that confers on each amino acid its unique properties and distinguishes one amino acid from another.

We have said that amino acids are linked, one to the other, to form proteins. How does this occur? The exact process by which living cells manufacture proteins will be discussed in Chapter 7. But here we want mainly to emphasize the nature of the chemical bond between amino acids. Figure 1.25 shows the condensation of two amino acid molecules into a *single molecule.* As indicated in the figure, the reaction occurs between the carboxyl group of one amino acid and the amino group of the other. In the

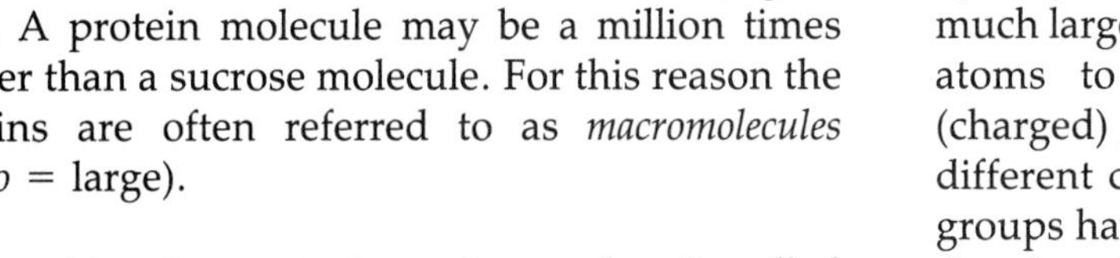

Figure 1.24
The structure of an amino acid.
(a) The general structure of an amino acid, with R indicating a variable group of chemicals. (b) The same general amino acid structure, with the amino group, variable side group, and carboxyl group clearly identified.

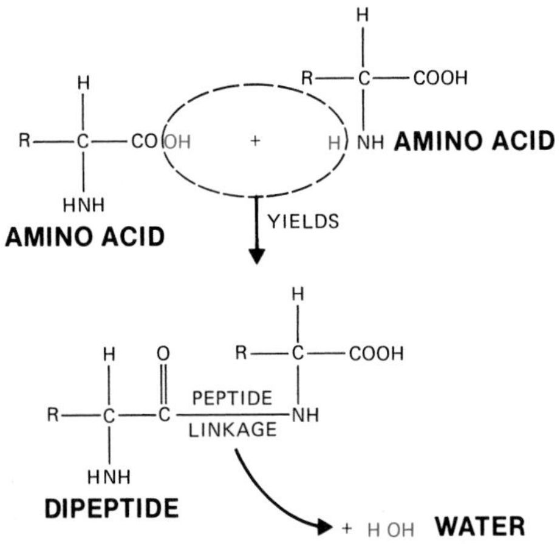

Figure 1.25
Condensation of two amino acids. The −OH (shown in color) of the carboxyl group of one amino acid and −H (also in color) of the amino group of a second amino acid are removed, forming water. The two amino acids are bound into one molecule, a *dipeptide,* through the resulting *peptide linkage.* There is still an available amino group at one end and a carboxyl group at the other end of the new molecule, and to either or both of these groups more amino acids can be attached. Thus long chains of amino acids can be built into *polypeptides.*

process, water (H_2O) is split out. Such a union results in the formation of a strong covalent bond between the two amino acids. In proteins this bond is referred to as the **peptide linkage.** You will note that there is still a carboxyl group at one end of the new molecule and an amino group at the other end. Both are available sites for reaction with additional amino acids. In this fashion the number of amino acids that can be linked one to another is enormous.

The levels of protein structure The specific sequence of amino acids in a protein constitutes the protein's *primary structure* (Figure 1.26). The primary structure determines the fundamental nature of the polypeptide. For example, a polypeptide consisting of 20 identical amino acids will have certain chemical and physical properties. However, a polypeptide made up of 20 different amino acids will have very different chemical and physical properties. Furthermore, if the 20 different amino acids are arranged in a different sequence, the polypeptide will have yet another set of chemical and physical properties. We can therefore see that the composition and sequence

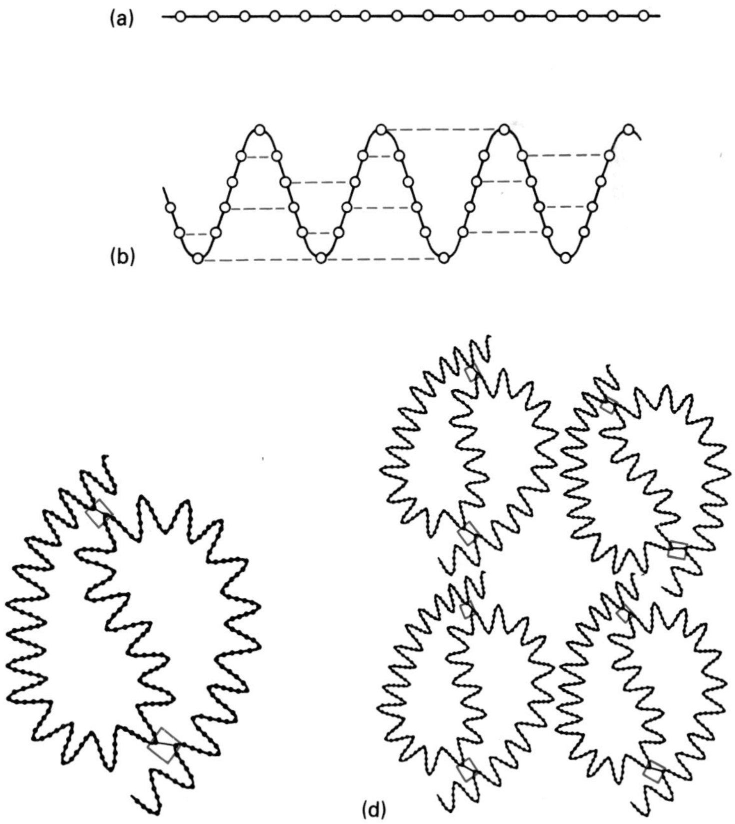

Figure 1.26
Levels of organization in proteins. (a) The *primary* structure of a protein (polypeptide) is determined by the sequence of amino acids in the polypeptide chain. Natural polypeptides contain from dozens to hundreds of amino acids. (b) The *secondary* structure of a polypeptide is brought about by the formation of hydrogen bonds between amino acids along the chain. One frequent result of such hydrogen bonding is the formation of a helix. (c) The *tertiary* structure of a protein appears when two cysteine (sulfur-containing amino acid) portions of a polypeptide chain come close together. A sulfur-to-sulfur bond (a disulfide bridge) forms, holding the polypeptide chain firmly at that point (colored rectangles). (d) The *quaternary* structure of some proteins is the result of the combination of several protein molecules into one supermolecule.

PROTEIN AND PERMANENT WAVES

When a hair shaft is produced by the cells of the human epidermis, its overall shape is determined by the shape of its protein subfibrils (a). The main constituent protein, *keratin,* contains a relatively large amount of the sulfur-containing amino acid cysteine. About one of every ten amino acids in keratin is cysteine. The sulfur of one cysteine can make an —S—S— bond, or disulfide bridge, with another cysteine, thereby binding the entire protein molecule into a firm shape (b). It follows that a high proportion of cysteine residues (component amino acids) makes a stable molecule and consequently a durable hair shape.

If the hair shaft is straight, it will remain straight because of the stability of the disulfide bridges, indicated in color in (1). Those bridges, however, can be broken by heat or chemically, and restored at will. These facts are applied in giving "permanent waves." A person who has straight hair and wants it curled or has curled hair and wants it straight can have the disulfide bridges broken by heat—unpleasant and hard on the hair—or by application of chemical solutions that accomplish the same effect (2). While the bridges are broken, the hair is bent or "set" in the desired form (3). Then it is either cooled or treated with a second solution that restores the bridges (4), and the hair will stay put. New hair pushed out from the hair follicle will have the old natural shape, because the artificial treatment affects only hair that has already grown.

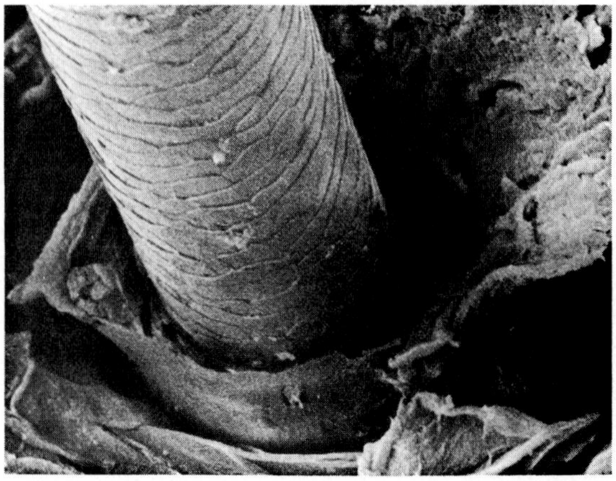

(a) × 1000

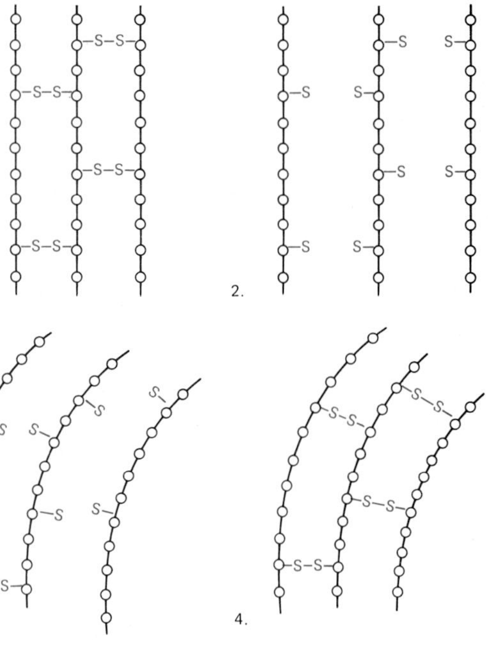

(b) 1. 2. 3. 4.

of amino acids in a protein are very important to its structure and function. When a chain of amino acids twists into a helix, the protein molecule is shaped roughly like a coiled spring. The coil is maintained by hydrogen bonds between the various amino acids. This coiling is referred to as the *secondary structure* of protein.

In very large proteins the coils tend to fold back on themselves. This folding gives the protein its *tertiary structure.* The particular chemical nature of the different R groups of amino acids plays a very important role in determining just how the helical regions fold to produce the tertiary structure. Proteins with different amino acid compositions will fold into different shapes and have different tertiary structures. Covalent, ionic, and hydrogen bondings between R groups are important in maintaining a stable tertiary structure.

The particular biological role of a protein depends on its three-dimensional tertiary structure. If that structure is altered, the appearance, properties, and biological activity of the protein will be destroyed. Consider the familiar "egg white" of an uncooked hen's egg, which contains the protein albumen. When an egg is boiled or otherwise subjected to heat, its three-dimensional structure changes profoundly. A cooked egg cannot be "uncooked." The change is due to the disruption of the secondary and tertiary structure of the albumen protein by the high heat.

Still another level of protein structure is possible, the *quaternary structure*. Several folded proteins, each with its own primary, secondary, and tertiary structure, may be loosely joined into a "superprotein." The molecule hemoglobin, responsible for the transport of oxygen in the blood, is such a "superprotein" and is composed of four subunits (see Figure 1.26).

Certain proteins may also join with ions, sugars, or vitamins to produce conjugated proteins. Hemoglobin is an important example of such a protein. It contains iron, which is important to the biological function of the protein. The protein component of hemoglobin cannot work without the iron, and the iron cannot work without the protein.

TO SUM UP PROTEINS

1. Amino acids make up proteins. Different proteins have different amino acid composition.
2. Amino acids are linked, one to the other, through covalent bonds (the peptide linkage).
3. What is linked together is then coiled. The coil is held together by hydrogen bonds. Such coiling is characteristic of many proteins, no matter what their amino acid composition.
4. What is coiled is then folded. Proteins can be folded into an enormous number of shapes. The particular folded structure is determined by the R groups on the amino acids.

Nucleic Acids

The last group of compounds in our "basic vocabulary" of biological materials, the **nucleic acids,** are such special molecules that they will be considered by themselves in Chapters 6 and 7.

BIOLOGICAL LEVELS OF ORGANIZATION

One of the lessons learned from the scientific study of nature is that there are levels of organization, from subatomic particles through atoms and molecules to cells, tissues, organs, whole organisms, societies, ecosystems, solar systems, and galaxies. In the study of biology we must relate these levels to one another (except the last two, with which we are scarcely concerned).

One point must be kept in mind throughout the consideration of all biological study: The whole is more than the simple sum of its parts.* For example, all the materials necessary to make up the human blood pigment, hemoglobin, are present in a rusty nail and a breath of air, but these atoms of iron, hydrogen, oxygen, carbon, and nitrogen cannot function as hemoglobin does unless they are organized into a hemoglobin molecule. Atoms that exist separately from one another act in a way that is totally different from what they do when they are part of a hemoglobin molecule. In passing from an atomic to a molecular state, they reach a new level of organization, and at that new level we find features that did not exist in the lower level.

To take the hemoglobin example a step further, we see that an isolated molecule of hemoglobin is not functional unless it is part of a blood cell. Just as a molecule is more than a collection of atoms, so a cell is more than a collection of molecules. You can kill a cell by squeezing it because you break up its organization; that is, you disrupt the level at which it operates. A dead cell has all the molecules it had before it was killed, but it does not function as a cell. Further, a mass of cells, perhaps growing artificially in a culture dish, cannot perform the duties of, say, a kidney, because a working organ such as a kidney is organized at an organ level, and has greater abilities than a mass of cells. On more complex levels, an entire organism can do things no organ can, and an organized community of organisms is more than simply a collection of individuals.

We started the study of biology by learning the main characteristics of the simpler particles, atoms and molecules. Now we study the more complex entities: cells, tissues, organs, organisms, and communities.

* British physicist Sir Arthur Eddington said in his book *The Nature of Physics*, "We often think that when we have completed our study of *one* we know all about *two*, because *two* is *one* and *one*. We forget that we still have to make a study of *and*."

SUMMARY

1. Everything in the universe is made up of *matter,* which is composed of *elements,* substances that cannot be broken down by ordinary chemical means into simpler particles. An *atom* is the smallest unit of an element that retains the chemical characteristics of that element.
2. Atoms are made up of a *nucleus,* with its electrically charged *protons* and uncharged *neutrons,* and *electrons* orbiting around the nucleus. The chemical and physical properties of an element are determined by the number of protons and neutrons, and by the number and arrangement of electrons.
3. *Isotopes* of an element are atoms with the same number of protons and electrons, but different numbers of neutrons.
4. In heavier elements the electrons are distributed as though they were in layers, known as *energy-level shells.* All biological actions involve changes in electron positions and consequent gains or losses of energy.
5. A *molecule* is the smallest unit into which a substance can be divided and still retain the properties of that substance. When two or more different atoms combine, they form a *compound.*
6. Atoms held together by the gaining, losing, or sharing of electrons are united by chemical *bonds.* The important chemical bonds are *covalent, ionic,* and *hydrogen* bonds. Many large biologically active molecules are held in proper shape by hydrogen bonds.
7. In a *chemical reaction,* bonds between atoms are broken or formed, and different combinations of atoms result. Some biologically important chemical reactions are *oxidation-reduction, hydrolysis,* and *condensation.*
8. An *ion* is a particle with an electrical charge; it may be a single atom or a group of atoms.
9. *Acids* are compounds that release hydrogen ions, and *bases* are compounds that accept hydrogen ions.
10. The separation of a molecule into ions in water is *dissociation;* the recombining of the ions to form an uncharged molecule is *association.* Dissociation cannot take place without water.
11. Acidity can be expressed numerically on a *pH scale,* which is an indication of the concentration of hydrogen ions free in water. A *buffer* is a compound that resists pH change even if a strong acid or base is added.
12. Some important biological molecules are *water, carbohydrates, lipids, proteins,* and *nucleic acids.* A protein is made up of units called *amino acids* that link together into a chain called a *polypeptide.*
13. Carbon is an especially important element because it is a part of all biological molecules and participates in all biological reactions. Compounds containing carbon are *organic,* and compounds not containing carbon are *inorganic.*
14. An *enzyme* is a protein that speeds up a chemical reaction without permanently entering into the reaction itself.

ASK YOURSELF

1. How do two isotopes of an element (for example C^{12} and C^{14}) differ in their atomic structure?
2. What environmental factors can affect the activity of an enzyme?
3. Explain how it is possible for a small quantity of an enzyme to act upon a large quantity of a substrate.
4. What energy change occurs when an electron in an atom moves from an outer to an inner energy-level shell?
5. What is the difference between a neutral atom and an ion?
6. What is the commonest type of chemical bond in biological compounds?
7. What one compound is always involved in a hydrolytic reaction?
8. If a compound is oxidized, what happens to it with respect to electrons? What if the compound is reduced?
9. Why is pH regulation important in a living system?
10. What are the main biological uses of carbohydrates?
11. What is a saturated fat? What is an unsaturated fat?
12. What is the difference in electrical charge between a neutral fat and a phospholipid?
13. What is meant by the "primary structure" of a protein? The "secondary structure"? The "tertiary structure"?

2
Energy for Life

SOME KEY POINTS

1. Life is impossible without a supply of energy. Our most important energy source is the sun, which makes photosynthesis possible and provides a suitable temperature range for life on earth.
2. Energy can be changed from one form to another, but it cannot be created or destroyed. Every time an energy-releasing or energy-requiring action occurs in an organism, some energy is converted to heat and is not available to do work.
3. Energy stored in ATP molecules is readily available as usable energy when it is needed by living organisms.
4. The main energy-transforming processes, photosynthesis and respiration, are made possible by the transfer of electrons from one compound to another.
5. Cells are able to rearrange chemical bonds to release stored energy.

ENERGY IS USUALLY DEFINED AS THE CAPACITY to do work. Biologically, we can translate this *capacity* to do work into its simplest terms: Energy is used in the operation of all life processes. No animal moves, no plant grows, no bacterium multiplies, no cell divides without a continual supply of energy.

Our most important source of energy is the sun, 150 million kilometers (93 million miles) away. The sun's energy enters the biological world by way of green plants, which use sunlight to create their food supply. This process, known as *photosynthesis,* will be discussed in detail in Chapter 4. Animals are unable to manufacture food by the use of light energy, and they depend on consumption of plants or other animals for their survival. But besides merely eating plants and animals, animals depend on the sun for their oxygen supply, which is obtained from plants as a by-product of photosynthesis.

The sun furnishes energy for more than photosynthesis. Its energy also provides the warmth that keeps the earth's temperature in a range suitable for life. When it is too cold, enzymes do not function. Equally important, if the sun did not yield heat, water would be frozen and would not be recycled by wind and rain.

Plants can change light, or *radiant,* energy from the sun into *chemical* energy in food (Figure 2.1). When a plant makes food, it stores energy in food molecules, specifically in the bonds that hold the atoms together. The stored energy is called *bond energy.* While plants are converting the light energy into the bond energy in chemical compounds, they are at the same time giving off the oxygen required by animals (and plants, too). Animals, and plants in the absence of light, give off carbon dioxide, which is taken in by plants in light. Thus the carbon is re-

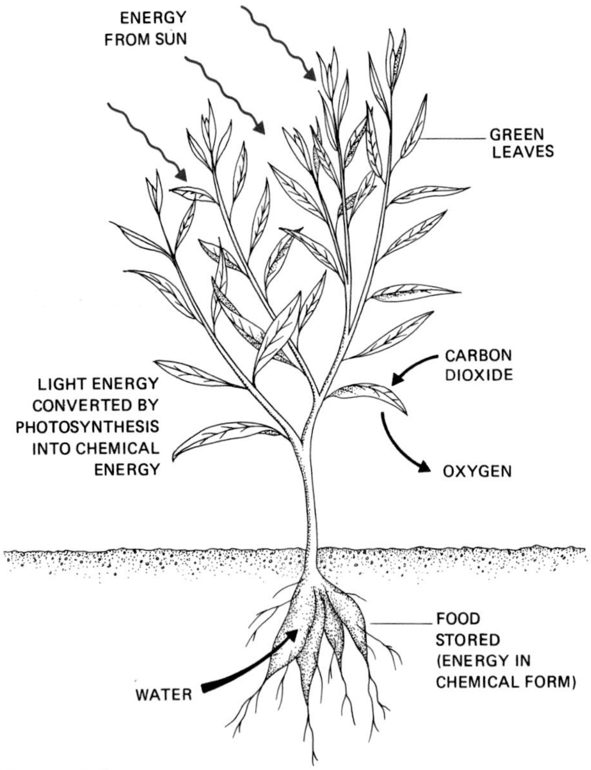

Figure 2.1
Diagram of the conversion of light energy from the sun to chemical energy of stored food. The green plant contains the necessary components to combine the elements of carbon dioxide and water to make food and give off gaseous oxygen.

cycled. The energy for the recycling is continually fed in from the sun. When energy contained in food is released, it can be converted into other forms such as heat, movement, electricity, chemical energy of new compounds, or even back to light once again.

SOME PRINCIPLES OF ENERGY CONVERSION

In general, and certainly as far as living systems are concerned, *energy can be changed from one form to another, but it cannot be created or destroyed.* This is a simple statement of the **First Law of Thermodynamics,** a "law" that biological systems adhere to with absolute strictness. If an insect eats a leaf, all the energy stored in that leaf still exists somewhere, even after the insect has digested the leaf and made some of it into insect tissue. Some of the energy is in the chemical bonds of molecules, some in heat dissipated into the air. Any example of energy consumption and conversion would follow the same rule.

Green plants do not create energy when they synthesize food during photosynthesis, nor do animals create energy when they use food for metabolic activities. They are only converting the energy of sunlight into different forms.

At the same time, any energy conversion involves some loss of energy as far as the converting system is concerned. (Actually, the "lost" energy has been converted to heat energy that dissipates into the surroundings in the process. Remember, energy is neither created nor destroyed.) The insect mentioned above, having eaten the leaf, ends up with less energy in its cellular components than there was originally in the leaf. Some heat is "lost" at every step along the way as the insect digests the leaf molecules and recombines their atoms into new forms. If the insect could capture all the potential energy of its food and turn it into useful work, it would be 100 percent efficient in anything it did (Figure 2.2).

Every time an energy-releasing or energy-requiring action occurs, some energy is "lost" as heat and is not available to do work. That is the **Second Law of Thermodynamics,** and it, too, is "obeyed" absolutely by living things. It is worth noting that, as with any natural "law," if some organism could "break" it, the law would simply cease to be, since the so-called natural laws are nothing more than our observations of how the universe operates.

The highly ordered system of cells, tissues, organisms, and populations of organisms that make up the living world is achieved only by the continued input of energy. It may seem at first glance that the living world does indeed break, or at least bypass, the Second Law of Thermodynamics, because as a plant grows, it obviously accumulates energy. However, it does not break any energy laws. The plant receives much more energy from the sun than it succeeds in storing.

Figure 2.2
Relative efficiencies of some machines compared with the main biological energy converters.

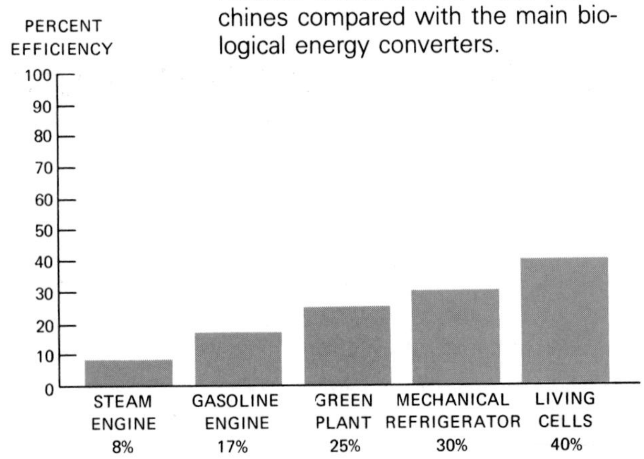

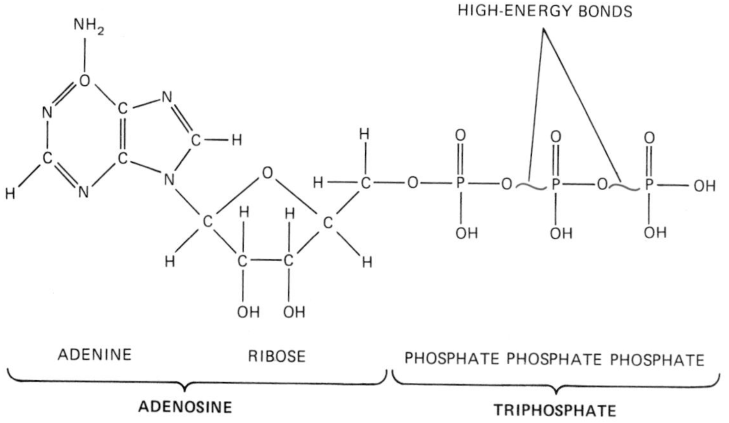

Figure 2.3
A structural diagram of an ATP molecule.

HOW IS ENERGY MEASURED?

Energy is measured in several ways. Some energy units are joules for electric current, ergs for mechanical devices, BTUs (British Thermal Units) for coal and oil, and calories for food. Since these units are all interconvertible, and since it is convenient to use a common term, biologists usually use the calorie as the basic energy unit. A **calorie** (from the Latin *calor*, heat) is the amount of energy required to raise the temperature of a gram of water one degree Celsius, specifically from 14.5° to 15.5°. A **kilocalorie,** written Kcal, or more commonly Calorie, is a thousand calories.

In popular literature, as in diet instructions and on food labels, a calorie is in fact almost always a kilocalorie. Thus, if a commercial breakfast cereal is said to contain 110 calories per ounce, it means 110,000 calories, or 110 Kcal. Biologists think and write in terms of Calories almost exclusively, whether they are concerned with energy from the sun, the formation of carbohydrates in a green plant, or the release of energy by a respiring cell.

In practice, the calorie content of materials, especially foods, is measured by burning a weighed sample in a well-insulated container called a calorimeter, and determining how much heat is released during a chemical reaction. This is done by measuring the temperature change of a given quantity of water. The energy requirements of many biological actions, however, are measured by a combination of experimental and theoretical means that do not always yield the same answers.

Since the energy values for various compounds are usually determined experimentally, they are not known absolutely. The most recent figures are those commonly given because it is assumed that new techniques are more accurate. At present, for example, the energy content of a mole* of glucose is considered to be 686 Kcal, and that of a mole of ATP to be 7.3 Kcal, but these figures, though reasonably close to correct, are subject to further refinement. Students need not be surprised to find that not all workers and all books use exactly the same numbers for the energy content of various biologically useful compounds.

THE RELEASE OF BIOLOGICAL ENERGY IS USUALLY SLOW AND STEADY

Living systems all convert energy in their own ways, and yet the basic chemical process is the same. When a burning piece of wood gives off heat, it is actually releasing energy that was contained within its chemical bonds. It would be simple if energy could be released from a cell by burning. But burning produces temperatures too high for a living cell to endure, and the burst of energy release is too short and too sudden. Organisms need a slow, steady energy source, and especially one that can be controlled and regulated, since the chemical reactions in organisms can use energy only in small packages. Cells have effective ways of regulating energy release.

Living things—from bacteria to orchids to humans—for the bulk of their energy requirements use a small organic molecule, adenosine triphosphate, customarily abbreviated as **ATP.** The ATP molecule stores energy in chemical bonds, where the energy remains locked up until it is released by rearrangement of those bonds (Figure 2.3). Al-

* A *mole* is the quantity of a compound, in grams, equal in number to the combined weight of all the atoms in a molecule of that compound. Moles are used because a mole contains a standard number (Avogadro's number, 6.023×10^{23}) of molecules, and moles of all compounds are comparable.

BIOLOGICAL LUMINESCENCE

One striking biological example of the conversion of energy from one form to another is the conversion of chemical energy to light energy. Many organisms, from bacteria to fishes that dwell in salt water, are luminous, but no luminous organisms are found among higher plants, amphibians, reptiles, birds, or mammals. This direct conversion of chemical energy to light energy is known as **biological luminescence,** or **bioluminescence.**

Aristotle and other early observers knew about bioluminescence and studied it, but it was not until 1887 that the French physiologist Raphael Dubois experimented with a luminous clam and isolated a substance he named *luciferin,* after Lucifer, the light-bearer. Luciferin emits light in hot water, and another substance, an enzyme that Dubois called *luciferase,* enables luciferin to emit light in cold water. Early in this century, E. Newton Harvey of Princeton University confirmed that bioluminescence is an enzyme-controlled process. He showed how to produce luminescence by adding water to the powder derived from a dried crustacean. Such powder, which contained both luciferin and luciferase, was later used by Japanese soldiers during World War II as a low-intensity light source when a stronger light might have revealed their field positions.

Although much information has been accumulated about the luciferin-luciferase combination, the principal unanswered question remains: Exactly how do so many diverse organisms produce their eerie and mysterious lights?

Some species use their light-producing ability as an alarm system, others use it to lure their prey, and still others, like fireflies, incorporate it into their mating rituals. One type of squid emits a radiant cloud instead of a screen of ink when it wants to elude a predator. An angler fish lures its deep-sea prey into its luminous jaws. Female marine worms near the Bermuda coast rise to the surface of the water exactly three days after a full moon and secrete a luminous circle that attracts the males, by then producing their own intermittent light. The nocturnal courtship ends with eggs and sperm being discharged into the glowing water. This relationship between mating periods, phases of the moon, and bioluminescence has been observed in many other organisms. In fact, bioluminescence can be predicted well enough to have become a scheduled tourist attraction in some Caribbean areas.

A "flashlight fish" that carries luminous bacteria under its eyes: Photoblepharon palpebratus. *About actual size.*

From "Flashlight Fishes" by John E. McCosker.

One deep-sea diver has observed that below depths of 400 meters (1300 ft), more than 95 percent of the fish are bioluminescent. Such findings suggest that some organisms in dark environments need to produce light to hunt their prey or simply to communicate. Simple organisms like fungi and bacteria can also be luminous, and several dramatic examples of symbiotic (Greek, "living together") luminescence have been cited, in which bacteria provide the light for their host. One fish carries around luminous bacteria near blood-rich vessels under its eyes (see the drawing). When the fish does not want the illumination, it can slide a darkening lid over the bacteria without obscuring its own vision.

Although much remains to be learned about the mechanism of bioluminescence, we know from studies of fireflies that two substances are always involved: oxygen and that ever-present energy producer, ATP.

---

though ATP stores energy as described, ATP itself is not stored and built up in ever-growing quantities. What is important is the ready *availability* and *transferability* of the energy in ATP.

The ATP molecule is built up from smaller subunits of adenine (which we will meet again when we discuss the nucleic acids in Chapter 6), a five-carbon sugar (ribose), and three phosphate groups.

Most of the energy in ATP available for biological systems is contained in the energy-rich bonds between the last two phosphate groups. These bonds are called *energy-rich* because they yield their energy readily. When ATP reacts with water (*hydrolysis*), the last of the three phosphate groups is separated. Such a reaction yields adenosine *di*phosphate (ADP) plus inorganic phosphate (P_i) and energy. Most of the energy released from ATP can be used for the cell's immediate needs.

Suppose that a synthesis is required—say, the joining of two small amino acid molecules into a double molecule called a *dipeptide* (see Chapter 1). Such a joining requires energy, which can be obtained from the breakdown of nearby ATP.

1. $\text{ATP} \rightarrow \text{ADP} + P_i + \text{Energy}$.
2. Amino acid + Amino acid + Energy → Dipeptide.

Two reactions such as these, starting with one set of reactants and ending with different ones, are called **coupled reactions** because they are related through the passage of a phosphate group to an intermediate compound not shown in the equations as written. The first reaction *releases* energy, passing a phosphate group to an intermediate, which can then react to give the final result. The second reaction *requires* energy, which is provided by ATP. Energy flow in most biological reactions is made possible by coupled reactions involving the passages of phosphate from ATP.

If the generation of ATP from ADP and P_i requires an energy source, the question arises: How does the energy get into the ATP molecule? We know that some reactions in cells release energy, and when these reactions occur, the energy made available is transferred by intermediates to the ADP + P_i reaction. The result is energy-rich ATP. The action is reversed when ATP is hydrolyzed. The energy released is transferred to whatever energy-requiring reaction the cell is carrying on. In most instances, neither the resultant ADP nor the inorganic phosphate becomes attached to the products of the reaction energized. The ATP merely provides the energy for the reaction. A complete explanation of how ATP releases energy is not yet available.

Activities that require energy, such as the lifting of a weight or the cellular synthesis of a sugar molecule, are known as *endothermic* (heat in). Activities that liberate energy, such as allowing a weight to drop or breaking down a sugar molecule, are called *exothermic* (heat out). One extremely important exothermic biological reaction is

$$\text{ATP} \rightarrow \text{ADP} + P_i + \text{Energy}.$$

The reverse reaction synthesizes* ATP:

$$\text{ADP} + P_i + \text{Energy} \xrightarrow{\text{enzyme}} \text{ATP}.$$

The synthesis of ATP is an endothermic reaction because energy from sunlight or food is required for the reaction to take place.

Cells have relatively few ways of generating ATP, and of course they must obtain energy to do so. In contrast, all biological activities require energy, and it is usually obtained from ATP. Indeed, ATP in biological jargon is called the energy "currency" of the cell, as though it were cash to pay for the energy needs of the cell.

Although ATP is a common energy source for biological activity, it is not the only one. A number of other compounds are available, some of which are used for specific functions. Guanosine triphosphate (GTP), for example, is used instead of ATP in certain steps of protein synthesis. The energy for muscle contraction is frequently supplied by another phosphorus-containing compound, creatine phosphate. Still a third compound, phosphoenolpyruvate (PEP) is used in sugar synthesis. Indeed, it would be surprising to find that any biological process used one and only one compound. Life does not proceed so.

TO SUM UP ENERGY RELEASE

1. The ATP molecule stores energy in chemical bonds. When energy is needed, it is readily available for release by the rearrangement of those bonds.
2. Most of the available energy in ATP is contained in its last two phosphate groups. When ATP reacts with water, a phosphate group is separated, and usable energy is released along with ADP and inorganic phosphate.
3. Coupled reactions are those that occur together, when one reaction provides energy for the next.
4. Energy gets into an ATP molecule through transfer of energy to an ADP molecule and inorganic phosphate.

* "Synthesize" means to form a new, more complex compound from its simpler parts. *Synthesis* comes from a Greek word meaning "to put together."

THE PHYSICAL TRANSFER OF ENERGY REQUIRES A TRANSFER OF ELECTRONS

The two main energy-transferring processes in biology are *photosynthesis* and *respiration*. They are so important to life that they will be given special treatment in Chapters 4 and 5, but we will briefly discuss here some features they share that are fundamental to all the activities of living cells. They are both essentially ATP-generating processes. Photosynthesis uses light energy to make ATP, which is then used to make organic compounds. In respiration, the energy of organic compounds is used to make ATP, which is then used for the various needs of the cell.

The essential action in these processes is the transfer of electrons from one compound to another. In photosynthesis, light energy brings about the release of electrons from the green pigment, chlorophyll, which pass to some electron acceptor. As the electrons pass, the energy is transferred with them. When an electron goes from a molecule, that molecule is said to be **oxidized.** The receiving molecule is said to be **reduced** (Figure 2.4). *For every oxidation there is a reduction.* The oxidized molecule gives up energy, and the reduced molecule receives it. In living cells, energy transfers involve oxidation-reduction reactions, and it is by means of them that energy is kept flowing.

Life on earth is a watery life, and all active cells are bathed in abundant water. The water is partially ionized, so that instead of existing as H_2O, some of the molecules dissociate into *hydroxyl ions*, OH^- (bearing an extra, unshared electron), and *hydrogen ions* or *protons*, H^+ (lacking an electron). This is an oversimplification, but it will suffice for practical purposes. The H^+ ions are ready electron acceptors, and since they are present and free to move around, they tend to go where the electrons go. Consequently, when an electron is passed from one molecule—say, chlorophyll—to another, which then has an extra electron, an H^+ ion is likely to go along, too. The receiving molecule thus gains not only an electron but a hydrogen ion as well (Figure 2.5).

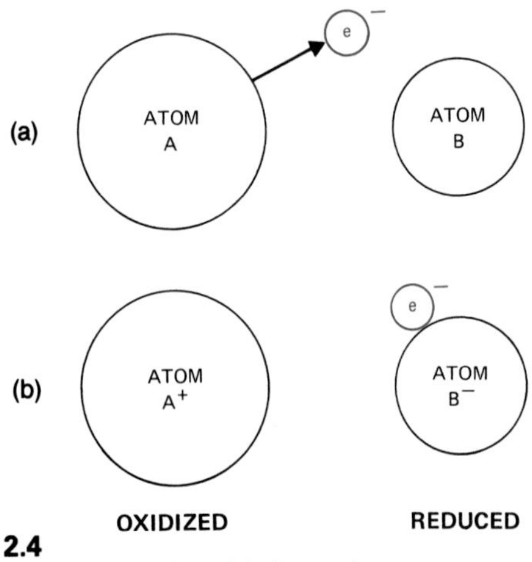

Figure 2.4
Oxidation and reduction. (a) Atom A loses an electron to atom B. (b) Atom A, having lost an electron, is said to be *oxidized,* and atom B, having gained an electron, is *reduced.* In this way entire molecules can be oxidized and reduced.

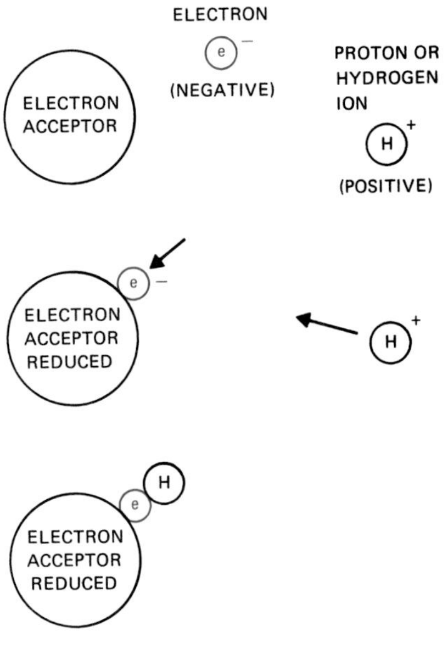

Figure 2.5
Diagram of transfer of an electron to an electron acceptor (which is reduced) and accompanying acceptance of a hydrogen ion.

Biologists frequently speak of hydrogen transfer, but what they mean is electron transfer accompanied by hydrogen pickup from dissociated water. The end result is the same regardless of the words used. The main point is that *the molecule that receives the electron and the hydrogen and is therefore reduced is the one that now has the energy.* Electron, or hydrogen, transfer is the main means of energy transfer in biological systems.

Energy in living systems manifests itself in several forms. *Movement* is apparent in muscle contraction, the upward push of trees growing, and the movements of chromosomes and other subcellular particles. *Heat* is released by all metabolic activities. *Light* is emitted by flashing fireflies, certain fishes, and several luminous microorganisms. *Electrical energy* is liberated by several animals, notably electric eels. No matter what the form of energy released, ATP or some similar energy source is involved.

All these uses of energy demonstrate how the cell manages to rearrange chemical bonds and release energy at exactly the moment it is needed. We still know little of how such a basic and vital request for energy is transmitted; in our study of biology we will see over and over again that natural functions are so complicated that we still do not understand all of them fully.

SUMMARY

1. *Energy* is usually defined as the capacity to do work. In biological terms, we can say that energy is used in the operation of all life processes.
2. The sun is our most important source of energy. Green plants use sunlight to create their food supply in a process called *photosynthesis.* Besides producing food for plants, photosynthesis liberates oxygen.
3. *Bond energy* refers to energy stored in the chemical bonds that hold molecules and compounds together.
4. The *First Law of Thermodynamics* states that energy can be changed from one form to another, but it cannot be created or destroyed. The *Second Law of Thermodynamics* states that every time an energy-releasing or energy-requiring action occurs, some energy is "lost"; that is, it is converted into heat energy and is not available to do work.
5. Biologists use the *calorie* as the basic energy unit. A calorie is the amount of energy required to raise the temperature of a gram of water from 14.5° to 15.5° Celsius. A *kilocalorie,* or *Calorie,* is a thousand calories.
6. All living things, for the bulk of their energy requirements, use *ATP.* ATP stores energy in chemical bonds, where the energy remains locked up until it is released by rearrangement of those bonds. The key feature of ATP is not simply the storage of energy in the chemical bonds of the molecule, because every molecule can store energy in its bonds. Of crucial importance is the *ready availability and transferability* of the energy in the ATP molecule.
7. When ATP reacts with water *(hydrolysis),* the last of its three phosphate groups is separated. Such a reaction yields ADP, inorganic phosphate, and energy.
8. *Coupled reactions,* though not using the same beginning and final products, are closely associated through energy-carrying intermediates.
9. Activities that require energy are called *endothermic* (heat in). Activities that liberate energy are called *exothermic* (heat out).
10. The two main energy-transferring processes in biology are photosynthesis and respiration. The essential action in these processes is the transfer of electrons from one compound to another.
11. When an electron goes from a molecule, that molecule is *oxidized.* The molecule that receives the electron is *reduced.*
12. The direct conversion of chemical energy to light energy is known as *biological luminescence,* or *bioluminescence.*

ASK YOURSELF

1. What is wrong with saying that an animal makes energy out of the food it eats?
2. What is a calorie?
3. Someone says, "I fed my laboratory mouse 100 calories worth of food and he produced 100 calories worth of movement from it." How do you know that he is either lying or has made some technical mistake?
4. Why is ATP called "the energy currency of the cell"?
5. Into what forms may an animal convert the energy of the food it eats?

PART TWO
THE LIFE OF CELLS

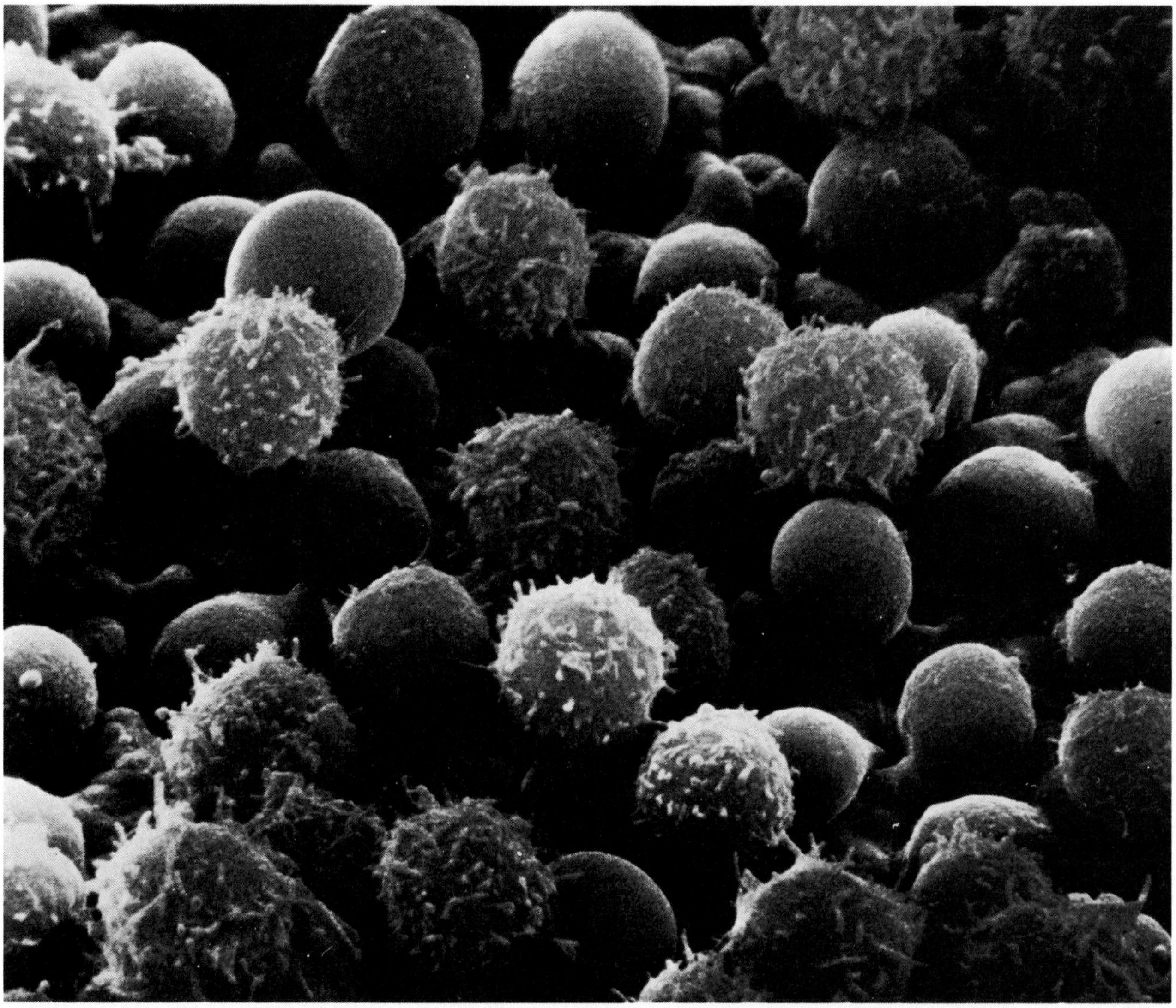

Scanning electron micrograph of human blood cells. The lymphocytes have rough outer surfaces with numerous microvilli. $\times 2000$

3
Cells and Cell Structure

SOME KEY POINTS

1. Cells are chemical factories that provide an environment suitable for all the processes of life.
2. Cytoplasm, the part of the cell outside of the nucleus, contains the many cell components known as organelles.
3. Cell membranes regulate the passage of substances into and out of cells, and they may take an active part in the transport of some substances.
4. Mitochondria specialize in producing the energy-rich ATP that is used to accomplish cellular work.
5. The nucleus contains chromosomes, which are the main carriers of the hereditary material of cells.

CELLS ARE THE CHEMICAL FACTORIES OF LIVING things. Compartmentalized but integrated, specialized, and precisely regulated, living cells provide an environment suitable for the processes of life. In this chapter we will examine the structure and function of a model cell. No single cell contains all possible subcellular structures, and no typical cell exists. For convenience, however, and for illustrative purposes, we will build a composite picture of an imaginary cell with everything possible in it.

WHAT ARE CELLS?

More than 300 years ago, in 1665, the English scientist Robert Hooke looked through one of the first microscopes and described the compartments in thin slices of cork (Figure 3.1). He called these com-

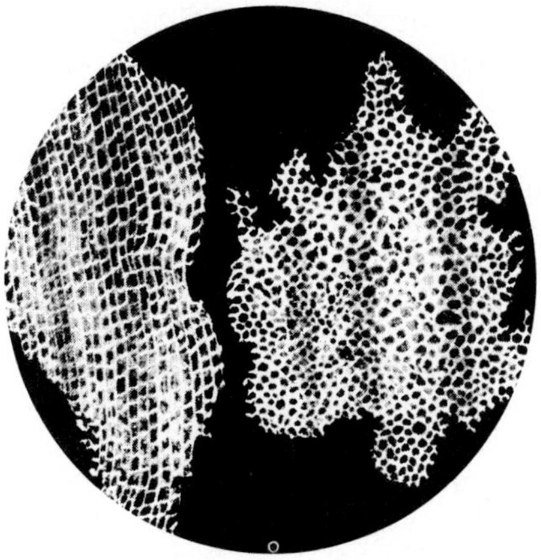

Figure 3.1
Robert Hooke saw the first biological cells ever viewed by human eyes. Because he chose to look at a thin sliver of cork, he saw only cell walls, not cell contents. His drawing of cork cells is regarded by biologists as one of the landmarks of biological history.

partments *cells* because they reminded him of empty chambers. By the early 1800s, the **cell theory** was generally accepted. The three most important postulates of the cell theory are these:

1. The cell is the unit of structure and function of all living things.
2. All cells come from former cells, as a result of cellular reproduction.
3. All living things are composed of a cell (or cells) and cell products.

Originally the emphasis was on the walls of cells, but there is now general recognition that the functional part of a cell is its contents. Microscopes have been refined continually since Hooke's early version, and by the 1950s scientists were able to use electron microscopes, with sharp and clear magnifications up to 50,000 times, to obtain unique information about the inner structure and workings of cells (Figure 3.2).

In the following description of cell components, each entity will be treated more or less separately and its form and activity described as if the information had been obtained from a series of still pictures selected from a strip of movie film. Indeed, much of what is known about cells has been gained in just that way. But one must remember that cells are in fact as busy as complex factories in full production, with many things going on at once. At a given moment, a fully functional cell may be replicating its DNA, digesting complex food molecules, building up chains of polypeptides by joining up residues at the rate of several hundred a minute, pouring out energy-rich ATP, engulfing external particles, packaging and expelling internal particles, and creating new cell parts. All this is going on while the cytoplasm is surging about with a speed, relative to the size of the suspended particles, like that of a river at full flood.

Cells vary enormously. They can range in size from rickettsias, bacteriumlike organisms less than a micrometer (1 millionth of a meter) in diameter (see Table 3.1) to nerve fibers more than a meter long (but too thin to be seen by the naked eye), to the biggest of all, ostrich egg yolks, which are usually the size of a small grapefruit. Most cells probably range from 10 to 100 micrometers. A quarter of a million liver cells could fit into a cube smaller than the small letters on this page.*

Figure 3.2
Three kinds of microscopy and the results they produce. All three photographs show liver cells. (a) A light microscope, drawn inverted for comparison with the other two instruments, and a photograph of liver cells as seen through such a microscope. (b) A transmission electron microscope, which uses electrons from a hot filament instead of light. The electrons, guided by magnets called "lenses," pass through the specimen (transmission), so that the image on a screen or photographic plate is a shadow of the specimen, with varying densities caused by differences in the electron absorption or scattering in the specimen. The electron micrograph is also of liver cells, but at a higher magnification. (c) A scanning electron microscope, which also uses electrons to form an image. However, instead of passing through the specimen, electrons are beamed *onto a surface* of a specimen. An image of the surface is picked up by a detector like a television receiver and made visible on a tube like a television screen. The image gives an illusion of three-dimensional depth, as shown in the scanning electron micrograph of the outer surfaces of liver cells. Electron microscopes are usually housed in special rooms with braced floors to support their massive weight. Temperature, humidity, and voltage are controlled.

The living stuff of cells is *protoplasm,* and the protoplasm of a single cell consists of a *nucleus* and the nonnuclear portion, the *cytoplasm.* Protoplasm is colorless unless a pigment is present. It is slimy and somewhat liquid, but it varies from almost watery to almost solid and has the ability to change its viscosity (resistance to flow) greatly and quickly. Protoplasm is usually slightly heavier than water.

* There is no direct relationship between the size of an animal and the size of its cells. The size of an animal is determined by the number of its cells rather than by their size. The cells of an elephant are barely larger than those of a mouse, but the elephant has many more cells than the mouse does.

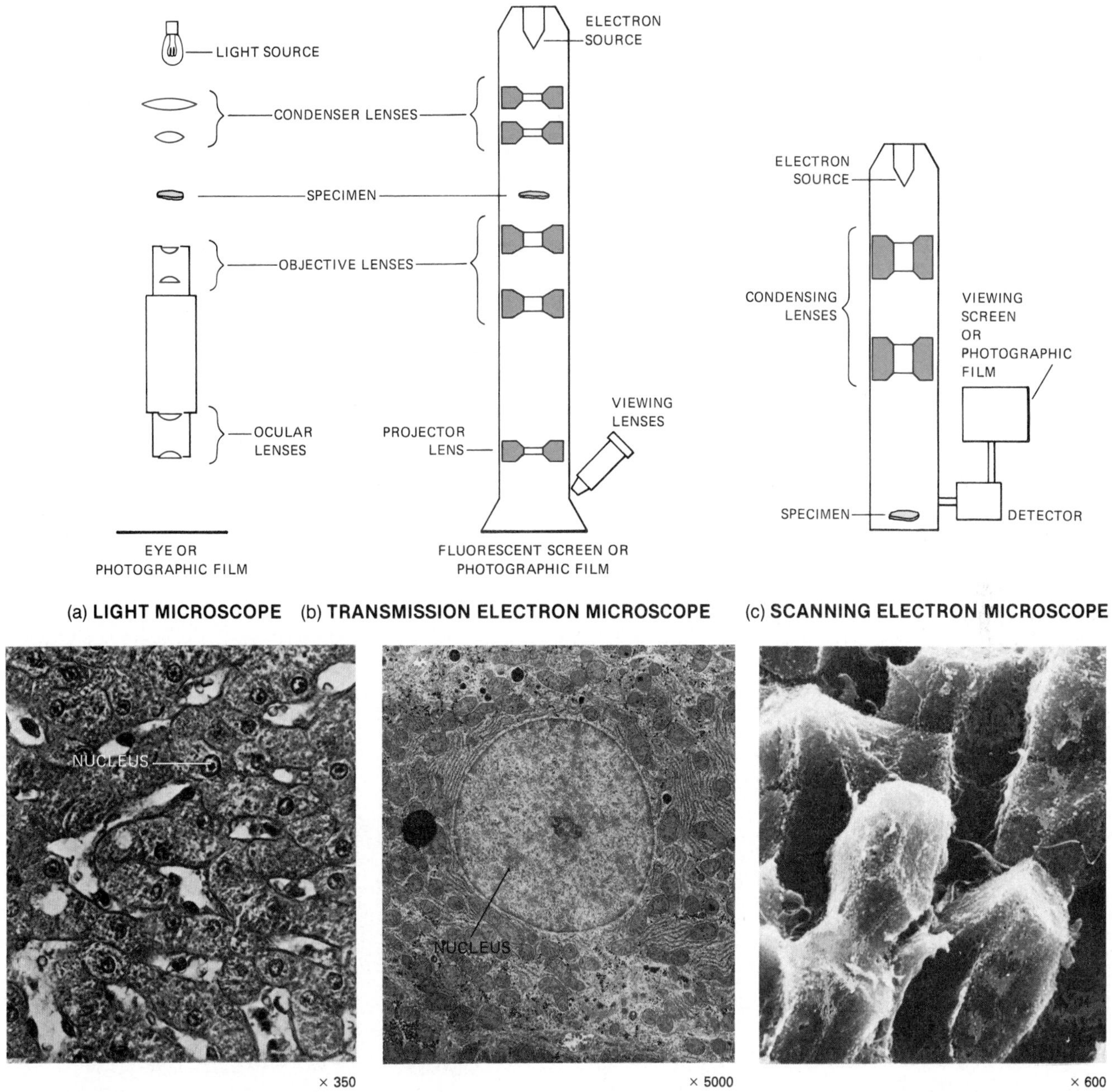

TABLE 3.1
Comparison of Units of Measurement in Cell Biology

MILLIMETER	MICRON (= MICROMETER)	MILLIMICRON (= NANOMETER)	ANGSTROM (= DECINANOMETER)
1	1,000	1,000,000	10,000,000
0.001 (1×10^{-3})	1	1,000	10,000
0.000001 (1×10^{-6})	0.001 (1×10^{-3})	1	10
0.0000001 (1×10^{-7})	0.0001 (1×10^{-4})	0.1	1

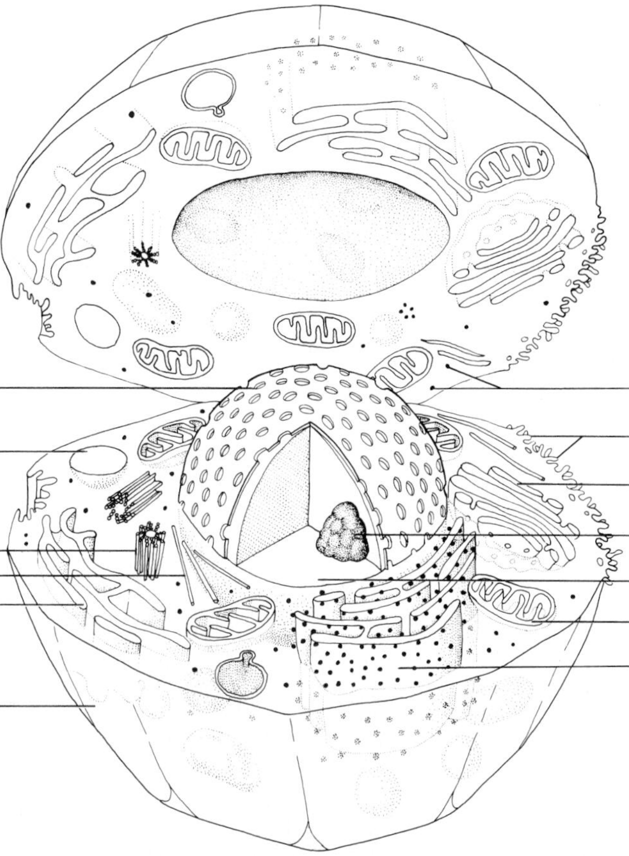

Figure 3.3
Diagrammatic view of an imaginary animal cell sliced open in such a way as to show some of the cytoplasmic inclusions. Such figures are made up from a number of sources, especially from thin sections and electron micrographs, and they represent what has been called a "textbook" cell, to indicate that no such cell exists.

Living cells are composed of many discrete parts (Figure 3.3), and in the rest of this chapter we will discuss the structure and function of the most important subcellular units, starting with plant cell walls and animal cell coats.

PLANT CELL WALLS AND ANIMAL CELL COATS

Because protoplasm is wet and runny, an organism that maintains any dependability of form must have some stabilizing structure. Most cells do have extracellular coats: soft envelopes in animals or stiff cell walls in plants.

Plant Cell Walls

Plants owe most of their rigidity to their walls. Walls of woody plant parts are thick and strong, and even those of soft parts, such as tender young leaves, are firm enough to keep the plant's shape. In land plants the commonest component of cell walls is cellulose, probably the commonest organic compound in the world. Cellulose, which is built into submicroscopic fibers in a meshwork around the outside of cells, is laid in place by the activity of the

× 20,000

Figure 3.4
Electron micrograph of crisscrossed cellulose fibers in the wall of a green alga. The orderly meshing of the fibers, so small that it is visible only in electron micrographs, gives plant cell walls their strength, elasticity, and flexibility.

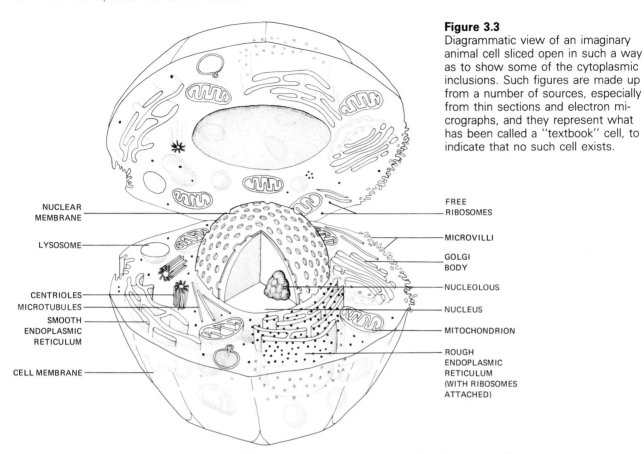

protoplasm inside (Figure 3.4). How the protoplasm inside the cell can synthesize cellulose and deposit it outside the cell is not known. It is as though a carpenter were to shingle the outside of a house while remaining inside with the doors and windows shut.

Each cell of a plant has its own cell wall. The walls of adjacent cells are held together (or held apart?) by a layer of jellylike pectin, the familiar material that causes jelly to gel. The pectic layer, the middle lamella, is responsible for the structural unity of plant parts. A crisp pickle has intact middle lamellas; a mushy pickle is one whose middle lamellas have been insufficiently salted or have been attacked by the enzymes of some invading mold.

Molds and mushrooms (as well as many animals, such as shrimp and insects) have coverings containing *chitin* (KYE-tin), a substance resembling cellulose, except that instead of being composed of simple glucose, it is composed of a nitrogen-containing sugar. Bacterial cell envelopes are different from all others, being made of special organic acids (Figure 3.5).

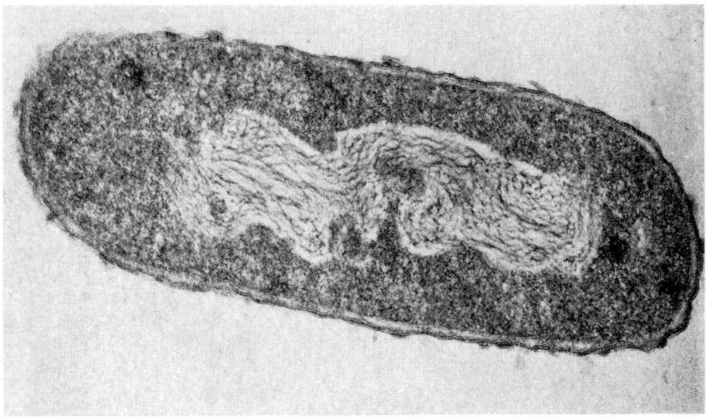

× 38,000

Figure 3.5
An electron micrograph of a thin section through a bacterial cell, *Escherichia coli*. The pale outer margin is the cell envelope.

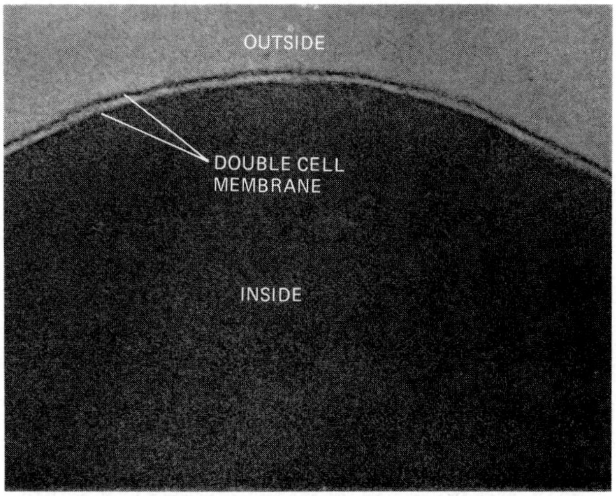

× 150,000

Figure 3.6
Electron micrograph of a section cut through a cell membrane, showing the usual double lines that appear in such preparations.

Animal Cell Coats

In contrast to the firm cellulose walls of plants, the cellular envelope of animals is flexible and elastic. For a simple comparison of the two, pat a tree and then pat your sweetheart. The animal cell coat, as distinct from the plasma membrane, is made of a conjugated protein, that is, a complex of protein-plus-carbohydrate, which not only gives the cell toughness and resilience but establishes its reactivity toward other cells.

Animal cells have sensitive outer surfaces that can distinguish subtle chemical differences in the extracellular environment. This sensitivity can be demonstrated, and experimentally altered, by using cells in tissue culture. If a few cells of kidney or liver are collected under germ-free conditions and grown in glass vessels with a suitable nourishing fluid, the cultured cells ooze about in a thin layer on the glass surface. If kidney cells and liver cells are mixed, each will find its own kind, so that there will be clumps of kidney "tissue" and clumps of liver "tissue." In fact, liver cells from human and mouse are more alike than cells from human liver and human muscle.

CYTOPLASM

The nonnuclear portion of a cell is the **cytoplasm.** Nuclei and cytoplasm have been traditionally treated separately because it is easier to work on and to talk about small, simpler parts than whole, complex cells, but nuclei and cytoplasm are in fact so closely associated that they are absolutely dependent on each other. In the following discussion, cytoplasm will be described first, including the various particles known as *organelles.*

Cell Membranes

The outermost protoplasmic layer of a cell is a membrane so thin that it cannot be seen with a light microscope. It can be photographed in electron microscopes, but because it is only about 100 angstrom units thick, it is below the limits of resolution of a light microscope (Figure 3.6). An angstrom unit

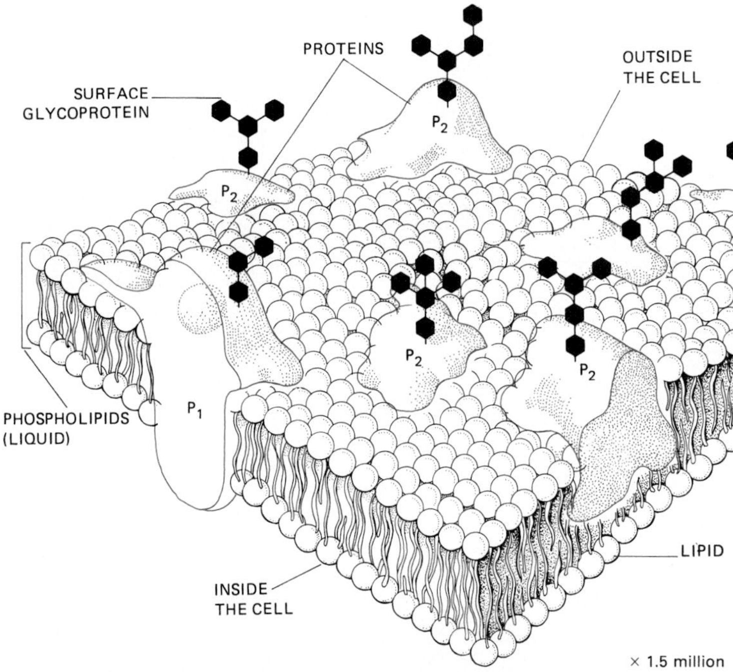

Figure 3.7
The fluid mosaic model of a cell membrane as envisioned by Singer and Nicolson. The double-layered membrane (see Figure 3.6) is composed of two layers of phospholipid (see Figure 1.20, p. 33). The sphere is the charged portion of a molecule, and the tail is the uncharged portion. Protein molecules float in the liquid membrane. The protein portions are shown extending completely through the membrane in one instance (P_1) and simply being embedded in the membrane in others (P_2). (The "fluid" part of the fluid mosaic model is the liquid membrane; the "mosaic" part is the protein.) Carbohydrates (surface glycoproteins) protrude from the proteins on the outside layer of the membrane. These glycoproteins enter into all phases of cell recognition, including immune reactions, and act as receptor sites for hormones such as insulin. The entire membrane is thought of as being in a constant state of flux, not a firm, immobile sheet.

(usually abbreviated Å) is one ten-billionth of a meter. It would require a stack of about a million cell membranes for them to become barely visible to an unaided human eye.

The **cell membrane** is also known as the *plasma membrane* or the *cytoplasmic membrane.* Cells have many membranes besides the outer one, and the more that is found out about cells, the more it becomes apparent that membranes are essential to most normal cell functions.

Most subcellular particles are surrounded by membranes. These membranes guarantee the separateness of the particles so that they do not interfere with each other. Membranes also control what substances pass in and out.

Cell membrane structure It is now thought that the membrane is composed of a double fatty layer that is studded with protein molecules. Some protein molecules may be on the outside of the membrane, some may be on the interior side, and others may extend through both fatty layers. If this description (or *model*) is correct, it is likely that the transport of most substances through the cell membrane takes place by way of the protein "passages."

The fluid mosaic model Cell membranes can be measured, and they can be separated from the rest of the protoplasm and analyzed chemically. The membranes consist mainly of phospholipids and proteins. Knowing the thickness of a membrane and the size of phospholipid molecules, one can guess

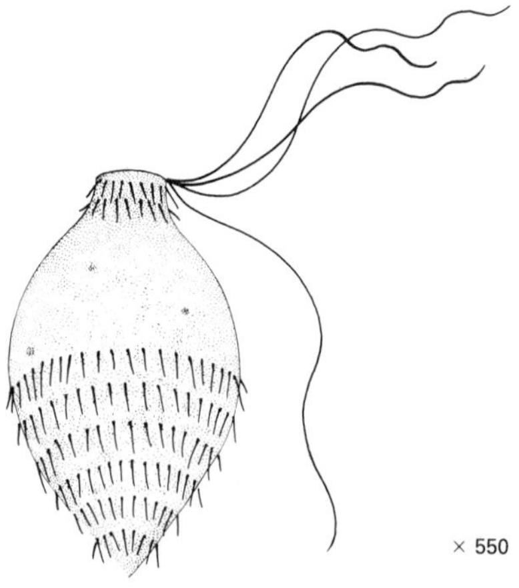

Figure 3.8
A protozoan, *Hyperdevescovina,* from a termite gut, showing the four flagella at the anterior end, as well as scattered bacteria attached to the "head" and "body," but absent from the "neck." With the head fixed against an object, the protozoan can slowly spin its body at a rate of one turn every one or two seconds. This is strong direct evidence of the fluidity of cell membranes.

that the phospholipids are arranged in a double layer (see Figure 3.6). This makes sense because phospholipids have one charged end and one uncharged end, and the charged ends can stick into the watery protoplasm inside and the similarly watery external environment while the uncharged ends face each other. Furthermore, electron micrographs of sections cut properly through the membranes show pairs of lines (see Figure 3.6). The problem lies in interpreting these pairs of thin parallel lines.

Any imaginary model of a membrane must take into account several demonstrable facts, especially concerning *permeability,* that is, the ability to allow passage of substances. Some substances, such as water and fat solvents, pass readily through membranes. Some, such as uncharged organic molecules, pass through with varying degrees of difficulty, depending in part on their size. Others, such as K^+ or Ca^{++} ions, pass only with considerable difficulty. The molecular architecture of membranes remains a puzzle, but the currently popular model, shown in Figure 3.7, includes the additional feature that the entire membrane is capable of constant rearrangement. This **fluid mosaic model** is substantiated by a number of theoretical and experimental studies, but like other membrane models, it is not universally accepted.

One observation that supports at least the fluid part of the fluid mosaic model comes from a study of the one-celled animals inhabiting the intestinal tracts of termites (Figure 3.8). These microscopic creatures swimming in the termite gut have a narrow neck near the front end, so that they seem to have a "head" and a "body." The body bristles with attached bacteria, and the head has four threadlike appendages, or *flagella* (see p. 68). Such an animal can stick its head against a firm object, hold it still, and rotate its body. It is difficult to imagine how such a spinning action could occur unless the membrane is indeed capable of unlimited molecular rearrangement, that is to say, is fluid.

TO SUM UP THE FLUID MOSAIC MODEL

Apparently, cell membranes are composed mainly of phospholipids and proteins, and are *fluid* rather than solid. Being fluid, the entire membrane is in a constant state of flux, shifting and changing, while at the same time retaining its basic structure and properties. Figure 3.7 shows how the proteins float like icebergs in the fluid phospholipids. The important points about the fluid mosaic model of cell membrane structure are as follows:

1. Phospholipids are represented as looking like balloons on a string. They are arranged in two layers, like a sandwich, with the spherical "balloon" heads as the "bread" of the sandwich, and the inward-facing tails as the "meat" (see Figure 3.7)

2. The "tails" (the *lipid* portion) of the phospholipid molecules are attracted to each other, and are repelled by water. As a result, the "balloons" (the *phosphate* portion) of the phospholipid molecules line up over the entire cell surface, and the fatty portions compose the center of the membrane between the two phosphate layers (see Figure 3.7).

3. The *protein* portions are embedded in the cell membrane like tiles in a mosaic. Sometimes they protrude only into the phospholipid sandwich, and sometimes they extend all the way to the cytoplasm. Proteins that stud the membrane (shown as P_1 and P_2 in Figure 3.7), even if they reach the cytoplasm, are enzymes. These enzymes are involved in the many chemical reactions that take place at their particular site on the membrane. All the proteins are capable of moving about the flexible double layer of phospholipid molecules. It is thought that the amount of protein in a cell membrane is a measure of that membrane's metabolic activity. The membranes of mitochondria, for example, contain more protein than the membranes of less active organelles.

4. Protruding from the floating proteins are antenna-like *surface glycoproteins,* proteins plus branched chains of polysaccharides composed of several types of sugar (see Figure 3.7). These glycoproteins are necessary for one cell to recognize or reject another. Apparently, glycoproteins of similar cells "fit together" chemically, whereas a foreign cell is not "recognized," and is repelled or even destroyed. (A mixed-up mass of liver cells and kidney cells, for example, will eventually separate into two distinct masses—one composed of liver cells, and one composed of kidney cells.) This principle undoubtedly helps blood cells accept transfusions of the same blood type and reject blood that is not a compatible type, and also applies to skin grafts, organ

transplants, and even mating on a cellular level; that is why sperm cells and egg cells of different species usually cannot unite. Without surface glycoproteins, cells would not form tissues, tissues would not form organs, organs would not form systems, and systems would not form a complete organism.

Membrane permeability and osmosis If it is to survive, a cell must maintain a proper balance between its internal and external environment. What is meant by "proper" depends on the cell in question, whether it is in a desert cactus, a freshwater worm, or a marine fish. The balance involves, among other things, regulation of the concentration of dissolved materials in water. In an enclosed space, such as a cell, the concentration of dissolved material, the **solute,** increases as the **solvent** (in cells, water) decreases, and conversely. One can think in terms of a high concentration of solute or a high concentration of solvent. A cell can raise the concentration of solute either by bringing in more solute or by expelling solvent. Or it can lower solute concentration by expelling solute or bringing in more solvent. Still another way of changing solute concentration is by changing undissolved particles such as starch into soluble sugar and back again. All these methods are used by cells to keep their water balance.

Unless a substance is at absolute zero, −273°C, its molecules are in constant random motion. As a result of this motion, molecules left to themselves will tend to spread from a region where they are crowded to become more evenly distributed. This phenomenon of **diffusion** (Figure 3.9) is usually described as the tendency of molecules to move from a region of greater concentration toward a region of lesser concentration.

The rate of diffusion depends on the state of matter of the diffusing molecules. Gases diffuse rapidly; that is how perfume works. Liquids diffuse more slowly, at a rate measurable in millimeters per week. Solids are the slowest of all, their rate being measurable in micrometers per year, but if, for example, a flat piece of gold is pressed against a flat piece of silver, each will move into the other. The movement is not detectable by simple inspection, but it happens. The diffusion rate is also affected by temperature: the higher the temperature, the faster the diffusion. Molecular size, too, affects the diffusion rate, with a small glycerol molecule moving faster than a large fatty acid.

Cells must work constantly to keep themselves in order, and the word "work" here means that there must be a continuous expenditure of energy. A cell works at keeping its membranes intact, since the membranes act as gatekeepers in determining what comes and goes, and how much, and how fast. One of the ways membranes do this is by maintaining a state of **selective permeability,** so that

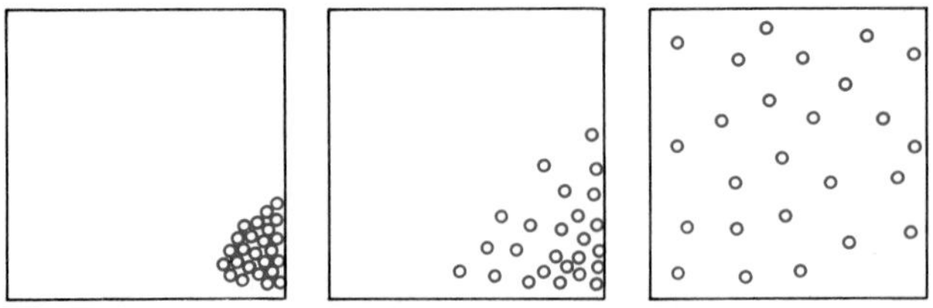

(a) (b) (c)

Figure 3.9
The phenomenon of *diffusion.* At the beginning (a), the soluble substance is concentrated in one small place; later (b) it spreads into the regions of lesser concentration; finally (c) it reaches a stable state when the substance is uniformly dispersed throughout the available space.

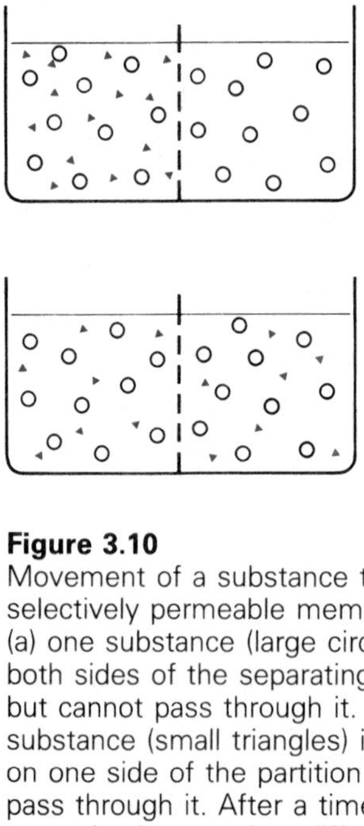

(a)

(b)

Figure 3.10
Movement of a substance through a selectively permeable membrane. In (a) one substance (large circles) is on both sides of the separating partition but cannot pass through it. The other substance (small triangles) is initially on one side of the partition but *can* pass through it. After a time (b) the second substance has diffused through the partition and is uniformly distributed throughout the system. The partition, permitting the passage of one substance but not another, represents a *selectively permeable membrane.*

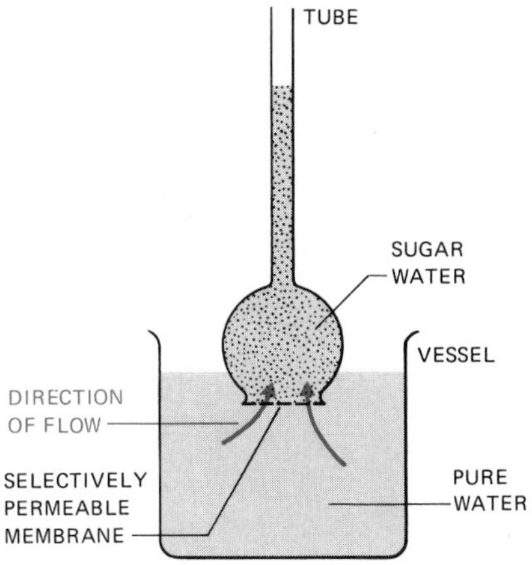

Figure 3.11
An *osmometer,* a device for measuring the amount of actual pressure that a solution can develop when separated from a pure solute by a selectively permeable membrane. The solution in the tube, a mixture of sugar and water, is separated from pure water in the vessel outside the tube. The water diffuses through the membrane, but since the sugar cannot pass the membrane, the total flow is from the vessel into the tube, and it builds up a pressure inside the tube. The fluid level rises in the tube.

some substances can pass through and others cannot (Figure 3.10). A gross example of a selectively permeable membrane is a window screen, which lets a breeze through and keeps flies out.

If a cell with some solute in its sap is immersed in pure water, there will be a tendency for the solute to diffuse out, but it cannot, because the cell membrane keeps it in. There is also a tendency for water to diffuse in because the pure water outside is more concentrated, in terms of water molecules, than it is inside, where it is diluted by the solute. The total movement, therefore, is inward, and the inward movement will continue until either the cell bursts or the pressure exerted by the cell wall in plants or the cell envelope in animals equals the inward diffusion pressure. Tough cells, like mature plant cells, swell and get plump, or *turgid,* like an inflated tire. Delicate cells, such as red blood cells, lacking a strong cover, burst when put in water. Before tender plant parts grow a stiffening tissue, they depend to some degree on being pumped full of water; otherwise they droop.

By means of an *osmometer,* one can measure the actual pressure developed when two substances, both potentially diffusible, are separated by a partition that allows the passage of only one (Figure 3.11). Since such pressure can be great enough to burst cells, they must either have strong walls or keep their internal and external environments in balance. Red blood cells, for example, exist only in the well-regulated blood plasma (Figure 3.12a). If the water concentration of the plasma were to go too high, the blood cells would explode (Figure 3.12b). If it went too low, water would diffuse out of the cells (there being relatively more water inside than out), and the cells would shrivel like raisins (Figure 3.12c).

Osmosis, the simple movement of a solvent through a selectively permeable membrane, is fre-

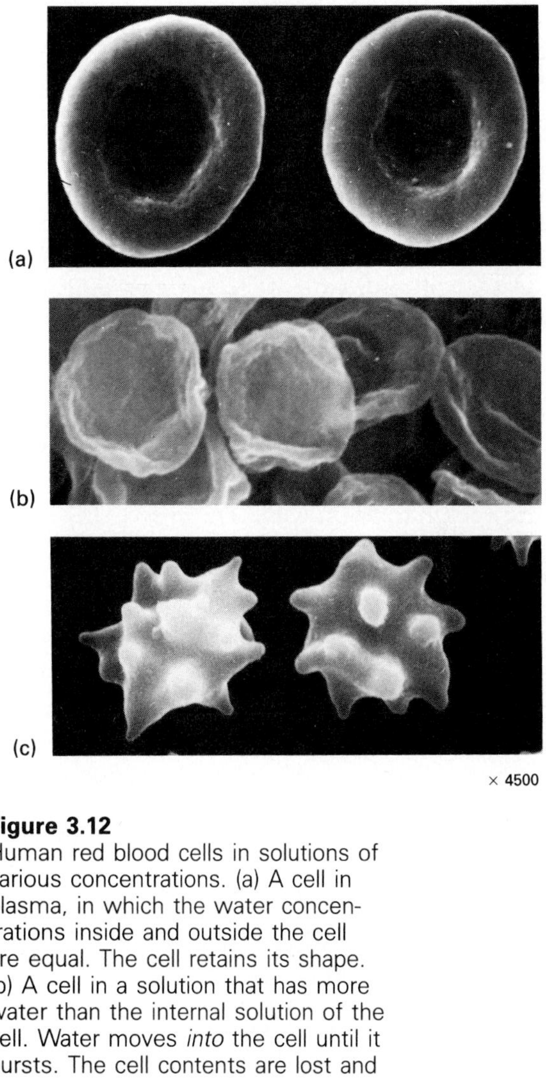

Figure 3.12
Human red blood cells in solutions of various concentrations. (a) A cell in plasma, in which the water concentrations inside and outside the cell are equal. The cell retains its shape. (b) A cell in a solution that has more water than the internal solution of the cell. Water moves *into* the cell until it bursts. The cell contents are lost and the empty plasma membranes are left as "ghosts." (c) A cell in a solution that has less water than the internal solution of the cell. Water moves *out* of the cell, leaving it shriveled.

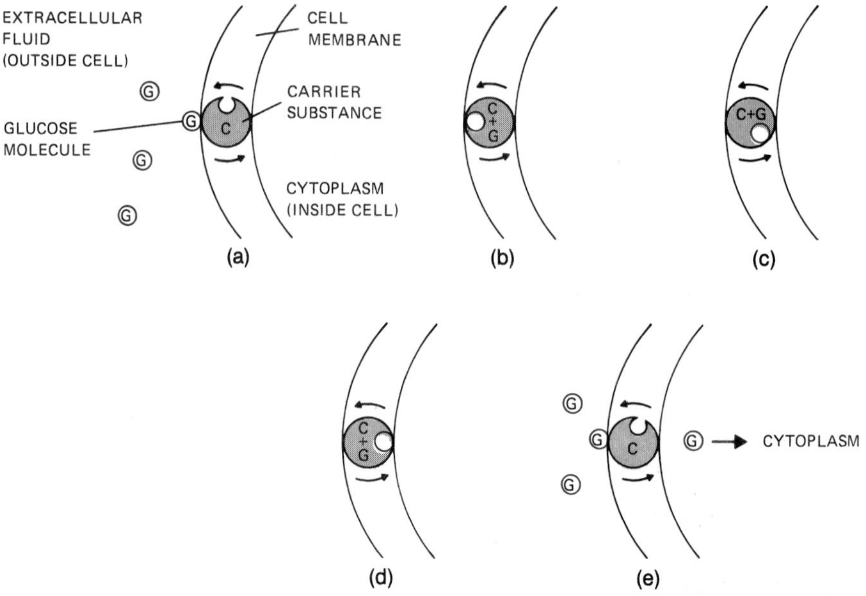

Figure 3.13
Facilitated transport. (a) Glucose molecules (color) in the fluid outside the cell reach the cell membrane. (b–d) A carrier substance inside the membrane combines with the glucose and carries it across the membrane into the cytoplasm. (e) After the carrier substance deposits the first glucose molecule inside the cell, it returns to the outside membrane to pick up another molecule of glucose. No energy is required from the cell.

quently invoked to explain water movement into cells. The rate at which water actually passes, however, is often too great to be explained merely by diffusion. Rather there must be massive movement of water, generally called *bulk flow*.

Transport across membranes In the examples given above of cells in water, there was always some solute, such as sugars or salts, in the cell sap. But if the membranes are impermeable to the solutes, how did they get in to begin with? The answer lies in the *selectivity* of the membranes, their ability to change that selectivity with the passage of time, and their ability to move molecules actively. Some molecules do cross membranes purely by diffusion, with the membrane having nothing to do with it. Such movement is **passive transport.** Passive transport requires no energy expenditure by a cell.

A second possibility, **facilitated transport,** does not require energy but does require a postulated carrier of some sort to "escort" a molecule across a membrane. The entry of glucose into red blood cells is the most intensively studied system of this sort. Glucose is insoluble in the double lipid layer of the cell membrane, but it is the body's main supply of usable energy, and it certainly must have a way of

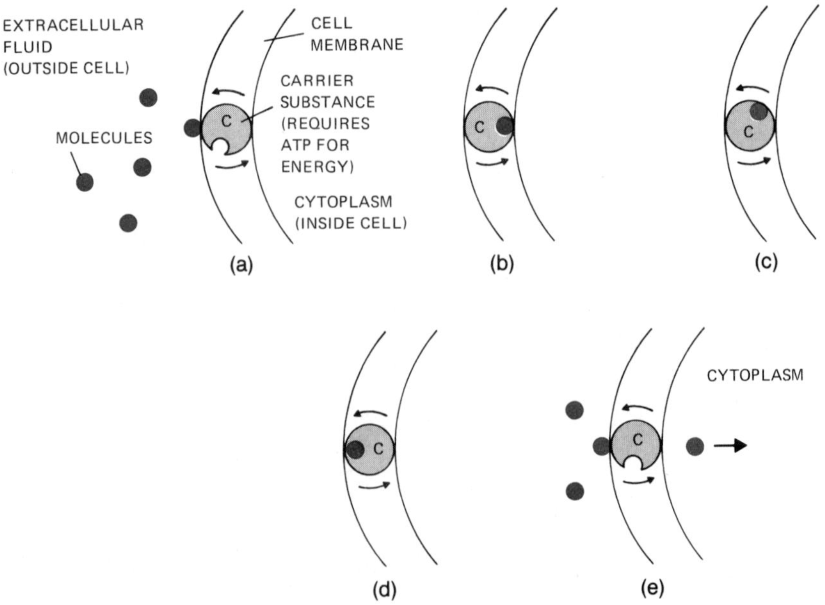

Figure 3.14
Active transport. The only difference between active transport and facilitated transport is that the carrier substance in active transport requires energy from the cell in the form of ATP. Active transport can be stopped by anything that interferes with a cell's ATP supply, such as cutting off its oxygen or adding a poison. Molecules can be carried out of the cell in the same manner as they are carried in. The mechanism of the carrier substance here and in Figure 3.13 is highly simplified.

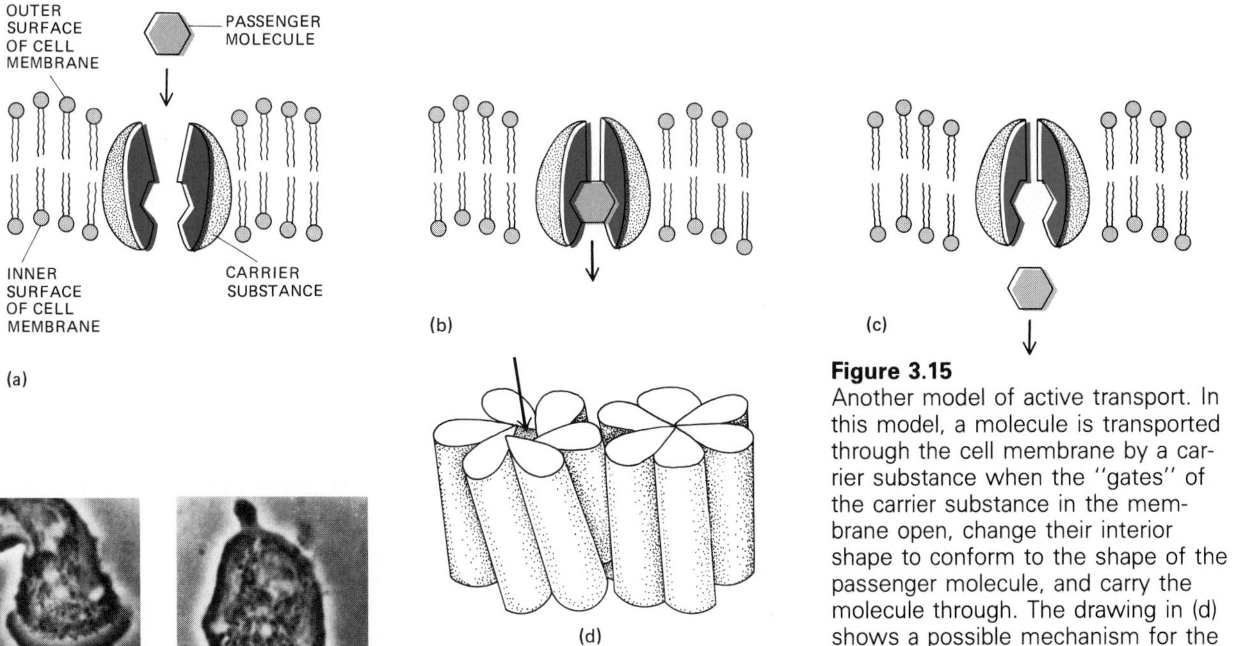

Figure 3.15
Another model of active transport. In this model, a molecule is transported through the cell membrane by a carrier substance when the "gates" of the carrier substance in the membrane open, change their interior shape to conform to the shape of the passenger molecule, and carry the molecule through. The drawing in (d) shows a possible mechanism for the tilting of the selective gates that creates a passageway (arrow).

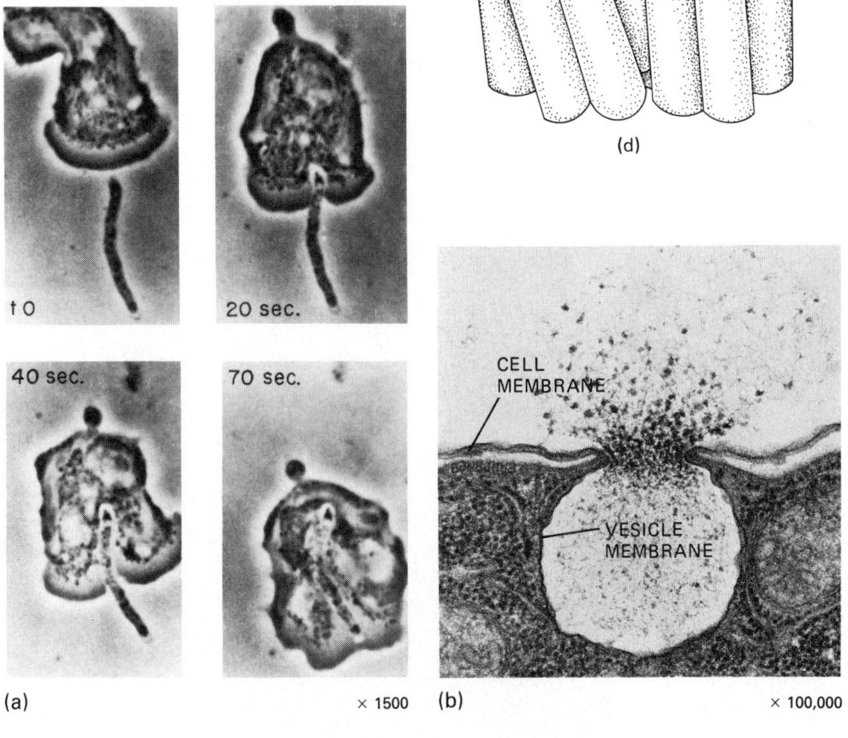

Figure 3.16
Phagocytosis and exocytosis. (a) A human white blood cell engulfing a bacterium, *Bacillus megaterium.* From the beginning, when the bacterium touches the blood cell, until the bacterium has been taken completely in, the action takes 70 seconds. (b) A secretory vesicle inside a cell discharges its contents through the cell membrane.

getting inside the cell. The passage of glucose through the membrane is accomplished with the help of facilitated transport (Figure 3.13).

Cell membranes, however, are usually far from passive. They can carry out a number of complex activities, which can be described better than they can be explained. Transport that requires energy is **active transport** (Figures 3.14 and 3.15). Anything interrupting a cell's energy source stops active transport. Indeed, no one has explained satisfactorily how a membrane can perform such feats as the following: (1) A white blood cell can approach a bacterium in the blood plasma, send out an encircling crater around it, cover it with the crested lips of the crater, pull crater, bacterium, and a bit of plasma down into the protoplast, and there pinch off a vesicle containing the whole thing; (2) A bacterium can distinguish between two molecules that differ in their atomic configurations, bring one into the protoplast, and ignore the other; (3) A marine algal cell can bring potassium into its protoplasm and keep bringing it in until there is a greater concentration of potassium inside than outside; this process is known as *building up against a gradient.*

Particle transport A white blood cell engulfing a bacterium or an amoeba engulfing a food particle is an example of what is called **phagocytosis** (Figure 3.16a), which means "cell eating." A similar phenomenon is **pinocytosis,** or the taking in of fluid by the same kind of membrane action. By a reversal of such action, a particle or a droplet in a cell—for example, a droplet containing a digestive enzyme—may be surrounded by an internal membrane in the protoplasm, carried to the cell boundary, and popped out through the cell membrane to the outside of the cell. This is called **exocytosis** (Figure 3.16b).

TO SUM UP CELL MEMBRANES

1. The outermost living layer of a cell is the cell membrane. Membranes help separate organelles within cells, and also control what substances pass in and out.
2. It is likely that substances pass in and out of cells through a double fatty layer that contains protein "passages."
3. One idea about cell membranes is that they are fluid, allowing for constant movement and rearrangement of their parts.
4. Cell membranes help to maintain a proper balance between the internal and external environment, including a regulation of dissolved materials in water.
5. Transport across membranes frequently takes place through simple diffusion, but in other instances it requires energy or a specialized carrier to escort molecules in and out.

Endoplasmic Reticulum

The cell membrane is only the outermost protoplasmic membrane. Inward from the cell membrane, and physically connected with it in places, there is the **endoplasmic reticulum** (Figure 3.17), frequently called simply the **ER.** It is a complex of double membranes branching and spreading throughout the cytoplasm.

The ER seems to act as a system of internal channels through which various materials move. It also serves as a point of attachment for ribosomes, structures that are active in assembling proteins (Chapter 7). In electron micrographs, the ER can be

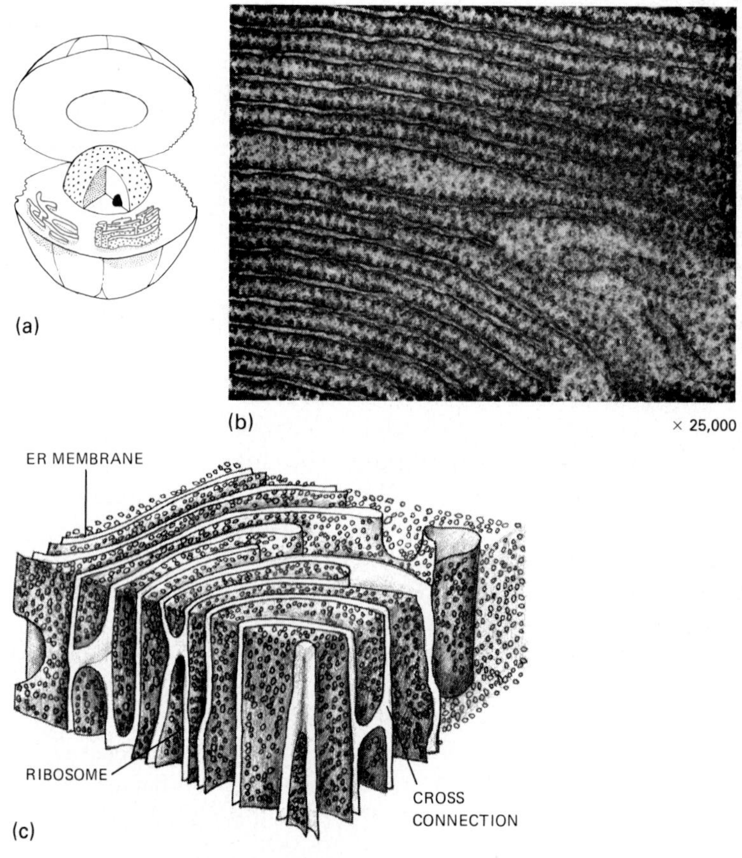

Figure 3.17
The endoplasmic reticulum, ER. (a) A diagrammatic cell cut open to show position and distribution of ER (color). (b) Electron micrograph of a thin section through cytoplasm, showing the rough ER studded with ribosomes on the cytoplasmic sides of the membranes. (c) Drawing showing the three-dimensional structure of ER.

seen studded with ribosomes (see Figure 3.17b and c). The terms "rough ER" and "smooth ER" refer to endoplasmic reticulum that does or does not have attached ribosomes. ("Rough ER" *does* have attached ribosomes.)

In plant cells, there is a tendency for the watery spaces between a pair of ER membranes to swell. One or more of the swellings enlarge until in old cells they may practically fill the cell, leaving only a thin layer of cytoplasm around the periphery of the cell. The central vesicle is the conspicuous **vacuole,** a poor name because it is not empty (Figure 3.18). It contains mostly water, but in the water are salts, proteins, crystals, and pigments. Vacuoles function as water reservoirs in maintaining turgor and apparently sometimes as a dumping ground for wastes.

Ribosomes

It is established that **ribosomes** are necessary for the synthesis of protein from amino acids (see Chapter 1). Ribosomes appear in electron microscope photographs as slightly flattened spheres with a slight constriction around the middle, showing that a ribosome is a double structure (Figure 3.19). During times of active protein synthesis, a string of five or more ribosomes may be linked together, presumably by a strand of messenger ribonucleic acid (mRNA), to form a polyribosome, or simply a *polysome.* (The activity of ribosomes is so important in the life of any cell that ribosomes will be given detailed treatment in Chapter 7.)

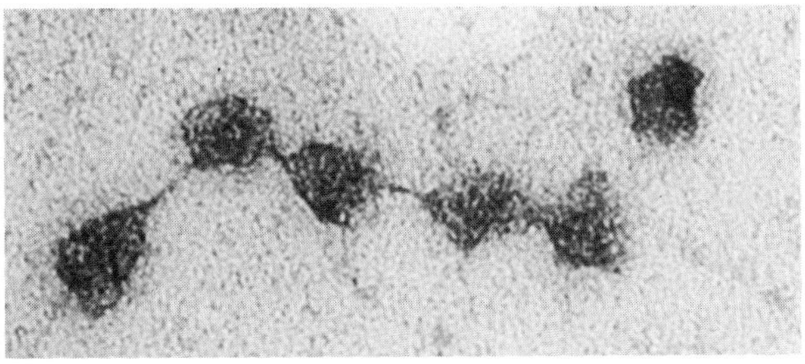

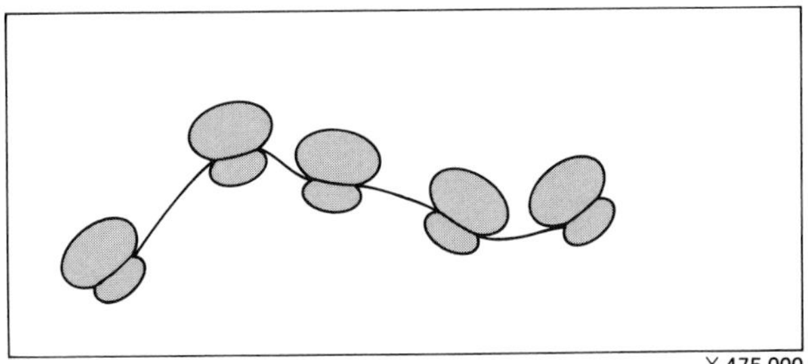

Figure 3.19
(a) Electron micrograph and (b) schematic drawing of *ribosomes* from rabbit cells. Five ribosomes, connected by a strand of a special nucleic acid, messenger ribonucleic acid (mRNA), constitute a *polysome.* The subunits of the ribosomes, shown in the drawing, are not apparent in this micrograph, but they are visible in some preparations and can be separated by physical means.

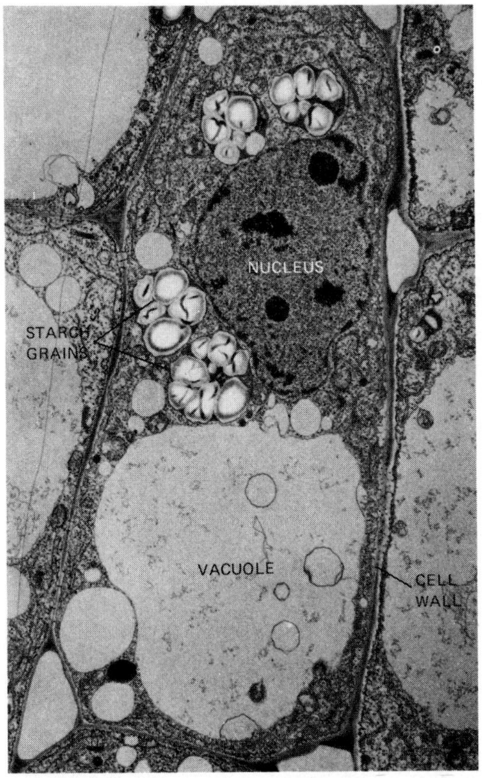

Figure 3.18
Electron micrograph of a thin section through a plant cell, showing the water-filled *vacuole,* produced by swelling of the cisternae between pairs of ER membranes.

Golgi Bodies and Lysosomes

Golgi bodies (named for their discoverer, Camillo Golgi) are collections of membranes associated with the ER. Golgi bodies have been described as stacks of saucers, but if the parts of a Golgi body are like saucers, they are saucers with irregular, frilly edges

Figure 3.20
Golgi bodies. (a) Schematic view of a cell, cut open to show the position of Golgi bodies (color). (b) Electron micrograph of a Golgi body, showing secretory vesicles budding off the edges. (c) Three-dimensional reconstruction of Golgi bodies to show how transfer vesicles from the endoplasmic reticulum compose the forming face of the Golgi body and how secretory vesicles bud off from the mature face of the Golgi body on their way to delivering secretions. (d) The route of material to be secreted from a cell, starting with the rough ER (1), proceeding over Golgi bodies (2), and finally being expelled from secretory vesicles (3).

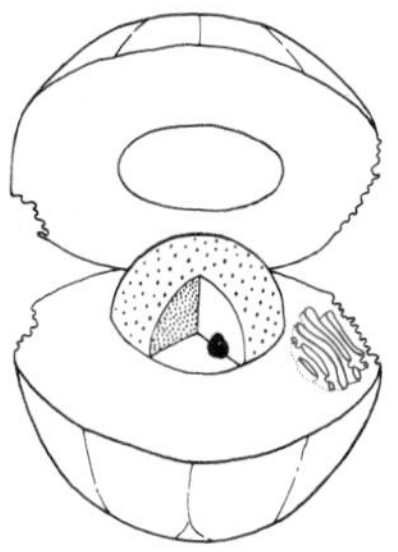
(a)

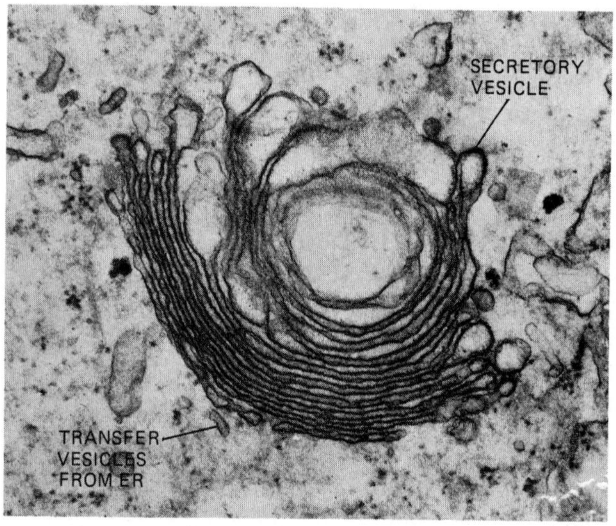

(b) × 50,000

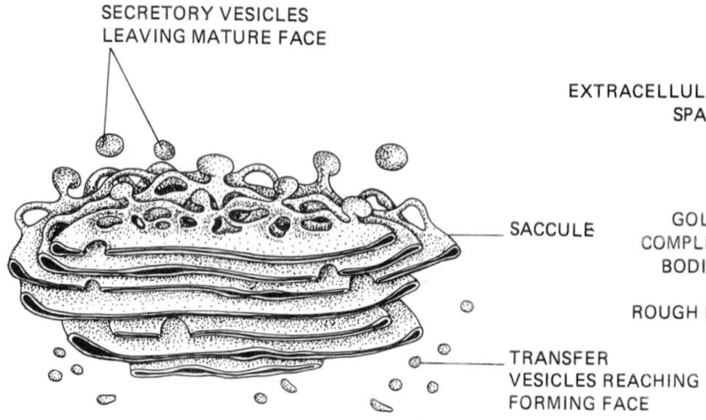

(c)

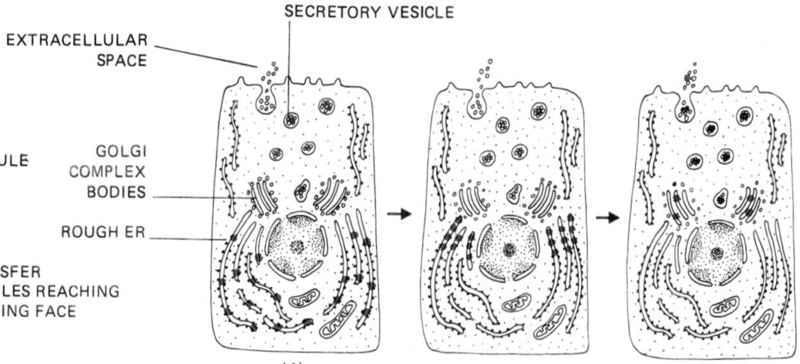

(d)

(Figure 3.20). Apparently, Golgi bodies are formed by the budding off of portions of the ER, and they in turn bud portions off themselves. Substances synthesized in the cytoplasm are delivered into the compartment between ER membranes and are passed on to the compartment defined by the membranes of the Golgi bodies. Droplets are pinched off from the Golgi bodies, and move to their final destination, usually outside the cell (Figure 3.20d).

Golgi bodies are most apparent and abundant in cells with secretory activity: pancreas cells secreting digestive enzymes, nerve cells secreting transmitter substances, plant cells laying down cell plates between newly divided cells. The Golgi bodies concentrate the substance to be secreted and deliver it to its destination. In some instances they actually engage in some synthetic work along the way.

Lysosomes can be seen easily in any microscope as droplets, but they are so characterless that they cannot be identified by simple inspection. They were discovered when homogenized cells had their components separated, and these spherical bodies, covered by a single membrane, proved to be extraordinarily rich in hydrolytic enzymes (Chapter 1). Lysosomes are widely distributed in all but bacterial cells, but they are most abundant in tissues that experience rapid changes, such as in lung tissue. They are loaded with digestive enzymes, and can act as scavengers, ingesting and digesting "worn out" cell parts. Or they can burst, releasing the concentrated enzymes, which can then digest the cell they are in, as happens to the cells in a tadpole's tail. When a tadpole is changing into an adult frog, the lysosomes in its tail cells cause the breakdown of structure, and the tail is absorbed into the frog's body. Lysosomes can also contribute their enzymes to digest invading bacteria. Lysosomes were called "suicide bags" by their discoverer, Nobel laureate Christian DeDuve, but other people have called them "garbage disposal units."

Plastids

Responsible for the conversion of light energy to chemical energy, **plastids** of plants occupy such a special place in the economy of life that they will be treated separately in Chapter 4.

Mitochondria

The main activity of the **mitochondria** is the production of ATP for use in cellular work. The conversion of organic molecules to usable energy occurs in the mitochondria, and it is for this reason that a mitochondrion is sometimes called the "powerhouse of the cell." Live mitochondria are vigorously active. They squirm, roll up, stretch, grow lumps, divide, flow together, swell, shrink, and even seem to swim against cytoplasmic currents (Figure 3.21a). Most cells of higher organisms contain several hundred mitochondria, with other quantities ranging from one or two in some protozoa to as many as 150,000 in the egg cells of amphibians.

An outer membrane, sometimes continuous with the ER, delimits a mitochondrion. Inside, another membranous structure is folded and doubled on itself so as to form incomplete partitions, the *cristae*, or crests. High-resolution micrographs sometimes show the inner membrane to be covered with pebbly bumps that seem to be stalked (Figure 3.21d and e). These bumps are suspected by some investigators to be the functional units of mitochondrial work, ATP production, but such an idea is not fully substantiated.

Besides being repositories for the enzymes of cellular respiration (Chapter 5), mitochondria have many other uses in cells, being concerned with calcium metabolism, rubber synthesis, antibody formation, protein storage, and flagellar movement. New functions of mitochondria continue to be found, and it has even been suggested that cancerous growth of cells is a result of mitochondrial dysfunction.

Plastids and mitochondria as symbiotic organelles

Because of their size and appearance, it was suggested long ago that mitochondria are something like cellular slaves that live in the cytoplasm of cells but are in fact bacteria. As more is learned about all these entities, the resemblances among bacteria, mitochondria, and plastids (see Chapter 4) increase, until it is becoming accepted that mitochondria and plastids may indeed be separate organisms, which have inhabited other organisms so long that now neither can live without the other.

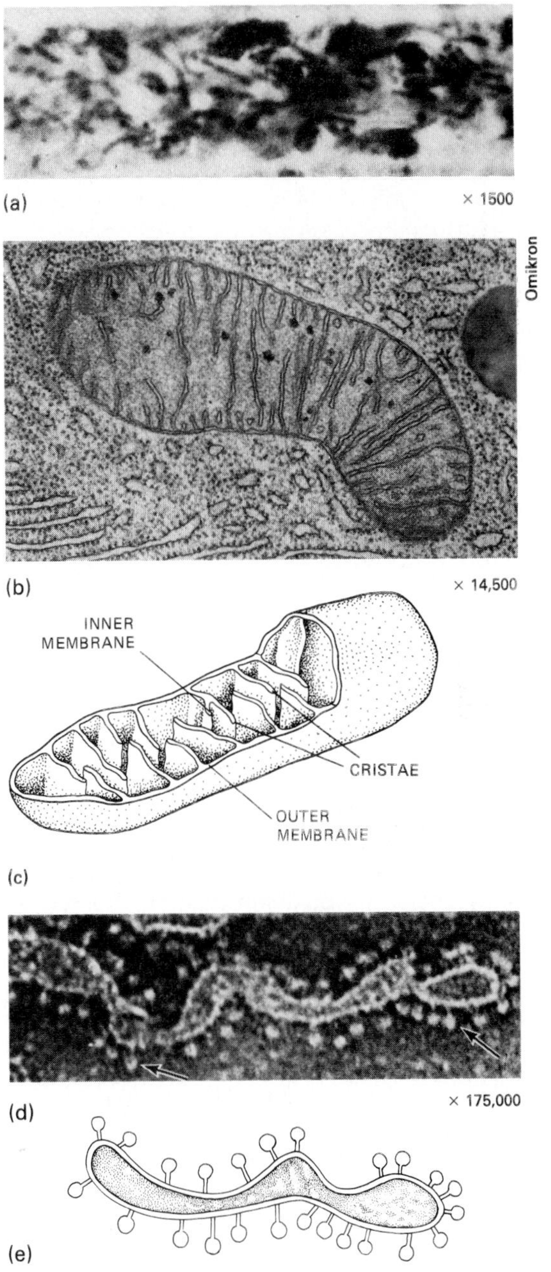

Figure 3.21
Mitochondria. (a) Living mitochondria in a filament of a fungus as seen in a light microscope. (b) Electron micrograph of a sectioned mitochondrion, showing covering membrane and folds of the inner membrane (cristae). (c) Schematic drawing of a reconstructed mitochondrion to show its three-dimensional structure. (d) Electron micrograph of the inner membrane at high magnification; arrows point to the functional units;
(e) Sketch interpreting the electron micrograph of the functional units.

TAKING CELLS APART TO PUT THEM BACK TOGETHER

In order to see most organelles in position in cells, one must slice the cells, perhaps stain them, and then look at them with a light microscope or photograph them with an electron microscope. Most such treatments are destructive, and the cells are dead practically from the start of the work. How can an experimenter know what an organelle does simply from inspecting a dead one?

The answer is that no single procedure can give a satisfactory picture. The experimenter usually must see the organelle but must also get it out of its cell, accumulate a sufficient quantity to experiment on, and get rid of foreign particles that might confuse the results. First, live cells must be broken up. The easiest and commonest way to do that is to drop a piece of tissue in a kitchen blender of the type that was originally made to stir drinks. The steel blades, whirling at 3000–4000 rpm (revolutions per minute), can disrupt cells without destroying their internal organelles, especially if everything is kept cold and a liquid medium is provided that does not upset the water balance. Common cane sugar in water is frequently used as the liquid medium. The result is a slurry mixture containing nuclei, mitochondria, and all the other subcellular particles.

The next step is to separate out the different pieces of the cells. This can be done by taking advantage of the different densities of the various organelles. Cell walls are heavier than nuclei; nuclei are heavier than mitochondria; mitochondria are heavier than ribosomes; and so on. If the complex mix is spun in a centrifuge at several thousand rpm, the effect is comparable to increasing the force of gravity from several hundred times to 100,000 or more. Particles that are only slightly heavier than the fluid medium will settle as a pellet at the bottom of the tube. By controlling the speed (and hence the force) and the time from a few minutes to several hours, one can harvest a series of different organelles selectively, starting with the heaviest. Once a pellet of mitochondria, for example, is collected, it can be tested for enzyme reactions or analyzed for chemical residues or checked to find where radioactive tracers have gone. It can even be made to carry out a particle's original functions.

Once a number of such investigations have been made, the problem of integrating the information begins. Some syntheses have been accomplished, but so far there is no machine that can function as an antiblender and put together what man has put asunder.

Besides the similarities in size and shape, bacteria and plastids and mitochondria are all bound by membranes, they all have their own DNA (see Chapter 6), which, unlike the DNA of higher organism chromosomes, is not associated with protein, and they have ribosomes that are like one another and smaller than the ribosomes of higher cells. Plastids and mitochondria have a partially independent genetic life, typical of organelles with their own DNA. One biologist has asked, "Are we really powered by bacteria?" Perhaps we are.

Microtrabeculae

Present-day electron microscopes reveal a skeletal network in the cytoplasm. A mesh of filaments, **microtrabeculae,** suspends the organelles in a cell, holding mitochondria, ribosomes, and other particles more or less firmly in place (Figure 3.22). At the same time, microtrabeculae are capable of quick changes in response to changes in the cell environment, and can determine the distribution of particles within the cell as well as cell shape.

Microtubules

Microtubules may be up to several micrometers long, but being only about 250 Å thick, they are far too slender to be seen with a light microscope. In electron micrographs, they are so straight that they seem to have been drawn with a ruler (Figure 3.23). When separated from a number of different cell types, they prove to be chemically identical, made of a protein appropriately called *tubulin*. The tubulin

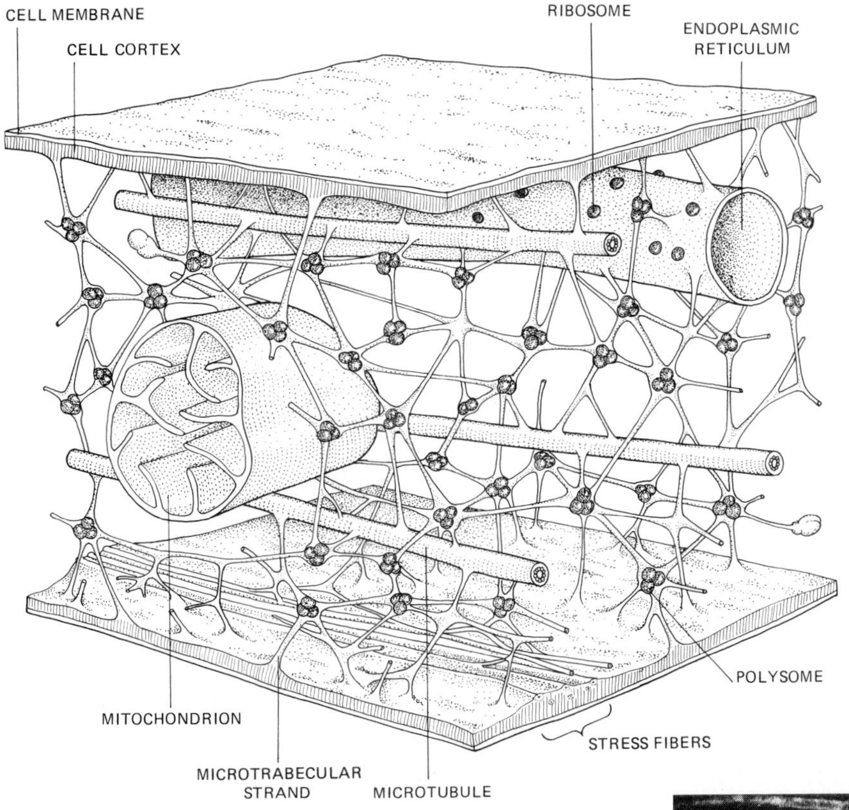

Figure 3.22
A model of the lattice formed by *microtrabeculae* within a cell. The model shows how the strands of microtrabeculae support other components of cytoplasm.

subunits can assemble into microtubules or disassemble and disperse in the cytoplasm, influenced by the cellular environment.

Microtubules are commonly associated with movement: in cilia and flagella, in pseudopods of amoebae, in cellulose deposition, and in chromosome movement during nuclear division (Chapter 8).

Centrioles

Although **centrioles** are barely visible under a light microscope, it has been known for nearly a hundred years that they are associated with nuclear division in animals and some lower plants. Flowering plants have no centrioles. Electron micrographs show that a centriole is a cylindrical organelle, about 0.2 micrometers thick and 0.5 micrometers long. When associated with nuclei, they occur in pairs and act as focal points at opposite ends of a cell during nuclear division. Centrioles from many organisms, from algae through mammals, show a standard structural

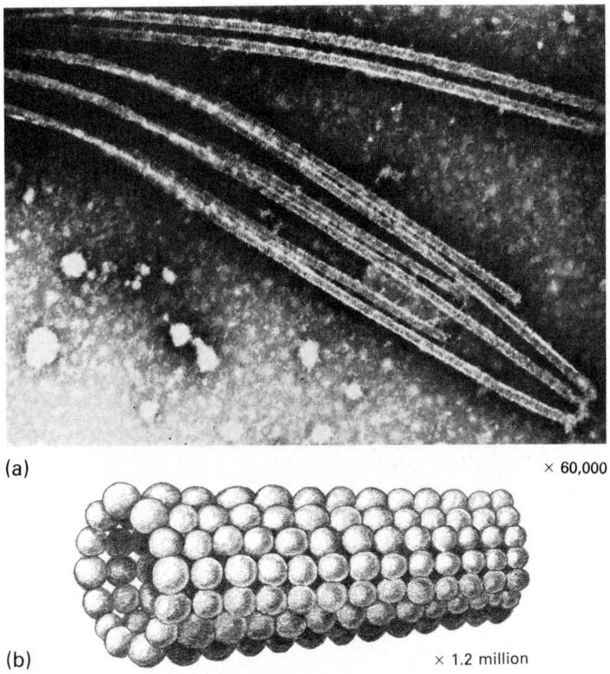

Figure 3.23
(a) Electron micrograph of long, straight, slender *microtubules* from the flagellum of a sperm. (b) Drawing of a portion of a microtubule to show the arrangement of protein units that make up the tubule.

Figure 3.24
Centrioles. (a) Position of centrioles (color) near the cell nucleus. (b) Schematic drawing of a pair of centrioles, showing their relation to each other, one at right angles to the other, and the nine triplets of tubules that form a roughly cylindrical body of the centriole. (c) Electron micrograph of a cross section through a centriole, showing the arrangement of the tubules and the central "spokes." (d) A drawing of the electron micrograph, emphasizing the nine rows (each row containing three tubules) of "cartwheeling" tubules.

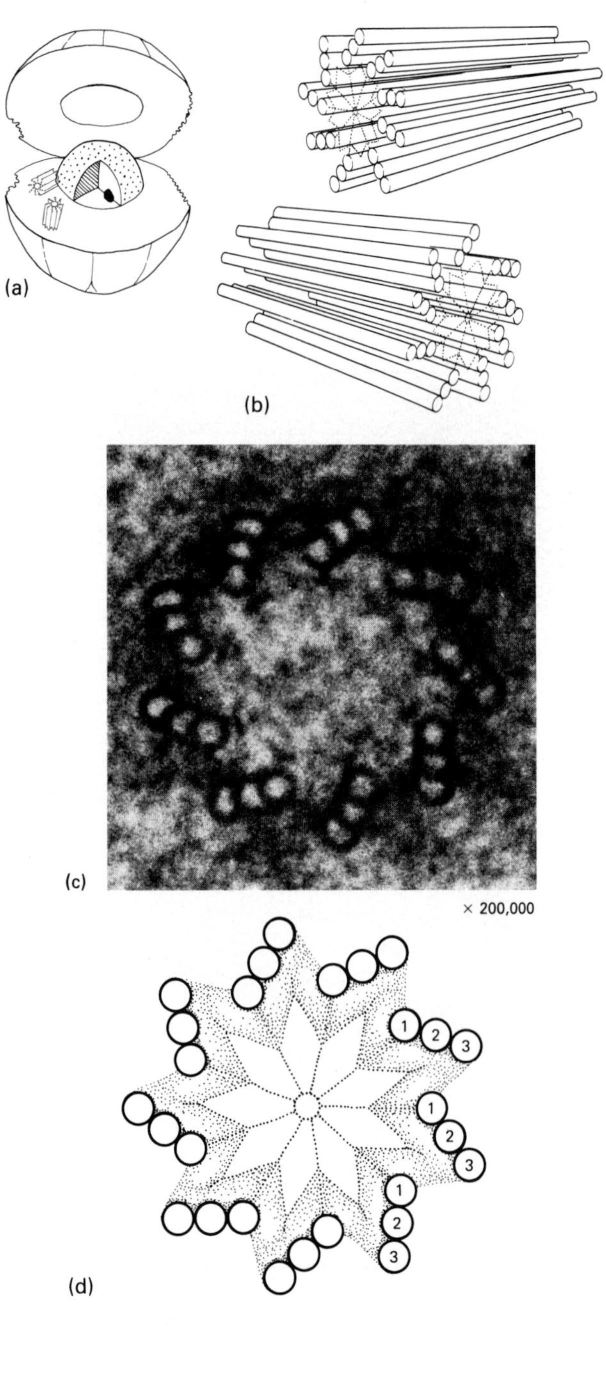

pattern of microtubules (Figure 3.24). They are also usually associated with other microtubules, such as the star-like asters that form during cell division (mitosis) (Chapter 8) or cilia (see the essay on p. 68), and are connected in some unknown way with cellular motility.

THE NUCLEUS AND NUCLEAR MEMBRANES

The word **nucleus** is derived from the Latin word for kernel, that is, the core or essential part. Indeed, the nucleus is the control center of the cell, and it contains the necessary information to direct the reproduction and heredity of new cells.

In all but the simplest organisms (bacteria and related forms), the nucleus is a well-defined region clearly delimited from the general surrounding cytoplasm by a membrane. It is roundish or flattened, usually colorless, and easily visible through a light microscope. The nuclear boundary appears to be something like the cell membrane, but with some important differences. It is not one double membrane, but two double membranes, with a discernible "space" (compartment) between them. Nor is the nuclear membrane continuous. It has numerous pores in it, so that it is something like the plastic balls children call "whiffle balls" (Figure 3.25). Thus the nuclear membrane separates the cytoplasm from the nuclear sap, but it does not do so completely. There must be two-way traffic between the nucleus and the cytoplasm, and all theoretical and experimental evidence says this is so. During nuclear division (Chapter 8), the nuclear membrane usually breaks up and remains "lost" (undetectable) in the cytoplasm until the end of the division period. After two new nuclei have been formed, the ER spreads new nuclear membranes over the chromosome masses.

Chromosomes

Chromosomes are the structures that contain deoxyribonucleic acid (DNA); they can be seen temporarily during cell division. Chromosomes (Greek, "color bodies") were so named because of their stainability with the textile dyes available to the early biologists. Between divisions the chromosomes are not readily visible; it is then that they are busy with their two main functions, replicating themselves and regulating the synthesis of proteins.

Nuclei of cells in all higher organisms have chromosomes, which vary in number from one per nucleus in a variety of intestinal worm to hundreds in some ferns and shrimp, and in size from less than a micrometer in some fungi to more than 100 mi-

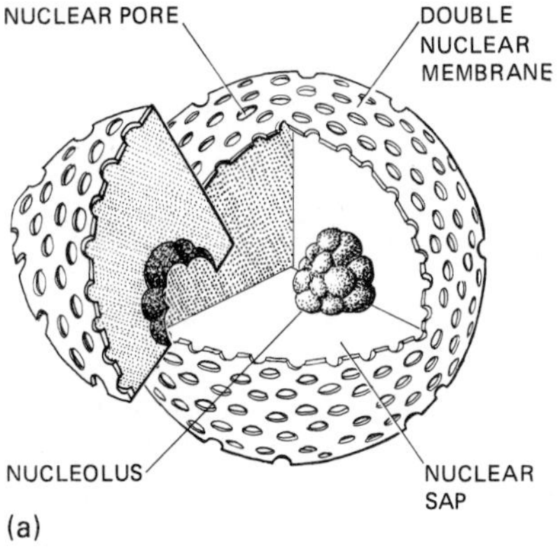

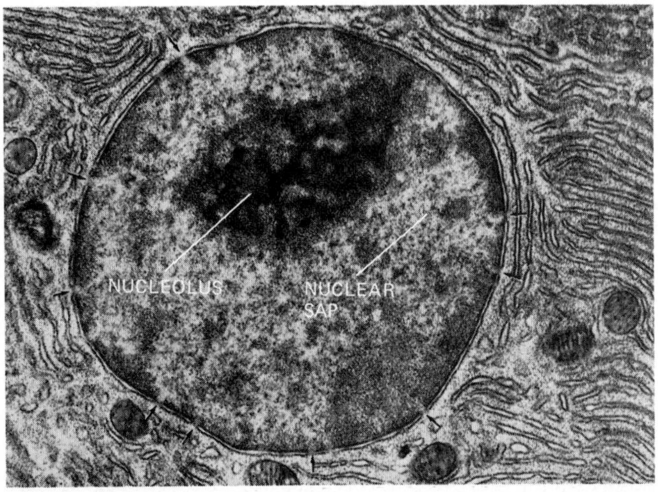

Figure 3.25
(a) A *nucleus,* as shown in a reconstruction from light and electron micrographs. Floating in the nuclear sap is a nucleolus, of which there may be one or several. Nuclei and nucleoli are variable in shape. (b) Electron micrograph of a section through a nucleus. The double nuclear membrane has many pores (arrows). The grainy appearance of the nuclear sap and the irregular, heterogeneous nature of the nucleolus are characteristic of such photographs. (c) A scanning electron micrograph of a nucleus from a cat pancreas shows nuclear pores (NP) clearly.

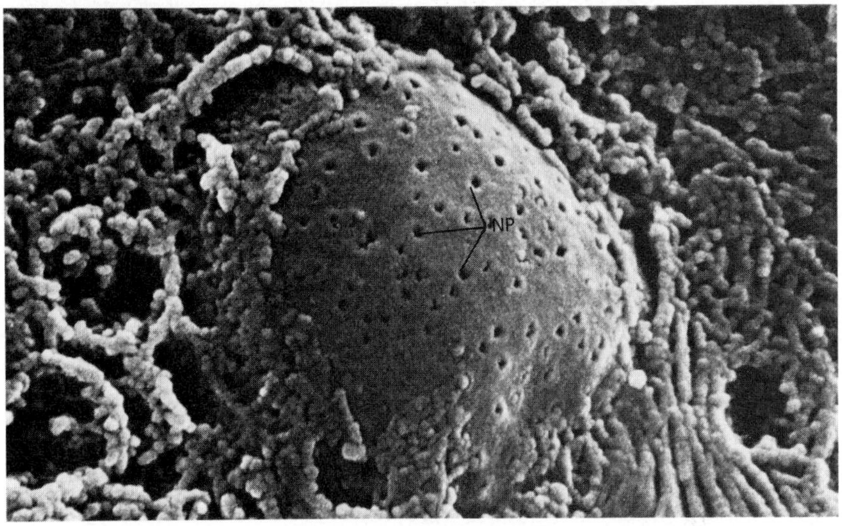

crometers in fly larvae. No membrane covers a chromosome; it lies uncovered in the nuclear sap. One or more strands of DNA run the length of a chromosome, coiled to varying degrees. During the time when a cell is metabolically active and not in the process of dividing, the coiling is minimal and the chromosome is so long and thin that it cannot be seen with a light microscope. Before and during nuclear division, the DNA coils more and more tightly. Coil upon coil is formed until a chromosome becomes a shorter, stubbier, recognizable body (Figure 3.26). The bulk of material in a chromosome is protein. Despite continuing attempts to find the use of chromosomal protein and the way it is physically built into the chromosome body, final answers are still lacking. The combined DNA plus protein is *chromatin*.

Most chromosomes have a special region somewhere along the DNA where microtubules

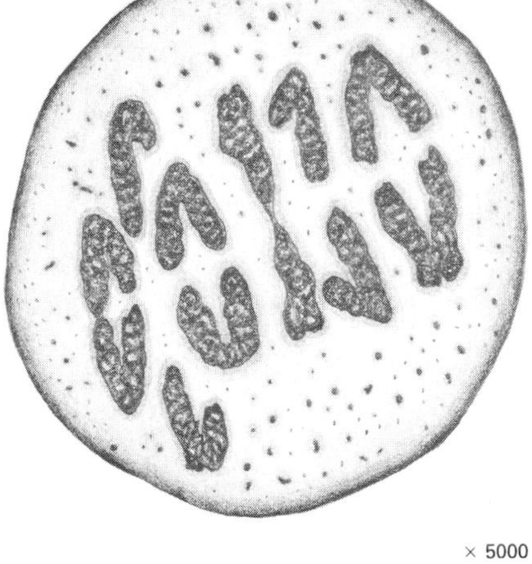

Figure 3.26
Tradescantia chromosomes in stained material, as seen through a light microscope.

CILIA AND FLAGELLA: MICROSCOPIC MOVEMENT

Cilia (Latin, "eyelid") and **flagella** (Latin, "whip") provide a means of locomotion for small, free-swimming organisms, or a means of forcing water past fixed cells. Cilia and flagella are alike in structure and function but differ in length. Cilia are about 25 times as long as they are thick, and flagella may be thousands of times as long as thick. The important feature of cilia and flagella is their ability to bend actively. A short cilium is something like a thin paddle, and a long flagellum can send waves of movement along its length like a snake fastened by its tail. Both effectively create water currents.

A flagellum is a membrane-bound tube full of cytoplasm and is provided with a set of microtubules anchored at the basal end and running parallel through or nearly through the length of the tube. There are of course some variations from species to species, but in general the fine structure of flagella is remarkably constant throughout the living world. The arrangement of microtubules is about the same in swimming algae, one-celled

× 1700

Scanning electron micrograph of cilia on the lining of a human trachea. In cells such as these, the beating of cilia causes fluid to move along the cell surface, but in free-floating cells, such as ciliated protozoa, the cilia make the cell move.

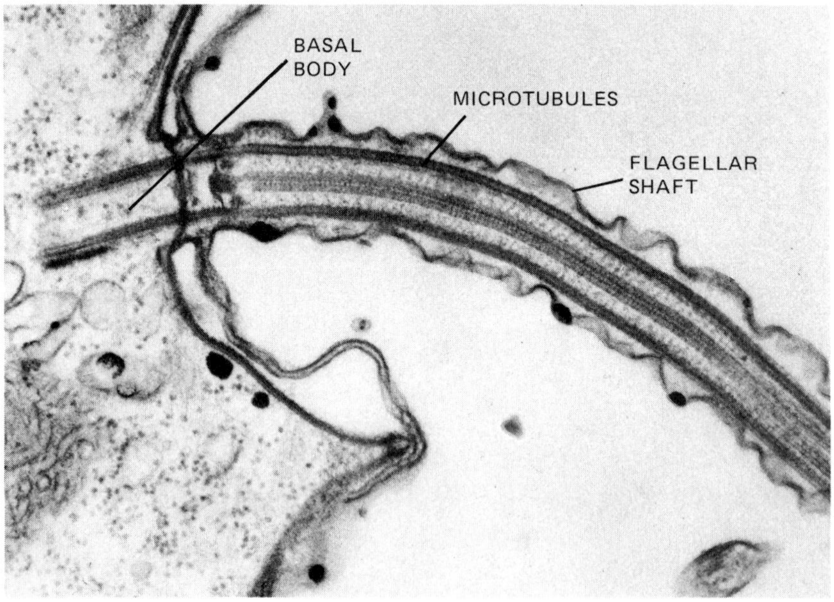

Omikron × 30,000

(a)

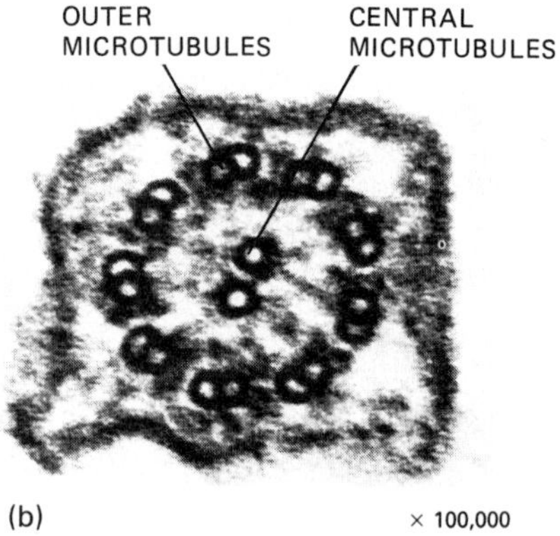

(b) × 100,000

(a) Longitudinal section of a flagellum from a frog sperm cell, as shown in an electron micrograph. The microtubules of the flagellar shaft are anchored in the basal body in the cytoplasm. (b) Cross section of a flagellum of a protozoan, showing the nine-plus-two arrangement of the microtubules, the "arms" protruding

An imagined dissection of a flagellum reconstructed from sections and broken flagella as seen in electron micrographs. The shaft of the flagellum contains the standard nine double peripheral tubules and two central ones. Rows of projections, the "arms," and T-shaped "hammerheads" are thought to act as rachets, somehow clicking past each other to make the tubules slide against one another, thus causing the shaft of the flagellum to bend. The base of the flagellum is anchored in a centriole or basal body in the cytoplasm of the cell.

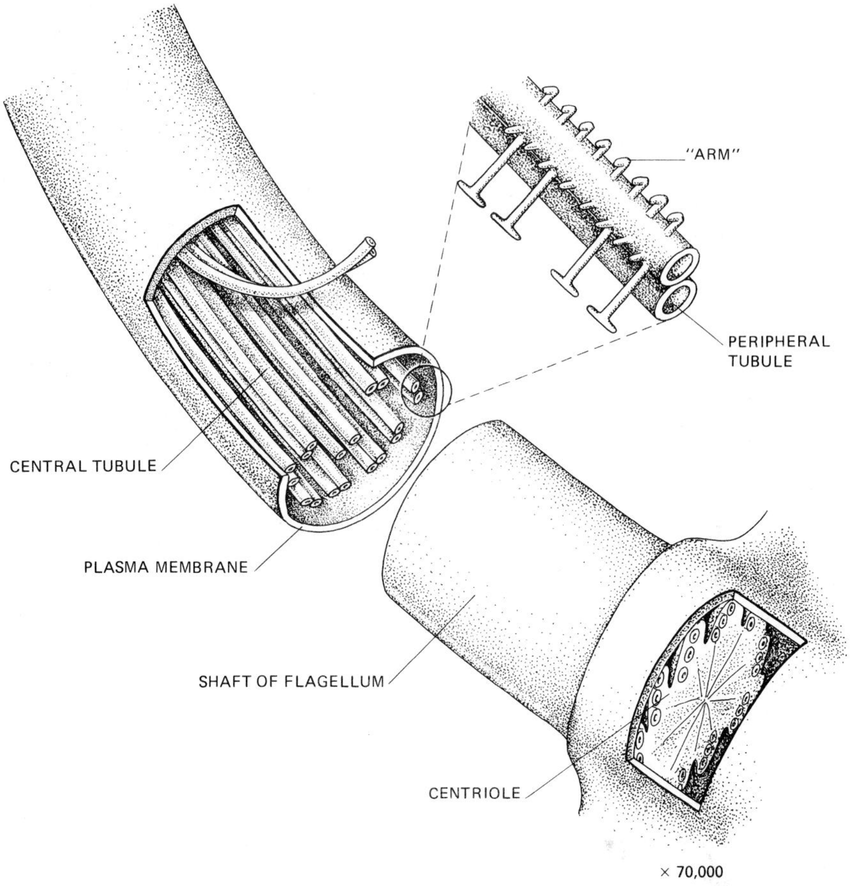

Diagram of power stroke and recovery of a flagellum.

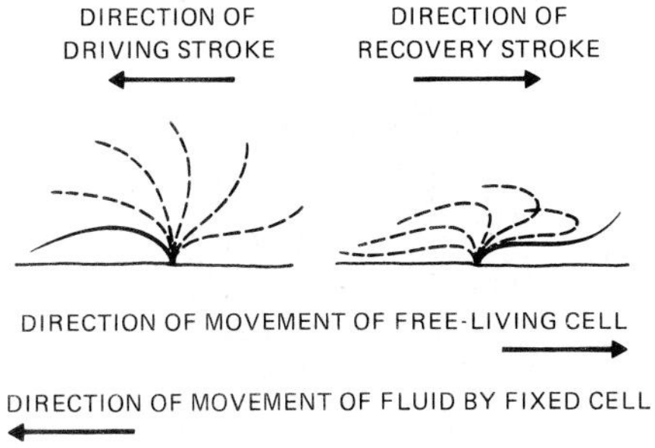

animals, mammalian sperm or fern sperm (where the special structure was first discovered), clam gills, and human mucus membranes. A section through a flagellum shows a ring of nine double microtubules around the periphery and two separate center ones. The unproved assumption is that the microtubules actively slide past one another to make the entire flagellum bend. They certainly can bend, even after being broken from cells, but only if they are provided with an energy source, such as ATP.

A ciliary beat consists of a sharp snap in one direction followed by a gentler, flexible recovery. What triggers the beat is not known, but it is evident that the impulse is controlled with great precision. In a protozoan or in a piece of ciliated epithelium, waves of movement sweep along as accurately synchronized as in a group of experienced handbell ringers.

The "invention" of flagella was probably one of the most important advances in the history of organic evolution. Surely an organism experiences an enormous advantage if it can move away from its fixed or passively floating competitors when overcrowding pollutes the environment or the food runs low.

become attached. This **centromere** (Figure 3.27) seems to be the hitching point for the connection of spindle fibers (microtubules), and it guarantees that a chromosome will proceed to its proper destination during cell division (see Chapter 8).

Nucleoli

Nuclei contain one or more roundish or lumpy **nucleoli** (new-KLEE-uh-lie), visible in living cells, as well as in stained cells, where they appear as dark balls (see Figure 3.25). Like chromosomes, nucleoli lack covering membranes. Nucleoli are composed of protein, presumably brought into the nucleus from the cytoplasm, and ribosomal RNA (see Chapter 7). At the onset of cell division, nucleoli disappear, to reappear after the division process is complete. Special chromosomal regions, the nucleolar organizers, start the process of forming ribosomes, so that a nucleolus can be thought of as a preassembly point for ribosomes, which move from the nucleolus out into the cytoplasm.

Figure 3.27
A chromosome reconstructed from electron micrographs, as it appears during nuclear division. The tangled chromatin threads and the centromere, the attachment point for the spindle fibers (microtubules), are conspicuous. Many chromosomes show a constriction at some point other than the centromere.

PROKARYOTES AND EUKARYOTES

Some organisms, such as the bacteria and blue-green algae, seem to have survived for several billion years without advancing beyond their original primitive state. These are the **prokaryotes,** meaning "before a nucleus." They differ from more advanced organisms in several structural features, the most obvious being that they lack distinct, membrane-bound nuclei. They are, indeed, characterized by their many structural deficiencies. It is noteworthy, however, that in spite of these seeming deficiencies, prokaryotes are quite effective. They compete successfully in a world filled with rivals.

Prokaryotes have no ER, no mitochondria, no plastids, no nucleoli, no Golgi bodies, and no true flagella. A prokaryotic "chromosome" is a naked strand of DNA without the proteins that are usually associated with chromosomes. Such a DNA strand is usually a continuous loop, inaccurately described as "circular." Prokaryotes do have ribosomes and extensions called flagella, but the so-called flagella of motile bacteria are more like single microtubules than like the complex true flagella of higher organisms.

The **eukaryotes,** meaning "true nuclei," are the organisms whose cells have all the features described in this chapter. Apparently the evolution of these diverse organelles has made possible the range of variation that is observable in all the plants and animals of the living world (Table 3.2).

SUMMARY

1. Robert Hooke first discovered and named cells in 1665. Almost 200 years later the *cell theory* was formulated.
2. *Cells* are the smallest independent units of life, and all life as we know it depends on the chemical activities of cells.
3. The living stuff of cells is *protoplasm;* the protoplasm of a single cell consists of a *nucleus* and the nonnuclear portion, the *cytoplasm.*
4. Plants owe most of their rigidity to their *walls.* In land plants, the commonest component of cell walls is *cellulose.* In contrast to the firm walls of

TABLE 3.2
Presence and Functions of Cell Organelles

STRUCTURE OR ORGANELLE	IN PROKARYOTIC CELLS	IN EUKARYOTIC CELLS	MAIN FUNCTION(S)
Cell covering	Envelope usually containing muramic acid	Walls in plants, cell coat in animals	Protection and support
Cell membrane	Present; site of major enzymatic activity	Present	Regulation of transport
Photosynthetic chromatophores	Present in some	Absent	Photosynthesis
Plastids	Absent	Present in most plants	Photosynthesis
Lysosomes	Absent	Usually present	Contain digestive enzymes
Golgi bodies	Absent	Usually present	Concentration and passage of materials
Mitochondria	Absent	Present	Aerobic respiration
Ribosomes	Present	Present	Active in protein synthesis
Nucleus	Absent	Present	Contains chromosomes and nuclear sap
Nuclear membrane	Absent	Present, with pores	Surrounds nucleus
Chromosomes	Continuous loop of naked DNA	Present, with centromeres and protein matrix	DNA replication and control of protein synthesis
Nucleoli	Absent	Present	Preparation of ribosomes
Endoplasmic reticulum	Absent	Present	Internal structure and transport
Microtubules	Present	Present	Movement
Flagella	True flagella lacking; microtubulelike structures	Present; consistent microtubule structure	Motility; water current
Centrioles	Absent	Present in all but higher plants	Mitotic activity and basal granules of cilia
Microtrabeculae	?	Present	Cytoplasmic structure

plants, the cell *coat* of animals is flexible and elastic. The animal cell coat is made of a complex of protein-plus-carbohydrate.

5. The *cell membrane* is essential to most normal cell functions. Cell membranes are not only selective but active. They allow some substances to pass through while refusing passage to others, and they can initiate and control the active transport of substances in and out of the cell.

6. In the *fluid mosaic model* it is postulated that the cell membrane is a double layer of phospholipids studded with protein molecules; in this model the entire membrane is capable of constant internal rearrangement.

7. Diffusion is one of the factors affecting water balance in cells. *Diffusion* is the tendency of molecules to move from a region of greater concentration of themselves toward a region of lesser concentration.

8. *Osmosis* is the movement of a substance through a selectively permeable membrane.

9. *Passive transport* occurs when some molecules cross membranes purely by diffusion, and the membrane has nothing to do with it. *Facilitated transport* does not require energy, but does require a molecular carrier of some sort. A cell membrane demonstrates definite, energy-requiring activity during *active transport.*

10. The *endoplasmic reticulum* is a complex of double membranes. It seems to act as a system of internal channels through which various materials move.

11. *Ribosomes* are necessary for the synthesis of protein from amino acids, but the exact function of the ribosomal RNA in the process is still unknown.

12. *Golgi bodies* are most apparent and abundant in cells with secretory activity. The Golgi bodies concentrate the substance to be secreted and deliver it to its destination, and in some instances they actually engage in some synthetic work along the way.

13. *Lysosomes* are loaded with digestive enzymes, and they can act as scavengers, ingesting and digesting "worn out" cell parts.

14. *Plastids* are responsible for the conversion of light energy to chemical energy.

15. *Mitochondria* are known as the powerhouses of the cell because their main work is the production of ATP in cellular respiration.

16. *Microtrabeculae* hold mitochondria, ribosomes, and other particles in place.

17. *Microtubules* are slender tubules that are commonly associated with cellular movement.

18. *Centrioles* are associated with nuclear division and with cilia and flagella.

19. The *nucleus* is the control center of the cell. It contains all the necessary information to direct the reproduction and heredity of new cells. The *nuclear membrane* is a double membrane with numerous pores in it, and there is two-way traffic between the nucleus and the cytoplasm.

20. *Chromosomes* are the main DNA-carrying structures, located within the nucleus in higher organisms. They have no limiting membranes, but they do have special regions, the *centromeres,* which are attachment centers for spindle fibers during nuclear division. Chromosomes are the cell's chief self-replicating organelles; they organize *nucleoli,* and they determine the structure of cell proteins.

21. *Nucleoli* are assembly centers in nuclei, functioning in the partial construction of ribosomes.

22. *Cilia* and *flagella* provide a means of locomotion for small, free-swimming organisms, or a means of moving water past fixed cells.

23. *Prokaryotic* cells lack distinct, membrane-bound nuclei. *Eukaryotic* cells have all the features described in this chapter.

ASK YOURSELF

1. What are the main ideas in the "cell theory" in biology?
2. What is the major chemical component of plant cell walls?
3. What cellular structures make the difference between the hardness of wood and the softness of animal muscle?
4. Why are the words "fluid" and "mosaic" used in describing cell membranes?
5. What are the main chemical components of cell membranes?
6. Why do biologists say that cell membranes are "selectively permeable"?
7. What is the difference between diffusion and active transport?
8. Describe phagocytosis and give an example.
9. What is the difference between smooth and rough endoplasmic reticulum?
10. Why have lysosomes been called "suicide bags" and "garbage disposal units"?
11. What is the general utility of Golgi bodies in animal cells?
12. What are the main chemical components of chromosomes?
13. What is the function of a nucleolus?
14. What components does a eukaryotic cell have that a prokaryote lacks?

4
Photosynthesis: Trapping and Storing Energy

SOME KEY POINTS

1. Through the process of photosynthesis light energy is converted into chemical energy and stored as carbohydrates that plants use as food. Water and carbon dioxide are combined to form sugar (glucose) and oxygen.
2. The requirements for photosynthesis are light, chlorophyll, chloroplasts, carbon dioxde, water, and oxygen.
3. The light and dark reactions of photosynthesis take place in light simultaneously and continuously. The light reactions trap light energy and convert it to chemical energy. The dark reactions use chemical energy to produce glucose.

LIFE RUNS ON SUNLIGHT. PLANTS CAN CONVERT the light energy of the sun into usable chemical energy, and this process is called **photosynthesis,** one of the most important chemical processes in nature. Plants store enough food with their light-trapping abilities to feed not only themselves but animals as well. Without photosynthesis there would be almost no usable energy for the cells of all living organisms. Not only does photosynthesis provide the food and energy to sustain plant life, it indirectly supplies the basic fuel for animals as well. Because animals cannot manufacture their own food, they must consume plants and other animals. Besides obtaining their nutritional needs from plants, animals receive their oxygen supply as a by-product of photosynthesis. Without photosynthesis, therefore, life as we know it could not exist.

WHAT IS PHOTOSYNTHESIS?

The essential facts about photosynthesis are these: Light energy is converted by means of the chlorophylls and accessory pigments in plants into chemical energy and stored as carbohydrates; the simple compounds used are water and carbon dioxide; the final products are organic molecules (sugars, starches, fats, etc.), and oxygen. The overall reaction is usually written

$$6CO_2 + 6H_2O \xrightarrow[\text{light}]{\text{chlorophyll}} C_6H_{12}O_6 + 6O_2.$$

Carbon dioxide	plus	Water	in the presence of chlorophyll and light yields	Glucose	plus	Oxygen

This process becomes more understandable if we analyze the word "photosynthesis." *Photo* means light, and *synthesis* means the process of building up a complex substance through the union of simpler substances. But the seemingly simple process of photosynthesis is aided by complex support mechanisms, as we shall see.

PRODUCERS AND CONSUMERS: AUTOTROPHS AND HETEROTROPHS

No life can exist without a constant source of energy. There are many variations in the methods used to extract and use energy, but the two major types of energy-requiring cells are called *autotrophic* and *heterotrophic*. When scientists first discovered that plants are able to manufacture their own food, organisms were reclassified as either **autotrophs,** organisms that can synthesize their food from inorganic sources, or **heterotrophs,** organisms that must acquire their food because they cannot manufacture it themselves.

The chief autotrophs are green plants, which acquire their energy by transforming the light energy of the sun into carbohydrate plus oxygen. Heterotrophic cells, such as those found in the human body and in other animals, in fungi, and in plants that do not contain chlorophyll, obtain their energy through the breakdown of food, such as sugar, in a process called *respiration*. Note that photosynthesis builds up compounds to convert and store energy, whereas respiration breaks down compounds to release energy. Cellular respiration will be discussed in detail in the next chapter.

Some autotrophic bacteria use the energy from oxidation of inorganic materials to make their own food, using a process called *chemosynthesis* rather than photosynthesis. These bacteria are not of great importance in the total economy of nature, but students should know that not *all* autotrophs are necessarily photosynthetic. This is one more example of the rarity of absolute statements in biology.

WHAT ARE THE REQUIREMENTS FOR PHOTOSYNTHESIS?

The principal requirements for photosynthesis are light, chlorophyll, chloroplasts, carbon dioxide, and water. Oxygen and some critical minerals must also be present. A further requirement is structural integrity of the functional units, the chloroplasts, which means that their pigments, membranes, and enzymes must be held in a proper physical relationship. If chloroplasts are damaged, they stop making food.

Light

The electromagnetic spectrum is composed of radiations of varying wavelengths, ranging from very short wavelengths of high frequency (gamma rays, X-rays) to very long ones of low frequency (infrared radiation, radio waves). That portion of the electromagnetic spectrum usable by plants for photosynthesis ranges from about 380 to 750 nanometers (a *nanometer*, abbreviated nm, is a billionth of a meter). This is approximately the same range that the human eye can detect, that is, the visible portion of the spectrum.

The energy in light is inversely proportional to the wavelength. The short (violet) end of the visi-

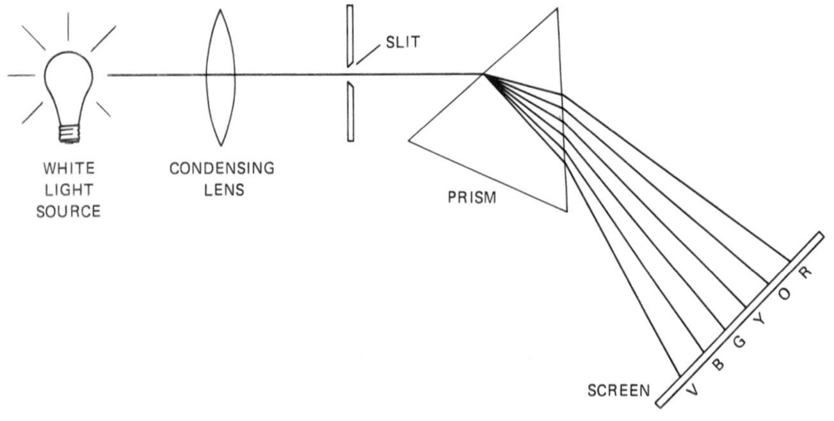

Figure 4.1
Breaking white light into its components. A source of white light, a mixture of all wavelengths and frequencies, has a beam focused through a slit and then through a prism. The glass of the prism changes the direction of the beam differently for different wavelengths, so that the shorter ones (violet, about 380 nm) are bent more than the longer ones (red, 700–750 nm). The other wavelengths are distributed between, with green about 550 nm and yellow about 600 nm. The resulting spectrum can be projected on a screen to produce a visible "rainbow."

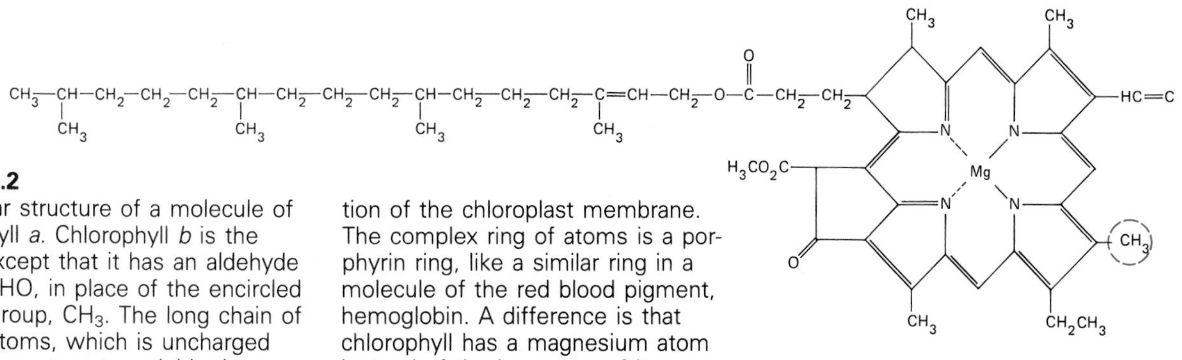

Figure 4.2
Molecular structure of a molecule of chlorophyll *a*. Chlorophyll *b* is the same, except that it has an aldehyde group, CHO, in place of the encircled methyl group, CH_3. The long chain of carbon atoms, which is uncharged and hence not water-soluble, is thought to extend into the lipid portion of the chloroplast membrane. The complex ring of atoms is a porphyrin ring, like a similar ring in a molecule of the red blood pigment, hemoglobin. A difference is that chlorophyll has a magnesium atom instead of the iron atom of hemoglobin.

ble spectrum contains about 1.4 times more energy than the long (red) end (Figure 4.1). Chlorophyll pigments absorb some light along the entire visible spectrum, but they absorb more effectively at the red end. Indeed, plants appear green because of their great absorption of red light and the reflection of the green portion of the spectrum.

The *intensity* of light, which is the amount of energy delivered on a given area per unit of time, is as important as the wavelength (or "quality") of the light. If the light intensity is too low, photosynthesis ceases. At low light intensity the plant uses up the photosynthetic food products as fast as they are produced. At the other extreme, light can reach a point of intensity so high that the plant can use no further light. Many plants, especially those adapted to low light intensities, as in deep water or in tropical rain forests, are actually harmed by light that is too intense, and chlorophyll is destroyed.

Chlorophyll, the Photosynthetic Pigment

In general, pigments are substances that absorb light. The most conspicuous photosynthetic pigments are the **chlorophylls.** The two common chlorophylls of higher plants are chlorophyll *a* and chlorophyll *b*, which are chemically similar. Both are complex molecules composed of carbon, hydrogen, oxygen, nitrogen, and magnesium (Figure 4.2).

The special feature of chlorophyll is its ability to be "excited" by certain wavelengths of light. During such excitement, which lasts about one-billionth of a second, pairs of electrons are pushed farther away from their nucleus and are thus raised to a higher, unstable energy level (Figure 4.3).

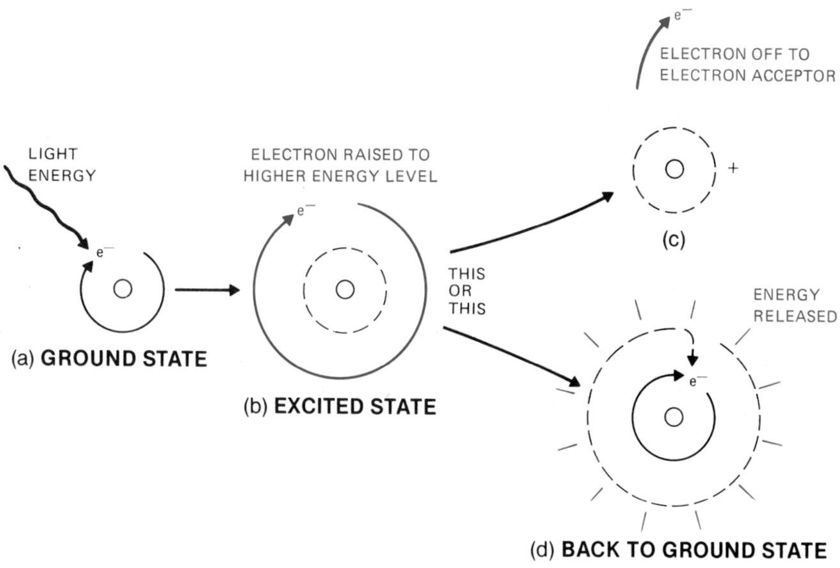

Figure 4.3
Schematic representation of activation of a molecule of chlorophyll by light. When a molecule in its ground state (or inactivated state) is energized by light (a), an electron is raised to a new, higher energy level (b), leaving an unoccupied place in the electron shell, frequently called a "hole." If an electron acceptor is available, the electron can pass from the original molecule (such as chlorophyll) to the acceptor, and the energy will be passed to the acceptor (c). If there is no acceptor, the energized electron will fall back into its original position, and in so doing will re-emit energy (d). The re-emitted energy may be in the form of heat, or it may be in the form of light.

Chlorophyll *a* seems to be most directly involved in photosynthesis. Chlorophyll *b* and the other related pigments are able to absorb energy, but instead of using it directly, they apparently pass it along to chlorophyll *a*.

Plastid Organization

The organelle of photosynthesis is the plastid. **Plastids** are relatively large, membrane-bound, colored organelles found only in plant cells. The special function of plastids is the synthesis and storage

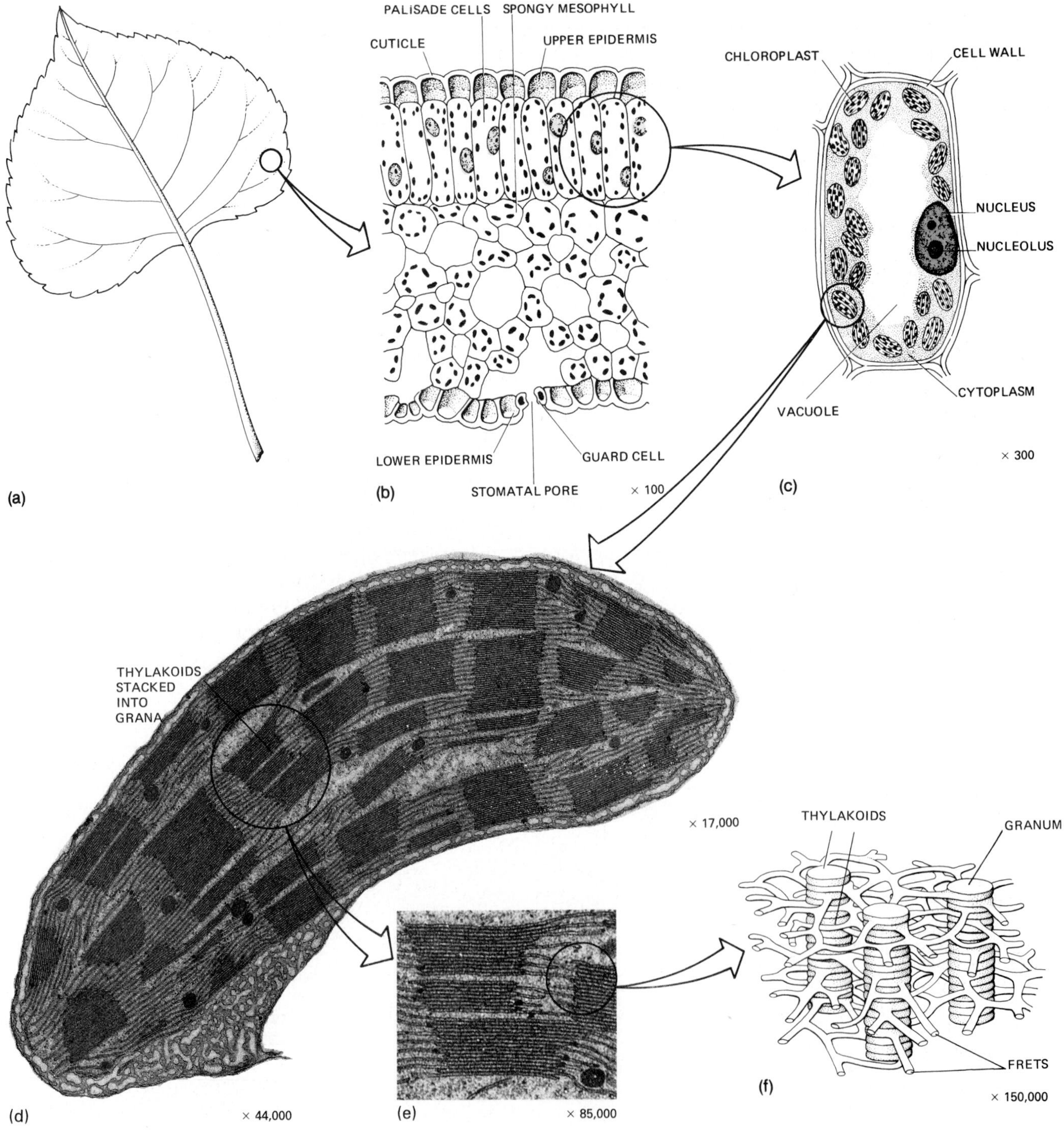

of food. Three types of plastids are known, and their names also indicate their colors. *Chromoplasts* can be yellow, orange, or red (Greek *chroma,* color). *Leucoplasts,* containing oil or starch, are almost colorless (Greek *leukos,* clear). The photosynthetic plastids, **chloroplasts,** contain the green chlorophyll (Greek *khloros,* greenish yellow) (Figure 4.4b and c).

With the use of the electron microscope, we can see the complex internal structure of a chloroplast (Figure 4.4d–f). Included are **grana,** which are composed of stacks of flattened, membrane-bound vesicles, the *thylakoids.* The membranes in the grana are like other cell membranes in that they contain proteins and phospholipids, but they are special because they contain the chlorophylls, the accessory pigments, and the various electron acceptors necessary for the "light reactions" of photosynthesis. Electron micrographs showing surface views of thylakoid membranes reveal regularly arranged bumps, each about 150 Å across. These bumps, called *quantasomes,* may be the actual photosynthetic units of the chloroplast. Non-grana regions of the chloroplasts, known as the **stroma,** are the sites of the so-called dark reactions of photosynthesis, to be discussed later in this chapter.

Figure 4.4
The photosynthetic machinery. (a) An intact leaf, natural size. (b) A section through a leaf. (c) A single chloroplast-bearing cell from the spongy mesophyll of a leaf. (d) Electron micrograph of a section through a single chloroplast, showing the covering membrane and *thylakoids* stacked into *grana.* (e) Electron micrograph of a section through four grana, showing neatly arrayed thylakoids making up the grana and connecting *frets* extending through the *stroma* from one granum to the next. (f) A reconstruction of several grana, shown in three dimensions. The drawing is a composite of electron micrographs made from sections cut at different angles. The grana are seen as stacks of thylakoids, connected by membranous frets between the grana.

Carbon Dioxide

There are three to four molecules of carbon dixoide in every 10,000 molecules of gases in the earth's atmosphere. This is a sufficiently high concentration so that plants can pick up the carbon necessary for the buildup of carbohydrate in photosynthesis. If the concentration goes too low, say to five parts per 100,000, ordinary photosynthesis ceases. If the concentration gets too high, carbon dioxide can become toxic. Plants could use more carbon dioxide than they normally get, up to five parts per 100 or more, but such a concentration is practically never found in nature. If it were economical to do so, crops might be "fertilized" by adding carbon dioxide to the air around leaves during daylight, but the high photosynthetic rates caused by these high carbon dioxide concentrations cannot be kept up indefinitely.

Water

One of the few absolute statements one can make about anything biological is that water is essential for all vital activities. It is the common cellular solvent, it helps regulate temperatures, it enters into chemical combination in countless reactions, it furnishes the essential protons for energy exchanges, it provides turgor pressure to create the stiffness of tender plant parts, it acts as a lubricant and as a medium of transport, and it even enters into the structural parts of cells. But it has a special value in photosynthesis because it is one of the two raw materials from which organic molecules are synthesized. The other is of course carbon dioxide. A molecule of water, broken apart in an active chloroplast, yields two protons (H^+), two electrons, and an atom of oxygen. The protons, or hydrogen ions, and the electrons become incorporated into a sugar or some other energy-containing molecule. The oxygen atom combines with another oxygen atom to form molecular oxygen, O_2, which will escape into the atmosphere.

Because water is essential to photosynthesis, plants must have some way to conserve it. The surfaces of plants are usually covered with a waxy cuticle called *cutin* that prevents excessive water loss, invasion by parasites, and damage such as bruising (see Figure 1.19). The waxy surface also shuts out carbon dioxide, and plants have had to evolve means of letting in the gas. The trick was to develop microscopic, controllable openings between cells in the leaf epidermis (outer layer). These openings are

stomata, each one bounded by a pair of *guard cells* (Figure 4.5).

Stomata are distributed on leaf surfaces in such a way that they allow adequate entrance of carbon dioxide and the liberation of oxygen during photosynthesis. Because the stomata cannot choose which gases can pass, they must also allow water vapor to escape. Water will necessarily escape because the internal surfaces of the leaf are wet with water brought up from the roots, and the air outside is usually drier than the air in the spaces of the leaf. Excess water loss can be fatal. As a compromise, leaves allow carbon dioxide, oxygen, and water vapor to move through the stomatal openings when the leaves are in light, but they close the openings in darkness. We can make stomata open or close and can watch them doing so, yet the exact mechanism is not known. The best that can be said is that light, water, and potassium movement are involved.

Other Photosynthetic Requirements

Besides the direct requirements of light, raw materials, and chloroplasts, a number of indirect requirements must be met if photosynthesis is to proceed. For example, there must be some oxygen present, and mineral elements, such as iron, copper, magnesium, sulfur, nitrogen, potassium, phosphorus, and boron are of critical importance.

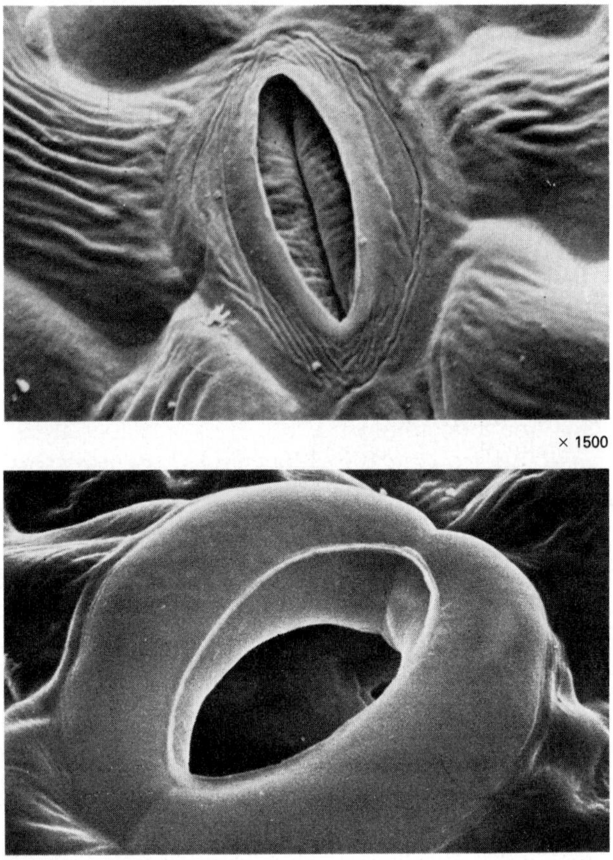

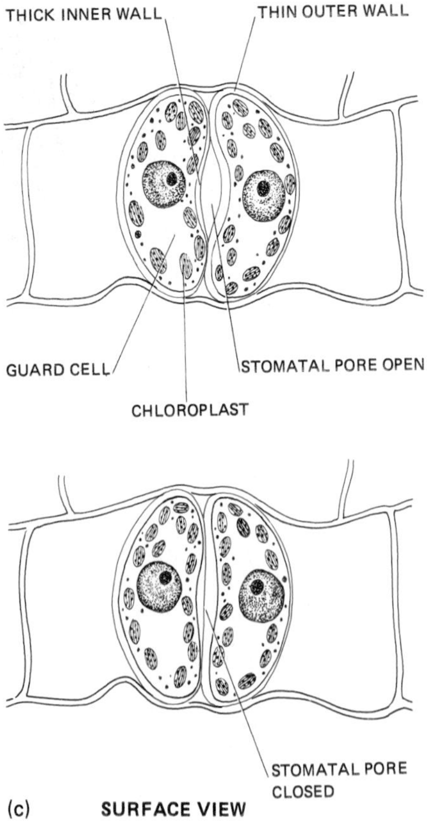

Figure 4.5
Stomata. (a) Closed stomata of a parsley leaf. (b) Open stoma of a cucumber leaf. (c) Movement of guard cells in stomata. When water enters guard cells and makes them firm, the pressure against the thin outer walls causes them to bulge outward, pulling the stomatal pore open (top drawing). When water is lost from guard cells they become soft, and the thin outer walls lose their bulge. The thick inner walls force the stomatal pore shut (bottom drawing).

UNDERSTANDING THE TWO PHASES OF PHOTOSYNTHESIS

It has been known for more than 50 years that there are at least two steps in the photosynthetic process, one requiring light and one not requiring light. Before discussing the "light" and "dark" reactions of photosynthesis, we should emphasize certain points to avoid confusion:

1. The light reactions *do* utilize light energy.
2. The dark reactions are not part of the light-trapping process. The use of the term "dark" *does not* mean that the dark reactions necessarily take place in the dark or at night. It means simply that these reactions may proceed without light as a requirement.
3. The light and dark reactions are consecutive steps in the overall process of photosynthesis, which starts with the trapping of light energy (the light reactions) and ends with the production of sugar (the dark reactions).
4. Photosynthesis is a continuous process. The light and dark reactions are discussed separately only to divide the light-absorbing phase from the "chemical" phase, which produces carbohydrate through complex chemical reactions. The light reactions *convert light energy to chemical energy;* the dark reactions *use chemical energy to produce complex organic molecules.*

THE LIGHT REACTIONS OF PHOTOSYNTHESIS

In the light reactions of photosynthesis, two products are formed, one an energy carrier (ATP) and the other a hydrogen carrier with the long name of nicotinamide adenosine dinucleotide phosphate, usually abbreviated to NADP (Figure 4.6).

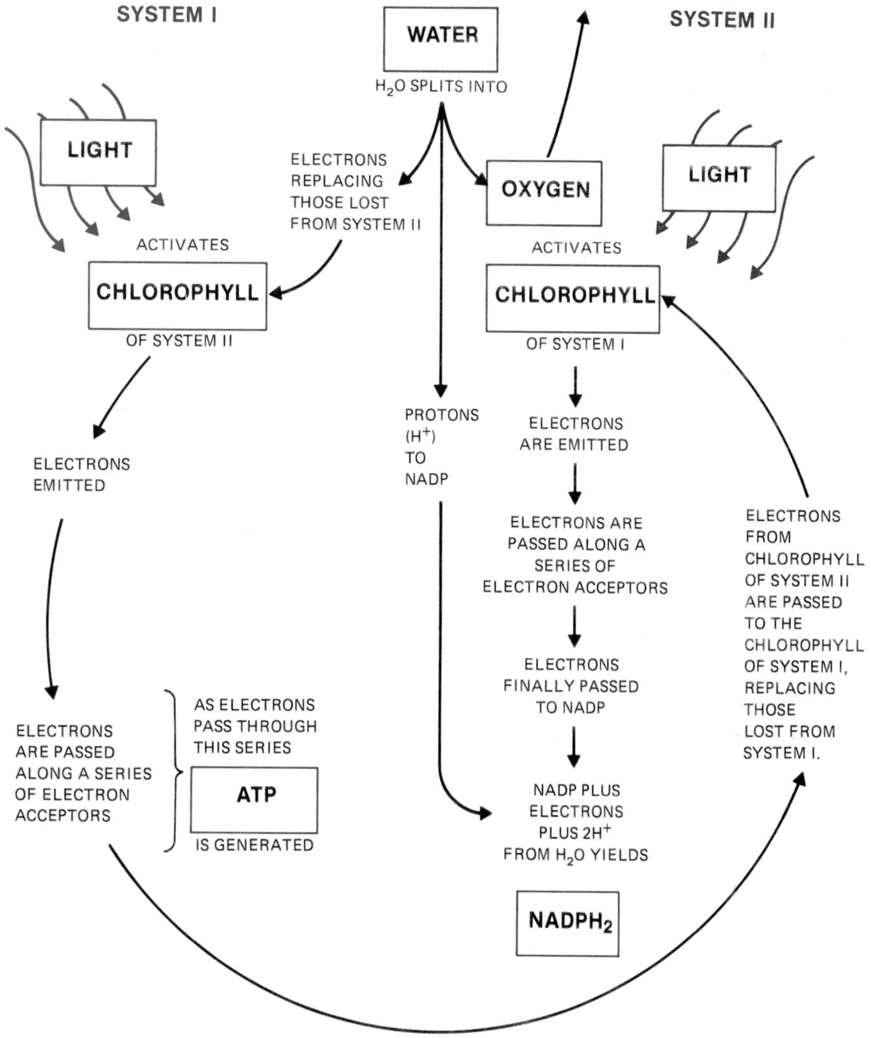

Figure 4.6
The light reactions of photosynthesis.

When light falls on a chloroplast, several simultaneous actions are started. There are two kinds of light-sensitive centers, designated *Pigment System I* and *Pigment System II,* which differ in their components and in the wavelengths of light they absorb. Pigment Systems I and II are alike in that they contain chlorophyll and that they can be activated by light energy.

When chlorophyll in Pigment System I is activated by light with a wavelength of 700 nm, pairs of electrons are raised to a high enough energy level to be released from the chlorophyll molecule. These energy-carrying electrons are passed along a series of acceptors. The electrons ultimately are transferred to NADP, which accepts not only the electrons but also a pair of protons, $2H^+$, reducing NADP to $NADPH_2$. The compound $NADPH_2$ is able to carry its load of hydrogen (H^+ and electron) to the next photosynthetic step, the dark reactions. In the dark reactions the hydrogen will be used in the buildup of carbohydrate.

Meanwhile, the electrons have left a "gap" in the chlorophyll from which they came, and that gap must be refilled before photosynthesis can continue. That refilling is accomplished by the action of Pigment System II.

While Pigment System I was working, System II was also being activated, but by light of wavelength of 680 nm. Its chlorophyll has pairs of electrons raised to high energy levels, and they are removed from the chlorophyll, accepted by a series of electron acceptors, and finally passed to the above mentioned "gaps" in the chlorophyll of Pigment System I. This process of course leaves gaps in the chlorophyll of Pigment System II, and they in turn must be refilled.

The electrons lost from Pigment System II are replaced by electrons from water molecules that are split into protons (H^+), oxygen, and electrons. The protons go to NADP, generating $NADPH_2$. The electrons fill the gaps left in the chlorophyll of Pigment System II, and they thus keep the process operating. The remaining oxygen atoms form molecular oxygen, O_2, which diffuses into the atmosphere.

Another important action is going on at the same time. As the electrons in Pigment System II are passed along their chain of carriers, they become less energized at each step, but not all their energy is lost to the plant. Some of it is caught and used in the formation of ATP from ADP and inorganic phosphate. This ATP will be used as an energy supply in the dark reactions to follow.

In some primitive plants, as well as in higher plants, there is a process called *cyclic phosphorylation.* It is called "cyclic" because electrons are recycled in Pigment System I and "phosphorylation" because phosphorus is transferred. In this process, no new electrons are received from water, no NADP is reduced, and the only product is ATP.

TO SUM UP THE LIGHT REACTIONS

The *main events in the light reactions* are: (1) the light activation of chlorophylls, with the release of energy-bearing electrons; (2) the splitting of water with the release of H^+, electrons, and oxygen; (3) the reduction of NADP to $NADPH_2$; and (4) the generation of ATP.

The Requirements of the Light Reactions	*The Products of the Light Reactions*
1. Light	1. Oxygen
2. Enzymes in the chloroplasts, structurally intact	2. ATP
3. Water	3. $NADPH_2$
4. NADP	
5. ADP and inorganic phosphate	

THE DARK REACTIONS OF PHOTOSYNTHESIS

The dark reactions of photosynthesis take place in the chloroplasts. During the dark reactions, complex molecules of sugar and starch, composed of carbon, hydrogen, and oxygen, are being made from the simpler molecules of carbon dioxide and the hydrogen of $NADPH_2$. The sequence of events here was postulated by Melvin Calvin.

The first identifiable carbohydrate in most land plants is a 3-carbon compound, phosphoglyceric acid (PGA). During the **Calvin cycle** (Figures 4.7 and 4.8), carbon dioxide from the atmosphere is enzymatically united with a 5-carbon sugar, ribulose diphosphate (RuDP), to form a 6-carbon compound, which immediately breaks down into two molecules of 3-carbon PGA.

Energy to drive the system comes from ATP, and hydrogen is brought in via $NADPH_2$, both of which have been produced in the light reactions. The PGA is reduced to phosphoglyceraldehyde (PGAL), a 3-carbon compound that living cells can use as a starter for the synthesis of all the uncounted substances of living matter.

After PGAL has been made, it has several alternatives available. Some of the 3-carbon PGAL can be

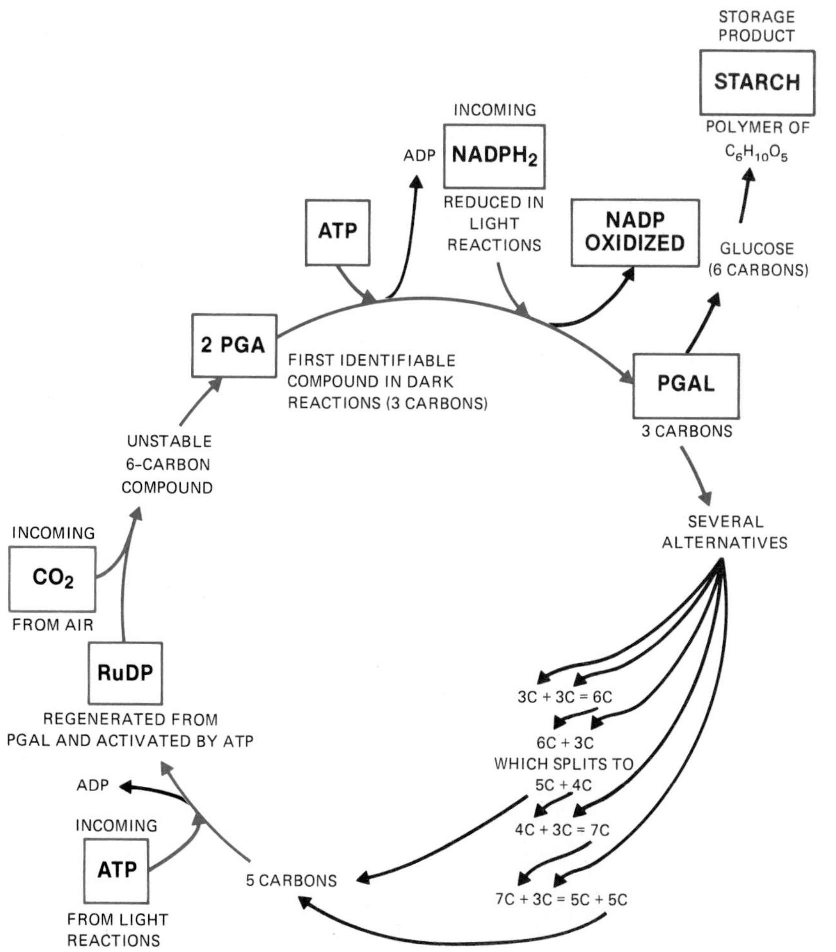

Figure 4.7
The Calvin cycle, the dark reactions of photosynthesis. When carbon dioxide enters the photosynthetic process, it joins the 5-carbon sugar ribulose diphosphate (RuDP) to form a temporary union, which quickly breaks to two 3-carbon molecules of phosphoglyceric acid (PGA). This is the first identifiable carbohydrate formed in photosynthesis. The PGA is reduced by hydrogen from NADPH, which itself was earlier reduced during the light reactions of photosynthesis. The NADP, now oxidized, is ready for another hydrogen-carrying trip, and the PGA is reduced to phosphoglyceraldehyde (PGAL).

The 3-carbon compound PGAL may be used in a number of ways. The most obvious use of PGAL is in the building of reserve food in the plant. Two molecules of PGAL can form one molecule of the 6-carbon sugar fructose, which in turn can be converted to glucose. Glucose can then be condensed to starch, the usual insoluble storage form of plant food.

Some of the PGAL must be reserved for use in the regeneration of more RuDP. This can be accomplished in several ways, as shown in the schematic diagram.

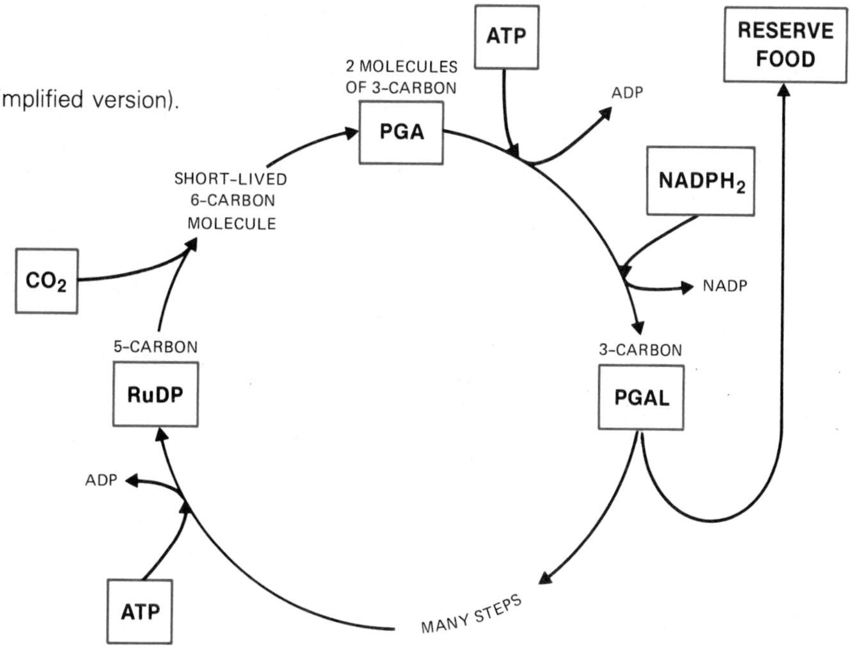

Figure 4.8
The Calvin cycle (simplified version).

condensed to 6-carbon sugars, such as fructose and glucose. These may be further condensed to a common storage product, starch, or may be enzymatically converted to fats or (with the addition of nitrogen) to amino acids. Such conversions are begun within minutes after a chloroplast is exposed to light. However, if all the PGAL were made into reserve food, there would be no new 5-carbon RuDP to pick up incoming new CO_2. Some of the PGAL, therefore, is used to generate new RuDP by means of several intermediates, including 4-carbon and 7-carbon compounds. The result is that although the bulk of PGAL synthesized during photosynthesis is used directly to make reserve food, enough is recycled to maintain an adequate supply of RuDP so that the process continues.

TO SUM UP THE DARK REACTIONS

The Requirements of the Dark Reactions	*The Immediate Products of the Dark Reactions*
1. ATP	1. Carbohydrate (PGAL)
2. $NADPH_2$	2. ADP
3. Carbon dioxide	3. Inorganic phosphate (P_i)
4. Ribulose diphosphate (RuDP)	4. Water
5. Enzymes in the chloroplasts	5. NADP

THE FATE OF WATER IN PHOTOSYNTHESIS

In general, photosynthesis can be thought of as the combination of carbon dioxide and water to yield oxygen and carbohydrate. As stated at the beginning of this chapter, the overall reaction is frequently written

$$6CO_2 + 6H_2O \xrightarrow[\text{light}]{\text{chloroplast}} C_6H_{12}O_6 + 6O_2.$$

However, if we recall that the resulting oxygen comes from the water and not from the carbon dioxide, we can see that the equation above is not quite correct. If the molecular oxygen on the right side of the equation comes from the water, then how can we account for 12 oxygen atoms from only 6 molecules of water? The answer is that additional water has entered into the reaction, and new water is formed as a result. Consequently, the equation is rewritten

$$6CO_2 + 12H_2O \xrightarrow[\text{light}]{\text{chloroplast}} C_6H_{12}O_6 + 6H_2O + 6O_2.$$

During photosynthesis, water and carbon dioxide are combined to form glucose and the important by-product oxygen. During respiration, glucose reacts with oxygen to release stored cellular energy, and carbon dioxide and water are by-products that are used again during photosynthesis. The energy released through respiration is distributed bit by bit in a controlled process that will be discussed in the next chapter.

ALTERNATIVE PHOTOSYNTHETIC SCHEMES

Much of what has been described in this chapter was determined from experiments on a green alga, *Chlorella*, or spinach, usually with an artificially high

Figure 4.9
A sugarcane field is a productive energy trap because in addition to using the 4-carbon method of photosynthesis, it thrives in conditions where the light is intense, the temperature is high, the water supply is constant, and the soil is rich.

concentration of carbon dioxide. Asking themselves whether the results from such special experimental conditions could be applied confidently to plants in general, some investigators looked at photosynthetic events in a number of plants under natural conditions. They found that some desert plants and some grasses, such as maize and sugarcane, do not use the Calvin cycle exclusively.

In sugarcane the first-formed compounds are not the expected 3-carbon PGA, but 4-carbon acids: malic, aspartic, and oxaloacetic acid. The Calvin scheme is consequently known as *3-carbon photosynthesis,* and the *Hatch-Slack scheme,* named for its discoverers, is the 4-carbon method. (No single species can shift back and forth under varying conditions between 3-carbon and 4-carbon photosynthesis. Any given land plant always uses *either* the 3-carbon system or a combination of both systems.) The Hatch-Slack method functions when 3-carbon photosynthesis is inhibited by too high a light intensity, too high a temperature, or too low a carbon dioxide concentration. It is therefore effective in tropical or desert conditions. It has been known for a long time that a sugarcane field is one of the most productive energy traps in all nature (Figure 4.9), and part of the explanation is the use by sugarcane of the 4-carbon method of photosynthesis, which allows the plant to take advantage of intense light and a hot climate. Thus photosynthesis, like many other biosynthetic phenomena, proves to be a process in which a given result can be achieved by more than one pathway.

SUMMARY

1. Plants can convert the light energy of the sun into chemical energy by the process called *photosynthesis,* one of the essential chemical processes in nature.
2. The essential fact about photosynthesis is that light energy, by means of the chlorophylls and accessory pigments, is converted into chemical energy and stored in carbohydrates.
3. Organisms are classified as either *autotrophs,* those that can synthesize their food from inorganic sources, or *heterotrophs,* those that must acquire their food because they cannot manufacture it themselves. The chief autotrophs are the green plants.
4. The principal requirements for photosynthesis are light, chlorophyll, chloroplasts, carbon dioxide, and water. The principal products are glucose, other organic molecules, and oxygen.
5. The *light reactions* of photosynthesis convert light energy to chemical energy. The *dark reactions* use chemical energy to produce complex organic molecules.
6. During the light reactions, chlorophyll is activated, releasing electrons. As the released electrons pass through a series of electron acceptors, ATP is generated.
7. Some of the electrons from chlorophyll reduce NADP to $NADPH_2$.
8. Electrons lost from activated chlorophyll are replaced by new ones from water. Water also supplies protons to NADP. The oxygen from water is released into the air.
9. In the dark reactions, carbon dioxide unites with a 5-carbon sugar to yield two molecules of 3-carbon PGA.
10. PGA is reduced by $NADPH_2$ to PGAL, which can either be made into a storage form of food or used to regenerate more 5-carbon sugar.
11. In some plants, the first-formed organic compounds are 4-carbon acids instead of the 3-carbon PGA.

ASK YOURSELF

1. What is the original source of energy for living things?
2. What two compounds are the raw materials for food production in green plants?
3. What is the source of renewed oxygen in the air?
4. What is an autotroph? A heterotroph?
5. If you illuminate a solution of chlorophyll, carbon dioxide, and water in a glass vessel, will the mixture produce sugar? Defend your answer.
6. What portion of the radiation from the sun is used by green plants for photosynthesis?

7. What are the end products of the light reactions of photosynthesis?
8. What is the immediate energy source for the dark reactions of photosynthesis?
9. What does the splitting of water in photosynthesis yield?
10. When electrons are lost from chlorophyll, how are they restored?
11. What is the function of NADP in photosynthesis?
12. What is the functional product of cyclic photophosphorylation?
13. When a molecule of carbon dioxide is joined to a molecule of five-carbon ribulose diphosphate, what is the immediate result?
14. If you could catch *all* the PGAL a green plant produces and remove it, what would happen to photosynthesis?
15. How does crab grass, a 4-carbon photosynthesizer, succeed in outgrowing other lawn grasses in midsummer but not in early spring?

5
Cellular Respiration: The Release of Energy

SOME KEY POINTS

1. Respiration releases the usable chemical energy that has been converted from light energy during photosynthesis.
2. Respiration involves four biochemical systems: glycolysis, fermentation, the Krebs cycle, and the electron transport system.
3. Most of the ATP generation and energy release in cells is made possible by a series of reactions in the Krebs cycle.
4. The electron transport system is the main producer of ATP.

DURING PHOTOSYNTHESIS, CHEMICAL ENERGY is stored in the bonds of organic compounds such as glucose. The energy is useless to cells, however, unless it can be released. During cellular respiration, as chemical bonds are rearranged, the stored energy is released. The process of cellular respiration is made up of many separate chemical reactions, each of which is controlled by a separate enzyme.

Respiration is a process that takes place inside cells, mainly in mitochondria.* It occurs in *all* living cells to produce the moment-to-moment energy required for basic life processes; because cells need energy all the time, respiration is a constant process. In contrast, photosynthesis takes place only at certain times in specialized plant cells (Table 5.1).

Although several organic compounds are involved in respiration, we usually use glucose as a starting material because much of what is known about cellular respiration was learned from studies on yeast, and yeast uses glucose as an energy source. The equation for respiration is usually written as

$$\underset{\text{Glucose}}{C_6H_{12}O_6} \underset{\text{plus}}{+} \underset{\text{Oxygen}}{6O_2} \underset{\text{yields}}{\longrightarrow} \underset{\text{Carbon dioxide}}{6CO_2} \underset{\text{plus}}{+} \underset{\text{Water}}{6H_2O} \underset{\text{plus}}{+} \underset{\text{Energy}}{277\ \text{Kcal}}$$

This is essentially the reverse of the equation for photosynthesis. Although the equation appears reasonably simple, it actually represents a complex se-

* The word "respiration" on its own can be taken to mean "breathing." In this chapter, when we say *respiration* we specifically mean *cellular respiration.*

TABLE 5.1
Photosynthesis and Respiration

	PHOTOSYNTHESIS	RESPIRATION
Where?	In chlorophyll-bearing cells	In all living cells
When?	In light only	All the time
Input?	CO_2 and H_2O	Reduced carbon compounds and O_2
Output?	Carbon compounds, O_2, and H_2O	CO_2 and H_2O
Energy source?	Light	Chemical bonds
Energy result?	Energy stored	Energy released
Chemical reaction?	Reduction of carbon compounds	Oxidation of carbon compounds

ries of chemical reactions. Remember that each step in the process is controlled by a separate enzyme; approximately 100 enzymes are used in the respiration of glucose.

THE STAGES OF RESPIRATION

Although the release of cellular energy is a continuous process that is carried on step by step from an organic molecule, such as glucose, to the final release of carbon dioxide, water, and energy, it is convenient to divide respiration into four distinct stages:

1. **Glycolysis** (which means "breakdown of sugar") is the initial breakdown of glucose into an intermediate compound. It takes place in the cytoplasm of a cell, and does not need oxygen.
2. **Fermentation** is the partial breakdown of pyruvic acid (produced during glycolysis) by enzyme action in the absence of free oxygen.
3. During the **Krebs cycle** all the remaining carbon is removed as carbon dioxide, and hydrogen ions are also given up as oxidation occurs.
4. The **electron transport system** is the main producer of ATP; it uses molecular oxygen in this process. The Krebs cycle and the electron transport system occur only in mitochondria. Glycolysis occurs in the cytoplasm of a cell.

Glycolysis, the first portion of the respiration process, takes place in the absence of oxygen, and is part of **anaerobic respiration** (*anaerobic* means "life without air"; the "an-" is a negative prefix that denotes "not" or "without"). Organisms, especially bacteria, that are able to obtain all their energy from glycolysis or fermentation reactions are called *anaerobes.* The Krebs cycle and the electron transport system are **aerobic** processes because they require free oxygen from the atmosphere. Most organisms obtain the bulk of their energy from the aerobic portion of the respiration process.

Glycolysis: The Initial Breakdown of Glucose

When a molecule of glucose is to be broken down into smaller units, its energy level must be *activated,* that is, have its energy level increased. This activation is achieved by the transfer of a phosphate group from a molecule of ATP. (See Figure 5.1, which reviews the structure of ATP and the location of high-energy bonds between the phosphate groups.) The activation energy supplied by ATP can be compared to the energy required to start a large boulder rolling down a hill when the boulder is blocked by a small pebble; all that is needed to start the process is to move the small pebble.

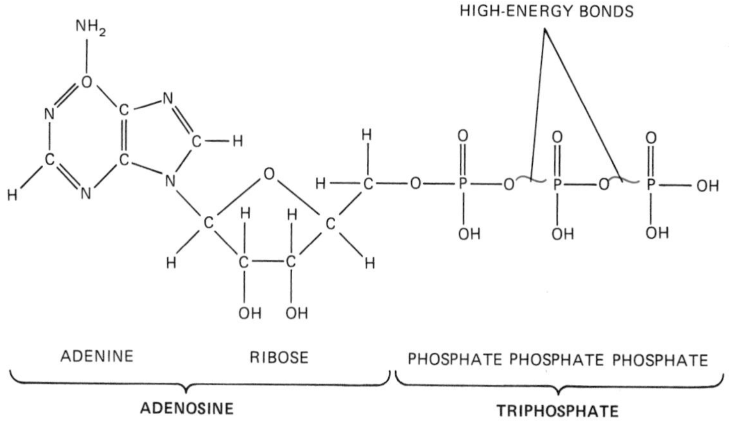

Figure 5.1
A structural diagram of an ATP molecule. Note the high-energy bonds in color between the phosphate groups. Energy is released when ATP, adenosine *tri*phosphate ("tri" means "three," referring to the three phosphate groups) loses its final phosphate group, forming ADP, adenosine *di*phosphate ("di" means "two," referring to the remaining two phosphate groups).

Step 1. The activation energy is supplied by ATP. In the reaction, ATP yields phosphate to glucose, which becomes glucose phosphate, a 6-carbon sugar.

Step 2. Another molecule of ATP is expended to form ADP and fructose diphosphate, another 6-carbon sugar.

Step 3. Fructose diphosphate is split into two 3-carbon molecules, which are called **PGAL** (phosphoglyceraldehyde).

Step 4. The PGAL molecules each pick up an inorganic phosphate group from the surrounding cytoplasm to become Di-PGA (diphosphoglyceric acid), which is oxidized by the removal of hydrogen. The hydrogen is picked up by two molecules of NAD (nicotinamide adenine dinucleotide) to form two molecules of $NADH_2$, the reduced form of NAD.

Steps 5 and 6. At this point, there are two molecules of a 3-carbon compound, Di-PGA. In a series of chemical reactions, the phosphate groups at both ends of the Di-PGA molecules are transferred to four ADP molecules, generating four ATP molecules. Since two ATP molecules were required to start the initial activation of a glucose molecule, and four have been formed, there is now a net gain of two molecules of ATP. With the release of the phosphate groups from the Di-PGA, two molecules of 3-carbon PGA (phosphoglyceric acid) are formed (Step 5). With the generation of the final two ATP molecules, two molecules of 3-carbon **pyruvic acid** are formed (Step 6).

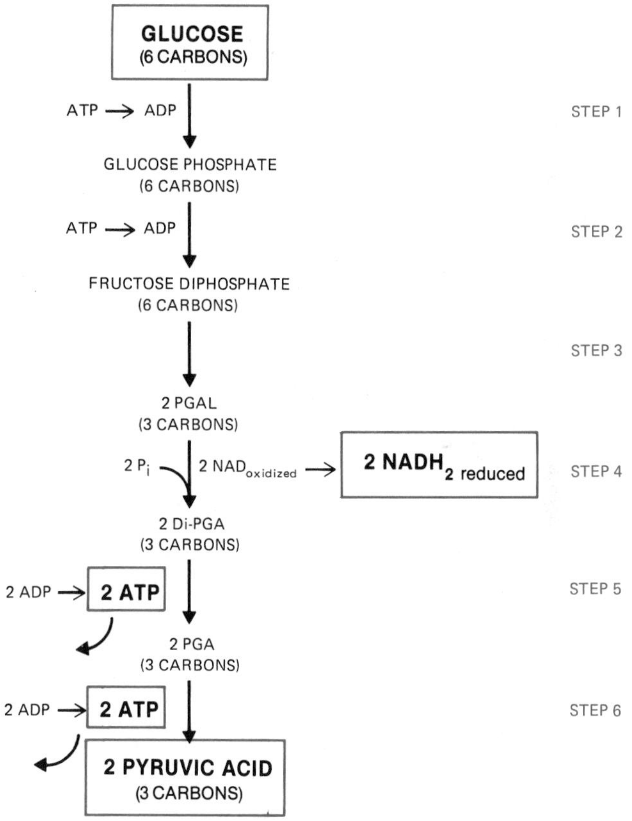

GLYCOLYSIS

Pyruvic acid is considered the end of glycolysis. Note that when a cell has broken down a glucose molecule as far as pyruvic acid, it has increased its ATP supply by two molecules, and it has on hand two molecules of reduced, energy-carrying $NADH_2$; the cell has not released any carbon dioxide or used any oxygen. Most of the chemical-bond energy originally present in the glucose is still present in the pyruvic acid.

Once a cell has some pyruvic acid, it may do various things with it. The pyruvic acid may be used to make amino acids, it may be partially respired anaerobically to make such compounds as alcohol or lactic acid, or it may be completely respired to form carbon dioxide and water. Because a sequence of biochemical reactions is called a *pathway,* pyruvic acid may be thought of as standing at a sort of biological crossroad (see Figure 5.2).

TO SUM UP GLYCOLYSIS

	Input		*Output*
1.	Glucose	1.	2 molecules of pyruvic acid
2.	2 molecules of ATP	2.	4 molecules of ATP (but net gain of only 2)

BEER, WINE, AND YOGURT

Long before they knew the biological reasons for what they were doing, people used the anaerobic activities of microbes. The curing of tobacco and tea leaves is a bacterial fermentation process, as are the preparatory steps in making vanilla extract from the fruits of orchid vines, known as "vanilla beans," or in making chocolate from the fruits of cacao trees.

Humans have been making alcoholic drinks longer than they have been writing. The usual active organism is one or another variety of the yeast *Saccharomyces cerevisiae.* Any sugar-containing, watery liquid can be used to make wine: plums, apples, blackberries, elderberries, oranges, diluted honey, dandelion flowers, and of course most commonly, grapes. Fermentation is allowed to continue until most or all (for dry wines) of the sugar has been broken down to carbon dioxide and ethyl alcohol, which can reach a concentration of about 12 percent. If the product is bottled before the fermentation is complete, additional carbon dioxide will be produced, resulting in bubbly wines such as champagne or sparkling burgundy.

Wine can be spoiled by the invasion of alcohol-digesting bacteria, which further ferment the alcohol to acetic acid, turning the wine into vinegar (*vinegar* literally means "sour wine"). Fortified wines, such as sherry, have their alcohol content artificially raised to about 20 percent by the addition of distilled alcohol. They do not spoil even when not refrigerated after they are opened.

Beer is similar to wine except that the source of the alcohol is not fruit or flowers but a grain, such as barley or corn. Originally beers, like champagne, were bottled before fermentation was complete in order to retain the bubbles of carbon dioxide, but now commercial beers are usually artificially carbonated. The method used in artificial carbonation was invented by the same Joseph Priestley who discovered oxygen and the process of photosynthesis in the eighteenth century.

The same species of microorganism that is called brewer's yeast in a brewery is called baker's yeast in a bakery. Brewers are mainly interested in the alcohol liberated by the yeast, whereas bakers want the bubbles of carbon dioxide to swell and make the bread light. The alcohol in yeast dough is boiled off during baking.

Fresh milk ordinarily deteriorates rapidly when bacteria grow in it, as they invariably do under natural conditions. Some bacteria, however, ferment the milk sugar, *lactose,* to lactic acid, creating an environment that eliminates the undesirable bacteria. Milk that receives a starter of a proper strain of *Lactobacillus* can be changed into any of a number of cheeses, of which yogurt is a well-known type. Yogurt was originally made because it has better keeping quality than fresh milk, but now it is made in quantity because some people like the taste of it, especially when fruit or flavoring is added.

3. 2 inorganic phosphate groups	3. 2 pairs of hydrogen atoms (4 hydrogen atoms)

Fermentation: One Possible Fate of Pyruvic Acid

The word "fermentation," as we will use it here, refers to the production of alcohol by microorganisms, as well as the formation of lactic acid in animal tissue. The critical point is that both alcohol and lactic acid are incompletely oxidized end products, formed without the use of atmospheric oxygen.

One possible product of fermentation is ethyl alcohol (Figure 5.2). In a yeast cell, for example, a molecule of carbon dioxide is lost from pyruvic acid, producing a 2-carbon molecule of acetaldehyde. The acetaldehyde is reduced by hydrogen from $NADH_2$, which was produced during glycolysis, and the final end product is *ethyl alcohol*. The process is known as **alcoholic fermentation,** and may be represented by the following equation:

$$C_6H_{12}O_6 \longrightarrow 2C_2H_5OH + 2CO_2 + 15 \text{ Kcal}$$

Glucose may be broken down into Ethyl Alcohol plus Carbon Dioxide plus Energy

Another possible anaerobic activity of pyruvic acid yields *lactic acid* (see Figure 5.2). In an active muscle, for instance, pyruvic acid can be reduced by the hydrogen from $NADH_2$ to form lactic acid. This can happen during strenuous exercise, when oxygen is not being delivered to the muscle fast enough for complete oxidation of the available food supply. Accumulation of lactic acid makes a muscle "feel tired," and can eventually stop muscle contraction until the lactic acid is removed when adequate oxygen is provided. The process that produces lactic acid from pyruvic acid is called **lactic acid fermentation,** and is shown in the following equation:

$$C_6H_{12}O_6 \longrightarrow 2C_3H_6O_3 + 15 \text{ Kcal}$$

Glucose may be broken down into Lactic Acid plus Energy

Note that fermentation produces 15 Kcal of energy, while aerobic respiration yields 277 Kcal (see the equation on page 85). Fermentation produces such a relatively low amount of energy because most of the energy in glucose remains in the chemical bonds of the lactic acid or ethyl alcohol.

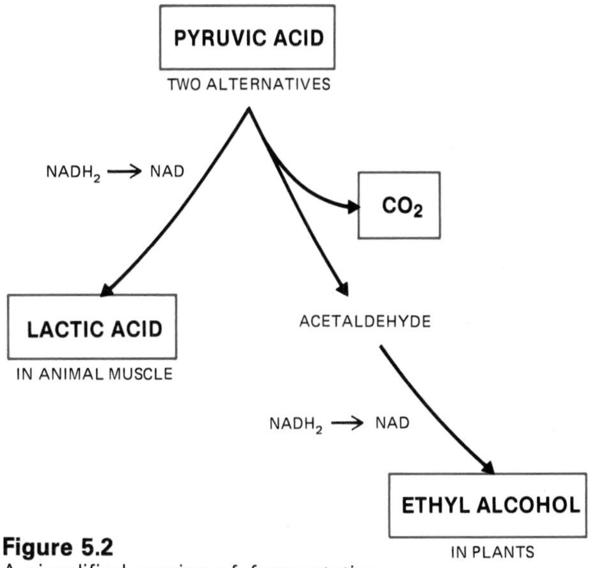

Figure 5.2
A simplified version of *fermentation,* or anaerobic respiration. Generally, fermentation yields lactic acid in animals and ethyl alcohol in plants and fungi. Note especially that carbon dioxide may be given out, but no oxygen is taken in.

TO SUM UP FERMENTATION

Input	*Output*
1. 2 molecules of pyruvic acid (from 1 molecule of glucose)	1. 2 molecules of an incompletely oxidized compound (for example, lactic acid or ethyl alcohol)
	2. 2 molecules of CO_2 (in alcoholic fermentation)

The Krebs Cycle: A Second Possible Fate of Pyruvic Acid

Most of the ATP that is generated and energy that is released in cells is made possible by a series of reactions called the **Krebs cycle,** also called the *tricarboxylic acid cycle* (TCA) or the **citric acid cycle.** During the Krebs cycle, molecules of acetyl-CoA (acetyl-coenzyme A) are taken up by mitochondria and altered in a series of enzyme-controlled steps.

The Krebs cycle is named for the biochemist Hans Krebs, who suggested in the 1930s the theo-

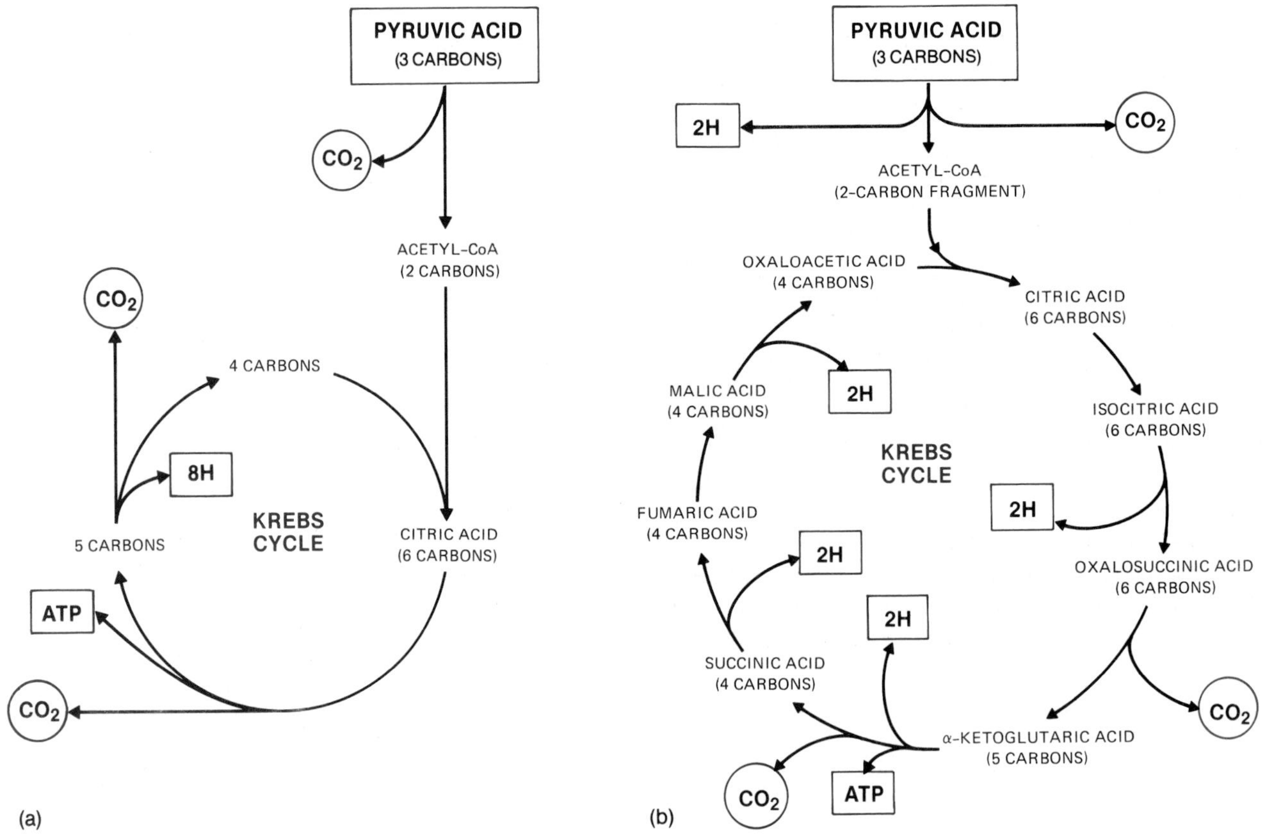

Figure 5.3
(a) A simplified representation of the Krebs cycle. The two carbons of acetyl-CoA combine with a 4-carbon compound to form 6-carbon citric acid. One carbon is removed as CO_2, leaving a 5-carbon compound; removal of another CO_2 leaves a 4-carbon compound, which combines with another 2-carbon acetyl-CoA to start the cycle all over again. (b) A slightly more detailed diagram of the Krebs cycle. See Figure 5.4 for an even more detailed representation.

retical possibility of a cycle of organic acids as a means of releasing carbon dioxide and hydrogen during respiration. Krebs received a Nobel prize in 1953 for his work in tracking down the complete breakdown of pyruvic acid.

The Krebs cycle (Figure 5.3), in its bare essentials, is as follows:

1. A molecule of pyruvic acid, with the help of an enzyme, gives up a molecule of carbon dioxide and hydrogen ions. The actual Krebs cycle now begins.
2. With one carbon atom gone from the 3-carbon pyruvic acid, two remain, forming an acetyl group, which combines with a molecule of coenzyme A (CoA) to form acetyl-CoA.
3. The two carbons of acetyl-CoA combine with the 4-carbon compound oxaloacetic acid to form a 6-carbon molecule of citric acid. From here on around the Krebs cycle, there is an ordered sequence of reactions that serve several purposes: Some are simply rearranging reactions that prepare a molecule for the next reaction, some release carbon dioxide, and some release hydrogen. It is the hydrogen-releasing reactions that are the important ones for energy transfer; the rest are means to that end.
4. The citric acid is rearranged to isocitric acid.
5. The isocitric acid *loses a pair of hydrogen atoms,* and the result is oxalosuccinic acid.
6. The oxalosuccinic acid *loses a molecule of carbon dioxide,* and the result is 5-carbon α-ketoglutaric acid.

7. The α-ketoglutaric acid *loses a pair of hydrogen atoms, and a molecule of carbon dioxide.* The result is 4-carbon succinic acid. A molecule of ATP is generated by the series of reactions.
8. The succinic acid *loses a pair of hydrogen atoms,* and the result is 4-carbon fumaric acid.
9. The fumaric acid is rearranged to yield 4-carbon malic acid, which *loses a final pair of hydrogen atoms,* yielding oxaloacetic acid, the same 4-carbon acid that accepted a 2-carbon acetyl group to begin the cycle.
10. The 4-carbon oxaloacetic acid combines with another 2-carbon acetyl group to start the cycle all over again.

Each of the reactions in the Krebs cycle is assisted by a specific enzyme. Without these enzymes the reactions would not take place fast enough to support life as we know it. The enzymes are spatially arranged in the membranes of mitochondria so that each is in its proper place to function when the time comes.

A more detailed representation of the Krebs cycle is shown in Figure 5.4.

TO SUM UP THE KREBS CYCLE

Input	*Output*
1. Acetyl-CoA (2-carbon) from pyruvic acid	1. 2 molecules of CO_2
2. 1 molecule of ADP and 1 phosphate ion	2. 4 pairs of hydrogen atoms (8 hydrogen atoms)
	3. 1 molecule of ATP

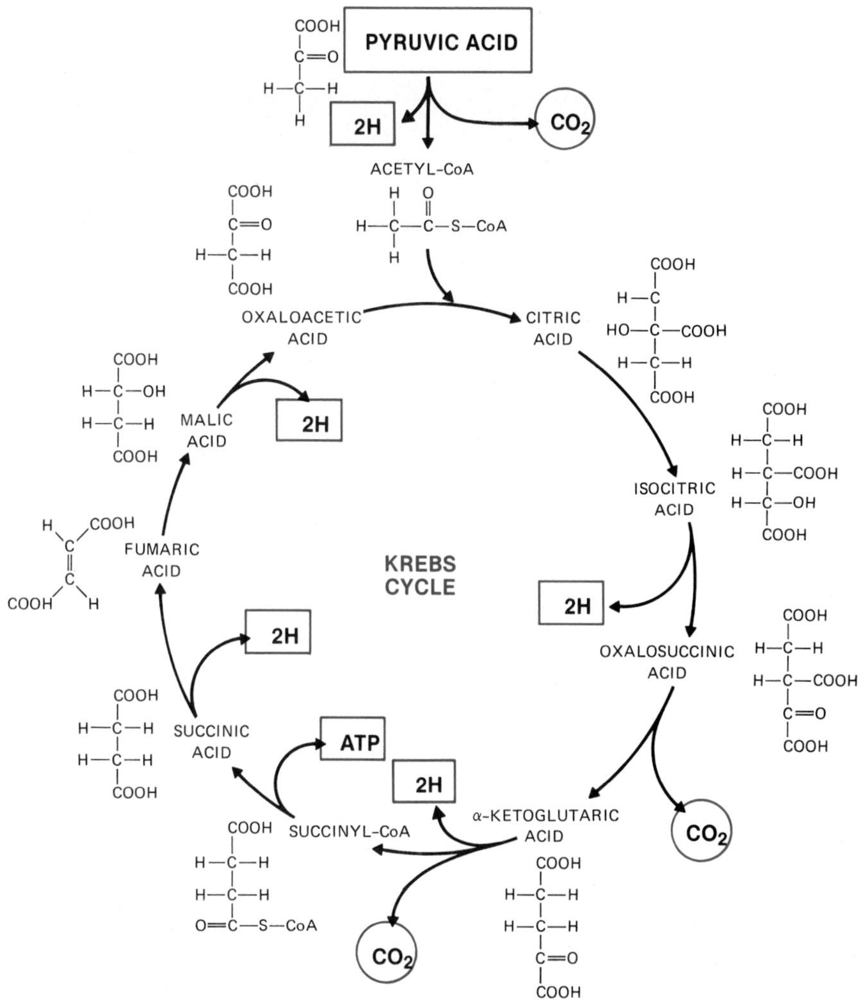

Figure 5.4
Schematic representation of the Krebs cycle. Simplified versions are shown in Figure 5.3.

CYANIDE, MUSHROOMS, AND OTHER FOES OF RESPIRATION

Most animals require a constant supply of oxygen, delivered all the time to all cells. The best-known cause of death from lack of oxygen is probably drowning. Without free air, a person loses consciousness within seconds, and death follows in minutes. Even if a rescue is effected, oxygen deprivation for several minutes can cause permanent brain damage, because the brain is more sensitive to loss of respiration than are other tissues.

But turning off the air supply is not the only way to stop respiration. Anything that interferes with cellular respiration can effectively cause a sort of "drowning." Some poisonous mushrooms contain a variety of protein toxins that interfere with the transmission of nerve impulses, thereby preventing the contraction of the rib and diaphragm muscles that causes the physical act of breathing. When these muscles become inoperative, breathing stops and death follows shortly. Other mushroom toxins break up the oxygen-carrying red blood cells. In either instance, a poisoned person dies because of lack of cellular respiration. Antidotes to mushroom poisoning are just beginning to be developed.

× 0.5

An Amanita, *one of the most poisonous of mushroom genera.*

Cyanide poisoning is specifically a respiratory blockage. A cyanide-treated cell cannot use oxygen because cyanide, HCN, inhibits the electron transport system and thus stops the activity of the terminal oxidations. A person dying of cyanide poisoning turns blue (*cyanide* means "blue" in Greek) and dies quickly. Contrary to the lore of spy stories, however, the dying is grimly painful.

Another foe of respiration is tetanus poisoning, or "lockjaw." The tetanus bacterium, *Clostridium tetani,* can exist only when oxygen is absent. It exudes a toxin 50 times as poisonous as cobra venom and 150 times stronger than another well-known poison, strychnine. Buried rusty nails are common dwelling places for such anaerobes, so that anyone stepping on a nail and driving it through the skin causes self-inoculation, by introducing the bacteria into an airless environment where they can thrive. Immunization consists of injecting a small amount of weakened toxin into the body to stimulate it to build up its own antitoxin. Anti-tetanus shots are a practical necessity for farmers, soldiers, children, and any others who have frequent skin breaks and are likely to be in contact with soil.

Ordinarily, individuals will not notice a lack of oxygen as long as they can get rid of carbon dioxide. The oxygen concentration in submarines sometimes gets so low that there is not enough oxygen to allow a person to light a cigarette; but such a shortage of air causes no feeling of discomfort. In some of the huge fires caused by bombings in World War II, the oxygen in entire city blocks was used up so completely by the flames that people died quietly and, judging from the expressions on the corpses, quite peacefully.

The Electron Transport System is the Main Producer of ATP

So far we have accounted for only four new molecules of ATP (two from glycolysis and two from *two* turns of the Krebs cycle; recall that glycolysis results in *two* molecules of pyruvic acid made available for the Krebs cycle). Where is the rest of the energy that is being released from glucose? Also, what has happened to all the hydrogen atoms that have left molecules?

When electrons from the hydrogen atoms leave a molecule to be taken on by another molecule, energy goes with them. The electrons are transferred to an orderly arrangement of hydrogen acceptors called *cytochromes* (protein-plus-pigment molecules containing iron). The cytochromes are accompanied by appropriate enzymes. The series of transfers is called the **electron transport system,** or *cytochrome system* (Figure 5.5). Most of the energy released in a cell results from the transfer of electrons (removal of hydrogen) in the electron transport system.

Cytochromes can be alternately reduced (receive electrons) and oxidized (give up electrons), with an energy loss or gain accompanying any electron transfer. Oxidation involves either the addition of oxygen or the removal of hydrogen.

As a pair of hydrogen atoms is passed from cytochrome to cytochrome, some energy is necessarily lost as heat, but some of it is caught up in coupled reactions that bring together ADP and inorganic phosphate, to form ATP.

The final hydrogen acceptor is oxygen, and when finally an oxygen atom accepts two hydrogen atoms, the result is water. This so-called water of respiration is the main source of body water in some animals, such as the larvae of clothes moths, which never drink liquid water. *The final, biologically useful product of respiration is ATP.* The carbon dioxide and the water can be considered waste.

If oxygen is present to act as the final hydrogen acceptor, each pair of hydrogen atoms released by respiratory activity contributes to ATP production. There are six reactions in the entire respiration process that liberate hydrogen: one during glycolysis, one at the formation of acetyl-CoA from pyruvic acid, and four during the Krebs cycle.

Hydrogen from PGAL and from pyruvic, isocitric, α-ketoglutaric, and malic acids reduces NAD to $NADH_2$. Passage of each pair of hydrogen atoms from $NADH_2$ along the electron transport system generates three molecules of ATP. Hydrogen from succinic acid, however, reduces FAD (flavin adenine diphosphate) to $FADH_2$, and hydrogen from this compound produces only two molecules of ATP per passage along the electron transport system (see Figure 5.5).

From one molecule of PGAL, there comes enough hydrogen to reduce five molecules of NAD to $NADH_2$ and one of FAD to $FADH_2$. Five $NADH_2$ yield 15 ATP and 1 $FADH_2$ yields 2 ATP, for a total of 17. Since a molecule of glucose results in two mol-

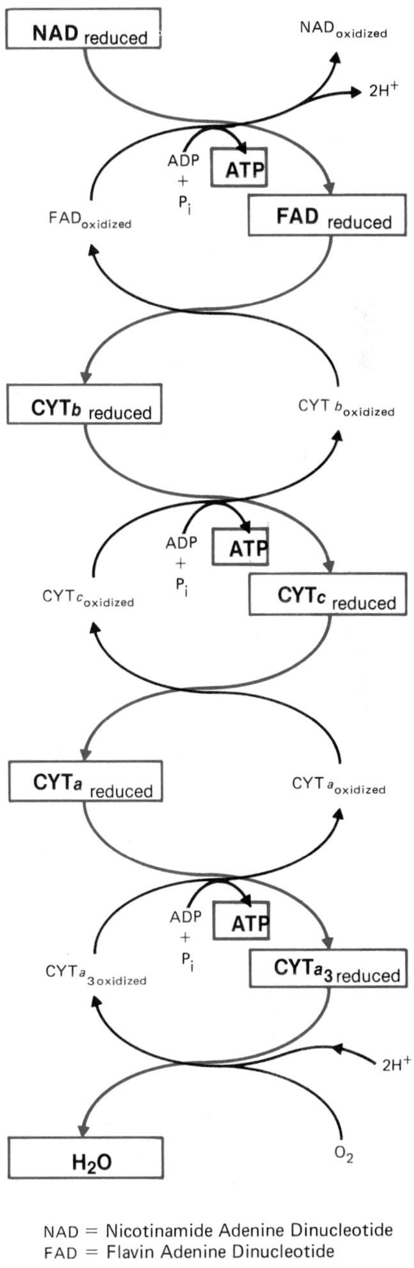

Figure 5.5
The electron transport system.

ecules of PGAL, this figure is doubled, to provide 34 molecules of ATP. Besides, each turn of the Krebs cycle also yields one ATP molecule without the use of the electron transport system. Thus, the total number of molecules of ATP produced from one molecule of glucose is 38, with 2 from glycolysis, 2 from two turns of the Krebs cycle, and 34 from the electron transport system.

Of the final 38 ATP molecules, according to the usual count, all but the two produced during glycolysis are the result of aerobic respiration, making aerobic respiration responsible for 36/38 or about 95 percent of the ATP energy made available from the breakdown of glucose. This figure makes it obvious why air-breathing, warm-blooded animals like human beings, with a large, constant demand for immediate energy, die so quickly when they cannot breathe and obtain oxygen.

If a molecule of ATP can yield 7.3 Kcal of energy, then respiration with its production of 38 molecules of ATP per molecule of glucose, which is "worth" 686 Kcal, is about 40 percent efficient:

686 Kcal = energy initially available
7.3 × 38 ATPs = 277 Kcal = energy obtained
277/686 = roughly 40 percent

Such efficiency compares very favorably with that of human-made machines, which commonly achieve between 5 percent and 30 percent efficiency, in terms of energy utilization.

TO SUM UP THE ELECTRON TRANSPORT SYSTEM (for one molecule of PGAL)

Input	*Output*
1. 6 pairs of hydrogen atoms (12 hydrogen atoms)	1. 5 $NADH_2$ reduced (from glycolysis and from pyruvic, isocitric, α-ketoglutaric, and malic acids): 5 × 3 ATP molecules = 15 ATP
2. 6 oxygen atoms	2. 1 $FADH_2$ reduced (from succinic acid): 1 × 2 ATP molecules = 2 ATP
3. 17 ADP	3. 6 molecules of water
4. 17 phosphate ions	

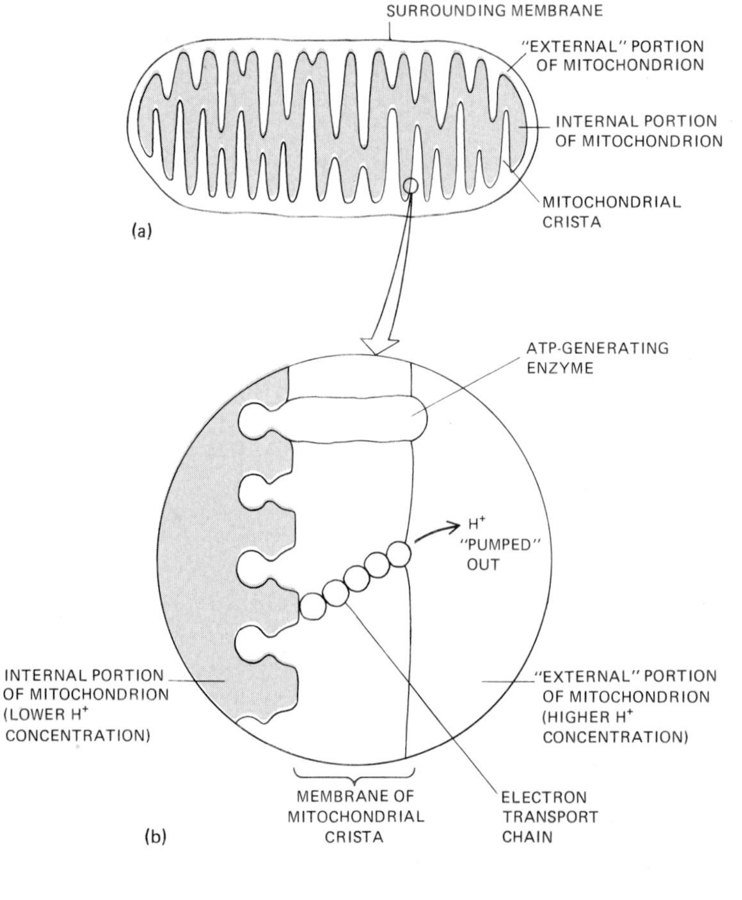

Figure 5.6
The chemiosmotic explanation of ATP generation. (a) A section of a mitochondrion, with a surrounding membrane and an internal membrane folded into *cristae.* The space inside the internal membrane is the internal portion of the mitochondrion. The space between the internal membrane and the surrounding membrane is the "external" portion. (b) A section through a membrane of a crista. The electron transport chain is shown diagrammatically as a series of rings, and an ATP-generating enzyme is shown as penetrating the membrane. The system uses the energy from electrons from the Krebs cycle to pump H^+ ions from the left (internal portion) to the right (external portion). The greater concentration of H^+ ions on the right, as compared to the lesser concentration on the left, is used to drive the generation of energy-rich ATP (from ADP and inorganic phosphate) by the mediation of an ATP-forming enzyme.

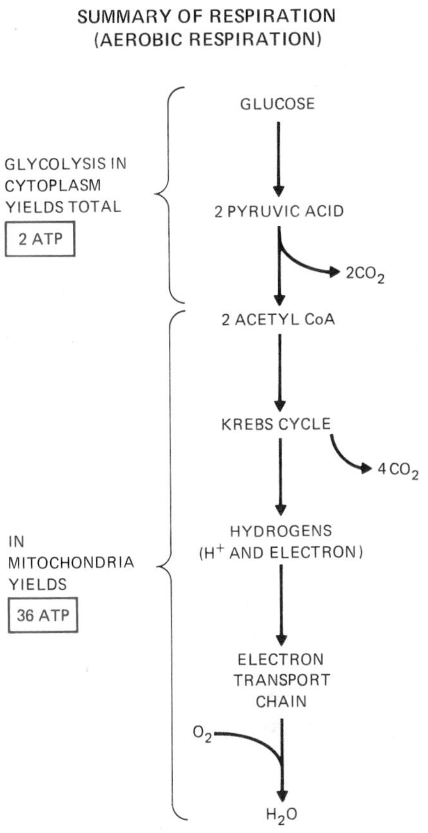

Figure 5.7
Generalized scheme of cellular respiration, from a beginning sugar to the production of ATP, carbon dixoide, and water. Note especially that most of the ATP is generated by the action of the Krebs cycle in mitochondria, and that oxygen is brought into the reactions only as the final step.

TO SUM UP ATP PRODUCTION IN RESPIRATION

Source	*Number of ATP Molecules*
From glycolysis	2
From Krebs cycle without electron transport system (two turns)	2
From electron transport system: 2×17	34
Total	38

THE CHEMIOSMOTIC EXPLANATION OF MITOCHONDRIAL ATP PRODUCTION

No one knows exactly how cells generate energy-rich ATP. But one explanation, known as the *chemiosmotic hypothesis,* is gaining acceptance. It is based on the idea that a membrane in a mitochondrion or chloroplast can use the energy of electrons (the "chemi-" part of the term) to build up a greater concentration of protons (H^+) on one side of the membrane than on the other (the "-osmotic" part of the term). The idea is carried further to assume that the resulting *difference* in H^+ concentration is used by ATP-generating enzymes as an energy source in forming ATP from ADP and inorganic phosphate (Figure 5.6).

This explanation fits the observation that ATP production stops if membranes of mitochondria or chloroplasts are damaged. Further, experiments on chloroplasts support the idea. If chloroplasts *in the dark* (no light energy available) are artificially made to have a higher H^+ concentration on one side of their membranes than on the other, those chloroplasts can make ATP.

RESPIRATION OF OTHER FOODS

Animals eat and plants make many kinds of food besides glucose: they include proteins, fats, and various starches (see Chapter 22). These respirable foods can be used to generate ATP, but they must be chemically changed so that they can be brought into a form suitable for providing hydrogen to the electron transport system (Figure 5.8). *Proteins* are digested to amino acids. An amino acid can be further metabolized to yield an amino group and an organic acid. These organic acids can enter the Krebs cycle and be dehydrogenated (oxidized). *Starches* are digested to glucose, which is readily respired via glycolysis and the Krebs cycle. Fats are digested to glycerol (which can be taken into the glycolytic scheme) and fatty acids. Fatty acids are broken down by the removal of two carbon atoms at

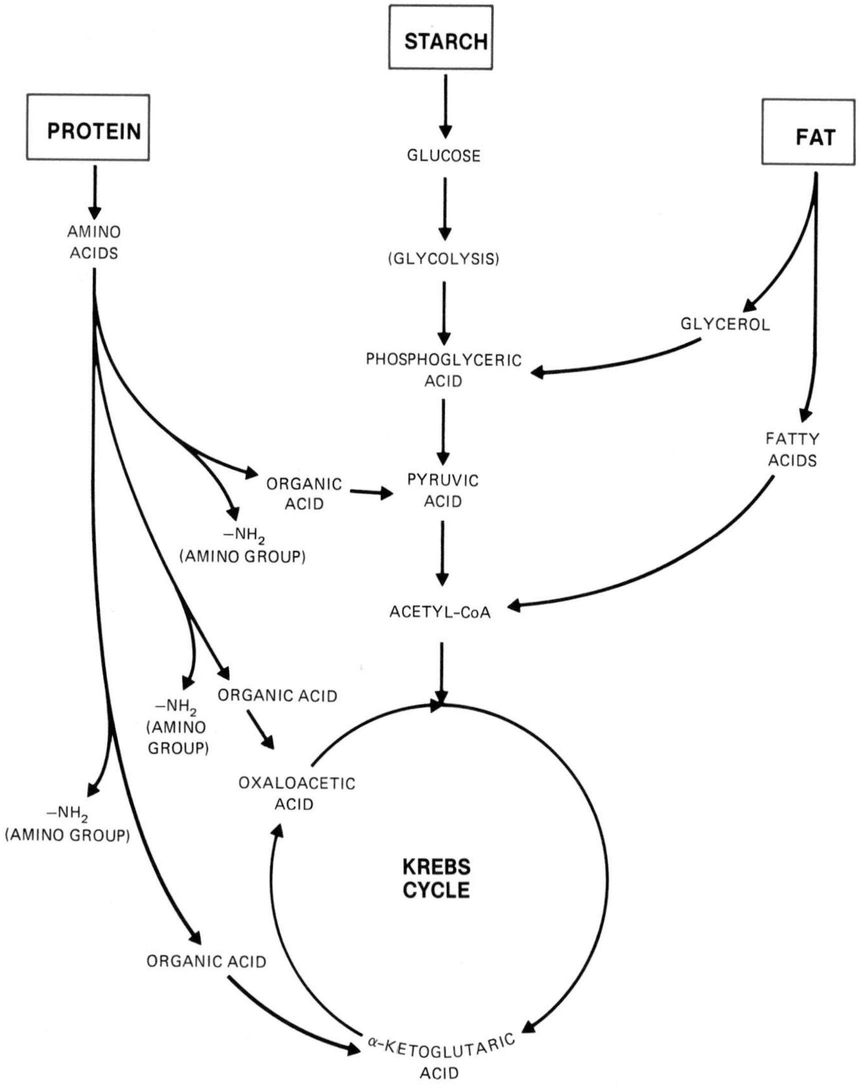

Figure 5.8
How proteins, starches, and fats are made available for the production of ATP. Proteins are fed into the Krebs cycle by way of amino acids. Starches enter the glycolytic pathway by way of simple sugars. Fats are digested to glycerol (which can enter the glycolytic pathway) and fatty acids (which can enter the Krebs cycle). Once in the Krebs cycle, any of these compounds can yield hydrogen and consequently be used to provide energy for the generation of ATP.

a time (acetyl groups), and these can be attached to coenzyme A and then brought into the Krebs cycle.

Not all organisms have the necessary enzymes to make all these conversions. Most animals, for example, lack a digestive enzyme for cellulose, the most common glucose-containing compound in the world. But if the enzymes are present, practically any organic molecule can be changed into a form that provides hydrogen, thereby making it suitable for energy production.

SUMMARY

1. In a biological sense, *respiration* is the release of the free energy stored in chemical bonds of certain organic compounds. Respiration makes available to cells the chemical energy that has been converted from light energy during photosynthesis.

2. Energy must be trapped within the molecules of ATP to be usable by organisms, and respiration is a highly efficient mechanism for regulating the systematic release of trapped energy.

3. The overall process of respiration is made up of many reactions or steps, and each step is controlled by a specific enzyme.

4. Respiration occurs in all living cells to produce the energy for basic life processes. In contrast, photo-

synthesis takes place only in specialized plant cells.

5. The general events of respiration are these: Every molecule of glucose, when combined with 6 molecules of oxygen, yields 6 molecules each of carbon dioxide and water and liberates usable energy in the form of 38 ATP molecules.

6. *Glycolysis* is the initial breakdown of glucose to an intermediate compound; it occurs in the cytoplasm of a cell. *Fermentation* is the partial breakdown of organic molecules in the absence of free oxygen. During the *Krebs cycle,* respirable molecules yield carbon dioxide and hydrogen. The *electron transport system* uses molecular oxygen and is the main producer of ATP. The Krebs cycle and the electron transport system take place in the mitochondrion.

7. *Aerobic* processes require free atmospheric oxygen to carry on their normal cellular activity. *Anaerobic* processes do not require free oxygen.

8. *Glycolysis* uses 1 molecule of glucose, 2 ATPs, and 2 phosphate groups, and it produces 2 molecules of pyruvic acid, 4 ATPs, and 2 pairs of hydrogen atoms.

9. *Fermentation* uses 2 molecules of pyruvic acid from 1 molecule of glucose and produces 2 molecules of an incompletely oxidized compound (such as lactic acid or ethyl alcohol), and 2 molecules of carbon dioxide.

10. The *Krebs cycle* uses acetyl-CoA from pyruvic acid and produces 2 molecules of carbon dioxide, 4 pairs of hydrogen atoms, and 1 molecule of ATP.

11. The *electron transport system* uses 12 hydrogen atoms, 6 oxygen atoms, 17 molecules of ADP, and 17 phosphate ions, and produces 17 molecules of ATP per molecule of PGAL.

12. The *chemiosmotic hypothesis* of mitochondrial ATP production proposes that as protons (H^+) are concentrated on one side of a mitochondrion's membrane the difference in H^+ concentration is used by ATP-generating enzymes as an energy source in forming ATP from ADP and P_i.

13. Animals eat and plants make many other kinds of foods besides glucose, and all respirable foods can be used to generate ATP after they have been changed to a form that provides hydrogen to the electron transport system.

ASK YOURSELF

1. What is the main *physical* result of cellular respiration?
2. What is the main *chemical* product of respiration?
3. What is the chemical end-product of the glycolysis portion of respiration?
4. What are some possible end products of fermentation?
5. Why is fermentation an inefficient method of respiration?
6. During what portion of the aerobic respiratory process is carbon dioxide eliminated?
7. Why are the hydrogen-releasing steps in the Krebs cycle important to a cell's energy supply?
8. Assuming that every chemical step in respiration requires a specific enzyme, estimate how many enzymes would be needed to break down a molecule of glucose to carbon dioxide and water.
9. During what portion of the aerobic respiratory process is oxygen taken in?
10. Where in the aerobic respiratory process is most of the ATP generated?
11. How can a cell use such noncarbohydrate foods as proteins and fats to release energy?

6
The Importance of DNA

SOME KEY POINTS

1. DNA is the genetic material, contained within the nucleus of a cell.
2. Nucleic acid containing the sugar deoxyribose is designated as DNA, and nucleic acid containing the sugar ribose is known as RNA.
3. Nucleic acids are made up of units called nucleotides, and each nucleotide is made up of a phosphate group, a sugar, and a nitrogen-containing base.
4. DNA is in the form of a double helix. A double-ringed adenine combines with a single-ringed thymine, and a double-ringed guanine combines with a single-ringed cytosine. These base pairs are held together by hydrogen bonds, and are attached to the sugar-phosphate backbones of the twisted ladder of the double helix.
5. When DNA replicates itself, the hydrogen bonds break, the old bases combine with new complementary bases, and new hydrogen bonds are formed. The result is two copies of the original strand.

MOST OF US HAVE WONDERED, AT SOME POINT in our lives, how it was possible for us to inherit our mother's blue eyes, our father's black hair and too-large nose, and our Uncle Charlie's winning smile. We may also have wondered how our cells know how to make identical copies of themselves. How do hereditary characteristics get passed on from generation to generation, and even more to the point, how do hereditary characteristics get passed on from cell to cell within our own bodies?

Scientists have agreed for a long time that the answer to these questions probably lies in the cellular material itself. In order to find the answer, they would first have to isolate the genetic *substance.* Then they would have to determine its *structure.* Only then could they hope to find out how the miracle of the continuation of life is performed. Although our basic knowledge of the genetic material dates back to the middle of the nineteenth century, it took the advanced technology of our own century, plus a group of persistent scientists, to unravel the threads of the intricate pattern of life.

Two scientists, James Watson and Francis Crick, received the credit for putting the pieces together in 1953. When they postulated the structure of a molecule called DNA, they provided enough clues so that many of the remaining questions could be answered within a decade or two. One of the riddles of life was about to be solved.

DNA was first discovered in 1869 by Friedrich Miescher, a Swiss physician. Although he had not actually set out to find the substance that determined our genetic heritage, Miescher inadvertently started one of the most dramatic searches in scientific history when he found the substance he called *nuclein.* Miescher gave it that name because it was located in the nuclei of the white blood cells he was studying. Nuclein was subsequently named *nucleic acid,* and then *DNA.*

During the century following Miescher's isolation of nuclein, a great deal of investigation was carried out on the substance. Its chemical composition was determined, and many chemical properties were identified. Nuclein (DNA) was shown to be the hereditary material, but its three-dimensional structure remained obscure. The knowledge of this structure was essential to explaining how DNA could function in heredity. In 1953, almost a hundred years after Miescher's discovery, the structure of DNA was revealed by Watson and Crick in their double helix model—a discovery that has been ranked with the contributions of Newton, Darwin, and Einstein. What *is* DNA, and why is it so important?

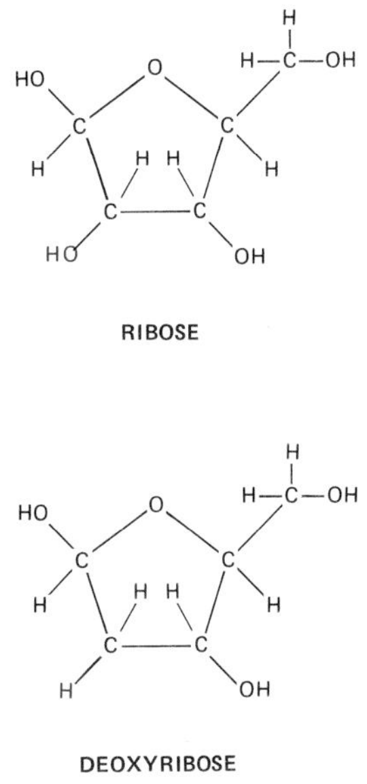

Figure 6.1
The sugars in nucleic acids. Ribose occurs in RNA and deoxyribose in DNA. The difference between the two kinds of molecules is the presence of one oxygen atom in ribose that is absent in deoxyribose. The colored atoms on the diagrams show the positions affected.

NUCLEIC ACIDS

Nucleic acids are extremely large molecules (macromolecules), probably the largest found in living cells. Their size results from the linking of small units that are repeated over and over again. Scientists are now certain that nucleic acids are found in all living things, and that they are the primary factors that control the genetic (hereditary) processes of life. We also know that nucleic acids are sometimes found outside the nuclei of cells, and that the nucleic acids of all living organisms, from the tiniest viruses to humans, differ in the sequence of the units. We will see later in this chapter how specific differences in the structure of nucleic acids give each organism its unique characteristics.

There are two main kinds of nucleic acids, each with its own sugar structure. One kind of nucleic acid contains the sugar ribose. Nucleic acids containing ribose sugar are known as **ribonucleic acids,** or **RNA.** The other type of nucleic acid contains a sugar with one less oxygen atom, which is therefore called *deoxy*ribose (Figure 6.1). The nucleic acid itself is designated as **deoxyribonucleic acid,** or **DNA.**

Nucleic acids are made up of units called **nucleotides,** and each nucleotide is made up of three components: phosphate, sugar, and nucleotide base (Figure 6.2). DNA contains four kinds of nucleotide bases: the double-ringed *purines* (adenine and guanine), and the single-ringed *pyrimidines* (thymine and cytosine). **Adenine, guanine, thy-**

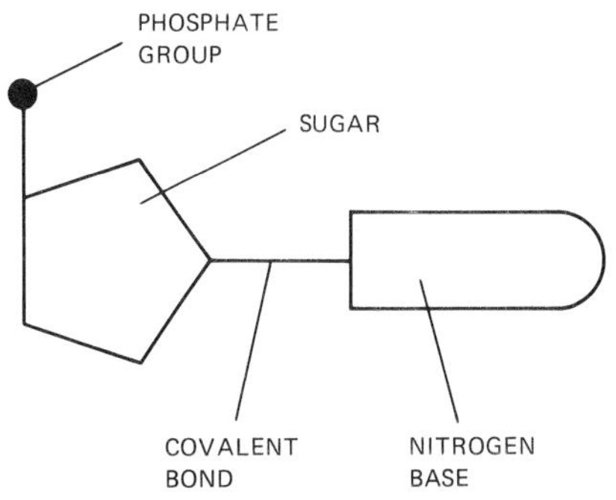

Figure 6.2
The structure of a nucleotide. Each nucleotide contains a phosphate group, a sugar, and a nitrogen base. Nucleotides bond together to form nucleic acids such as DNA. Note the covalent bond between the sugar and nitrogen base.

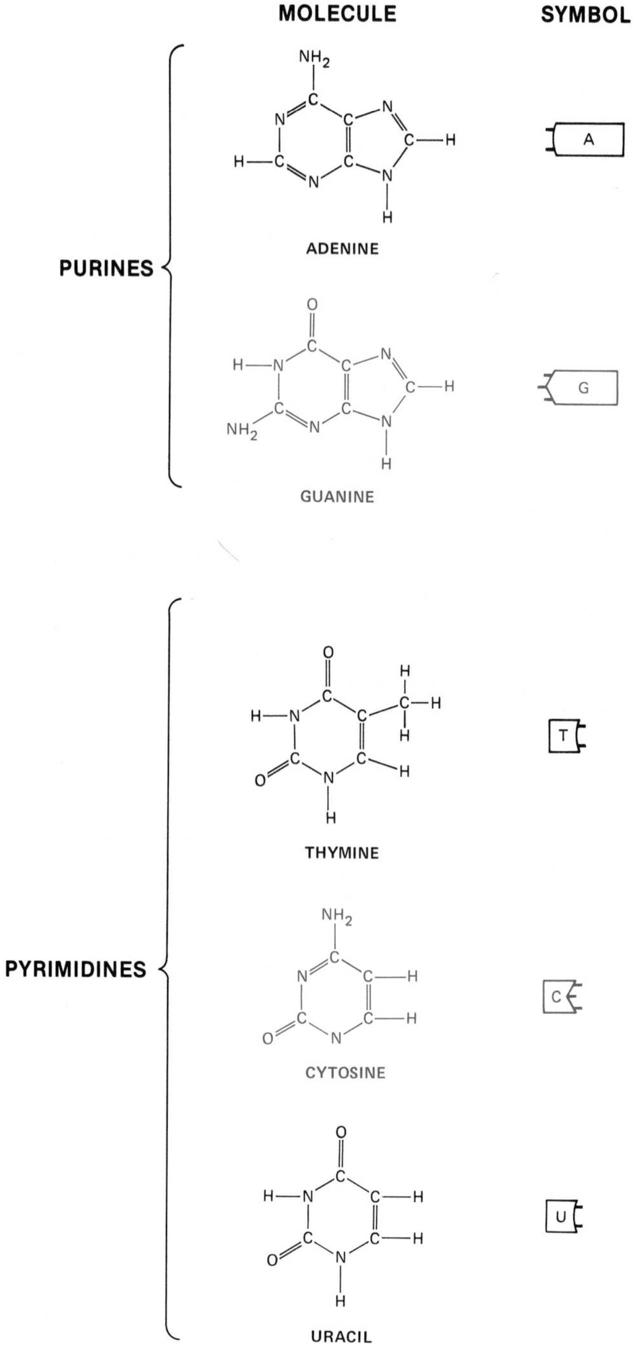

Figure 6.3
The nitrogen bases that commonly make up nucleic acids. The *purines* are *adenine* and *guanine*, and the *pyrimidines* are *thymine*, *cytosine* (in DNA), and *uracil* (in RNA). Note that the purines have a double-ring structure and the pyrimidines have a single-ring structure. Guanine and cytosine are able to form three hydrogen bonds each, whereas adenine, thymine, and uracil can form only two each.

mine, and **cytosine** are usually referred to in a shorthand version as A, G, T, and C (Figure 6.3).

RNA molecules differ from DNA molecules in three ways besides their sugar makeup: (1) RNA contains uracil instead of thymine; (2) RNA is usually single-stranded, whereas DNA is double-stranded; and (3) RNA occurs in cytoplasm, nuclei, and several organelles in cells, but DNA is usually restricted to chromosomes or chromosomelike structures. (Small but important quantities of DNA are found in chloroplasts and mitochondria.)

In summary, nucleic acids contain sugar, phosphate groups, and nitrogen-containing bases. A combination of one of each of these components makes up a nucleotide, a self-contained unit within the overall molecule of nucleic acid.

PROOF OF THE GENETIC IMPORTANCE OF DNA

In 1869 Friedrich Miescher was interested in determining the content of the nuclei of cells, and he began his experiments with the white blood cells in the pus that adhered to discarded bandages. He was able to isolate a nuclear material, and from this he further isolated an unfamiliar acidic substance that was also rich in phosphorus. When Miescher announced his results in 1871, he named his new substance *nuclein,* and in later experiments with the nuclei of salmon sperm he isolated a similar substance.

Early in the twentieth century researchers discovered that chromosomes carry the hereditary information, and that chromosomes contain protein as well as DNA and some RNA. Because the protein seemed more complex than the nucleic acids, scientists generally accepted the idea that only the protein was capable of carrying the hereditary information necessary to maintain life. We now know that it is DNA, not protein, that controls heredity, but it was not until the 1940s that any substantive proof became available.

In 1924 the German biochemist Robert Feulgen found that a decolorized dye, basic fuchsin, turned a bright purple in the presence of DNA. By means of this staining technique it was relatively easy to show that DNA was present in the nucleus of a cell. When it was clearly shown that there was DNA in the chromosomes of cells, scientists began to believe that DNA might be connected to heredity.

Later, Alfred Mirsky and his colleagues at the Rockefeller Institute found that the cells of any given species all contain the same amount of DNA. Sex cells contain exactly half the amount of DNA in

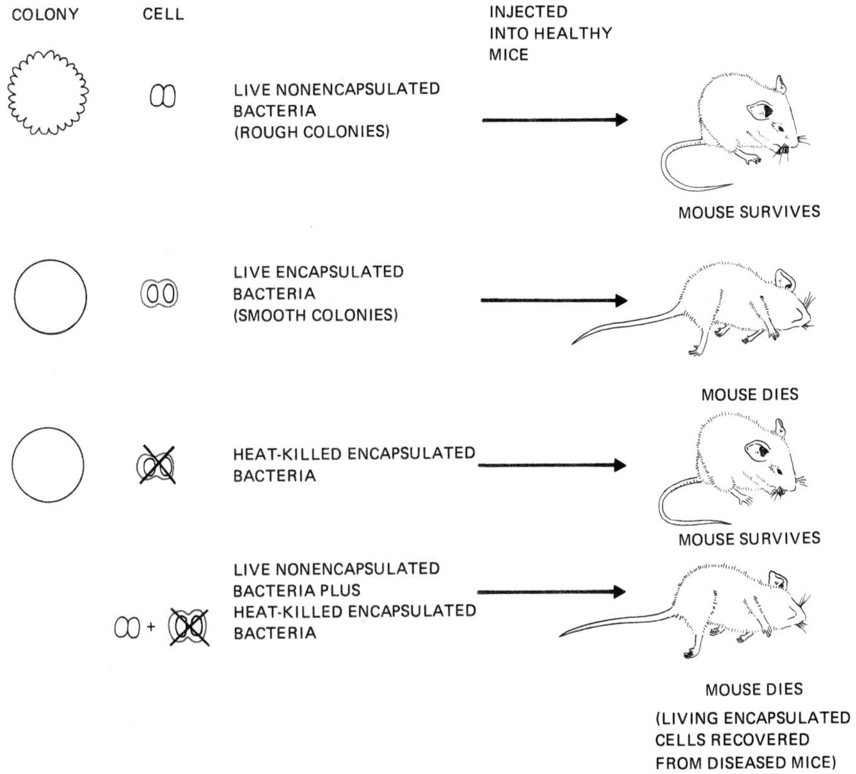

Figure 6.4
Griffith's experiments with encapsulated and nonencapsulated *Diplococcus pneumoniae* on mice, the work that started the idea of bacterial transformation and eventually led to the demonstration of DNA as the genetic material of cells.

normal body cells, but when the male and female sex cells unite in fertilization, the single resultant cell carries the same amount of DNA as other body cells.

By the time Fred Griffith began his experiments with the pneumococcus bacterium, scientists considered it possible that DNA was the substance that controlled heredity. Definite proof was still necessary, and in 1928 Griffith began the series of experiments that would ultimately bring acceptable proof in the 1940s and 1950s.

While seeking a cure for pneumonia, Griffith infected mice with two kinds of pneumococcus bacteria. Type I (or R, for rough) was a harmless strain without a covering capsule, and Type III (or S, for smooth) was the infective strain, with each cell covered by a protective coat (Figure 6.4). When mice were infected with the R strain, nothing happened. When they were exposed to the virulent S strain, the mice contracted pneumonia and died. These results were expected.

The unexpected happened when Griffith tried another approach. Mice were injected with the nonvirulent, uncoated R strain, and at the same time, the mice received a dose of coated S strain that had been heated and killed. To Griffith's astonishment, the mice subsequently died of pneumonia. To add to the puzzle, living coated bacteria were obtained from the dead mice. If the uncoated bacteria were harmless, and the coated ones had been killed by heating, what caused the mice to die? And how did *live* coated bacteria get into the blood of the dead mice?

What had happened was a process called **transformation,** which allowed the harmless bacteria to acquire the genetic characteristics of the virulent bacteria. But how had this happened? What substances were involved?

In 1944, after much prior experimentation, Oswald T. Avery, Colin MacLeod, and Maclyn McCarty finally managed to isolate the factor that had caused the transformation in Griffith's experiments. First they removed the coats from the S strain bacteria to determine if the coats were crucial to the hereditary transfer. But heated S bacteria without coats still transformed the R strain into a virulent form. The scientists then removed the protein from the S strain and left only nucleic acid in the bacteria. Again the R strain bacteria were transformed. Finally, when the nucleic acid (DNA) was separated from the S strain bacteria and only protein remained, the bacteria could no longer affect the harmless R strain.

Without a doubt, Avery had identified DNA as the effective agent of heredity. So it was DNA after all, not protein, that held the clue to life itself. That most important and thrilling discovery was considered "revolutionary" by leading scientists, who saw the consequences that would follow from it.

Another classic experiment by Alfred D. Hershey and Martha Chase in 1952 confirmed the results of Avery, MacLeod, and McCarty. Hershey and Chase were aided by the separate findings of two researchers. Thomas Anderson used electron micrographs to show that the shells of viruses become empty after the viruses come into contact with bacteria. Roger Herriott found that when a virus attaches itself to a bacterium, the virus releases DNA into the host cell.

Bacteriophages, or *phages* for short, are viruses that infect bacteria by attaching themselves to the host bacterial cell and injecting their single molecule of DNA into the host. Once inside the bacterium, the viral DNA begins to give instructions for the multiplication of new viruses. The new viruses soon burst from the host cell, able to infect other bacteria (Figure 6.5). Bacteriophages have been used extensively by scientists in their search for the genetic properties of DNA. By using radioactive tracers, Hershey and Chase verified that only the DNA from the head of the virus was injected into the host bacterium; the protein of the virus shell remained outside. Subsequently, new viruses, complete with protein coats, were reproduced. These new viruses were identical to their parent virus.

Hershey and Chase reasoned that if new viruses were being created that were identical to their parent, a genetic (that is, a hereditary) force must be at work. If only viral DNA was being injected into the host cell, then it was sensible to assume that

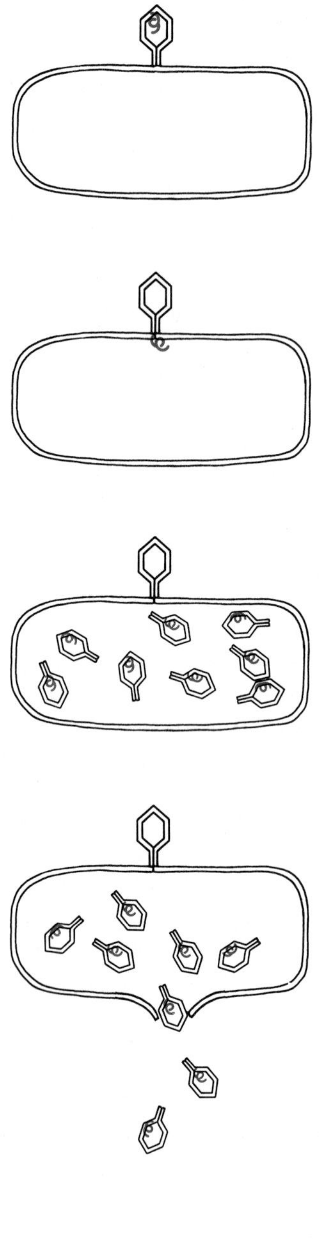

(a) (b) (c) (d)

Figure 6.5
Viruses consist of a head of protein surrounding a single strand of nucleic acid (usually DNA, but sometimes RNA). Completing the simple viral structure is a protein tail. Viruses cannot reproduce on their own. To reproduce they must infect a living cell, where they make use of the cell's enzymes and nourishment. As a result of viral infection and reproduction, the cell is usually killed. (a) When a virus infects a host bacterial cell, it attaches itself by its tail to the host cell and injects its single molecule of DNA into the bacterium (b). (c) The viral DNA inside the host cell directs the multiplication of new viruses. (d) Within 20 minutes, approximately 100 new viruses, complete with protein coats, have been reproduced, and they burst from the cell ready to infect other hosts. The original host has been totally sacrificed to the parasitic intruders. (e) Electron micrograph of a section through a bacterial cell containing replicated virus particles. Other viruses can be seen outside the cell. (f) A closer view of the edge of a cell with four viruses attached to the cell surface.

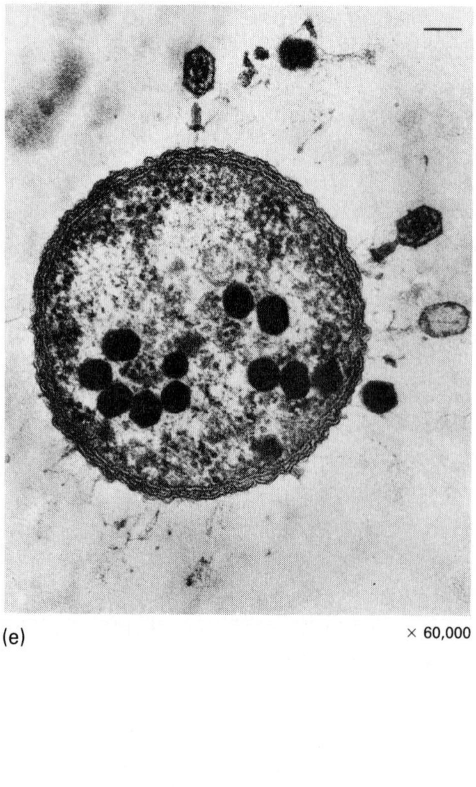

(e) × 60,000

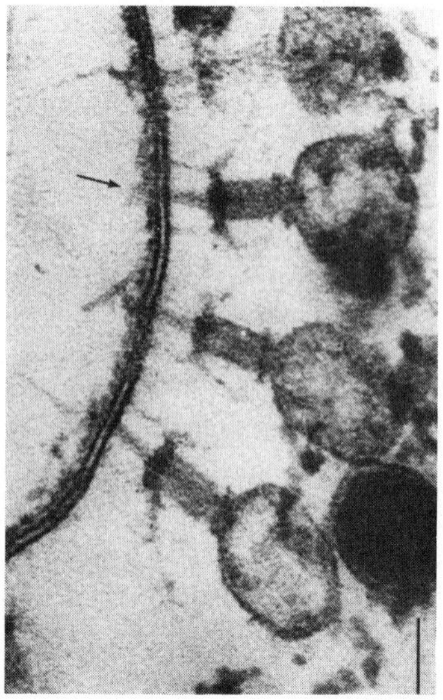

(f) × 130,000

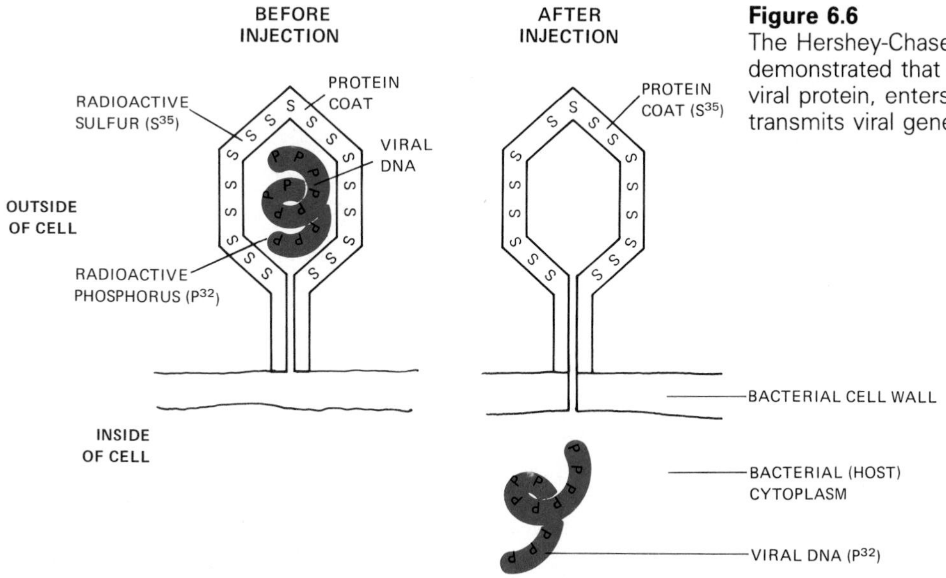

Figure 6.6
The Hershey-Chase experient, which demonstrated that only viral DNA, not viral protein, enters a host cell and transmits viral genetic control.

DNA was the genetic material. It was known that proteins contain sulfur but do not contain phosphorus, and that the content of DNA is the opposite—phosphorus but no sulfur. So when the viral sulfur and phosphorus were made radioactive, they could be traced to determine which substance entered the host cell and was able to direct the genetic activity. Of course, the results showed that phosphorus was on the inside and sulfur on the outside, proving that DNA alone had entered the cell (Figure 6.6).

This final confirmation of the work of Griffith and Avery left no further doubt that DNA was the genetic material, and it cleared the way for the frantic search for the *structure* of DNA.

The reason the structure of the DNA molecule was considered so important was the belief that by knowing its structure, we could explain how DNA could do the two things that any hereditary material must do: (1) make replicas of itself, and (2) regulate the activities of cells.

THE RACE FOR THE DOUBLE HELIX

After the conclusive experiments of Hershey and Chase, intensive research to discover the structure of the DNA molecule began. The chemical composition of DNA had been known for 50 years (see p. 100), but now the important questions were: How is the DNA molecule constructed? And how does DNA accomplish its work? It was these questions that would be answered in 1953 by James D. Watson, an American, and Francis H. C. Crick, an Englishman.

Probably the most famous entrant in the race for the double helix was chemist Linus Pauling. Working at the California Institute of Technology, Pauling had done outstanding work in chemistry, including a description of the three-dimensional helical structure of protein molecules—the work for which he would be awarded the Nobel prize in 1954. Confident from his success with proteins, Pauling took aim on DNA.

Most scientists agreed that DNA would be shaped like a *helix* (similar to the winding of a spiral staircase), and Crick recalled several years later that many of the most influential researchers (including Pauling) were thinking of a helical structure for DNA. "They were all building helical models," he said. "I would say, you would be eccentric, looking back, if you didn't think DNA was helical."*

Watson and Crick, searching for the structure of DNA, relied heavily on intuition and the published and unpublished research findings of others. They did no formal "research" during their attempts to build a model of DNA, and they agreed with other scientists that the answer to how DNA worked would be found in its three-dimensional *structure.* Watson arrived in Cambridge, England, in 1951 to work on this problem. He became acquainted with Crick, and their famous collaboration began.

Besides the assumption that DNA would be a helix and the rather complete knowledge of its chemical components, there was additional information about DNA available through X-ray diffraction photographs. Maurice H. F. Wilkins and later Rosalind Franklin were producing excellent X-ray

* From recorded interviews with Robert Olby in Cambridge, England, March 8, 1968, and August 7, 1972. Robert Olby, *The Path to the Double Helix* (Seattle: University of Washington Press, 1974).

diffraction photographs at King's College in London, and their work would be one of the most important aids to Watson and Crick (Figure 6.7).

Recall that a DNA molecule consists of sugar, phosphates, and four nitrogen bases—adenine, guanine, thymine, and cytosine. Watson and Crick knew that. They also knew from the X-ray photographs by Wilkins and Franklin that DNA was most likely a coiled structure, with the twists in the coil occurring at every tenth nucleotide. This was a consistent pattern, no matter how many twists there were or how long the entire strand of DNA was. What Watson and Crick still did not know was how the bases and sugar-phosphate backbones were arranged in three-dimensional space, or how the entire mechanism could duplicate itself, with all the necessary information being supplied from within the DNA.

By 1952 Linus Pauling and his co-workers at Caltech had become increasingly interested in DNA, and in December 1952 Watson and Crick learned that Pauling and R. R. Corey had built a structure for DNA. Although Pauling gave no details, Watson was convinced that Pauling had beaten them. But when Pauling and Corey published their findings in February 1953 they showed a *triple* helix structure. Most surprising for a chemist of Pauling's stature, the model contained some obvious chemical errors, and Watson calculated that he and Crick had four to six weeks before Pauling himself realized his mistakes and built the correct model.

Watson and Crick knew that in any given molecule of DNA, adenine and thymine appear in equal amounts, and guanine and cytosine also are present in equal amounts. This principle had been presented in 1949 by Erwin Chargaff, but it was not until 1951 that Watson and Crick were able to see the obvious one-to-one relationship between the bases. In his book *The Double Helix,* Watson tells how important the Chargaff relationships were in making the connection that would solve the puzzle:

> *Suddenly I became aware that an adenine-thymine pair held together by two hydrogen bonds was identical in shape to a guanine-cytosine pair held*

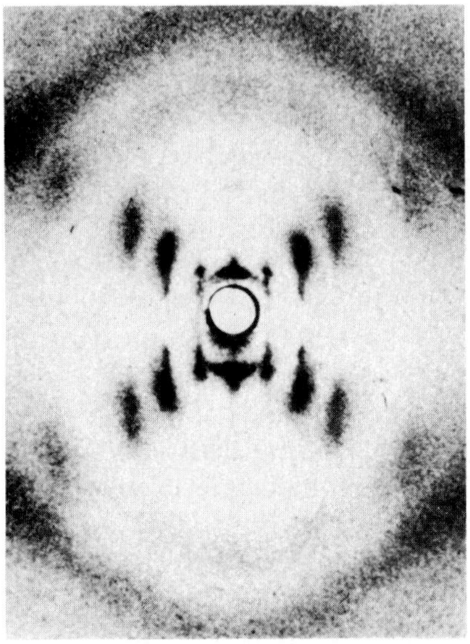

Figure 6.7
An X-ray photograph of DNA. To obtain such a photograph, the crystallographer sends an X-ray beam through a small hole in a lead sheet, allowing the beam to strike a sample. If the sample is constructed in a regular pattern of repeating particles, the X-ray beam will be changed in direction in a regular way. If the resulting "diffracted beams" are caught on a photographic plate, one can develop a pattern of light and dark spots whose spacing and position can be measured. From such measurements, it is possible to calculate the spacing and position of the particles in the sample. From this particular photograph, several dimensions of spacing and arrangement of atomic groups in DNA were deduced.

(a)

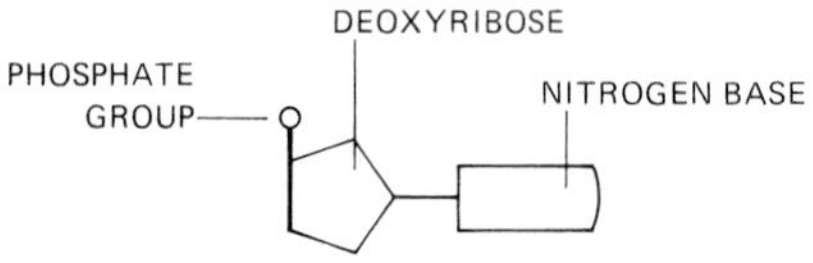

(b)

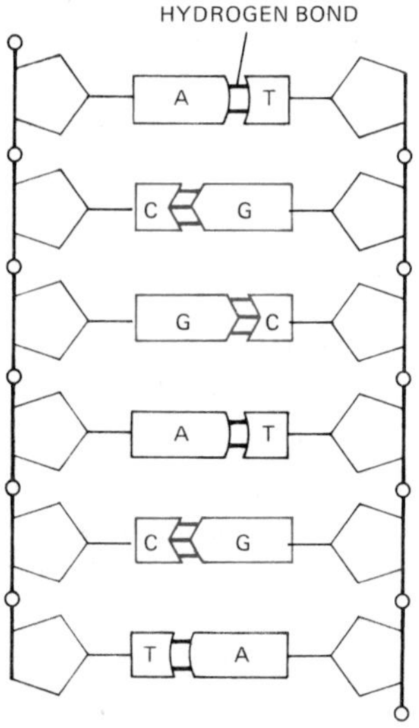

Figure 6.8
(a) A single nucleotide, with a phosphate group, a five-membered ring of deoxyribose, and a nitrogen base. Two such nucleotides, joined by hydrogen bonds, make a *base pair.*
(b) A strand of DNA consisting of six base pairs. Note that the space can be adequately filled only when a purine, such as guanine (with its double-ring structure), is matched with a pyrimidine, such as cytosine (with its single ring). The number of atoms available for hydrogen bonding is important in determining which purine is joined with which pyrimidine. With three hydrogen bonds possible between guanine and cytosine and only two between adenine and thymine, the proper bases normally pair up accurately—guanine only with cytosine and adenine only with thymine.

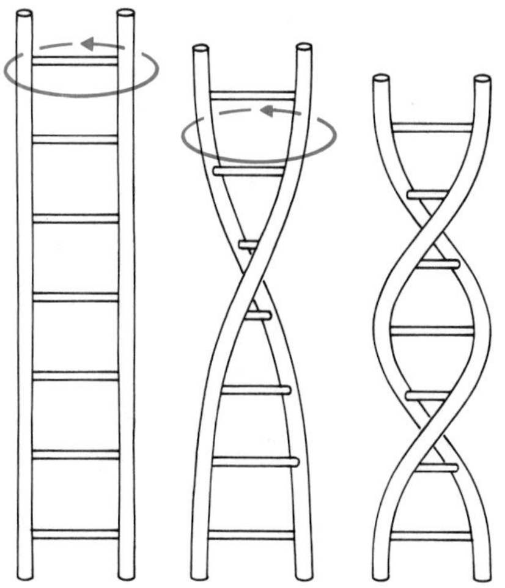

(a)

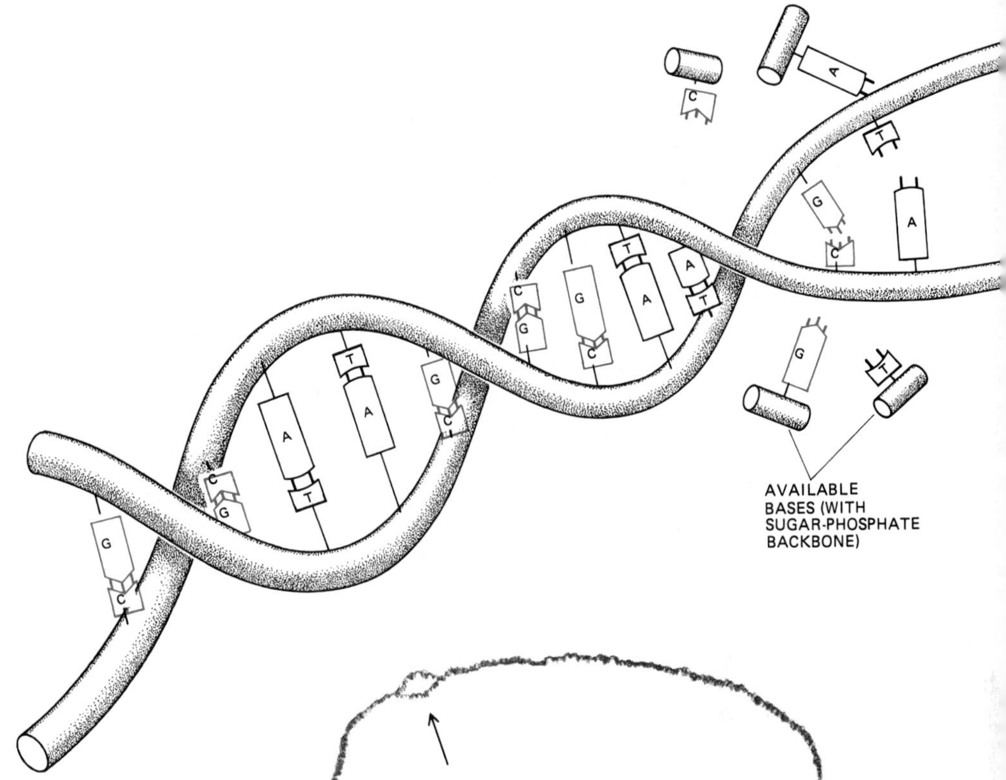

(b)

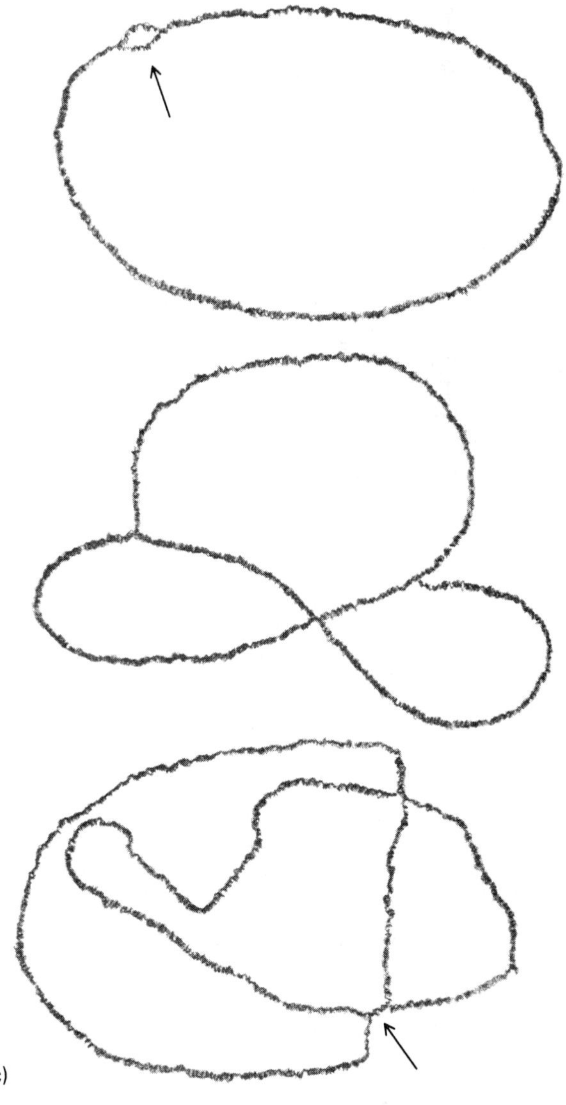

(c)

Figure 6.9
(a) The double helix form. If a ladder could be twisted, the two uprights would form a double helix. In a DNA molecule, the repeating strands of sugars and phosphate groups can be thought of as the uprights and the nitrogen bases as rungs. (b) A portion of a DNA molecule, with its double helix of sugars and phosphate and cross-connectors of bases. A complete DNA molecule has millions of base pairs, and an actual molecule would be long enough to see with the unaided eye if the strand were not so thin. (c) Replication of a bacterial chromosome, which is a naked DNA strand in one continuous loop. In the first picture, one small strip has begun to separate at the upper left (arrow). In the second, almost half the chromosome has become visibly doubled, and in the third, the loop has separated except at one final sticking point (arrow).

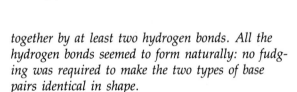

together by at least two hydrogen bonds. All the hydrogen bonds seemed to form naturally: no fudging was required to make the two types of base pairs identical in shape.

Now Watson and Crick could visualize how the double-ringed adenine could combine with the single-ringed thymine, and likewise the double-ringed guanine with single-ringed cytosine without deforming the spaces between the sugar-phosphate backbones (Figure 6.8). And, as Watson said, the weak hydrogen bonds fell into place between the bases naturally without any forcing or "fudging."

The structure was a *double helix*, twisted like a vertical ladder (Figure 6.9). Watson and Crick had won the race.

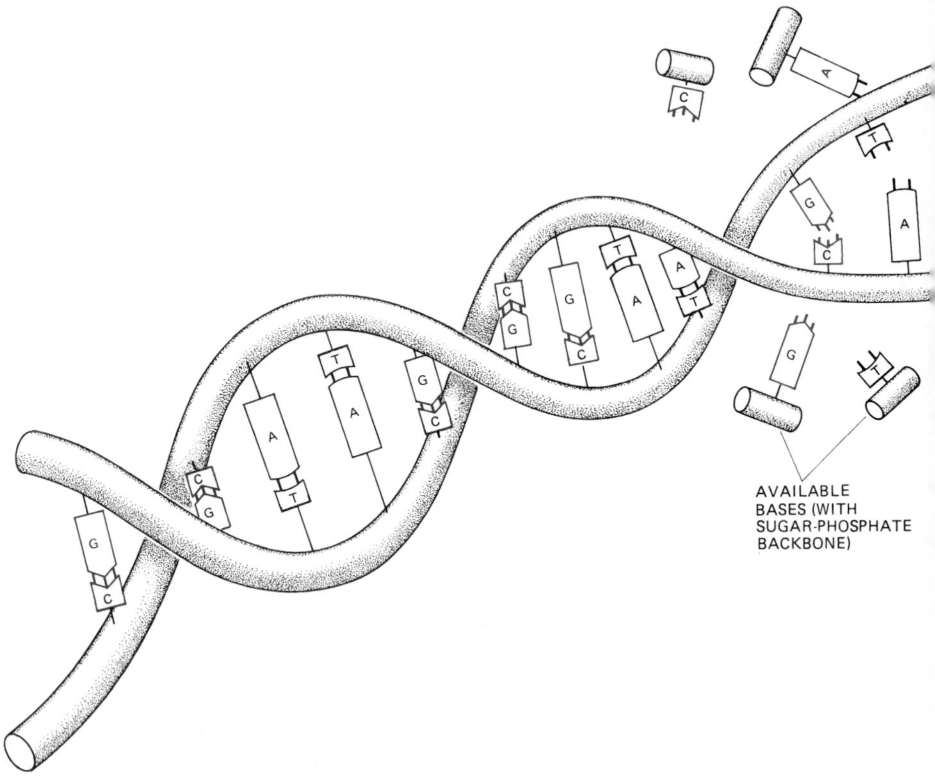

Figure 6.10
The replication of DNA. Scientists believe that the double helix starts the replication process by untwisting. The weak hydrogen bonds joining the purine and pyrimidine bases are somehow released. The molecule "unzips" itself. Each adenine-thymine pair separates and each guanine-cytosine pair separates. This state of separation does not exist for long. From within the cell's cytoplasm, available adenine (complete with sugar-phosphate backbone) finds its way to the newly exposed thymine, and with the help of appropriate enzymes, new hydrogen bonds are created.

On the other side of the unzipped backbone, the partnerless adenine will be joined by a free-floating thymine. Guanines will link up with stationary cytosines, and floating cytosines will form bonds with guanines still attached to the sugar-phosphate backbone. Each nitrogen base has achieved a coupling with its complementary base partner, and the single unzipped strand of DNA has now become two exact copies of the original. It is this principle of *complementary bonding* that ensures that an exact copy will be made.

Crick has proposed that the untwisted strand begins to twist again as soon as the new bonds are formed. If so, the complicated activity of untwisting, unzipping, rezipping, and retwisting is all taking place at the same time.

In the April 25, 1953, issue of *Nature,* Watson and Crick published their ideas about the DNA molecule in a brief article. Later in 1953 they published a follow-up article discussing how this structure could permit the reproduction of identical molecules. Their work was almost unanimously accepted, and it has since been confirmed by numerous experiments. As hoped, the structure of DNA was the key to the replicating process, and Watson and Crick modestly understated that supposition in their original article: "It has not escaped our notice that the specific pairing we have postulated immediately suggests a possible copying mechanism for the genetic material."

Pauling, informed of the content of the article, declared graciously:

Although it is only two months since Professor Corey and I published our proposed structure for nucleic acid, I think we must admit that it is probably wrong. . . .

Although some refinement might still be made, I feel that it is very likely that the Watson-Crick structure is essentially correct. . . .

I think that the formulation of their structure by Watson and Crick may turn out to be the greatest development in the field of molecular genetics in recent years.

In 1962 Watson, Crick, and Wilkins were awarded the Nobel prize for their assessment of the structure of DNA. Franklin died in 1958. Although her work with Wilkins was not cited by the Nobel prize committee, her contribution to the Watson-Crick model has since been fully acknowledged.

Now that the structure of the DNA molecule had been revealed, it was possible to explain its method of replication.

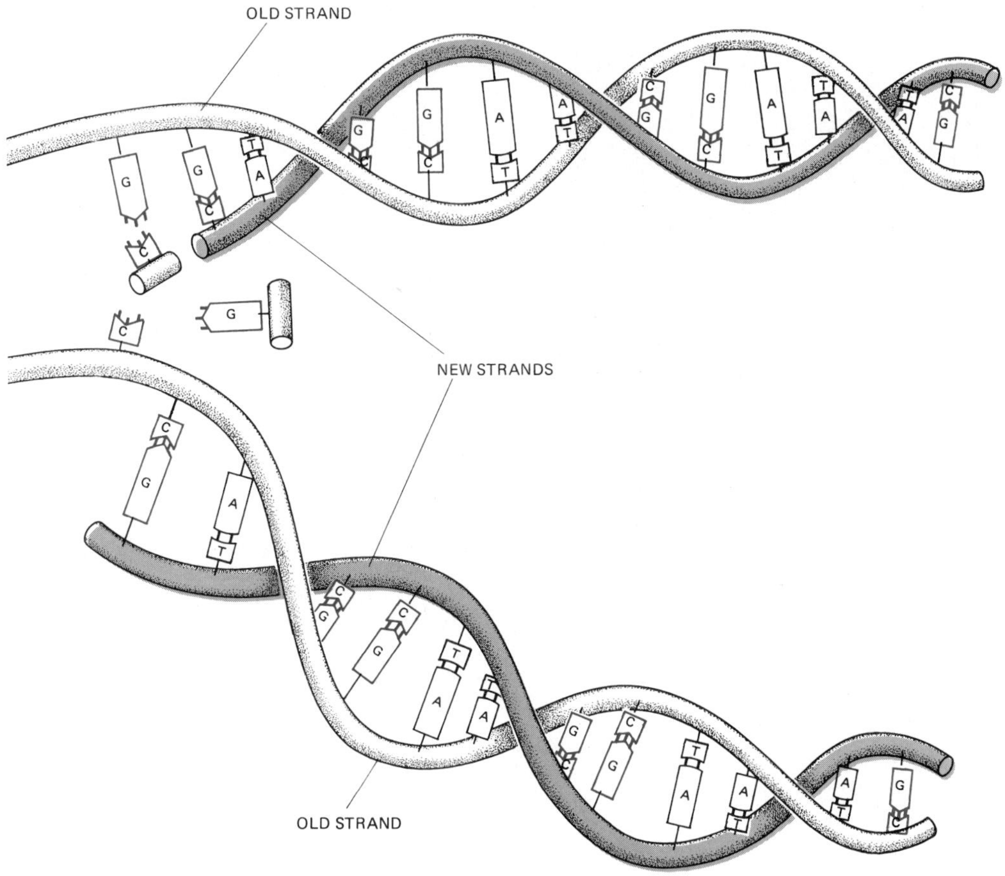

HOW DNA REPRODUCES ITSELF

Let us now visualize the DNA molecule in more specific terms. If we continue to think of the molecule as a twisted vertical ladder, the side uprights consist of alternating sugars and phosphates, and the rungs consist of the base pairs of A—T, G—C, T—A, or C—G. There is no restriction on the overall length of the ladder (although each organism will have its own consistent lengths of DNA strand), and the pairings of purines with pyrimidines may occur in any sequence. *It is the sequence of nucleotides that specifies the genetic code.*

It has been estimated that the total number of paired bases in a human DNA molecule may be counted in the millions. If all the DNA in a human cell were in one long strand, it might be 2.5 cm (1 in.) long, with a molecular weight of perhaps 50 million. Such a molecule could contain the equivalent of the information in a large library of books. How could it be possible to duplicate so much information, over and over again, without numerous errors? (When errors are made, as they are sometimes, they are called *mutations.* We will discuss them in Chapter 12.) And how could so much information be contained in a single molecule?

The Watson-Crick model seems to provide suitable answers to these questions. Because the paired bases of adenine-thymine and guanine-cytosine can be arranged in any sequence as long as purines are joined with pyrimidines, the possible number of combinations is enormous. Therefore it is possible to provide enough chemical instruction to regulate the manufacture of proteins from amino acids.

In order for organisms to go on living, their cells must be able to reproduce, and because DNA is essential to life, *it* must be able to reproduce. The Watson-Crick model permits us to understand *how DNA replicates itself* (Figure 6.10). A simplified, un-

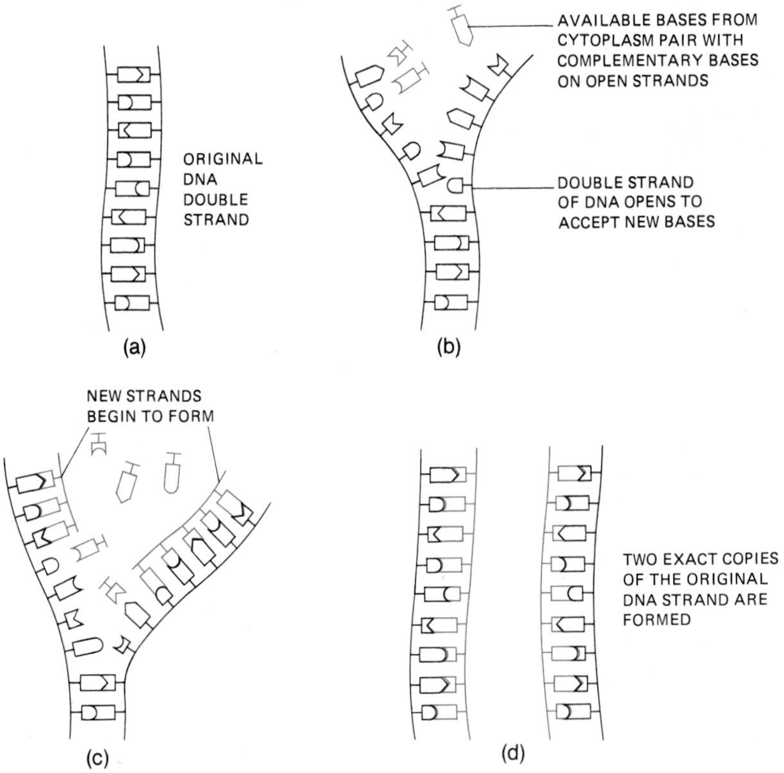
AVAILABLE BASES FROM CYTOPLASM PAIR WITH COMPLEMENTARY BASES ON OPEN STRANDS
ORIGINAL DNA DOUBLE STRAND
DOUBLE STRAND OF DNA OPENS TO ACCEPT NEW BASES
(a)
(b)
NEW STRANDS BEGIN TO FORM
TWO EXACT COPIES OF THE ORIGINAL DNA STRAND ARE FORMED
(c)
(d)

Figure 6.11
The complicated replication of DNA has been simplified in these drawings, which do not attempt to show the strands twisting and untwisting.

twisted model of DNA replication is shown in Figure 6.11.

The connection between the replication of DNA and the making of proteins, the genetic code that holds the key to the production of life-giving proteins, and the function of RNA in synthesizing proteins will be the concerns of the next chapter.

SUMMARY

1. Nucleic acids are extremely large molecules (macromolecules), probably the largest in living cells. Scientists are now certain that nucleic acids are found in all living things, and that they are the primary factors that control the hereditary processes of life.

2. There are two main kinds of nucleic acids, each with its own sugar structure. *Ribonucleic acid,* or *RNA,* contains ribose. *Deoxyribonucleic acid,* or *DNA,* contains a sugar with one less oxygen atom. RNA contains *uracil* instead of thymine. It can also be found in the cytoplasm outside the cell's nucleus.

3. DNA is made up of units called *nucleotides,* and each nucleotide is made up of three components: phosphate, sugar, and nitrogen base. DNA contains four kinds of nucleotide bases: *adenine, guanine, thymine,* and *cytosine.* Adenine and guanine are *purines,* and thymine and cytosine are *pyrimidines.*

4. Friedrich Miescher discovered DNA, which he called *nuclein,* in 1869, but he did not determine its function. It was not until the 1940s that any proof became available that DNA, not protein, controls heredity.

5. In 1928 Fred Griffith discovered the process of *transformation,* whereby DNA transfers its genetic characteristics from one organism to another. Sixteen years later, Avery, MacLeod, and McCarty confirmed that it was the DNA in Griffth's experiments that was the effective agent of heredity.

6. In 1952 Alfred D. Hershey and Martha Chase used radioactive tracers to show how viruses injected their DNA into host cells and thereby manufactured new, identical viruses. This final confirmation that DNA was the genetic material cleared the way for the search for the three-dimensional structure of DNA.

7. The three-dimensional structure of the DNA molecule was important to explain how DNA could do the two things that any hereditary material must do—make replicas of itself and regulate the activities of cells.

8. In 1953 James D. Watson and Francis Crick, with the assistance of X-ray diffraction photographs by Maurice Wilkins and Rosalind Franklin, identified the structure of the DNA molecule as a *double helix.* Their discovery permitted an acceptable explanation of the *replication* of DNA; it has also been shown that the *sequence of nucleotides* specifies the genetic code.

ASK YOURSELF

1. In what ways does DNA differ from RNA?
2. What is a nucleotide?
3. How does a purine differ from a pyrimidine?
4. Why is the difference between a purine and a pyrimidine important in a living cell?
5. How did experiments on bacteria show the genetic importance of DNA?
6. Why were proteins rather than DNA long suspected of being *the* important genetic constituents of cells?
7. In a DNA molecule, what nucleotide base can pair with a thymine base? With a guanine base?
8. If a DNA molecule is compared to a twisted ladder, what components of the molecule are the uprights? The cross pieces?

7
The Genetic Code and Protein Synthesis

SOME KEY POINTS

1. A gene is a specific segment of DNA that controls specific cellular functions.
2. The production of each protein is controlled by a specific gene.
3. Messenger RNA (mRNA) is composed of chains of nucleotides, which act three at a time to code for specific amino acids. The triplets of mRNA are called codons.
4. The complete set of triplet nucleotide symbols that indicate specific amino acids is called the genetic code.
5. Nucleotides of transfer RNA (tRNA) also act three at a time to match the placement of mRNA. Triplets of tRNA are called anticodons.
6. Protein assembly starts when single-stranded RNA is synthesized under the direction of DNA. The process is called transcription.
7. Bacterial cells apparently regulate their own internal activities through a complex interaction of specialized genes.
8. Eukaryotes may have a system of gene regulation that provides for greater flexibility than prokaryotes have in producing new and different genes.

WE SAW IN THE PREVIOUS CHAPTER HOW DNA molecules can make copies of themselves. That is something most molecules cannot do. Proteins, for example, are not capable of duplicating themselves. But some proteins are enzymes, and enzymes control all cellular reactions. These reactions ultimately create organic molecules such as fats, sugars, and even other proteins. If proteins are critical to life, and if they cannot reproduce, how can life go on?

Once Watson and Crick had described the structure of DNA, the next big question was: How can we use what we know about the replication of DNA to discover how proteins are made? Scientists had known since the 1940s that DNA is involved in the production of proteins, and just a few years after the double helix of DNA was revealed, scientists began to understand the function of the *other* nucleic acid, RNA. The "central dogma" of modern biology was formulated: **DNA makes** (*codes for*) **RNA, which makes** (*codes for*) **proteins.** In this chapter we will find out *how.*

WHAT IS A GENE?

The double-helix model of DNA opened the door to a staggering amount of genetic research. Between 1958 and 1965, *14 out of the 24* Nobel prizes awarded in chemistry and medicine/physiology went to scientists who had studied nucleic acids, proteins, or genetics. New discoveries were followed by new definitions. Textbooks of the pre-DNA explosion might have called a gene simply "a unit that carries hereditary traits." Now, our reason-

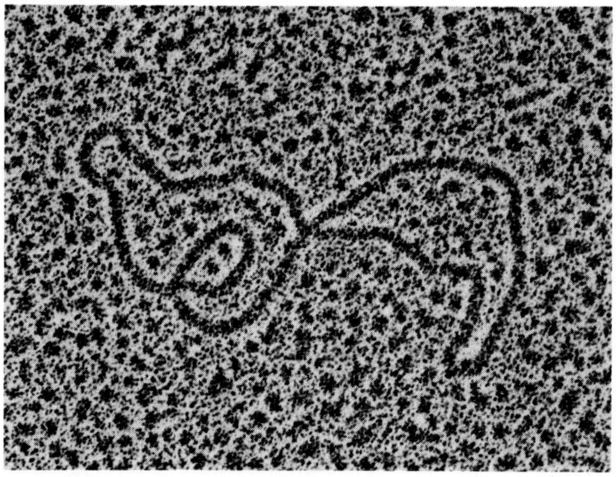
× 70,000

Figure 7.1
Electron micrograph of a gene isolated from the bacterium *Escherichia coli*. This particular gene, estimated to be about 1.4 micrometers long, is concerned with the production of an enzyme that acts on the sugar galactose.

ing about genes takes this more complete approach: Every cell contains DNA. Each DNA strand contains specific segments called **genes** that control specific cellular functions, either by synthesizing enzymes or other proteins, or by regulating the action of other genes (Figure 7.1).

The definition of a gene has changed many times since Gregor Mendel first introduced the idea of "factors" that influence hereditary traits. At present we accept the concept described above, but in years to come new information may bring us closer to a more nearly complete and accurate description of a gene.

THE ONE GENE–ONE ENZYME HYPOTHESIS

In 1941, George W. Beadle and Edward L. Tatum began their research with the bread mold *Neurospora*, an organism that normally synthesizes the amino acids and most of the vitamins it needs. Beadle and Tatum decided to "begin with known chemical reactions and then look for the genes that control them." They induced changes (mutations) in the genes in *Neurospora* DNA by using X-rays; the mutated gene could no longer produce the enzyme that controlled the particular synthesis being observed. The researchers found that in almost every case they could trace a single enzyme malfunction to a single gene.

The idea was developed that each gene controls a specific enzyme; it was further proposed that because enzymes are proteins, and because enzymes are involved in the making of other proteins, *genes actually control protein synthesis*. With this line of reasoning, now universally accepted, Beadle and Tatum originated the "one gene–one enzyme" hypothesis. The theory is now referred to as the **one gene–one polypeptide** hypothesis, because we have since learned that genes control the formation and activity of *all* proteins, not just enzymes.

THE TRIPLET CODE

Proteins are the working molecules of cells. In Chapter 1 we saw that proteins, including all enzymes, are made up of structural units called *amino acids*. Only 20 different amino acids are used by cells to make up the countless varieties of proteins needed by the human body; enzymes alone account for 10,000 to 100,000 of these different combinations of amino acids, each structurally specialized to do its own particular job. (See the essay on enzymes on p. 28.) How can only 20 amino acids make the thousands of proteins an organism needs? Each enzymatic function is determined by the three-dimensional shape of the protein molecule, and that shape is in turn determined by the arrangement of amino acids in the molecule.

There are only four kinds of nucleotides available in the coding material of RNA (adenine, cytosine, guanine, and uracil), but these four must somehow code for the 20 different amino acids important to living things. If one nucleotide coded for one amino acid, four nucleotides could code for only four amino acids. If two nucleotides at a time were used, there could be 16 possible combinations (4^2 or 4×4), but that is still not enough. With three nucleotides at a time, however, the 64 possible combinations (4^3 or $4 \times 4 \times 4$) are more than adequate to produce the 20 amino acids we need. We can say that the language of DNA-RNA contains only four letters, from which 64 three-letter words may be written. (Similarly, the 26 letters of our alphabet can be combined in various ways to make an entire dictionary of words.) The combination of three nucleotides **(triplets)** in a specific sequence that codes for a specific amino acid is called a **codon.** (We will come back to codons shortly.)

THE GENETIC CODE

The complete set of triplet nucleotide symbols that indicate specific amino acids is called the **genetic code** (Table 7.1). Experiments with different organisms have shown that a given codon, say AUG, dependably specifies one particular amino acid (AUG "means" methionine, CAG "means" glutamine; what amino acid does GGG code for?). Table 7.1 shows that 64 possible codons can be derived from the four RNA nucleotide symbols (U, C, A, G); 61 of these codons correspond to amino acids, and the remaining three codons are "stop" signals, any one of which will cause the protein synthesis to stop. Note that some amino acids have only one codon, but others have as many as six. Such repetition is thought to allow for a certain amount of error without destructive results for an organism.

Strange as it may seem, a bird, a bean, and a human being—practically all living organisms—have the same genetic code. The difference among species is not in the arrangement of letters within a codon, but in the *sequence* of codons.

THE TRANSCRIPTION OF RNA FROM DNA

Protein assembly starts with a process known as **transcription,** in which single-stranded RNA is synthesized under the direction of DNA. The unwinding of a double-stranded DNA molecule during replication was described in the previous chapter. A similar separation of hydrogen bonds between the members of a pair of DNA bases opens up the molecule before the transcription begins (see Figure 7.3a). However, when RNA instead of more DNA is to be synthesized, the process is slightly different in three important ways. First, the bases in RNA differ, in that RNA contains uracil instead of thymine. Second, the newly assembled chain of nucleotides, that is, the new RNA molecule, separates from its DNA "mold" and moves off as a single-stranded compound. Third, the RNA nucleotide contains the sugar ribose instead of the sugar deoxyribose (Figure 7.2).

During transcription, each new nucleotide must match a proper nucleotide base on the opened DNA strand. Where a DNA nucleotide contains

TABLE 7.1
The Genetic Code (RNA)*

FIRST LETTER IN THE CODON	**SECOND** LETTER IN THE CODON: U	C	A	G	**THIRD** LETTER IN THE CODON
U	Phenylalanine	Serine	Tyrosine	Cysteine	U
	Phenylalanine	Serine	Tyrosine	Cysteine	C
	Leucine	Serine	Stop†	Stop†	A
	Leucine	Serine	Stop†	Tryptophan	G
C	Leucine	Proline	Histidine	Arginine	U
	Leucine	Proline	Histidine	Arginine	C
	Leucine	Proline	Glutamine	Arginine	A
	Leucine	Proline	Glutamine	Arginine	G
A	Isoleucine	Threonine	Asparagine	Serine	U
	Isoleucine	Threonine	Asparagine	Serine	C
	Isoleucine	Threonine	Lysine	Arginine	A
	Methionine	Threonine	Lysine	Arginine	G
G	Valine	Alanine	Aspartic acid	Glycine	U
	Valine	Alanine	Aspartic acid	Glycine	C
	Valine	Alanine	Glutamic acid	Glycine	A
	Valine	Alanine	Glutamic acid	Glycine	G

*The amino acid coded for appears under the second letter of the triplet. For example, UGG codes for the amino acid tryptophan.

†UAA, UAG, and UGA do not correspond to particular amino acids; any one of these three codons will cause protein synthesis to stop.

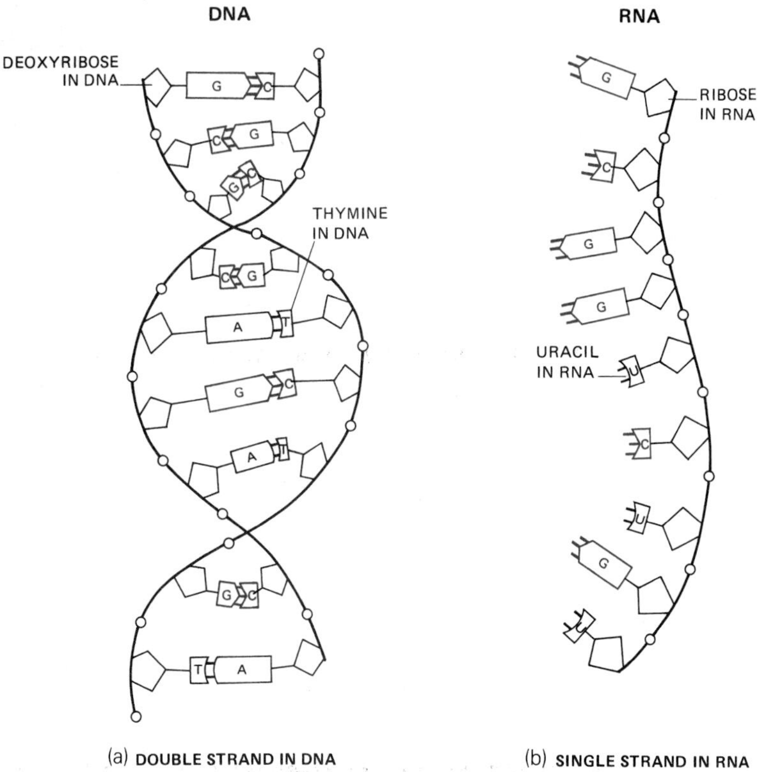

Figure 7.2
The differences between DNA and RNA: (a) DNA is double-stranded, has deoxyribose as the sugar portion of each nucleotide, and incorporates thymine as one of its nitrogen bases; (b) RNA is single-stranded, has ribose as its sugar, and incorporates uracil instead of thymine. (RNA also contains a number of nitrogen bases besides the common ones, but their exact functions are not understood.)

HOW THE GENETIC CODE WAS DECIPHERED

To guess that a triplet, or codon, designates a specific amino acid is one thing; to show that this is actually so is something else again. The genetic code began to be deciphered when Severo Ochoa prepared a nucleotide chain in a laboratory in 1959. He made his nucleotide chain of the same nucleotide throughout. Ochoa's work was followed up in 1961 by that of Marshall Nirenberg.

Ochoa's synthetic mRNA was composed of uridines only, and so it was polyuridine, or *poly-U.* If poly-U in a test tube is supplied with all 20 tRNAs, all 20 amino acids, ribosomes, and an energy source, it works to form proteins without the presence of entire cells. The only amino acid selected out of the amino acid pool by the poly-U is phenylalanine, and the polypeptide chain becomes a chain of phenylalanines. In other words, UUU is the mRNA code symbol for the amino acid phenylalanine.

The breaking of the code of mRNA triplets had begun. If a synthetic mRNA is made only from adenine instead of uracil (polyadenine or poly-A), a polypeptide containing only the amino acid lysine is produced, and consequently AAA is said to code for lysine. Similarly, polyguanine, poly-G, was shown to code for glycine, and polycytosine, poly-C, for proline.

When mRNAs with varied nucleotide residues are considered, such straightforward methods are not possible. The necessary indirect experiments have been performed, however, and meanings have been assigned to all possible combinations of nucleotides. For example, CUU "means" leucine, GCA "means" alanine, and so on. Not all triplets, however, code for an amino acid. For example, UAG and UGA seem not to code for anything, although there is speculation they may serve as "punctuation marks." Further, some amino acids are coded for by several triplets, as glycine is by GGU, GGC, GGA, and GGG. Such repetition is thought to reduce the number of serious errors in copying.

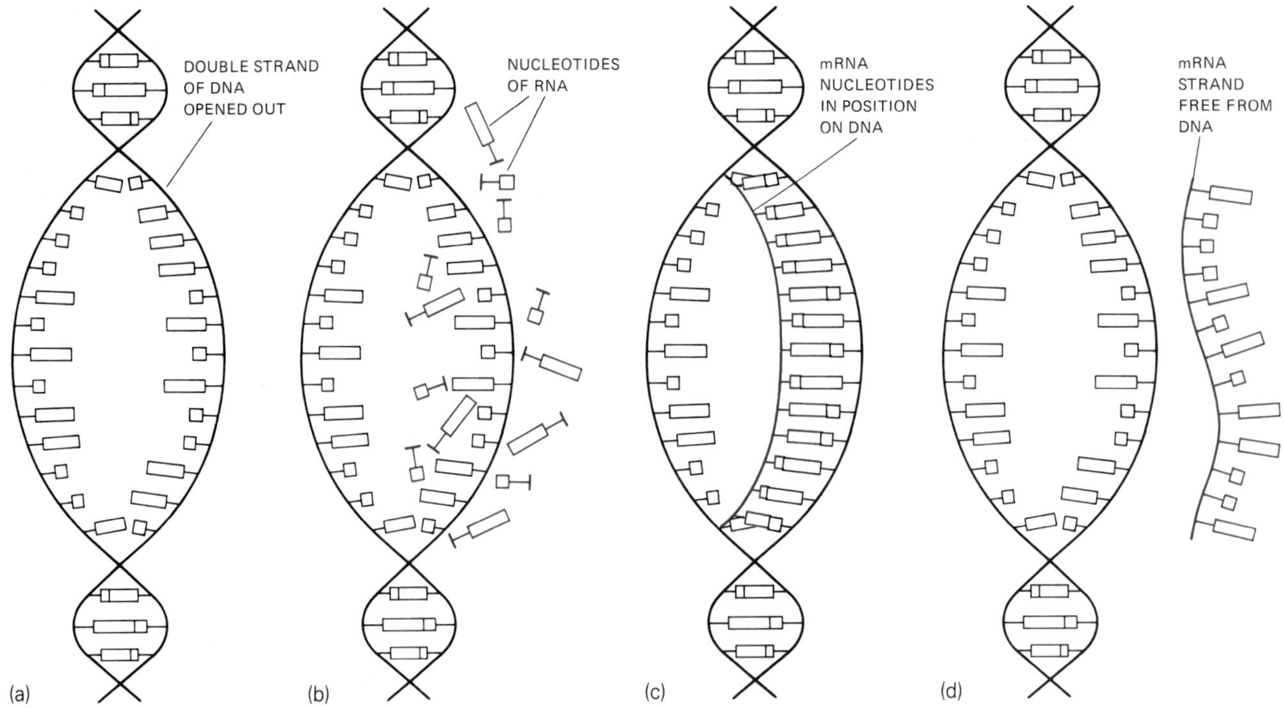

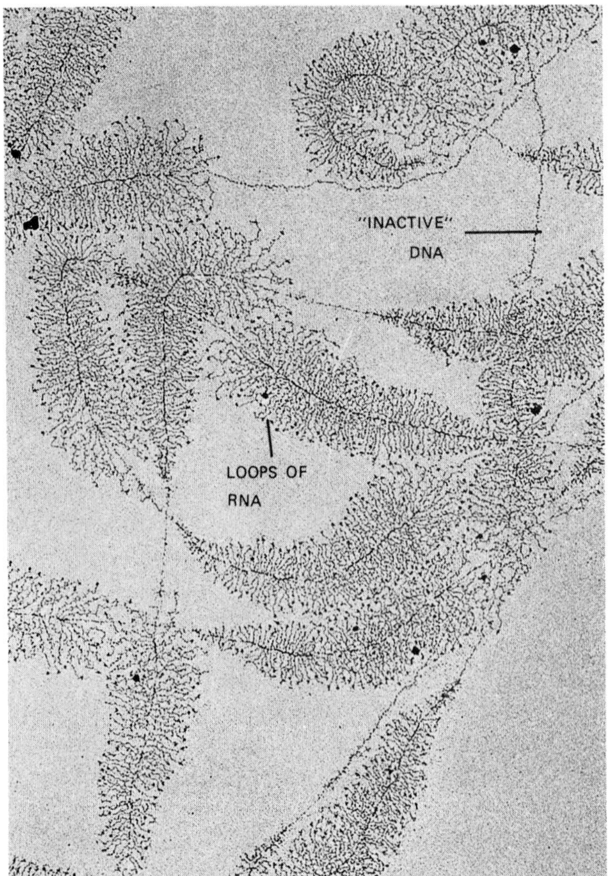

(e)

× 11,000

Figure. 7.3
The formation of RNA from DNA. (a) The double strand of DNA in a chromosome opens, and (b) one of the strands accumulates the proper complementary bases (in color), which have come from the cytoplasm of the cell into the nucleus. (c) The complementary bases are in position along the opened DNA molecule, and (d) the string of newly assembled bases, made into an RNA strand, leaves the chromosome. The DNA is now available to produce more copies of the same RNA. Only a few bases are shown in this simplified figure, but in reality a strand of mRNA could have hundreds or thousands of bases. (e) Electron micrograph of DNA strands in an amphibian oocyte, caught in the act of transcribing RNA (in this instance, ribosomal RNA). The long, thin lines are DNA, and the feathery strands extending from the DNA are RNA strands whose lengths vary with the amount of RNA that had been produced at the moment the cell was killed for study. The short strands have just begun transcription. The naked stretches of DNA were not transcribing when the cell was killed.

guanine, the opposite RNA nucleotide will contain cytosine; conversely, a DNA cytosine will attract an RNA guanine. A DNA thymine will attract adenine. Then the irregularity: A DNA adenine, instead of attracting an RNA thymine, will attract a uracil. (The linking of the nucleotides into a molecule of RNA is dependent on the action of specific enzymes, especially RNA polymerase.) With the exceptions noted, the new RNA is an accurately copied imprint, a kind of negative of the DNA that made it, with proper nucleotides faithfully positioned (Figure 7.3). Three kinds of RNA are transcribed: ribosomal RNA, messenger RNA, and transfer RNA.

Ribosomal RNA

Certain regions of chromosomes transcribe **ribosomal RNA** (rRNA), so called because it will eventually become part of ribosomes. As described in Chapter 3, ribosomes are cytoplasmic organelles composed of about half protein and half RNA. They are the sites of protein synthesis. Although we do not know the use of rRNA, we do know that the final stages of protein synthesis cannot occur without it.

Messenger RNA

A second RNA is **messenger RNA** (mRNA), the molecules of high molecular weight that are directly responsible for the sequence of amino acids in proteins. Messenger RNA is well named, since it does indeed act as though it were carrying a message from DNA, the control center in the nucleus, to the ribosomes in the cytoplasm, telling them which amino acids are to be included and where they are to be placed.

Transfer RNA

The third RNA is **transfer RNA** (tRNA). It is the smallest of the RNA molecules, with a molecular weight of about 25,000 and composed of about 75 nucleotides (mRNA is composed of 1000 to 10,000 nucleotides). Transfer RNA is described as having the shape of a cloverleaf or hairpin (Figure 7.4). One or more types of tRNA exist for each of the 20 amino acids. Each tRNA attracts to itself, and carries a specific amino acid. A triplet of nucleotides in one loop of the tRNA molecule is arranged so that it determines how the particular tRNA will place itself on the mRNA (see the description of protein synthesis in the next section). Because it is the match of the mRNA codon, the triplet on the tRNA is called an **anticodon** (see Figure 7.4b). In the next section we will see how codons and anticodons work together in the synthesis of proteins.

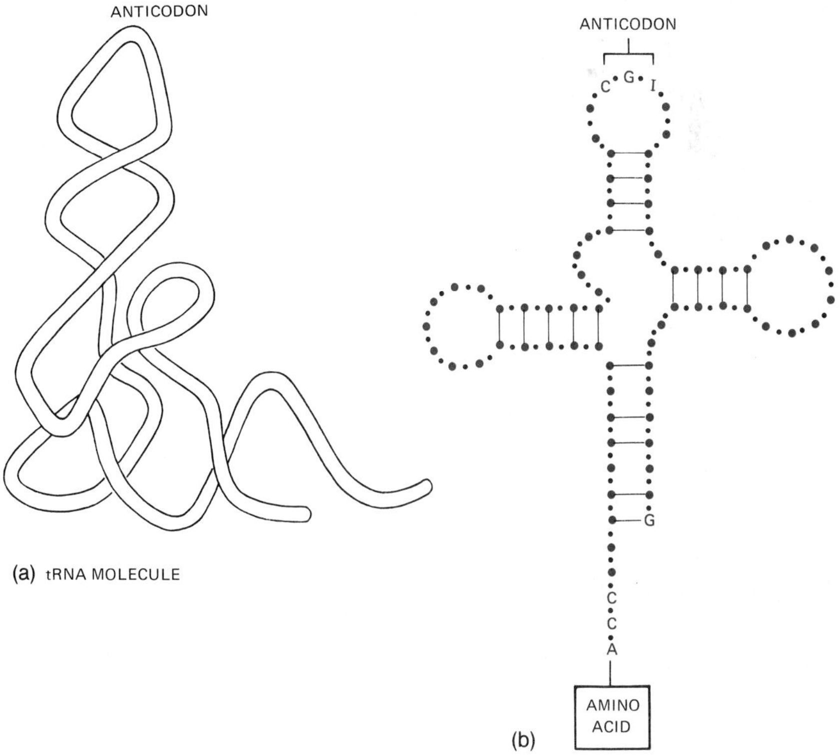

Figure. 7.4
(a) A schematic view of a molecule of transfer RNA (tRNA) carrying an amino acid. A *codon* in a molecule of mRNA will match the complementary bases of the *anticodon* at the end of one of the loops. (b) The sequence of bases (colored dots) in a particular tRNA molecule. Some of the base pairs are complementary, allowing for hydrogen bonding between them, and such bonding keeps the tRNA molecule in shape. The anticodon in this example is C-G-I. Every amino acid has its own special tRNA, yet all RNAs terminate in C-C-A at one end and in G at the other (color).

HOW ARE PROTEINS SYNTHESIZED?

The many proteins the body needs are produced from amino acids that must be assembled according to each organism's own genetic blueprint. This genetic program for the specific linking sequence of amino acids is contained in DNA located in the cell's nucleus. The DNA itself does not assemble the amino acids in the correct sequence. Instead, DNA works together with RNA, which is present in the cytoplasm as well as in the nucleus; that is an important difference—*the RNA is not restricted to the nucleus, but is free to move about the cytoplasm.* The following drawings and description show the process of protein synthesis.

1. The first step in the synthesis of a protein takes place in the nucleus. The genetic message contained in the DNA in the nucleus is copied from one of the strands of the DNA to produce a single-stranded RNA molecule; this RNA molecule is appropriately called **messenger RNA** (mRNA), because it is carrying the genetic message from the DNA.

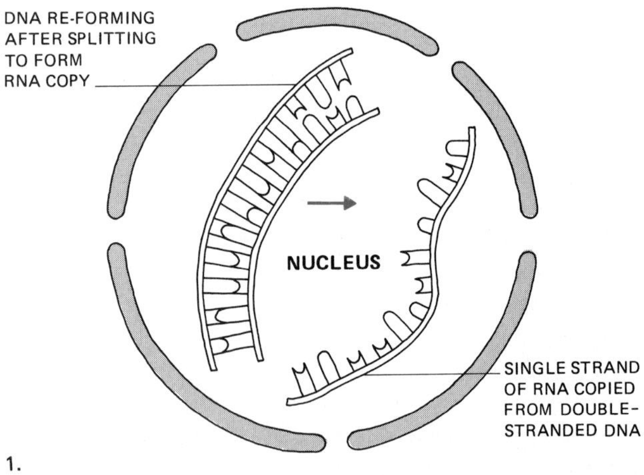

1.

2. The messenger RNA carries the message out of the nucleus to the ribosomes in the cytoplasm. (Only a tiny portion of the strand of messenger RNA is shown.)

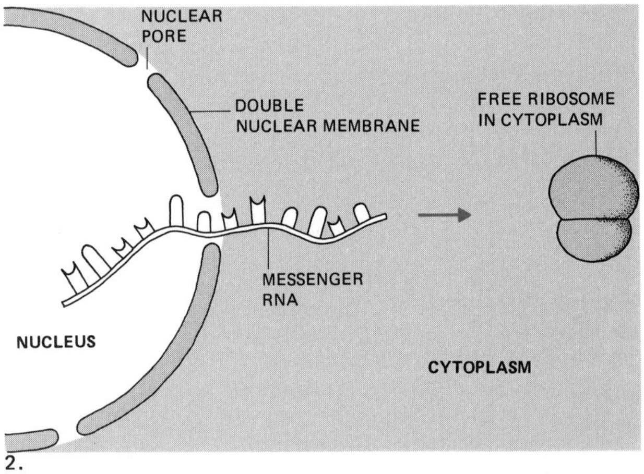

2.

3. The messenger RNA attaches itself to a ribosome, and the ribosome begins to move from one end of the strand of messenger RNA to the other. (Actually, a string of five or more ribosomes is usually involved; see photo below.)

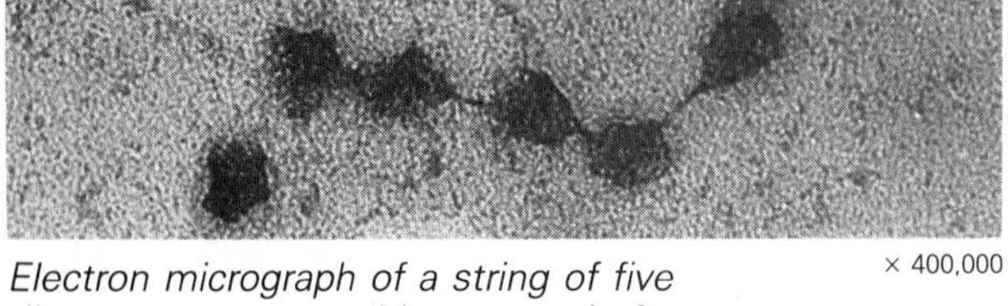

× 400,000

Electron micrograph of a string of five ribosomes connected by a strand of mRNA.

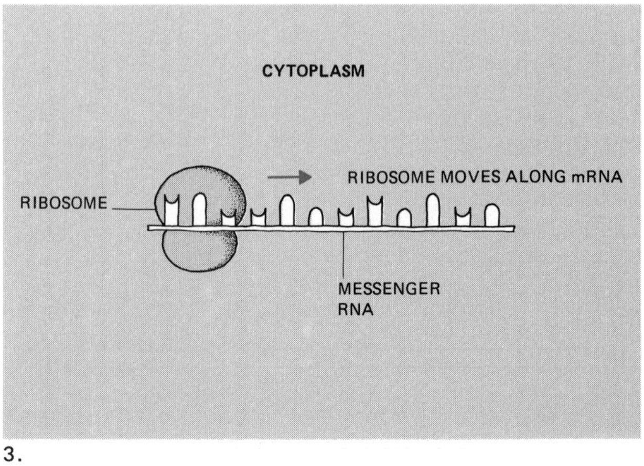

3.

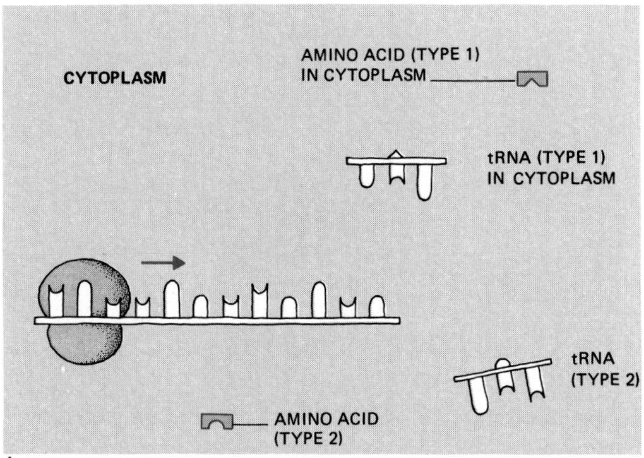

4.

4. Another type of RNA, **transfer RNA** (tRNA), is also in the cytoplasm; amino acids are in the cytoplasm as well. There are about 60 or so different types of transfer RNA molecules, one or more for each type of amino acid. Complexes of transfer RNA and specific amino acids become activated. Each transfer RNA molecule contains two parts, both of which operate on a "lock-and-key" basis: (1) the "mold" part enables it to fit its specific amino acid and (2) the "triple mold" part pairs it with a specific set of three bases (triplet codon) on the strand of messenger RNA (the "triple mold" on the transfer RNA is called a *triplet anticodon*). Each correctly matching transfer RNA molecule and amino acid are linked by a specific enzyme, with energy from ATP.

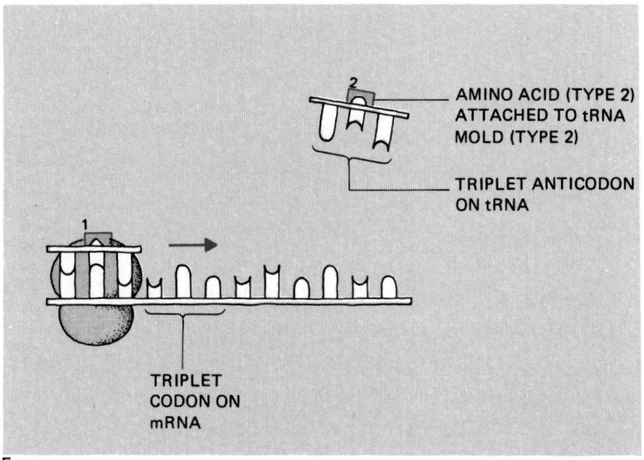

5.

5. Each triplet codon on the messenger RNA is linked with its matching triplet anticodon on the transfer RNA, which is also carrying an attached specific amino acid.

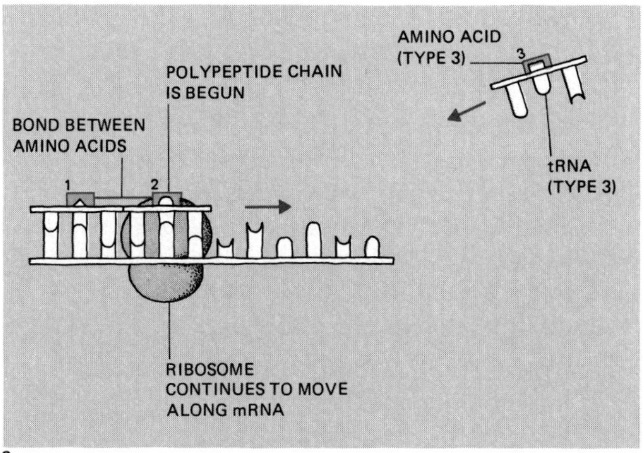

6.

6. As the ribosome continues to move along the strand of messenger RNA to the next triplet of bases, another transfer RNA molecule, with its attached specific amino acid, links up with that second triplet on the messenger RNA. The second amino acid bonds with the first, starting a *polypeptide chain*. Additional transfer RNA molecules, with matching amino acids, approach the strand of messenger RNA.

(Continued on next page.)

7. The ribosome moves steadily onward, and a third link is made between triplets on the messenger RNA and a third transfer RNA. Amino acids continue to bond together in a polypeptide chain. The transfer RNA molecules move off the strand of messenger RNA after their amino acids become part of the growing polypeptide chain. The transfer RNA molecules can be used again after they have deposited their amino acid "passengers."

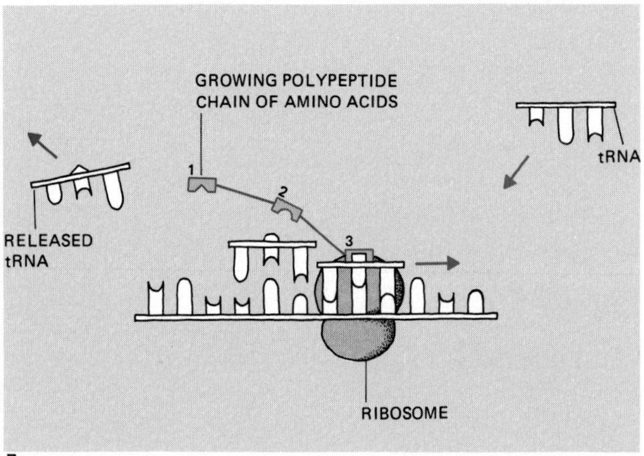

7.

8. One after another, amino acids are bonded together in an ever-lengthening chain until a complete protein molecule is built up according to the coding order on the strand of messenger RNA. The last codon on a complete strand of messenger RNA, instead of coding for an amino acid, signals that the chain is complete. The polypeptide chain is released (the chain shown here is greatly shortened for simplicity), and may combine with other polypeptide chains if a more complicated protein molecule is to be formed. The protein then performs its job. Some proteins remain inside the cell, and others, such as hormones, leave the cell for their assigned destinations.

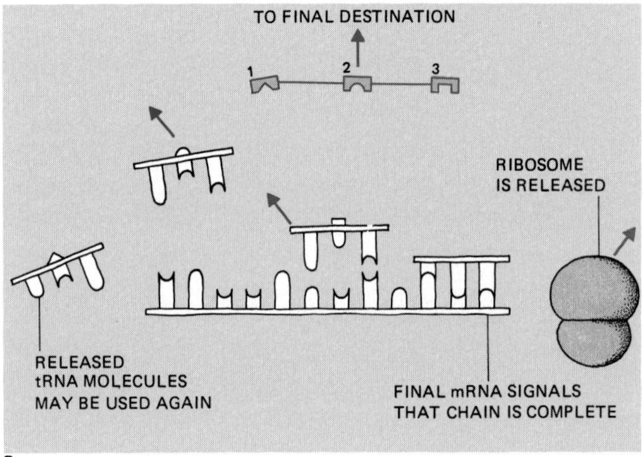

8.

The formation of mRNA from DNA was referred to as *transcription;* the formation of a polypeptide by mRNA is **translation.** Thus, the central dogma can now be rewritten as

$$\text{DNA} \xrightarrow{\text{transcription}} \text{RNA} \xrightarrow{\text{translation}} \text{Protein}$$

where the arrows indicate the direction of flow of information (see Figure 7.5).

TO SUM UP PROTEIN SYNTHESIS

1. The nucleus contains the DNA code message for the synthesis of proteins. The message is transcribed from one of the strands of the DNA to a single-stranded messenger RNA molecule. The DNA molecule serves as a template, or mold, so that the RNA strand is complementary to the DNA strand.
2. The messenger RNA carries the message out of the nucleus to the ribosomes in the cytoplasm. The mRNA attaches itself to the ribosomes.
3. Each of the 20 amino acids is activated by a specific enzyme. Each amino acid attaches itself to a tRNA molecule. After an amino acid is attached to its specific tRNA molecule, the amino acid can be attached to another amino acid at a site on a ribosome.
4. Each triplet codon on the mRNA links with a complementary anticodon on a tRNA, with its attached amino acid.
5. A tRNA and its specific amino acid are attached to mRNA at a ribosome as the mRNA moves along the ribsomes. One after another, amino acids are bonded together in an ever-lengthening chain until a protein molecule is built up according to the coding order on the mRNA molecule.
6. The transfer RNA can be used again to carry a new amino acid after it has deposited its former amino acid "passenger" on the growing polypeptide chain.

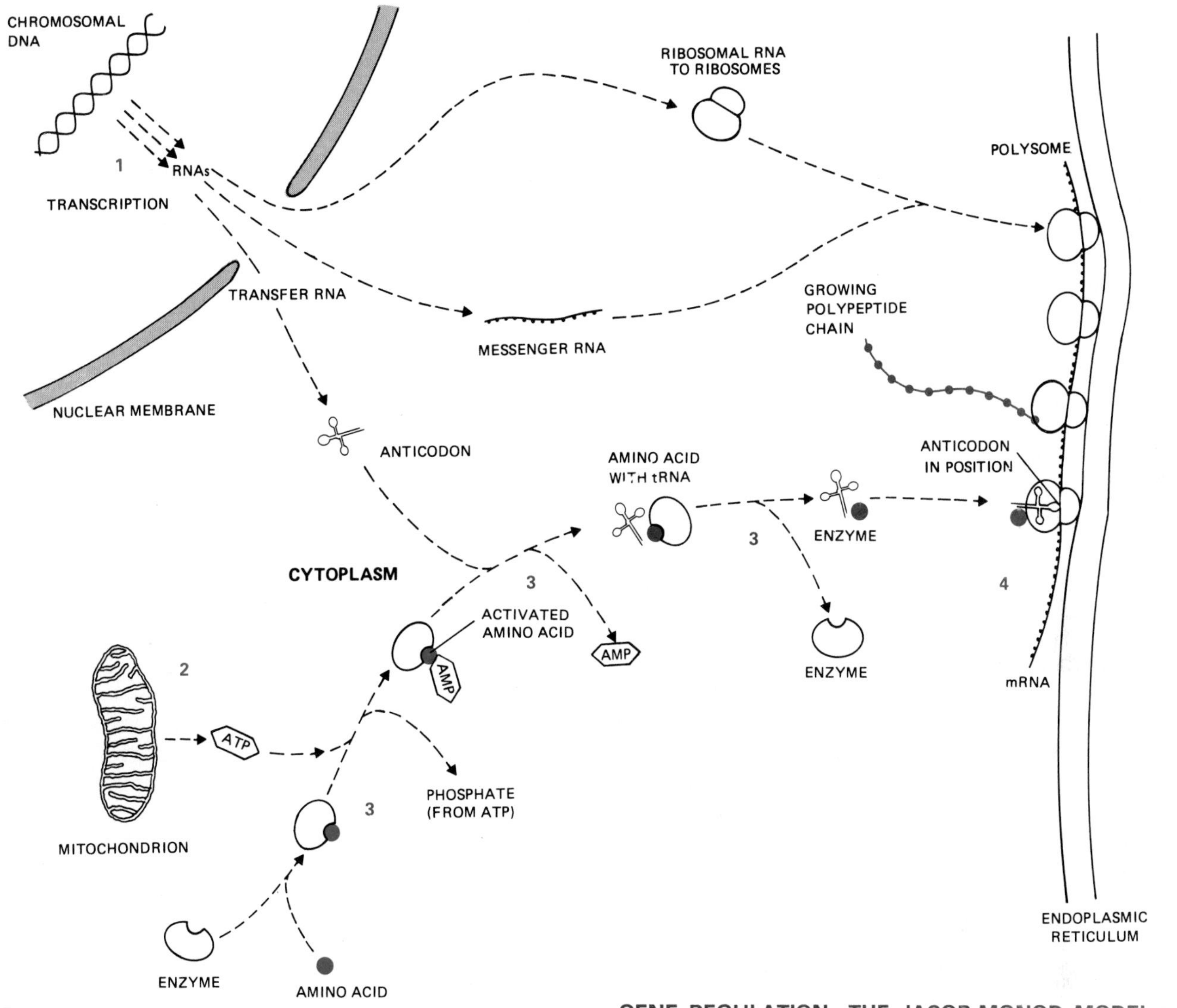

Figure 7.5
Assembling amino acids in the buildup of a *polypeptide.* The diagram shows (1) a DNA section of a chromosome transcribing three kinds of RNA: ribosomal, messenger, and transfer; (2) a mitochondrion producing ATP, which will be used to provide activation energy for an amino acid; adenosine monophosphate (AMP) is released; (3) the sequence of events during which an activated amino-acid–tRNA complex is formed; and (4) a portion of endoplasmic reticulum with an attached polysome (mRNA and ribosomes) and a chain of amino acids being built into a polypeptide (protein).

GENE REGULATION: THE JACOB-MONOD MODEL

Every cell nucleus contains genes that regulate *every* function of the whole organism. But individual cells are specialized to perform only a *few* special functions (liver cells and red blood cells, for instance, do different things), so most of the genes in the nucleus of any given cell are not actually doing anything. What makes some genes in a cell "turn on" while most of the others are "turned off"?

The best current theory about the mechanism of gene activation and inactivation comes from studies of mutants of the bacillus *Escherichia coli* performed by François Jacob and Jacques Monod. The scheme is called the **Jacob-Monod model,** and it describes how *structural* and *operator* genes help regulate the everyday activities of cells, including the production of enzymes and other proteins. (Structural genes are the ones that produce mRNA, and are instrumental in the production of proteins, including enzymes. A

TURNING GENES ON AND OFF

One can directly observe gene control in some cells, especially those that have large chromosomes and that are active in producing proteins. Fruit flies are particularly suited to experimentation because some of the chromosomes in their secretory glands are the so-called giant chromosomes. These chromosomes are large enough to be seen in detail with a light microscope, and they change visibly during the changes from one activity to another.

During the larval stage (1a), before the flies become adult (1b), their secretory glands are producing large amounts of digestive enzymes and other proteins to be used in growth. Cells from such glands contain the large chromosomes (2), which are several hundred times as large as those of the usual body cells. A single chromosome, seen under a microscope, appears banded, with cross stripes definitely and recognizably placed (3). Unlike ordinary chromosomes, these giant chromosomes are made of multiple identical strands, with each strand very tightly coiled at specific places. The thickened regions, lined up side by side, give the appearance of bands, as shown diagrammatically in (4). At certain places along such a chromosome, enlargements appear as rather formless swellings. The enlargements are now more commonly known as **puffs** (5).

Chromosome puffs are believed to be produced in the following way. The many strands of DNA that make up the bulk of the chromosome can be thought of as a bundle of parallel fibers (6), which may uncoil, thus increasing their length. When this happens, the lengthened strands are pushed so that they bend outward (7) and, after still greater uncoiling, develop into loops (8). The hundreds of loops are uncoiled DNA strands, but the entire puff contains more than DNA, especially protein and RNA (9 and 10).

The discovery that certain regions of chromosomes make puffs at certain times led to the supposition that a puff is an expression of genes in action. The supposition was verified in a series of experiments that were simple in their conception but highly imaginative and technically clever. (One of the main experimenters was Ulrich Clever.)

When a larva was injected with a hormone that induces molting, a puff developed on a special place on a chromosome; without the hormone, the puff did not develop. This was good evidence that that particular puff was associated with molting. The nature of the puff itself could be found. A reasonable guess was that it contained RNA, because RNA is required in active protein synthesis. An experiment was devised to test that guess. Cells convert uridine into uracil, and uridine is therefore a precursor of uracil. Since uracil is incorporated into RNA but not into DNA, a uracil precursor, labeled with radioactive tritium (H^3), was injected into larvae. The larvae were killed, the secretory glands were removed, and the giant chromosomes from the glands were placed on photographic film. The developed film showed darkened areas directly over the puffs, indicating that the radioactive tracer was concentrated in the puffs. The fact that the tracer was in a uracil precursor showed that the puffs contained RNA. This finding was further confirmed by staining procedures. Additionally, treatment of larvae with an antibiotic that inhibits RNA transcription (Actinomycin D) prevented the buildup of radioactivity at the sites of the puffs. Puffs can thus be produced by hormone treatment or prevented by antibiotic treatment.

The puffing of polytene chromosomes shows that specific places on chromosomes (and therefore on DNA strands) become active when called on, and they produce RNA at such times. When those places are not called on, they relapse into inactivity. Genes can indeed be turned on and turned off, and they can be seen at work.

row of structural genes on one DNA strand is called an *operon*. Operator or "control" genes act like switches to stop or start the structural genes.)

Jacob and Monod used one of *E. coli's* 2000 genes in their study, a gene that, when active, produces an enzyme called *beta-galactosidase*. This enzyme helps break down the sugar lactose, which provides energy for cells. Each operator gene is controlled by yet another gene called a **regulator gene,** which codes for the production of a protein called

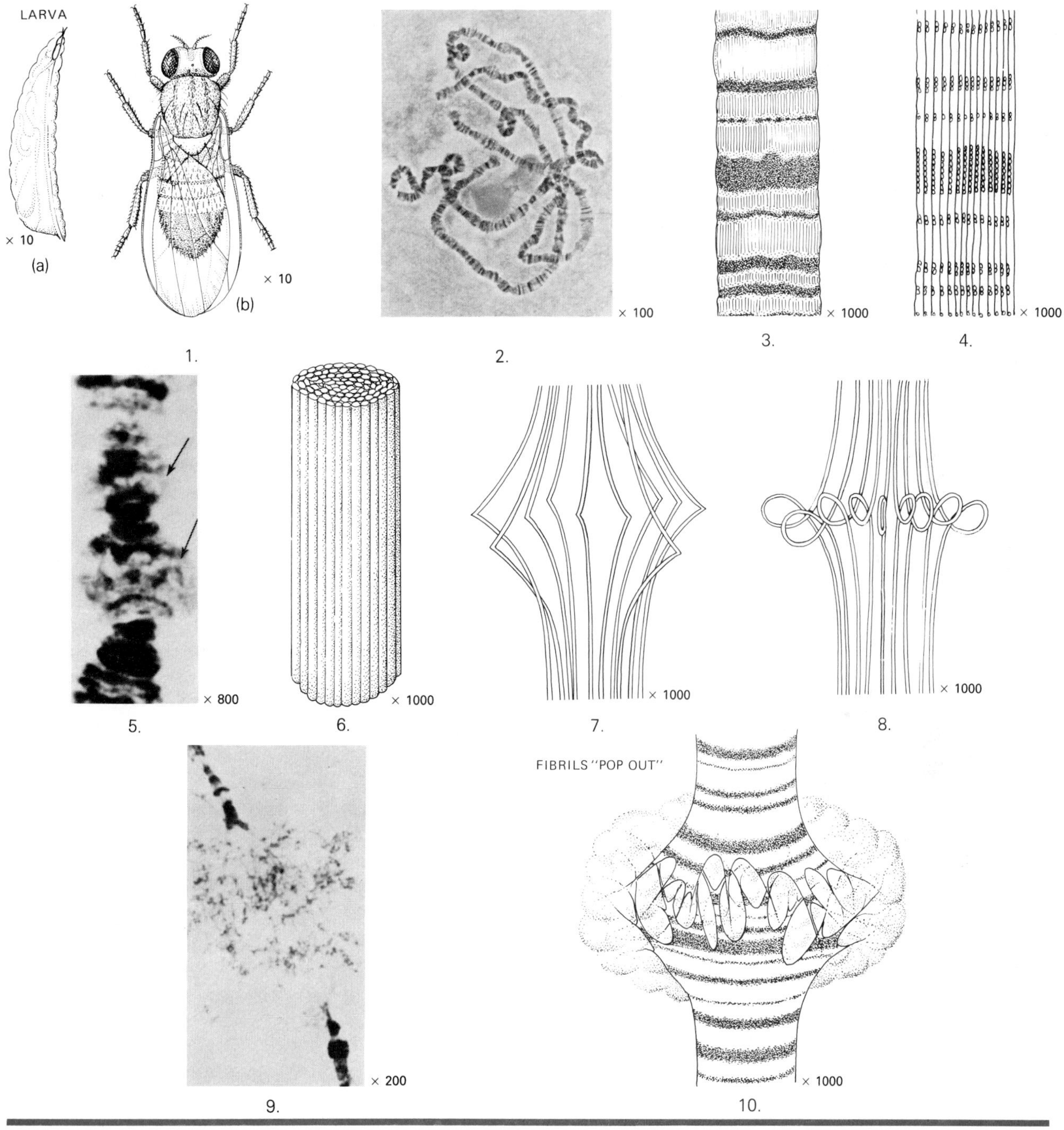

the **repressor.** The regulator gene produces a steady supply of the repressor protein, which usually rests on a small section of the DNA strand next to the beta-galactosidase *gene.* In this position, the repressor protein prevents the production of beta-galactosidase, and as a result, the cell is unable to make use of lactose. *E. coli* can grow immediately in a glucose solution, but in a lactose solution there is a period of delay before it starts to grow. It is as if the cells are at first unable to use lactose, but seem to "learn" to do

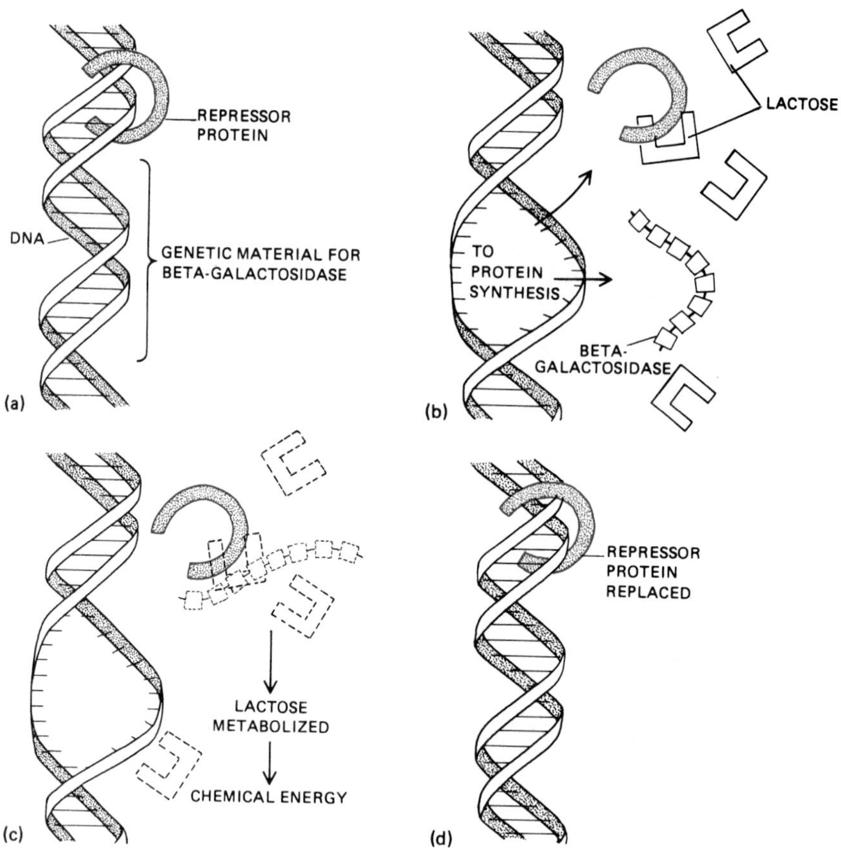

Figure 7.6
A highly stylized representation of the *induction* portion of the *Jacob-Monod model* of gene regulation in a cell. (a) The repressor protein located on the DNA strand, prevents the adjacent gene portion from producing the enzyme beta-galactosidase. (b) When the sugar lactose enters the cell, it binds to the repressor protein, and the beta-galactosidase gene is able to start the synthesis of beta-galactosidase. (c) The enzyme breaks down the lactose, releasing stored energy that the cell needs to carry on its functions. (d) If all the lactose has been digested, more repressor protein attaches itself to the original repressor site on the DNA, and the production of beta-galactosidase is blocked once more. The presence of additional lactose will start the process all over again, providing the cell with a steady energy supply as long as food (lactose in this case) is available for digestion by enzymes.

so, given time. Why are they not able to use lactose immediately, but are then able to use it after a while?

Jacob and Monod found the answer to this puzzle in a complex feedback system. (A portion of the Jacob-Monod model, much simplified, is shown in Figure 7.6.) When a certain amount of lactose enters the *E. coli* cell, the repressor protein is bound by lactose and ceases to act as a repressor. With the removal of the inhibiting repressor protein, the operator gene turns on the adjacent beta-galactosidase gene, which is then free to synthesize the enzyme beta-galactosidase. The enzyme breaks down the lactose in the cell, and energy is released for the cell's use. Once the lactose is used up, more repressor protein goes to its site on the DNA strand, and the production of beta-galactosidase is stopped.

The system is economical. Because of the action of the repressor protein, the cell needs to produce beta-galactosidase only when lactose is present. Note that the repressor gene and the enzyme gene are always present; it is the presence or absence of the sugar lactose that starts the machinery that turns the genes on or off. Repressor proteins can be inactivated either by some substance from outside the cell or by some product synthesized inside the cell. Either way, the structural genes can be turned on or off. With such a system, it is possible for a cell to use those genes that are appropriate at a given time. For example, if a certain food molecule is not available, it would be a useless waste of energy to make the enzyme needed to digest it. Without the certain food molecule to digest, the enzyme-making machinery is turned off. Such a food molecule (lactose or any other) is called an **inducer,** because it induces, or initiates, the production of an enzyme. The process of gene regulation up to this point is called **induction.**

In the **repression** portion of the Jacob-Monod model, if the structural genes are producing enzymes that are working in a cell to yield internally synthesized end products, those end products can bind to a repressor that has been inactive up to that point. The *inactive* repressor is then changed into an *active* repressor, which keeps the cell from building up an excess of any given substance.

The previous description of gene regulation has stressed the process rather than the participants in the process. The following text and drawings present the entire Jacob-Monod model, including the terms used to describe the several participants:

Gene regulation according to the scheme of Jacob and Monod:

(a) The row of structural genes on one DNA strand, all regulated by one control, is an **operon.** *Structural genes* are those that produce the messenger RNA responsible for the working proteins of the cell. Structural genes are made to produce or are kept from producing by a nearby *operator gene,* which acts as an on-off switch. The operator gene may be on or off, depending on whether or not a *repressor protein* is present and active. The repressor protein comes from still a third kind of gene, a *regulator.*

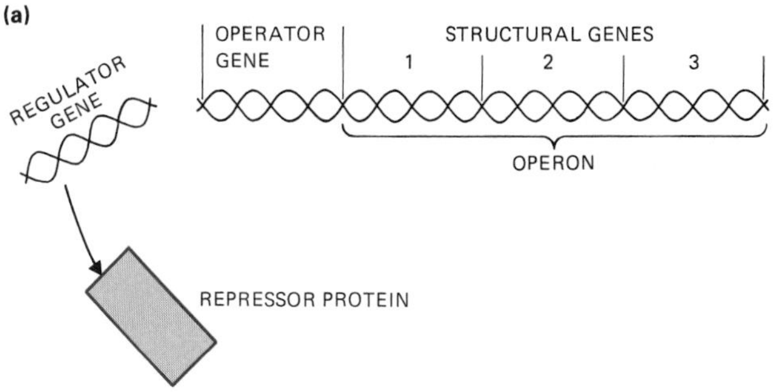

(b) In the process of **induction,**

1. the regulator sends out an *active* repressor, which turns off the structural genes by binding with the operator gene, so no mRNA and consequently no protein are made from those genes.

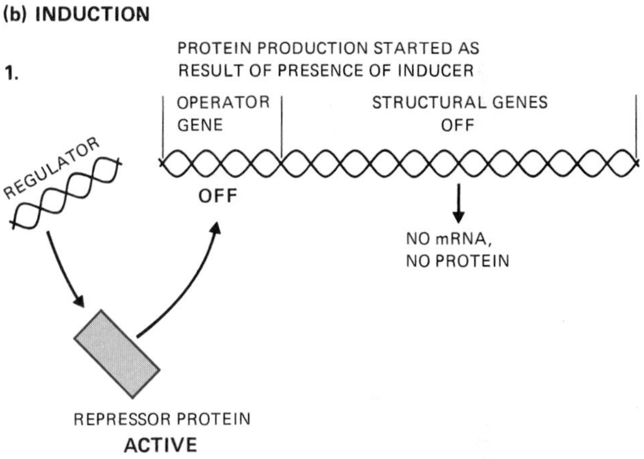

---

2. If a usable outside material (such as lactose in *E. coli*) enters the cell, it can bind itself to the repressor protein, thereby keeping it from holding the operator in the off position. When that happens, the operator turns the structural genes on, and they start producing mRNA and then protein.

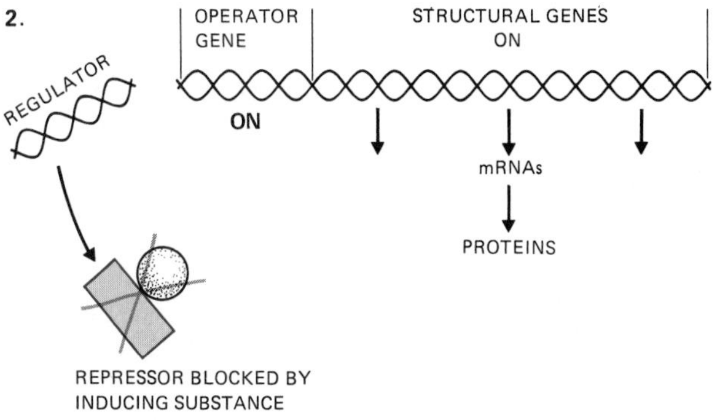

(Continued on next page.)

(c) In **repression,**

1. an *inactive* repressor is assumed to be present, so that the operator and the structural genes are on and functioning.

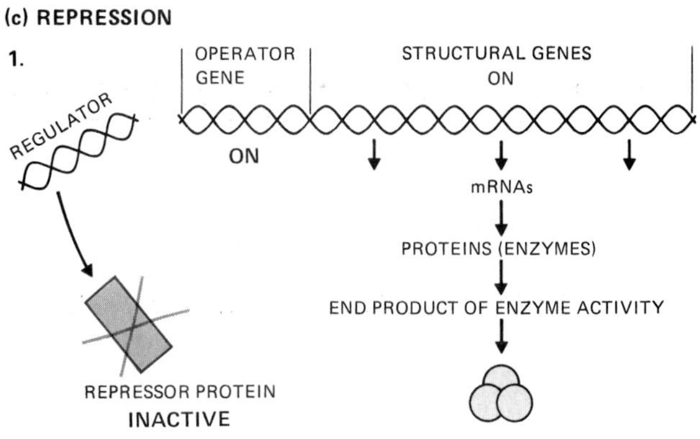

2. If the structural genes are producing enzymes that are working in a cell to yield end products, those end products can bind themselves to the *inactive* repressor, thereby making it an *active* repressor, which can turn off the operator and structural genes and thus prevent excess buildup of the end product. When the inactive repressor is made active by a product within the cell, that product is called a *co-repressor*.

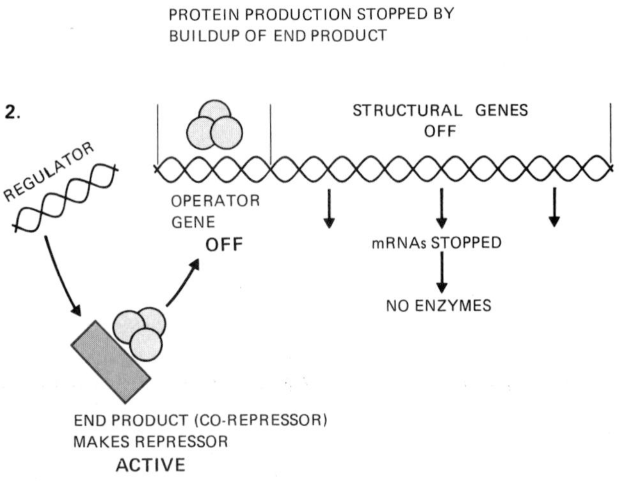

TO SUM UP GENE REGULATION

1. *Structural genes* produce mRNA, and are instrumental in the production of polypeptides. They are the genes that function in the synthesis of enzymes.
2. An *operon* is a row of structural genes on one DNA strand. The controller of an operon is an *operator gene,* which acts as a switch to "turn on" the structural genes. "Turned on" structural genes begin to form mRNA.
3. Operator genes are "turned off" by a *repressor protein,* which is produced by a *regulator gene.* When the repressor protein is inactivated, the operator "turns on" the structural genes.
4. A substance from outside a cell can cause the cell to start making enzymes that will act on that substance. Thus, external influences can direct the cell's specific activity.
5. A feedback system within the cell can regulate the amount of any given substance that a cell manufactures. Thus, a cell is in control of its own internal affairs.

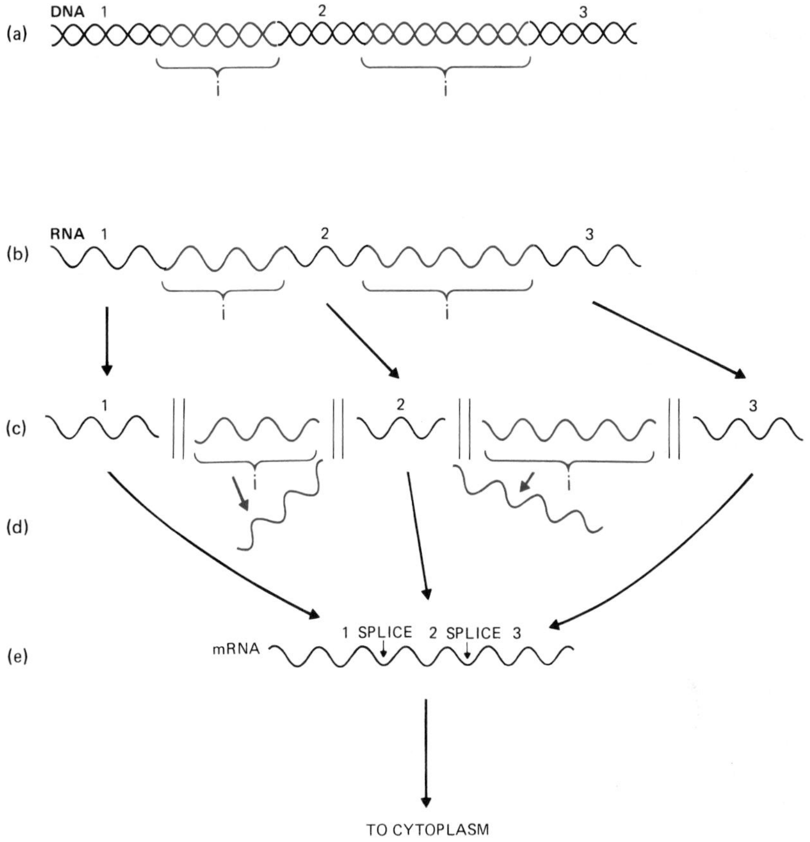

Figure 7.7
One way in which a *split gene* may operate. (a) A length of DNA with three portions coding for amino acids in black, and two noncoding portions, or *introns* (i) in color. (b) The entire length of the DNA strand is transcribed onto a length of RNA. (c) The RNA is cut by enzymes at the places where coding and noncoding nucleotides meet. (d) The noncoding introns are eliminated. (e) The coding portions of the now-finished mRNA are rejoined, and the mRNA is ready to move from the nucleus out into the cytoplasm, where it will join up with ribosomes and form a polypeptide. (In this drawing, chromosomal protein is omitted, and proportional lengths of DNA are much reduced for simplicity.)

The Jacob-Monod model applies best to bacterial cells. It is not known how the Jacob-Monod model applies to plants and animal cells, because they have much more complex systems, and they certainly have different chromosomes. Besides, many of the so-called genes of eukaryotic organisms exist not just in one or two copies of each gene per nucleus, but in many copies, perhaps hundreds. The following section presents some recent ideas about protein production in eukaryotes.

SPLIT GENES

The Jacob-Monod scheme of protein production is well documented for prokaryotes, with their relatively simple, naked strands of DNA. In eukaryotes, however, chromosomes are more complex, and some adjustments in the story must be made. First, eukaryotic chromosomes have proteins closely bound to their DNA. The function of these proteins is not known, but presumably they affect the physical shape of the DNA and influence the way in which RNA is made. Second, a gene in a eukaryotic chromosome is not a continuous coding stretch of DNA. The DNA in such a chromosome is interrupted, sometimes by 50 or more stretches of nucleotide strands that do not code for any amino acids. The interruptions are called **introns.**

Genes with interrupting introns, seemingly the common sort in eukaryotic chromosomes, are called **split genes.** How split genes function in living cells is not yet understood, but a sensible guess is that they piece together the required parts of the resultant mRNA and thus finally produce working proteins (Figure 7.7).

One assumption is that by using split genes, cells can more easily recombine various lengths of DNA and have greater flexibility in making new and different enzymes. With greater flexibility, cells would have more chances for "experimentation" with new kinds of proteins, and more chances for variation in organisms. Certainly the rate of evolutionary advance has been much greater in the centuries following the advent of eukaryotes than it was before (see Table 31.1 on p. 642).

HOW TO MAKE A GENE

In August of 1976, after nine years of research, biochemist Har Gobind Khorana and his team at the Massachusetts Institute of Technology produced the first synthetic gene that is able to function within a living cell, an accomplishment that is considered as important as the discovery of the structure of DNA in 1953 by Watson and Crick. Part of the technique of deciphering the genetic code involved the synthesis of many different nucleotide base sequences into artificial messenger RNA. Khorana received a Nobel prize in 1968 for his part in preparing such compounds. A strand of messenger RNA, however, is not a gene; concisely, a gene is a strand of DNA responsible for RNA assembly.

A first requirement for the test-tube manufacture of a gene is a knowledge of the base sequence of DNA, and that cannot be determined without knowing the base sequence of a specific RNA. Such a sequence has been known since 1965, when Robert Holley (corecipient of the 1968 Nobel prize) worked it out for the transfer RNA that is associated with the amino acid alanine in yeast: alanine-tRNA. Once a transfer RNA has been mapped, the necessary bases in DNA are known, since they are complementary to the mRNA bases. Furthermore, methods for incorporating genes into chromosomes have been developed (the dramatic technology of recombinant DNA, by which genetic material from one organism is inserted into another), and suitable tests have been devised for determining whether a functional gene has indeed been made.

When Khorana set out to synthesize a gene that coded for a known sequence of transfer RNA, he decided not to construct the gene by adding on one nucleotide at a time. Such a technique would have required huge quantities of material to allow for the proper chemical purification of each step and the expected loss of material at each stage of the many reactions involved. Instead, Khorana chose to construct small pieces of the molecule separately and attach them in their proper sequence later. Each piece of the chain was a single strand 10 to 15 nucleotides long. The single-stranded pieces were then joined into a double-stranded molecule, with each strand overlapping its complementary partner to provide a "sticky end" for the next strand to recognize. The separate strands were attached with the help of an enzyme called *DNA ligase,* which was discovered only a couple of years before Khorana used it.

The first gene to be made by such chemical methods was one that is responsible for a highly specialized RNA, a mutant transfer RNA for the amino acid tyrosine. Khorana and his collaborators

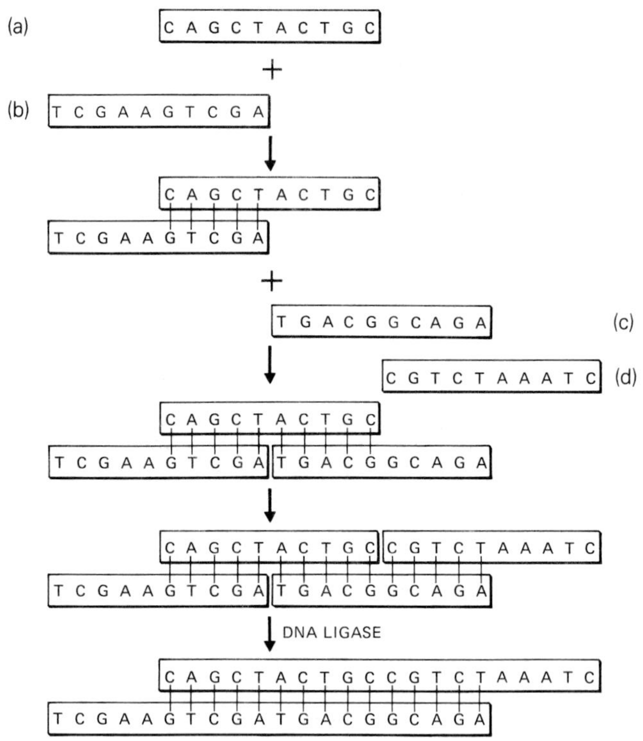

(1)

The "sticky end" technique used to build the gene from single-stranded pieces: The sticky end of fragment (a) left protruding after base-pairing with fragment (b) is complementary to the first few bases in fragment (c). That in turn leaves a sticky end recognizable by fragment (d). The individual single-stranded fragments are welded into continuous chains with the aid of the joining enzyme, DNA ligase.

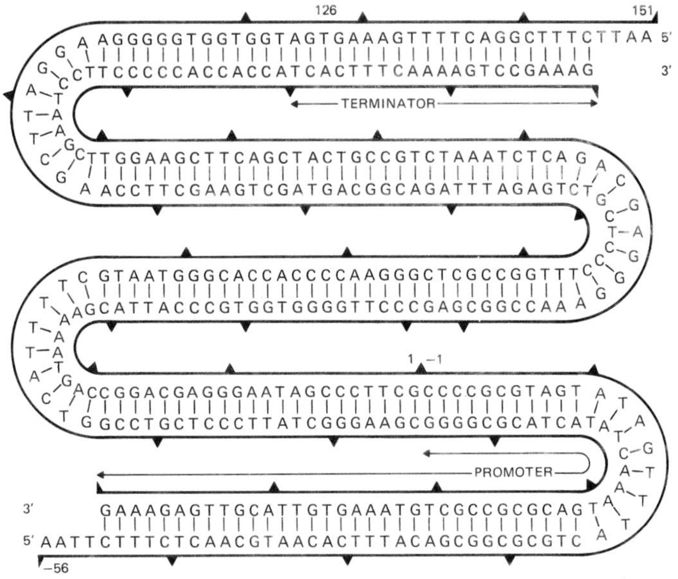

(2)

Structure of the synthetic E. coli *tyrosine transfer* RNA *gene synthesized by the Khorana group. Segments between points were linked chemically, then joined enzymatically to form the entire* DNA *double helix.*

(From the original drawings, Massachusetts Institute of Technology)

chose it because it can be effectively tested in certain strains of *E. coli,* the organism most favored for experiments in microbial genetics laboratories. The working part of such a transfer RNA contains only 126 base pairs, but the DNA responsible for its formation must contain about 200. Part of the reason for the extra required length is the need for two "ends," one indicating the start of the copying and the other for the end. These are the *promoter* (the first 52 nucleotide pairs) and the *terminator* (the final 21 nucleotide pairs).

The first gene to be introduced into *E. coli* cells by Khorana and his group failed to function. But another brand-new discovery guided these researchers to a bacteriophage that will not grow in *E. coli* unless a certain suppressor mutation is present. Khorana's synthetically prepared gene is such a suppressor, and after the gene was introduced into the DNA of the specific phage in *E. coli,* the phage was able to grow. Thus researchers had made a fully functioning gene, and the magnitude of the discovery was diluted only because the latest technology had made the synthesis of a gene inevitable. Nevertheless, it was a gigantic scientific achievement.

As usually happens in research, each success brings up new questions; the questions are especially pressing in the instance of synthetic genes. What can and will be done with synthetic genes? Plainly, one of the main uses of such knowledge will be the making of synthetic promoters and terminators in an effort to find out how cells succeed in controlling their own genes. Such techniques may be important in correcting genetic defects such as sickle-cell anemia. Researchers will also concentrate on creating "controlled mutations" in order to test their effect on both the gene and the transfer RNA. But with the development of applied aspects of such techniques come even more difficult questions. Now that genes can be made artificially and can be incorporated into whole chromosomes, will we be able to control our own genes? A human gene is 1000 to 3000 base pairs long. How soon will it be possible to synthesize such a complex gene? And who will decide which genes—and whose—are to be tampered with?

SUMMARY

1. A *gene* is a specific segment of a DNA strand that controls specific cellular functions, either by synthesizing proteins or by regulating the action of other genes.

2. The "one gene–one enzyme" hypothesis was formulated by Beadle and Tatum to indicate how each gene controls a specific enzyme. The hypothesis is currently referred to as the "one gene–one polypeptide" hypothesis, because genes control the formation and activity of all proteins, not just enzymes.

3. Besides forming the structural elements that hold cells firmly, proteins are responsible for the enzymatic action that makes life dynamic and continuing.

4. The combination of three nucleotides (triplets) in a specific sequence that codes for a specific amino acid is a *codon.* There are 64 possible triplet combinations, more than enough to code for the 20 amino acids needed by living organisms.

5. The complete set of triplet nucleotide symbols that indicate specific amino acids is called the *genetic code.*

6. Protein assembly starts with *transcription,* in which single-stranded RNA is synthesized under the direction of DNA. Three kinds of RNA are transcribed: ribosomal RNA, messenger RNA, and transfer RNA.

7. *Messenger RNA* (mRNA) carries a coded message from the nucleus to the ribosomes in the cytoplasm, telling them which amino acids are to be included and where they are to be placed.

8. *Transfer RNA* (tRNA) codes for a specific amino acid. After attaching itself to the amino acid, the triplet of tRNA then attaches to its matching triplet on the mRNA. The matching tRNA is an *anticodon.*

9. A protein molecule is synthesized from amino acids according to the specific coding order of the mRNA molecule.

10. The formation of a polypeptide by mRNA is *translation.* The "central dogma" of modern biology states that DNA transcribes (codes for) RNA, which translates into (codes for) protein.

11. The *Jacob-Monod model* proposes that a bacterial cell governs its own internal affairs through a complex interaction of specialized genes.

12. Eukaryotes may have a system of gene regulation that has "split genes," with interruptions called *introns.* It is speculated that split genes give eukaryotes additional flexibility in making new and different enzymes.

ASK YOURSELF

1. Give a definition of a gene.
2. What common microorganism was used to develop the idea of one gene–one polypeptide?
3. How is it possible to make the uncounted numbers of proteins from only 20 amino acids?
4. What is a codon? An anticodon?
5. What are the three different kinds of RNA?
6. What is meant by "transcription" of a nucleic acid? What is "translation"?
7. What is the function of mRNA? Of tRNA?
8. What is the function of a polysome?
9. In the Jacob-Monod scheme of gene regulation, what is a structural gene? An operator gene? A regulator gene?
10. What does a repressor protein do in gene regulation?
11. In bacterial physiology, what is an inducing substance?
12. What is meant by a "split gene"?

8 Cellular Growth, Reproduction, and Specialization

SOME KEY POINTS

1. In order for cells to do their proper jobs, they must mature, differentiate, and specialize.
2. The process of mitosis reproduces cells and distributes equal DNA to each new cell.
3. Meiosis is the pair of cellular divisions required to reduce the number of chromosomes from the double to the single condition, in preparation for sexual union of male and female cells.
4. Cellular differentiation usually follows mitosis. Once a group of cells has become specialized to perform a particular function, it is called a tissue.
5. A clone is a group of individuals all with identical genetic makeup.
6. Homeostasis is the maintenance of an organism's inner stability, even though the outside environment changes. Biological feedback provides controls for stability. Cancer occurs when feedback control is missing.

THE FINAL FORM OF AN ORGANISM DEPENDS DIrectly on its cellular arrangement. Multicellular plants and animals start life as single cells, and they achieve maturity as a result of repeated cell divisions and changes in the form and function of individual cells. The first event in the development of complex organisms is the division of the original cell; then subsequent cells divide until a sufficient mass of cells has developed. Despite intensive efforts to find the initiating drive, no one knows what causes a cell to undergo division. Cells can be held back from division by cold and by chemical means, but experimental, predictable, controllable cell division is not yet possible. We can only provide conditions that favor growth in general and allow cell divisions to occur.

If a cell mass is to achieve the complex form of a well-developed animal, such as a human being, the cells must be directed when to divide, how often, and in what direction. The direction in which a cell divides is important to an organism. After an initial cell divides, if subsequent divisions are always in the same direction, the result will be a single strand of cells, end to end. Such organisms exist, but they are severely limited in what they can do.

Cell reproduction and placement are not enough, either, to make a functional organism. If a human egg divided the proper number of times and made the divisions in the correct directions, the result might be something that looked like a human, with arms and fingers and the rest, but if all the cells were alike, that mass of cells would be no human. It is the maturation, the differentiation, the specialization of cells that give them their final working form. Functional specialization is more striking than spe-

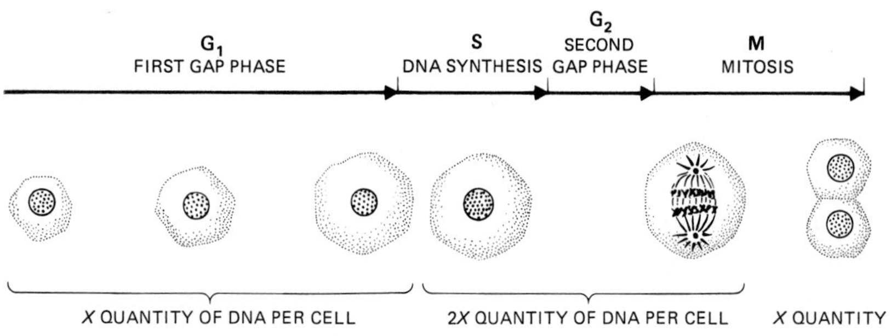

Figure 8.1
The life of a cell. Once produced as a result of a cell division, a new cell, with its chromosomes and a standard amount of DNA, grows in size during the first gap phase (G_1). Having reached full size, it goes into a phase of DNA synthesis, the S phase, during which the amount of DNA in the chromosomes is doubled. The cell then enters a second gap phase (G_2), which ends when the cell divides, parceling out its DNA equally to the new daughter cells.

cies specialization; a human liver cell is less like a human eye cell than it is like a pig liver cell.

In this chapter, we will describe some of the observable events that lead to a complete organism and give some of the current ideas on cellular control, but we cannot explain completely the forces that lead to cell division, directional growth, or cellular differentiation because those forces are not known. As this book repeatedly points out, the amount that we know about life is undoubtedly much less than the amount we do not know.

THE MECHANICS OF CELL DIVISION

The story of cell division, as seen with a light microscope, has been known since the 1860s. Since then, details and variations have been found and facts added, especially for that period between divisions known as the *interphase.*

After a cell has divided into two, each of the new cells is about half as large as the original one. Immediately after a small new cell is produced, it enters a period of growth known as the first "gap phase," or G_1 (Figure 8.1). The G_1 lasts from a few minutes to years, or it may even be lacking in some young embryos. During that time, there is synthetic activity, mostly protein synthesis, in all cell parts except the chromosomes, and the cell increases in size until it is usually as large as its parent cell was before division.

Some time during interphase, if the cell is going to divide, it enters a new "synthetic phase," or S phase, during which the DNA of the chromosomes doubles. The important action here is the exact replication of the DNA, which is not simply a doubling of the quantity, but a doubling that provides the cell with two complete, identical sets of chromosomes. During the S phase, the chromosomes are stretched out so thin that they cannot be seen with a light microscope. That fact made it impossible to learn when duplication occurs until such new techniques as radioactive tracing were available.

Mitosis Distributes Equal DNA to Each New Cell

With the S phase past and the chromosomes doubled, the cell goes into a second gap phase, G_2, in preparation for nuclear division. Further general synthetic activity continues, but the G_2 phase seems to be less variable in time than G_1. It seems that, having once "decided" to divide, a cell goes ahead and does so. At the end of G_2, the cell begins the actual nuclear division process, generally known as **mitosis,** or the *M phase*. The word "mitosis" comes from the Greek *mitos,* which means "thread," and refers to the threadlike appearance of chromosomes during this period.

The key structures that contain the replicated DNA are the chromosomes in the nucleus, and they are seen prominently in the drawings and photographs that follow. The drawings show only four chromosomes, to make the details of cell division clearly visible. The principles of mitosis are the same no matter how many chromosomes are involved.

For convenience, mitosis is conventionally divided into stages, but these stages flow smoothly from one to the next and must not be thought of as coming in jumps, as selected still pictures may seem to indicate. The stages of mitosis are usually separated as follows: (1) interphase (*inter* = between), (2) prophase (*pro* = before), (3) metaphase (*meta* = after), (4) anaphase (*ana* = again), and (5) telophase (*telos* = end).

INTERPHASE

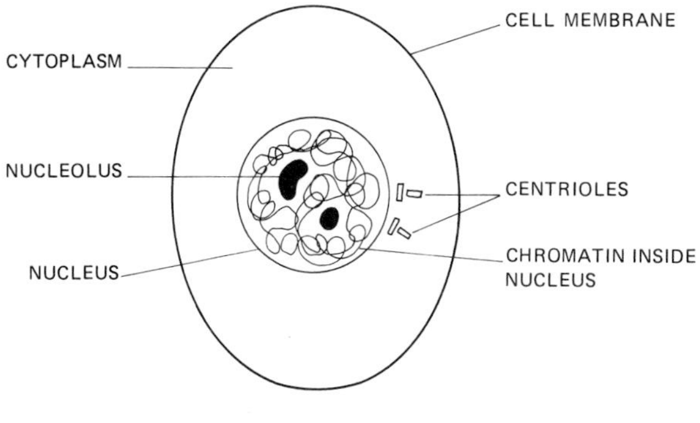

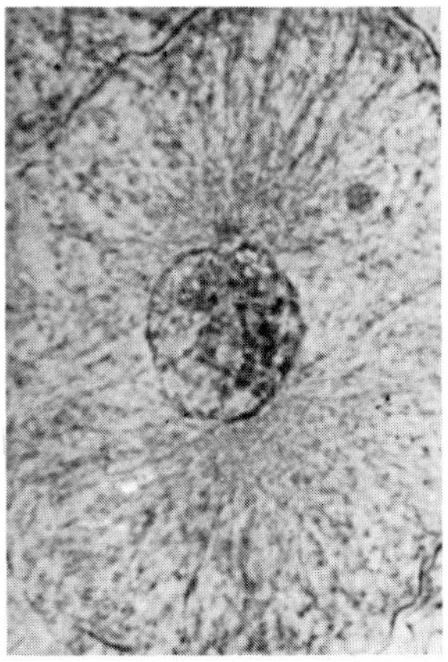

Interphase

Interphase is not actually a part of mitosis, but is the phase *between* cell divisions. During interphase, normal cell functions other than reproduction take place. The chromosomes consist of a complex of protein and DNA called chromatin. At this stage it resembles a jumbled mass of fine threads. One or more nucleoli are present in the nucleus. In animals, two pairs of centrioles appear just outside the nucleus. *DNA replication occurs during interphase;* before mitosis actually begins, the DNA in each chromosome has already been duplicated. The synthesis of RNA and proteins also occurs during interphase.

PROPHASE

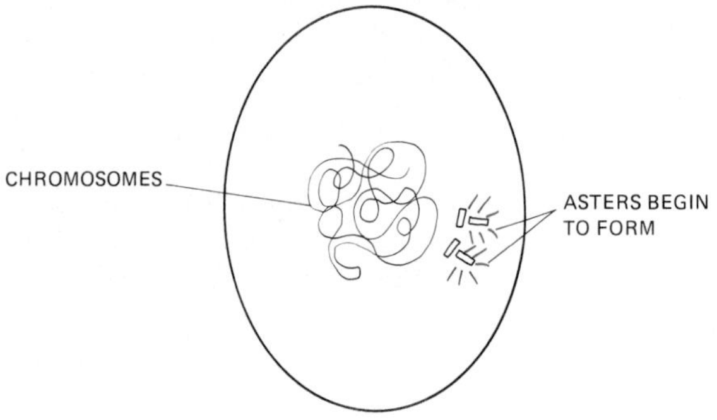

EARLY PROPHASE

Prophase

At the onset of **prophase,** the chromosomes begin to shorten and thicken, coiling upon themselves. Before the coiling begins, the total length of the DNA strands in a single human nucleus is greater than the length of the entire human body, but the strands are too thin to see. When they reach their minimum length of only a few micrometers during late prophase, they are thick enough to be readily visible with a light microscope. By late prophase, each chromosome appears double. The two halves are called *chromatids.* Each pair of chromatids has a *centromere,* to which spindle fibers are attached (see drawing and photo below).

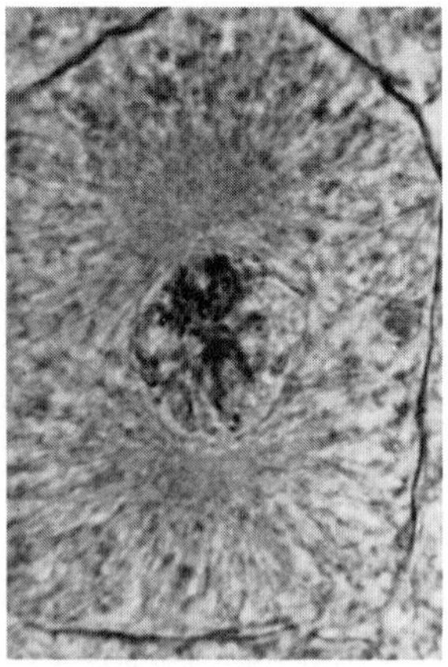

CHROMOSOME

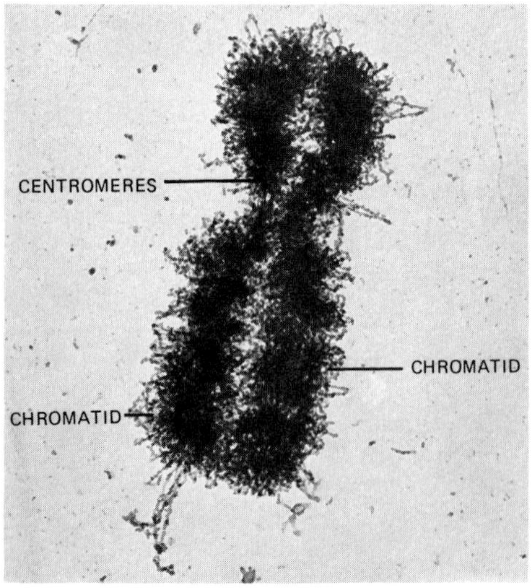

× 1100

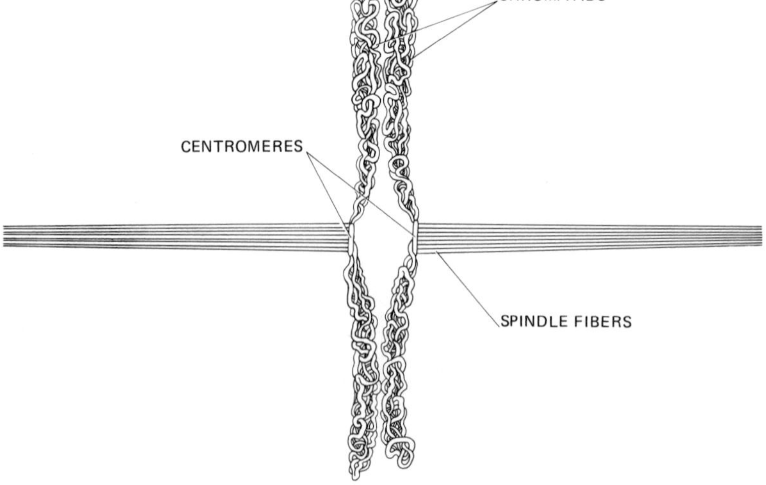

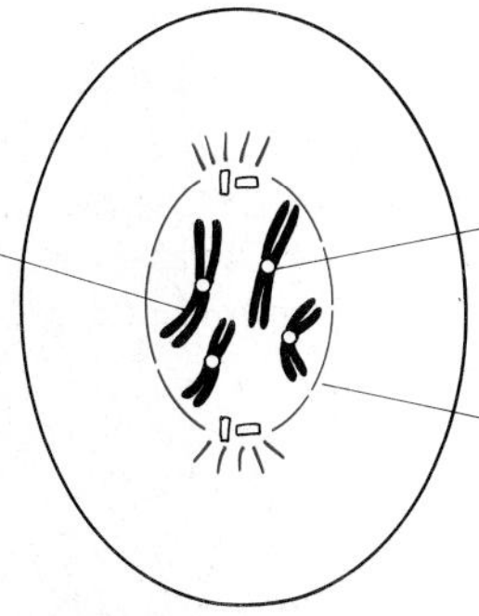

LATE PROPHASE

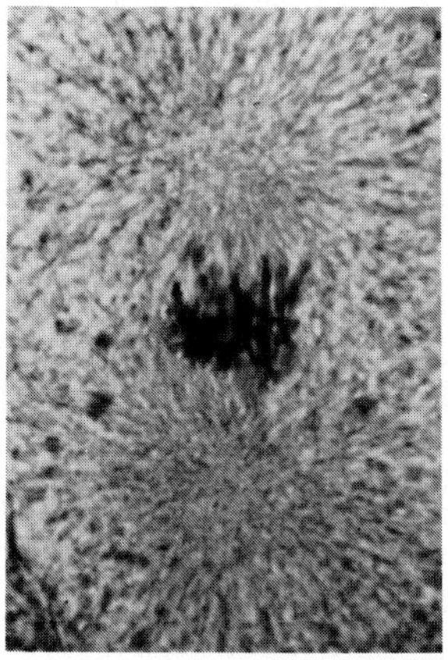

While the chromosomes are preparing for separation of their chromatids, other activities are progressing. Outside the nucleus, the two pairs of centrioles have begun to move apart, and by the end of prophase they are at opposite sides of the nucleus at the *poles* of the cell. The position of the centrioles at the poles determines the direction in which the cell will divide. Meanwhile, the nucleoli break up and disappear, as does the nuclear membrane. Between the centrioles, strands of microtubules constitute the *spindle*. In animals, a sunburst of microtubules radiates out from the centrioles, forming the *asters*. (The asters, spindle, centrioles, and microtubules are collectively called the *mitotic apparatus;* see the drawing below.) Higher plants have neither centrioles nor asters, but they do have spindles and the other mitotic components.

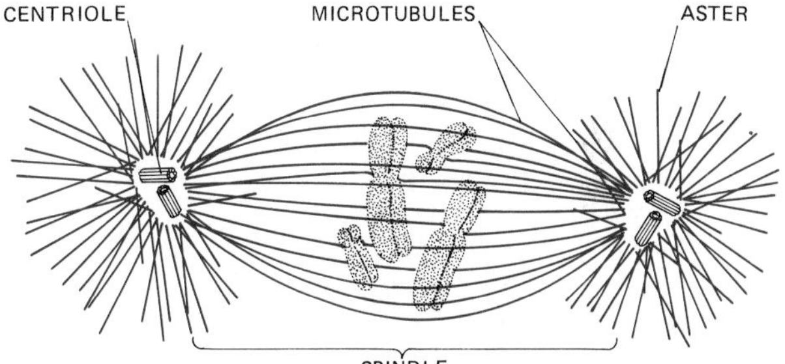

THE MITOTIC APPARATUS

METAPHASE

METAPHASE

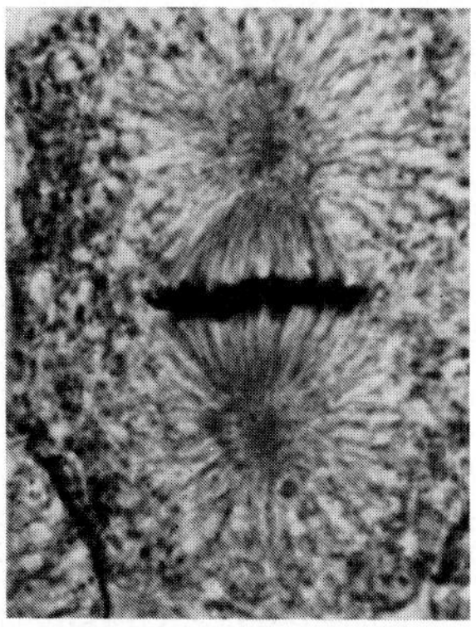

Metaphase

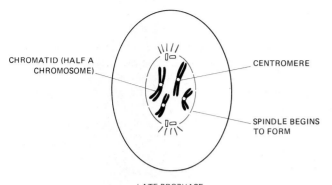

When the chromosomes have shortened to their minimum length, they move into a somewhat flattened plate across the cell and come into **metaphase.** Some spindle fibers extend from pole to pole, but others extend only from one pole and attach to the centromere of a chromosome. In early metaphase the pairs of chromatids that make up a chromosome are held together by a single centromere, but toward the end of metaphase the centromeres become double, so that each chromatid has one. It is then appropriate to call each chromatid a full-grown chromosome. Metaphase is short, lasting only a few minutes in some cells.

Anaphase

The beginning of **anaphase** is marked by the separation of the chromosomes, with one member of the pair of chromosomes starting toward one pole and the other moving in the opposite direction. Meanwhile, the poles themselves are moving farther apart, and the microtubules of the spindle are changing. The pole-to-pole microtubules are lengthening while the pole-to-chromosome microtubules are shortening. The force causing anaphase movement of chromosomes is not known, although the microtubules are essential.

ANAPHASE

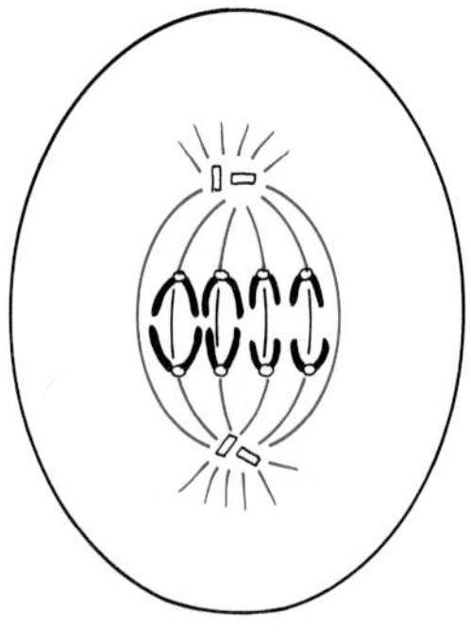
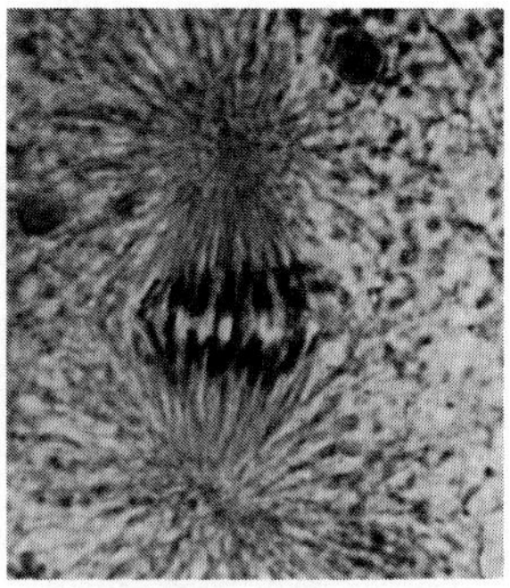

EARLY ANAPHASE

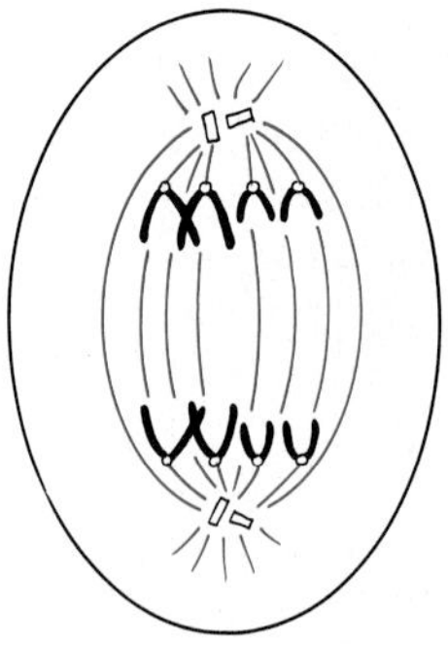
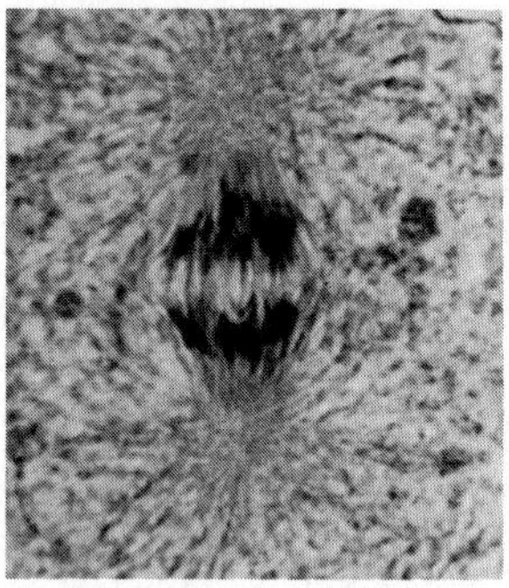

LATE ANAPHASE

TELOPHASE

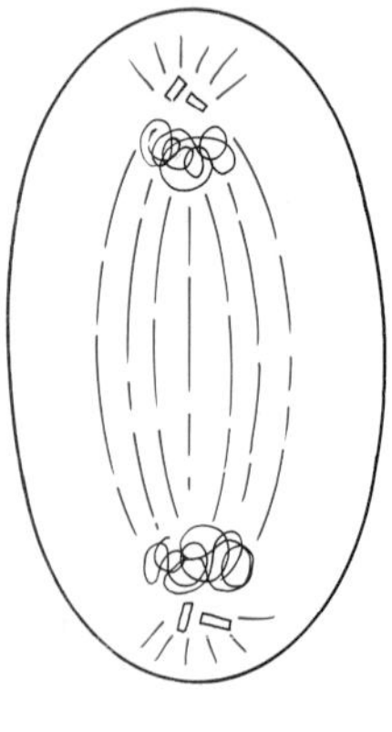
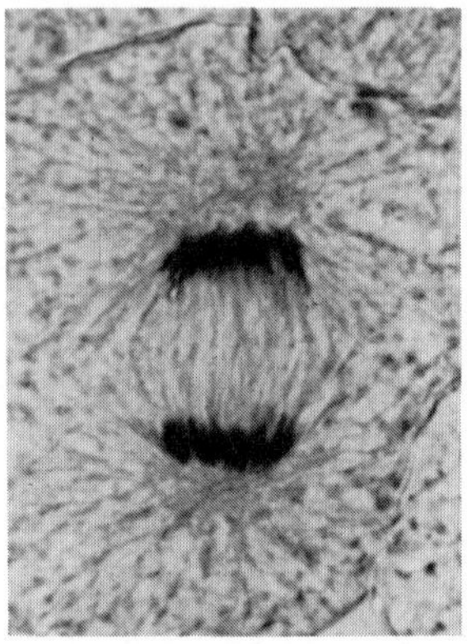

EARLY TELOPHASE

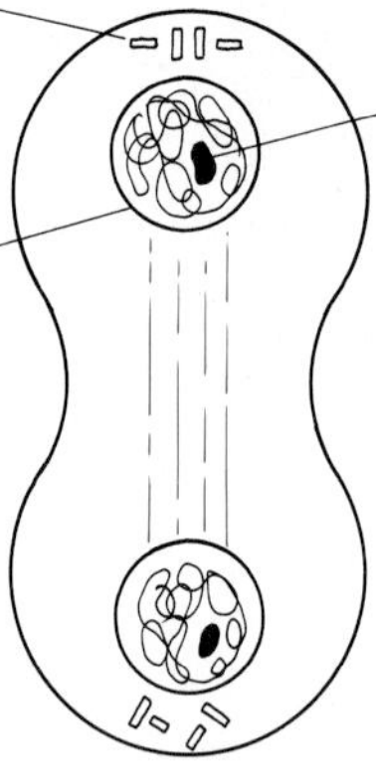

LATE TELOPHASE

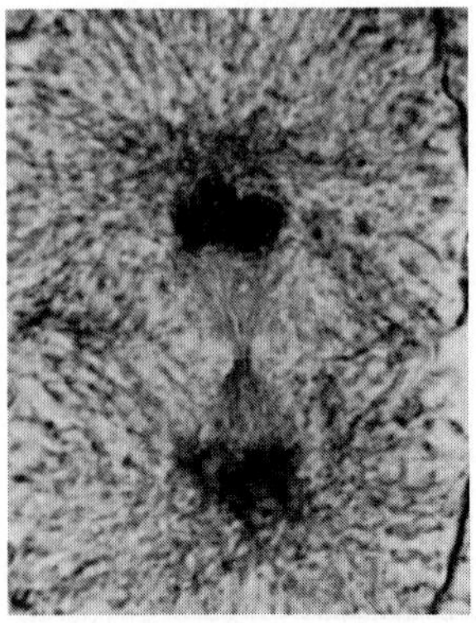

Telophase

Telophase is marked by the arrival of the chromosomes at the poles. Once there, they lose their distinctness, begin to lengthen, and resume their interphase condition. About this time, too, the centriole at each pole duplicates as if preparing for the next mitosis to come. The endoplasmic reticulum spreads out around the chromosome mass, making a new nuclear membrane, and the nucleolar regions of the chromosomes build new nucleoli. Nuclear division is over.

CYTOKINESIS

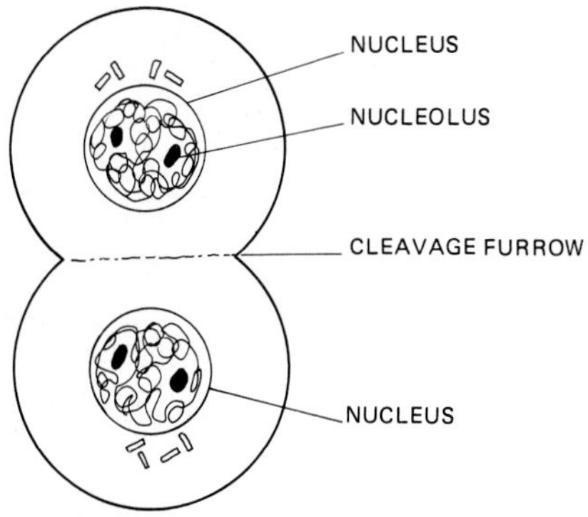

CYTOKINESIS (TWO CELLS)

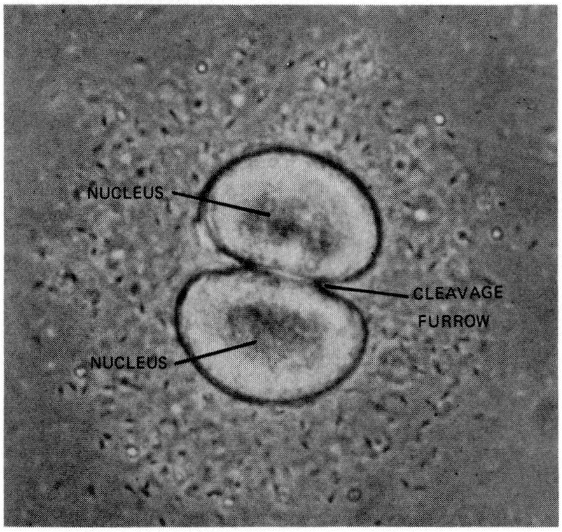

Cytokinesis

Nuclear division is over, but *cell division* is not. The new nuclei are still in the same cytoplasmic unit. Commonly, nuclear division is followed by separation of the cytoplasm into two parts. In animals this separation is accomplished by a pinching of the cell membrane, as though someone had looped a cord around the equator of the cell and pulled it tight. The pinching near the middle of an animal cell forms a *cleavage furrow* (see drawing and photo). The process of cytoplasmic division is called **cytokinesis** (Gr. *cyto*, cell; *kinesis*, movement). In plants, the cytoplasmic division is achieved by the formation of a partition, the *cell plate*, between the two portions of the cell (see drawings below). Cytoplasmic division is not necessarily a part of nuclear division, however. Many cells allow a normal mitosis without subsequent cell division, and the result is multinucleate cells.

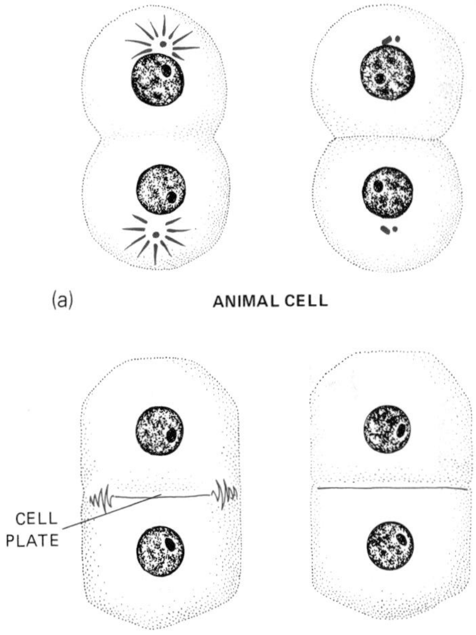

(a) ANIMAL CELL

(b) PLANT CELL

Mitosis accomplishes two things: (1) the reproduction of cells and (2) the equal distribution of DNA to each. Cells do not deviate from this pattern unless they are behaving abnormally or are undergoing those special divisions called *meiosis* (discussed in the next section).

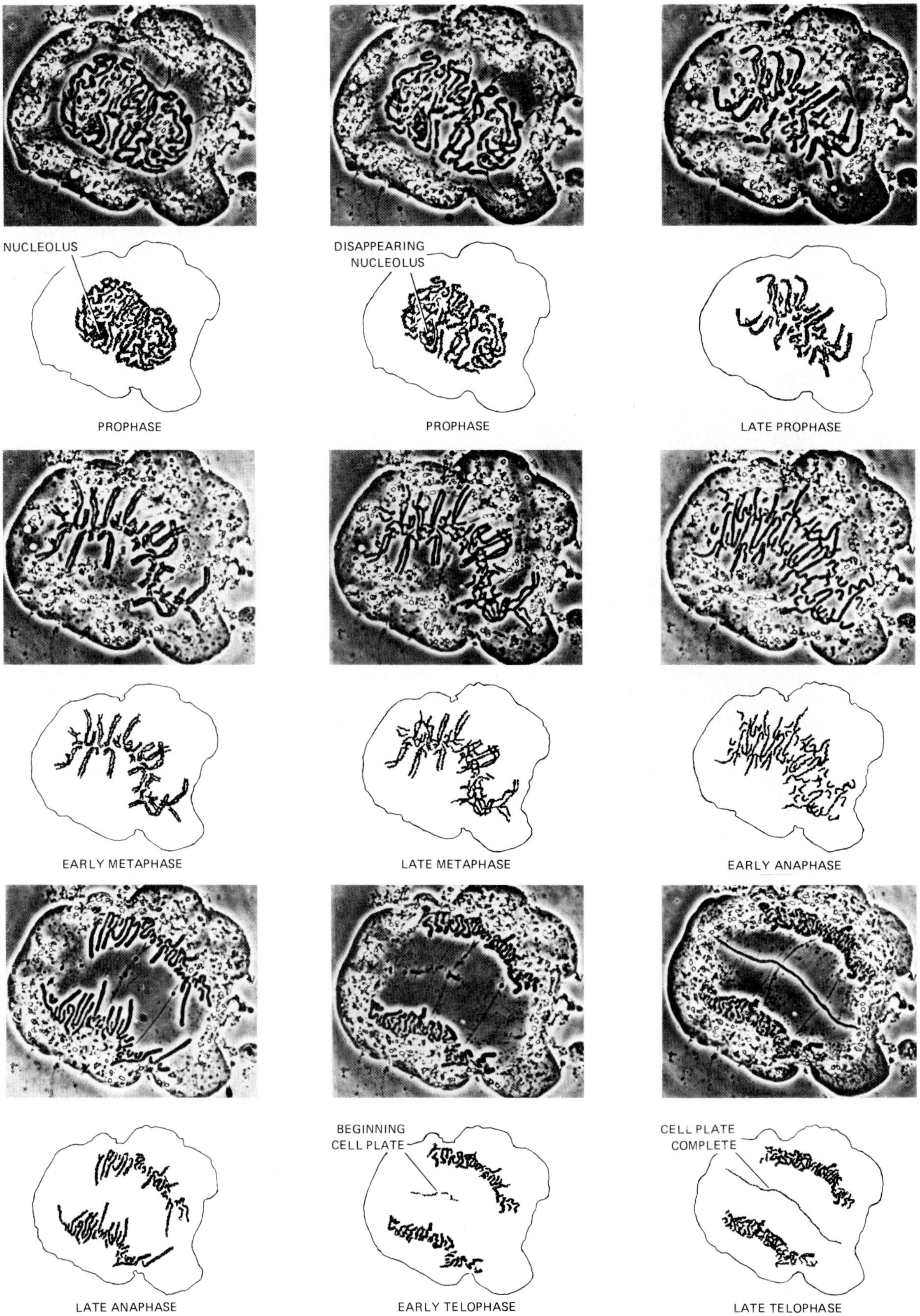
NUCLEOLUS
PROPHASE
DISAPPEARING NUCLEOLUS
PROPHASE
LATE PROPHASE
EARLY METAPHASE
LATE METAPHASE
EARLY ANAPHASE
LATE ANAPHASE
BEGINNING CELL PLATE
EARLY TELOPHASE
CELL PLATE COMPLETE
LATE TELOPHASE

Figure 8.2
Selected photographs and drawings from a series showing *mitosis* in a living cell of an African blood lily in culture. The granules in the cytoplasm are mitochondria and other organelles. The drawings emphasize details within the nucleus. The disappearance of the nucleolus, the doubleness of the chromosomes in metaphase, and the buildup of the cell plate in telophase are notably clear.

TO SUM UP MITOSIS

1. The nuclei of cells divide during mitosis. Mitosis accomplishes the reproduction of nuclei and the equal distribution of DNA to each.
2. The phases of mitosis are prophase, metaphase, anaphase, and telophase.
3. During *Early Prophase* the chromosomes thicken, the nucleoli and nuclear membrane begin to disintegrate, centrioles separate, and asters begin to form.
4. In *Late Prophase* chromatids are formed, the chromosomes continue to thicken, poles are established, and the nucleoli disappear.
5. During *Metaphase* spindle fibers are complete and the chromosomes move to the equator of the cell.
6. *Anaphase* occurs when the chromosomes split, and each member of a pair of chromosomes moves to an opposite pole.
7. During *Telophase* the centrioles double, the nucleoli and nuclear membrane reappear, and the chromosomes "unwind" and disappear from view.
8. After telophase, two new cells are usually formed.

Meiosis Reduces the Chromosome Number to Half

Mitosis guarantees that daughter* cells receive exactly the same genetic information as the parent cell. The parent cell and daughter cells each have the same number of chromosomes. New cells can continue to be formed with this method as long as they are not sex cells. For example, a cat cell has 38 chromosomes. If a cat *sperm* cell contained 38 chromosomes and a cat *egg* cell contained 38 chromosomes, their combination during fertilization would produce a cell with 76 chromosomes. Such a cell would not be a normal cat cell.

As soon as researchers discovered that chromosome numbers in organisms are constant, it became clear that there must be a time in the life of a sexually reproducing plant or animal when the chromosome number is cut in half. The problem is solved by another process of cell division, called **meiosis** (Greek "to diminish"). Meiosis reduces the number of chromosomes in sex cells (*gametes*) to half. When a cell has the full number of chromosomes (38 in cats, 18 in carrots) the cell is said to be in the **diploid** (double) condition. When sex cells receive half the number of chromosomes, the cells are in the **haploid** (single) condition.

In the following simplified drawings, the diploid number is again shown as four, as it was in the drawings of mitosis. This number is chosen simply for clarity.

Meiosis involves *two* sequences of cell division, called Meiosis I and Meiosis II. Meiosis I and II produce *four* haploid cells, no matter what the chromosome number. (Recall that mitosis produced *two* cells.) Each sequence of meiosis contains the same stages as mitosis: prophase, metaphase, anaphase, and telophase. Superficially, meiosis resembles mitosis, but there are two major differences:

1. The chromosome number is *reduced to half* during meiosis.
2. In meiosis there is a close pairing and subsequent *exchanging of parts of chromosomes.*

Both these points deserve special consideration, and will be explained further in the following discussion of meiosis.

* The use of the word "daughter" has nothing to do with gender. A "daughter" cell is not necessarily a *female* cell, but merely an *offspring,* a new cell formed from a parent cell or cells.

MEIOSIS I

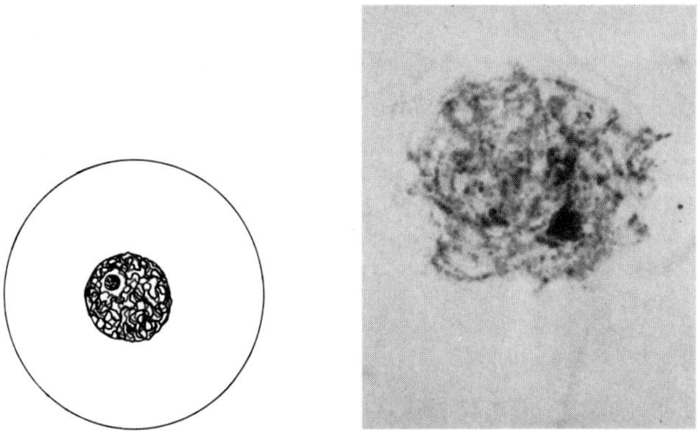

(1) A diploid cell, about to undergo meiosis.

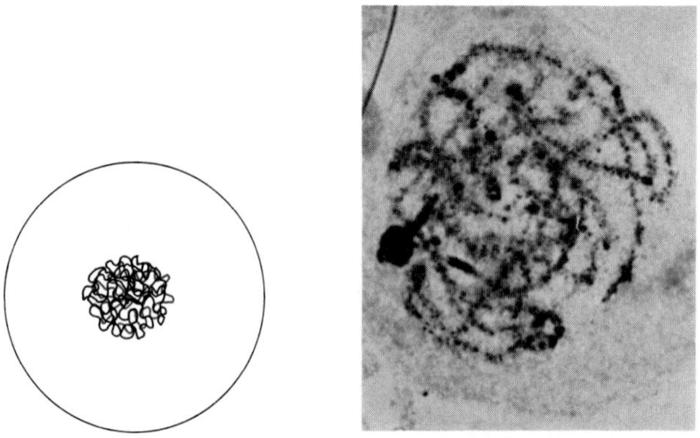

(2) Threads become somewhat thicker, forming visible lines.

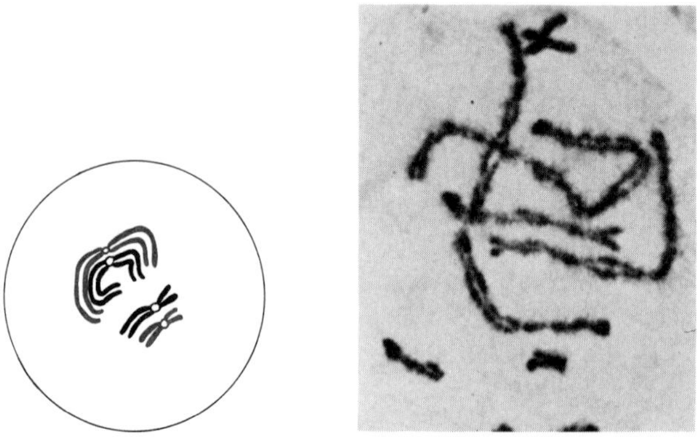

(3) Chromosomes become apparent. In this scheme there are four chromosomes, one long pair and a shorter pair. The chromosomes that were derived from the male parent (of the organism whose cells are illustrated) are black, and the chromosomes from the female parent are colored. The chromosomes line up exactly. Each chromosome is itself doubled, so that the *bivalent* is four-stranded.

Prophase I

In animals at least, most of the body cells are *diploid;* that is, their nuclei contain two full sets of chromosomes, one set having come from the animal's female parent and one from its male parent. The chromosomes (with some exceptions to be described later) are present in matching pairs. The two members of a matching pair are *homologous chromosomes,* or simply **homologs** (see the drawing below). During the working life of the animal, homologs behave almost independently, ignoring each other during the mitotic cycles. But during the first prophase of meiosis, homologs approach one another, two by two. Early in Prophase I (the first prophase), when the chromosomes are long and thin, each member of a homologous pair comes to lie alongside its opposite member, normally matching point for point with great accuracy.

The picture is complicated by the fact that each homolog itself is double, consisting of two chromatids, (a chromatid is a new chromosome that has not yet separated from its twin), so that what may appear as a single strand is really four strands: two homologs, each with two chromatids. Such a complex is called a **bivalent** (see the drawing below). The two chromatids belonging to one chromosome are *sisters.* Still during Prophase I, the chromatids undergo a series of breaks. Instead of healing themselves and returning to the way they were, however, nonsisters break at corresponding points. When healing does occur, one chromatid becomes joined up with a nonsister, and vice versa, so that there is an actual exchange of chromatid parts.

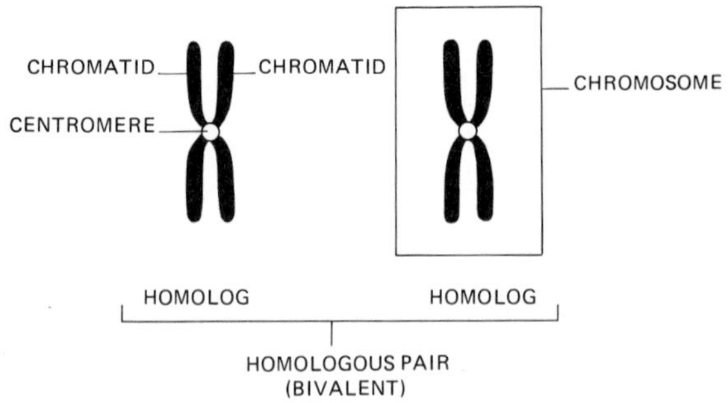

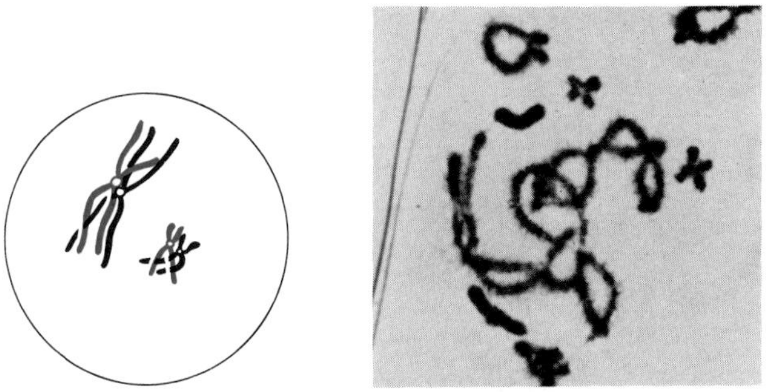

(4) Strands of the bivalents (*chromatids*) come to lie across one another.

After homologs begin to separate, the points where exchanges took place cause a tangling of the chromatids, with strands crossing over one another, frequently looking like Xs when seen in a microscope. These crossover regions are consequently called *chiasmata,* from the Greek letter chi (χ). Not only are chiasmata observable, but their genetic effects can be detected in the next generation of organisms. It can be demonstrated experimentally that the Xs visible in meiotic cells do indeed represent physical exchanges of DNA, and they can be correlated with specific genetic exchanges.

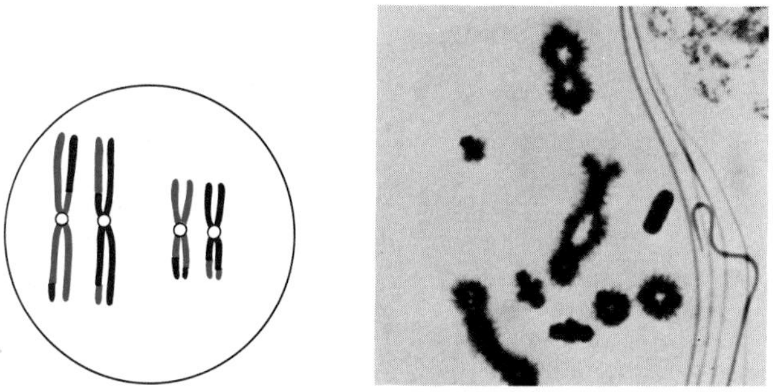

(5) Parts of maternal and paternal chromatids are exchanged. All the events shown in (1) through (5) occur during the first prophase.

Metaphase I

At **Metaphase I** the homologous pairs line up along the equator. Each pair moves as a unit.

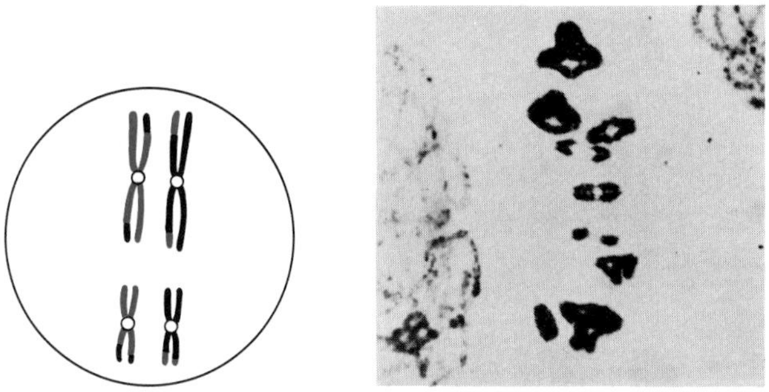

(6) The bivalents line up on the equator of the cell, bringing the cell to Metaphase I.

Anaphase I

At Anaphase I, the centromeres of a pair of homologs separate and the chromosomes move to opposite poles, carrying with them the newly recombined pieces of DNA that have been exchanged with nonsister chromatids. Whether the original centromere came from a mother's egg or a father's sperm makes no difference; the diploid set of chromosomes is effectively shuffled, rearranged, and separated out. Even without chiasmata, the genetic material can be well shaken up during meiosis, but when the chiasmata are taken into account, it becomes easy to understand how among the billions

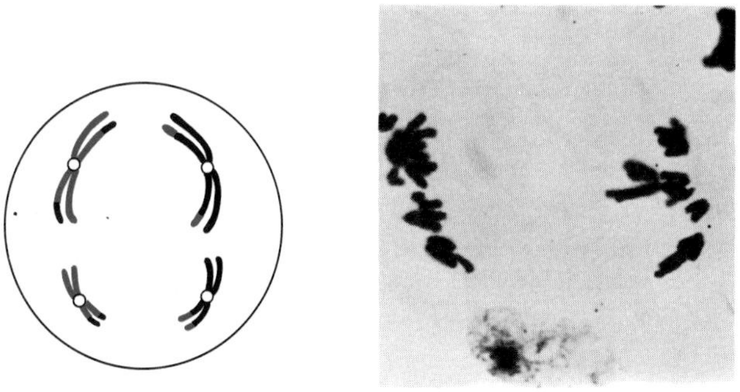

(7) During Anaphase I the centromeres of the bivalents separate and start toward the poles of the cell, taking the reshuffled chromosomes with them.

of people, to take an example of only one species, no two are genetically alike. Identical twins are of course an exception, but they arise after meiosis. Humans, with their 23 pairs of homologous chromosomes could theoretically produce 2^{23} (8,388,608) kinds of gametes: If chromosome exchanges ("crossing over") between chromatids occur, the number becomes too large to think about. Meiosis is an effective method of redistributing genes.

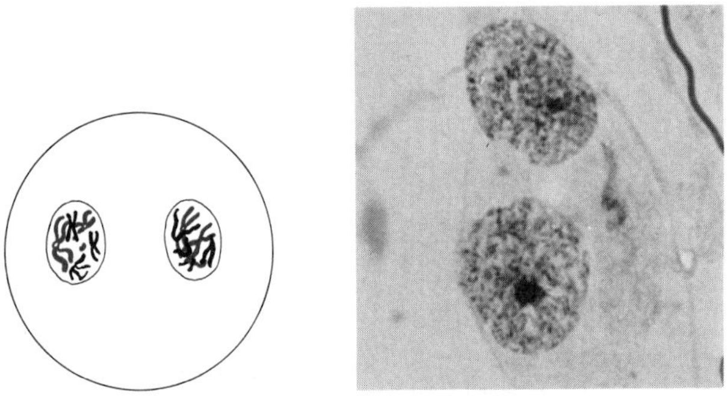

(8) At Telophase I, which is of short duration, two nuclei are complete.

Telophase I

After Anaphase I, there is a short Telophase I, followed by a second division: Prophase II, Metaphase II, and so on.

MEIOSIS II

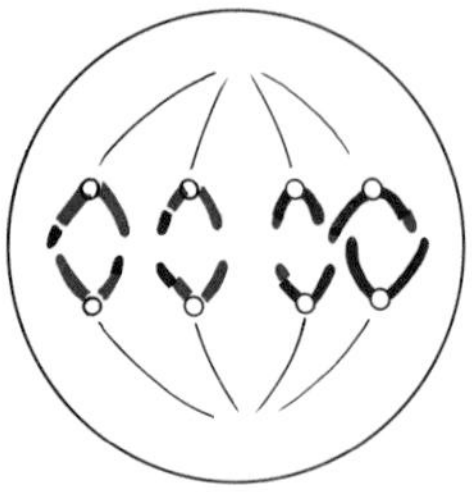

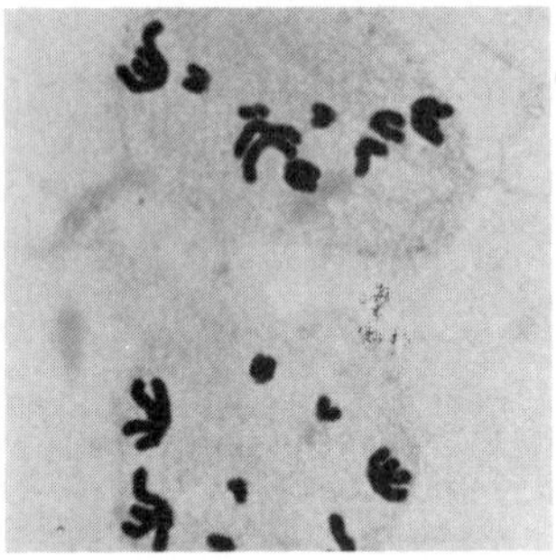

(9) At Metaphase II there are two sets of spindle fibers.

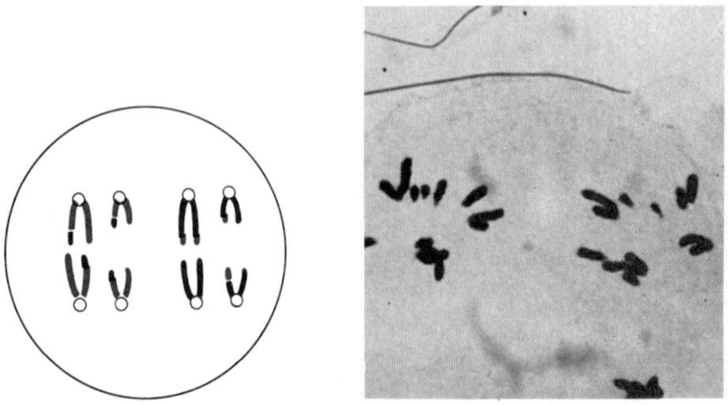

(10) During Anaphase II the chromatids separate, and they can now be called chromosomes. Each pole receives one long and one short chromosome, but these are different from the chromosomes the cell started with, having undergone chromatid exchanges, or crossovers.

At Anaphase II the final separation of chromatids occurs, and by Telophase II there are four nuclei, each with a full set of chromosomes, with one representative of each chromosome (that is, the nuclei are haploid) but with the DNA redistributed in a way that is practically unique. It is theoretically possible that two human sperm cells could have exactly the same set of nucleotide sequences in their DNA, but the chances are about as good as your winning a national lottery every day all your betting life.

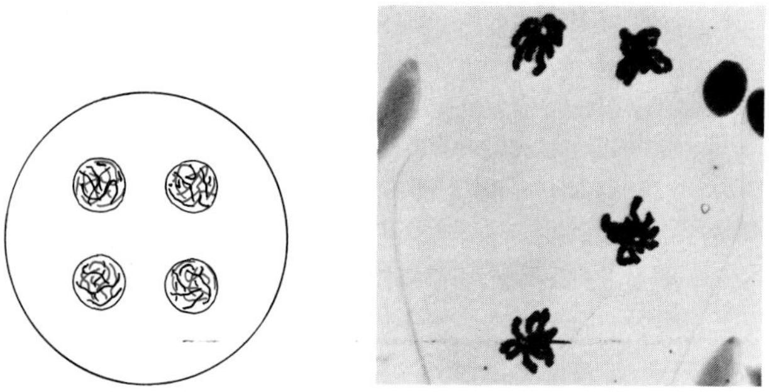

(11) At Telophase II, four nuclei, now *haploid*, are re-formed.

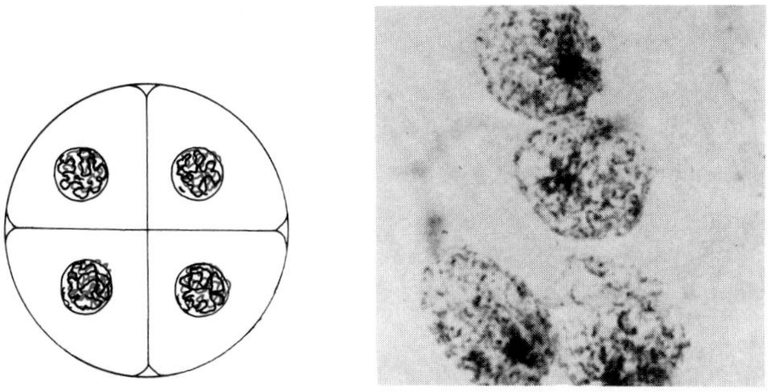

(12) Each haploid nucleus is separated from the others. Four cells with a reduced number of chromosomes have been produced.

TO SUM UP MEIOSIS

1. *Meiosis* is the process that reduces the chromosome number in the nucleus from the double (diploid) to the single (haploid) condition.
2. Meiosis requires two divisions. The phases of meiosis are similar to the phases of mitosis.
3. During *Prophase I,* chromosomes become apparent. Each chromosome, having two chromatids, lines up with a matching chromosome (homolog). A matched pair of doubled chromosomes is a bivalent.
4. Chromatids lie across one another, and parts of the nonsister chromatids are exchanged.
5. The bivalents line up on the equator of the cell during *Metaphase I.*
6. During *Anaphase I* the centromeres of the bivalents separate and move toward the poles of the cell, taking the reshuffled chromosomes with them.
7. Two nuclei are complete at *Telophase I.*
8. *Prophase II* follows Telophase I immediately.
9. At *Metaphase II* there are two sets of spindle fibers.
10. During *Anaphase II* the chromatids separate, and can now be called chromosomes. Each pole receives one haploid set of chromosomes, which have undergone genetic exchanges.
11. At *Telophase II,* four haploid *nuclei* are formed.
12. After Telophase II, four haploid *cells* are formed.

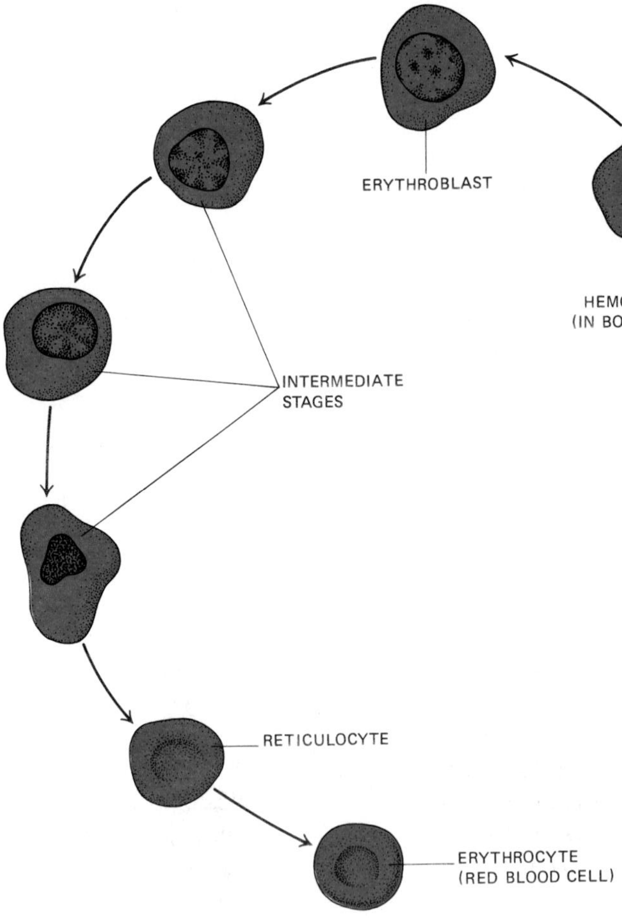

Figure 8.3
Formation of an erythrocyte from a bone marrow cell (hemocytoblast). As differentiation proceeds, the generalized cell loses its nucleus and other cell components, synthesizes hemoglobin, and assumes the typical shape of a red blood cell.

Once a group of cells has become specialized to perform some particular function, it is called a **tissue.** The study of tissues is sometimes pursued as an end in itself, but in this book we will discuss each tissue in relation to its function. For example, we will describe heart tissue in connection with blood circulation, plant vascular tissue with water movement, and nerve tissue with responsiveness.

Specialization of function and specialization of structure go together. A cell with a highly specific duty, as for example, a nerve cell, a cartilage cell, or a wood fiber, may be so modified in structure as to be scarcely recognizable as a cell. However, an unspecialized cell, such as a pith cell in a stem or a liver cell (which has so many jobs that it must not become too peculiar), looks like a textbook cell: more or less roundish and furnished with the usual organelles.

COMMON TYPES OF CELLULAR SPECIALIZATION

We have considered the two kinds of cell division: mitosis, which results in cell multiplication, and meiosis, which results in a reduction of chromosome number in a nucleus. We now turn to the events following mitosis. Cellular reproduction and subsequent increase in size are not enough to make a working organism. A cell must undergo the process of **differentiation,** meaning that it develops in some special way to become fitted for a particular function. Take for example a newly formed bone marrow cell that is destined to become a red blood cell (Figure 8.3). It separates from its associated cells, synthesizes millions of molecules of a special protein, hemoglobin, digests its own nucleus and mitochondria, and flattens itself into a disk. When all that is done, it is a fully differentiated red blood cell. Cell differentiation accounts for such diverse structures as hooves, feathers, wood, brains, and mushrooms.

BIOLOGICAL FEEDBACK PROVIDES HOMEOSTASIS FOR THE SYSTEM

A dynamic, functioning organism must make an unending series of compromises between two opposing requirements. It must keep changing to eat, to breathe, to mate, and to escape destruction. At the same time it must keep itself somewhat the same to maintain its strength and to retain its accumulation of experience. A balance between these two contrasting demands is attained at the cellular level and is made possible by the intake and outflow of energy and material. The result is that cells seem to remain stable while working. In a functional sense, they do remain stable; some individual cells remain alive for years, but they do so by virtue of a constant turnover of cellular material. The maintenance of a measure of stability during change is called **homeostasis,** meaning "standing the same."

By direct observation, a given cell can be seen to maintain its integrity, carrying out its normal activities for an extended length of time. It is known to have stability. But experimental treatment of cells proves that they are not stable in the same sense that a building is stable. If an animal eats amino acids with radioactive, labeled tracers, the tracers will soon show up in the structural proteins of the animal's cells. The animal has replaced some of its old proteins with new ones, and cells can make such substitutions rapidly. It is difficult to understand how a molecule can "wear out," yet some liver enzymes are taken out of circulation and replaced within a few hours. Some muscle proteins are 50 percent replaced after six months. A human being can expect to be rebuilt, molecule for molecule, about every year and a half. It is as though a building has a few stones removed from its structure every day and has new ones slipped into their places, so that after a time the entire building is renewed. It still has the same look and the same rooms, and it has not been out of service during the renovation.

You might ask, "If I am not the same person I was five or ten years ago, how can I remember what happened then?" A provocative answer is, "Because your cells are able to maintain homeostasis."

Feedback Systems Affect Stability

Biological controls and indeed all self-regulating controls depend on a **negative feedback system** (Figure 8.4). A familiar example is a thermostatically controlled furnace. An action (burning of a fire in the furnace) causes a change in the system (warming of the house), which acts on a responsive sensor (the thermostat), which then causes a change in the original action (the fire is cut off) in a direction opposite to the original action. Any dynamic system that remains stable must have an adequate control, either imposed from without (radiators are turned on and off, windows are opened and shut) or by negative feedback in the system itself (the thermostat turns the system on and off automatically). The principle is in use in the chemistry of cells, the actions of organs, the behavior of whole organisms, and even in the regulation of populations.

Positive feedback, sometimes called *runaway feedback,* occurs rarely and is usually disastrous. The alcoholic drinks more and more until he gets so drunk he passes out. The victim of heatstroke gets hotter and hotter as he continues to generate more internal heat until he dies. Both are suffering from positive feedback. In the example of a thermostat, if the thermostat was wired wrong, an already hot room would call for heat from the furnace and would keep calling for heat until either the house burned or the fuel gave out. Positive feedback is *not* normal in biological systems.

Cellular processes are generally dependent on a feedback system. When feedback is mentioned, it is understood to mean *negative* feedback since positive feedback is too uncommon to be considered. For

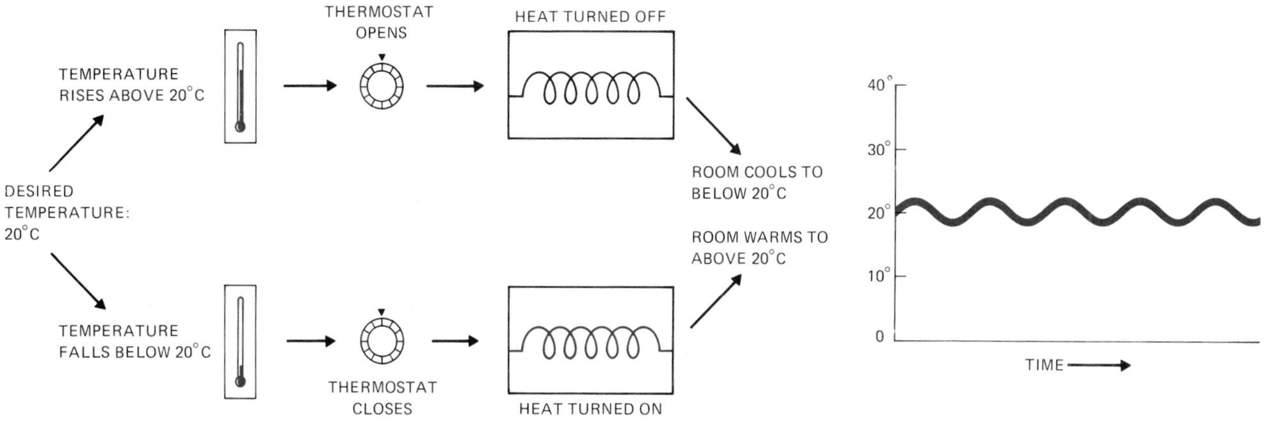

Figure. 8.4
Example of a *negative feedback system.* When a room with a thermostat *cools* below the set point, the system calls for *heat,* and the converse also applies. Such a system does not produce an absolutely even temperature, but it does keep the temperature close to the set point. The colored line in the graph shows how a negative feedback system produces a wobbly effect which means that any variations are slight, and conditions proceed almost in a straight line. With sufficiently precise sensors, the ripple can be made so slight as to be unimportant.

NUCLEAR TRANSPLANTS, EMBRYOS, AND CLONES

Every organism in a sexually reproducing population is genetically unique. (Exceptions, such as identical twins, are too rare to consider.) Even such "mass-produced" organisms as the plants in a field of wheat or the fish in a school of herring are all unique. Each one is the result of the fusion of an egg that was unique and a sperm that was also unique.

A group of individuals all with identical genetic makeup—a **clone**—can be made only by avoiding the usual sexual events of meiosis and fertilization. That is possible with many plants because asexual propagation by cuttings or grafting is relatively easy and has been common practice for thousands of years. ("Clone" comes from the Greek word *klon,* which means "twig.") Plants such as grains and legumes are not readily grown from cuttings and cannot be grafted in quantities, but even these plants should prove reproducible by cell cultures and somatic embryos (see the essay on p. 363).

Animals are generally different. The only way at present to get a chicken is to grow one from a hen's egg. A snippet of muscle tissue from a frog's leg will not regenerate a whole new frog. Still, the idea of being able to clone animals is one that has fascinated biologists and animal husbandry specialists more and more in recent years. Theoretically, there is the challenge of doing something simply because it is hard to do. Biologists are like mountain climbers: they want to see how much they can accomplish and find what is on "the other side of the mountain." Practically, it would be economical to have herds of cattle or flocks of fowl that were all alike in growth rate, feeding requirements, and productivity.

In their efforts to produce cloned animals, biologists work mainly with eggs or with cells taken from very young embryos, and practice the microsurgical art of **nuclear transplantation.** The procedure for cloning a frog is as follows: An egg of a frog is large enough to handle readily, and with a fine hollow needle one can suck out its nucleus, making an enucleated egg (an egg without a nucleus). One can then insert into the egg a new nucleus, and encourage the egg to grow into an embryo and eventually into an adult animal. If the source of the transplanted nucleus is a multicellular embryo (all of whose cells are genetically identical), then it should be possible to transplant a number of such nuclei into a number of enucleated eggs. The resulting animals would all be one clone.

Nuclear transplantation has been successful in the relatively large eggs of frogs (see Figure 8.5) and in mice, although mouse eggs are much smaller and harder to obtain. The removal of mouse eggs and early embryos, and the transplantation of mouse nuclei have been achieved by a few people with unusual patience and skill.

Karl Immensee of the University of Geneva and Peter Hoppe of the Jackson Laboratories in Bar Harbor, Maine, transplanted nuclei from the cells of an early mouse embryo into enucleated eggs. Three of the eggs so treated and then im-

example, when a cell is making a new DNA molecule, it must have on hand in the cytoplasm balanced amounts of adenine and thymine, and these materials must be synthesized enzymatically by the cell. Too much of one or the other is wasteful, and cells cannot afford to be wasteful. If there is too little of either one, however, DNA synthesis cannot proceed. The problem for the cell is to provide the proper amounts of both adenine and thymine , and it does so by a feedback loop that shuts down adenine production if the adenine-synthesizing mechanism produces more than is being used, or that turns on the adenine producer when adenine is in short supply. The same kind of control works for the thymine producer. The result is that the cell keeps itself in order, with adequate amounts of the necessary building products. Many such examples are known, and the supposition is that cell regulation is usually if not always based on some similar arrangement.

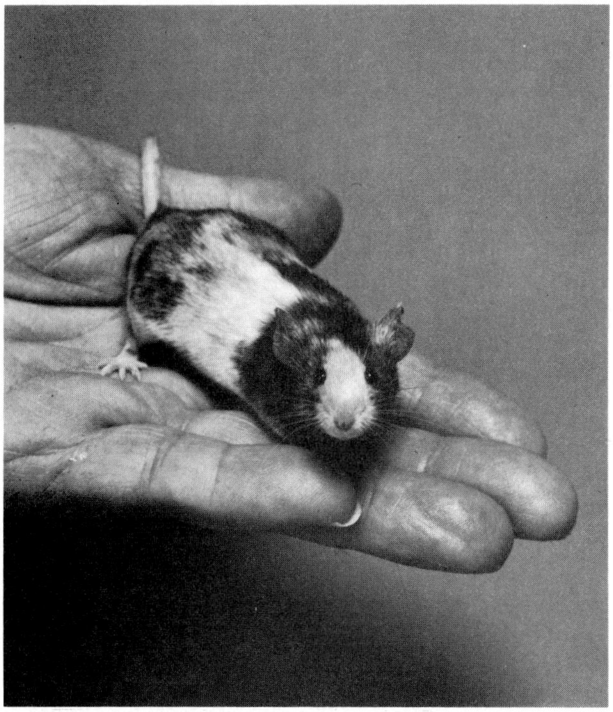

A mouse with six parents.

planted in female mice survived in their foster mothers and were born normally. As far as their nuclear genes are concerned, the three little mice were a clone.

Clement Markert and Robert Petters of Yale University grew to maturity a mouse with six parents (see photo). They did this by removing three immature embryos, each of which had two parents, putting together a few cells from each, and implanting the resulting mixed cell mass into a female mouse. Of the original three embryos, each was genetically set to have fur of a different color—one black, one yellow, and one white. The mixed embryo was born normally, and, as the picture shows, it had patches of the three expected colors of fur, showing that it had genes from the three original embryos. (Subsequently, a mouse with eight parents has been produced under laboratory conditions.)

The scientific and technical problems of cloning seem to be nearing solution. The history of biology shows that today's advanced procedures are tomorrow's standard clinical practice or even routine laboratory exercises. But when the scientific and technical questions are answered, we come to the social and ethical ones. Suppose we learn to clone human beings. What would it be like to have a thousand identical factory workers or half a million equally capable and disciplined soldiers all in the same-sized uniform? How would the cloned people *like* being unindividual? Who would rear them? Other cloned people? Finally comes the hardest question of all: Who would be empowered to decide which ones are to be cloned?

The students of today may have to make these decisions; if and when they do, they will need to know why they are deciding what they decide.

TOTIPOTENCY: EVEN MATURE CELLS CONTAIN ALL THE GENETIC CONTENT OF THE ORGANISM

Totipotency is the ability of any nucleus in an individual to perform any of the thousands of functions it might be required to do. The term, meaning "with all power," is believed to be applicable to nuclei in general.

For many years biologists wondered: After a cell has become fully differentiated, does it still have all its original genetic abilities, or are they lost along with the process of specialization? We now know that the first supposition is the correct one. From persuasive demonstrations in vertebrate animals and in flowering plants, we know that cells do retain all their synthetic power.

In 1952 Robert W. Briggs and Thomas J. King transplanted living cell nuclei into frog eggs and successfully produced genetically identical offspring, which were also genetically identical to their

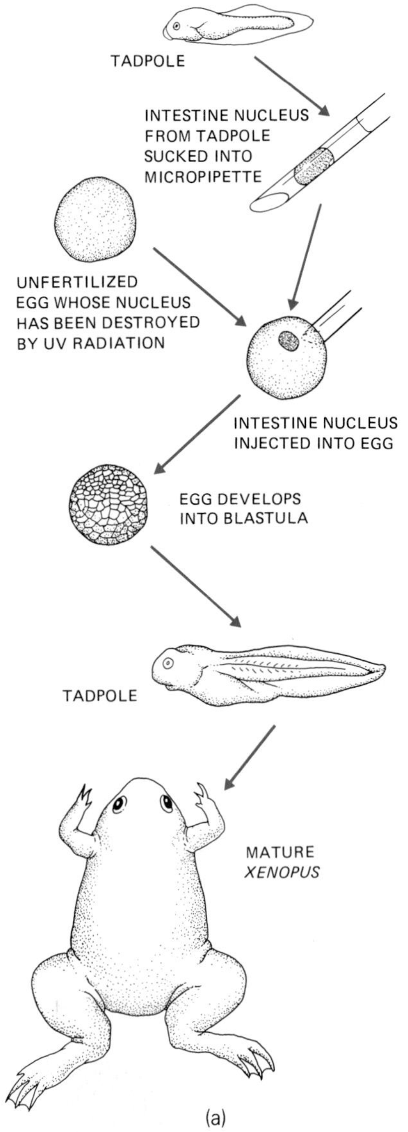

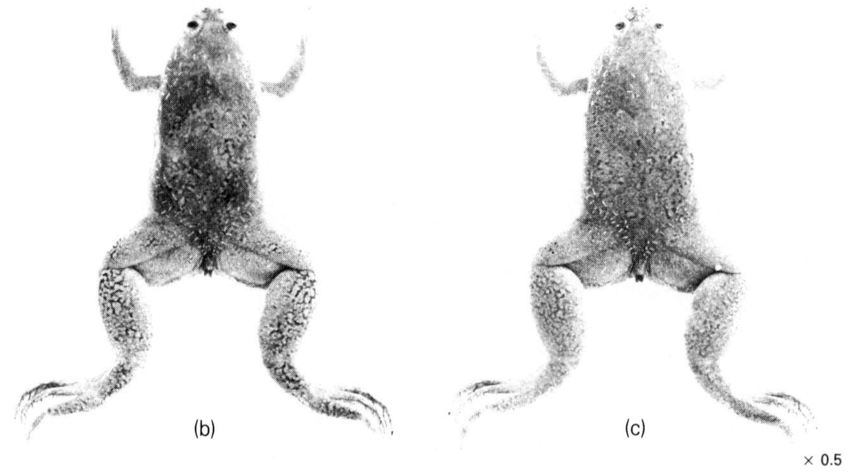

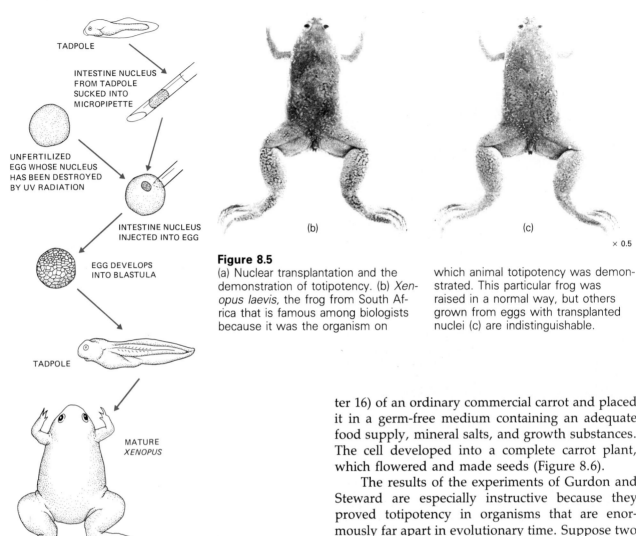

Figure 8.5
(a) Nuclear transplantation and the demonstration of totipotency. (b) *Xenopus laevis*, the frog from South Africa that is famous among biologists because it was the organism on which animal totipotency was demonstrated. This particular frog was raised in a normal way, but others grown from eggs with transplanted nuclei (c) are indistinguishable.

single parent donor. Shortly thereafter, J. B. Gurdon at Oxford University dissected the nucleus from an egg of the African clawed toad, *Xenopus*, and replaced it with a nucleus taken from a completely differentiated intestinal cell (Figure 8.5). The egg with the transplanted nucleus grew into a normal animal, complete with all the variety of cells an adult toad requires. This process, in which the nucleus from a cell of an organism is inserted into another cell, is technically known as **transplantation.** (See the essay on page 144.)

F. C. Steward of Cornell University, using a different technique and a different organism, proved the same point Gurdon had proved. Steward and his co-workers removed tissue called *phloem* (Chapter 16) of an ordinary commercial carrot and placed it in a germ-free medium containing an adequate food supply, mineral salts, and growth substances. The cell developed into a complete carrot plant, which flowered and made seeds (Figure 8.6).

The results of the experiments of Gurdon and Steward are especially instructive because they proved totipotency in organisms that are enormously far apart in evolutionary time. Suppose two demonstrations had been made on two closely related animals, say, a toad and a frog, or on two closely related plants, such as carrot and parsley. The original question would remain: Is totipotency limited to some special types? When two such distantly related organisms as toads (vertebrate animals) and carrots (flowering plants) can be shown to have totipotent nuclei, it seems safe to assume that the phenomenon is a general biological one, applicable to nucleated cells across a wide range of living things.

CANCER OCCURS WHEN FEEDBACK CONTROL IS MISSING

As long as cells remain under proper control, they grow normally and perform their normal functions, but if anything breaks into the feedback system, they may die or go into uncontrolled growth. Skin

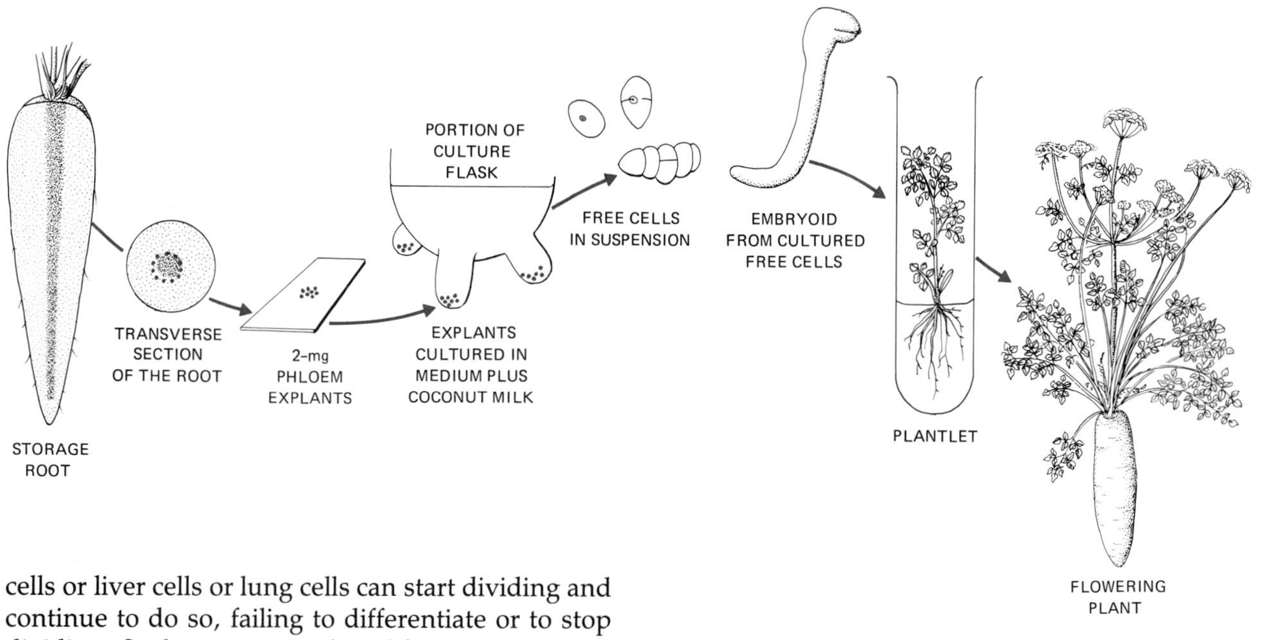

Figure 8.6
Steward's demonstration that a complete plant can be regenerated from a single differentiated cell. The ability of a phloem cell to develop into a complete plant with roots, flowers, and all other normal plant parts is taken as proof of the totipotency of plant cells.

cells or liver cells or lung cells can start dividing and continue to do so, failing to differentiate or to stop dividing. Such unrestrained proliferation is **cancer,** meaning "crab," something that nibbles at you (Figure 8.7); it can occur in any active tissue, including that of plants.

Among the characteristics of cancer in higher animals is an abnormal, seemingly uncontrolled growth of body cells, which invade and destroy

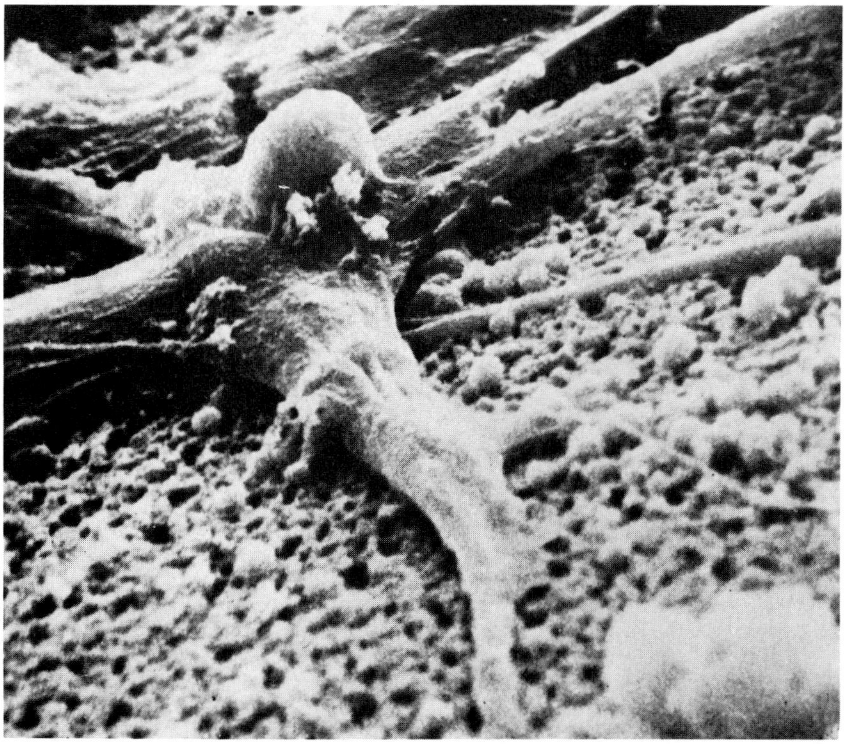

Figure 8.7
Scanning electron micrograph of a cancer cell in a tissue culture.

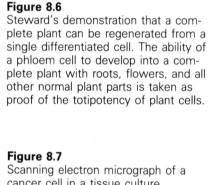

ENVIRONMENTAL CAUSES OF CANCER

We do not know as much about cancer as we would like to. We don't know how to prevent it or cure it. We *do* know that cancer is the second leading fatal disease in the United States, behind circulatory diseases, and it accounts for 20 percent of all deaths in the United States. (Every two minutes someone in the United States dies of cancer.)

The current belief is that many forms of cancer are caused by environmental factors. Statistically, at least, the evidence for environmental causes is substantial. For example, Jews who emigrate to Israel from the United States or Europe contract cancers representative of the countries they leave, but their children who are born in Israel show a lower incidence of cancer in general.

Lung cancer in women has only recently been appearing, whereas lung cancer in men began to be prevalent about 50 years ago. The reason for this difference may be simple: Men started smoking long before women, and cancers attributed to smoking appear on the average about 20 years after the first incidence of smoking. (Cigarette smokers are about four times more likely than nonsmokers to die of lung cancer.)

Cancer of the large intestine may be related to diet. Richer countries, where red meat is consumed regularly, generate more cancer deaths than countries where more cereal than meat is eaten. Studies of women in various countries indicate that deaths due to cancer of the large intestine are 40 times more prevalent among women in New Zealand, where red meat is plentiful, than in Nigeria and 30 times more prevalent among American women than among Japanese women. (Japanese women who move to the United States and adopt a typical American diet show an increasing incidence of cancer of the large intestine. The same is true of Japanese men.)

(a)

× 2500

(b)

× 2500

Lung cancer usually begins in the bronchi, the treelike tubes that transport gases in and out of the lungs. (a) Scanning electron micrograph of normal tissue lining a bronchus, showing healthy cilia, and (b) a cancerous bronchus, with the cilia destroyed. The sweeping cilia help to protect the lungs from foreign particles.

Vaccines have virtually eliminated some viral diseases such as polio. Unfortunately, there is still no simple connection between viruses and human cancer, so that a preventive vaccine is impossible, at least for the present.

neighboring normal tissue (Figure 8.8). Cancer cells may also break off from the original mass and enter the blood or lymph channels, to be circulated throughout the body. These cells may then take hold and destroy the organ where they have settled. Besides having a somewhat independent growth, cancerous cells also appear irregular under a microscope. They may vary so much in size and structure that no recognizable formations remain.

Almost half of all cancer deaths are caused by malignancies in the lung, large intestine, and breast. Cancer of the lung is the leading killer, accounting for almost 20 percent of all cancer deaths. The major areas of fatal cancer in American men are the lungs, the digestive system, and the prostate gland. The most vulnerable areas in women are the breasts, the digestive system, and the lungs.

Elderly people are the most susceptible to cancer, and the occurrence of cancer of the large intestine, for instance, increases about 1000 times between the ages of 20 and 80. Current thinking is that certain genes help prevent cancer until they are made ineffective by a mutation. The chances of a successful mutation are naturally increased after several unsuccessful attempts. Therefore the probability increases in direct proportion to age. Further, if it is true that several mutations are required to finally cause a cancer, then the cancer may have actually begun 10 or even 30 years before it becomes evident.

The World Health Organization has reported that approximately 85 percent of all cancer cases result from such environmental causes as smoking, overeating, overexposure to sunlight, overdrinking, and exposure to dangerous chemicals like asbestos and polyvinyl chloride, a widely used plastic. It could very well be, then, that humans are actively elevating cancer to epidemic status. We do know for certain that high doses of X-rays and similar radiation can cause leukemia and other cancers. The extremely high incidence of leukemia and cancer of the breast, bowels, and brain that have occurred in survivors of the bombings of Hiroshima and Nagasaki during World War II are a clear example of human-provoked cancers. At least one possibility for cancer prevention is clear: Humans can prevent the cancers they cause.

Only a tiny fraction of cells that go out of control actually progress far enough to become harmful. Most abnormal cells die simply because they *are* abnormal, and abnormal cells are not equipped to survive. Other abnormal cells die early because the

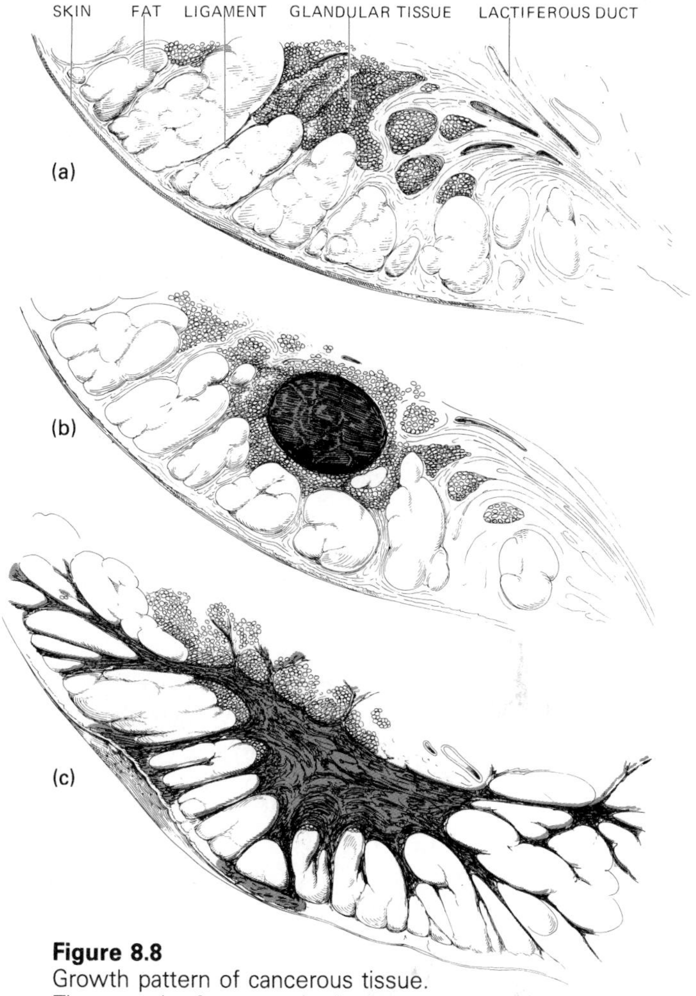

Figure 8.8
Growth pattern of cancerous tissue. The growth of a tumor in the breast ordinarily threatens life only when the tumor can spread to distant parts of the body. (a) The normal breast is organized into a glandular tissue, fat, and other structures. Tumors arise almost exclusively in the glandular tissue. They are composed of cells in which the normal restraints on growth and reproduction have been removed. (b) A benign tumor can grow rapidly and become quite large, but it cannot escape the tissue in which it develops. (c) A cancerous (malignant) tumor can spread throughout the glandular tissue, can often involve ligaments and skin, and can sometimes penetrate the muscle underlying the breast. In addition, cancers can in some cases migrate through the blood or the lymphatic system to establish new colonies of cells in distant, unrelated organs. This is the process known as *metastasis*. It is the metastatic spread of cancers that is responsible for their lethality.

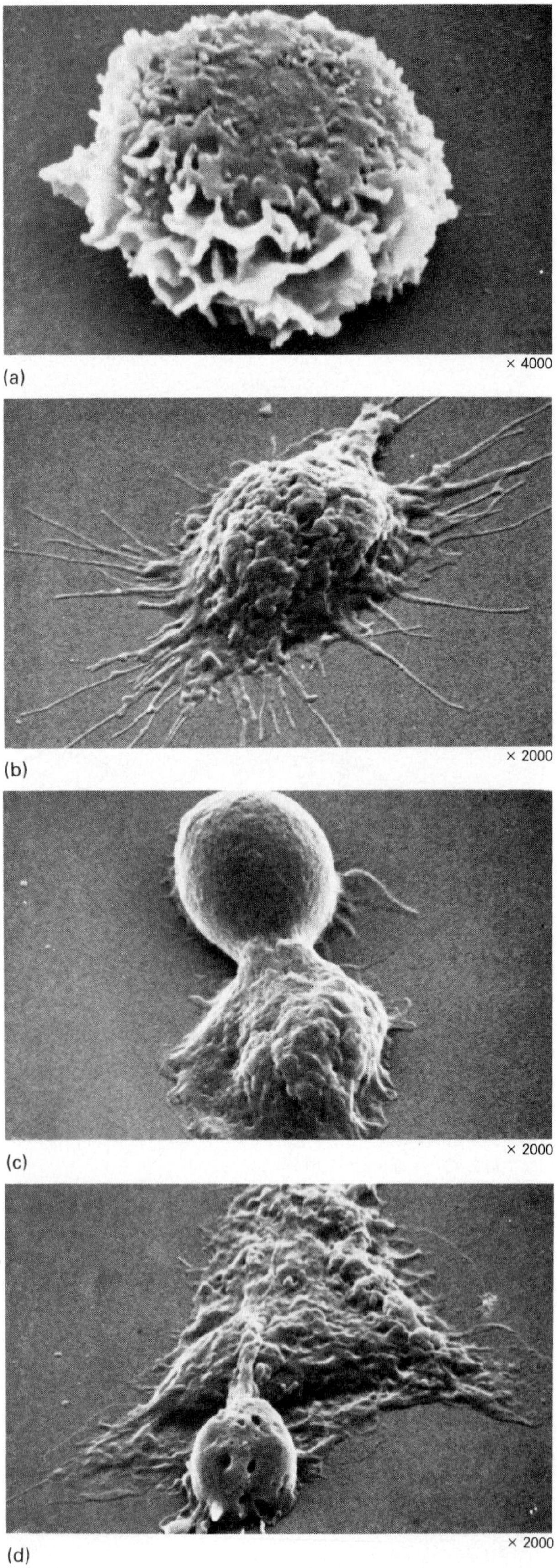

Figure 8.9
The death of a cancer cell. This series of scanning electron micrographs shows (a) a mouse cancer cell (shown disproportionately large), (b) a defense cell called a macrophage, anchored by its tendril-like filopodia, (c) the macrophage moving toward the cancer cell and attacking it; the typically wrinkled cancer cell begins to smooth out as it weakens, and (d) the cancer cell develops holes in its surface after about six hours of attack by the macrophage; the holes indicate that the cancer cell is dead or close to death.

AGING AND DEATH

Aging and death have been the subjects of philosophical speculation for centuries, but only within recent years have scientific investigations been started on the physiology of aging. At present, there is little firm information, and there are no broadly accepted general principles.

Concerning aging, some facts have been obtained from studies on cell cultures. Experiments on cell cultures do not necessarily prove that what happens under artificial conditions is the same as what happens in a live animal, but they are instructive. If normal embryonic human cells are grown in culture, the culture can be maintained through about 50 generations. Cells from children survive about 30 generations and cultures from adults last only about 20 generations. These facts have prompted a supposition that the number of times cells can divide in a lifetime is preset and can be modified only to a minor extent by the life an individual leads. Other guesses as to the reason for aging include the hypothesis that after repeated divisions, cells come to possess more destructive mutations than they can stand, or that an increasing accumulation of toxic metabolic products leads to an eventual stoppage of activity. Experimental evidence supports all these views.

A puzzle is added by the ability of cultured cancer cells to grow and divide indefinitely. A cel-

body's immune system recognizes them as "different" and forms antibodies that destroy them (Figure 8.9). The cancer cells that survive, despite the body's built-in defenses against them, are hearty.

The causes of cancer are many, probably working in different ways to produce a similar result. Similarly, an automobile may "run away" because it was left on a hill with the brakes off or because the brakes failed or because the driver went to sleep. Regardless of the reason for the lack of control, the end result is the same. In the case of cellular growth, control can be destroyed by viruses; by certain chemical compounds; by radiation, which disrupts molecules in cells; by mechanical abrasion; and probably by other, still unidentified forces. Cancer-causing agents are called *carcinogens*.

Two of the most important theories concerning the causes of cancer emphasize the genetic material and changes in that material. One hypothesis suggests that mutations occurring in the body cells may result in the uncontrollable cellular growth characteristic of cancer. Another hypothesis suggests that cancer results when the DNA of an infectious virus is inserted into the genetic material of otherwise normal cells. So far, research has not been able to choose between these two hypotheses. We *do* know that DNA extracted from cancerous cells is different in a number of ways from DNA in normal cells. Indeed, the DNA in cancerous cells is visibly changed, and it has been known for more than 65 years that one type of cancer in chickens is caused by a virus. Although researchers have identified no virus that

---

ebrated instance is the culture of HeLa cells, started in 1952 from a cancerous growth in a woman named Henrietta Lacks. Cultures of HeLa cells have been used in so many experiments in hundreds of laboratories that their total quantity far exceeds the number of cancerous cells the original donor ever possessed. Yet HeLa cells are growing today as vigorously as ever.

If an individual's cells are "programmed" to last only so long, how does a transformation from a normal to a cancerous condition upset the program? As with so many fundamental biological questions, no one knows.

If the biochemical attack on the problems of aging proves to be as effective as such attacks have been on metabolic and genetic problems, then we may come to understand the process and possibly even control it. If we do, a new question poses itself, this time a question that cannot be answered by biology or chemistry or any science: How desirable would it be for a person to live indefinitely?

Average life expectancy has increased statistically because of medical and other improvements, but a rise in the average age at death merely shows that more people have escaped an early death by disease or accident and have thereby come closer to living to the maximum. The maximum itself has not risen. A time comes when every animal with finite size dies.

But what is death? As discussed in the introduction to this book, life eludes exact definition. It is therefore not especially instructive to define death as the absence of life, but that is what it is, an absence of the functional organization that enables a plant or animal to do things. Death has traditionally been regarded as a Thing, something positive and real, even though it cannot be measured or even satisfactorily defined in medical, legal, or biological terms. Death, like cold and darkness, has terrified human beings for thousands of years, but death and cold and darkness are all negative. As physicists consider cold to be the absence of heat, and darkness the absence of light, so biologists can regard death as the absence of life.

The only whole organisms that are potentially immortal are those that reproduce by fission. A single-celled alga, on dividing, ceases to be, yet the two products of the division live on, and their offspring live on, limited in succeeding generations only by the restrictions of nourishment and space. Most plants and animals, as individuals, die, but if they leave offspring, they transmit a living cell to the next generation. In a way, any organism that reproduces has a chance at a tiny bit of biological immortality.

causes cancer in humans, leukemia is one form of human cancer suspected of being virally induced. Since more than 100 cancer-causing viruses have already been isolated in other species, it is likely that investigators will some day identify a virus that causes cancer in humans. At this time, no single virus has been identified as the *sole* cause of any type of cancer in humans, and it is believed that neither mutations nor viral infections cause *all* cancers. (See the essay on p. 148.)

A cell without a functional feedback control for mitosis does not "know" when to stop growing. In a living body, all the requirements for growth are available: warmth, water, nutrients, vitamins. A cancerous tissue therefore goes on growing without restraint until it impairs some vital function of the organism, whereupon both organism and cancer die.

As unexplained forces may cause cells to lose their self-control, so other forces may reverse the damage. Perhaps some individuals develop specific cancer-controlling mechanisms, or perhaps a reversal of the original defect occurs by chance. Regardless of the cause, there are rare and occasional *spontaneous remissions*. The reasons for such remissions are still as mysterious as the reasons for the start of the cancer.

SUMMARY

1. The final form of an organism depends directly on its cellular arrangement. Multicellular plants and animals start life as single cells and achieve maturity as a result of repeated cell divisions and changes in the form of individual cells.
2. It is the maturation, the differentiation, the specialization of cells that gives them their final working form. Functional specialization is more striking than species specialization.
3. During the initial steps in cell division there is an exact replication of the DNA, which provides the cell with two complete, identical sets of chromosomes.
4. The actual nuclear division process is known as *mitosis*. The phases of mitosis are the *prophase*, the *metaphase*, the *anaphase*, and the *telophase*.
5. Mitosis accomplishes the reproduction of cells and the equal distribution of DNA to each.
6. *Meiosis* is the term applied to the pair of divisions required for reduction of the chromosome number from *diploid* to *haploid*.
7. Cellular *differentiation* following mitosis achieves specialization of cells.
8. Any dynamic system depends on a *negative feedback system*, which helps maintain homeostasis by compensating for actions that may be harmful to the system. Cells are regulated by such a system of biological feedback.
9. *Totipotency* is the principle that states that all the cells in an organism possess all the genetic capabilities of that organism, even after the cells have undergone differentiation.
10. In nuclear *transplantation* the nucleus of a cell is replaced by a nucleus from another cell. A *clone* is a group of individuals all with identical genetic makeup.
11. *Cancer* occurs when the feedback systems of cells break down, causing cells to die or grow uncontrollably. The causes of cancer are many, probably working in different ways to produce a similar result.

ASK YOURSELF

1. What is the main result of mitosis?
2. When mitosis is complete, how does the number of chromosomes in a daughter cell compare with the number in a parent cell?
3. What are the four recognized stages of mitosis?
4. After nuclear division is over at the end of telophase, how is *cell* division accomplished in animals?
5. What is the difference between a cell in the haploid condition and one in the diploid condition?
6. Why is the exchanging of parts of chromosomes during meiosis important?
7. What is a homolog? A chromatid? A bivalent?
8. What is homeostasis, and why is it important to all living organisms?
9. What is totipotency?
10. How does cancer actually kill an organism?

PART THREE
THE DIVERSITY OF LIFE

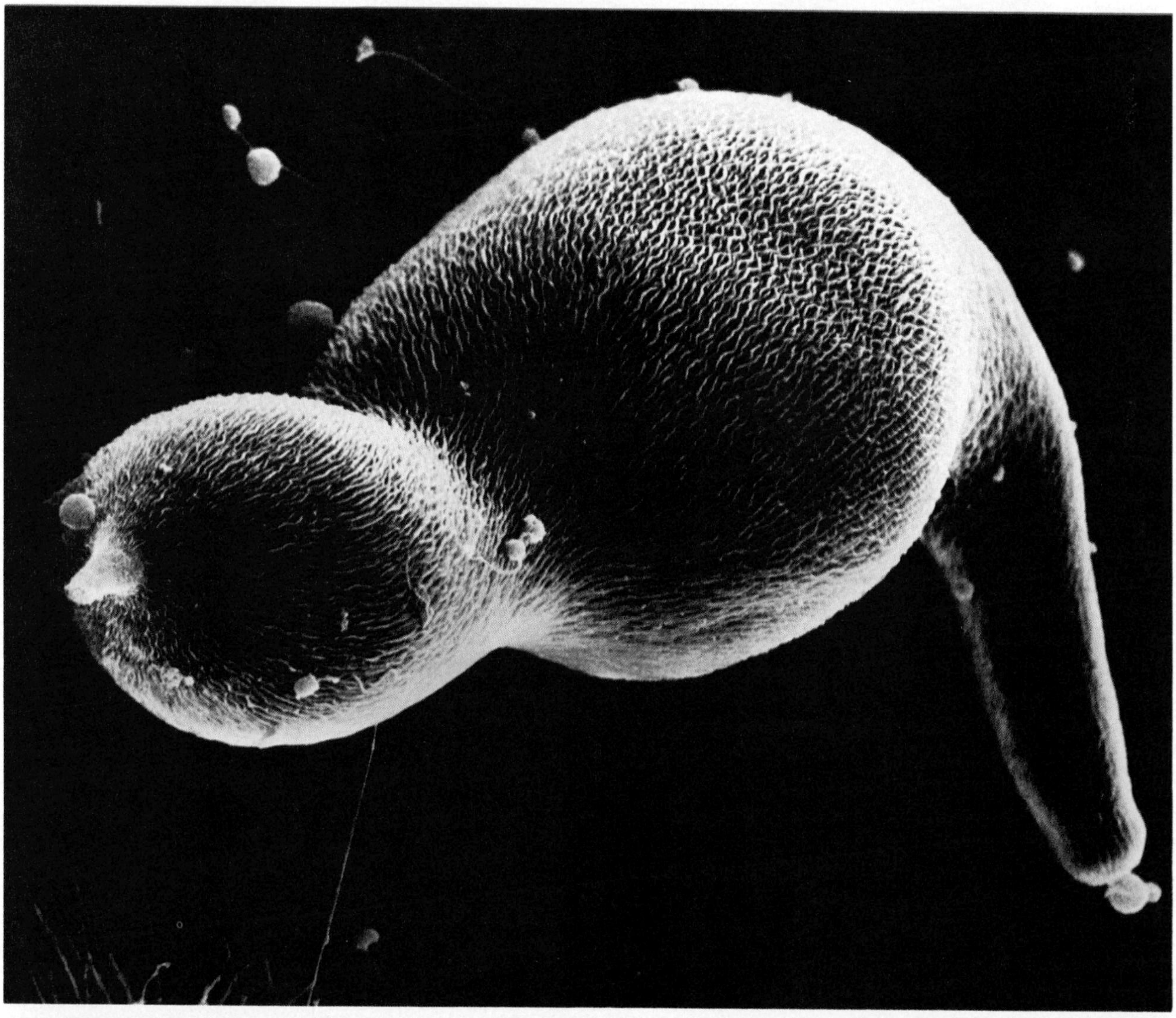

Scanning electron micrograph of a miniature parasite, Scyzogregori, that is found inside clams. This protozoan moves about by contracting and stretching its body. × 750

9
The Living Kingdoms: The Lower Groups and Plants

SOME KEY POINTS

1. The number of different kinds of plants is enormous, and some organisms have not yet even been discovered.
2. Organisms are typically classified according to species, genus, family, order, class, phylum, and kingdom.
3. Kingdoms may be classified as Monera, Protista, Fungi, Plants, and Animals.
4. Plants that have a system of conducting vessels give the earth its main green covering. They include the ferns, conifers, and flowers. These vascular plants are the most recent, the most complex, and the most successful plants.
5. The flowering plants, or angiosperms, first evolved nearly 200 million years ago, and have become the most conspicuous feature of the landscape. They are either monocots, with one seed leaf in the embryo, or dicots, whose embryos have two seed leaves.

ANYONE WHO LOOKS AT THE EARTH'S ORganisms sees a contrast between constancy and diversity. Living things have many features in common, such as cellular arrangement and chemical controls. Until now we have stressed general structures and functions that are common to most organisms, and we have done so for the very reason that they *are* common and therefore must be fundamental to the activities of life. Now, however, we will emphasize differences. In this chapter and the next we will give some idea of the major kinds of organisms alive today, with occasional mention of some that lived once but are with us no more.

The sheer number of different organisms is impressive. We do not even know how many have been collected and described because the literature on the subject is so vast that it has never been completely catalogued. In addition, unknown numbers of organisms have never been systematically recorded. There is not even complete agreement as to what a species is. Species is frequently defined as "a population that can produce fertile offspring," but not all biologists accept that definition.

Despite such difficulties, rough guesses about the numbers and kinds of organisms can be made, and these numbers are usually greater than nonbiologists expect. Some groups of organisms are relatively small. There are, for example, only about 1500 recognized species of bacteria and about 4500 species of grasses. Orchids, with about 15,000 species, make a larger group. The numbers of animal species are overwhelming: 23,000 species of fishes and 8500 species of birds, for example. But the beetles are the winners, with close to 300,000 species in the world, a tenth of them living in North America.

Obviously, no one person can know more than a small fraction of the living species, although a good field biologist can reasonably expect to recognize most of the specimens in restricted groups. With experience, one can come to know the birds, trees, and mammals in an area. To know the mosses or fishes takes more time, and even an expert never learns all the crustacea or insects, even in a limited region.

More striking than simply the number of kinds of plants and animals is the way in which modifications can occur. From types that are considered ancient or primitive, there has evolved a bewildering array of alterations. Some well-known examples are the mouth parts of insects, the leaves of ferns, the teeth of mammals, and the appendages of marine crustacea. Anyone who looks at a number of species or even at the individuals within a species is impressed by the endless variety. There is, to be sure, an underlying unity. Otherwise there could be no biological generalizations worth paying attention to. But there is, at the same time, this diversity, which is one of the main sources of biological excitement.

In this chapter and the next, we shall describe some aspects of a few selected kinds of organisms. We have chosen to include those that are in some way peculiarly instructive or bizarre or those that affect human health or wealth. In some groups of organisms, the usual trend toward complexity seems to be reversed, as they exhibit a return to simplicity. In the groups thought to be the oldest and most primitive, structural differentiation is minimal, though even the simplest types are biochemically beyond our present understanding. You can follow a small history of life by examining a bacterium, a moss, a fern, and finally a flowering plant. The increasing complexity in animals parallels that in plants, and you can repeat history by learning first of animals so small as to be invisible, progressing through intermediate examples, and finally arriving at the so-called pinnacles of evolution, a few examples of which might be a squid, a honeybee, a hummingbird, and a sculptor.

SCHEMES OF CLASSIFICATION

Just as there is disagreement among biologists as to species, so there is disagreement over the larger units of classification. The aim of the taxonomist, that is, the student of classification, is to make a system that shows the evolutionary relationships of organisms. However, we do not know what the relationships are because most of the organisms that could clarify those relationships are dead and lost forever. The so-called family tree of living things could better be compared to a bush that is almost submerged in a murky swamp. Only the tips of some of the branches stick out above the surface of the water, and all the connecting branches are hidden. We can make guesses as to which twigs are growing on branches close to other twigs, but the guesses are speculative, regardless of the care and logic we apply to them.

The main standard of relationship is similarity. The more points of similarity that can be found between two specimens, the closer they are thought to be related. As many similarities as possible are taken into account: physical structure, embryological development, ratios of DNA base pairs, chromosome structure, reproductive methods, respiratory processes and pigments, and biochemical components. A single point of similarity may be striking but superficial, and it may lead to wrong conclusions. The tropical cecilians, for instance, look like snakes, but they are not snakes, they are more like legless frogs. Organisms must be taken as whole organisms when evolutionary comparisons are made, and *all* characteristics must be considered.

Individuals that fit a taxonomist's definition of a species are placed in a **species.** A group of species that seem so similar that they are believed to be closely related are placed in a **genus** (plural, genera). The system of naming an organism by giving its genus and species, established by the so-called Father of Classification, Carolus Linnaeus (1707–1778), is still followed. The genus name, always starting with a capital letter, comes first, followed by the species name. Both words are italicized or underlined. The words are in the Latin form, even when the root words are Greek or some other language. A number of genera may be considered as belonging to a **family.** Families are grouped into **orders,** orders into **classes,** classes into **phyla** (singular, phylum), and phyla into **kingdoms.** Any category of classification is a **taxon** (plural, taxa). A species is a taxon, as is a genus, a family, or an order.

Here are some examples of a simplified classification of two familiar organisms, a plant (carrot) and an animal (horse):

Kingdom Plantae	**Kingdom Animalia**
Phylum Tracheophyta	Phylum Chordata
Class Angiospermae	Subphylum Vertebrata
Subclass Dicotyledonae	Class Mammalia
Order Umbellales	Order Perissodactyla
Family Umbelliferae	Family Equidae
Genus *Daucus*	Genus *Equus*
Species *carota*	Species *caballus*

TO SUM UP CLASSIFICATION

1. The smallest classification group is a *species;* the largest classification group is a *kingdom.*
2. Closely related species are placed in a *genus* (plural, genera).
3. A number of genera are included in a *family.*
4. Families are grouped into *orders.*
5. Orders are grouped into *classes,* and classes into the still-larger *phyla* (singular, phylum).
6. Phyla are grouped into *kingdoms.*

THE LIVING KINGDOMS

Until recently, all organisms were classified in one of two kingdoms: Animals and Plants. Plants are green and stationary; animals eat and move about. Such a simple scheme is adequate for grass and horses, but so many living species do not fit into either a plant or an animal category that alternative systems were devised. The system used in this book is one currently favored by many biologists, but it must be remembered that many others exist, and even this one is not completely satisfactory for many kinds of organisms. In this system, the following five kingdoms are recognized:

Monera: the prokaryotes, including bacteria, blue-green algae, actinomycetes, viruses, and viroids.

Protista: some unicellular algae, protozoa, and some intermediate forms.

Fungi: the "true" fungi (phycomycetes, ascomycetes, basidiomycetes).

Plants: red, brown, and green algae, mosses and some similar forms, and the vascular plants (ferns, a few fernlike nonseed plants, and seed plants).

Animals: all the familiar types, except protozoa.

THE KINGDOM OF THE MONERA

The **Monera** are the simplest, at least structurally, of all living things, and fossils resembling present-day Monera can be found in the oldest fossil-bearing rocks. The Monera are characterized more by what they lack than by what they possess. They have no nuclear membrane, no well-defined nucleus with chromosomes, no protein associated with their DNA, no flagella with microtubules, and no membrane-bound organelles, such as plastids or mitochondria. They all have some kind of external covering, but the materials are different from those of eukaryotic cells. There are ribosomes, but they are smaller than eukaryotic ribosomes. Perhaps present-day prokaryotes are organisms that have retained their primitive features and may therefore resemble the ancestors of eukaryotes. Being primitive does not mean being unsuccessful, however. The Monera are eminently successful in the biological sense of surviving in great numbers.

Viruses

Viruses were known by their activities long before actual virus particles were demonstrated. They are usually far too small to see directly with a light microscope, although a few outsized ones are as large as the smallest bacteria. These parasitic organisms were known to be capable of reproduction because infection, in tobacco plants, could be demonstrated to result from the transmission from plant to plant of the disease known as tobacco mosaic (which causes spotted leaves). By recognizing that the causative agent had reproduced, scientists could explain the spread of the disease.

A major advance in knowledge of viruses came in 1933, when Wendell Stanley concentrated enough tobacco mosaic virus to be able to purify a small quantity of it. To the surprise of the biological world, the purified virus could be crystallized. Only molecules with precise three-dimensional shape can be crystallized. When suspended in water, the crystallized virus again became infective and would produce the mosaic disease in tobacco plants. Since the ability to reproduce is one of the prime characteristics of living material, and since crystallization is a feature of a nonliving chemical compound, viruses seem to occupy a borderline position between the living and nonliving worlds.

During the 1950s, electron microscopes were developed that had the ability to resolve particles about one-thousandth the size of those visible with light microscopes. Virus particles, called *virions,* could be photographed to show their size, shape, and fine structure. Electron micrographs reveal that virions may range from about 0.02 to 0.3 micrometer. Thus a row of about 25,000 of the small virions or about 1500 of the largest ones would stretch across the period at the end of this sentence. Shapes, too, are various. The tobacco mosaic virus is

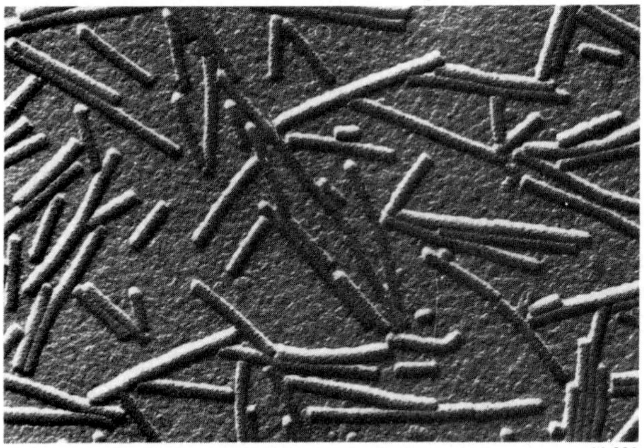

(a) × 11,000

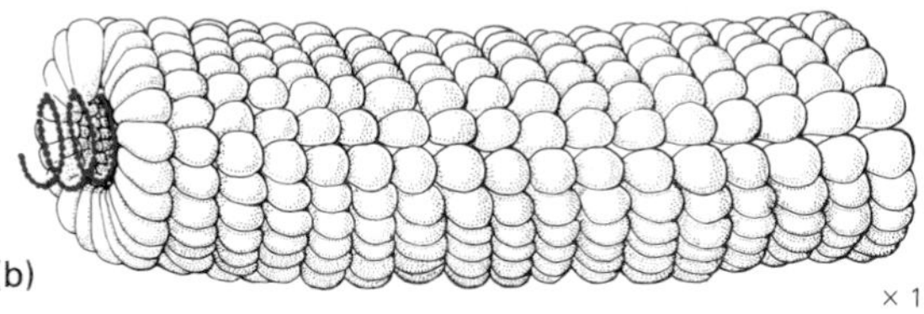

(b) × 110,000

Figure 9.1
(a) Electron micrograph of tobacco mosaic virus. (b) A drawing interpreting the electron micrograph, showing the nucleic acid core (color) of the virus particle with protein units wound around it.

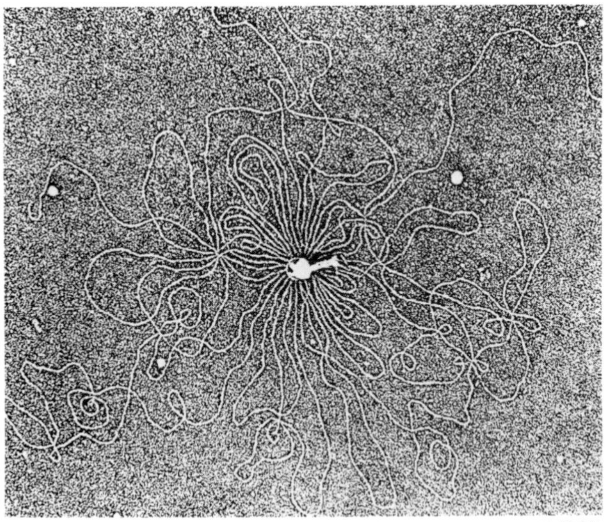

× 40,000

Figure 9.2
Viral DNA streaming from a virus particle, having been exploded by treatment with water and photographed with an electron microscope. One free end of the viral DNA can be seen at the right of the picture and the other at the left.

a slender cylinder (Figure 9.1), but many viruses are roundish with flat-sided facets like those on a cut gem. In high-resolution pictures, individual protein units of the virus coat are visible, and when a virus particle is burst open, its internal strand of DNA can be seen like a spread tangle of thread (Figure 9.2).

A virus particle has no life of its own, no respiration or known synthetic ability. It does have its nucleic acid strand—DNA in some viruses and RNA in others—and it has a covering of protein. Several different proteins may be incorporated into the virus coat. We know that a virus must have some enzyme that can weaken the outer coat of a cell it is attacking. We know this because viruses grow only in host cells, and host cells have outer coats.

The method of attack of a cell by a virus is known best for the viruses that infect bacteria, known as **bacteriophages** (Greek, "bacterium-eaters"), or simply as **phages.** One phage (Figure 9.3) resembles a tadpole; it has a large body and a slender tail with six fibers. These fibers attach the tip of the tail to the outer wall of the victim and hold the phage body in place while the nucleic acid is injected through the host wall and into the cell. The protein coat of the virus remains outside. Inside the cell, the viral DNA causes the host to carry out the genetic directions of the viral DNA. The end result of this process is the synthesis by the cell of virus nucleic acid and virus protein and their assembly into complete virus particles. In an active virus infection, as many as 200 new particles may be assembled within 20 minutes. When the new particles break out of the destroyed cell, they are ready for another round of infection.

Under some circumstances, a virus in a cell remains inactive and does little or no harm to the host cell until the host cell undergoes a change, when the virus may become destructive. For example, the virus that causes human cold sores, herpes simplex virus, seems to be almost perpetually present, yet it fails to cause the typical symptom of ruptured cells unless the host cells are weakened, such as by another infection.

Possibly all living cells harbor viruses, most of them so specific that they can inhabit only one kind or a few kinds of host cells. Although many viruses seem to do little or no harm, many others are the causes of plant and animal diseases. These range from minor disturbances (such as the variegated color of Rembrandt tulips and the cold sores on human lips) to lethal disruptions (such as virus infection of elm trees and rabies in mammals). The

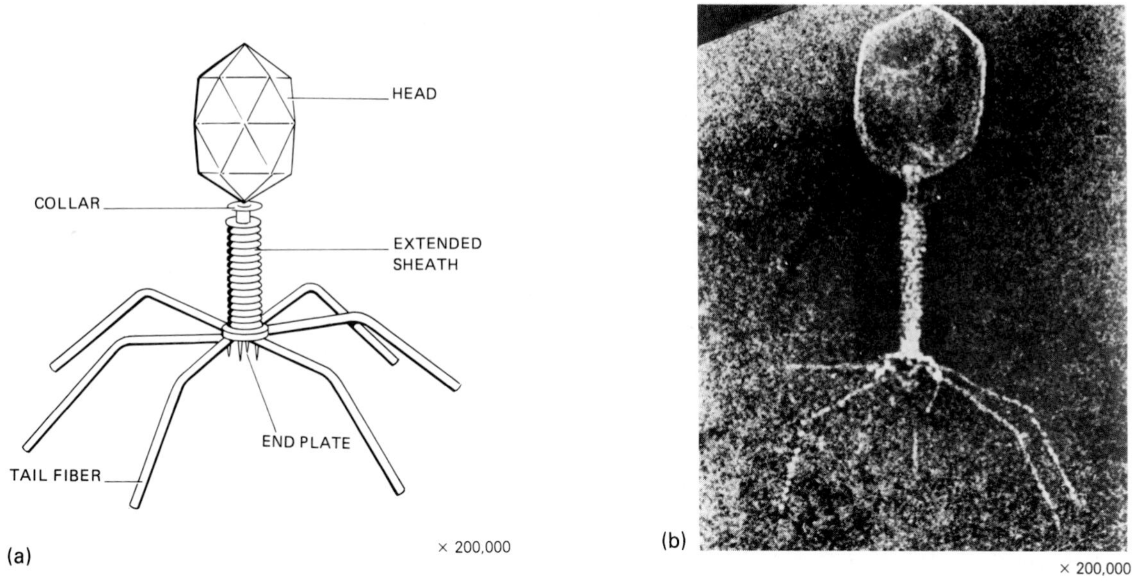

Figure 9.3
Bacteriophages (viruses). (a) A diagram of a virus. The protein-coated head encloses the DNA. The protein sheath around the tail is contractile, allowing the virus to inject its DNA into a host cell. The tail fibers and end plate provide a firm attachment to the surface of the host cell. (b) Electron micrograph of a virus, showing the shape of the head, the tail, and the tail fibers.

virus of yellow fever is carried from person to person by mosquitoes; some virus-induced tumors of plants are transmitted by tiny insects called leaf hoppers. In our human-oriented way, we think only of the human sufferer as having a disease, but the insect is as much afflicted as the larger host. Other human virus diseases are measles, common colds, the various types of influenza, poliomyelitis, mumps, and the once-dreaded but now essentially conquered smallpox. A growing suspicion that some human cancers are virus-induced is strengthened by the fact that some tumors in other animals and in plants are definitely viral in origin.

The question of the origin of viruses is another of the unanswered biological questions. Viruses were once thought to be examples of an extremely primitive form of life. Their specificity of infection, however, is scarcely a primitive feature, and because viruses are so selective, they are now regarded as *not* primitive.

Bacteria

A varied group of organisms is included in the general category of **bacteria.** They were for a long time considered a single taxonomic group because they are all prokaryotic, having the usual characteristics of prokaryotes. Increasing evidence, however, shows that there are two kinds of bacteria, each group with its own features. The group more familiar to biologists is the *Eubacteria,* the one we will be mainly concerned with in this book. A second group, the *Archaebacteria,* differs from the Eubacteria in a number of fundamental ways.

Archaebacteria do unusual things. They generate methane (marsh gas) or live in unusual habitats (at least for bacteria) such as hot sulfur springs, acid waters, or very salty places. They have their own kind of nucleotide sequences in their ribosomal RNA, special wall-forming compounds, peculiar fatty compounds, a unique method of building amino acids into proteins, and reactions to antibiotics that differ from those of Eubacteria. For these reasons, microbiologists are coming to think that the Archaebacteria should be considered as organisms distinct from all others on earth, and some would regard them as members of a separate kingdom, to be added to the Monera, Protista, Fungi, Plants, and Animals. In this book, we will treat the Archaebacteria as distinct, extremely ancient organisms still provisionally included in the Kingdom Monera.

Distribution of bacteria Bacteria have become adapted to more different living conditions than have any other organisms. They are inhabitants of soil and water, and they exist in enormous numbers on the surfaces of most, probably all, plants and animals. They can be found in Antarctic ice, in hot springs, in the upper atmosphere, in petroleum, in the digestive tracts of animals, and even inside developing seeds.

Some species can withstand the high pH of alkaline springs, the crushing pressures deep in the sea, the enzymatic action of animal digestive systems, lack of oxygen, long and severe drying, and even in some instances a short time in boiling water. Generally, however, bacteria cannot stand long periods at high temperatures, are intolerant of low pH, and cannot grow actively (although they are not killed) without an adequate water supply. With such adaptability, bacteria have so completely invaded practically every bit of the earth's surface, as well as portions well above and below the surface, that the mass of bacterial cells is estimated to outweigh the mass of all the plants and animals. For a convincing demonstration of the ability of bacteria to get into things and grow, try keeping them out of some place where they are not wanted, such as food, wounds, or laboratory materials.

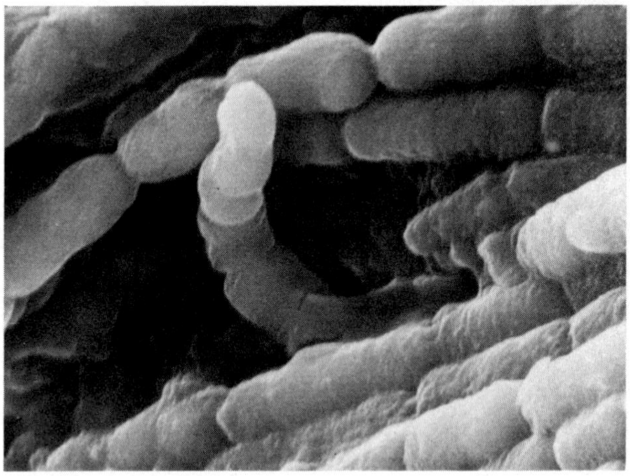

Figure 9.4
Scanning electron micrograph of rod-shaped bacterial cells, or bacilli. About 80 such cells would stretch across the period at the end of this sentence.

Activities of bacteria Covering the metabolic activities of bacteria is impossible in a short space, but we can consider a few functions. Bacteria generally obtain their food by bringing it into their cells through their outer membranes. The food may be in a diffusible form already, as in a sugar solution, or it may be digested by enzymes released from the bacteria. The solvent must be water, and that is why dehydrated foods do not spoil and why salted meats generally, but not invariably, do not suffer bacterial decay. Most bacteria require oxygen, but some can live, though inefficiently, without it. Oxygen inhibits the growth of certain bacteria that live deep in soil or in wounds that do not allow entrance of air. The best known is *Clostridium tetani,* the cause of tetanus, or "lockjaw."

Bacteria are usually *heterotrophic,* requiring an outside food supply, but a few species are *autotrophic,* making their own food. Some of the autotrophs are *chemosynthetic,* obtaining the energy necessary for food manufacture from the oxidation of such materials as sulfur and iron compounds, nitrates, and hydrogen. A few photosynthetic species, such as the purple bacteria (e.g., *Rhodospirillum rubrum*), obtain their energy from light, but their chlorophylls are different from those in higher plants. Further, they may use some substance other than water as a hydrogen source (Chapter 5).

Metabolic products, especially proteins, are exuded from bacteria, and these compounds can make important changes in the bacterial environment. Enzymes can break down materials outside the bacterial cells, and exuded toxins can affect a human or other animal body. Some bacteria exude *antibiotics,* which inhibit the growth of other microorganisms. Antibiotics are used to regulate selectively the effects of bacteria on human and animal health, as well as in many experimental processes in biology.

Identification of bacteria Bacteria are all small. The largest are only about 10 micrometers long, far below the limits of human vision. The smallest are about 0.3 micrometer, which is about at the lower limit of light microscopy (Figure 9.4). Most bacteria are about 1 to 4 micrometers across. They may be spherical *(cocci)* or long and rod-shaped *(bacilli).* Some of the corkscrew-shaped bacteria, called "spiral," may be seen readily with a low-power microscope.

Besides size and shape, one identifying feature of some bacterial cells is the presence of *flagella,* which are extremely slender filaments (Figure 9.5). The position and number of flagella are useful in identifying bacterial types. Bacteriologists also need to know what energy sources a bacterium can use; whether, for example, a given species can digest

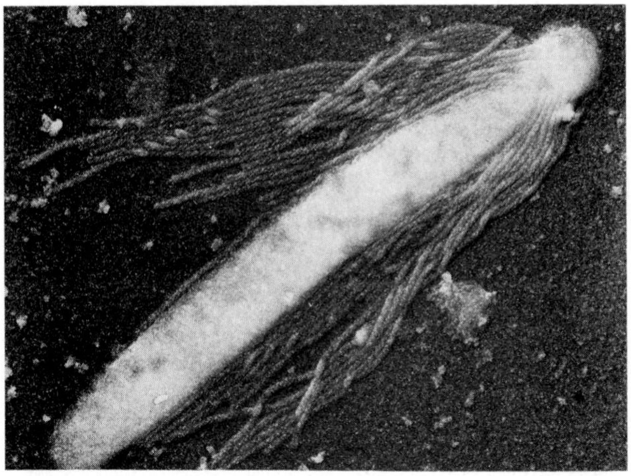

× 21,500

Figure 9.5
Electron micrograph of a bacterium isolated from an African snail. The flagella, clustered at one end, are remarkable in having the appearance of twisted yarn.

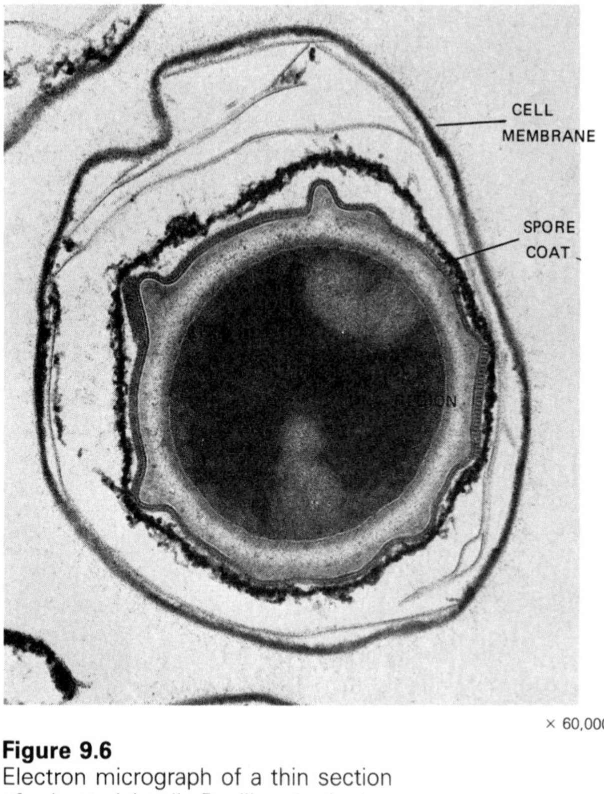

× 60,000

Figure 9.6
Electron micrograph of a thin section of a bacterial cell, *Bacillus thuringiensis,* with a spore formed inside it. The wavy inner layer is the protective spore coat. The clear area inside the spore is the region containing the cell's DNA.

special sugars or acids, or alcohols; and whether or not it produces gases.

Still another identifying feature is a type of cell wall that is peculiar to bacteria. For example, if a cell has a relatively heavy wall, it will absorb and hold a purple stain, even when washed with alcohol. If the wall is thin, the color will rinse out. The procedure, known as the *Gram's stain,* can be used to determine whether a bacterium is Gram-positive (that is, retains the stain because it has a heavy wall) or Gram-negative. Let us say that a given species is described as a motile (having flagella), Gram-negative rod. Then by finding which compounds the species is capable of using for growth, a microbiologist can identify the organism.

Other features used in identification are the shape and texture of a mass of cells known as a **colony,** the presence or absence of pigments, temperature tolerances, and the production of various gases. Unlike most other organisms, bacteria are classified not so much on a basis of how they are constructed as on what they can do.

Structure of bacteria The internal structure of a bacterial cell is that of a usual prokaryote, without mitochondria, plastids, endoplasmic reticulum, or nuclear membrane. Ribosomes are present—otherwise protein synthesis would not be possible—and there is a strand of DNA without accompanying chromosomal protein. Various granules of fatty or carbohydrate reserve materials may be scattered throughout the cytoplasm, and pigments are concentrated in *chromatophores,* which are simpler than the organized plastids of eukaryotic plants. Respiratory activity is carried on in the plasma membrane, which may be infolded into complex masses, remindful of the internal cristae of mitochondria in eukaryotes. Some bacteria have *pili* (singular, pilus), which resemble flagella but are shorter and seem to function in the connection of bacterial cells to surfaces or to other bacteria during mating. Further, some bacteria can build an internal, firm body known as a **spore,** which is much more resistant to heat and drying than the usual vegetative cell (Figure 9.6). Spore-forming species are harder to kill than non-spore-forming species. They may persist for years in dried soil or in clothes that have been in contact with soil.

Reproduction of bacteria Bacterial reproduction is mostly asexual. The so-called circular chromosome duplicates itself, and then the two strands separate. This move is followed by a constriction around the cell as though a thread were drawn tight around it,

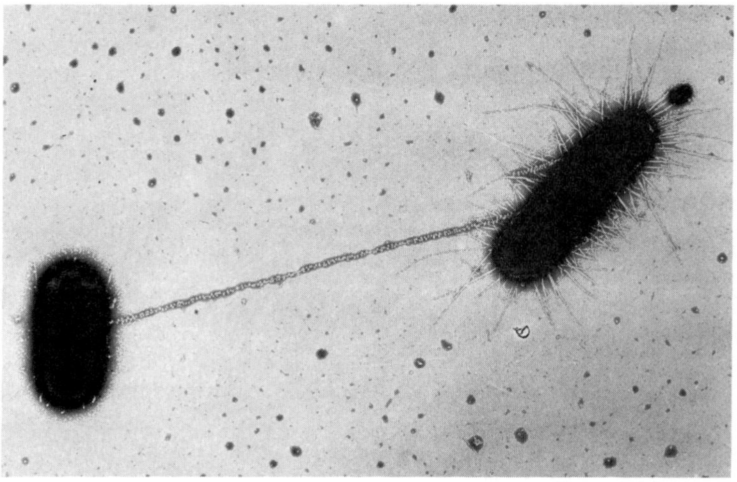

Figure 9.7
Mating in bacteria. Instead of fusing two cells into one, bacteria form a protoplasmic bridge from one cell to another, and bacterial nuclear material is sent through the connection, called a *pilus*. Since the mating time of a bacterium can be about an hour and a half, and the life span of a vegetative (nonreproducing) bacterium in good growing conditions is only about half an hour, mating lasts three times as long as the ordinary life of the individual.

and two cells result. Under optimal conditions of temperature, space, and nutrients, a bacterium can divide in about 20 minutes. Sometimes bacteria mate. A protoplasmic bridge is formed between a pair of cells, and the DNA of one passes into the other (Figure 9.7). It was the discovery of this action of bacteria that led to our realization that DNA is the important genetic material.

Destructive action of bacteria Bacteria are best known to most people because of their ability to cause disease, and indeed some of the most destructive scourges in human history have been the result of bacteria, or as they are popularly called, "germs." By far the most deadly scourge was the Black Death, or simply the *plague,* which is estimated to have killed perhaps half the population of continental Europe during the fourteenth century. In England, where as many as two-thirds of the people died of plague in the seventeenth century, the action of a microbe, *Yersinia pestis,* brought about social, economic, and political change. The shortage of laborers resulting from the plague made ordinary workers really valuable for the first time in human history, and their social status rose in a way that has never been reversed. Another effective killer was *Rickettsia prowazekii,* the causal agent of typhus. Both typhus and plague are dangerous when people must live in crowded, unsanitary conditions, where rats and their accompanying fleas can carry the bacteria from victim to victim. The most effective way to prevent these diseases is to understand how they are contracted and clean out the rats with their fleas and rickettsias.

Other bacterial scourges of mankind have been diphtheria, caused by *Corynebacterium diphtheriae;* tuberculosis, caused mainly by *Mycobacterium tuberculosis;* and typhoid fever, caused by a *Salmonella* species. Other *Salmonella* species cause a food poisoning, sometimes lethal, that is popularly but inaccurately known as ptomaine poisoning. The really dangerous food poisoning is that caused by the toxin of *Clostridium botulinum.* The botulism poison is so fantastically potent that a person could be killed by as little as one fifteen-millionth of a gram, certainly not enough to see and so little that the residue on a clean-looking spoon could be lethal. Fortunately, it is uncommon, and it can be rendered harmless by heat.

Plants as well as animals are susceptible to bacterial disease. One of the first identified and most serious is a destructive malady of pear trees known as fire blight. Despite the fact that several hundred bacterial diseases of plants are known, however, bacteria are not the real threat to plants. As we shall see, fungi are to plants what viruses and bacteria are to animals.

Protection against bacteria Means of combating bacteria are available. Living animals have special ways of preventing or overcoming bacterial infection through a chemical method of developing *immunity.* The subject of immunity is so important that it deserves special treatment, and it will be discussed in detail in Chapter 24.

Sulfa drugs and antibiotics are used to protect animals against bacteria. These drugs inhibit bacterial growth, although in different ways.

In processing food, bacteria and fungi (see p. 170) must be kept from digesting what we would like to preserve. Common methods are freezing (keeping the temperature so low that bacterial enzymes cannot function), canning (killing bacteria by heat and then keeping live ones from entering the product), salting or candying (lowering the water content of the product by addition of salt or sugar), pickling (lowering the pH to a point below the tolerance limits of bacteria), or dehydration. When human consumption of a product is not expected, preservation can be ensured by the use of a poison such as alcohol, phenol, or any of a number of synthetic chemicals.

Usefulness of bacteria On balance, the usefulness of bacteria far outweighs the damage they do in causing disease and in destroying valuable materials. Without bacteria, the life of this world as we know it could not go on. The biological importance of bacteria lies in their ability to release enzymes. All biological products can be broken down by bacterial enzymes, with the result that bacteria, with assistance from fungi, maintain the cycling of materials in the living world. The enormous amounts of cellulose, lignin in wood, proteins, and fats that are produced by the higher plants could make life on earth impossible if bacterial enzymes were not available to return the oxygen, carbon, nitrogen, phosphorus, and sulfur to the air, water, and soil, where they can be used over and over. Only certain manufactured substances, such as plastics and synthetic fibers, are not *biodegradable* (capable of being decomposed by natural biological processes), but there is a fair likelihood that sooner or later, given the versatility of bacterial evolution, even such refractory substances as nylon may be attacked by some enzyme. When that happens, chances are that the enzyme will be of bacterial or fungal origin.

Bacteria are used commercially in many fermentations: butter and cheese making; the curing of tea, tobacco, and cocoa; the cleaning of flax fibers for linen; the production of a long list of organic chemicals; and the commercial production of such antibiotics as bacitracin and streptomycin. One feature of bacterial activity that helps in human economy as well as in the general economy of nature is the ability of bacteria, especially in the Genus *Rhizobium,* to take nitrogen (N_2) from the atmosphere, reduce it to ammonia (NH_3), and incorporate the ammonia into amino acids and then into protein. The whole process is called **nitrogen fixation.** Nitrogen gas is not usable by most organisms, and yet nitrogen is essential to all cells. Consequently, the nitrogen fixation by rhizobia in the roots of leguminous plants gives these plants an advantage (Figure 9.8). Rhizobia cannot use atmospheric nitrogen as long as they live free in soil, but they can work their way into the root hairs of susceptible plants, such as beans, clovers, alfalfa, and locust trees. Once rhizobia are in a root, a colony develops, and the host plant grows small tumors a millimeter or two thick. In those tumors the bacteria begin a new life, trapping atmospheric nitrogen and reducing it to ammonia. Nitrogen thus fixed can be used by both the bacteria and the host plant to make any of the nitrogen-containing compounds that are essential to life.

Actinomycetes One subgroup of bacteria, the actinomycetes, deserves mention because the group is the source of many useful antibiotics (streptomycin, chloramphenicol, the tetracyclines) and because members of the group are so universally distributed in soil. Actinomycetes give off a volatile substance with a characteristic, familiar odor that gives soil its distinctive smell. Any gardener—indeed, any child who has "played in the dirt"—knows the smell of fresh, moist soil. That smell is not soil; it is actinomycetes.

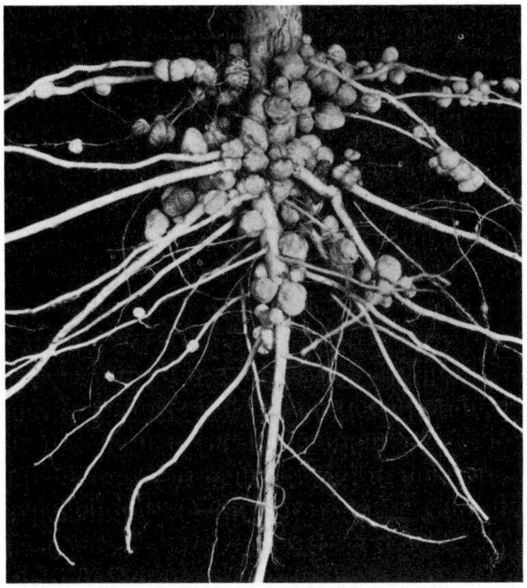

× 3

Figure 9.8
Nodules on the roots of a soybean plant. Each nodule contains millions of bacteria that can use atmospheric nitrogen for the building of organic nitrogen compounds. If the nodules are left in the ground after the plant is harvested, the soil can have its usable nitrogen content increased.

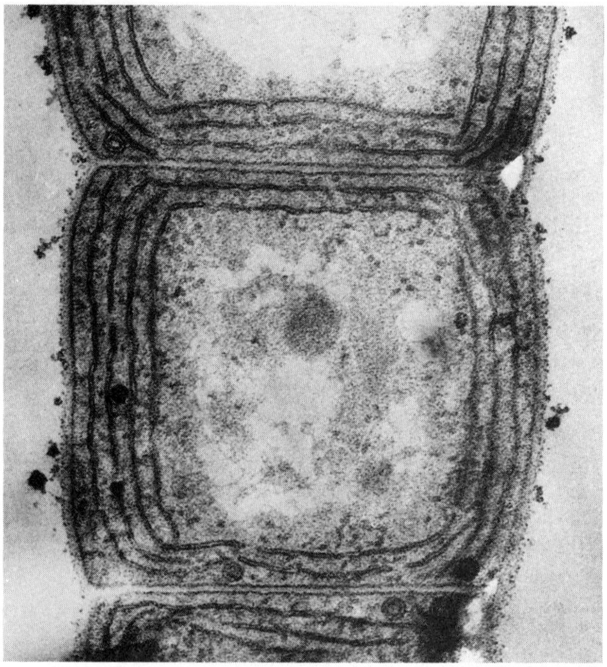

Figure 9.9
Electron micrograph of a thin section of a cell of the blue-green alga *Anabaena.* There is an outermost plasma membrane. Inward from that are other membranes, which bear the photosynthetic pigments. The DNA of the cell is only faintly visible in the center.

Blue-green Algae

The **blue-green algae*** (cyanobacteria) are prokaryotic in their lack of membrane-bound nuclei and of most other cell components typical of eukaryotes, but they differ most from bacteria in their ability to carry on photosynthesis with the release of oxygen from water, as higher plants do (Figure 9.9). The blue-greens frequently grow into chains of cells. Each cell is physically separate from the others, not united by protoplasmic connections as those of typical plants are. For this reason, some biologists believe a blue-green filament is merely an aggregation of individual cells, not a multicellular organism. Nevertheless, some blue-greens produce specialized cells at specific intervals along the filament, thus showing a simple but definite power of differentiation—a characteristic of multicellular organisms (Figure 9.10). The blue-greens may be more than just a cluster of cells.

Although they are called blue-green, many members of the group are in fact red, because in addition to chlorophyll *a* and a blue pigment, they have a red pigment. Some blue-greens have pigments also in their outer coverings, giving the organisms a range of colors that can become strikingly visible. The Red Sea, for example, is so named for the occasional masses of red-pigmented "blue-green" algae that accumulate there, and the sulfur-bottomed whale has a yellow belly because of the growth of these algae.

Blue-green algae are almost as adaptable as bacteria to different habitats. They can stand temperature ranges from ice to hot springs. Much of the colored deposits around the Yellowstone hot springs is an accumulation of blue-greens. Although they are so common in soil as to be practically ubiquitous, blue-green algae are typically aquatic. Besides being difficult to control or to filter out of water, they can give drinking water a vile taste and smell, and some are toxic enough to make people and animals sick. On the positive side, blue-greens are like the rhizobia in their ability to make atmospheric nitrogen available for amino acid and protein production. In some crops, especially rice, blue-greens provide a free source of nitrogen, in part replacing artificial fertilizer.

Coupled with their adaptability, the asexual reproductive effectiveness of the blue-greens has made them biologically successful. They can live almost anywhere, and they can proliferate at such a rate as to produce enormous masses of cells in a short time. Although some recombination of characteristics has been reported, indicating an exchange of genetic material between strains, actual mating between cells of blue-greens has not been seen. Blue-greens seem to lack true sexuality. Mating may take place, but no one has yet observed it.

* The word "algae" has no technical meaning, but it is commonly used by biologists and laymen to refer to small aquatic plants of many sorts, as well as to the large seaweeds. Any photosynthetic organism that is not a mosslike, fernlike, or seed-bearing plant is likely to be called an alga, just as insects are called "bugs" and whales are called "fishes" without regard to taxonomic niceties.

TO SUM UP MONERA

1. The Monera have the simplest structures of all living things.

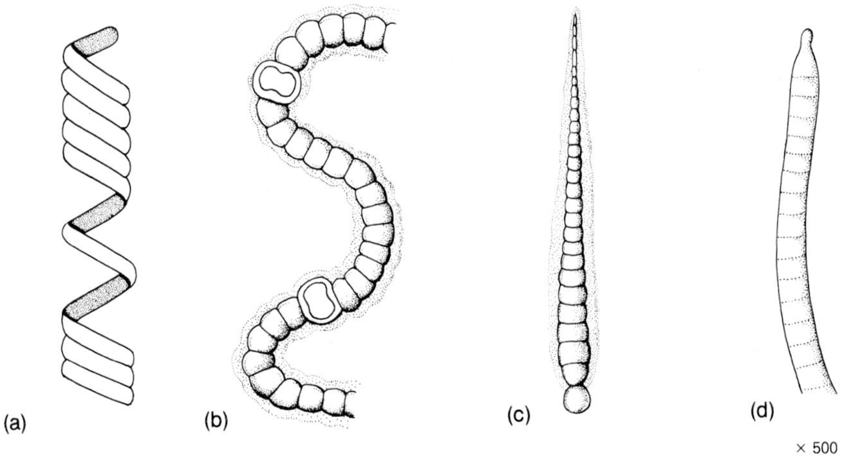

Figure 9.10
Representative blue-green algae. (a) *Spirulina,* a single coiled cell. (b) *Anabaena,* with enlarged, empty cells. (c) *Rivularia,* one filament of a mass of filaments. (d) *Oscillatoria,* the commonest of the blue-greens, capable of a swaying movement.

2. *Viruses* and *viroids* have no life of their own, but can inject their DNA or RNA into living host cells.
3. *Bacteria* are prokaryotic and usually are unable to manufacture their own food. Reproduction is mostly asexual. Many bacteria cause disease in human beings and other animals.
4. *Blue-green algae* are capable of photosynthesis and are almost as adaptable as bacteria. They reproduce asexually at extremely high rates.

THE KINGDOM OF THE PROTISTA

The **Protista** are structurally more complex than the Monera, and we can find features that show how more advanced organisms evolved. We also find greater diversity of form and can therefore make reasonable guesses as to how they diverged during their development from primitive ancestors.

As is true of the Monera, the Protista consist of a heterogeneous assemblage. They are eukaryotic. As we will see, Protista exhibit a bewildering diversity of shapes, habits, and chemical components. Many are photosynthetic; others strictly heterotrophic. Some Protista form aggregations of loosely connected cells called *colonies.* Some of these colonies show very simple specialization and integrated control of some cells and thus constitute a primitive multicellular organism. But most Protista live their lives as single-celled individuals. Some of the phyla of the Protista are groups popularly called algae. The word "algae" (singular, alga) has no technical significance but is commonly applied to any plantlike organisms that are photosynthetic but structurally too different from ordinary land plants to be included with them. The various phyla grouped as Protista are probably not closely related in evolutionary history. They are thought to have diverged billions of years ago and to have developed along independent lines. Today they are far enough apart to be well-established, peculiar groups.

Phylum Protozoa

The single cell of a **protozoan** such as a *Paramecium* is a complete, functioning organism. Like most microorganisms, protozoa exist in any habitat that provides a source of nourishment and liquid water: in soil, fresh and salt water, in and on plants and animals, even floating (in dried form) in air. Most protozoa are microscopic, ranging from about 5 to 100 micrometers, but some large ones are several millimeters across and could pose a threat to a small beetle. Four classes of protozoa are recognized:

1. The *Flagellata,* so called because they have one or more flagella. These organelles, constructed like those of higher animals, have a membranous covering and microtubules. Many members of this group are photosynthetic, and they are as much like plants

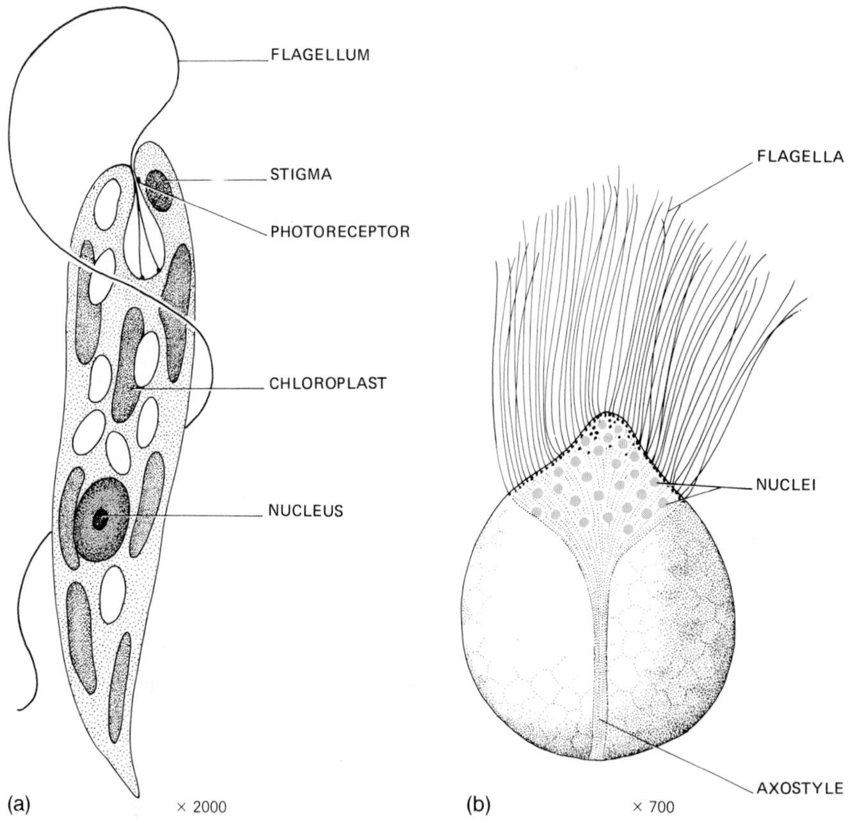

Figure 9.11
Examples of flagellated protozoans. (a) *Euglena*, a photosynthetic, free-living cell with a red stigma, or "eye spot," close to a light-sensitive photoreceptor. The flagellum is anchored at the base of a gulletlike sac. Green chloroplasts manufacture a special kind of starch. (b) *Calonympha* from the gut of a termite, where, with the help of bacteria, it lives on the cellulose eaten by the termite. There are many flagella and many nuclei, as well as a central stiffening rod, the axostyle.

as like animals (Figure 9.11). Members of the genera *Euglena, Pandorina,* and *Volvox* are frequently considered to be plants, but nongreen flagellates are clearly heterotrophic. One flagellate, *Peranema,* could be called a colorless *Euglena.* Most flagellates are free-living in soil or water, but some, the trypanosomes, are parasites, causing, among other diseases, African sleeping sickness. The parasite is passed from one person to another via the biting tsetse fly.

2. The *Sarcodina.* The members of this class have no specific organelles for locomotion. They extend blunt or slender cytoplasmic points *(pseudopods),* and then the cell body follows the pseudopods, which can be pushed out and retracted repeatedly. Amoebas are familiar to all biology students as free-living dwellers in water. Some amoebas can cause human dysentery. Marine shelled sarcodinians through geological time have deposited billions of skeletons, which now make layers on the bottom of the sea (Figure 9.12). The "radiolarian ooze" or "Globigerina ooze" is studied by oil prospectors because the presence of certain species gives clues to possible petroleum deposits.

3. The *Sporozoa.* The cells of sporozoans are usually nonmotile and parasitic (Figure 9.13). The best-known members of the group are those that cause malarial fever. The word *malaria,* which means "bad air," refers to the prevalence of the disease in swampy regions, where mosquitoes thrive. Once it was known that mosquitoes can carry the parasites and inject them into people, prevention of the disease became easier. When mosquito control is impossible, the fever can still be controlled by the use of an ancient folk remedy: quinine from the bark of cinchonas, the Peruvian fever trees. Modern substitutes, such as quinacrine (Atabrine), are synthetic.

4. The *Ciliata.* Protozoa with numerous short cilia, so named because of the resemblance of cilia to eyelashes (the word really means "eyelid"), are so different from other protozoa that some biologists classify them as a distinct subphylum. They use their accurately synchronized cilia either to move themselves through water or, if they are attached, to bring water containing food particles near or into the protist's body. Many ciliates have a groove or depression called a *gullet* into which food can be brought, and although no mouth opening is pres-

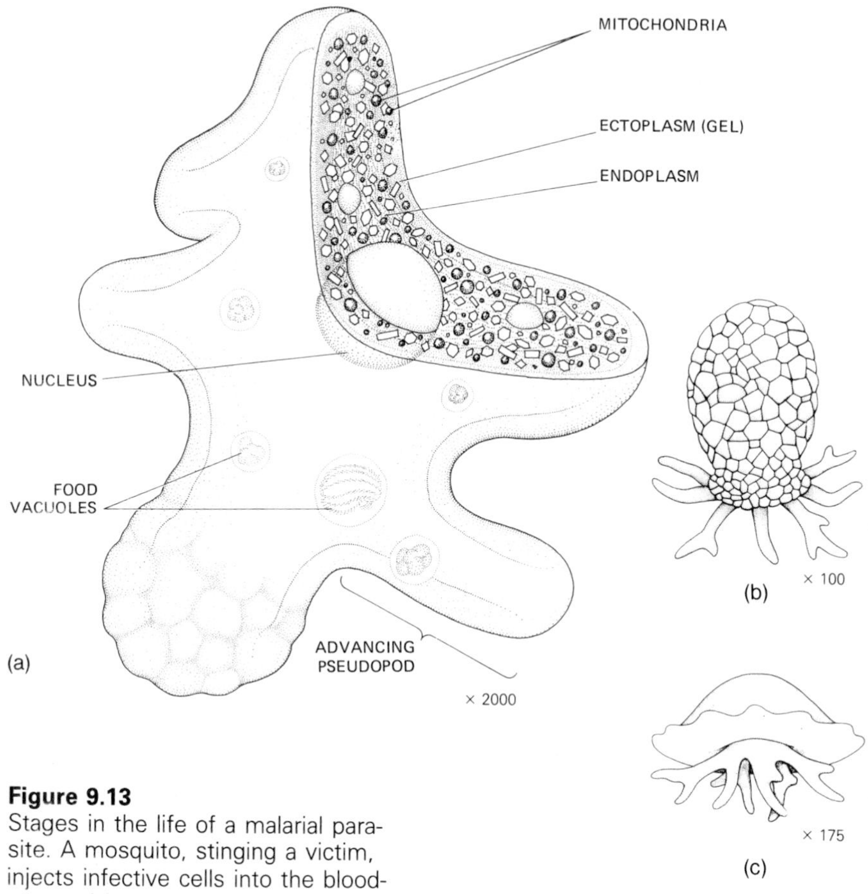

Figure 9.12
Some representative Sarcodina.
(a) *Amoeba proteus,* the common free-living amoeba. (b) *Difflugia,* with a covering shell of sand particles stuck to the outside of the animal's cell. (c) *Arcella,* a sarcodinian that secretes its own shell. (d) Shells of various sarcodinians dredged from deep-sea sediments.

Figure 9.13
Stages in the life of a malarial parasite. A mosquito, stinging a victim, injects infective cells into the bloodstream (a). The infective cells are circulated in the blood until they reach the liver. After penetrating the liver cells (b), the malaria cells multiply (c), and can repeat this multiplication again and again. Parasite cells escape from liver cells (d), and then reenter the blood, where they attack red blood cells (e) and multiply enormously (f). When great numbers of malaria cells break out of red blood cells (g), they cause the periodic chills and fever typical of the disease. This cycle of entry, reproduction, and breakout is repeated at definite intervals, depending on the species of parasite. Some of the parasites become potential gametes, which can be sucked up by a female mosquito taking blood (h, i). In the mosquito's stomach, gametes (j, k) fuse to form a zygote (l), which moves (m) and bores into the mosquito's gut wall (n), where it multiplies (o). When the cyst containing numerous malaria cells ruptures, the parasite cells float in the mosquito's body fluid (p). Some of them find their way to the salivary glands (q), from which they can enter a new warm-blooded host when the mosquito bites it.

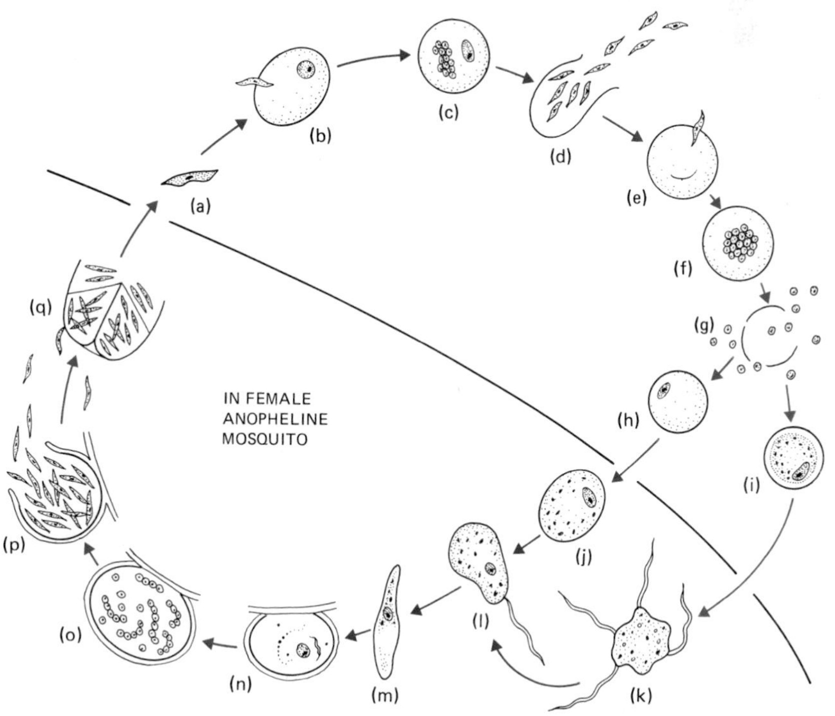

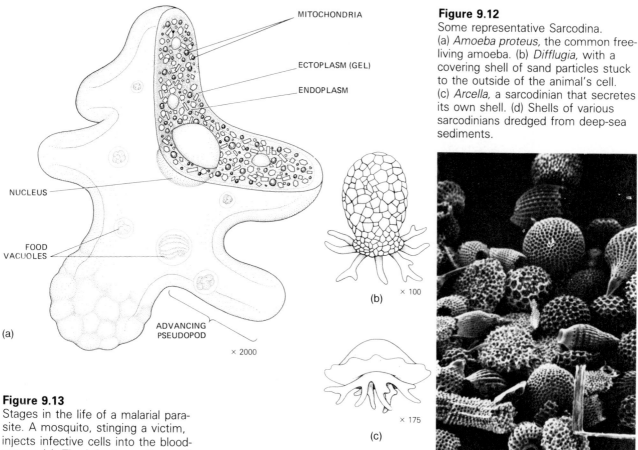

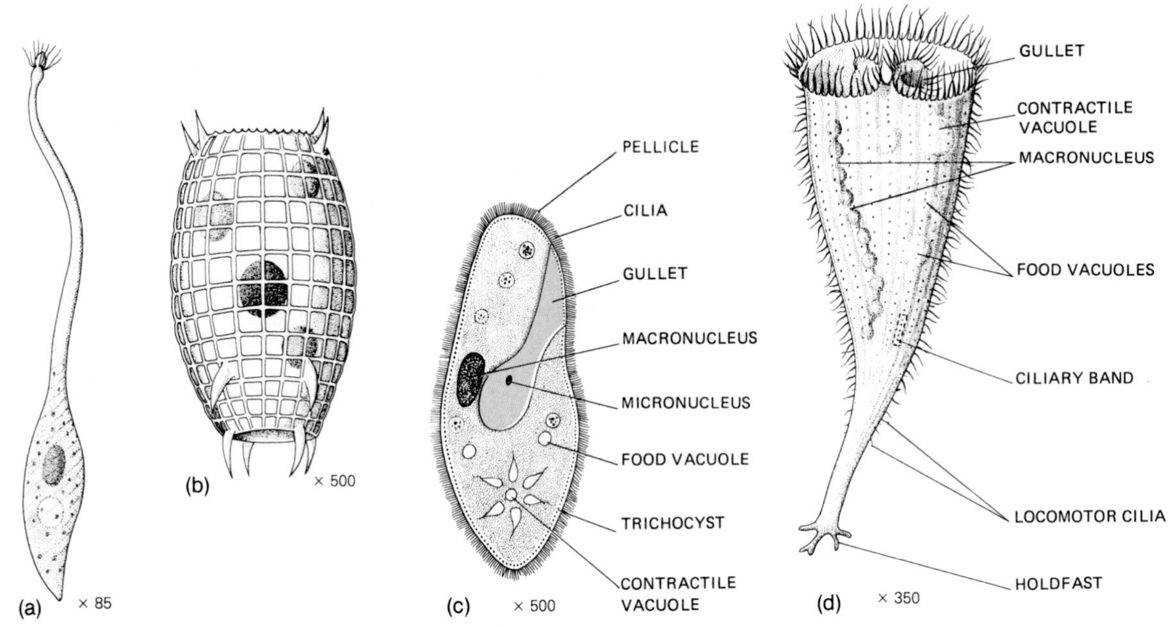

Figure 9.14
Some representative ciliated protozoans. (a) The teardrop-shaped *Lachrymaria.* (b) A spiny, firm-bodied *Coleps* from fresh water. (c) The common "slipper animalcule," *Paramecium,* one of the most complex cells. Food is trapped in the gullet, from which it is taken into the cell. Small particles are circulated in the cytoplasm in food vacuoles. There is a larger macronucleus and one or more smaller micronuclei. Water content within the cell is regulated by a pair of contractile vacuoles, which alternately fill by way of star-shaped canals. One such vacuole is shown at the posterior end being filled. Under the exterior covering, the pellicle, are thousands of trichocysts, like microscopic darts that can be shot out to aid in the capture of prey. (d) *Stentor coeruleus,* a blue ciliate.

ent, solid bits can be brought inside the body by a process of engulfing (Figure 9.14).

Most reproduction is asexual, but mating between compatible strains of ciliates occurs, with a resulting exchange of genetic material. Mating does not include increase in numbers. Mated individuals come together, go through an elaborate set of nuclear exchanges, and then separate. However, the main genetic effect of sex is there: the possibility of recombination. Their ready availability in almost any drop of pond or gutter water, their unexpected shapes, their colors (usually rather gray, but occasionally pink or blue), their liveliness, their elaborate and controlled behavior, their unparalleled cytoplasmic complexity, and their nutritional requirements have served to make the ciliates subjects for investigation by biologists for several centuries.

Phylum Chrysophyta

Chrysophyta comes from the Greek meaning "golden plants." These organisms are named after their conspicuous yellow pigments, the carotenoids. However, they do possess two forms of chlorophyll, *a* and *c*, and can photosynthesize effectively. One group of the Chrysophyta, the **diatoms** (Greek, "two cuts," from the double nature of the walls), have been for millions of years some of the commonest inhabitants of lakes and seas. The diatoms are unique in the structure of their shells, called valves, which fit like the halves of a Petri dish, and in their gliding manner of moving (Figure 9.15). Their cell walls, which are impregnated with glass-like silicon, are practically indestructible, and they pile up in thick layers on ocean bottoms, where they build up diatomaceous earth. Diatoms are commercially mined, and the diatom-rich earth is used as insulation, polishing powder, or filtering materials. The walls are so minutely and delicately sculptured that they have long been used as test objects in determining the quality of microscope lenses. Because they are very common and because they store reserve food in the form of fat, diatoms are one of the main sources of food in seas. The fat has a peculiar

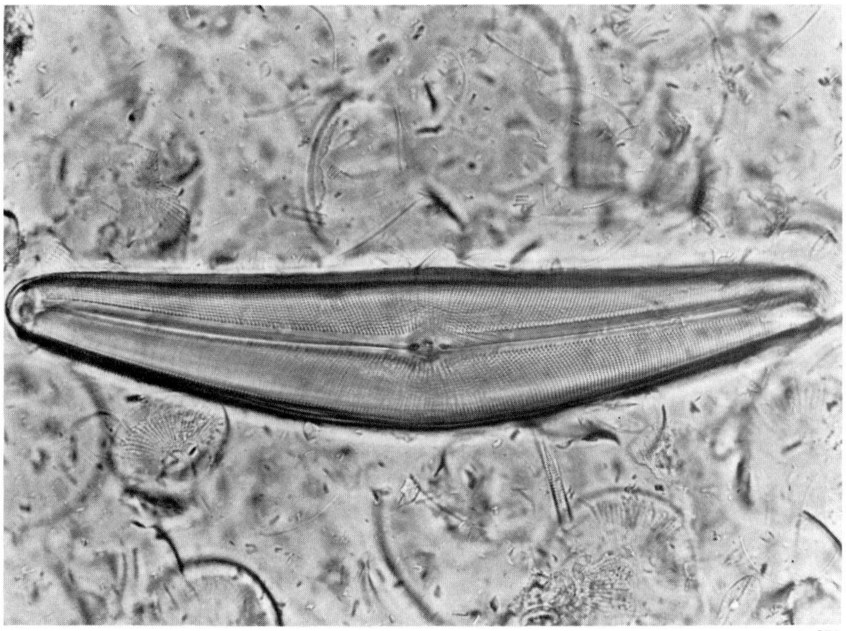

× 850

Figure 9.15
Representative diatom. Diatom shells are so nearly indestructible that they remain in soil and water indefinitely. When alive, many of the diatoms are capable of moving along like microscopic boats. *Navicula,* a common boat-shaped diatom.

flavor and odor that it retains even after it is eaten by animals. It can be accumulated in such quantity that it gives fish their "fishy" taste. Compare the flavor of a freshwater trout, whose ultimate food source is *not* diatoms, with that of a marine mackerel, whose food supply *is* largely diatomaceous.

Phylum Pyrrophyta

The **Pyrrophyta** (meaning "flame-colored plants") are microscopic aquatic algae, mostly with two tinsel-type flagella attached in a characteristic manner (Figure 9.16). They are important in the ecology of the sea because, like several other protists, they furnish food for small animals, which are in turn eaten by larger animals. Pyrrophyta thus serve as a base supply for commercial fishes. Having chlorophylls *a* and *c* plus carotenoids, they are photosynthetic. A dramatic effect of one class of Pyrrophyta, the dinoflagellates, is produced when these organisms are exposed to air. A churning boat propeller, the swish of an oar, or the breaking of a wave will cause a dinoflagellate to emit a flash of light. When millions of these creatures flash at night, they can make the sea glitter with a luminosity that sailors have long called "the burning of the sea." (See the essay on bioluminescence on p. 42.)

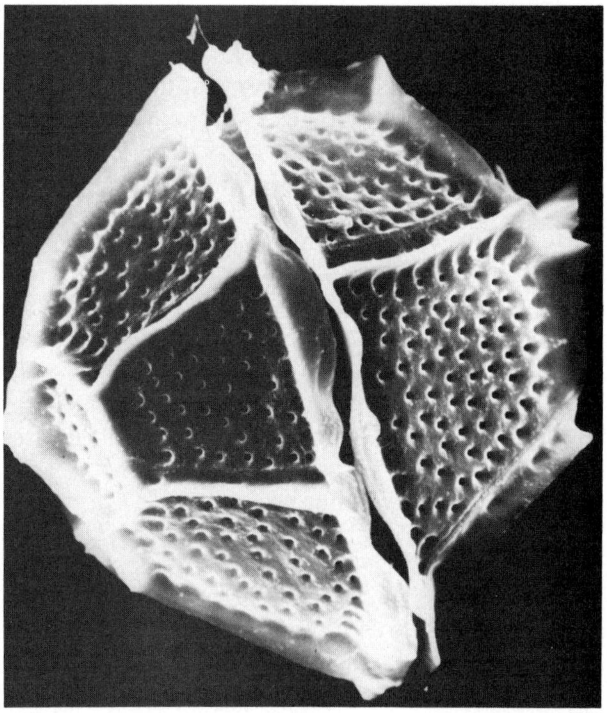

× 2000

Figure 9.16
An example of the Pyrrophyta, *Gonyaulax,* as seen in a scanning electron micrograph. The wall plates are conspicuous, but the flagella do not show.

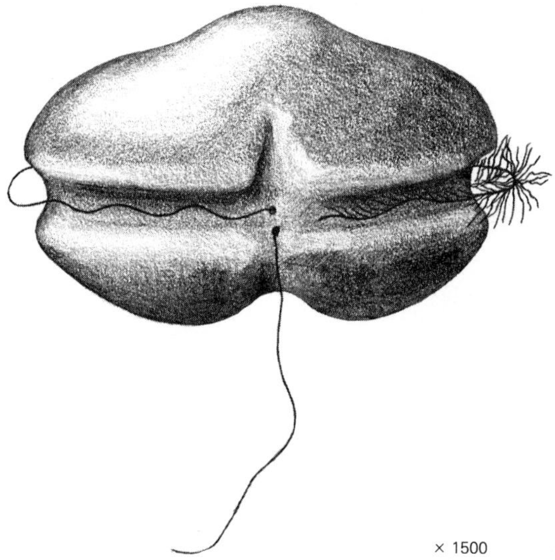

Figure 9.17
One of the toxic Pyrrophyta, *Gymnodinium brevis.* This is an organism whose poison can kill fish by the million in Florida, where they are sometimes numerous. It is responsible for the massive blooms known as red tides.

Another striking activity of dinoflagellates is the production of the so-called red tides by infestations of red-pigmented species. Under certain still undetermined conditions, some species of dinoflagellates multiply enormously until the very water becomes visibly colored (Figure 9.17). The trouble, from a human view, is that these minute creatures synthesize a powerful nerve poison that can kill vertebrates. With a burst of reproductive speed, the dinoflagellates become so numerous that they can furnish the main food supply for small crustaceans (shrimp and water fleas) and molluscs (clams and oysters), which feed by filtering the dinoflagellates from the water. These primary feeders are not harmed by the toxin, but people or fish can be poisoned if they eat smaller animals that have fed upon such dinoflagellates. During a severe red tide, countless dead fish may be cast up on the beach, as they sometimes are on the Florida Gulf Coast. California has endured many red tides, and occasionally Atlantic infestations are carried as far north as New England. American Indians in precolonial days were aware of the danger and used to alert their contemporaries by lighting warning fires along beaches when the unseen dinoflagellates contaminated the sea and sea life.

TO SUM UP PROTISTA

1. The Protista show a greater diversity of form than the Monera.
2. *Protozoa* are single-celled, complete, functioning organisms. Protozoa may be classified as *Flagellata, Sarcodina, Sporozoa,* or *Ciliata.*
3. *Chrysophyta* have two forms of chlorophyll, and photosynthesize effectively. One common group is the diatoms.
4. *Pyrrophyta* are microscopic aquatic algae that provide food for small sea animals and cause red tides.

THE KINGDOM OF THE FUNGI

Fungi look more like plants than like anything else, and they do have two definitely plantlike features: firm cell walls and reproductive bodies called *spores.* But the cell walls usually do not contain cellulose, being composed largely of *chitin.* Chitin, a polymer of an amino sugar, is more characteristic of insect bodies than of plants. Further, the spores of fungi, especially the sexually produced spores, are different from the spores of familiar higher plants in appearance and method of production. The texture of such a fungus as a mushroom is pithlike, but the cells of mushroom tissue are formed by the massing of filaments rather than by the kinds of cell division used by either plants or animals. The starchlike glycogen commonly stored by fungi as a reserve food is like the glycogen of animals, but unlike most animals, fungi do not eat. Like bacteria and tapeworms, fungi obtain nourishment by absorbing predigested food from the environment. Even the chromosomes and nuclei of fungi are peculiar to the group. The chromosomes lack the proteins characteristic of other eukaryotes, and mitoses are completed within a nuclear membrane. Like animals, fungi have centrioles, but unlike animals, they do not separate new cells by constriction of the cell coating. For all these reasons, current opinion favors the classification of fungi in a separate kingdom.

The Phycomycetes

One heterogeneous class of fungi with cellulose walls and *coenocytic hyphae,* the **Phycomycetes,** contains several loosely related groups of fungi. *Hyphae* (singular, hypha) are the filamentous strands that fungi generally have instead of the compacted cells found in the tissues of higher organisms; *coenocytic* means that instead of having distinct cells, each with its own nucleus, the hypha is a continuous strand without cross walls. A mass of fungus hyphae is a *mycelium.*

Phycomycetes commonly reproduce asexually by producing cells that can swim or that are blown about, later to germinate into a new fungus. Sexual reproduction involves fusion of nuclei, but in many species, nuclei will join only if they are of proper "mating types." Since they are neither male nor female, they are called "plus" or "minus" types. A mycelium derived from a single spore of such a species is incapable of sexual reproduction.

Several Phycomycetes have been devastating to plants, causing enormous difficulties. The downy mildew of grapes, *Plasmopara,* practically destroyed the French wine industry for a while in the 1880s. The blight that struck potatoes in Ireland during the late 1840s was caused by the Phycomycete *Phytopthora infestans.* It was so destructive that widespread famine resulted. The virtual elimination of the main staple food and the poverty and loss of energy that followed, coupled with poor sanitation and limited knowledge, created a double hazard of starvation and disease, much of it cholera and typhus, caused by other microorganisms. Because many were dying and those who could escape the country did, the Irish population was halved in about 10 years, and it has not fully recovered even yet. The emigrants made their presence felt all over the world.

The Ascomycetes

The distinctive feature of the **Ascomycetes** is the production of spores within a single cell, with a sexual fusion of nuclei followed by immediate meiosis and rounding up of usually eight spores with haploid nuclei. The spore-carrying cell, known as an *ascus* (plural, asci), is the only characteristic common to all the Ascomycetes. The structure, distribution, size, and metabolic habits of Ascomycetes are so varied that no simple description can be adequate (Figure 9.18). Unlike the Phycomycetes, the Ascomycetes have cells of their hyphae compartmented, with one or two nuclei in each cell. Their walls are impregnated with chitin. Besides sexually producing spores (the *ascospores,* in asci), most Ascomycetes also form *conidia* by pinching off a piece of a

Figure 9.18
Fruiting bodies of Ascomycetes. (a) Scanning electron micrograph of fruiting bodies of a fungus (*Peziza*). (b) Diagram of a section through a fruiting body, showing five asci (in a real one there are dozens or even hundreds), each with eight ascospores. In different species, asci contain anywhere from four to hundreds of spores.

(a)
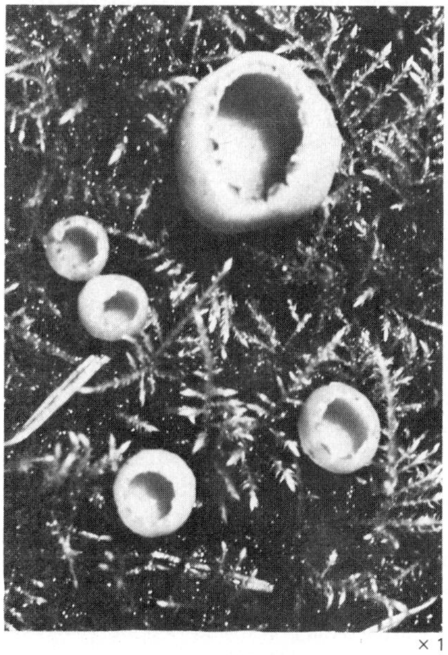

(b)
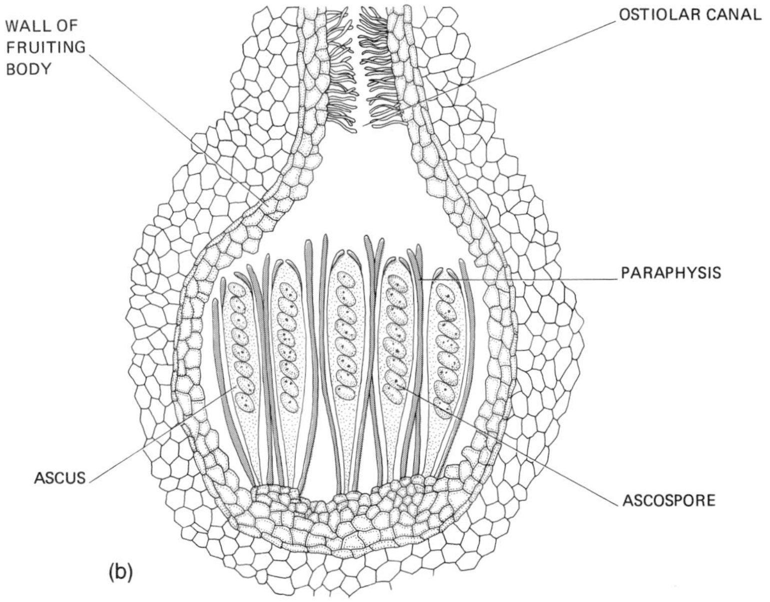

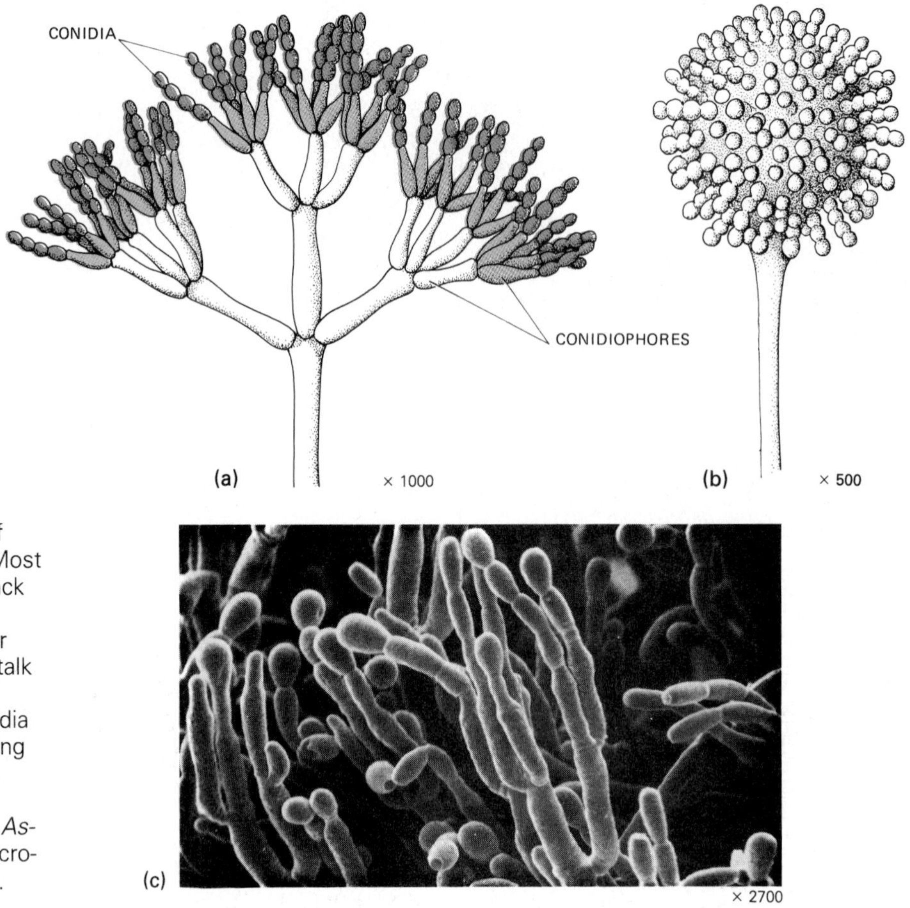

Figure 9.19
Penicillium and *Aspergillus,* two of the commonest imperfect fungi. Most of the blue, green, brown, and black molds that grow on fruits, paper, leather, and cloth belong to one or the other of these genera. (a) A stalk (conidiophore) of *Penicillium* with chains of *conidia.* Only a few conidia are shown, but a vigorously growing mold has thousands of conidia on such a "head," with hundreds in each chain. (b) A conidial head of *Aspergillus.* (c) Scanning electron micrograph of a few *Penicillium* conidia.

hypha containing a nucleus and enough cytoplasm to make an asexual reproductive body. Indeed, some Ascomycetes have gone so far in the use of conidia that they seem to have lost the ability to form ascospores.

Failure to find ascospores in a growing fungus is no guarantee that the sexual phase is actually lacking; it may be that no biologist has found the proper mating types. But extensive searches have frequently failed to yield any ascospores, so that some strains of fungi under study appear to remain consistently asexual. Such strains are called *fungi imperfecti,* the imperfect fungi (Figure 9.19). Enough imperfect fungi have produced ascospores on occasion to warrant the guess that most of them are in fact Ascomycetes. Conidia of imperfect fungi are so widely distributed that they are probably second only to bacteria in being present everywhere. There is hardly a gram of soil or square centimeter of surface, whether of plant, animal, or inanimate object, that does not carry a load of conidia, ready to germinate into a vigorous mycelium. These are the common blue, green, brown, pink, violet, and yellow "molds" that flourish wherever moisture and a few molecules of nourishment are available.

Yeasts are microscopic Ascomycetes. Of the hundreds of wild yeasts that live wherever fungi can grow, including oceans and rivers, most are of no economic use. However, one species of inestimable value is *Saccharomyces cerevisiae,* baker's or brewer's yeast (Figure 9.20). Mixed with flour and a little sugar, the yeast cells can respire anaerobically, yielding carbon dioxide that lightens bread. In fruit juice, the same respiratory activity produces alcohol (Chapter 5).

On the negative side, Ascomycetes are responsible for some diseases. Very few women live a whole life without ever having a vaginal infection by some yeastlike fungus. And various kinds of itch, athlete's foot, and "ringworms" are due to fungi, most of them Ascomycetes. Wild and domestic animals have their share of fungus infections, too, but it is in plants that these parasites cause their greatest disasters. Ascomycetes and their relatives, the im-

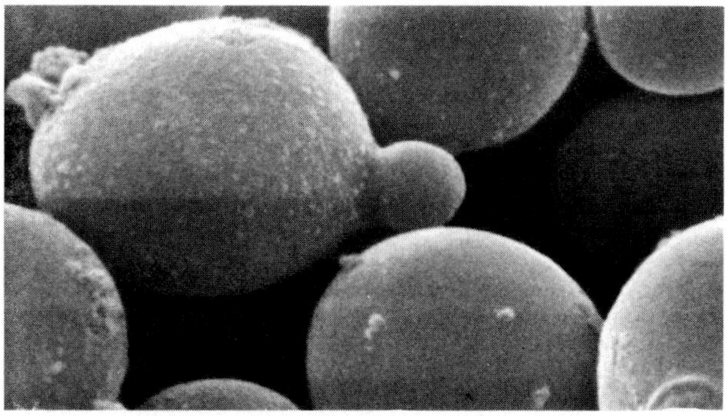

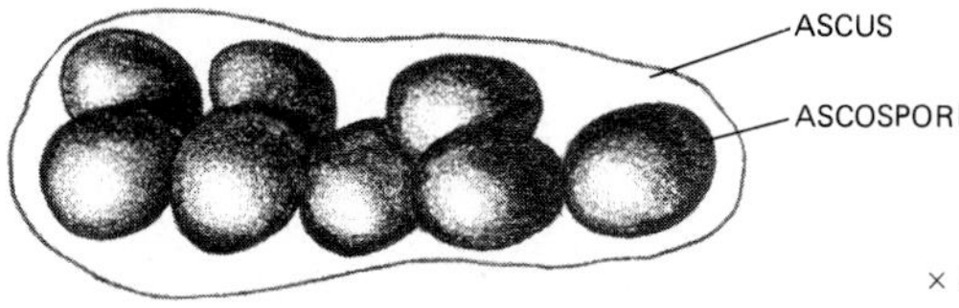

Figure 9.20
Growth and reproduction of yeasts. (a) Scanning electron micrograph of cells of baker's yeast, *Saccharomyces cerevisiae.* Most yeasts grow new cells by putting out a *bud,* which increases in size until it is as large as the parent, and then breaks away. (b) Spore formation in a yeast. In this species (*Schizosaccharomyces octosporus*) there are eight spores. The yeasts, superficially unlike most fungi, are classified with the Ascomycetes because of the way they form spores. Compare this picture with the asci and ascospores in 9.18(b), and note the similarity.

perfect fungi, produce such an array of mildews, blights, cankers, wilts, spots, necroses, and rots that it is a wonder there is anything left for the insects, birds, and rodents, let alone people. The great American chestnuts, which filled the forests of the eastern United States at the beginning of this century, are reduced to a few sickly suckers, all doomed to an early death because of the chestnut blight fungus, *Endothia parasitica.* Row after row of the stately elms that graced the towns of New England are dead, killed by another Ascomycete, *Ceratostomella ulmi.* Rye can be infected with ergot, *Claviceps purpurea,* which not only damages the rye grains but produces unfortunate effects in consumers of infected rye flour. They undergo fantastic and horrifying delusions, see their families as monsters, and may have blood supplies to their extremities so restricted that their ears or fingers rot off. In severe enough cases they die. Strangely enough, the same class of fungi that causes so much misery contains the two classic fungus delights of gourmets, morels and truffles (Figure 9.21), and it is also the source of the workhorse of the medical doctor's array of antibiotics, penicillin from *Penicillium.*

Figure 9.21
A morel, *Morchella,* one of the most desirable of edible fungi. The honeycombed upper part of the fruiting body is lined with millions of asci containing ascospores, but the growing conditions necessary for successful development of the fungus are so special that only about one spore from each crop of morels germinates and grows to maturity (else the species would be increasing in numbers, and it is not).

The Basidiomycetes

The fungi included in the **Basidiomycetes** are so various that at first sight one would not expect them to be even remotely related. There are mushrooms, stinkhorns (looking remarkably like vegetable phalluses), the messy black pustules of corn smut, the orange incrustations of blackberry rust, shelves on decaying birch trees, wads of jelly on soil or wood, the rainbow-colored coral fungi, and the tiny golden fingers of *Dacrymyces.* They all have one feature in common: the formation of *basidia* (singular, basidium), hyphal ends with horn-shaped extensions, *sterigmata* that bear spores at their tips. These spores are the *basidiospores* (Figure 9.22).

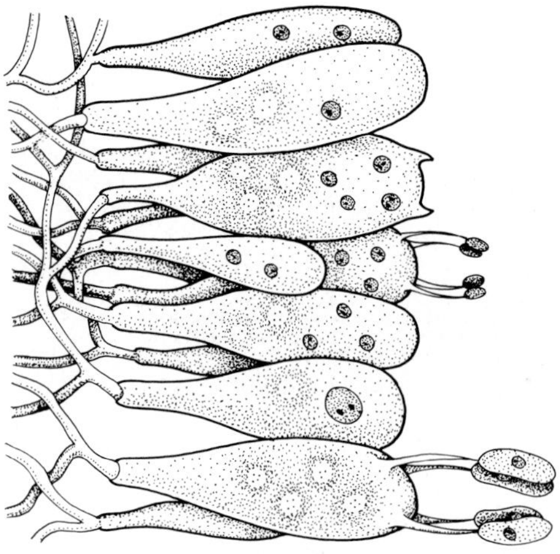

Figure 9.22
Basidia on gills of a mushroom, with spores in several stages of development.

The most popularly known Basidiomycetes are mushrooms, such as the commercially grown *Agaricus* sold in grocery stores. Basidiomycetes are synthesizers of some unique chemical compounds, some of them deadly to humans (Figure 9.23). Poisons from certain mushrooms may destroy red blood cells or interfere with nerve transmission. In either case the victim dies of lack of oxygen. No general rules for determining safety can be given; one simply has to know personally which species have been proved to be killers. The sacred mushrooms of the Mexican Indians, teonanacatl, or "God's flesh," are mostly of the genus *Psilocybe.* Eating them makes people see and hear things that no one else sees or hears.

A widespread and economically destructive genus is *Puccinia,* the red rust of wheat. This parasitic fungus is like a number of rusts in that it normally requires two plant hosts: cultivated wheat and barberry bushes. During summer, the invader lives on wheat, sometimes drastically reducing the yield of grain. Spores pass from field to field in the wind, spreading the disease, which looks like red rust. Other spores live through the winter. In the following spring, they infect a different plant, or *alternate host,* the barberry. Still other spores from the barberry can infect the new wheat crop. For complexity of spore forms in such a vegetatively simple fungus, wheat rust is without equal anywhere in the living world, and it is not surprising that the rusting of wheat remained mysterious to farmers for thousands of years. One reason that the rust has not been eliminated from wheat-growing regions is that as fast as wheat geneticists breed new resistant strains of wheat, new infective strains of rust appear. Some perhaps develop by mutation, but most come from genetic recombinations arising from sexual matings. The competition between the crop breeders and the parasite appears endless.

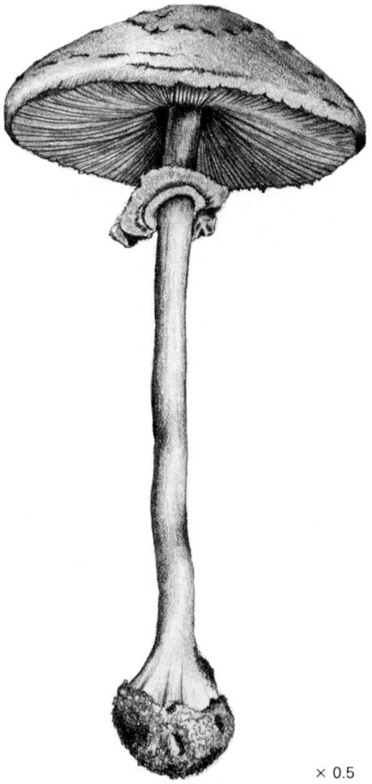

Figure 9.23
An example of a mushroom. The parasol mushroom, *Lepiota procera,* so called because the ring is so loose around the stalk that it can usually be slid up and down.

Lichens

Like most of the lower organisms, **lichens** puzzled early biologists. They were variously considered fungi, mosses, nonliving growths, or "queer" algae. However, the Swiss botanist S. Schwendener dis-

×1

Figure 9.24
A lichen, *Parmelia*, on tree bark.

covered in 1867 that a lichen is an association of fungus and alga, each so intimately associated with the other that together they make essentially a whole new organism. Later the word *symbiosis* (Greek, "living together") was coined to express the relationship, and that term has since been broadened in meaning to include any example in which organisms of more than one species live together in some special physical or nutritional relationship (Figure 9.24). Lichens have little economic value, but they are important in the ecology of some parts of the earth (Chapter 28). They also fascinate biologists because they are unique in their ability to make what looks like one organism, with the apparent attributes of a real species, while being composed of two distinct organisms.

Lichens are so sensitive to air pollution that they are killed in cities, probably by the products of combustion in furnaces and cars. Strictly urban people never see lichens, but in uncontaminated regions, whole forests or whole landscapes may be gray-green with them.

TO SUM UP FUNGI

1. The so-called *true fungi* are Phycomycetes, Ascomycetes, and Basidiomycetes.
2. The *Phycomycetes* may reproduce asexually or sexually. They cause diseases of plants, including grapes and potatoes.
3. The *Ascomycetes* are distinctive because of their production of spores within a single cell. Ascomycetes include many molds, some of which cause infections in plants and animals.
4. The *Basidiomycetes* include mushrooms. Their common feature is the formation of spore-bearing basidia.
5. *Lichens* are an intimate association of fungus and alga that forms a whole new organism.

THE KINGDOM OF THE PLANTS

Included here are photosynthetic organisms, most having cellulose walls covering the cells, usually multicellular with well-developed vegetative bodies, and a life cycle in which sexual reproduction alternates with asexual. The red and brown algae are somewhat different in certain details from the rest of the organisms considered here, but they are in most characteristics so definitely plantlike that it seems sensible to keep them here rather than to designate more kingdoms. The green algae are like some of the Protista in certain respects, but their chemical similarity to the higher plants has convinced most biologists that there is a line of evolutionary development leading from the green algae to the more advanced groups.

Figure 9.25
Irish moss, or carrageen, an edible red alga (*Chondrus crispus*). The extract *carrageenin* is used in ice cream, jellies, hand lotions, and toothpaste to give smoothness and body. It can be boiled with milk, sugar, and flavoring to make pudding, and it has been so used for centuries in maritime countries. Soaked in whiskey and chewed, it is said to be useful as a cough medication.

× 0.5

Phylum Rhodophyta, the Red Algae

The **Rhodophyta** (meaning "rose-colored plants") are usually red because in addition to chlorophylls *a* and *d*, they have a red pigment. All but a few of the red algae are marine, and most are tropical, thriving especially in deep waters, where they can make use of what little light energy can penetrate sea water (Figure 9.25). Like fungi, red algae may have solid-looking bodies, but they do not have cellular tissues like those of other plants. Instead, red algae have masses of filaments compacted into well-defined leaflike or stemlike structures.

They are of little economic importance, although some calcium-rich ones are like coral animals in that they are tropical reef builders; some, such as Irish moss, *Chondrus crispus*, are edible; and some furnish the agar used in microbiological work. The use of agar as a solidifying agent for bacterial cultures (discovered by a Jersey City housewife) is so widespread that if the supply of agar were cut off, thousands of hospital, university, and research laboratories would have to change their ways of working considerably.

Phylum Phaeophyta, the Brown Algae

The **Phaeophyta** (meaning "dusky plants") are those seaweeds popularly known as *kelps*. Anyone who knows the beaches of North America is familiar with the rockweeds. Despite the similarity of their parts to real leaves and stems of higher plants, the brown algae have simpler tissue structures, and their pigments, besides the usual chlorophyll *a*, include chlorophyll *c* and the brown *fucoxanthin* that gives the plants their characteristic color.

Some brown algae are among the most conspicuous plants in the sea. The rockweed, *Fucus*, covers rocks of the cooler shores of the northern hemisphere with thousands of tons of growth. The giant kelps of the American Pacific coast (*Macrocystis* and *Nereocystis*) are among the world's largest plants, growing up to 100 meters (330 ft) in length. *Sargassum* loosely covers more than five million square kilometers (two million square miles) of the Atlantic Ocean just north of the equator and gives that region its name, the Sargasso Sea. One brown alga, *Postelsia*, standing on rocky shores and splashed by breaking waves, looks like a small, sloppy palm tree (Figure 9.26). Alginates from brown algae are so

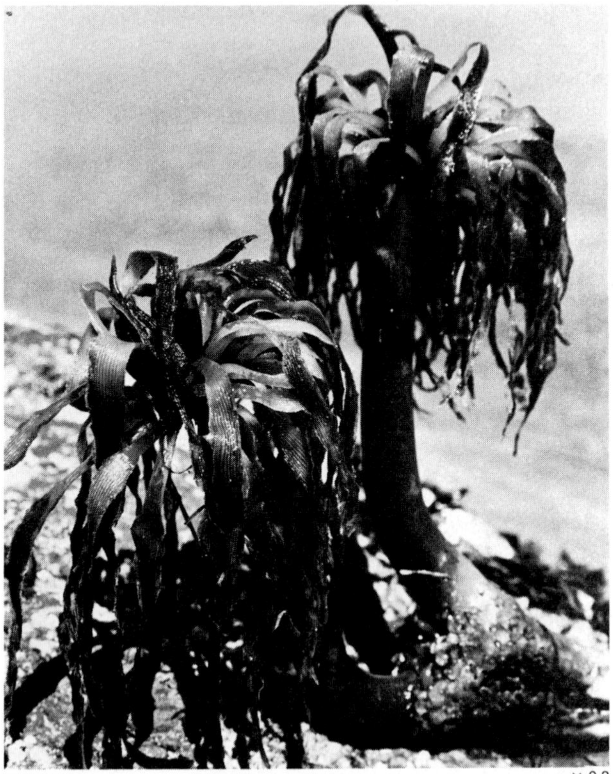

× 0.2

Figure 9.26
The brown alga *Postelsia* grows on the rocky Pacific coast of the United States. With a definite stalk and long "leaves," it has a general resemblance to a small tree, but it has no vascular tissue.

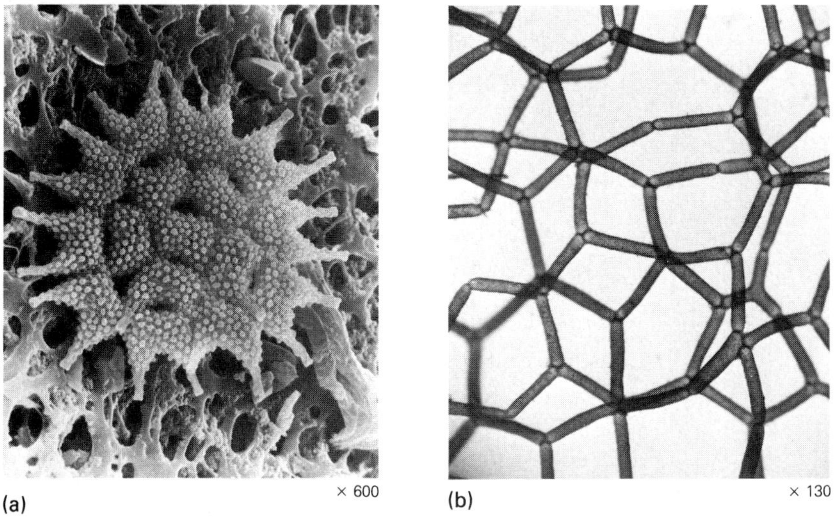

Figure 9.27
Green algae. (a) Scanning electron micrograph of *Pediastrum.* (b) A portion of water net, *Hydrodictyon,* which grows in the form of delicate little sacs, readily visible to the unaided eye. Each segment, usually meeting at a three-way point, is a separate cell.

useful commercially in applications requiring an inert, gelatinlike material that they are harvested, and attempts are being made to cultivate them artificially.

Phylum Chlorophyta, the Green Algae

Green algae are placed with the higher plants because they share with mosses, ferns, and seed plants three kinds of chemical compounds: (1) the photosynthetic pigments cholorphyll *a* and chlorophyll *b*, (2) the storage carbohydrate starch, and (3) cellulose as the main constituent of the cell wall. The common ability of all these plants to synthesize these substances points to a common ancestry, and it seems therefore reasonable to include in one taxonomic unit those organisms that have such striking similarity. This cannot be said of any of those algal groups that are included with the Protista in this chapter.

Green algae have long been favorite objects of study both by students and by professional biologists. At first, people were fascinated by the structural elegance of these microscopically delicate organisms. Then when green algae seemed to offer clues to the evolution of the vascular plants, interest heightened (Figure 9.27).

To a giant 15 kilometers (9 mi) tall, a forest would seem like nothing more than a thin film on the earth; this is how pond scums may seem to the average human observer. But to a microscopist, those slimy messes become worlds of beauty, with a brilliance of color and a fine precision whose esthetic appeal has never been adequately captured in pictures. In the green algae, chloroplasts have evolved in many ways, so that living examples are formed into balls, stars, nets, loops, rings, and helices. Plant bodies range from single cells through branched or unbranched filaments, to sheets and even, in *Fritschiella,* to thickened structures that resemble parts of higher plants.

Green algae are of little direct economic importance, although they have been repeatedly suggested as food sources for people and domestic animals because they are rapid growers and can be manipulated to form proteins and fats in high concentrations. Green algae have also been studied as possible suppliers of oxygen in submarines, but so far the scheme has met technical difficulties. Much of the research on photosynthesis has been carried out using the unicellular genus *Chlorella.*

Algae are the main photosynthetic organisms in aquatic environments. Just as grasses on land serve as a prime food supply for animals, so algae in streams, lakes, and seas serve to feed small animals, fish, and eventually even the great marine mammals, such as the whales and dolphins. Green algae, along with diatoms and some of the Pyrrophyta, are so necessary to the survival of aquatic animals that they have been called the "grasses of the sea." The reproduction of green algae is discussed in Chapter 15.

Phylum Bryophyta, the Moss Plants

Mosses and their relatives are typically plants of moist, shady places, but some species have enough tolerance of dryness to allow growth on sunbaked rocks and in sidewalk cracks. Indeed, one species, *Bryum argenteum,* meaning "silver moss," finds its way into urban cracks, and can be found easily in any of our cities. In contrast, the peat mosses,

× 4.5

Figure 9.28
Peat moss, *Sphagnum*. The feathery growths are masses of leaves, each with a mix of photosynthetic and empty cells. The latter are capable of absorbing large amounts of water.

× 1

Figure 9.29
A thick stand of moss, *Catherinea*, with "leaves" so small they are scarcely visible. Here "leaves" is written in quotation marks because the photosynthetic organs of mosses are not structurally comparable to the true leaves of ferns or flowering plants.

Sphagnum, live only in marshy places, where they are always soaking wet (Figure 9.28). In fact, all mosses must have liquid water if eggs are to be fertilized because moss sperm are motile, flagellated cells that swim to the eggs.

Mosses seem to be intermediate between green algae and the higher vascular plants (Figures 9.29 and 9.30), but are not believed to be ancestral to the higher plants because fossil vascular plants can be found that are older than any fossil mosses. Consequently, the current opinion is that mosses arose from an algal ancestor, and that the vascular plants have evolved independently.

Besides the ordinary mosses and the specialized sphagnums, the bryophytes include the liverworts and hornworts. These are either mosslike (the leafy liverworts) or ribbonlike, lying flat on soil or rocks or frequently in water. Liverworts are thought to be an evolutionary dead end, having given rise to no more advanced plants.

Bryophytes are important colonizers of bare areas (Chapter 28), but they have little direct effect on human affairs. Decayed peat moss is ground and sold for garden mulch, and compressed blocks of dead peat are cut for fuel. The "turf" of Ireland is burned extensively there for home heating and even for public electrical energy. The smoke from a turf fire has a strange, sweet smell, and it has been said of the Irish in the United States that if a whiff of such smoke were wafted across the country, 10 percent of the population would burst into tears.

Phylum Tracheophyta, the Vascular Plants

The **Tracheophytes** are the plants that give the earth its main green covering. This group, including the ferns, the conifers, and the flowers, contains the most recent, the most complex, and the most successful plants ever to have evolved. The feature that gives the Tracheophytes their name is the system of vessels and associated cells that serve for conducting water and dissolved foods and minerals throughout the plant. The *tracheae* (singular, trachea) of plants are the water vessels, and they are so called because under a microscope they look like the tracheae or breathing tubes of animals. When plants during their evolution from aquatic to terrestrial life began to live on land, they had to have several features if they were to survive. They needed stiffening if they were to stand up and obtain sunlight, an anchoring structure to hold them firmly in the soil, a way to absorb water and minerals, a system for conducting water up to the aerial parts and transporting food down to the underground parts (which were

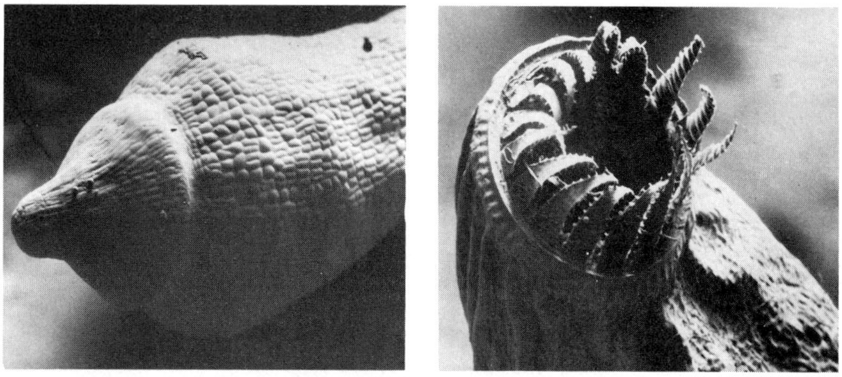

(a) × 27 (b) × 33

Figure 9.30
Spore dispersal in a moss, as seen by scanning electron microscopy. (a) The tip of a moss capsule is covered by a cap until the spores are mature. (b) The cap falls off. The opening of the capsule is surrounded by a ring of *peristome teeth,* which change position with changes in moisture. They can curl down into the spore cavity or straighten, lifting spores out into the air. The spores thus freed can blow to new growing places.

necessarily in the dark), a waterproofing cover to prevent excess drying, and some means of allowing photosynthetic and respiratory gas exchange. These needs were met by the development of fibers and vessels of the vascular system, roots or rootlike parts, a waxy cuticle over stems and leaves, and stomata with regulatable openings. No plant can live on land and stretch its leaves far into light and air without these components.

The first vascular plants were probably somewhat fernlike and still needed liquid water as a medium for transport of swimming sperm in fertilization. Eventually, however, even that need was overcome, and now the most advanced plants (the conifers and flowering plants) are free from the necessity for liquid water when seeds are produced.

Primitive Tracheophytes Of the small subphyla of the Tracheophyta that are not commonly known, three should be mentioned even in a brief survey. They are the Psilopsids, the Lycopsids, and the Sphenopsids. The Psilopsids are represented possibly by the Florida whisk fern, *Psilotum nudum.* The Sphenopsids now living are the horsetails of the genus *Equisetum* (Figure 9.31). The Lycopsids are the club mosses, *Lycopodium,* and the little club

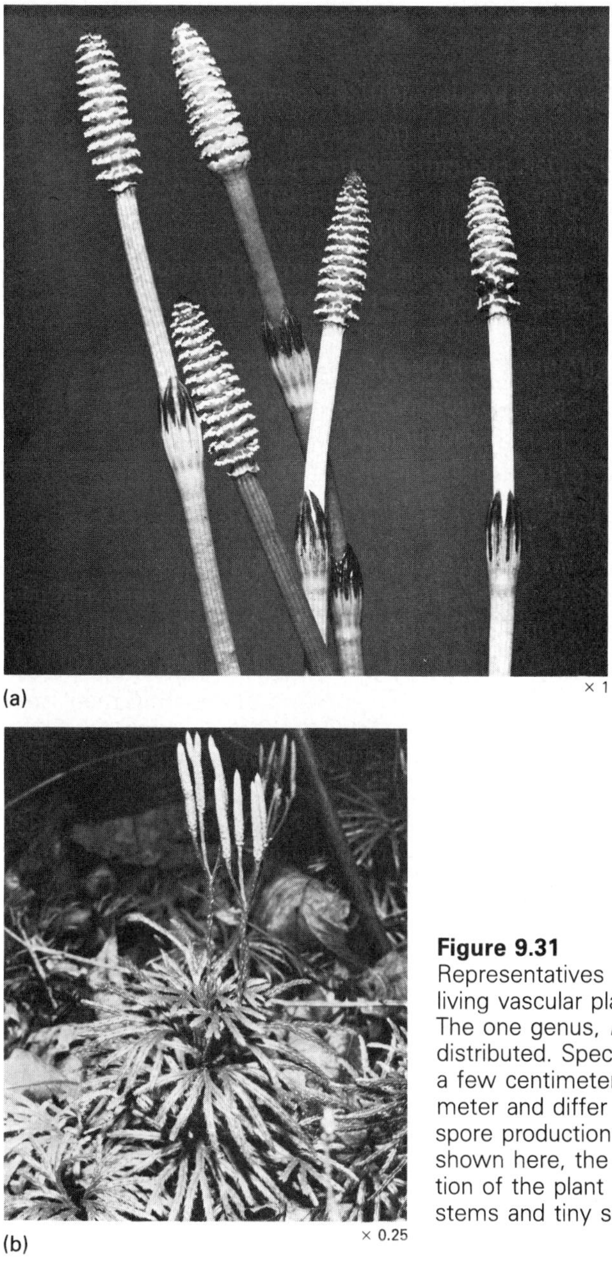

(a) × 1 (b) × 0.25

Figure 9.31
Representatives of some primitive, living vascular plants. (a) Horsetails. The one genus, *Equisetum,* is widely distributed. Species vary in size from a few centimeters to more than a meter and differ in the method of spore production. In the species shown here, the photosynthetic portion of the plant has whorls of green stems and tiny scales (photosynthetically useless) for leaves. The reproductive portion of this species, which is brown rather than green, is unbranched. Spores are produced in cones, each consisting of a hundred or so *sporophylls,* visible in this photograph as small, pale lumps. (b) The club mosses of the genus *Lycopodium,* commonly called ground pine or ground cedar.

(a) × 0.15 (b) × 1.25

Figure 9.32
Ferns. (a) A mosaic of leaves of the bracken *Pteridium*, widely distributed throughout the north temperate part of the world in fields and forests. (b) Instead of unfolding as leaves of flowering plants do, fern leaves unroll in characteristic fiddleheads, or *croziers*, shown in the cinnamon fern, *Osmunda*.

mosses, *Selaginella*. All these are slender remnants of plant groups that were once the important plants on earth. During the Carboniferous Period some 300 million years ago, the giant Lycopsids and Psilopsids were forest trees, whose dead bodies yielded the coal for which the period was named.

Subphylum Pteropsida

Most of the familiar plants of the world are in the subphylum **Pteropsida.** These are the ferns and the seed-bearing plants. Despite the wide variety of forms in the group, all members have two features constant enough to make inclusion in one subphylum possible: (1) All make spores on special spore-bearing leaves, even though the spores are microscopic and not easily seen (see Chapter 15), and (2) all have complex leaves based essentially on one structural plan and with complex connections to a stem. The similarities in all this large mass of plants make it reasonable to think that they belong in one great evolutionary category.

Class Filicineae, the ferns In the algae it was chloroplasts with which evolution played a game of theme-and-variations. In the ferns it was leaves. Fern leaves range in size from tiny cliff brakes only a few centimeters tall to the lacy sheets of tree ferns three meters (10 ft) long (Figure 9.32). They may range from the simple flat frond of a hart's tongue fern through the once-divided Christmas fern and the twice-divided lady fern to the filigree of five-times-divided *Davallias.* The stems of ferns are frequently horizontal, growing along under the soil, following the contours of the surface, branching at the tips, and spreading away from their old beginnings. After many years a fern may expand into a circle of advancing stem tips, each with a cluster of leaves coming up, to make a "fairy ring." Internally, fern stems have vessels for the conduction of water and food, plus an abundance of thick-walled strengthening fibers, but different species of ferns have such different arrangements of these tissues that no general description is possible. For the first time in this survey, we meet in ferns real roots with well-defined conducting tissues, not the hairlike structures of mosses.

Ferns produce their spores in sporangia on the leaves (Chapter 15). Sometimes the sporangia are in

× 4

Figure 9.33
Spore production in a fern. The underside of a leaf in some genera (as here in the wood fern, *Dryopteris*) has clusters of sporangia, or *sori* (singular, sorus), covered with thin scales, the *indusia*. In this specimen, each sorus with its indusium looks like a raised bump. Sporangia are too small to show at this magnification, but they are in fact crowded under the indusia.

clusters on the lower surfaces (Figure 9.33) or along margins, at the tips of ordinary leaves, or on special leaves whose only function is spore formation.

People find ferns useful as ornaments because of the elegance of their leaves. On the minus side, the tough, inedible bracken is responsible for occasional infestations of pasture land. Ferns are of little importance in the human economy, although one minor value is that well-rotted fern roots are the medium of choice for growing orchids.

Class Coniferae, the conifers The **Coniferae** are variously called the conifers, the evergreens, or the softwoods, but all these names are inaccurate in part. Coniferae do not all bear cones; yews have red "fruits" that look like berries. They are not all evergreen; larches and bald cypresses are deciduous (leaf shedding). And the wood is not necessarily soft; yellow pine is considerably harder than the so-called hardwood of basswood.

All conifers are woody plants. (How many such absolute statements can you find in this book?) Their general manner of growth by means of extension at stem and root tips and expansion of a growth layer under the bark is similar to that of flowering plants (Chapter 16), but the cellular structure of the wood in one important respect is not like that of flowering plants. The water-conducting cells, called *tracheids*, are thickened and elongated, but they are not open-ended like sections of water pipe. The open-ended water vessels, called *tracheae*, did not appear until the evolution of the flowering habit.

Leaves of conifers are frequently needle-shaped, as they are in pines, firs, and spruces, but they may be scalelike, as in cypresses, arbor vitae, and white cedar, or they may be broad-bladed. Most conifer leaves remain on the tree for more than one season, over-wintering in a green condition and thus appearing "evergreen." Any single leaf usually lasts only two summers, but certain types of trees keep leaves for many years, with some of the ancient bristlecone pines having leaves that are 30 years old. The leaves are generally adapted to dry conditions, being well waxed, being impregnated with such antidrying agents as resins, having stomata in depressed pits, and having relatively little surface exposed to the air. In pines and some others, needles are borne in bundles of two, three, five, or—rarely—one, at the ends of stubby branches that are so short as to be inconspicuous.

Reproduction in conifers Conifers are nonflowering seed plants (Chapter 15), but they are not the only ones. They are loosely known as *gymnosperms* (Greek, "naked seeds") because the seeds are not enclosed in a covering tissue but are borne on the surfaces of spore-bearing leaves, *sporophylls*. In pinecones, to be sure, the ovules that develop into seeds are effectively tucked between the cone scales, but they are still not encased in a complete

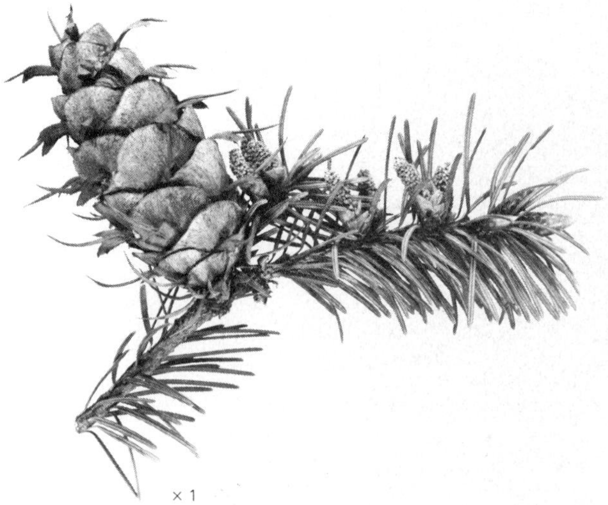

Figure 9.34
Branch and ovulate cone of Douglas fir, *Pseudotsuga.*

ovary, such as can be found in flowering plants. The seed in a gymnosperm is like a coin grasped in a fist: out of sight and protected, but not structurally inside the hand in the same sense that a finger bone is inside.

As its name suggests, the conifer literally bears cones, of which pinecones are familiar examples. Those usually seen are the ovulate cones, or seed cones, which bear the embryos (Figure 9.34). Cones vary in size from small hemlock cones, no bigger than a fingernail, to massive cones of the sugar pine, *Pinus lambertiana,* the size of a small watermelon. They also vary in structure from thin, long, and dry, as in eastern white pine, *Pinus strobus,* to the berrylike balls of red cedar, *Juniperus virginiana,* whose cone scales are so plump they mask their really conelike nature. The *staminate cones,* which produce pollen, are much smaller, usually only a few millimeters long, and short-lived; they appear for a few weeks in spring and then dry and fall.

Distribution of conifers Unlike many ancient plants that were crowded off the earth when more advanced species evolved, the conifers are thriving. After first appearing in the Permian Period of the Paleozoic Era about 300 million years ago, they multiplied and spread until by the Triassic Period they had formed large forests, as they still do throughout the northern hemisphere. Although conifers are poorly represented in lands south of the equator, they are so numerous in parts of the three northern continents that they give the landscapes their characteristic appearance.

Among trees, conifers are record breakers. They are the tallest, with redwoods, *Sequoia sempervirens,* reaching 117 meters (385 ft). (There are unconfirmed reports of taller eucalyptus trees in Australia.) They are the thickest. The Big Tree of Tule in Oaxaca, Mexico (*Taxodium mucronatum*), about 40 meters (132 ft) in diameter, was already revered by the Indians for its size when Europeans first saw it. Conifers are also the oldest, with some bristlecone pines, *Pinus aristata,* in the White Mountains of California having lived for 4600 years.

Uses of conifers The wood of the pines, spruces, firs, the various woods known as cedar, the redwoods, hemlocks, cypresses, and yews furnish the bulk of the lumber cut for general construction and other commercial purposes. In addition, conifers provide the 40 million tons of pulpwood used annually in the United States in the manufacture of paper. No other material has ever been found to have the resonant quality of spruce wood in the soundboards of pianos and the tops of violins.

Conifers and fungi Pines—and indeed many other kinds of plants up and down the evolutionary scale—enter into symbiotic relations with fungi. In the soil, a fungal mycelium penetrates a root and may grow between the cells of the host root or actually probe inside root cells. Such a fungus is called a *mycorrhiza,* from the Greek meaning "fungus root." It obtains food and protection from the host and in return makes easier, or even possible, the entry of minerals into the root. Some pines are so dependent on their mycorrhizas that they cannot thrive or survive without them, and emigrating peoples have been known to carry with them a little home soil to add at the bases of transplanted trees because they had learned from experience that something essential was contained in it.

Conifers have impressed people in a way that no other plants have. Visitors have been moved emotionally by the great white pine forests of New England—all gone now, destroyed by the same people who admired them—by the huge Douglas firs of the Pacific Northwest, the incomparable California redwoods, the swamp cypresses, dark and draped with gray Spanish moss, and the strange wilderness of the Jersey pine barrens (Figure 9.35).

Class Angiospermae, the flowering plants To most people, "a plant" means "a flowering plant." Since

(a) (b)

Figure 9.35
Two representative conifers. (a) A giant sequoia, *Sequoiadendron giganteum,* in the Sierra Nevada Mountains of California. For sheer bulk, giant sequoias surpass any organism that has ever lived (unless larger ones left no fossil trace). (b) A cypress swamp in Arkansas, with buttressed trees *(Taxodium).* They commonly grow in water and send up "knees" from their submerged roots.

they first evolved in the Cretaceous Period nearly 200 million years ago, flowering plants have become increasingly dominant as the main flora of the earth, until they now make the most conspicuous feature of the landscape. Except for seas, deserts, some northern forests, and a few artificial spots, such as major cities, the *look* of the world is determined by the angiosperms (Greek, "vessel seeds"). They are the basis of practically all human economy, furnishing the main requisites for survival in the form of food, clothing, and shelter. Even if we were not dependent on them for necessities, a world without flowering plants would be a poor place to live. Therefore the emphasis in such a book as this is on angiosperms, so much so that instead of treating them in a general survey of the Plant Kingdom, we give them special treatment, particularly with respect to their structure and development (Chapter 16) and their physiological functions (Chapter 17). Their reproductive activities will be described in Chapter 15, and their photosynthetic ability was covered in Chapter 4.

The **angiosperms** have evolved in two lines, the Monocotyledon and the Dicotyledon line. These groups are named for the presence of one or two seed leaves, cotyledons, in the embryos (Chapter 15). The number of cotyledons is distinctive, but it is only one of a number of features that characterize the two plant groups. The names are commonly shortened to "monocot" and "dicot." In addition to having only one cotyledon, monocots usually have

(a)

(b)

Figure 9.36
Representative flowers. (a) A lily, *Lilium*, a monocotyledonous flower, showing the threefold nature of the flower, with six stamens, three sepals, and three petals (the sepals and petals look alike). (b) A dicotyledonous flower of the rose mallow, *Hibiscus moscheutos*. The five-part corolla, typical of dicots, is evident. The central mass of stamens surrounds the style, which terminates in five stigmas.

the veins in their leaves more or less parallel along the long axis. Their vascular bundles are distributed as small, isolated strands of cells running up the stems. They usually grow little or not at all in girth, because they lack a lateral mitotic layer of cells (cambium). And their flower parts occur in threes or multiples thereof (Figure 9.36). In contrast, dicots have networks of veins in their leaves. Their vascular system consists of a central cylinder. They do have a lateral cambium, which contributes to increase in girth. And their flower parts come in fours or fives or multiples thereof.

The angiosperms are the most numerous (in number of species) and most diverse plants on earth. Of the several hundred families of angiosperms, only a few are of economic importance or are well known as ornamental plants. In the monocot group, the grass family, the Gramineae, is not only the source of most of the human food in the world (wheat, rice, corn, rye, barley, millet, sorghum, sugarcane) but covers vast areas of the earth in grasslands that feed uncounted wild and domestic grazing animals. Other well-known monocots are the lilies, palms (including coconut, date, and oil palms), yams, and orchids. Among dicots are the oaks, maples, legumes (beans, peas, clovers, locusts), roses (not only the flowers, but apples, pears, peaches, almonds, plums, prunes, strawberries, blackberries), heaths (heather, blueberries, cranberries), asters in the broad sense (sunflowers, artichokes, goldenrods, dandelions), mustards, and mints.

TO SUM UP PLANTS

1. The *Red Algae* (Rhodophyta) live mostly in deep marine waters, where their red pigment traps light energy for photosynthesis.
2. The *Brown Algae* (Phaeophyta) are seaweeds known as kelps. Although they appear similar to land plants, they are not.

3. The *Green Algae* (Chlorophyta) are placed with the higher plants because they have the same photosynthetic pigments, starch, and cell walls consisting mostly of cellulose. These similarities point to a common ancestry with the higher plants.
4. The *Moss Plants* (Bryophyta) are intermediate in complexity between green algae and the higher plants.
5. The *Vascular Plants* (Tracheophyta) give the earth its main green covering. The group includes ferns, conifers, and flowers. These plants have a system of vessels for conducting water, dissolved food, and minerals. Flowering plants, or *angiosperms,* are the main flora of the earth, and are classified as monocots or dicots.

ASK YOURSELF

1. What is the correct order of the following groups, starting with the largest: genus, family, class, phylum, order, species?
2. What is the simplest kingdom, in terms of structure?
3. What kingdom includes the viruses and bacteria?
4. What are the simplest organisms that are able to carry on photosynthesis?
5. What are some of the reasons that current opinion is in favor of placing fungi in a separate kingdom?
6. What makes the ascomycetes distinctive among fungi?
7. Where are the red algae usually found?
8. Why are green algae classified with the higher plants?
9. What are some distinguishing characteristics of the vascular plants?
10. What are some of the classes of plants in the subphylum Pteropsida?
11. Why are conifers also called gymnosperms, which means "naked seeds"?
12. What are the differences between "monocot" and "dicot" plants?

10
The Living Kingdoms: The Animals

SOME KEY POINTS

1. Animals may be characterized as organisms that breathe, move, and eat.
2. Several characteristics help biologists classify animals: (a) cellular nature, (b) symmetry of the animal body, (c) gross structure of the body, and (d) embryological patterns.
3. Protostomes are animals that in early embryos develop a mouth first and an anus later as an opening at the posterior end of the digestive tract. Deuterostomes develop an anus first, and later the mouth opening breaks through at the anterior end. Deuterostomes are the starfishes and their relatives (Phylum Echinodermata) and the vertebrates and their relatives (Phylum Chordata). The protostomes are all the other animals, including most of the invertebrates.
4. Human beings (*Homo sapiens*) may be several million years old. Some ancient human-like beings were the *Dryopithecus, Australopithecus, Homo habilis,* and *Homo erectus.*
5. Typical human activities such as the domestication of animals, the use of metals, the settlement of towns, and writing are less than 10,000 years old.

ANIMALS ARE GENERALLY THOUGHT OF AS those creatures that breathe, move, and eat, as indeed most animals do. Yet we cannot define animals so simply, because not all animals inhale and exhale, even though we know that all living cells respire. Not all animals move about, nor do they all have mouths and digestive systems. In spite of our lack of a completely satisfactory definition of animals, and in spite of living examples that do not fit neatly into any formal definition, we can still categorize animals roughly as those organisms that breathe, move, and eat.

The system of classification of animals used in this book is a simplified one, which omits some of the smaller and less well known phyla. The animals to be considered here will be described according to the following outline, in which phyla, some subphyla, and classes will be named, with English names of examples given in parentheses.

Phylum Porifera (sponges)
- Class Calcarea (calcareous sponges)
- Class Hexactinellida (glass sponges)
- Class Demospongiae (bath sponges)

Phylum Coelenterata
- Class Hydrozoa (polyps)
- Class Scyphozoa (jellyfishes)
- Class Anthozoa (corals and anemones)

Phylum Platyhelminthes (flatworms)
- Class Turbellaria (planaria)
- Class Trematoda (flukes)
- Class Cestoda (tapeworms)

Phylum Aschelminthes (roundworms)

Phylum Mollusca (molluscs)
  Class Amphineura (chitons)
  Class Scaphopoda (tooth shells)
  Class Pelecypoda (clams)
  Class Gastropoda (snails)
  Class Cephalopoda (squids)

Phylum Annelida (segmented worms)
  Class Oligochaeta (earthworms)
  Class Polychaeta (clam worms)
  Class Hirudinea (leeches)

Phylum Arthropoda (arthropods)
 Subphylum Trilobita (extinct trilobites)
 Subphylum Chelicerata
  Class Merostomata (horseshoe crabs)
  Class Arachnida (spiders)
 Subphylum Mandibulata
  Class Crustacea (shrimps)
  Class Chilopoda (centipedes)
  Class Diplopoda (millipedes)
  Class Insecta (insects)

Phylum Echinodermata
  Class Crinoidea (sea lilies)
  Class Asteroidea (starfishes)
  Class Ophiuroidea (brittle stars)
  Class Echinoidea (sea urchins)
  Class Holothuroidea (sea cucumbers)

Phylum Hemichordata (acorn worms)

Phylum Chordata (chordates)
 Subphylum Urochordata (sea squirts)
 Subphylum Cephalochordata (lancelets)
 Subphylum Vertebrata (vertebrates)
  Class Agnatha (lampreys)
  Class Chondrichthyes (sharks)
  Class Osteichthyes (bony fishes)
  Class Amphibia (frogs)
  Class Reptilia (snakes)
  Class Aves (birds)
  Class Mammalia (mammals)
   Subclass Prototheria (platypus)
   Subclass Metatheria (opossums)
   Subclass Eutheria (the "beasts" and human beings)

BASES OF ANIMAL CLASSIFICATION

Biologists can work directly only with individual organisms, but after they have observed thousands of individuals, patterns of similarity become apparent. The similarities serve as bases for the general classification schemes. In the arrangement of animal phyla and the lesser subdivisions, several characteristics are used as indicators of relationships. These characteristics are basic enough to enable us to tell which groups of animals are closely related.

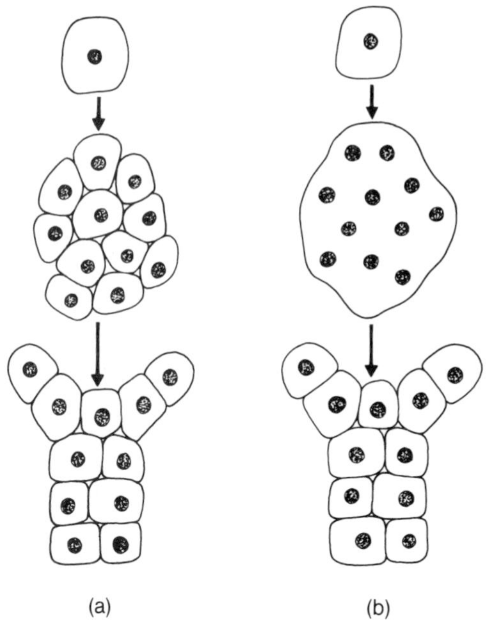

Figure 10.1
Two ways in which metazoan (multicellular) animals may have arisen. (a) One-celled animals remained in contact. (b) One-celled animals underwent repeated nuclear divisions without cytoplasmic division.

1. The first characteristic is the *cellular nature* of the organism. If an animal is one-celled it belongs with the protozoa, which we examined in Chapter 9. If it is a multicellular animal, it belongs with the metazoa (see Figure 10.1). The evolutionary origin of the metazoa is uncertain. One possibility is that one-celled animals remained in contact after dividing, forming a mass of cells that began to function as a single entity, finally becoming an integrated, single organism (Figure 10.1a). A second possibility is that a one-celled animal underwent repeated nuclear divisions without cytoplasmic division, so that a large, multinucleate protoplasmic mass began to function as a single organism and subsequently interposed cell membranes between nuclei, thus becoming a multicellular organism (Figure 10.1b). The question of the origin of the metazoa remains unanswered.

2. A second feature of fundamental importance is the *symmetry* of the animal body. As metazoans evolved, they might have grown without any symmetry, simply becoming irregular masses. They might have developed into spheres, a form that has not been successful. A third "choice" was radial symmetry, that is, the symmetry of a wheel. Radial symmetry has not been very successful either. When animals became more motile, radially symmetrical animals could not glide through water. The final possibility was bilateral symmetry, in which an animal is two-sided, and the sides are mirror images of each other, or nearly so. With bilateral symmetry came differentiation along other planes than the one down the middle. Swimming animals came to have a head end and a tail end, as well as a top (or dorsal) side and a bottom (or ventral) side. Once these axes were established, there was also a right and a left. In our synopsis of the animal kingdom, everything above the Phylum Coelenterata has bilateral symmetry.

3. A third characteristic of animal bodies that is useful in delimiting large groups is the *gross structure* of the body itself. In the familiar higher animals, there is a body cavity called the *coelom,* which contains many of the main organs (Figure 10.2). Some of the primitive animals (the flatworms, for example) have no such cavity; they have an outer epidermal covering, a mass of tissue beneath that, and an inner digestive tract. In our synopsis, all animals above Phylum Platyhelminthes have some sort of coelom. It is necessary to say "some sort of coelom" because some animals have a variation of the typical coelom found in vertebrates.

4. A fourth pattern, mainly embryological, serves to separate the coelomate animals into two great groups. During development, the first opening into the embryo becomes the mouth in the *protostomes*

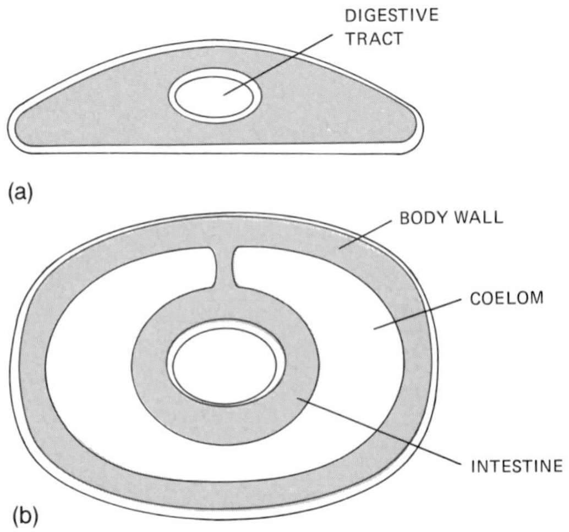

Figure 10.2
Body cavity, or lack of it, in animals. (a) In flatworms, ancient and simple animals, there is an outer skin, an inner lining of the digestive system, and a loosely formed tissue between. There is no "empty space" in the body, that is, no coelom or body cavity. (b) In most animals, the coelom is a "space" between the outer body wall and the internal organs.

Figure 10.3
Methods of alimentary tract formation in different animals. (a) After cleavage, when an animal embryo undergoes the infolding process known as *gastrulation,* it forms a hole where the infolding occurred, the *blastopore.* (b) In some animals—for example, molluscs and arthropods—when the alimentary tract is completed the mouth is formed where the old blastopore was, and the anus breaks through at the opposite end. This is the *protostome* method of development. (c) In the echinoderms and chordates, the anus is formed where the old blastopore was, and the mouth breaks through at the other end of the alimentary tract. This is the *deuterostome* method of development.

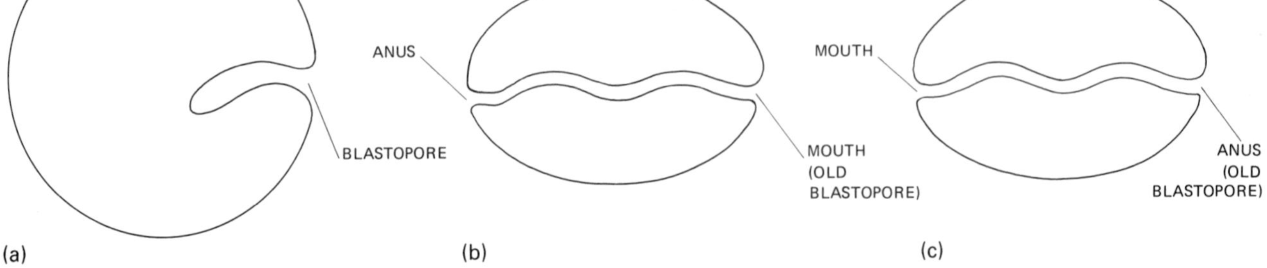

(Greek, "first mouth") and becomes the anus in the *deuterostomes* (Greek, "second mouth") (Figure 10.3). The deuterostomes are Phylum Echinodermata, the starfishes and their kin, and the Chordata, the phylum in which humans are included. The protostomes are all the rest. People are deuterostomes and so are starfishes, regardless of how different they seem.

There are further differences between the protostomes and the deuterostomes, especially in the larval forms of those animals that have larvae, but the differences given here suffice to show that there is a fundamental separation between the two groups. Biologists believe that such differences and such constancy of differences indicate an evolutionary relation of long duration. Consequently the protostomes and the deuterostomes are represented in schemes of classification as being separate branches of the evolutionary tree.

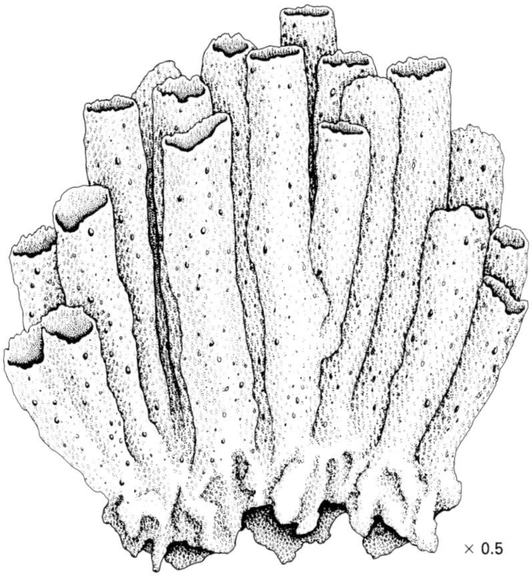

Figure 10.4
A tube sponge, *Callyspongia vaginalis*. Water containing bits of food is drawn in through the body walls by the action of flagellated cells. The usable parts are caught, and excess water and unused material are pushed out through the openings at the top.

Section 1: Sponges through Insects

PHYLUM PORIFERA, THE SPONGES

The **Porifera,** meaning "pore bearers," are among the simplest of animals (Figure 10.4). They do form definite bodies, some with complex systems of water channels, and there is some cellular differentiation. They can build some fantastically intricate forms, but the elementary nature of their cellular relations is shown by the fact that an entire sponge body, after being broken up and passed through a fine sieve, can reassemble itself into a functional animal. (This demonstration was one of the classic experiments in developmental biology.) In obtaining food, a sponge drives water in through pores in the sides of the body wall, creating currents by the beating of flagella. As water passes through the sponge pores, particles of food are captured and taken into the feeding cells.

There are three classes of sponges. The **Calcarea** have needle-shaped lime crystals scattered throughout the middle cell layers. The **Hexactinellida** also make crystalline skeletal elements, *spicules,* but of siliceous material (hence the common name for the group, glass sponges) and usually with six-part geometry (Figure 10.5). Some of the Hexactinellida, such as *Euplectella,* the Venus flower basket of Asian oceans, make such delicately woven glassy skeletons that when they were first brought to Eu-

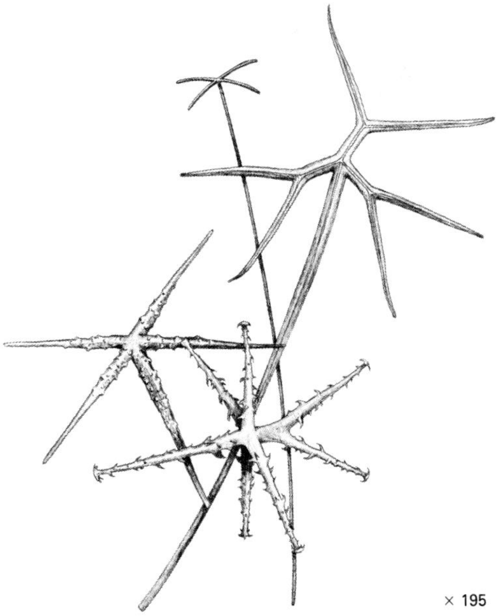

Figure 10.5
Sponge spicules, the siliceous or glassy particles that help support the sponge body.

rope, people did not believe they were natural objects. The **Demospongiae** have a proteinaceous skeleton as well as spicules. The skeleton is the spongy part that is preserved for human use as the commercial natural sponge. Sponge fishers pull the live sponges from the sea, let them die and decay on the beach, and wash away everything but the fibrous, flexible protein, spongin.

Sponges have little organization of function. Most metabolic activities are carried on by individual cells, without any visible coordinating nerve network or internal circulation. Reproduction is by asexual fragmentation or by the sexual union of eggs and sperm.

PHYLUM COELENTERATA

The **coelenterates** (Greek, "hollow intestine") are the polyps, jellyfishes, sea anemones, and corals. They are characterized by their radial body symmetry, the presence of stinging cells, a simple network of nerve cells, tentacles, and a saclike *gastrovascular cavity,* in which one opening serves both for intake of food and outflow of waste. Some coelenterates exhibit a phenomenon called *dimorphism* (Greek, "two forms") or *polymorphism* (Greek, "many forms"), with individuals of a particular species occurring in two, or several, forms.

Coelenterates that have a conspicuous polyp (tree-shaped) stage are in the **Class Hydrozoa.** One of the best known in the class, *Hydra,* has only a polyp stage (Figure 10.6). Hydras, common in fresh water (not a usual place for a coelenterate), are used in experiments on regeneration because they can grow new parts when dismembered or mutilated. Also, because of the primitiveness of their nervous systems, they are instructive in experiments on feeding, light and temperature reactions, and tolerance of stresses. They are favorites in biology laboratory courses for observations on their ability to capture moving prey with the stinging cells, *nematocysts,* with which their tentacles are equipped. The **Class Scyphozoa** contains the free-swimming jellyfishes, marine creatures ranging from a few centimeters to a meter or more across (Figure 10.7). In

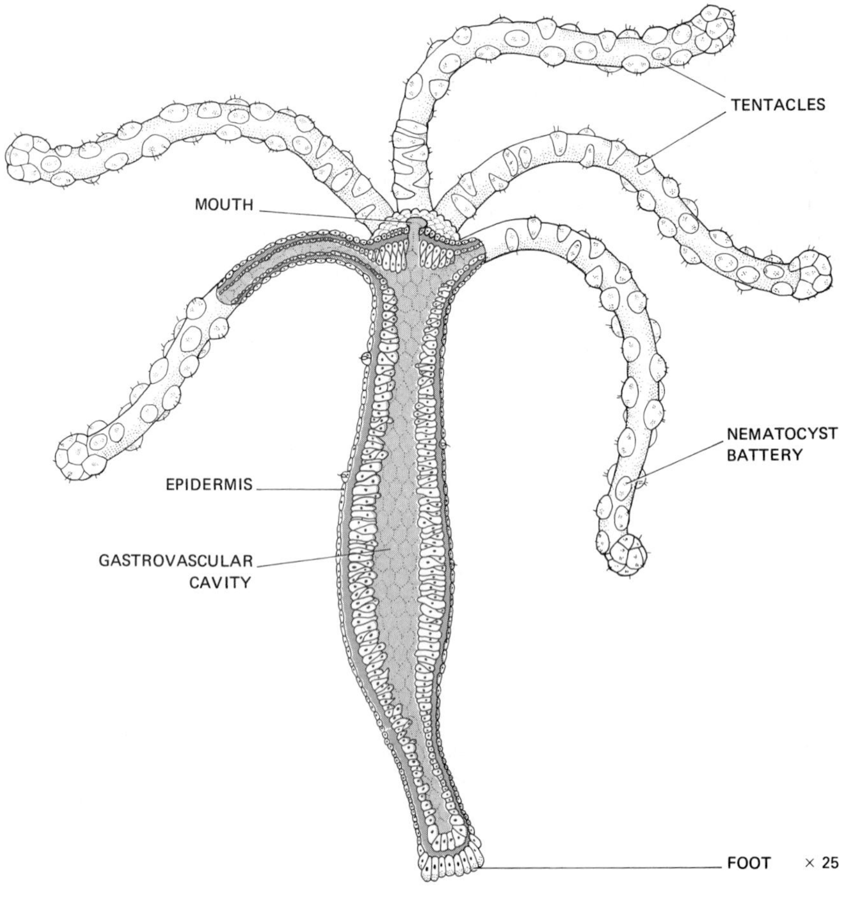

Figure 10.6
The common freshwater coelenterate, *Hydra.* The body and tentacles are soft and extensible, so that the animal can shrink down to a length of 3 or 4 millimeters or stretch out to 50 millimeters. Stinging cells in the tentacles can paralyze small animals swimming by, after which the prey is brought to the mouth and ingested. *Hydra* can remain attached by the basal foot to a solid surface, or it can move about by turning a series of somersaults.

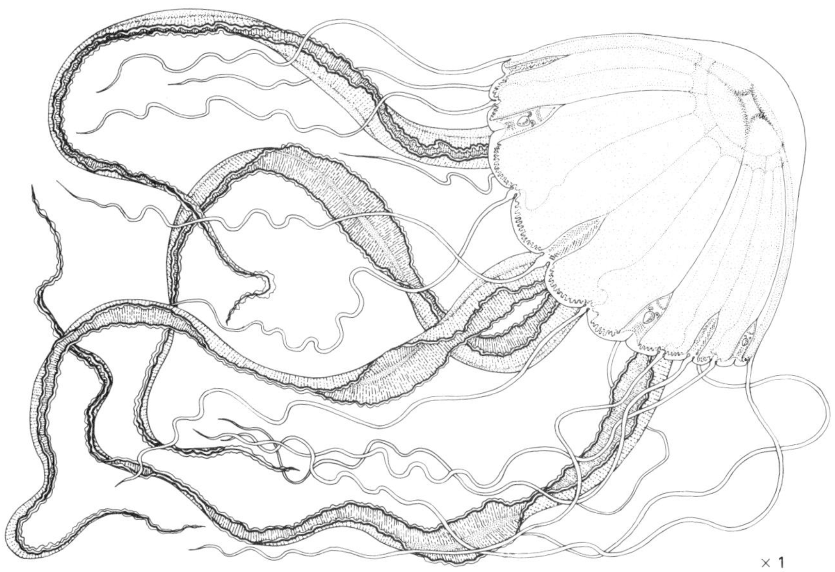

× 1

Figure 10.7
A free-swimming jellyfish, *Chrysaora*. The mouth is in the center of the underside of the umbrella, shown here as a square at the top of the picture. Digestive canals radiate out from the mouth to the circumference of the body. Like *Hydra*, jellyfishes have stinging cells in the tentacles.

× 1

Figure 10.8
Skeleton of a reef-building coral, *Oculina diffusa*. Each of the flowerlike structures is an individual animal, forming part of a large colony. The living parts of the corals have been washed away, leaving only the calcareous hard skeleton.

the **Class Anthozoa** are the sea anemones and corals (Figure 10.8). Corals differ from other coelenterates in secreting hard, limy supporting structures that can be built over a period of centuries into such massive piles as to form habitable islands, the coral atolls of the tropical seas.

Other interesting coelenterates are the pink corals of the Mediterranean Sea, which are made into jewelry, and the Portuguese man-of-war. This last-named "animal" (*Physalia physalis*) is not an individual but an aggregation of individuals living as a colony, floating about on the surface of the sea, buoyed up and kept moving in the wind by an inflated air float, which shimmers with iridescent colors (Figure 10.9). Drifting below are meter-long streamers provided with deadly nematocysts, which can sting,

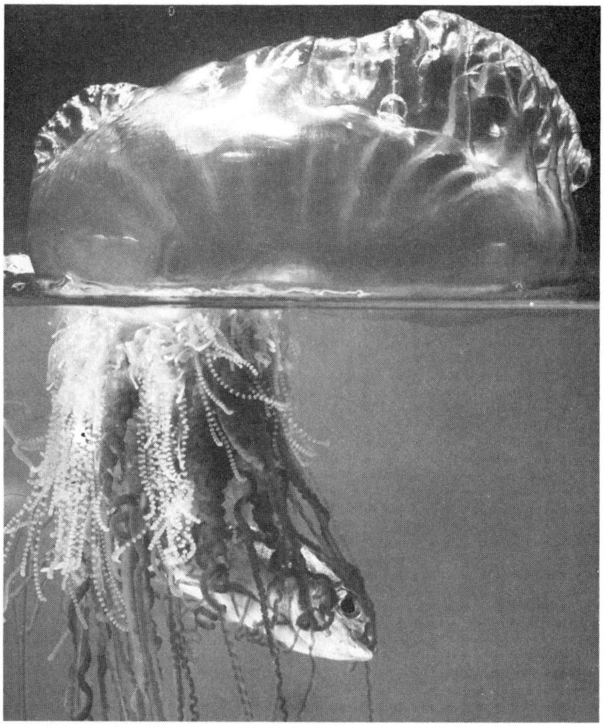

× 0.3

Figure 10.9
A coelenterate: the Portuguese man-of-war, *Physalia*, with a captured fish. These colonial coelenterates live in the warm waters of tropical seas, where they drift about, blown by wind on the sail-like crest.

Figure 10.10
A flatworm. The dark rectangular structure in the middle of the body is the pharynx with a mouth at its end. The highly branched digestive system is visible, almost filling the body, with one main branch toward the head end (right) and two less conspicuous ones toward the tail.

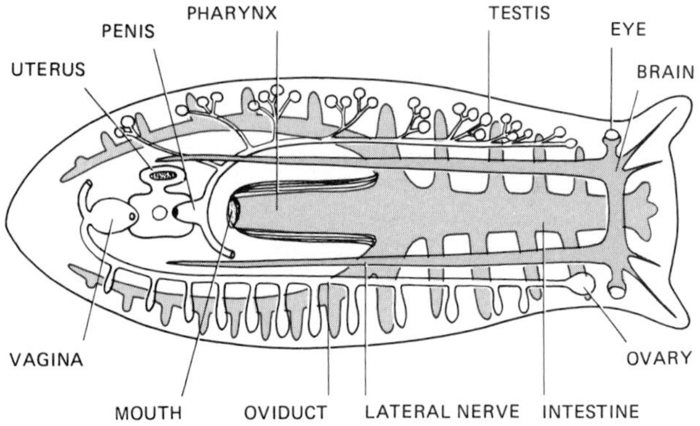

Figure 10.11
A tapeworm *(Moniezia expansa)* from a sheep's intestine. The slender, tapered end is the head end, with hundreds of units growing larger and larger toward the older end. The oldest units are filled with masses of tapeworm eggs.

paralyze, and kill any unwary creature they touch. They can be dangerous even to as large an animal as a human being. Since *Physalia* is not one animal but a collection, or colony, of animals, one separate tentacle with all its stingers can remain active after being washed up on a beach. Sitting bare-legged on a man-of-war tentacle is an experience long to be remembered.

PHYLUM PLATYHELMINTHES, THE FLATWORMS

The **flatworms** are the simplest, in evolutionary terms, of the bilaterally symmetrical animals. The body has an outer epidermis, a digestive tract (except in some parasitic species), and a mass of tissue between (Figure 10.10). As the tissue is practically full of material, mostly cellular, there is no coelom. Most flatworms have contractile muscle cells and a definite nervous system, with a pair of ganglia (nerve masses) at the end, a nerve down each side, and cross-connections like the rungs of a ladder.

The **Class Turbellaria** includes such free-living flatworms as the planarian, *Dugesia,* familiar in many biological laboratories as a subject for experimentation on feeding, regeneration, and learning. On being cut into pieces or variously mutilated, planarias can make whole new worms from small parts or heal in bizarre ways, but they are at the same time well-organized animals and can provide information on problems of cellular differentiation. They have even been trained to learn simple mazes, gliding along on their epidermal cilia, and turning right or left as they have been conditioned to do. To a person accustomed to the idea of a worm with a mouth at the head end, a planarian, which has a mouth in the middle of its underside (but no anus), is an improbable animal, especially since it has eye spots that give it a comical cross-eyed appearance.

A second group of flatworms, the **Class Trematoda,** includes the parasitic flukes that live in the lungs, livers, and other organs of many animals of many phyla. One of them, the sheep liver fluke, shows to what complex lengths parasites can go in completing their life cycles. In requiring more than one host and in its multiplicity of forms, a liver fluke rivals the red rust of wheat (Chapter 9). The life cycle of the liver fluke includes, in addition to a planarianlike form that inhabits livers, a second host (an aquatic snail), and several larval stages. The human lung fluke infests still a third host, a crab.

The parasitic tapeworms are in the **Class Cestoda** (Figure 10.11). A human tapeworm is an example of the way in which a parasite can undergo an evolutionary simplification, or, as it is frequently called, degeneration. The body of a tapeworm consists of a head, the *scolex,* provided with suckers for clinging to the host's intestinal wall, and back of the head, a string of units that are little more than bags of eggs. There is an outer covering and a pair of nerves that extend down each side of the flattened row of units, but there is no mouth, since the animal never eats, and no digestive tract. All metabolic requirements are met by passage of nutrients inward and waste products outward through the tough epidermis and cuticle.

Tapeworms, which infest many animals besides humans, get into their hosts by being eaten, usually while the tapeworms are in an embryonic or larval stage. They can be avoided by practicing clean personal habits, and by not eating meat that can bear worms, especially beef, or by ensuring adequate cooking. A human tapeworm five meters (16 ft) long, with thousands of units, can weaken a human, but such a worm is dwarfed by the broad fish tapeworm, which inhabits humans, fishes, and small crustaceans, and which may carry many more thousands of units in its 20-meter (65-ft) length.

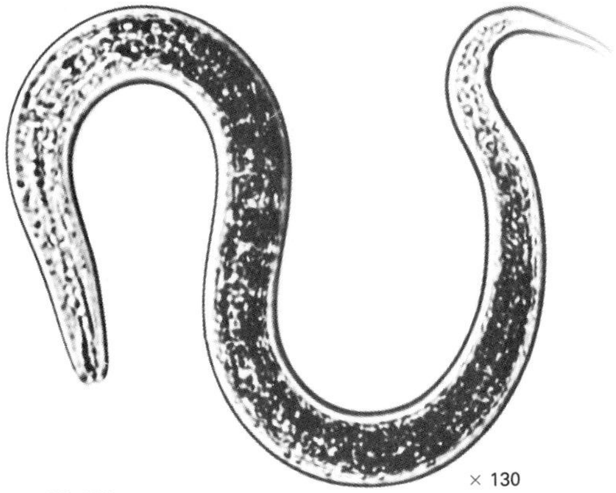

Figure 10.12
A roundworm that lives on the roots of plants. These worms have no effective means of locomotion, but they whip about violently, sometimes managing to progress forward. Because they are transparent, their internal organs are readily visible.

Figure 10.13
The shell of a clam, *Tridacna squamosa,* in the genus with the giant clams of the South Pacific, which are frequently used for fountains and garden ornaments.

PHYLUM ASCHELMINTHES, THE ROUNDWORMS

The **roundworms** are more complex than flatworms in body structure (Figure 10.12). Flatworms have either no mouth or a mouth located centrally on the underside of the body, but roundworms have a mouth at the anterior end, a tubular alimentary canal, and an anus at the posterior end.

The most notorious roundworms are human parasites: hookworms, which inhabit intestines; trichina worms, which infest pig muscles and are passed to human hosts via incompletely cooked pork; filaria worms, which are carried by mosquitoes and cause the monstrous deformities of elephantiasis; pinworms, which at one time or another infect almost every human being; and the guinea worm. The last is a slender, meter-long parasite that has earned its fame not so much from the damage it does as from the bizarre cure once devised for those afflicted with it. Before modern chemical methods were available, guinea worms were carefully tweezed out from under the victim's skin and slowly wound up on a stick to ease the attacker out.

The roundworms noted above are known from their destructive effects on people, but most roundworms are microscopic, free-living, harmless animals, present in and on animals, plants, soil, and water. They occur in such enormous numbers that it is impossible to avoid eating them.

Some roundworms, commonly called nematodes, are destructive to plants. Root-eating nematodes can so thoroughly fill the roots of plants that nothing is left but a squirming mass of worms.

PHYLUM MOLLUSCA, THE MOLLUSCS

The **Mollusca** (Latin, *mollis,* soft) have soft bodies, which in most molluscs are encased in hard shells. Although molluscs vary in size and shape, many have some common features: a mass of muscle on one side of the body known as the *foot* (easily seen in snails), a fold of tissue over the body called a *mantle,* and a rasping organ, the *radula,* or tongue, with a filelike surface. Molluscs have a well-developed set of functional systems: a digestive tract, effective sense organs coupled with an integrating nervous system, circulation of blood pumped by a heart, excretory organs *(nephridia),* respiratory organs (gills or functional lungs), and gonads. They range in size from almost microscopic snails to the *Tridacna* clams of the South Pacific Ocean, which grow up to a meter long and weigh 150 kilograms (330 lb) (Figure 10.13). The huge squids of the North Pacific, the

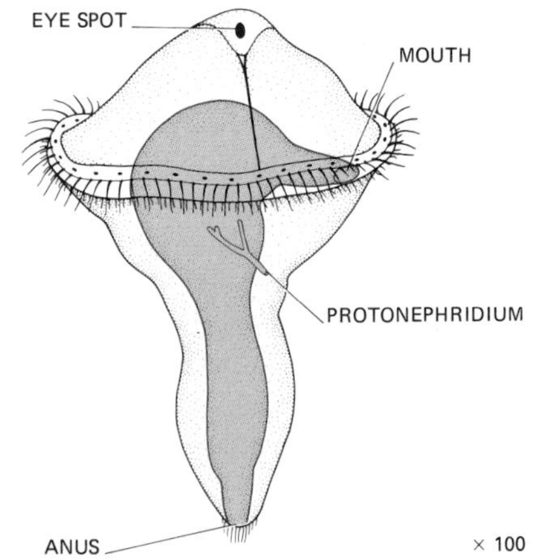

Figure 10.14
A *Molluscan* larva, a characteristic immature form common to both molluscs and segmented worms. The mouth is at one side, and the anus is at the posterior end. Although molluscs and segmented worms are quite different during adult life, the appearance of similar larval forms is taken as an indication of the evolutionary relationship of the two phyla.

largest of the invertebrates, in 10 years can reach a length of 16 meters (52 ft).

Molluscs are important to humans in many ways. We eat clams, oysters, snails, mussels, squids, octopuses, and periwinkles. Many snails are hosts for disease-producing parasites. Snails and slugs are devastating pests to crop plants in moist regions (see Figure 10.17). Molluscs known as shipworms bore into wood so effectively that they can scuttle a wooden ship within months or reduce massive pilings to a fragile lacy net.

With innumerable individuals and about 100,000 species now known, the Mollusca are among the most successful of animal phyla. They have been in existence for at least 500 million years and remain one of the three or four most important phyla. Their embryological development shows that they belong with the evolutionary line of the protostomes. Some primitive molluscs have characteristics in common with the segmented worms, the next phylum to be described. Such similarities indicate a relationship between molluscs and segmented worms (Figure 10.14).

Classes of Mollusca

Two classes of the **Mollusca** that are not generally familiar are the **Class Amphineura,** the segmented chitons, and the **Class Scaphopoda,** the slender, conical tooth shells. The remaining three classes are better known, largely because they are edible or because they are the producers of most of the seashells. The **Class Pelecypoda** (Greek, "hatchet foot") are the bivalves: clams, mussels, oysters, scallops, and thousands of similar kinds of animals. The body of a bivalve, such as a clam, is covered with two fleshy lobes called the *mantle,* which secretes the matched pair of shells and can close over the body, leaving two openings at the posterior end (Figure 10.15). If the hinge side of the shells is the dorsal side, then the mantle opening that allows

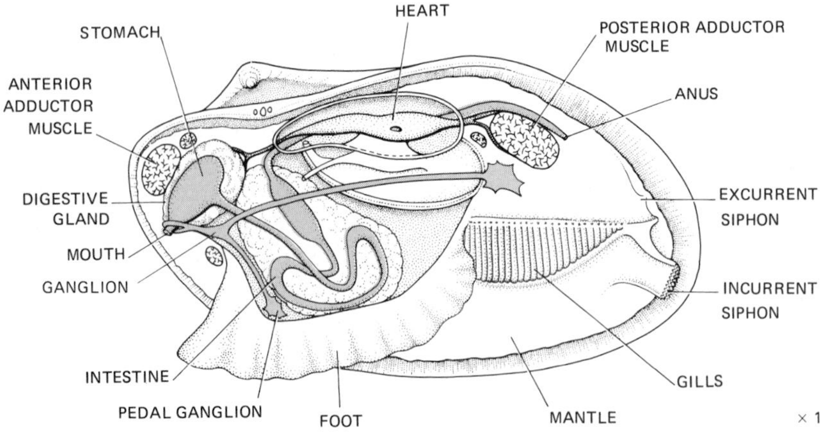

Figure 10.15
The internal structure of a clam, shown with one valve and portions of the body removed. The mass of the body is the foot, by means of which the clam can plow its way along the ocean bottom. A digestive tract leads from the mouth to a coiled intestine, which winds through the foot past a digestive gland and directly through the heart, finally ending in the anus. The whole soft body of the clam is covered by a pair of mantles, one of which has been cut away for this figure. A clam has no brain but does have collections of nerve cells, especially in ganglia below the esophagus, in the foot, and behind the foot. The valves can be held tightly shut by two stout muscles, the anterior and posterior adductor muscles.

water to enter the mantle cavity, the *incurrent siphon*, is ventral, and the other, which allows water out, the *excurrent siphon*, is dorsal. Water is moved over two pairs of *gills* by the action of beating cilia and made to flow in the incurrent siphon, through and past the gills, and past the *mouth*, which picks up tiny particles of food. The water then flows out the excurrent siphon, carrying with it dissolved carbon dioxide, alimentary wastes, and sometimes sex cells.

The bulk of the body is the *foot*, a mass of muscle a clam can use to plow slowly along through the mud at the ocean bottom. In the body are an alimentary canal; a pulsing heart, which sluggishly pumps body fluid through a system of cavities; and a nervous system, which consists of a pair of ganglia above the mouth and connecting nerves. There are sensory organs, which are associated with water testing (a mollusc can taste and smell), balance, and light perception. Scallops have a row of bright blue eyes along the edge of the mantle lobes. Clams produce eggs, which collect in the gills, where they are fertilized by sperm brought in through the incurrent siphon.

Although many bivalves are both male and female, they usually do not produce both eggs and sperm at the same time, so that self-fertilization is avoided. The shells of bivalves are hinged by a tough ligament that tends to spring the shells apart, but a set of muscles, especially the two large anterior and posterior *adductor muscles*, can hold the shells together. The scallops sold for human food are the adductor muscles only; the rest of the scallop body, except the gonads (which are considered a special delicacy in some regions), is discarded. Some unscrupulous restaurateurs have been known to stamp out cylindrical chunks of shark meat to sell as "scallops," and it is not easy to tell the difference.

The **Class Gastropoda** (Greek, "stomach foot") are the whelks, various coiled seashell animals, snails, and slugs (shell-less gastropods) (Figure 10.16). The land-dwelling snails are the only gastropods in which lungs have evolved, but even they must remain moist. The foot is conspicuous in gastropods, especially in the land snails, which use the foot for locomotion (Figure 10.17). A slick secretion is oozed out of the foot, and the animal travels along the bed of slime, rippling its muscles to propel itself forward. Hundreds of millions of garden snails, *Helix pomatia*, are prepared annually in French kitchens. The "sound of the sea," heard when a large conch shell is held to the ear, is only the resonance of sounds already in the air, but they are too faint to

× 0.5

Figure 10.16
A whelk. Whelks are marine gastropods and as such are essentially oversized, fancy snails. When the animal is alive, its body almost fills the shell, but it can extend its massive foot and move around, or it can withdraw into its shell. Whelks are similar to, but smaller than, conchs (pronounced "conks"), which are also marine gastropods.

× 1

Figure 10.17
A garden snail. Snails can be destructive to cultivated plants because their tongues, like microscopic files, can rasp away plant tissues. They have the ability, rare in animals, to digest cellulose.

be noticed without the help of the air column inside the shell.

The **Class Cephalopoda** (Greek, "head foot") are the squids and octopuses. Squids have ten tentacles and octopuses have eight, as their name indicates (Figure 10.18). Cephalopods are highly developed animals with sensitive perception of light and touch. The eyes of squids are of special interest to biologists because they are structurally similar to vertebrate eyes. Since squids and vertebrates are far removed from each other in evolutionary history, the fact that they have developed a similar set of light-sensitive organs furnishes an instructive example of convergent evolution (Chapter 26). One feature of the squid eye that could teach humans a lesson in humility is the position of the retina. In human eyes, light must pass through a layer of nerve cells before arriving at the light-sensitive retina, and it is therefore less precisely focusable than it might be. In squid eyes, the retina is in front of the retinal nerves and receives light directly from the lens, clearly a superior arrangement. Besides being sensitive, squids are probably the most readily trainable of the invertebrates. They can learn and remember, and therefore, like rats and pigeons, they have been used experimentally by students of behavior to study how learning is accomplished.

× 0.4

Figure 10.18
An octopus. Octopuses, squids, and cuttlefishes are almost shell-less molluscs. Despite their reputation for ferocity in folklore, these animals are usually rather small, timid creatures, and they are exceptionally sensitive and intelligent for invertebrates.

PHYLUM ANNELIDA, THE SEGMENTED WORMS

Many animals, including humans, show evidence of the serial repetition called **segmentation,** but in few is segmentation as noticeable as it is in the **Annelida.** Not only do the annelids illustrate segmentation, but they also have simple, clearly identifiable organs for most animal functions. Consequently annelids (Latin, "little ring"), especially earthworms (angleworms, fishing worms, night crawlers), are much used in elementary biology courses for dissection and instruction. In vertebrates and arthropods, segmentation is demonstrable, but in annelids it is obvious. An earthworm, even externally, looks like an animated stack of coins, and the external segmentation is continued inside the animal's body. Segmented organization interests biologists because they think it is a fundamental body plan for animals that evolved later from the first segmented types.

The **Class Oligochaeta** (Greek, "few hairs") is the best-known class of annelids, because it contains the earthworms *(Lumbricus terrestris).* They are so common in soil almost all over the world that they are familiar even to city dwellers, who may see them squirming on sidewalks after rain. Any small patch of soil is likely to contain thousands of earthworms during moist weather, when they make little mounds of earth by casting up pellets outside their burrows. With the arrival of hot, dry summers, earthworms disappear deep in the soil, giving rise to the myth that they turn into June bugs. With their perpetual eating of the soil, passing it through their alimentary tracts as they go, they stir it up, aerate it, turn it over, and keep it soft. In a set of classic observations, Charles Darwin, who put his mind to many a biological subject, found earthworms to be a major factor in maintaining the fertility of the earth, casting up as much as 18 tons of soil per acre per year.

Besides segmentation, the body of an earthworm shows many features of organization like those of more advanced animals. The symmetry is bilateral, with a definite head end where the nerve

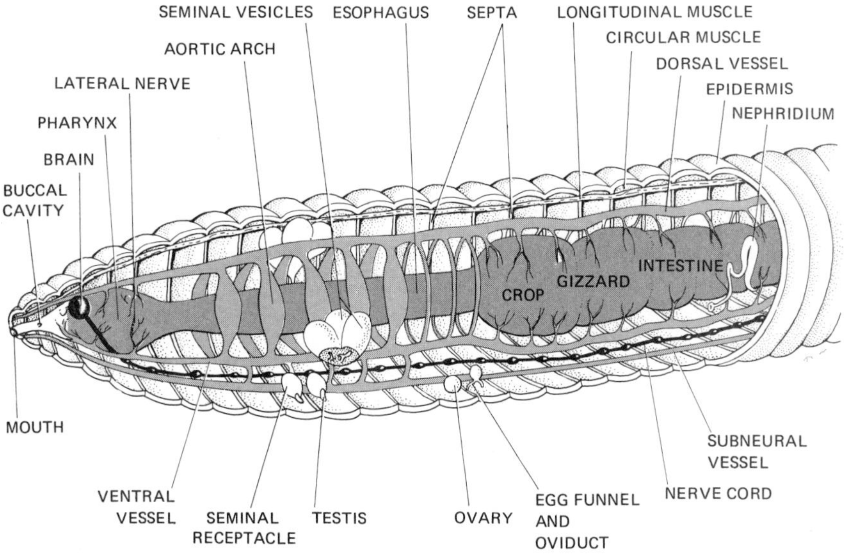

Figure 10.19
Internal structure of an earthworm. Segmentation, visible on the outside of the body, is striking internally because each segment is divided from the others by a partition, or septum, through which nerves and other structures pass. Many of the organs are repeated in more than one segment: hearts (aortic arches), ganglia, and nephridia, for example.

ganglia are, a true coelom lined with mesoderm, and all the functional organs that a well-developed, independent animal needs (Figure 10.19). The body wall, covered with a protective cuticle, has longitudinal muscles that can shorten the worm and circular muscles that can squeeze it out long, plus four pairs of whiskerlike *setae* (singular, seta) per segment, which can be pushed out or pulled in. With this arrangement, the worm can extend itself with its rear setae anchored, thus pushing forward. Then it releases its rear setae, anchors its forward ones, and shortens its body, drawing the tail end up. Coordinated and speedy, an earthworm can elude all but the fastest predators.

The alimentary tract of an earthworm is a straight tube through the length of the body. It extends from a muscular *pharynx* back of the mouth through a slender *esophagus* to a thin-walled storage *crop,* followed by a tough, muscular grinding *gizzard,* and finally a long *intestine,* whose dorsal wall has its surface increased by a longitudinal fold, the *typhlosole.*

The nervous system consists of a pair of ganglia above the pharynx, connected by a nerve ring to a ventral nerve cord that innervates the whole body, with a ganglion in each segment and branch nerves from that (see Figure 10.19). Without eyes, the earthworm uses its epidermis as a light-sensitive organ. It is sensitive to touch and to soil vibrations. There are testes and ovaries in each individual, but cross-fertilization is accomplished by transfer of sperm from one worm to seminal receptacles of another (Figure 10.20). Later, a band around the body,

× 1

Figure 10.20
A pair of earthworms (fishing worms, night crawlers) mating. Like some molluscs, these annelid worms have gonads of both sexes, and when they mate they exchange sperm. Fertilization of the eggs occurs when the eggs are laid.

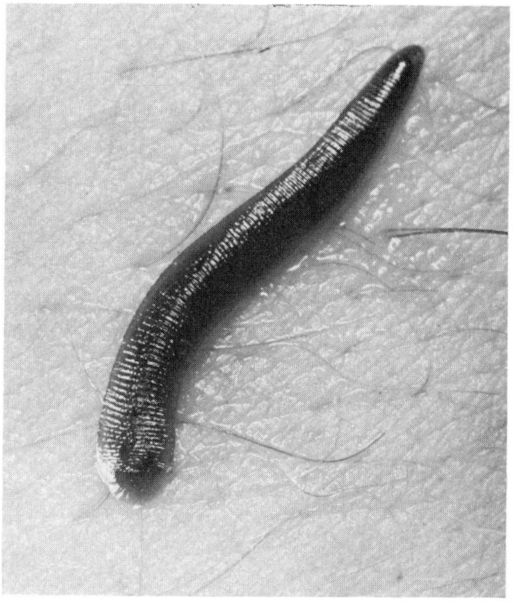
× 2

Figure 10.21
A leech, *Hirudo medicinalis.* These animals can draw up almost into a ball or stretch out to a length of many centimeters. When a problem of "too much blood" was considered a medical matter, leeches were attached to patients and allowed to suck out blood.

the *clitellum,* secretes a cocoon, which is slipped along the body, like a ring being drawn off a finger. Eggs are deposited in the cocoon, to be fertilized as the band passes the seminal receptacles. Embryological development occurs in the cocoon in the soil.

The excretory system is made up of a number of *nephridia,* two in most segments. A single nephridium is a sinuous tube with a ciliated funnel at its free end, capable of picking up wastes and taking them to the outside of the animal. Earthworms have a *closed circulatory system,* that is, a system in which blood (red in earthworms) is pumped through a set of vessels and stays in the vessels, as opposed to an *open circulatory system,* in which blood oozes about through various fluid-filled spaces. No special respiratory system is needed in such small animals when they live in moist environments where sufficient oxygen and carbon dioxide can be exchanged directly through the epidermis.

Other annelids, members of the **Class Polychaeta,** are equipped with pairs of appendages at the sides of the segments and with various head appendages, such as eyes, jaws, and tentacles. Common polychaetes are the clam worm *(Nereis)* and lugworm *(Arenicola),* which live in sandy beaches, and the "palolo worm" of Samoa, whose annual swarming makes feasts for the islanders. Leeches are in the **Class Hirudinea** (Figure 10.21). Leeches are such famous bloodsuckers that the word, which originally meant physician or healer because doctors used leeches to bleed patients, has now come to be used metaphorically to refer to any person who lives off another. At the head end, a leech, such as *Hirudo medicinalis,* has a sucker equipped with three sharp teeth. It can attach itself firmly to skin, puncture a blood vessel of the victim, secrete an anticoagulant, and suck blood until the body of the leech is bloated. In an old romantic movie, "The African Queen," Humphrey Bogart played the archetype of the rough and tough adventurer, who carelessly braves bullets, jungles, disasters, and threats of disasters, but who is reduced to shivering horrors on climbing out of the river to find himself dripping with leeches.

PHYLUM ARTHROPODA

Although the human species has made such an obvious impact on the earth that the present age is sometimes called the Age of Man, the arthropods at the same time have such a number of species and such a number of individuals that calling this the Age of Arthropods could more easily be defended. Except for microorganisms, there are more arthropods, any way you count them, than there are members of any other phylum of animals. Insects have taken the land as their province, and crustaceans have taken the sea. Every imaginable habitat and every imaginable behavior (including habitats and behaviors that are practically unimaginable) have been tried out successfully by arthropods.

The **arthropods** are named for their paired, jointed appendages (Greek, "joint foot"), which are covered by a horny *exoskeleton* composed largely of *chitin.* Arthropods are protostomes. They have a true body cavity, lined on both sides with mesodermal tissue. Beyond these few main characteristics, the arthropods are such a diverse lot of animals that they can be considered further only by examining the subphyla and classes separately.

Subphylum Trilobita

In our discussion, most animals known only from fossils will not be considered, but the **trilobites** (Latin, "three lobes") deserve special attention because they were so numerous (with millions of fossilized individuals still present), because they endured for such a long time (a period of more than 400 million years), and finally because their Precambrian ancestors, or something like them, must have given rise to the present-day arthropods (Figure 10.22). Before they became extinct in the Permian Period, about 200 million years ago, trilobites were common. They existed in a number of forms, all with the same general, fundamental plan. Two longitudinal grooves, running lengthwise, divided the body into three regions. Trilobites had eyes, antennae, and down each side of the body a row of forked, plumy appendages. They generally ranged from about 5 to 50 centimeters (2 to 20 in.) in length. Trilobites lasted well past the time of arrival of such major arthropod classes as crustaceans and insects, and for their time they were immensely successful.

× 1

Figure 10.22
A trilobite fossil, showing the two longitudinal furrows that divide the animal's body into three parts, giving it the name. Although now all extinct, trilobites were once among the commonest of animals, living all over the earth and evolving thousands of species. They endured for millions of years and left abundant fossils, especially in Cambrian rocks. They had well-developed compound eyes like those of modern insects and, like arthropods generally, pairs of jointed appendages.

Subphylum Chelicerata

The **Chelicerata** (Greek, "claw horn") differ from other arthropods in having claws or fangs as the first pair of appendages, no antennae, and a body with an unsegmented *cephalothorax* (head and thorax in one section) and an abdomen. There are usually four pairs of legs.

One small, ancient class, the **Merostomata,** contains the king crabs, or horseshoe crabs (Figure 10.23). They have remained on earth almost unchanged since the Triassic Period, 180 million years

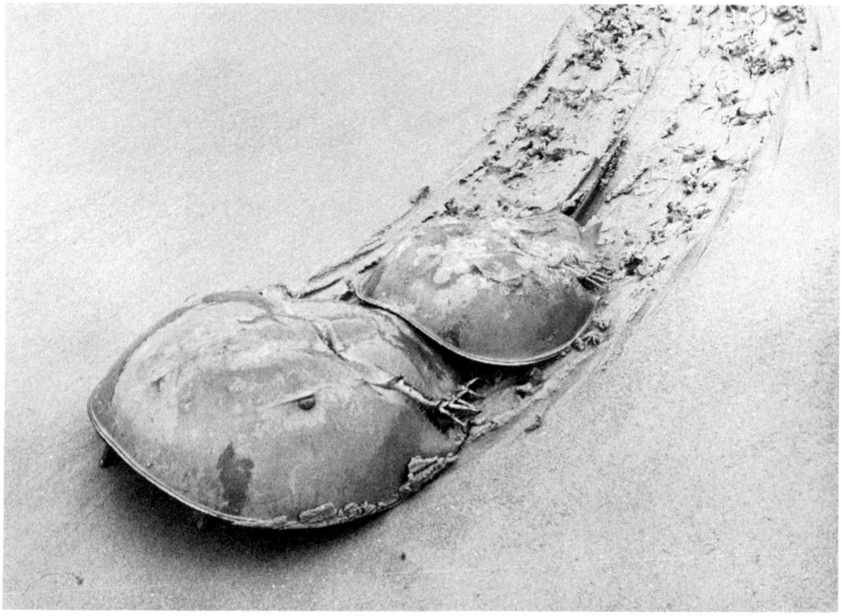

× 0.2

Figure 10.23
A pair of horseshoe crabs mating. The male, the smaller of the two, fertilizes the eggs after the female has laid them in the sand, and the female then covers the eggs with sand. In a couple of weeks they hatch, long after the couple has returned to the water. Horseshoe crabs, also called king crabs, look so similar to crustaceans that they were for a long time classified with them. Actually more nearly related to spiders, they are now usually considered to be in a class of their own.

ago or more, and somewhat similar animals have existed as far back as the Cambrian Period, more than 500 million years ago. Only a few species have survived, but those are locally abundant. Resembling crustaceans, horseshoe crabs have a tough *carapace,* jointed to a smaller abdomen, and a slender, sharp, tail-like *telson.* The horsehoe crabs of the Atlantic coast are used to furnish body fluids for immunological research.

Another class of Chelicerata is the **Arachnida,** the spiders, scorpions, mites, ticks, and harvestmen, or daddy longlegs (Figure 10.24). The garden spider, *Argiope,* can serve as an example of a familiar spider. It has a two-part body: an anterior *cephalothorax* and a posterior *abdomen.* There are four pairs of legs, a pair of poison fangs, and a pair of *palps,* which are a combination of sensory and chewing appendages. There are eight simple eyes but no antennae. At the tip of the abdomen, the *spinnerets* are capable of extruding a silky filament used for web building. Gas exchange is accomplished by means of thin-layered folds of tissue, called *book lungs.*

Mating in spiders is frequently preceded by elaborate species-specific courtship activities by the small male. He goes through stereotyped movements of waving his palps or assuming special positions, presumably to ensure recognition by the larger female, who might otherwise take him for food. She may eat him later anyway, but only after her eggs have been fertilized. When she lays the eggs, usually in a cocoon of spun web material, up to several hundred small spiders hatch out, and some of them eat others before breaking out of the cocoon.

Despite their dark reputation in popular folklore, spiders are generally harmless creatures, more our allies than our enemies because they feed largely on insects. They rarely bite anyone, and then only when pressed. Even the infamous black widow (*Latrodectus mactans*), with a pair of red triangles on the underside of the abdomen, is so peaceable that inducing her to bite for experimental reasons is difficult. Black widows will on occasion bite, but the

(a) × 3

(b) × 2

Figure 10.24
Arachnids. (a) A spider, *Dolomedes,* sucking the blood from a small fish it has caught. The four pairs of legs and the two-part body are evident. Spiders are generally carnivorous, but live mainly on insects. (b) A scorpion in defensive position, with its pincers extended and its tail raised over its back. At the tip of the tail is a venomous sting attached to a poison bulb. A scorpion's sting is not dangerous but can be painful and can kill small animals. A mother scorpion carries her young for a while on her back. The scorpion's shape is so characteristic that it has been recognized for centuries as special, even giving its name to the constellation Scorpio.

(a) × 1

(c) × 0.7

(b) × 0.5

Figure 10.25
Representative crustaceans. (a) A shrimp, *Palaemonetes.* (b) An edible crab. In crabs the abdomen is reduced and tucked under the thorax so completely that it is invisible from the top side. (c) A cluster of barnacles. These active barnacles have extended their feathery *cirri* (singular, cirrus), with which they agitate the water and obtain food. Except during larval life, barnacles are firmly attached to some solid surface, such as rocks, driftwood, or ship bottoms.

number of instances in the United States has dropped since the almost universal installation of plumbing in houses. The other potentially dangerous American spider, with the sinister name of brown recluse (*Loxosceles reclusa*), lives in the lower midwestern and southeastern states. In contrast to spiders, scorpions do not bite. They sting with their tails, and though the sting is painful, it is usually not lethal.

The other arachnids are either completely harmless, as the daddy longlegs is, or do their damage by transmitting disease. The most dangerous ones are ticks, which not only carry Rocky Mountain spotted fever but can cause painful allergic reactions in humans and domestic animals.

Subphylum Mandibulata

The **Mandibulata** (Latin, "jaws") differ from the Chelicerata in having the front appendages in the form of antennae and chitinous jaws that work from side to side (instead of up and down as vertebrate jaws do). Containing as it does the crustaceans and the insects, the subphylum has the greatest number of species and individuals of any group of present-day multicellular organisms.

Class Crustacea The **crustaceans**—crabs, shrimps, lobsters, crayfishes, barnacles, and a host of smaller animals—are mainly aquatic, and they occupy a position roughly comparable to that of the insects on land (Figure 10.25). A crustacean typically has

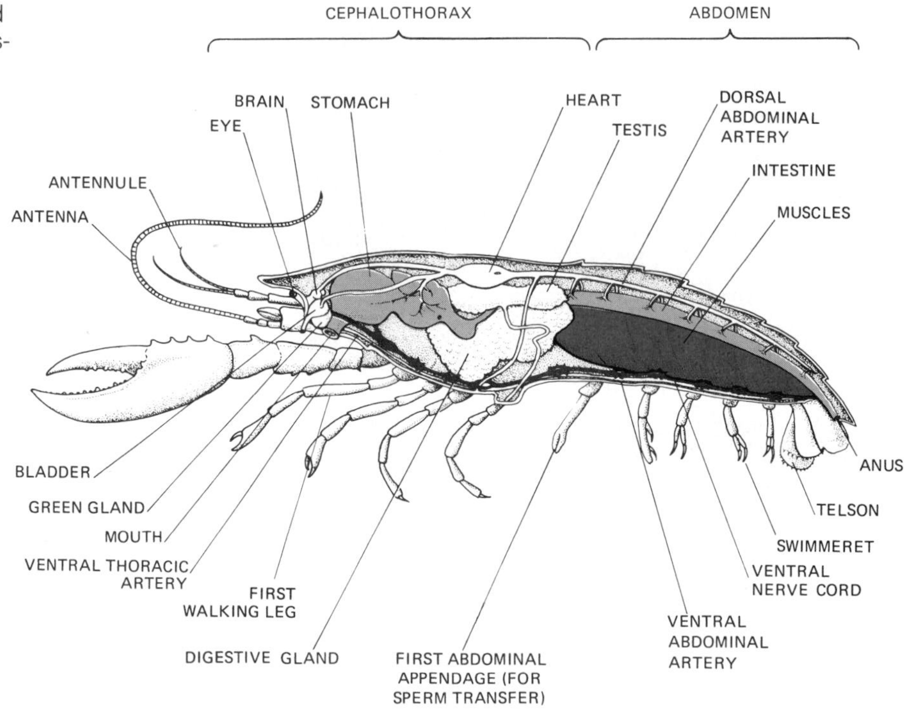

Figure 10.26
Internal anatomy of a lobster, a representative crustacean. Notable organs are the small heart, a stomach with internal grinders, a nerve cord, and the large, powerful abdominal muscles.

two pairs of antennae; a pair of mandibles, or jaws; and two pairs of paddlelike appendages, or maxillae. Other appendages, attached to segments along the body, are variously modified for grasping, walking, swimming, or sexual activities. The body is covered with an exoskeleton of calcium-impregnated chitin. The general body plan is segmented, but the segmentation is obscure in some parts. There is usually a head and thorax, although these two parts may be more or less fused into a cephalothorax, and there is usually a clearly segmented abdomen. The evolutionary source of the Crustacea is unknown, but one generally accepted guess is that early crustaceans were like trilobites.

Crayfish, which are structurally like lobsters, are frequently used as examples in descriptions of Crustacea. The body of a crayfish is divided into two recognizable parts: a cephalothorax, with a pointed *rostrum* at the anterior end; and a jointed, flexible abdomen (see Figure 10.26). The cephalothorax is covered by a firm *carapace* that covers the sides of the body like overhanging eaves of a roof, protecting the feathery gills through which oxygen and carbon dioxide are exchanged. The abdomen has six segments, each with a firm exoskeleton but united to the next segment by a flexible membranous joint, as are the joints of the appendages. At the tail end there is a paddle-shaped extension, the *telson,* which the animal uses to escape danger. When a crayfish is startled, it can curl its abdomen, which is mostly filled with muscle, thus giving its telson a sudden pull forward, and that movement jerks the whole animal backward. There are five pairs of *walking legs,* the first of which is tipped with a strong pinching claw, all attached below the cephalothorax. Five pairs of *swimmerets,* like hairy forked paddles, grow from the underside of the abdomen.

Crayfish feed by grasping large particles with their pincers or by fanning small ones to the mouth. A pair of antennae and a smaller pair of antennules help it find food by smelling or tasting. The *mandibles* can tear food but not chew it. Food is chewed in the stomach, where three horny teeth form a *gastric mill,* which grinds coarse food. Fine particles are digested by a *digestive gland* that secretes hydrolyzing enzymes; the coarse material is passed on through the intestine and out the anus. Other body wastes are collected by a *green gland,* which functions as a kidney and passes the waste out through an opening near the base of the antennae. The circulatory system is open; that is, the yellowish blood is pumped by a dorsal heart into vessels that empty into the tissues. Channels through various tissues allow the blood, with its copper-containing pigment (in contrast to the iron-containing pigment of red-blooded animals), to flow past the gills and eventually to make its way back through the heart. A ganglion above the esophagus, serving as a brain, is connected by a ring of nerves to another ganglion below the esophagus, then by a ventral nerve cord along the body, with a ganglion in each segment and branching nerves from them to the rest of the body (Figure 10.26).

When crayfish mate, the male uses his first pair of swimmerets to transfer sperm from his body to *seminal receptacles* in the female's body. The sperm remain in the receptacles until the female lays eggs, several hundred at a time, and the eggs are then fertilized. The eggs are attached to the mother's swimmerets until they hatch out as miniature crayfish. When a female crayfish or lobster is carrying eggs, she is said to be "in berry."

When crustaceans or any animals with exoskeletons grow, they swell inside their exoskeletons until something must give. When hormonal conditions so direct, the outer covering splits, and the animal pulls its soft inner parts completely out of the old shell, down to the tips of claws, antennae, and even eyes, like pulling a hand out of a glove. After this process of **molting,** the body is temporarily vulnerable, but it quickly hardens a new and larger covering. "Soft-shelled crabs" are crabs caught while in an early postmolt condition. Most of the increase in size occurs just after the molt, when the crustacean swells with the intake of water. As long as it lives and grows, it must continue to molt at intervals (see Chapter 18).

Body form is as variable in the Crustacea as one would expect in so large and ancient a group. There are the familiar lobsters; Alaskan crabs with broad cephalothoraxes, little abdomens tucked under, and legs with a spread of three meters; tiny ostracods, swimming like minute bivalves; copepods with one eye, named *Cyclops* for the mythical Homeric monster; brine shrimps, popularly sold in dry dust to hatch out in home aquaria as "sea monkeys"; and barnacles with calcareous plates, long thought to be molluscs. But if body form varies, it is with the appendages that evolution has been most prodigal in its diversity. From the rather simple antennules of lobsters and the two-branched swimmerets, there have developed uncounted differences in length, shape, and ornamentation. Some of the deep-sea crustaceans have such feathery appendages, several times body length, that they seem to be all frills.

Crustaceans are important to humans in a number of ways. Aside from the obvious direct consumption of crabs, crayfish, shrimps, prawns, and lobsters as human food, there is the indirect feeding of most marine fishes, which live largely on small crustaceans. The cool waters of the Antarctic and adjacent oceans are rich in small shrimplike krill, the main food source of the baleen whales and other large marine animals. Barnacles, attaching themselves to the hulls of ships, can create enough drag to reduce speed considerably, and they have to be scraped off. Crustaceans are also indirectly responsible for disease in that they are hosts for parasites, such as the broad tapeworm, that also infect humans.

Classes Chilopoda and Diplopoda The **Chilopoda** and **Diplopoda** are small classes of Arthropoda with little biological or economic importance. The first are the **centipedes** (Latin, "hundred feet"), segmented animals with one pair of legs on most segments (Figure 10.27). Though some tropical centipedes can bite painfully, they are mostly harmless. The temperate ones live on insects, but their help in killing cockroaches and bedbugs is not generally appreciated, because they are too slithery and too much like small dragons to appeal to nonbiologists. The Diplopoda are the **millipedes** (Latin, "thousand feet"),

× 2

Figure 10.27
A representative chilopod: a centipede, also called a hundred-legger. Some centipedes have a slightly poisonous bite, but their relatives, the millipedes, are quite harmless.

differing from the Chilopoda most conspicuously in having two pairs rather than one pair of legs on each segment of the body. They are mostly vegetarian scavengers, living in moist, dim places, rarely inside houses. (There are several other minor classes of Mandibulata that are not treated here.)

Class Insecta The **insects** (Latin, "cut into," in reference to the thin neck and wasp waist) are sometimes known as the Hexapoda (Greek, "six feet"). They represent the culmination of terrestrial evolution in the arthropod line, with a bewildering variety of forms and activities. With some 900,000 described species, and with new species being found all the time, insects occur in more forms than do the members of any other biological group.

Insects affect people in every aspect of daily living. They eat our crops and flowers, destroy our stored grains, gnaw holes in clothes and carpets, tunnel through the woodwork, pulverize books, bite our skins, pester our livestock, parasitize our pets, and infest our food. They carry plant diseases, such as yellows, mosaics, witches'-brooms, rots, knots, cankers, scabs, stains, mildews, wilts, blights, streaks, and stunting. They also transmit animal diseases, including human diseases: tapeworm, lung fluke, and guinea worm infections, malaria, sleeping sickness, cattle fever, blacklegs, dysentery, plague, anthrax, typhoid fever, cholera, typhus, spotted fever, yellow fever, and encephalitis. Yet the same bees that may sting painfully—even to death if the victim is allergic to the stings—provide honey. Bees are responsible for the pollination of most of the fruits that people eat and most of the flowers that people enjoy (the staple grains are wind pollinated).

The general activities of nature and the economy of humans depend on the insects. Insects give silk, serve as fish bait, make shellac and dyes, aerate soil, help keep down undesirable weeds, feed songbirds, and even provide medicines. Butterflies, moths, and beetles are sources of artistic inspiration, and they offer pleasure to many amateur and professional *entomologists* (students of insects). Ants, grasshoppers, grubs, and caterpillars are eaten by humans in many parts of the world. In our competition with insects, we find insects themselves our allies against their own kind. Ladybugs eat scale insects, and mantises eat almost anything they can catch. Wasps parasitize caterpillars, and aquatic bugs eat mosquito larvae. Finally, insects are great scavengers, helping clean up the dung, feathers, and dead carcasses of animals. On balance, in spite of the trouble they can cause, insects probably do more good than harm, and certainly without them life as we know it would change enormously—and probably for the worse.

As evidence of the intimacy with which we regard insects, we have applied a wealth of names to thousands of them, many of them picturesquely illuminating. Heading the list are insects named for animals: toad bugs, lizard beetles; cuckoo wasps; hawk moths; elephant beetles; tiger, rhinoceros, and ermine beetles; leopard, owlet, and virgin tiger moths; zebra swallowtails; and orange dogs. Some insects are named for other insects: flea beetles; spider beetles (although spiders are not insects); and bee moths. Some are considered outlaws: assassin bugs; robber flies; ambush bugs; and masked hunters, which earned their title by accumulating under-the-bed dust on their heads. Less sinister are the checkered skippers, the pinching bugs, and the kissing bugs, which tend to bite people around the mouth. Some seem military, such as the army worms and army ants, the bombardier beetles, and the sword-bearing crickets. Glamour comes with the handsome earwigs, bog damsels, painted beauties, and darling underwings; and the supernatural world is represented by fairy moths, ghost moths, deathwatch beetles, phantom craneflies, and black witches. Mental states are applied in thick-headed flies and confused flour beetles. Even modern technology enters the picture, with wheel bugs, which have an obvious piece of cogwheel growing out of the thorax; railroad bugs, with green lights in front and red lights behind; and the short-circuit beetles, which chew their way into lead-covered electrical cables. These are all living insects, and these names can be found in standard entomological works.

In the some 300 million years that insects have inhabited the earth, they have come to occupy practically every available habitat except the sea. One genus, *Halobates,* does ride the surface of salt water, even far out at sea, and many live in salt or brackish marshes, but almost all insects are terrestrial or live in fresh water. They are not usually noticed because most of them are small, but even in apparently inhospitable regions, the soil is teeming with insects. (There are more insects in 10 square miles of Arctic tundra than there are mammals in all of North America.) They live all over animals, penetrating their skins, clinging to fur and feathers, and crawling into any available opening. They eat plant roots, hollow out stems, suck or chew leaves, or tunnel along between upper and lower epidermises of leaves, and eat their way into fruits and seeds. Peo-

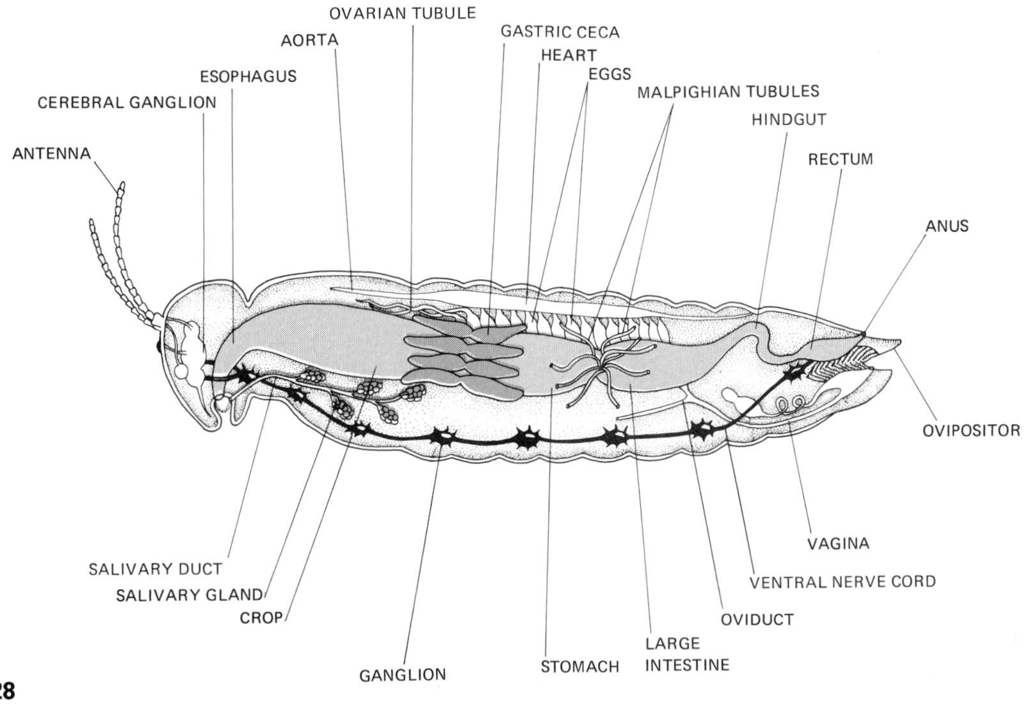

Figure 10.28
Internal structure of a grasshopper as a representative insect. Note especially the cerebral ganglion, or "brain," the ventral nerve cord with repeated ganglia, the large digestive tract with outpockets (gastric ceca), the large crop for storage of food, and the digestive glands (salivary). The Malpighian tubules are excretory organs. The ovipositor delivers eggs.

ple and their houses offer special if unwilling protection to a host of insects, the more obnoxious ones being fleas, lice, bedbugs, cockroaches, silverfish, earwigs, and houseflies. Insects may thrive in icy streams, hot springs, snowfields, oil slicks, caves, manure heaps, haystacks, and cadaver vats in medical schools. Unless the larger scavengers (vultures, opossums, crows, jackals, hyenas, eagles, etc.) get to them first, dead animals are consumed by a succession of insects from the fur and skin eaters to the carrion beetles and fly larvae, or maggots. By far the most delicate and accurate way to prepare skeletons for study is to let beetles clean the bones.

The large insects attract attention by their very size, since a 12-centimeter (5-in.) beetle, a 30-centimeter (1-ft) stick insect, or a 25-centimeter (10-in.) moth may be spectacular. However, the great majority of insects are so small that they pass unnoticed by most people. The average insect is less than 5 or 6 millimeters (¼ in.) long, and some tiny ones are recognizable only with a lens and identifiable only with a microscope.

Communication among insects is as varied and complex as the other insect activities. Much communication is visual; males especially have specific shapes or colors, or they are capable of stylized movements that make them recognizable to females. Nocturnal insects can recognize and find one another in the dark. Some do so by means of flashing lights, others by means of characteristic sounds, usually produced by scraping one body part (such as a rasping leg) across another part. The most sensitive communication method is through smell; insects have much more accurate perception of smell than vertebrates do (Chapters 29 and 30).

The structure of insects The usual body plan of an insect consists of a head attached by a more or less flexible neck to a thick thorax, which continues as a segmented abdomen (Figure 10.28). The head bears a pair of antennae (which range from short and stubby to long and plumy, with many variations in between), a pair of compound eyes and several simple eyes, and a set of mouth parts. The compound eyes are made up of a few to several thousand individual components, the **ommatidia,** each one a slender cone with a light-sensitive receiver at the base. Such eyes are not as effective in perceiving still im-

× 10

Figure 10.29
The face of an insect. The most conspicuous features are the long antennae, the bulging compound eyes with hundreds of units (ommatidia), and the heavy mandibles that work from side to side. This specimen is a yellow jacket, *Vespula*.

ages as they are in perceiving motion, but they have wide angles of vision in that they point in many directions. A dragonfly can see in almost all directions at once because its bulging eyes point practically every way except inward. The mouth parts of insects are adapted to chewing, piercing, lapping, or sucking (Figure 10.29). In some insects, such as mayflies, they are so reduced that the adult form does not eat. A primitive set of mouth parts, from which more specialized types are thought to have evolved, consists of a hinged upper lip overlying a pair of side-to-side *mandibles,* or jaws, and other appendages. From this basic arrangement, mosquitoes have developed stilettoes for piercing skin (although mosquitoes live mostly on plants and only females bite animals), flies have a bulbous set of appendages for sponging, and butterflies have a structure that can suck up nectar or coil up like a watch spring when not in use.

The insect thorax bears the locomotor organs: three pairs of walking legs and in most insects two pairs of wings. Wings may be leathery, horny, membranous, scaly, feathery—or lacking—but they are usually characteristic of members within a group. Separation of insects into orders is usually based on wing structure. Wings have been of enormous importance to insects because flight has enabled them to escape predators, to take advantage of otherwise inaccessible habitats, and to be carried on wind currents wherever wind blows. An occasional insect spatters against the windshield of an airplane flying at a height of six to eight kilometers (four to five miles). Walking or running is a more stable means of moving for an insect than it is for a two- or four-legged animal, because the insect always has three feet on the ground.

The insect abdomen is without appendages except for the genitalia and, in females, egg-laying apparatus. Along the sides of the abdomen, breathing holes, **spiracles,** provide access for oxygen and carbon dioxide exchange, leading into a system of branched tubes known as **tracheae.** Adult insects have no lungs, but they do pump air in and out of the tracheae through the spiracles, contracting and relaxing the abdominal muscles in a way that resembles vertebrate breathing. Respiratory gases are delivered directly to the tissues, in which colorless blood oozes by means of an open circulatory system. There is a heart, but movement of the blood is accomplished more by body movement than by forceful heart action.

The digestive tract in insects leads from the mouth to an enlargement of the esophagus called a crop, then to a gizzardlike grinding sac, and finally to the hindgut. Along the way, digestive enzymes work on the food, and nutrients and water are absorbed through the gut wall until only a compressed, rather dry pellet of fecal matter is passed out.

The diet of insects is so various that no brief description of the enzymes can be made. Termites, "digest" cellulose, not by their own enzymes, however, but by the enzymes of the bacteria that accompany the protozoa in the termite gut. The larvae of some flies spit out enzymes that digest the meat on which they are growing, and then all the larvae have to do is lap up the digested protein. Mosquitoes inject a droplet of saliva into the puncture of an animal skin, thus preventing clotting of the blood and making the characteristic itching irritation. Mayflies do not feed during the adult stage; they mate, lay eggs, live a few days, and die.

Reproduction of insects The life of social insects is so specialized and complex that it will receive treat-

(a)

(b)

(c)

(d)

Figure 10.30
Complete metamorphosis in the monarch butterfly, *Danaus plexippus.* (a) A tiny caterpillar hatches from an egg and eats voraciously until it increases as much as ten thousand times in weight. (b) When it reaches its full size, it forms a protective covering about itself and becomes a pupa. The pupal coat is formed inside the skin of the caterpillar. During the pupal stage, the animal seems inactive, but inside the case the larva is completely reorganizing to form a new kind of body, that of the adult, or *imago.* When the imago is mature and external conditions are suitable, the pupa case splits down the back (c) and the soft, moist butterfly crawls out. (d) Blood is pumped into the delicate wings, and as soon as they dry and harden, the butterfly is ready to fly.

ment on its own in Chapter 30. The sexes are separate in insects, and fertilization is internal. Beyond that, few generalizations can be made. In the simplest types, eggs are laid, and they hatch into small copies of adults. In many orders of insects the newly hatched individual is somewhat like an adult but has juvenile features. The young, called **nymphs,** go through several stages of development, the *instars,* becoming larger and more mature with each molt, and finally arriving at adulthood. The series of changes is *gradual* or **incomplete metamorphosis.** In the orders that contain flies, beetles, and butterflies, there is **complete metamorphosis** (Figure 10.30). From the egg comes a **larva** (Latin, "ghost"), which may be a maggot, grub, caterpillar, or other rather wormlike, actively feeding form. After a series of molts, the larva covers itself with a protective coat, either of chitinous or of spun fibrous material (silk), and goes into a period of apparent inactivity. In fact, it is during this **pupal stage** that major changes occur, eventually giving rise to the **adult,** or **imago.** When the pupa opens, the imago climbs out, soft and moist, leaving the pupal shell behind. The new imago stretches and plumps up its body and appendages, dries, and is ready for adult life. (Further discussion will be found in Chapter 18.)

× 3

Figure 10.31
A representative insect: a cockroach, *Blatta orientalis,* producing an egg case. Relatively primitive insects, cockroaches have been and remain enormously successful animals in terms of survival and sheer numbers. Three species of roaches, some called "water bugs" or Croton bugs, infest houses—this one, *Blattella germanica,* and *Periplaneta americana.*

× 20

Figure 10.32
A weevil on a hairy petal of an iris flower. The long snout on this weevil, which is a kind of beetle, has a pair of jaws at the tip with which it can gnaw at plant tissues. It can eat the tissues or make holes in which to lay eggs. The eggs hatch into larvae (grubs), and they, too, eat plant tissues. Weevils are well adapted to live on hard, dry foods, and they can be especially destructive to stored grains, peas, and beans.

The orders of insects Of the 27 orders of insects recognized by most entomologists, many are small or not commonly known, or they consist of such minute animals that they are seen only by specialists, but half a dozen are so conspicuous and widely distributed that they are known to some extent by almost everyone. One relatively primitive order is the **Order Orthoptera** (Greek, "straight wings"), which includes grasshoppers, locusts, katydids, mantises, cockroaches, walking-sticks, and crickets (Figure 10.31). The **Order Coleoptera** are the beetles and weevils (Figure 10.32). There are more beetles than any other group of organisms on earth, and nearly 30,000 species of them live in the United States.

In the **Order Lepidoptera,** (Greek, "scale wings") the butterflies, moths, and skippers have four wings, which, like much of the body, are covered with loosely set, microscopic scales (Figure 10.33). As a result they have a dusty appearance.

In the **Order Diptera** (Greek, "two wings"), the hind wings are modified into *halteres,* or balancers (Figure 10.34). These insects are the true flies, mosquitoes, gnats, and midges.

To an entomologist, the only true *bugs* are in the **Order Hemiptera**; all other "bugs," despite popular terminology, are technically not really bugs. Bedbugs, stinkbugs, lace bugs, and giant water bugs are true bugs, with the typical thickened forewings

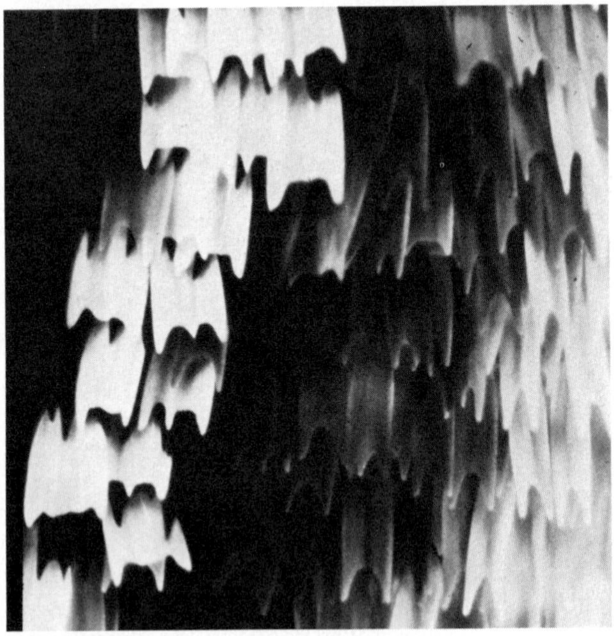

× 50

Figure 10.33
Photomicrograph of the wing scales of a large moth, *Hyalophora cecropia.* Scales give the wings of butterflies and moths a powdery appearance and also create distinct color patterns.

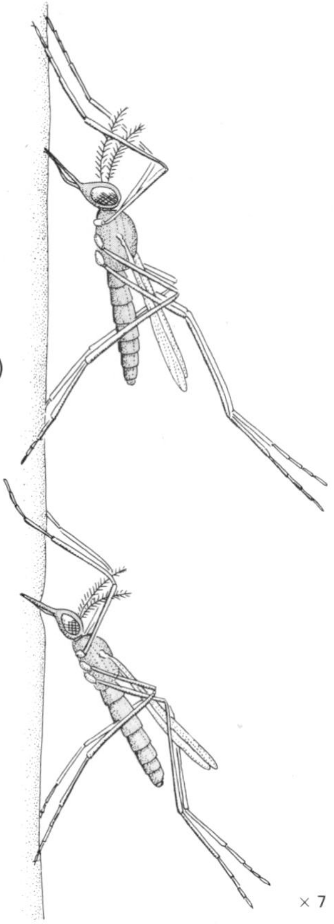

Figure 10.34
Mosquitoes, members of the same order of insects as flies. (a) A *Culex* mosquito in a biting stance, with her body parallel to the surface. These mosquitoes are annoying but do not carry human disease. (b) An *Anopheles* mosquito in a biting stance, with her body tilted head downward. *Anopheles* is the malaria carrier.

crossed over the delicate hindwings to form a triangle on the top of the thorax.

The **Order Hymenoptera** (Greek, "membrane wings") are the bees, wasps, and ants. One final order of insects, the **Order Thysanura,** should be mentioned because they are believed to represent the most primitive of living insects. They are exemplified by the common household pest, the silverfish (Figure 10.35).

Figure 10.35
Wingless insects of the Order Thysanura. These are young and mature examples of the silverfish, *Lepisma saccharina,* which lives commonly in houses, eating the glue and paper-filler in books.

TO SUM UP SPONGES THROUGH INSECTS

1. *Porifera,* the sponges, are primitive animals, with little organization of function.
2. *Coelenterata* include corals and jellyfishes. They have a radial body symmetry, stinging cells, simple nerve cells, tentacles, and a simple sac for food intake and elimination.
3. *Platyhelminthes* are the flatworms, the simplest bilaterally symmetrical animals. Included in this phylum are planaria and the parasitic flukes and tapeworms.
4. *Aschelminthes,* the roundworms, have a mouth, a tubular alimentary canal, and an anus. They include parasitic hookworms and trichina worms, in addition to many harmless worms.
5. *Mollusca,* the molluscs, usually have soft bodies encased in hard shells. Some well-known molluscs are clams, snails, octopuses, and periwinkles.
6. *Annelida,* the segmented worms, have clearly identifiable organs for most body functions. They include earthworms and leeches.
7. *Arthropoda* include spiders, shrimps, centipedes, and insects. There are more arthropods than members of any other animal phylum.

Section 2: Echinoderms through Mammals

PHYLUM ECHINODERMATA

The **echinoderms** (Greek, "hedgehog skin") are almost radially symmetrical animals with a trace of bilaterality in the adult forms and clear bilaterality in the larvae. They are therefore thought to have evolved from a bilateral ancestor, with their radial shape as a secondary development. In their embryological development, echinoderms are different from the previously described phyla, but they are like the chordates, which will be treated later. Echinoderms are deuterostomes; that is, the formation of the body during early development is characterized by the growth of a digestive tract with a mouth breaking through at the anterior end, away from the embryonic blastopore. This general and fundamental event led biologists to classify the echinoderms and chordates as one major branch of evolutionary development, under the assumption that these two groups had a common ancestor and that they are different from the other phyla, such as the Mollusca and Arthropoda.

Echinoderms are strikingly different in their adult forms from any other animals. Aside from their apparent radial symmetry, they have embedded in the body wall skeletal elements of calcareous plates or spiny spikes. Some echinoderms (sea urchins) have shell-like *tests*. One special feature, a *water-vascular system*, can be used as a hydraulic pump to extend the soft, extensible *tube feet*, which have muscular contractility and carry terminal suckers (see Figure 10.36). Echinoderms use their tube feet for movement, capturing food, respiration, and sensory perception. Another peculiarity of echinoderms is the possession in some species of *pedicellariae* (singular, pedicellaria), hard, jawlike pairs of pincers that act like tiny pliers with crossed jaws and that are useful for cleaning the skin surface. Echinoderms are widely distributed in practically all marine habitats, from the shallowest to the deepest seas. They are a peculiar side branch of animal evolution and have given rise to no more advanced types.

Classes of Echinoderms

Five classes of echinoderms are living. The **Class Crinoidea** contains the stalked sea lilies. The **Class Asteroidea** is familiar to anyone frequenting temperate or tropical shores. Its members are the star-

× 0.35

Figure 10.36
The underside of a starfish. The white spots on the body are calcareous plates that give the outer skin its hardness. Tube feet are visible along the groove in each arm. The mouth is at the center of the star, and a light-sensitive organ is at the end of each arm.

× 2.5

Figure 10.37
The underside of a starfish (*Asterias*). The grooves under the arms are filled with waving *tube feet*, which the animal can push out by filling them with water or can retract by contracting the muscles in each tube. At the end of a tube foot, there is a sucker that works like a vacuum pad. Using many of those suckers, a starfish can exert a strong and continuous pull and open a tightly shut mussel, as this one is doing.

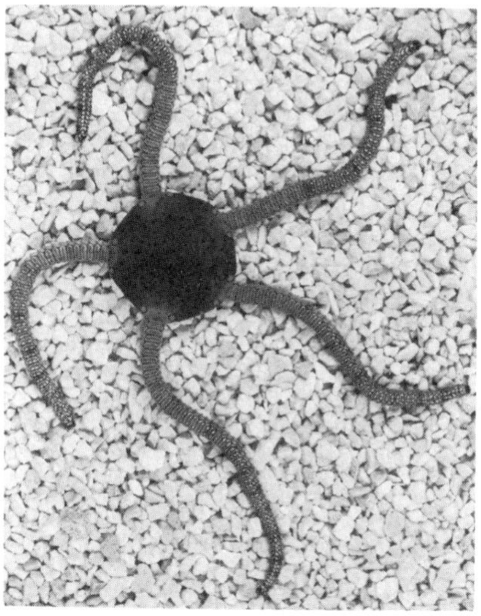

× 1

Figure 10.38
Brittle star of the Class Ophiuroidea. They are called brittle stars because their arms break easily; they can grow new ones within a few weeks.

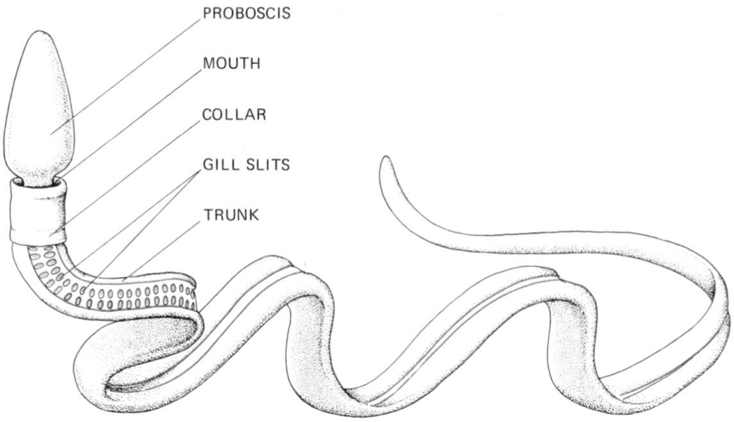

Figure 10.39
A hemichordate, *Dolichoglossus,* (closely related to *Balanoglossus*), a wormlike dweller in New England sea water.

× 1

fishes (Figures 10.36 and 10.37). The five-rayed body is stiffened by endoskeletal plates, protecting both the internal water-vascular system and the extensive digestive system. The *ambulacral groove* along the underside of each arm is lined with tube feet, with which the starfish can creep and capture prey. The **Class Ophiuroidea** contains the brittle stars, which are like starfishes but have slender or much-branched arms (Figure 10.38). The **Class Echinoidea** is characterized by the presence of a firm globular or flattened skeleton or test, from which movable spines radiate. Some common echinoids are sea urchins, sea biscuits, and sand dollars, whose bleached tests are cast up by the millions on beaches all over the world. The **Class Holothuroidea** contains the sea cucumbers, which are soft-bodied animals like flabby, wet sacks with a ring of tentacles at the oral end. When disturbed, a sea cucumber can throw up its insides and cast away the whole mass, later regenerating a new set of organs. Some Asian people consider the sea cucumber a gastronomic delicacy, calling it "trepang."

PHYLUM HEMICHORDATA

The **hemichordates** are soft-bodied, usually wormlike marine animals, little known except to biologists, who regard them with interest because they are thought to occupy a unique position in the evolutionary development of the vertebrates. Their deuterostome embryonic development shows that they belong in the echinoderm-chordate line of evolution. They are like chordates in having gill slits (although they use them mainly for feeding) and a hollow, dorsal nerve cord. They also have a rodlike structure that for many years was thought to be comparable to the notochord of vertebrate animals. The *notochord* is a cellular rod, extending almost the entire length of the animal just ventral to the dorsal nerve cord. It acts as a stiffening support in those animals that retain it throughout life and in the embryos of those animals that replace it with a backbone.

One of the better-known hemichordate examples is the acorn worm, *Balanoglossus,* a secretive mud-dweller whose main claim to fame is its intermediate position in evolution between the echinoderms and the chordates (Figure 10.39).

PHYLUM CHORDATA

Although **chordates** vary from tiny, immobile sea squirts, through such unlikely phylum-mates as hagfishes, to chickens, whales, and humans themselves, they are all classified in one great biological group because they have several features in common and are therefore thought to have evolved from a common ancestral group. No one knows what manner of animals those ancestors were, or even if there was one common ancestral group. The evidence seems reasonable that the chordates share a beginning with the echinoderms. Both phyla are deuterostomes, and current opinion has it that echinoderms and chordates represent one important, two-forked line of evolutionary development.

In addition to the characteristics they share with echinoderms, the chordates have a number of features uniquely their own. (1) Every chordate has a *notochord* at some stage of individual development, although it is a temporary embryological organ in most species. (2) A *tubular nerve cord* runs the length of the body, near the dorsal suface, and is enlarged and complicated at the head end into a brain. (3) *Gill slits* are openings in the region of the pharynx that lead from the internal alimentary canal to the outside of the animal. Gill slits function in respiration in fishes, but they are closed early in embryological development in air-breathing vertebrates. The slits are clearly evident in young human embryos. (4) A *tail*, extending beyond the anus, is always present at some stage of development, even though it may be absorbed or overgrown in such animals as frogs and humans. Other characteristic features of chordates are shared with animals of other phyla: bilateral symmetry; evidence of segmentation in the body; a well-defined head end; the usual three embryonic germ layers of ectoderm, mesoderm, and endoderm; a well-developed coelom; paired appendages in most groups; and a firm internal skeleton, or endoskeleton. A few have external skeletons, or exoskeletons.

× 6.5

Figure 10.40
An ascidian, or sea squirt. During the adult stage of a sea squirt's life, shown here, the animal is not recognizable as a chordate or perhaps even as an animal.

Subphylum Urochordata

Little known except to biologists and beachcombers, the sea squirts (Figure 10.40) are so different in general appearance from most chordates that they would not be recognizable as such, except for their larval stages. Larval sea squirts, however, have the essential features of chordates: the notochord, the hollow dorsal nerve cord, and the pharyngeal gill slits. After a free-swimming stage, the larva attaches itself to a solid substrate and undergoes such a series of changes that by the time it has become an adult it is a different-looking animal. The adult is a *sessile* (attached) animal, usually growing in colonies on marine vegetation, where it feeds by filtering particles out of the water and forcibly ejecting spurts of water from its excurrent siphons—hence the name "sea squirt."

The resemblance of the larva to a hypothetical chordate ancestor and the lack of resemblance of the adult to anything else have prompted the suggestion that the entire vertebrate line of evolution has been from a larva that gained the power of reproduction. Some larvae in flatworms, insects, and amphibia can indeed reproduce, so such a suggestion is not as farfetched as it might seem at first inspection.

Subphylum Cephalochordata

The most famous **cephalochordate** to biologists is a small fishlike creature known as the lancelet (or amphioxus), *Branchiostoma* (Figure 10.41). Lancelets, each only about 5–7 centimeters (2–3 in.) long, burrow backward in the sandy shores of warm seas, where they feed by filtering the water. They resemble larval sea squirts, but they remain permanently free-living. They have the notochord, dorsal nerve cord, gill slits, and postanal tail characteristic of chordates, and they are so like what zoologists imagine the prototype chordate to have been that

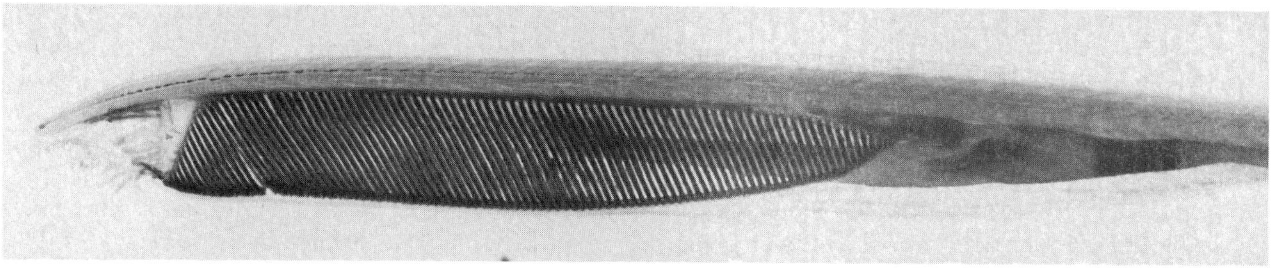

Figure 10.41
A lancelet or amphioxus. This tiny fishlike animal, a sand-dweller along sea beaches, serves as a prototype chordate. It has a dorsal nerve cord, and it retains its notochord and gill slits throughout its life. In this photograph, the V-shaped, segmented muscles are faintly visible, but most of the internal organs cannot be seen. Amphioxus may well represent the kind of animal from which the more advanced chordates, the vertebrates, evolved. However, amphioxus is still thriving. In some places, such as in the Pacific Ocean along the China coast, they are abundant enough to be caught and eaten.

× 4

they have been called "the blueprint of the phylum."

Subphylum Vertebrata

To most nonbiologists, the **vertebrates** are *the* animals. In addition to the usual chordate features, vertebrates have a number of structures all their own (Figure 10.42). These include a special kind of covering, generally called skin, consisting of at least two layers, one of ectodermal and one of mesodermal origin, with capacity for forming a number of accessory parts, such as glands, hair, nails, claws, hooves, feathers, and scales. There is a spinal column of vertebrae, from which the entire subphylum takes its name, and that column usually has two pairs of bone-supported appendages. In the closed circulatory system, the ventral heart pumps red blood throughout the body. A pair of kidneys removes waste from the blood. The brain, at the head end, is encased in some kind of skull, is connected to the typical dorsal nerve cord, and is provided with 10 or 12 cranial nerves. An endocrine glandular system, distributed throughout the body, helps regulate physiology and behavior by means of hormones. The typical body has a head, trunk, and tail, usually fore and hind appendages, and sometimes a neck. Sexes are usually separate. This, then, is the vertebrate plan, and it has been one of the great evolutionary successes, along with the arthropod plan, in the earth's history.

The taxonomic arrangement of vertebrates, especially fishes, is always in a state of change. As in most systems of classification, there is never complete agreement among zoologists as to which scheme is most satisfactory. The arrangement given here is a conservative one, simplified by the omission of some of the smaller and less well known groups.

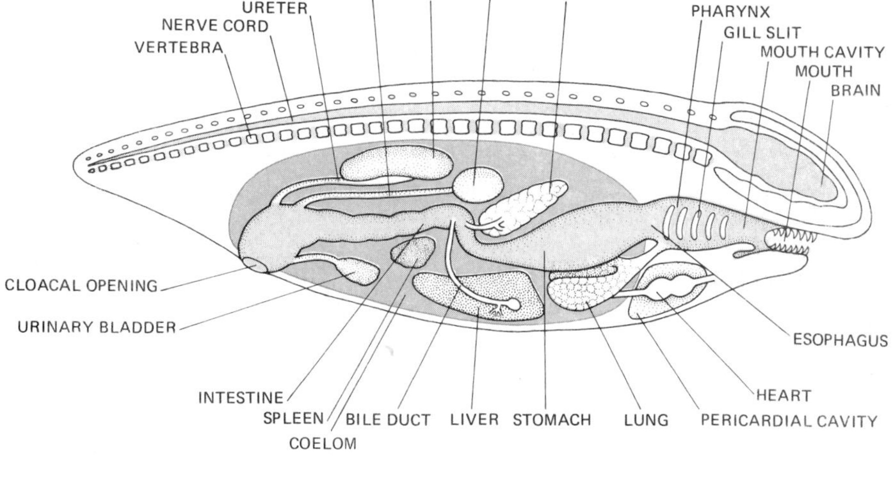

Figure 10.42
A longitudinal section through a stylized vertebrate animal. Only two features are common and lasting throughout the life of most vertebrates: the row of *vertebral bones* and the *dorsal nerve cord* with a brain at the anterior end.

Class Agnatha In the **Class Agnatha** (Greek, "no jaws") are the lampreys and hagfishes. As the name indicates, these animals have no jaws but have a sucker, some with teeth, by means of which they attach themselves to living or dead fishes. Lampreys are such serious predators that once they began to live in the Great Lakes of the United States during the 1950s, they almost destroyed the commercial fisheries there. The parasites attach themselves to a fish and suck out its juices until it dies. Attacking the larval stages, which develop in small tributaries of the lakes, has brought them under partial control. Lampreys have cartilaginous skeletons, eyes, brains, and either male or female gonads, but living as they do on fish juice, they do not need and do not have stomachs. Hagfishes, in contrast, are not parasites, only scavengers. Agnatha are regarded as primitive examples of vertebrate development.

Class Chondrichthyes The **Class Chondrichthyes** (Greek, "cartilage fishes") contains the sharks, skates, and rays (Figure 10.43). Their skeletons are of cartilage instead of bone. The typical shark body is streamlined into a spindle shape, but in rays the basic shape is obscured by the outgrowth of lateral "wings," which give the fishes something of a butterfly appearance. The mouth on the underside of the head is equipped with rows of teeth, which are continually replaced as they are lost. Five gill slits allow the passage of water from the pharynx over the gills to the outside. Nostrils, indentations in the underside of the head, do not open to the inside because they are functional only as smelling organs. A pair of anterior *pectoral fins* and a pair of posterior *pelvic fins,* supplemented by two median dorsal fins and an asymmetrical tail, provide movement. The skin feels rough because of the presence of tiny enamel-covered scales, formed in the same way as the teeth in the mouth. Along each side of the body a *lateral line organ* acts as a sensitive receptor of pressure changes. With that sensory organ the fish can detect currents, prey, and mates. Sharks also have functional eyes, but they have no eyelids, so it is impossible to know whether or not they sleep. Sharks, however, do go into periods of immobility, during which they act as if they must be asleep.

Figure 10.43
A sand tiger shark.

Sharks are neither so ferocious as fictional stories imply nor so mild-tempered as some skin divers insist. They are generally rather innocuous creatures, but well-authenticated instances of shark attacks on swimmers have been recorded, especially attacks by great white sharks, mako sharks, and tiger sharks. Rays, too, can be dangerous because of their rasping tails, but they are more impressive than actually injurious, with their enormous winglike flaps. Some rays can leap out of the water and wave themselves ponderously, and because of their great size they seem to be flying in slow motion. Other rays have the ability to build up enough electrical potential to give a stunning shock even to a human adult.

Blood circulates through a closed system, which carries it past the gills, through the body, and back to a two-chambered heart with an atrium and ventricle in tandem. The increase in complexity of the hearts and circulatory systems of more advanced animals is important to watch for in the descriptions coming next, because the different animal groups show a steady progression from the simple shark heart to the double pump of "warm-blooded" animals. In the Chondrichthyes, sexes are separate and fertilization is internal. Some female sharks retain eggs within their bodies until young sharks are partially developed. Other species nourish the young within a uterus until they are born alive.

Class Osteichthyes The bony fishes of the **Class Osteichthyes** (Greek, "bone fishes") are the dominant animals in water as birds and mammals dominate terrestrial environments (Figure 10.44). The bony fishes have evolved into some 18,000 modern species (as compared with only about 3000 Chondrichthyes) inhabiting the waters of the earth,

Figure 10.44
A representative of the bony fishes, or Osteichthyes. This one is a salmon, *Salmo*, swimming up an Alaskan waterfall on its way to the spawning territory where it will lay (or fertilize) eggs.

Figure 10.45
The internal structure of a fish. The bulk of the body is muscle, with a small space along the ventral side containing the digestive, excretory, and reproductive organs. The external nares, or nostrils, are pits that do not open into the pharynx, but are supplied with sensory nerves. The gill slits allow water taken in by the mouth to pass over the gills and out, allowing the gill arteries to give off carbon dioxide and take in oxygen.

from shallow ponds to the darkness of marine deeps. They have evolved into many shapes, from the pencil-like slenderness of pipefishes to the squatty lumpiness of toadfishes. Lungfishes can survive for months out of water, and some catfishes can migrate across stretches of land, struggling along the ground with their fins.

Some mature minnows may be only a few centimeters long, whereas swordfishes may be 4 meters (12 ft) long, but the bony fishes do not rival in size the great sharks, with their 10-meter (35-ft) length. Most fishes are shaped to glide easily through the water, though some of the sluggish ones have various bulbous shapes. Their skins may be scaleless, or they may be furnished with any of several kinds of scales, some of which, like growth rings in a tree trunk, indicate the age of the animal. The mouth is at the end of the head, as contrasted with the under-the-head shark mouth, and the gill slits are covered by a bone-supported flap, the *operculum*. Both paired and median fins are strengthened by ribs of bone or cartilage. In bony fishes, the tail fin is typically symmetrical.

Like Chondrichthyes, Osteichthyes have a two-chambered heart, with one atrium and one ventricle, that pumps red blood with nucleated red blood cells throughout the body. Oxygenation of the blood occurs when it is pumped past the gills. Some fishes also have an air chamber, or *swim bladder*, which can be filled with gas or emptied, keeping the buoyancy of the body within necessary limits (Figure 10.45). The sexes are separate and fertilization is usually external.

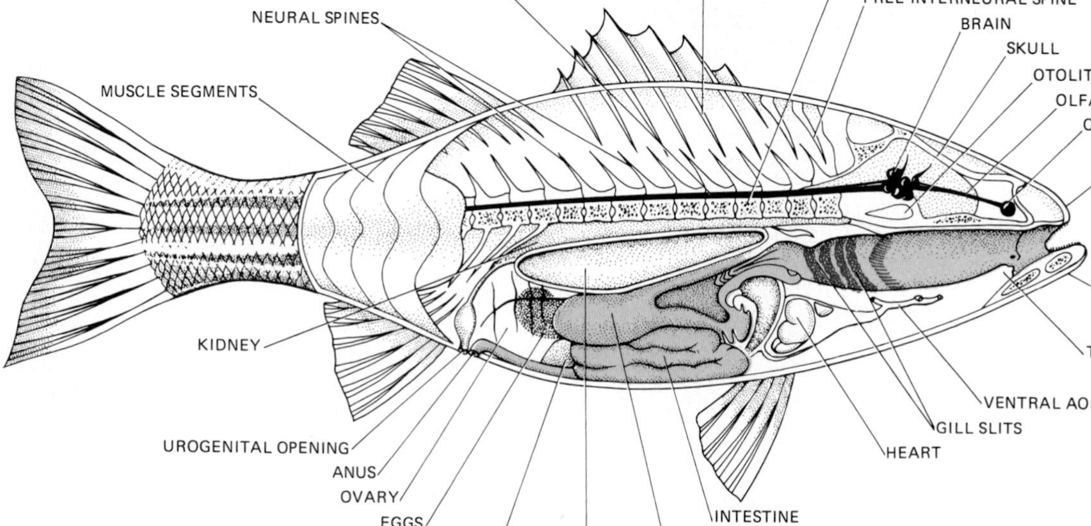

One rare fish deserves special mention. In 1938, commercial fisherman caught a specimen of a group of lobe-finned fishes, or coelacanths, of the genus *Latimeria* (Figure 10.46). Lobe-finned fishes had been long known from fossils, but they were believed to be extinct. The discovery of this "living fossil" stirred up interest among zoologists because the paired, lobed fins resemble the fore and hind limbs of amphibians. It is thought that from such fins could have come the tetrapod (four-footed) body plan that became established as typical of land-dwelling vertebrates. Over the years, several dozen more latimerias have been hauled up from the deep seas off the east coast of Africa, and they have been intensively studied, as is fitting for a relic that has survived almost unchanged for millions of years.

× 0.1

Figure 10.46
A coelacanth fish, *Latimeria chalumnae.* This rare animal, caught only a few times off the east coast of Africa, belongs to the group known as lobe-finned fishes. They must have been fairly common some 700 million years ago, lasting for 400 million years and then becoming rare.

Fishes have been a main source of human food, probably as long as people have been on earth. With the recent enormous increase in human population, there has been greater and greater emphasis on catching seafood, and some unrealistic optimists have thought that the oceans would be our salvation. But with ever-greater numbers of fishers and evermore efficient methods of capturing fishes, the fish population is being depleted rapidly and some estimates indicate that it is being depleted faster than it can recover. Some schemes for increasing the food supply from the sea include processing trash fishes for protein flour; developing intensive fish farms, using algae as fish food; and inducing the upwelling of mineral nutrients from the depths of the sea, where they are now not available for algal growth and therefore for the subsequent nourishment of organisms higher on the feeding hierarchy. Each of these methods presents technical problems, and all

(a) × 2

(b) × 1

Figure 10.47
Two representative amphibians. (a) A tree frog. Frogs are typically tailless, and they have unusually strong hind legs. Tree frogs have in addition a specially adapted set of suction disks on the digits of their front feet and can therefore cling effectively to twigs or even to smooth surfaces. Being amphibians, they may spend most of their lives on land (or on vegetation), but they must pass their larval, or tadpole, stages in water, even if that water is only a few milliliters trapped in a leaf cup. (b) A salamander. In European folklore, salamanders were thought to live in fire, probably because they lived in moist places, such as old logs, and if a log was added to a fire, the salamanders would be driven out.

must for now be regarded as possibilities rather than probabilities.

Class Amphibia In the **Class Amphibia** (Greek, "double life") are the animals that came out of the water and established a successful life on land. Getting onto land offered some advantages, especially escape from competition, availability of a greater number of places to live, and a freer oxygen supply. But with the advantages there were problems. Temperature fluctuations in air are greater than they are in water. Air has great drying power, which can be dangerous to any organism. And air lacks the great buoyancy of water, so that a land animal needs stouter support than an aquatic one. Amphibians took advantage of the improved possibilities and adapted themselves to the difficulties, mainly by remaining close to water, by keeping a soft, moist skin, by developing lungs, and by evolving a sturdy, bony skeleton with a strong vertebral column and four legs.

Of the three orders of amphibians, two are familiar. In the **Anura** (Greek, "no tail") are frogs and toads. In the **Urodela** (Greek, "with a tail") are the salamanders, efts, newts, mud puppies, and congo eels (Figure 10.47). The third order, the **Apoda** (Greek, "no feet"), contains a few rare, tropical, legless amphibians, which are mainly biological curiosities. All three orders share a number of characteristics. The bony endoskeleton is the main body support; the notochord is absorbed during development. Breathing is mostly by means of skin and lungs, although external gills are present in larvae. The circulatory system shows an advance over that in fishes. There is a three-chambered heart, with two atria and one ventricle, and a double circulation—that is, one set of blood vessels to the body as a whole and a separate one to the lungs. The two systems, however, are not entirely separated as they are in birds and mammals, and the amphibia are "cold-blooded," that is having internal temperatures that vary with the environment.

Amphibia usually have four limbs, with powerful hind limbs in frogs, but weak, rather ineffective legs in salamanders. Salamanders and other urodeles can scarcely be said to walk. Rather they slither along on their bellies, pushing themselves forward by a combination of wiggling and flapping their legs. The original use of lobed fins by prototype amphibians, and later of legs, was probably not so much to get *onto* land as to get *across* land in an effort to get back to water from a drying mud hole.

Modern amphibians still require water for the early development of embryos and larvae. Most frogs, toads, and salamanders lay eggs in ponds, where the young begin to grow, but a number of methods enable some species to bypass such obvious nurseries. Some lay eggs in small pools of water in cupped leaves, and some carry the young in folds of parental skin or even in the mouth of the father.

Class Reptilia The **Class Reptilia** (Latin, "crawl"), containing the snakes, lizards, turtles, and crocodilians, is probably second only to the Arachnida (spiders) in its association in popular mythology with a mysterious and sinister world (Figure 10.48).

Figure 10.48
Two of the main living types of reptiles. (a) A painted turtle. Turtles are special in their chunky body form and their possession of a top and bottom shell. (b) A green python, one of the world's large snakes. Because they are legless, snakes have had to make some adaptations in their body structure, including hinged jaws and a streamlining of internal organs that fit in the slender tubular body.

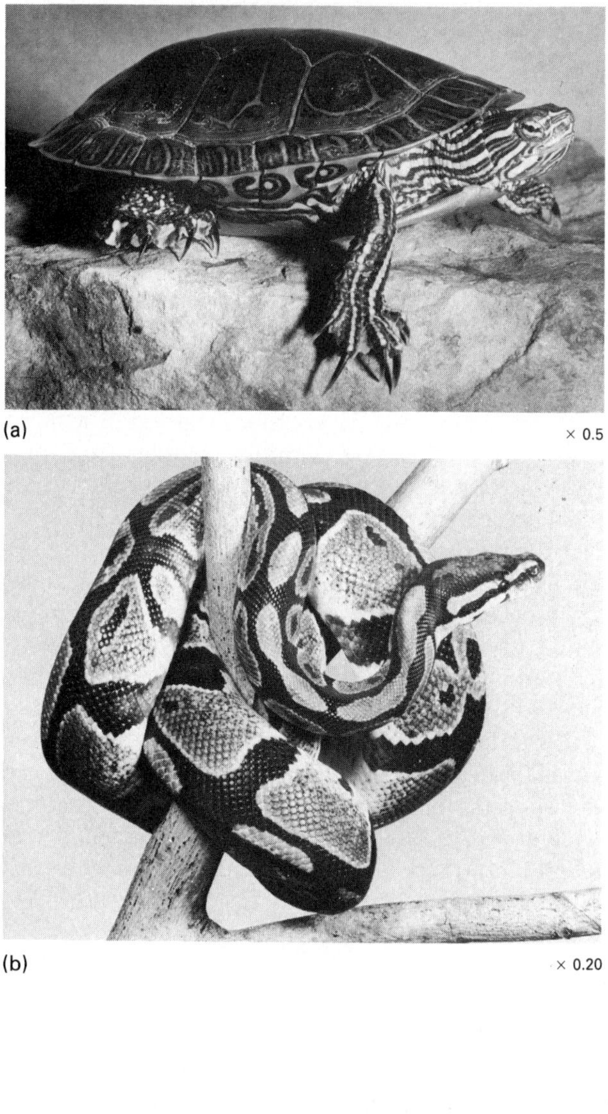

(a) × 0.5

(b) × 0.20

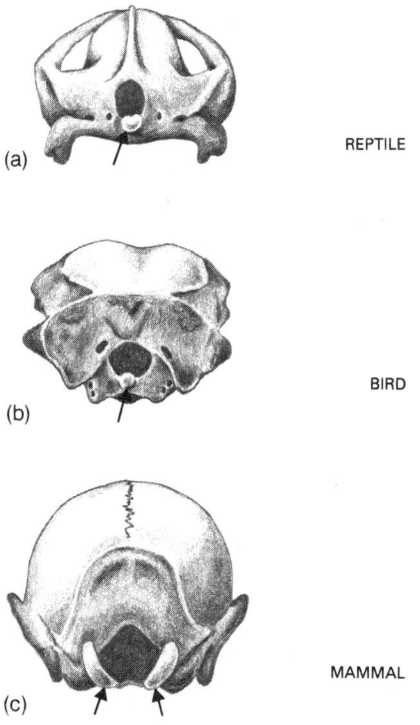

Figure 10.49
One indication of the relationship that exists between reptiles and birds: the presence of a single *occipital condyle,* the bearing surface of a skull on the first vertebra. In (a), which shows a reptile (a turtle), and in (b), which shows a bird (a gull), the arrows point to the single occipital condyle. In contrast, (c) shows the skull of a mammal (an opossum), with two occipital condyles.

Modern reptiles are in fact a lingering remnant of a once dominant land fauna that included dinosaurs and various flying and swimming reptiles. To dispel a few myths about reptiles, they are no colder than most objects in nature, and indeed most maintain a warmer body temperature than the surrounding air. They are not slimy unless they have slimy algae growing on them (as some do). They are no more ferocious than any other animal that must eat to live, and they are no more secretive than any other animal that must escape its hungry predators.

Reptiles probably evolved from an amphibian ancestor nearly 300 million years ago, developing the special characteristics that make them distinctively reptilian. The skeleton is that of the usual vertebrate, but there is an additional covering of horny plates or scales, and the skull has a peculiar feature: one *occipital condyle* (Figure 10.49), the bearing surface where the skull rests on the first *cervical* vertebra, or neck vertebra. Other classes of vertebrates, except for birds, have two occipital condyles. Breathing is accomplished by means of lungs, gills having been left permanently behind in evolutionary development. In most reptiles, the heart is three-chambered, but in alligators and crocodiles it is four-chambered, so that these animals are the first (as treated in this evolutionary sequence) to have such an organ. They are still not "warm-blooded," however.

Reptiles were the first vertebrates to become true land-dwellers. Their ability to breathe air, their possession (in most cases) of effective walking legs with five toes each, and the strength to stand on those legs all contributed to their terrestrial success. None of these features, however, would have been enough to free the reptiles completely from some aquatic stage without one additional development: the ability to reproduce without free water. That problem was overcome with the evolution of the ability to lay *amniotic eggs.* Such eggs are water-filled and covered with a water-retaining sheath, which in reptilian eggs is leathery.

Four orders of reptiles are recognized. The **Order Rhynchocephalia** contains only one species, *Sphenodon punctatum,* the tuatara. This is another "living fossil," a lone survivor of an ancient reptilian group, now alive only in a few islands of the Cook Strait in New Zealand. One of its remarkable features is a scale-covered third eye.

The **Order Testudines** is the order of the turtles and tortoises. These vary from the small box tortoises of North American woods to the great sea turtles and the giant land tortoises of the Galapagos and Seychelles Islands. In the Testudines, the vertebrae are fused to a dorsal shell, or *carapace,* and the ribs are fused to a ventral *plastron.* Testudines have no teeth, but they have horny jaws. The sea turtles are known for their ability to find their way year after year across the oceans to specific spawning grounds, where they mate and lay eggs on sandy beaches. Like many edible marine animals, these turtles have been hunted by humans until their numbers have been dangerously reduced.

In the **Order Squamata** (Latin, "scaly") are the snakes and lizards. Snakes can be thought of as essentially legless lizards, although boa constrictors sometimes even yet grow a few vestigial bones where the hind limbs of ancestors used to be. Lacking chewing teeth, snakes either swallow their prey alive, crush it in their coils until it ceases to breathe, or kill it by poison. Having a lower jaw that can drop far down from the upper jaw, and having no breast bone to restrict the rib cage, a snake can swallow an object larger than itself.

Snake poisons act on the nervous system or the blood (or both) of the victim, in either way causing death from lack of oxygen. In the United States, rattlesnakes, copperheads, cottonmouth moccasins (water moccasins), and coral snakes are the main venomous reptiles, and hundreds of bites are reported annually. Nonetheless, snakes are in general timid animals, which will escape if they can and will bite humans only when alarmed. Even nonvenomous snakes can bite painfully, but their most effective defense is the voiding of foul-smelling excreta onto an attacker. Snakes eat mainly small animals, especially rodents and insects, though some of the large tropical snakes, such as the great pythons and anacondas, can handle prey as large as a pig. Snakes in turn are eaten by other snakes and by predaceous birds and mammals.

Snakes can see, they can smell, mainly with the help of the extrudable tongue, and they are sensitive to temperature changes, but they have no ears and can perceive sound only if tremors shake the ground. Snakes can move by coiling up and lunging forward, but they progress mainly by muscles that move the backward-directed scales on their bellies, which they use to anchor against rough surfaces. A snake on a polished floor is practically helpless. Lizards, in contrast, have achieved true terrestrial locomotion, and many are fleet of foot. Some can rise on their hind legs and run, even skittering along over the surface of water when they are in a hurry. (See the picture essay in Chapter 21.)

The **Order Crocodilia** contains the broad-snouted alligators and the slender-snouted crocodiles, both inhabiting the United States (Figure 10.50). For all their carnivorous ways—some aggressive crocodiles are capable of capturing deer—they have a remarkable fondness for marshmallows.

Figure 10.50
An American alligator, *Alligator mississippiensis.* By making water holes in the swamps of Florida and the Gulf Coast states, alligators provide living (and sometimes dying) quarters for many other animals that would otherwise not be able to inhabit the areas during the dry seasons.

× 0.2

Figure 10.51
A bird is easy to define: an animal with feathers. Feathers, developing on reptilian animals during the Jurassic Period some 150 million years ago, were probably at first useful as an effective covering to keep the animals warm. Only later were they used to help in flying. The example here is a pelican in Florida.

Class Aves All birds are in the **Class Aves** (Figure 10.51). They make a rather homogeneous grouping, apparently having evolved as one branch from an earlier reptilian stock. Their scaly feet and the possession of a single occipital condyle are good indications of their relation to reptiles, but even more convincing is the evidence from fossils. The famous archaeopteryx fossil, discovered in Bavaria in 1877, is the kind of fossil that gladdens the evolutionary taxonomist, because it has obvious features common to both the old and the new. Archaeopteryx had the teeth, the long reptilian tail, and the clawed wing digits that show its reptilian nature, but it also had the two distinguishing characteristics of birds: wings and feathers.

Many of the structural and behavioral features of birds are associated with their ability to fly. The streamlined body shape, the light, hollow bones, the air sacs in the body, and the smooth covering of feathers are all part of the pattern of a bird. (See the picture essay in Chapter 21.) Being able to fly away from danger enabled birds to survive without many of the elaborate defenses that earthbound animals developed.

Birds can learn, they are fantastically agile, and their perceptions (especially sight) are acute. Birds have four-chambered hearts with lung and body circulation completely separated. With the effective insulation of feathers, they maintain a higher body temperature than do most other animals, usually 40–42°C (104–107°F). Another avian development is the egg with a hard, calcareous shell, which must

Figure 10.52
The duck-billed platypus, *Ornithorhynchus anatinus.* The platypus survives only in Australia and Tasmania. It is a queer enough animal in shape and habit, with its ducklike snout and aquatic preference, but its more unusual features include webbed hind feet with poison glands in the claws, its habit of laying eggs, and the feeding of its young by having them lap milk from the mother's teatless underside. This cat-sized animal eats small aquatic creatures.

× 0.2

be kept warm during incubation. The requirement for parental care has led to elaborate behavior patterns of mating, nesting, care for the young after hatching, and even extensive migration (Chapter 29).

Occasionally birds become nuisances, as when they get into airplane jet engines or accumulate, as starlings do, in flocks of millions where they are unwanted, but in general birds are regarded as friendly to humans. They are useful scavengers, and they eat unimaginable numbers of harmful insects. Some are themselves edible, and some effect pollination of flowers. They are frequent subjects of biological experimentation, and with their pleasing calls and colorful feathers, they add to our esthetic enjoyment of what would be a less attractive world without them.

Class Mammalia The **Class Mammalia** has been divided into three subclasses. The most primitive, **Subclass Prototheria** (Greek, "first beasts"), contains the egg-laying mammals of Australia: the duck-billed platypus (*Ornithorhynchus*) and the spiny anteater (*Tachyglossus*) (Figure 10.52). The name Mammalia is derived from the Latin word for breast, and the presence of breasts for feeding the young is one of the most distinctive features of the class.* Although the Prototheria do not possess breasts like those of other mammals, they do have milk glands, modified sweat glands that secrete milk into the fur along the mother's belly, where the young can lick it. The **Subclass Metatheria** (Greek, roughly translatable as "next beasts") are the marsupial mammals, the opossums, kangaroos, wombats, wallabies, and koala bears, which bear young in a poorly developed condition, then keep them in an abdominal pouch, the *marsupium,* where each embryonic baby attaches itself to a nipple and nurses until it achieves a measure of independence (Figure 10.53). In the United States, opossums are the only marsupials, but in Australia marsupials were the dominant mammals before Europeans introduced other animals. The third subclass, the **Eutheria** (Greek, "true beasts"), is made up of the placental mammals, those whose young are developed in the maternal uterus and are fed after birth by nursing.

Mammals and insects now dominate the earth. Mammals inhabit practically every available space, from the Arctic ice to the forests, grasslands, and deserts and to the depths of the sea. They have been increasing in numbers (but not necessarily in numbers of species) and importance since the dinosaurs went into decline toward the end of the Mesozoic Era, about 70 million years ago, but fossil animals from the Jurassic Period, 150 million years ago, show a combination of reptile and mammal characteristics.

The early mammal-like animals were small, many no larger than rats, but they stood up on their four legs, developed a greater agility than the reptiles had, and along with that agility had an effectively coordinated nervous system. Being smarter than the reptiles they were displacing, they took advantage of all their other emerging adaptive features to increase their domination of the earth. As

× 0.3

Figure 10.53
An American opossum and her family of half-grown young. When born, baby opossums are naked, ill-formed objects, scarcely more than embryos, about the size of honeybees. They have nonfunctional eyes and ears, and their hind legs are poorly developed. But they have mouths and vigorous front feet, and they can find their way into their mother's pouch, where each baby finds a nipple and attaches itself. There it hangs, sucking milk until it develops the completed organs that most mammalian babies have at birth.

* "Mama" is an easily articulated, repetitious sound, uttered by babies in many countries and in many languages, including Indo-European, Greek, Russian, Welsh, Irish, and the Romance languages. It has apparently been taken to mean both the mother herself and the mother's nourishing breasts.

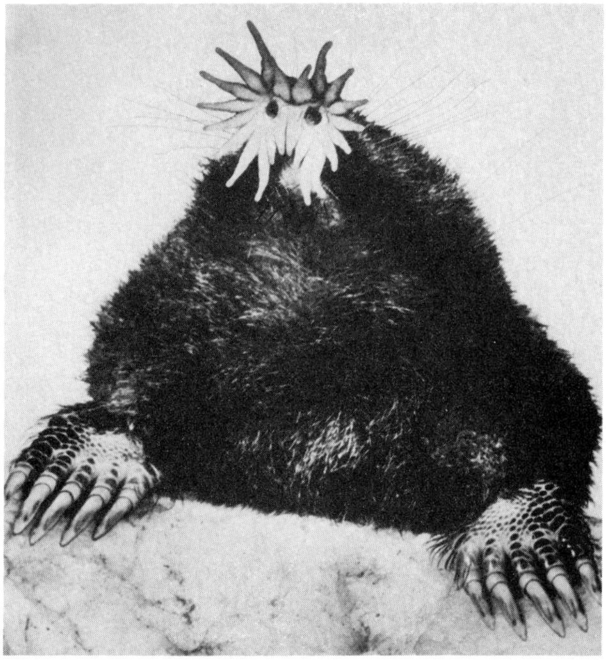

× 1.5

Figure 10.54
An insectivore, the star-nosed mole, *Condylura cristata.* Its unusual face has a ring of sensitive feelers around the snout. Powerful diggers, moles can practically swim through the soil, pushing ahead with their noses and throwing the soil back with their front feet. They live on insects and other small animals, and they eat almost constantly.

× 0.6

Figure 10.55
A rodent, the tiny European harvest mouse, *Micromys minutus.* Only about 6 centimeters long, harvest mice have so great a surface-to-volume ratio that they lose heat rapidly and must eat during most of their waking hours.

more and more truly mammalian features evolved, the animals became warm-blooded and thus were able to enter and remain in places that were effectively closed to the cold-blooded reptiles.

Associated with increased intelligence was a series of improvements in sensory receptors: movable eyelids and directional external ears. Their feeding habits were also helpful. With teeth in both jaws, jaw joints that could work sideways as well as up and down, and a mouth-and-nose combination that allowed breathing during eating, the evolving mammals could nourish themselves better than any of the previous vertebrates. Finally, the reproductive method of keeping the young in a uterus and feeding them milk after birth got the mammals away from the hazards of laying eggs in nests. Once established, mammals replaced the next higher group, the reptiles, relatively fast. Within a space of some 25 million years (which is "relatively fast" for such a biological development), they reached a position of major importance about 50 million years ago and have kept it ever since.

Since the human species is mammalian, more research has been done on mammals than on other kinds of animals; consequently, we know more about them, and most people are more interested in them. In this book, major emphasis will be on the nutrition, physiology, and reproduction of mammals, especially humans, and the student is referred to those chapters that treat specific topics. Here we will briefly note the ten major orders of mammals. Some of the smaller and less well known orders are not included.

The orders of mammals The **Insectivora** (Latin, "insect eaters") are moles and shrews, small mouselike animals that live close to or under the ground, eating grubs, beetles, and ants (Figure 10.54). The **Rodentia** (Latin, "gnaw") are squirrels, rats, mice, beavers, and muskrats (but not rabbits), with their front, or *incisor,* teeth adapted to a chiseling action (Figure 10.55). Tough, aggressive, and adaptable,

rats compete with people for food and carry several dangerous diseases; they are probably our most formidable biological enemy. **Carnivora** (Latin, "meat eater") are cats and their wild kin, dogs, bears, otters, foxes, and most of the animals used for their fur. Carnivores have long, sharp *canine* (Latin, "dog") teeth, but small, weak incisors. Their teeth are thus well adapted to tearing flesh but not very good for grinding and chewing (Figure 10.56). The **Artiodactyla** (Greek, "even toes"), including the pigs, cattle, deer, goats, sheep, camels, hippopotamuses, and giraffes, walk on the tips of two toes on each foot, the other three toes of ancestral types having been reduced to a functionless condition (Figure 10.57). The hooves are comparable to the claws of carnivores or the toenails of primates. Most artiodactyls are grazing animals, and they have a special digestive system (see Chapter 22) that allows them to eat quickly and later "chew their cud" in safety at leisure. The **Perissodactyla** (Greek, "odd toes") are the animals that run and walk on the hoof of a modified middle digit: horses, zebras, the tapirs

Figure 10.56
A carnivore, the African leopard, *Panthera pardus,* eating a Thomson gazelle. Many big cats carry their prey to high branches before eating them.

Figure 10.57
A member of the Artiodactyla, animals with an even number of hooves on each foot. This is a female deer with her new fawn.

of Central America, and rhinoceroses (Figure 10.58). The **Cetacea** (Greek, "whale") are mammals that have returned to the sea (Figure 10.59). Besides whales, there are other cetaceans: the dolphins, porpoises, and narwhals. The last-named bear a long, bony spike in the head, which used to be sold commercially as a "unicorn horn." The **Chiroptera** (Greek, "hand wing") are the bats, more helpful than not, in spite of the actual transmission of rabies (carried by an occasional bat) and some vampirism of animals by tropical bats (Figure 10.60). Many plants are pollinated by bats, and many insects are eaten by them. They are the only mammals that fly actively; other animals, such as "flying" squirrels, merely glide on outstretched folds of skin between limbs. The ability of bats to use echolocation in navigating and finding food is discussed in Chapter 29.

The **Lagomorpha** (Greek, "rabbit form") are the rabbits, hares, and pikas. They are somewhat like rodents, but their teeth are different enough from those of rodents to justify placing these mammals in a different order. The **Proboscidea** (Greek, "front feeder") are the elephants, the only living animals in which the nose is elongated into a trunk and the

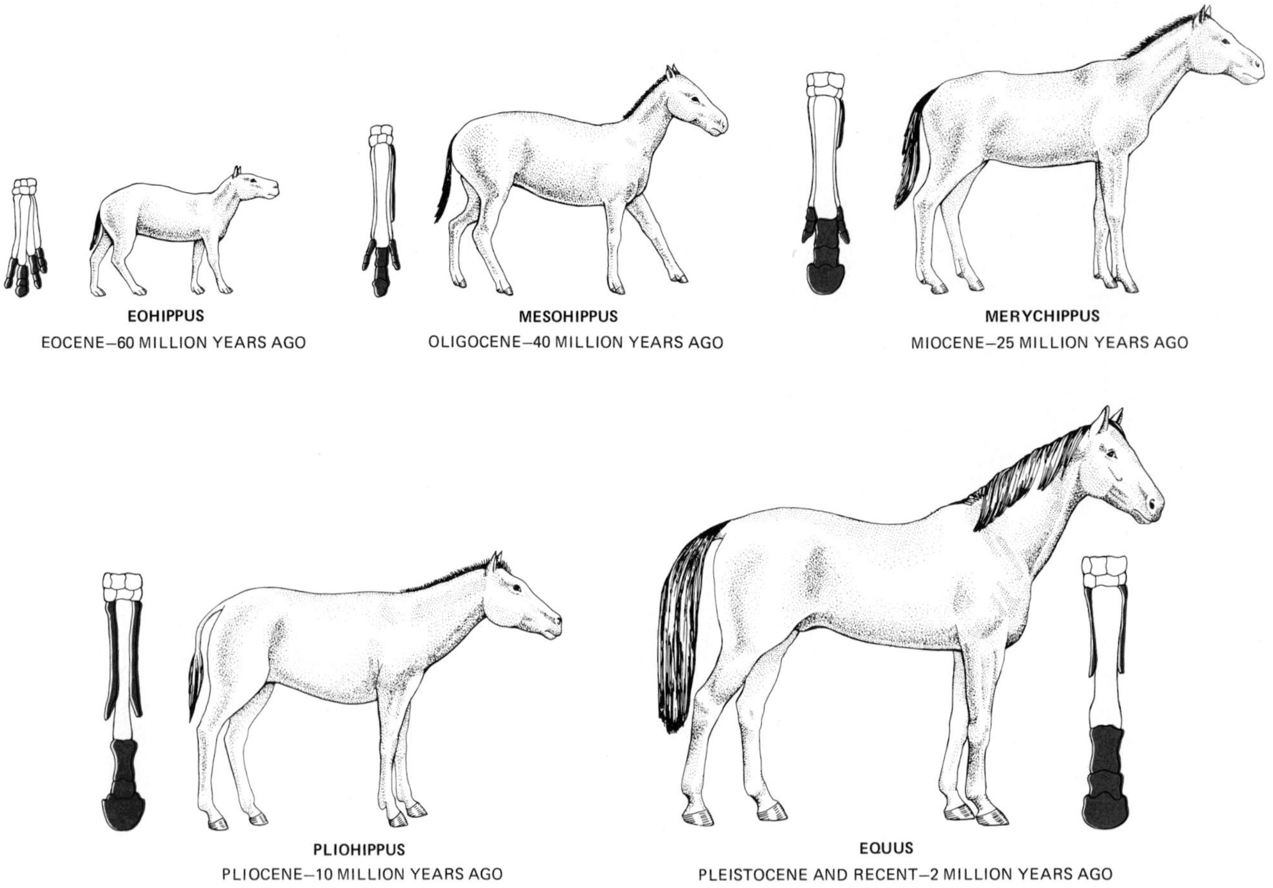

Figure 10.58
The development of the modern horse. By comparing older and older fossils, going from the obviously horselike bones of recent types to older, smaller, and less horselike ones, paleontologists have been able to reconstruct a convincing series. One ancestor was *Eohippus*, a dog-sized beast with four toes. It lived in North America when the climate was warm, some 60 million years ago. It was followed by a somewhat larger descendant, *Mesohippus*, which had several toes but put the mass of its weight mainly on the middle toe. *Merychippus* was an obviously horse-like animal with a definite hoof on its middle toe, although the side toes were still present. *Pliohippus*, living about ten million years ago, had lost the side toes and used the middle toe exclusively. Its feet were like those of a present-day horse, but its body was small and stocky. The modern horse, *Equus*, has been on earth only about two million years. It disappeared from the region of its ancestors, so there were no representatives in North America when Europeans reintroduced the horse in the sixteenth century. The wild mustangs now inhabiting the western plains of the United States are escaped descendants of the horses brought to Mexico by the Spanish explorers.

Figure 10.59
A cetacean, the humpback whale, *Megaptera novaeangliae.* Humpbacks are relatively large whales (up to about 15 meters) with unusually long flippers that are plainly visible when the animal leaps out of the water. They are vocal animals, communicating with one another under the water by a series of complex, repeatable sounds, popularly known as the "Song of the Humpback Whale."

incisor teeth are in the form of tusks. In the **Primates** (Greek, "first") are a number of familiar animals, such as the Old World monkeys, the New World monkeys, chimpanzees, gorillas, baboons, orangutans, and human beings, as well as some lesser-known, more primitive primates: tree shrews (the Etruscan shrew is the smallest mammal, weighing less than 3 grams), tarsiers, and lemurs. They have grasping digits with flat nails, five to a limb.

TO SUM UP ECHINODERMS THROUGH MAMMALS

1. *Echinodermata* include starfishes, sea urchins, and sea cucumbers. They have a water-pumping system that helps extend their tube feet, and most classes have rough or spiny outer coverings.

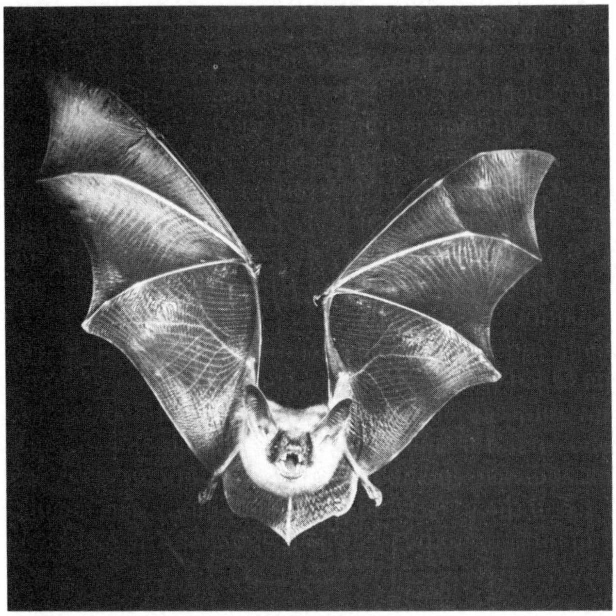

× 0.5

Figure 10.60
A bat, of the Order Chiroptera. Aside from being the only flying mammals, bats show a number of unusual features. They mate a couple of months before fertilization of the eggs occurs, and the female keeps the sperm alive in her uterus during the waiting period. The female has two functional teats from which the young nurse, but she also has an additional pair that does not give milk, serving only as something for the young to cling to while the mother hangs or flies. Male and female are alike except for the sexual organs, and the male has a bone in his penis. Females are impatient with the young, biting them vigorously when the babies are annoying. Contrary to popular myth, they do not entangle themselves in women's hair, and they can see quite well.

2. *Hemichordata* is a phylum of soft-bodied animals that usually live in water. The acorn worm is an example. It has gill slits, a hollow dorsal nerve cord, and a stiffening support similar to a notochord.
3. Every member of the Phylum *Chordata,* which has a wide range from sea squirts to human beings, has a notochord at some stage of development, a tubular dorsal nerve cord terminating in a brain, and at some developmental stage, gill slits and a tail.
4. A subphylum of the Phylum Chordata is the *Vertebrata,* or vertebrates. Included are eels, sharks, bony fishes, frogs, snakes, birds, and mammals. Examples of animals in subclasses of mammals are the platypuses, opossums, and human beings.

THE EVOLUTION OF HUMAN BEINGS

We have relatively little factual material to help complete the story of human historical development, with only an occasional fragment of bone that escaped destruction, or the refuse in ancient campsites, or bits of shaped rock that look as if they might have been used as tools. Not much remains, but geologists, anthropologists, and archeologists have picked painstakingly through tons of rubble on all the continents looking for evidence of our own forebears. Alone among animals, we possess the curiosity that makes us look back into the past to ask, "Where did we come from?"

We are finding that humanlike creatures have been inhabiting the earth for a much longer time than we had suspected. Instead of a presumed age of perhaps 10,000 years, which was a common estimate 50 years ago, new discoveries of fossils and tools have pushed the early dates back hundreds of thousands of years, and the current guess is that the living species, *Homo sapiens,* may be a million years old. Humanlike species, similar enough to modern humans to be thought of as members of genus *Homo,* lived in East Africa as much as five million years ago. These figures are still being continually revised, so that final dates have surely not been determined.

Although the age of humans has been revised from a few thousand to several million years—and that is certainly a great change—we are still recent arrivals. We must note that the oldest dated rocks are about 3.8 billion years old, and that the earliest fossils of any kind are also more than three billion years old. On a time scale measured in billions of years, a million or two one way or another begins to seem less substantial. One common way to emphasize the shortness of human history is to compare biological time to a 24-hour day. If we assume that life began more than three billion years ago and call that time span 24 hours, then humans came on the scene a few *seconds* before the end of the 24-hour period.

Some 20 to 25 million years ago in Europe, Asia, and Africa, there were animals that lived in trees, ate fruit, and were somewhat apelike. From their fossils we can deduce something of their structure and habits, since the kinds of limbs and teeth an animal has can give clues to its diet and means of moving. These old beasts, now thought to be the forerunners of both present-day apes and modern humans, have been named *Dryopithecus,* meaning "ape of the woods." Remains of dryopithecines are not rare, but there are no clearly intermediate fossils to tell us what kinds of development led from them to later fossils. The story is still incomplete and consequently speculative, but more old fossils are being found, and we can hope that some of the blank spaces in our knowledge may be filled in.

Instructive fossils, along with indications of social activities, have been found in Africa and Eurasia. Between one and five million years ago, several species of prehuman animals were already walking on their hind limbs. Called *Australopithecus,* these animals were small (under 50 kg, or 110 lb) and had brain cases only one-third to one-half that of modern humans. A famous partial African skeleton, nicknamed Lucy, was an *Australopithecus.* The presence of shaped pebbles near their remains is taken as an indication that they had a rudimentary idea of making tools.

A now-famous fragment of a skull known as ER 1470 belonged to a creature worthy of being called a man of genus *Homo* (Figure 10.61). Modern dating techniques have shown ER 1470, found in East Africa, to be about two million years old. The possessor of this old skull was small, only about 1.3 meters (4 ft) tall. He had a brain capacity of about 800 cubic centimeters, which was large for his time, and he knew how to shape and use pieces of rock as tools. Because he was too humanlike to be excluded from the genus *Homo* but was at the same time not a fully evolved modern human, he has been provisionally put in a different species, *H. habilis,* which means

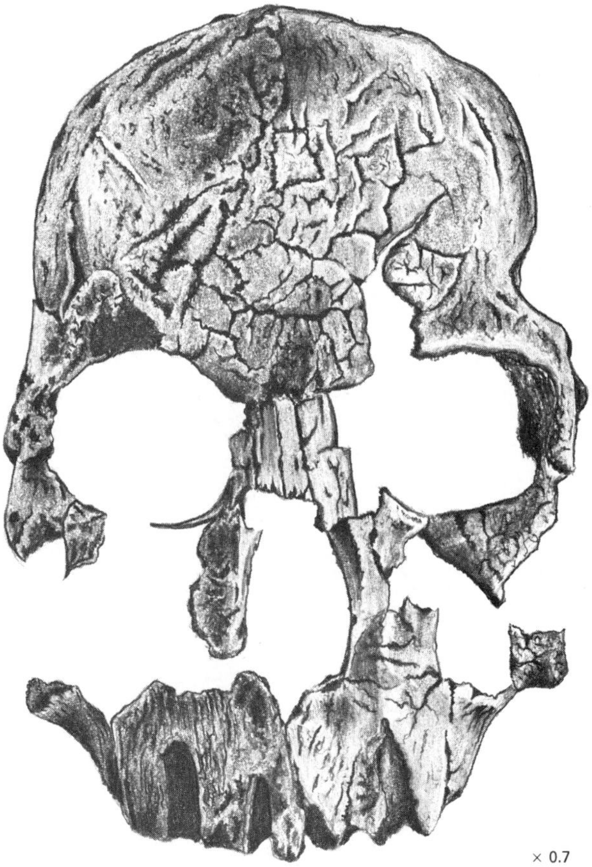

Figure 10.61
A portion of the oldest known human skull, ER 1470, which is thought to be about two million years old. Its relatively large size is notable.

clever or dexterous man, to distinguish him from *H. sapiens*, the wise man—a subtle but definite distinction. At the present writing, ER 1470 is the oldest identifiable *Homo* known.

Another early human, called *H. erectus*, spread successfully over the Old World. Fossils of this species have been found with rock tools that are more advanced than the flaked tools of earlier human-like species. The simple broken rocks that were in use practically unchanged for perhaps a million or two years were called Oldowan tools, named for the Olduvai Gorge in southeast Africa, where some of them were found. *Homo erectus*, living until about 300,000 years ago, improved on Oldowan-type tools by chipping them on two sides. To a modern student, that may not seem much of an improvement, but when one considers how long those dim minds took to achieve it, it becomes more impressive.

A better-known early human was the **Neanderthal man.** Neanderthals, named for specimens from the Neander Valley in Germany, actually lived in many places. They had sloping foreheads, relatively large brains, and rather prominent supraorbital ridges, those lumps of bone over the eyes that are characteristic of modern apes. The last of the Neanderthals lived about 40,000 years ago.

The last in the series is of course *H. sapiens* (Figure 10.62). Existing for at least a million years, probably longer, this species spread all over the surface of the earth while land connections still existed between Asia and Australia and between North America and Asia. When those connections were later covered by the water of rising seas, the various populations were isolated. Once geographically separated, the diverging populations developed some differences, but never enough to cause loss of interfertility. All the races of present-day humans are interfertile, and mixing of genes has occurred whenever people have come back together, even after thousands of years of separation.

The closest living relatives of the human species include the apes and monkeys, with the chimpanzees of equatorial Africa being the most nearly human. The ape-monkey, or anthropoid, line of

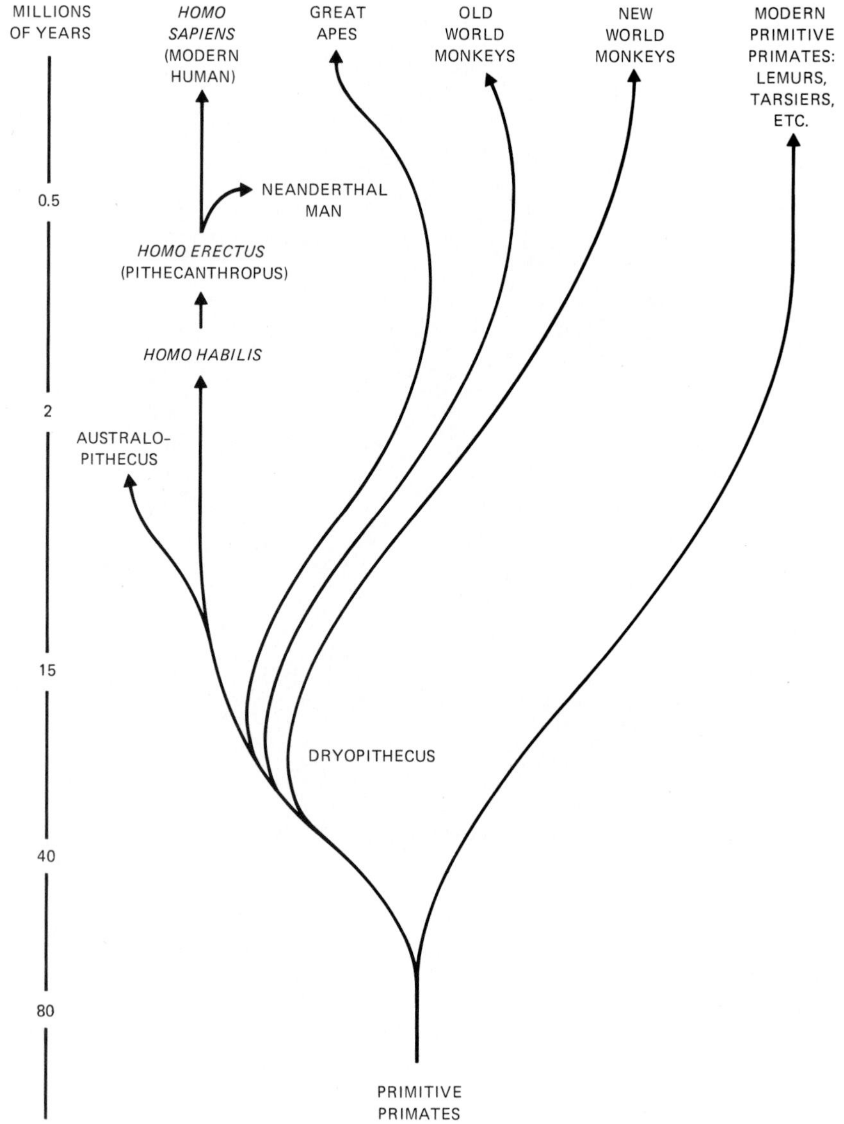

Figure 10.62
A diagram of a possible line of descent for *Homo sapiens,* with skulls, reconstructions (except for modern human) of entire heads, and representative tools.

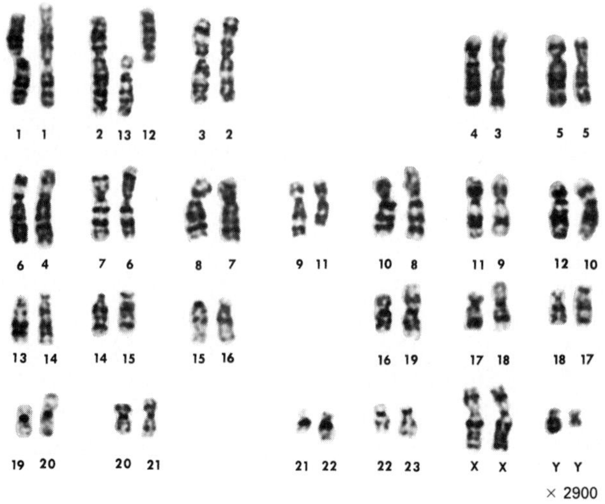

Figure 10.63
Comparison of human and chimpanzee chromosomes. In these *karyotypes* of the two animals, the human chromosomes are shown paired with those of the chimpanzee. In each pair, the human is on the left, numbered consecutively from 1 to 22, plus X and Y. Some differences of structural detail can be seen, but there is generally consistent similarity in the number, size, and banding patterns of the two animals.

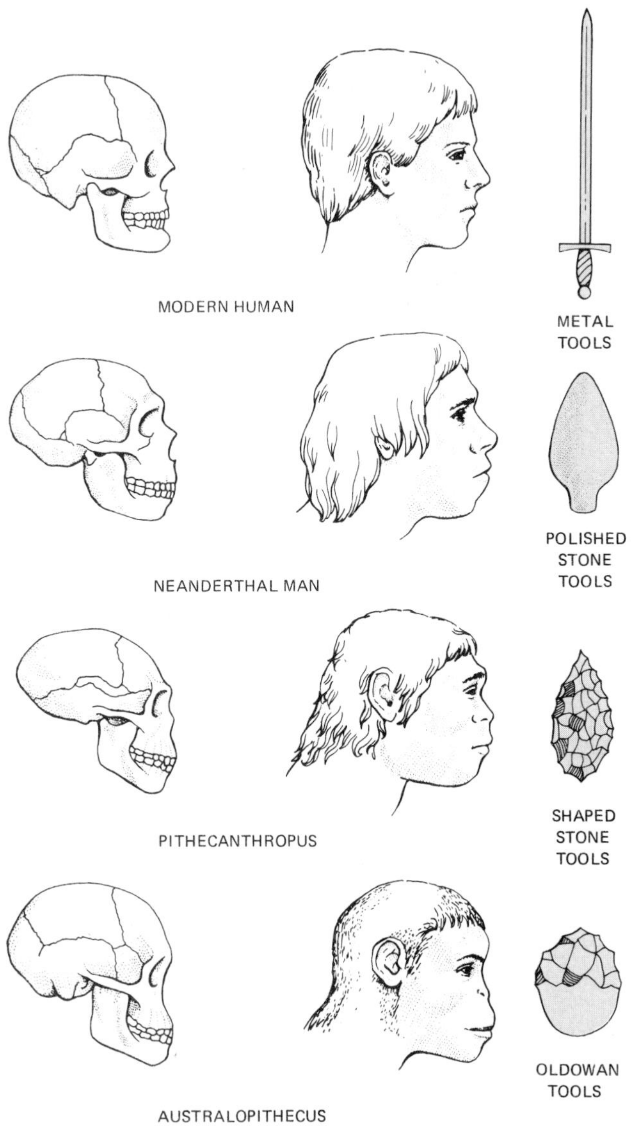

evolution seems to have become separated from the hominid line (the line leading toward modern humans) about 15 million years ago. Besides the easily observed similarities between humans and chimpanzees, most of the chromosomes of the two animals can be matched with considerable accuracy. There are only three chimpanzee and two human chromosomes that do not match (Figure 10.63).

Some of the features and abilities leading from prehuman forebears to modern human beings are biological; some are cultural. Among the structural changes are those that allowed upright locomotion or brought about a reduction in the canine teeth, the long, sharp, tearing teeth that are prominent in dogs and in such anthropoids as baboons. The hominid brain reached its present size of about 1200 cubic centimeters sometime between 100,000 and 500,000 years ago. The use of stones as tools may go as far back as five million years, but speech, in the sense of grammatical sound communication, is possibly only a million years old. Prehumans could not have known the subtlety, variety, and complexity of speech as we know it. Certain activities that are characteristic of today's humans are recent: the cultivation of plants and domestication of animals, the use of metals, the settlement of towns, and finally, the practice of writing. These are all less than 10,000 years old, a mere flicker of time in comparison with the length of life on earth. With the invention of writing, *Homo sapiens* came into the period of "history."

ASK YOURSELF

1. What animals are members of the coelenterate phylum?
2. What are the simplest bilaterally symmetrical animals?
3. What is an important structural difference between flatworms and roundworms?
4. Why do you think molluscs are considered to be one of the three or four most important phyla?
5. Why is it appropriate to use segmented worms for dissection in biology laboratories?
6. What phylum includes insects and crustacea? Why would you consider it to be a major phylum in terms of evolution?
7. What is molting, and why is it an important feature for some animals?
8. What are at least five major orders of insects?
9. What makes echinoderms different from other animals? What classes of echinoderms are still alive?
10. What is the difference between a protostome and a deuterostome?
11. What features do the chordates have that echinoderms do not have?
12. What distinctive characteristics do vertebrates have beyond the general characteristics of chordates?
13. What is the basic structural difference between the Class Chondrichthyes and the Class Osteichthyes?
14. What are the three classes of mammals?

PART FOUR
GENETICS AND ANIMAL DEVELOPMENT

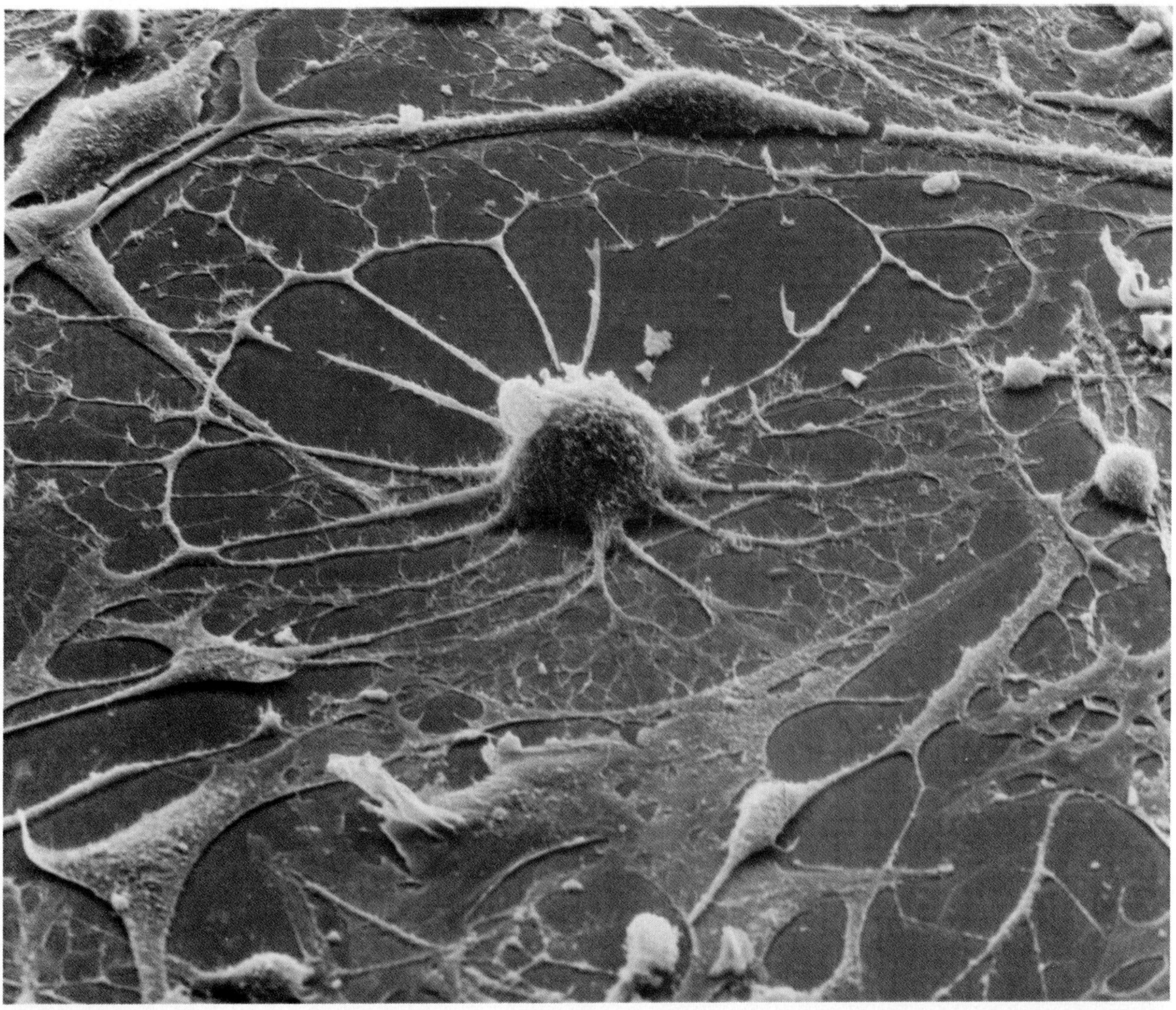

Scanning electron micrograph of a mouse embryo cell infected with a tumor virus called polyoma. $\times 700$

11
Mendelian Genetics

SOME KEY POINTS

1. Mendel recognized dominant and recessive characteristics in the hereditary traits of organisms.
2. The Law of Segregation states that factors determining contrasting characteristics in an individual do not blend or contaminate one another; rather, they segregate (separate) when gametes are formed, and any single gamete receives only one factor.
3. Alleles are alternative forms of a gene, and they determine different expressions of a trait such as eye color.
4. In a monohybrid cross, inheritance of one trait is considered; in a dihybrid cross, two traits are considered.
5. Every body cell has a pair of alleles of each gene. When both members of the pair of alleles are the same (aa), the individual is homozygous; when they are different (Aa), the individual is heterozygous.
6. A genotype is an individual's genetic makeup. The phenotype is the physical expression of the genetic trait.
7. The Law of Independent Assortment states that when one pair of alleles segregates during the formation of gametes, it is not affected by the segregation of another pair of alleles.
8. The chromosome theory of inheritance states that genes, the determinants of heredity, are located on the chromosomes. This idea makes it clear that there is a definite connection between the reproductive activities of cells (meiosis) and inheritance in whole organisms.

THAT PLANTS AND ANIMALS, INCLUDING HUmans, breed their own kind is such an obvious fact that it has been recognized as long as we have any records. But knowing that something is so is different from knowing *why* it is so, and only recently, in terms of human history, have we begun to approach a detailed explanation of what has been accepted for uncounted generations as a mystery. In the days before experimental science was applied to the study of inheritance, there was only untestable guesswork, which didn't lead very far.

With the rise of experimental biology in the eighteenth and nineteenth centuries, many attempts were made to work on problems of inheritance, but they were too vaguely descriptive, and they made the mistake of looking at too much at once. That is, early investigations on genetics took into account whole plants and animals, and the results were so complicated that no one could draw conclusions from them. The very word "genetics" didn't exist until the English biologist William Bateson coined it in 1906, although the Book of Genesis ("beginnings") is several thousand years older. The term "gene" was not introduced until 1909 by Wilhelm Johannsen, a full three years after Bateson introduced the word "genetics."

MENDEL AND THE BEGINNING OF STATISTICAL GENETICS

Gregor Mendel (Figure 11.1) was born to peasant parents in 1822 in Moravia, now part of Czechoslovakia. At the age of 21, he joined the Augustinian monastery in Brunn, Austria, as a novice. His entrance into the monastery placed him among many

Figure 11.1
Johann Mendel, better known under his monastic name of Gregor (1822–1884). He is now recognized as the man who started the science of genetics.

teachers who provided a genuine cultural and scientific environment suitable for Mendel's own intellectual development.

In his early years as a priest, Mendel could not successfully complete his duties as an attendant to the sick, but he subsequently found contentment as a temporary teacher at the local high school, and he was graciously relieved of all pastoral duties. In the summer of 1850, Mendel took an examination in an attempt to qualify as a permanent teacher, but he failed the final portion of the exam. This momentary setback, however, was to lead to one of the most important developmental stages of his life, because the following year it was recommended that he enter the University of Vienna to enrich his apparently inadequate background in the natural sciences.

From 1851 to 1853 Mendel was a student at the University of Vienna, where he studied botany, mathematics, and physics with some of the leading scientists of that time. It was surely during this stimulating period that Mendel learned of Schleiden's work with the cellular composition of plants and of the hybridization experiments of Kolreuter and Gartner, and developed the basic statistical concepts he was to use so fruitfully in his pea experiments. Indeed, Mendel's knowledge and practical application of mathematical statistics was one of the reasons for his unique success in an experimental area where so many others had preceded him without noteworthy scientific accomplishment.

When Mendel began growing peas in 1856 in his monastery garden, he started a whole new science, one that was to change all biological thinking. The 1850s and 1860s were decades of enormous development in biological knowledge. Darwin's revolutionary ideas on evolution were born, and there was a dazzling burst of information on cell structure, the chemistry of living material, internal anatomy, and geographical distribution.

Mendel was at once the luckiest and unluckiest of men. He was lucky because he had enough imagination and insight to invent questions and to devise experiments that could provide answers to them, and he had enough training and perception to recognize his results as answers. He also had predecessors who understood the sexual nature of plants and who had provided him with agriculturally useful peas, which had been bred by trial and error until they were genetically dependable, that is, until they "bred true," or produced offspring like themselves. (Mendel would not have spoken of them in those terms.) Finally, he was lucky in that he chose to investigate seven visible characteristics in peas, and peas have seven chromosomes (Table 11.1). What is most improbable is that all his chosen characteristics had their genetic locations on *different* chromosomes. As will become apparent later, if it had not happened that way, he might have been hopelessly misled, although with a mind like Mendel's he might have worked his way through those difficulties as well.

He was unlucky in that after working with peas he turned to honeybees, one of the most unlikely animals for genetic studies because male honeybees are hatched from unfertilized eggs, and that fact seriously alters the possibilities of parental combinations of genes. Mendel's misfortune was compounded when he also attempted his crosses with hawkweed. He did not know that hawkweed produces seeds asexually, and so there could be no genetic combinations of parental traits.

Finally, Mendel was unlucky in that he was born at the wrong time. In 1865 he presented the results of his pea experiments to the Brunn Society of Natural Science and published his conclusions in the Proceedings of the Society the following year. Nothing happened. No one agreed, no one disagreed, no one did anything. Why? Volumes have been written trying to explain why clear, convincing explanations of fundamental, thoroughly documented work should have been passed over by the sharpest scientists of the time. Perhaps if Mendel had been an established scientist instead of an unknown "hobbyist" his experimental results would have awakened the scientific community. Perhaps Charles Darwin had completely captured the attention of world scientists when he published his highly controversial ideas about evolution in *The Origin of Species* in 1859. But most likely, the scientists of Mendel's time were simply not ready to understand the significance of his work. By 1900, chromosomes and cell division had been described, and Mendel's observations could be directly linked to currently acceptable principles of cell biology. As is so common in the history of science, the world was now ready for the idea, and the rush was on. Almost simultaneously, the Dutch botanist Hugo De Vries, the German botanist Carl Correns, and the Austrian botanist Erich von Tschermak independently discovered Mendel's genetic laws and had little trouble establishing their importance.

Mendel made two decisions that brought him experimental success where others had failed. First, he narrowed his interest to specific details. He paid attention to only one or two characteristics at a time, not allowing himself to be confused by the overwhelming complexities of an entire plant. Second, he counted his plants, keeping accurate records of

TABLE 11.1
Mendel's Seven Pairs of Contrasting Traits in Garden Peas

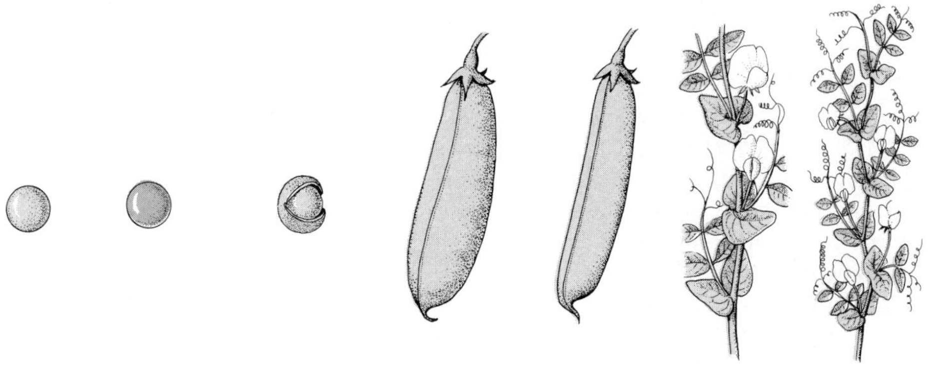

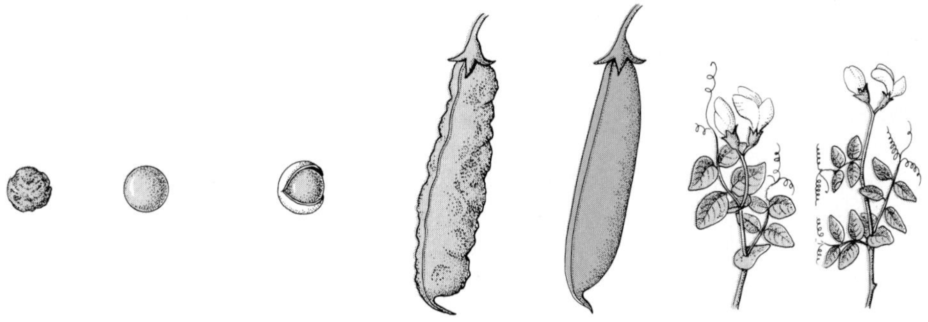

TRAIT STUDIED	SEED SHAPE	COTYLEDON COLOR	SEED COAT COLOR	POD SHAPE	POD COLOR	FLOWER POSITION	STEM LENGTH
DOMINANT	Round	Yellow	Gray	Inflated	Green	Axial	Tall
RECESSIVE	wrinkled	green	white	constricted	yellow	terminal	dwarf

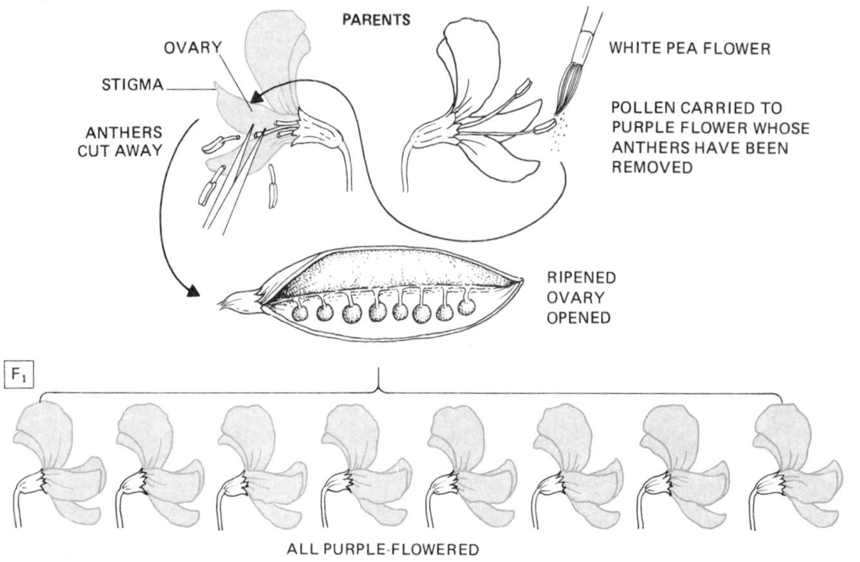

Figure 11.2
Method of making a hybrid by placing pollen from a white-flowered pea on the stigma of a purple-flowered one.

everything he grew. He started the practice of statistical treatment of biological data. Once these two technical necessities were established, the way was open to all that was to follow.

THE MENDELIAN LAW OF SEGREGATION

One of Mendel's varieties of peas grew about a meter (3 ft) tall, as contrasted with another that grew as a dwarf. He removed the pollen-bearing anthers from one before they ripened. This prevented self-fertilization. Then he dusted pollen from the other on the virgin flowers and covered the flowers with a sack to prevent the entrance of other pollen, just as plant breeders do today (Figure 11.2). When the seeds ripened, he kept them for planting the next season.

When the plants grew, Mendel found that they were all tall, whether the seeds had come from tall plants pollinated from dwarf ones or vice versa. These new plants were the F_1, or the *first filial generation.* He allowed the F_1 plants to pollinate themselves, as peas do, and again he kept the seeds. The following season, he grew 1064 plants (the F_2 generation) and found that 787 of them were tall and 277 were dwarf. The numbers struck him. They showed a ratio of almost 3 to 1. It was certainly close: 798 to 266 would have been exactly 3 to 1 (Table 11.2).

From the numbers, Mendel deduced what had happened. First, he realized that a characteristic (tall) could appear in all individuals in the first generation, to the exclusion of another characteristic (dwarf). However, the masked characteristic was still present in the F_1, as was shown by its reappearance in the F_2 (second filial generation). The characteristic that showed he called **dominant,** and the one that remained temporarily hidden he called **recessive.** (Mendel did not know what genes were; he called them "factors.") The recognition of dominant and recessive characteristics was his first contribution.

He noted further that when one individual has two alternative factors, the two can *segregate* in the next generation. That was the basis for his **Law of Segregation,** one of the fundamental principles of modern genetics, which states that contrasting factors in an individual do not blend or contaminate one another; rather, they segregate when gametes are formed, and any single gamete receives only one factor (Figure 11.3).

Alternative forms of a gene are called **alleles.** For instance, *dwarf* may be the alternative form of the dominant *tall.* In modern notation, letters are used to designate alleles, with capital letters indicating dominance and lowercase letters indicating recessiveness. Thus an allele for tallness is T, and for dwarfness is t. Similarly, the yellow-seed allele is Y, and the green-seed allele is y. We would now say that an organism having two identical factors is **homozygous,** and that it has a pair of identical alleles. An organism whose alleles are not identical is **heterozygous** for that pair of alleles. A homozygous tall pea plant is designated TT, and a heterozygous one is Tt. A homozygous tt is necessarily a dwarf (see Figure 11.3).

In that first series of experiments, Mendel had performed a **monohybrid cross,** in which the mated

TABLE 11.2
Results of Mendel's Monohybrid Crosses on Seven Pairs of Characteristics in the Garden Pea

PARENTAL CROSS	F_1 GENERATION	F_2 GENERATION		ACTUAL RATIO
Round × wrinkled seeds	All Round	5474 Round	1850 wrinkled	2.96:1
Yellow × green cotyledons	All Yellow	6022 Yellow	2001 green	3.01:1
Red × white flowers	All Red	705 Red	224 white	3.15:1
Inflated × constricted pods	All Inflated	882 Inflated	299 constricted	2.95:1
Green × yellow pods	All Green	428 Green	152 yellow	2.82:1
Axial × terminal flowers	All Axial	651 Axial	207 terminal	3.14:1
Tall × dwarf stem	All Tall	787 Tall	277 dwarf	2.84:1
All characteristics combined	All Dominant	14,949 Dominant	5010 recessive	2.98:1

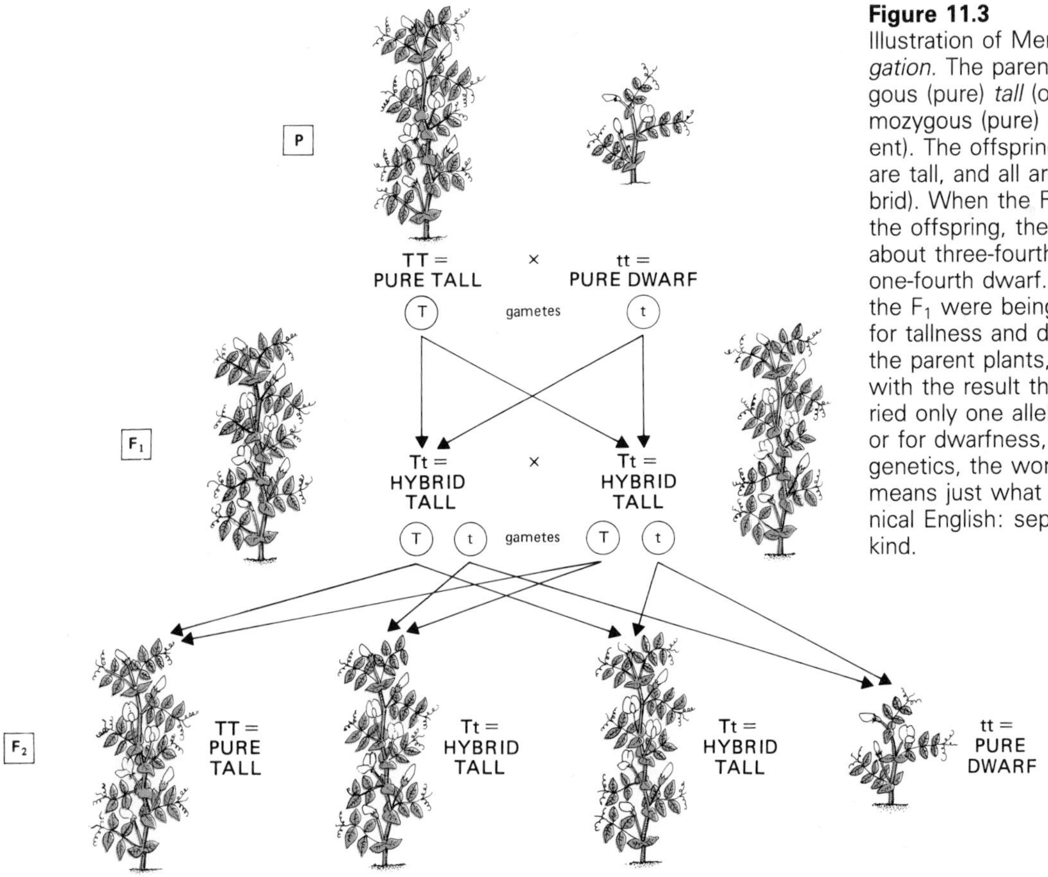

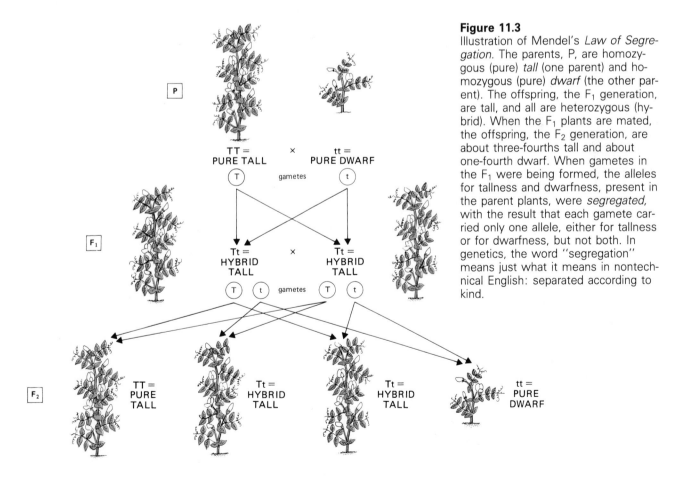

Figure 11.3
Illustration of Mendel's *Law of Segregation.* The parents, P, are homozygous (pure) *tall* (one parent) and homozygous (pure) *dwarf* (the other parent). The offspring, the F_1 generation, are tall, and all are heterozygous (hybrid). When the F_1 plants are mated, the offspring, the F_2 generation, are about three-fourths tall and about one-fourth dwarf. When gametes in the F_1 were being formed, the alleles for tallness and dwarfness, present in the parent plants, were *segregated,* with the result that each gamete carried only one allele, either for tallness or for dwarfness, but not both. In genetics, the word "segregation" means just what it means in nontechnical English: separated according to kind.

Figure 11.4
A "checkerboard," or *Punnett square,* named for the geneticist R. C. Punnett, can be used to show the predictable outcome of a genetic cross. All possible kinds of sperm from the male parent are displayed across the top, and all kinds of eggs down one side. The boxes can be filled in to indicate the kind of fertilized egg produced by each pair of gametes. Then the visible characteristics (phenotypes) can be determined and counted. This method gives the numbers of kinds of offspring that can be expected from any cross in which the genetic makeup (genotypes) of the parents is known. In this example, a heterozygous pea plant produces two kinds of pollen, T and t, which are written across the top of the square. The eggs, also T and t, are written down the left side, and the resulting genotypes are written as letters (TT, Tt, or tt). The phenotypes are drawn as images, and the three tall to one dwarf is evident. In actual crosses, in which dozens or hundreds or more individuals are used, the numbers only approach the theoretical ratios obtained from Punnett squares or from calculations.

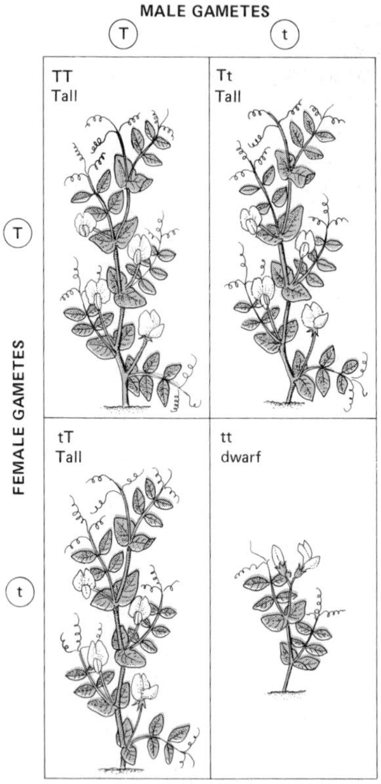

individuals differed in the expression of a single trait (Figure 11.4). (Mendel crossed a *tall* plant with a *dwarf* plant, for example.)

Genotypes and Phenotypes

If an individual is homozygous for a recessive characteristic, then that characteristic will be visibly expressed, and if one knows that the characteristic is recessive, one also knows that the individual must be homozygous for that characteristic. However, if the dominant characteristic is expressed, one cannot tell by inspection whether the individual is heterozygous or homozygous (see Figure 11.5). An individual's genetic makeup, sometimes hidden, is that individual's **genotype,** and it may be either homozygous or heterozygous for any one characteristic. The gene that is visibly expressed determines how the individual looks or acts and determines the **phenotype,** meaning "that which shows."

Mendel did not use the words *phenotype* and *genotype,* but he obviously understood the basic concept of visible and hidden characteristics. Without using the exact words, Mendel described the nature of phenotypes and genotypes in peas: "It can be seen how rash it may be to draw from external resemblances [phenotype] conclusions as to their internal nature [genotype]."

Figure 11.5
A test cross. If an organism shows a dominant characteristic, its genotype cannot be determined simply by inspection. In order to determine its genotype, it can be crossed with a phenotypic recessive, whose genotype is obviously that of a homozygote. In the example given here, a tall pea plant of unknown genotype, T?, is crossed with a short one, tt.

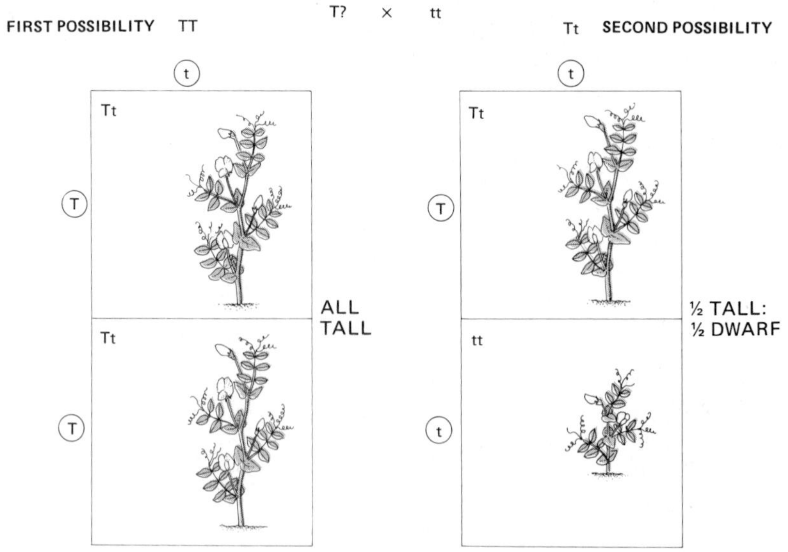

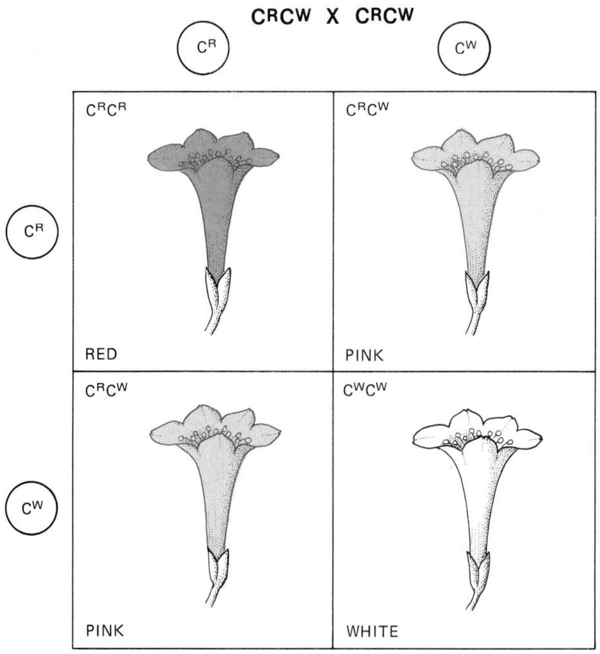

Figure 11.6
A Punnett square showing the results of a cross between heterozygotes with *incomplete dominance* in four-o'clock flowers.

The Test Cross

When an organism shows a dominant characteristic, one cannot know by inspection whether it is homozygous or heterozygous. One can, however, cross such an organism with a phenotypically recessive one, and from the cross count the offspring that exhibit the dominant and recessive characteristic. For example, a tall pea plant may be either TT or Tt. The unknown genotype is written T?, and the recessive (dwarf) is tt. Such a mating is a **test cross** because the offspring will reveal the genotype of the tall parent. In this test cross, if all the offspring come out tall, then the genotype of the tall parent must be TT; but if half the offspring come out tall and half short, then the genotype of the tall parent must be Tt. (Figure 11.5).

Incomplete Dominance

Some alleles fail to act either as dominants or recessives, but are expressed as intermediates. This **incomplete dominance** is shown in some flower colors, one well-known example being the garden plants called four-o'clocks. If a red-flowered plant is mated to a white-flowered one, the F_1 generation has pink flowers. If those pink-flowered plants are self-fertilized, the F_2 is one-fourth red, two-fourths pink, and one-fourth white. This 1:2:1 ratio is the same as the genotypic ratio in a typical Mendelian cross. If we call the allele for red color C^R and for white C^W, then the genotypic ratios from an inbred C^RC^W are 1 C^RC^R: 2 C^RC^W: 1 C^WC^W. In such an instance, the genotype can be recognized by simply looking at the phenotype (Figure 11.6).

A SIMPLE GENETICS GLOSSARY

ALLELE One form of a gene, responsible for one of two or more contrasting characteristics.

DIHYBRID CROSS A cross between individuals differing in two inheritable characteristics, or in which only two such different characteristics are considered by an experimenter.

DOMINANT Pertaining to an allele that expresses itself to the exclusion of its recessive allele.

GENE A specific segment of DNA that controls a specific cellular function; the foundation of inheritable characteristics.

GENOTYPE An individual's genetic makeup.

HETEROZYGOUS Pertaining to an organism in which the paired alleles for a particular trait are different.

HOMOZYGOUS Pertaining to an organism in which the paired alleles for a particular trait are identical.

LAW OF INDEPENDENT ASSORTMENT A principle of Mendel, which states that when one pair of alleles segregates during gamete formation, its manner of segregation is not affected by the manner of segregation of a second, different pair of alleles.

LAW OF SEGREGATION Mendel's first law, which states that a pair of alleles is segregated, or separated, during the formation of gametes, so that a gamete has only one of each pair of alleles.

MONOHYBRID CROSS A cross between individuals differing in only one inheritable trait, or in which only one trait is considered by an experimenter.

PHENOTYPE The appearance or discernible characteristic of an individual, as determined by the genetic makeup and environmental influences.

RECESSIVE Pertaining to an allele that is masked by a dominant allele in a heterozygote.

TEST CROSS A cross between an organism of unknown genotype and one with a recessive phenotype, to determine the genotype of the one that is unknown.

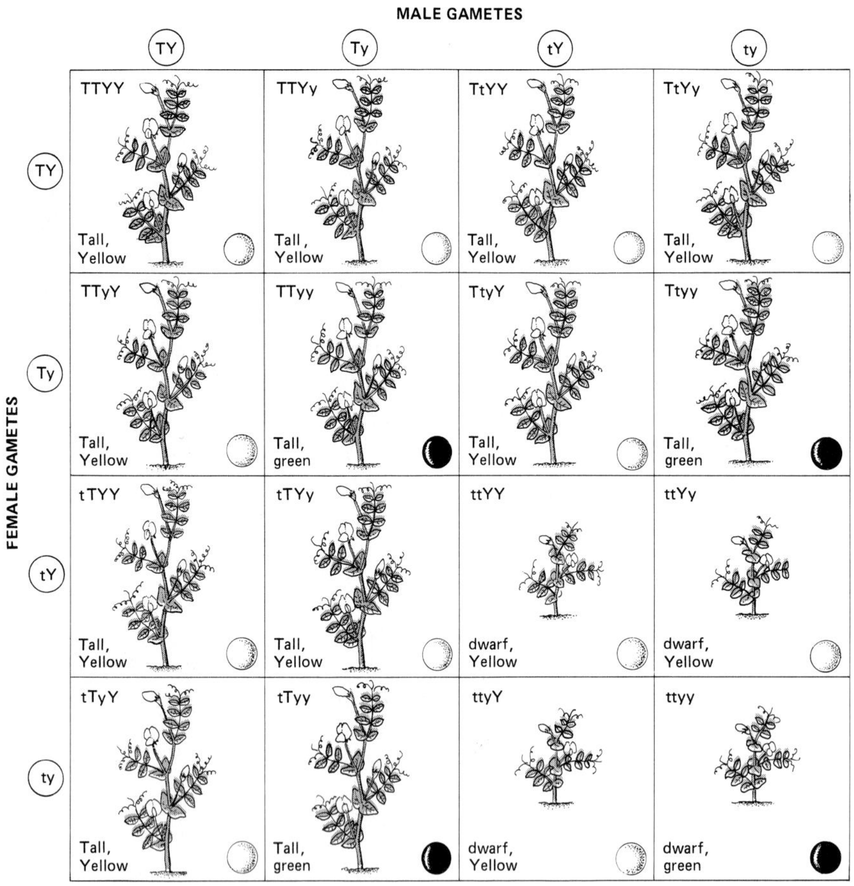

Figure 11.7
A Punnett square showing the expected results of crossing two individuals, both heterozygous for two pairs of alleles: a *dihybrid cross*. The parents are tall and have yellow seeds (white ball), so the kinds of sperm that can be produced are TY, Ty, tY, and ty, and the kinds of eggs are the same. Filling the boxes appropriately shows that of the 16 squares, only one has the genotype TTYY (phenotypically Tall, Yellow), only one is ttyy (phenotypically dwarf, green; black ball), nine are Tall and Yellow (but not genotypically identical), and so on. Both parents were heterozygous for both pairs of alleles, but showed only the dominant allele, and appeared Tall with Yellow seeds. Since some of the offspring were dwarfs with Yellow seeds, or Tall with green seeds, *recombination* occurred and became visible. In a monohybrid cross, one heterozygous pair of alleles gives two kinds of gametes and a matrix with four boxes. In a dihybrid cross, two heterozygous pairs of alleles give four kinds of gametes and a matrix with 16 boxes. In a trihybrid cross, three heterozygous pairs of alleles would give eight kinds of gametes and a matrix of 64 boxes.

MENDEL'S LAW OF INDEPENDENT ASSORTMENT

After completing his monohybrid crosses, Mendel next considered the pattern of inheritance of *two* traits in a single mating; that is, he completed a **dihybrid cross** (Figure 11.7). (When writing of monohybrid or dihybrid crosses, the geneticist means that attention is being paid to only one or two pairs of alleles, respectively, not that the organisms actually differ in only one or two features. In reality, it is unlikely that a pair of organisms would differ in only one pair of alleles.) Mendel crossed a plant that produced *round, green* seeds with a plant that produced *wrinkled, yellow* seeds, for example. Some of his peas had yellow seeds, and some had green. Further, some were round when ripe, and some were wrinkled. When the two features were considered separately, the yellow color proved to be dominant over green, and round shape was dominant over wrinkled. Therefore, if each dominant characteristic was crossed with its contrasting recessive, all offspring of the F_1 showed the dominant trait.

If the two pairs of characteristics were treated simultaneously, a new phenomenon appeared. Each pair of alleles behaved as if it were unaffected by the other pair of alleles. The number of plants with dominant seed *color* was three times that with recessive color, and the same was true of seed *shape*. But if Mendel crossed two pea plants that were heterozygous for the two pairs of alleles, then new combinations of alleles appeared in the next generation (see Figure 11.7). Although the parents were round-and-yellow or wrinkled-and-green, some of the offspring were round-and-green or wrinkled-and-yellow. *Recombination* had occurred. Recombination indicates new gene combinations.

These observations led to Mendel's second generalization, the **Law of Independent Assortment,** which states that when one pair of alleles segregates during the formation of gametes, its manner of segregation is not affected by the manner of segregation of a second, different pair of alleles. Note that the simple English phrase *independent assortment* means just what it says.

In present-day genetic research, whether on humans, bacteria, or viruses, recombination is one of the bases on which progress depends because it is mainly by observing recombination, visibly expressed, that geneticists can determine what genes have done.

HOW IS MEIOSIS RELATED TO MENDELIAN PRINCIPLES?

Mendel's observations are especially remarkable when we remember that he knew less about chromosomes and cell division than you do. Mendel was able to offer an explanation of the results of his crosses without knowing the cellular mechanism that made them possible. But very soon after Mendel's work was rediscovered in 1900, his principles could be reinterpreted in terms of cell biology. It was the time when cell biology had reached a point of maturity, when all the pieces were being put together in a comprehensive way. Cell biologists recognized that Mendelian principles about "factors" could also be applied to chromosomes during meiosis.

In 1903 Walter S. Sutton, while still a graduate student at Columbia University, demonstrated a parallel between the behavior of chromosomes during meiosis and the behavior of pairs of Mendel's "factors." Sutton called attention to "the probability

PROBABILITY AND CHANCE IN MENDELIAN GENETICS

Carrying the idea of independent assortment further, Mendel saw that the number of individuals with every possible combination of characteristics was close to what one would expect if the pairs of alleles were distributed by chance. For example, if you toss two pennies at a time and count the heads and tails, you will find that after a hundred or so pitches, you have about 25 percent both heads, 50 percent heads-and-tails, and 25 percent both tails. If you think of heads as being dominant in a Mendelian sense, the 25:50:25 can be expressed as 75:25, or 3:1—or what peas demonstrate in the F_2. Next, if you toss pairs of pennies and pairs of dimes simultaneously many times, you can obtain 16 combinations. Again taking into account the idea of dominance, you will find that 9 out of the 16 possibilities will show two heads, 3 will show one head, 3 the other head, and 1 all tails. The proportions are the same ones that peas exhibit when two pairs of alleles are considered at once, 9:3:3:1. Since the outcomes of pitching coins are the result of chance, and since the ratios obtained in making crosses give the same numbers that chance gives, the conclusion is that segregation and assortment of genetic features must also be a matter of chance. Distribution of chromosomes at meiosis is a matter of chance, and so is the meeting of gametes in fertilization. Random events give statistically predictable results, whether we are dealing with inanimate coins, or living, reproducing organisms.

Two basic principles of probability should be considered in the overall discussion of probability and chance in Mendelian genetics:

1. If a coin is tossed 100 times, and it comes up heads every time, the chance of its coming up heads again on the next toss still remains 50 percent, the same probability of coming up heads on the first toss, or the seventeenth, or on any given toss. *The result of one chance event does not affect the results of later occurrences of the same chance event.* In human terms, parents who already have two daughters are just as likely to have another daughter as a son on the third attempt.

2. The probability of one die coming up with any of its six numbers is $\frac{1}{6}$. Therefore, the probability of two dice each producing the same number during one toss is $\frac{1}{6} \times \frac{1}{6} = \frac{1}{36}$. *The probability that two independent events will occur together is the product of their probability of occurring separately.* Again, in human terms, the chance that a woman's first child will be a boy is $\frac{1}{2}$, and the chance that her sister's first child will be a boy is also $\frac{1}{2}$. The chance that both sisters will have first children who are boys is $\frac{1}{2} \times \frac{1}{2} = \frac{1}{4}$. In other words, the chances are 3 to 1 against the occurrence of such an event.

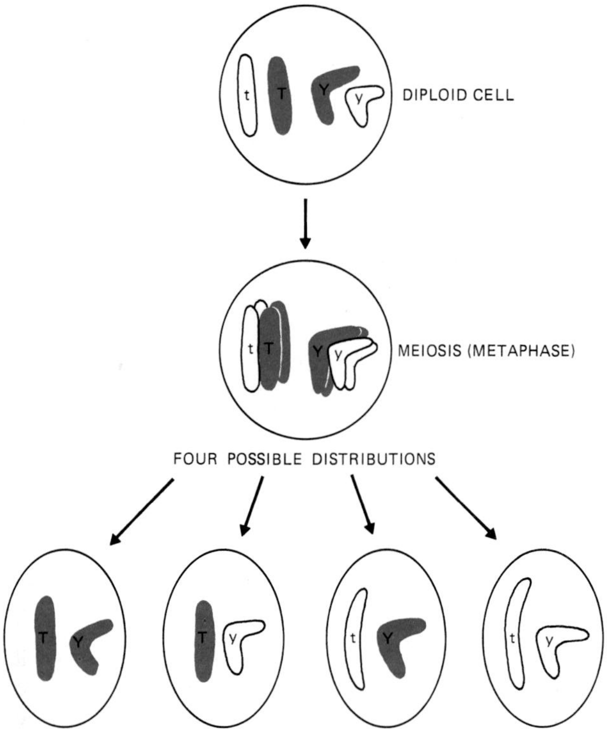

Figure 11.8
The parallel between the physical distribution of chromosomes and the phenotypic development of organisms. In these diagrams, chromosomes bearing a dominant allele for tallness in peas (T) and those bearing the dominant allele for yellow (Y) are in color, and those bearing the recessive alleles, t and y, are white. To make the distinction more readily visible, one pair of homologs (T and t) are straight, and the other (Y and y) are bent. At meiosis, the chromosomes are distributed randomly, with the restriction that each gamete must have a full set of chromosomes. The result is that one-fourth of the gametes carry the alleles T and Y, one-fourth carry T and y, one-fourth carry t and Y, and one-fourth carry t and y. The distribution of *chromosomes* themselves can be followed by observing them with a microscope. The distribution of the *genes* can be followed by making a test cross, which shows that the active genetic factors function in a manner comparable to that of chromosomes. The similarity of behavior of chromosomes in meiosis to the distribution of genes in inheritance gives strong support to the idea that genes are carried by chromosomes.

that the association of paternal and maternal chromosomes in pairs and their subsequent separation during [meiosis] may constitute the physical basis of the Mendelian law of heredity."

Without disturbing either the Mendelian principles or the new information about how chromosomes behave during meiosis, the word "chromosome" could, in fact, be substituted for the word "factor":

1. Chromosomes in parents are paired.
2. Chromosomes separate, and only one from each parent goes to an offspring.
3. Chromosomes are paired again in the offspring, one chromosome coming from each parent during fertilization.
4. It is a matter of chance which member of a parental pair of chromosomes goes to an offspring.

Both the *Law of Segregation* and the *Law of Independent Assortment* could now be interpreted in terms of Sutton's **chromosome theory of inheritance,** which proposed that genes, the determinants of heredity, are located on the chromosomes. This was the first specific reference to genes as the basic units of inheritance. Just as Mendel's "factors" segregated during gamete formation, so members of a pair of chromosomes segregate into different gametes during meiosis. Also, during meiosis chromosome pairs align randomly, providing a cellular parallel for the principle of independent assortment (Figure 11.8). Thus it was shown that there is a clear, understandable relation between the reproductive activities of cells and inheritance in whole organisms. The new science of genetics had come into being.

TO SUM UP MENDELIAN GENETICS

(Modern genetics terms are used here; Mendel would not have used such words as gene, allele, or chromosome.)

1. A gene is a segment of a DNA strand that controls a specific cellular function.
2. An allele is an alternative form of a gene at a specific place on a chromosome.

3. Inherited traits are controlled by alleles, which occur in pairs in sexually reproducing organisms.
4. When alleles in a pair are different, one may be dominant over the other, which will be recessive.
5. A pair of alleles is separated during the formation of gametes, and distributed unchanged into the gametes with equal probability.
6. When one pair of alleles on a chromosome segregates during gamete formation, they are distributed to gametes independently of how other alleles on a chromosome are distributed.

GENETICS PROBLEMS

Using the symbols T (for the "tall allele" in peas) and t (for dwarf), Y (for yellow seeds) and y (for green), complete the following exercises.

1. A phenotypically tall pea plant is pollinated by a dwarf plant, and the seeds of the first generation hybrid produce 327 tall plants and 321 dwarfs. Give the genotypes of all the plants.
2. Two tall pea plants, both heterozygous for height, were crossed, and 860 seeds were collected. What number of what phenotypes would be expected to grow from the seeds?
3. A tall, yellow-seeded plant, heterozygous for both characteristics, was pollinated by a dwarf, green-seeded one. From the 680 seeds collected, the plants were: 172 tall, yellow; 168 tall, green; 167 dwarf, yellow; 173 dwarf, green. Give the genotypes of all plants.
4. A plant of genotype TTYy was crossed with one of genotype ttYy. If the cross yielded 800 seeds, what number of what phenotypes probably resulted? What were their genotypes?
5. Two tall, yellow-seeded plants were crossed, and some dwarf, green-seeded plants resulted.
   a) What were the genotypes of the parents?
   b) What possible genotypes might there be among the tall, yellow-seeded offspring?
6. A TtYy pea plant self-pollinates, and one seed is picked at random for planting.
   a) What is the chance that the seed will produce a tall, green-seeded plant?
   b) If the plant actually turns out to be tall and yellow-seeded, what is the chance that its genotype is TTYY?
7. If you have a phenotypically tall plant, describe several ways in which you could learn its genotype.
8. A dominant gene (S) in snapdragons produces bilaterally symmetrical ("irregular") flowers; the recessive allele (s) produces radially symmetrical flowers. Another pair of alleles controls color, yielding red (A^1A^1), pink (A^1A^2), or white (A^2A^2). If a cross is made between two snapdragons of genotype SsA^1A^2, what will be the phenotypes of the F_1? Give the proportions of each type.

ANSWERS TO GENETICS PROBLEMS

1. Parents, Tt and tt. All the tall offspring were Tt, the dwarfs tt.
2. 645 tall, 215 dwarf.
3. Parents TtYy, ttyy; the offspring, TtYy, Ttyy, ttYy, and ttyy.
4. 600 phenotypically tall and yellow, of which there were 200 TtYY, 400 TtYy; and 200 tall and green, Ttyy.
5. a) Both TtYy.
   b) TTYY, TtYY, TTYy, and TtYy.
6. a) $3/16$. b) $1/9$.
7. You might find the genotypes of its parents, but a surer way would be to cross it with a short plant, tt—that is, make a test cross. You might also use self-pollination if the plant is self-fertile.
8. Irregular red, $3/16$; irregular pink, $6/16$; irregular white, $3/16$; regular red, $1/16$; regular pink, $2/16$; regular white, $1/16$.

SUMMARY

1. Before Mendel, work on problems of inheritance was vaguely descriptive, and because it tended to look at too much at once, the results were too complicated to understand.

2. Mendel made two decisions that brought him experimental success where others had failed. First, he paid attention to only one or two characteristics at a time. Second, he started the practice of statistical treatment of biological data.

3. Mendel's *Law of Segregation* states that contrasting factors in an individual do not blend or contaminate one another; they segregate when gametes are formed, and any single gamete receives only one factor.

4. The *Law of Independent Assortment* states that when one pair of alleles segregates during the formation of gametes, the way it separates is not influenced by the action of another pair of alleles.

5. Mendel's hypotheses may be summarized (in current genetic terminology) as follows:
   a) Indivisible units called *alleles* (Mendel called them "factors") function as pairs to determine hereditary traits. One member of the pair is contributed by each parent.
   b) Any diploid cell contains two genes for any particular trait; these genes may be alike *(homozygous)* or different *(heterozygous).*
   c) Alternative versions of genes, such as tall versus dwarf, are called *alleles.*
   d) If in one organism the two alleles are different, the *dominant* allele will be expressed phenotypically while the *recessive* allele remains hidden.
   e) One member of each pair of alleles is segregated into a gamete at the time of gamete formation. Each member of the pair is unaffected by the other member of the pair. One allele is just as likely as the other to go into a given gamete.
   f) Gametes are united randomly during fertilization, and the distribution of traits among offspring is predictable if the genotypes of the parents are known. Inherited traits are therefore varied but predictable.

6. Soon after Mendel's work was rediscovered in 1900, both the Law of Segregation and the Law of Independent Assortment could be interpreted in terms of Sutton's *chromosome theory of inheritance.* Sutton showed a clear, understandable relation between the reproductive activities of cells and inheritance in whole organisms.

12
Genetics and Chromosomes

SOME KEY POINTS

1. Mutations are changes in genes or chromosomes. When they occur in sex cells they can be passed on to future generations.
2. Gender in humans is determined by parental chromosomes. Females have an XX pair, and males have an XY pair. If an X of the male parent combines with an X of the female parent, the offspring is female; when a Y chromosome from the male parent combines with a female parent's X, the offspring is male.
3. When chromosomes do not separate normally during meiosis, the result is called nondisjunction.
4. Genes that do not segregate and tend to be inherited together are said to be linked.
5. Chromosomes are capable of breaking and exchanging parts during meiosis, making genetic crossovers that associate genes from the mother with genes from the father.
6. Chromosome abnormalities can occur when a chromosome or a piece of a chromosome is lost, unnecessarily repeated, placed out of sequence, or transferred to another chromosome.
7. A mutation occurs at a certain chromosome position, or locus. When in a population there are more than two possible alleles at a specific locus, they are multiple alleles.
8. Certain human traits are determined by genes that are linked to the sex-determining genes of the possessor.
9. Some visible characteristics, such as skin color, are affected by more than one gene or pair of alleles.

ONE OF THE IRONIC TWISTS OF HISTORY IS that the original work of Mendel lay practically unnoticed for 35 years. When it was at last disinterred, a rush of experimenters set about proving or disproving his results. Although peas were ideal for the kind of work Mendel did, they are too large and reproduce too slowly to produce the masses of statistical information geneticists need to obtain new information. In their search for a more useful organism, investigators chose fruit flies, of the genus *Drosophila,* those small insects that gather in swarms around decaying fruit. Fruit flies offer some special advantages. They are so small that a laboratory can raise millions of them inexpensively, they produce a generation in less than two weeks, they have only four (haploid) chromosomes, and during the larval stages their glands have relatively huge chromosomes that can be seen in detail with a light microscope (Figure 12.1). (Still later, geneticists turned to the fungi, then to the smaller and more-rapidly reproducing bacteria, and finally to the smallest and most rapid of all, viruses.) Analysis of human beings has also contributed to our understanding of genetics. Even though breeding experiments in humans are not possible, a great deal has been learned from family pedigrees and from studies of human chromosomes.

MUTATIONS MUST BE EXPRESSED VISIBLY

Before one tries to understand how experiments on heredity were performed, it is useful to know how differences in organisms can be used by biologists and how their significance came to be appreciated. Differences, sometimes trivial and sometimes great,

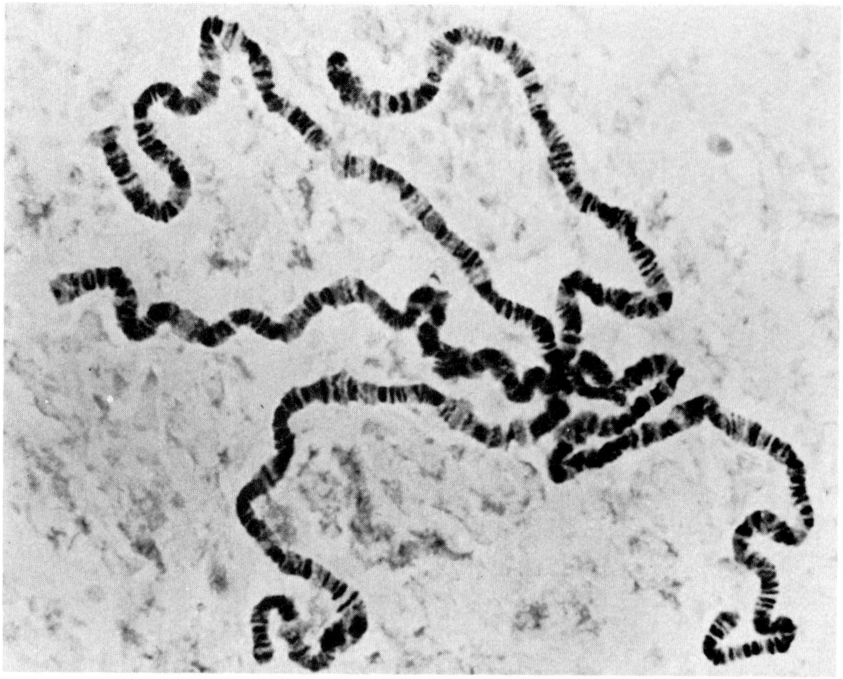

× 200

Figure 12.1
"Giant" chromosomes from the salivary glands of a *Drosophila* larva are about 100 times larger than other body cells. These chromosomes, able to be seen with an ordinary light microscope, were invaluable in early genetic research.

Figure 12.2
Hugo De Vries, holding two evening primroses. He is remembered not only for his part in the rediscovery of Mendel's work but for his enunciation of the idea of *mutations*. Either contribution alone would have been enough to establish him as one of the main founders of the modern science of genetics.

are frequently the result of hereditary forces in an organism, and their visible expression is affected little or not at all by environmental forces. DNA changes in an organism, whether they are expressed visibly or not, are **mutations,** and the story of how mutations came to be recognized starts in the Dutch garden of a botanist, Hugo De Vries (Figure 12.2).

De Vries was interested in breeding plants known in English as evening primroses. Among the results of his breeding experiments, De Vries noticed that at unexpected intervals he obtained different kinds of plants, one of the most famous being an unusually large type that he appropriately called *gigas,* the giant. De Vries found that his different plants *bred true;* that is, they produced offspring identical to themselves. "Here," said De Vries, "is the way Darwin's variations come into being. They occur suddenly, as large changes, and the changes are inherited."

De Vries called these changes mutations, from a Latin verb that means "to change." He wondered if anyone else had ever described such changes, and while searching through the published literature, he came upon Mendel's forgotten paper on inheritance in peas. De Vries recognized Mendel's paper for what it was: a simple, clear, original, satisfactory explanation of the principles of inheritance. The obscure monk had been rediscovered, and as the twentieth century began, evolution and genetics met, to grow more and more intimate with the passing years.

Ironically, De Vries may be more important for having found Mendel than for having explained the basics of mutations. As it turned out, the "mutations" in evening primroses are not really mutations in the present-day sense. They are instead only chromosome rearrangements of a sort peculiar to evening primroses and a few other plants. Stable, heritable changes, however, were sought and found in many other plants and animals, until we now use mutations as the basis for practically all genetic research. The "mutations" found by De Vries and the true gene mutations found by practical geneticists have both been used to gain information on patterns of inheritance and interactions among genes. Now when a biologist speaks of a mutation, he means a gene mutation, which is a change in DNA. If he merely means a chromosome rearrangement, he explicitly calls it a "chromosome mutation."

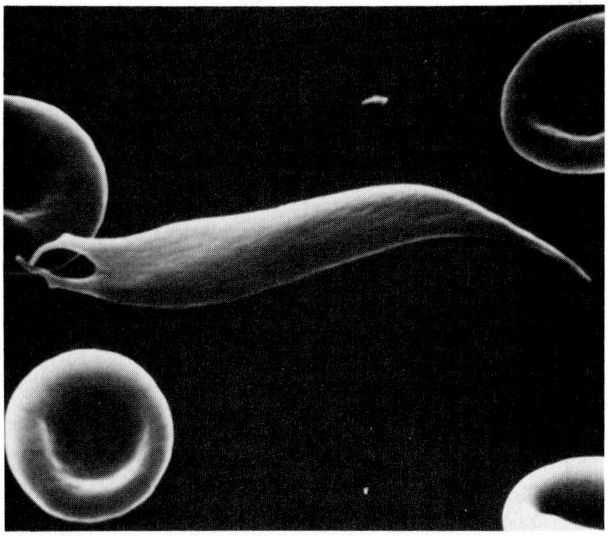

× 3000

Figure 12.3
The effect on an organism of a change in a single DNA nucleotide. The scanning electron micrograph shows normal red blood cells from a person suffering from sickle-cell anemia and one sickled cell, enormously stretched out. Besides not carrying oxygen, sickled cells tend to pile up in small blood vessels, impairing normal blood circulation.

The Frequency of Mutations

Because of the speed of reproduction, mutation rates are most readily determined with experiments on bacteria and viruses, in which one mutation in a billion can be detected. These experiments show that not all genes mutate at the same frequency. If a gene is chemically stable, it will mutate less frequently than will an "unstable gene." It is not unusual for one mutation to occur in a million DNA replications, but some unstable genes may mutate once in 2000 replications. On the other hand, an exceptionally stable gene may remain unaltered through billions of replications.

Causes and Effects of Mutations

Gene mutations occur when nucleotides in a DNA molecule are added, left out, or changed. Before a cell division can take place, there must be replication of DNA. The process almost always works accurately, but occasionally something goes wrong, and the DNA nucleotide sequence is altered. That is a mutation. Sometimes the mutation is not phenotypically detectable, and no further change takes place. At the other extreme, one change may result in the death of an organism. Such a mutation, regardless of how it occurs, is known as a *lethal mutation.* Ordinarily, geneticists are most concerned with mutations whose effect is between the undetectable and the lethal.

A single "misplaced" amino acid can make the difference between normality and malfunction, or even between life and death. For example, if one of the 300 amino acids in hemoglobin* is changed, the action of the whole hemoglobin molecule is changed. A number of such changes are known, a famous one in human biology being the substitution of valine for glutamic acid in sickle-cell anemia (Figure 12.3). That seemingly small change makes an enormous difference to the life of an individual in whose blood it occurs. This example indicates the importance of the specific sequence of amino acids in a protein.

Remember that Table 7.1, which illustrated the genetic code, showed that some amino acids can be produced by more than one codon. For example, leucine can be synthesized by UAA, UUG, CUU, CUC, CUA, and CUG. This repetitious ability, called the *degeneracy* of the genetic code, is thought to allow for a certain amount of error in the codons without destructive results for an organism.

* Hemoglobin is a protein in red blood cells responsible for carrying oxygen.

Artificially induced mutations We still do not know what causes natural gene mutations, but mutations may be induced artificially by any radiation that has sufficient energy to cause chemical changes of molecules in cells. The most effective agents of radiation have been X-rays, gamma rays, neutrons, and ultraviolet light. We have also learned that radiation from radioactive elements, such as that from the dust of nuclear explosions, will cause mutations. The first person to demonstrate the effectiveness of radiation to induce genetic mutations was Hermann J. Muller, who in 1927 developed a technique for proving the presence of lethal mutations caused by X-rays in *Drosophila.* He subsequently won a Nobel prize in 1946. By that time, the first atomic bombs had been dropped on Hiroshima and Nagasaki, and the possibility of mutations in human beings caused by human action began to be realized.

Mutations in sex cells may be accelerated by increasing the temperature or by introducing "chemical mutagens," such as the oxides of nitrogen from automobile exhausts, mercury, DDT and other pesticides, food additives, and a wide variety of organic compounds, such as dyes and the ingredients of tobacco smoke. Most cancer experts agree that the demographic distribution of cancer types supports the hypothesis that many, perhaps the majority, of human cancers are induced by such environmental "carcinogens."

The Fate of Mutations

Most mutations are detrimental. The effect is somewhat as if you made a thoughtless change in the circuitry of your finely tuned stereo sound system. The chance of your improving its performance by such a change is slim indeed. Existing organisms must be already well adapted to their environment, as proved by the very fact of their survival. Any change is more likely than not to be a change for the worse, that is, one that may make an organism less well suited to its environment. It is obvious, however, that over a long period of time, occasional mutations have been advantageous, because favorable mutations are the basis for the genetic variability that makes evolution work.

Besides being usually harmful, mutations are also usually recessive. Therefore, most new mutations are not expressed phenotypically in sexually reproducing organisms, although they can be expressed if by chance mating they happen to occur as a homozygous pair of alleles. In view of the relative rarity of mutations and the improbability of bringing together an identical pair of mutant alleles, it is apparent that many mutations may exist in a population generation after generation without being expressed. Also, if the possessor of a gene mutation dies without having passed it on to offspring, the mutation may be eliminated without ever having been expressed at all.

When a mutation occurs in a cell, it will be passed on to the descendants of that cell, but what eventually happens to it depends on the kind of cell in which the mutation originally appeared. Only those mutations that occur in sex cells in animals have any real chance of being passed on to later generations. A mutation in a liver cell (which is a body cell, or *somatic* cell), for example, will be lost when the animal dies, but a mutation in an ovary may be included in an egg and then in offspring.

TO SUM UP MUTATIONS

1. Mutations are changes in genes (DNA). A chromosome rearrangement is explicitly called a *chromosome* mutation.
2. Gene mutations occur when nucleotides in a DNA molecule are added, left out, or changed. The cause of natural gene mutations is not known, but mutations can be induced artificially by radiation and other means.
3. Mutations are usually not beneficial, and are usually recessive.
4. In animals, only mutations in sex cells can be passed on to future generations.

SEX-LINKED TRAITS IN DROSOPHILA

In general, the reality of Mendelian ratios was confirmed by new experiments in the early 1900s, but soon inexplicable irregularities appeared. In 1910, Thomas Hunt Morgan and his associates at Columbia University began a series of classic experiments with the fruit fly, *Drosophila.* Morgan noticed among his stock of fruit flies a fly with white eyes instead of the normal red eyes (Figure 12.4). The mutant fly, a male, was mated with a normal red-eyed female, and all their F_1 offspring had normal red eyes. Morgan correctly concluded that red eyes are dominant, and these results were therefore not unusual. When red-eyed F_1 males were mated with red-eyed F_1

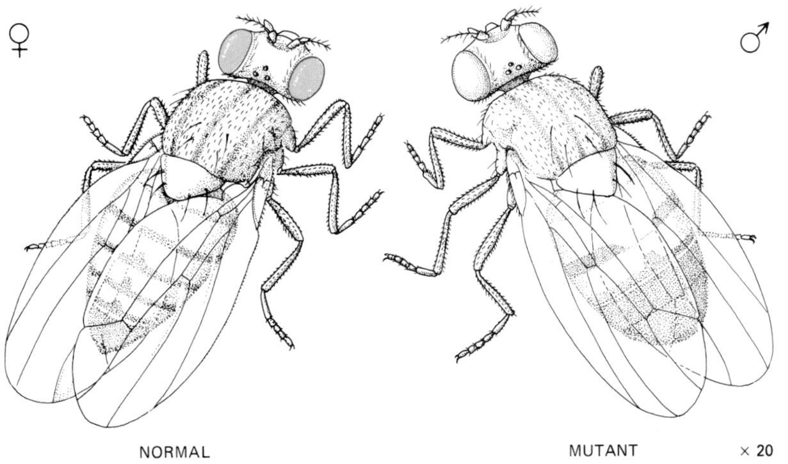

Figure 12.4
Eye color in fruit flies. The usual, or *wild-type,* eye of *Drosophila melanogaster* is red, but unpigmented white eyes occur in some mutants.

females, the F_2 generation contained 3470 red-eyed flies to 782 white-eyed.

At this point Morgan made a crucial discovery. *All* 782 of the white-eyed flies were *males.* This was a puzzle, and Morgan realized it could not be a coincidence. He deduced the connection between white eyes and sex (specifically, the male sex) and thereby made the first identification of a trait in *Drosophila* that was linked to the sex of the fly. A **sex-linked trait** is a visible feature that is specifically associated with the sex of its possessor, that is, one that is carried on a sex chromosome.

Even before Morgan began his observations of *Drosophila* chromosomes, two types of chromosomal sex determination had been postulated: the XO–XX and XY–XX types: In some insect species the male has a single X (sex) chromosome and the female has two X chromosomes. The male is symbolized as XO, with the O indicating the absence of a homolog of X, and the female is XX. Both humans and *Drosophila* have the **XY–XX** type of sex chromosomes (Fig-

Chromosomal Sex Determination in Drosophila

Once an inherited bodily trait was known to be linked to sex inheritance in fruit flies, Morgan proceeded to examine *Drosophila* chromosomes to see if chromosomes were related to the determination of sex, and if W. S. Sutton (Chapter 11) had been correct in hypothesizing that genes are located on chromosomes. (They are.) Another proposal by Sutton could also be tested: Are several genes usually located on one chromosome? (They are.)

In *Drosophila,* there are four pairs of chromosomes, or a diploid number of eight. In females there are four matching pairs, but in males only three pairs match, and they have counterparts in females. One chromosome pair in males is mismatched. The mismatched pair, called the XY pair, has one chromosome (the X chromosome) like those in females, and one odd, hook-shaped one (the Y chromosome) (Figure 12.5). The XX pair in females and the XY pair in males are the **sex chromosomes;** the other three pairs, identical in both male and female, are **autosomes.**

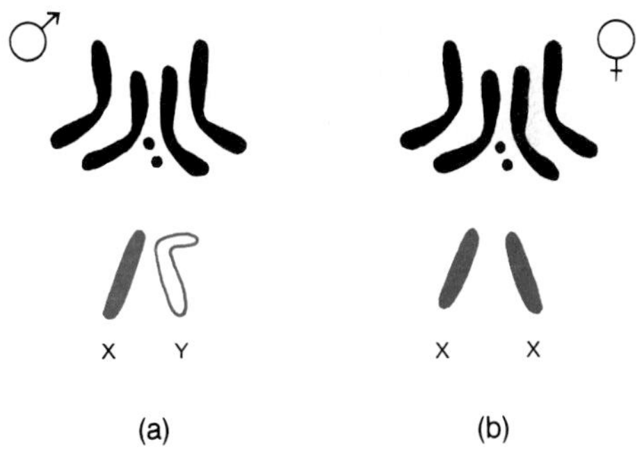

Figure 12.5
Fruit fly chromosomes. (a) The chromosomes of a male fly. The cell has three pairs of autosomes in matching pairs, shown in black, and one pair of unmatched sex chromosomes, shown in color. The colored *straight* sex chromosome is designated X; the *bent* one is designated Y. (b) The chromosomes of a female fly. Again the autosomes are shown in black and the sex chromosomes in color. In females, all the chromosomes occur in matching pairs.

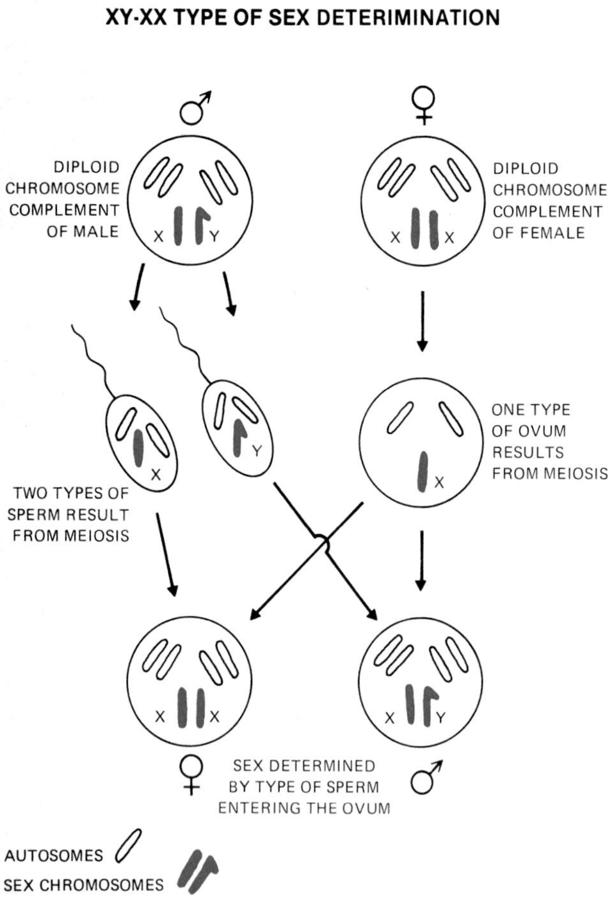

Figure 12.6
Diagrams of one kind of sex determination, the XY–XX type. There are two different "sex chromosomes." (Compare Figure 12.5.) There are two kinds of sperm, those carrying an X chromosome and those that carry a Y chromosome. A Y-type sperm, fertilizing an X-carrying egg, produces an XY zygote, which develops into a male. An X-type sperm, fertilizing an egg (all eggs have an X chromosome), produces an XX zygote, which develops into a female.

ure 12.6). The female has two chromosomes, **XX,** and the male has one chromosome identical to the female's, plus one of the shorter variety mentioned earlier. The male is designated as **XY.**

NONDISJUNCTION IN DROSOPHILA

Calvin Bridges, one of Morgan's co-workers, guessed that at meiosis, when homologous chromosomes normally separate, occasionally some chromosomes simply failed to separate. The result is that

RECOMBINANT DNA

What is recombinant DNA? It is the same as gene cloning, gene splicing, genetic engineering, or gene manipulation, and involves the transfer of genes from one organism to another, followed by unlimited reproduction of the transferred genes.

Once the general outlines of the DNA-to-protein story were understood, and techniques for detailed handling of DNA became available, biologists began to ask, "Why can't we make any organism do anything any other organism can do?" If, for example, one could obtain a length of DNA that is capable of coding for the production of a specific human protein, could one then get that DNA into a nonhuman cell and let that cell take over the production of that protein?

One of the first practical ideas was to obtain the bit of DNA, the gene, for human insulin synthesis, incorporate it into a bacterial cell, and have bacteria synthesize human insulin. Instead of the old difficult and expensive way of obtaining insulin from animals, one could culture masses of insulin-making bacteria and harvest the product as one would culture yeast in the manufacture of beer.

One promising experimental organism is the colon bacillus, *Escherichia coli,* which has been worked on so intensively that a store of information about it is available. (Similarly, a number of viruses are known in genetic detail.) It is known that in addition to its main chromosome, *E. coli* cells may contain smaller loops of DNA, known as plasmids, which replicate when the bacterial cell divides. And techniques are known by means of which plasmids can be introduced into cells. Further, enzymes can be used to break DNA molecules at specific places.

As the diagrams show, a plasmid can be taken from a bacterium and "cut" to open the ring. With the ring opened, again by enzyme action, a length of foreign DNA can be inserted into the opening. The foreign DNA might be from some other organism, even from a human cell, or it might be artificially synthesized so as to have a desired nucleotide sequence. Once the plasmid has accepted the foreign DNA—that is, has un-

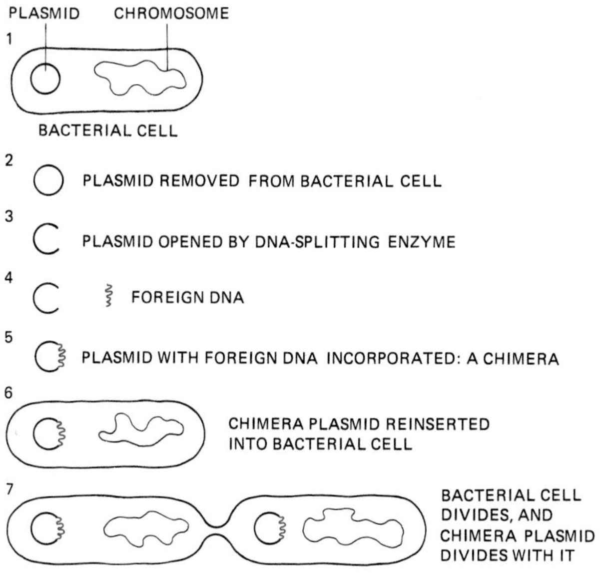

dergone recombination—it then has a mixture of two kinds of DNA and is a *chimera* (kye-MEER-uh). The altered plasmid, or chimera, is reintroduced into a bacterium. When the bacterium divides, so does the plasmid with its new length of DNA, and all subsequent divisions will carry it, producing a clone of identical cells. If the artificially added DNA codes for the production of some desired product, the ensuing mass of cells will produce as much of this product as the bioengineers want.

At present, some substances can be gotten only from animals, and they cost far more than their weight in gold, or cannot be had commercially at any price. Bioengineers have already made insulin-producing bacteria and yeast that makes antivirus material (interferon). They are looking forward to the production of other substances such as scarce hormones or various blood components. Such once-expensive substances, if synthesized in large vats by bacteria, could be made cheaply.

Medical supplies are only one possible application of the theory behind recombinant DNA. The list of desired materials that might be made cheaply and in abundant quantities includes tasty and nutritious proteins, fats and oils for eating, burning, or soap-making, and alcohols for commercial chemistry or for vehicular fuel.

Extending the idea from bacteria to other organisms, bioengineers talk of new crop plants. Why not corn that fixes its own nitrogen as legumes do without expensive fertilizer? Why not beans that carry all, instead of just some, of the amino acids needed for complete human nutrition? Why not wheat that resists rust fungi? The list seems to be limited only by human imagination. Huge amounts of money could be made and medical supplies enormously increased.

If the new biotechnology promises such fortunes and such benefits to humanity, why have citizens' groups, scientists (singly and in associations), and government forces made so much noise against it, or at least insisted on regulating it so closely? Two reasons appear. One reason is ethical and theoretical. Some argue that human beings do not have the right to make new forms of life, perhaps eventually trying to "engineer" a human body. A second reason is practical. Some scientists fear that a bacterium or some other recombined-DNA organism may escape from its home laboratory, and, if it proved uncontrollable, might bring on a disastrous epidemic of disease. Current regulations are intended to make such a catastrophe highly unlikely. But, say the opponents of recombinant-DNA work, laboratories are run by people, and people are fallible.

In spite of opposition, ranging from limited to absolute, research continues, and over a hundred plant and animal genes have been inserted into vectors—viruses that can carry a bit of recombined DNA from one cell to another. Progress, both in laboratory research and in commercial development, is so rapid that it is impossible to do more than give a general idea of the present state of knowledge and activity in this explosively fast-growing field.

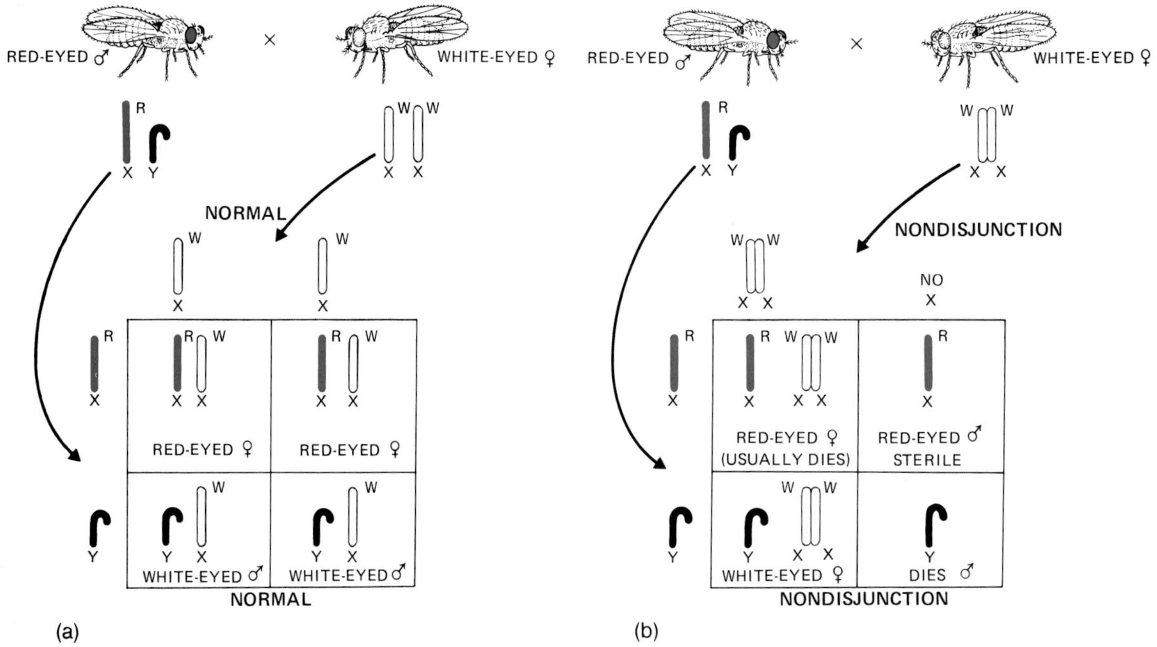

an egg might contain two Xs instead of the usual one X. It was also possible that both X chromosomes might go into the polar bodies, leaving the egg with no X chromosome. Microscopic examination of cells from certain female flies with unusual sex ratios among their offspring verified that their eggs did indeed have either an extra X chromosome or no X chromosome at all. This failure of chromosomes to separate during meiosis is called **nondisjunction** (Figure 12.7).

Figure 12.7
Normal and abnormal gamete production in fruit flies. (a) In the normal distribution of sex chromosomes (and of other chromosomes as well), one member of each homologous pair goes into a gamete. Thus every egg has a set of *autosomes* and an X chromosome, and every sperm has a set of autosomes and either an X or a Y chromosome. (b) If meiosis fails to proceed properly, the two X chromosomes in the ovary of the female may both go into one egg, or they may both go into a polar body, leaving an egg without an X chromosome. This is *nondisjunction.* If a female fruit fly suffers a nondisjunction, her daughters will have either three X chromosomes (and usually die), or the XXY condition. Her sons will lack any X chromosome, and will die.

Nondisjunction and Sex Determination

In fruit flies, if a two-X egg is fertilized by an X-carrying sperm, the XXX zygote develops into a sterile female, which usually dies. If the two-X egg is fertilized by a Y-carrying sperm, the XXY zygote develops into a fertile female. This fact shows that in flies *the Y chromosome itself does not determine maleness.* This is *not* true of human sex determination. In humans the Y chromosome is a male determiner, since an XXY zygote develops into a fertile male, and an XO (having no Y) becomes a sterile female.

Regardless of the difference between flies and humans, the important fact to come out of all this work is that chromosomes do indeed determine sex. An even more important conclusion was that if chromosomes could control sex, then they might be able to control everything an organism has or does. Thus it was established that of all the things that might imaginably control any biological activity, it is the chromosomes that are the chief regulators.

The ancient riddle of the determination of sex turned out to be a matter of Mendelian assortment. Morgan could say: "The old view that sex is determined by external conditions is entirely disproved."

GENE LINKAGE AND CROSSING OVER

Morgan discovered many other irregularities in the fruit fly besides the white-eyed mutant. For example, there are mutant black bodies as opposed to the normal wild-type gray bodies, and there are mutant reduced (vestigial) wings as opposed to normal wings (Figure 12.8). Gray bodies and normal wings are both dominant. If a black-bodied, vestigial-winged fly is crossed with a gray-bodied, normal-winged one, the F_1 flies are all gray-bodied and normal-winged, as one would expect. Next, if the F_1 hybrids are inbred, one would expect 9/16 to show both dominant characters (Gray, Normal), 3/16 one dominant character (black, Normal), 3/16 the other dominant character (Gray, vestigial), and 1/16 both recessives (black, vestigial). However, the expected ratios do not occur. Instead, the F_2 generation has too many individuals resembling one grandparent or the other and too few recombinants.

In practice, geneticists do not typically inbreed the F_1 progeny as Mendel did. They use instead a **test cross,** in which an experimental organism is mated with a phenotypic recessive. The phenotypic recessive is of course homozygous, since otherwise the recessive phenotype would not be apparent. A test cross is informative because the results can be assessed for genotype by direct inspection. To make a test cross of those flies that do not show Mendelian assortment, the geneticist mates one of the F_1 females (Gray-bodied and Normal-winged, but heterozygous for both features) with a double-recessive male (black-vestigial). The expected results from such a mating are:

Gray, Normal	25%	50%	Parental types
black, vestigial	25%		
black, Normal	25%	50%	Recombinant types
Gray, vestigial	25%		

But the actual kinds and numbers of flies among the offspring, puzzling at first, are as follows:

Gray, Normal	41.5%	83%	Parental types
black, vestigial	41.5%		
black, Normal	8.5%	17%	Recombinant types
Gray, vestigial	8.5%		

There are more gray-normals and black-vestigials than were expected and not enough recombinants, and it is especially notable that the majority of the flies are like one parent or the other. Instead of assorting themselves as the genes did in Mendel's

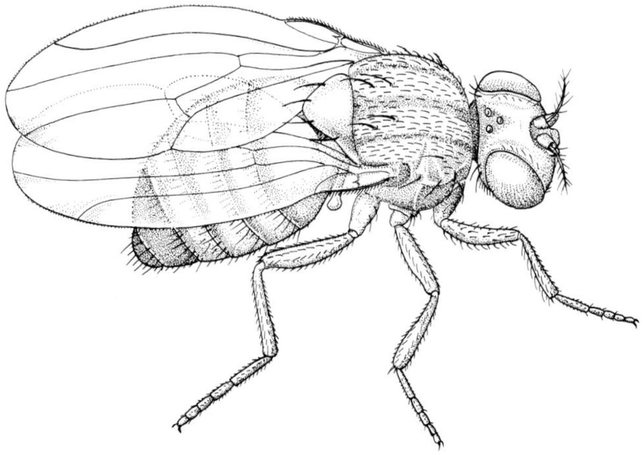

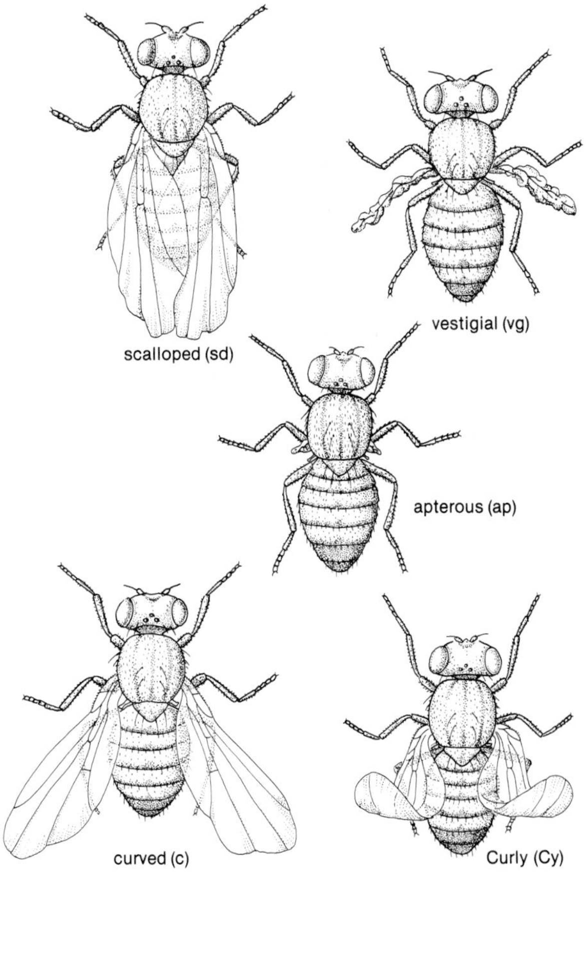

Figure 12.8
Some mutations of wing structure in *Drosophila.* A normal or *wild-type* fly is shown directly above, about 25 times natural size, and the deviations are to the right, each with its name and genetic abbreviation. The *apterous* mutant (Greek, "without wings") is designated ap; the use of lowercase letters indicates that it is determined by a recessive allele. Curly, in contrast, is Cy; the upper-case initial indicates a dominant allele. Dozens of wing mutations have been described, as have hundreds of mutations involving various fly structures and colors.

peas, the genes for body color and wing shape seem to be stuck together in most, but not all, of the flies. This observation set the stage for still other important genetic discoveries.

Gene Linkage

When genes tend to be inherited together, they are said to be **linked.** With body color and wing form in fruit flies, linkage results in 83 percent of the flies having the same genetic characteristics as the parents; only 17 percent are unlike either parent (see above).

When further crosses were made with flies and with a number of other organisms, more and more examples of linkage were discovered, until lists of linked genes were built up. Some genes were closely linked, with low percentages of recombinants, and other genes were less closely linked. *Linkage lists* were made for fruit flies after hundreds of trial matings until it became apparent that four linkage groups exist in fruit flies. In corn, another favorite experimental organism, ten linkage groups were established. The revealing fact about these experimental organisms is that fruit flies have *four* linkage groups and *four* chromosomes, and corn has *ten* linkage groups and *ten* chromosomes.

Crossing Over

The numerical equality of linkage groups and chromosomes indicated that chromosomes are indeed the genetically important parts of cells, and other evidence in support of this notion was—or soon

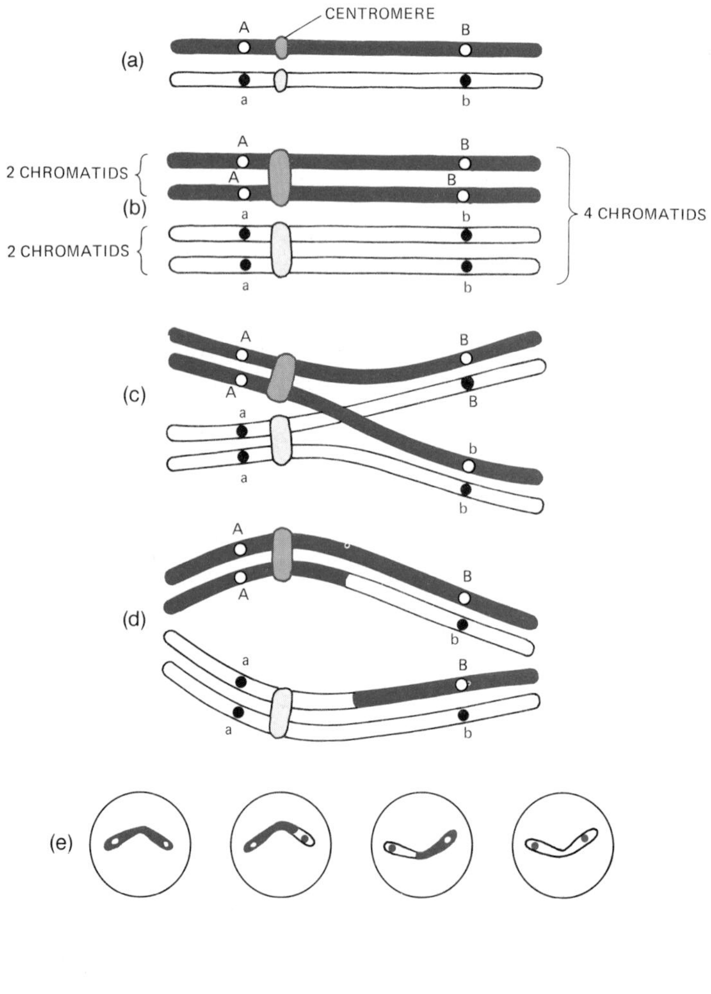

Figure 12.9
Crossing over. (a) A pair of homologous chromosomes is represented, one in color and one in outline, each with two "genes," shown as circles.

(b) Each of the chromosomes duplicates, but the two *chromatids* of each homolog are held together at the *centromere*.

(c) During meiosis, two chromatids, each from a different homolog, cross over each other. At the point of contact they both break and rejoin, but not in the original way.

(d) As the chromosomes move apart during anaphase I, the chromatids, which have now exchanged parts, begin to separate.

(e) After meiosis is complete, each resulting reduced cell has a complete chromatid (now a chromosome), but the two that have experienced a crossover are unlike anything that went into the system before meiosis. The new chromosomes are physically and genetically different. In this example only one crossover is shown, but in a real cell there may be several crossovers, making complex rearrangements possible.

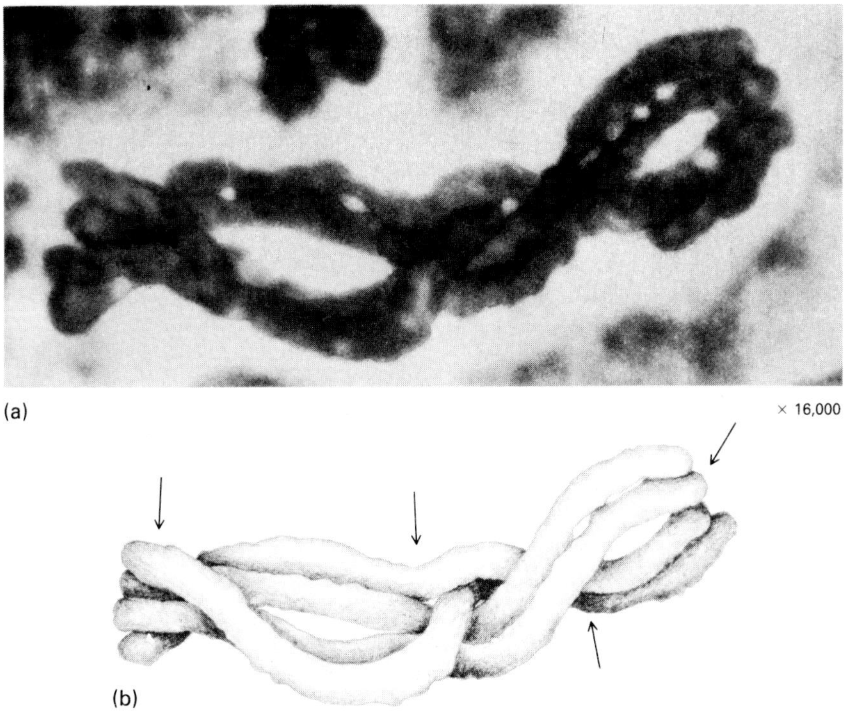

Figure 12.10
Crossing over. (a) Photomicrograph of chromosomes in grasshopper testis during meiosis. Four chromatids are twisted about one another, but they are still in contact at four places, where genetic crossovers occur during chromatid exchange. (b) Drawing made from the photograph, emphasizing the distinctness of the chromatids. The arrows show points of crossovers.

became—available. Segregation and independent assortment of alleles parallel the activity of chromosomes during meiosis and fertilization.

If one assumes that a chromosome is a string of genes, one can explain segregation of alleles at meiosis, as well as linkage of genes. But if genes are linked together on a chromosome, how do they succeed in making any recombinations at all? The supposition was that even linked genes do not always remain together as a group, and that homologous chromosomes can break and exchange parts during meiosis, making genetic **crossovers** (Figures 12.9 and 12.10). These crossovers are expressed as recombinants. During a crossover, genes from a *maternally* derived chromosome become associated with genes from a *paternally* derived chromosome.

The physical appearance of chromosomes during meiosis supported the idea of crossing over, although it is usually impossible to know whether actual chromosome exchange has occurred simply by looking. But by the use of specially marked chromosomes or radioactive tracers, it has now been convincingly demonstrated that chromosomes can break apart during meiosis, and when chromosomes exchange pieces to mend the break, blocks of linked genes are exchanged. Morgan and Bridges had guessed correctly.

Once the principles of nondisjunction, gene linkage, and crossing over were basically understood by Morgan and his colleagues, there remained little doubt that genes located on chromosomes are the carriers of hereditary traits.

Chromosome Mapping

The idea of crossing over was brand-new and still unproved when another Morgan colleague, Alfred H. Sturtevant, speculated that genes were arranged in a linear sequence. He attempted to find out if the frequency of recombinant crossovers was somehow related to the positions of the genes along a chromosome. Sturtevant supposed that the closer together the genes were, the less would be the chance for their crossing over and subsequent recombination.

Repeated matings of fruit flies showed that when two genes are linked, they produce recombinants in dependable proportions. For example, the "forked bristles" trait is linked to "fused veins" in fruit flies, with recombination in 2.8 percent of the offspring. However, the proportions differ when different pairs of linked genes are studied. The linked traits of "fused veins" and "carnation eyes" give a 3 percent recombination. Sturtevant began constructing **chromosome maps,** which located

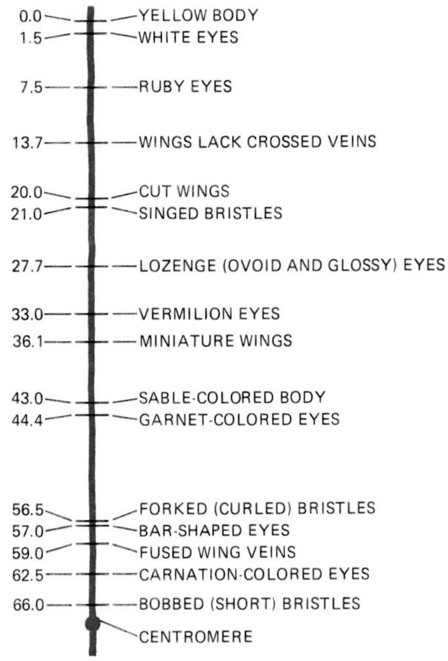

Figure 12.11
A chromosome map representing the sequence of gene positions along the X chromosome of the fruit fly, *Drosophila.* The numbers indicate the distance from one end of a chromosome as determined by crossing experiments, and they are based on the frequency of crossing over. These *map distances* are numerically equal to percent crossovers between loci (positions on a chromosome), and show the order of the loci, but they do not necessarily indicate actual physical distances on a chromosome.

genes along a chromosome based on the percentage of crossovers (Figure 12.11). He translated the recombination percentage into distance, letting one percent equal one "map unit." In 1915, only two years after he began his research, Sturtevant was able to map the loci of more than 85 mutant genes in *Drosophila.*

CHROMOSOMAL ABERRATIONS SOMETIMES OCCUR

Chromosomal replication must be a dependable process. Otherwise the millions of kinds of plants and animals would not be able to maintain their stability. Chromosomal aberrations are less common than gene mutations, but nevertheless, mistakes do occur. The best known of the physical rearrangements in chromosomes are *deletions, duplications, inversions,* and *translocations* (Figure 12.12).

A **deletion** occurs when a whole chromosome or a piece of a chromosome is lost. In general, such losses are likely to be detrimental. **Duplications** are less dangerous, but they can have effects on organisms. An **inversion** occurs when a chromosome becomes looped around itself, breaks, and rejoins with a portion of the chromosome reinserted backward. The frequent result is that some of the products of meiosis are sterile. A **translocation** is the transfer of a part of one chromosome to a new place, usually to some other chromosome. Many plants live successful lives in spite of translocations. In other cases, a translocation can disrupt the functioning of the whole organism, making it essentially unworkable.

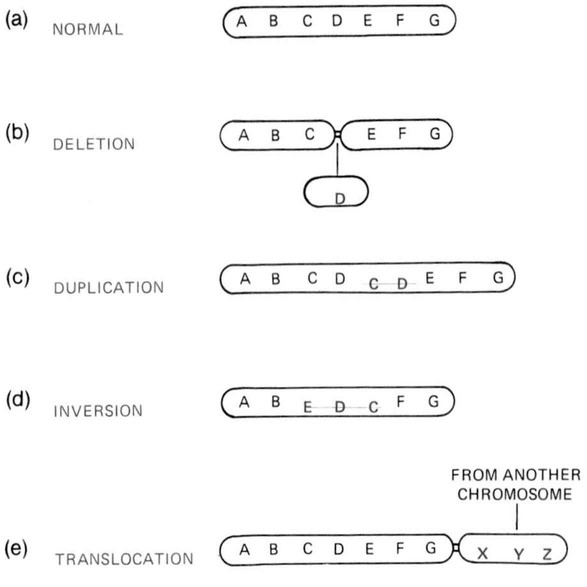

Figure 12.12
Chromosomal aberrations. (a) A normal chromosome. (b) *Deletion.* A segment of a chromosome is lost from a midsection, and broken ends heal together. (c) *Duplication.* A segment of a chromosome is repeated one or more times. (d) *Inversion.* A segment of a chromosome is removed and reinserted so that a portion of the gene sequence is reversed. (e) *Translocation.* A segment of a chromosome is detached from one chromosome and becomes attached to another one, either to its homolog (in which case the receiver would also have a duplication) or to a nonhomologous chromosome.

TABLE 12.1
Blood Types and Their Reactions

				THEORETICALLY	
BLOOD TYPE	CELL ANTIGEN	PLASMA ANTIBODY	GENOTYPE	CAN GIVE BLOOD TO	CAN RECEIVE BLOOD FROM
A	A	Anti-B	I^AI^A or I^Ai	A, AB	A, O
B	B	Anti-A	I^BI^B or I^Bi	B, AB	B, O
AB	A and B	None	I^AI^B	AB	A, B, AB, O
O	None	Anti-A and Anti-B	ii	A, B, AB, O	O

MULTIPLE ALLELES AND BLOOD GROUPS

So far, alleles have been considered as if only two possibilities exist—for example, red eyes or white eyes in *Drosophila.* Of course, any diploid organism normally does have only two alleles. Suppose, however, that in a separate organism a mutation occurs at the "red eye" position *(locus)* on a chromosome. Then instead of two possible alleles in the population, there are three. Indeed, a whole series of alleles for eye color has been found in fruit flies, ranging from deep red through a variety of shades of pink to white. When a population carries more than two alleles at a specific locus, the alleles are called **multiple alleles.** Any one diploid individual can have only two alleles at a specific locus.

Multiple Alleles and Blood Grouping

A well-known example of multiple alleles is the ABO blood-grouping system in humans. In this system, at least three alleles are available to determine the blood type of any person. Table 12.1 shows that each blood type may contain one or more antigens in the red blood cells, and it also contains in the plasma those antibodies *for which it has no antigen.* (An *antigen* is usually a foreign substance that causes a body reaction that subsequently stimulates the production of an *antibody,* a protein that attempts to neutralize the antigen and its effect on the body.) As an example, blood type A carries antigen A in its cells and antibody anti-B in the plasma. This means that mixing blood types A and B will cause *agglutination,* a clumping of blood cells (Figure 12.13), because the anti-A antibody in the B-type blood will affect the A antigen, and vice versa.

Obviously, blood grouping is important in transfusions. For example, if a person with blood type A gives blood to a person with blood type B, the anti-A antibody in the recipient's blood will cause a clumping of the A-type cells in the donor's blood, and the result will probably kill the recipient. In contrast, a person with blood type AB has neither anti-A nor anti-B plasma antibody and can safely receive any blood of the ABO series. A person with blood type O has no cell antigens and can safely

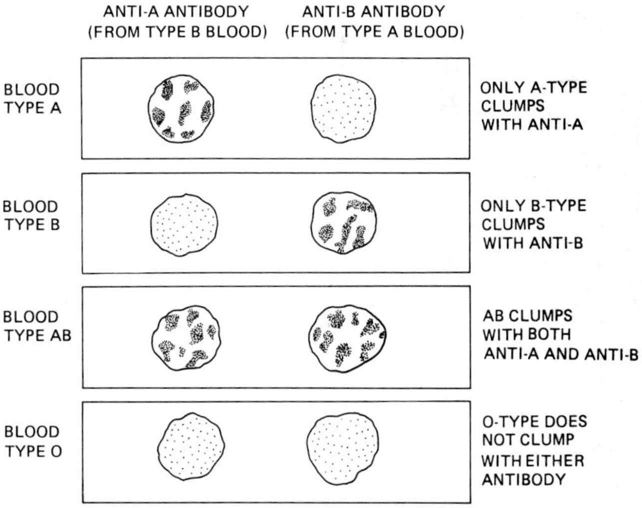

Figure 12.13
Determination of blood groups. Two separate drops of blood of unknown type are put on a glass slide. A drop of anti-A is added to one and a drop of anti-B to the other. The resulting clumping or lack of clumping tells the blood type.

give blood to any of the other types. The person with blood type O is therefore known as a *universal donor*.

From the genotypes and phenotypes (blood types) in Table 12.1, and by application of simple Mendelian rules, the inheritance of ABO blood types can be determined (Table 12.2). For example, a woman with type A blood may be I^AI^A or I^Ai. If she marries a man with type B blood (I^BI^B or I^Bi), she may have children with type A (I^Ai), B (I^Bi), AB (I^AI^B), or O (ii). If she did indeed have four children with all the possible blood types, one would know that both she and her husband were heterozygous for the ABO series.

It is necessary to state "for the ABO series" explicitly because the ABO system is not the only blood group. Another blood antigen system involves the Rh factor, so called because it was first discovered in *Rh*esus monkeys in the 1930s. Of the several genes responsible for different blood antigens, this one is of clinical interest because it sometimes causes Rh disease.

Rh disease can kill a young infant if the blood of the newborn is not replaced with fresh blood. In general, the circulation of a fetus is separate from the mother's circulation, but sometimes some fetal blood cells may leak into the mother's blood during late pregnancy or at birth. If the fetal cells carry an Rh antigen inherited from the father, and if the mother has no such cells, she will produce anti-Rh antibody (Figure 12.14). This series of events hap-

TABLE 12.2
Inheritance of Human ABO Blood Types

BLOOD TYPES OF PARENTS	CHILDREN'S BLOOD TYPE POSSIBLE	CHILDREN'S BLOOD TYPE NOT POSSIBLE
A + A	A, O	AB, B
A + B	A, B, AB, O	
A + AB	A, B, AB	O
A + O	A, O	AB, B
B + B	B, O	A, AB
B + AB	A, B, AB	O
B + O	B, O	A, AB
AB + AB	A, B, AB	O
AB + O	A, B	AB, O
O + O	O	A, B, AB

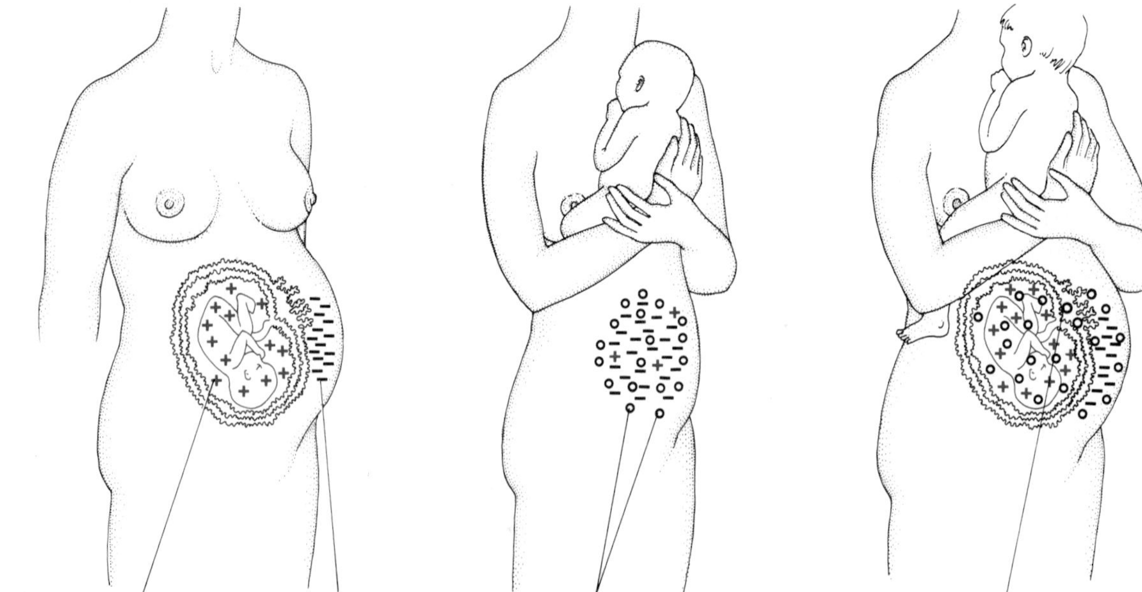

Figure 12.14
The development of Rh disease in an infant. (a) The Rh-negative mother carrying an Rh-positive fetus may develop anti-Rh antibody. (b) A first child does not have time to be affected, but a second child may receive some of its mother's antibody (c), which may destroy that second child's red blood cells unless appropriate countermeasures are taken.

MARY LYON AND CALICO CAT

Normal female mammalian cells contain two X chromosomes. And in every female cell there is a small object that rests against the wall of the nucleus. This inert body was discovered by Murray Barr, and so it is called a *Barr body,* or more descriptively, a *sex chromatin body,* stainable and microscopically visible. Barr found the sex chromatin body in female cats (male cats do not have such a structure), but all he knew about it was that it was there. He did not connect it with chromosomes.

Later, Barr bodies were found in many mammals, including human beings, but always in females only. It was also discovered that each female cell contained one less Barr body than the number of X chromosomes present. Ordinarily this meant one Barr body for each cell, but abnormal XXX cells, for instance, have two Barr bodies. Apparently Barr bodies had something to do with sex, and since they reacted to stains that usually work for DNA, it seemed likely that they contained genes.

Appropriately, it was a woman, Mary Lyon, who discovered the significance of the sex chromatin bodies. She proposed the "single active X hypothesis," which stated that the sex chromatin body is an X chromosome that did not uncoil during mitosis. It remains as a nonfunctioning mass, clumped against the nuclear membrane. Why? Probably to maintain the proportion of the sex chromosome with the autosomes. The inactivation of one X chromosome in a female restores the same balance as that in a normal male cell, which has only one X chromosome.

Lyon postulated that both X chromosomes function during an animal's early prenatal development, but that only one of them remains permanently active, the second one retiring and becoming a metabolically inert Barr body, or sex chromatin body. An instructive example of such a development is found in cats called calico or tortoiseshell, which have patchy markings of black and orange-yellow. A calico cat, typically a female, has two X chromosomes, one carrying a gene for black fur and the other carrying its allele for orange-yellow. On some parts of such a cat, the black gene goes dormant during embryonic life, making an orange-yellow patch, but on other parts, it is the colored gene that goes dormant, leaving the black to be expressed. The result is the varied coat.

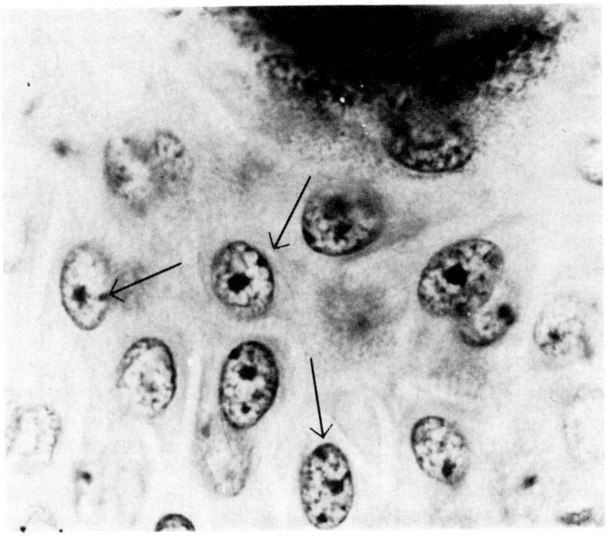

× 600

Isolated nuclei showing Barr bodies (arrows).

Male cats normally have an XY chromosome pair. The Y chromosome carries no color gene, and the X carries either black or orange-yellow but not both. Since the black-orange-yellow locus is not the only one affecting fur color, some males *can* be mottled, but they cannot be true calicos unless they happen to have the abnormal XXY complement, and then they are sterile. Now, after centuries of guessing, we know something of why calico cats are females, but we still do not know why one X chromosome or the other can be effective to the exclusion of the opposing one.

pens only when an Rh-negative woman becomes pregnant with an Rh-positive fetus and usually does not affect the firstborn child, which is born before it has time to receive a dangerous amount of maternal antibody. Once sensitized to the Rh antigen, however, a woman carrying anti-Rh antibody may give some of that antibody to a second child. Therefore, when an Rh-negative woman has an Rh-positive pregnancy, she should be alerted to the danger to possible later children and receive injections that prevent her from making anti-Rh antibody.

Although only the ABO and the Rh antigen systems have been included here, there are in fact about 80 described blood antigen systems. When one considers the number of combinations of blood antigens possible, it appears likely that there are enough different arrangements to provide every human being with an individual blood type. Since only a few types seem to be medically important, those few are well known, but the rest are present and functioning.

SEX-LINKED GENES IN HUMAN BEINGS

In animals that, like humans and fruit flies, have an XY–XX type of sex determination, any gene on an X chromosome can be heterozygous only in a female (with her XX pair of homologs). Since the Y chromosome does not carry alleles to match anything in the X chromosome, a male cannot be heterozygous for any sex-linked locus. He cannot be homozygous, either. His condition is called *hemizygous*. Any gene he has on his X chromosome will be expressed phenotypically, even if it is recessive, because there will be no matching dominant allele on the Y chromosome. Because of this fact, sex-linked characteristics in humans tend to be expressed mainly in males, who *do not* pass such characteristics to their sons. They may well pass it to their grandsons through their daughters, however, so this kind of inheritance is sometimes called "skip-generation" inheritance.

All the many genes located on X chromosomes give inheritance patterns like those in white-eyed flies, provided the species is of the XY–XX type. The Y chromosome has few genes, but when a mutation occurs in a Y chromosome, it is passed from fathers to sons only, and daughters cannot even be carriers (Figure 12.15).

The most famous example of sex-linked inheritance in humans is hemophilia, the bleeder's disease. Queen Victoria of England was apparently a mutant heterozygous carrier for the hemophilia gene, which she transmitted to some of her children (Figure 12.16). She herself, of course, was not a hemophiliac. The gene spread through several royal families in Europe, sometimes with serious political and personal results. Victoria's granddaughter Alexandra, a carrier, gave birth to Alexis, the son of Czar Nicholas II of Russia. Alexis was a hemophiliac. The best Russian doctors could not cure him, but the opportunist monk Rasputin claimed he could erase the disease, and the royal family fell under Rasputin's influence. Rasputin added to the country's political corruption, and the Russian Revolution, which foreshadowed the current Soviet

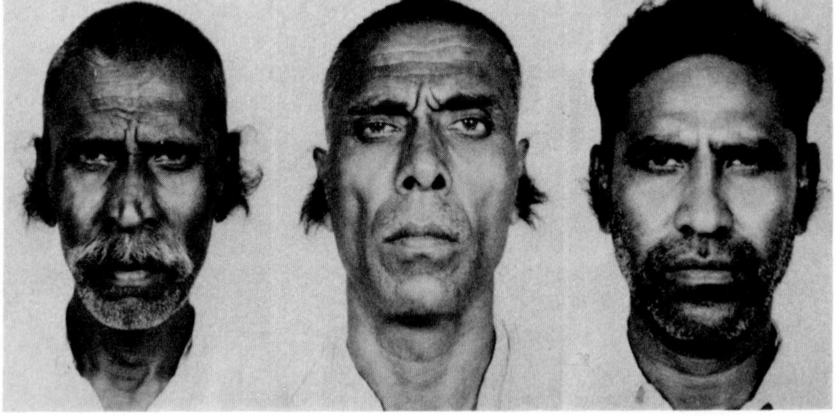

Figure 12.15
A human trait inherited on the Y chromosome: hairy ears. These three Indian brothers show the feature. A characteristic is thought to be carried on the Y chromosome if it is passed from father to sons, is inherited by all the sons of a father, and is not inherited by any daughters or children of daughters.

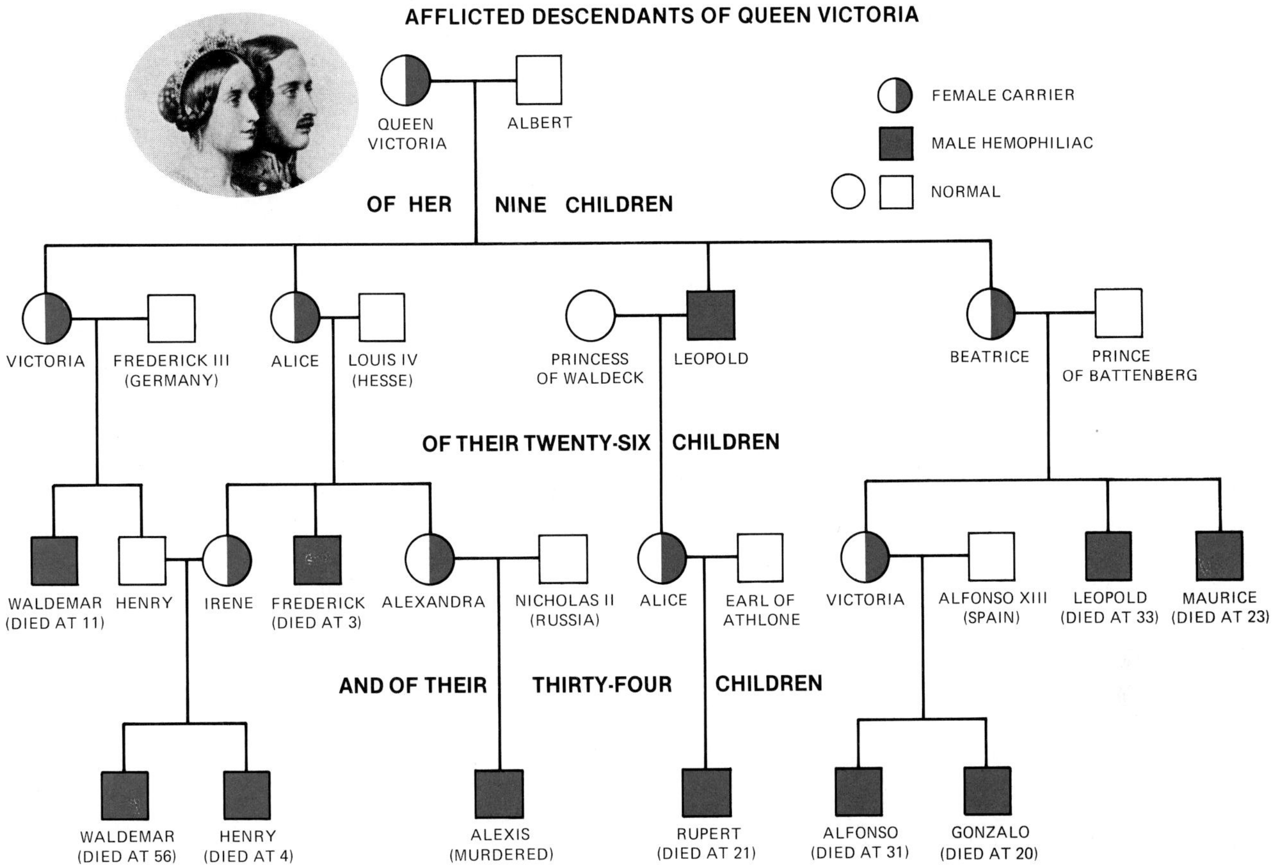

doctrines, erupted into a bloody overthrow of the Czarist regime. Alexis would have become Czar of Russia if the 1914 Revolution had not led to his death. Because he was a hemophiliac, the course of world history was dramatically affected.

Several kinds of **color blindness** are sex-linked. The alleles for color blindness, like those for hemophilia, are on the X chromosome. Color blindness is more likely to occur in men than in women. In fact, there are 5 to 10 times more color-blind men than color-blind women. A recessive allele on the X chromosome carries the color blindness trait, and there is no allele on the male Y chromosome to counteract it. In order for a man to be color-blind, he need only receive one X chromosome with an allele for color blindness from his mother, whereas a woman must receive the color blindness trait from both mother and father. Therefore, if a male receives the color blindness allele from his mother's X chromosome, he will be color-blind. (It is as though the recessive allele on the X chromosome suddenly became dominant, and in a sense it *is* dominant to the neutral Y chromosome.)

Figure 12.16
A famous example of a human mutation: hemophilia in the descendants of Queen Victoria. The chart shows the relation of the queen to several royal houses into which her children married, and the results of those marriages. Note that four of Victoria's nine children were either carriers or actually afflicted, and that of her other descendants, ten males were hemophiliacs and four females were carriers. Because it has no afflicted members, the British royal family of today is not represented on the chart.

SOME EFFECTS OF HUMAN CHROMOSOMAL ABERRATIONS

A number of human deficiencies and deformities are caused by chromosomal deletions, duplications, or translocations. They can be recognized by microscopic examination of the chromosomes and correlated with the set of symptoms that characterize the

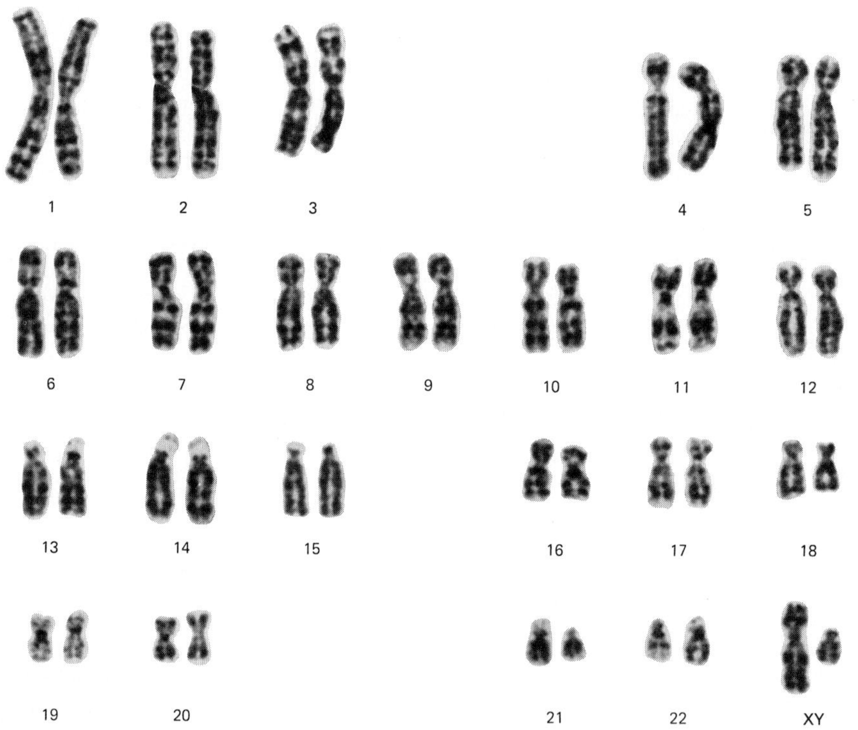

Figure 12.17
The chromosomes of a man. Preparation of a *karyotype* involves taking a tissue sample, usually blood, and culturing it. Cells are killed and stained. A cell is photographed, and each chromosome is cut out of the photograph. All the chromosomes of the cell are arranged in order from the longest, number 1, to the shortest, number 22, followed by the sex chromosome—XX for a female, or XY, as here, for a male. The X chromosome is about the size of a number 10, but the Y is about like a number 22.

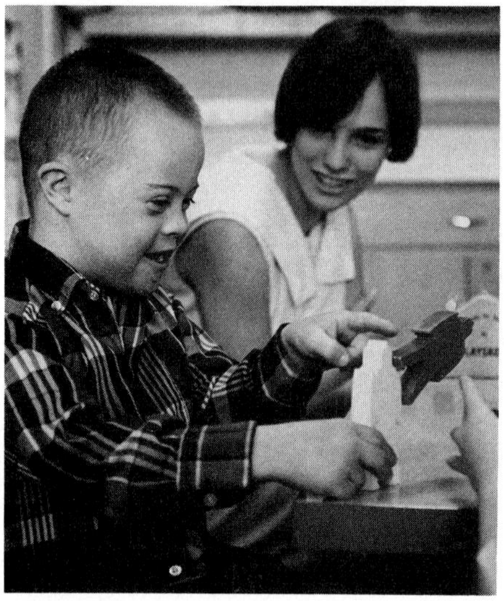

Figure 12.18
A boy with Down's syndrome. In the general population, about 14 live births per 10,000 have Down's syndrome, but the rate is much higher when the mother is over 35 years old. Apparently chromosomes suffer increasing tendency toward nondisjunction with advancing maternal age. Many potential sufferers die before birth, so the actual number of conceptions of trisomics of this kind is much higher than the live birth figures indicate.

phenotype. A **karyotype,** or chromosome set of a male, has 22 matched pairs of chromosomes plus one unmatched pair, the XY pair (Figure 12.17). A female has 23 matched pairs, including the XX pair. Any deviation from the normal is likely to have a harmful effect on the development of the individual suffering from the deviation.

Nondisjunction in Human Chromosomes

If an egg undergoes an inaccurate meiosis, two members of the same homologous pair of chromosomes may go into the final egg, and if that egg is then fertilized by a normal sperm with one homolog, the zygote will have three copies. Such a zygote is a *trisomic,* caused by nondisjunction, the failure of a pair of homologs to separate at meiosis.

An abnormality known as **Down's syndrome** is caused by the presence of a third chromosome 22. (It is still called "trisomy 21" because for years it was thought to be a result of an extra chromosome 21.) People with Down's syndrome are mentally retarded and have lowered resistance to common diseases (Figure 12.18).

Aberrations in Human Sex Chromosomes

A number of examples of sex chromosome aberrations have been described. An individual with only one sex chromosome, the XO condition, is female but has permanently infantile sex organs, fails to develop normal secondary sex characteristics, and has abnormal body structure (**Turner's syndrome**). An individual with **Klinefelter's syndrome** has an extra X chromosome, which produces a genotypic arrangement of XXY and a total of 47 chromosomes. Such a person is usually phenotypically male but has underdeveloped male genitalia and some evidence of female breasts. The body is usually somewhat female in appearance, with long legs, sparse body hair, and some hip development. The individual is sterile and may be mentally retarded. It was not until 1956 that the cause of Klinefelter's syndrome was discovered.

After a woman athlete with an XXY configuration was disqualified from a track meet, all female athletes competing in the Olympics and other international meets are required to take a chromosome test.

MULTIPLE FACTOR INHERITANCE

In many instances, a visible characteristic can be influenced by more than one pair of genes. When several pairs of genes act on a single characteristic, they are called *multiple genes,* and they are usually concerned with some quantitative aspect of inheritance, whether simply size, or perhaps amount of a pigment synthesized. Consider, for example, the overall growth of a human body, in which there is no such distinct division into tallness or dwarfness as occurs in garden peas. Most humans are neither giants nor dwarfs. A few are very short, many are of medium height, and a few are very tall. A range, with many intermediates, is common in a variety of features, such as skin color or eye color in humans, grain color in wheat, and size of cob in corn.

Skin color in humans is an example of the expression of multiple genes. Just as people are generally neither giants nor dwarfs, so is their skin color neither pure black nor pure white. The degree of lightness or darkness depends on the thickness of the skin and the amount of melanin present. Melanin is a pigment in the outer layers of the skin. The more melanin, the darker the skin. Current thought is that there are probably four pairs of genes involved in the expression of skin color, with the resulting possibility of nine different skin shades. Figure 12.19 shows how medium-skinned parents may produce children with lighter or darker skin than their own.

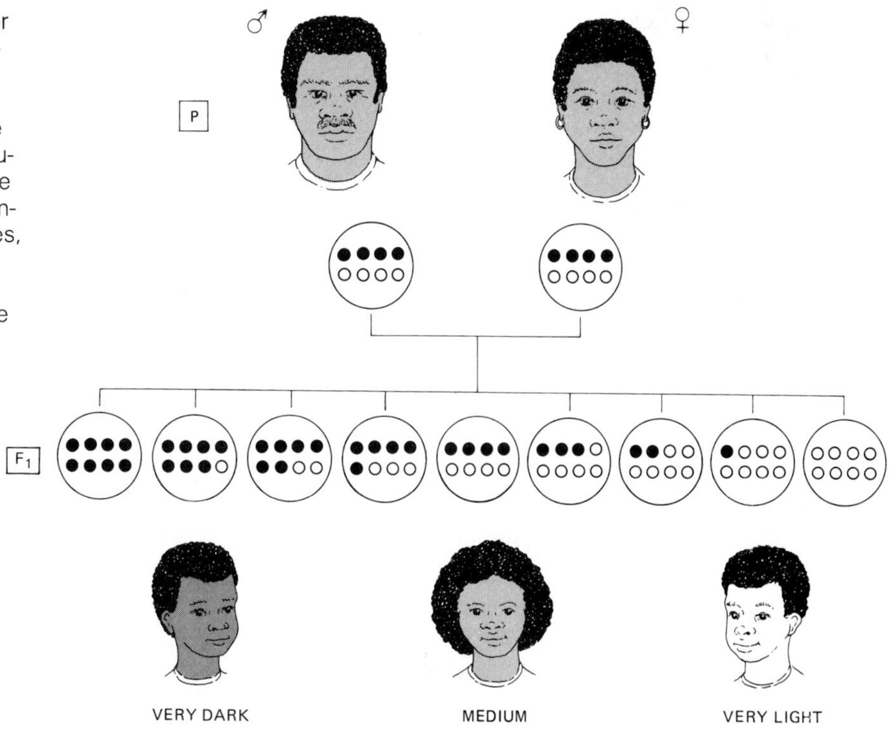

Figure 12.19
Quantitative inheritance of skin color in humans. If pigmentation is determined by four pairs of melanin-producing genes, then a couple with medium coloration might each have four melanin-producing and four neutral genes. Their children might have anything from none to eight melanin-producing genes, or nine possibilities, as shown in the diagram. It is thus possible for moderately pigmented parents to produce children who are very dark or very light or any of seven intermediate shades.

GENETICS PROBLEMS

1. In fruit flies, males have XY sex chromosomes, females have XX, and white eye color is sex-linked. What are the expected sexes and eye colors from the following crosses?
   a) red-eyed (homozygous) female × white-eyed male
   b) red-eyed (heterozygous) female × white-eyed male
   c) white-eyed female × red-eyed male
2. In pigeons, as in birds generally, males have a matching pair of sex chromosomes, and females have an unmatched pair (the ZZ–ZW type). A sex-linked gene in pigeons affects the color of the head feathers. Gray-headed females, mated with cream-headed males, give equal numbers of cream-headed males, gray-headed males, and gray-headed females.
   a) Which allele is dominant?
   b) Offer an explanation for the lack of cream-headed females.
3. From the pedigree shown below, answer the following questions. (Circles indicate females; squares, males. Open figures indicate normal phenotypes; black, color-blind individuals.)

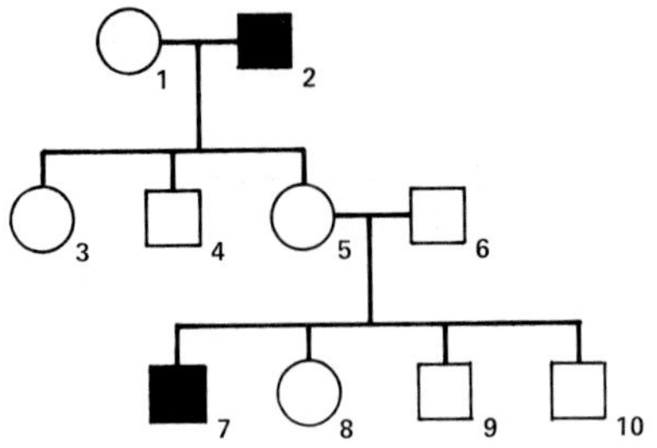

   a) What is the probable genotype of 1?
   b) What is the genotype of 5? Of 9?
   c) How certain are the answers to (b)?
   d) If 8 marries a normal man, what are the chances that she will have a color-blind son? If she marries a color-blind man?
4. The Y chromosome in a human male carries a few known genes, such as the one for ichthyosis hystrix gravior (the "porcupine men" of eighteenth-century England). If a porcupine man has children, what can you expect with regard to genotypes and phenotypes and the distribution of sexes?
5. If an organism is heterozygous for genes at four loci, all on different chromosomes, what is the possible number of kinds of gametes, at least with respect to those loci?
6. In the chromosomes of male fruit flies, crossing over does not occur. Genes at loci A and B are linked, with a distance of 18 map units between them. What kinds of gametes can be produced by male flies with genotype AaBb?
7. In a disputed paternity case, a woman with blood type B had a child with blood type O, and she claimed that it had been fathered by a man with blood type A. What can be proved from these facts?
8. What are the possible parental genotypes of a person with blood type AB?

ANSWERS TO GENETICS PROBLEMS

1. a) All red-eyed, half male and half female.
   b) One-quarter red-eyed female, one-quarter white-eyed female, one-quarter red-eyed male, one-quarter white-eyed male.
   c) Half female, all red-eyed, and half male, all white-eyed.
2. a) Cream. b) "Cream" must be lethal. It can be tolerated only if counteracted by a "gray gene" on a homologous chromosome. Female pigeons have the ZW chromosome pair, and there is no chance for such a counteraction.
3. a) Number 1 may seem to be homozygous normal because she had three normal children by a color-blind man, but she still might be a carrier.
   b) Number 5 is a carrier, and number 9 is a normal male.
   c) Both genotypes are certain.
   d) Number 8 has a 50% chance of being a carrier, since her mother is definitely a carrier and her father is normal. She also has a 50% chance of passing the color-blind gene to a child, and a further 50% chance that a child will be male. Thus the chance of producing a color-blind son is $\frac{1}{2} \times \frac{1}{2} \times \frac{1}{2} = \frac{1}{8}$. The chance is the same whether her husband is normal or color-blind, since he will pass only a Y chromosome to any son.
4. All sons will be phenotypic porcupine men. Daughters will not be affected either genotypically or phenotypically, because they receive no Y chromosomes.
5. One locus could give two kinds of gametes, or 2^1, and two loci give four kinds, 2^2. Therefore four loci could give 2^4, or 16 kinds.
6. If A and B are on one homolog and a and b on the other, then the only kinds of gametes are AB

and ab. If A and b are on one homolog and a and B on the other, the only kinds of gametes are Ab and aB.

7. Since the man could have been genotypically AO and the woman BO, he *could* have been the father, but this is no proof that he actually was or was not the father.

8. $I^AI^A \times I^BI^B$; $I^AI^A \times I^AI^B$; $I^AI^A \times I^Bi$; $I^Ai \times I^BI^B$; $I^Ai \times I^AI^B$; $I^Ai \times I^Bi$; $I^AI^B \times I^BI^B$; $I^AI^B \times I^Bi$; $I^AI^B \times I^AI^B$.

SUMMARY

1. An explanation of both Mendelism and Darwinian variations was proposed by Hugo De Vries during the period 1901–1903. De Vries called inherited changes *mutations.*

2. The so-called mutations studied by De Vries and other geneticists provided some useful information on patterns of inheritance and interactions among genes, but the chemical explanation for mutations had to wait for the demonstration of DNA as the genetic material.

3. Although the causes of natural mutations are not known, biologists can induce them artificially by a number of methods. Mutations can be induced by X-rays and other radiation that has sufficient energy to cause chemical changes of molecules in cells. Mutations can be induced by increasing the temperature and by introducing chemical mutagens, such as atmospheric pollutants and pesticides.

4. Most mutations are not only recessive but detrimental. However, occasional mutations have been advantageous; in fact, favorable mutations are the basis for the genetic variability that makes evolution work.

5. Only those mutations that occur in sex cells in animals have any chance of being passed on to later generations.

6. In the early 1900s researchers discovered that in fruit flies chromosomes are responsible for the determination of sex.

7. *Nondisjunction* in fruit flies occurs when a pair of homologous chromosomes fails to separate during meiosis.

8. In humans, the Y chromosome determines maleness.

9. When genes tend to be inherited together, they are said to be *linked.* Genes that are linked compose a *linkage group,* and the specific genetic traits carried by those genes are inherited as a group, not separately.

10. Chromosomes can break and exchange parts during meiosis, making genetic *crossovers.* These crossovers are expressed as recombinants. With information about nondisjunction, gene linkage, and crossing over, geneticists concluded that genes located on chromosomes are the carriers of hereditary traits.

11. Geneticists are able to construct *chromosome maps,* which locate genes along a chromosome, based on the percentage of crossovers.

12. *Recombinant DNA* involves the transfer of genes from one organism to another, followed by unlimited reproduction of the transferred genes.

13. Physical rearrangements in chromosomes are *deletions, duplications, inversions,* and *translocations.*

14. When there are more than two possible types of allele at a specific locus in a population, the alleles are called *multiple alleles.* A well-known example of multiple alleles is the ABO blood-grouping system in humans.

15. A *sex-linked characteristic* is one regulated by a gene on a sex chromosome. The best-known examples of sex-linked traits in humans are hemophilia and color blindness.

16. One effect of nondisjunction in human chromosomes is *Down's syndrome.* Examples of aberrations in human sex chromosomes are *Turner's syndrome* and *Klinefelter's syndrome.*

17. Some characteristics, such as body size or skin color, can be influenced by more than one pair of genes. When several pairs of genes act on a single characteristic, they are called *multiple genes.*

13 Human Reproduction

SOME KEY POINTS

1. Whether animals reproduce sexually or asexually, they pass on genetic traits to future generations through DNA. Sexual reproduction allows for greater variation and diversity than asexual reproduction does.
2. The reproductive function of the human male is the production of sperm cells and their delivery to the female. Human female sexuality is more complex. Females produce eggs, nourish, carry, and protect the developing embryo, and care for the newborn young.
3. Sex is determined in human beings at the time of fertilization. An X-carrying sperm makes a female, and a Y-carrying sperm makes a male.
4. A rhythmic menstrual cycle takes place in the human female as long as pregnancy does not occur. If pregnancy does occur, a new hormonal program goes into effect in preparation for the development and birth of the baby.
5. When fertilization occurs in human beings, the nucleus of the egg, with 23 chromosomes, fuses with the nucleus of the sperm, with 23 chromosomes, to form a zygote, a single cell with 46 chromosomes.

ANIMALS REPRODUCE BY SEXUAL AND ASEXUAL means. In either case, hereditary traits are passed on from generation to generation through the genetic code of DNA. The advantage of sexual reproduction is that it allows for variation and diversity through a *recombination* of parental genes. Animals that reproduce asexually can produce only offspring identical to themselves, and only through infrequent mutations can variations develop that might help the species adapt to a changing environment.

ASEXUAL REPRODUCTION IS USUALLY LIMITED TO THE SIMPLER ANIMALS

Asexual reproduction in higher multicellular animals is rare; it is usually limited to the simpler types. One-celled animals, like one-celled plants, frequently undergo fission, although most of them use some sexual methods as well. Corals bud off small replicas of themselves, and many parasitic worms simply break into pieces, each of which is capable of developing into a complete worm.

Some female insects, such as aphids and bees, can lay viable eggs without fertilization. Although such reproduction is literally sexless, it is plainly derivative of a sexual method and usually serves only as a partial substitute for a sexual method.

VARIATIONS IN ANIMAL REPRODUCTIVE PROCESSES

The essential feature of sexuality is the union of unlike chromosome sets. In the unicellular (or as they are sometimes called, the noncellular) animals, the protozoa, sexual mating involves the fusion of two nuclei, but there is no increase in the number of individuals. Thus there is sex without actual reproduction, but *genetic recombinations* do take place, and that is the *essence of sexuality*. Table 13.1 outlines the variations of animal reproduction.

Many animals have fascinating methods of mating, and some of the more interesting reproductive

TABLE 13.1
Variations of Animal Reproduction

METHOD	TYPICAL ANIMALS USING METHOD	CHARACTERISTICS OF REPRODUCTION	NATURE OF OFFSPRING
Asexual			
Fission	Protozoans, other one-celled organisms	Body of parent divides into two equal parts. May be transverse or longitudinal.	Exact replicas of parent.
Budding	Hydras: external budding; Freshwater sponges, bryozoans: internal budding	Unequal division. Bud develops organs like parent's and detaches itself. Buds are usually external; if internal they are gemmules.	The new individual arises as an outgrowth (bud) from the parent. Replicas of parent.
Fragmentation	Flatworms, and echinoderms	Organism breaks into two or more parts.	Each new part capable of becoming a complete animal. Replicas of parent.
Sporulation	Protozoans	Multiple fission. Many cells are formed, enclosed together in a cystlike structure.	Replicas of parent.
Sexual Modification			
Parthenogenesis	Rotifers, plant lice, some ants, bees, and crustaceans	Modification of sexual reproduction. Unfertilized egg develops into complete individual. Usually occurs for several generations, followed by a biparental generation in which the egg is fertilized.	Fertilized bee eggs become females (queens or workers); unfertilized bee eggs become males (drones). Other examples various and complex.
Sexual			
Conjugation	Protozoans	Two individuals fuse together temporarily, exchange micronuclear material. Union may occur between like cells (isogametes) or different cells (anisogametes). May be difficult to distinguish sex.	Contain nuclear material from both parents. Morphologically but not genetically like parents.
Biparental reproduction	Most vertebrates and invertebrates	Familiar method involving separate and distinct male and female individuals. Each individual produces either sperm or eggs, rarely both. Separate sexes (dioecious).	Similar to parents but not identical.
Hermaphroditism	Flatworms, hydras, tapeworms, and some molluscs	Individuals have both male and female organs (monoecious condition). Self-fertilization usually prevented by producing gametes at different times.	Similar to parents but not identical. Contain equal amounts of DNA from each gamete.

Adapted from Cleveland P. Hickman, Sr., Cleveland P. Hickman, Jr., and Frances M. Hickman, *Integrated Principles of Zoology,* Sixth Edition (St. Louis: The C. V. Mosby Company, 1979) pp. 798–800. Used with permission.

variations are presented here. One of the eight arms of a male octopus is actually a *hectocotylus,* a chamber for storing sperm. At the time of mating, the male inserts this special arm into the breathing cavity of the female, where eggs are already ripening. The arm breaks off inside the female (another arm will grow in time for the next mating season), and when the eggs are ripe, the arm ruptures and releases the sperm to meet the eggs. Female bristle worms eat the sexual parts of the males in order to fertilize their eggs. The sperm are liberated from their sacs when they reach the female's stomach. From there the sperm can travel to fertilize the eggs. The starworm, *Bonellia,* has an even more unusual reproductive pattern. The dwarf males spend their entire lives as parasites within the female's body, where they continue to fertilize eggs as the eggs become ripe. Honeybees have a female-oriented society, as we will see in Chapter 30. After the queen bee mates with many male bees, their male reproductive organs break off, remaining inside the queen to prevent the loss of sperm. The male honeybees, having fulfilled their prime function, proceed to bleed to death.

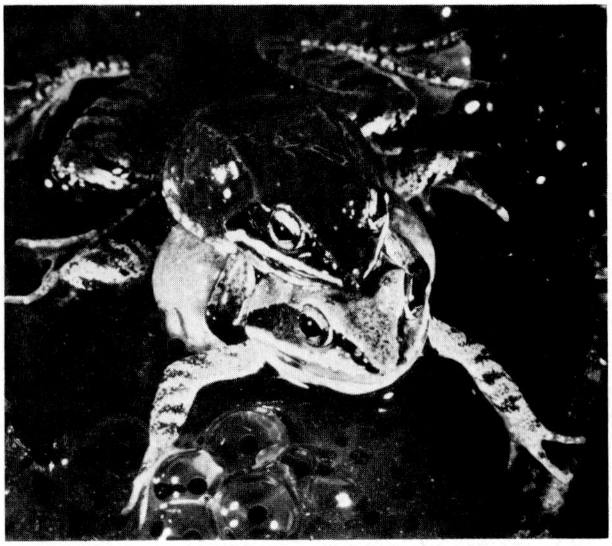

Figure 13.1
Mating between frogs. Close association of the male and the female is involved, but fertilization is external. As the female lays eggs, the male deposits sperm over them.

Many aquatic animals simply shed gametes into the water they live in, allowing sperm to find eggs as best they can. Such a dependence on fertilization of free-floating gametes (mature sperm and egg cells) may seem to be a risky way to start new generations, but several factors increase the chances of fertilization and survival: (1) The eggs contain enough stored food to give the young animals a start in life until they can feed themselves. (2) The chances for successful fertilization are improved by the synchronized release of sperm and eggs, as well as by the huge numbers of gametes released. (3) Specialized behavior patterns tend to bring sperm close to eggs. (4) Some method of timing ensures that eggs and sperm have a good chance to meet; phases of the moon that affect the flow of tides are well correlated with spawning times, and some animals correlate their spawning times with the length of days, thus liberating their gametes, each in its proper season. Thus aquatic invertebrates and fishes do breed successfully. The factors that contribute to successful fertilization and survival are the products of long evolutionary trends that ensure the survival of the species.

Almost all land-dwelling animals have adopted some form of mating that allows fertilization of eggs inside the body of the female. Even such diverse groups as insects, birds, and snails practice internal fertilization. In the lower vertebrates, some (sharks) use internal fertilization, but most fishes and amphibians (frogs, toads, salamanders) have specialized behavior that guarantees that sperm will be deposited close to the eggs. In frogs, for example, the male clasps the female at egg-laying time and ejects sperm over the egg masses as the female liberates them (Figure 13.1). Fertilization and embryo formation occur in water, usually with no parental attention, although in some species of toads the female carries the eggs in pouches in the skin of her back. The wet environment there allows the embryos to grow until they can live independently. Some fishes have ritualized behavior that increases the chance for fertilization (Chapter 29).

Sex cells can survive and perform their task of fertilization only in a wet medium. This requirement provided no insurmountable problems for aquatic animals, but the animals that migrated to the land developed special reproductive structures that retained a wet environment for fertilization, even though mating itself took place on the dry land. In humans and most other land-dwelling vertebrates, the penis is inserted into the vagina during copulation to accomplish a wet, internal fertilization. Fertilization is *always* internal in mammals. Although the human species is not typical of mammals, it is understandably of special interest to students, so human reproductive structures and functions will be emphasized in this book.

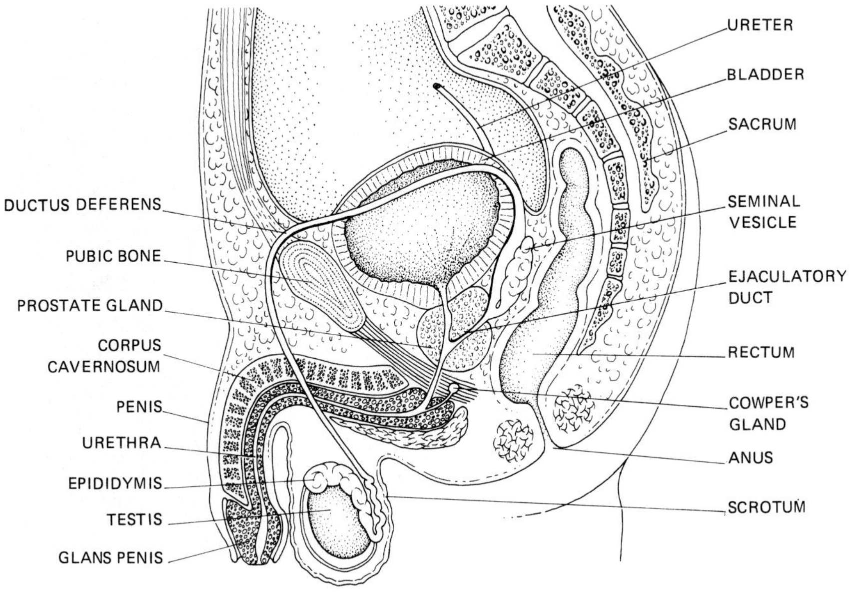

Figure 13.2
A cutaway side view of a male human pelvis.

HUMAN SEXUAL ANATOMY AND FUNCTION: MALE

The reproductive function of the male is the production of sperm cells and their delivery to the female (Figure 13.2). The organs that produce the sperm cells, *testes,* also produce the primary male hormone, *testosterone,* which aids sperm production. These duties, simple in comparison with the duties of the female, require only the *gonads* (testes), *accessory glands* to furnish a carrying fluid for the gametes, *ducts* for their transmission, and the *penis.*

The Testes

The paired **testes** are formed inside the body cavity, but they descend shortly before birth into an external sac between the thighs, the **scrotum.** The lower temperature outside the body is necessary for active sperm production; the sperm remain infertile if they are retained inside. For the same reason, a fever is capable of killing hundreds of thousands of sperm cells. In warm temperatures the skin of the scrotum hangs loosely, and the testes are held in a low position. In cold temperatures the muscles under the skin of the scrotum contract, and the testes are pulled closer to the body. In this way, the temperature of the testes remains somewhat constant. Sweat glands also help to cool the testes.

The testes contain about 15 to 20 coiled **seminiferous tubules,** which produce the **sperm cells,** or **spermatozoa,** at a daily rate of a billion or so in young men. The actual combined length of the seminiferous tubules in both testes is about 225 meters (750 ft) (Figure 13.3).

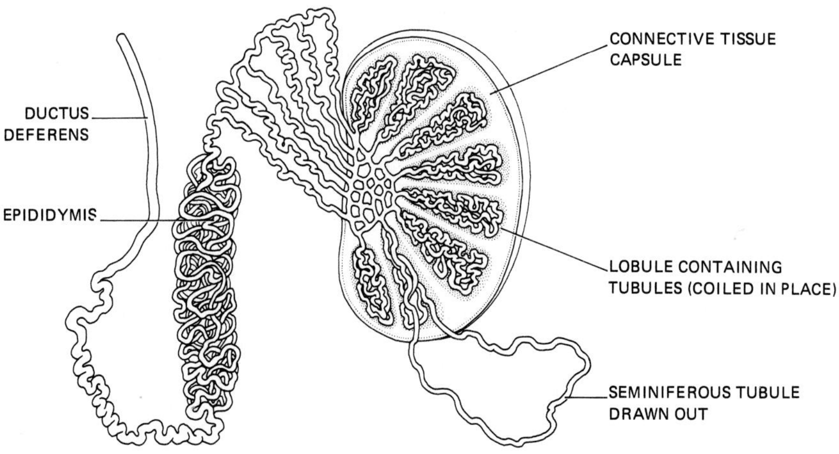

Figure 13.3
Structure of a mammalian testis, dissected to show the seminiferous tubules, within which sperm are produced. The sperm cells pass along the *seminiferous tubules* and are stored in the coiled tubules of the *epididymis,* from which they are passed to the outside.

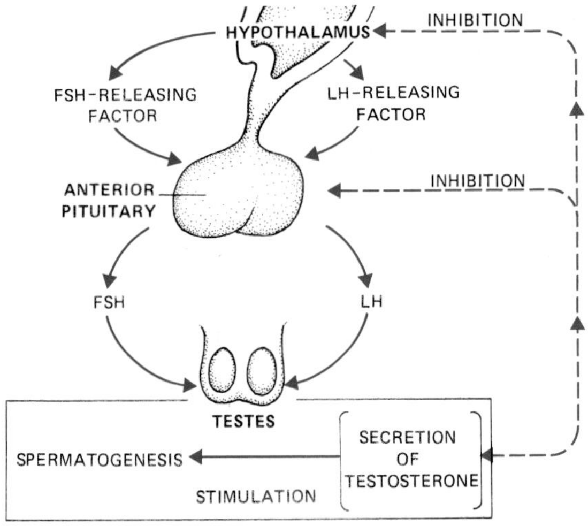

Figure 13.4
Scheme of sperm-production control by hormones. Releasing factors sent from the hypothalamus stimulate the anterior pituitary to release hormones that stimulate the secretion of testosterone. Testosterone stimulates sperm production but also acts as an inhibitor to both hypothalamus and pituitary in a negative feedback system. (Compare Figure 13.14.)

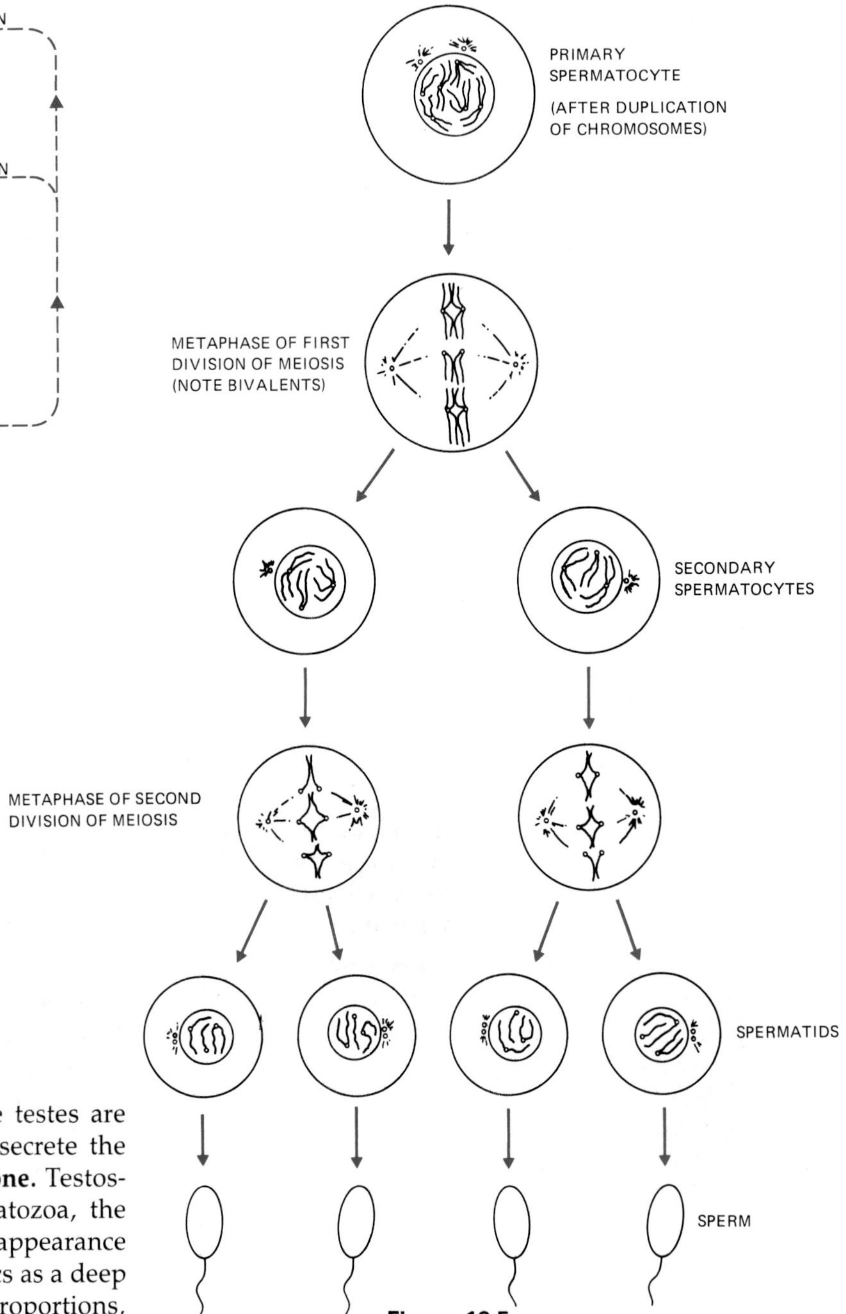

Figure 13.5
The cellular activity that leads to sperm production. A cell that is about to have its chromosome number reduced is a *primary spermatocyte.* It has a diploid chromosome number, and each chromosome itself is double-stranded. At the end of the first meiotic division, the cells resulting are *secondary spermatocytes.* After the second meiotic division, the cells are *spermatids.* Spermatids must undergo a reorganization in order to become sperm; this includes formation of a penetrating point, or *acrosome,* and a tail, or flagellum.

Hormonal Regulation in the Male

Among the seminiferous tubules in the testes are small masses of *interstitial cells,* which secrete the male sex hormones, especially **testosterone.** Testosterone affects the production of spermatozoa, the development of the sex organs, and the appearance of such secondary male sex characteristics as a deep voice, facial and body hair, skeletal proportions, and distribution of body fat.

If there is a deficiency of testosterone, all the accessory male reproductive organs will decrease in size and activity. In most cases, both penile erection and volume of ejaculation will diminish markedly also. It is not yet certain how (or if) testosterone affects male behavior in general, but it is recognized that deficiencies of testosterone will produce an overall decrease in male sex drive and behavior. It is also known that secondary male sex characteristics will be affected by testosterone deficiencies. A male who is castrated before puberty will gradually acquire such female characteristics as fatty deposits in the breasts and hips, lack of facial hair, and smooth skin texture. If an adult male is castrated, he will usually lose his facial hair, and his bones and muscles will probably diminish in thickness and size, but other feminine or childlike characteristics will be minimal.

Besides testosterone, males also secrete the sex hormones **FSH** (follicle-stimulating hormone) and **LH** (luteinizing hormone), also known as **ICSH** (interstitial cell-stimulating hormone). Both FSH and LH are produced by the pituitary gland and are chiefly responsible for the stimulation of spermatogenesis and testosterone secretion in the testes (Figure 13.4). As we shall see in Table 13.2, FSH and LH are also major female hormones.

Puberty and "Menopause"

A young male usually matures sexually when he is about 14 years old, although sexual maturation may take place a couple of years earlier or later. This period of sexual maturation is called **puberty.** Apparently puberty is initiated by unknown factors that cause the hypothalamic portion of the brain to stimulate the pituitary gland to secrete LH and FSH. These hormones, in turn, stimulate testosterone secretions from the testes and steroid secretions from the adrenal glands.

Although males usually experience a gradual decrease in testosterone secretion after they reach 40 or 50, there are no such drastic changes as occur with women at that age. (Those changes will be described later.) Indeed, any psychological problems during this period are probably caused not by a lack of testosterone but by a self-imposed fear of impotency and old age. Even with a decrease in testosterone, normal males may retain their sexual potency well into their eighties.

Spermatogenesis

Sperm cells are formed in the testes after a complex, precisely controlled series of events occurs in the seminiferous tubules. After repeated mitotic divisions, which increase the number of cells (the *spermatogonia*), meiosis results in quartets of haploid cells, the *spermatids* (Figure 13.5). Changes in shape that each spermatid undergoes include the development of an *acrosome* at the front tip, which contains several enzymes that aid the sperm in penetrating an egg; a compact *nucleus* containing the chromosomes (and therefore all the genetic material, DNA); a *middle piece* containing mitochondria, which supply the energy for movement; and an undulating *tail,* which drives the swimming cell forward (Figure 13.6).

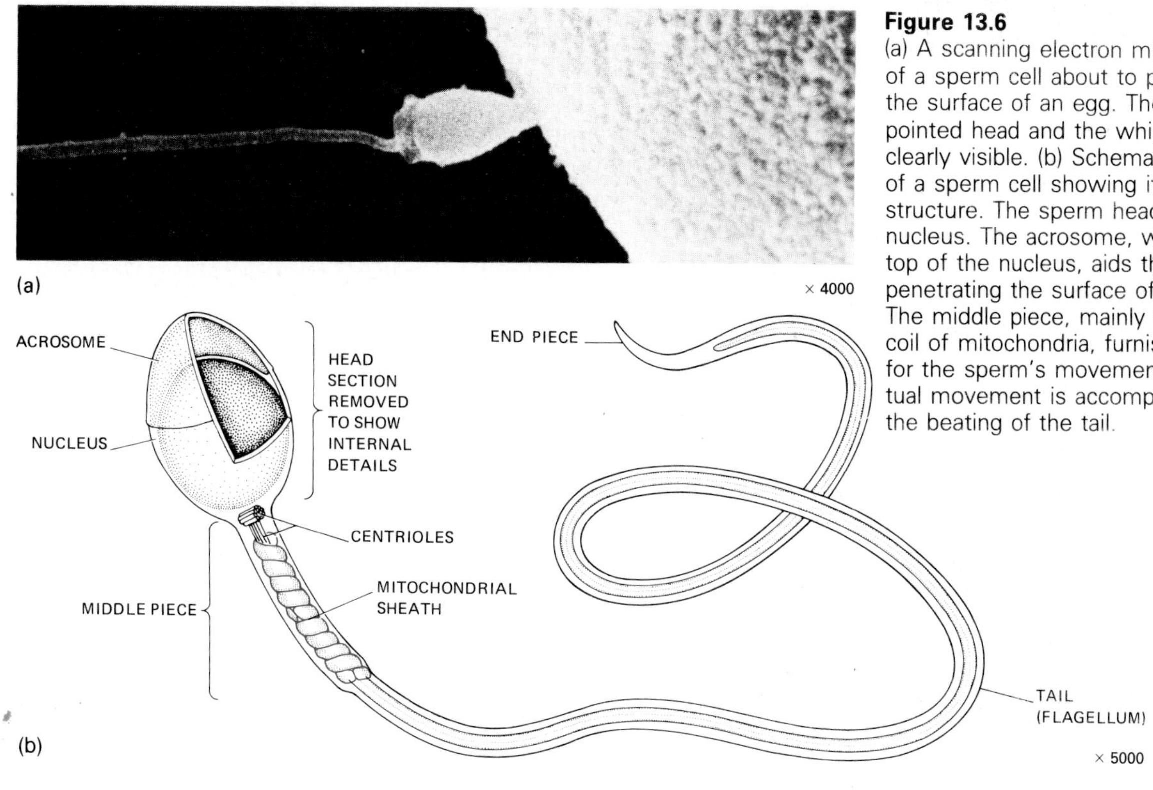

Figure 13.6
(a) A scanning electron micrograph of a sperm cell about to penetrate the surface of an egg. The compact, pointed head and the whiplike tail are clearly visible. (b) Schematic drawing of a sperm cell showing its internal structure. The sperm head is mostly nucleus. The acrosome, which sits on top of the nucleus, aids the sperm in penetrating the surface of the egg. The middle piece, mainly made of a coil of mitochondria, furnishes energy for the sperm's movement. The actual movement is accomplished by the beating of the tail.

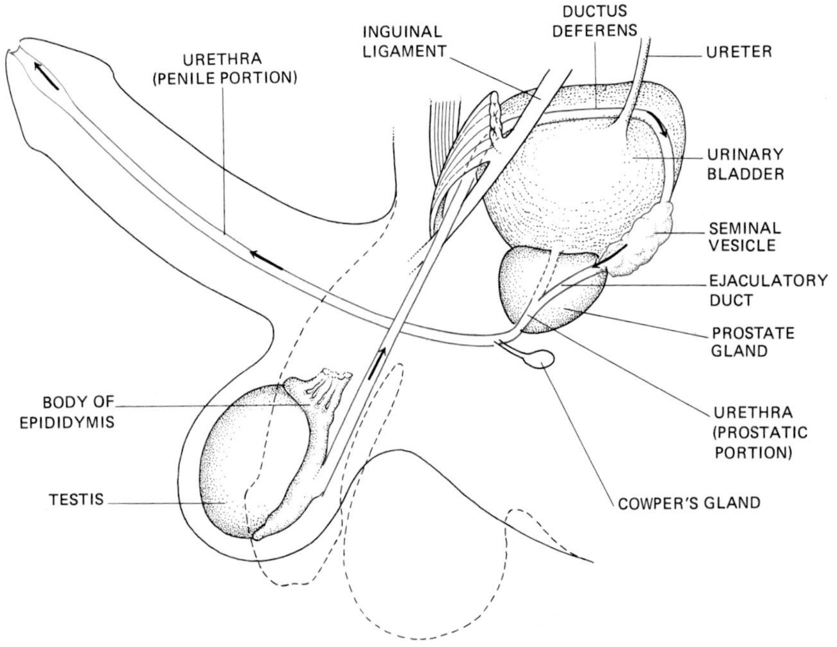

Figure 13.7
Dissection of the male human sexual organs, with the penis in an erect state. Compare this figure in which only the testes, penis, and accessory parts are emphasized, with Figure 13.2, in which the general anatomy of the whole pelvic region is shown. The arrows trace the looped path of sperm ejaculated from the testes through the urethra to the outside.

The production of sperm is periodic in most mammals, coinciding with seasonal breeding activity. But humans do not have a specific breeding season, and spermatogenesis proceeds on a continuous basis. Sperm cells at varying degrees of maturation are always present in the seminiferous tubules.

A sperm cell is one of the smallest cells in the body (the head of a sperm is about 0.005 mm long), and it is basically a nucleus with a tail.* Yet its development requires almost three months for completion. If a small thimble were filled with sperm cells, it would contain more sperm than there are people in the world. (The population of the world is about five billion.)

Accessory Ducts and Glands

Several structures are associated with the testes (Figure 13.7). Applied directly to each testis is an **epididymis** (Greek, "upon the twins"; the "twins" are of course the testes), a coiled tube that would be about six meters (20 ft) long if it were stretched out. The haploid sperm cells are stored in the epididymis. All the seminiferous tubules of the testis lead into the epididymis, which then fuses into the major seminal duct, the **ductus deferens.** The ductus deferens receives secretions from the **seminal vesicles** and the **prostate gland** as it passes alongside them to the urethra. The **urethra** is the final tube, which leads through the penis and transports sperm outside the male body. During ejaculation, fluids from the seminal vesicles and prostate gland are secreted into the ductus deferens, and these secretions plus the sperm cells make up the **semen.** With the onset of sexual excitement, the *bulbourethral* (Cowper's) glands secrete alkaline fluids into the urethra to neutralize the acidity of any remaining urine. These fluids also act as a lubricant within the urethra to facilitate the ejaculation of semen.†

Sperm cells make up only about 5–10 percent of human semen. The remainder is from the diluting medium of the accessory glands, which provides nourishing fructose for the sperm, a suitable alkaline medium to help neutralize urethral and vaginal acidity that might otherwise inactivate the sperm, and buffering salts and phospholipids that make the sperm motile.

The Penis

The function of the penis is twofold: It carries urine through the urethra to the outside, and it transports semen through the urethra during ejaculation.‡ In

* Placed end to end, about 600 sperm would equal 2.5 cm (1 in.), and if placed side by side, with heads touching, 6000 sperm would equal 2.5 cm. It would take about 1000 sperm cells to cover the period at the end of this sentence.

† Secretions from the Cowper's glands sometimes contain some sperm cells, so it is possible for a woman to become pregnant after sexual intercourse even if the man has not ejaculated.

‡ It is physically impossible for a man to urinate and ejaculate at the same time, and in fact, a man cannot urinate if he has an erection *and* is sufficiently sexually aroused. This happens because just prior to ejaculation a valve closes off the opening from the bladder, where urine is stored. The muscle that controls this valve does not relax until the ejaculation is completed, which explains why a man can urinate soon *after* an ejaculation.

addition to the urethra, the penis contains three cylindrical strands of *erectile tissue,* mainly the two spongy **corpora cavernosa.** During erotic stimulation, which may be either tactile or mental, the spaces of the corpora cavernosa become engorged with blood under high pressure. As part of the overall response to stimuli from the nervous system, the arteries leading to the penis are dilated (enlarged), and the veins leading away are constricted. This dual action prevents the blood from escaping, and the penis becomes enlarged and firm in an **erection** (Figure 13.8; see also Figure 13.7). The erect penis can be inserted into the vagina during sexual intercourse. The tip of the penis is the *glans,* a sensitive area containing many nerve endings, and therefore an important source of sexual arousal. The word "glans" is from the Latin for "acorn," which is the basic shape of the glans.

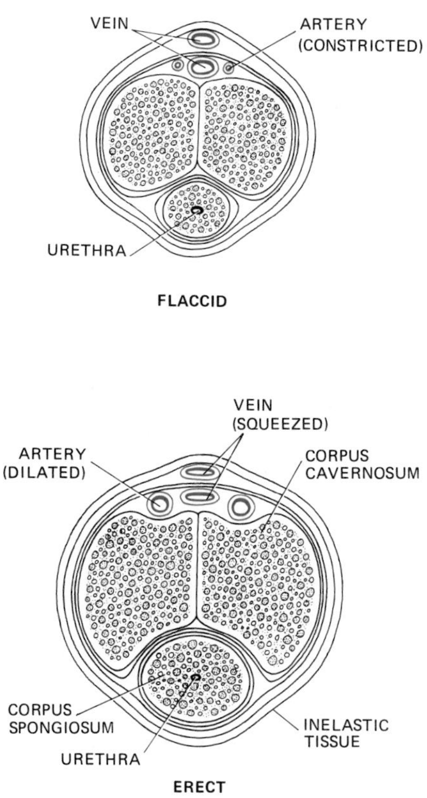

Figure 13.8
Cross sections of a human penis, showing the difference between the flaccid and erect conditions. Erection is produced when blood under pressure enters the penis via the artery. As the *corpus cavernosum* enlarges, it squeezes shut the vein that normally carries blood back to the body. The resulting buildup of pressure creates a stiffening of the organ.

SOME CLINICAL CONSIDERATIONS

Ordinarily, there are 300–500 million sperm cells released during a normal ejaculation. If fewer than about 150 million normal spermatozoa are released per ejaculation, the semen is likely to be ineffective. A man who produces an inadequate number of sperm cells is said to be **sterile,** but sterility may also be caused by other abnormalities. The removal of the testes, *castration,* used to be practiced on some young boys to guarantee that they would remain sopranos to take female roles in musical productions or to make certain that harem guards were harmless. A castrated male will not produce testosterone, and so after a while there will be little or no male sex drive; however, erection of the penis may still be possible after castration. Castration is still used to produce tender roosters *(capons)* or steer steaks.

Sterility may also be caused by venereal diseases, which may interfere with sperm production in many ways, or by such diseases as mumps.

Impotence, which is simply the inability to produce or maintain an erection of the penis, is more likely to be a psychological problem than an anatomical one.

Circumcision is the removal, for religious or health reasons, of the *prepuce,* or *foreskin,* a fold of skin over the end of the penis. Today the principal reason most circumcisions are performed is the belief that the operation practically ensures freedom from cancer of the penis.

HUMAN SEXUAL ANATOMY AND FUNCTION: FEMALE

Sexuality in females is more complex than it is in males. Not only do the females have to produce eggs, but after fertilization they must nourish, carry, and protect the developing embryo. They must also nourish it for a time after it is born. Besides all of this, females have the added complication of a monthly rhythmicity imposed on the entire system.

The Ovaries and Oviducts

The egg-producing organs are the paired **ovaries,** elongated bodies about 5 centimeters (2 in.) long at

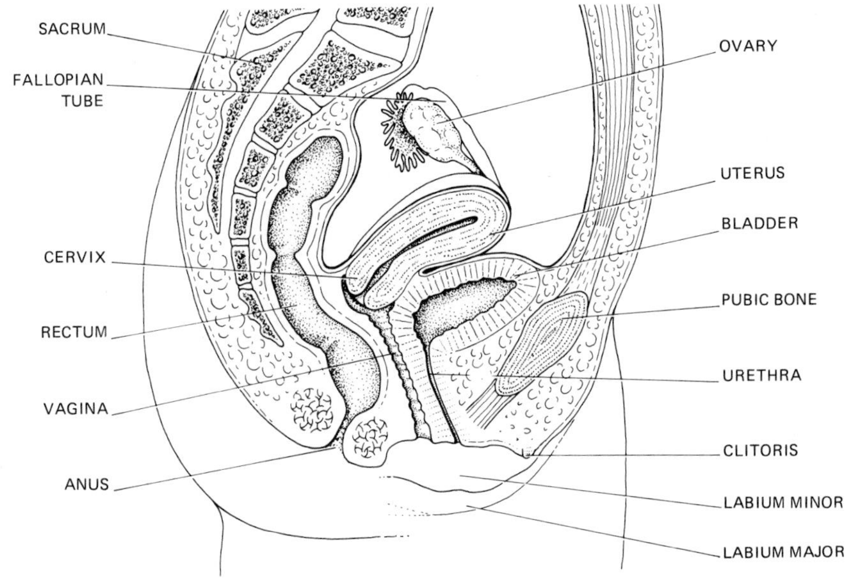

Figure 13.9
A cutaway side view of a human female pelvis. This figure is the female counterpart of Figure 13.2, which shows a male pelvis. The uterus and vagina, which are central and single, are shown in section, but the ovary and oviduct (Fallopian tube), both of which are bilateral and double, are shown with only the left-hand one in place.

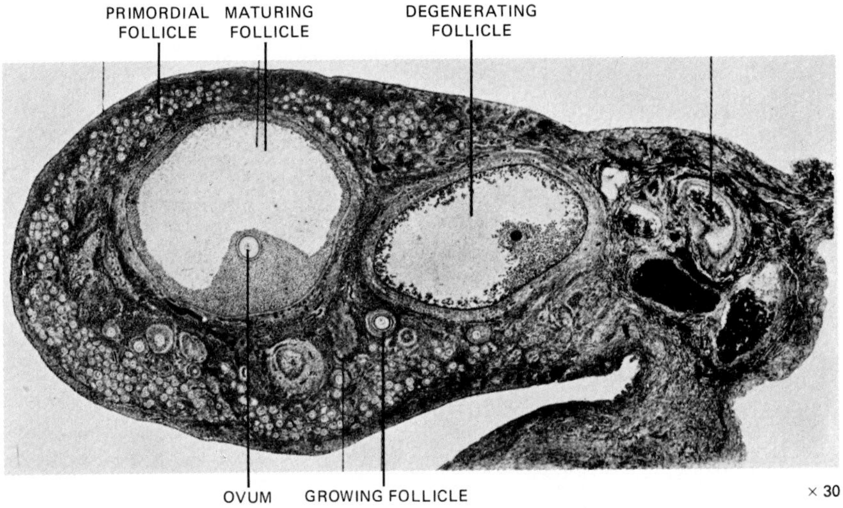

Figure 13.10
Photomicrograph of a section through a monkey ovary, showing various stages in the development of eggs. The *primordial follicle* is one that might have developed into a mature egg-containing follicle. It is necessary to say "might have" because only a few of the potential follicles actually do ripen. A *growing follicle* is just what it says. The *maturing follicle,* which is practically full size, shows the enormous difference in size between it and the primordial follicles. After the egg has been discharged from the follicle, the follicle is absorbed (unless pregnancy has resulted) as the *degenerating follicle,* smaller and irregular in shape, shows.

maturity (Figure 13.9). Like testes, ovaries have a dual purpose. They release eggs (ova) on a fairly regular monthly schedule, and they produce female sex hormones. At birth a human female has about 400,000 potential eggs, but most of them never mature. In fact, no more than 400 of these potential eggs will mature. Beginning at puberty, the eggs mature one at a time (rarely more) in keeping with the monthly reproductive cycle, and they continue maturing during the reproductive years. Most women cease to be fertile during their forties, although some women in their late fifties may still bear children.

The ovaries contain **follicles,** the actual centers of egg production. Each follicle contains a potential ovum, and follicles are always present in several stages of development (Figure 13.10).

Each ovary is partially covered by a funnel-shaped *infundibulum,* into which the egg passes when it is released from the ovary. Apparently there is no particular pattern that selects one ovary over the other each month. The egg is effectively swept across a tiny gap into the infundibulum by cilia (Figure 13.11), and few ova are ever lost in the abdominal cavity. From the infundibulum, an egg can pass down the **Fallopian tube,** or oviduct, which leads to the uterus, or womb.

Oogenesis

The maturation of eggs, **oogenesis,** differs from the maturation of sperm (see Figure 13.5). Not only are the eggs matured one at a time under the rhythmic direction of nervous and hormonal control, but the cellular procedure is unlike that of sperm production. Where one primary spermatocyte yields four spermatozoa after meiosis, a primary oocyte yields only one egg (Figure 13.12). After the first meiotic nuclear division, which produces daughter cells

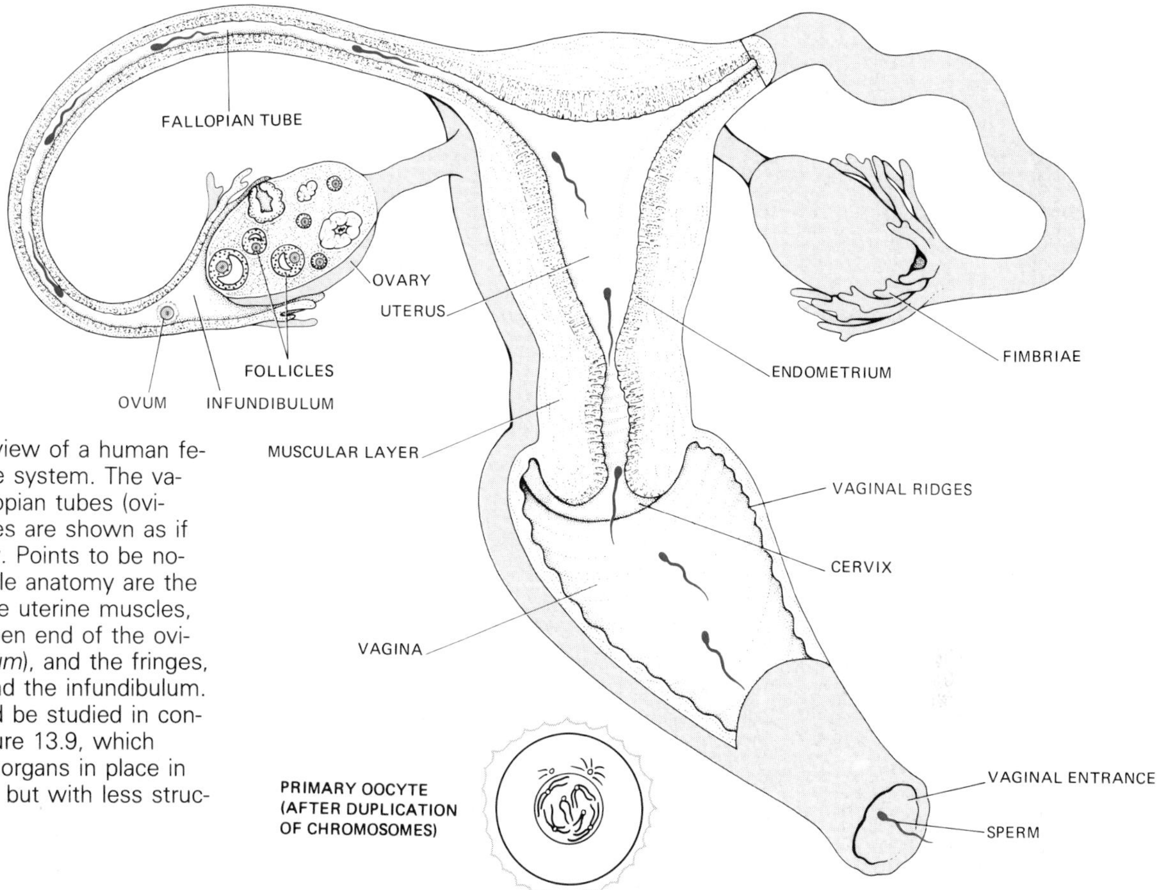

Figure 13.11
Frontal cutaway view of a human female reproductive system. The vagina, uterus, Fallopian tubes (oviducts), and ovaries are shown as if partially cut away. Points to be noticed in the female anatomy are the vaginal ridges, the uterine muscles, the funnel-like open end of the oviducts (*infundibulum*), and the fringes, or *fimbriae*, around the infundibulum. This figure should be studied in conjunction with Figure 13.9, which shows the same organs in place in the pelvic region, but with less structural detail.

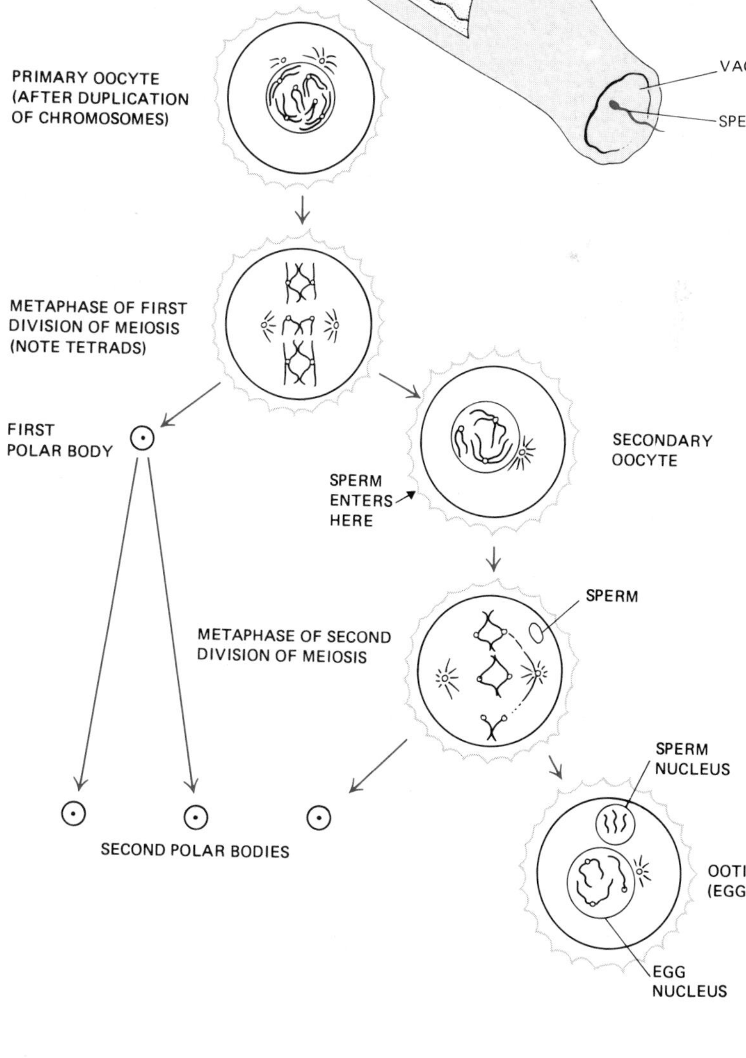

Figure 13.12
The cellular events in the maturation of an egg. The diploid *primary oocyte* (pronounced OH-oh-site) has two complete chromosome sets, and in every pair of chromosomes there are four *chromatids;* that is, each chromosome is double. As meiosis proceeds, a *secondary oocyte* is produced after the first meiotic division, but the two cells resulting from that division are not equal. One is the functional cell (the secondary oocyte), and the other is a *polar body,* destined for destruction. After the second meiotic division, which follows the entrance of the sperm, reduction of the chromosome number is complete, and the result is a haploid *ootid* (pronounced OH-oh-tid), which will develop into a mature egg. The first polar body may or may not divide, so there may be only two final polar bodies instead of the three shown in the diagram. The nonfunctional polar bodies actually remain in contact with the egg. The bulk of the cytoplasm from the primary oocyte goes into the one functional egg.

with 23 chromosomes, the egg-to-be divides into two cells of unequal size. One is perhaps a thousand times as large as the other, and it contains almost all of the food-rich cytoplasm. The small one, a *polar body,* may divide again or disintegrate, but it never functions in reproduction. The large cell goes through a second meiotic division, and another tiny polar body is "pinched off." It, too, is destined to die. The larger cell becomes the egg.

The Uterus, Vagina, and External Genitalia

The Fallopian tubes terminate in the **uterus,** a pear-shaped organ located directly behind the urinary bladder. Not only is the uterus pear-shaped, but it is pear-*sized* as well. However, it increases from three to six times in size during the nine months of pregnancy.

Every month, in response to the secretion of the hormone estrogen, the lining of the uterus *(endometrium)* is built up in preparation for the possible implantation of a fertilized egg (the phenomenon of *pregnancy*). Secretions of another hormone, progesterone, aid in the development of the endometrium into an active gland rich in nutrients and ready to receive a fertilized egg. If pregnancy does not occur, the endometrium is sloughed off, and another monthly, or *menstrual,* cycle begins.

If fertilization and implantation do occur, the uterus houses, nourishes, and protects the developing fetus within its muscular walls. As the pregnancy continues, estrogen secretions develop the smooth muscle in the uterine walls, in preparation for the expulsive action of childbirth.

The uterus leads downward to the **vagina,** a muscle-lined tube about 8–10 centimeters (3–4 in.) long where sperm from the penis are deposited during sexual intercourse and through which the fetus passes down from the uterus during childbirth. An inner mucous membrane secretes acids that help prevent infection (but also create an environment hostile to sperm). A fold of skin called the *hymen* partially blocks the vaginal entrance. The hymen is usually ruptured during the female's first sexual intercourse, but it may be broken earlier through other physical activities. The vaginal canal ends at the *cervix,* or neck of the uterus. The cervix is a frequent location of infection or cancer that should be checked regularly by a physician.

The external genital organs are the labia major, the labia minor, vestibular glands, the clitoris, and the opening of the vagina (Figure 13.13). These organs are collectively called the **vulva.** The *labia* are folds of skin and mucous membrane, fatter in the major than in the minor, that surround and protect the vaginal and urethral openings. The primary lubrication during sexual arousal comes from a "sweating" action of the vaginal walls and labia minor. The *clitoris,* just anterior to the vaginal opening, is structurally comparable to the penis; like the penis, it is composed almost exclusively of erectile tissue. During sexual excitement the clitoris becomes engorged with blood. However, although it is one of the major sources of sexual arousal for

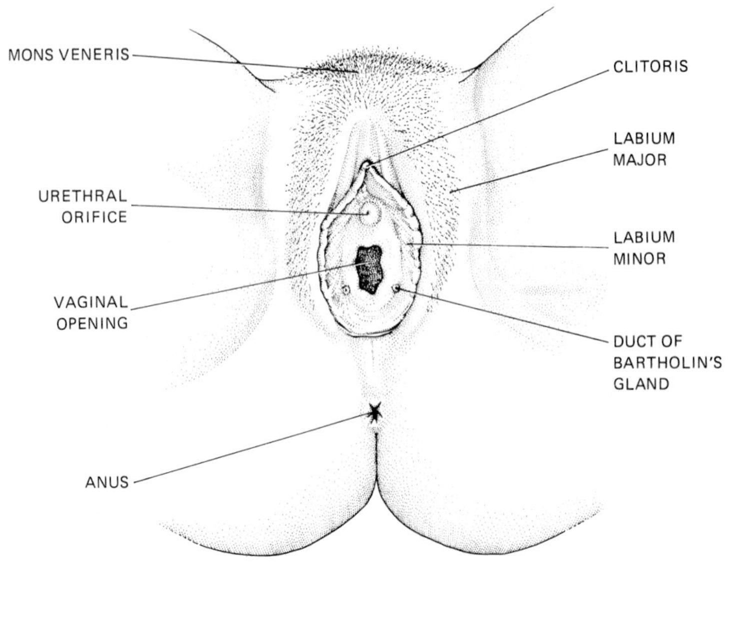

Figure 13.13
External view of the human female genitalia.

TABLE 13.2
Major Human Reproductive Hormones

HORMONE	FUNCTION OF HORMONE	SOURCE OF HORMONE
Female		
FSH (follicle-stimulating hormone)	Causes immature egg and follicle to develop; increases estrogen; stimulates new gamete formation and development of uterine wall after menstruation.	Pituitary gland (controlled by hypothalamus)
LH (luteinizing hormone)	Stimulates further development of follicle and egg; stimulates ovulation; increases progesterone; aids development of corpus luteum.	Pituitary gland (controlled by hypothalamus)
Progesterone	Stimulates thickening of uterine wall.	Corpus luteum (controlled by LH)
Estrogen	Stimulates thickening of uterine wall; stimulates development of female characteristics; decreases FSH; increases LH.	Follicle, corpus luteum (controlled by FSH)
CG (chorionic gonadotropin)	Prevents corpus luteum from disintegrating; stimulates estrogen and progesterone secretion from corpus luteum.	Embryonic membranes, placenta
Prolactin	Allows mammary glands to secrete milk after childbirth.	Pituitary gland (controlled by hypothalamus)
Oxytocin	Stimulates uterine contractions during labor.	Pituitary gland (controlled by hypothalamus)
Male		
Testosterone	Increases sperm production; stimulates development of male characteristics; inhibits LH secretion.	Testes (controlled by LH)
LH (ICSH) (interstitial cell–stimulating hormone)	Stimulates secretion of testosterone.	Pituitary gland (controlled by hypothalamus)
FSH	Increases testosterone production; aids sperm maturation.	Pituitary gland (controlled by hypothalamus)

women, the clitoris plays no direct role in reproduction.

The Mammary Glands

Breasts are present in an undeveloped form in children and men, but during puberty in the adolescent female the breasts begin their development as mammary, or milk-producing, organs. During pregnancy, secretions of estrogen and progesterone cause the milk-secreting glands within the breasts to develop; the actual milk-producing hormone is *prolactin*.

The breasts contain an extensive drainage system comprising many lymph nodes. These nodes may be susceptible to breast cancer (the most common form of cancer in females), and frequent self-examination of the breasts for lumps should be a routine practice.

Hormonal Regulation in the Nonpregnant Female

The male is continuously fertile from puberty to old age, and throughout that period sex hormones are secreted at a steady rate. The female, however, is fertile only during a few days of her monthly uterine cycle, and the complicated pattern of hormone secretion is intricately related to the cyclical release of an egg cell from the ovary (Table 13.2).

The maturation of an egg in an ovary is timed by a cyclical production of hormones in the hypo-

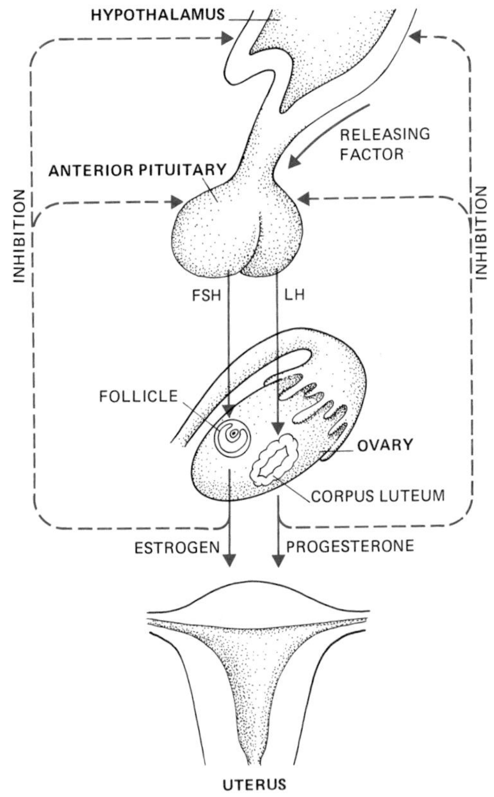

Figure 13.14
Scheme of follicle and ovum (egg) development, as influenced by the follicle-stimulating and luteinizing hormones (FSH and LH). The presence of LH (from the pituitary) in the ovary enhances follicle and egg development. It also increases ovarian output of estrogen. Estrogen stimulates follicle and egg development and at the same time inhibits the output of both the hypothalamus and pituitary. Thus gamete formation in females somewhat parallels that in males (see Figure 13.4).

thalamic region of the brain (Figure 13.14). The timing is determined in an unknown manner in every individual even before birth. The action is indirect, in that the *gonadotropin-releasing hormones* from the hypothalamus act on the pituitary gland. The pituitary gland then releases two additional hormones, FSH and LH, to bring about the egg's maturation and release from the ovary, a process known as **ovulation.**

One way to understand the hormonal alterations of a normal monthly cycle is to follow the development of the ovum. The approximate sequence (the *exact* sequence cannot be precisely outlined) is as follows:

TO SUM UP HORMONES AND THE MENSTRUAL CYCLE

1. The follicle-stimulating hormone, **FSH,** promotes the development of the ovum and one of the immature follicles, the egg-producing centers in the ovary.
2. An ovarian hormone, **estrogen,** is produced by the follicle and causes a buildup and enrichment of the endometrium, as well as the inhibition of further production of FSH.
3. The elevated estrogen level triggers the pituitary secretion of **LH,** the luteinizing hormone, which causes the release of the ovum and also brings about the conversion of the collapsed follicle into an endocrine body, the **corpus luteum.**
4. The corpus luteum secretes another hormone, **progesterone,** which completes the development of the endometrium and maintains this uterine lining for 10–14 days.
5. If the ovum is not fertilized and implanted in the endometrium, there is a disintegration of the corpus luteum, and progesterone production ceases (Figure 13.15b).
6. Without progesterone, the endometrium breaks down, and **menstruation** occurs. (The Latin word for "month" is *mensis.*) During menstruation, the dead cells and blood released by hemorrhage of the uterine arteries are drained from the body through the vagina.
7. As the progesterone level decreases, the level of LH drops off also. This causes the pituitary to renew active secretion of FSH, which stimulates the development of another ovum. The monthly cycle begins again (Figure 13.15d).

After ovulation, estrogen and progesterone act in the bloodstream to inhibit the release of LH and FSH by the pituitary. This is a feedback control that keeps more than one follicle at a time from maturing. There will be no more maturation of a follicle and no more ovulation until the next turn of the cycle, which will usually begin approximately 28 days after the start of the previous cycle *unless the egg is fertilized and pregnancy ensues.* Then the story changes and requires a separate treatment.

Figure 13.15
Correlation of hormonal and physical events in the human menstrual cycle. (a) Concentrations of LH and FSH in the blood. (b) Concentrations of the steroid hormones estrogen and progesterone in the blood. (c) Maturation of the follicle and egg, release of the egg, and degeneration of the follicle (no pregnancy resulting after ovulation). (d) Diagram of uterine activity through two menstrual periods and one intermenstrual period. (e) The menstrual cycle, showing the menstrual flow during the first five days of the 28-day cycle, and the most fertile phase occurring between the eleventh and seventeenth days. The menstrual cycle varies from person to person, and should not be thought of as being as regular as these drawings indicate.

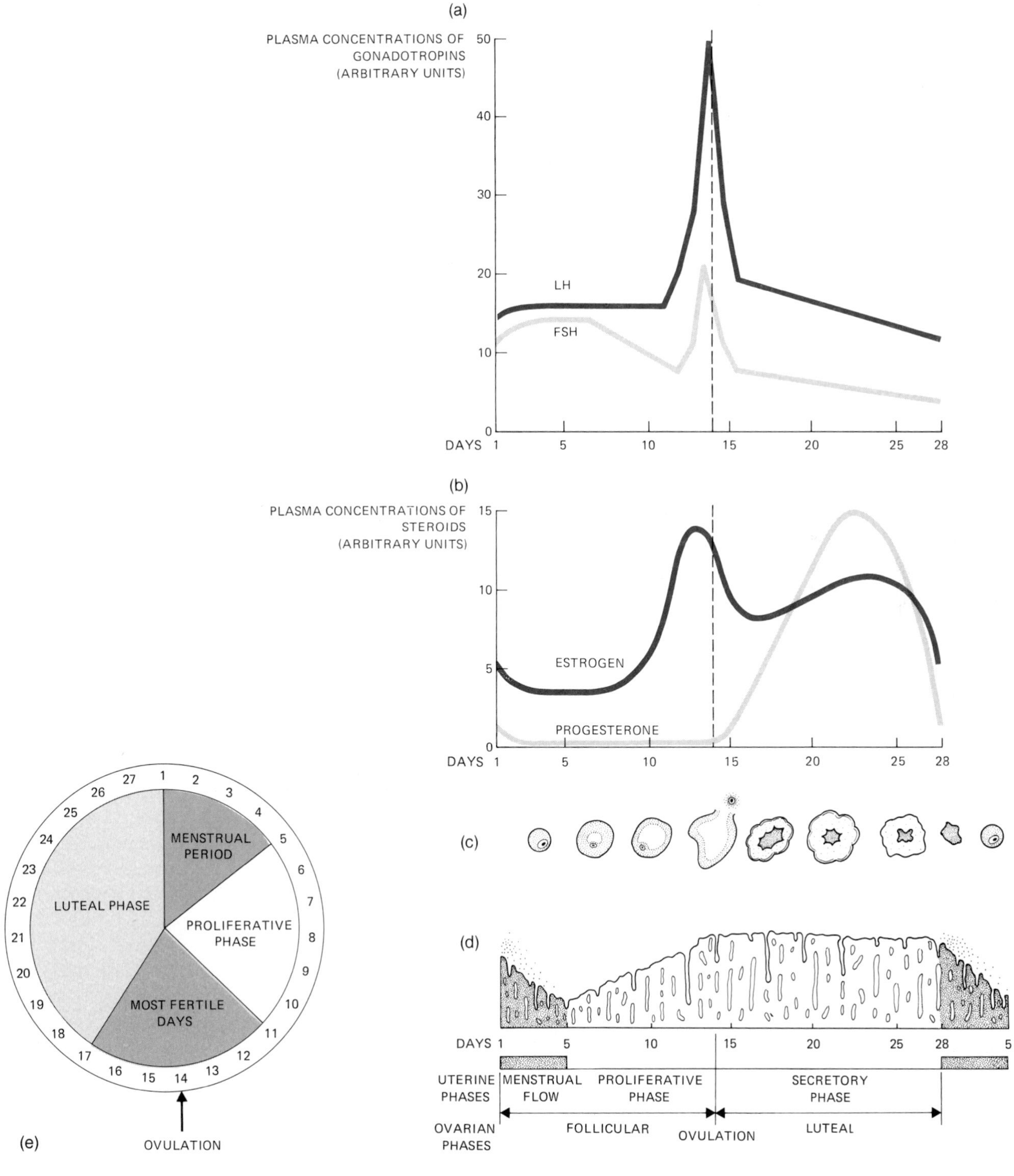

Hormonal Regulation in the Pregnant Female

Pregnancy sets a new series of events in motion. The ovaries themselves are affected because as the embryo develops, its covering membranes and the placenta release a hormone, *chorionic gonadotropin* (CG), that keeps the corpus luteum from disintegrating. Remember that if there is no pregnancy, the corpus luteum does disintegrate. If there *is* pregnancy, the implantation of the embryo triggers a feedback that causes the corpus luteum to remain during most of the period of pregnancy. This scheme ensures that the corpus luteum will continue to produce progesterone. If the amount of progesterone does not remain especially large, the uterine wall will break down, and the embryo will become separated from the uterus. This is one cause of *miscarriage*.

As the embryo develops, other hormones are secreted. *Prolactin* and *oxytocin* induce the mammary glands in the breasts to secrete milk after childbirth, and oxytocin also stimulates the uterine contractions that expel the baby from the uterus during childbirth.

Lest all this sound more certain than it really is, we should point out that the entire mammalian reproductive method is subject to great variability, not only from species to species, but even within the human species. For example, most women living together in a dormitory will usually menstruate within the same two-week segment of the month. How can we explain such a synchronization of menstrual periods? (Apparently they are influenced by body odors that change.)

Puberty and Menopause

The onset of puberty in young girls usually occurs between the ages of 10 and 14 years. During puberty, the external genitalia, uterus, vagina, and breasts grow and mature under the influence of pituitary gonadotropins and estrogen from the ovaries. Hair begins to grow in the pubic region and the armpits, the pelvis widens and accumulates fatty deposits, and an active sex drive commences. Estrogen also stimulates a thickening of the skin and an increase in its water content, which tends to counteract somewhat the excessive secretion of adolescent oils and fats that may lead to acne. Finally, estrogen brings about an increase in calcium deposits at the ends of the long bones and causes the bone cells to stop dividing; the combination of these events ordinarily signals the end of skeletal growth.

The first menstrual period, the *menarche* (Greek, "beginning the monthly"), occurs during the latter stages of puberty, although actual ovulation may not occur until a year or so later. The natural cessation of menstrual periods, the *menopause* ("ceasing the monthly") usually occurs sometime between the ages of 40 and 50, signaling the end of reproductive ability. Menopause does not happen suddenly, and menstrual periods and ovulations will most likely begin to be irregular a few years before the final menstrual period.

After menopause, the breasts, uterus, and external genitalia begin to atrophy, but the sex drive does not necessarily decrease, and in fact it may increase. Mental and physical symptoms may occur during menopause. Hot flashes of the skin, which are typical, are due to estrogen deficiency that somehow causes dilation of blood vessels in the skin. Some of the symptoms of menopause may be relieved by small doses of estrogen.

Prior to menopause, diseases such as hardening of the arteries occur infrequently in women, probably because estrogen reduces cholesterol in the blood. But as estrogen secretions continue to diminish after menopause, the incidence of cardiovascular disease becomes almost equal in men and women.

HUMAN MATING

Many animals have elaborate courtship rituals (Chapter 30), but few have anything like the variety of sexual practices that humans have. Men are sexually potent from the onset of puberty until late in age, and they have nothing comparable to the rutting season of deer or the nesting season of birds. Women are affected by the cycle of ovulation and menstruation, but they are fertile at monthly intervals throughout the year, and they remain so from about the first menstruation until the last.

In humans, the act of sexual intercourse, **coitus** (Latin, "to come together"), is usually preceded by a period of preliminary sexual excitement, in which physiological and psychological changes occur in both sexes as a preparation for actual *copulation*, or coupling. Until fairly recently, it was thought that female sexual response was different from and probably less intense than the sexual response of a male. Fortunately, an objective approach to the study of sexual response during recent years has shown that previous ideas about sexual response were not only narrow-minded but incorrect as well. We know now that males and females can experience relatively similar feelings during sexual intercourse.

Orgasm, the climax of sexual excitement, is accompanied in males by *ejaculation* of the semen, which is expelled through the penis via the urethra by massive contractions of the ductus deferens, the seminal vesicles, and the other accessory glands, all of which contribute their secretions to the semen. Each normal ejaculation produces 3–5 milliliters (about a teaspoon) of semen. The seminal fluid enhances the sperm's motility, and they begin actively swimming at speeds up to 5 centimeters (2 in.) an hour, wiggling their tails erratically at 14–16 cycles per second. Some sperm may reach the uterus a few minutes after ejaculation and may reach the Fallopian tubes as soon as 30 minutes after ejaculation. (Sperm cells seem remarkably adept at progressing through the uterus into the Fallopian tubes, in spite of opposing currents generated by cilia.) Female orgasm does not necessarily accompany every copulation, and in humans, at least, female orgasm is not necessary for fertilization to occur. In the female, sexual climax is accompanied by the usual psychological effect of orgasm, with contractions of the uterine muscle, high breathing and heart rate, and flushing of the skin. Although multiple orgasms may occur, there is nothing in the female climax that corresponds to ejaculation.

Figure 13.16
Scanning electron micrograph of the inner lining of a Fallopian tube. The irregular sphere is an egg. After it is released from the ovary, the mature egg is swept toward the uterus by the many protruding cilia (Ci) that line the Fallopian tube.

CONCEPTION

Once every 28 days or so, a single human egg oozes through a particularly fragile portion of the ovary wall and is carried by ciliary action through the Fallopian tube toward the uterus (Figure 13.16). The egg, moving much more slowly than the sperm, takes four to seven days to travel the 10 centimeters (4 in.) from the ovary to the uterus. During the first third of the journey, where fertilization is most advantageous, the egg slows its pace, as if awaiting the sperm. Fertilization must occur no more than 24 hours after ovulation, since the egg will remain viable for only that period of time. Sperm cells may remain viable in the female reproductive tract for about 72 hours. (If semen is stored in a laboratory at very cold temperatures the sperm will remain viable for many months.)

During mating, millions of sperm cells enter the female's vaginal canal.* If mating takes place at about the same time as ovulation, these millions of spermatozoa enter the vagina, and some may travel toward the opposite-moving ovum, but only one sperm cell will eventually enter and fertilize the egg. If only one sperm may enter, why are millions dispatched? Most likely, more than enough sperm are discharged to ensure that at least one will successfully complete its mission. Not only must the sperm travel from the vagina all the way to the Fallopian tubes (a trip of 15–20 centimeters that may take as much as a few hours); they must also resist the spermicidal acidity of the vagina and overcome opposing fluid currents in the uterus and oviducts. So it is not surprising that the mortality rate of the sperm is enormously high. It is also postulated that the quantity of sperm discharged is large so that at least several thousand will reach the egg and act in unison to provide sufficient amounts of hyaluronidase. This enzyme, which is found in the head of the sperm

* Allowing for enormous differences in size, a mature sperm swims about 50 percent faster than an Olympic swimmer. Put another way, a sperm can swim about 600 times its own body length in eight minutes, but the male Olympic swimmer travels about 400 times his body length in eight minutes. (And the Olympic swimmer does not have all the obstacles and unfriendly current to negotiate.)

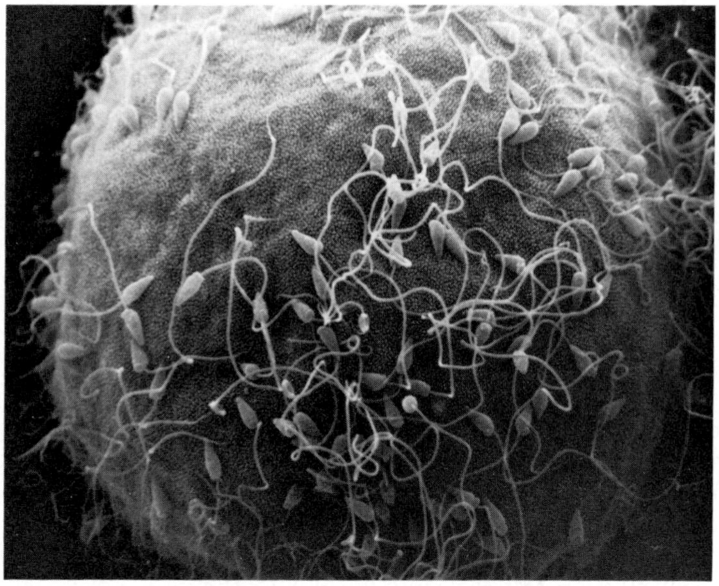

× 250

Figure 13.17
Scanning electron micrograph of the meeting of sperm and egg. A sea urchin egg, the large sphere, is practically surrounded by hundreds of sperm. The outer covering of the egg is not its cell membrane but a special coating, the *vitelline membrane,* on which only certain areas are receptive to sperm. Immediately after the entry of a sperm, the membrane thickens. It is then called a *fertilization membrane,* which prevents other sperm from penetrating the egg, ensuring that each egg is fertilized by only one sperm.

Figure 13.18
Scanning electron micrographs of the penetration of a sperm into an egg. The acrosome of the sperm touches the tiny projections (microvilli) on the egg surface (a). The microvilli in the region of contact rise up to meet the sperm (b). The vitelline membrane rises and thickens, becoming the fertilization membrane, through which it is impossible to see. The fertilization membrane has been peeled away, disclosing the sperm, almost buried in the egg's cell membrane after three minutes (c). The sperm rapidly sinks into the egg cytoplasm, and a minute after the sperm body enters the cell membrane only the tail remains outside (d).

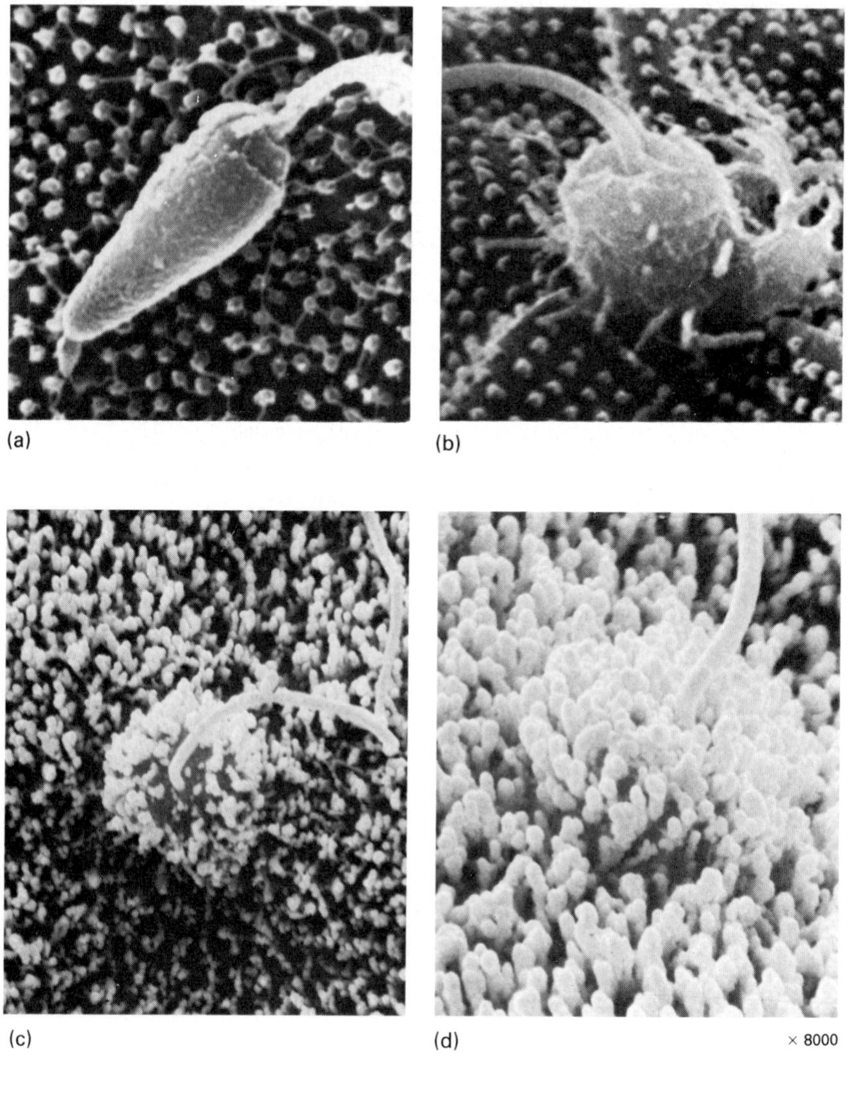

(a) (b) (c) (d)

× 8000

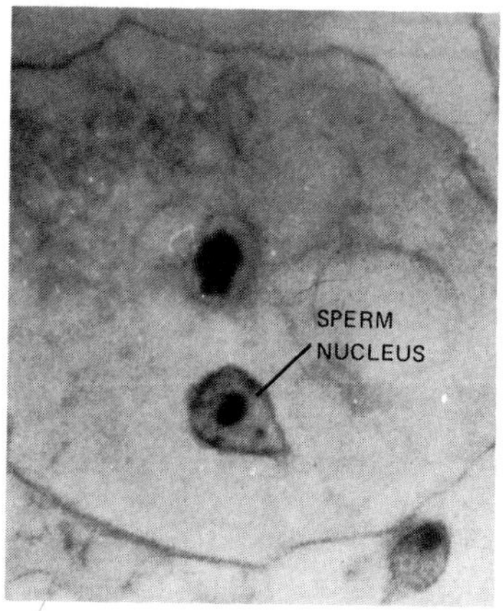

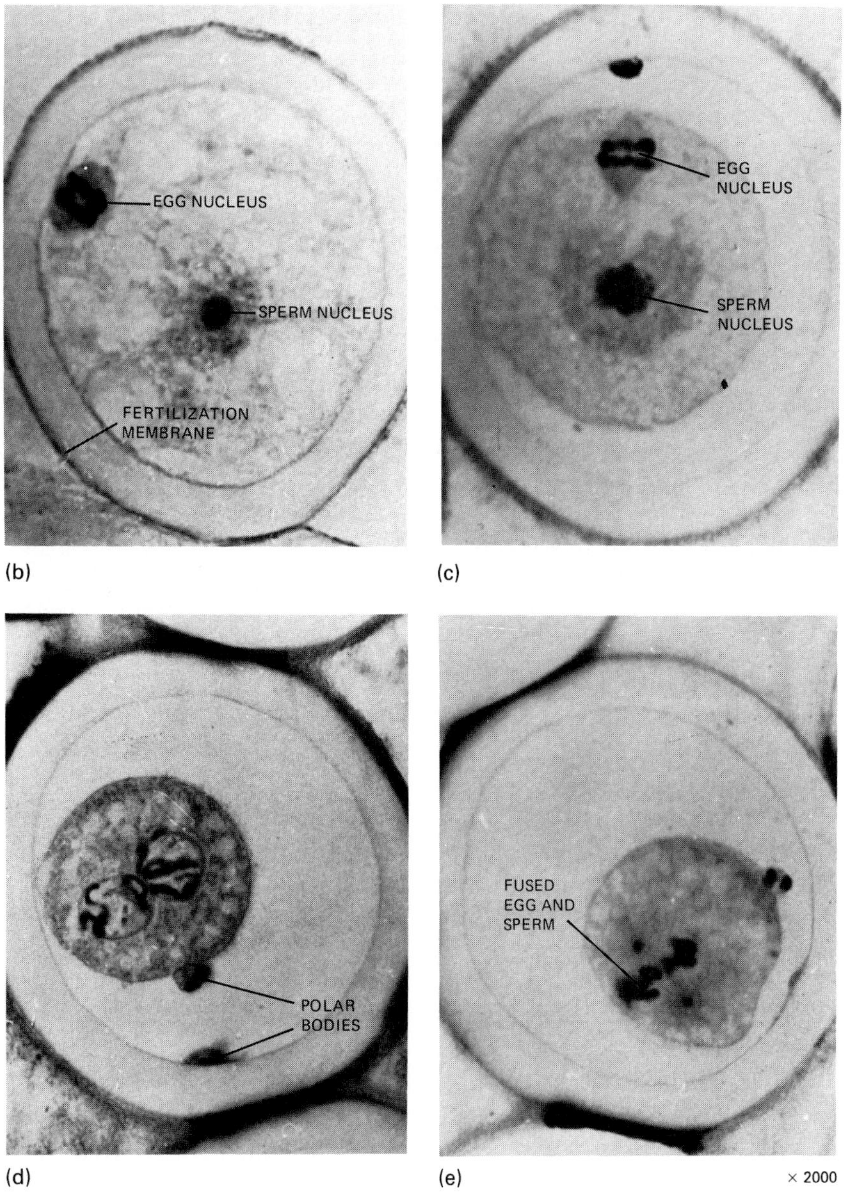

Figure 13.19
Fertilization as seen inside an egg. The specimen is *Ascaris,* a parasitic intestinal worm of swine and horses, which has long been a favorite subject for studies on fertilization. (a) The sperm has just penetrated the egg. The heavy fertilization membrane surrounds the egg. (b) The sperm loses some of its compactness as it approaches the egg nucleus in the center of the egg cytoplasm. (c) The egg nucleus, not having undergone meiosis until the sperm is present, is undergoing its first reduction division. (d) The egg nucleus has completed meiosis, and the two disintegrating polar bodies are outside the egg cytoplasm, visible as dark spots on the top of the figure. The egg and sperm nuclei are touching, but which is male and which is female is impossible to tell. (e) The egg and sperm nuclei have fused, and the zygote nucleus, the first diploid nucleus of the new generation, is in its first metaphase, on its way to producing an embryo.

cells, chemically dissolves just enough of the outer wall of the ovum to allow a single sperm to enter. Immediately after fertilization, a *fertilization membrane,* which sperm cannot penetrate, forms in an unknown manner around the fertilized egg.* In this way, only one sperm enters the ovum. Now embryonic development begins (Chapter 14). (If more than one sperm somehow enters the ovum, the development of the fertilized egg is almost always abnormal.)

Some aspects of fertilization are illustrated in Figures 13.17 through 13.19. Fertilization usually takes place in the Fallopian tube. (If a fertilized egg should become implanted in the Fallopian tube, for instance, instead of properly in the uterus, an *ectopic pregnancy* ensues, and surgery is usually performed to remove the implanted embryo.) About an hour after fertilization, the haploid (23 chromosomes) nuclei of the egg and sperm fuse to form a single diploid cell with 46 chromosomes, the **zygote.** (Not surprisingly, it is in this context that the egg and sperm are called *gametes,* derived from the Greek word for "to marry.")

* Recent studies with sea urchin eggs have revealed that an egg's outer membrane has a negative electrical charge before penetration by a sperm, and a positive charge after penetration. Apparently, the reversal of the membrane charge blocks the entry of any sperm after the first one enters the egg. The causal mechanism for the change is not known. After a minute or so, the egg's outer membrane switches back to its original negative charge, but by that time the membrane has been permanently strengthened and is impenetrable.

"Test-tube babies" are conceived outside a woman's body when she has had a history of unsuccessful pregnancies because of blocked Fallopian tubes or some other problem. An egg is surgically removed from one of the prospective mother's ovaries at the time of ovulation and fertilized in a laboratory with sperm from the husband. Two or three days later, after cell division has begun, the embryo is implanted in the woman's uterus through her cervix. Rabbits, mice, and cattle test-tube babies have

TABLE 13.3
Methods of Contraception (in decreasing order of effectiveness)

METHOD	MODE OF ACTION	EFFECTIVENESS IF USED CORRECTLY	ACTION NEEDED AT TIME OF INTERCOURSE	POSSIBLE SIDE EFFECTS OR INCONVENIENCES
Vasectomy*	Prevents release of sperm	Very high	None	Usually irreversible
Tubal ligation†	Prevents egg cell from entering uterus	Very high	None	Usually irreversible
Oral pill, 21-day administration	Prevents follicle maturation and ovulation	Very high	None	Early: some water retention, breast tenderness, nausea; Late: possible blood clots, hypertension
Intrauterine device (coil, loop)	Prevents implantation	High	None	Some women do not retain device; possible menstrual discomfort, abnormal bleeding, infection
Diaphragm with jelly	Prevents sperm from entering uterus; jelly kills sperm	High	Must be inserted before intercourse	None, but may cause overlubrication; cannot be fitted to all women
Condom (worn by male)	Prevents sperm from entering vagina	High	Must be put on prior to intercourse	Some reduction of sensation in male, may interrupt foreplay
Temperature rhythm	Determines ovulation time by noting body temperature at ovulation	Medium	None	Requires abstinence during part of cycle
Calendar rhythm	Abstinence during parts of cycle	Medium to low	None	Requires abstinence during part of cycle
Vaginal foam	Kills sperm	Medium to low	Requires application before intercourse	None usually, may irritate
Withdrawal	Remove penis from vagina before ejaculation	Low	Withdrawal	Frustration in some
Douche	Washes out sperm	Lowest	Immediately after	None, but must be done immediately after intercourse

*Vasectomy prevents the release of sperm but does not alter the production of male hormones.
†Tubal ligation prevents the passage of the mature egg cell to the uterus but does not alter the production of female hormones.

TABLE 13.4
Other Methods of Contraception under Investigation

METHOD	MODE OF ACTION	EFFECTIVENESS IF USED CORRECTLY	ACTION NEEDED AT TIME OF INTERCOURSE	POSSIBLE SIDE EFFECTS OR INCONVENIENCES
"Minipill' (very low content of progesterone)	Inhibits follicle development	High	None	Irregular cycles and bleeding
"Morning-after pill"	Arrests pregnancy probably by preventing implantation; fifty times normal dose of estrogen	By currently available data, high	None	Breast swelling, nausea, water retention; possible cause of vaginal cancer in female offspring
Vaginal ring (inserted in vagina, contains progesteroid)	"Leaks" progesteroid into bloodstream through vagina at constant rate; thereby inhibits follicle maturation	Studies are "promising"	None	Spotting, some discomfort
Once-a-month "pill"	Injected in oil base into muscle; slow passage of birth control drug into circulation inhibits follicle maturation	Said to be 100 percent	None	Similar to oral pill
Medroxyprogesterone acetate (Depo-Provera; DMPA) 3-month injection	Injected; inhibits follicle development	By currently available data, high	None	Similar to oral pill

been produced successfully for several years, but it was not until 1977 that a human test-tube baby was successfully conceived, implanted, developed, and born (in 1978). (Compare the methods described in the essay on page 144.) We now think toward the next development—growing an embryo, then a fetus, and finally a fully developed baby independent of a uterus.

CONTRACEPTION

All contraceptive methods have one aim: prevention of pregnancy. This aim can be achieved by preventing the production of eggs or sperm, by keeping eggs and sperm from meeting, or by preventing the implantation of an embryo in the uterus. All these methods are possible, but some are difficult to achieve, some are fallible, some are dangerous to the female, and some may be considered unacceptable to some people because of psychological factors or religious beliefs. Table 13.3 describes the many methods of contraception now available, and Table 13.4 lists several methods under investigation.

VENEREAL DISEASES

Some parasites are transferred from host to host mainly by sexual contacts. Consequently, they are known as **venereal** parasites (from Venus, Greek goddess of love). Contrary to popular myths, venereal diseases (VD) are rarely contracted by casual, dry contact with persons or objects. The most common venereal diseases in humans are type 2 herpes simplex, caused by a viral infection, and gonorrhea and syphilis, both caused by bacteria.

Type 2 herpes simplex (also called herpes genitalis) is usually (but not always) transmitted through

HUMAN SEX DETERMINATION

The male sex cell, the sperm, contains the haploid number of 23 chromosomes. The female sex cell, the ovum, or egg, also contains 23 chromosomes. When a sperm fertilizes an egg, the two sets of chromosomes enter into a single nucleus, and the diploid number of 46 chromosomes results. A new individual, with its full complement of hereditary material, begins its development. But what determines whether this new organism will be a boy or a girl? The answer is: the nature of the chromosomes. In both males and females, 22 pairs of chromosomes always match. The 22 dependably matching pairs (44 chromosomes) are called **autosomes.** The remaining two chromosomes are the **sex chromosomes.** The pair of sex chromosomes look alike in females, and they are designated **XX**. Sex chromosomes do not look alike in males, and they are designated **XY.**

After meiosis in males, half of the sperm contain 22 autosomes and an X chromosome, and the other half contain 22 autosomes and a Y chromosome. If an egg (always containing an X chromosome) is fertilized by the X-chromosome-bearing sperm, an XX zygote results and will develop into a female; an egg fertilized by a Y-bearing sperm produces an XY zygote and will develop into a male. So the father actually determines the sex of the child, since only his sperm cells contain the variable, the Y chromosome.

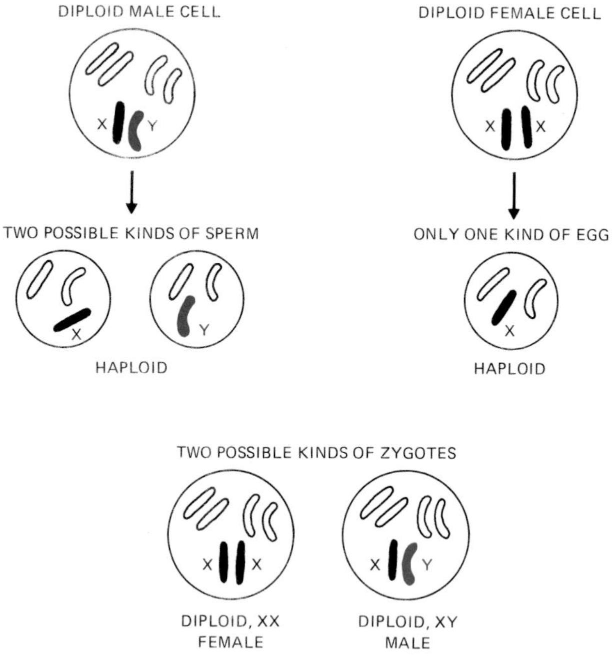

sexual intercourse, and may be a permanently recurring problem, even if no further sexual contact is made. It ordinarily appears as blisterlike sores on or near the external genitalia about a week after intercourse with an infected partner. Fever, muscle aches, and swollen lymph nodes may also be apparent. When the blisters rupture they produce shallow ulcers that can be very painful, especially for a woman during urination. When the blisters burst they release millions of infectious virus particles. The blisters usually heal after a week or two, but the virus particles retreat to nerves near the lower spinal cord, where they remain dormant until the next attack. Some people never have a "next attack," but most do, harboring the infectious viruses for months, years, or a lifetime. Perhaps the most serious complication of type 2 herpes simplex is the infection of a newborn infant as it passes through the birth canal. For this reason, if a pregnant woman is known to have type 2 herpes simplex, the baby is usually delivered by cesarean section. As of this writing, there is no known cure.

Gonorrhea is primarily an infection of the reproductive and urinary tracts, but any moist part of the body, especially the eyes, may suffer. The causative organism, *Neisseria gonorrhoeae,* can be recognized by microscopic examination. If left untreated, gonorrhea becomes difficult to cure, causing painful inflammation of any mucous membrane, but it can be cured by antibiotics, especially if it is caught early. Before routine treatment of eyes in the newborn was established, thousands of babies were blinded by gonorrheal infection at the time of passage through the vaginal canal. Gonorrhea is commonly called *clap.*

Syphilis is a more dangerous disease, caused by a motile, corkscrew-shaped bacterium, *Treponema pallidum.* Its early symptom is a sore, the hard chancre, at the place where infection occurred. After perhaps a few months, various other symptoms develop, including fever and general body pain. The symptoms frequently disappear, leaving the victim with the false impression that the disease is gone. But later, circulatory or nervous tissue may degenerate, so that paralysis, insanity, and death follow. Some individuals develop a tolerance for the parasites and can therefore carry and transmit the disease, even though they do not seem to have it. One of the most pitiful aspects of venereal infection is the transmission of the parasites from a mother to a baby. The syphilis bacterium is able to cross the placenta during pregnancy, whereas the gonorrhea bacterium seems unable to do so. Thus the developing child can contract syphilis early in its development and exhibit some of the terrible manifestations of the disease at birth. If the baby contracts syphilis during the actual birth process, it is not likely to exhibit any symptoms at the time of birth. However, a baby infected with syphilis will grow poorly, be mentally retarded, and die early.

Like gonorrhea, syphilis can be cured by antibiotics if treatment is started in time, and it seems likely that a syphilis vaccine for humans will be developed in the near future. Rabbits injected with dead syphilis bacteria have shown a partial resistance to subsequent injections of live syphilis bacteria, indicating that protective antibodies had been formed. But many more tests with animals such as chimpanzees must be concluded successfully before a vaccine can be used with humans. In the meantime, it is well to remember that the occurrence of venereal disease has reached epidemic proportions in this country, especially among teenagers and young adults, and false feelings of security about quick antibiotic cures should not be encouraged. It is true that venereal diseases can be cured with greater ease than ever before, especially if they are reported and treated early, but antibiotic-resistant strains of the gonorrhea bacterium are known. The most effective treatment is intelligent prevention.

SUMMARY

1. A number of methods can be used in sexual reproduction, but the end result is similar in most animals, with one animal producing small, motile gametes (spermatozoa), and the other producing larger, nonmotile food-filled gametes (eggs, or ova).

2. In protozoa there may be sex without actual reproduction, but genetic recombinations do take place, and that is the essence of sexuality. Almost all land-dwelling animals have adopted some form of mating that allows fertilization of eggs inside the body of the female. Most fishes and amphibians have specialized behavior that guarantees that sperm will be deposited close to the eggs.

3. The reproductive function of male mammals is the production and delivery of sperm cells to the female. These activities require the *testes, accessory glands* and *ducts,* and the *penis.*

4. Among the *seminiferous tubules* in the testes are small masses of *interstitial cells,* which secrete the male sex hormones, especially *testosterone.*

5. The formation of sperm cells in the testes *(spermatogenesis)* takes place after a complex, precisely controlled series of events occurs in the seminiferous tubules. After repeated mitotic divisions, which increase the number of cells (the *spermatogonia*), meiosis results in quartets of haploid cells, the *spermatids.*

6. Haploid sperm cells are formed in the testis and stored in the *epididymis,* which leads into the *ductus deferens.* The ductus deferens passes alongside the *seminal vesicles* and the *prostate gland* (and receives secretions from them) to the urethra. During ejaculation, fluids from the seminal vesicles and prostate gland are secreted into the ductus deferens, and these secretions plus the sperm cells make up the *semen.*

7. The *penis* contains three cylindrical strands of erectile tissue, mainly the *corpora cavernosa,* which facilitate an erection during sexual excitement.

8. A man who produces an inadequate number of sperm cells for reproduction is *sterile.* Sterility may also be caused by castration, venereal diseases, and other diseases, such as mumps.

9. Sexuality in female mammals is more complex than it is in males. Females produce eggs, nourish, carry, and protect the embryo, and care for the newborn offspring. Hormonal controls in females are more varied and intricate than they are in males.

10. The egg-producing organs in humans are the *ovaries,* which release eggs *(ova)* on a fairly regular monthly schedule. They also produce the female sex hormones *estrogen* and *progesterone.* The ovaries contain many structures called *follicles,* which are the centers of egg production.

11. In the maturation of eggs *(oogenesis)* a primary *oocyte* yields only one egg. After the first meiotic nuclear division, the egg-to-be divides into two cells of unequal size; the smaller one is a *polar body,* which never functions in reproduction. After the second meiotic division, another polar body is pinched off and dies. The remaining cell is the egg.

12. The *vagina* is the muscle-lined tube where sperm are deposited during sexual intercourse and through which the fetus passes from the uterus during childbirth. The external female genital organs are collectively called the *vulva.* During pregnancy, the *uterus* houses, nourishes, and protects the developing fetus within its muscular walls.

13. The maturation of an egg in an ovary is timed by a cyclical production of *gonadotropin-releasing hormones* in the *hypothalamus.* These hormones cause the *pituitary gland* to release *FSH* and *LH,* which stimulate *ovulation.* The *corpus luteum* produces *progesterone,* and the ovarian *follicles* secrete *estrogen.*

14. *Menstruation* occurs if the egg is not fertilized and implanted. The uterine wall breaks down, and the dead blood cells and tissues of the uterus are discharged through the vagina.

15. If the egg is fertilized, another hormone, *chorionic gonadotropin* (CG), prevents the corpus luteum from disintegrating. The secretion of progesterone continues, and other hormones are secreted that aid pregnancy, childbirth, and milk production.

16. Few animals have the variety of sexual practices that humans have. Men are sexually potent from the onset of puberty until late in life, and women are fertile at monthly intervals throughout the year and remain so from about the first menstruation until the last. In humans, the act of *coitus* is usually preceded by a period of preliminary sexual excitement, in which physiological and psychological changes occur in both sexes.

17. *Fertilization* usually takes place in the Fallopian tube, where the nuclei of the haploid sex cells fuse to form a single diploid cell, the *zygote.* Immediately after fertilization, a *fertilization membrane* forms around the zygote, preventing further penetration of sperm cells. An X-carrying sperm determines that the zygote will develop into a female; a Y-carrying sperm determines that the zygote will develop into a male.

18. All *contraceptive* methods have one aim: prevention of pregnancy. This aim can be achieved by preventing the production of eggs or sperm, by keeping eggs and sperm from meeting, or by preventing the implantation of an embryo in the uterus.

19. *Venereal* parasites are transferred from host to host mainly by sexual contacts. The common venereal diseases in humans are *type 2 herpes simplex,* caused by a viral infection, and *gonorrhea* and *syphilis,* both caused by bacteria.

ASK YOURSELF

1. What is probably the most important advantage of sexual reproduction over asexual reproduction?
2. How do land-dwelling animals manage to have fertilization take place in a wet medium?
3. What is the role of the testes?
4. What is the difference between sperm and semen?
5. What is the function of an ovarian follicle?
6. What is a basic difference in the processes of spermatogenesis and oogenesis?
7. What is the vulva?
8. What is a corpus luteum?
9. What roles do the hormones FSH and LH play in ovulation?
10. Where does human fertilization normally occur?
11. How is it possible for only one sperm to enter an egg?

14
Human Development

SOME KEY POINTS

1. After a sperm cell fertilizes an egg, a series of cell divisions called cleavage begins.
2. About six to eight days following fertilization in humans, the embryo becomes implanted in the inner wall of the mother's uterus. This is the start of pregnancy.
3. Nutrients are passed from the mother to the embryo through blood vessels in the umbilical cord and placenta, and the embryo's wastes are sent out through opposite blood vessels. There is normally no direct connection between the maternal and fetal circulatory systems.
4. After the second month of development the embryo is called a fetus. The baby is usually born about 280 days after the last menstrual cycle before conception.
5. Immediately after the baby is born it begins to breathe on its own, no longer depending on the umbilical cord and placenta.

THE ADULT HUMAN BODY CONSISTS OF ABOUT 50 trillion (50 million million) cells, and each type of cell is a specialist with a specific job to do. Cells that are programmed to be heart cells, for instance, will never be anything else. In fact, cells are so specialized that the liver cells of a human and a horse are more alike than are human liver cells and other types of human cells. Cells are functional. They *do* something special, and if they are healthy and normal they will never do anything else. Organs do something specific, too. For instance, *all* vertebrate livers do basically the same things. Structure and function are closely related.

Such specialization may sound simple, but no one has yet determined how it occurs (see Chapter 8). We do know that all the many specialized cells of an adult body are derived from a single fertilized egg cell. As the fertilized ovum divides again and again, the resultant cells begin to differ more and more from their original forms. They also become increasingly different from one another. The information coded in each cell originates in DNA. Further, certain cells may secrete substances akin to enzymes or hormones that somehow provoke other cells to grow and differentiate. Even environmental conditions may be important stimuli. But still, how does a cell "know" when and how to differentiate? And if mitosis creates daughter cells genetically identical to the parent cell, how is their identity altered to produce specialized, different cells?

For thousands of years it has been understood that the act of sexual intercourse is a prerequisite for a mammal's conception and birth. But until the late nineteenth century it was not realized that both parents contribute genetic material (chromosomes) to the zygote through egg and sperm. With this knowledge it was relatively easy to understand that the embryonic development of the new organism was programmed by the genetic input of the parents.

A REVIEW OF MITOSIS AND MEIOSIS

Before we begin a discussion of embryonic development, it may be helpful to review the basic principles of cell division, originally presented in Chapter 8. In Figure 14.1 we can observe **mitosis,** the repro-

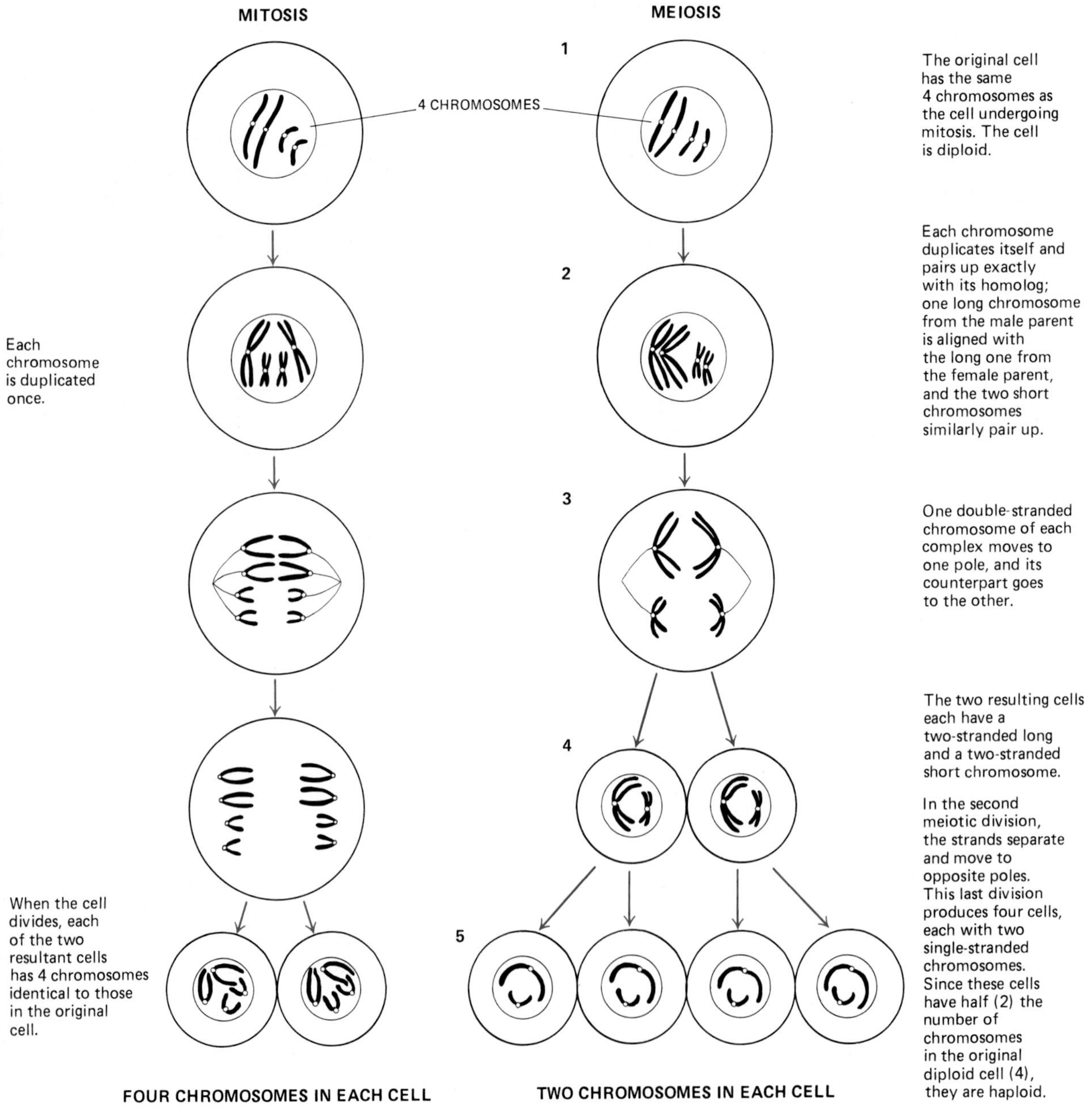
MITOSIS
MEIOSIS
4 CHROMOSOMES
1
The original cell has the same 4 chromosomes as the cell undergoing mitosis. The cell is diploid.
2
Each chromosome is duplicated once.
Each chromosome duplicates itself and pairs up exactly with its homolog; one long chromosome from the male parent is aligned with the long one from the female parent, and the two short chromosomes similarly pair up.
3
One double-stranded chromosome of each complex moves to one pole, and its counterpart goes to the other.
4
The two resulting cells each have a two-stranded long and a two-stranded short chromosome.
In the second meiotic division, the strands separate and move to opposite poles. This last division produces four cells, each with two single-stranded chromosomes. Since these cells have half (2) the number of chromosomes in the original diploid cell (4), they are haploid.
5
When the cell divides, each of the two resultant cells has 4 chromosomes identical to those in the original cell.
FOUR CHROMOSOMES IN EACH CELL
TWO CHROMOSOMES IN EACH CELL

Figure 14.1
A comparison of mitosis and meiosis. In the example shown here, there are four chromosomes: two long ones (one from the organism's male parent and one from the female parent), and two short ones (one from the male parent and one from the female parent). Mitosis, with one division, results in two cells identical to the original one, with four chromosomes in each cell. Meiosis, with two divisions, produces four cells from the original one, with *two* chromosomes in each cell.

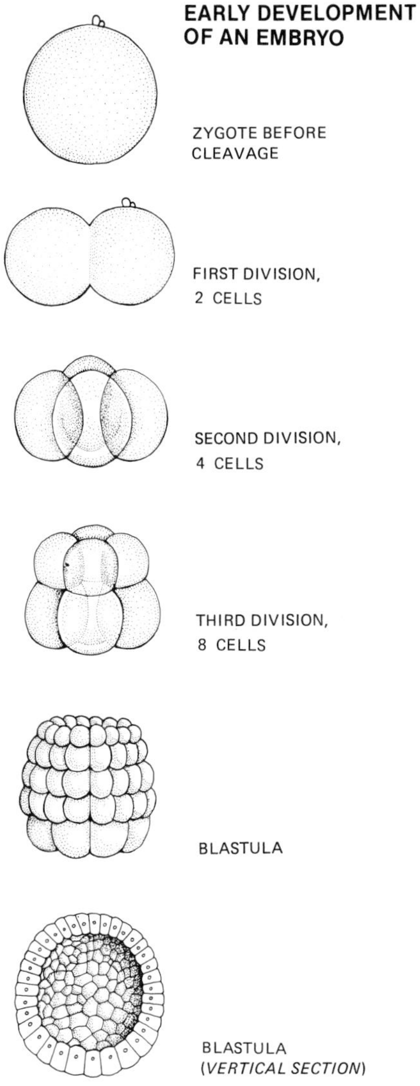

(a)

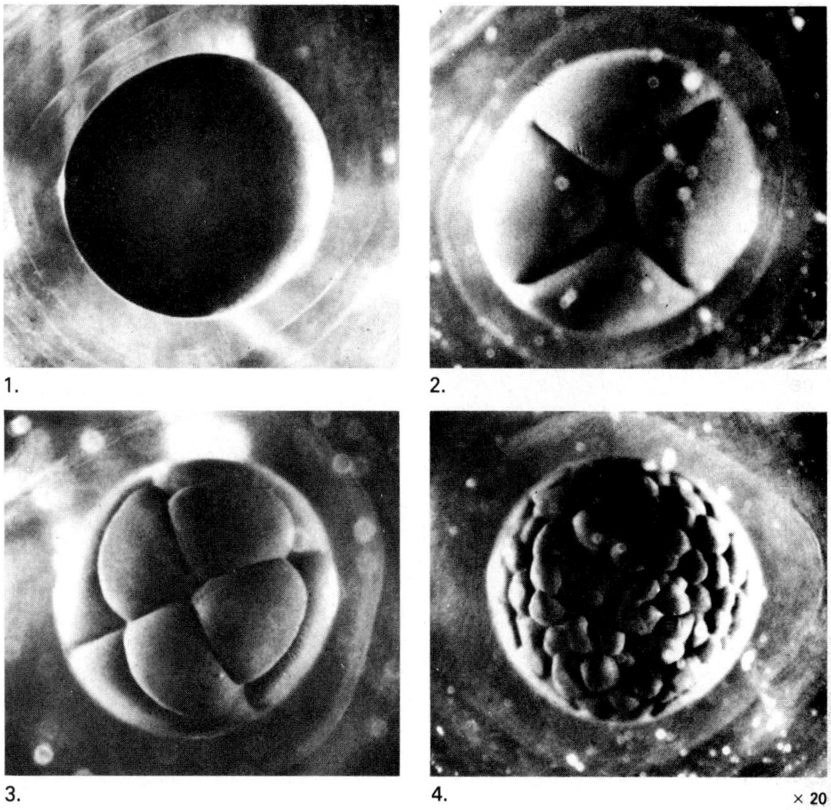

(b)

Figure 14.2
(a) The first divisions (*cleavage*) in an idealized embryo. The single-celled zygote still has two polar bodies attached to it. Successive divisions result in two, four, eight, sixteen cells, and so on, each new cell smaller than the zygote cell. These divisions continue until a *blastula*, a hollow ball of cells, is formed. The whole blastula, consisting of many small cells, is scarcely larger than the one original zygote cell. (b) Photographs of cleavage in a frog embryo. Compare the reality of these pictures with the diagrammatic quality of the drawings in (a).

duction of nuclei and the equal distribution of DNA to each, and **meiosis,** the pair of nuclear and cellular divisions that reduce the chromosome number from diploid to haploid.

PRINCIPLES OF EARLY EMBRYONIC DEVELOPMENT

When a sperm meets an egg, it digests its way into the egg cytoplasm and sheds its tail. Meiosis in the egg, which had stopped at metaphase of the second division, is resumed after entry of the sperm. The second polar body is pinched off, leaving the egg nucleus with a haploid set of chromosomes. The egg and sperm nuclei then fuse to form the diploid *zygote* nucleus. This nucleus contains all the genetic material, DNA, which subsequent mitotic divisions will distribute equally to all cells of the embryo. After one sperm has gained entrance to the egg, changes in the egg's surface take place to prevent additional sperm from entering.

After a sperm penetrates an egg, the egg quickly surrounds itself with an enveloping coat, the *fertilization membrane,* and a series of cell divisions called **cleavage** begins (Figure 14.2). Cleavage

Figure 14.3
The process of *gastrulation* in the lancelet or amphioxus, a primitive chordate. (a) A blastula has developed. (b) One side of the growing embryo changes in such a way as to make the wall bend inward (invaginate). The result is a double-walled cup, a *gastrula.* (c) The hole where invagination occurs is the *blastopore.* In the amphioxus, as in all deuterostome animals, the blastopore is at the posterior end of the embryo. The blastopore will become the anus of the mature animal. Later the mouth will break through the opposite (anterior) end.

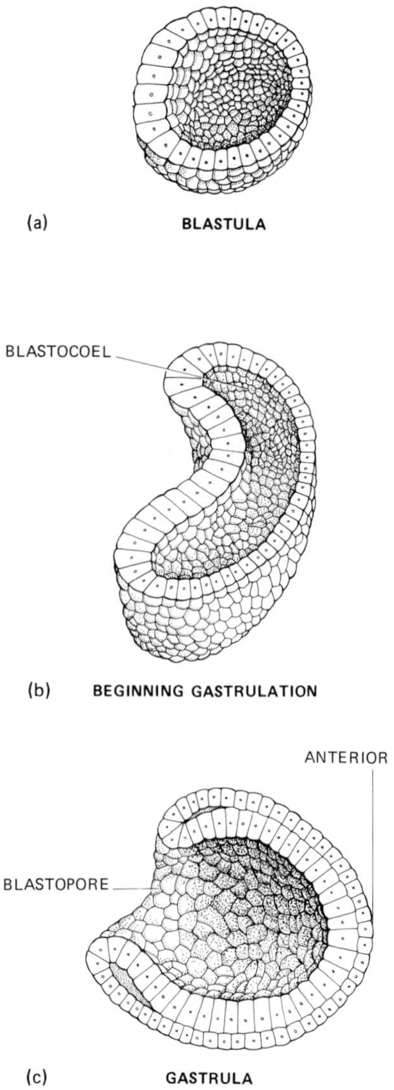

divisions, unlike later growth divisions, result in smaller and more numerous cells but not in increased size of the early embryo. Human cleavage is relatively slow. The first cleavage occurs about one day after fertilization, and subsequent cleavages take place about twice a day. In comparison, frog eggs may undergo cleavage every hour.

Early Cleavage in Vertebrates

Because early stages in human development are difficult to obtain for observation and almost impossible to experiment on, most descriptive and experimental information comes from other animals, such as frogs, but the essential principles hold for all vertebrates. Regardless of what kind of egg is undergoing cleavage, the result is the formation of a number of cells that will start the embryo toward increasing developmental specialization. In many animals, an early stage in embryo formation is a hollow ball of cells called a *blastula* (see Figure 14.2).

Gastrulation

Cell divisions continue, but since the whole embryo remains the same size at this early stage, the cells are becoming smaller as cleavage progresses. In animals with small, free-floating eggs, the cells begin an organized movement on the surface of the young embryo, entering a phase called **gastrulation,** during which the rearrangement of cells and growth of new ones result in an embryo with a developing body. The essential feature is the formation of layers of cells, with each layer capable of developing into special tissues (Figure 14.3).

Once gastrulation has occurred, the whole embryo has an outer coating, the **ectoderm** (*ecto,* outside; *derm,* skin), and an inner lining, the **endoderm** (*endo,* inside) (Figure 14.4). The ectoderm cells will give rise to the outer layers of skin, the hair, the fingernails, the enamel of teeth, and the brain with its associated nervous tissue. The endoderm will form the lining of the digestive tract, digestive glands, and lungs. Between the ectoderm and endoderm, new cells proliferate, filling the part between the two original layers and eventually building the bulk of the embryo. This is the **mesoderm** (*meso,* middle), the progenitor of connective tissue in lower skin layers, bone, muscle, blood, and the covering tissues of the internal organs.

Early Human Development

In mammals, including of course humans, early development is somewhat different from that in ani-

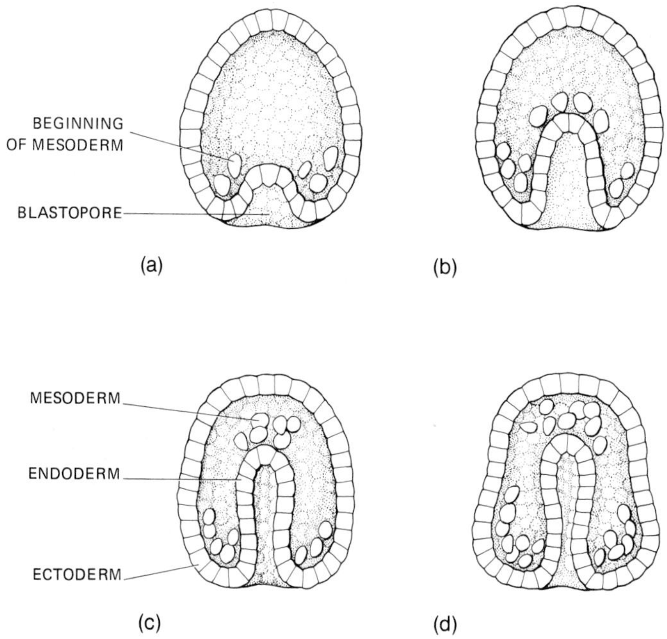

Figure 14.4
Origin of the three *germ layers* in a sea urchin. The *ectoderm* is the outermost layer. The *endoderm* is started when gastrulation starts. By the time gastrulation is complete, the endoderm is well established as the inner layer of the gastrula. At certain places within the embryo, cells begin to grow in the space between ectoderm and endoderm. The resulting tissue is *mesoderm*, which in most animals makes up the bulk of the mature body.

mals with small eggs that float freely. About 4½ to 5 days after fertilization in humans, the cells of the beginning embryo form a sphere called a **blastocyst.** The central cavity of the blastocyst contains fluid and a mass of cells from which the embryo proper will grow. The inner mass becomes a two-layered plate, the germinal disk. One of the layers is ectoderm, and the other is endoderm. They function the same as the ectoderm and endoderm in vertebrates in general. During the third week of development, the disk grows a thickened strip of cells called the **primitive streak,** which gives off migrating cells. Those cells move in between the ectoderm and endoderm, and thus add a third layer, the mesoderm. Once the three-layered germinal disk is established, the embryo is ready to go ahead with the formation of all the parts of a complete animal.

DEVELOPMENT OF HUMAN EMBRYOS

The following description is directed specifically at human development. As soon as a single sperm cell penetrates an egg cell, the egg undergoes its final meiotic division, and the polar body containing very little cytoplasm is formed. The haploid nuclei of the sperm and the egg unite in the act of fertilization, and the egg, now a diploid zygote, continues its journey toward the uterus. The process of ovulation has already caused the corpus luteum to initiate the monthly development of the uterine walls, and the fertilized egg will be accepted by the uterus, as an unfertilized egg would not be.

Implantation in the Uterus

The fertilized egg travels through the Fallopian tube until it reaches the uterus four or five days later. While it has been moving toward the uterus, the fertilized egg has been dividing. After two or three more days of floating freely in the uterus, where it is nourished by uterine fluids, the embryo is ready to be received by the enriched uterine walls. By now the embryo is a fluid-filled hollow sphere of about 100 cells plus a covering mass of cells, the *trophoblast,* which will develop into a system of membranes outside the embryo. These membranes will transport nutrients to the embryo and will remove wastes from it. The whole mass, including trophoblast and the contained embryo, is a **blastocyst.**

The uterine wall and the blastocyst have continued to develop in the few days since fertilization occurred. When the uterus and the blastocyst have reached compatible levels of development, the blastocyst becomes enveloped by the nutritious inner

THE DEVELOPMENT OF THE SPINAL CORD AND EYE

As an embryo develops, it undergoes differentiation of its parts. In studying differentiation, biologists have asked what, where, when, how, and why. The questions of how and why are just beginning to be answered and the answers are still not satisfactory. Such questions are the stuff of developmental biology, one of the most active of present-day scientific fields. Once the ectoderm, endoderm, and mesoderm are formed, other organs and tissues of the body can begin to differentiate. The nervous system, for example, develops very early and actually plays a part in regulating the development of other parts of an animal body. The development of the vertebrate spinal cord and eye are presented here.

This series of pictures of developing frog embryos shows the origin and differentiation of the spinal cord and some other major body structures. Even though frog eggs, and consequently frog embryos, have features that are conspicuously different from their human counterparts, the general way that development proceeds is similar in both species.

In (a-1) gastrulation has occurred. A section through the midregion shows a lump at the top: the neural plate, destined to become part of the central nervous system. In (a-2) the ridge, which is on the dorsal side of the embryo, has risen more sharply. The section shows the neural plate thickened into two folds so heaped up that a narrow furrow has formed between them. Below these neural folds there is a rounded body (seen in cross section), the notochord, which is flanked on each side by developing masses of mesoderm. These masses will become muscle. In (a-3) the neural folds have grown together along the upper edges to form a hollow tube that will be the spinal cord. In (a-4) the dorsal ridge is more pronounced than it will ever be again. The body cavity, the coelom, shows as two slits in the mesoderm, right and left of the notochord. At its anterior end, the spinal cord will enlarge, retaining its essential hollowness, to become the brain. Nerves grow out from the brain and spinal cord, growing and branching until they form the nervous system. In this way, one entire set of internal structures has an ectodermal origin.

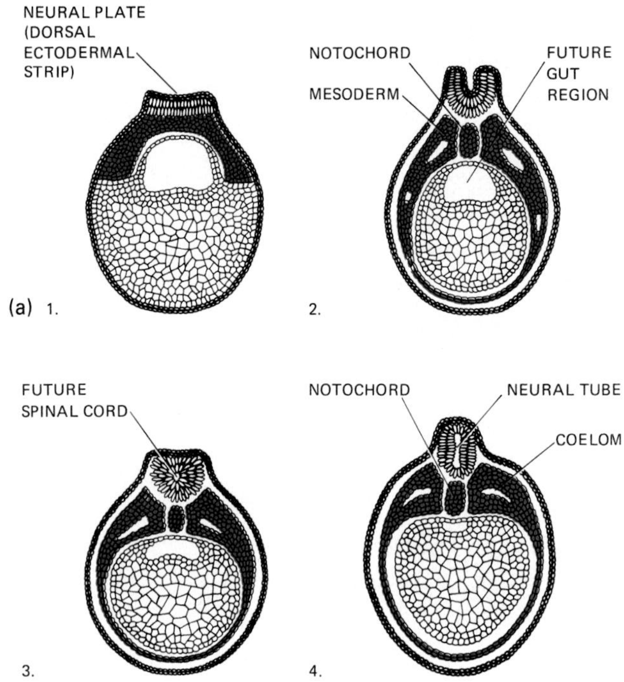

The series of events leading to the formation of a vertebrate eye is instructive because it illustrates an embryological principle of general applicability. That is the principle of **induction,** in which one tissue is caused, or *induced,* to change its developmental pattern as a result of the influence of a different tissue. The physical events are relatively simple. At the anterior end of the embryo, swellings of the enlarging brain grow out right and left toward the outer covering of ectoderm. They are the *optic vesicles,* which are the forerunners of eyeballs (b-1). As an optic vesicle approaches the outer covering, the ectoderm, it causes the ectoderm nearest the vesicle to thicken (b-2), then dimple inward toward the vesicle while the vesicle is folding in on itself (b-3). The dimple becomes a deep invagination, which sinks beneath the surface of the ectoderm and pinches off like a submerged bubble (b-4). This bubble, now surrounded by the collapsed vesicle, becomes the eye lens, and the vesicle becomes the outer wall of the eyeball, while the old ectoderm, healed over, becomes the outer layer, the cornea or sclerotic layer, of the eye (b-5).

In the induction of eye lens formation, it is obvious that at least two factors are at work. One can be generally called a genetic factor, since a vertebrate eye could not develop in, say, a spider. The proper genes must be present. The other factor is one of time and place, since the eye lens develops only when the optic vesicle is close by. Both these factors are contained within the embryo itself, but development can also be influenced by forces from outside the embryo. During the 1950s, many European women took a tranquilizer, thalidomide, during the early stages of pregnancy. The drug crossed the placenta into the embryonic blood and interfered with normal growth of the developing limbs. Many babies were born with deformed arms or legs, or even without limbs. Another dramatic cause of birth defects is the virus of German measles, or rubella, which can penetrate the unborn baby's defenses during the first three months of pregnancy, and last even past birth. Rubella virus can damage the heart, eyes, ears, and the blood.

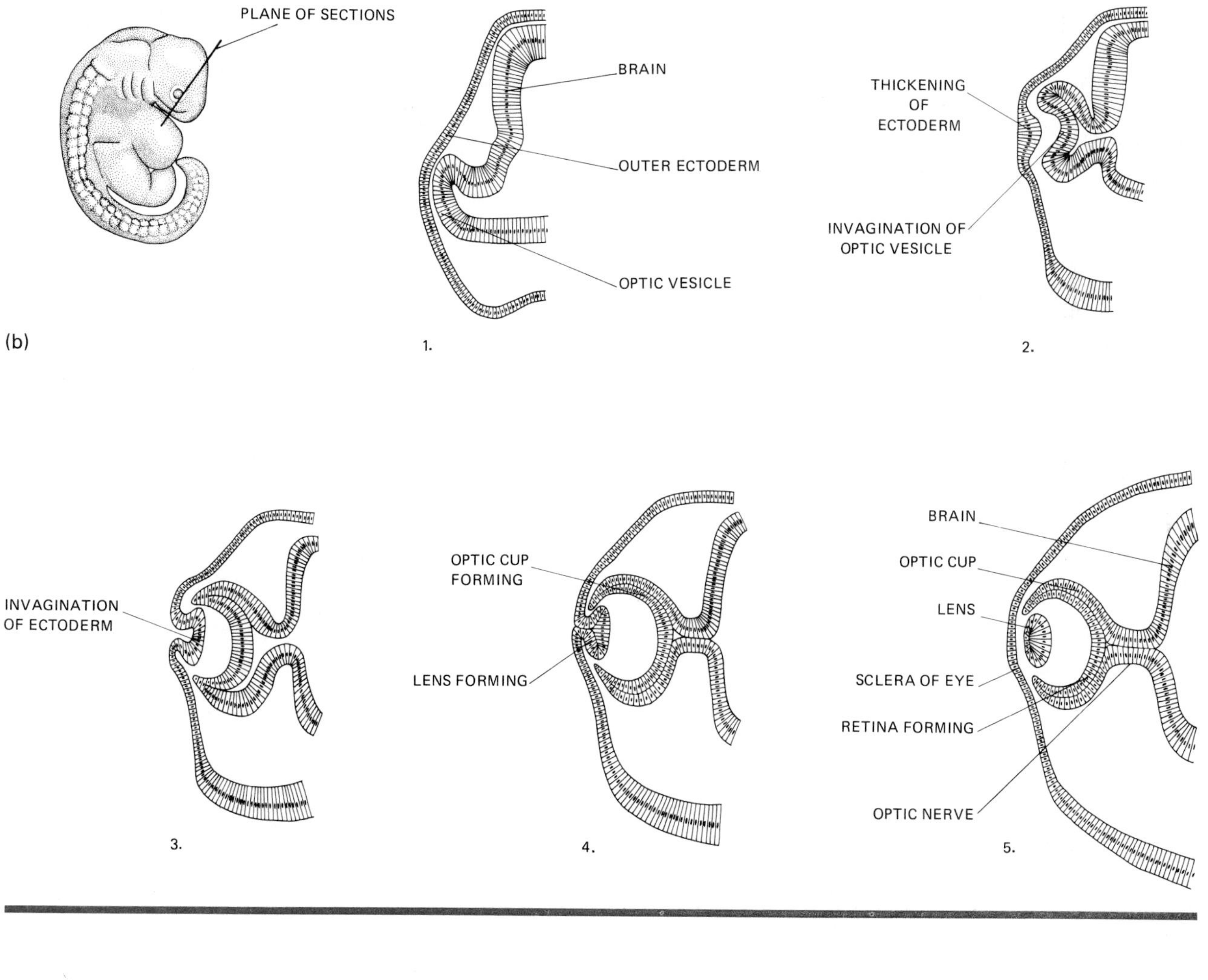

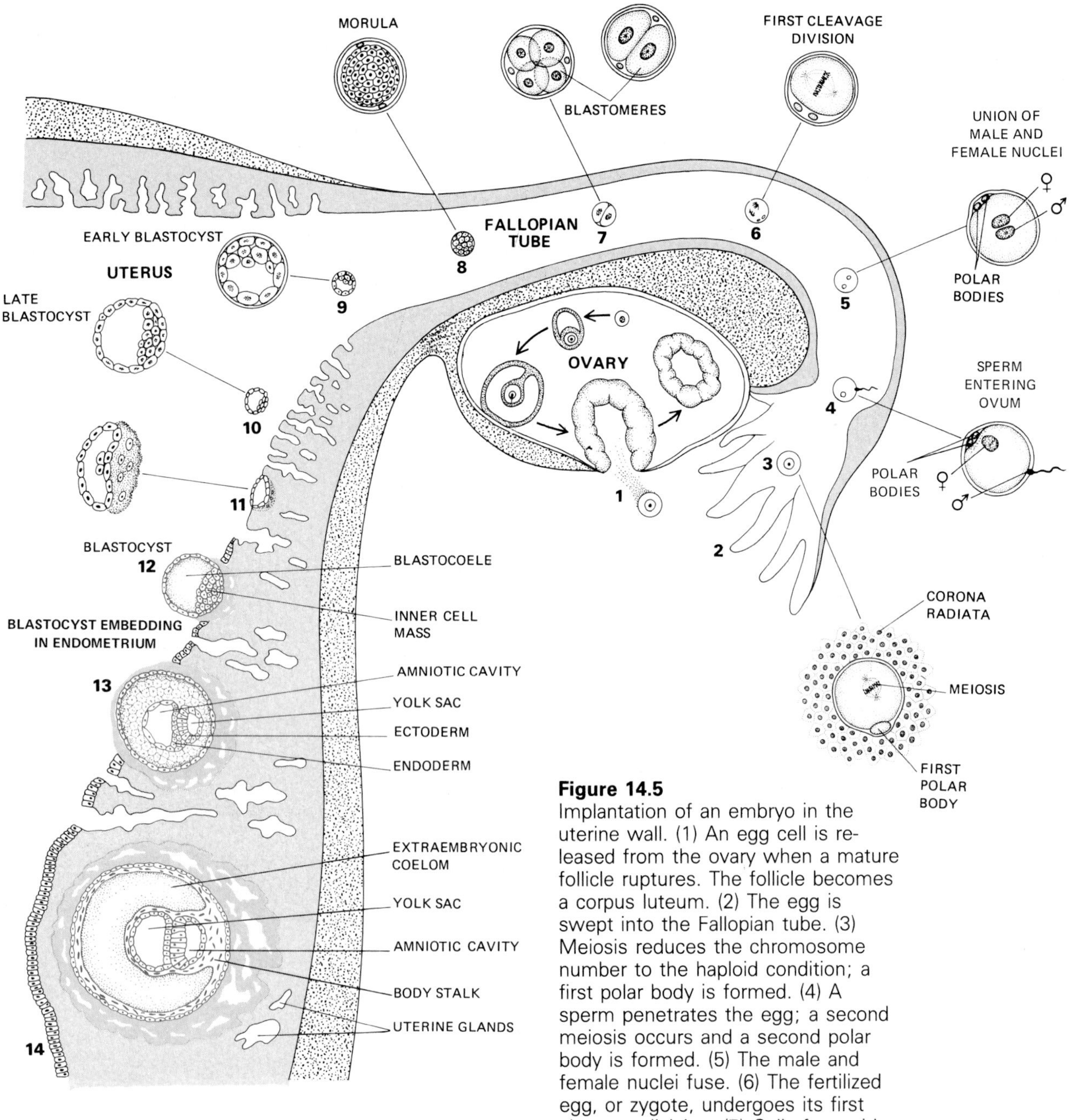

Figure 14.5
Implantation of an embryo in the uterine wall. (1) An egg cell is released from the ovary when a mature follicle ruptures. The follicle becomes a corpus luteum. (2) The egg is swept into the Fallopian tube. (3) Meiosis reduces the chromosome number to the haploid condition; a first polar body is formed. (4) A sperm penetrates the egg; a second meiosis occurs and a second polar body is formed. (5) The male and female nuclei fuse. (6) The fertilized egg, or zygote, undergoes its first cleavage division. (7) Cells formed by cleavage divisions continue to divide until a cluster of cells called a *morula* is formed (8). (9) An early blastocyst is formed. (10) About four days after fertilization, a late blastocyst is formed. (11) About a week after fertilization, the blastocyst begins the process of implantation into the uterine wall (12–14). Once implantation has been accomplished, the uterine wall grows over the embryo and starts contributing the outer portion of the *placenta;* the embryo itself contributes the inner part.

layer of the uterine wall, the *endometrium*. This process, known as **implantation** (Figure 14.5), is the point at which pregnancy begins.

The Fetal Membranes and the Placenta

As the trophoblast grows, it branches and extends into the tissues of the uterus, which increase along with it. The two kinds of tissue, the embryonic and maternal, grow until there is sufficient surface contact to ensure adequate passage of nourishment and oxygen from the mother, as well as removal of metabolic waste, including carbon dioxide, from the embryo. The embryo is soon completely covered by the envelope of *extraembryonic membranes,* one of which is the *amnion* filled with fluid (the *amniotic fluid*). The embryo floats in the amniotic fluid inside the amnion (Figure 14.6). At one side, the developed trophoblastic tissue becomes that part of the embryonic mass, the *chorion,* which absorbs nutrients from the mother. At the site of implantation, the chorion joins intimately and intricately with uterine tissue to develop into the **placenta.** A full-term human placenta is about 2.5 centimeters (1 in.) thick, and 22 centimeters (9 in.) across, and it weighs about 0.45 kilogram (1 lb).

The extensive mass of blood vessels and connective tissue that composes the chorion makes an inner lining of the placenta. These blood vessels are formed from the embryo and are connected to the embryo by way of the **umbilical cord.** The cord contains two arteries, which carry carbon dioxide and nitrogen wastes from the embryo to the placenta, and a vein, which carries oxygen and nutrients from the placenta to the embryo. A gelatinous cushion surrounds the vessels of the umbilical cord. This resilient pad, together with the pressure of blood and other liquids gushing through the cord, prevents the cord from twisting shut when the fetus becomes active enough to turn around in the womb.

There is normally no direct connection between the embryonic and the maternal tissue, at least no actual blood flow and no nerve connection (Figure 14.7). Sugars, water, oxygen, and hormones can

Figure 14.6
Two important membranes form around the developing embryo at the time of implantation. The outermost membrane is the *chorion,* which makes up most of the placenta. The chorionic membrane contains many fingerlike projections (*villi*), which allow the exchange of nutrients, gases, and metabolic wastes between mother and embryo. Because of the intricately folded system of chorionic villi, the total surface of the chorion is about 50 times the surface area of the skin of the newborn baby. The second membrane is the *amnion,* a fluid-filled sac that enables the developing embryo to float suspended in a relatively injury-free environment during pregnancy.

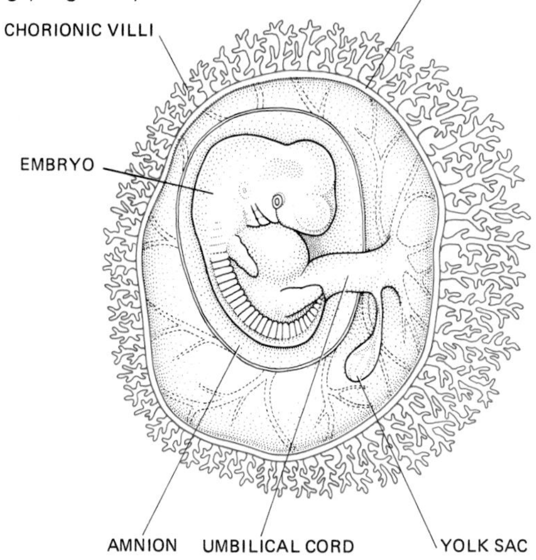

Figure 14.7
Circulation in the placenta. A section of uterine wall is shown at the place where the umbilical cord is connected. Two umbilical arteries in the cord carry blood, poor in oxygen and rich in carbon dioxide, from the fetus to the placenta (downward arrows). The connection between the fetal part of the placenta and the maternal part is close, with thousands of villi on the fetal part embedded in the maternal part, increasing the contact surface enormously. The fetal capillaries come close to the maternal capillaries and exchange their load of carbon dioxide for oxygen. In addition to exchange of respiratory gases, the placenta makes possible the elimination of metabolic wastes from the fetus (downward arrows) and allows the entry of foods, vitamins, and salts (upward arrow).

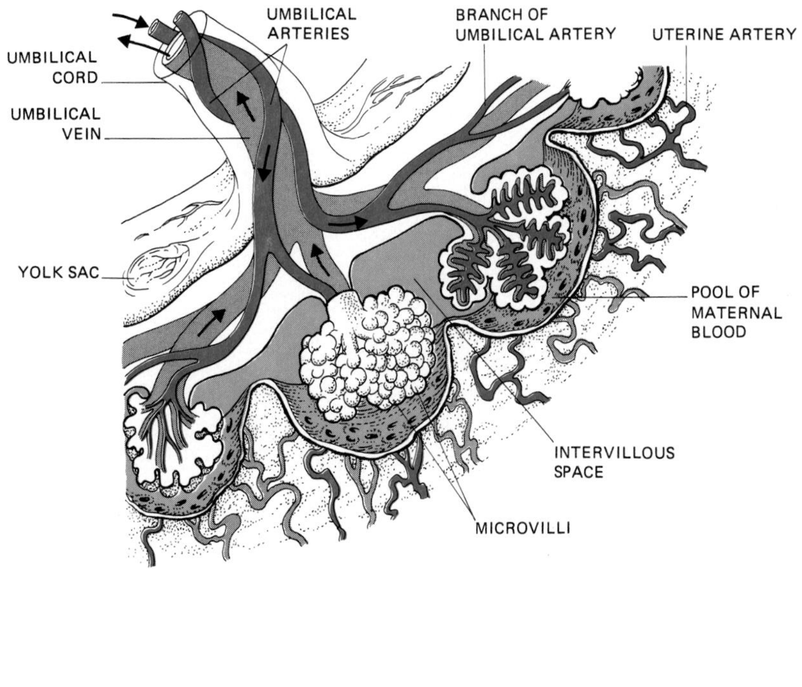

MULTIPLE BIRTHS IN HUMAN BEINGS

Multiple births are always an exception in human beings. Twins are born once in 86 births, triplets once in 7400 births, and quadruplets once in 635,000 births. The chances of quintuplets being born are astronomically low—about one in 55 million. In most cases, one or more of the children born as quintuplets or sextuplets do not survive beyond infancy. Women who have taken "fertility drugs" may have more multiple births than usual. Also, women who are older than 35 or who have had children previously may produce multiple offspring at an abnormally high rate. This increase

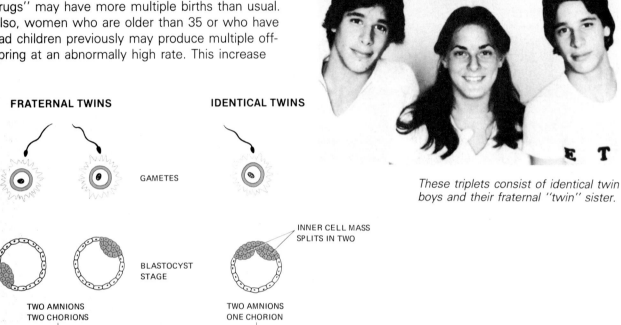

These triplets consist of identical twin boys and their fraternal "twin" sister.

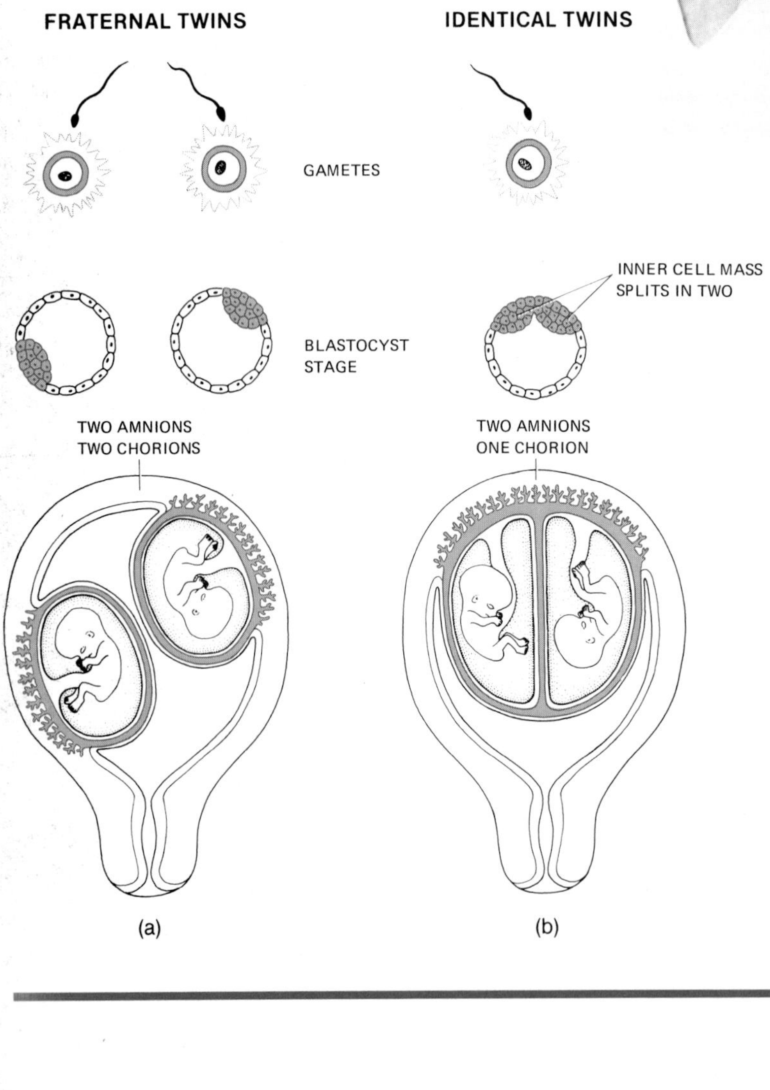

may be caused by irregular ovulation patterns in older women and the resultant release of more than one egg at a time. Apparently the tendency to release more than one mature egg at a time is an inherited trait. However, the cell separation that produces *identical* twins is not due to a hereditary factor, and its cause is not known.

Twins may be identical or fraternal. *Fraternal twins* (see drawing a) are formed when more than one egg is released from the ovary or ovaries and two eggs are fertilized. Each fraternal human twin has its own placenta, umbilical cord, chorionic sac, and amniotic sac. Fraternal twins are the commonest of multiple births; 70 percent of all twins are fraternal, and 30 percent are identical. Aside from being the same age, fraternal twins resemble one another no more than any other brothers or sisters. Drawing (b) shows the formation of *identical twins,* who are developed from one fertilized egg, and are therefore genetically identical. Because of their common inheritance, identical twins are always the same sex. Identical twins are created when the embryo from a single fertilized egg breaks in two at a very early stage of development. Note that identical twins have two amniotic sacs but share the same chorionic sac and umbilical cord.

The set of triplets shown in the photograph consists of identical twin boys and their fraternal "twin" sister, who bears no more than a family resemblance to the identical-looking boys. In this case, more than one egg was fertilized, creating fraternal male and female twin embryos. Then one of the fertilized eggs (the male) split apart into two identical embryos, forming the identical boys. Triplets may also be formed in other ways. One fertilized egg could divide more than once, forming four identical embryos. If one of them died at the embryo stage, only three identical triplets would be born. Or three separate eggs could be fertilized, producing three "fraternal" triplets. Numerous other combinations are possible.

cross the placental barrier, as can such poisons as lead, insecticides, and drugs. A newborn baby can show drug withdrawal symptoms if its mother used heroin during pregnancy. Nevertheless, the growing embryo is well insulated from most of the possibly harmful influences to which the mother is subjected. The placenta is later shed as the *afterbirth.*

It has long been thought that oxygen passed through the placenta from the maternal blood to the fetus by *passive diffusion* (the random movement of highly concentrated gas molecules to areas of low concentration). Actually, molecules of oxygen may be shuttled across the placenta at an unusually fast rate by means of an unidentified carrier molecule. It also seems that certain chemicals, such as anesthetic gases, insecticides, tranquilizers, and the carbon monoxide in cigarette smoke, may interfere with the efficiency of the carrier. Fetuses that are deprived of this facilitated oxygen transport may develop into underweight children with a susceptibility to developmental defects. Severe oxygen deprivation may also cause miscarriages. It has been known for some time that female anesthetists and women who smoke heavily during pregnancy have more miscarriages and underweight children than do other pregnant women.

TO SUM UP HUMAN DEVELOPMENT

1. The fertilized egg undergoes cell division as it travels through the oviduct toward the uterus.
2. After two or three days in the uterus, the embryo (now about 100 cells) and its covering are called a *blastocyst.*
3. The blastocyst becomes implanted in the nutrient-rich wall of the uterus.
4. Now floating in a protective amniotic sac, the embryo receives its nutrients and discharges its wastes through the *umbilical cord,* which is intimately joined with uterine tissue in the *placenta.*
5. The placenta is shed as the afterbirth.

LATE EMBRYONIC DEVELOPMENT AND BIRTH

The human ovum is one of the largest cells in the body, just big enough to be seen with the unaided eye. After a week's growth, the embryo is a formless group of cells, about half a millimeter in diameter.

AMNIOCENTESIS: "VISITING" THE BABY BEFORE BIRTH

Amniocentesis is the technique of obtaining cells from an unborn infant. The procedure is to locate the fetus by bouncing high-frequency sounds off it and recording the echoes. This is a safe method, unlike the use of X-rays, which are potentially dangerous to embryos. Then a hypodermic needle is inserted into the amnion, usually directly through the abdominal wall. The embryo is surrounded by the amnion, and the amnion is filled with fluid in which a number of loose cells float. Some of the cells are from the embryo itself and some from the amnion, but they are all derived from the original zygote and are therefore genetically alike. Some 20 milliliters (0.6 fl oz) of the amniotic fluid is drawn into a syringe and may be grown as a tissue culture. The cells in the fluid may be studied directly, but cultured cells can be made to yield more information. However, nearly a month is needed to establish a satisfactory culture, so the delay may be undesirable.

By amniocentesis the sex of the fetus can be determined, but outside of satisfying the possible curiosity of the parents, it has no medical importance. What is important is to find whether or not there are chromosome abnormalities, which might indicate possible deformities. The chromosome situation can be read with fair accuracy. For example, if three chromosomes of chromosome number 13 are present instead of the usual pair, the prediction can be made that the embryo, if allowed to come to term, will not survive infancy. The parents may decide that the best action is to terminate the pregnancy by having an abortion. On the other hand, if there is a history of abnormality in the family and consequent doubt concerning the unborn child, amniocentesis may provide reassurance by showing that the chromosomes are normal and that the child should be healthy.

Since the earlier an abortion is performed, the easier it is on the mother, amniocentesis is best done as soon as any question arises. The procedure of amniocentesis is reasonably safe for both embryo and mother, and it is advisable when there is any likelihood of serious fetal abnormality. This would be true if a mother is known to carry a chromosome aberration, or if she is over 40. Women over 40 have a disproportionately high number of babies with chromosome defects. It is judicious for older expectant mothers to have the cells of their unborn babies checked.

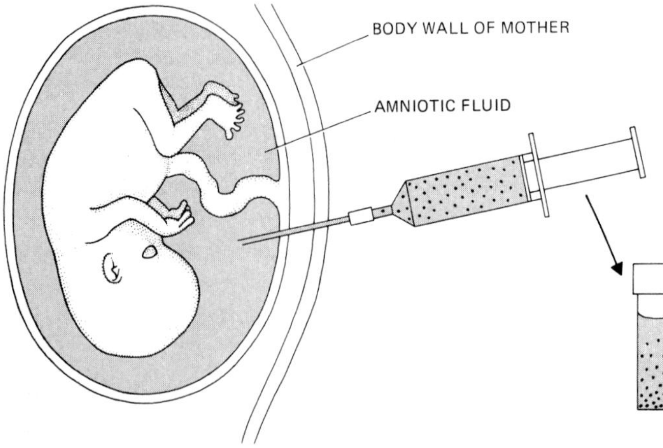

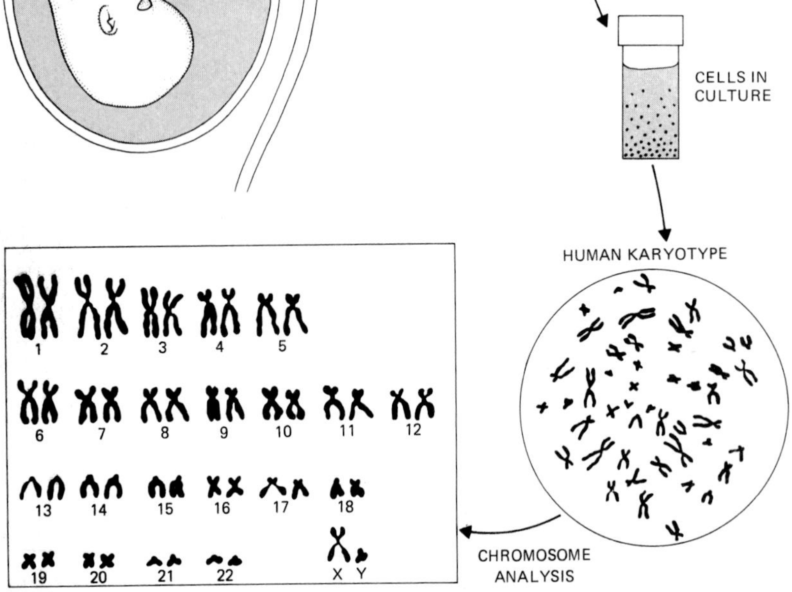

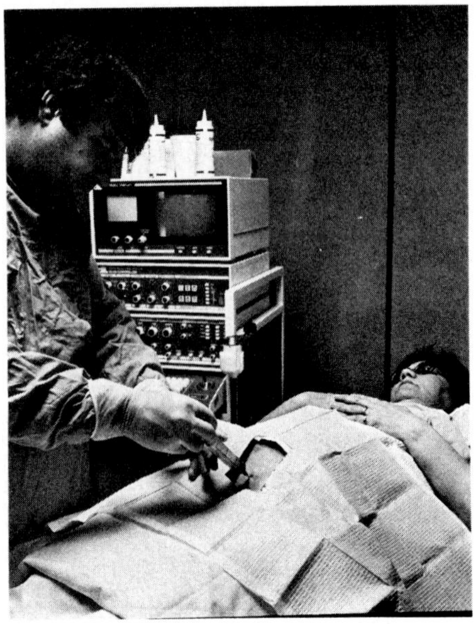

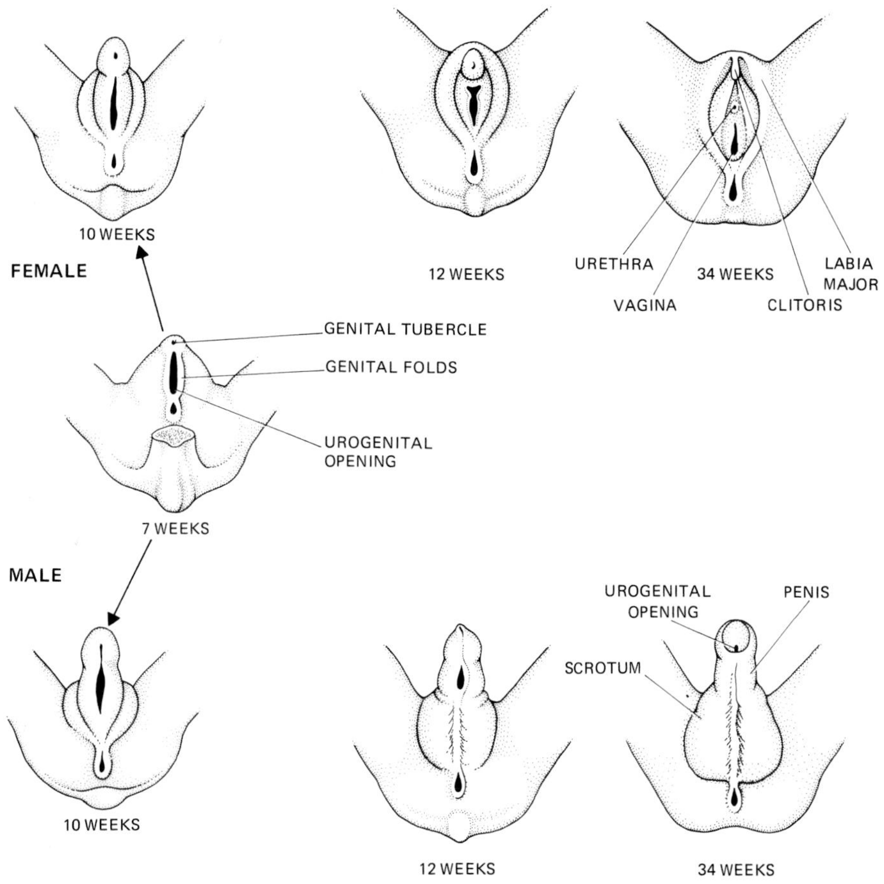

Figure 14.8
Embryonic human sex organs appear the same for males and females until the embryo is about seven weeks old. After 10 weeks, the female genitalia (top row) develop a distinct bud at the top that will become the clitoris. Swellings in the middle area will develop into the labia, and the vertical slit that is evident from the earliest stages will remain separated to become the openings for the urethra and vagina. The genitalia of the male fetus (bottom row) do not differ greatly from the female organs at 10 weeks. By the twelfth week it is evident that the bud is developing into the penis, and the swellings fuse together over the vertical slit to become the scrotum. The testicles will descend into the scrotum when the fetus is about seven months old. After 34 weeks, both the male and female sex organs look very much as they will at birth.

At the end of two weeks, it has begun tissue differentiation, but it is still mostly made up of extraembryonic cells essential for its nourishment. (See the full-color portfolio, How a Human Embryo Develops, following p. 302.) By three weeks, a 2.5-millimeter (0.1-in.) embryo has begun organ formation, including the primitive gut and nervous system. An incomplete but beating heart is present in a month, and the embryo reaches a diameter of about 5 millimeters. During the second month, organ building continues, with discernible features appearing, including limbs, facial contours, and sex organs (Figure 14.8). Once this stage is reached, the embryo is referred to as a **fetus.**

After three months in the uterus, a fetus is recognizably human. Although only about 50 to 60 millimeters (2–2.4 in.) long, it contains all the organ systems characteristic of the adult. The last six months of pregnancy are devoted to increase in size and maturation of the organs developed during the first three months. By the time the fetus is 100 millimeters (4 in.) long, it can move and be felt by the mother, but it is thin, wrinkled, hairy, and watery. As it ages, the fetus loses most of its hair, its bones begin to harden, it picks up fat, and it becomes mature enough to be born. It is said to have come *to term.*

The Process of Childbirth

At term, about 280 days after the last menstrual cycle before conception, the muscles of the uterus

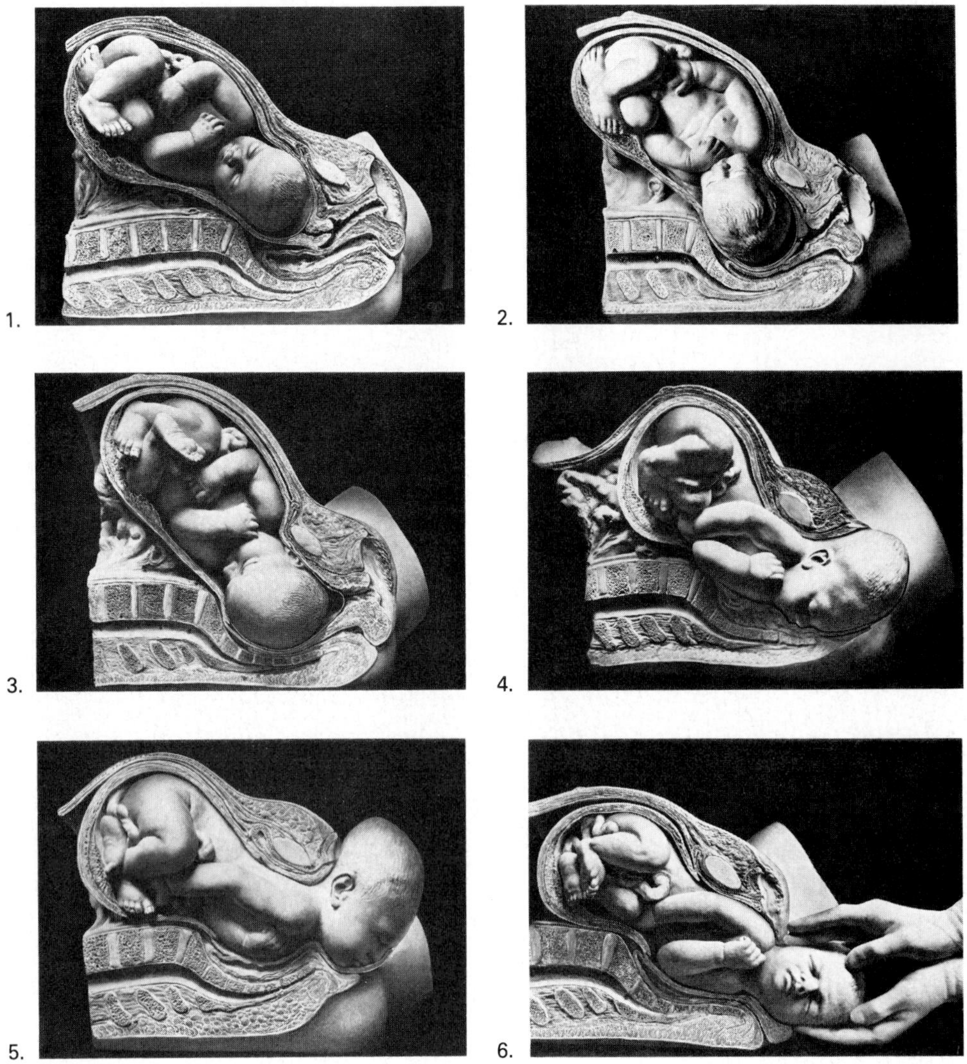

begin to contract rhythmically (Figure 14.9). The exact cause of these initial contractions is not known, but several different causes have been postulated, including hormones and the size of the fetus.

At any rate, the initial contractions stimulate the liberation of oxytocin from the pituitary gland, which further stimulates even more powerful uterine contractions. Waves of muscular contractions spread down the walls of the uterus, forcing the fetus toward the cervix. By now, the cervix is dilated close to its maximum diameter of about 10 centimeters (4 in.). Early contractions occur about every half hour and last about one minute. Later, the contractions become increasingly stronger, finally occurring every minute or so just prior to childbirth.

The amniotic sac may burst at any time during labor, or it may have to be ruptured by the attending physician. Either way, the ruptured sac releases the amniotic fluid. This phenomenon is known as "los-

Figure 14.9
The birth of a human baby. The internal events, as shown in a series of models. (Photographs of live births show only external events.) When the time has come for a baby to be born, strong, rhythmic contractions of abdominal and uterine muscles force the baby out through the birth canal. The amniotic sac breaks, releasing the amniotic fluid in which the baby has floated for most of its life, and the muscles of the cervix and of the vagina relax. Most babies are born head first. If the head (the baby's largest part) does not come first, the baby can sometimes be turned around before delivery. However, it may be delivered breech (hindquarters) first, although that makes for a difficult delivery.

HOW A HUMAN EMBRYO DEVELOPS

The First Week
Initial cell divisions take place. The embryo in the oviduct is in a four-cell stage; by the time it reaches the uterus four or five days later, it has grown to about 32 cells. Blastocyst and trophoblast form, and implantation in the uterine wall takes place.

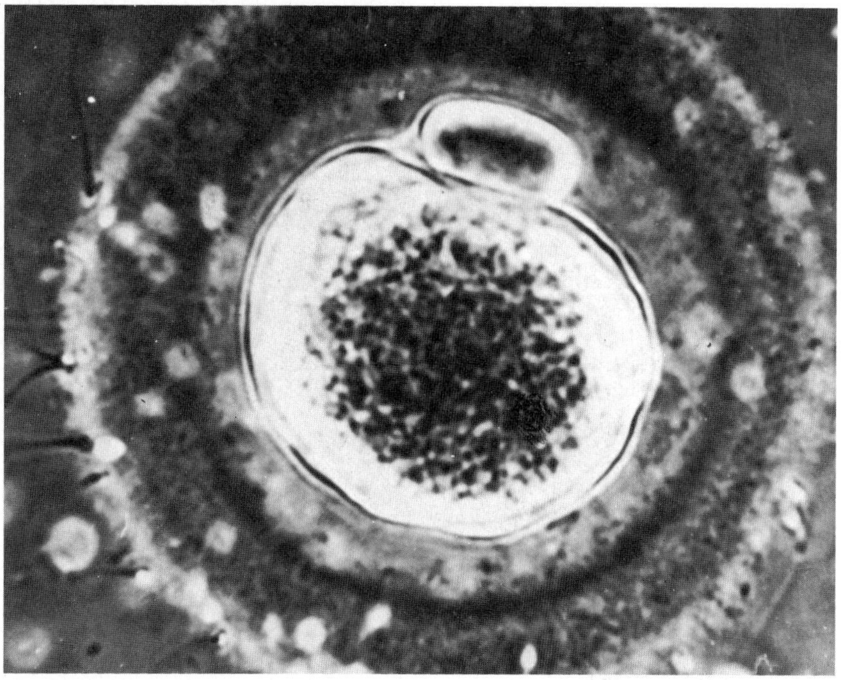

Sperm and egg fuse in fertilization.

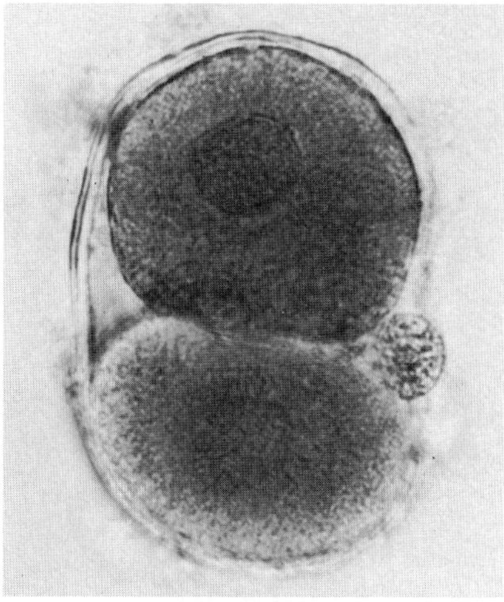

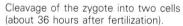

Cleavage of the zygote into two cells (about 36 hours after fertilization).

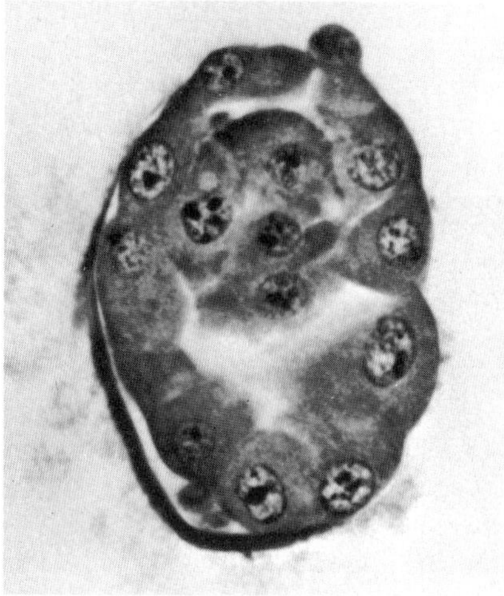

Cleavage and morula formation (about the third day after fertilization).

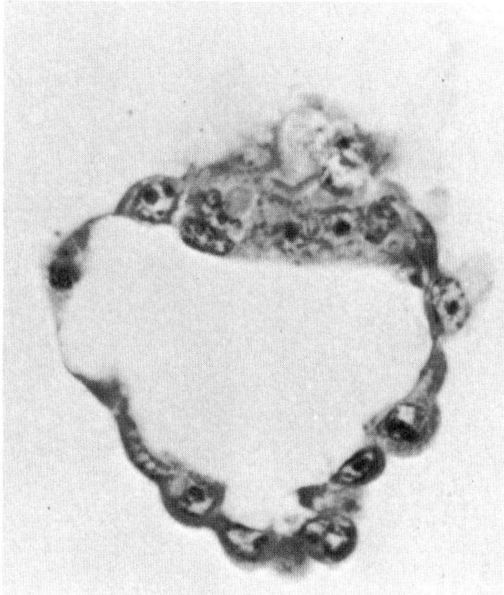

The blastocyst (about four days after fertilization).

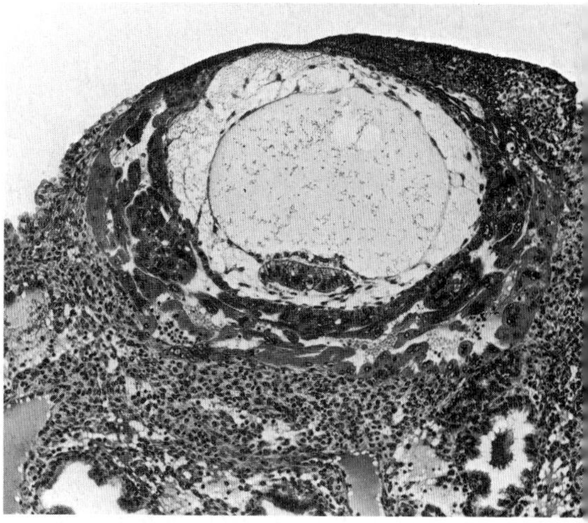

Implantation of the blastocyst (about 6–8 days after fertilization).

Second Week
The extra-embryonic membranes, the chorion and the amnion, begin to develop; blood vessels grow within developing chorionic villi. The placenta is formed from villi and uterine tissues. The connection between placenta and embryo is the umbilical cord. Formed at this time are the three embryonic tissues from which all other mature tissues derive. (about 1.5 mm; all measurements from crown to rump)

Third Week
Body, head, and tail are distinguishable. The neural tube begins to form and will eventually develop into the brain, the spinal cord, and the rest of the nervous system. Primitive blood cells and vessels are present. (about 2.5 mm)

Fourth Week
The U-shaped heart pumps blood through a simple system of vessels. Placenta is functioning, and definitive villi appear. Limb buds are visible. Umbilical cord begins to develop. Body is tubular and C-shaped. Intestine is a simple tube. Liver, gall bladder, pancreas, and lungs begin to form. Eyes, nose, and brain begin to form. (about 5.0 mm)

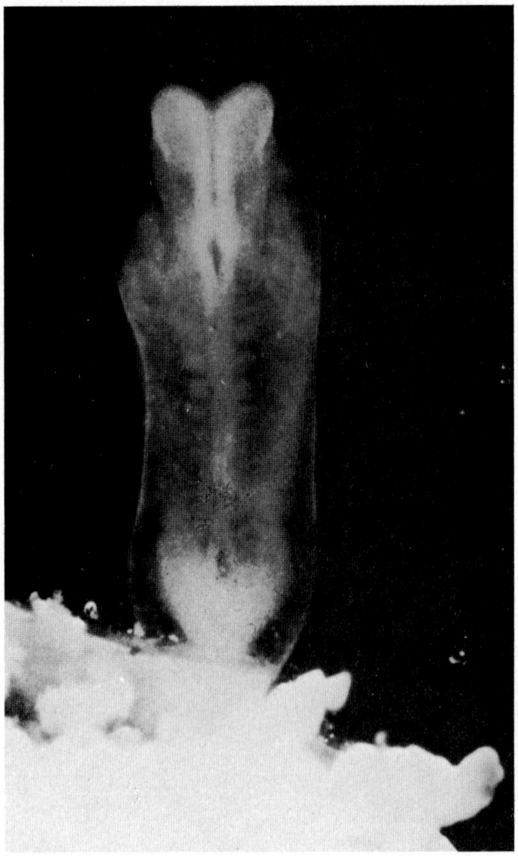

A three-layered embryo after about three weeks.

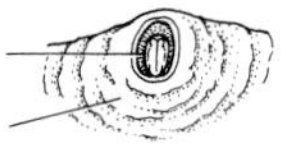

Actual size

THE DEVELOPMENT OF A HUMAN HAND

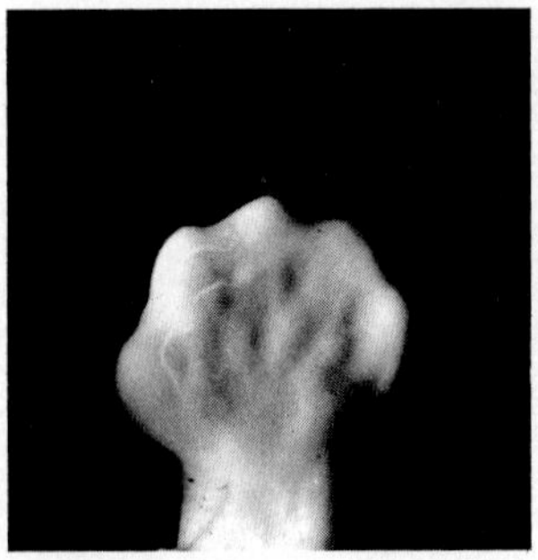

At about five weeks the forearm is shorter than the hand, and finger ridges begin to form.

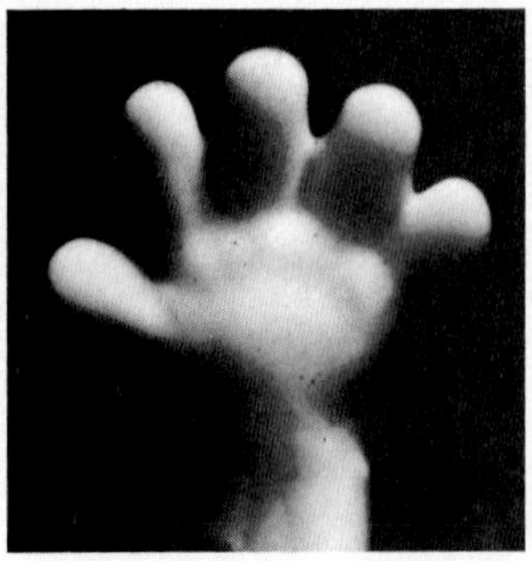

At six weeks the hand has begun to develop clearly delineated fingers.

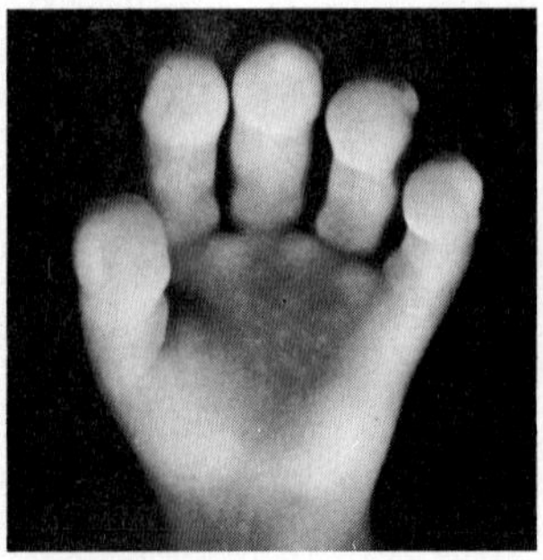

At 11 weeks fingertip buds are developed, and the muscles are active.

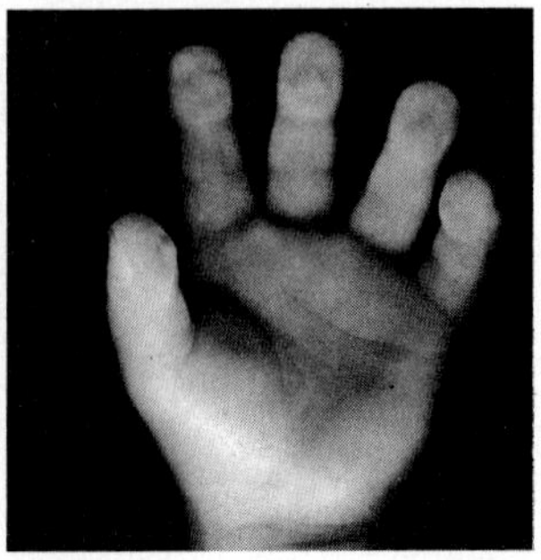

At five months the hand is practically complete, and thumb sucking may start.

Fifth Week

Primitive nostrils are present. Tail is prominent. Heart, liver, and primitive kidney begin to form. Jaws are outlined. Intestine elongates into a loop. Genital ridge bulges. Primitive blood vessels extend into head and limbs. Spleen is indicated. Spinal nerves are formed, and cranial nerves are developing. Stomach begins to form. Premuscle masses appear in head, trunk, limbs. Epidermis is gaining a second layer. Cerebral hemispheres are bulging. It is at this stage that drugs taken by the mother may affect the embryo; also such diseases as German measles may be transmitted from the mother to the embryo. (about 8.0 mm)

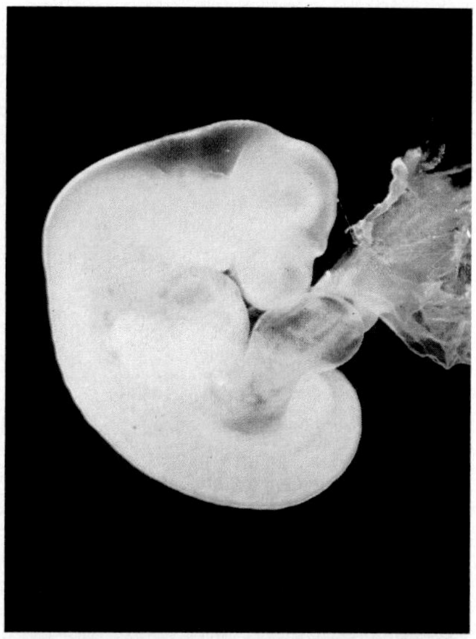

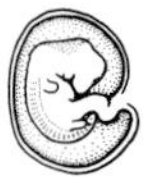

Actual size

The embryo at about five weeks.

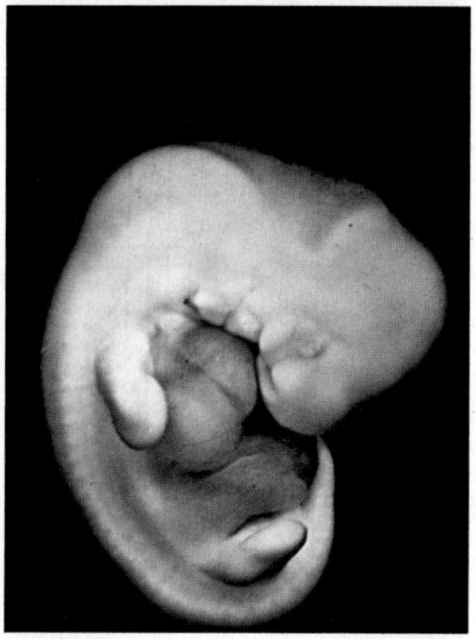

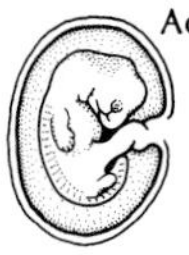

Actual size

The embryo at the end of five weeks.

Sixth Week

Upper jaw components are prominent but separate; lower jaw halves are fused. Head becomes dominant in size. External ear appears. Arms and legs are recognizable. Simple nerve reflexes are established. Heart and lungs acquire definitive shapes. Eyes continue to develop and become colored. (about 12.0 mm; $\frac{1}{2}$ in.)

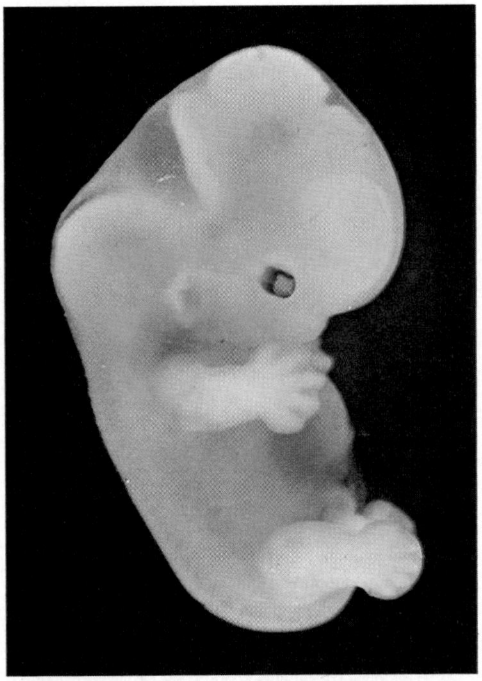

The embryo at about six weeks.

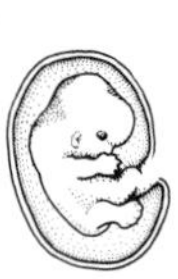

Actual size

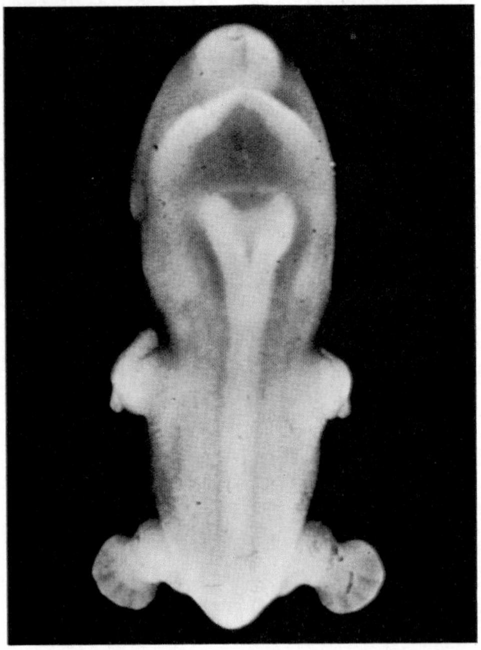

A rear view of the embryo at about six weeks.

Seventh Week

Face and neck begin to form. Fingers and toes are differentiated. Back straightens. Tail is regressing. Three segments of upper limbs are evident. Jaws are formed and begin to ossify. Stomach is attaining final shape and position. Muscles are differentiating rapidly throughout the body and assuming final shapes and relations. Brain is becoming large. Eyelids are forming. (about 17.0 mm; $\frac{3}{4}$ in.)

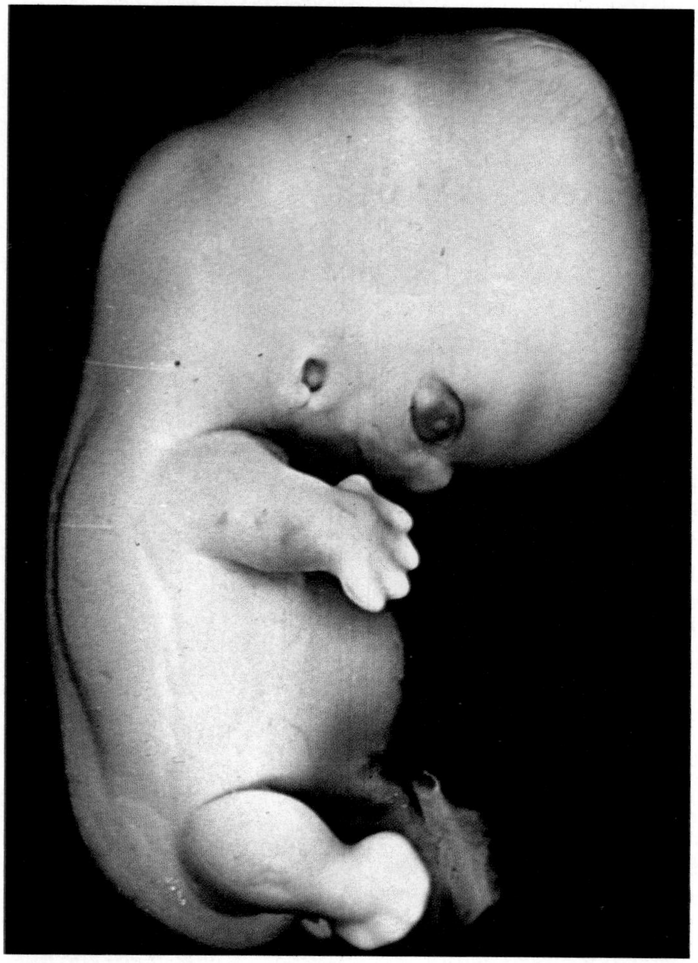

The seventh week.

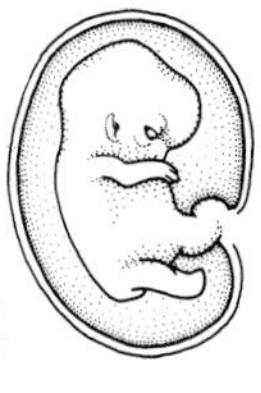

Actual size

Eighth to Ninth Week
The embryo is now called a *fetus*. Nose is flat; eyes are far apart. Fingers and toes are well formed. Growth of intestine makes body evenly round. Head is elevating. Tongue muscles are well differentiated; earliest taste buds form. Ear canals are distinguishable. Small intestine is coiling within umbilical cord; intestinal villi are developing. Liver is relatively large. Testes or ovaries are distinguishable as such. Main blood vessels assume final plan. First indications of skeletal bone formation. Muscles of trunk, limbs, and head are well represented, and the fetus is capable of some movement. All five major subdivisions of brain are formed, but brain lacks convolutions that are characteristic of later stages. External, middle, and internal ear are assuming final form. (about 23.0 mm; 1 in.)

Tenth to Eleventh Week
Head is erect as in later life. Limbs are nicely modeled. Fingernail and toenail folds are indicated. Lips are separate from jaws. Intestines withdraw from umbilical cord and assume characteristic positions. Anal canal is formed. Nasal passages are partitioned. Kidneys begin to function. Bladder expands as sac. Vaginal sac is forming. Rudimentary sex ducts form. Early lymph glands appear. Enucleated red cells predominate in blood. Earliest hair follicles begin developing on face. Spinal cord attains definitive internal structure. Eyelids are fused. Fetal pulse is detectable. Tear glands are budding.(about 40.0 mm; $1\frac{1}{2}$ in.)

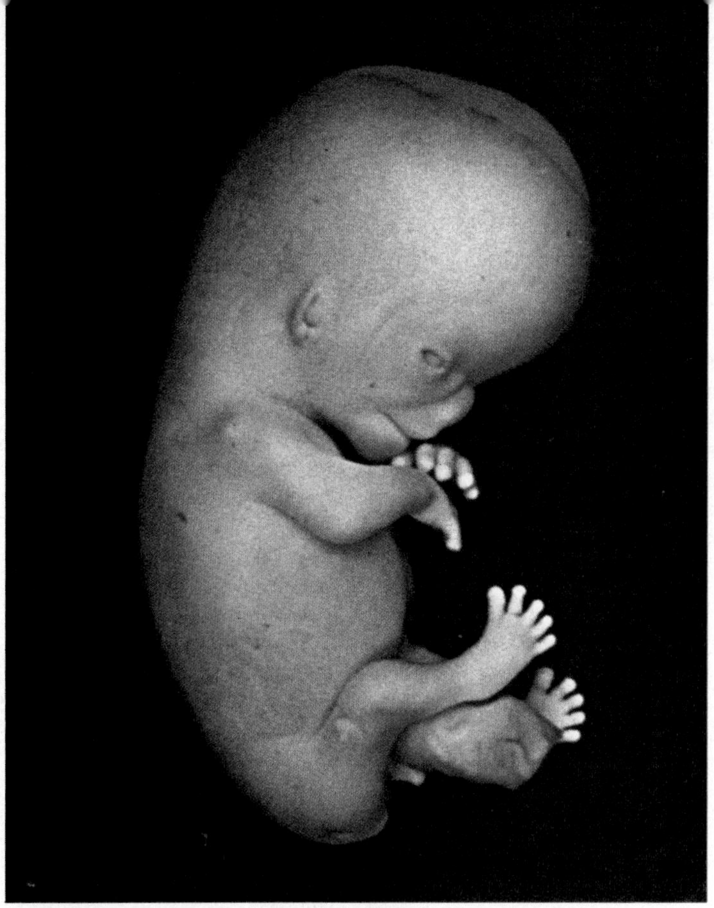

The fetus at about eight weeks.

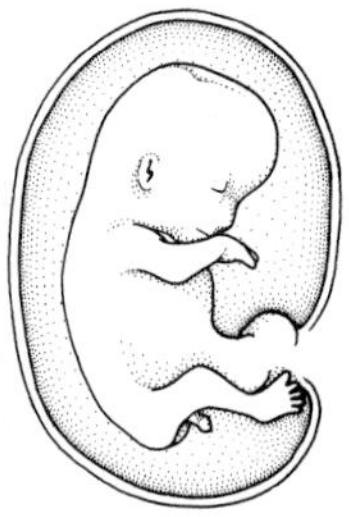

Actual size

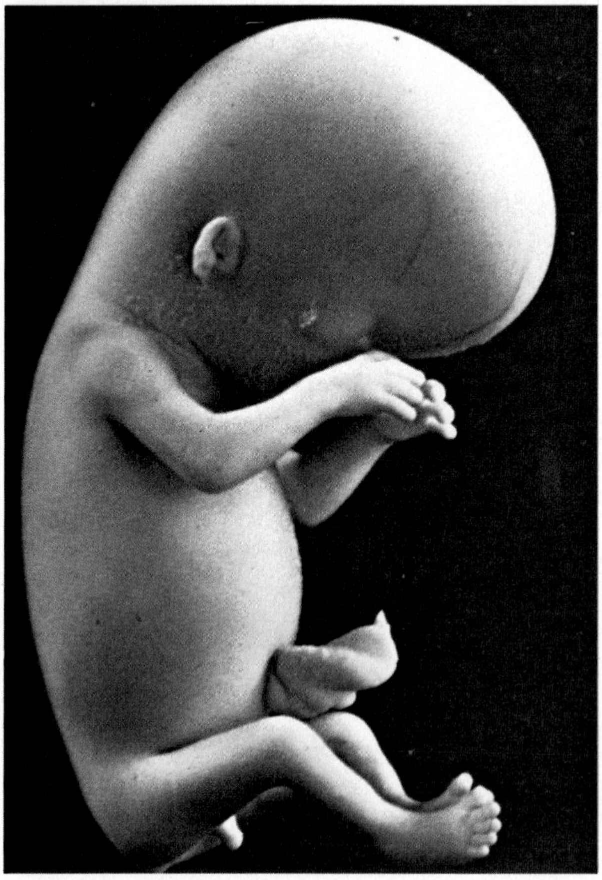

The fetus at about nine weeks.

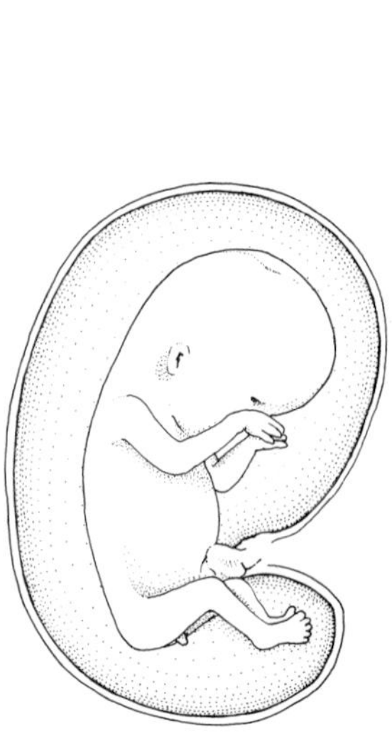

Actual size

Twelfth to Fifteenth Week
Head is still dominant. Nose gains bridge. External genitalia attain distinctive features. Tooth buds and bones form. Cheeks are represented. Nasal glands form. Lungs acquire definitive shape. Prostate and seminal vesicles appear. Blood formation begins in bone marrow. Some bones are well outlined. Epidermis is three-layered. Brain attains its general structural features. Characteristic organization of eye is attained. Mother feels uterus enlarging. All vital organs are formed. (about 56.0 mm; $2\frac{1}{4}$ in.)

Sixteenth to Nineteenth Week
Face looks "human." Hair appears on head. Muscles become spontaneously active. Body grows faster than head. Hard and soft palates are differentiating. Gastric and intestinal glands begin to develop. Kidneys attain typical shape and plan. Testes are in position for later descent into scrotum. Uterus and vagina are recognizable as such. Blood formation is active in spleen. Most bones are distinctly indicated throughout body. Stretching movements, not felt by mother. Epidermis begins adding other layers to form skin. Body hair starts developing. Sweat glands appear. Eyes, ears, and nose approach appearance more characteristic of adult human form. General sense organs begin to differentiate. (about 112.0 mm; $4\frac{1}{2}$ in.)

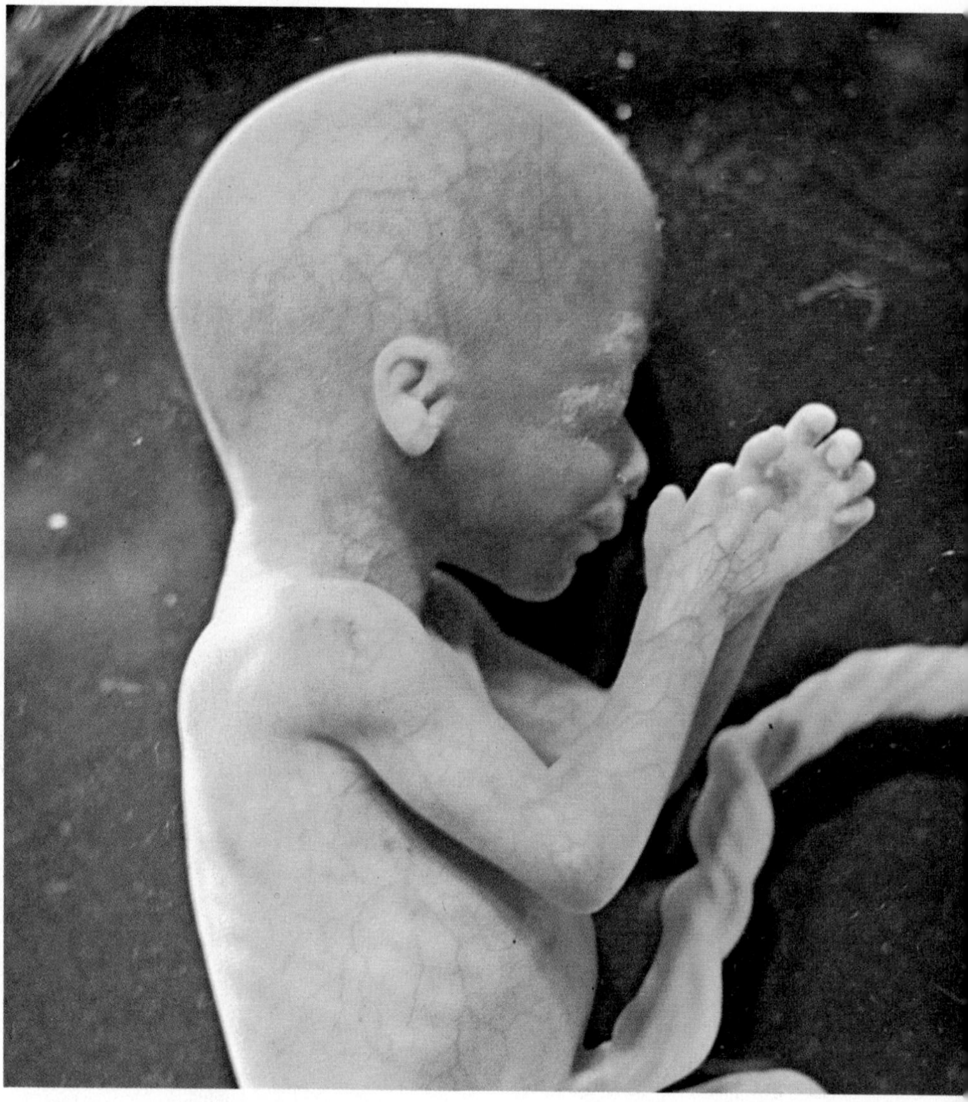
The fetus at about seven lunar months.

Fifth Lunar Month *
Body is covered by downy hair called lanugo. Nose bones begin to harden. Vaginal passageway begins to develop. Until birth, blood formation continues to increase in bone marrow. Fetal heart sounds are audible with stethoscope. Heart beats at twice adult rate. Lungs are formed but do not function. Gripping reflex of the hand begins to develop. Kicking movements and hiccuping may be felt by the mother. (about 160.0 mm; $6\frac{1}{2}$ in.)

Sixth Lunar Month
Body is lean but better proportioned. Colon becomes recognizable. Nostrils reopen. Hairs emerge. Cerebral cortex of the brain is layered, as is more typical of adult. Internal organs occupy normal positions. Thumb sucking may begin.

Seventh Lunar Month
Fetus is lean, wrinkled, and red. Eyelids reopen; eyebrows and eyelashes form. Nervous system is developed sufficiently so that fetus practices controlled breathing and swallowing movements. Uterine glands appear. Scrotal sacs develop, and testes descend, from seventh to ninth month. Lanugo hair is prominent. Brain enlarges, cerebral fissures and convolutions appearing rapidly. Retinal layers of the eye are completed, and light can be perceived. If delivered at this stage, the baby has at least a 10 percent chance of survival.

Eighth Lunar Month
Testes settle into scrotum. Fat is collecting, wrinkles are smoothing, and body is rounding from eighth to tenth month. Taste sense is present. Weight increase slows down. If delivered at this stage, the baby has about a 70 percent chance of survival.

Ninth Lunar Month
Nails reach fingertips. Fetus still cannot hear.

* Because women's reproductive cycles are closer to the lunar month of 28 days than to the calendar month, fetal development is conventionally described in terms of lunar rather than calendar months.

Tenth Lunar Month
Pulmonary branching is only two-thirds complete; lungs do not function until birth. Some fetal blood passages discontinue. Lanugo hair is shed. Digestive tract is immature. Lack of space in mother's uterus causes decrease in fetal activity. Fetus turns to head-down position. Maternal blood supplies protective antibodies. Placenta regresses, placental blood vessels degenerate. (about 350.0 mm; 14 in.) (Measurements in the final months can vary greatly.)

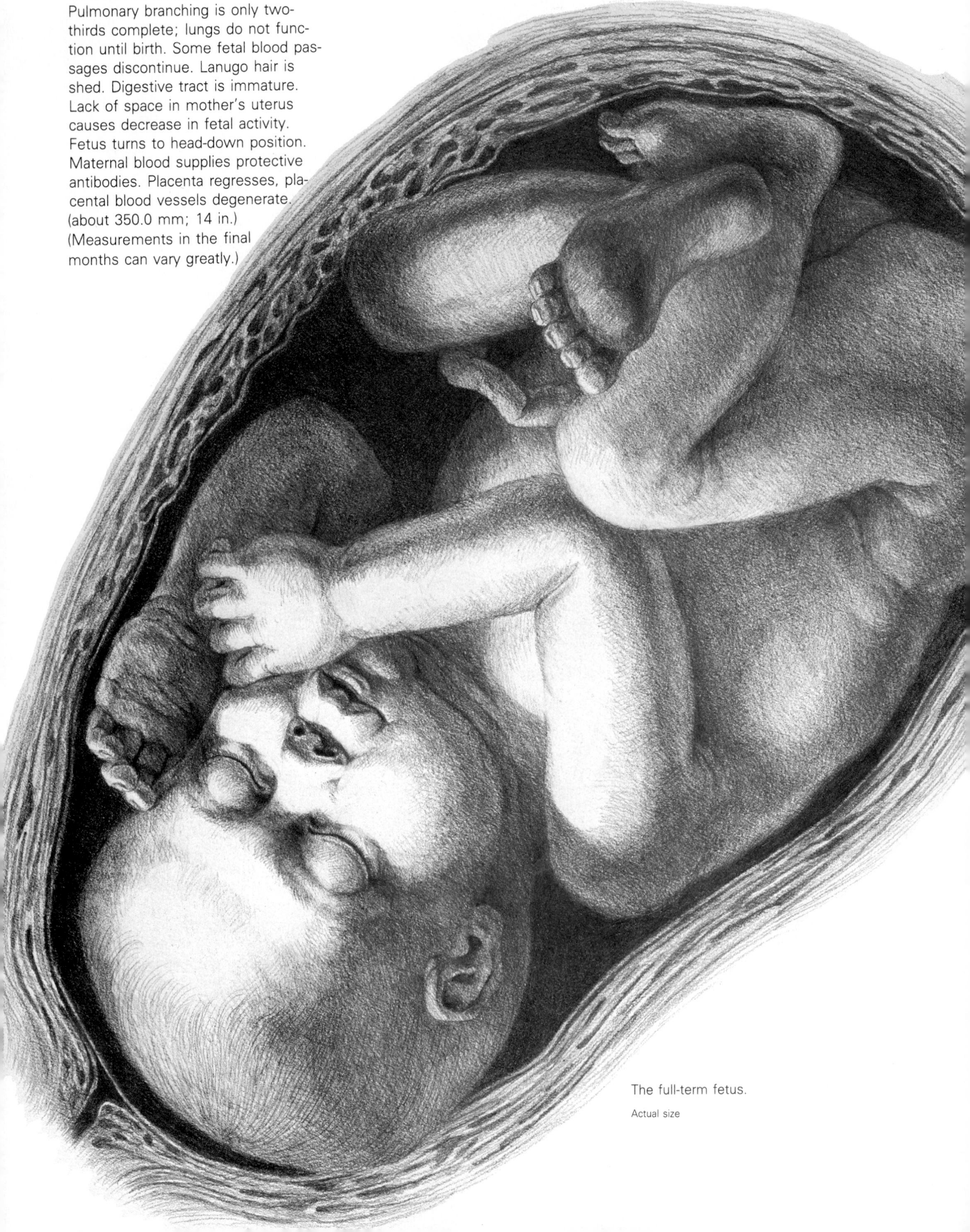

The full-term fetus.

Actual size

ing the waters," and if it occurs very early, it may signal the onset of labor. Another possible indication that labor has begun is the release of the cervical plug of mucus from the vagina. Ordinarily, either the bursting of the amniotic sac or the release of the cervical plug will occur early enough to provide ample time to prepare for childbirth, but great variations exist among pregnant women.

During pregnancy, the muscle cells of the uterus grow to as much as 40 times their former size, and the number of muscle cells increases also, transforming the uterus into an enormously powerful muscle. The sturdy walls of the uterus harbor and nourish the fetus during pregnancy, and finally, through the muscular contractions during labor, the uterus forcefully expels the fetus outward through the vaginal canal. (The fetus does not assist its own birth passage at all.) Normally babies are born head first, with the soft bones allowing some temporary deformation as the head is squeezed through the birth canal.

TABLE 14.1
Gestation Periods and Litter Sizes in Mammals

ANIMAL	GESTATION PERIOD IN DAYS (MEAN)	LITTER SIZE (MEAN)	LITTER SIZE (RANGE)
Opossum*	13	8	4–13
Mouse	20	4	1–9
Rat	22	8	6–12
Rabbit	31	8	1–13
Kangaroo*	33	1	1
Dog	60	6	1–12
Cat	63	4	2–6
Tiger	100	3	2–4
Pig	114	8	2–14
Lion	115	3	1–6
Sheep	148	2	1–3
Chimpanzee	227	1	1–2
Dolphin	255	1	1
Gorilla	260	1	1–2
Human being	280	1	1–5
Cow	280	1	1–2
Horse	340	1	1–2
Humpback whale	360	1	1
Elephant	625	1	1

*Marsupial animals are born at a very immature stage. Immediately after birth each offspring makes its way to the mother's pouch, where it finds and clings to a nourishing nipple for about three months. During this nursing period the newborn's development continues to a more mature stage.

After it is born, the baby is still attached to the placenta by the umbilical cord. Immediately after the baby is expelled from the uterus, the attending physician clamps off the umbilical cord and cuts it. (In other mammals the mother severs this cord with her teeth.) Now the newborn is on its own, to some degree.

A few minutes after the baby is born, further uterine contractions expel the placenta, now called the *afterbirth.* These final uterine contractions also help to close off the blood vessels that were ruptured when the placenta was torn away from the uterine wall. Then the uterus gradually returns to its original size.

The three stages of labor may be summarized as follows:

1. The first stage begins with the initial contraction of the uterus and ends when the cervix is dilated to its maximum.
2. The second stage is the delivery of the child.
3. The third stage is the delivery of the placenta (afterbirth).

The Breathing Process of the Newborn

Naturally, a fetus does not breathe air during its intrauterine life; it gets its oxygen from the placenta and delivers its carbon dioxide there. At birth, the baby must adapt quickly to life in air, and in order to do that, it must make several rapid physiological and anatomical adjustments. It must shift from a fetal-type circulation, which depends on the placenta, to an essentially adult circulation, which depends on the lungs for gas exchange. Before birth, the fetal heart has an opening (the "oval window") between the atria that allows blood from the venous and arterial systems to mix. Also, the *ductus arteriosus* carries blood from the pulmonary artery to the aorta (Chapter 24), thus bypassing the lungs (Figure 14.10). Both the atrial opening and the pulmonary bypass close at birth, allowing the blood to pass through the pulmonary artery to the newly functioning lungs. If either the "oval window" or ductus arteriosus does not close off, the oxygen content of the blood will be low, and the baby's skin will appear slightly blue. A newborn infant with this condition is therefore called a "blue baby." Usually a defective ductus arteriosus must be corrected surgically, but the surgery usually is not performed until a few days after birth.

The start of actual breathing can be helped if it does not begin immediately at birth. Fluid can be cleaned from the nose and mouth of the *neonate*

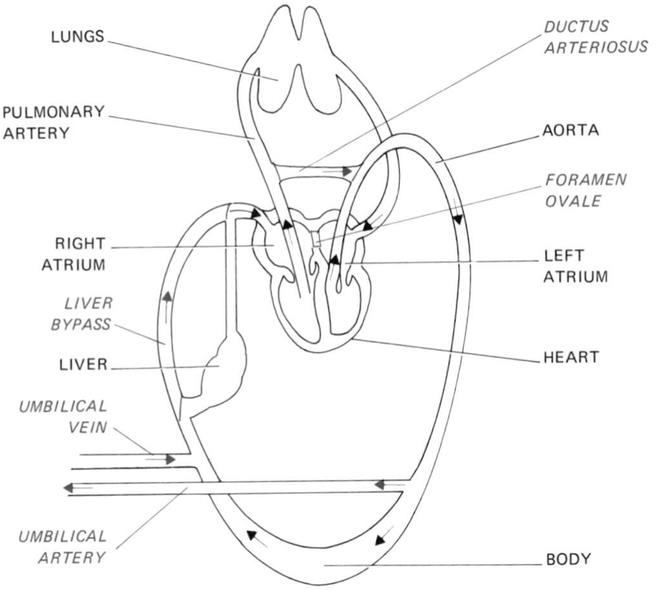

Figure 14.10
Human fetal circulation. The lungs of an unborn baby, being airless, cannot function in gas exchange. The respiratory exchange takes place in the placenta instead. Since the lungs of the fetus are useless until it is born, they receive only enough blood to allow them to grow, and most of the fetal circulation is detoured. One detour is the *ductus arteriosus*, a connecting vessel between the pulmonary artery and the aorta. A second is the *foramen ovale* ("oval window"), an opening between the right and left atria of the heart, which allows some of the blood from the body to bypass the lung entirely. There is also a vessel that temporarily bypasses the liver. In a fetus, the liver is not completely functional, and the fetus depends on the maternal liver. The colored arrows and colored labels in the figure show which parts of the circulatory system are strictly fetal; the others will continue to function after birth as they did before. At or soon after birth, the fetal detours are closed: The umbilical vessels are cut, and the other bypasses are blocked by growth processes.

(newborn), the baby can be held upside down to help drain the breathing passage, and if necessary, it can be given a smack on its bottom to shock it into crying, its first gasp of air. More commonly these days, breathing is stimulated by a small whiff of carbon dioxide.

Lactation

Lactation includes both milk secretion and milk removal from the breasts. Throughout pregnancy the hormone *prolactin* stimulates milk production in the mammary glands, but it is not until after the baby is born that milk is actually secreted from the mother's breasts. It appears that placental secretions of estrogen and progesterone inhibit the release of milk from the breasts, but after the placenta has left the uterus there remains no deterrent to milk production, and nursing can begin.

Although several other hormones, including oxytocin, play a part in milk production and secretion, the most important agent is prolactin. It is not known what causes the secretion of prolactin during pregnancy, but it is clear that the continued production of prolactin during breast-feeding is stimulated by the infant's sucking action on the nipple; breasts that are not emptied soon cease to produce milk. It also appears that nursing of the child aids in the return of the uterus to normal size and shape by the continued production of oxytocin, which is also responsible for the release of milk from the breast. Because of this stimulation by the suckling infant, lactation may be prolonged for two years or more if nursing continues.

On the average, nursing mothers produce about one liter of milk each day. Multiple births result in larger secretions, usually about one liter for each baby engaged in suckling.

Although menstruation may not take place while the mother is nursing her baby, ovulation may occur anyway. Despite tales to the contrary, a woman may indeed become pregnant during the lactation period. Nursing is certainly not an effective method of birth control.

SUMMARY

1. Each type of cell is a specialist with a specific job to do. But it is not yet totally clear how such specialization occurs.
2. Not until the late nineteenth century was it realized that both parents contribute genetic material to the zygote through egg and sperm.

3. *Mitosis* is the reproduction of nuclei and the equal distribution of DNA to each. *Meiosis* is the pair of nuclear and cellular divisions that reduce the chromosome number from diploid to haploid. Haploid cells develop into eggs or sperm.

4. When a sperm meets an egg, it digests its way into the egg cytoplasm and sheds its tail, and its nucleus fuses with the egg nucleus. The egg quickly surrounds itself with the *fertilization membrane*. The resultant diploid cell is the *zygote*.

5. The series of cell divisions initiated after fertilization is *cleavage*.

6. The earliest stages of embryo development are different in mammals from those in animals such as frogs with small, free-floating eggs.

7. In frog development, once a hollow ball of microscopically small cells, a *blastula*, is formed, a new growth phase ensues. The cells begin to move about on the surface of the blastula, next entering a reorganizing phase of *gastrulation*. The shifting of cells includes an invagination (bending in) at one place on the sphere of cells. The region where invagination occurs is the *blastopore*.

8. Once gastrulation has occurred, the embryo has an outer coating, an *ectoderm*, and an inner lining, an *endoderm*. The *mesoderm* fills the part between the two original layers of ectoderm and endoderm, and it eventually builds the bulk of the embryo.

9. The ectoderm is destined to develop into outer skin and nervous tissue, the endoderm into the lining of some internal organs, and the mesoderm into blood, bone, various other connective tissues, and muscle.

10. The human embryo begins to develop in the oviduct, but it passes down toward the uterus, cleaving as it goes, until after six to eight days it is ready for implantation in the uterine wall. By then the embryo is a ball of some 100 or more cells, which will become the embryo proper, plus a covering mass of cells, the *trophoblast*, which will develop into a system of membranes outside the embryo. The whole mass, including trophoblast and embryo, is a *blastocyst*.

11. The *placenta* consists of the absorbing part of the embryonic mass (the chorion) and uterine tissue; it is the embryo's only contact with the outside world. The *umbilical cord* carries wastes from the embryo to the placenta, as well as oxygen and nutrients from the placenta to the embryo.

12. As an embryo develops, it undergoes *differentiation* of its parts. We know that cellular specialization and growth can result in functional organs, even though the exact regulating forces are not understood.

13. The series of events that lead to the formation of a vertebrate eye illustrates the principle of *induction*, in which one tissue is caused or *induced* to change its developmental pattern as a result of the influence of a different tissue.

14. After two months of development, the human baby is called a *fetus*. After three months in the uterus, a fetus is recognizably human. The last six months of pregnancy are devoted to increase in size and maturation of the organs developed earlier. When the fetus is mature enough to be born, it has come *to term*.

15. At term, about 280 days after the last menstrual cycle, the muscles of the uterus begin to contract rhythmically, and the "labor" preceding actual childbirth begins. The exact cause of these initial uterine contractions is not known.

16. When a baby is born, it must make changes in its blood circulation to enable the baby to adapt to breathing air.

17. *Lactation* includes both milk secretion and milk removal from the breasts. Several hormones, including *prolactin*, play a part in milk production and secretion.

ASK YOURSELF

1. What is cleavage? What important processes are set up by gastrulation?
2. What is a blastocyst?
3. What is the event that signals the actual start of pregnancy?
4. What is the function of the placenta?
5. What is the difference between an embryo and a fetus?
6. Which hormone stimulates uterine contractions during childbirth?
7. What is the significance of the ductus arteriosus?
8. Prolactin and oxytocin are both involved with lactation. How do their roles differ?

PART FIVE
THE LIFE OF PLANTS

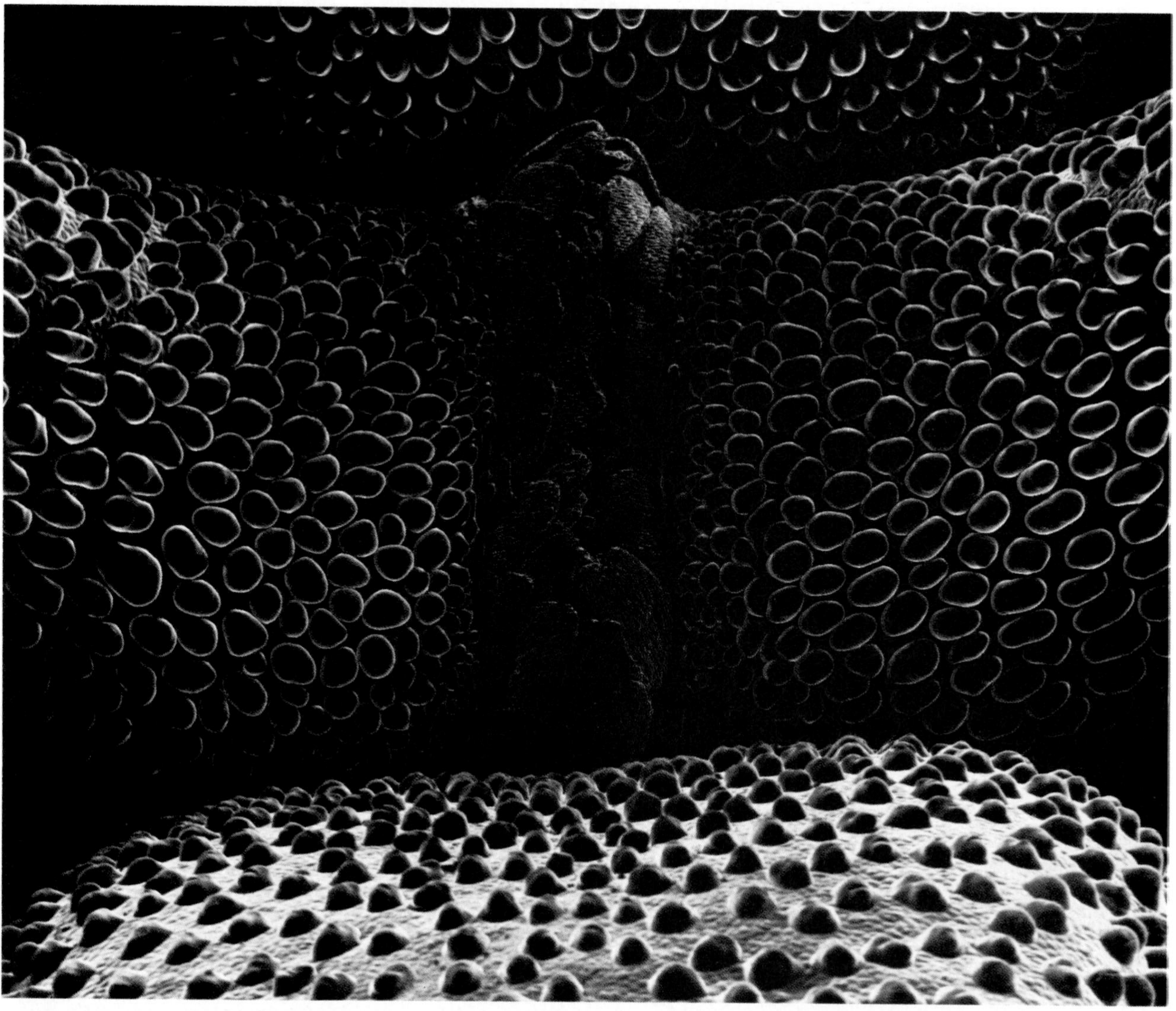

Scanning electron micrograph of a common house plant known as baby's-tears. A tiny bud is surrounded by larger, mature leaves. × 30

15
Reproduction in Plants

SOME KEY POINTS

1. Asexual reproduction is used by some plants. It is slower than sexual reproduction in distributing variations, but it does not need sexually compatible partners, it can be fast, and it usually produces genetically identical offspring. The simplest example of asexual reproduction is fission.
2. Spores are nonsexual reproductive cells.
3. The essential feature of sexual reproduction is the union of two nuclei, with the formation of a completely new nucleus. The result is usually a new set of gene combinations.
4. Most sexually reproducing organisms have male and female gametes—sperm and egg—but some primitive types have gametes that are alike.
5. Algae frequently reproduce sexually, but the commonest algal reproduction is asexual.
6. Fungi are nonphotosynthetic and in a kingdom of their own, and they have evolved special reproductive methods.
7. Mosses, ferns, and seed plants all have an alternation of a sexual, haploid, gamete-producing generation with an asexual, diploid, spore-producing generation.
8. In mosses, the sexual generation is the more conspicuous; in ferns, it is the asexual generation.
9. Conifers are nonflowering seed plants, in which the sexual generation is reduced to a microscopic, parasitic state.
10. A flowering plant is an asexual, spore-producing plant, and contains within itself a sexual phase. A seed is a ripened ovule containing an embryonic plant. A fruit is a ripened ovary containing seeds.

THE CONTINUITY OF LIFE ON EARTH DEPENDS on the ability of organisms to perpetuate their own kind with enough accuracy to maintain their identity, while providing enough variability and flexibility to allow for changing conditions. During the course of organic history, organisms have developed such a range of reproductive methods that no general scheme can be made to include them all. But in every method, the end result of reproduction is the production of new individuals, all more or less like their parents.

Patterns of reproduction have been evolving ever since the first prokaryotes were formed several billion years ago. At first, there was presumably no sex, no recombination of genes through mating; there was a mere pinching in two of a single cell. Then came simple cellular fusions. Later, specialization led to distinction between male and female gametes. In time, certain parts of an organism were set apart as being peculiarly adapted to gamete production, and these parts evolved ever greater specialization until the highly differentiated sex organs of multicellular plants and animals were formed.

In this chapter, we will consider plants and such plantlike organisms as the fungi and lower algae. The plants currently inhabiting the earth can be arranged in a series in which the structurally and functionally simple ones are placed at the beginning and the complex ones are at the end.

The most commonly accepted idea is that algae are the simplest and most ancient plants. Then, omitting some minor groups of plants, come the mosses, the ferns, the conifers, and finally the flowering plants. The last are the most complex and the most recent. Keep in mind that when complexity is mentioned, it is the entire life history of the plant that is being considered, not any particular part. Indeed, as we describe the several types of plants, it will become apparent that some parts of one type of plant may be more complex than those of a supposedly more primitive type, but that other parts may actually be simpler. This is notable in a comparison of the relatively more primitive mosses and the more highly developed ferns. The sexual generation of a fern is structurally simpler, smaller, and shorter-lived than the sexual generation of a moss, but the simplification is more than offset by the fern's enormously greater complexity in the asexual generation.

The examples presented in this chapter will illustrate a progression from the simplest asexual types of reproduction to the specialized reproductive methods of the higher plants. The evolution of the spore-producing generation is especially instructive. In those ancient plants that have undergone limited evolutionary change, the gamete-producing phase is more prominent than the spore-producing phase (see Figure 15.13). However, in plants of more recent origin, the situation is reversed (see Figure 15.14).

ASEXUAL REPRODUCTION IS USED BY SOME SPECIES OF PLANTS

Reproduction without sex, that is, without union of two sets of chromosomes, is simpler than sexual reproduction. Asexual reproduction suffers a disadvantage in comparison with sexual reproduction in that it results in a slower distribution of variation through a population. Asexual reproduction does, however, have several advantages, and many organisms, especially microscopic ones, commonly use it. Asexual reproduction bypasses the need for an organism to find a mate, an advantage of considerable importance for survival when the population is sparse or when new territory is being invaded. It can also be exceedingly fast, as one can see when a bacterial population dramatically increases in an infected host. Furthermore, except for an occasional mutation, all the offspring from an asexually reproducing organism are genetically identical. Having a whole population of individuals all genetically alike can be useful, especially in a restricted environment, such as in one animal body or a piece of decaying matter. The asexual method is so efficient that it is the common method in many plants.

Fission and Fragmentation

The simplest example of asexual reproduction is **fission,** the division of a single cell and the separation of the two resultant cells. Many small multicellular plants break readily into smaller pieces, and the pieces grow independently. The massive "blooms" of green algae in ponds are the result of such fragmentation.

× 3000

Figure 15.1
Scanning electron micrograph of the asexual spores *(conidia)* of black bread mold, *Rhizopus.*

Spores

Most plants produce special cells that are reproductive but involve no nuclear fusion. These asexual reproductive cells are usually called **spores.** Enormously variable in form and manner of distribution, spores are common reproductive cells in algae and molds, which are capable of giving them off in astronomical numbers (Figure 15.1). Spores of seed plants are produced inside cones or flowers.

The air, the earth's waters, and most exposed surfaces carry a load of microscopic spores. Spores can survive in harsh environments for long periods by remaining dormant while conditions such as temperature and moisture remain at their extremes. When moisture and temperature are suitable, spores can germinate and grow into new organisms.

Vegetative Reproduction

Higher plants may reproduce asexually, or *vegetatively,* through processes in which a new complete plant is produced from a root, stem, or leaf of the parent plant. This process may occur naturally or through human interference by cutting, grafting, or layering. In higher plants, prostrate stems or even masses of tissue can be spread beneath or on the

Figure 15.2
Asexual (vegetative) reproduction in a flowering plant. A lateral stem, or runner, branches out from a parent strawberry plant, and when it touches earth at its far end, it sprouts roots and makes an entire new plant.

surface of the soil; branching at the advancing ends and decaying at the back, these cuttings will eventually take root, with many separate individuals resulting. Strawberry plants, for instance, may produce *runners,* or *stolons,* which grow along the ground away from the parent plant. Runners form roots and shoots that develop into new plants (Figure 15.2). Ferns and mushrooms may also spread, forming conspicuous circles of plants, which may be several meters in diameter (Figure 15.3).

Humans sometimes desire genetically identical copies of plants in species that have special combinations of characteristics: apples, for example, or roses. These plants do not "breed true" from seeds, because seeds involve sexual fusions and are therefore genetically different from the parent plants. To guarantee biological uniformity, people use **clones,** or groups of individuals all having the same genetic background, obtained without an intervening sexual fusion. Obtaining clones is easy with plants,

Figure 15.3
Multiplication of mushrooms. From a central starting point, an ever-expanding underground mass of mushroom tissue sends up fruiting bodies around the outer edge, making a "fairy ring."

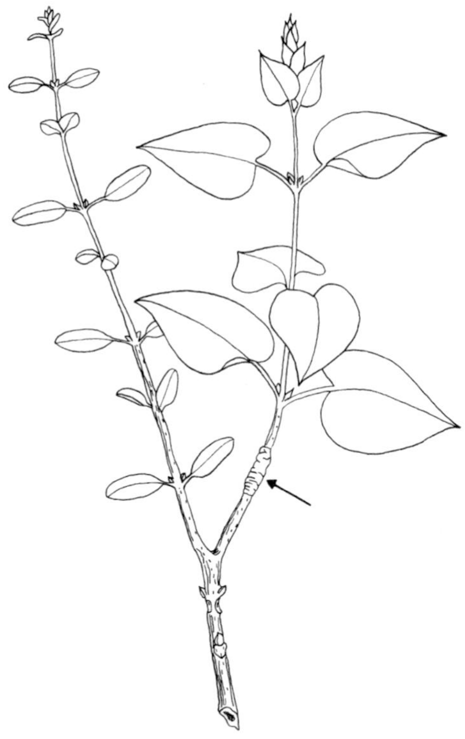

Figure 15.4
Asexual reproduction by grafting. A branch of a lilac bush has been connected to a stem of privet hedge. Care is taken to have the growing tissue of the *scion* (the lilac) connected to the growing tissue of the *stock* (the privet), and the joint is covered with wax and wrapped (arrow) to hold the graft in place and prevent drying. Grafts survive only if stock and scion are closely related. Lilac and privet, both in the olive family, are similar enough to ensure a successful graft.

because new plants can be grown from slips or cuttings. A piece of stem may be cut from a desirable variety and grafted onto a stem of any closely related seedling (Figure 15.4). With this procedure, one can have an orchard of uniform peaches or cherries or mangoes, a garden of identical lilac bushes, or even several kinds of apples on a single tree.

The principle of cloning has been known for centuries, and practical gardeners have known how to produce multiple copies of plants almost as long as they have been cultivating them.

TO SUM UP ASEXUAL REPRODUCTION

1. Asexual reproduction takes place without the union of two sets of chromosomes.
2. *Fission* is the simplest example of asexual reproduction. It is the division of a single cell and the separation of the two resultant cells.
3. Most plants produce *spores,* special asexual reproductive cells that do not involve nuclear fusion.
4. Higher plants may use *vegetative reproduction,* which produces a new complete plant from a root, stem, or leaf of the parent plant.
5. A *clone* is a group of individuals genetically identical to each other and to their parent organism.

THE ORIGIN OF COMPLEX SEXUALITY

The essential feature of any sexual reproduction is the union of two nuclei, usually of different genetic makeup, resulting in the formation of a single new nucleus. In the simplest and presumably most ancient organisms, mating involves the complete fusion of two unicellular individuals (Figure 15.5).

By observing a number of *simple* organisms, biologists have found a series of sexual types that may illustrate the way in which *higher* organisms came to have special sexual procedures. The simplest sexual process is the joining of *any* two free-living cells, with cytoplasmic union followed by nuclear fusion. A slight advance over such a primitive procedure occurred when special mating types evolved—a foreshadowing of differentiation into maleness and femaleness. This evolution of mating types happens in some species in which certain individuals, not visibly different, can mate *only* with compatible individuals. This phenomenon is true of many algae (see p. 177).

The next step in sexual complexity came with a visible difference between the sex cells, a slight difference in size or pigmentation. Still another advance came with a change in motility. One sex cell, the larger one, lost its ability to move, while the smaller remained active. Since in higher organisms the larger, nonmotile cell is the female gamete, and the smaller, motile one is the male, the terms "male" and "female" are sometimes used even in simple, microscopic species.

All the sexual methods described above can be observed in one genus of one-celled green algae, *Chlamydomonas* (Figure 15.6). But additional innovations have resulted in still more complex organisms. One was the development of multicellular bodies in which only a special portion of the organism is de-

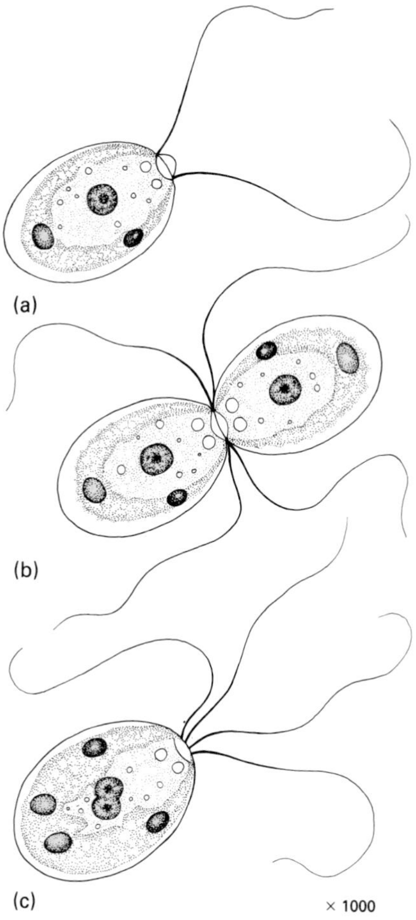

Figure 15.5
Sexual fusion of entire organisms. (a) A unicellular green alga, *Chlamydomonas.* (b) Two compatible cells meet. Each is haploid (N number of chromosomes). (c) A single cell, a *zygote,* results from the fusion of the two separate cells. The zygote is diploid (2N chromosomes). After a period of time, meiosis occurs, and four haploid cells are formed.

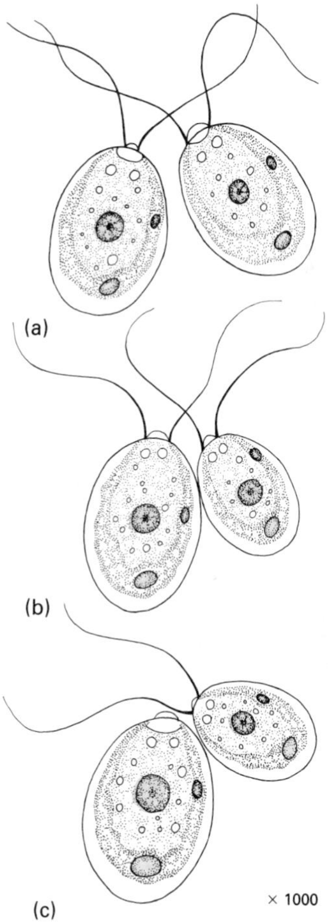

Figure 15.6
Three gradations of sexuality in one genus, *Chlamydomonas.* (a) Two cells, structurally alike, meet and mate. This is the simplest kind of sexuality imaginable. (b) In a different species, one partner is larger than the other, making a sexual differentiation. (c) In still a third species, one partner not only is larger than the other but has lost its motility, although the smaller one is still able to swim. Having a larger, nonmotile cell (usually called an egg) mate with a smaller, motile cell (usually called a sperm) is typical of the sexual methods of most plants and animals.

voted to gamete production. Such a development must have occurred early in organic history, as shown by the fact that some relatively primitive species have it, and it is now used by most living things.

The most recent complication was the provision for parental care of the new generation. Either the egg is provided with enough reserve food to ensure a reasonable chance for survival of the new individual, or the new individual is kept inside the parent during its early life.

SEXUAL REPRODUCTION IN LOWER PLANTS

Algae

In the heterogeneous assemblage of plants popularly known as algae, reproductive methods are as various as the plants themselves. Indeed, algae are so diverse that the current opinion among biologists is that they are not a single evolutionary group, but that they instead represent several independent lines of development (Chapter 9).

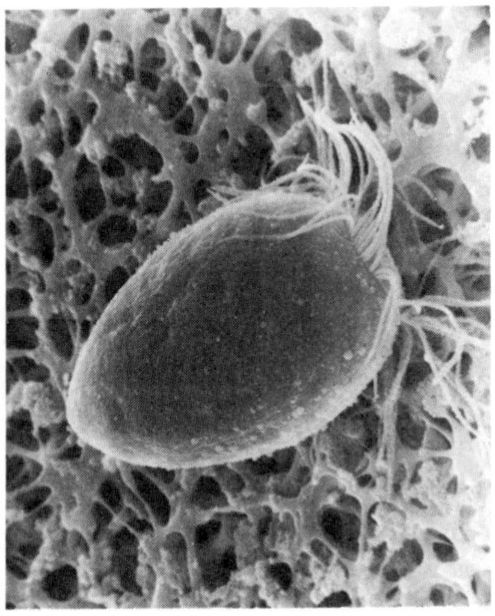

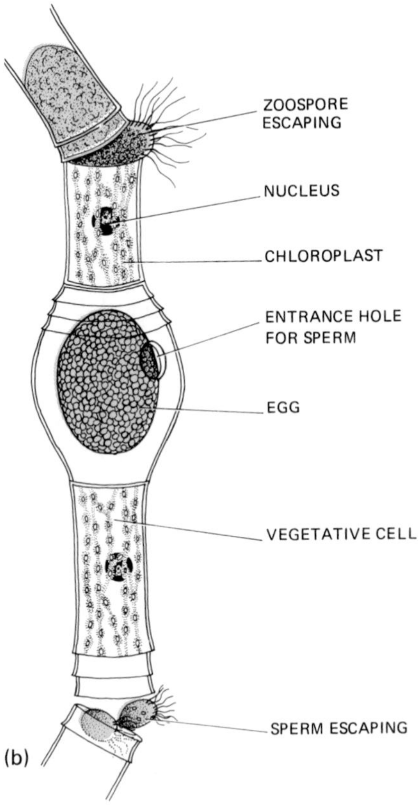

Figure 15.7
Reproduction by both asexual and sexual methods in the green alga *Oedogonium.* (a) Scanning electron micrograph of a zoospore of Oedogonium. It bears a conspicuous ring of flagella at the anterior end. (b) Asexual zoospores break out of the parent filament, swim away, and later settle down to produce new individuals. When *Oedogonium* reproduces sexually, one cell grows into an enlarged egg while others are producing sperm cells. The sperm escape, swim to a hole in the cell containing the egg, and fertilize the egg. Later, the fertilized eggs undergo meiosis and germinate to produce new haploid individuals.

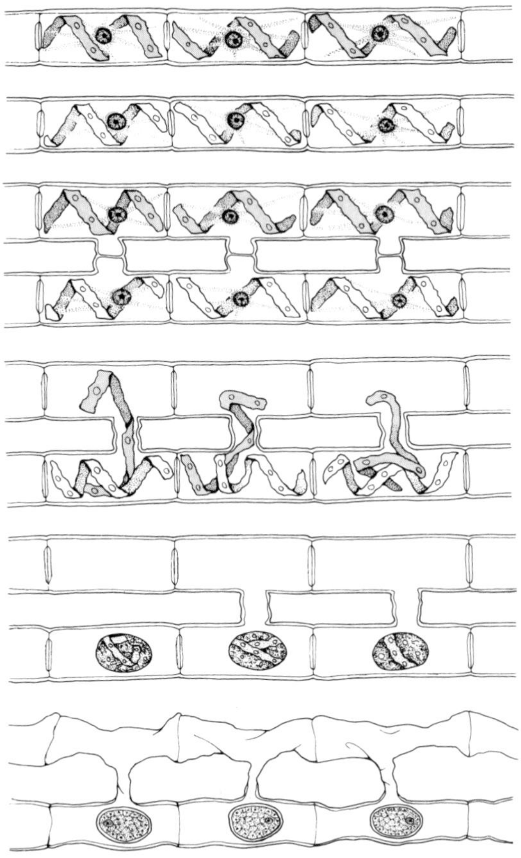

Figure 15.8
In the green alga *Spirogyra,* mating procedure is by conjugation. (a) Two filaments lie side by side. (b) Tubes are sent out from each cell to a neighboring cell. (c) All the cellular material passes from one cell into the mating partner. (d) The cells fuse into a single *zygospore.* (e) The zygospores go into a period of dormancy, later to undergo meiosis and give rise to new filaments.

Representative examples of algal reproduction illustrate some common procedures in microscopic, freshwater types. *Oedogonium* (Figure 15.7), and *Spirogyra* (Figure 15.8) are common freshwater genera, each with its characteristic reproductive method.

Fungi

Fungi are nonphotosynthetic organisms. Because of structural similarities to plants, they have been traditionally considered as plants, and for convenience they will be treated along with plants in this book. Reproduction in fungi is as varied as it is in algae. In fact, some fungi, such as those causing rust diseases in plants, exhibit the most complex reproductive patterns to be found anywhere in nature.

Some molds, such as the black bread mold, *Rhizopus,* not only have asexual spores but practice a kind of conjugation, in which adjacent cells meet, fuse their cytoplasm and nuclei, and grow a heavy protective wall around the resultant zygospore (Figure 15.9). Later, meiosis occurs and the zygospore germinates to produce a new fungus.

Mushrooms, the most familiar fungi, produce sexual spores. The structural and functional features of mushroom spore production are shown in Figures 15.10 and 15.11. As might be expected, many

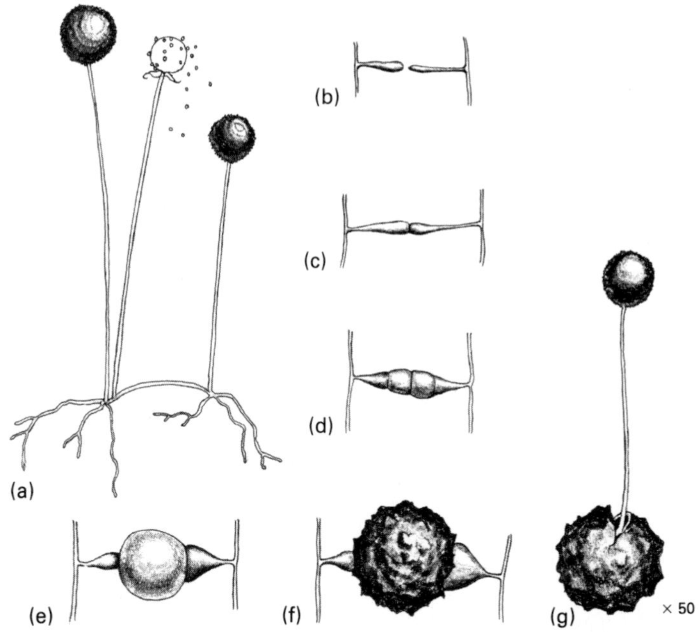

Figure 15.9
Reproduction in the black bread mold, *Rhizopus*. (a) The fungus threads send up hyphae, *sporangiophores,* bearing spore cases, *sporangia,* full of spores (see Figure 15.1). These asexual spores are the usual means of reproduction unless compatible mating types meet. If they do, mating hyphae grow toward one another (b), meet (c), and form a bridging pair of cells (d), which fuse to make a single zygospore (e) containing hundreds of compatible nuclei. The nuclei mate, two by two, and the zygospore develops a thick covering (f). After a period of dormancy, the zygospore undergoes meiosis and then germinates, to send out a sporangiophore (g) with spores that are haploid.

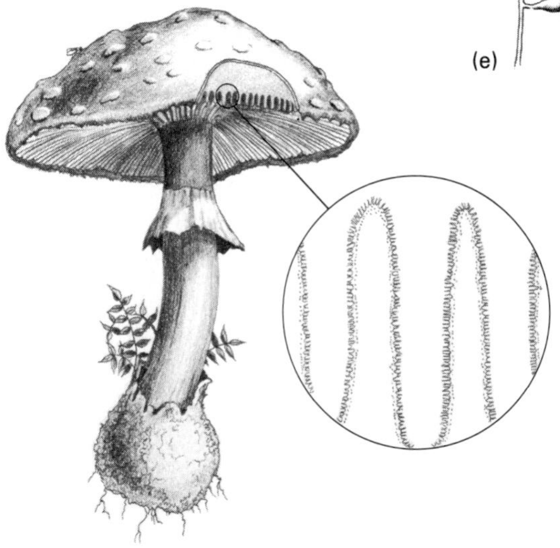

Figure 15.10
Reproduction in a mushroom. The layers of tissue under the cap of the mushroom are the gills (shown enlarged in the inset), which bear the spore-producing cells. (See Figure 15.11.)

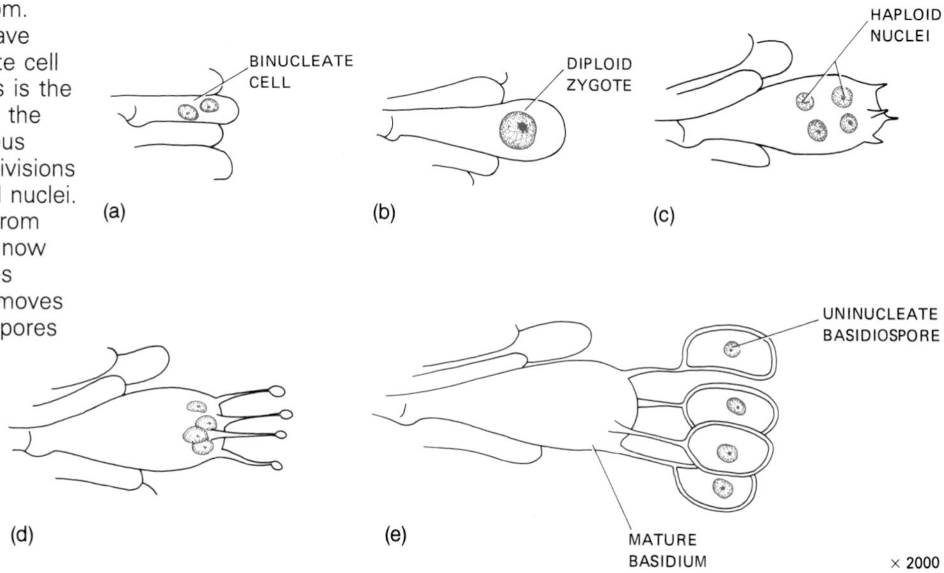

Figure 15.11
Spore production in a mushroom. (a) The cells of a mushroom have two nuclei each. One binucleate cell is shown. The nuclei fuse. This is the act of mating. (b) The result of the fusion is a diploid zygote nucleus (2N). (c) Meiotic or reduction divisions follow, resulting in four haploid nuclei. Four slender processes grow from the end of the cell, which can now be called a *basidium*. (d) Spores begin to swell, and a nucleus moves into each one. (e) The basidiospores grow to full size.

variants are known, but this procedure is a common one, and it is used by hundreds of species of mushrooms.

Mosses and Ferns

Sexual reproduction in mosses and ferns illustrates the general direction of plant sexuality, which is distinctly different from the sexual methods of animals. In practically all animals, a diploid organism possesses specialized cells that undergo meiosis, and the resulting haploid cells become eggs or sperm cells. Then the union of an egg and a sperm makes a diploid zygote. Plants rarely proceed so. Instead, plants have a whole haploid phase during which gametes are produced without meiotic divisions (Figure 15.12). That phase is the *gametophyte generation,* or the gamete-producing plant (*gamete,* Greek, "marry"; *phyte,* Greek, "plant"). The union of gametes produces a diploid zygote, which may grow mitotically into a diploid plant.

At maturity some of the diploid plant's cells undergo meiosis, and the resultant haploid cells are spores. This phase is the *sporophyte generation.* Since the haploid, gametophyte (or sexual) phase, is followed by the diploid, sporophyte (asexual) phase and then by the gametophyte phase again, the whole phenomenon is called **"alternation of generations."**

The moss plants usually seen on earth or rocks are sexual plants. Flask-shaped **archegonia** produce eggs on this moss, and egg-shaped **antheridia** produce sperm, all by mitotic nuclear divisions, that is, without meiosis (Figure 15.13). Since the sperm are motile, they can swim from the antheridia to the archegonia, where fertilization occurs. The diploid zygote grows first into a few-celled embryo inside the archegonium. This new sporophyte bursts the archegonial walls to become a long, thin body with its lower end embedded in the old gametophyte and with a swollen bulbous capsule at the upper end. In the capsule, some cells divide meiotically, yielding quartets of haploid cells that become spores. The haploid spores are released from the capsule, and if they fall into a proper environment, they germinate into new haploid, gametophyte moss plants.

Mosses are the simplest living plants in which the gametes are produced in special multicellular organs, the antheridia and archegonia. Algae and fungi, in contrast, produce their gametes inside single cells. The mosses are also notable in that they nourish embryonic sporophyte plants in the archegonia. That arrangement or some other that is similarly protective is typical of all plants from mosses on up through the evolutionary scale of plants.

In mosses the gametophyte is the larger and more actively photosynthetic generation, but in ferns the arrangement is reversed. A haploid fern spore germinates to develop into a flat, heart-shaped plant about five millimeters across, with rootlike hairs growing into the soil, archegonia with eggs, and antheridia with sperm (Figure 15.14). The haploid, motile sperm can swim from its antheridium to an archegonium and there fertilize a haploid egg. The resultant diploid zygote first grows into an embryo, then sends a root into the soil, and later sends an upward-thrusting leaf into the air. This is the young stage of the familiar fern plant, and it grows to become a mature fern plant. At maturity, fern leaves produce spore capsules, known as **sporangia** (singular, sporangium), in which meiosis occurs, followed by the formation of spherical quartets of haploid spores. After release from a sporangium, a spore can germinate to yield a new fern gametophyte.

TO SUM UP SEXUAL REPRODUCTION IN LOWER PLANTS

1. Sexual reproductive methods in algae and fungi are diverse. Some fungi use a kind of conjugation, in which cytoplasm and nuclei of adjacent cells fuse and produce a new fungus, and other fungi, such as mushrooms, produce sexual spores.
2. Mosses and ferns reproduce through a phenomenon called *alternation of generations.* In these plants, sexual and asexual stages alternate. The sexual stage is the haploid, gamete-producing gametophyte, and the asexual stage is the diploid, spore-producing sporophyte.
3. In mosses, the gametophyte is more conspicuous than the sporophyte; in ferns, that is reversed.

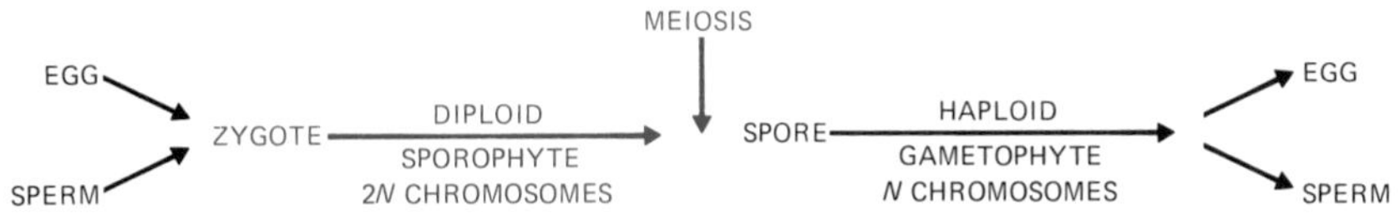

Figure 15.12
Generalized scheme of *alternation of generations*. The diploid, sporophyte generation begins with fertilization and the formation of a zygote, and it ends with meiosis at the time of spore production. The haploid, gametophyte generation begins when meiosis results in spore production, and it lasts until egg and sperm meet in fertilization.

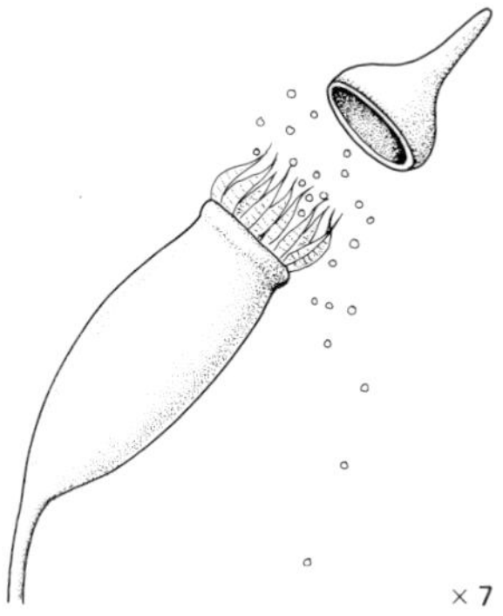

(b)

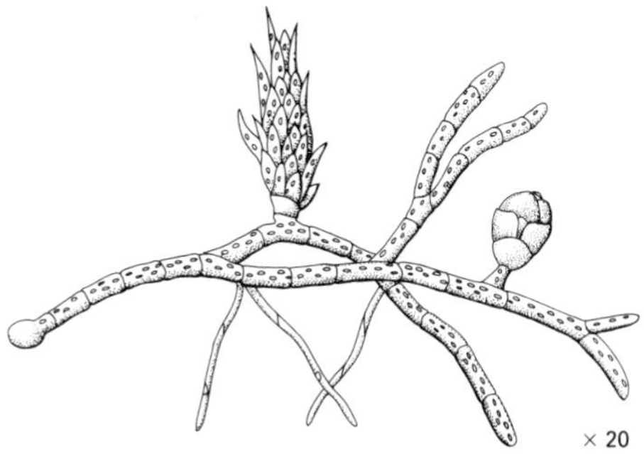

(c)

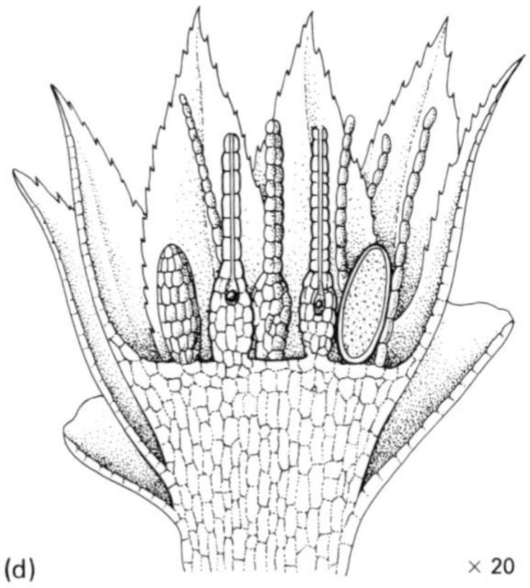

(d)

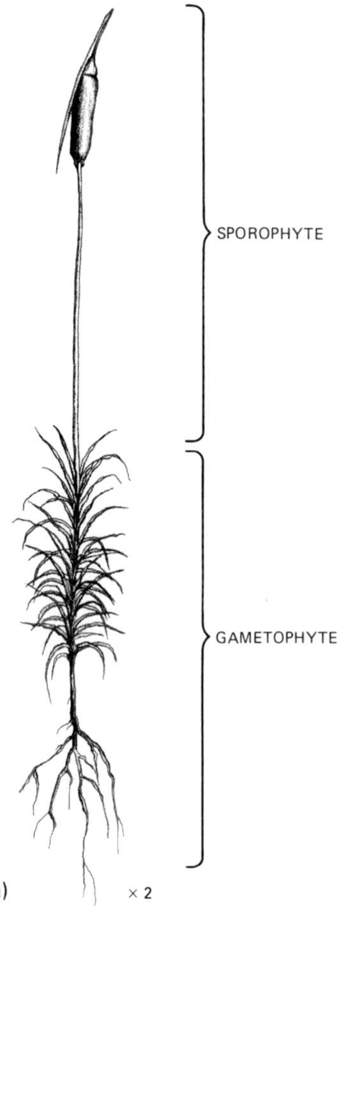

(a)

Figure 15.13
Alternation of generations in a moss *(Catherinea)*. Note that the chromosome count is reduced from diploid to haploid in the spore capsule when spores are produced, and the count is restored to diploid when the egg in the archegonium is fertilized. (a) The leafy gametophyte, the sexual generation, is shown with the sporophyte (2N) growing out of its top. For a while, the spore capsule carries the remains of the old *archegonium* (see f below). The capsule is closed with a lid when young. (b) When the spores are ripe, the lid falls off, and spores (N) are released. (c) When spores germinate, they produce green filaments resembling algae. These form buds that develop into tiny, leafy plants (N). (d) The moss plant, on reaching sexual maturity, produces flask-shaped *archegonia* containing eggs and egg-shaped *antheridia* containing sperm. (e) A single antheridium, split open to show the flagellated, swimming sperm. (f) An archegonium, cut open to show the egg and the canal down the neck through which sperm can swim. After fertilization, the zygote develops into an embryo, which then grows up to become a mature sporophyte with its capsule.

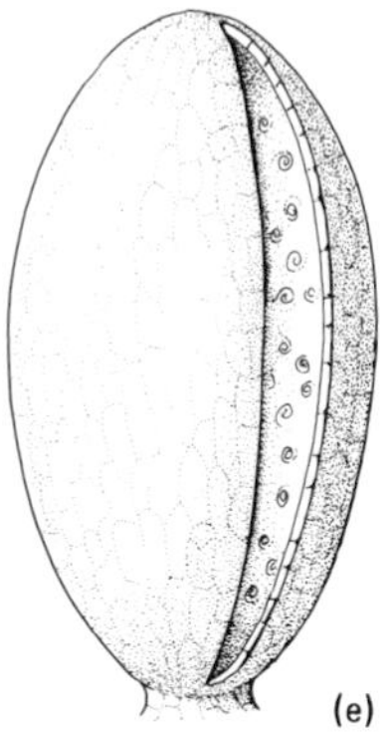

(e) × 90

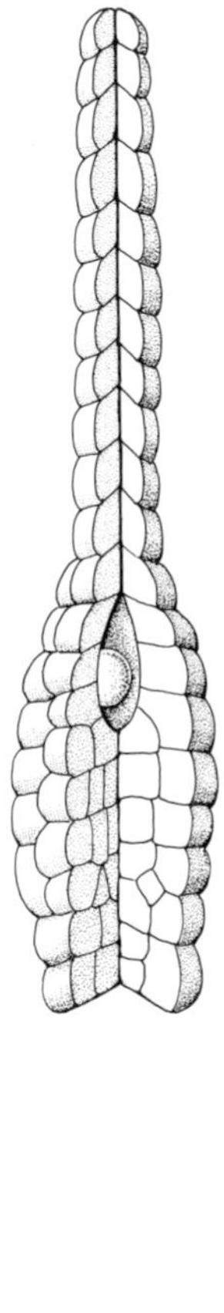

(f) × 90

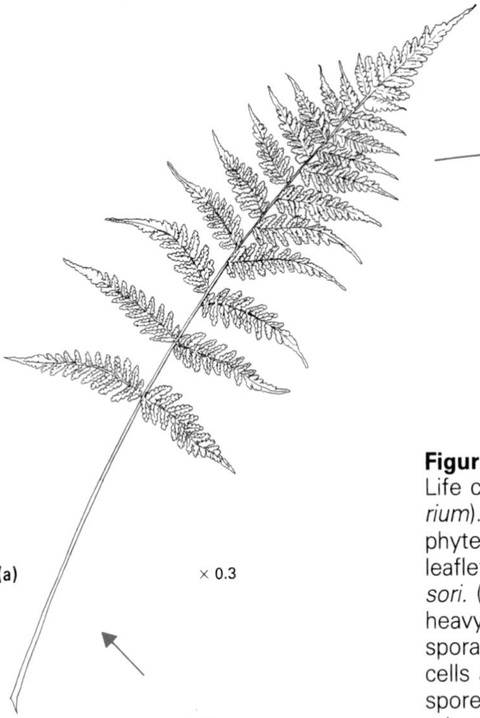

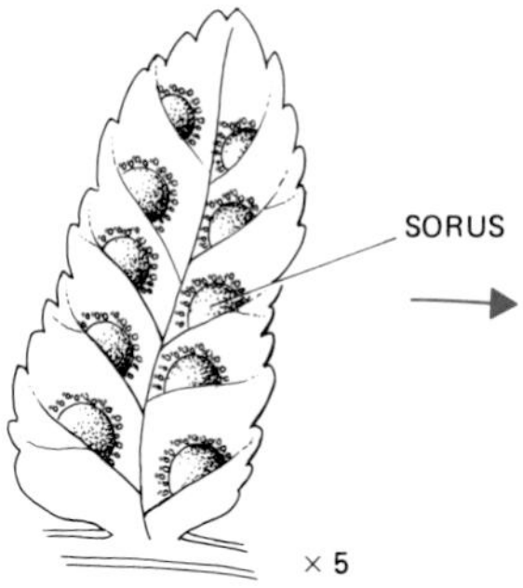

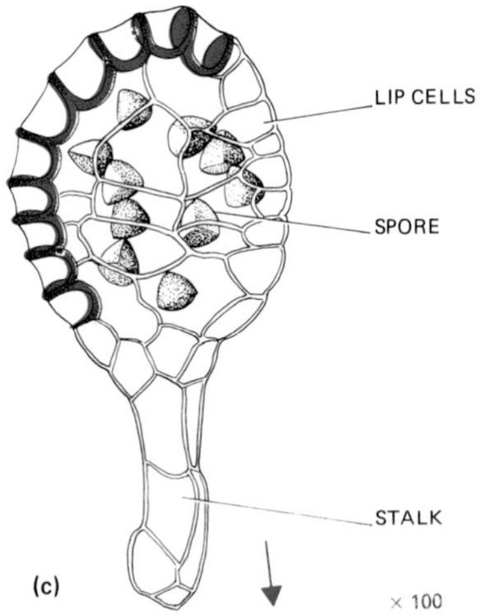

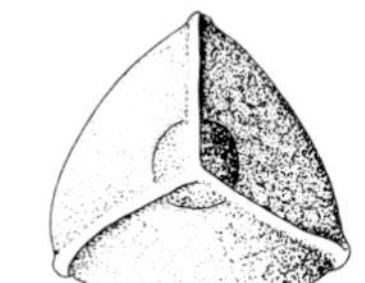

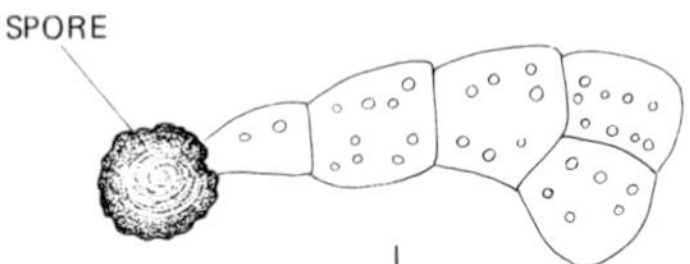

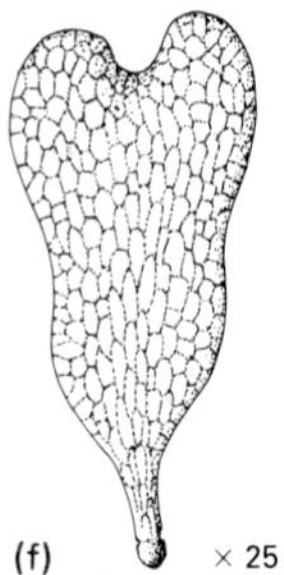

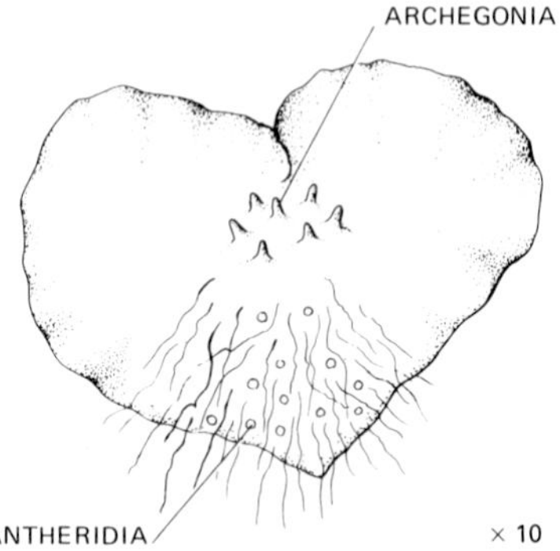

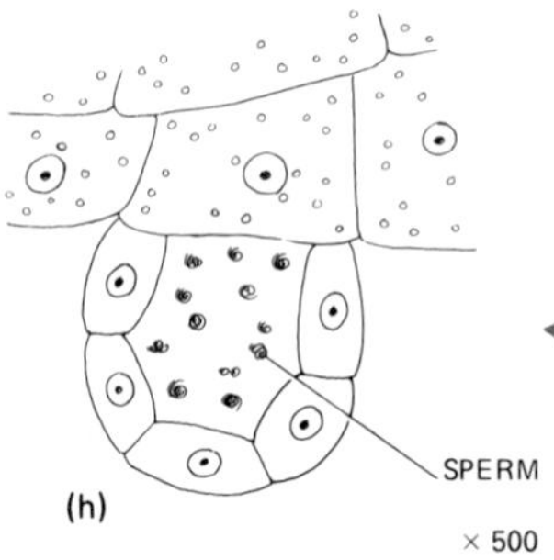

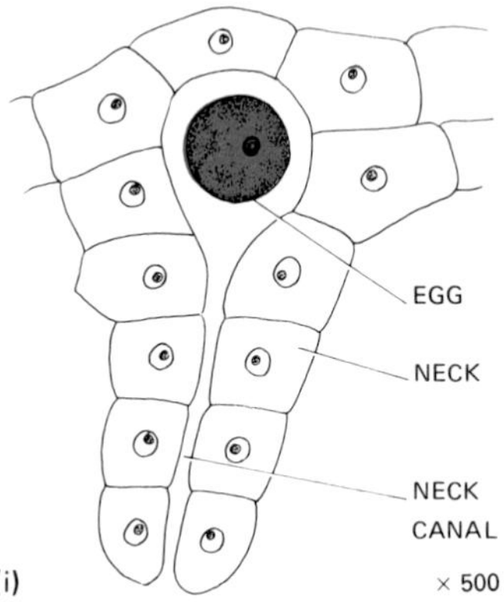

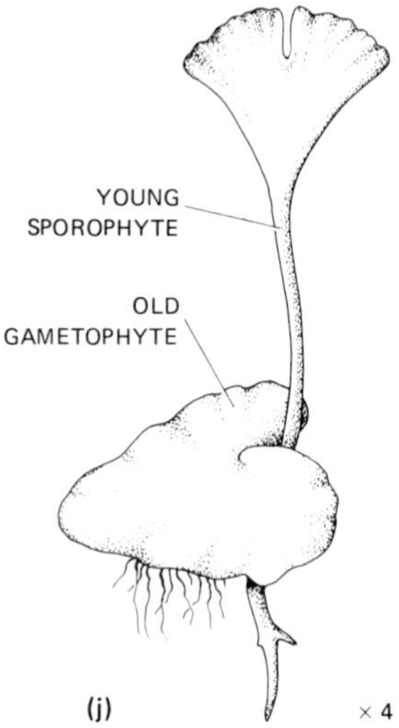

Figure 15.14
Life cycle of a fern (Lady fern, *Athyrium*). (a) The fern plant is the sporophyte (2N). (b) The underside of a leaflet bears clusters of sporangia in *sori*. (c) A sporangium has a ring of heavy-walled cells, which pulls the sporangium open, splitting the lip cells and liberating the haploid spores. (d) A spore. (e) A spore germinates to produce a green, filamentous body (N). (f) The body, a young gametophyte, grows photosynthetically into a flat plate. (g) When mature, the gametophyte is a *prothallium.* It bears *archegonia* with eggs and *antheridia* with sperm. (h) A section through an antheridium containing sperm. (i) A section through an archegonium showing the egg, neck, and neck canal. (j) A young sporophyte (2N) growing out of an archegonium. Although still attached to the parent gametophyte, it has its own root and green leaf, and it will become a completely independent plant.

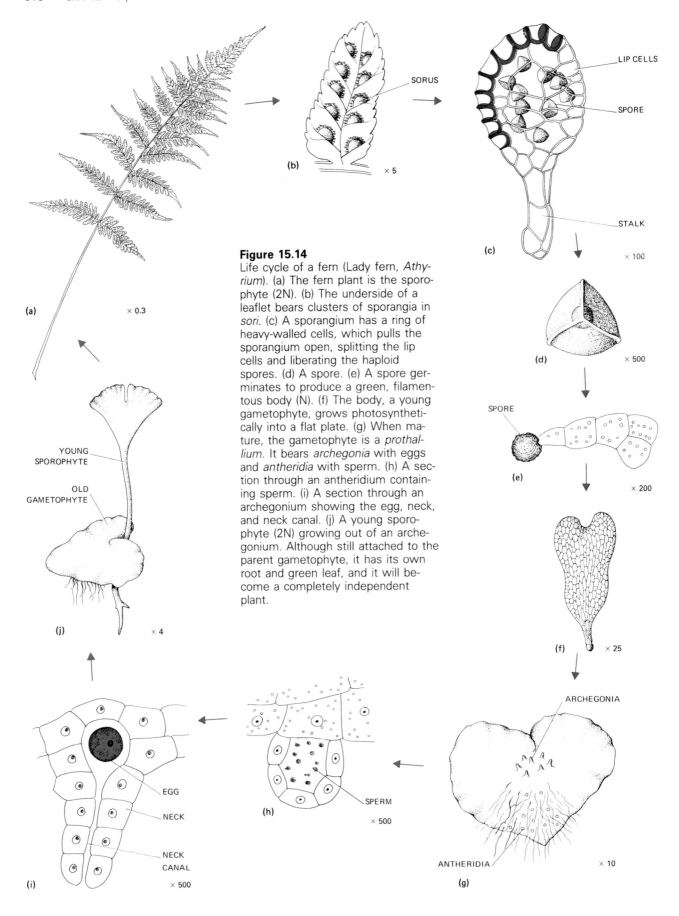

SPORE AND SEED PRODUCTION

By the time plants had developed the habit of reproducing by means of seeds, they had increased the complexity and relative size of their diploid, asexual, sporophyte stage. They had also reduced the complexity and relative size of their haploid, gametophyte, sexual stage so much that biologists would probably not know what these stages mean in seed-bearing plants if it were not for the more readily understood models provided by the lower plants. Microscopic observations of cells during all the phases of a plant's life reveal the time and place when the chromosome number is reduced by half, thus establishing when meiosis takes place. (Figure 15.15). Further investigation shows which structures are spores and where and when sexual union occurs. By comparing the chromosome counts and the manner of spore and gamete production in seed plants with those phenomena in ferns and mosses, one can learn how cones, flowers, fruits, and seeds function.

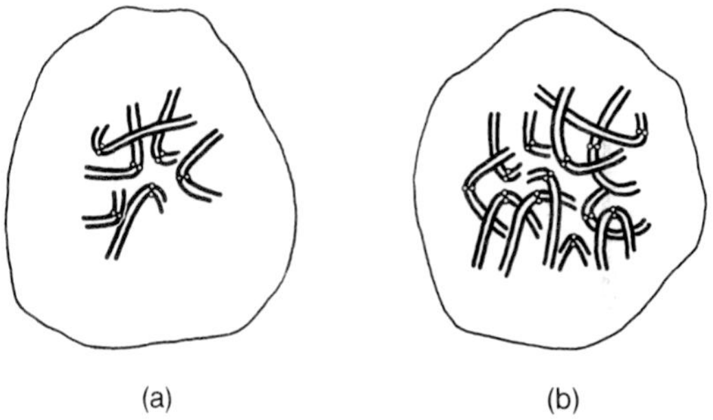

Figure 15.15
Comparison of haploid and diploid mitoses. In the haploid mitosis (a), characteristic of the gametophyte generation of a plant, there is only one representative of each chromosome type; in this species there are six types. In the diploid mitosis (b), there are two representatives of each chromosome type, as there would be in the sporophyte generation of the plant. By counting chromosomes, one can determine where the meiotic divisions occur in the life of a species.

Heterospory

Ferns and fernlike plants make spores that, having undergone meiosis, grow into haploid, gametophyte plants, usually capable of yielding both eggs and sperm. At many times and places, however, plants have evolved a method of producing two kinds of spores: those that germinate to form female gametophytes only or male gametophytes only. Since spores that are destined to produce female gametophytes usually are relatively large, they are called *megaspores* (Greek, "large seeds"). The smaller ones, which will develop into male gametophytes, are *microspores* (Greek, "small seeds"). The phenomenon of making two kinds of spores is **heterospory** (Greek, "different seeds"), in contrast to *homospory* (Greek, "same seeds"), in which only one kind of spore is produced. Heterospory is typical of seed-bearing plants.

In seed plants, spores are formed in a sporangium following meiosis, but they are not immediately released from the parent plant to begin an independent life as they would be in mosses and ferns. Instead, the spores begin developing while they are still in the sporangium, and the kind of development depends on the kind of spore. Megaspores in seed plants are produced in megasporangia, which in seed plants are commonly called **ovules** (see Figure 15.22). Megaspores develop into female gametophytes, which are haploid, egg-producing plants. They are typically nongreen, and they live parasitically on the parent sporophyte. Microspores are produced in microsporangia, which are called **anthers** in flowers. Microspores develop into male gametophytes. They, too, are parasitic, but they remain small, undergo only a few mitotic nuclear divisions, and when shed from the microsporangium are called **pollen grains.** When pollen grains are transported to the vicinity of an egg nucleus, they germinate, forming a long pollen tube that carries sperm nuclei directly to the egg nuclei.

The familiar seed plants of the world are *diploid sporophytes*. Since they produce haploid megaspores and microspores, they are essentially asexual. The sexual generation is the microscopic mega- or microgametophyte. When the gametophytes have produced eggs and sperm, and the resulting zygotes have begun to be embryos, new sporophyte generations have begun. These, with their surrounding tissues, are **seeds.** The embryo sporophyte in a seed can be thought of as the grandchild of the plant that bears it because the immediate parents of the embryo were the gametophytes.

Conifers

Coniferous plants are seed plants, but they have no flowers, and they are less advanced in an evolutionary sense than flowering plants. They are woody, and most bear needle-shaped evergreen leaves.

Pine trees are representative conifers, showing one common method of reproduction. They produce two kinds of cones: ovulate ones, which are

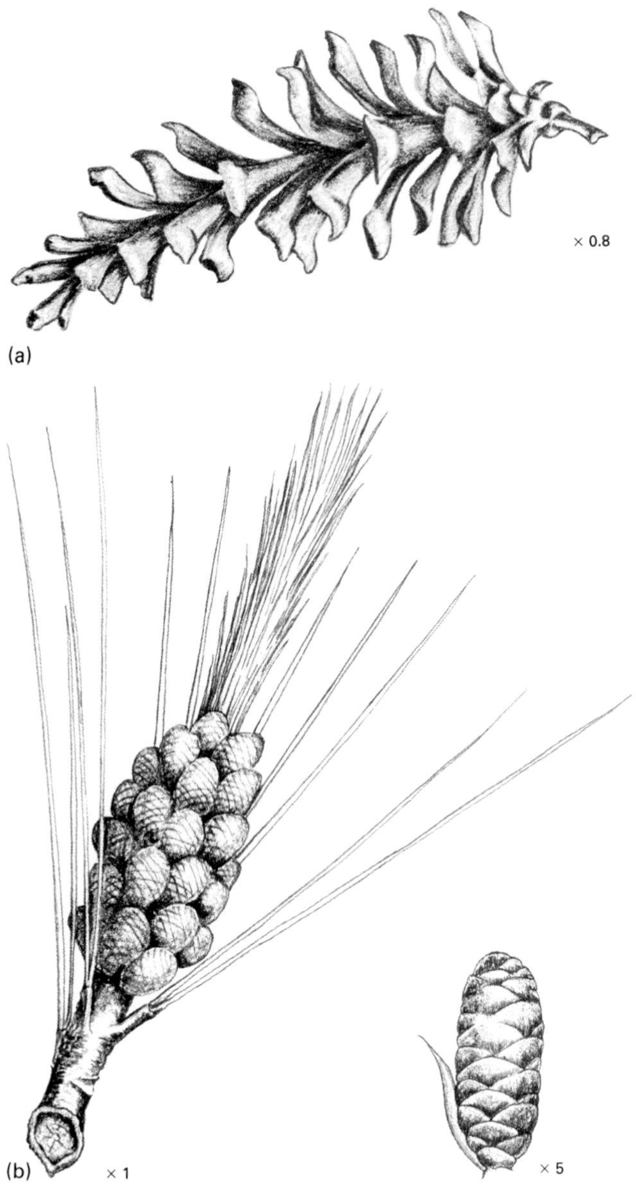

Figure 15.16
(a) An ovulate cone of white pine, *Pinus strobus.* (b) A cluster of staminate cones of pine, as seen in early spring, with old needles below and young needles at the end of the stem. One cone is shown enlarged.

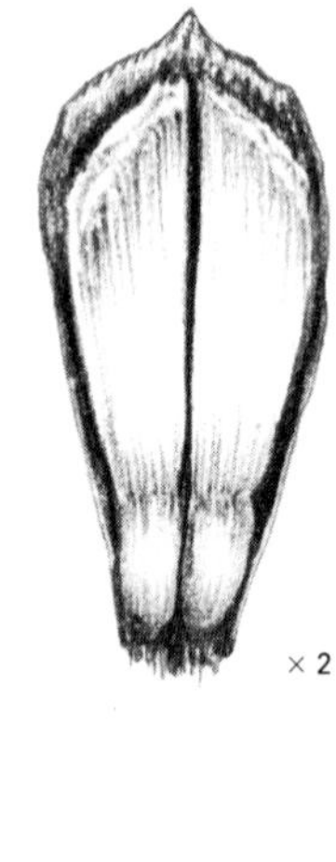

Figure 15.17
The upper surface of a single scale from an ovulate cone with two seeds loosely attached to it. The roundish lumps are the ripened ovules containing embryos, and the flat wing attached to the seed helps in wind dispersal.

the conspicuous familiar pinecones, and the small staminate cones, which appear for a few weeks in spring (Figure 15.16).

The ovulate, or female, cone, produced at the apex of a branch, is a cluster of scales arranged spirally around a central axis. Each scale bears on its upper surface a pair of *ovules* (Figure 15.17). In the tissue of the ovule, one cell is destined to undergo meiosis. This is the *megaspore mother cell.* As is typical of spore formation, this megaspore mother cell undergoes meiosis, producing four haploid megaspores, three of which abort. The surviving one follows the pattern of any megaspore: It grows by mitotic divisions into a haploid female gametophyte. When mature, the female gametophyte is a tiny mass of tissue with several archegonia (Figure 15.18). The archegonia are not such definitely constructed organs as those of mosses; rather, they consist only of a vesicle with a few neck cells and a conspicuous egg nucleus.

In early spring, while the ovulate cones are being formed, staminate ones are growing on other branches (Figure 15.19). Each scale of a staminate cone bears a pair of microsporangia that produce hundreds of diploid microspore mother cells. Each microspore mother cell divides meiotically into four

SEEDS AND CIVILIZATION

Early humans wandered about the earth in small groups, killing such animals as they could overpower, finding fruits in season, digging up starchy rhizomes, and even eating leaves when they could find nothing better. Because large societies are impossible to feed when everything must be hunted down or collected by chance, the tribal size was limited, probably to a few dozen at most. The very smallness of hunting tribes made protection difficult. Wild animals and marauding human enemies made life risky. When societies were tiny and the bulk of a human's life was taken up in self-protection and scrounging for a bit of food, there was neither the power nor the technique for building large, permanent settlements.

Change came when it was discovered that fruits and seeds could be stored and then planted to produce food crops that could be harvested. Discoveries came slowly and independently in

Figure 15.18
A long section through a young ovulate cone. The insert shows a portion of a single cone scale. On the upper surface of the cone scale there is a pair of ovules, one of which is shown sectioned. The ovule is covered with *integuments,* part of the cone which will form the seed coat. The female gametophyte (haploid) has two archegonia, each with an egg. An opening in the integuments allows a pollen grain to come in close to the eggs. The pollen grain germinates, sending a pollen tube into the ovule, where haploid sperm nuclei are deposited in the archegonia and the haploid eggs are fertilized. The diploid embryo then develops in the ovule.

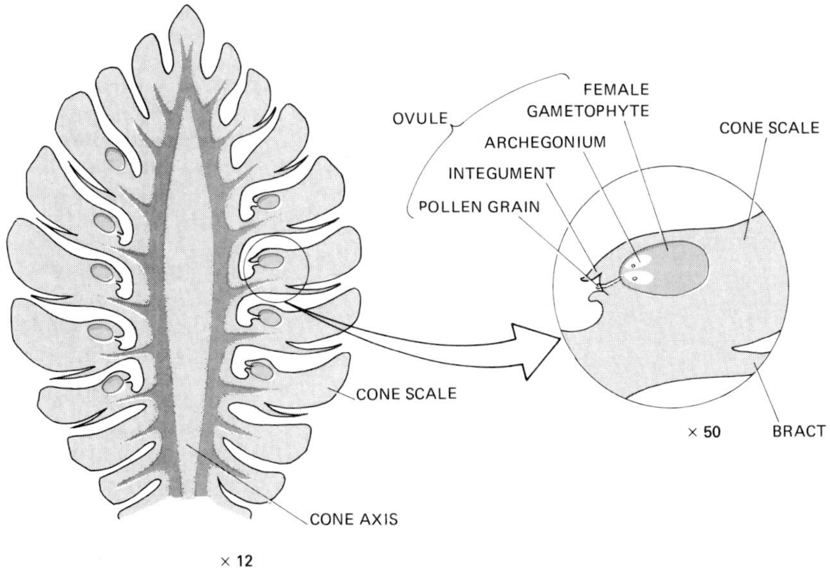

Figure 15.19
A long section through a staminate cone. The mature pollen sacs are shown full of mature pollen grains.

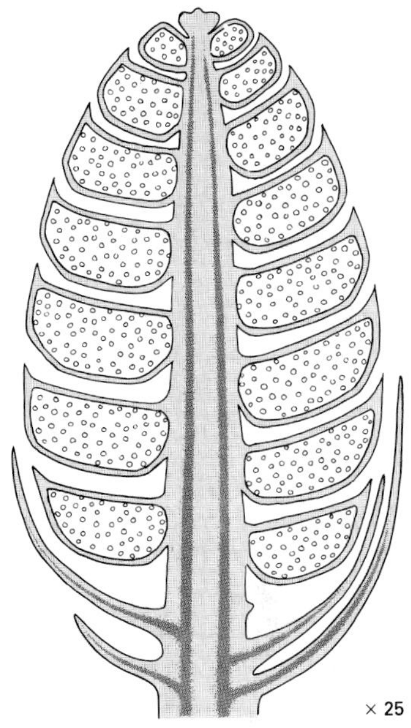

scattered places. In Jordan and Israel flint sickles about 10,000 years old indicate that grains, probably wheat and barley, were then being harvested. Knowledge of farming seemed to have spread westward to Europe about 5000 B.C., with wheat and barley as the main crops. Meanwhile, skill in cultivation moved eastward to China, where there was wheat, but where another grass, rice, became the staple grain. From eastern Asia, people came through what is now Alaska and migrated to South America. The hunters in the cold territories of the north remained hunters, but in the gentler climates where rich valleys offered the opportunities, humans found wild grasses, which they learned to grow, to keep, to select, and to improve. By about 4000 B.C., the Mexicans were cultivating varieties of maize so specialized that their wild counterparts are still unknown.

An occasional genius may have painted animals on rock walls or carved figures in stone or bone, but cities and armies were possible only when there was an assured food supply. When humans made seeds their servants, they began to make buildings that were more than simply shelters. These structures were built with elegance and ornamented with carving and color, and they served for governmental or religious functions or for personal ostentation.

The great civilizations of the world have depended on the ability of seeds to store and concentrate a durable source of human food. The arts, the sciences, literature, architecture, politics, philosophy, even wars—all became possible because of grains. Figuratively speaking, the temples of Canton were built on rice, the cathedrals of Paris on wheat, and the pyramids of Chichén-Itzá and the fortresses of Machu Picchu on maize. The greatest achievements of humans have developed from seeds of grasses.

microspores, which develop into *pollen grains.* These are male gametophytes, which form sperm nuclei.

Pollen drifts in air to the young ovulate cones, which open their scales and temporarily allow pollen to come close to the ovules. Pollen lies quiet in the ovulate cone for a year until the eggs in the archegonia mature. In the second spring in the life of the ovulate cone, a pollen grain sends out a germ tube containing sperm nuclei, one of which enters an archegonium to fertilize an egg nucleus (Figure 15.20). From the resulting zygote there grows an embryo with an axis and several embryonic leaves, the *cotyledons.* The embryo is contained in and nourished by the female gametophyte.

The complete pine seed consists of the embryo embedded in the old female gametophyte, which is in turn surrounded by the integument, or seed coat (Figure 15.21). When the seed is mature at the end of the second summer, the cone scales spread, releasing the seed along with a piece of the scale, which peels off, giving the seed a wing that helps it drift in moving air.

Figure 15.20
A germinated pine pollen grain. This is the complete male gametophyte (haploid). It has the *body cell,* which will form two sperm nuclei; the *tube nucleus;* and the *pollen tube,* which will penetrate into the ovule.

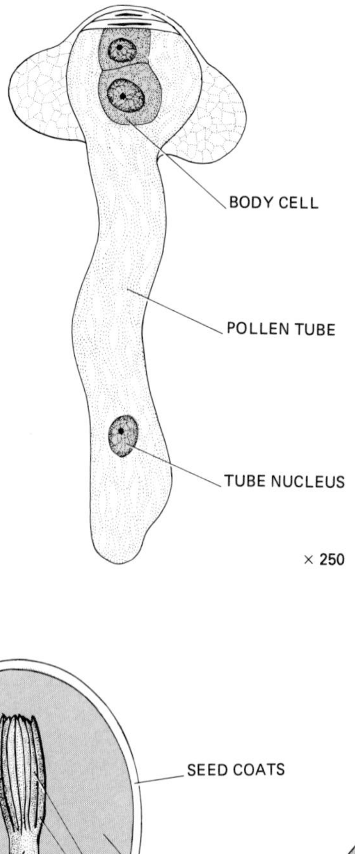

TO SUM UP SPORES AND SEEDS

1. Seed plants and a few others produce two different kinds of spores that germinate to form either male or female gametophytes. The phenomenon is called *heterospory.*
2. In seed plants, megaspores or microspores are formed following meiosis; a microspore eventually develops into a *pollen grain,* and a megaspore produces a female gametophyte.
3. *Seeds* are mature ovules of a seed plant, containing an embryo.
4. Conifers are seed plants that produce ovulate and staminate cones.
5. A complete pine seed consists of the embryo embedded in the old female gametophyte, which is surrounded by the seed coat.

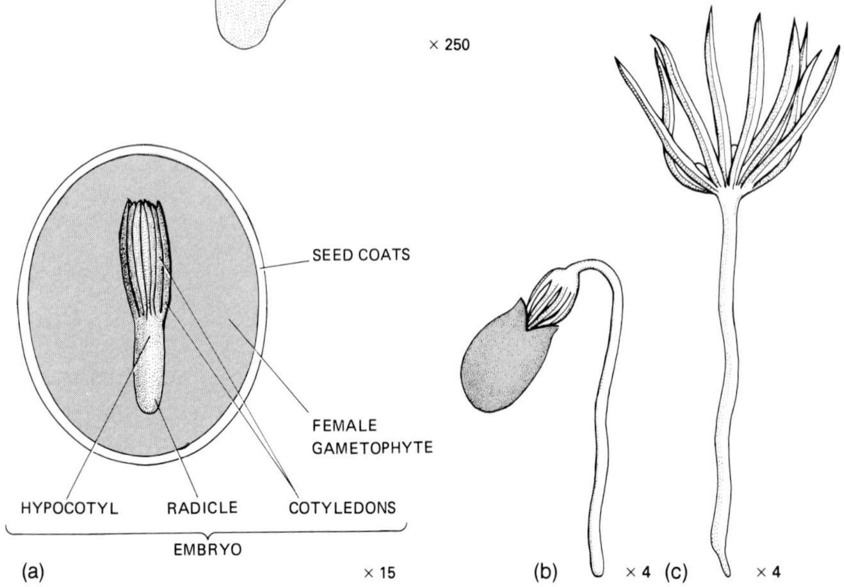

Figure 15.21
Pine seed and seedlings. (a) A long section through a pine seed. The embryo is nourished by the old haploid female gametophyte. The diploid embryo is a small sporophyte; it has an axis, with a growing point at each end and a ring of seed leaves, or *cotyledons.* (b) When the seed germinates, the *radicle* of the embryo stretches, and new cells at the tip contribute to the growing root. As the *hypocotyl* elongates, it drags the seed above ground. (c) When the top of the seedling reaches light, the seed coat and old gametophyte fall off, the hypocotyl straightens and the cotyledons expand. The little pine tree is established and ready to grow on its own.

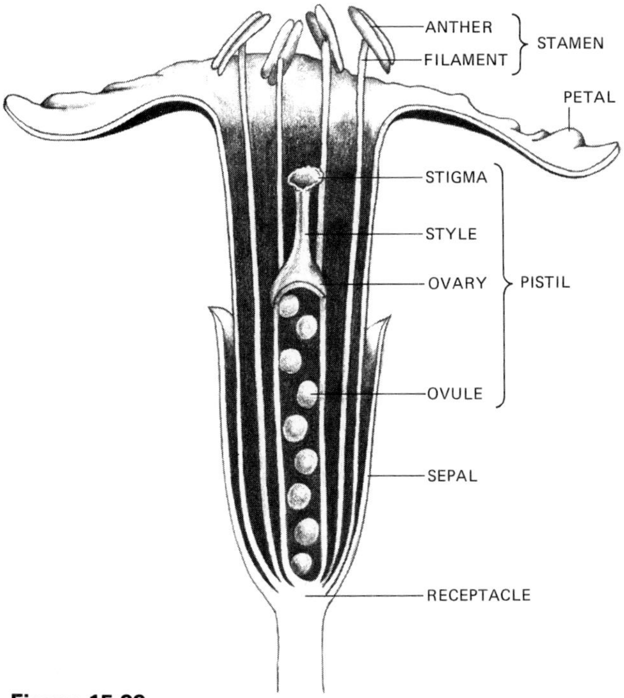

Figure 15.22
A section through a representative flower. The stem end bearing the flower is the *receptacle.* The sepals, usually green, collectively make the *calyx.* The petals, frequently brightly colored, collectively make the *corolla.* The *pistil,* containing the *ovules,* has a more or less elongated *style,* topped by a sticky *stigma* on which pollen lands and germinates.

FLOWERING PLANTS

Flowering plants are the most recent and most complex plants on earth. Since the first flowers on earth apparently were never fossilized, the transition forms that led to modern flowers are not known. One possibility is that flowers evolved in rather dry uplands, where conditions were not favorable for preservation, with the result that the links are indeed missing. Once having developed, flowering plants came to dominate the land, and they are now the most conspicuous plants on the surface of the earth.

Flowers: Structure and Variation

A complete flower has all floral parts present: the usually green *sepals* under the usually colored *petals,* plus the reproductive parts (Figure 15.22). A ring of sepals makes a *calyx,* and a ring of petals makes a *corolla.* In most flowers there is a ring of *stamens* inside the ring of petals, and each stamen has a pollen-bearing *anther* at the end of a supporting *filament.* In the central part of the flower, a *pistil* containing *ovules* carries a columnar *style,* at the top of which is a *stigma,* which receives pollen.

"Textbook flowers," as described above, seldom occur, although poppies and wild roses approach the theoretical type. Most real flowers vary in one or more details, and every imaginable possibility can be found somewhere in nature (Figure 15.23).

Figure 15.23
Representatives of diverse families of flowering plants. (a) A jack-in-the-pulpit (Arum Family). (b) Oats (Grass Family). (c) Apple flowers (Rose Family). (d) Dog-tooth violet (Lily Family).

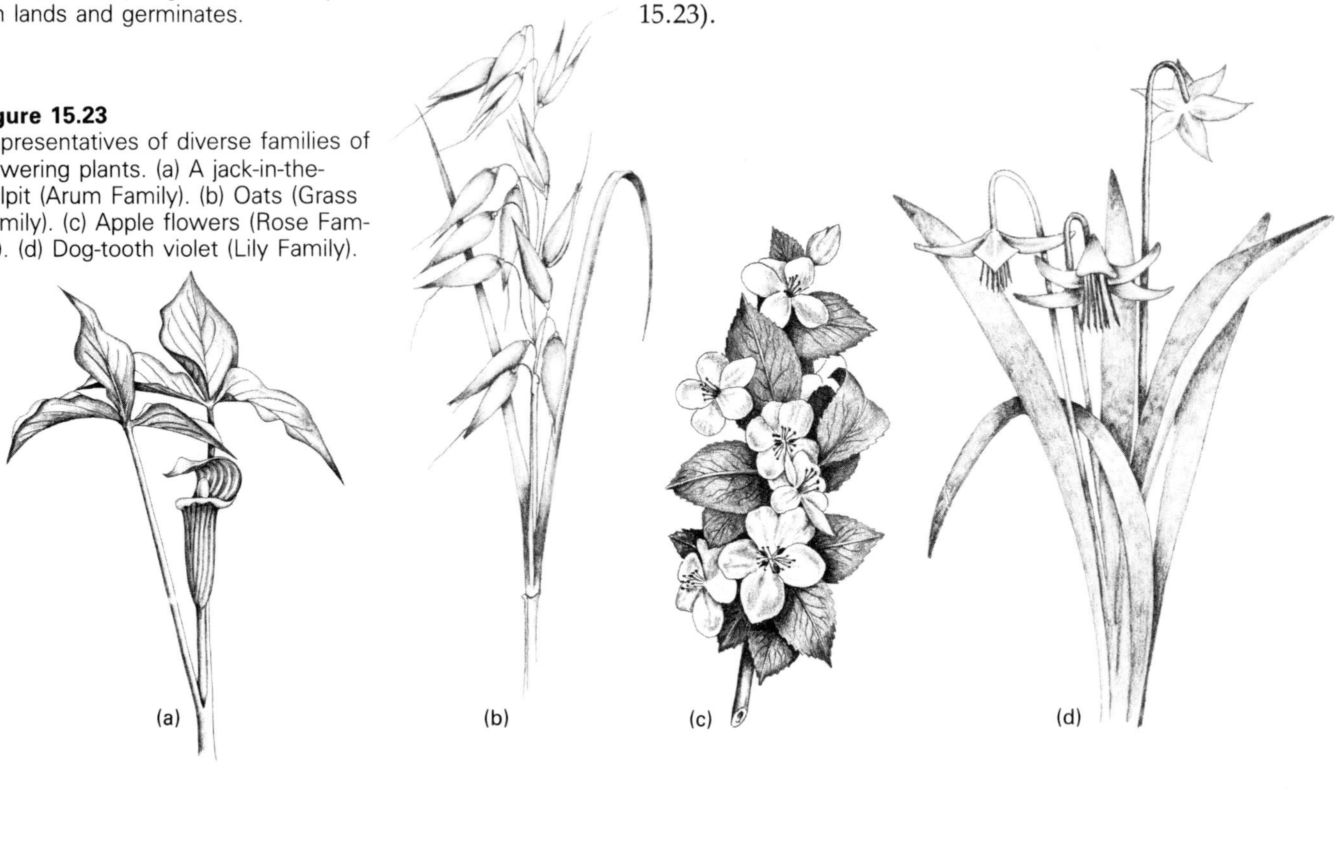

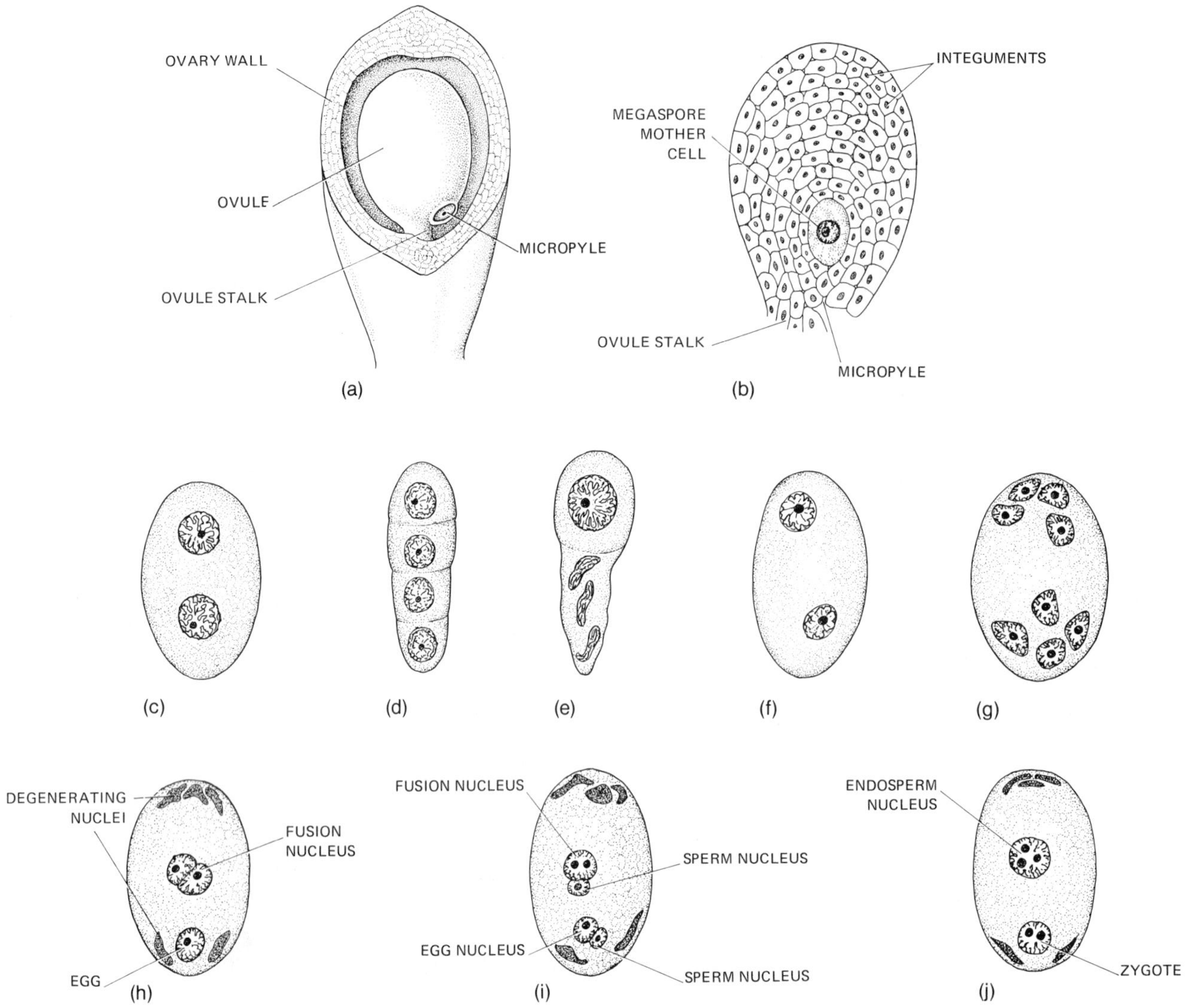

Figure 15.24
Development of the embryo sac in the ovule of a flower. (a) A sectional view into a cut ovary, showing one ovule. (b) A section through an ovule, attached to its placenta. The *micropyle,* through which the pollen tube will pass later, and the *integuments,* which will form the seed coats, are visible but not conspicuous. The largest cell in the ovule is the diploid *megaspore mother cell.* The megaspore mother cell undergoes the first of the meiotic divisions, resulting in a binucleate cell (c). The second meiotic division results in a row of four haploid megaspores (d). One megaspore survives; the other three die (e). The surviving megaspore nucleus divides mitotically to form a binucleate cell (f). Two more mitotic divisions produce an eight-nucleate *embryo sac* (g). Two nuclei, one from each end of the embryo sac, fuse to make a *fusion nucleus.* One nucleus near the micropyle becomes the egg. The other nuclei (three at one end of the embryo sac and two near the egg) disintegrate. The embryo sac is ready for fertilization (h). The pollen grain has deposited two sperm nuclei in the embryo sac (i). One is fusing with the egg nucleus and the other with the (now diploid) fusion nucleus. The act of *double fertilization* is a special feature of flowering plant reproduction. In (j) a haploid sperm and a haploid egg fuse to make a diploid *zygote,* which is ready to develop into an embryo. The fusion of a haploid sperm with a diploid fusion nucleus makes a *triploid endosperm nucleus,* which will produce the triploid endosperm tissue, a nourishing mass surrounding the developing embryo.

Reproductive parts of flowers may even be lacking. An *imperfect flower* lacks one of the reproductive organs; that is, such a flower is either *staminate* or *pistillate.* Many trees (hollies and willows) have imperfect flowers. (If petals or sepals are lacking the flower is simply called *incomplete.*)

In size, flowers range from the almost microscopic blooms of the floating *Wolffias* to the great, heavy magnolias. They may be solitary, as in tulips, or collected into *inflorescences* that contain a few flowers (cherries, for example) to hundreds or even thousands of individual flowers in a single head (sunflowers).

Spore Production in Flowers

A flower is not primarily a sexual structure, nor are pollen grains comparable to sperm cells. The main plant, the part that is conspicuous and does the main metabolic work, is the diploid, asexual sporophyte, and the primary function of the flower is to produce spores. The mechanism of spore production is basically similar in all flowers, although enormous variation occurs in the external appearance of different families of flowering plants.

Like conifers, flowers make two kinds of spores: **microspores,** which can develop into male gametophytes, and **megaspores,** which can develop into female gametophytes. Megaspores are produced in ovules. Ovaries of flowers are not truly comparable to the ovaries of animals, because they do not produce eggs directly following meiosis. The ovule in an ovary of a flower is a sporangium, in which meiosis results in spores, specifically four megaspores. One of these haploid megaspores grows into a female gametophyte inside the ovule, but even at maturity it is microscopic and hidden from external view. Since the ovules are enclosed within the ovaries, the entire group of flowering plants is called the **angiosperms,** which means "seeds in a vessel."

In most flowers, the megaspores divide mitotically, and this division is followed by two more mitoses, so that the mature female gametophyte finally comes to have eight nuclei. It is then called an **embryo sac** (Figure 15.24). Three of the eight nuclei are of special importance: one that will be the egg nucleus, or female gamete, and two that fuse to make a new kind of nucleus, the diploid *fusion nucleus,* which is to function during the development of the seed. At this stage the embryo sac is complete and ready for fertilization.

While the ovules are producing megaspores and eventually embryo sacs, the anthers are in the process of forming pollen. An anther when young usually has four elongated lobes, each filled with tissue composed of diploid cells. Each diploid cell is a *microspore mother cell.* When these cells undergo meiosis, each microspore mother cell yields four haploid microspores. The microspores, which become almost spherical free-floating cells, represent the first stage of a short-lived gametophyte. The microspore nuclei divide mitotically to form a binucleate cell that is soon covered by a firm double coating. This is the pollen grain, and at this stage, many plants split open their anthers to release the pollen. Pollen grains are variously shaped and covered with such distinctive markings that they can be identified, frequently with specific accuracy (Figure 15.25). Once the anthers are open, the pollen is ready to be carried to the stigma of the flower.

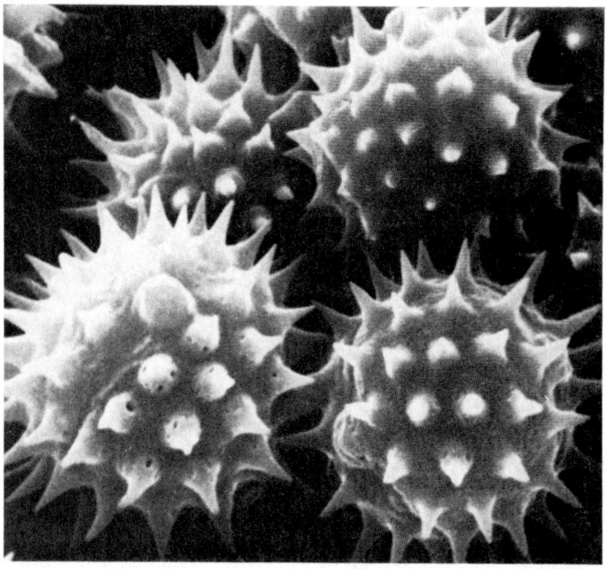

(a) × 4000

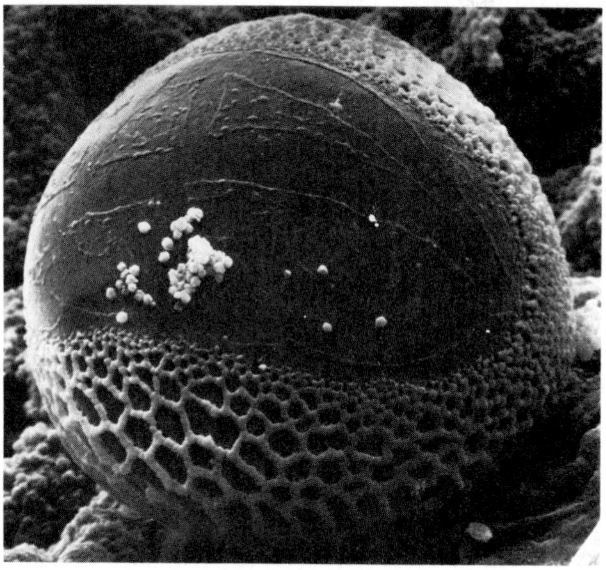

(b) × 2500

Figure 15.25
Scanning electron micrographs of pollen grains. (a) Pollen of *Cosmos.* (b) Pollen of a lily.

Pollination

Pollination is not fertilization; it is only the movement of pollen from anther to stigma. Actual fertilization occurs later. Pollination, however, must precede fertilization, and it is a requirement for initiation of seed growth. Flowers have evolved many methods of ensuring pollination, some of them complicated and involving structural and behavioral modifications of the pollinating agent. For example, flowers pollinated by hummingbirds produce sweet nectar in glands deep in the throat of the flower, and hummingbirds have the hovering flight and the long, thin beaks needed for obtaining the nectar (Figure 15.26). Each species needs the other.

Most flowers, perhaps the earliest insect-pollinated ones, are pollinated by beetles, but other insects are important. Each species is likely to be

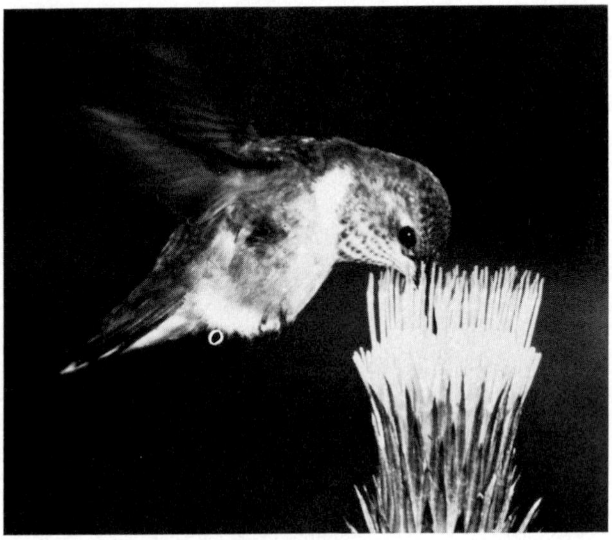

Figure 15.26
Flowers of a thistle being pollinated by a hummingbird. Flowers pollinated by hummingbirds usually have nectar deep in long, narrow floral tubes, where most animals cannot reach it. The long slender beaks of the hummingbirds are well suited to obtaining the nectar, and as the birds feed, they also pollinate the flowers.

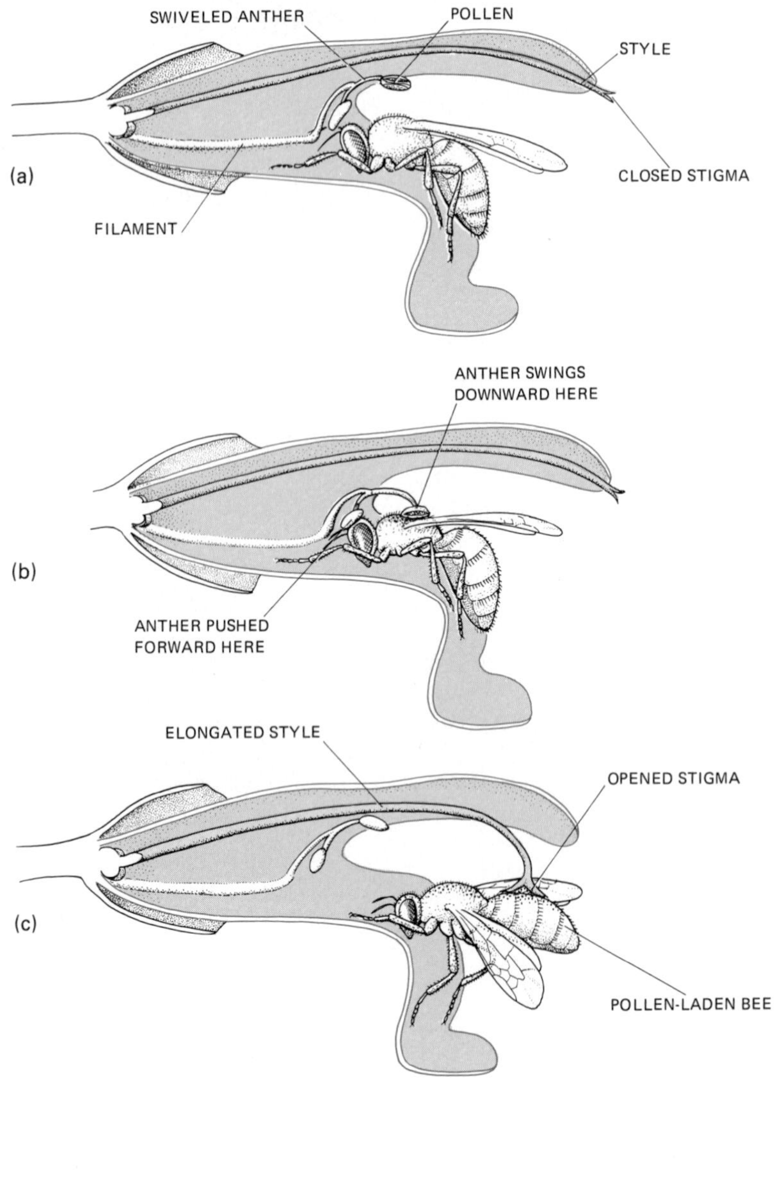

Figure 15.27
Pollination of the sage, *Salvia.* (a) A bee, entering the flower for the nectar at the base of the tube, runs into an anther blocking the tube entrance. The swiveled anther can seesaw like a rocker, so when the bee's head strikes the lower end, the upper end is pushed down against her back, dusting it with pollen, as shown in (b). Note that the stigma is at first high and closed. Later the style elongates and the stigma spreads, as in (c). When a bee, with her head and back loaded with pollen, enters an older flower, she brushes her back against the opened stigma, rubbing pollen onto it. Not only is pollination accomplished, but self-pollination is avoided, and cross-pollination is practically ensured.

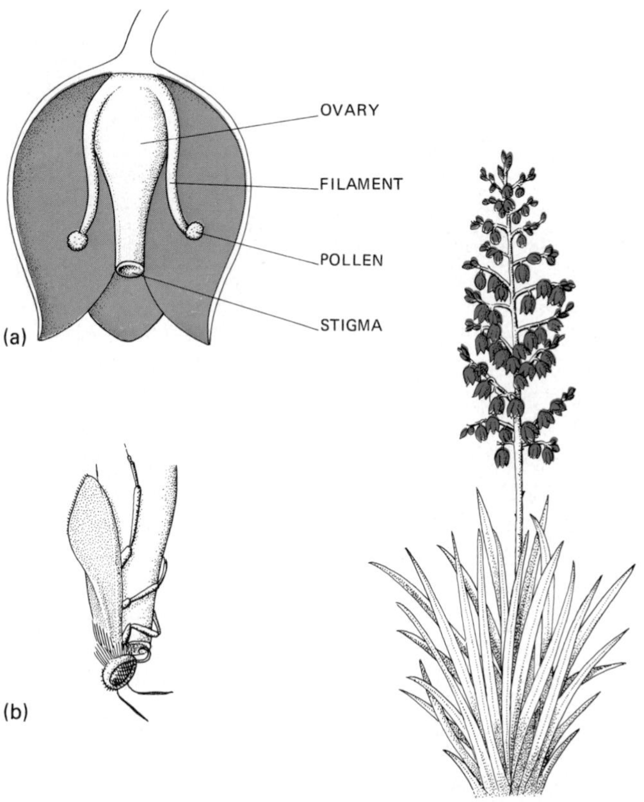

Figure 15.28
Pollination in the rock lily, *Yucca.* (a) The flower has its stigmatic surface on the underside of a concave cup. The pronuba moth drills several holes in the flower ovary and lays eggs among the ovules. (b) Then she gathers balls of pollen and crams them up into the concave stigmatic surface. This action ensures fertilization of the *Yucca* eggs and provides ripening ovules for moth larvae to eat. After the larvae are grown, they eat their way out of the ovary, let themselves down to earth by a spun thread, and pass their pupal stages below ground. The following summer, the adult moths emerge in time for the flowering of the *Yucca,* and the story is repeated. Everything must be integrated: the structural adaptations of moth and flower, the behavior of the moth, and the timing of both organisms. This is coevolution and mutual dependence.

adapted to the visits of some particular insect, whether ants, bees, wasps, butterflies, or moths.

Several instances are known in which the evolution of both flower and pollinator has resulted in carefully controlled mutual benefits. In wild sage, *Salvia,* hinged anthers with movable levers dust the hairy backs of feeding bees with pollen, and the downward pointing stigmas catch the pollen as bees enter the blossoms (Figure 15.27). *Yucca* flowers provide a hatching and feeding ground in their ovaries for pronuba moths, in compensation for which the egg-laying female moth collects pollen and applies it to the inverted, cuplike stigma (Figure 15.28). Both the insect and the plant benefit from the association, and indeed neither could survive without the other. Some orchids (*Ophrys*) have flowers that resemble female bees both in appearance and odor. Male bees, in seeming to attempt mating with a female bee, will bring about the pollination of the orchid plant (Figure 15.29).

It cannot be inferred that either the plant or the pollinator has evolved such elaborate symbioses by an effort of will or intelligent design. There is no evidence that any biological change has ever been so made, regardless of our human habit of ascribing purposeful action to plants and animals. The explanation most satisfactory to contemporary biologists

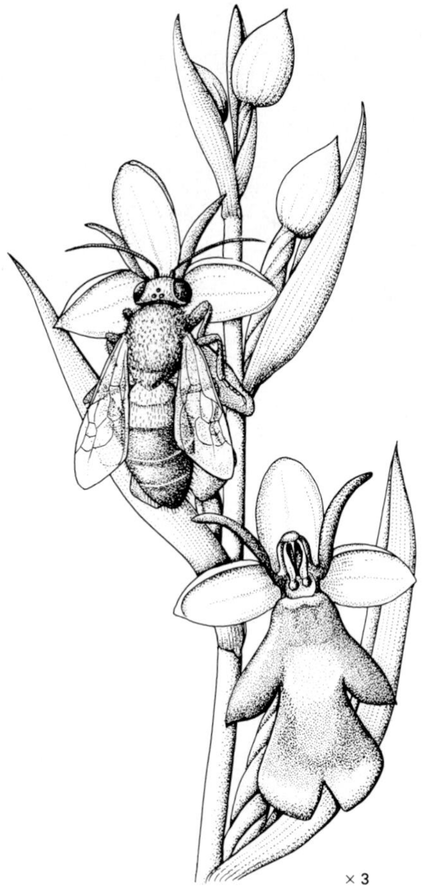

Figure 15.29
The fly orchid, *Ophrys,* has flowers that look and smell so like a female bee that male bees try to mate with them. This action effects pollination.

is that adaptations that happen to function are selected in the competition for survival.

One aspect of pollination is its side effect on humans: It causes hay fever in susceptible victims. Wind-pollinated plants yield great quantities of light, airborne pollen, which can be inhaled and can cause human allergies. Ragweeds, grasses, conifers, and catkin-bearing trees are the most troublesome (Figure 15.30). In general, such plants have small, inconspicuous flowers. Brightly colored flowers like goldenrods, even though they are popularly accused of causing hay fever, are insect-pollinated and do not make enough pollen to affect people.

Seed and Fruit Formation

When a pollen grain arrives on a receptive stigma, it sends out a tube. This is the *pollen tube,* which grows down through the style of the flower, until it comes to an ovule. There it finds its way into the ovule and penetrates the embryo sac. By this time, pollen tubes generally have three nuclei: a tube nucleus and two sperm nuclei. In most flowers, the nucleus that results from the fusion of the fusion nucleus and the second sperm nucleus is a triploid nucleus, that is, with three sets of chromosomes (see Figure 15.24i). The union of one sperm nucleus with the fusion nucleus and the other with the egg constitutes the phenomenon of *double fertilization,* which is characteristic of flowering plants but of no other plant group.

After double fertilization has occurred, the zygote nucleus and the endosperm nucleus divide repeatedly. From the zygote comes the embryo, an immature plant of the new sporophyte generation (Figure 15.31). An embryo in a seed, even when the seed is ripe, is an incompletely formed plant, but it usually has a rudimentary stem with an apical growing point at the "upper" end and a root growing point at the other, along with one or two special structures called seed leaves, or **cotyledons.** Meanwhile, the endosperm nucleus produces a formless mass of triploid tissue, the **endosperm.** The endosperm functions as a nourishing tissue for the growing embryo, sometimes becoming larger than the embryo itself, as in coconuts or the cereal grains.

Figure 15.30
Wind-pollinated flowers. (a) The *catkins* of the aspen (*Populus*). (b) Ragweed *(Ambrosia)* grows in waste places, producing great quantities of pollen from the small, green flowers. Ragweed is one of the worst hay fever plants in the northern temperate zone.

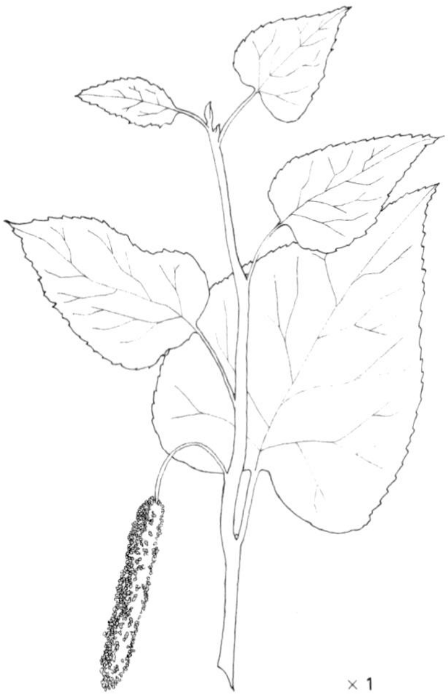

(a) × 1

(b) × 0.75

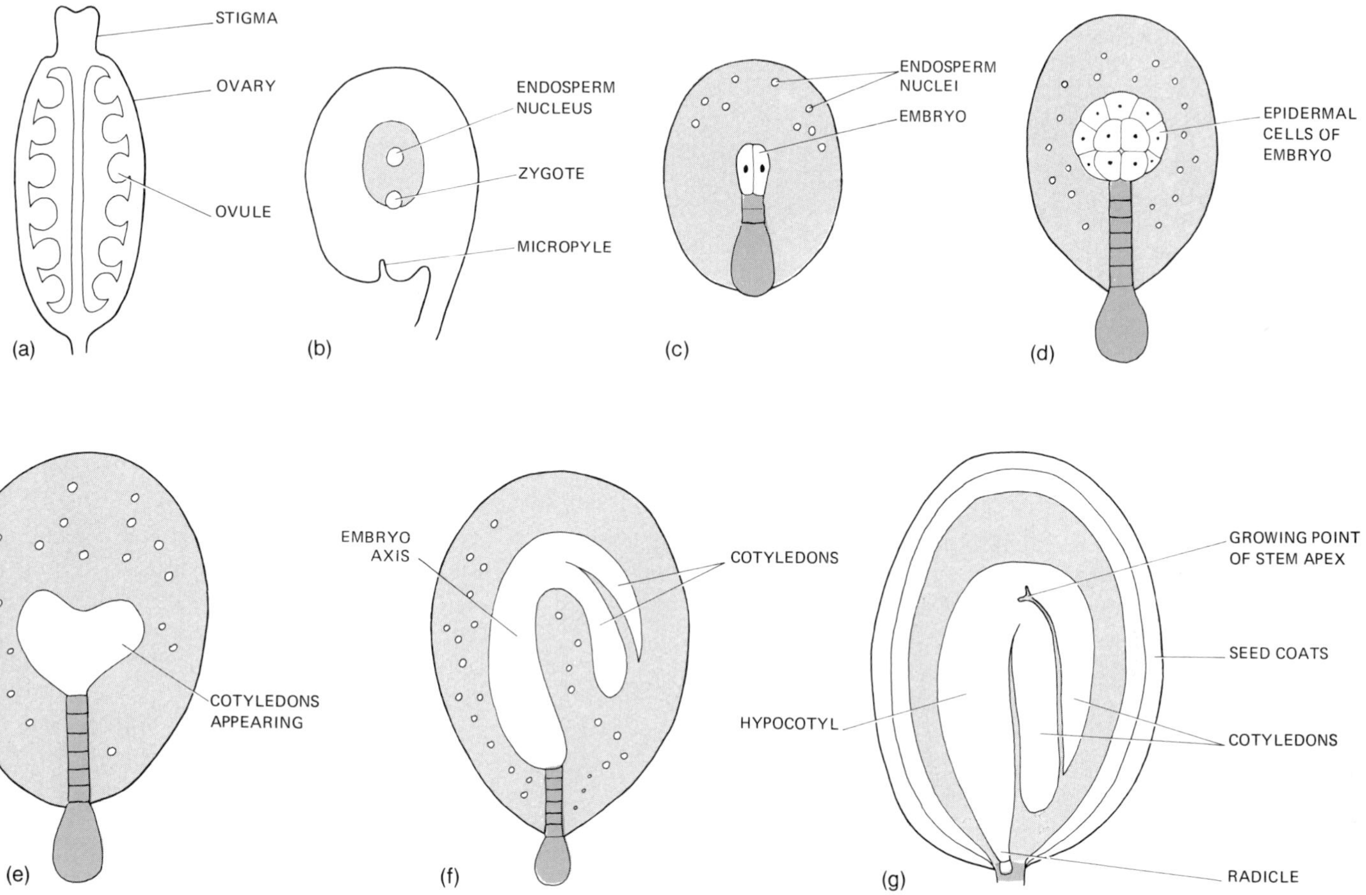

Figure 15.31
Development of a seed. (a) A long section of an ovary with young ovules. (b) An ovule, sectioned longitudinally. Fertilization has just taken place. There is a diploid zygote and a triploid endosperm nucleus. (c) The endosperm nuclei are dividing and making a loose mass of tissue around the growing embryo. The embryo has one enlarged cell at the base, a short row of cells, and the embryo proper, which at this stage has only two cells. (d) Further cell divisions in the embryo have produced an outer layer of cells that will form the epidermis of the future plant. (e) The endosperm continues to grow into a diffuse covering around the embryo, which has grown into a heart-shaped mass, indicating the future cotyledons. (f) Although there is still some endosperm, most of the space in the ovule is taken up by the growing embryo, which has an axis and two cotyledons that have been bent around. (g) The mature seed has seed coats derived from the integuments of the ovule. The endosperm has been absorbed, and the embryo practically fills the entire seed. The axis of the embryo consists of a *hypocotyl* with a root growing point (the *radicle*) at the lower end and a large pair of cotyledons filling the ovule. An apical growing point between the cotyledons will make the stem apex of the future plant.

However, the endosperm may be used up by the time the seed is mature, so that the finished seed contains an embryo but no endosperm. Seeds of peanuts and beans are examples of seeds without endosperm.

Since the bulk of the world's food supply consists of flour or meal from wheat, rice, corn, millet, barley, and oats, and since the bulk of each of these grains is made of endosperm tissue, it is evident that humans live largely on triploid tissue (Figure

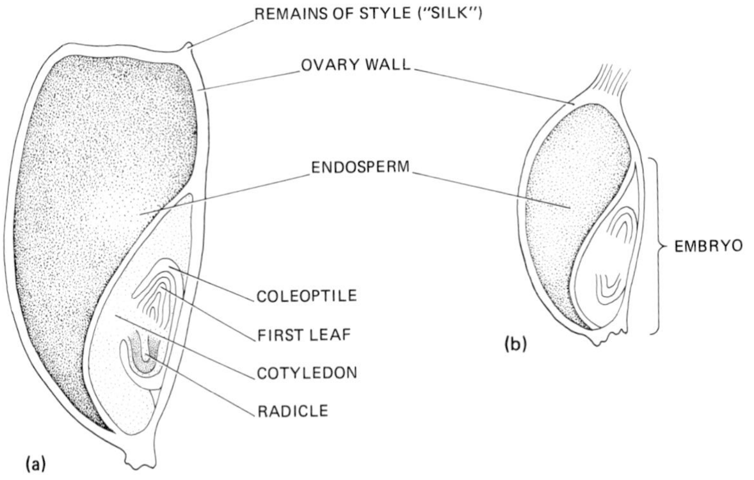

Figure 15.32
Sections through fruits of two of the world's most important food plants. (a) Corn, or maize (*Zea*). The most conspicuous part is the starchy endosperm. Its outermost cells are heavily impregnated with protein. The embryo, too, contains protein, as well as quantities of fat, the source of corn oil. The first leaf has a special covering, the coleoptile. (b) Grain of wheat, generally similar to that of corn but with a relatively larger embryo. The endosperm is the source of wheat flour. Note that a grain is not simply a seed but a complete fruit covered by the ovary wall.

15.32). In all these grains, the embryo is a relatively small object, popularly known as the "germ," and it is usually removed from the grain during the process of making flour or when the grain is being finished for commercial sale.

As the endosperm and embryo grow, so do the integuments of the ovule and the walls of the ovary. At maturity the ovule has become a seed, the integuments have become seed coats, and the ovary wall has become the outer covering of the fruit. A fruit is in fact a ripened ovary, although it may include some nonovarian tissue in some plants (Figure 15.33). A pea pod is simply an ovary with seeds in it, but apples and strawberries are quite different. The ovary of an apple flower is only the core, and the edible part of the apple consists essentially of stem. In a strawberry, the ovaries are what are usually called "seeds" on the berry, and the juicy part of the berry is, once more, a swollen, sweet, juicy stem.

FRUITS WITHOUT SEEDS

Bananas do not have seeds; at least those commonly sold in markets do not. Neither do pineapples and some varieties of grapes, oranges, and grapefruit. Where do such fruits come from? And how is it possible to produce them in commercial quantities?

The answer to the first question is: They come from seeded parents that underwent some mutation that prevented ripening of the ovules but allowed the development of ovary walls and accessory parts. Most fruits abort when seeds are not formed, but some, known as *parthenocarpic* (Greek, "virgin fruit"), grow almost normally even without seeds. Most parthenocarpic fruits sold commercially were found by chance, appreciated by their discoverers, and saved from extinction by human interference.

Once obtained, a seedless fruit may be propagated in several ways. Seedless grapes and citrus fruits, such as seedless oranges, are propagated by grafting seedless branches on ordinary rootstocks. After a few plants begin growing and branching, practically unlimited numbers of grafted seedless plants can be produced, so that extensive vineyards or orange groves are possible. Bananas and pineapples are propagated by cutting away and transplanting side shoots.

Even seedless fruits have ovules, but they do not mature. Ovules can be seen as black flecks in bananas or as pinpoint-sized white granules in pineapples. In some varieties of both fruits, real seeds do mature. Banana seeds are shiny, brown, and flattened, like small beans. Pineapple seeds, when ripened, look like apple seeds. Seed-bearing individuals may be used in plant-breeding stations to yield new varieties, adding to the stock of variability and offering a chance to combine mutations in novel, possibly desirable, ways. In this manner, many seedless varieties can be produced.

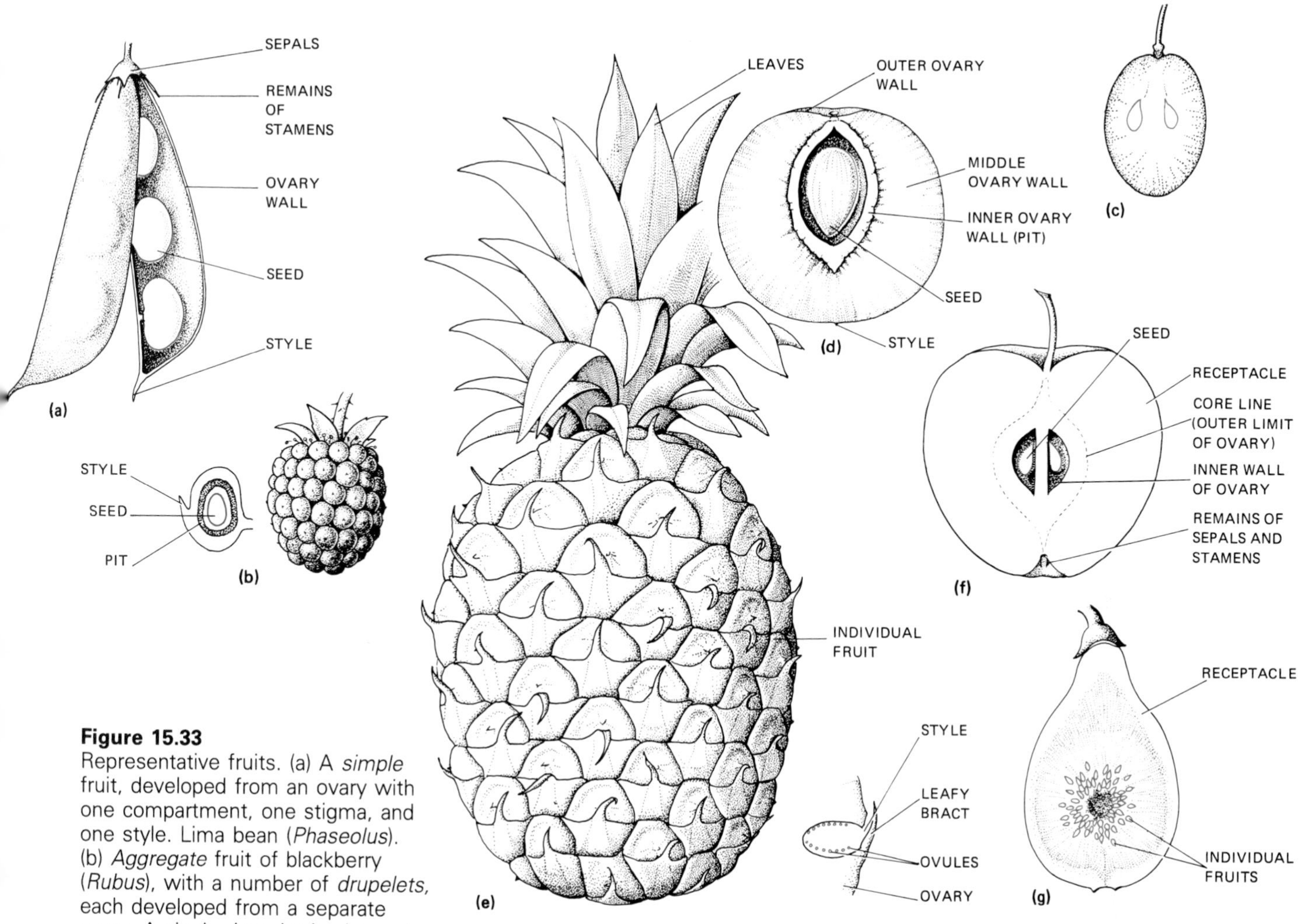

Figure 15.33
Representative fruits. (a) A *simple* fruit, developed from an ovary with one compartment, one stigma, and one style. Lima bean (*Phaseolus*). (b) *Aggregate* fruit of blackberry (*Rubus*), with a number of *drupelets,* each developed from a separate ovary. A single drupelet is shown in section with its pit inside. (c) *Berry* of grape (*Vitis*). The fleshy part is the ovary wall, and the seed coats are hard. (d) *Drupe* of peach (*Prunus*). The skin is from ovarian epidermis, the edible part is the middle ovary wall, and the hard pit is the inner ovary wall. The seed, covered with brown seed coats, is inside the pit. Almonds, nectarines, cherries, plums, and prunes are similar. (e) *Multiple fruit* of pineapple (*Ananas*). This is a mass of fruits, each derived from a separate flower. A single flower is shown in section, with tiny aborted ovules inside. (f) An apple (*Pyrus*) is mostly receptacle (stem) tissue. The ovary makes the core of the apple, and the papery inner layer of the ovary covers the seeds. (g) A fig (*Ficus*) is a special case of a multiple fruit, with the so-called seeds, structurally individual fruits, each derived from a separate flower.

POINTS TO REMEMBER

1. The embryo in a seed is not the direct offspring of the plant that bears it. It is rather the offspring of a pair of gametophytes contained in the parent plant, or more accurately, the grandparent plant.
2. A seed contains tissues of three generations: (a) The seed coats belong to the original plant. (b) The pollen tube and embryo sac belong to the next, or gametophyte, generation although by the time a seed is mature, those two parts are usually destroyed. (c) The embryo belongs to the new sporophyte generation.

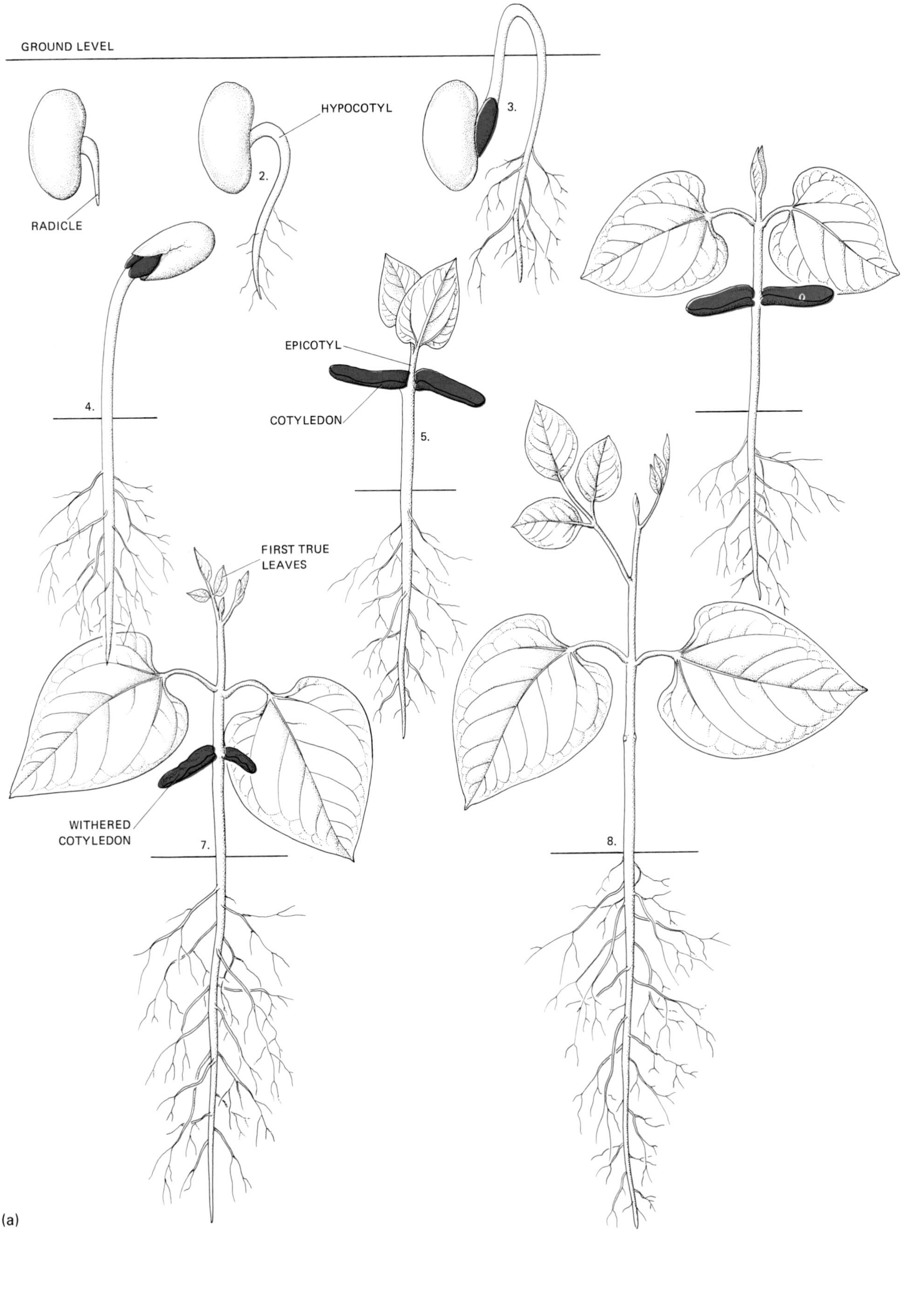
GROUND LEVEL
HYPOCOTYL
RADICLE
2.
3.
4.
EPICOTYL
COTYLEDON
5.
FIRST TRUE
LEAVES
WITHERED
COTYLEDON
7.
8.

(a)

Figure 15.34
Germination of seeds. (a) Like all seeds, beans start germination by putting forth a root by extension of the *hypocotyl* and growth of its tip, the *radicle* (1). As the hypocotyl stretches (2), it arches up, shoves its way through the soil, and pulls the *cotyledons* along (3). Once into the light aboveground, the arch straightens (4), and the cotyledons expand (5). The stem above the cotyledons is the *epicotyl.* The growing point puts forth a pair of *juvenile leaves,* unlike any that will develop later (6), and then produces the first typical bean leaves (7). Meanwhile secondary roots branch from the first root, or primary root, and the plant is established (8). (b) Corn grains also start germination by sending out a root (1). The first part of the upward growing shoot to emerge is a slender, thimble-shaped *coleoptile,* which covers the first leaf (2). The *cotyledon* remains inside the grain, acting as an absorbing organ that transports food from the bulky endosperm to the growing seedling. After the first leaf breaks through the coleoptile and uncurls, other leaves follow, and the seedling becomes independently photosynthetic (3-5). The root system consists not only of the primary and secondary roots but of *adventitious* roots, which sprout from the base of the stem and will eventually become the main root system of the mature plant.

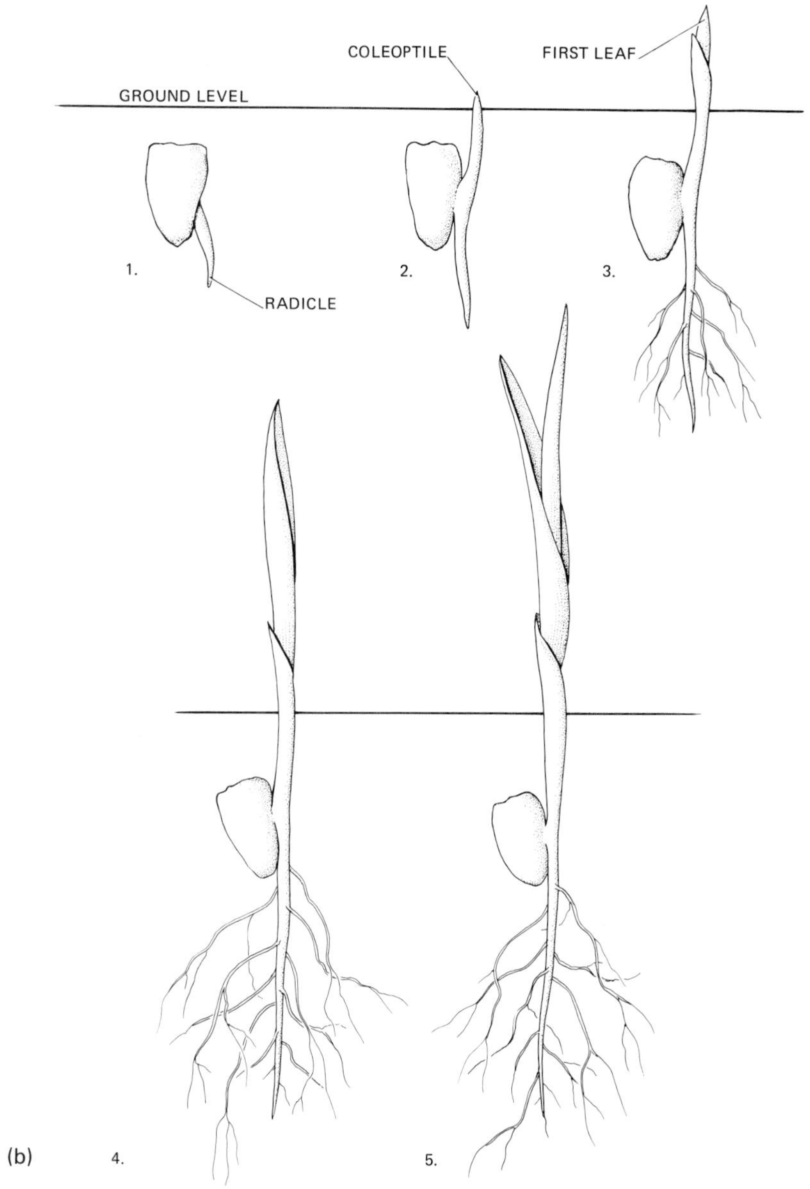

A seed is essentially an embryo in a protective coat, waiting in suspended animation until proper warmth, moisture, and sometimes sufficient aging make germination possible. Seed production is an effective way for plants to wait out cold or dry periods safely. The higher animals have no such dormancy period; when a mammalian embryo starts life, it must proceed to be born and grow up. The only alternative is death.

Flowering plant seeds are of two possible kinds: those with one embryonic leaf (**monocotyledons,** such as grasses, lilies, and palms) and those with two embryonic leaves (**dicotyledons,** such as beans, walnuts, and oaks). When a seed sprouts, it first grows a root downward (Chapter 17) and then a shoot upward (Figure 15.34). Once the shoot reaches the light, it develops chlorophyll and begins manufacturing its own food, but until that time it depends for its energy and material supply on nutrients stored either in the embryo itself or in the endosperm.

The ability of plants to pack a rich supply of carbohydrates, fats, proteins, vitamins, and minerals into a small volume of a seed has been important to animals as well as to the seeds themselves. Animals find seeds such a nourishing, concentrated food supply that in many instances animal survival depends on an adequate seed crop. It is probably no coincidence that the evolution of the warm-blooded animals came only after the evolution of seeds, since the availability of a constant high-energy source is essential to furnishing the warmth that such animals require.

TO SUM UP FLOWERS

1. A complete flower has sepals, petals, anthers, and pistil.
2. Flowers make microspores in anthers and megaspores in ovules. The megaspores divide mitotically to become *embryo sacs.*
3. Pollination is the movement of pollen from anther to stigma.
4. In *double fertilization,* one sperm nucleus unites with a diploid endosperm nucleus and a second sperm nucleus unites with the egg.
5. *Cotyledons* are seed leaves in an embryo. Flowering plants may be *monocotyledons* (having one embryonic leaf) or *dicotyledons* (having two embryonic leaves).

SUMMARY

1. Most multicellular animals use strictly sexual reproduction, but plants commonly use both sexual and asexual methods, varying from species to species and occasionally within a single species.
2. Asexual reproduction results in a slower distribution of variation through a population than does sexual reproduction, but it has several advantages. It bypasses the need for an organism to find a mate, it can be exceedingly fast, and all the offspring are genetically identical, a situation that often is desirable.
3. The simplest example of asexual reproduction is *fission,* the division of a single cell and the separation of the two resultant cells.
4. Most plants produce special reproductive cells called *spores* that do not involve nuclear fusion.
5. Higher plants may reproduce asexually, or *vegetatively,* through processes in which a new complete plant is produced from a root, stem, or leaf of the parent plant, sometimes with human interference by cutting, grafting, or layering.
6. To guarantee biological uniformity, people use *clones,* or groups of individuals all having the same genetic background, obtained without an intervening sexual fusion and identical to the parent from which they came.
7. The simplest sexual process is the joining of two free-living cells, with cytoplasmic union followed by nuclear fusion. The next step in sexual complexity came with a visible difference between the sex cells, a slight difference in size or pigmentation. Still another advance came with a change in motility. The larger sex cell lost its ability to move while the smaller remained active. Another innovation was the development of multicellular bodies in which only a special portion of the organism is devoted to gamete production. The most recent complication was the provision for parental care of the new generation.
8. In the heterogeneous assemblage of plants popularly known as *algae,* reproductive methods are as various as the plants themselves.
9. Reproduction in *fungi* is as varied as it is in algae, and some fungi exhibit the most complex reproductive patterns to be found anywhere in nature.
10. Most plants have a whole haploid phase during which gametes are produced without meiotic divisions; that phase is the *gametophyte generation.* The union of gametes produces a diploid zygote, which may grow mitotically into a diploid plant. At maturity, some of the diploid plant's cells undergo meiosis, and the resultant haploid cells are spores; this is the *sporophyte generation.* The whole phenomenon is called *"alternation of generations."*
11. *Mosses* are the simplest living plants in which the gametes are produced in special multicellular organs, the *antheridia* and *archegonia.* Algae and fungi, in contrast, produce their gametes inside single cells. The mosses are also notable in that they nourish embryonic sporophyte plants in the archegonia.
12. In mosses the gametophyte is the larger and more actively photosynthetic generation, but in *ferns* the arrangement is reversed.
13. Spores that are destined to produce female gametophytes are usually relatively large, and they are

therefore called *megaspores*. The smaller spores, which will develop into male gametophytes, are *microspores*. The phenomenon of making two kinds of spores, called *heterospory*, is typical of seed-bearing plants. Megaspores are produced in megasporangia, which in seed plants are called *ovules*. Microspores are produced in microsporangia, which are called *anthers*. When the gametophytes have produced eggs and sperm, and the resulting zygotes have begun to be embryos, new sporophyte generations have begun.

14. *Coniferous plants* are seed plants, but they have no flowers, and they are less advanced in an evolutionary sense than flowering plants. They are woody, and most bear needle-shaped evergreen leaves.

15. *Flowering plants* are the most recent and most complex plants on earth. Flowering plants are now the most conspicuous plants on the surface of the earth.

16. An ideal complete flower has all floral parts present: the usually green *sepals* under the usually colored *petals*, plus the reproductive parts. A ring of sepals makes a *calyx*, and a ring of petals makes a *corolla*. In most flowers there is a ring of *stamens* inside the ring of petals, and each stamen has a pollen-bearing *anther* at the end of a supporting *filament*. In the central part of the flower, a *pistil* containing *ovules* carries a columnar *style*, at the top of which is a *stigma*, which receives pollen.

17. A flower is not primarily a sexual structure, nor are pollen grains comparable to sperm cells. The main plant, the part that is conspicuous and does the main metabolic work, is the diploid, asexual sporophyte, and the primary function of the flower is to produce spores. After spore production, the sporophyte protects and nourishes the gametophyte and later the fruit and seed. Like conifers, flowers make two kinds of spores, microspores and megaspores. The entire group of flowering plants is called the *angiosperms*.

18. *Pollination* is not fertilization; it is only the movement of pollen from anther to stigma. Actual fertilization occurs later. Pollination, however, must precede fertilization, and it is a requirement for seed setting.

19. A seed contains tissues of three generations: The *seed coats* belong to the original plant, the *pollen tube* and *embryo sac* belong to the next, or gametophyte, generation, and the *embryo* belongs to the new sporophyte generation. A seed is essentially an embryo in a protective coat, waiting in suspended animation until proper warmth, moisture, and sometimes sufficient aging make germination possible.

20. Flowering plant seeds are of two possible kinds: those with one embryonic leaf, *monocotyledons*, and those with two embryonic leaves, *dicotyledons*.

ASK YOURSELF

1. Give some examples of plant evolution in which later plants showed greater complexity than earlier ones. Give other examples in which simplification rather than complication evolved.
2. What is the major difference between a spore and a seed?
3. What is a clone?
4. What is the simplest possible method of sexual reproduction? Give an example.
5. How are spores formed in mushrooms?
6. What are the distinguishing features of a gametophyte plant (chromosome condition, reproductive product)? Of a sporophyte?
7. Where are the spores of mosses formed?
8. How is it known that the large, familiar stage of fern plants is the sporophyte generation?
9. What do megaspores produce? Microspores?
10. What do archegonia in a fern produce? Antheridia?
11. What is a cotyledon?
12. What is the difference between a microspore and a pollen grain?
13. Justify the statement that the main plant in a flowering species is the nonsexual phase.
14. What is "double fertilization" in a flowering plant?
15. What is the difference between pollination and fertilization?
16. What is the function of endosperm in the seed of a flowering plant?
17. What is the chromosome condition (ploidy) in endosperm?
18. Why are monocots so named?

16
Development of Higher Plants

SOME KEY POINTS

1. The usual vascular plant has a shoot with a growing point at the tip, leaves produced at nodes, branches from side buds, and roots with growing tips.
2. Roots are anchoring and absorbing organs.
3. Stems are supporting and conducting organs.
4. Thickening of stems and growth of wood is due to cell divisions in a cambial layer beneath the bark.
5. Leaves are adapted for maximum photosynthetic efficiency.

WHEN PEOPLE SAY "PLANTS," THEY ALMOST invariably mean vascular plants, those with woody conducting tissues, including the ferns and fernlike plants, the cone-bearing trees, and the woody and herbaceous flowers. In the present chapter, we are concerned with the general body plan of vascular plants, with the developmental processes by which mature form is achieved, and with some of the variations that have evolved from a presumably common ancestor.

There is one basic plan for all vascular plants. It consists of a *stem,* with a terminal growing point, bearing leaves along its length, and a *root,* also with a terminal growing point. Both stem and root contain internal tissues with conducting ability, the downward-flowing *phloem* and a more central, upward-flowing stiffening tissue, the *xylem.* Two major variations on that theme have proved successful. One is that of the nonflowering vascular plants and the dicotyledons, exemplified by most common herbs and woody trees, and the other is that of the monocotyledons, exemplified by grasses, lilies, and palms. (For their evolutionary position, see Chapter 9.) In spite of numerous and sometimes bizarre variations in some plant families, these two main types are so fundamental to the general scheme of higher plant structure that if they are understood, then the structure of almost any plant encountered in ordinary experience will be recognizable. Figures 16.1 and 16.2 offer a schematic comparison of the monocotyledonous and dicotyledonous plant bodies. Botanists usually shorten these terms to **monocot** and **dicot.**

Superficially a head of cabbage is enormously different from a tropical vine stretching almost into

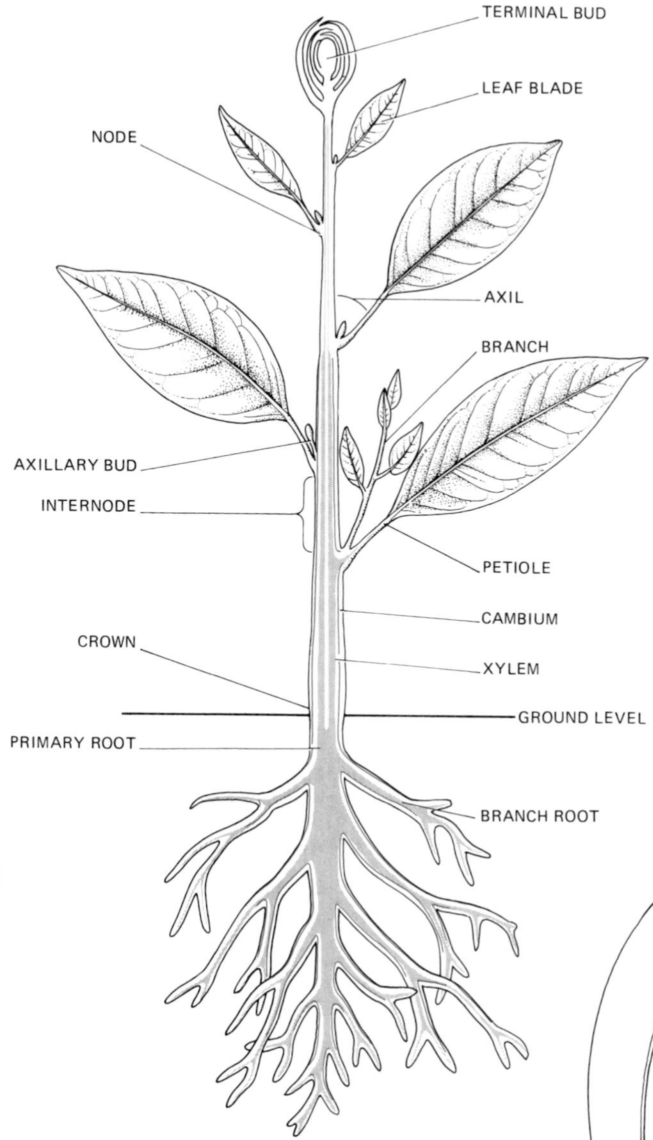

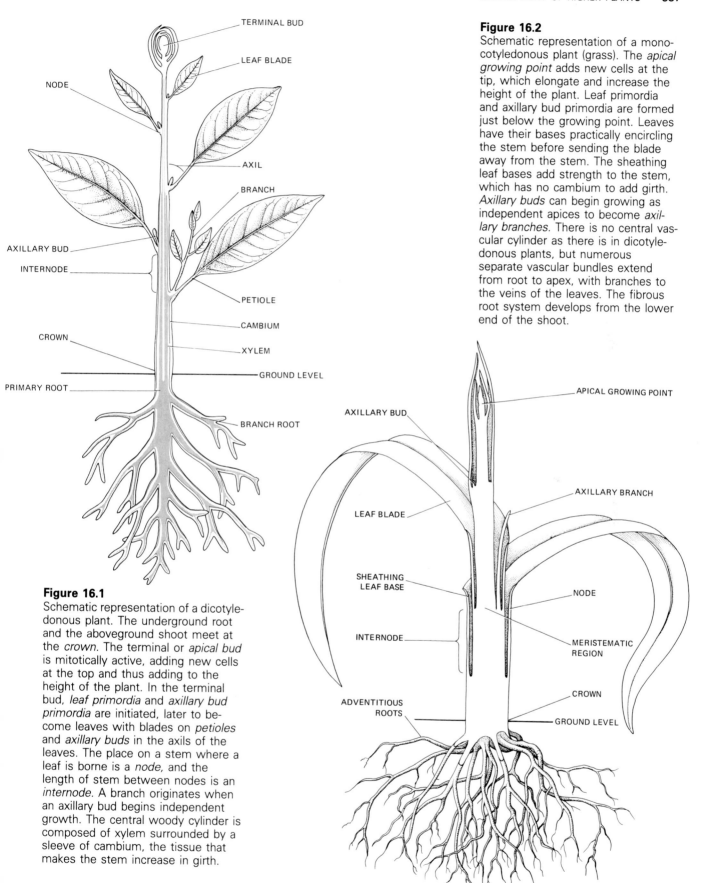

Figure 16.1
Schematic representation of a dicotyledonous plant. The underground root and the aboveground shoot meet at the *crown.* The terminal or *apical bud* is mitotically active, adding new cells at the top and thus adding to the height of the plant. In the terminal bud, *leaf primordia* and *axillary bud primordia* are initiated, later to become leaves with blades on *petioles* and *axillary buds* in the axils of the leaves. The place on a stem where a leaf is borne is a *node,* and the length of stem between nodes is an *internode.* A branch originates when an axillary bud begins independent growth. The central woody cylinder is composed of xylem surrounded by a sleeve of cambium, the tissue that makes the stem increase in girth.

Figure 16.2
Schematic representation of a monocotyledonous plant (grass). The *apical growing point* adds new cells at the tip, which elongate and increase the height of the plant. Leaf primordia and axillary bud primordia are formed just below the growing point. Leaves have their bases practically encircling the stem before sending the blade away from the stem. The sheathing leaf bases add strength to the stem, which has no cambium to add girth. *Axillary buds* can begin growing as independent apices to become *axillary branches.* There is no central vascular cylinder as there is in dicotyledonous plants, but numerous separate vascular bundles extend from root to apex, with branches to the veins of the leaves. The fibrous root system develops from the lower end of the shoot.

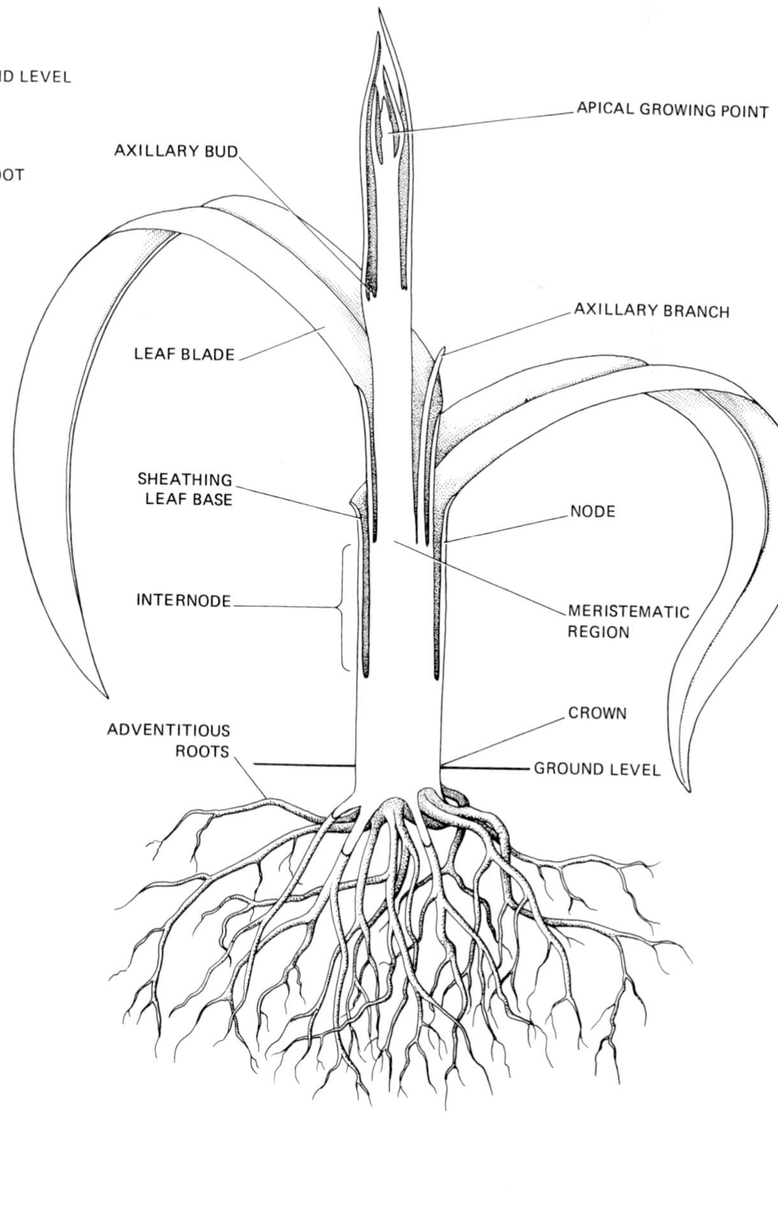

invisibility to the top of a mahogany tree, and the fine turf of a putting green seems to have little in common with a giant bamboo, but the way these dissimilar plants grow and their structural similarities, once discovered, are clearly fundamental.

GERMINATION IS THE RECOMMENCEMENT OF THE LIFE OF A SEED

In order to start growing, a seed must be physiologically prepared. This may simply mean that there is a temperature above freezing, an adequate oxygen supply (because a germinating seed has a high respiratory rate), and enough water. Some seeds have additional special requirements, such as light, a ripening time to allow enzymes to act, the decay of a hard seed coat, or the presence of specific external stimulating compounds. But *water, warmth,* and *oxygen* are universally needed, and indeed the best way to preserve living, dormant seeds is to keep them dry, cold, and airless.

A mature, viable seed, ready to germinate, provides a reasonable starting point for a discussion of vascular plant development and anatomy.

When the embryonic plant in a seed begins to grow, the first activity is in the **radicle,** the lower tip of the embryonic axis. Cells expand, and mitotic divisions begin at the extreme tip, causing the first or **primary root** to push through the seed coat. In a dicot seed such as a radish, the two embryonic leaves, the cotyledons, begin to expand. With the primary root growing down and that portion of the shoot below the cotyledons (the **hypocotyl**) expanding, the cotyledons are pushed up until they are above the ground surface. They then open out, turn green, and expose the apex of the stem. As the primary root grows, it sends out branches, **secondary roots,** which firmly anchor the seedling. At the same time the stem apex is mitotically active, growing upward and beginning to produce the first true leaves. The length of stem above the attachment of the cotyledons, the **epicotyl,** elongates, and the seedling is on its way to becoming a self-sustaining plant.

Monocot seeds, such as grasses, begin germination by sending out a primary root, but the single cotyledon acts as an absorbing organ that swells and pushes itself into the seed's main food supply in the endosperm (Chapter 15). The apical growing point is covered by a *coleoptile,* inside which the first true leaf, like a dagger in a sheath, grows upward. The leaf pushes through the coleoptile, splitting it and leaving it as a temporary basal collar. With a primary root established and a true leaf extending into the light, a monocot seedling is established.

ROOTS ARE THE ANCHORING AND ABSORBING ORGANS OF A PLANT

When an expanding radicle breaks out from a germinating seed, part of the increase in size is simply the swelling of cells, but such increase is limited, and soon mitosis commences at the growing root tip. As new cells are formed, some remain as a permanent part of the root, but others are pushed forward (that is, usually downward) beyond the actual root tip, and they form a loose covering over the tip. The thimble-shaped cover is the **root cap,** whose wet, slimy cells make it possible for the elongating root to ease its way among the soil particles. As roots keep growing, root caps are continually worn away and replaced. At the very apex of the root tip is a *quiescent zone,* in which mitosis is minimal. Most mitotic division takes place in the tissue just behind the quiescent zone.

Once a cell has divided into two cells, the new cells increase in size, not only taking in water but synthesizing new proteins and other protoplasmic compounds. Simple increase in size is accompanied by little or no apparent increase in complexity, but once full cell size is reached, the cells begin to *differentiate.* Some round up to become storage cells, some elongate and thicken to become conducting cells, and some send out extensions to become **root hairs.** The regions where these events occur can be seen in thin sections of roots: (1) mitoses in the region of cell division, (2) an obvious increase in cell size in the region of elongation, and (3) an increase in diversity of shape and wall thickness in the region of differentiation. Long sections of root tips are frequently used in studies of differentiation because they show plainly the events through which all cells pass from first formation to maturity (Figure 16.3).

Some of the epidermal cells back of the region of elongation are capable of pushing out extensions as long tubular growths, the root hairs. These structures, which grow thickly enough to give a young root a fuzzy appearance, increase the absorbing surface of the root by about twentyfold. They squeeze their way among the soil particles, coming into such close contact that they can absorb almost all the water. Root hairs generally last only a few days, dying as new ones are grown near the growing point of the ever-elongating root (Figure 16.4).

By the time a root reaches functional maturity, it has a number of differentiated tissues, each with its own work to do. The outermost layer of epidermis not only bears root hairs but does some absorbing of water and mineral salts. Inward from the epidermis a usually voluminous cylinder of cortex is

Figure 16.3
Diagrammatic view of a median long section through a root tip. In the *procambium* just behind the root cap, mitoses increase the number of cells, which elongate as they age, pushing the root tip forward through the soil and driving the root cap ahead of the tip. Within the space of a few millimeters, cells can be found in all stages of development, from the perpetually embryonic mitotic region, through the region of cell enlargement, and into the region of final differentiation.

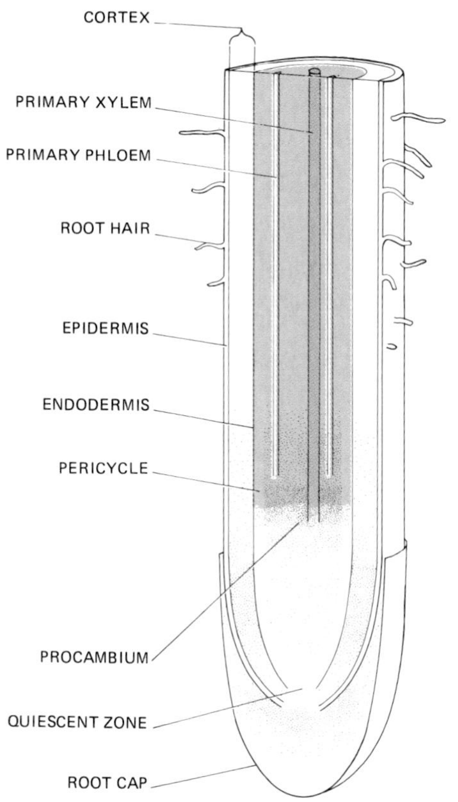

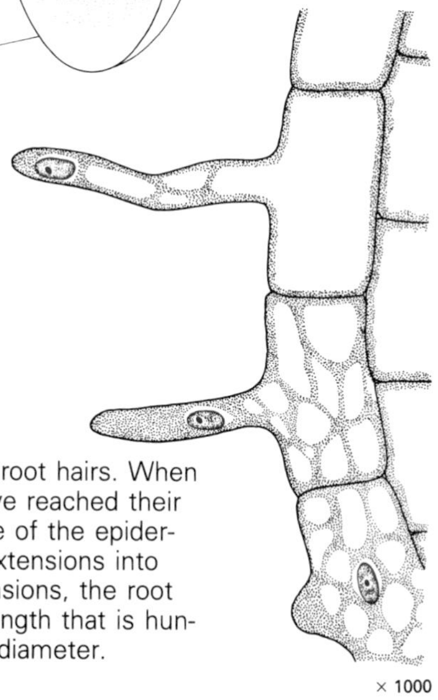

Figure 16.4
The development of root hairs. When cells in a root tip have reached their maximum size, some of the epidermal cells send out extensions into the soil. These extensions, the root hairs, may reach a length that is hundreds of times their diameter.

made up of rather large, simple cells, which pass water from the epidermis to the conducting tissues and also store reserve food, usually starch. The central conducting cylinder is sheathed in a sleeve of specialized cells, the *endodermis* (Greek, "inside skin"). The endodermis is lined with still another layer of structurally simple cells, the *pericycle*. The bulk of the conducting tissue is **xylem,** whose elongated, heavy-walled cells are open-ended. Running lengthwise in the fluted grooves of the xylem cylinder are strips of **phloem,** whose elongated, thin-walled cells are mainly transporters of dissolved food. In most roots, the cells between the xylem and phloem, the **cambium** (Latin, "change"), which retains its *meristematic* (mitotic) ability, is capable of increasing the girth of the root by producing new xylem cells inward and new phloem cells outward (Figure 16.5).

Figure 16.5
Diagrammatic cross section of a young dicot root. All the tissues here are primary, having come from the apical growing point. The solid central core of xylem in this example is like a fluted column, appearing somewhat star-shaped in cross section. In the "arms" of the xylem are phloem strands, separated from the xylem by a layer of cambium that will eventually contribute secondary xylem and phloem. A sleeve (a ring in cross section) of undifferentiated pericycle surrounds the entire vascular cylinder, or *stele.* Outside that is a second sleeve of specialized endodermis. A voluminous cortex surrounds the stele, usually making up the bulk of the young root. The outermost layer is the epidermis, some of whose cells grow into root hairs.

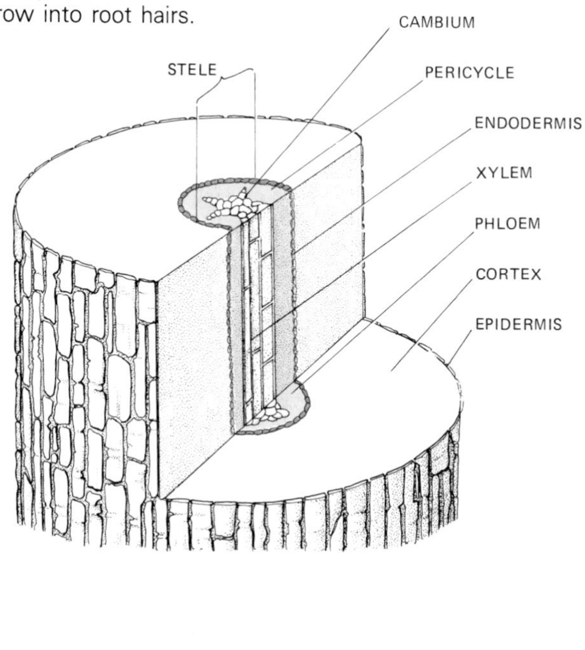

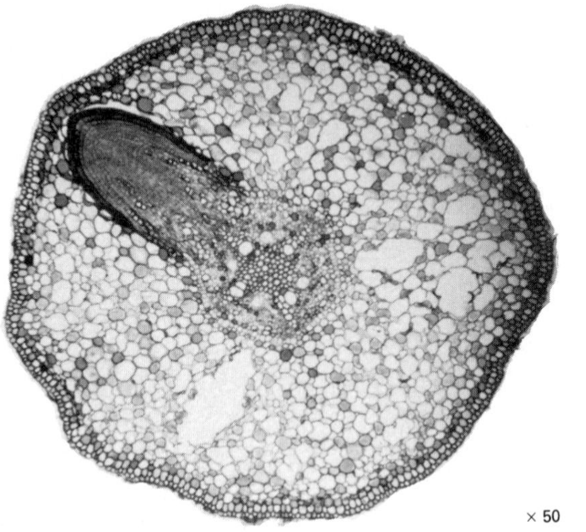

× 50

Figure 16.6
The origin of branch roots. In this cross section of an old root, a branch can be seen pushing its way from the central stele out through the cortex and bursting through the epidermis. The first cells to begin dividing when a branch starts are in the pericycle of the old root. They become meristematic and start a growing point, which becomes the apex of the branch root. Branches originate at points where xylem strands are close to the pericycle, but there do not seem to be any predestined places where such renewed mitotic activity commences. In stems, in contrast, branches originate superficially from branch buds in the axils of leaves, and their distribution is predetermined by the position of the leaves.

Root branches may originate anywhere along a root except at the growing tip. Some pericycle cells may become mitotically active, and by divisions they begin to form a new root apex, which swells and pushes outward through the endodermis, then through the cortex, finally bursting through the epidermis. Meanwhile, the cells just back of this growing tip are becoming like those behind the growing point of any root, with a region of mitosis, one of cell elongation, and finally one of differentiation. After a branch root has grown out from the parent root, xylem and phloem are differentiated in the new root, and they form a functional connection with the old xylem and phloem. Thus there is a continuous connection of the vascular system throughout the entire root system (Figure 16.6).

In woody plants, as roots grow older, the cambial cells come to make a complete cylinder around the xylem, and they keep adding more xylem year after year until an old root is mostly xylem, with only a thin covering of phloem and its associated tissues. Since the original epidermis can cover only a small root, it splits when a root swells in growth and is soon lost, but it is replaced by new, corky cells, which make an effective covering over the phloem. Old roots support large plants and conduct water, minerals, and food, but they are not absorptive.

Roots are more extensive than most people think. Because they are out of sight and appear small when plants are pulled out of the soil, root systems are not generally appreciated, but they are in fact more extensive than the aboveground plant parts. Tree roots generally extend well beyond the tips of branches. One of the best-known inventories of root growth is that of the botanist H. J. Dittmer, who measured all the roots of a single rye plant and reported a total of more than 600 kilometers (380 mi) of roots, with a surface area of 230 square meters (2500 ft^2), not counting the extra area provided by root hairs. With the leaves measuring only 4.7 square meters (51 ft^2), the roots had more than 50 times as much surface as the leaves. Although grasses (rye is a grass) may have relatively more root than woody trees do, the comparison is probably fair for most plants.

The original primary root grows out from the radicle of the seed, but in mature plants, roots may grow from any part. A root produced from an organ other than the radicle or another root is an **adventitious root.** Most of the roots of grasses, with maize (corn) as a clear example, are adventitious, coming from the base of the stem. Stems readily produce new roots, as anyone knows who has made cuttings or "slips" of garden plants. In such plants as African violets, even leaves are capable of generating roots.

Since they are in an underground environment, which is generally more stable than the aerial environment, roots are less variable than either stems or leaves. Botanists say that roots are "conservative." Nevertheless, considerable variation can be found. Grass roots, for example, are numerous, thin, and fibrous, but sweet potato roots may become swollen with reserve starch. Aerial roots of climbing vines grow out from stems and attach themselves to trees or buildings. A few orchids have green roots that carry on the photosynthetic work of the plant.

TO SUM UP ROOTS

1. As root cells develop they differentiate into storage cells, conducting cells, epidermal cells, or root hairs.

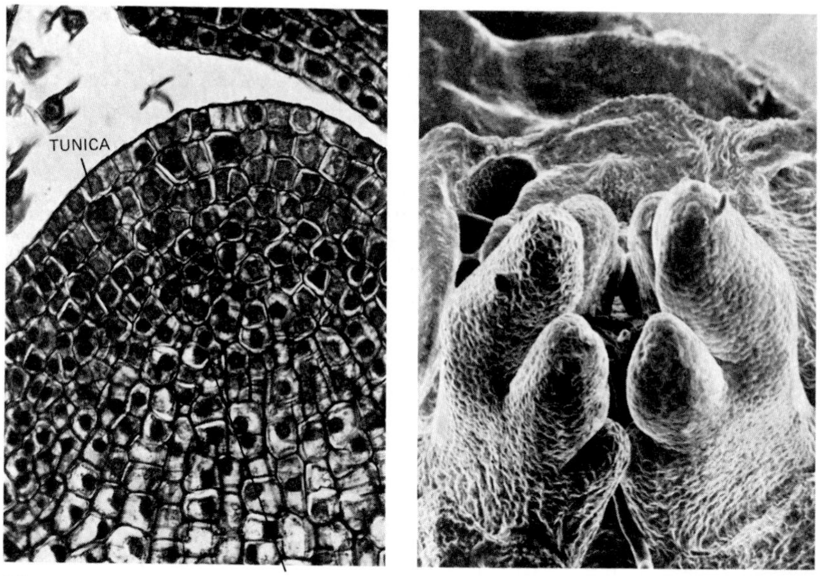

Figure 16.7
Apical growing points of stems. (a) In this median long section through the apex of a *Chrysanthemum* stem, the outer layer of tunica (which will give rise to later epidermal cells) covers the surface. Under the tunica, the corpus consists of a mass of small cells with thin walls, relatively large nuclei, and little cytoplasm. The corpus will give rise to all the other primary tissues of the leaves and stem. (b) Scanning electron micrograph of the shoot apex of a buckeye (*Aesculus californica*). The apex is just visible in the center of the picture. The points sticking up above the apex are leaf primordia.

2. The bulk of conducting tissue is *xylem,* which carries water and minerals. *Phloem* transports dissolved food.
3. The vascular system is continuous throughout the root system and the entire plant.
4. Besides absorbing and conducting water, minerals, and food, roots support the plant.
5. *Adventitious roots* are produced from an organ other than the radicle or another root.

STEMS SERVE MAINLY FOR SUPPORT AND CONDUCTION

The epicotyl of a germinating seed, with a meristematic apex at its tip, gives rise to the shoot. Stem growth in length is dependent on mitotic activity in the apex. A **stem apex** is a soft, dome-shaped mass of tissue, covered with a layer of meristematic cells known as the **tunica.** Under the tunica is a mass of cells, the *corpus,* that divide in such a way as to make the stem increase in length (Figure 16.7).

Just below the apical growing point, usually only a fraction of a millimeter, bumps grow out at precisely regulated places. These are the incipient leaves, known as **leaf primordia.** Inside the young stem, events are similar to those back of a root growing point. Once produced, new cells first become enlarged and then begin to differentiate into conducting xylem and phloem. As leaf primordia grow, they extend in length and become separated from one another by the elongation of the stem. The place on a stem that bears a leaf is a **node,** and the intervals between nodes are **internodes.** Nodes elongate little, but internodes elongate greatly, especially in dim light, causing a separation of the nodes and consequently a separation of leaves along the stem. As leaves mature, strands of xylem in them differentiate, and a meeting of the conducting tissues of leaf and stem results (Figure 16.8).

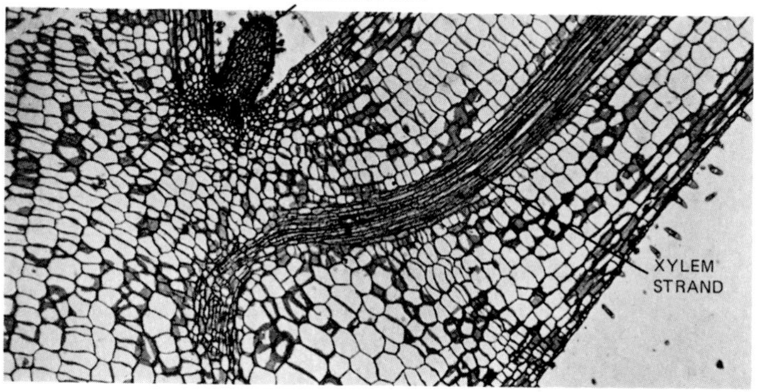

Figure 16.8
A median petiole section through a node of a *Coleus* stem. The xylem strand in the petiole of the leaf, stretching toward the upper right side of the picture, connects to the xylem of the stem. The bump in the angle formed by the stem and the petiole is a young axillary bud.

Soon after a stem apex produces a leaf primordium, it begins to produce a second, smaller mass of cells just above the leaf primordium. The new growth is a *lateral* or *axillary bud primordium*. When the leaf is mature, the axillary bud will have developed in the axil of the leaf. The axillary bud is a copy of the terminal apex, and given the proper stimulation, it can begin to grow just as the terminal apex did; that is, it can expand, grow out, and become a side branch of the old shoot. Stem branches in general originate in this way.

The manner of stem branching is different from that of root branching. Stem branches are produced specifically at nodes and come from superficial cells that originate in the stem apex. Root branches, in contrast, may appear anywhere along a root and are the result of meristematic cells originating deep inside the root, in the pericycle. Roots do not have nodes and internodes, and this is one of the surest ways of distinguishing between roots and stems, which sometimes may closely resemble one another.

The patterns of maturation in monocot and dicot stems are somewhat different. We will follow first the dicot pattern (Figure 16.9). As the stem cells mature and the vascular tissues differentiate, a number of xylem strands form either a continuous cylinder inside the stem or a ring of separate strands. Then, outward from the xylem cylinder (or strands), phloem cells begin to differentiate. Anywhere from a few millimeters to several centimeters below the apex, depending on the species and environmental conditions, a fully functional stem develops. At such a place, before a cambium has begun to make new growth, all the cells are initially formed in the apex. Tissues composed of such cells are **primary tissues,** in contrast to secondary tissues, which will be derived later from cambium. In monocot stems, the early xylem strands are not developed in a ring or cylinder around a central pith. They are numerous and scattered throughout the "background" tissue.

As the apex grows upward and leaves the older part of the stem farther and farther behind, the

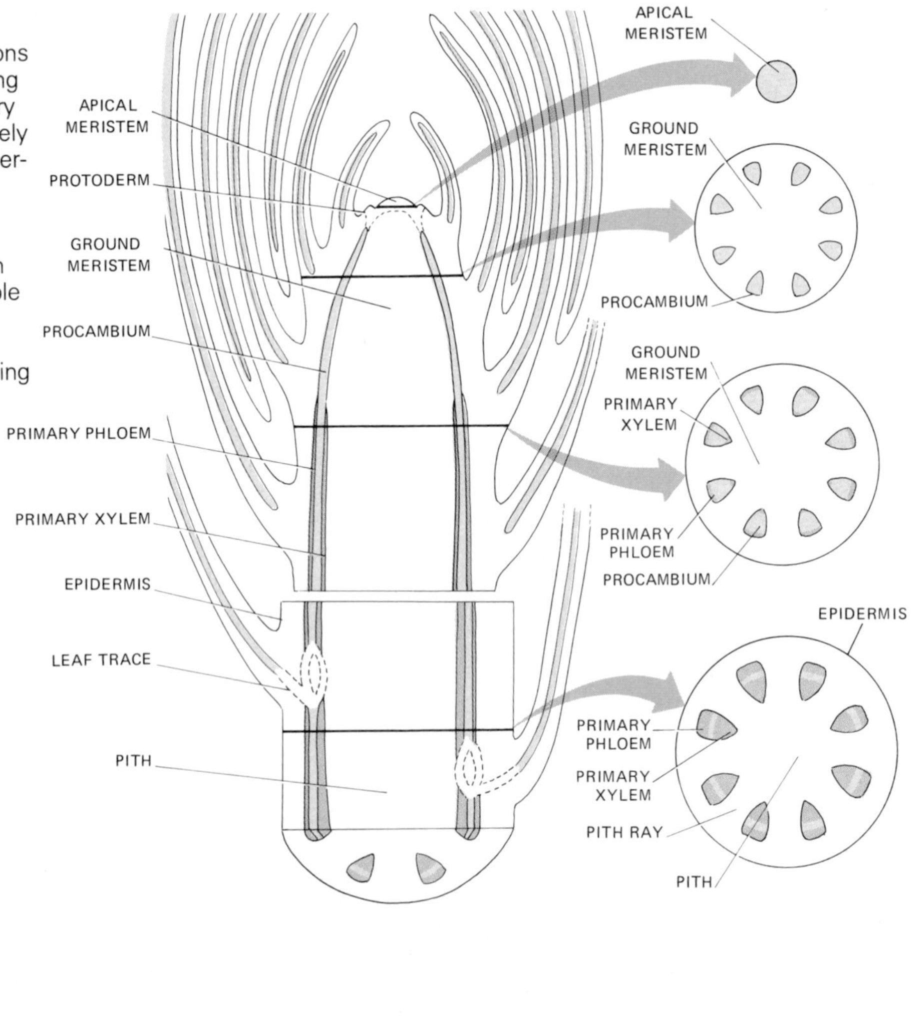

Figure 16.9
Diagrams of cross and long sections through a dicot stem apex showing maturation of the tissues. The very tip of the stem, which is completely embryonic, is composed of undifferentiated cells. However, as new cells enlarge, they push the apex higher and higher, leaving behind the cells that differentiate to form the *vascular tissues* and the simple *ground tissues: cortex* and *pith.* Later, when the *vascular cambium* has become the main growing tissue, the bundles will enlarge until they make a complete cylinder of wood.

older part of the stem begins secondary growth. The cells between a xylem strand and a phloem strand become meristematic, dividing so as to produce more xylem inward and new phloem outward. Thus the xylem is increased and the stem grows in girth. By the time a stem is a few years old, it will have a remnant of the old primary cells down the center (the pith) and a small amount of primary xylem, surrounded by a thick cylinder of secondary xylem, which is in turn surrounded by a thin cylinder of cambium. Outside the cambium is the bark, containing phloem and a replacement for the long lost epidermis. This replacement is the **cork,** produced by a second meristematic cylinder, the **cork cambium.** Several kinds of cells make up the phloem. There are undifferentiated cells, strengthening fibers, conducting cells with holes in the end walls called *sieve tubes,* and slender cells called *companion cells* whose function is not known (Figures 16.10 and 16.11).

The cambium, only one or two cell layers thick, has two kinds of cells: long, slender ones, which are to yield long, slender xylem or phloem cells, and

Figure 16.10
A cross section through a woody stem: American sycamore (*Platanus occidentalis*). The central pith is surrounded by a cylinder of xylem, through which rays radiate out to the encircling bark. The innermost layer of cells of the bark, just outside the xylem, is the cambium. The pale splotches in the bark are masses of phloem, outside which is the cortex. In a young stem such as this (one year old), the outer epidermis is still complete.

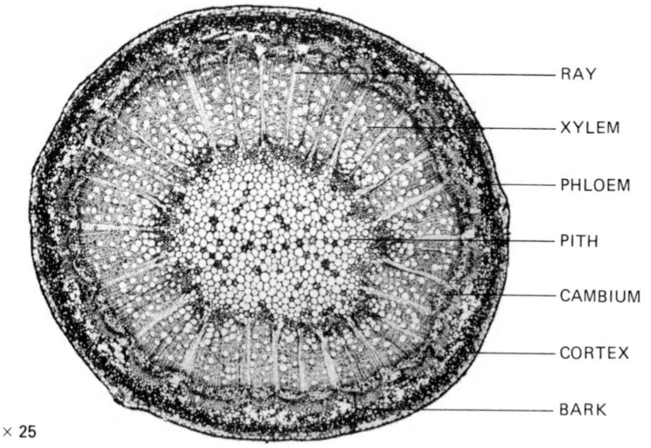

Figure 16.11
Enlarged portion of a cross section through an old woody stem: linden (*Tilia*). A corky layer at the left is produced by a cork cambium, which keeps making more cork as the stem grows. Inward from the cork is a layer of undifferentiated cortex. The vascular tissues, phloem and xylem, are complex tissues, each consisting of several kinds of cells. In the phloem are phloem tubes, rays, and fibers. The cambium between phloem and xylem, which is the main growing tissue, produces phloem toward the outside and xylem toward the inside. In the xylem are open-ended water-conducting cells (vessels), fibers, rays, and parenchyma. (Parenchyma is a term applied to any essentially undifferentiated plant tissue.)

clusters of smaller and rounder cells, which are to yield rays (Figure 16.12). The cambium is especially active at the beginning of each growing season in the spring of the year, making a new ring of large xylem cells around last year's old xylem. As the season progresses, the cambium produces smaller and smaller xylem cells until by late summer, when it ceases growth, it is making relatively small ones. Each year, then, a new cylinder of xylem is laid down around last year's cylinder, and when a tree is cut across, these *annual rings* can be easily seen. Since normally only one ring is laid down each year, the number of rings equals the number of years the tree has lived. As the xylem increases in girth, the cambium expands, leaving behind it not only the annual rings but also the strips of ray tissue, which appear in cross cuts of wood as radial lines running from the central pith toward the bark (see Figure 16.11).

Annual rings vary in thickness with weather; a dry year gives a thin ring and a wet year a thicker one. By making careful measurements of successive rings in trees of known age and overlapping a number of samples, we have been able to draw a number

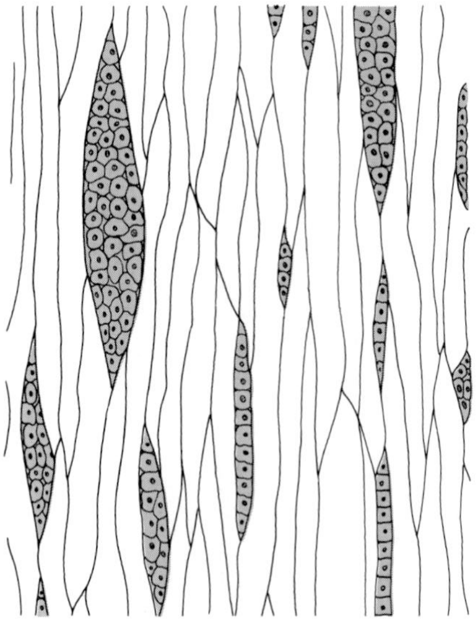

Figure 16.12
A tangential view of cambium of black walnut, *Julgans nigra,* as it would appear if a piece of bark were stripped down to the wood. The long, thin cells divide lengthwise to yield long xylem cells toward the inner side or phloem cells toward the outer side. The short, rounder cells in groups (color) divide to produce rays.

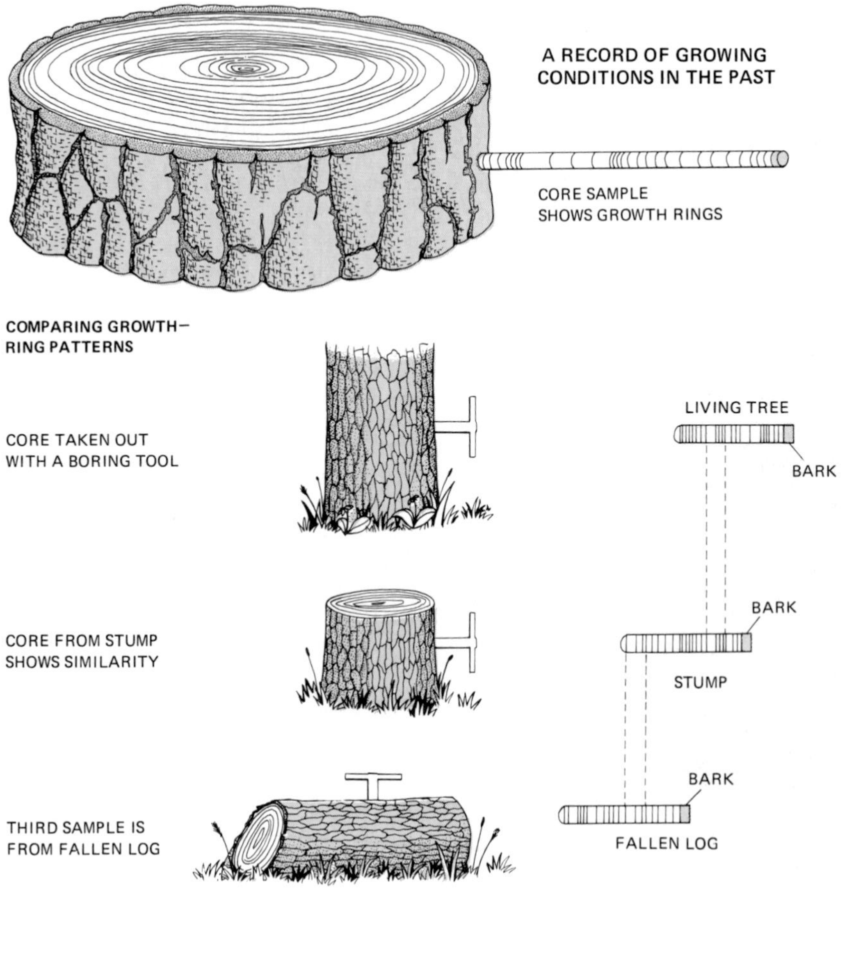

Figure 16.13
Tree-ring dating. A slender cylinder of wood, called a core, is cut from a tree (or log or sawed piece of lumber) with an increment borer. The spacing between rings is carefully measured. A sequence of spacings is established and can be compared with that of other samples. In any particular locality, all the trees will have thin rings in poor growing years and thicker rings in better years. A sample from a living tree provides a starting date. The older part of that sample will overlap an older sample of unknown date. More and more overlapping samples have made it possible to date wood hundreds of years old with great accuracy.

of conclusions about the past from the science of *dendrochronology* (Greek, "tree timing"). Tree-ring dating has made possible the establishment of accurate dates for many historical edifices, especially the buildings of the Indians of the American Southwest.

Also from tree-ring measurements, climates can be described as far back as 8000 years. One unexpected development of dendrochronology is the demonstration that dating by the use of carbon-14 is not accurate without a correction (see the essay, p. 16). Carbon-14 dating, based on the assumption that the amount of carbon-14 has remained constant through the years, does not agree with tree-ring dating, but tree-ring dating is the more convincing method, and therefore the carbon-14 method has to be adjusted (Figure 16.13).

The differences among different kinds of woods are determined by the species of trees, the climates in which they grew, the part of the trunk from which they were taken, the angle of the saw cut, and individual peculiarities. The hardness of wood is mainly determined by the species. *Softwood* is officially derived from conifers, whether the wood is actually soft or not, and *hardwood* comes from flowering trees. Although linden wood from a flowering tree is softer than yellow pine wood from a conifer, linden is sold as a hardwood and yellow pine as a softwood, but in general the distinction is well founded. There is a difference, too, in woods of different ages, because some trees deposit pigments in the older parts of the trunk more than in the younger parts. The central core (the *heartwood*) of walnut, cherry, and red cedar (*Juniperus virginiana*), for example, is deeply colored, but the outer rim (the *sapwood*) is pale. (See Figure 16.17.)

The *grain* of wood is determined by the way the cambial cells lay down xylem and by the final differentiation of the xylem cells. Cambium may produce uniform masses of cells, all neatly lined up parallel to each other and with little difference in cell size or wall thickness, and the result, as in white pine, is a homogeneous, smooth, almost characterless wood. Or the cambium may be twisted and give rise to xylem cells contorted in various ways, which may mature into tissues with many sizes of cells with different wall thicknesses, as happens in curly or tiger-striped or fiddleback maple. If there is a striking difference between the wood made early in the season (*spring wood*) and that made later (*summer wood*), the annual rings will be conspicuous, as they are in yellow pine (Figure 16.14).

Some woods produce a few large water vessels in the spring but few or none later on, as in oak. When oak wood is seen in cross section, the spring wood shows a ring of holes, and it is called *ring-porous*. In contrast, linden wood has vessels formed throughout the year and is *diffuse-porous* (Figure 16.15). Since softwoods have no water vessels, only

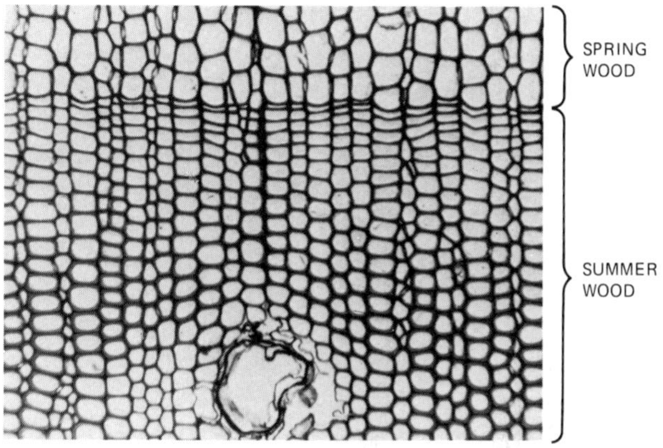

Figure 16.14
The difference between spring wood and summer wood in white pine, *Pinus strobus*. In this cross section the spring wood with its larger cells is at the left. The alternation of cell size makes rings of softer tissue (large cells) and harder tissue (smaller cells).

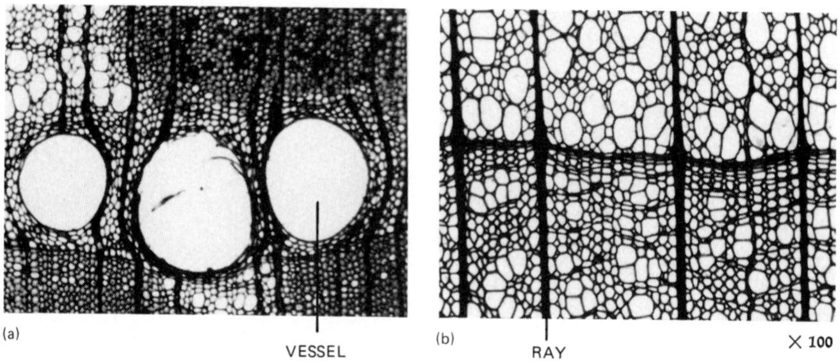

Figure 16.15
Cross sections of hardwoods. (a) In oak (*Quercus*), the vessels produced in spring are conspicuously larger than those produced later in the season. (b) In linden (*Tilia*), the vessels are about the same size throughout any annual ring.

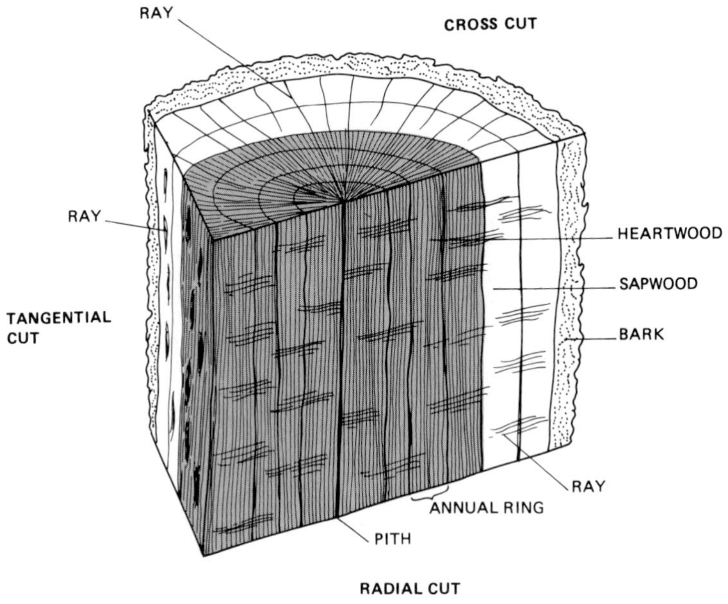

Figure 16.16
The effect of the direction of cut on the appearance of wood grain. In a cross section, the annual rings appear actually as rings, and the rays as narrow lines crossing the annual rings. In a radial section, which passes directly through the center of the stem, the annual rings appear as long streaks, and the rays as cross-hatching across the annual rings. In a tangential section, which passes anywhere except through the center, the annual rings appear as long streaks, and rays as short lines.

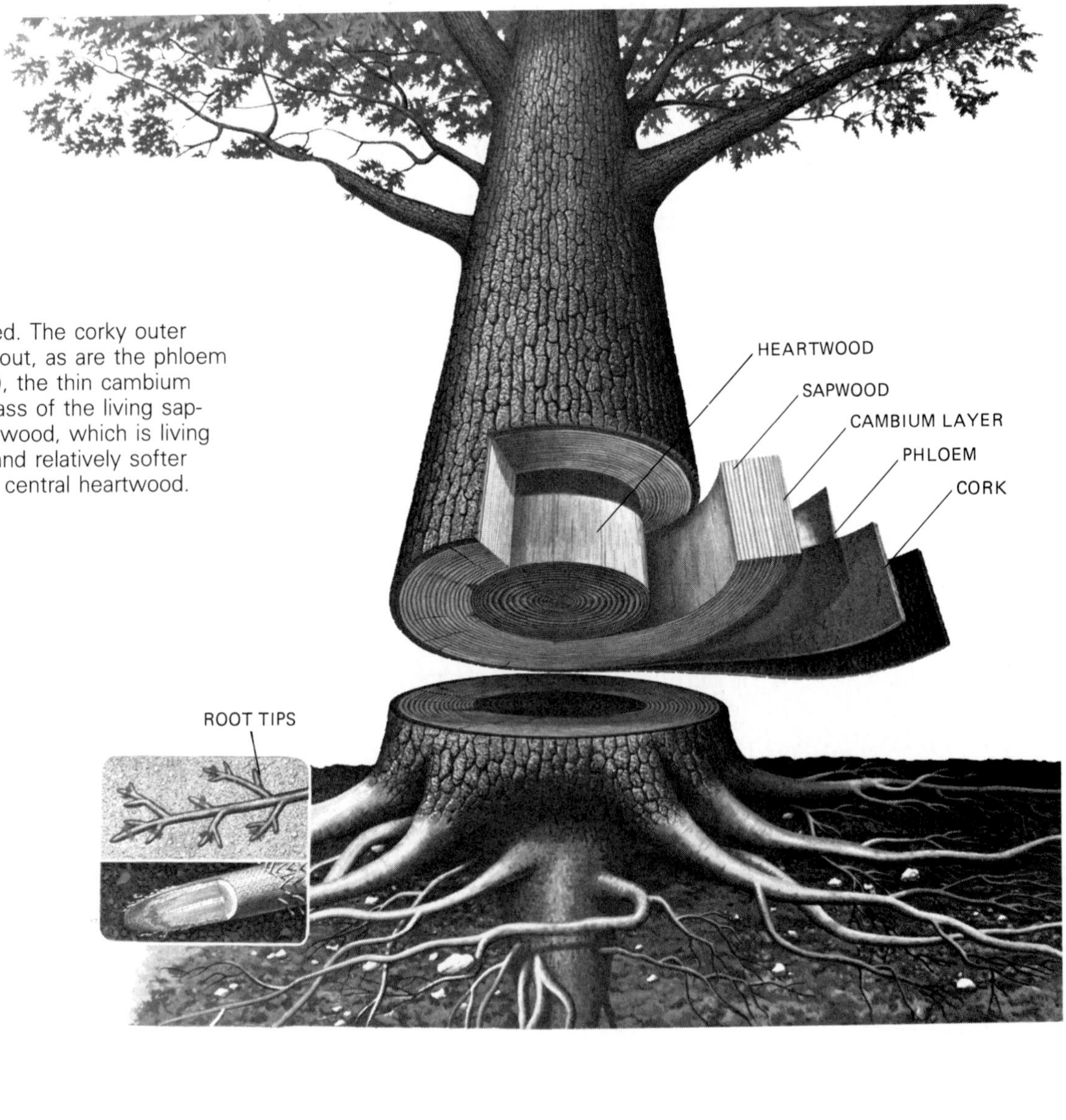

Figure 16.17
A tree dissected. The corky outer bark is peeled out, as are the phloem (the inner bark), the thin cambium layer, and a mass of the living sapwood. The sapwood, which is living wood, is wet and relatively softer than the dead, central heartwood.

tracheids, they are nonporous. Rays, too, affect the appearance of wood. If they are vertically shallow, they will not be visible to the naked eye, as in birch. If they are extensive, as in oak, cherry, and maple, they may be conspicuous.

The way in which wood is cut makes a great difference in the appearance of the cut surface. If it is cut across (a *cross section*), the annual rings appear as circles or parts of circles, with rays running radially. If a cut is directly through the pith, that is, along a radius of the cylinder of the tree trunk (a *radial section*), the annual rings appear as more or less parallel lines, provided that the tree was reasonably straight. The rays then appear as shining strips of various widths, depending on the species, running across the lines of the annual rings. If the cut is through a log anywhere else, as it usually is (a *tangential section*), the annual rings will show up as wavy, irregular streaks, and the rays will usually show poorly or not at all (Figure 16.16).

With wood becoming ever more scarce and expensive, fine woods are commonly cut in thin sheets and glued over a base of less desirable wood to make *veneer*. *Plywood* is made by gluing sandwiches of three to seven thin wooden sheets together with the grains alternately crossing each other. Since plywood can be made with the inner layers of flawed wood, it can be made cheaply. Besides, if made with weatherproof glue, it is not affected by moisture, and it is unsplittable.

Woods are prized for many special qualities: red cedar for its aromatic resins, mahogany for its color, lignum vitae for its hardness, hickory for its resilient flexibility and for the flavor of its smoke, rosewood for its color patterns, white pine for its homogeneity, spruce for its ability to resonate in musical instruments, pernambuco for its springiness in violin bows, teak for its durability, beech for its resistance to water damage, white oak for its retention of liquid in wine casks, redwood for its resistance to decay, dogwood for its resistance to wear. In recent times, the cost of fine wood has driven wood technologists to make partially synthetic substitutes, mold boards from wood chips bonded in plastic, or print photographs of wood onto plastic sheets, as in wall panel boards and table tops.

The general structure of stems is subject to as much variability as is their tissue structure (Figure 16.18). The differences depend on the relative amounts of growth of the apices and cambium. Just

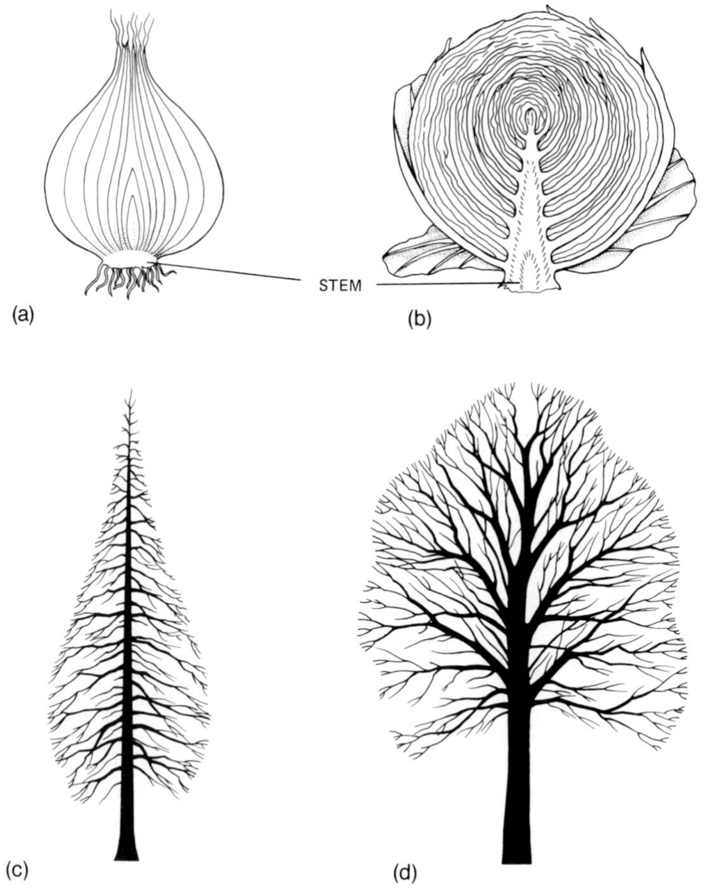

Figure 16.18
Variation in the form of stems. (a) An onion stem is a short, broad lump at the base of the onion bulb, with an apex, no cambium, and roots growing from the bottom. The bulk of the onion bulb is made up of succulent leaf bases. (b) A cabbage head has a longer but still stubby stem. (c) A tree with one main stem and minor side branches. (d) A tree with no one main stem but many almost equal side branches.

as some things that do not look like stems (potato tubers) really are, so some that look like stems are not. The apparent stem of a banana plant is a mass of leaf bases, and the morphological stem, except when the plant is flowering, is a short nub of tissue near the base, deeply hidden in the leaf bases.

TO SUM UP STEMS

1. New cells in a stem apex differentiate into xylem, phloem, ground tissue, and epidermis. Leaf primordia grow and become separated as the stem lengthens.
2. Stems have *nodes,* where leaves appear; stem branches are produced at nodes. Roots do not have nodes.
3. A vascular cylinder in a dicot stem develops around a central pith. Vascular bundles in monocot stems are numerous and scattered.
4. The cambium has two kinds of cells: long, slender ones, which are to yield long, slender xylem or phloem cells, and clusters of smaller and rounder cells, which are to yield rays.
5. Annual rings are formed when new layers of xylem are laid down around the previous year's xylem.
6. The tissue structure and general structure of stems are varied, but the principle functions of any stem are conduction and support.

LEAVES ARE THE MAIN PHOTOSYNTHETIC ORGANS OF PLANTS

Leaves arise as outgrowths of tissue just below the growing point of the stem apex. As a young leaf increases in size, it begins to flatten out laterally and to undergo differentiation of its internal tissues. While still in the terminal bud, it is folded in a neat, compact package, but when it opens, its cells swell and some of them continue for a while to divide until the leaf achieves its mature form.

The tissues of leaves vary from species to species. The amount of cuticle ranges from none in aquatic leaves to a heavy covering in such succulent leaves as those of *Sedum.* Stomata occur on both sides of some leaves, or sometimes only on the upper epidermis, as in floating water lily leaves. The epidermis may be one thin layer, or it may have

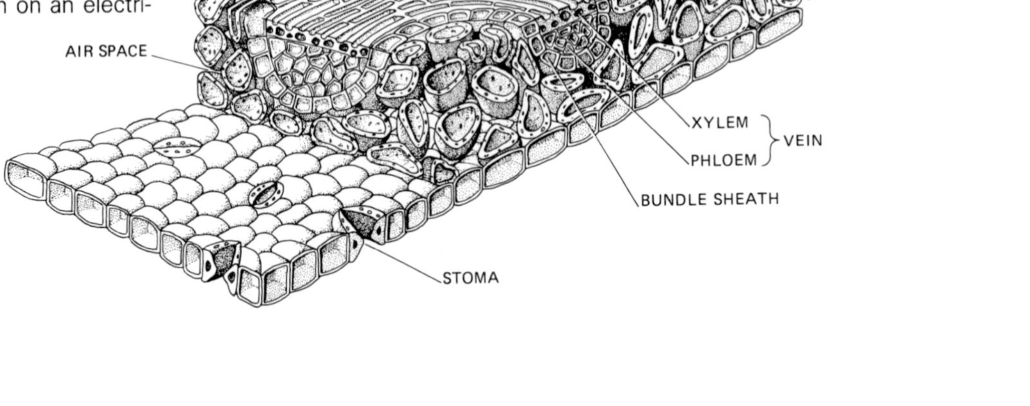

Figure 16.19
A dissected leaf. A mature leaf has a layer of *epidermal cells* over its surface. On the outside, these cells have a waterproof coating of waxy *cuticle. Stomata* are usually more common in the lower than in the upper epidermis. The stomatal pore leads into an air space inside the leaf. The *spongy mesophyll* (Greek, "middle leaf"), located above the lower epidermis, consists of chloroplast-bearing cells that are partially pulled away from one another as the growing leaf expands, with the result that they are loosely connected and have a labyrinth of air channels throughout the tissue. Above the spongy mesophyll the cells of the *palisade parenchyma,* also chloroplast-bearing, are neatly compacted like the nap of velvet. A system of *veins* extends through leaves. Veins are coated with cells, a *bundle sheath,* making a covering like the insulation on an electrical wire.

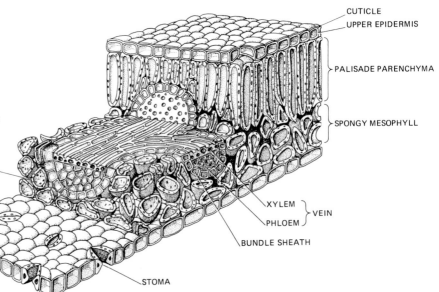

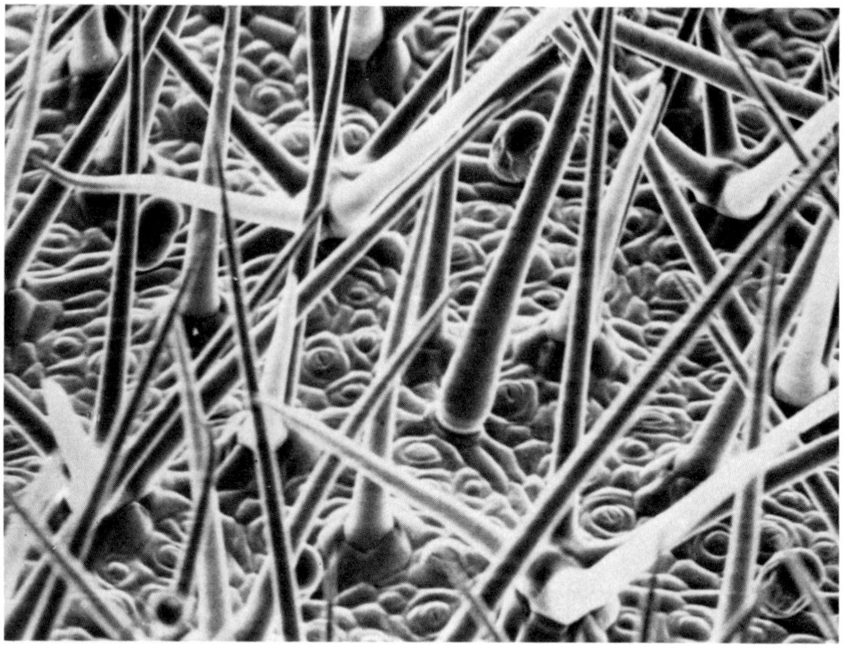

× 125

Figure 16.20
A scanning electron micrograph of a leaf surface. Spiky extensions of epidermal cells give the leaf a bristly appearance. Their function is unknown.

several thick layers of cells. The palisade parenchyma may be one, two, or three layers deep, or, as in many ferns, not present at all. The mesophyll may be so thin that the leaves wilt on any warm, dry day, as squash leaves do, or it may be so thick, juicy, and filled with mucilaginous materials that the leaves resist drying for weeks, as in the succulent "air plants" (*Kalanchoe*) (Figures 16.20 and 16.21).

The arrangement of leaves on stems and their gross structure are determined by genetic forces that are completely unknown. In the apical bud of a mint plant, leaf primordia grow opposite each other in pairs, and when the leaves mature, they occur two at a node, one on each side of the stem. A few plants, such as bedstraw (*Galium*), produce a circlet of leaves at a node, but most plants bear only one.

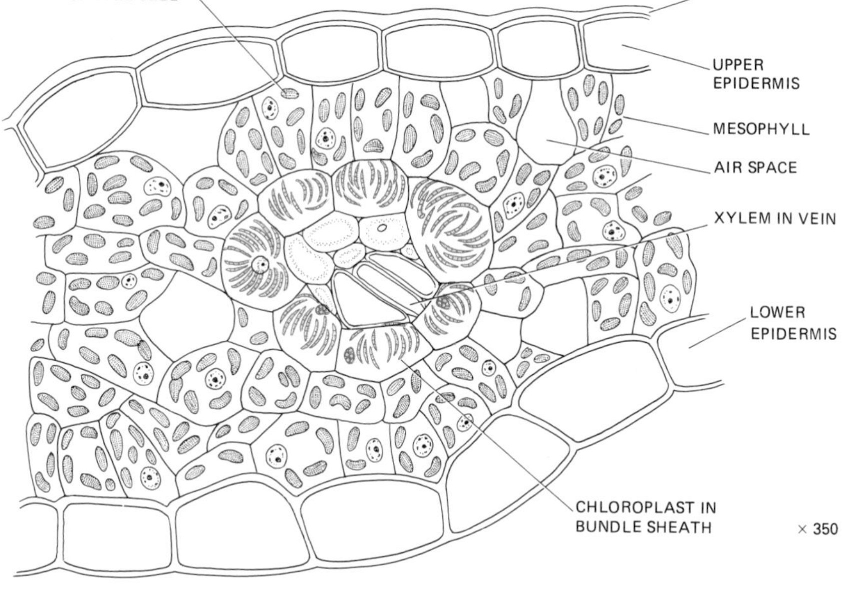

Figure 16.21
A portion of a cross section through a corn leaf (*Zea mays*), a C-4 plant. The chloroplasts in the ordinary mesophyll cells do not build up starch, but they do have grana. Carbon dioxide is fixed in the mesophyll. In the bundle sheath cells, those that surround the vascular bundles, the chloroplasts *do* form starch grains (indicating carbohydrate synthesis) but have atypical grana.

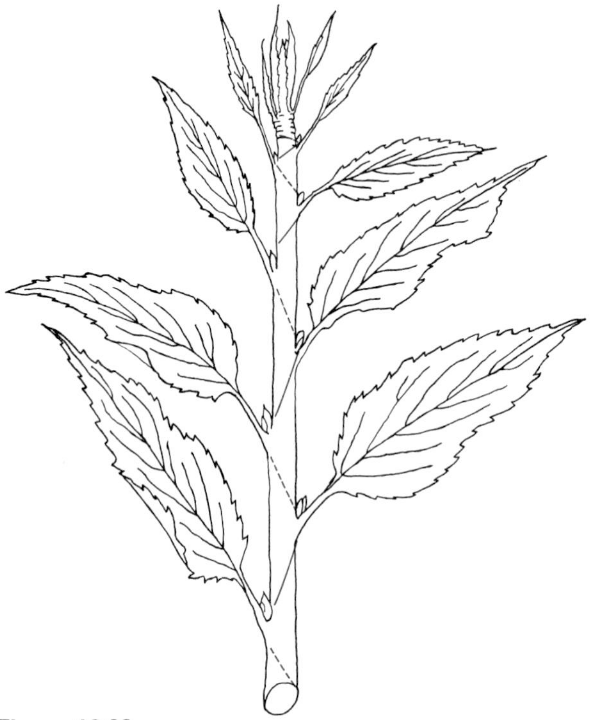

Figure 16.22
A spiral arrangement of leaves on a stem. In this example, every leaf is directly above the second one below. The arrangement, which is dependable for a species, is determined by the way the apical growing point starts its leaf primordia.

Solitary leaves, however, are so placed that one can follow an imaginary spiral twining up a stem and passing through the base of each leaf stalk, or *petiole.* The geometry of leaf arrangement (Figure 16.22) is so precise and so predictable in most species that it has excited the curiosity of botanists and mathematicians for many years. It is possible to express these arrangements with formulas that can be applied as well to other natural structures or phenomena, such as the distances between planets of the solar system, or the vibrations of musical scales.

Many leaves consist of a simple blade, but they may be lobed or deeply incised or so reduced in surface that they are little more than veiny skeletons. In asparagus, butcher's-broom (*Ruscus*), and most cacti, leaves are nothing more than tiny, colorless scales that achieve nothing, and the green stems do all the photosynthetic work (Figure 16.23). Mulberry and sassafras leaves are not all alike even on one plant. When a leaf blade is all in one piece, it is said to be *simple,* but when it is subdivided into *leaflets,* it is *compound.* The difference becomes apparent when one finds the position of the axillary bud, which is produced at the base of the petiole of a *leaf* but not at the base of a *leaflet* (Figure 16.24). Leaflets themselves may be compounded, making a *doubly compound* leaf, and many ferns are triply compounded. One genus, *Davallia,* which is five times compounded, has leaves like a fine filigree (Figure 16.25). Leaf form helps in determining species of plants, but it is of little use when applied to larger taxa, such as genera and families, because leaves are the least stable in form of all plant parts.

In all but a few rare species, leaves are temporary organs. At the end of the growing season in herbaceous plants and deciduous trees, leaves are shed because of a weakening of the cells across the base of the petiole. The timing of leaf drop, or abscission, is regulated by the amount of the growth-control compounds, *auxin* and *abscisic acid.* In most instances, abscission is facilitated by the alteration of a few cells that make an *abscission layer* across the petiole (Figure 16.26). The cells are impregnated with a corky, waterproofing, waxy material, which seals the scar where the leaf was. Evergreen leaves remain in place at least through a winter, but many are cast off at the end of the second growing season.

Figure 16.23
Stems that look like leaves. In the butcher's-broom (*Ruscus*), the broad, flattened branches are green and have such a leafy aspect that one could easily mistake them for leaves. But real leaves have buds in their axils, and these "leaves" have no buds. Instead, *they* occur in the axils of tiny, colorless scales, which in this instance are the morphological but functionless true leaves. Another indication that these leaflike branches are actually branches is the fact that they bear flowers in the middle of the surface.

Figure 16.24
Simple and compound leaves. (a) A simple leaf of privet (*Ligustrum*), in which the leaf blade is not divided. (b) A compound leaf of mountain ash (*Sorbus*), in which the leaf blade is divided into two rows of leaflets, with the midrib of the leaf naked between the leaflets. (c) The doubly compound leaf of the Kentucky coffee tree (*Gymnocladus*), in which each leaflet is in its turn subdivided into still smaller leaflets. In all these examples, the whole leaf grows from a stem at a node, and it has an axillary bud just above the point of attachment of the leaf. There is no bud in the axil of a leaflet. The vascular structure in a leaf stalk, or petiole, is quite different from that of a stem.

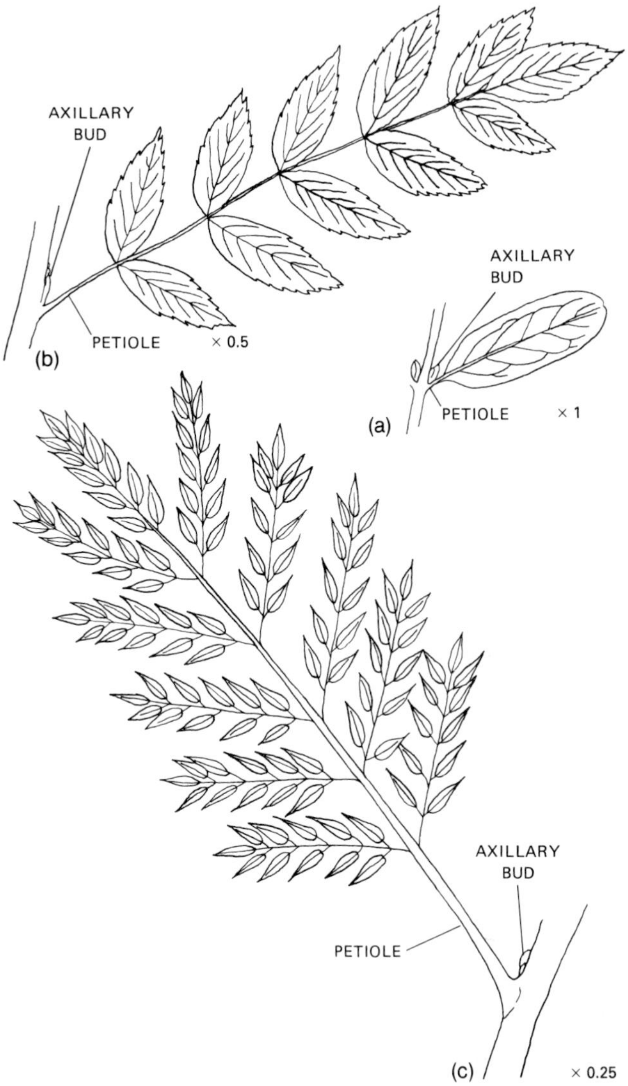

Figure 16.25
A leaf of the rabbit's foot fern, *Davallia bullata*. Such a lacy structure may be useful to a plant in allowing light to penetrate through the upper leaves and shine down to the lower ones.

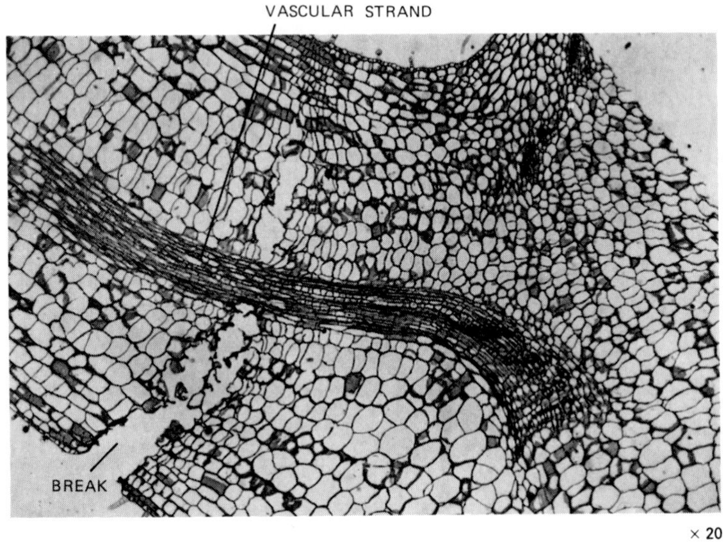

Figure 16.26
How a leaf is cut off from a branch: the *abscission layer*. At the end of a growing season, or when a leaf grows old or is damaged, the cells across the base of the petiole are weakened so that a slight movement of the leaf will break it off. This section cut through a node of a *Coleus* stem just before leaf fall shows the crack appearing across the petiole. The vascular strand, the dark streak in the photograph, is the last tissue to break. Abscission layers are generally formed in temperate climates at the approach of cool weather, or in tropical climates at the beginning of a dry season.

In some unusual species, such as the bristlecone pine of the California mountains, individual leaves have been found to remain in place as long as 30 years.

TO SUM UP LEAVES

1. Most photosynthetic activity takes place in leaves.
2. Mature leaves have a surface layer of *epidermal cells,* which are coated with a protective *cuticle. Stomata* permit the passage of gases into and out of the leaves.
3. Photosynthetic *chloroplasts* are located in the mesophyll and palisade parenchyma.
4. Leaves are usually temporary organs. The timing of leaf drop is regulated by the growth-control compounds, auxin and abscisic acid.

SUMMARY

1. Water, warmth, and oxygen are universally needed for *seed germination.*
2. When the embryonic plant in a seed begins to grow, the first activity is in the *radicle.* Then mitotic divisions cause the *primary root* to push through the seed coat. As the primary root grows, it sends out *secondary roots.* The *epicotyl* elongates further, and the seedling is on its way to becoming a self-sustaining plant.
3. All seeds begin germination by sending out a *primary root.* In monocot seeds, a single cotyledon acts as an absorbing organ that presses into the seed's main food supply in the endosperm. The first true *leaf* pushes through the *coleoptile,* splitting it. With a primary root established and a true leaf extending into the light, a monocot seedling is established. Variations exist in both dicot and monocot seed germination.
4. When an expanding radicle breaks out from a germinating seed, mitosis soon commences at the growing root tip. A *root cap* makes it possible for the elongating root to ease its way among the soil particles. Once elongation has ceased, the cells *differentiate,* with epidermal cells sending out extensions to form *root hairs.*
5. By the time a root reaches functional maturity, it has a number of differentiated tissues: *epidermis, cortex, endodermis, pericycle, xylem, phloem, cambium,* and in some species *pith.*
6. After a branch root has grown out from the pericycle of the parent root, xylem and phloem are differentiated in the new root, and they form a functional connection with the old xylem and phloem. There is a continuous connection of the vascular system throughout the entire root system.
7. A root produced from an organ other than the radicle or another root is an *adventitious root.*
8. Since they are in an underground environment, roots are less variable than either stems or leaves. Nevertheless, considerable variation can be found.
9. The epicotyl of a germinating seed gives rise to the shoot. Practically all stem growth in length in dicots is due to mitotic activity in the *stem apex.* No one knows how the orientation of mitosis is regulated, but the result is a continuing growth of the apex.
10. Incipient leaves are *leaf primordia.* Soon after a stem apex produces a leaf primordium, it begins to produce a second, smaller mass of cells, a *lateral bud primordium,* just above the leaf primordium.
11. Both phloem and xylem are complex tissues, each consisting of several kinds of cells. In the phloem are *sieve cells, companion cells,* and *phloem fibers.* In the xylem are water-conducting cells, either open-ended *tracheae* or closed *tracheids, xylem fibers, rays,* and *xylem parenchyma.*
12. The *cambium* has two kinds of cells: those that result in elongated xylem or phloem cells, and those that produce rays.
13. Each year a new cylinder of xylem is laid down around last year's cylinder, producing *annual rings.* Tree-ring dating has made possible the establishment of accurate dates for historical edifices and the description of past climates.
14. *Softwood* comes from conifers and *hardwood* from angiosperms. The central core of wood, the *heartwood,* may be more dense and pigmented than the outer *sapwood. Spring wood* has larger cells and is softer than *summer wood. Rays* are strips of cells streaking across the grain of wood in a radial direction toward the *bark,* which contains all tissues outside the cambium.

15. The *leaves* of vascular plants arise as outgrowths of tissue just below the growing point of the stem apex.

16. A completed leaf has a layer of *epidermal cells* over its surface. These cells are coated on the outside with a waxy *cuticle*. *Stomata* are usually more common in the lower than in the upper epidermis; the stomatal pore leads into an air space inside the leaf. The internal tissues of leaves are the *spongy mesophyll*, the *palisade parenchyma*, and the *veins*, consisting of *xylem* and *phloem* and sometimes *cambium*, covered by a *bundle sheath*.

17. In all but a few rare species, leaves are temporary organs. Leaf drop at the end of the growing season, *abscission*, is regulated by the amount of the growth-control compounds, *auxin* and *abscisic acid*.

ASK YOURSELF

1. What is the first part of a plant embryo to emerge from a germinating seed?
2. What are the functions of a root cap?
3. What is the function of root hairs?
4. What is the difference in the method of branching in roots and stems?
5. What is the difference between primary and secondary tissues in plants?
6. Why does a cross-section of a tree show rings in the wood?
7. How did tree-ring dating reveal errors in carbon-14 dating methods?
8. What is the difference between hardwoods and softwoods?
9. What is the difference between a tracheid and a water vessel in wood?
10. If you drive a nail into a tree three feet above ground level, where will the nail be fifty years from now?
11. Why does a plant need stomata?

17
Nutrition, Growth, and Regulation in Plants

SOME KEY POINTS

1. Soil and soil water are the sources of all the elements in a plant except carbon.
2. Soils are made of rock particles, water, gases, and living and dead organic matter.
3. Plants require seven mineral elements in relatively large quantities and seven others in small quantities.
4. Mineral elements are used over and over by generations of plants and animals.
5. Water and minerals rise in plants through wood vessels in xylem.
6. Water is pulled up plant stems by evaporation from leaves.
7. Food and water move in stems in the phloem tissue in the bark.
8. Plants establish and maintain position by growth regulators, especially auxin.
9. Phytochrome is a light-sensitive pigment involved in flowering, seed germination, pigment formation, and other plant activities.

PRACTICAL BIOLOGISTS—FOOD HUNTERS AND farmers—have known for centuries that warmth, water, and soil are necessary for plant growth, but it was only with the advent of experimental procedures, beginning in the nineteenth century, that we began to understand in some detail how plants use the environment and how they regulate their own internal affairs. Energy capture by photosynthesis (Chapter 4) and energy release by respiration (Chapter 5) have already been described. The input of materials other than carbon, hydrogen, and oxygen will be treated in the present chapter, along with those biologically synthesized materials with which plants determine their own rate, direction, and form of growth.

MINERAL NUTRITION OF PLANTS

Soil is the source of all the mineral nutrients taken in by plants and consequently (with some negligible exceptions) by animals. Soil is therefore of prime importance in the continuation of life. Far from being just an aggregation of rock particles, soil is such a changing and variable medium that students of soil frequently think of it as something almost living. Besides the rock particles that make up the bulk of most soils, there are water, gases, organic matter, and living organisms.

Soils

The chemical composition of the parent rock from which any soil was derived naturally determines much of the nature of the soil (Figure 17.1). Even though all rock particles are soluble to some extent

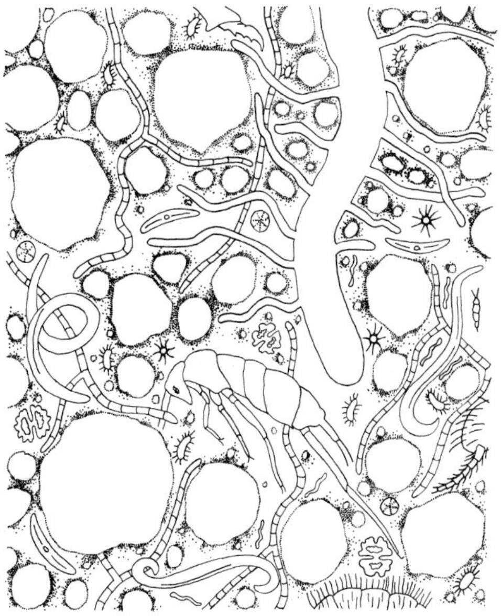

Figure 17.1
A worm's-eye view of a pinch of soil. In soil are roots with root hairs, diatoms, various green and blue-green algae, protozoa, roundworms, earthworms (too large to show in such an enlarged view), bacteria, unicellular and filamentous fungi, water, dissolved minerals, air that is rich in carbon dioxide, such insects as springtails, and rock particles of all sizes.

in water, some are more resistant than others. Silicates, such as granite rocks, are less soluble than carbonates such as limestone, or phosphates. Some soils are rich in magnesium, others in iron, others in calcium.

The size of soil particles is as important as their composition. Large particles fail to hold much water, but extremely small particles can become hard and compact. The most satisfactory soils for plant growth contain particles of mixed size, such as occurs in a **sandy loam** (Table 17.1).

Soils are not just loose mixes of particles, like the free-flowing sand of a dry sandy beach. Rather, they are made up of loose aggregates of particles, called **crumbs,** held in place by organic material. The crumb structure of a soil affects many of its qualities as a medium for plant roots, especially its water-holding power, its softness, and the amount of air it can contain.

The acid-base reaction of soil is important for soil fertility. If a soil solution is too acid, the crumb-producing ability of calcium is damaged, organisms in the soil are inhibited from growing, and general productivity is impaired. If the soil solution is too alkaline, iron becomes insoluble, and organisms cannot live without iron. The "acid-loving" plants, such as the heaths (blueberries, azaleas, rhododendrons, heather), are not so much lovers of acid as lovers of iron. On the pH scale (Chapter 1), with neutrality at 7, soils are generally most productive at pH 5 to 6, or slightly on the acid side. A few crops, such as legumes (clover, peas, alfalfa), do best in slightly alkaline soils.

Water may simply trickle through soil after rain without remaining long enough to be of much use to plants. Or it may be present, stuck to soil particles as a thin layer. Such water is too tightly bound to soil to be available to roots. In clay soils, with their high proportion of extremely small particles, there may be as much as 20 percent water, but it is so firmly attached that it is useless to plants. **Capillary water,** however, is water that is present in liquid form, not stuck to soil particles yet held firmly enough so that it does not drain away to the *water table* below. Capillary water is the important source of water to growing plants, and it must be renewed periodically, or roots must grow to it. In life, roots do indeed grow into the moist places. As a root tip advances, followed by its region of root hairs (Chapter 16), it removes the water from the soil particles it can reach, making a dry cylinder of soil around the root. This does not imply that roots actively seek water, for they do not, but it does mean that much water uptake is the result of active growth. If one recalls that Dittmer's rye plant produced nearly 600 kilometers of roots (Chapter 16) in one season, it is easy to understand that daily growth of a root system can be considerable.

Soil must be provided with abundant air spaces, or most plants cannot grow in it. If air spaces are eliminated by trampling, as on paths, or by heavy machinery, if they are filled with water by flooding, or if gas exchange with the free atmosphere is prevented by piling on of soil (as happens in construction projects), plants usually die. Corn in a flooded field or a tree whose roots are covered by a bulldozer in a building lot is doomed because the

TABLE 17.1
Classification of Soil by Particle Size

SOIL CLASSIFICATION	DIAMETER OF PARTICLES (IN MM)
Sand	2 to 0.02 mm
Silt	0.02 to 0.002 mm
Clay	Less than 0.002 mm

roots, which are alive and require an adequate oxygen supply, will be killed. Even under the best of conditions, air pockets in soil contain higher concentrations of carbon dioxide than are in the atmosphere because of the respiratory action of roots and soil organisms, with no compensating photosynthesis. (It is, after all, dark down there.)

Organic material makes up varying amounts of soils. In the black soils of swampy regions, the soil may be almost entirely organic and will burn if ignited, but drifting sand dunes may be practically devoid of organic material. Good growing soils have a mixture of inorganic matter, which provides lightness and crumb texture, and organic material, which holds water and helps maintain crumbs. The organic material comes from dead and partially decomposed plant and animal parts and from the excreta of animals, worked on by soil animals, bacteria, and fungi. The familiar dark color of a good organic soil is due to the presence of **humus,** a decay-resistant mix of protein and lignin derived largely from woody plant parts.

Because its inhabitants are mostly too small to see, soil organisms are rarely noticed, but actually the soil is alive with bacteria, protozoa, fungi, various worms, insects, plant roots, and even large burrowing animals. A handful of a rich agricultural soil can contain as many bacteria as there are people living on the earth. As a result of the biological activity in the upper levels of the soil (living things are not common below the top meter or so), the soil is being constantly stirred up, acidified by carbon dioxide production, and altered by the living, dying, eating, and excreting of its inhabitants.

Mineral Requirements of Plants

Some elements are needed in rather large quantities by plants and animals (Figure 17.2). Excluding oxygen, carbon, and hydrogen, which are obtained from air and water, all the required elements must be absorbed by plants from soil. Those needed in greater amounts are calcium, nitrogen, phosphorus, potassium, sulfur, magnesium, and iron. These are the **macronutrients** (Table 17.2). To get into the living world, they must be absorbed into plant roots, usually via root hairs, and then incorporated into plant protoplasm, where they remain until the plant dies and its components are returned to the soil by the bacteria and fungi of decay or until the plant is eaten by some animal. Eventually, the animal itself is eaten or dies. In either instance, the elements originally incorporated into the plant will find their way back to the soil to be used over and over.

Every element is recycled, even those that are not metabolically useful. If an element is in the soil, it will dissolve to some extent, and a plant will take it in. Traces of such useless elements as aluminum, lead, and gold can be found in plants, and their concentration is an indication of the concentration of the elements in the soil. Gold analysis of plants, for example, may eventually be used in an activity called geobotanical prospecting, by which people hope to be able to locate precious metals. Biologists, however, are mainly interested in biologically useful elements and have traced the recycling of these elements through many possible pathways (see Chapter 27).

Besides the macronutrients, there are the **micronutrients,** sometimes called **trace elements.**

Figure 17.2
Determination of the need for mineral elements. Tobacco seedlings were grown in a solution containing all the elements needed (on the right) and other solutions lacking some elements. The differences in growth are apparent.

TABLE 17.2
Mineral Elements Necessary for Growth of Plants, Listed in the Order of Quantities Needed

MINERAL ELEMENT	LOCATION, FUNCTION
Macronutrients	
Nitrogen	Part of almost every organic molecule except neutral fats, carbohydrates, and hydrocarbons.
Potassium	Required by many plant activities but with poorly understood function.
Sulfur	Part of amino acids (cystine, cysteine) and some enzymes.
Calcium	Required for firm intercellular middle lamellas; presence affects permeability of cell membranes; used in enzymes.
Magnesium	Part of chlorophyll molecules; cofactor in enzyme actions.
Phosphorus	Part of DNA, RNAs, membrane lipids, ATP and related compounds, and coenzymes.
Iron	Part of molecule in enzymes, essential in cytochromes.
Micronutrients (trace elements)	
Boron	Probably concerned with sugar transport.
Chlorine	Unknown.
Sodium	Unknown.
Manganese	Coenzyme action in photosynthesis, respiration, and nitrogen activities.
Zinc	Amino acid synthesis; coenzymatic activity.
Copper	Part of respiratory enzymes.
Molybdenum	Part of nitrate-nitrite reduction in bacteria.

They are required in such small concentrations that they were slow to be recognized, because efforts to demonstrate their necessity were hindered by the difficulty of obtaining pure enough chemicals.

TRANSLOCATION IN PLANTS

Plants do not have a circulatory system in the sense that animals do, with fluid continually going and returning to a starting place. Instead, water and dissolved salts or ions usually enter plants through the roots. The water mostly passes through the plant and is evaporated from the leaves, and the nonvolatile materials are left behind in the plant. The minerals are incorporated into protoplasm or into foods, and some of the water is used in photosynthesis. The resulting foods are either stored in leaves or transported to other storage organs or to active regions where they are used metabolically. There is, then, upward and downward movement of materials, but the transporting system, especially as far as water is concerned, is an open system, and it is better compared to a pipeline than to a circulating system.

Water enters roots mainly through root hairs, the driving force being primarily the diffusion pressure of the water molecules. Living roots contain many dissolved materials: sugars and other organic compounds and salts. The water in the soil is more nearly pure water than the water inside the root cells, and there is a difference in water concentration between the soil water and cell sap. Since the cell membranes are freely permeable to water but not to the solutes in the cell sap, they allow water to diffuse in. Once in a root, water can diffuse from cell to cell in the cortex of the root, always following a "downhill gradient," that is, moving continuously from the place where there is more water (less solute) to where there is less water (more solute). Once across the cortex, the water can enter the water vessels of the root xylem. It is then free to pass up the roots and to the leaves by way of the stems, carrying some dissolved minerals with it.

Rise of Water

Water pours through stems with such speed and volume and to such heights as cannot be explained by the usual mechanical methods of water rise. Neither capillarity nor suction, aided by atmospheric pressure, could raise water to the top of even a 25-meter (80-ft) tree or vine, much less one of the tall eucalyptus or Douglas fir trees. The most widely

Figure 17.3
Demonstration of the pulling power of evaporation on a water column. Mercury cannot be pushed up by air pressure (we usually say "sucked up by a vacuum") higher than 76 centimeters (30 in.). A continuous water column, however, has enough tensile strength to allow it to pull a mercury column higher than air pressure can push it. With an apparatus like the one shown here, a mercury column can be pulled up over 100 centimeters (40 in.). A plug of some porous material, such as plaster or wood, is made thoroughly wet, and the glass tube is filled with water and dipped in a mercury reservoir. As evaporation proceeds the mercury column is pulled up until it passes the limit imposed by the weight of air on earth (atmospheric pressure). This experiment establishes the principle of "transpiration pull," by means of which evaporation from leaves can pull water from the soil to the tops of tall trees.

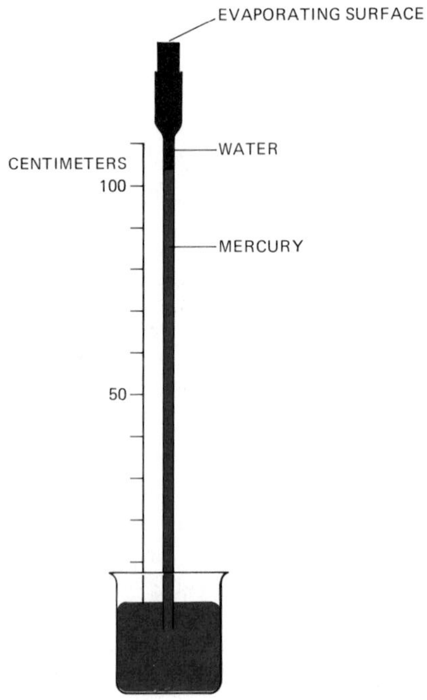

accepted explanation for the flow of water up trees is the tension-cohesion theory of Dixon and Joly, which holds that transpiration, which is the evaporation from the leaves, pulls the water up from the roots, taking advantage of the tensile strength* of water in a closed tube (Figure 17.3). Some people are surprised to learn that water does indeed have tensile strength, but the theory is supported by many observations and experiments on living plants, as well as by mechanical models that can raise water higher than atmospheric pressure can account for. So long as leaves are actually transpiring, there is a flow of water from soil to roots to stems to leaves, pulled along by a water deficit in the leaves. Other suggestions, such as the active pumping of water by xylem along the way, have not stood up to experimental checking.

The transpiration pull that exists when a tree is in full leaf, however, could not exist in a deciduous tree in early spring before leaves have appeared. Yet water does rise in leafless trees, and demonstrations of root pressure indicate a pushing force that could help water move upward.

Phloem Transport

The movement of materials in the xylem can be followed with a fair degree of certainty. Both the usual transport of water and mineral salts upward during times of active transpiration and the transport during the spring of dissolved sugars, especially notable in sugar maple trees, can be readily verified. The movement of substances in the phloem cannot be so easily followed. Phloem tissues are thin, and they are protected by outer cells that are frequently thick and corky (see Chapter 16). The phloem cells themselves are so delicate that any experimental treatment is likely to interfere so seriously with their functioning that little has been learned from such direct interference.

Radioactive tracers that leave the living cells intact have been used to show that phloem can carry water and dissolved foods both up and down stems, but the forces that determine the direction of flow and provide the driving power remain unknown. Unlike water-conducting vessels in xylem, conducting tubes in the phloem are alive, although the main conducting cells, the **sieve tubes,** are strangely lacking in nuclei. The fact that the sieve tubes are perforated at the ends and sides but are provided with active cell content has prompted the suggestion that sieve cells somehow act as living pumps. The smaller, closed companion cells, which accompany sieve tubes, may have some effect on translocation by the phloem, but their actual function is not known.

One ingenious approach to the study of phloem transport is the use of aphids (Figure 17.4). These tiny insects can drive their suckers through the epi-

* The tensile strength of a material is the amount of stretching force it can stand without breaking.

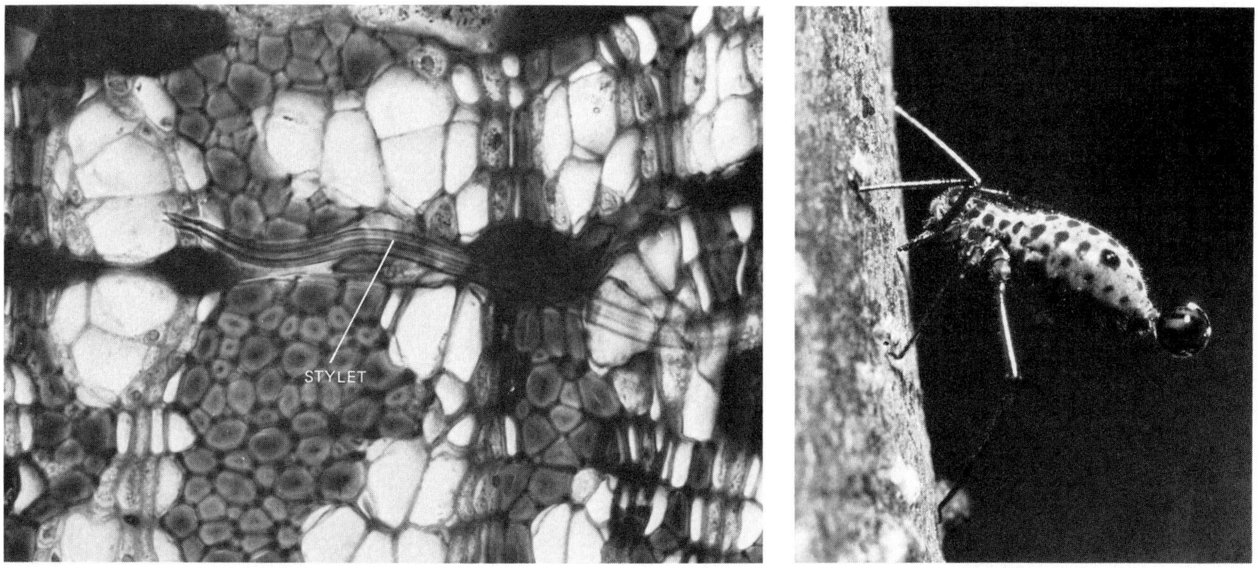

(a) × 200 (b) × 40

Figure 17.4
Using aphids to collect samples of plant sap. An experimenter trying to obtain sap from a living tree would have to use such a gross needle that the phloem would be damaged, thus invalidating the experiment. But the slender sucker of an aphid can enter a phloem cell apparently without harming it. If the aphid's sucker is cut off with a razor, phloem fluid oozes out of the cut stump, where it can be collected for analysis. Thus the content of phloem fluid can be determined. In the picture on the left, a section through the bark of a basswood twig shows an aphid's stylet in the phloem cells. The picture on the right shows an aphid sucking on a twig. A droplet of exudate, called honeydew, is hanging from the aphid's abdomen.

dermis of a stem and into a living phloem cell, then suck out the juice, as if through a microscopic hypodermic syringe. The puncture does not seem to harm the plant cell. The aphid beak, called the stylets, can be neatly severed with a razor blade, and phloem exudate will continue to flow from the cut stump, forming droplets that can be collected for analysis. A series of such analyses shows that dissolved sugars, amino acids, and various other substances can be transported by phloem either from leaves or from roots, but the mechanisms of transport remain obscure.

GROWTH REGULATION IN PLANTS

Plants have no recognizable glands like the glands of animals; nor do they have a circulatory system. They do, however, have the ability to synthesize compounds that are transported from one part of the organism to another and that can result in specific physiological reactions. Such compounds in animals are called **hormones,** and when regulatory compounds were found in plants, they were also called hormones. The term hormone is still used for some plant substances, but many botanists prefer to call them **growth-regulating substances,** of which a number of natural and synthetic examples are known.

Tropisms and Auxin

The early work on plant growth-regulating substances was done in an effort to understand tropisms. **Tropisms** are growth movements whose direction is determined by the direction from which the stimulus comes. They are *positive* when growth is toward the stimulus and *negative* when growth is away from the stimulus. A plant that is illuminated from one side and grows toward the light is showing *positive phototropism.* If a shoot grows away from the gravitational pull of the earth, it is showing *negative geotropism.* Other tropisms are *thigmotropism* (response to touch) and *chemotropism* (response to chemicals).

The original observations on phototropism were made by Charles Darwin, who investigated plant movement, earthworm habits, barnacles, and

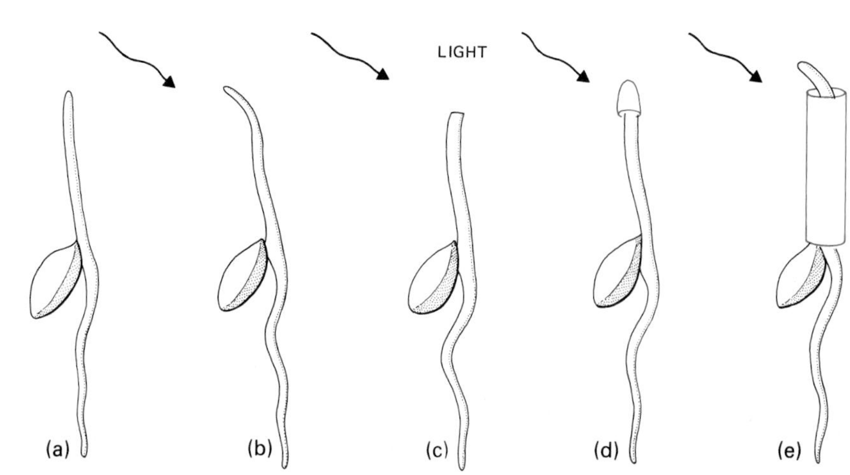

Figure 17.5
Darwin's demonstration of the light receptor site in a seedling. (a) A grass seedling grown in the dark or in uniform illumination sends its coleoptile straight up. (b) If illuminated from one side, the coleoptile bends toward the light source. (c) If the tip is cut off, the coleoptile will not bend, even when unilaterally illuminated. (d) If a cap is put over the tip, the coleoptile is insensitive to unilateral illumination. (e) If the shaft of the coleoptile is kept dark by a sleeve, the same part of the coleoptile that bent in (b) will bend. Therefore the tip is the sensitive part, but the shaft of the coleoptile is the responding part.

animal behavior, to name some of his lesser contributions (Figure 17.5). He used grass seedlings to show that the light-sensitive part of the seedling is the tip of the coleoptile (Chapter 16), but that the active part is *below* the coleoptile. This finding indicated that something in the tip was affected by light, but that the "something," whatever it was, moved down lower to cause a curvature at some distance from the receiving source. Later experiments showed that if the tip was removed and the seedling was illuminated from one side, there was no response, but if the tip was removed and replaced *off center,* the seedling would bend even when uniformly illuminated all around or when not illuminated at all. Thus was born the idea of a plant hormone.

The actual substance responsible for the observed curvatures was found after techniques for its separation were developed. First it was discovered that if coleoptile tips were cut off and placed on agar (an inert jelly) sheets, then small portions of the agar, stuck to the side of a growing coleoptile, could cause bending (Figure 17.6). The active substance was extracted, concentrated, and identified as indoleacetic acid, or IAA, now commonly called **auxin.** It can be shown experimentally that auxin is synthesized in the coleoptile tip of grass seedlings and that it moves away from its source under several influences: (1) It moves geographically downward under the influence of gravity; (2) it moves away from a one-sided light source; and (3) it moves morphologically downward, that is, "downward" with respect to the plant axis. Just as animals have a head end and a tail end, so plants have an essentially upward-growing shoot and a downward-growing root, and that *polarity* seems to be irreversible. It can also be shown that in shoots, higher concentrations of auxin can cause cellular enlargement.

With this much information, we can begin to understand how a plant turns toward light: (1) Auxin is present in a coleoptile tip, where it is synthesized. (2) It moves downward. (3) A light source at one side drives the auxin toward the dim side. (4) The auxin, which is more concentrated on the dim side, causes the cells on that side to elongate. (5) The cells on the illuminated side do not elongate as much. (6) The shoot bends toward the light. The plant is not "trying to get to the light"; it simply cannot do anything else.

A similar series of events "explains" upward growth away from gravitational pull. (The word "explains" is in quotation marks because this is only a partial explanation, and no full explanation is yet available.) (1) Auxin moves away from the tip, but this time it moves *morphologically downward,* regardless of the position of the seedling; that is, it moves toward the base of the plant. (2) If the seedling happens to be lying on its side, the auxin will move not only morphologically downward but also toward the actual downward side of the shoot, and it will accumulate along the lower side. (3) Cells on the lower side will elongate faster than those on the upper side. (4) The shoot will bend upward.

These explanations seem acceptable until we look at the action of roots. When a primary root emerges, it shows *positive geotropism* and turns down. If auxin causes cells to elongate, a root on its side should grow upward, as a shoot does. The answer is that auxin causes cell enlargement only in the proper concentrations, and above a certain concentration, it can act as an inhibitor. Roots are much more sensitive to auxin than shoots are, and a concentration that can cause elongation in a shoot can cause inhibition in a root. Besides, roots have root caps, which seem to be receptors. Root caps in some way receive the gravitational stimulus and cause hormonal regulation of directional growth, involving auxin and other regulators, such as abscisic acid, to be discussed later.

With this much understood, the ability of a whole seedling to right itself becomes clear. Neither in nature nor in agriculture are seeds planted so that their radicles point down and their epicotyls up, yet when they germinate they align themselves properly. The auxin, moving from the shoot apex, determines that the shoot will grow upward. Auxin plus other regulators make the root grow downward until the seedling is established.

The amount of auxin used in one geotropic response is too small to measure directly. Only a millionth of a gram can be extracted from about 10,000 coleoptile tips. Under such circumstances, biologists use a technique that is effective in many instances when vanishingly small amounts of an active substance have to be determined: **bioassay.** In a bioassay, some organism of known ability is tested. Bioassays are routinely used in the measurement of quantities of vitamins, hormones, antibiotics, and of course, plant growth-regulating substances. To perform a bioassay on auxin, one starts with a known amount of auxin. Then after successive dilutions, the auxin is applied to the side of a decapitated oat seedling in the dark. The degree of curvature resulting is measured. When an unknown quantity of auxin is to be measured, the curvature of a new set of seedlings is measured and compared to that of the first set. The amount of auxin in the unknown can be determined by such comparison.

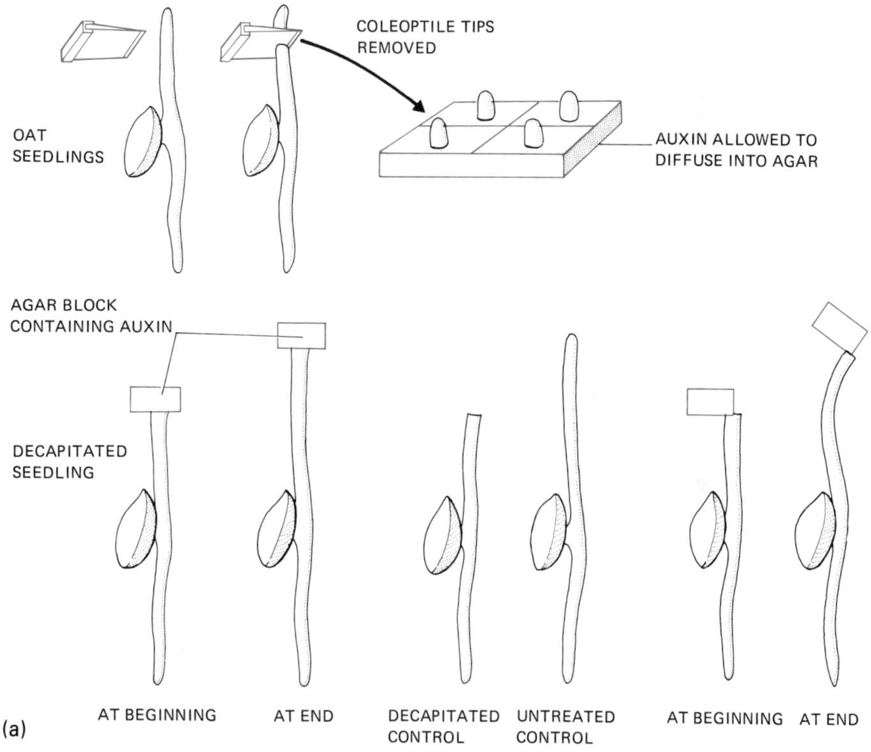

(a)

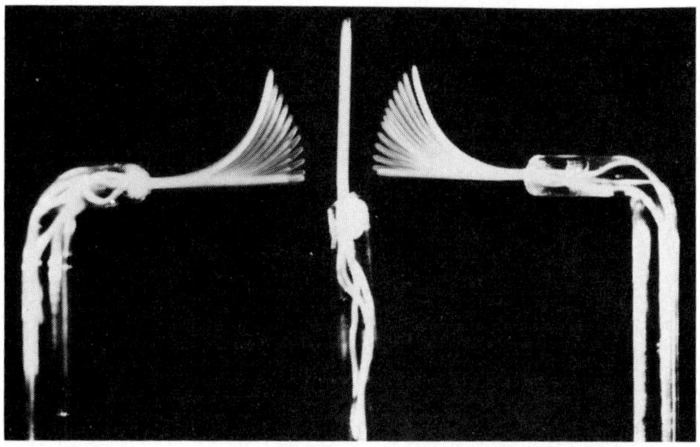
(b)

× 0.67

Figure 17.6
Hormonal control of tropisms. (a) Auxin and coleoptile bending. Coleoptile tips are cut and placed on small squares of agar to allow diffusion of auxin from the tips into the agar. If an auxin-containing agar block is placed on top of a decapitated coleoptile, the coleoptile will grow at a rate equal to that of an untreated seedling, but if a coleoptile is decapitated and no auxin is provided, growth stops. For still another proof, when an auxin-treated agar block is placed off center atop a decapitated coleoptile, the auxin diffuses down one side of the coleoptile, causing it to grow faster on that side and making the shaft bend. (b) One seedling was placed upright in the center, and other seedlings, one on each side, were fastened horizontally. They were photographed repeatedly on the same film for two hours. The images show the upward bending of the horizontally placed seedlings. Since auxin is pulled downward by gravity, it accumulates on the lower side of the horizontal seedlings and causes faster growth there. The result is the upward bending of the shaft.

Once IAA was found to be *the* auxin, several chemically related compounds were found to be capable of producing similar effects. Naphthalene acetic acid, indolepropionic acid, and indolebutyric acid, all synthetically produced, can act as auxins. Another related compound, 2,4-dichlorophenoxy-acetic acid (2,4-D), is such a potent growth stimulator that, when applied in concentrated form, it can cause a plant to grow itself to death. Consequently, it is useful as a cheap weed killer. In spite of this information, however, we still do not know what there is about the nature of some molecules that gives them an auxinlike quality or how, at a molecular level, auxin affects cells.

Other Actions of Auxin

Auxin is involved in more than tropisms. It enters into the regulation of plant life from seed to fruit. It stimulates root initiation. Application of auxin to the outside of a stem can cause it to put forth a veritable beard of roots, and synthetic auxins are commercially used to make roots grow on such woody cuttings as holly. Auxin can inhibit growth as well, as is shown when the auxin produced in an apical bud moves down a stem and keeps nearby axillary buds from breaking into growth. This phenomenon is called **apical dominance** (Figure 17.7). If a strong apical bud is cut off, the next lower side buds will soon send out branches and become apical buds, *inhibiting* in turn the buds below it. All gardeners know that nipping off terminal buds will make a bush bushier.

Leaf abscission is prevented by auxin. If most of a leaf is cut away, an abscission layer will form at the base of the petiole and the leaf will fall off, but if auxin is artificially applied to the cut surface, the leaf will remain. A similar action occurs on fruit stalks. Frequently, fruit drops off before ripening. Auxin sprays on apple orchards can make the trees hold the fruit instead of allowing the massive loss known as "June drop," when many little apples are likely to fall prematurely. Auxin influences the differentiation of cambium cells or cells in tissue cultures into xylem cells.

A last action of auxin is the stimulation of fruit development. When a pollen grain germinates on a stigma of a flower, it brings enough auxin with it to start growth of the ovary wall, making the fruit grow. Most fruits will not grow if the ovules are not provided with pollen-borne auxin, although bananas and pineapples are exceptions. As we will see, however, auxin frequently acts together with other growth-regulating substances, and it is not the only one or even the most potent one.

Figure 17.7
Apical dominance and the effect of removing the apex. In the intact plant (left) the main growing point is growing steadily, and the lateral buds in the axils of the leaves are held in check by the auxin draining down from the apex. When the apex is cut off, the auxin source is removed and the lateral buds begin to grow. They, in turn, now exert an inhibiting influence.

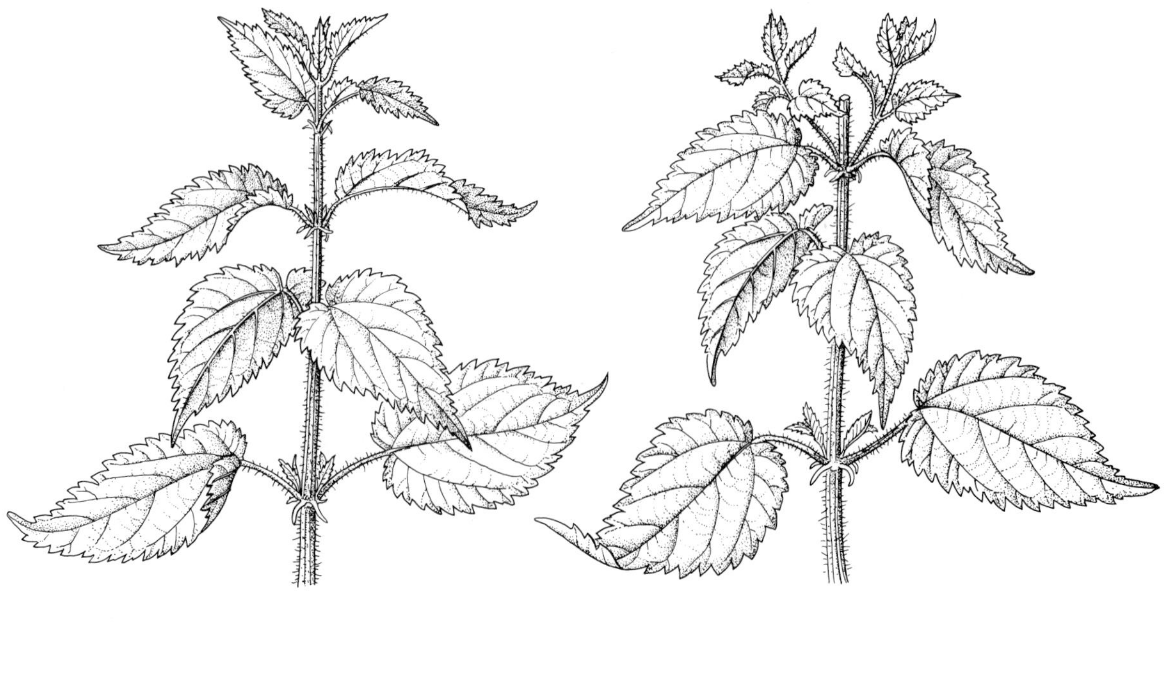

MANY FROM ONE OR TWO

When crop plants are grown, the more uniform the crop is, the more efficient the handling of the crop and its product. At present, however, making a large number of genetically uniform plants is either slow and expensive, as in grafted fruit trees, or impossible, as in cereal grains.

Some techniques, still limited to laboratory pilot production, promise to make possible masses of cloned plants. One way is to grow plant embryos from vegetative parts of a desirable kind of plant. Mature cells are cleanly removed from a plant organ, such as a root or leaf, and placed in a sterile solution in which a particular blend of plant hormones induces cell division. Soon a mass of rapidly dividing, undifferentiated cells floats in the liquid medium. When some of the cells are moved to another medium with a different hormone mixture, they begin to grow in an organized fashion and eventually become embryos. They look like zygotic embryos, but since they come from somatic (nonreproductive) tissues, they are called **somatic embryos.**

The growth of somatic embryos shows that it is possible to bypass the usual sexual cycle and still generate functional embryos. It also demonstrates that mature cells retain a full complement of genetic information; that is, they are *totipotent* (see p. 145). The mature cells have everything necessary to form an entire new plant.

Somatic embryos can be grown quickly and in large numbers. Over 100,000 carrot embryos can be grown in a single Petri dish in about four weeks. These embryos are useful for studying the processes of growth and differentiation because, unlike ordinary zygotic embryos, they are not buried in maternal tissue.

Somatic embryos of the caraway plant (*Carum carvi*) are shown in the photo floating in a medium containing sugar, mineral salts, vitamins, and equal concentrations of the plant growth regulators zeatin, gibberellic acid, and abscisic acid. Each Y-shaped body has a hypocotyl and two cotyledons and looks and acts like an embryo from a typical seed.

Because there is no genetic segregation or recombination, each somatic embryo, as well as the plant that grows from it, has exactly the same genetic makeup as the plant that originally provided the cells. Growing multiple somatic embryos therefore provides a quick and efficient method of cloning.

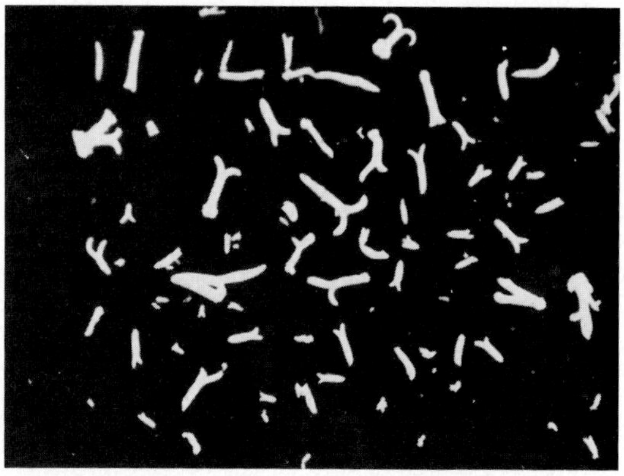

Somatic embryos of caraway. ×1

A different method uses the technique of hybridizing *protoplasts*. A plant protoplast is a cell that has had its covering wall of cellulose digested away by enzymes, leaving the plasma membrane as its outermost boundary. Plant tissues may be broken apart, yielding individual cells, but the cells keep their walls and are so heavily encased therein that they cannot unite with other cells. But naked protoplasts do fuse with one another, even if the cells are of unrelated species. Fusions of protoplasts are therefore unlike grafts, which must be made with attention to evolutionary relationships.

Hybridization between various animal cells was accomplished before it could be done with plants; animal cells have no interfering walls. Since methods have been developed for ridding plant cells of their walls, and some novel hybrids have been produced, the way seems open for manufacturing plants almost to order. Heretofore, plant breeding programs were limited to the mating of strains that were amenable to cross-pollination. Sperm from wheat pollen, for example, cannot fertilize potato ovules. But with protoplasts, the hybridization of entirely new kinds of plants is theoretically possible.

By following the use of protoplast hybrids with the growth of unlimited numbers of somatic embryos, a new era in food production seems possible.

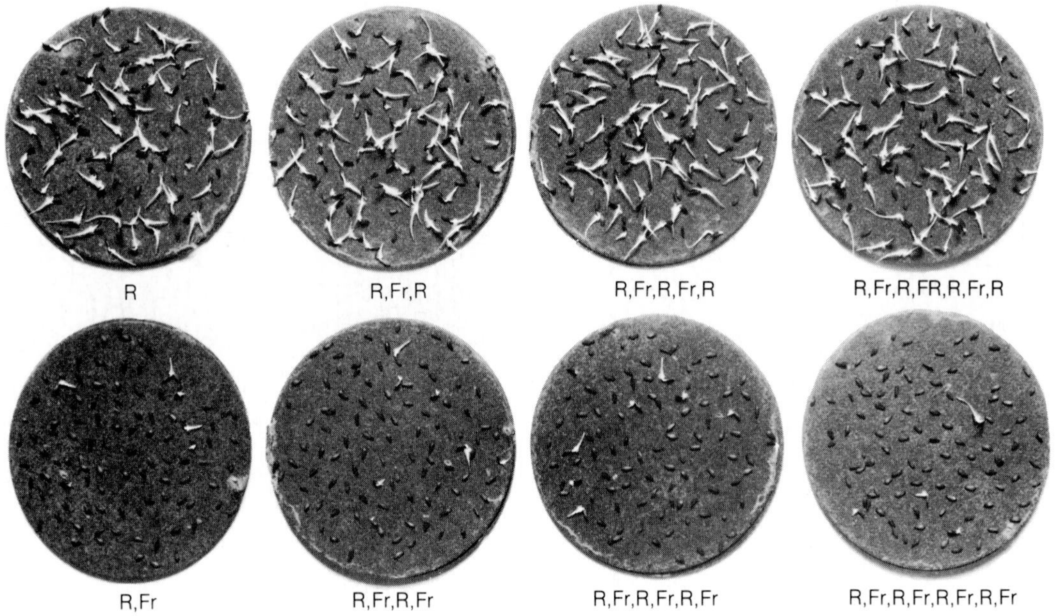

Figure 17.8
Demonstration of the effects of red and far-red light on lettuce seed germination. The top row of Petri dishes shows most of the seeds sprouting, but the bottom row shows almost no seedlings. The letters under each dish tell the treatment given, with R indicating exposure to red light and Fr indicating far-red. Those seeds that had red light as their last exposure germinated; those that had far-red did not. The critical point is that in every instance the final treatment was the one determining whether or not the seeds germinated.

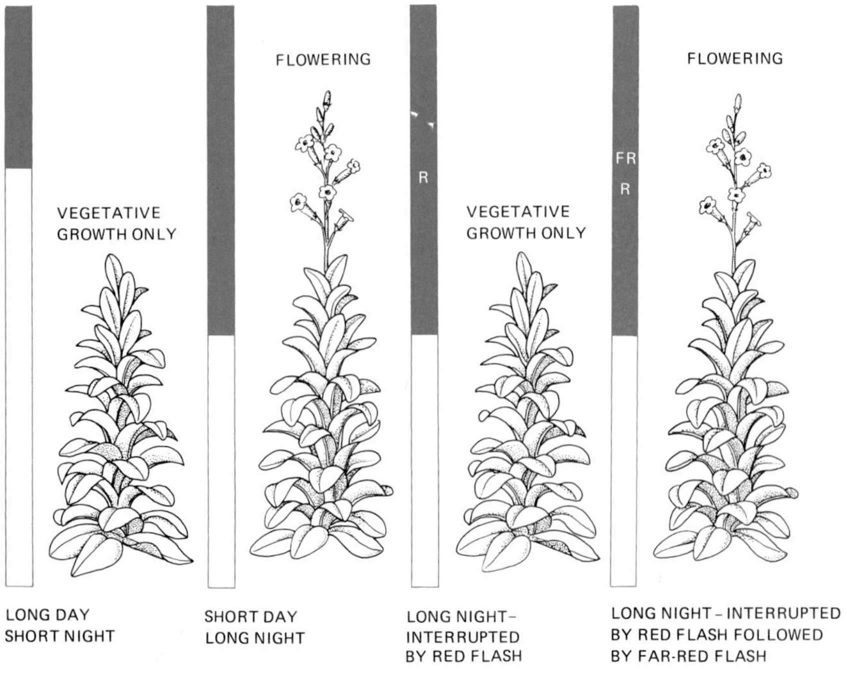

Figure 17.9
The red–far-red reversal effect in the plant on which much of the original photoperiodism work was done: Maryland Mammoth tobacco. The bars indicate a 24-hour day, with the white portion representing the relative amount of light and the colored portion representing darkness. When grown in long days and short nights, the plant remains vegetative. When grown in short days and long nights, the plant flowers. However, if the long night is interrupted by a flash, only a few minutes long, of red light (or white light containing red), the plant behaves as if it had had only a short night and stays vegetative. But the effect of the red light can be obliterated by a flash of far-red light. If that is provided, the plant reverts to its short-day behavior and flowers.

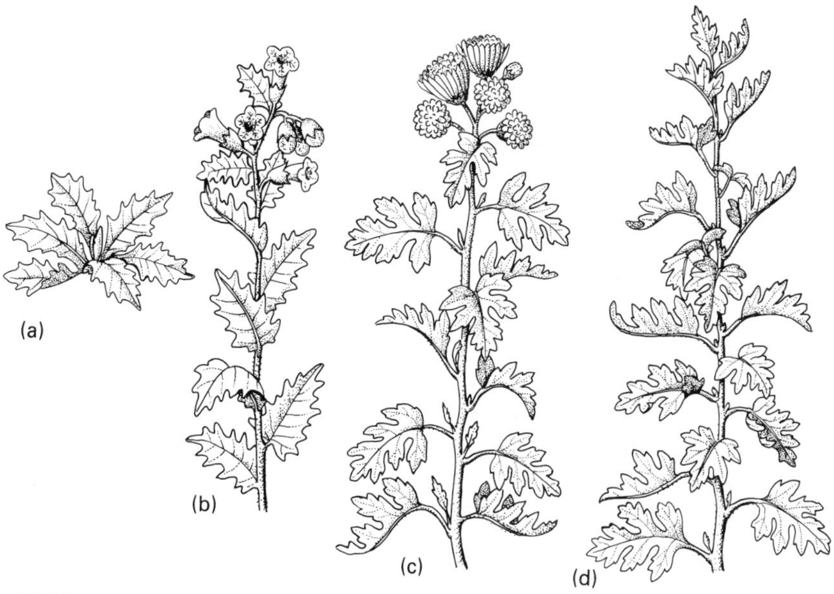

Figure 17.10
The effects of day length on flowering. Black henbane (*Hyoscyamus niger*) is a long-day plant, which flowers only when days are longer than a critical period. (a) A plant of black henbane kept artificially on a regime of short days remains vegetative. (b) The same species flowers when provided with long days and short nights. Chrysanthemum varieties, however, are frequently short-day (or long-night) plants. (c) A chrysanthemum plant kept on a short-day (long-night) regimen flowers. (d) A similar plant kept on a long-day schedule remains vegetative.

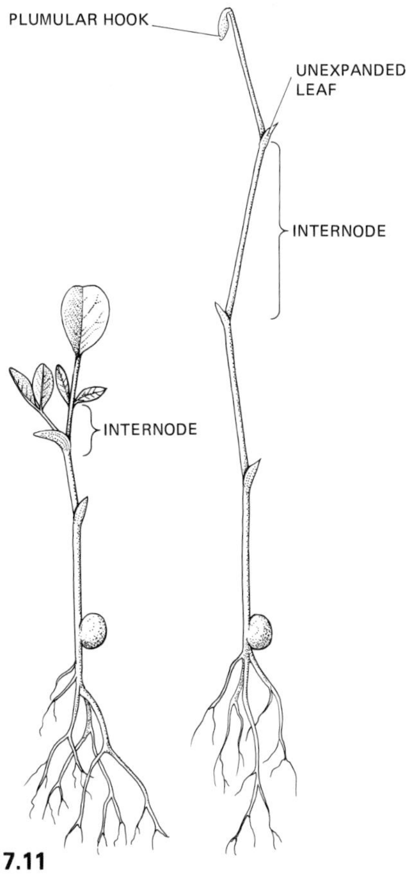

Figure 17.11
The red–far-red reaction in a seedling. When illuminated with red light (or white light containing red), the seedling grows normally as shown at the left. The internodes are restricted in length, the leaves expand, and the plumular hook at the apex straightens. When illuminated with far-red light only, the seedling has much longer internodes, the leaves remain unexpanded, and the plumular hook stays bent over as if it were in the dark.

Phytochrome and Photoperiodism

Phytochrome is a greenish-blue plant pigment involved in a number of activities. Among other things, it affects the time of germination of seeds, the timing of blooming of flowers, and the shape of seedlings.

All of these plant activities have a feature in common, and that is the plant's reaction to special portions of the spectrum. Those special portions are in the long wavelengths, specifically the red at 660 nm (nanometers) and the far-red 730 nm. For example, Grand Rapids lettuce seeds germinate when illuminated with red light at 660 nm, but remain dormant in far-red light (Figure 17.8). Under natural conditions, such seeds would not germinate when buried in dark soil, but would germinate if brought to the surface where they could be illuminated by daylight, which includes red light.

The flowering habit of certain plants is regulated by the same wavelengths. A tobacco variety, Maryland Mammoth, will flower if it is grown under conditions of short days and long nights. But it can be kept from flowering even under a short-day regime if the long night is interrupted by a burst of red light. If the burst of red light is followed by a burst of far-red, however, the plant will flower. The *last* treatment is the effective one (Figure 17.9). Other plants are known to flower only during long days (Figure 17.10). The phenomenon of responding to day length is **photoperiodism.**

Another activity controlled by the red–far-red light system is the form of germinating seeds (Figure 17.11). A pea seedling grown in white light

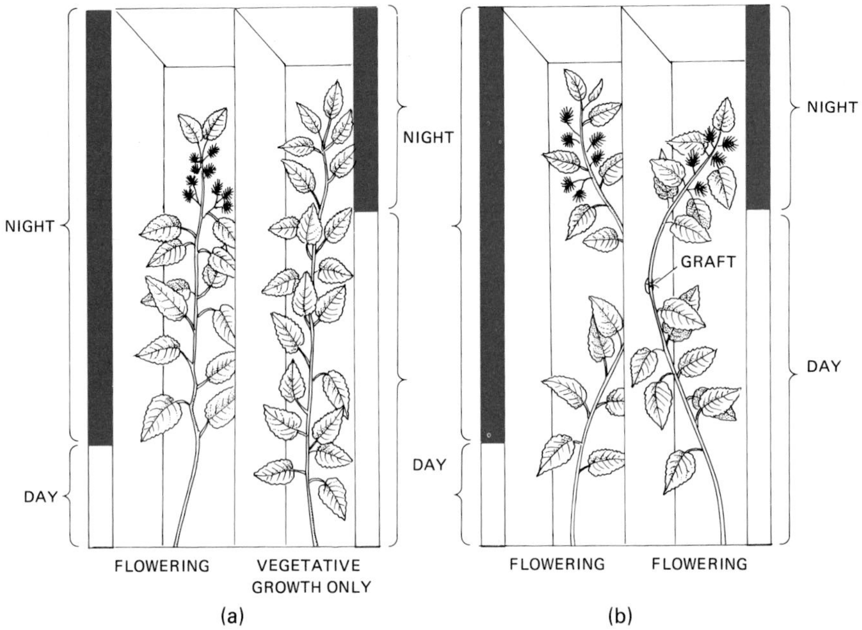

Figure 17.12
Grafting and flowering. The cocklebur *Xanthium*, is a short-day plant. (a) If plants are grown in light-tight boxes and provided with short days, they flower; if given long days, they do not. (b) If a plant on a short-day regimen is grafted through a hole in its box to another plant that is kept on a long-day regimen, the second plant will flower even though the light it is receiving would normally prevent flowering. This indicates that some diffusible substance is passing through the point of the graft and causing the second plant to overcome the inhibition that its own light regimen should dictate.

(containing red light at 660 nm) or in red light has short internodes, expanded leaves, and an erect apex. A seedling grown in far-red light grows long and thin, keeps its leaves unexpanded and has a bend in the stem near the apex. The bend, or plumular hook, is useful to a plant that is pushing its way up through soil. If a seedling comes up through dark soil and reaches daylight, the red light absorbed will cause the leaves to expand and the plumular hook to straighten.

The germination of the lettuce, the flowering of the tobacco, and the action of the pea seedlings are all controllable by the same red–far-red system. The absorbing pigment in all of these actions is thought to be phytochrome, because phytochrome absorbs light preferentially in a way that is like the light-absorbing systems in living plants. Just as germination and other activities are affected by red or far-red light and can change in response to light changes, so phytochrome can change. When a solution of phytochrome is illuminated by red light at 660 nm, the pigment absorbs the light and undergoes a change. After the change, it absorbs far-red light preferentially. Phytochrome in the red-absorbing state, called P_r, can be shifted to the far-red–absorbing state, P_{fr}, and back again, over and over. In living plants, phytochrome is thought to revert spontaneously to the P_r form during periods of darkness.

"Florigen"

Phytochrome itself is not thought to initiate the setting of flower buds directly. Instead, it causes some as yet undiscovered metabolic change, probably by altering the DNA activity of meristematic cells, with resultant changes in enzyme (protein) content. Grafting experiments indicate that some diffusible substance is directly responsible, with the phytochrome mainly in leaves acting as the receptor (Figure 17.12). If a short-day plant, such as a cocklebur, is exposed to short days, so that it is stimulated to flower, and is grafted through a hole in an opaque partition to another plant that is kept on a nonflowering schedule, the plant on the nonflowering schedule will flower. This result is taken to mean that something like a flower-inducing hormone migrates from the short-day side to the long-day side and makes the long-day side bloom, regardless of the light duration. Repeated efforts to isolate this postulated hormone have failed, but it already has a name: **florigen** (Latin, "flower maker"). Recently, other complex grafts between short-day and long-day plants have produced evidence of still another hormone, this one an antiflorigen that inhibits flowering. The switch from a leaf-producing to a flower-producing bud is a complex change, but progress is constantly being made toward understanding it.

Other Growth-Regulating Substances

Continuing investigation has revealed the presence of several additional growth-regulating substances. As was true of auxin and phytochrome, the discoveries came as a result of research on something other than hormones. Studies on leaf fall, on tissue cultures, on plant diseases, on dormancy of buds, and on ripening of fruits have shown that all these activities, plus a number of additional unexpected ones, are influenced by an array of compounds that can act either alone or in concert to regulate the general form and behavior of plants.

Rice suffers from a disease characterized by spindly growth. A fungus, *Gibberella fujikuroi,* causes the disease. When a substance, named **gibberellin,** was extracted from cultures of the fungus and applied to healthy rice plants, they developed typical symptoms of the disease. Later, about forty similar compounds were isolated from many parts of normal plants of various species; these substances were found to affect the growth habits of plants in a number of ways. A gibberellin can, for example, make a genetically dwarf plant, such as Mendel's dwarf peas, grow as tall as a nondwarf. Or a lettuce plant, typically formed into a compact head, can be made to "bolt," that is, to stretch its stem upward and separate the leaves (Figure 17.13). When a barley seed begins to germinate, a gibberellin influences the digestion of stored starch to soluble sugar. Gibberellins also affect the formation of flowers and the opening of buds. They seem to act in concert with auxins in determining the form and action of plants.

In the 1950s, F. Skoog and C. O. Miller found a new class of growth-regulating substances, which have been called **cytokinins** (Greek, "cell movers") because they stimulate cell divisions. Auxin can make cells enlarge, but it has no direct ability to promote mitosis. The most striking characteristic of the cytokinins is just that. The cytokinins are chemically similar to a component of DNA—the nitrogen base adenine. Several natural cytokinins have been found, including zeatin and kinetin, besides some forty adeninelike compounds (see the essay on p. 368). These substances not only can affect mitotic rates but are important in the normal development of embryos, and they have the surprising ability to prolong the durability of chlorophyll in isolated bits of leaf. Like the gibberellins, the cytokinins work in collaboration with auxin in maintaining normal plant growth.

Ethylene, C_2H_4, is the simplest and the longest known of growth regulators. One of its most readily observable features is its ability to hasten the ripening of fruits. In fact, people have known for centuries that hard pear fruits would soften if they were individually wrapped in paper. Each fruit, so enclosed, kept its self-generated ethylene in the package and so brought about the desired ripening. Now ethylene is used commercially to encourage artificial softening of hard fruits. On the other hand, if softening is not wanted in stored fruits, the ethylene can be washed away by a flow of inert gas, keeping fruit firm longer. Ethylene hastens leaf abscission, inhibits internode elongation, causes increase in stem thickness, and interferes with geotropism. These actions are antagonistic to those of auxin, and it is likely that ethylene and auxin work together, as other growth-regulating substances do, to make plants conform to workable patterns. With one force working to speed up an activity and another force working to slow it down, the activity can be kept at a functionally successful rate.

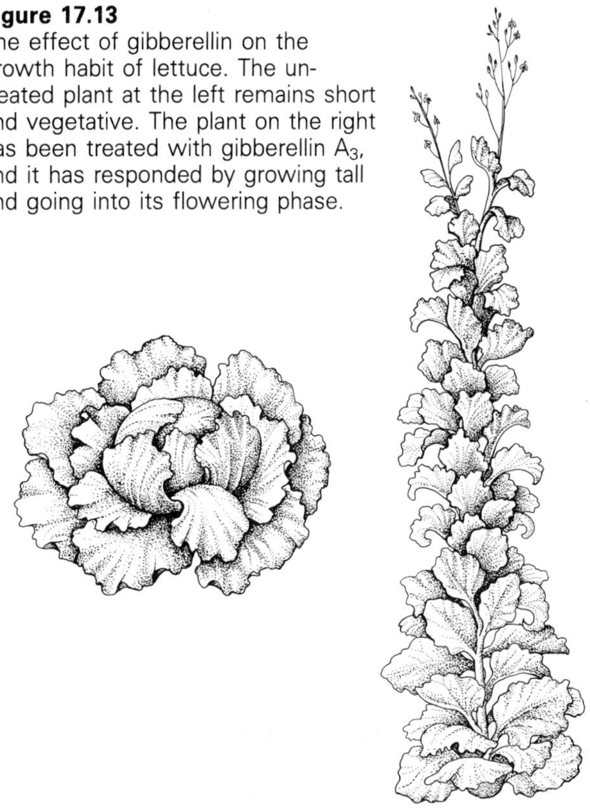

Figure 17.13
The effect of gibberellin on the growth habit of lettuce. The untreated plant at the left remains short and vegetative. The plant on the right has been treated with gibberellin A_3, and it has responded by growing tall and going into its flowering phase.

The most recent plant growth regulator to be discovered is **abscisic acid,** or ABA. It is involved in leaf abscission and in the dormancy of such woody plant buds as those of maples. It is well known that oak or maple branches brought into a warm place in midwinter will not open. Abscisic acid holds them tightly shut. Abscisic acid also acts as an antagonist to auxin in such matters as cell elongation. Another feature of abscisic acid, its ability to cause stomata of leaves to close, suggests that it plays a part in controlling water loss in times of stress.

The actual or supposed uses of the various plant hormones or growth regulators have been briefly indicated but not their distribution within a plant. Auxin occurs throughout a plant, but it is

WITCHWEED

A North Carolina farmer plants corn. It comes up and grows, but along with it is a small, red-flowered weed, and the corn crop that year is so poor that the farmer decides not to plant corn again next year but to try peanuts instead. During the next season, a peanut crop is weed-free, so the encouraged farmer in the third year goes back to corn. There is the troublesome weed again. It attaches its roots to the roots of corn and saps the strength of the corn, seriously lowering the productivity of the crop. Its seeds lie in the soil ungerminated until the proper host begins to grow. Then the weed seeds sprout, and the seedlings send their parasitic roots to meet the host roots. After that, the weeds thrive at the expense of the host.

The parasite is one of the witchweeds of the genus *Striga,* from a Greek word for "a bird that screams in the night." Our species is *Striga asiatica,* but the other species attack other grasses, such as sorghum, and they are equally destructive and specific.

The question of how witchweed seeds are able to remain dormant until corn grows nearby prompted one listener to this story to ask, "Do they have little eyes?" The answer is that a substance required for germination is diffused into the soil by corn roots but not by roots of other crops. When that substance reaches a witchweed seed, then and only then does the seed begin growth. By testing witchweed seeds in culture dishes and adding various compounds isolated from corn roots, experimenters identified the critical substance as a cytokinin (6-furfurylaminopurine):

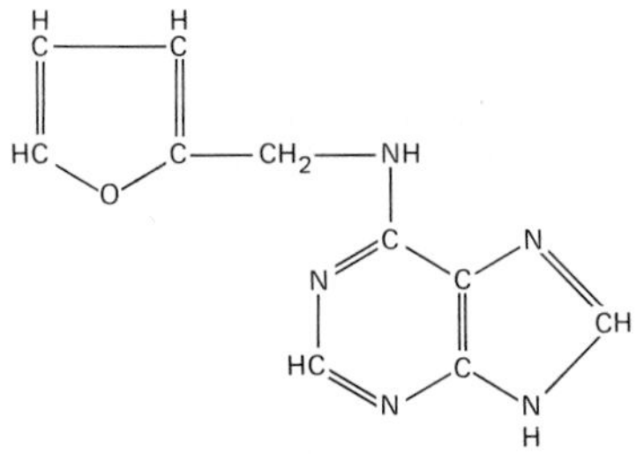

Cytokinin.

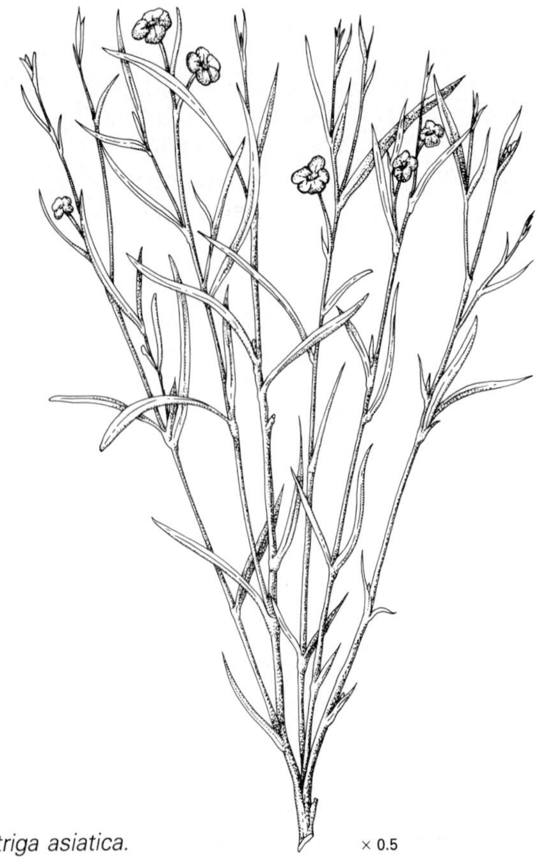

Striga asiatica. × 0.5

Witchweed seeds can be made to germinate by adding kinetin to the water in which the seeds are soaked or by growing corn roots near them, but most crops apparently do not exude kinetin into the soil, and witchweed seeds can lie dormant for years until a suitable host grows nearby.

The original problem of how to control witchweeds is still unsolved, although we know at least how the seeds regulate their own germination time. As so frequently happens, however, in seeking one answer, experimenters found a new question. Using radioactive carbon and phosphorus tracers, biologists found that water and minerals pass more readily from corn to witchweeds than in the other direction, as though a one-way valve were operating, but manufactured food does not pass readily in either direction. How does the parasite determine the direction of flow? The answer, when found, will be a straightforward physical or chemical one, *mysterious* when unknown, but no more *mystical* than the once secret reason for the dormancy.

The practical aspect of the parasitism of witchweeds still poses a problem: How can you get rid of the parasite and leave the host undamaged?

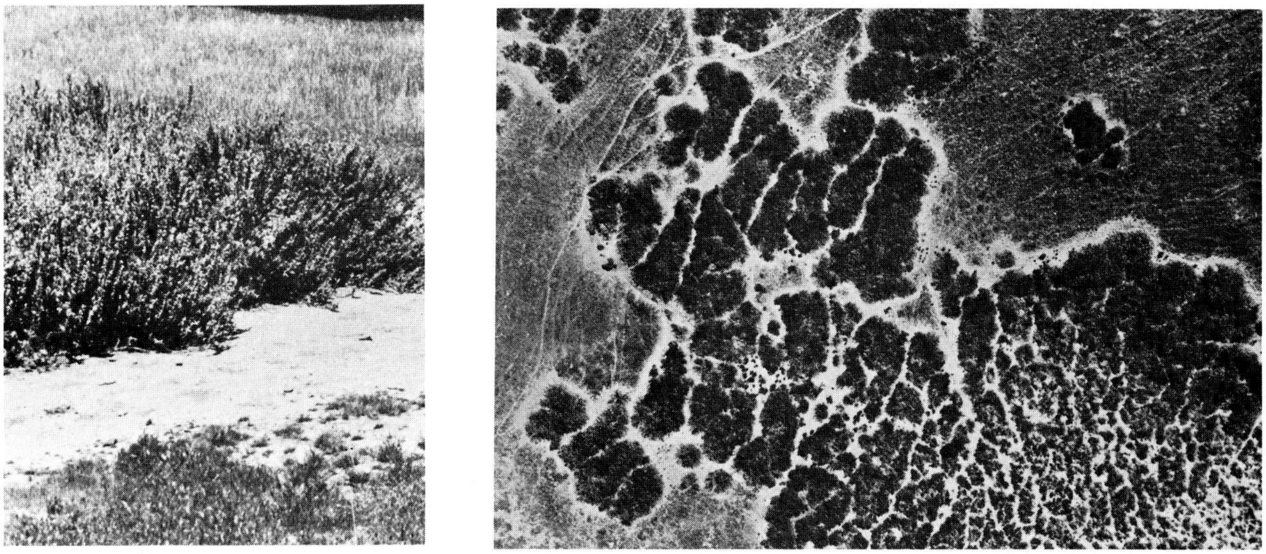

(a) (b)

Figure 17.14
Spacing in plants. The purple sage (*Salvia leucophylla*) of the southwestern United States exudes volatile oils that are toxic to other plants. When a sage bush is established and gives off its plant-repellant compounds, other plants cannot grow close to it. (a) A sage bush (left) is surrounded by an area of bare sand in which plant growth is inhibited. (b) An aerial view of patches of sage, each clearly surrounded by white margins devoid of plants.

most concentrated in the apex, young leaves, and root tips. Gibberellins are like auxins in their distribution. Cytokinins are restricted to actively growing regions where mitosis is occurring. Ethylene is most actively produced in maturing fruits, and abscisic acid in mature leaves. In our state of incomplete knowledge of these compounds, we can only say that there appears to be a balance, achieved through ages of evolution, that keeps plants working effectively through the seasons and through the vicissitudes of cold, drought, and disease.

Growth Retardants

Besides growth stimulators, there are a number of known compounds, both natural and synthetic, that inhibit vegetative growth of plants. One of these, *maleic hydrazide,* is used commercially to keep tobacco plants within desired limits, and it has been tried as a control for lawn and roadside grasses to reduce the need for mowing.

Some plants regulate not only their own lives but those of others. The ability of black walnut trees to hold down the growth of plants near them has been in the folklore for centuries. A compound called *juglone* that is present in walnut bark and hulls is toxic to other plants. In regions where survival for plants is precarious, as in the arid American Southwest, competition for water is intense, and some plants have evolved means of killing off competitors by exuding inhibitors and thereby saving the little precious moisture for themselves. One plant especially effective at protecting its own territory in this way is a shrub, *Adenostoma,* whose exudates include a number of phenolic compounds (related to carbolic acid). Another is a desert sage, *Salvia leucophylla,* which exudes camphor and other turpentinelike products that are poisonous to seedlings. The regular spacing of these plants, obvious even to casual observers, is the result of such exudates and helps guarantee adequate living space for a necessarily limited number of individuals (Figure 17.14). For examples of territorial protection in animals, see Chapter 30, and for competition among plants, see Chapter 27.

TO SUM UP GROWTH REGULATION

1. Plants contain *growth-regulating substances* that are transported from one part of the plant to another and that can produce specific physiological reactions.
2. *Tropisms* are growth movements whose direction is determined by the direction from which the stimulus comes. The active sub-

stance involved in tropisms is auxin, which also stimulates root initiation and can prevent the dropping of leaves or fruits.

3. *Phytochrome* is a light-sensitive pigment that affects the shape of seedlings, the time of germination of seeds, and the timing of flowering as determined by day length.
4. The habit of some organisms to respond to certain lengths of illumination time is called *photoperiodism.*
5. *Florigen* is a plant hormone that influences the flowering schedule.
6. *Gibberellins* seem to act together with auxins in determining plant growth and other actions, including starch digestion in seeds.
7. *Cytokinins* stimulate cell divisions; *ethylene* hastens fruit ripening; *abscisic acid* regulates leaf fall and other activities.

SUMMARY

1. *Soil* is the source of all the mineral nutrients taken in by plants and consequently by most animals. Soil is therefore of prime importance in the continuation of life. Besides the rock particles that make up the bulk of most soils, there are water, gases, organic matter, and living organisms.

2. The acid-base reaction of soil is important for soil fertility. If a soil solution is too acid, the crumb-producing ability of calcium is damaged, soil organisms are inhibited from growing, and general productivity is impaired. If the soil solution is too alkaline, iron becomes insoluble, and organisms cannot live without iron.

3. *Capillary water* is present in the soil in liquid form, but it is held firmly enough so that it does not drain away to the *water table* below. Capillary water is the water available to plants.

4. Soil must be provided with abundant air spaces, or most plants cannot grow in it.

5. Good growing soils have a mixture of inorganic matter, which provides lightness and crumb texture, and organic material, which holds water and helps maintain crumbs. *Humus* is a decay-resistant mix of protein and lignin derived largely from woody plant parts.

6. The soil is alive with bacteria, protozoa, fungi, various worms, insects, plant roots, and burrowing animals that constantly stir up the soil, acidifying it by carbon dioxide production and altering it by living, dying, eating, and excreting.

7. Some elements are needed in large quantities by plants and animals—these are the *macronutrients.* Excluding oxygen, carbon, and hydrogen, which are obtained from air and water, all the required elements must be absorbed by plants from soil. Besides the macronutrients, there are the *micronutrients,* sometimes called *trace elements,* which are required in very small amounts.

8. Plants do not have a circulatory system in the sense that animals do, but there is upward and downward movement of materials in an open transporting system. Water and dissolved salts or ions enter plants usually through the roots. The water mostly passes through the plant and is evaporated from the leaves, and the nonvolatile materials are left behind in the plant.

9. Water enters roots mainly through root hairs, the driving force being primarily the diffusion pressure of the water molecules.

10. The widely accepted tension-cohesion theory of Dixon and Joly states that water is able to flow up trees because the evaporation from the leaves pulls the water up from the roots, taking advantage of the tensile strength of water in a closed tube. The mechanism of transport in the phloem, however, remains obscure.

11. Plants have the ability to synthesize compounds called *growth-regulating substances,* which stimulate specific physiological actions.

12. *Tropisms* are growth movements whose direction is determined by the direction from which the stimulus comes. The substance responsible for tropisms in plants is *auxin.* In grass seedlings auxin is synthesized in the coleoptile tip, and it moves away from its source under several influences. Besides causing tropisms, auxin enters into the regulation of plant life from seed to fruit, inhibiting as well as stimulating.

13. *Phytochrome* is a light-sensitive pigment involved in seed germination and flowering. The action of phytochrome is influenced by its ability to change from a red-light-absorbing form (P_r) to a far-red-light-absorbing form (P_{fr}).

14. *Photoperiodism* is the response of organisms to specific lengths of illumination time.
15. The name "florigen" has been given to a hormone not yet isolated that seems to be responsible for inducing flower buds.
16. Other growth-regulating substances besides auxin and phytochrome have been discovered. *Gibberellins* are hormones that affect growth in height and starch digestion in seeds; *cytokinins* stimulate cell divisions; *ethylene* hastens the ripening of fruits and is generally antagonistic to auxin; and *abscisic acid* regulates leaf fall and other plant activities.
17. Besides the growth regulators, several known natural and synthetic compounds inhibit vegetative growth of plants.

ASK YOURSELF

1. What are a plant's sources for the following elements used in food production: carbon, hydrogen, oxygen, nitrogen, calcium, sulfur, phosphorus?
2. Why do most plants fail in very acid soils?
3. Why is humus in soil helpful to plant growth?
4. How do we know that water in the tops of tall trees is not pushed up by atmospheric pressure?
5. What materials are transported mainly in the bark of trees?
6. How is auxin movement affected by light? By gravity?
7. Criticize the statement: "a plant seeks light."
8. Explain the upward growth of stems in terms of auxin control.
9. What is a positive tropism?
10. What is photoperiodism?
11. What is the difference between the two forms that phytochrome can assume?
12. Why is the chemical structure of florigen not known?
13. What are gibberellins?
14. What is the effect of cytokinins on cells?
15. Why are commercial pears frequently wrapped individually in paper?

PART SIX
THE PHYSIOLOGY OF ANIMALS

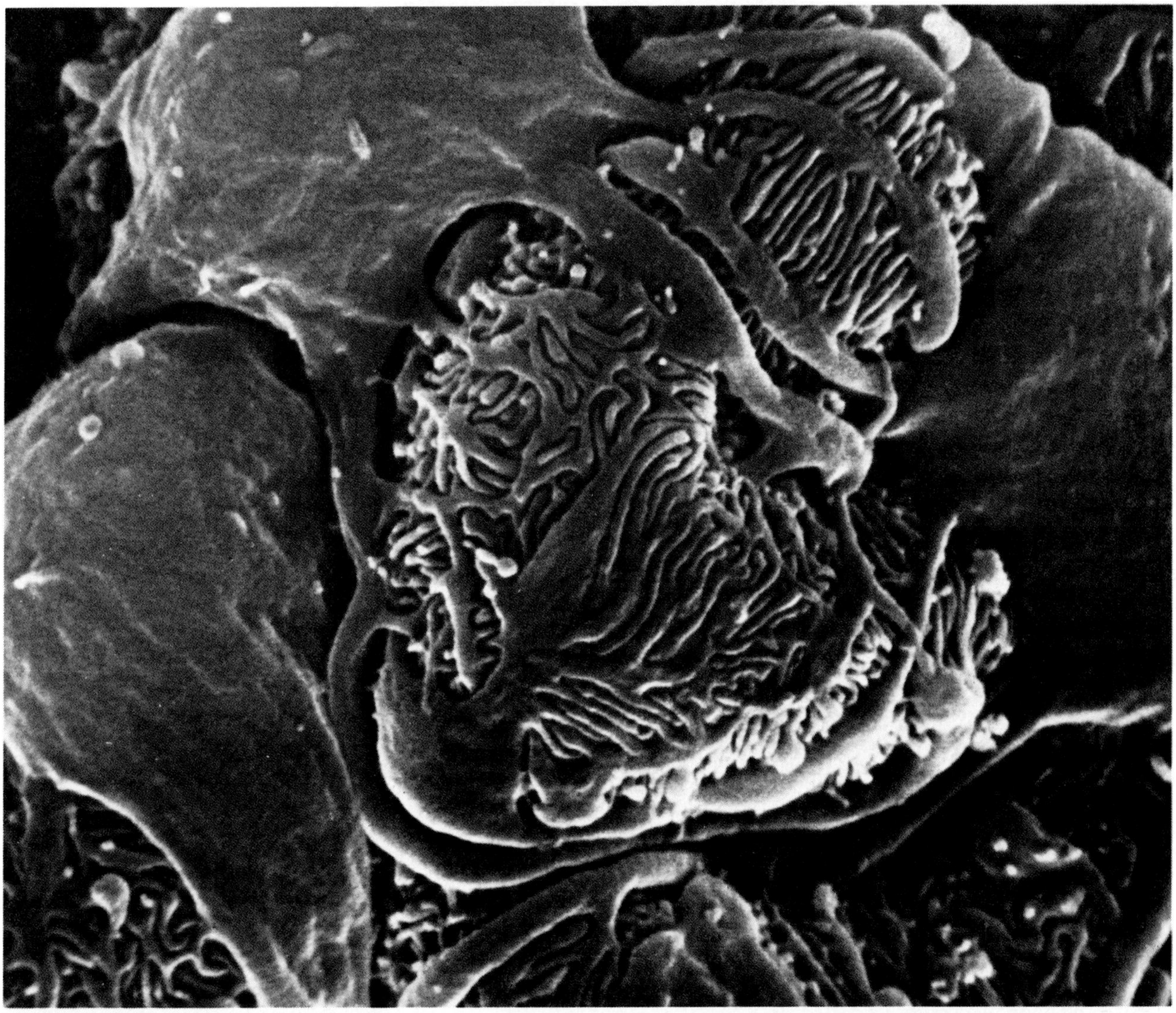

Scanning electron micrograph of a kidney glomerulus. × **2000**

18 Animal Hormones and the Endocrine System

SOME KEY POINTS

1. Any biological system's ability to maintain a constant internal environment, even when external conditions change, is called homeostasis.
2. Hormones, the secretions of endocrine glands in mammals, are discharged directly into the bloodstream, and affect specific target areas in the body.
3. Hormones are generally part of a feedback system that works with other hormones and the nervous system to maintain the body's inner stability.
4. The pituitary gland has some control over most of the other endocrine glands.
5. The endocrine system probably functions by a mechanism called the second-messenger concept.

ABOUT A HUNDRED YEARS AGO, THE GREAT French physiologist Claude Bernard said, "All the vital mechanisms, however varied they may be, have only one object, that of preserving constant the conditions of life in the internal environment." Bernard was speaking of the body's ability to maintain an inner stability (**homeostasis**), no matter what the external conditions are. Bernard compared our cells to flowers in a greenhouse. The conditions outside the greenhouse may be sweltering or frigid, and yet the environment of the flowers remains constant.

We tend to take homeostasis for granted. If we feel well today, we expect to feel the same way tomorrow. We don't think about homeostasis until we lose it. The word homeostasis literally means "standing the same," and as long as our body systems are functioning properly, our bodies will "stand the same." No matter what the outside conditions may be—cold in the winter or hot in the summer, dry in the desert or wet in a tropical rain forest—the internal conditions of our bodies will remain constant. We know clearly enough when "the conditions of life in the internal environment" have gone awry. We simply do not feel well. If homeostasis does not return soon, we feel worse, and if our regulatory mechanisms break down beyond the point of repair, we die.

The next few chapters will concentrate on how the bodies of animals work: physiology. No system of the body can work alone; they are all connected in some way, and it is not necessary to repeat this point every time another body system is discussed. As Claude Bernard said, "All the vital mechanisms, however varied they may be, have only one object, that of preserving constant the conditions of life in the internal environment." Put another way, "All the body systems work together to maintain homeostasis."

THE ENDOCRINE SYSTEM DEPENDS ON A FEEDBACK SYSTEM

Glands are tissues or organs that secrete chemical substances. They are usually classified according to their system of secretion. Ductless glands, or **endocrine** glands, secrete their substances directly into the bloodstream, and *exocrine* glands, such as salivary and sweat glands, secrete into ducts that lead to specific sites. The endocrine system is one of two great regulating systems in the body (Figure 18.1); the other is the nervous system. The endocrine system typically regulates such long-term processes as growth and reproductive ability, whereas the nervous system controls short, quick responses to stimuli. In fact, these systems are so interrelated that they might very well be considered a single regulatory agency. For example, the hypothalamic portion of the brain controls the secretions of many endocrine glands, and the secretions of the endocrine glands, called **hormones** (Greek, "to arouse"), have a direct effect on many aspects of nervous control. Hormones regulate body metabolism, growth and development, stress responses, and many other critical functions. In general, hormones are indispensable for the maintenance of homeostasis.

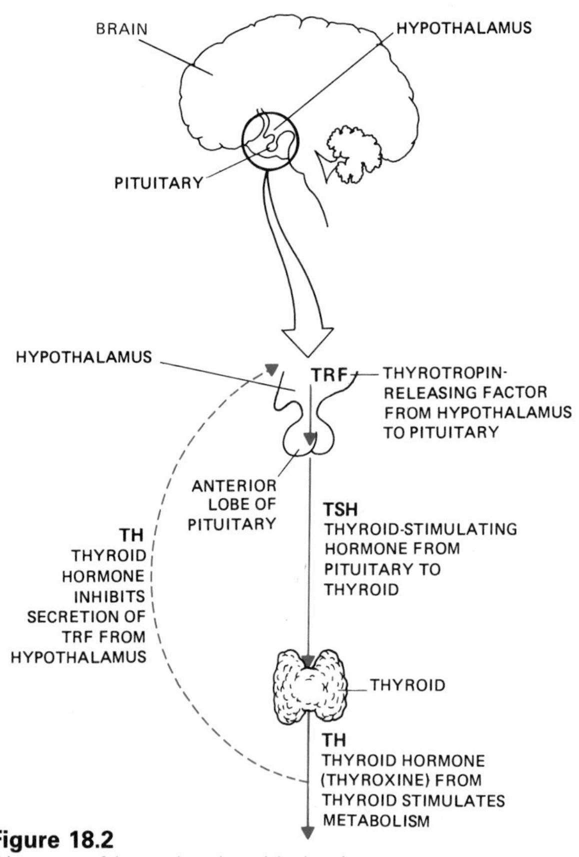

Figure 18.2
Diagram of how the thyroid gland controls metabolism and is itself controlled. The hypothalamus produces a hormone-releasing substance, thyrotropin-releasing factor (TRF), that causes the anterior pituitary to release a hormone, thyroid-stimulating hormone (TSH), or thyrotropin, into the blood. When the hormone reaches the thyroid gland, that gland releases a thyroid hormone (TH), which brings about an increase in metabolic rate. High metabolic rates cause a decrease in the production of the thyrotropin-releasing substance by the hypothalamus. This self-balancing system keeps the metabolic rate of the body at a constantly fluctuating but safe level.

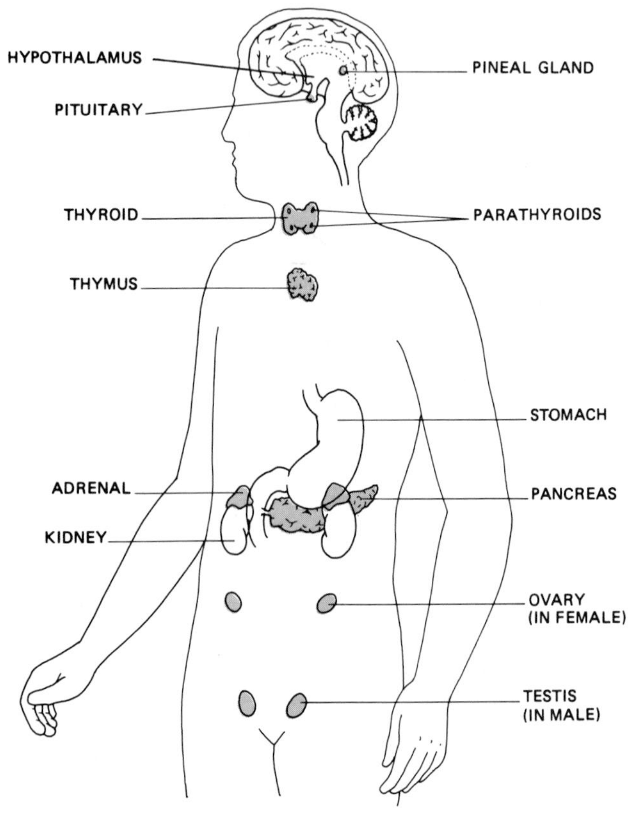

Figure 18.1
The endocrine glands of the human body.

Hormones are chemical compounds, usually either steroids (a type of lipid) or proteins. They are synthesized from raw materials in the cells and secreted into the bloodstream. When liberated into the blood, hormones are carried throughout the entire body, but they are effective only in those "target areas" where specific cells are capable of responding to certain hormones. Hormones are effective in extremely small amounts. Only a few molecules of a given hormone may produce a dramatic response in a target cell.

Hormones generally work on a negative *feedback system* (Figure 18.2). For example, the metabolic rate of an animal body must be kept steady, and a hormonally regulated feedback system ensures steadiness. If the metabolic rate drops, a region of the brain called the hypothalamus stimulates the release of a pituitary hormone, the *thyroid-stimulating hormone* (TSH). TSH causes the thyroid gland to secrete another hormone, *thyroxine*. Thyroxine causes an increase in metabolic rate. On the other hand, if the metabolic rate rises, production of thyroid-stimulating hormone by the pituitary is slowed down or stopped, the thyroid secretes less thyroxine, and the metabolic rate is lowered. In normal individuals, the thyroid-pituitary feedback system maintains a steady state of thyroxine secretion and consequently a constant metabolic rate.

The hypothalamus also operates together with the pituitary gland in a negative feedback system to affect the secretions of the adrenal glands and gonads. Besides regulating the release of pituitary hormones, the hypothalamus is involved in the regulation of body temperature and the ion-water balance in the body. Throughout this chapter we will see again and again how the endocrine glands harmonize not only with each other but with the nervous system as well.

THE VERTEBRATE PITUITARY GLAND AND ITS RELATIONSHIP TO THE HYPOTHALAMUS

The **pituitary** gland is sometimes called the *hypophysis* (Greek, "undergrowth") because it is located directly under the hypothalamus portion of the brain. It is about the size of a bean, and consists of two distinct lobes, the anterior lobe and the posterior lobe. The pituitary is connected by a stalk of nerve cells and blood vessels to the hypothalamus, providing a direct link between the nervous system and the endocrine system (Figure 18.3).

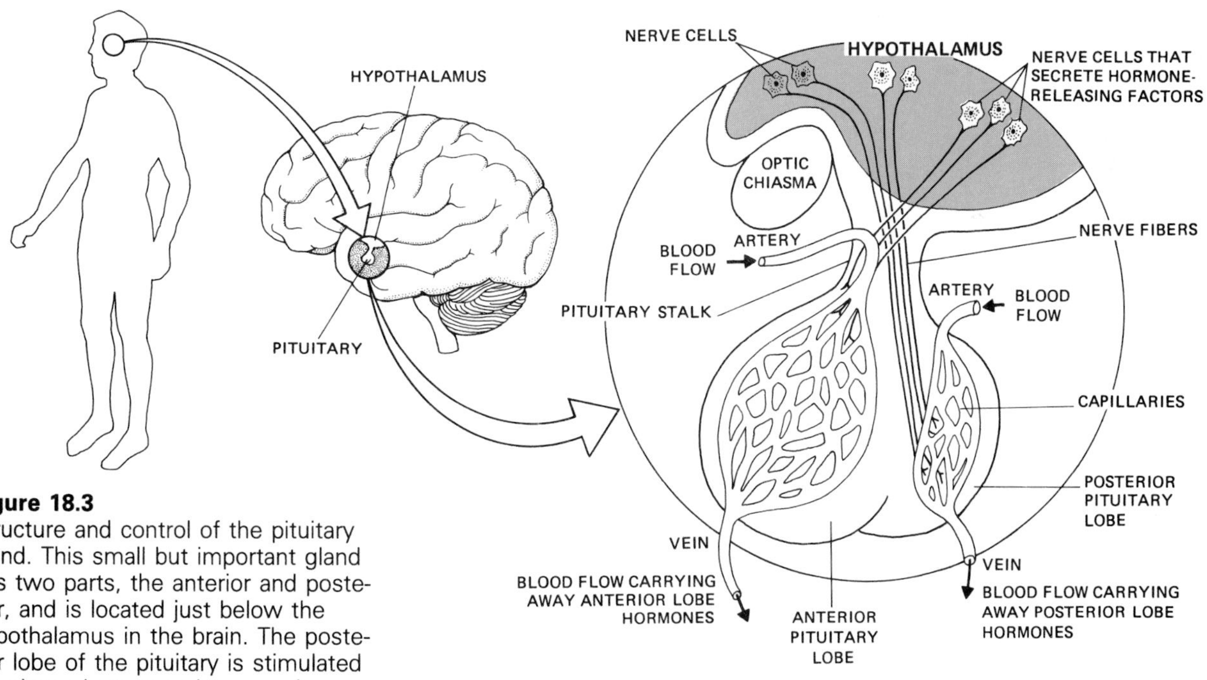

Figure 18.3
Structure and control of the pituitary gland. This small but important gland has two parts, the anterior and posterior, and is located just below the hypothalamus in the brain. The posterior lobe of the pituitary is stimulated to release hormones by nerve impulses from the hypothalamus, and the anterior lobe is stimulated by hormone-releasing factors that are secreted by nerve cells in the hypothalamus and transmitted by the blood.

TABLE 18.1
Major Human Endocrine Glands and Their Functions

GLAND	HORMONE	FUNCTION OF HORMONE	MEANS OF CONTROL	TARGET AREA
Anterior pituitary	Growth hormone (GH) (somatotropic hormone, somatotropin, STH)	Stimulates growth, increases protein synthesis, affects metabolism	Hypothalamic growth hormone-releasing factor	Bone and muscle
	Thyroid-stimulating hormone (thyrotropin, TSH)	Stimulates production and secretion of thyroxine by the thyroid gland	Hypothalamic TSH-releasing factor and thyroxine	Thyroid gland
	Prolactin	Stimulates development of mammary ducts and milk production	Hypothalamic prolactin-inhibiting factor	Mammary glands
	Adrenocorticotropic hormone (ACTH)	Stimulates secretion by adrenal cortex	Hypothalamic ACTH-releasing factor and cortisol	Adrenal cortex, skin, liver, mammary glands
	Luteinizing hormone (LH)	Stimulates development of corpus luteum, release of mature ovum, and production of progesterone and estrogen in female, secretion of testosterone in male	Hypothalamic LH-releasing factor and testosterone in male, progesterone in female	Ovaries, testes
	Follicle-stimulating hormone (FSH)	Stimulates ovulation in female, increases sperm production in male	Hypothalamic FSH-releasing factor, estrogen in female	Ovaries, testes
Posterior pituitary*	Oxytocin	Stimulates uterine contractions, milk ejection	Action potentials in hypothalamic secretory neurons	Uterus, mammary glands
	Vasopressin (antidiuretic hormone, ADH)	Inhibits urine formation by controlling water reabsorption from kidneys, contracts some blood vessels and causes a rise in blood pressure	Action potentials in hypothalamic secretory neurons	Kidneys, smooth muscles
Thyroid	Thyroxine	Controls rate of metabolism and growth	Thyroid-stimulating hormone (TSH) of pituitary	Most cell types
	Thyrocalcitonin	Lowers calcium and phosphate level in blood	Blood calcium concentration	Bone, kidneys, other cells
Parathyroid	Parathormone (parathyroid hormone, PTH)	Increases calcium level and decreases phosphate level in blood	Blood calcium concentration	Bone, intestine, kidneys, other cells
Adrenal cortex	Cortisol, cortisone, and related hormones	Regulates storage of glycogen by liver and conversion of protein into glucose; anti-inflammatory	ACTH	Many cell types
	Aldosterone	Controls retention of sodium and loss of potassium in urine	Angiotensin (a blood vessel-constricting substance) and blood potassium concentration	Kidneys

*Oxytocin and vasopressin are actually synthesized in the hypothalamus and released from the posterior pituitary.

GLAND	HORMONE	FUNCTION OF HORMONE	MEANS OF CONTROL	TARGET AREA
Adrenal medulla	Adrenalin (epinephrine)	Increases pulse and blood pressure	Sympathetic nervous system	Heart and smooth muscle
		Controls breakdown of glycogen, increases blood sugar		Liver and muscle
	Norepinephrine	Constricts blood vessels, increases metabolic rate	Sympathetic nervous system	Artery walls
Pancreas (islets of Langerhans)	Insulin	Lowers blood sugar, increases liver and muscle glycogen	Blood glucose concentration	Muscle, liver and fat tissue
	Glucagon	Decreases liver glycogen, increases blood sugar	Blood glucose concentration	Liver and fat tissue
Ovaries (follicle)	Estrogen	Development of sex organs and female characteristics	FSH	Reproductive tract and other parts of the body
Ovaries (corpus luteum)	Progesterone, estrogen	Influences menstrual cycle, stimulates growth of uterine wall, maintains pregnancy	LH	Uterus
Placenta	Estrogen, progesterone, chorionic gonadotropin (CG)	Maintains pregnancy	Unknown	Ovaries, mammary glands, uterus
Testes	Testosterone (androgens)	Development of sex organs and male characteristics	LH	Reproductive tract and other parts of the body
Digestive tract	Secretin	Stimulates release of pancreatic juice to neutralize stomach acid	Acid in small intestine	Cells of pancreas
	Gastrin	Produces digestive enzymes and hydrochloric acid in stomach	Food entering stomach	Stomach lining
	Cholecystokinin (CCK)	Stimulates release of pancreatic enzymes and gallbladder contraction	Food in small intestine	Pancreas, gallbladder

The pituitary is sometimes referred to as the "master gland" (even though its daily secretions are less than one-millionth of a gram) because of its control over most of the other endocrine glands. (In truth, the hypothalamus might justifiably be called the "master gland," since hypothalamic hormones control the secretions of the anterior pituitary.) This master control system is derived from the anterior lobe of the pituitary, which in turn is controlled by the hypothalamus. Some of the regulatory hormones secreted by the anterior lobe are the thyroid-stimulating hormone (TSH), which stimulates the thyroid to secrete *thyroxine;* the adrenocorticotropic hormone (ACTH), which means "the hormone that stimulates the cortex (outer part) of the adrenal glands"; and the gonadotropic hormones, follicle-stimulating hormone (FSH) and luteinizing hormone (LH), which control the secretions of the ovaries and testes.* The anterior pituitary also secretes a growth hormone, somatotropic hormone (STH), or *somatotropin,* which regulates the growth of the skeleton; and *prolactin,* a hormone that stimulates milk production in the mammary glands (Table 18.1).

* We can understand how one hormone is able to stimulate the same kind of action in two very different glands (ovaries and testes) if we remember that both the male and the female sex organs have the same embryonic origin.

INVERTEBRATE HORMONES

In the previous chapter we saw how the hormones of a plant regulate its growth. The growth and development of invertebrates (insects, for example) are also regulated by hormones, as are other activities, such as molting, coloration, and reproductive ability. Most of the hormones produced by invertebrates originate in cells of the nervous system. Although hormones have recently been identified in such invertebrates as crustaceans and molluscs, more explicit information is available on insect hormones, especially *ecdysone* and *juvenile hormone.*

In insects, a set of organs produces hormones that activate the genetic program of precisely timed and regulated developmental stages. Many insects, such as butterflies, ants, bees, and wasps, undergo a developmental transformation from egg to larva to pupa to adult. (The word "larva" comes from the Latin word for ghost, and "pupa" derives from the Latin word for doll.) This progression of developmental stages is called **complete metamorphosis.** Each stage requires the formation of a new exoskeleton as the previous one is shed, or **molted.** Molting and metamorphosis are both controlled by hormones.

Ecdysone, the molting hormone, is a steroid produced by a prothoracic gland of the larva. Ecdysone, under the control of a hormone from the brain, apparently acts directly on chromosomes to stimulate molting. Ecdysone is balanced by the second insect hormone, **juvenile hormone,** which is produced by two small organs near the brain. Ecdysone and juvenile hormone together control insect growth and development. Ecdysone favors the formation of the adult, and juvenile hormone prolongs the larval stages through several molts. As a result of decreasing production of juvenile hormone throughout the larval and pupal stages, the adult insect finally emerges.

Promising experiments with juvenile hormone have indicated that it may be effective as an insecticide spray because of its ability to prolong the larval stage. If juvenile hormone could be used in this way to inhibit insect development, it would be a welcome alternative to such environmentally harmful chemicals as DDT.

Hormone control of insect metamorphosis.

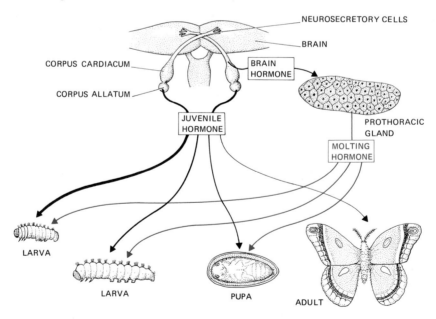

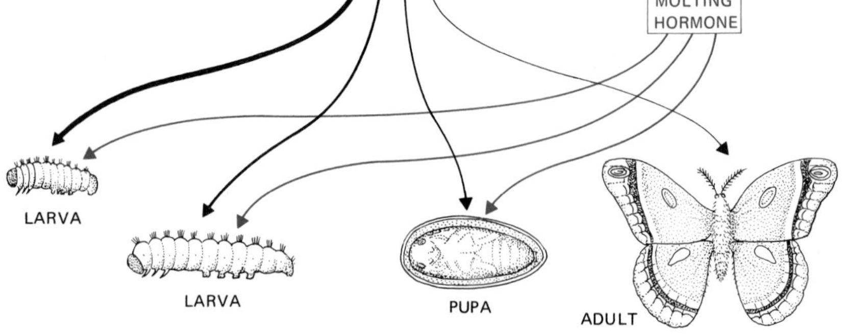

NEW IDEAS IN BIOLOGICAL CONTROL: RADIOIMMUNOASSAY (RIA)

A baby born with insufficient thyroid hormone secretion will seem normal for several months. Then, when the damage is too severe to repair, cretinism (infantile hypothyroidism) appears: The baby fails to develop normal bones and has suffered brain damage (see p. 384). We can imagine the expense and suffering of the parents and affected children, magnified by the number of people involved: 1 out of every 4000 children born, in a population of well over two hundred million in the United States.

Until the development of radioimmunoassay (RIA) techniques, no one could tell whether or not a newborn baby had a sufficiently high level of thyroxine in its blood. That has changed, mainly as a result of the research projects of Rosalyn S. Yalow (Nobel prize, 1977). Now a drop of blood from a newborn baby can be tested for thyroxine content, and for a couple of dollars an answer is available. An incipient cretin, otherwise doomed to an inferior life, can receive extra thyroxine and grow to become a normal child.

Detecting vanishingly low concentrations of biologically active compounds can be done readily by means of RIA. The technique is highly specific, working on only particular substances such as hormones or vitamins that would be undetectable by other methods. RIA uses radioactively labeled compounds that react with natural substances in the body. Radiation from the compound is monitored by radiation counters, and unimaginably small quantities can be measured.

Think of the composition of a human body: mostly water, a fair quantity of bone, carbohydrate, and fat, a collection of proteins, and finally a few molecules of hormones. Finding and identifying hormones in natural body fluids was formerly an almost impossible job. By RIA, however, those scarce hormone molecules can be detected. This feat is like detecting one *grain* of salt in a shipload of 5000 *tons* of sugar. Within less than 20 years, the ability of scientists to measure such minute amounts of biologically active substances has increased by several orders of magnitude.

As soon as the accuracy, efficiency, and sensitivity of RIA were shown, a whole range of substances and human disorders were studied. In addition to a long list of hormones, other substances that can now be measured by RIA are enzymes and other proteins, vitamins, drugs, and viruses.

One particularly nasty virus that RIA eliminated was "hepatitis B antigen." It is not detectable in blood (except by RIA), but when blood carrying the virus was used in transfusions after accidents or surgery, the recipient got the liver ailment, hepatitis. With routine screening of donor blood by RIA before it is given to a patient, hepatitis following transfusions is no longer a danger.

These examples demonstrate that the best medical care is the cheapest in the long run, because it offers prevention rather than cure. The virus causing smallpox, formerly a major killer, has been so nearly destroyed on this earth that even vaccination against it is no longer necessary. Vigorous vaccination programs have also drastically reduced the incidence of poliomyelitis, diphtheria, and typhoid fever, once dreaded diseases whose cost in lives, social burden, and human suffering were beyond estimation. Hypothyroidism seems to be under control. Diabetes may be next. Who knows when cancers—not just one disease, but several—will be preventable? If we learn the basic reasons for many of our ills, we can sometimes prevent them for pennies.

The posterior lobe of the pituitary secretes *oxytocin,* which stimulates uterine contractions during childbirth and milk release after childbirth; and antidiuretic hormone (ADH), or *vasopressin,* which aids water reabsorption in the kidneys and stimulates the smooth muscles of the arteries. Emotional strain can cause increased secretion of vasopressin, which in turn produces a rise in blood pressure and an inhibition of urine formation. Drugs such as nicotine and ether stimulate the secretion of vasopressin, but alcohol has the opposite effect. For example, if you consume large quantities of beer, you will stimulate your urine production, not only because you increase the liquid content of your body, but also because the alcohol in the beer inhibits vasopressin secretion, which then stimulates urine production. Besides, another component of beer, lupulin from hops, is a *diuretic,* or urine stimulator.

Figure 18.4
Growth control by pituitary hormone. A normal-sized individual (left) is compared with a "giant," whose pituitary was overactive, and a pituitary dwarf (midget), whose pituitary was underactive.

Some Disorders of the Pituitary

One of the most familiar results of a disorder of the pituitary gland is *giantism,* caused by oversecretion of growth hormone during the period of skeletal development (Figure 18.4). In this case, the body grows beyond the normal size range, and it may reach more than 2½ meters (8 ft) and 180 kilograms (400 lb). Death usually occurs before the individual is 30 years old. *Acromegaly* is a form of giantism that usually occurs in adults after skeletal development is complete. If too much growth hormone is secreted after maturity, the skeleton does not lengthen any further, but some cartilage and bone will thicken. Such thickening makes the face, hands, and feet widen considerably (Figure 18.5). Acromegaly may be treated with some success by using radiation therapy.

Figure 18.5
The effects of overproduction of pituitary growth hormone after maturity. If the pituitary recommences secretion of growth hormone after full growth is reached, only the bony extremities are affected, so that hands, feet, and jaws increase. The onset of the affliction, called acromegaly (Greek, "large peaks"), is shown in this series of photographs taken of an individual who was normal until she reached adulthood but then developed the acromegalic symptoms apparent in the last picture.

Insufficient secretion of growth hormone produces *pituitary dwarfs*, or *midgets*, persons who have normal intelligence and body proportions, but whose overall height does not progress beyond that of a normal six-year-old child (see Figure 18.4). Premature senility is common, however, and pituitary dwarfs usually die before they reach 50. Although sexual development is usually minimal and reproductive ability is probably impaired, some pituitary dwarfs are capable of producing normal children. As we will see in the next section, some dwarfism is caused by deficiencies of thyroid hormone rather than by deficiencies of pituitary growth hormone.

THE THYROID GLAND REGULATES METABOLIC FUNCTIONS

The **thyroid** gland has two lobes, one on each side of the junction between the larynx (voice box) and trachea (windpipe). The two lobes of the thyroid are connected by a bridge of thyroid tissue called an *isthmus* (Figure 18.6). The thyroid hormone is *thyroxine*, an amino acid containing iodine. Thyroxine controls *metabolism*, the rate at which food and oxygen are used to generate heat and energy throughout the body. The synthesis of thyroxine and its release are controlled by the thyroid-stimulating hormone (TSH) of the pituitary gland.

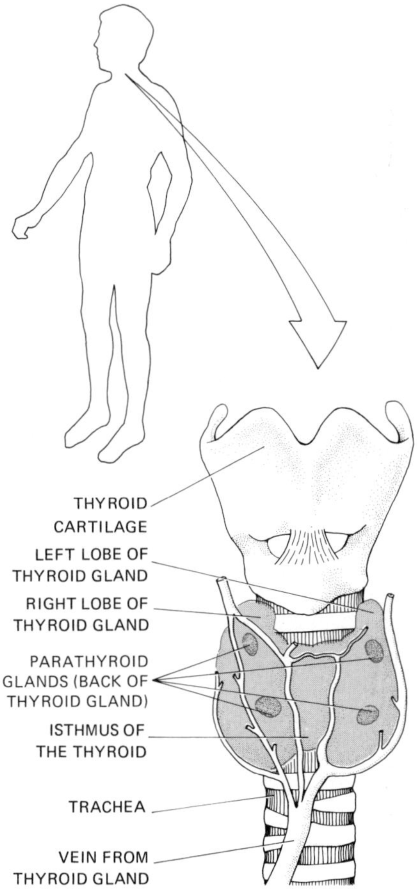

Figure 18.6
Position and anatomy of the thyroid and parathyroid glands.

Some Disorders of the Thyroid

The thyroid is concerned mainly with such metabolic processes as growth and cellular respiration, and any deviation from the normal rates of secretion will affect many body functions. This is especially true because thyroxine is not directed at a particular target area but is effective throughout the body.

Overactivity of the thyroid gland results in *hyperthyroidism* (Graves' disease), a condition caused by an oversecretion of TSH by the pituitary or by a thyroid tumor. Symptoms include nervousness, irritability, increased heart rate and blood pressure, weakness, weight loss, elevated use of oxygen at rest, and bulging eyes (exophthalmos), caused in part by an increase in the fluid behind the eyes (Figure 18.7). Drug therapy that inhibits thyroxine production and the administration of radioactive iodine have successfully replaced surgery as treatments for hyperthyroidism.

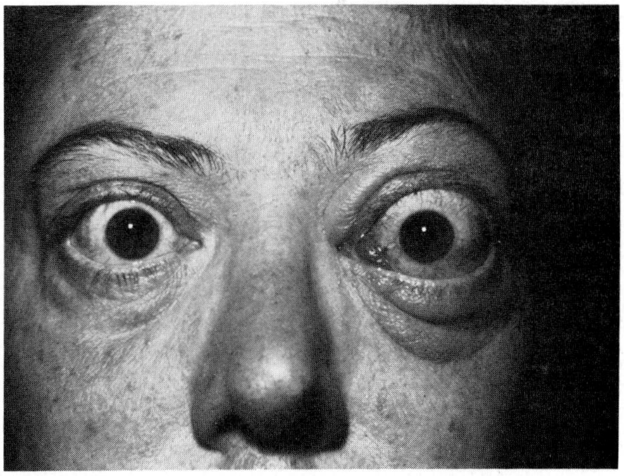

Figure 18.7
External effect of thyroxine overproduction. One of the results of heightened metabolism and increased blood pressure is a bulging of the eyes, which gives the sufferer a startled expression.

Underactivity of the thyroid is *hypothyroidism*, usually associated with *goiter* (Latin, "throat"), an enlarged thyroid. Such a swelling in the neck is caused when insufficient iodine in the diet forces

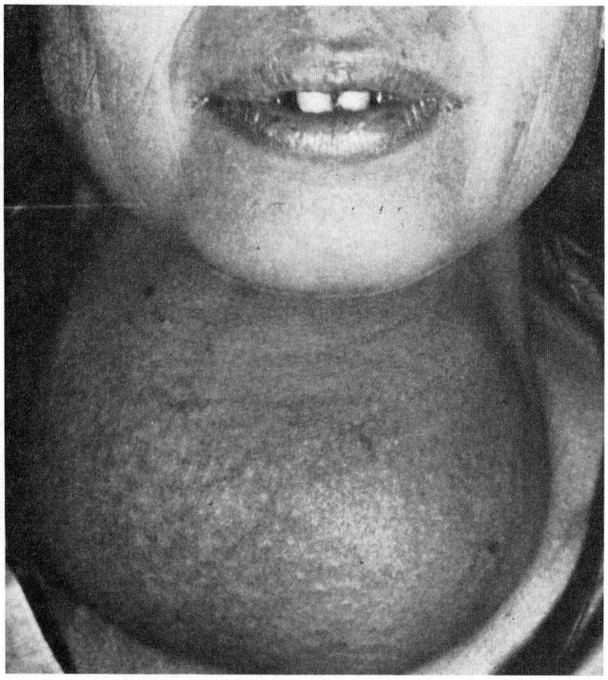

Figure 18.8
The effect of thyroxine underproduction: a *goiter.* If the thyroid gland does not secrete sufficient thyroxine, usually because of iodine deficiency in the diet, the gland becomes enlarged, although not so much in most cases as that shown here. The condition of *hypothyroidism* causes a whole series of difficulties, including mental retardation. It was formerly thought that people suffering from infantile hypothyroidism were in some way supernatural. Hence they were called *cretins,* a variant pronunciation of the French word *chrétien,* or Christian.

the thyroid to expand in an "effort" to produce more thyroxine (Figure 18.8). Most cases of goiter used to be found in areas away from the ocean, where iodine content in the soil and water supply is low. With the addition of minute amounts of iodine to ordinary table salt and drinking water in recent years, goiter has practically disappeared as a common ailment.*

Symptoms of hypothyroidism include decreased heart rate, blood pressure, and body temperature; lowered basal metabolism rate; and underactivity of the nervous system.

Underactivity of the thyroid during infancy causes *cretinism* (CREH-tin-ism), which results in mental retardation and irregular development of bones and muscles. The skin is dry, eyelids are puffy, hair is brittle, and the shoulders sag. Because of similar physical characteristics, cretinism is often confused with Down's syndrome (mongolism), a genetic disease (Chapter 12). Ordinarily, cretinism cannot be cured, but early diagnosis and treatment with thyroxine may arrest the disease before the nervous system is damaged. If the thyroid becomes underactive during adulthood, *myxedema* results, producing swollen facial features, dry skin, low basal metabolism rate, tiredness, possible mental retardation, and intolerance to cold in spite of increased body weight. Like cretinism, myxedema may be corrected if it is treated early.

The familiar "circus dwarf" is the irreversible result of an underactive thyroid gland that interferes with the normal growth of cartilage (Figure 18.9). Such dwarfs are usually intelligent and active, and have stubby arms and legs, a relatively large chest and head, and flattened facial features. It has been suggested by some scientists that Attila the Hun (406?–453) was such a dwarf, but no firm evidence exists for this captivating claim.

THE PARATHYROID GLANDS

The four **parathyroids** are tiny glands embedded in the thyroid gland (see Figure 18.6). The parathyroid hormone (PTH), or *parathormone,* regulates the use of several chemicals, especially calcium. Improper balance of calcium in the blood may cause faulty transmission of nerve impulses, destruction of bone tissue, and hampered bone growth, and it may produce a paralysis of the muscles (tetany). Small as the parathyroid glands are, their removal invariably causes death. Before their existence and location became known in 1850, the parathyroids were frequently excised during goiter surgery, and the patients died "mysteriously."

THE ADRENAL GLANDS

The two **adrenal** (Latin, "upon kidneys") glands rest like berets on top of the kidneys. Each adrenal is composed of an inner portion, the *medulla* (Latin, "marrow," meaning inside), and an outer portion, the *cortex* (Latin, "bark," referring to the outer portion of something, just as the bark of a tree is the outer covering) (Figure 18.10).

* Almost 4000 years ago, the Chinese treated goiters (they did not call them that) by adding the ashes of burned seaweed and sponge to the diet of the afflicted individual. Although the cause of the disease was not known, and iodine in the seaweed and sponge was not consciously being prescribed, the "cure" was usually at least partially successful.

Figure 18.9
One of the causes of dwarfism is underactivity of the thyroid gland. These brothers and sisters of the Owitch family are an example of that phenomenon.

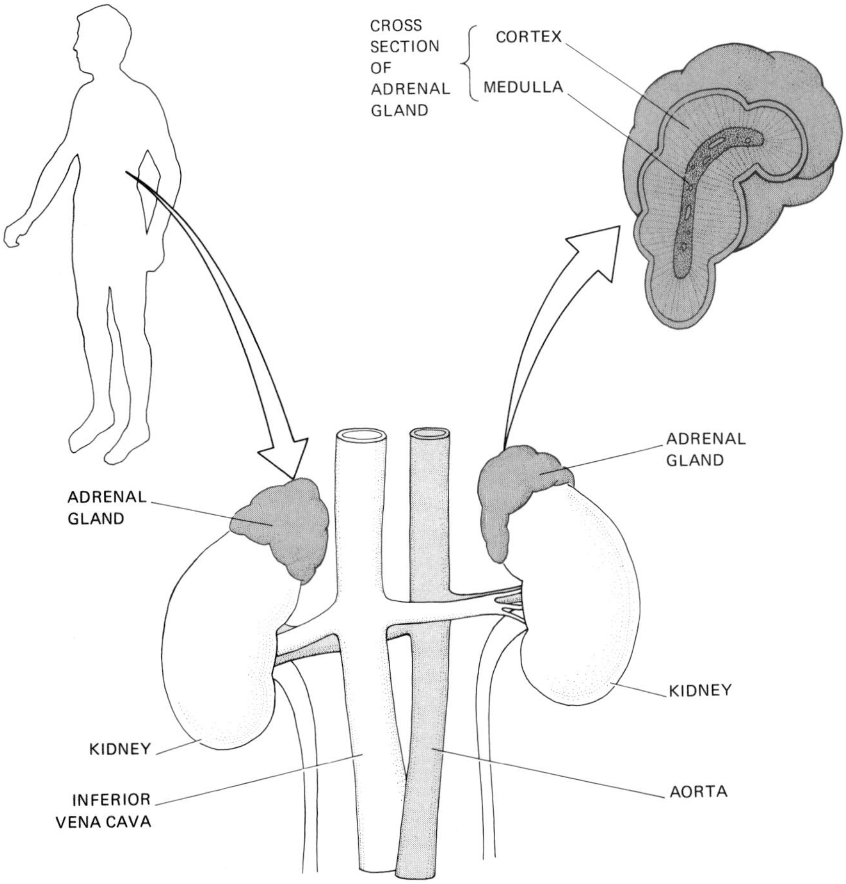

Figure 18.10
Position and anatomy of the adrenal glands, the source of a number of hormones.

JFK'S "SUNTAN" AND NAPOLEON'S NAPS

It is unlikely that the president of the United States, especially such a vigorous man as John Fitzgerald Kennedy, would want to admit to having an incurable disease. But during the campaign of 1960 it was "leaked" that Kennedy had Addison's disease, caused by an underactivity of the adrenal cortex. Ironically, the symptoms of Addison's disease include tiredness and an inability to deal with stress—characteristics not usually associated with the public JFK.

Basically, Addison's disease results from the failure of the adrenal cortex to produce its hormones. Although the full extent of Kennedy's case was never disclosed, it was known that he received oral doses of cortisol frequently. Without cortisol, Kennedy probably could not have synthesized enough blood glucose between meals, and

Napoleon at 34.

John Fitzgerald Kennedy.

Napoleon at about 50.

The **adrenal cortex** is divided into three specific zones, each providing a different type of steroid hormone. More than 50 adrenal cortical hormones have been identified, but the most widely known are the mineralocorticoid group (including aldosterone), which affect sodium metabolism, glucocorticoids (such as cortisol), which affect metabolism of foods, especially carbohydrates, and the cortical sex hormones (mostly male androgens). Cortisone and cortisol work as anti-inflammatory drugs for such diseases as arthritis, bursitis, and allergies by temporarily shifting the balance from the inflammatory mineralocorticoid hormones toward the anti-inflammatory glucocorticoids.

The **adrenal medulla** chiefly secretes *epinephrine,* commonly known as *adrenalin.* The secretion of his metabolism might have been sluggish because of an inability to mobilize proteins and fats from the tissues. Even with special doses of cortisol, it is likely that President Kennedy's muscles still would have been weak.

But Kennedy looked healthy. His constant suntan, for instance, made some envious of his life. He did sail and play touch football, but his "suntan" may have been partially due to Addison's disease, since one of its characteristics is a bronzing of the skin, apparently caused when excess ACTH produces a darkening of the skin pigment.

An even more dramatic example of a public figure with a hormonal defect is Napoleon, who supposedly suffered from a disorder of the hypothalamus that affected his pituitary, adrenals, thyroid, and sex organs. (And as if that wasn't enough, he contracted syphilis in later life.) As a young man, he was thin, virile, and so energetic that he rarely needed sleep. Napoleon was only 52 when he died, and by the time he was 46, his mind and body had deteriorated so much that, according to a report of the time, he actually fell asleep during the Battle of Waterloo, which he lost.

It is easy to speculate that Napoleon's indecision and fits of hysteria caused his late military failures, but there is disagreement over the report that his autopsy provided definite proof of his brain disease. One of the forms of a hypothalamic hormonal imbalance produces feminine characteristics in a male. When Napoleon died in 1821, he had acquired many of the secondary sex characteristics of a woman, even to the widening of hips and narrowing of shoulders. He was reputed to have been proud of his full breasts.

adrenalin is stimulated by the nervous system to produce a condition sometimes called the "fight or flight" effect, which permits the body to react quickly and strongly to emergencies. This extreme effect lasts a very short time; in fact, enzymes in the liver inactivate adrenalin in about three minutes. After adrenalin has been secreted, some blood vessels constrict and others dilate, redistributing blood flow to such organs as the brain and muscles, where it is most needed. Blood is moved away from the skin, reducing the danger from a possible surface wound, blood pressure rises, and the time needed for blood clotting is reduced. Circulation increases, and enzymes are activated in the liver to release sugar from glycogen for increased energy. Digestion, an energy-consuming function, is halted during the emergency by the diversion of blood from the stomach and intestines to the muscles, where extra energy may be needed. In each case, the secretion of adrenalin favors the body's survival in times of stress.

Some Disorders of the Adrenal Glands

Apparently, the adrenal medulla is never underactive, and though overactivity is very dangerous, it is quite rare. In contrast, disorders of the adrenal cortex do occur occasionally. Underactivity of the cortex produces *Addison's disease,* whose symptoms include anemia (deficiency of red blood cells), weakness and fatigue, increased blood potassium, and decreased blood sodium. Skin color becomes bronzed because excess ACTH, produced by the pituitary in an effort to restore a normal cortical hormone level, induces abnormal deposition of skin pigment (see the essay on p. 386). Until recently, Addison's disease was usually fatal, but now it can be controlled with regular doses of cortisol and aldosterone.

Two diseases caused by overactivity of the adrenal cortex are *Cushing's disease* and *adrenogenital syndrome.* Cushing's disease is usually caused by a cortical tumor that overproduces glucocorticoids. Symptoms include fattening of the face, chest, and abdomen (the limbs remain normal), accompanied by abdominal striations (stripes on the skin) and a tendency toward diabetes, caused by increased blood sugar. Protein is lost, and the muscles become weak. Adrenogenital syndrome is also caused by an overactive adrenal tumor, which stimulates excessive production of the cortical male sex hormones. These androgens cause male characteristics to appear in a female, and they will accelerate sexual development in a male. Hormonal disturbances during development of a female fetus may cause a distortion of the genitals, so that the clitoris and labia become enlarged and resemble a penis and scrotum. In an adult woman, an extreme case of adrenogenital syndrome may produce a bearded lady, so frequently found in a circus sideshow. Such pronounced male characteristics in a female may be caused by other hormonal malfunctions besides defects in the adrenal cortex.

THE PANCREAS

The **pancreas** is tucked directly under the stomach (Figure 18.11). About 99 percent of its weight is involved in producing digestive enzymes, which are transported through a duct into the small intestine. The true endocrine portion of the pancreas is found in its remaining one percent. This portion of the pancreas secretes two hormones from a cluster of cells called the **islets of Langerhans.** The islets contain two special groups of cells, designated as alpha and beta. Alpha cells secrete **glucagon,** and beta cells secrete **insulin.** Insulin is secreted when the level of blood sugar rises, such as right after a meal. The most important effect of insulin is to facilitate glucose transport across cellular membranes. Insulin also enhances the conversion of glucose to glycogen, which then can be stored in the liver as a source of sugar.

The liver acts as a regulator of blood sugar by taking up excess glucose when insulin and blood sugar are high, and by releasing sugar back into the blood when insulin and blood sugar are low (Figure 18.12). Glucagon has functions opposite to those of insulin. Glucagon's most important job is to increase blood glucose concentration by stimulating the liver to convert glycogen to glucose.

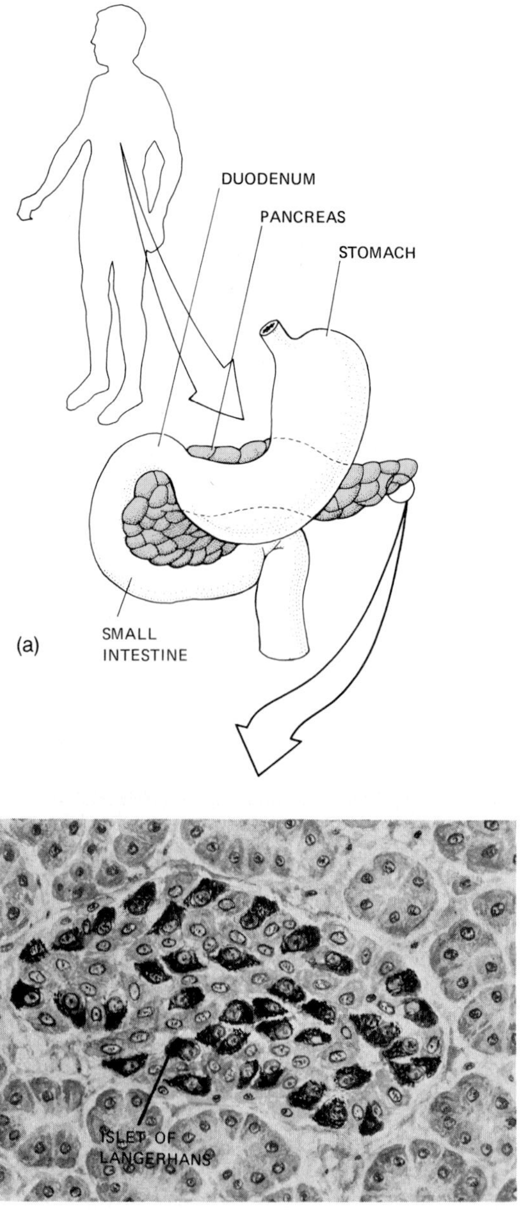

Figure 18.11
(a) Location and anatomy of the pancreas, which contains the islets of Langerhans. (b) Photomicrograph of a section of pancreatic tissue, showing an islet (the central mass) containing the darker-stained alpha cells and lighter beta cells.

Some Disorders of the Pancreas

About four percent of the United States population will develop *diabetes mellitus* at some time in their lives. Diabetes, an inherited disease, is caused when beta cells do not produce enough insulin. When this happens, glucose accumulates in the blood and is deposited in the kidneys, but cannot enter the cells. Because the cells are unable to use the accumulated glucose (the most readily available energy source in the body), the body actually begins to starve. Appetite may increase, but eventually the body consumes its own tissues anyway, literally eating itself up.

Because the removal of glucose from the kidneys requires large amounts of water, the diabetic person produces excessive, sugary urine. In response to the increased urine production, with the possibility of serious body dehydration, diabetics become extremely thirsty, drink huge amounts of liquids, and may excrete as much as 20 liters of urine per day.*

Diabetes is incurable in any form, but mild diabetes can usually be controlled by strict dietary regulation. More serious cases may require treatment with regular injections of insulin. If the disease is untreated, almost every part of the diabetic's body

* The word "diabetes" comes from the Greek word for "siphon," or "to pass through," referring to the speedy elimination of liquids. The word was actually used by the Greeks as early as the first century. In the seventeenth century, the sweetness of diabetic urine was discovered, and the name of the disease was lengthened to diabetes mellitus. "Mellitus" comes from the Greek word *meli,* which means honey. In England, the disease had already been called "the pissing evil."

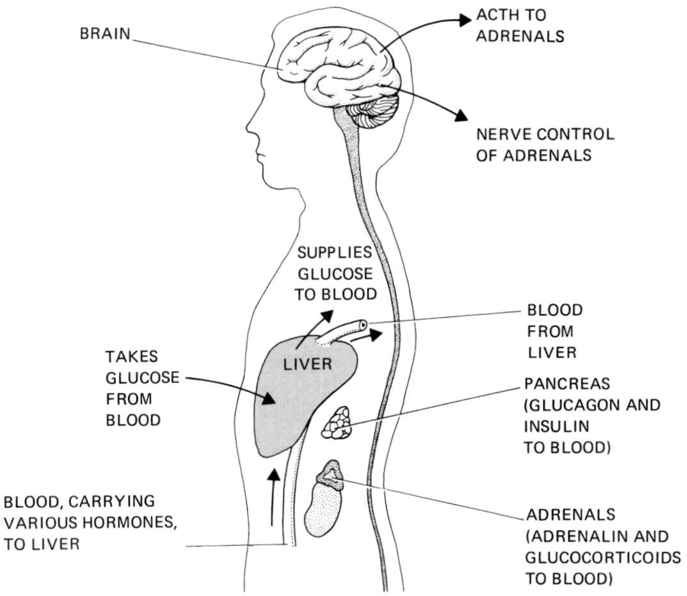

Figure 18.12
General scheme of blood sugar regulation. The liver is the center for storage of carbohydrate, as well as the center for the change of soluble glucose to stored glycogen, and vice versa. Whether glucose is changed to glycogen or glycogen is changed to glucose is determined by the relative concentrations of the hormones insulin and glucagon in the blood. Insulin makes the liver convert glucose to glycogen, and glucagon makes it convert glycogen to glucose. These hormones are regulated by adrenal hormones, which in turn are regulated by the adrenocorticotropic hormones (ACTH), which in their turn are regulated by the brain.

will be affected, and gangrene, hardening of the arteries, other circulatory problems, and further complications may occur.

Medicinal insulin is generally obtained from animals, but experimentally produced insulin can be extracted from bacteria (see the essay on p. 250). Bacteria-produced insulin is not yet available commercially, but rat genes for insulin production have been implanted in the bacterium *E. coli*, and at least a precursor of insulin has been harvested. Genetic engineers hope that by the time you are reading this text, techniques for bulk production of therapeutic insulin may be available.

Low blood sugar, *hypoglycemia*, may be caused by excessive secretions of insulin. Sugar is not released from the liver, and the brain is deprived of its necessary glucose. Hypoglycemia can be controlled by regulating the diet, especially the carbohydrate intake. On a short-term basis, the sugar in a glass of orange juice may restore the normal glucose balance in the blood, but a long-term treatment would usually consist of a reduction of carbohydrates in the diet. Carbohydrates tend to stimulate large secretions of insulin, thereby removing sugar from the bloodstream too quickly.

THE GONADS

In Chapter 13 we saw how the gonads secrete hormones that control reproductive functions. Besides the ovaries and testes, other sex structures, such as the placenta and corpus luteum, also contribute to the complex feedback mechanism of hormonal control, especially in the female. We refer the reader to Chapter 13 for a review of the intricate interrelationships of hormones that regulate reproductive mechanisms in human beings.

OTHER SOURCES OF HORMONES

Besides the major endocrine organs already described, other glands and organs carry on hormonal activity. We will consider briefly the pineal gland, the thymus gland, the digestive system, and the prostaglandins. The last are recently discovered hormones that are found in many parts of animal bodies.

The Pineal Gland

The pineal gland is a pea-sized body located in the midbrain. Several chemicals have been isolated from the pineal, especially melatonin. Melatonin affects skin pigmentation and seems to react to changes of light through the eye. Lizards and some amphibians appear to be able to camouflage themselves through this interrelationship of light and skin color. Although the function of the pineal gland in humans is uncertain, it may act as a "biological clock" that affects secretions of sex hormones.

The Thymus Gland

The thymus is located under the breastbone, in the center of the chest. It is large and active only during childhood, and it reaches its maximum effectiveness during early adolescence. After that time, the thymus atrophies because of the action of sex hormones. It finally ceases activity altogether after 50 years or so, and it may therefore play an important role in the process of aging. The main function of the thymus in humans seems to be the formation of a permanent immunity system of antibodies. There is speculation that the thymus may be a true endocrine gland that secretes a hormone (thymosin) into the bloodstream to affect other regions of antibody production, such as the lymph nodes and the spleen. There is also a possibility that the thymus may influence the secretion of reproductive hormones from the pituitary gland, and investigations continue.

The Digestive System

The best known of the several digestive hormones are the polypeptides gastrin and secretin. *Gastrin* stimulates the production of hydrochloric acid when food enters the stomach; thus gastrin is produced by the same organ, the stomach, that is its target organ. *Secretin,* the first hormone to be discovered (by British scientists Sir William Bayliss and Ernest Starling in 1902) and the first substance to actually be called a "hormone," stimulates a pancreatic secretion that neutralizes stomach acid that passes to the small intestine.

Prostaglandins

Prostaglandins are fatty-acid hormones that can be found in tissues of the intestines, liver, kidneys, pancreas, heart, lungs, thymus, brain, and both male and female reproductive organs. Human semen contains the highest concentration of prostaglandins in the body; however, in spite of their name, prostaglandins are mostly secreted from the seminal vesicles, not the prostate gland, as was once thought.

Prostaglandins can raise or lower blood pressure, regulate digestive secretions, inhibit progesterone secretion by the corpus luteum, reduce infection by stimulating production of bacterium-destroying blood cells, and regulate normal blood clotting. In addition, prostaglandins cause the contraction of smooth muscle in the uterus, and they dilate (open) air passages to the lungs. Prostaglandin drugs may be useful in inducing labor, and in relieving asthma, arthritis, ulcers, and hypertension (high blood pressure). There is also some evidence that prostaglandin drugs may inhibit the growth of viruses, and may be useful in treating diseases such as multiple sclerosis.

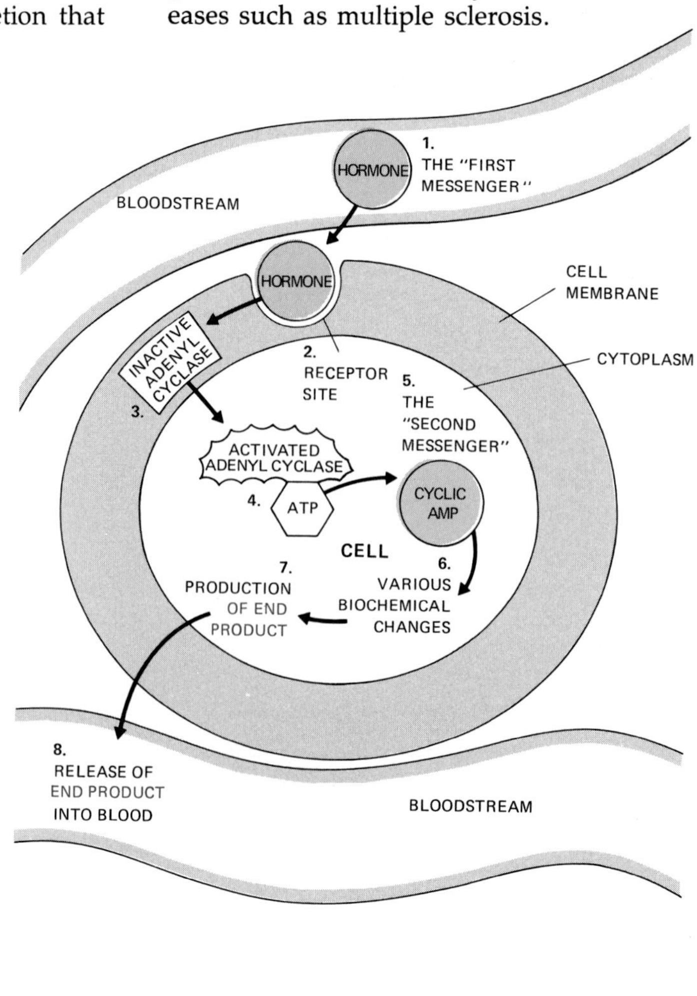

Figure 18.13
Hormonal action at the cellular level. A hormone is carried by the blood all over the body (1), but it is effective only in certain places, known as "target organs." Adrenalin, for example, can cause the liver to respond, but not retinas or spleens. Once the hormone has arrived at a target organ that recognizes it (2), it activates an enzyme, adenyl cyclase (3). Adenyl cyclase then brings about the production of cyclic adenosine monophosphate, cAMP, from ATP (4-5). Cyclic AMP is called the "second messenger," as if the hormones were working as a relay team. The cyclic AMP then induces intracellular changes (6-7), such as the release of glucagon, with the resultant release of glucose from glycogen in the liver (8).

Little is known about how prostaglandins work. These "super hormones" may work inside cells to turn chemical reactions on and off, or they may do their job by changing the composition of cellular membranes. Prostaglandins are being studied intensively.

WHAT MECHANISM CONTROLS THE ENDOCRINE SYSTEM?

In the late 1950s, Earl W. Sutherland, Jr., and his colleagues at Case Western Reserve University isolated a substance called **cyclic AMP.** Because Sutherland found a rise in cyclic AMP following increased hormone levels, he proposed that cyclic AMP played a major role in regulating hormonal activity, and in 1971 he received the Nobel prize in recognition of his **second-messenger concept.**

According to Sutherland's hypothesis, a hormone acts as a "first messenger," carrying its chemical message from the cells of the endocrine gland to the entire body. At the cells of the target organ, the first messenger binds to receptor sites on the cell membranes and reacts with an enzyme (adenyl cyclase), which then acts on ATP inside the cell. ATP is converted by the enzyme into cyclic AMP, which becomes the "second messenger." The second messenger then diffuses throughout the cell and finally causes the cell to respond with its specific endocrine function (Figures 18.13 and 18.14).

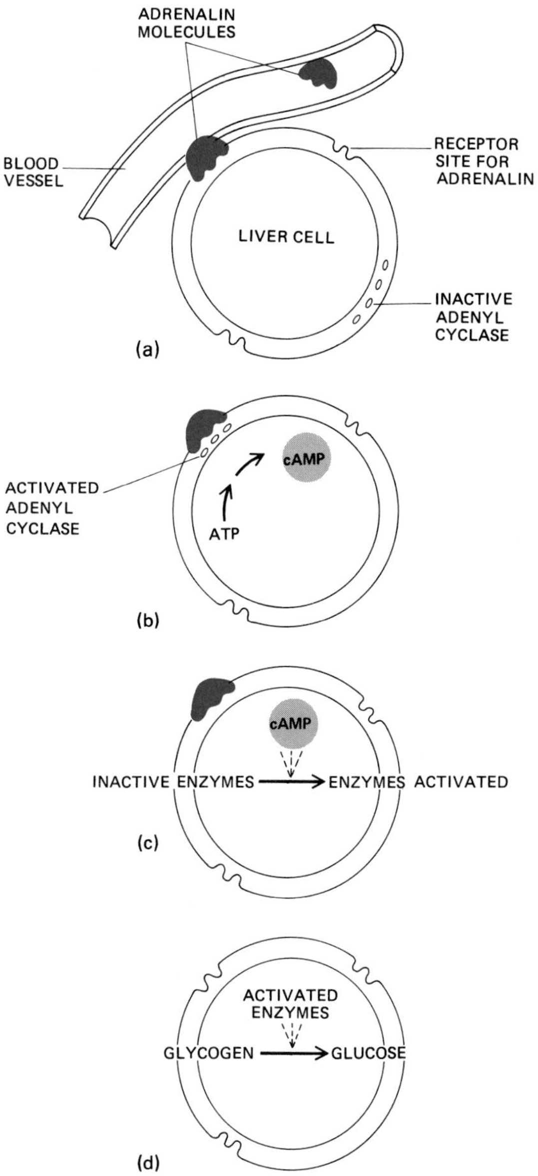

Figure 18.14
Step-by-step action of adrenalin and a "second messenger" on a specific target organ, the liver. Adrenal glands may be stimulated by a sudden fright to secrete the "first messenger," adrenalin. The hormone, released into the bloodstream, is carried throughout the entire body.
(a) When the hormone reaches a specific organ (in this instance, the liver), it passes through the wall of the blood vessel to the cells of the target organ. It is then attached to special receptor sites on the cell membranes.

(b) Adrenalin at a receptor site activates the enzyme, adenyl cyclase, in the cell membrane. The activated enzyme brings about the production of cyclic adenosine monophosphate (cAMP) from ATP.

(c) The cAMP, the "second messenger," causes a series of enzymes to become active.

(d) The activated series of enzymes converts insoluble glycogen in a liver cell to soluble glucose. The glucose can go out of the cell into the bloodstream, which can carry it to organs such as muscles, where it can be used.

The second-messenger concept has been shown to apply to some hormones (notably the protein hormones) but not all. Another possible mechanism for the regulation of hormonal activity concerns the stimulation of protein synthesis. Several hormones, such as the thyroid hormones (and ecdysone, the molting hormone of insects and other invertebrates), seem to act directly on specific genes by stimulating the synthesis of proteins (enzymes) and causing the transcription of certain kinds of messenger RNA. It is possible that the hormone inhibits the repressor substance, so that the DNA of structural genes is turned on, allowing mRNA to start building enzymes. (See the discussion of the Jacob-Monod scheme in Chapter 7.) The enzymes or their end products are then available for a specific activity, such as the release of glucose into the blood.

For the time being, the second-messenger concept and the stimulation of protein synthesis are considered the only credible mechanisms. Much remains to be learned about how hormones work.

SUMMARY

1. Ductless glands, or *endocrine* glands, secrete *hormones* directly into the bloodstream, whereas exocrine glands secrete into ducts that lead to specific sites. The nervous system and endocrine system are interrelated. Hormones generally work on a *feedback system*, and they are crucial for the maintenance of *homeostasis*, the internal stability of the body.
2. The *pituitary* gland is connected by a stalk of nerve cells and blood vessels to the hypothalamus, facilitating an integrated coordination of nervous and hormonal activities.
3. The pituitary is considered an especially important endocrine gland because together with the hypothalamus it controls most of the other glands. Some hormones of the anterior lobe are TSH, ACTH, FSH, LH, STH (the growth hormone), and prolactin. The posterior lobe secretes oxytocin and vasopressin. Some disorders of the pituitary are *gigantism*, *dwarfism*, and *acromegaly*.
4. The *thyroid* gland secretes *thyroxine*, which controls the rate at which food and oxygen are converted into heat and energy. Thyroxine contains the greatest concentration of iodine in the body. Some disorders of the thyroid are *hyperthyroidism*, *hypothyroidism* (goiter), *cretinism*, *myxedema*, and *dwarfism*.
5. The *parathyroids* secrete *parathormone*, which mainly regulates the balance of calcium in the blood.
6. Each *adrenal* gland is composed of the *medulla* and the *cortex*. The medulla chiefly secretes *adrenalin*, and the cortex secretes many hormones, especially *aldosterone* and *cortisol*, which affect metabolism. Some disorders of the adrenal cortex include *Addison's disease*, *Cushing's disease*, and *adrenogenital syndrome*.
7. The islets of Langerhans in the *pancreas* produce *glucagon* and *insulin*, which respectively increase and decrease blood sugar. A common disorder of the pancreas is *diabetes*.
8. The *gonads*, especially the female gonads and secondary sex organs, secrete steroid hormones, which are involved in complex and integrated feedback systems.
9. The *pineal* gland secretes *melatonin*, which affects skin color in amphibians and may affect sex hormones in humans.
10. The *thymus* gland atrophies after adolescence. It probably produces a permanent immunity system.
11. The *digestive system* produces several hormones, including *gastrin*, which stimulates the production of hydrochloric acid when food enters the stomach, and *secretin*, which influences the neutralization of acid in the small intestine.
12. *Prostaglandins* are distributed throughout the body. The effects of prostaglandins are many, including contraction of uterine muscle, inhibition of progesterone secretion by the corpus luteum, and lowering or raising of blood pressure. Prostaglandins may prove useful in relieving asthma, arthritis, ulcers, and hypertension.
13. Two mechanisms have been proposed for the control of the endocrine system. According to the second-messenger concept, a hormone causes the conversion of ATP to cyclic AMP inside a target cell. This action causes the cell to respond in its specific way. Another possibility is that some hormones may inhibit the repressor substance, thereby allowing the buildup of enzymes (proteins) that cause the target cells to begin their activity.

ASK YOURSELF

1. What advantage does radioimmunoassay (RIA) have over other biochemical analytic techniques?
2. Explain how the body temperature of a warm-blooded animal is controlled by a negative feedback system.
3. Why is the pituitary called the "master gland"?
4. Why does drinking a pint of beer cause greater urine production than drinking a pint of water?
5. What is a goiter?
6. What is cretinism?
7. What physiological effects follow a surge of epinephrine into the blood, causing the "fight or flight" condition?
8. Criticize the statement: "Diabetes is a kidney disorder."
9. How can hypoglycemia be quickly (but temporarily) relieved?
10. During what years of a human life is the thymus gland most active?
11. What tissues secrete prostaglandins?
12. Why is cyclic AMP called a "second messenger"?

19
The Action of Nerve Cells

SOME KEY POINTS

1. The basic unit of the nervous system is the nerve cell, or neuron.
2. There are neurons that carry impulses toward the central nervous system (brain and spinal cord), neurons that carry impulses away from the central nervous system, and neurons wholly within the central nervous system that are involved with processing incoming information and helping the body react to it.
3. A nerve impulse is carried along a nerve fiber as electrical and chemical changes occur inside and outside the fiber.
4. Nerve impulses travel from one neuron to another by way of chemical bridges (synapses) between them.

ONCE ANIMALS EVOLVED BEYOND THE SIMPLE one-celled stage, it became necessary to coordinate the actions of all parts of a single animal. Two methods developed, one chemical and one electrical, or nervous. As we saw in the last chapter, the chemical secretions of the endocrine glands go directly into the bloodstream and to all parts of the body. In contrast, the nervous system works by means of electrochemical impulses conducted by specialized nerve cells from one specific part of the body to another.

In an organism that must find food and shelter and avoid danger, an internal information system must be dependable enough to ensure survival. At the same time, the effects of this information system must be short-lived to allow for rapid adjustment to change. A chemical control system does not meet these requirements, but an electrical one does. (The endocrine system and the nervous system have been compared to a large city's postal system and telephone system, respectively.) But neither system is really in charge. Nerves work closely with hormones to keep the body functioning properly.

NEURONS ARE THE FUNCTIONAL UNITS OF THE NERVOUS SYSTEM

The fundamental unit of the nervous system is the nerve cell, or **neuron.** The human nervous system is an organized network of about 100 billion conducting cells, which are neurons. Most of the cells in the nervous system are not neurons, but *glia,* nonconducting cells that protect and nourish the neurons. In fact, glial cells protect the neurons from foreign substances so efficiently that antibiotics usually distribute poorly in the brain tissue. Similarly, white blood cells that enter the brain to fight a brain infection may merely group together in ineffective clumps.

The two basic properties of neurons are *excitability*, the capacity to respond to stimuli, and *conductivity*, the ability to conduct a current along a nerve cell. These properties combine chemical and electrical characteristics. We know much about the electrical characteristics of nerve impulses, but substantial information about the chemical action of nerve cells is lacking. However, we do know that when our neurons are operating, they release certain chemicals, known as neurotransmitters, that affect glands, muscles, or even other neurons. These chemicals (which will be described later) are similar in all vertebrates.

Parts of the Neuron

Like other cells, a neuron contains a nucleus and cytoplasm, but its cytoplasm extends far beyond the main body of the cell (Figure 19.1). A vertebrate

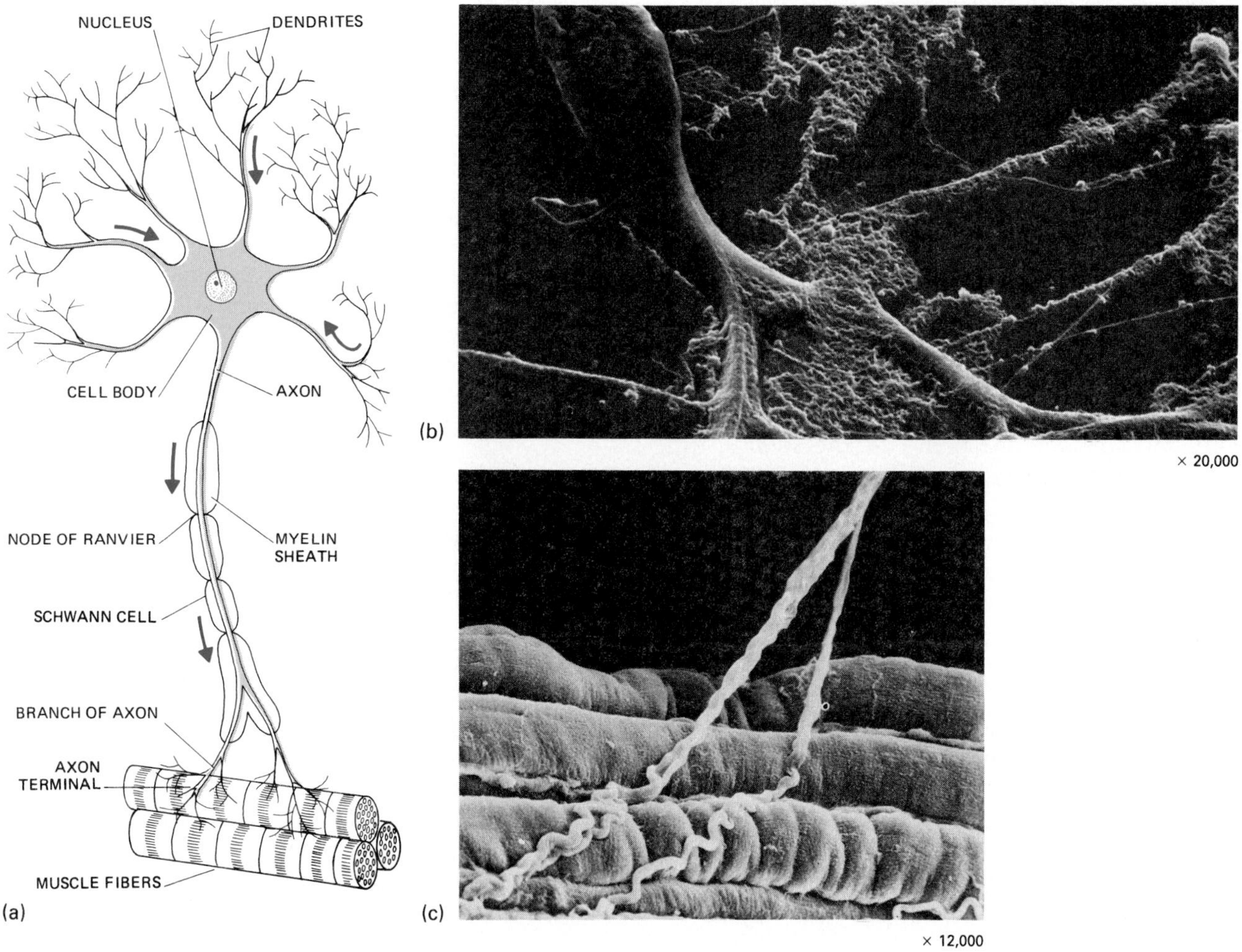

Figure 19.1
(a) Diagram of a vertebrate nerve cell, in this case a motor neuron, which carries an impulse to a muscle. The impulse is usually received by a dendrite and transmitted from the cell body along an axon. (The arrows show the direction of the nerve impulse). The nerve endings of the axon stimulate the muscle and cause it to contract. (See the photo in c.) An insulating sheath of fatty myelin surrounds the axon. Nodes of Ranvier are constrictions in the myelin sheath that occur between the Schwann cells along the axon. In this drawing, the axon is longer than the dendrites, but that is not always so. A nerve cell and its branches may be only a few micrometers from one end to the other, or the branches may be meters long, as in a giraffe's leg. (b) Scanning electron micrograph of a nerve cell, showing its branching fibers. (c) Scanning electron micrograph of nerve endings on muscle fibers.

neuron is made up of three identifiable parts, each associated with a specific function:

1. The **cell body** (soma) contains a nucleus and several organelles responsible for maintaining the metabolism, growth, and repair of the neuron. Some of the organelles in the neural cytoplasm are Nissl bodies, which contain RNA and apparently synthesize the protein necessary for such physiological purposes as memory tracing. Although a neuron may grow and be repaired to some degree, it is not capable of mitosis after very early childhood, and therefore it cannot reproduce after that time. Once a mature nerve cell of the brain or spinal cord dies, it cannot be replaced.

2. The **dendrites** (Greek, "tree"), are numerous, short, threadlike branches (processes) extending out from the cell body. A dendrite may be excited or inhibited. Dendrites carry impulses *toward* the cell body. A cell may have as many as 200 dendrites.

3. Each nerve cell generally has only a single **axon,** a long process extending as much as one or two meters from the cell body. (The diameter of the axon determines the velocity of the impulse conduction; thick axons conduct faster than thin ones.) An axon usually carries impulses *away* from the cell body, and it usually ends in a spray of numerous axon terminals. Unlike a dendrite, which may be excited or inhibited, an axon is usually only excited, passing its stimulation on to the next cell. An axon may be enclosed by a fatty insulation sheath called *myelin,* which aids in the conduction of electrical nerve impulses. Another covering, the *Schwann cells,* may also be present surrounding nerve cells outside the brain and spinal cord (the central nervous system); they enhance the regeneration of peripheral nerve tissue, which does not include the brain and spinal cord. Multiple sclerosis is the degeneration of myelin sheaths in the brain and spinal cord; the cause is unknown, and there is no cure.

Types of Neurons

There are three types of neurons, classified according to their functions:

1. **Afferent** neurons carry impulses *toward* the central nervous system (Figure 19.2). Some afferent neurons are also known as *sensory* neurons because they carry sensory information to the brain and spinal cord from distant parts of the body where actual stimuli (such as a stubbed toe) are received. There is no sensation until the message has been relayed to the central nervous system. Dendrites of afferent neurons receive impulses from *receptor cells,* special nerve cells that pick up stimuli (environmental changes).

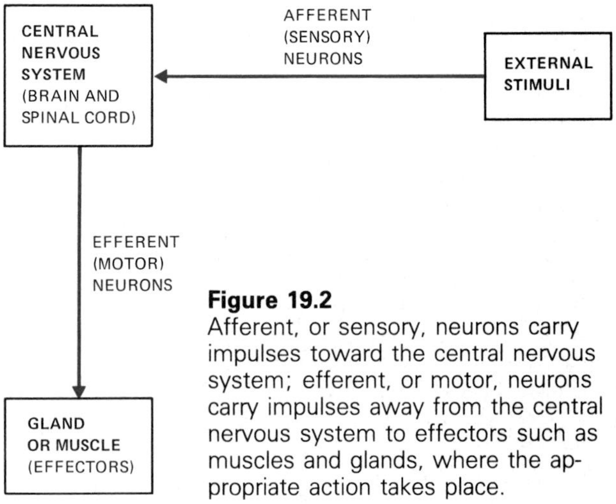

Figure 19.2
Afferent, or sensory, neurons carry impulses toward the central nervous system; efferent, or motor, neurons carry impulses away from the central nervous system to effectors such as muscles and glands, where the appropriate action takes place.

2. **Efferent,** or *motor* neurons, carry impulses *away* from the central nervous system to the glands or muscles where the physical activity actually takes place (see Figures 19.1 and 19.2). Axons of motor neurons terminate in *effectors,* such as muscles or glands, which bring about the relevant action in the effector organ. Your bare foot jerks away from a sharp carpet tack after effectors, in the form of leg muscles, receive an impulse from efferent neurons.

Poliomyelitis (infantile paralysis), which is a viral infection, sometimes causes the destruction of the neurons leading to skeletal muscle. The result is paralysis of muscles, sometimes including those needed for breathing.

3. **Interneurons,** or connector neurons, lie entirely within the brain and spinal cord and constitute the bulk of all nerve cells. Interneurons carry impulses from sensory neurons to motor neurons, and they are involved in the processing of input information and in such complex activities as learning, emotions, and language. The number of interneurons between afferent and efferent neurons depends on the complexity of the activity. Such intricate actions as learning and memory may require thousands of interneurons. In other, less complicated actions (such as a simple reflex), sensory and motor neurons may be directly connected, with few or no interneurons involved.

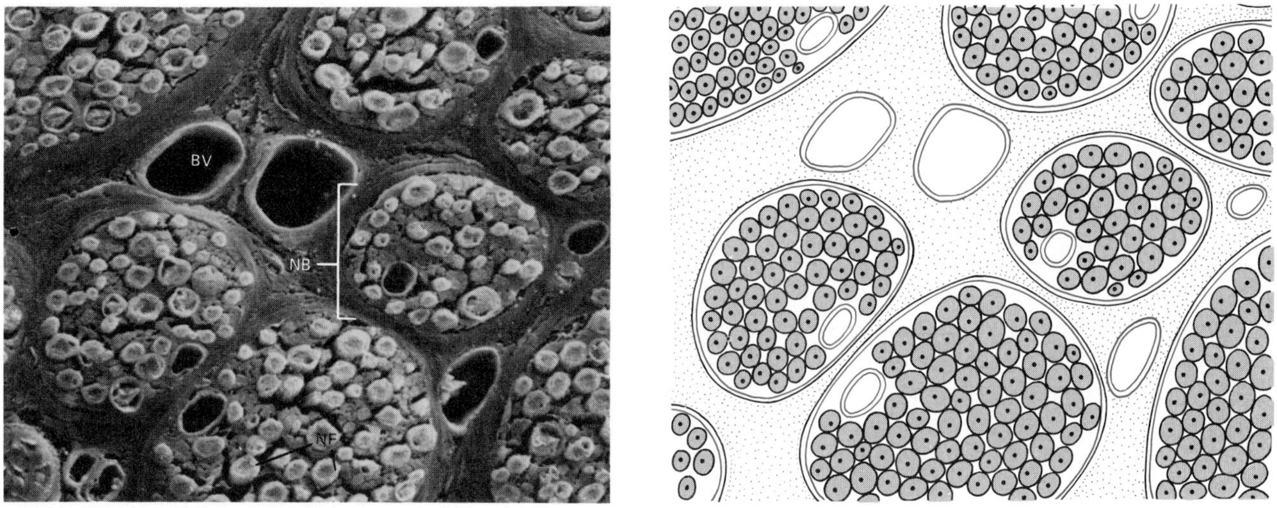

(a) (b)

Figure 19.3
A nerve in cross section. (a) A photomicrograph, showing bundles of nerve fibers (NB), blood vessels (BV), and single nerve fibers (NF) within a bundle. Of the thousands of fibers in a nerve, some may be carrying impulses in one direction while others are carrying impulses the other way. (b) A simplified diagram of a nerve, drawn to show the individual fibers more clearly.

A **nerve** is simply a bundle of fibers enclosed in connective tissue like many telephone wires in a cable (Figure 19.3). Nerves composed of sensory fibers from sensory organs that detect environmental changes are *sensory nerves,* and those made up of only motor fibers that stimulate muscles and glands are *motor nerves.* Nerves that contain both sensory and motor fibers, as most body nerves do, are called *mixed nerves.*

HOW IS A NERVE IMPULSE TRANSMITTED?

A **nerve impulse** is not the same as a flow of electricity. Instead, it is an electrochemical impulse, and it travels at a much slower rate than electricity. The actual rate is about 100 meters (330 ft) a second, as compared with about 300,000 *kilo*meters a second (the speed of light) for an electrical current. Another difference between a nerve impulse and an electrical current is that the nerve impulse does not diminish in power as it moves, as an electrical current does.

A minimum stimulus (*threshold*) is required to instigate an impulse, but an increase in intensity of the stimulus does not increase the strength of the impulse. The phenomenon is like firing a gun. If insufficient force is applied to the trigger, nothing happens. But once the sufficient minimum force is applied, pulling the trigger harder will not make the gun fire harder. The phenomenon of having a nerve cell fire at full power or not at all is an example of the **all-or-none law.**

The all-or-none law applies only to axons carrying impulses away from a nerve cell. Dendrites, carrying impulses toward a nerve cell, may carry *graded* impulses. These impulses may be excitatory or inhibitory. Once an impulse has passed through the cell body and goes out along an axon, then the all-or-none law operates.

To a person who can perceive the difference between a light touch and a strong one, or between a soft sound and a loud one, the all-or-none law may seem wrong. Actually, such differences can be perceived when the *frequency* of the impulse on the tissue is changed. Fibers may conduct impulses at a rate of a few a second or as many as 100 a second. The more frequent the impulses, the higher the level of excitation. Also, the *number of neurons* involved makes a difference in the way the intensity of the stimulus is perceived. For example, consider the difference you feel between getting a strong push and getting a light touch. Even though the same hand may be in contact with the same part of your body, more of your neurons are affected by the strong push. But in an experimental situation, one can show that the all-or-none law holds for axons.

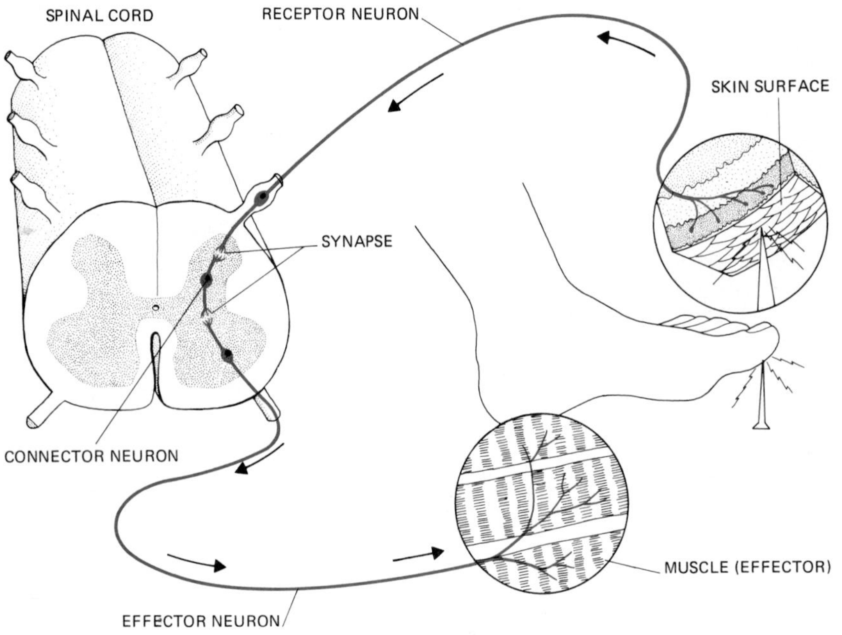

Figure 19.4
Diagram of a simple reflex arc, starting with a tack pricking the skin surface of a big toe. In the scheme as shown here, only three neurons are involved: a receptor, a connector, and an effector. An actual instance as simple as this is rare in such a complex animal as a human being, but it does illustrate the principle on which a reflex works. The arrows indicate the direction of the impulse, from the toe to the spinal cord to the muscle that jerks the foot away from the tack.

The Reflex Arc

If you hurt a finger, you pull back your hand before making a conscious decision to move. Or if a doctor taps the tendon just below your kneecap, your leg jerks up without your conscious control. Why? What has happened is a fairly simple reflex activity, conducted by two or three neurons connected in a **reflex arc** (Figure 19.4). A simple reflex arc makes possible an automatic reaction to a stimulus. It bypasses the brain, processing the incoming information in the spinal cord. In this way, the muscles can react faster than if they had to wait for instructions from the brain. Such reflex actions explain how a chicken whose head has been cut off can still run around awhile before it falls down dead.

A simple reflex arc usually involves (1) the reception of a stimulus, (2) transmission of the nerve impulse to a sensory nerve cell body in a cluster of nerve cells (a *ganglion*) in a spinal nerve, (3) transmission of the impulse to an interneuron in the gray matter of the spinal cord, (4) transmission of the impulse to a motor nerve cell body in the spinal cord, (5) transmission of the impulse out an efferent fiber to an effector, usually a muscle, and (6) contraction of the muscle.

Even in most of the simplest organisms there occurs specialization of cells that results in *sensory cells* that receive outside stimuli, *nerve cells* that conduct impulses, and *effector cells* that contract and move the organism. Such a simple system is all that is necessary to make up the mechanism of a reflex arc, but such simple reflexes are rare in reality, and a typical reflex arc in humans is certainly much more complex than the simple reflex shown in Figure 19.4. In higher animals and humans the more usual complicated reflexes are actually chains of reflexes, with the possibility of several muscles being activated almost simultaneously.

The Mechanism of Nerve Action

Little is known of how stimuli such as sound and touch can be changed into a nerve impulse, but some information is available about the way impulses are transmitted along nerve fibers. Although the nerve impulse is basically *electrical,* its transmission along the nerve fiber is caused by *chemical* changes that involve potassium and sodium ions on the inside and outside of the fiber membrane. The chemical changes that allow the nerve fiber to conduct an electrical current occur when the sodium and potassium ions move across the fiber membrane as follows:

1. In a *resting* nerve cell, potassium ions (black) are concentrated inside the cell, and sodium ions (color) are concentrated outside the cell membrane. Although potassium ions (K^+) and sodium ions (Na^+) both have positive electrical charges, the differing amounts of ions set up a negative charge inside the cell *relative to* the outside of the cell; this relative electrical difference is called the **resting potential,** and the resting cell is considered to be **polarized** (the inside is more negative than the outside). The cell remains polarized as long as the flow of ions into and out of the cell is selectively controlled:

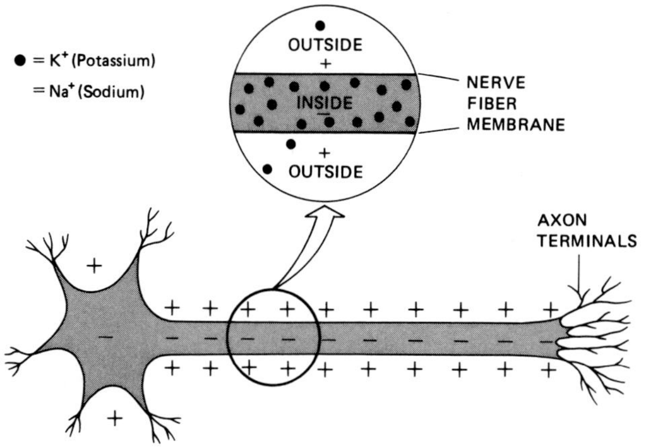

2. When the nerve cell is stimulated by an outside stimulus, the permeability of the membrane is changed, and the relative electrical charges are reversed at the point of stimulus. During the short period that the permeability change lasts (about $\frac{1}{1000}$ of a second), sodium ions rush into the cell at the point of stimulus. The change in the membrane's permeability, and the reversal of charges that makes the outside negative relative to the inside, is called **depolarization.** When a stimulus is strong enough to cause depolarization, the nerve cell is said to *fire*:

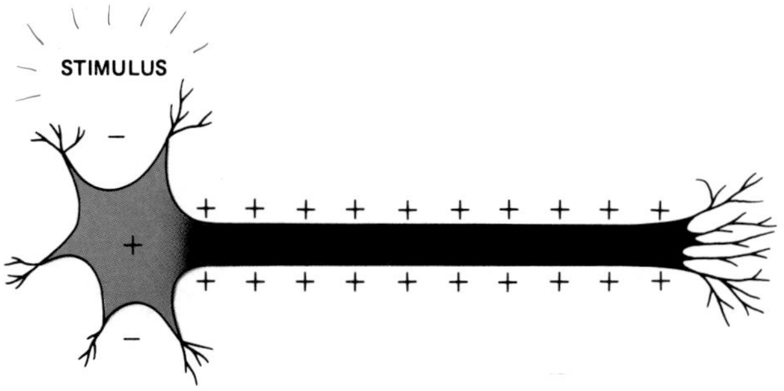

3. Depolarization at one place on the membrane stimulates an adjacent point, and sodium ions rush in; the electrical charges become reversed as they did at the original point of stimulus. The electrical change at each point is called the **action potential.** The action potential may be thought of as the nerve impulse traveling along the fiber membrane as a wave of depolarization that moves from one point on the fiber membrane to the next. The action potential occurs when the depolarization of the cell reaches a critical level of voltage intensity called the **threshold.** The events of a nerve impulse, as recorded on an oscilloscope, are shown below:

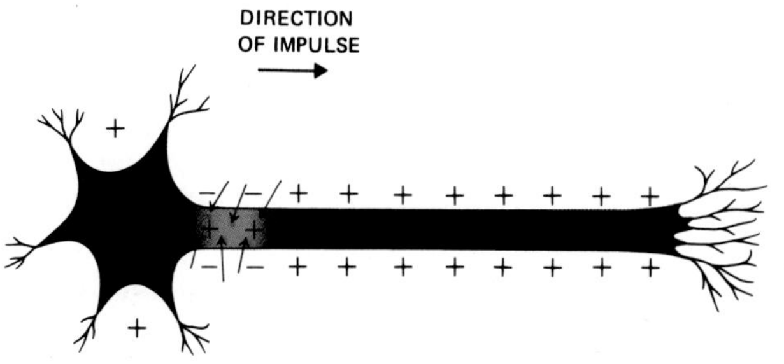

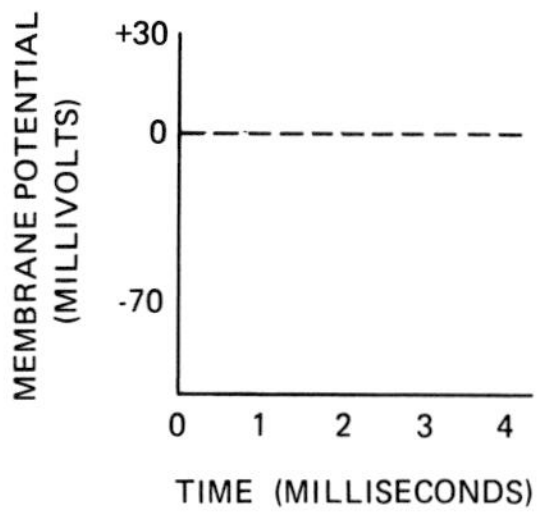

NO POTENTIAL

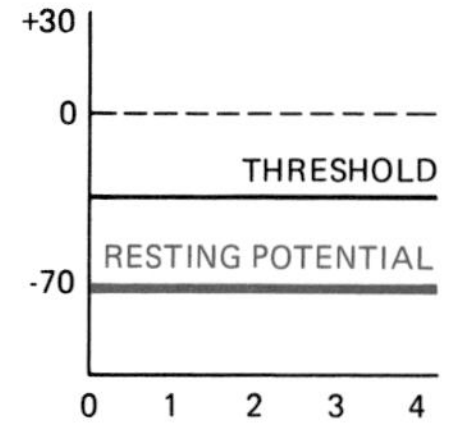

RESTING POTENTIAL

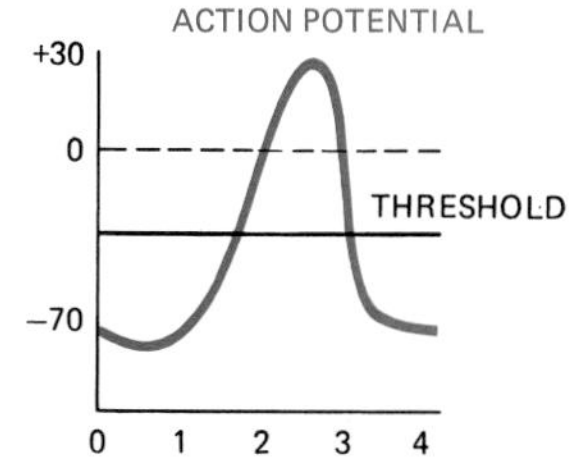

ACTION POTENTIAL

4. Immediately after the action potential occurs, the membrane at the original point of stimulus becomes permeable to potassium ions, and a large number of them rush to the outside of the cell. The outward rush of potassium ions restores the original condition (the resting potential), and the relative negative charge inside the cell and relative positive charge outside the cell are also restored at the point where potassium ions rush out. The membrane is said to be **repolarized,** and the impulse is ended momentarily at that point:

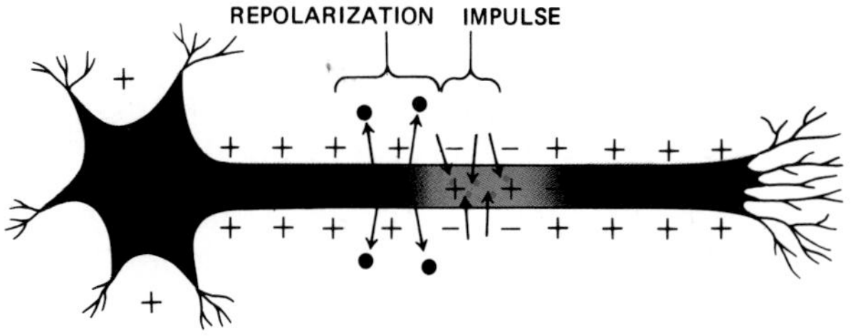

5. After each firing, the nerve cell membrane must rest for anywhere from $\frac{1}{2000}$ to $\frac{1}{500}$ of a second. During this **refractory period** the fiber membrane is being repolarized at that part of the membrane where the impulse has just passed. Afterward, the fiber is "recharged" and is ready to transmit another impulse. Human beings are limited to about 300 impulses per second:

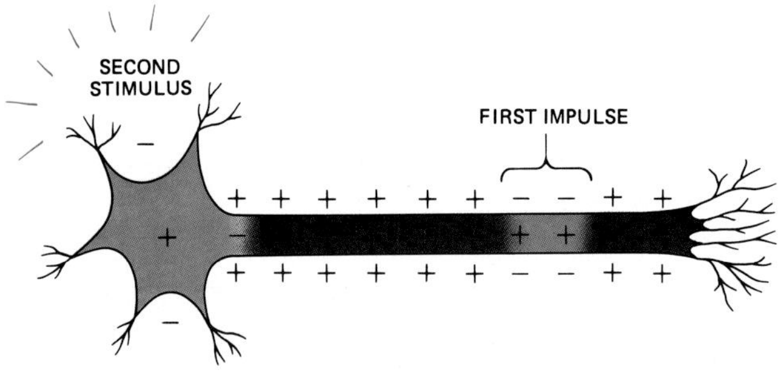

The sodium-potassium pump and ion channels The concentration of sodium and potassium ions inside and outside the membrane of a nerve cell is remarkably constant, even though ions of both kinds "leak" back and forth through the membrane. The fluid outside a neuron contains about 10 sodium ions for every potassium ion, and the ratio inside the cell is reversed, with potassium ions outnumbering sodium ions 10 to 1. Such a regular distribution of ions is necessary if a neuron is to retain its ability to conduct nerve impulses, but how is this homeostasis maintained if potassium and sodium ions are constantly passing back and forth through the semipermeable axon membrane? The answer lies in a series of ATP-powered **sodium-potassium pumps** and selective **ion channels** in the membrane that regulate the flow of ions during a nerve impulse and in the resting state (Figure 19.5).

Each revolution of the sodium-potassium pump exchanges three sodium ions on the inside of the cell for two potassium ions on the outside, and in this way it retains the cell's resting value of about −70 millivolts, slightly more negative inside the cell than outside. A small neuron may have as many as a million pumps. The ion channels open and close selectively to allow sodium and potassium ions to pass through the membrane in response to changes in voltage differences across the cell membrane. As described above, these changes occur when the neuron is stimulated. The initial impulse would not be able to travel along the nerve fiber without the repeated action of sodium ions moving into the cell and potassium ions moving out.

How Fast Does a Nerve Impulse Travel?

The speed of passage of an impulse along a fiber is related to the thickness of the axon. The larger the diameter of the axon, the faster is its rate of conduc-

THE GIANT AXONS OF SQUIDS

The diversity of living types does more than provide biologists with entertainment and a source of speculative philosophy. Knowing that organisms have an enormous variety of structures and ways of doing things, experimenters are always looking for special creatures that may be particularly useful as research tools. For the work Morgan wanted to do, he could scarcely have invented a better animal than the fruit fly. For biochemical genetics, Beadle and Tatum found an almost ideal organism in the fungus *Neurospora.* An object of choice in neurological research in 1936 by J. Z. Young was a squid.

A startled squid can shoot itself backward through the water by a sudden contraction of the muscular cover over its body, the mantle. The spurt of water coming out of its siphon tubes provides jet propulsion, accompanied by a cloud of black fluid. (It has been facetiously suggested that biologists, like squids, progress by going backward as fast as possible and covering their route with ink.) Squids need and have a rapid system for sending impulses to the mantle muscles, and that system includes a so-called giant axon. Most nerve fibers are visible to the naked eye, but they are too slender to study using the electrodes and micropipettes that were available to neurobiologists of the 1940s. However, the big squid axon is several centimeters long, even in a small specimen, and it is up to a millimeter thick, easily thick enough to receive a thin electrode. The axon, which is formed from the fusion of many axons, is physically a tubular sheath filled with axoplasm. With measuring electrodes placed both inside and outside the axon, one can determine changes in potential as impulses pass along the axon. The giant axon can be emptied and then refilled with different solutions. When Young

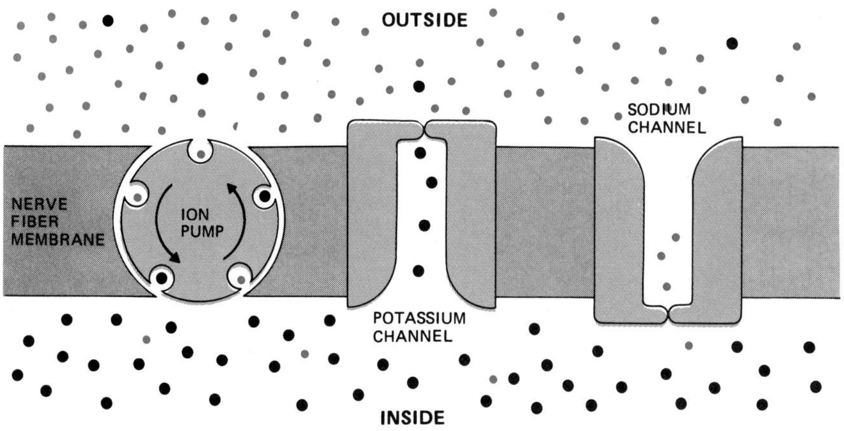

Figure 19.5
A sodium-potassium pump requires ATP to maintain the ionic balance between sodium (color) and potassium (black) on both sides of an axon membrane. The pump ensures a slightly negative electrical charge inside a nerve cell when the cell is at rest. The ion channels, shown here in the resting state, open and close in sequence when a nerve impulse is being transmitted along the axon.

tion. Also, thickly myelinated fibers will conduct more rapidly than fibers with only a thin sheath. The maximum speed the human nerve impulse travels is still relatively slow—about 100 meters a second (300 km/hr), which is three times faster than a car on a highway, but slower than a jet plane.

Fishes usually conduct impulses about half as fast as mammals, and snakes and earthworms may be one-third as fast as mammals, but no other animals are even close to the mammalian rate of conduction in nerves. Of course, *reaction time* to a stimulus is something else again. Some other animals—cockroaches, for example (see the essay on p. 511)—react faster than mammals mainly because the impulse has to travel a shorter distance in a small animal.

did so, he showed that an experimental axon can continue to transmit impulses as long as the inside solution contains potassium and the outside is bathed in a sodium solution. Further, he discovered that the speed of transmission is affected by the size of the axon. An ordinary squid axon, one of the small ones, carries impulses at only about 4.3 meters per second, but the giant axons are five to ten times faster.

Since the days of the classic experiments of A. L. Hodgkin and H. F. Huxley at Plymouth, England, finer electrodes have been developed and researchers can now work on ordinary nerve fibers, but the fundamental demonstrations of nerve action were made with squid axons. Giant axons are important to squids because a rapid transmission system is useful in escaping danger, and to us because we can take advantage of their specialized structure to learn biological principles.

The Synapse

There is always a narrow gap (about $\frac{1}{500}$ as thick as a strand of your hair) between one neuron and the next. How do nerve impulses travel from one neuron to another? The junction between the axon terminal of one neuron and the dendrite or cell body of the next neuron is a **synapse** (Greek, "union") (Figure 19.6). If an impulse is to reach a functional site, such as the brain or a muscle, it must pass across the chemical bridge of the synapse. The passage is accomplished by the tips of axons releasing a chemical substance that crosses the synaptic gap into the dendrite or body of the next cell (Figure 19.7). One such substance is *acetylcholine* (about 10 other such "transmitters" are known).

Acetylcholine, when passed to the adjacent neuron, causes changes in the cell membrane that start the same kind of depolarization that the previous cell experienced, and a wave of excitation passes along the second nerve fiber. The impulse is on its way again after a temporary delay at the synapse. The acetylcholine, after crossing the synaptic junction, is quickly destroyed by an enzyme, *acetylcholinesterase*. Some chemical substances, such as certain insecticides and "nerve gases," work by inhibiting the chemical action of acetylcholinesterase. Since acetylcholine is not destroyed, nerve impulses occur continuously. Muscle cells remain in a state of

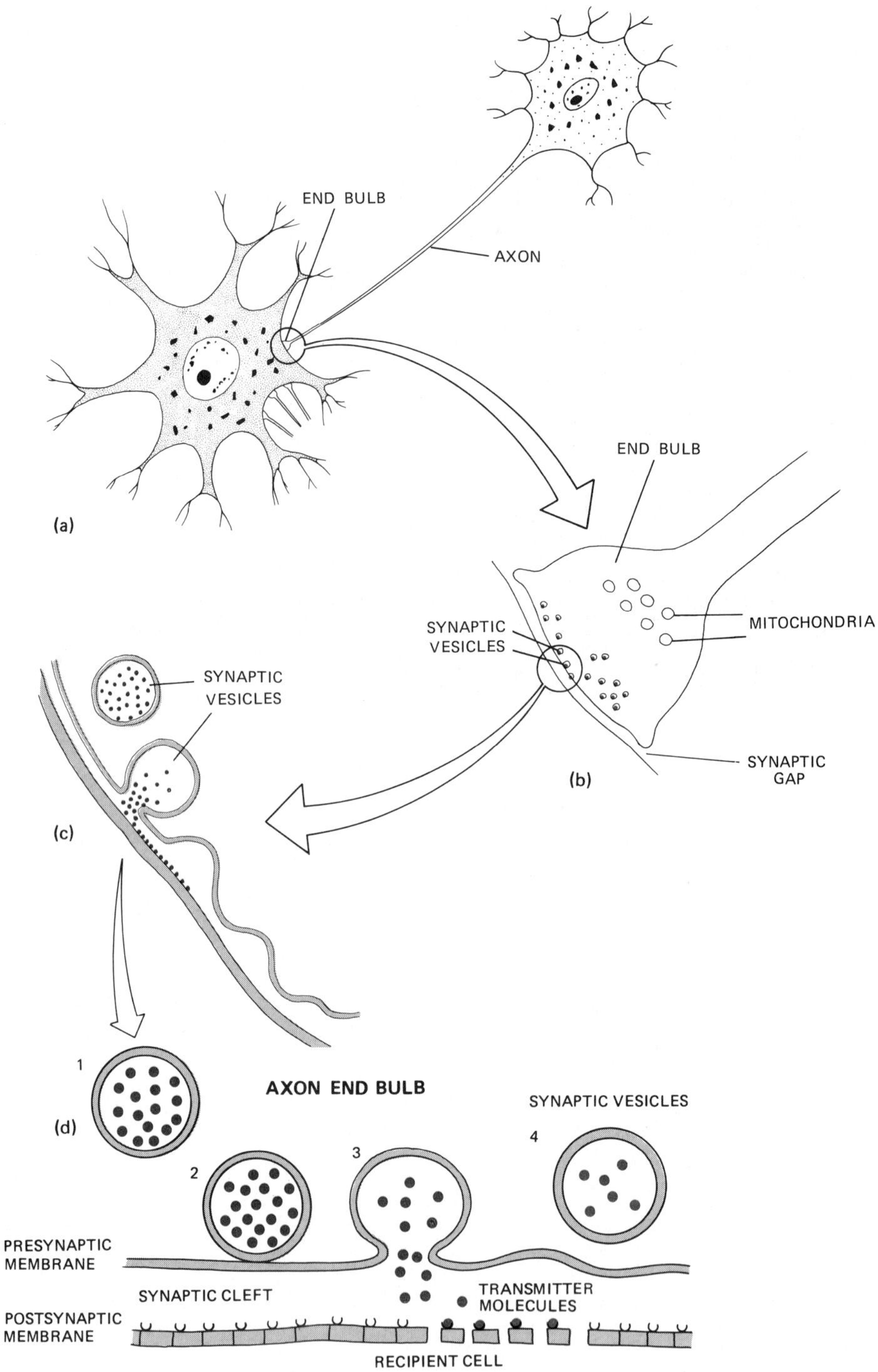

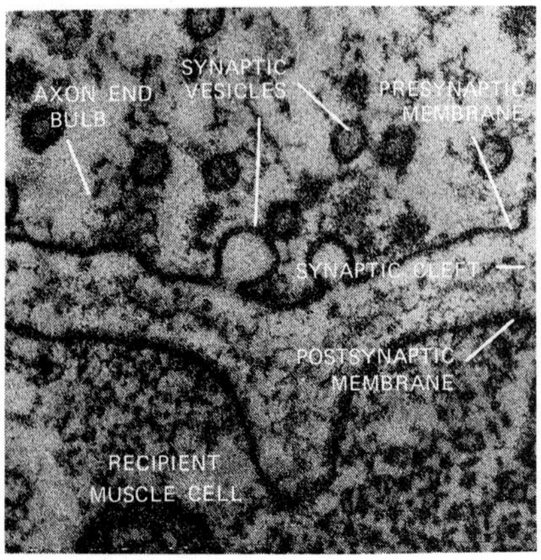

(e)

× 110,000

Figure 19.6
Transmission of an impulse across a synapse. (a) A low-magnification view of a nerve cell, touched by an end bulb of an axon ending (encircled on the drawing). (b) An enlarged view of the end bulb, containing tiny droplets called synaptic vesicles. (c) At still greater enlargement, only a portion of one side of the end bulb is shown. The sequence of events, from left to right: A synaptic vesicle approaches the membrane, makes contact with and breaks through the membrane, and empties its contents of transmitter molecules (color) into the cytoplasm of the nerve cell. The neurotransmitter carries the message across the synaptic gap to the recipient cell. (d) An enlargement of the activity in (c), showing how the transmitter molecules are expelled through the membrane of the axon end bulb by exocytosis, and how the transmitter molecules (color) cross the synaptic cleft and bind to the membrane of the recipient cell. (e) An electron micrograph catches the moment when transmitter molecules of acetylcholine are discharged into the synaptic cleft of a frog, between the axon end bulb and a muscle cell.

contraction, and the breathing muscles are paralyzed. Death by asphyxiation results.

The synapse acts as a one-way valve, with acetylcholine accumulating in the axon ends only, not in the dendrites. Impulses can pass from axon ends across the synapse, over to dendrites of other neu-

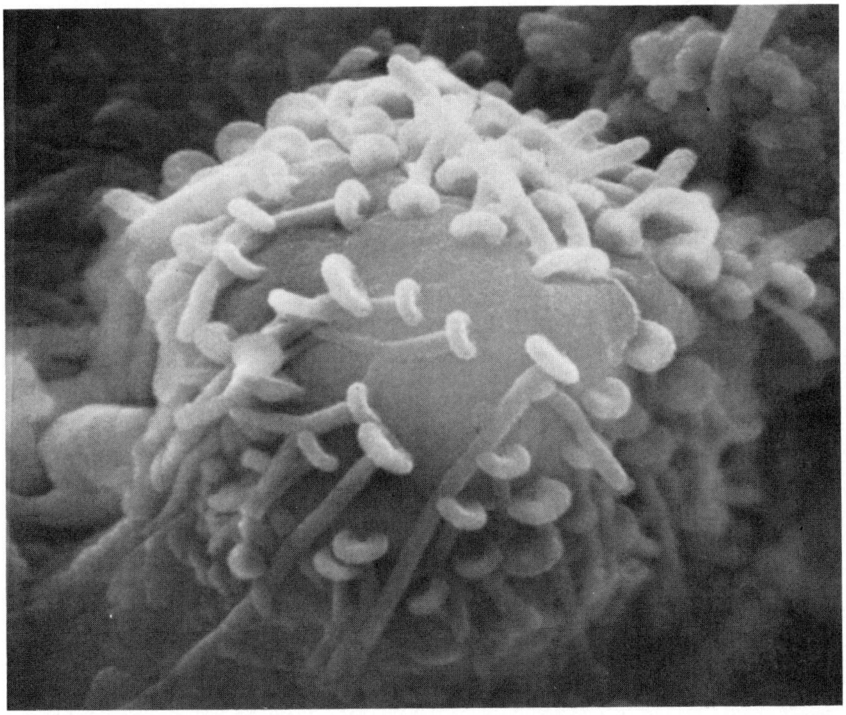

Figure 19.7
A scanning electron micrograph of synaptic end bulbs from several axons of a sea hare, *Aplysia californica,* in contact with the cell body of a postsynaptic neuron.

rons, but not the other way. Because of this built-in directional control by synapses, impulses do not travel in all directions at once. If they did, an organism with a complex nervous system would be thrown into uncontrollable spasms by every strong impulse.

Transmission across a synapse, like transmission along a nerve fiber, requires the input of energy from ATP. Details of how the transmitter chemicals work are incomplete, and even the nature of some of the compounds is uncertain, especially in the central nervous system.

It is possible for one neuron to have as many as 200,000 synapses. Such a great flexibility of possible connections and pathways makes possible the brain's unique complexity and unpredictability. Of course, the pathways to be followed for any given impulse are not chosen at random. They have been carefully programmed to avoid crossed circuits and garbled messages. In the next chapter we will see how nerve impulses are utilized to stimulate muscular and glandular activity.

TO SUM UP NERVE ACTION

1. Nerve impulses are electrochemical in nature.
2. A nerve cell fiber develops a *resting potential* when the electrical charge inside the cell is negative relative to the charge outside the cell.
3. When a nerve cell is stimulated beyond a point called the *threshold,* the charges at the point of stimulation are reversed.
4. Sodium channels in the membrane open, and sodium ions rush into the cell.
5. Potassium channels open at the point preceding the moving impulse, and potassium ions rush to the outside of the cell. The original resting potential is restored, and the impulse ends momentarily as both channels close.
6. The stimulated point on the axon membrane stimulates the adjacent point. The movement of the impulse along the fiber is the *action potential.*
7. After a brief refractory period, a new impulse may begin.
8. The resting potential is kept in balance by the active transport of sodium and potassium ions across the membrane by a *sodium-potassium pump.*
9. Nerve impulses travel from one neuron to another, or from a neuron to an effector, across chemical bridges called *synapses.*

SUMMARY

1. The fundamental unit of the nervous system is the nerve cell, or *neuron.* The two basic properties of neurons are *excitability* and *conductivity.*
2. A neuron is made up of three parts, each associated with a specific function: the *cell body,* the *dendrites,* and the *axon.*
3. There are three types of neurons, classified according to their functions: (1) *afferent* (sensory) neurons carry impulses *toward* the central nervous system; (2) *efferent* (motor) neurons carry impulses *away* from the central nervous system to effector organs (muscles and glands); (3) *interneurons* (connector neurons) carry impulses from sensory neurons to motor neurons and they are involved in the processing of input information and in such complex activities as learning, emotions, and language.
4. A *nerve impulse* is an electrochemical impulse, not a flow of electricity. The axon of a nerve cell is either conducting an impulse at full power or not conducting at all, a phenomenon known as the *all-or-none law.* The dendrites may carry graded impulses of an inhibitory or excitatory nature toward the axon.
5. A *reflex arc* involves a group of neurons regulating a fairly simple activity. It includes a receptor, at least one afferent neuron, usually an interneuron, at least one efferent neuron, and an effector. Reflex arcs can permit body movements without conscious participation of the brain.
6. A nerve impulse is a wave of electrical change that passes along the membrane of a nerve fiber, with inward movement of sodium ions and outward movement of potassium ions. The nerve fiber potential is restored to its resting state by the action of a "sodium-potassium pump," which moves ions out of the axon. The pump maintains the resting potential.
7. A neuron conducts impulses at speeds proportional to the diameter of the axon. Human nerve impulses are conducted at maximum speeds of about 100 meters per second.
8. A *synapse* is the junction between the axon terminal of one neuron and the dendrite or cell body of the next neuron. "Transmitter" chemicals such as acetylcholine cross the synaptic gap and excite or inhibit the next nerve fiber.

ASK YOURSELF

1. In what direction, with respect to a nerve cell body, do dendrites carry nerve impulses?
2. What is an axon?
3. What is the function of a myelin sheath?
4. In what direction, with respect to the central nervous system, do motor neurons carry nerve impulses?
5. What is a nerve?
6. How can nerve impulses allow a brain to distinguish between a loud noise and a soft one?
7. What is the minimum number of neurons required by a simple reflex?
8. What is the "refractory period" in nerve action?
9. What ions are involved in the transmission of a nerve impulse?
10. What is a synapse?
11. How does a "message" pass from one nerve cell to another?
12. How does such a poison as a nerve gas kill its victim?

20
The Nervous System

SOME KEY POINTS

1. The central nervous system is the brain and spinal cord. It is the body's main control center, and it monitors all the activities of the body.
2. The peripheral nervous system consists of nerve cells and fibers that lie outside the brain and spinal cord. It enables the brain and spinal cord to communicate with the entire body.
3. The main parts of the brain are the brainstem, cerebellum, and cerebrum.
4. Receptors are structures that can receive stimuli. The body's sensory system depends on receptors to change stimuli into nerve impulses.
5. Vertebrates have color-sensitive, camera-type eyes, and insects and other arthropods have compound eyes. Most animals are sensitive to light.
6. Little is known about the processes that regulate learning, memory, sleep, and dreams.

THERE IS PROBABLY NOTHING IN THE WORLD more complex and more beautifully designed than the human nervous system. Comparisons with computers, space machines, and communications systems fail (even though we will succumb to such comparisons in this chapter). The human nervous system, and especially the human brain, is just too far ahead of the competition.

The nervous system is a single communications network, but for the convenience of explaining and learning, it is usually divided into two main parts. (1) The *central nervous system* is the brain and spinal cord, and (2) the *peripheral nervous system* consists of those nerve cells and fibers that lie outside the brain and spinal cord. The peripheral nervous system includes the cranial and spinal nerves, and it is concerned with the transmission of sensory and motor information to all parts of the body. The *sensory system* is involved with vision, hearing, and the other senses.

Although the nervous system of humans differs from that of other animals in the development of the brain, it is basically similar to that of the other vertebrates (Figure 20.1), and the main subject of this chapter will be the human nervous system.

THE CENTRAL NERVOUS SYSTEM

The **central nervous system,** consisting of the spinal cord and the brain, is protected within the backbone (vertebrae) and skull. It may be likened to a central control system, a home base for a far-reaching communications network.

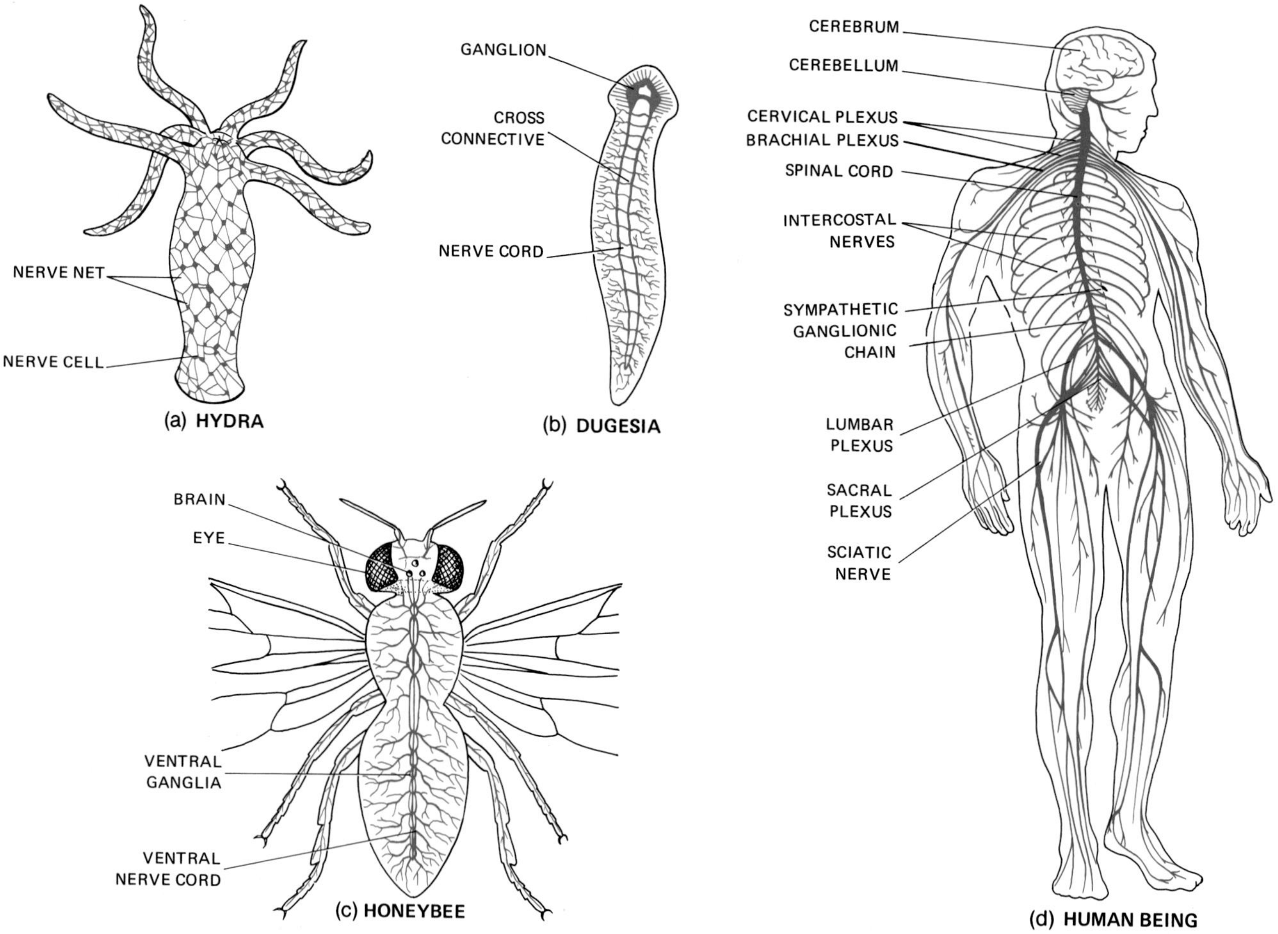

Figure 20.1
Nervous systems of four animals. (a) A coelenterate animal, *Hydra,* is simple compared with a mammal, but it does have tissue differentiation, including a network of specialized cells that function as a primitive nervous system. (b) A flatworm, *Dugesia,* has a definite head end, containing ganglia, and a pair of readily visible nerves, one running down each side of the body, with branches to all parts. (c) An insect, such as a honeybee, *Apis,* has a complex, well-developed nervous system, with ganglia and a central nerve cord running the length of the animal on the ventral side, having branches to all sensory and moving parts. (d) Mammals, especially the human species, have the most highly specialized nervous systems of all animals.

The Spinal Cord

We have already seen how the spinal cord operates in reflex actions. It is also the connecting link between the brain and most of the body. The spinal cord is about 45 centimeters (18 in.) long in adults and about as thick as your index finger, and it extends through the protective vertebrae of the spinal column from the base of the brain to the second lumbar vertebra (about waist level). In an infant, the spinal cord is about the same length as the spinal column (the vertebrae), but the column eventually outgrows the cord.

The space in the spinal column not occupied by the cord is filled by connective tissue and the *meninges,* three protective membranes enveloping the cord. The meninges, from the outside in, are the dura mater, arachnoid layer, and pia mater.

When cut across, the spinal cord shows a tiny central canal, and a dark portion of H-shaped "gray matter," surrounded by "white matter." The gray matter consists mostly of cell bodies and synapses, and the white matter consists mainly of white myelinated nerve fibers, which extend up and down the

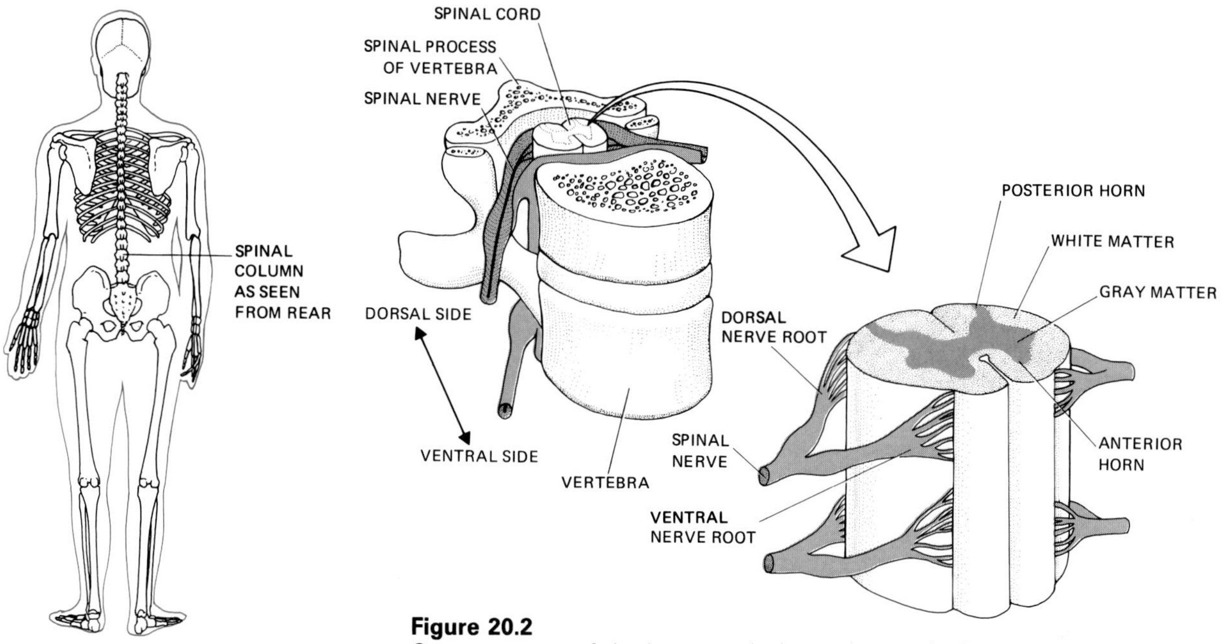

Figure 20.2
Gross anatomy of the human spinal cord and the bones that protect it. Each vertebra, precisely fitted to the ones on either side of it, has a passageway for the spinal cord and openings between the spinal processes through which pass the pairs of *spinal nerves.* As the figure shows, spinal nerves do not leave the spinal cord as single, solid outgrowths; instead, they have two parts, a *ventral* and a *dorsal root.* A tiny canal, too small to see in a picture of this size, passes the length of the cord. The central region of the cord contains the *gray matter,* which in cross section is shaped somewhat like the letter H. The *white matter* surrounds the gray matter. The entire cord is covered by the *meninges,* a familiar term to many people because infection of the meninges is spinal meningitis. The bumps that can be felt in the midline of a human back are *spinal processes* sticking out from the vertebrae.

cord (Figure 20.2). Inside the white matter are several pathways, or tracts. The descending tracts carry motor impulses to skeletal muscle fibers, and the ascending tracts carry sensory impulses to the brain. If the spinal cord is severed, there will be no movement or sensation below the cut.

The 31 pairs of spinal nerves branching to each side of the body, from the neck all the way down the column, are mixed, containing both sensory and motor fibers. The spinal nerves and the cranial nerves form the peripheral nervous system.

The Brain

The adult human brain weighs about 1.4 kilograms (3 lb),* its outermost portion contains more water than any other organ in the body, and it has the consistency of scrambled eggs. Yet it is the most complex and organized instrument ever evolved.

The brain consumes about one-quarter of the oxygen being used in the entire body, and it is so sensitive to shortages of either oxygen or glucose that lack of these substances will cause extremely rapid damage to brain tissue. Proper and continuous nourishment is so importnat to the brain that it has built-in regulating devices that make *impossible* a constriction of blood vessels that would reduce the incoming blood supply, even though it is relatively easy to dilate (enlarge) these same vessels through the use of certain drugs.

The brain is covered by the same three meninges that protect the spinal cord. Between the innermost layer and the middle membrane is a space filled with cerebrospinal fluid. The main function of the fluid and meninges is to cushion and protect the brain. The cerebrospinal fluid supplies a special "floating" environment, so the very soft brain is relieved of gravitational pull and is cushioned against hard blows on the skull and sudden movements

* The human brain is almost completely grown by the time we are seven years old, but soon after birth, nerve cells begin to die. They continue to die throughout our lives, never again equaling the number at birth.

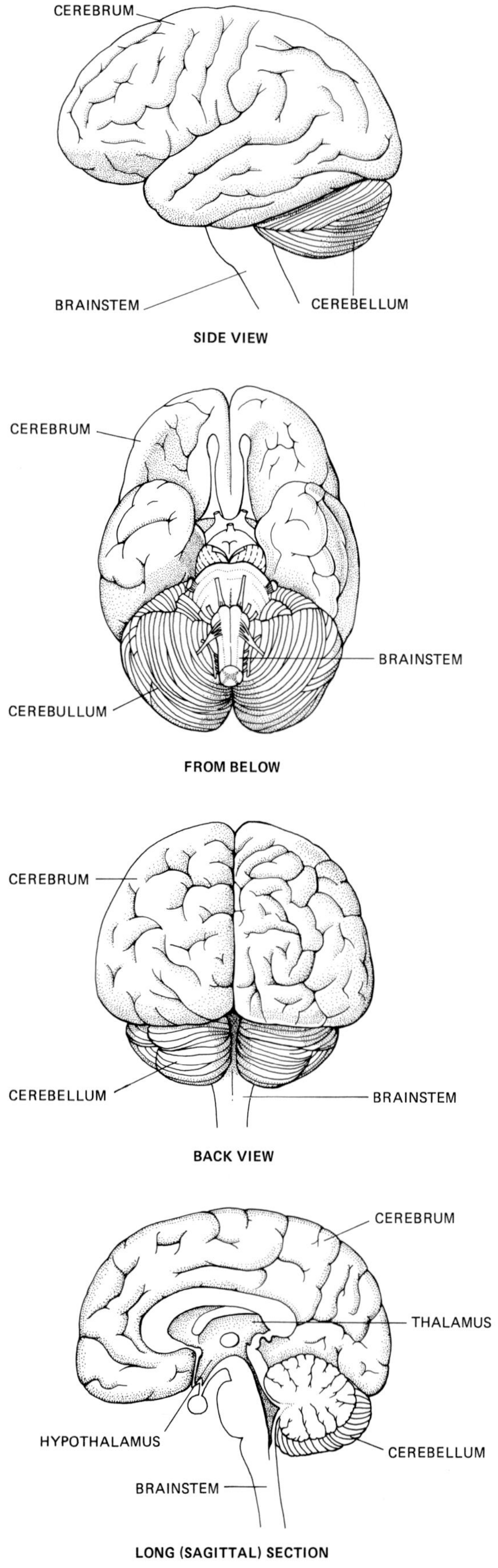

Figure 20.3
Side, bottom, back, and sectional views of the human brain.

that might force the brain against the skull. Such damage to the brain may cause concussion, which means "shake violently" in its Latin form.*

The three main parts of the brain are the brainstem, cerebellum, and cerebrum (Figure 20.3). The **brainstem** is the stalk of the brain, relaying messages between the spinal cord and the brain. It is further divided into the medulla, pons, and midbrain. Several groups of neurons in the *medulla* control such vital functions as heart rate, breathing rate, and blood pressure. The medulla also regulates vomiting, sneezing, coughing, and swallowing. The *pons* forms a bulge at the forward portion of the brainstem. It consists mainly of fibers from the medulla to the cerebrum, and of connections to the cerebellum. The pons controls certain respiratory functions. The *midbrain,* at the anterior end of the brainstem, is directly connected to the pons. Reflex centers are located toward the rear of the midbrain. Visual, auditory, and postural reflexes are monitored in the midbrain.

The **cerebellum** is the second largest portion of the human brain (the cerebrum is the largest). It is a coordinating center for muscular movement. The cerebellum transforms impulses from the cerebrum into actual, coordinated muscular movements. The cerebellum does not initiate muscular movement, but impulses affecting muscular action from the cerebrum would be meaningless without the cerebellum to carry them out. Balance, precision, timing, and body positions are concerns of the cerebellum in humans. In birds, which are generally more agile than mammals, the cerebellum is relatively more important and seems to be one of the major centers of learning (Figure 20.4).

The **cerebrum** is the largest portion of the human brain. It is divided by a deep fissure into two lateral halves, or hemispheres, and each hemisphere is further separated into four lobes, the fron-

* Woodpeckers would seem to be prime candidates for brain damage or concussion. They may bang their beaks against trees 40–45 times in less than three seconds, and they usually drum several hundred times a day, more during the mating season. Concussion and other injury are avoided, however, because woodpeckers have a narrow air space between the brain and its outer membrane; this air space probably reduces the fluid transmission of shock waves. In addition, a woodpecker's brain is cushioned with spongy bone and thick muscles that act as shock absorbers.

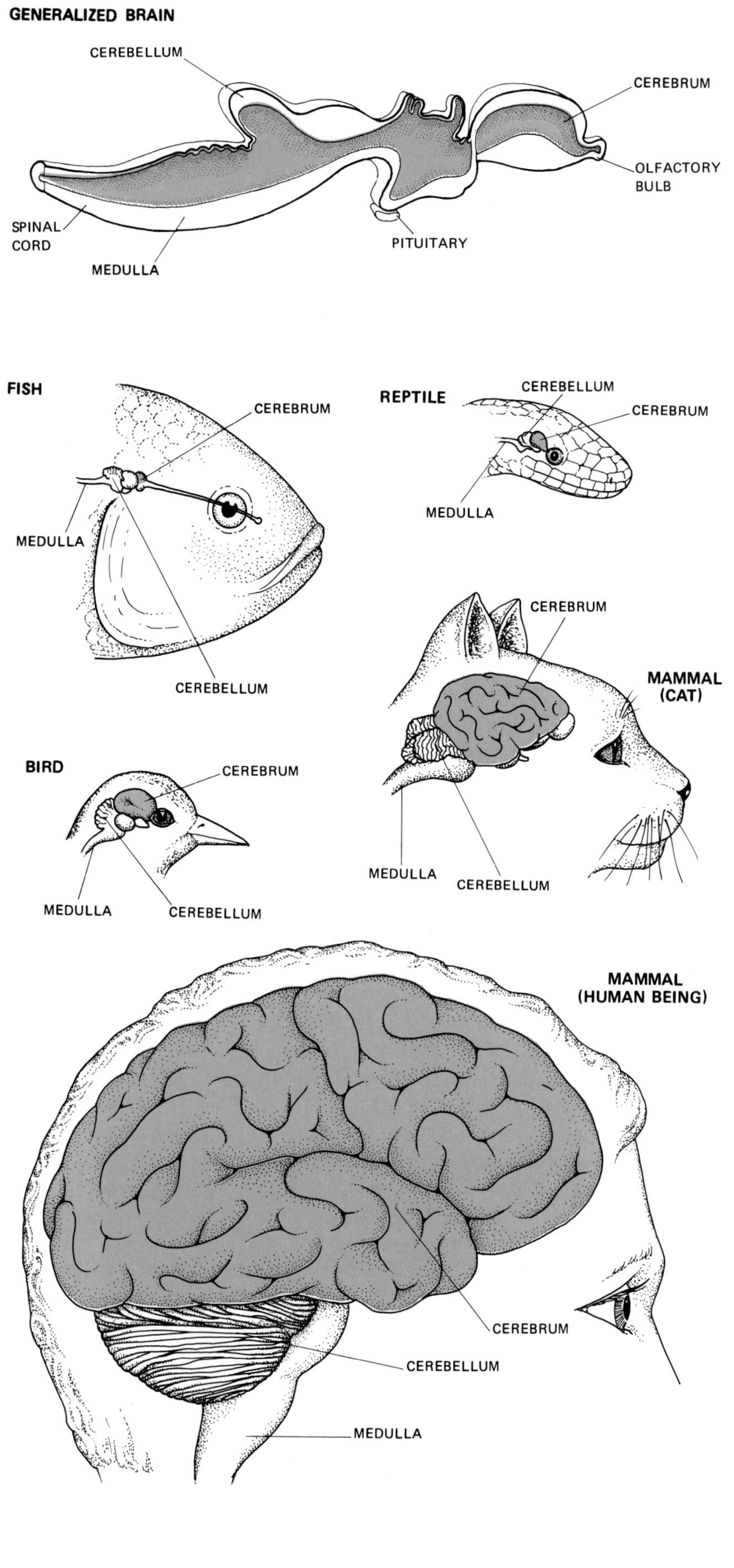

Figure 20.4
A generalized brain, compared with brains of selected vertebrate animals. The cerebrum is made conspicuous in all examples because it is used as a measure of the behavioral complexity of the animal. From fish through reptile to human, the relative size of the entire brain has increased, but the part of the brain taken up by the cerebrum increased proportionately even more. Compare the size of the cerebrum and cerebellum in the human brain with the same parts in the other animals.

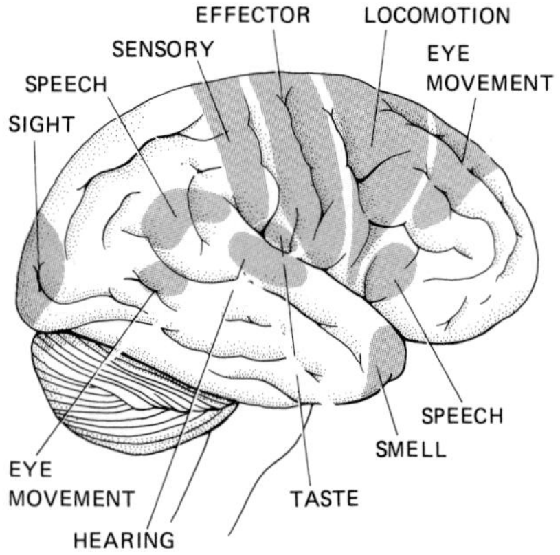

Figure 20.5
Areas of the human brain in which specific functions are localized. The information on which such a map is based has come largely from accidents in which loss of a particular sense or function can be correlated with damage to a known part of the brain. In experiments with animals other than humans, selected places in the brain can be electrically stimulated, and the resulting activity or lack of it can be observed.

tal, parietal, temporal, and occipital. Each lobe contains special functional areas, including speech, vision, movement, learning, and memory (Figure 20.5). A tough bridge of nerve fibers, the *corpus callosum,* connects the hemispheres and relays nerve impulses between them (see the essay below).

In humans, the cerebrum is essential for survival, although exactly which parts or functions are critically necessary is not known. Popularly, the cerebrum is considered the region where thinking is done, but no localization of consciousness or intellectual learning has been established. The cerebrum has been mapped for specific regions that perceive sensory stimuli from specific parts of the body, such as fingers, lips, trunk, and genitalia. It has also been mapped for the localization of regions that control specific parts of the body (see Figure 20.5).

If the cerebrum is damaged, there will be loss of function, depending on the specific area of the injury. For instance, damage to the speech area in a temporal lobe may produce *aphasia* (inability to understand or use written or spoken words).

The outer portion of the cerebrum, the *cortex,* is a thin, convoluted covering containing about 12 bil-

THE SPLIT BRAIN IN HUMAN BEINGS

Recently the left and right hemispheres of the cerebrum have come to be thought of as two separate brains. Such an idea was proposed intuitively in the past, but now there are enough experimental data to substantiate a claim for the split-brain theory in humans. Apparently the left hemisphere (the "left brain") dominates such activities as speech, writing, and logical reasoning. The right brain seems to dominate short-term memory, intuition, spatial perception, creativity, and athletic functions. In normal instances the right brain controls the left side of the body, and the left brain controls the right side. Connecting the two brains is the *corpus callosum,* a bridge of 200 million nerve fibers carrying nerve impulses from one hemisphere to the other.

Experiments with the corpus callosum solidified the case for the split-brain theory. In the early 1950s Ronald E. Myers and Roger W. Sperry found that when the corpus callosum of a cat was severed, both hemispheres functioned independently, almost as if they were two complete brains. Subsequently, the corpus callosum operation was performed on several patients suffering from epilepsy, and these individuals were usually improved as a result of the separation of their cerebral hemispheres. (In 1981, Sperry shared the Nobel prize in Physiology and Medicine with David H. Hubel and Torsten N. Wiesel for his split-brain research. The work of Hubel and Wiesel concentrated on the brain's role in vision.)

The speech center is located in the left hemisphere. When split-brain patients were asked to view objects across their total left-right field of vision, they could point to the objects they saw in their total field, but could *orally describe* only the objects in the left side of the field. Obviously, the patients could "see" with both sides of their brain, but if they had to talk about what they saw, their left brains were "blind." Remember, lion neurons. The folded convolutions of the cerebrum increase the surface area about three times. The gray color of cortex nerve cells gives the outer brain its common designation as "gray matter." The inner, medullary portion of the cerebrum shows the whiteness of myelinated efferent, afferent, or connective fibers. Also buried within the cerebrum are the cell bodies of cell clusters called basal ganglia, which help to coordinate muscle movements. A malfunctioning of the basal ganglia causes Parkinson's disease, which is characterized by involuntary tremors.

The **thalamus,** below the cerebrum (see Figure 20.3), is involved in all the activities of the central nervous system. It acts as a monitor to all incoming and outgoing impulses and modifies some of them. The thalamus contains part of a net of nerve cells, the *reticular formation,* which filters all incoming stimuli, discarding unimportant ones and sending others to the proper decoding centers in the brain.

The **hypothalamus,** below the thalamus (see Figure 20.3), controls many physiological and endocrine activities, and it contains centers that regulate basic behavioral patterns, such as those involved in feeding, fighting, reproduction, or escape. It is in the hypothalamus that experimentation has revealed "pleasure centers," "hunger centers," and "fighting centers."

the speech center is in the left brain, which relates to the *right* side of body.

Can the right brain learn to "speak" when the speech center in the left brain is damaged? Apparently it can, and apparently other functions can be transferred ("relearned," if you will) from one hemisphere to the other, especially in young children before the activity becomes totally specialized. But many adults have demonstrated the ability for fully developed language skills in their right hemispheres.

Western society has downgraded left-handedness and consequently the right brain for many centuries. ("Sinister" and "gauche" are both derived from words meaning "left," whereas words like "adroit" and "dextrous" are related to "right.") Now we must finally give the right brain its due. We still do not know why the left brain is usually dominant, but we do know that the right brain can be just as "adroit" as the left brain when it has to be.

TO SUM UP THE CENTRAL NERVOUS SYSTEM

1. The central nervous system is made up of the brain and spinal cord.
2. The spinal cord is the connecting link between the brain and most of the body. Descending tracts in the cord carry motor impulses to muscle fibers, and ascending tracts carry sensory impulses to the brain.
3. The brain consists of the brainstem, cerebellum, and cerebrum. The brainstem relays messages between the spinal cord and the brain and controls such vital functions as heart rate and breathing rate. The cerebellum is a coordinating center for muscular movement. The cerebrum controls speech, vision, and other functions, including learning and memory.
4. The thalamus acts as a monitor to all incoming and outgoing impulses, and modifies some of them. The hypothalamus controls many physiological and endocrine activities.

THE PERIPHERAL NERVOUS SYSTEM

The **peripheral nervous system** enables the brain and spinal cord to communicate with the entire body. It may be further divided into the *somatic* nervous system and the *autonomic* nervous system. The somatic system involves "voluntary" movements of the skeletal muscles, and the autonomic system controls such "involuntary" activities as breathing and glandular secretion.

The Somatic Nervous System

The **somatic nervous system** is composed of nerve fibers that lead from the brain through the spinal cord to the skeletal muscles. These fibers pass from the brain directly to the cells of the skeletal muscles without any synapses. The action of such motor neurons is always to contract muscles.

BIOFEEDBACK: THE BRAIN CONTROLS THE BODY

The so-called involuntary activities of the body, such as breathing and regulation of heart rate, are in fact not completely involuntary. The method of control of such "automatic" body functions is popularly called **biofeedback,** and it can be learned. One important factor in biofeedback training is knowing what is happening in one's cerebral and visceral functions. For instance, some migraine-headache sufferers, aware that their headaches are associated with certain dilated blood vessels, have learned how to prevent the vessels from expanding when they felt that a headache seemed to be approaching.

Recently, with a growing popular interest in Yoga and other Eastern mysticism, biofeedback training of one kind or another has developed rapidly. Soothing alpha waves are called up at will, heartbeats are slowed or steadied, blood is diverted into the hands away from a potentially aching head, and skin temperature is changed. Of course, some chicanery accompanies most fads, especially profitable ones. Prescribed body positions, incantations, special costumes, smells, sounds, and decorative environments are all unnecessary, though some may aid a faltering beginner by strengthening his resolve. The one factor that all the religious and other methods have in common is an insistence on unswerving thought, whether it is called prayer, meditation, concentration, or scheming. Directed thought is more difficult to achieve than most people expect; it is rare for anyone not disciplined in concentration to attend strictly to a single idea for more than a few seconds. Legitimate efforts with biofeedback training may provide some long-sought relief for such diseases as epilepsy and other brain-related problems. Indeed, perhaps illness need not be in the brain to be treated with biofeedback techniques.

The Autonomic Nervous System

The main differences between the somatic nervous system and the autonomic nervous system are shown in Table 20.1.

The **autonomic nervous system** helps the somatic system to regulate normal body functions and to control the use of body resources (Figure 20.6). It is composed of two parts, the *sympathetic* and the *parasympathetic systems,* each complementing the other. The **sympathetic** system, including nerves from the thoracic (chest) and lumbar (midback) regions of the spinal cord, can cause relaxation of intestinal wall muscles and an increase in sweating, heart rate, and blood flow to voluntary muscles. The **parasympathetic** system, containing nerves from both the brain and sacral (low) end of the spinal cord, acts in opposition to the sympathetic, causing contraction of intestinal wall muscles and a decrease in sweating, heart rate, and blood flow to voluntary muscles.

TABLE 20.1
Major Differences Between the Somatic and Autonomic Nervous Systems

Somatic

1. Has axons that go from the central nervous system to the effector muscle without synapse.
2. Activates skeletal muscles only.
3. Always excites the effector organ.

Autonomic

1. Has fibers that synapse in ganglia outside the central nervous system.
2. Affects smooth muscle, cardiac muscle, and glands.
3. May inhibit *or* excite effector organ.

Both systems include paired ganglia, to the right and left of the spinal cord. The ganglia of the parasympathetic system are in the effector organs, somewhat removed from the spinal cord; the ganglia of the sympathetic system are in two rows near the spinal cord. All the ganglia are connected to the spinal cord by preganglionic fibers, and to effectors by postganglionic fibers.

The postganglionic fibers secrete *norepinephrine* at their endings to transmit impulses from the fiber to the effector organ. Both the sympathetic and parasympathetic systems release acetylcholine as the chemical transmitter for ganglionic synapses.

The sympathetic system is the only nerve supply to the adrenal medulla, and it is capable of stimulating the flow of adrenalin from the adrenal gland in response to a typical stress situation, commonly called a "fight or flight" condition.

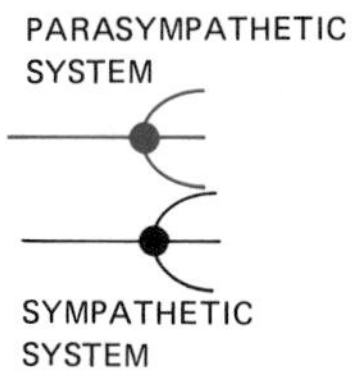

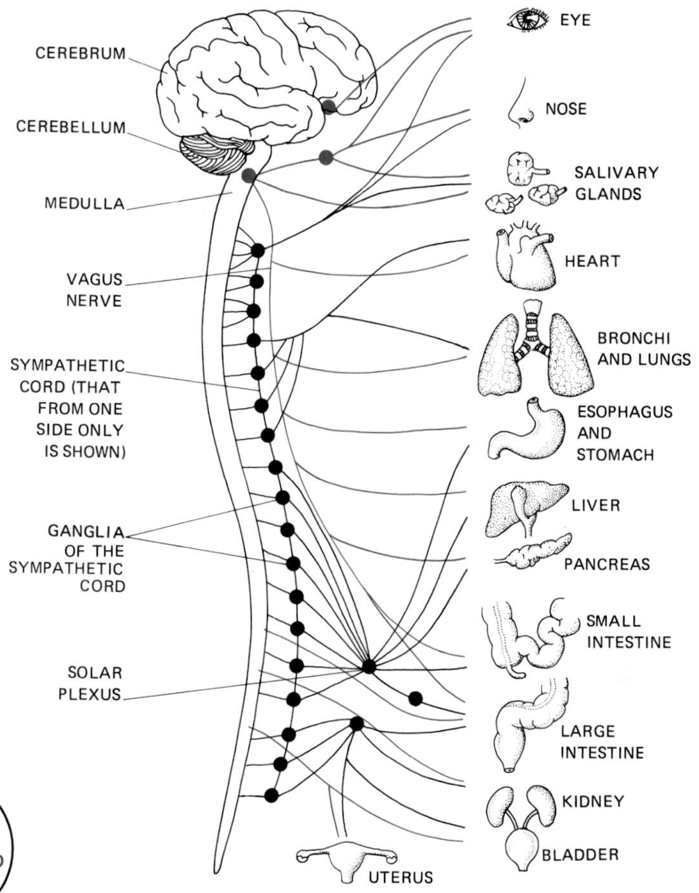

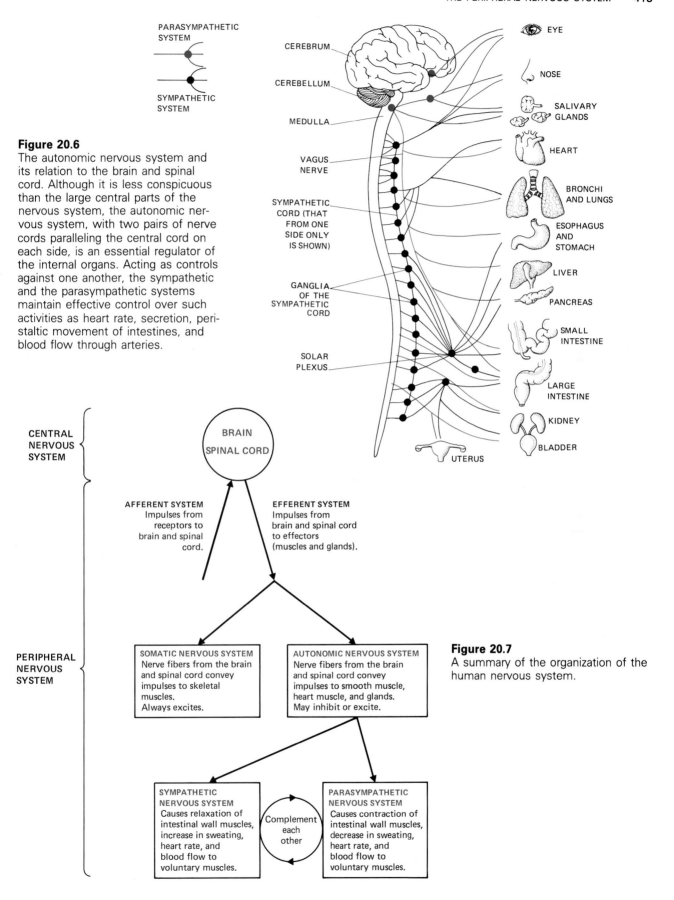

Figure 20.6
The autonomic nervous system and its relation to the brain and spinal cord. Although it is less conspicuous than the large central parts of the nervous system, the autonomic nervous system, with two pairs of nerve cords paralleling the central cord on each side, is an essential regulator of the internal organs. Acting as controls against one another, the sympathetic and the parasympathetic systems maintain effective control over such activities as heart rate, secretion, peristaltic movement of intestines, and blood flow through arteries.

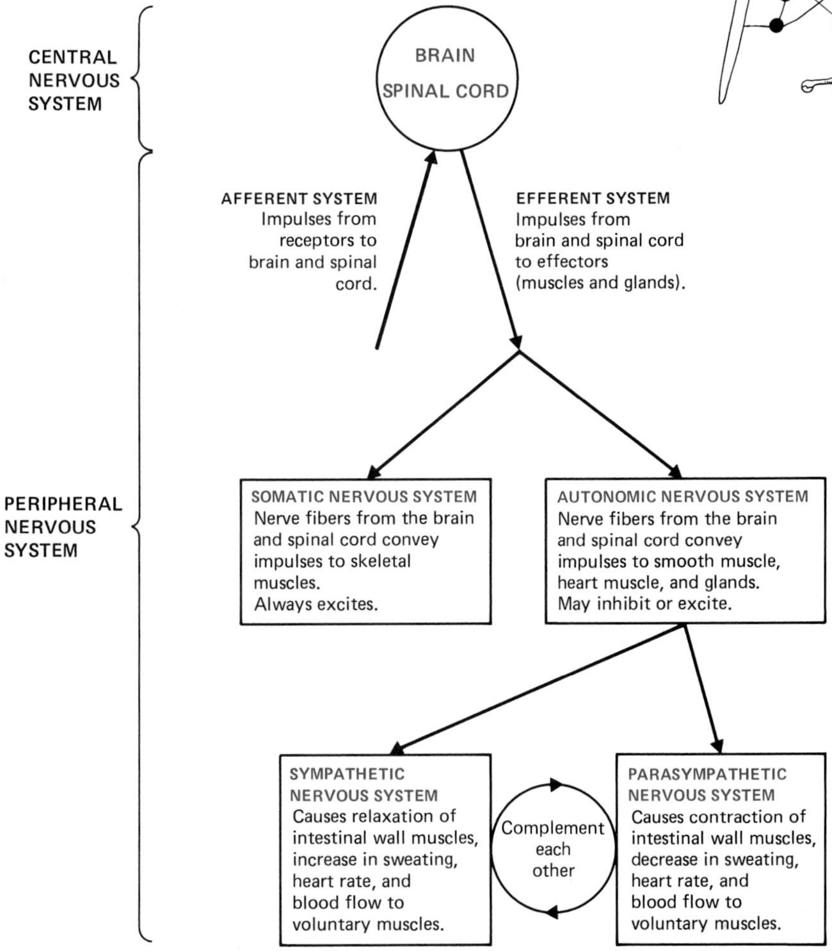

Figure 20.7
A summary of the organization of the human nervous system.

Although the autonomic system regulates many important body functions almost independently of the brain and spinal cord, it is basically incorrect to refer to it as an "involuntary" system. It is becoming increasing obvious to experimenters that animals (including humans) can learn to control such activities as heart rate and breathing rate if there is a motivation to do so (see the essay on p. 412).

TO SUM UP THE PERIPHERAL NERVOUS SYSTEM

1. The peripheral nervous system may be divided into the somatic nervous system and the autonomic nervous system.
2. Nerve fibers from the somatic nervous system connect the brain to skeletal muscles, and nerve impulses cause the muscles to contract.
3. The autonomic nervous system is further divided into the sympathetic and parasympathetic systems, which complement each other. It affects smooth muscle, heart muscle, and glands, and helps regulate many important body functions that are typically considered involuntary.

THE SENSES: PERCEPTION OF STIMULI

All knowledge and awareness depends on the reception and decoding of stimuli from the outside world. Such a process of assimilating afferent information is called *perception.* Practically all organisms, except perhaps bacteria, have some perceptive mechanisms, and there is some evidence that bacteria do also.

Organisms must maintain homeostasis, and one of the ways they do so is through a sensory system that permits the body to adjust to changing environmental conditions (including internal, visceral conditions). Structures that are capable of perceiving such changing stimuli are **receptors.** One method of classifying receptors is according to the kinds of energy they respond to, as follows: chemoreceptors (taste, smell), mechanoreceptors (touch, pain, hearing), photoreceptors (vision), thermoreceptors (heat, cold), and proprioceptors (balance and other information about one's own body). In terms of the physiology of the nervous system, a receptor may be thought of as the peripheral end of a sensory neuron. All stimuated receptors generate the same type of nerve impulse, and the different sensations are brought about when nerve fibers connect with specialized portions of the central nervous system. As with nerve impulses in general, the intensity of "sensory" stimuli may be altered by increasing or decreasing the frequency of the stimulus, or by stimulating more or fewer receptors at one time.

In summary, it appears that all receptors are capable of changing (*transducing*) stimuli into nerve impulses. Subsequently, localized regions of the brain translate unspecialized impulses into specific, conscious sensations. In this way, even mechanical pressure on an eye stimulates impulses in the optic nerves that will ultimately be interpreted as light. If you jar your eyes hard enough, even in total darkness, you will "see stars" because the brain recognizes impulses coming through optic nerves only as light.

The Eye and Vision

Visible light is a portion of the electromagnetic spectrum within the range of about 300 to 760 nanometers (a nanometer is one-billionth of a meter). The human eye is sensitive to wavelengths within this range, and so we all "see" objects that emit or are illuminated by waves in our receptive range. Other organisms (insects, for example) are sensitive to shorter wavelengths in the range of ultraviolet, and some organisms can "see" longer wavelengths, such as infrared waves (Figure 20.8). Of course, no eye actually *sees;* it merely receives light. It is the brain that "sees" by converting nerve impulses into information. If you were to shine a light directly onto the optic center of the brain, no light would be "seen," and in fact, no other response would be recorded either.

To go one step further, if you could switch the nerves that carry light impulses to the optic centers of the brain with the ones that carry sound impulses to the auditory centers, it would actually be possible for the brain to "hear" light and "see" sound. The brain is impartial. It will convert *all* sufficiently strong action potentials into nerve impulses. The brain doesn't "know" the differences between light, sound, touch, and the other senses. If the sight center of the brain is stimulated, the brain "sees," even if the external stimulus is actually the loud backfire of a truck.

Practically all animals respond to light in some way, and light receptors vary from the image-forming, color-sensitive **camera-type eyes** of vertebrates (Figure 20.9) to the simple photoreceptors of one-celled animals, which can distinguish only intensity. The nocturnal earthworm has no eyes, but its

WHAT ANIMALS SEE

Light and color are such common features in the world of human experience that they are almost taken for granted. But even so, we should consider two basic principles that make vision possible: absorption and reflection of light. Although sunlight has a white appearance, it actually consists of many different colors. A glass prism will separate sunlight into reds, oranges, yellows, greens, blues, and violets, the full spectrum of colors visible to the human eye. Sunlight also contains other "colors" not visible to the human eye, infrared and ultraviolet. When sunlight falls on an object, the color components of the light may be reflected or absorbed. If all the light is absorbed, the object will appear as black to the color-sensitive eye. "Color" is seen when light is reflected from an object. Green plants are perceived as green because their green pigment, chlorophyll, reflects the green portion of the spectrum to our eyes; the green plant absorbs the other colors in the light. To "see" a color, the light receptor of the organism must be sensitive to the particular colors being reflected from the object to the eye. On the following pages we will see how several different animals perceive color.

Predatory birds, such as hawks, falcons, and eagles, are superb hunters because they fly faster than any other animal can move. But to complement their flying skill, birds need exceptional vision so they can see prey at a distance. In fact, birds not only fly faster than any other animal moves; they also see more sharply. The resolution of some birds' eyes is about eight times that of human eyes because birds have about 1.5 million cone cells per fovea compared with our 200,000. Because of the abundance of cone cells, birds can resolve a distant object well enough to identify it clearly. A human would see the same object as a blur. The pictures here demonstrate how a ground squirrel at a distance would look to a hawk and to a human.

It should not be surprising that birds can also see close objects clearer than humans can. Birds are able to change their focus instantly, and they can see close objects as well as they can see distant ones. Visual acuity is so important to birds that many have eyes that are larger than their brains.

A distant ground squirrel as seen by a hawk.

The same ground squirrel as seen by a human.

If animals could talk to us and tell us what they see, we would know more about their color vision than we do now. But we still could not know everything, because there is no guarantee that one animal could successfully communicate its view of the world to another animal. Even a human being does not know exactly what another person sees. The senses are so subjective that we can only describe them approximately to other people. It is fairly easy to know if someone is seeing *no* color, or if a person mistakes one end of the spectrum for the other, but we cannot say that we know absolutely what another person or another animal sees. The photographs on these pages come as close as possible to showing the world as seen by a human, a dog, a horse, and an insect. Likewise, the color spectra on the following page are at best approximations of the world of color as seen by several animals.

Adapted from an idea in *Life Before Man,* a Time-Life Book.

A human being sees the world in full color, from violet to red, but certain colors, such as ultraviolet, are beyond the human's normal range of visible light. The field of vision is about 180 degrees, and the overall picture is exceptionally clear.

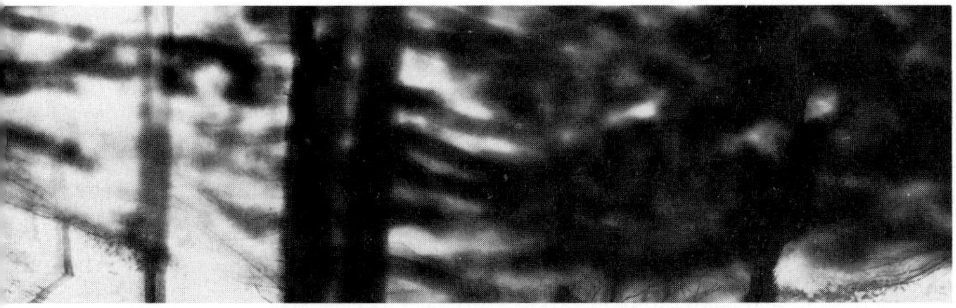

A dog does not see color; it sees various shades of gray. The field of vision is wider than a human's, but objects in the background tend to be out of focus because animals such as dogs and cats are nearsighted. Note that the foreground is sharper than in the picture above of the human's world.

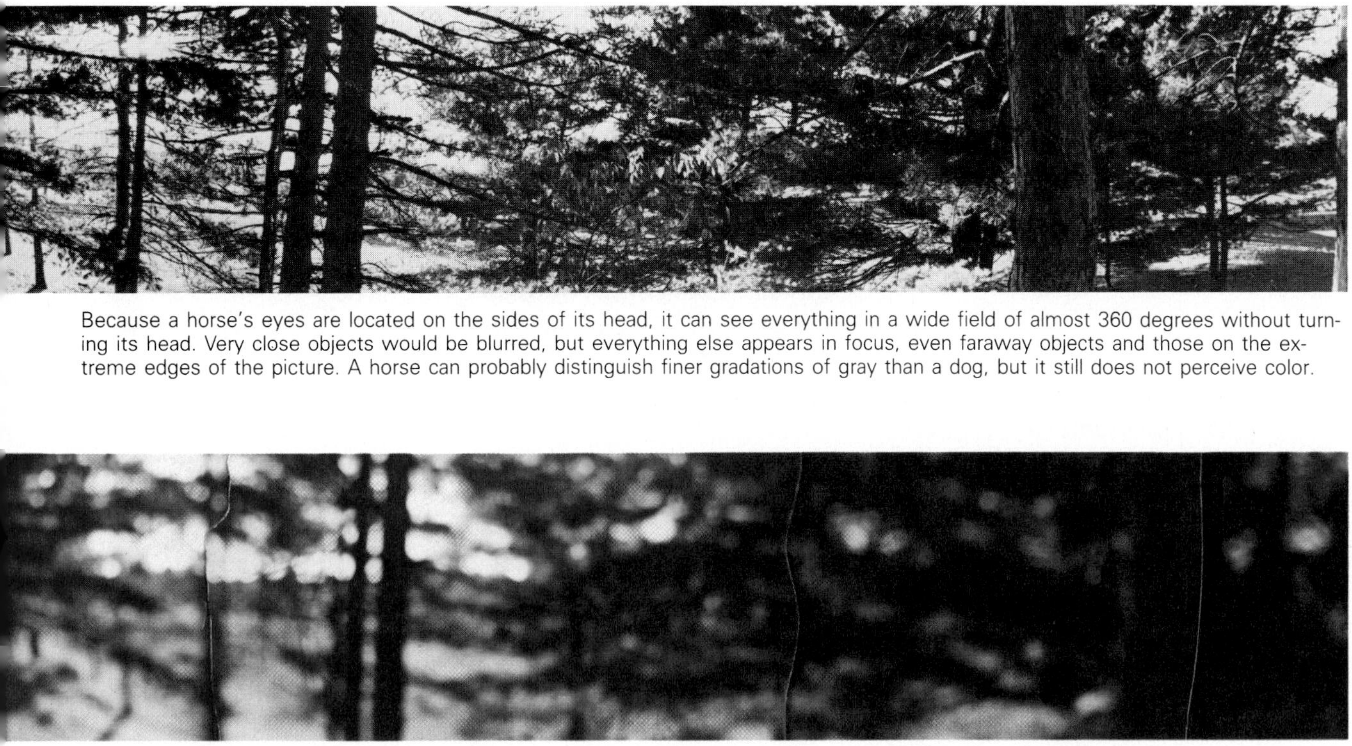

Because a horse's eyes are located on the sides of its head, it can see everything in a wide field of almost 360 degrees without turning its head. Very close objects would be blurred, but everything else appears in focus, even faraway objects and those on the extreme edges of the picture. A horse can probably distinguish finer gradations of gray than a dog, but it still does not perceive color.

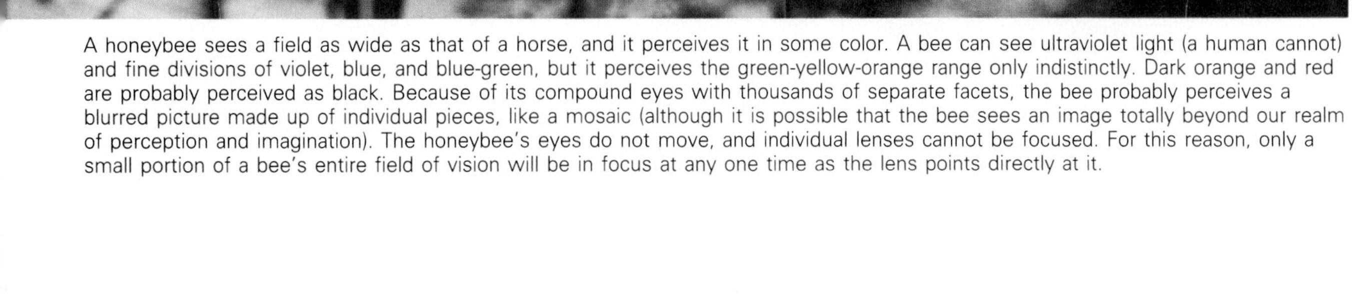

A honeybee sees a field as wide as that of a horse, and it perceives it in some color. A bee can see ultraviolet light (a human cannot) and fine divisions of violet, blue, and blue-green, but it perceives the green-yellow-orange range only indistinctly. Dark orange and red are probably perceived as black. Because of its compound eyes with thousands of separate facets, the bee probably perceives a blurred picture made up of individual pieces, like a mosaic (although it is possible that the bee sees an image totally beyond our realm of perception and imagination). The honeybee's eyes do not move, and individual lenses cannot be focused. For this reason, only a small portion of a bee's entire field of vision will be in focus at any one time as the lens points directly at it.

Color vision exists in insects, fishes, reptiles, and birds, but it is absent in most mammals. The primate mammals, including humans, can perceive color, and it is possible that among mammals *only* the primates are able to distinguish color. Other mammals, such as cats, dogs, horses, cows, lions, and elephants, are almost certainly color blind. Color perception is highly developed in fishes, reptiles, and birds, and in such insects as dragonflies and bees. Amphibians probably lack a discriminating sense of color vision, although frogs seem to be able to distinguish between blue and green. If a trapped frog is given a choice, it will jump toward blue and avoid green, presumably heading for the open areas of sky and water rather than the more confining environment of grass.

The different visual abilities of animals and the specific visual perceptions of each species were not randomly evolved. Rather, they are biologically sensible and appear to be related to specific problems and conditions in the environment. Bees, for example, have developed color sensitivities in areas where such sensitivities are useful. Bees usually do not pollinate red flowers because they cannot distinguish red. In contrast, bees *do* pollinate blue flowers and some flowers that reflect ultraviolet light. In these wavelengths, bees have a highly developed color sense. The natural selection of the evolutionary process has been effectively at work.

HUMAN BEING

Humans see a full range of color from violet to red but are unable to see ultraviolet and infrared light. As far as we know, humans have the most discriminating sense of color perception of all animals.

BIRD

A bird's eye is almost ten times as acute as a human eye and responsive to the same portion of the spectrum. Apparently, birds are partially blind to blue but very sensitive to red. Some birds of prey can detect infrared rays emanating at night from the bodies of mice and other prey. Some birds may not see violet. Many birds show no interest in green, and in fact they may not see green at all.

FISH

It is not known with certainty what colors fishes perceive, but they do seem to react to the bright colors exhibited during mating. One group of biologists presents the spectrum above, with the full range of colors, including ultraviolet light. Another group believes that fishes are attracted to yellow and green, and there is some evidence that they do not see the blue and violet areas at all. This group of biologists also thinks that fishes react to orange and red light as if it were black, as shown below. In either spectrum, the colors are probably blurred by the fish's murky environment.

FISH

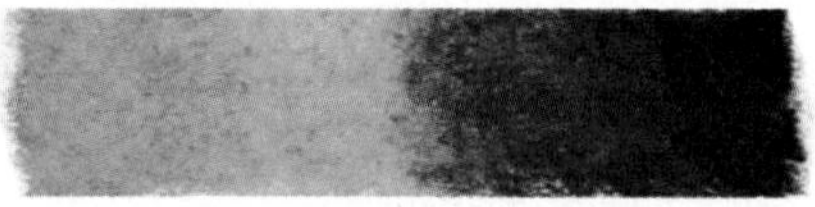

TURTLE

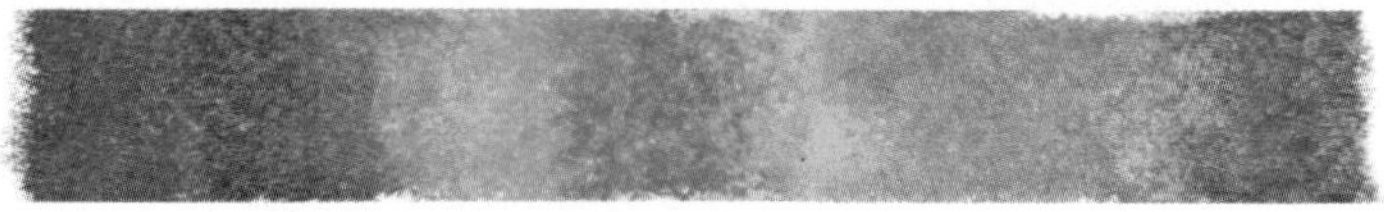

Turtles and most reptiles have a well-developed color vision. The most important colors seem to be orange, green, and violet. Yellow and yellow-green are apparently seen as orange. Red seems to appear as violet, thereby closing the color circle. It is not certain that snakes can distinguish color, but apparently lizards can.

HONEYBEE

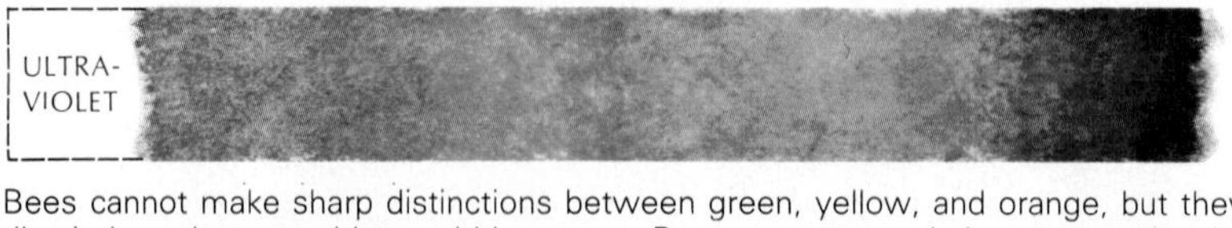

Bees cannot make sharp distinctions between green, yellow, and orange, but they can discriminate between blue and blue-green. Bees cannot see dark orange and red (red probably appears as black), but they do see ultraviolet light, which humans do not see.

CAT

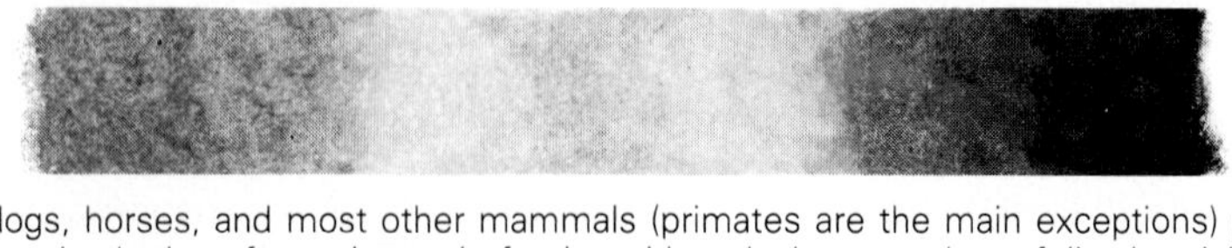

Cats, dogs, horses, and most other mammals (primates are the main exceptions) probably see only shades of gray instead of color, although they may have full-color vision.

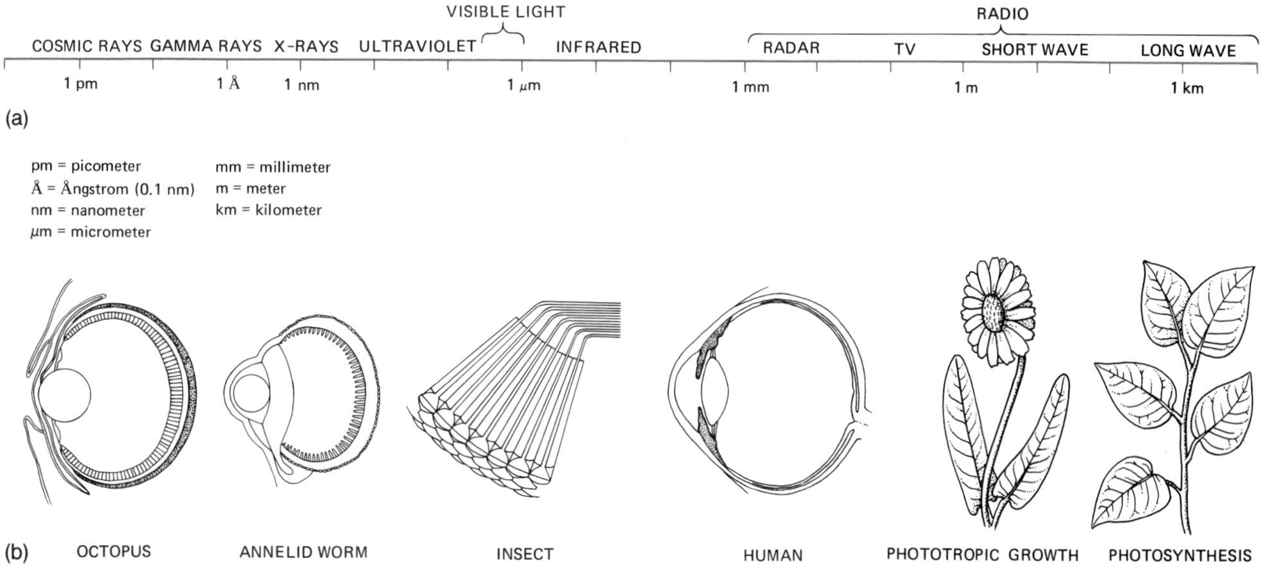

Figure 20.8
The electromagnetic spectrum and some biological radiation receptors. (a) A scale showing the wavelengths of electromagnetic radiation, from the shortest cosmic rays to long radio broadcast waves. That part of the spectrum commonly absorbed and used by living things is about midway between the extremes. It is called "visible light" because it is the part that affects human retinas. The limits vary from species to species, and in human eyes from individual to individual, but the range is generally bounded by the near ultraviolet at about 300 nm and the near infrared at about 700 nm. (b) Some representative light-absorbing parts: eyes of animals and leaves (generally) of plants. Regardless of the differences in form, all depend on having a light-gathering structure and a selective light-absorbing pigment. The octopus eye and the human eye, eons apart in evolutionary history, are notably similar, but the insect eye is built on quite a different plan. Plants do not usually have any means of concentrating light, although some mosses and flowering plants have lenslike thickenings that do effectively achieve such concentration.

Figure 20.9
Comparison of a human eye with a camera. There are several structural similarities. Light from an illuminated object passes through a hole in the front of the apparatus (pupil of an eye, aperture of a camera) and is focused by a lens on a sensitive field (retina of an eye, film in a camera). In both the image is reversed, but one can turn a photographic film over to look at it, and in much the same way a brain can interpret the image to understand which way is up. In an eye, the actual image is both upside down and reversed. One important difference between the camera and the human eye is that the eye focuses on nearer or more distant objects by changing the shape of the lens, but the camera (like some fish eyes) focuses by moving the lens toward or away from the film.

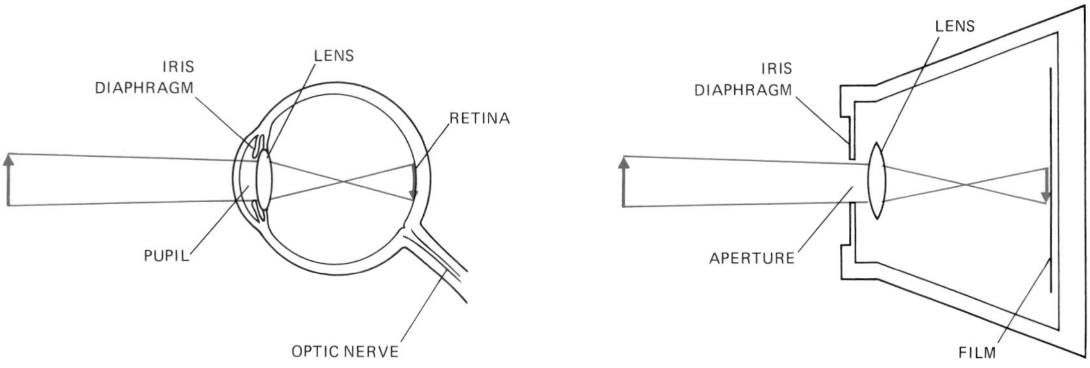

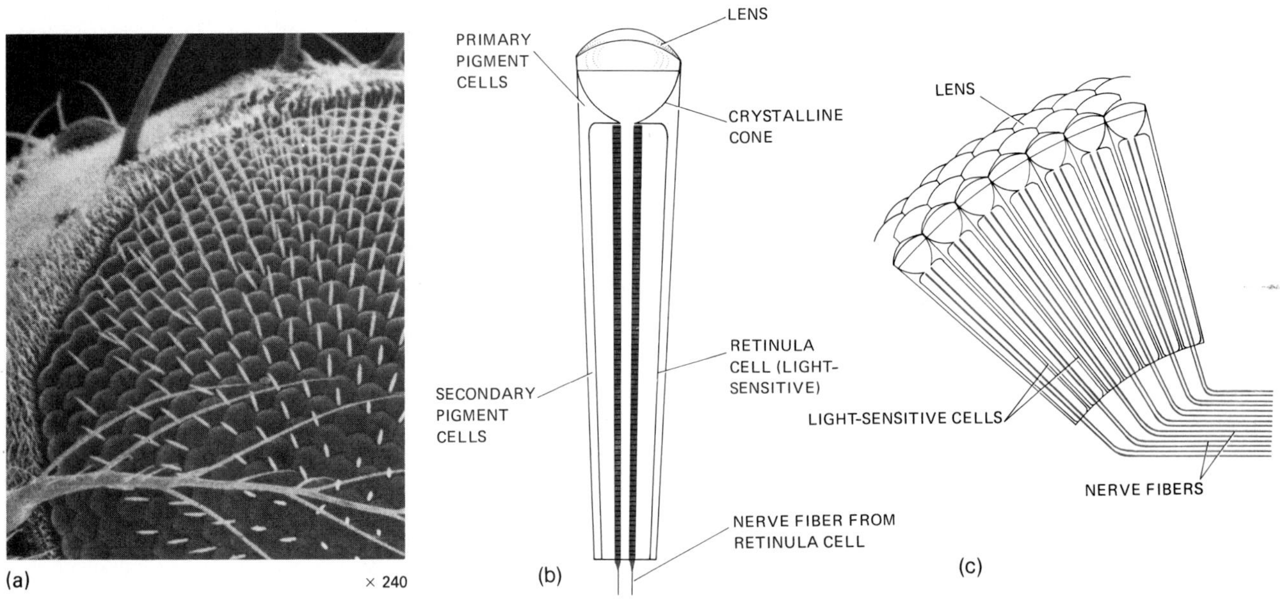

Figure 20.10
The compound eye. (a) A scanning electron micrograph of part of a fly's eye, showing the hundreds of simple units, *ommatidia,* that constitute the whole eye. (b) A diagram of a long section of a single ommatidium. Light is collected by a front lens and passed down a long tunnel to the receptor cell at the bottom. The field of view of an ommatidium is narrow. (c) A diagram of a compound eye, cut away to show the arrangement of the ommatidia. An insect or other animal with such compound eyes probably does not form a picturelike image; rather, it must have some sort of mosaic perception. Although some insects obviously have effective vision, we cannot imagine how an insect brain interprets the stimuli it receives. No one knows how the world looks to a dragonfly, even though it is possible to take a photograph through the myriad lenses of such an animal and look at the picture.

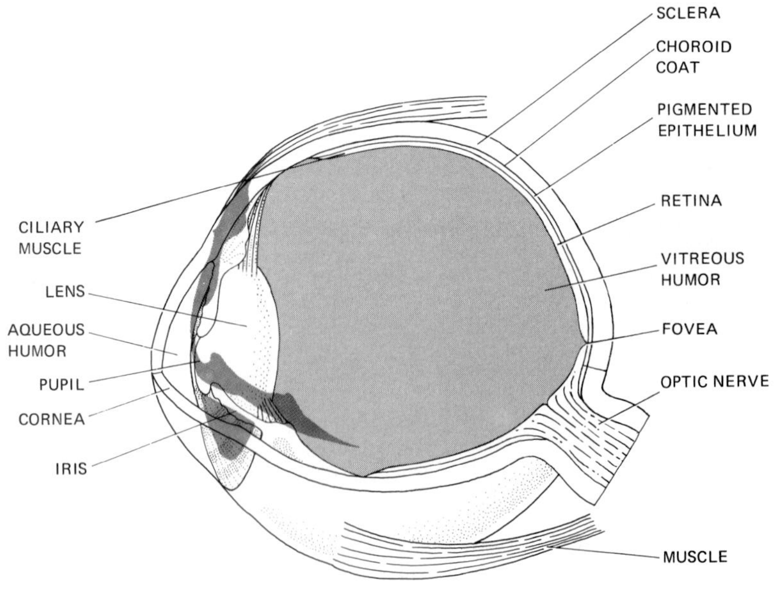

Figure 20.11
Diagram of a section through a human eye. The outer coat, the tough *sclera,* is transparent across the front, where it is called the *cornea.* Back of the cornea is a chamber filled with clear liquid, the *aqueous humor.* The *iris* is a muscular disk, in the center of which is a hole, the *pupil.* The *lens* not only helps form a visual image but can change to accommodate the eyes to near or distant objects. The lens is connected to radiating fibers from the *ciliary body,* the active focusing organ. The inner layer of the eye is the dark-pigmented *choroid coat,* which cuts down on reflection and glare by absorbing stray light. The bulk of the eye is filled with a semisolid *vitreous humor,* which sometimes has harmless particles floating in it. If you look at a neutral, featureless background, such as a uniform gray sky, you can see them drifting about. They are called muscae volitantes (Latin, "flying flies"). The sensitive layer of the eye, the *retina,* has a slight depression, the *fovea,* where the retinal elements are so distributed as to give maximum sharpness of vision. An eye can be moved in its socket by a set of six muscles.

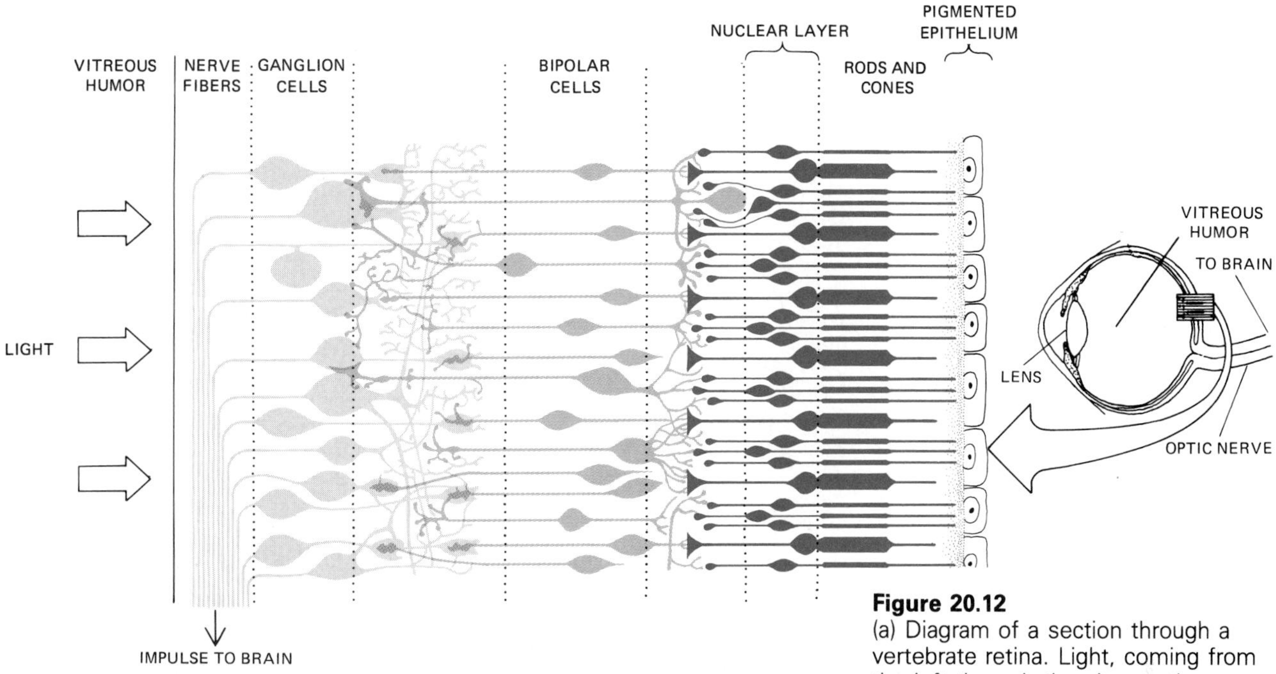

(a)

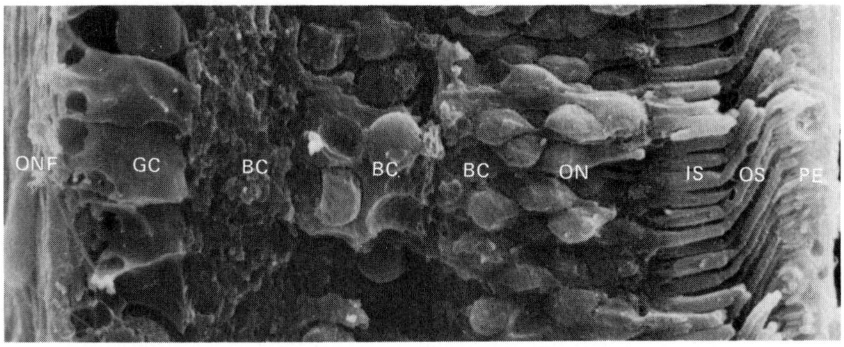

(b)

Figure 20.12
(a) Diagram of a section through a vertebrate retina. Light, coming from the left through the vitreous humor, passes through three layers of cells before reaching the light-sensitive *rods* and *cones*. Beyond the rods and cones there is a pigmented epithelial layer, which prevents reflection from the back of the retina. When light energy stimulates a rod or cone, that energy is transduced into the electrical energy of a nerve impulse. The impulse is sent from the receptor cells through an intermediate set of *bipolar cells* and finally to ganglion cells before passing on to the optic nerve. The function of the eye is to receive light stimuli, change the stimuli into nerve impulses, and send those impulses to the brain; interpreting the impulses is the work of the brain. (b) Scanning electron micrograph of a rabbit retina, showing pigmented epithelium (PE), outer segments (OS) and inner segments of rods (IS), outer nuclear layer (ON), bipolar cells (BC), ganglion cells (GC), and optic nerve fibers (ONF).

body is covered with light-receptor cells that cause it to move away from bright light. In contrast, predator birds like eagles have developed extremely efficient eyes that outweigh their brains. And a frog, which has perfectly good eyesight, will nevertheless respond only to moving objects. A frog surrounded by insects will starve to death if the insects do not move. (The subject of perception in animals is discussed in detail in Chapter 29.)

Insects and other arthropods have **compound eyes,** with dozens or even thousands of separate, tubular *ommatidia* (Figure 20.10). Each of these units has its own lens and light-sensitive receptor cells, and each ommatidium views a single section of the total picture. The compound eye is the most usual animal eye. It occurs in arthropods, and there are more arthropods than any other kind of animal. The compound eye is probably better suited to detecting motion than to perceiving a clear image of a visual field. The greater the number of ommatidia, the more distinct is the image. The eyes of bees contain about 15,000 ommatidia, and some predatory insects have twice that number. (A dragonfly has about 30,000 separate ommatidia.)

Within a human eyeball, about 2.5 centimeters (1 in.) in diameter, are over one million neurons that convert light waves into electrochemical impulses that are decoded by the brain (Figure 20.11). As we have seen, the eyes of vertebrates (including humans) work like a camera, the image being brought to a focus on the sensitive **retina** (Figure 20.12). The

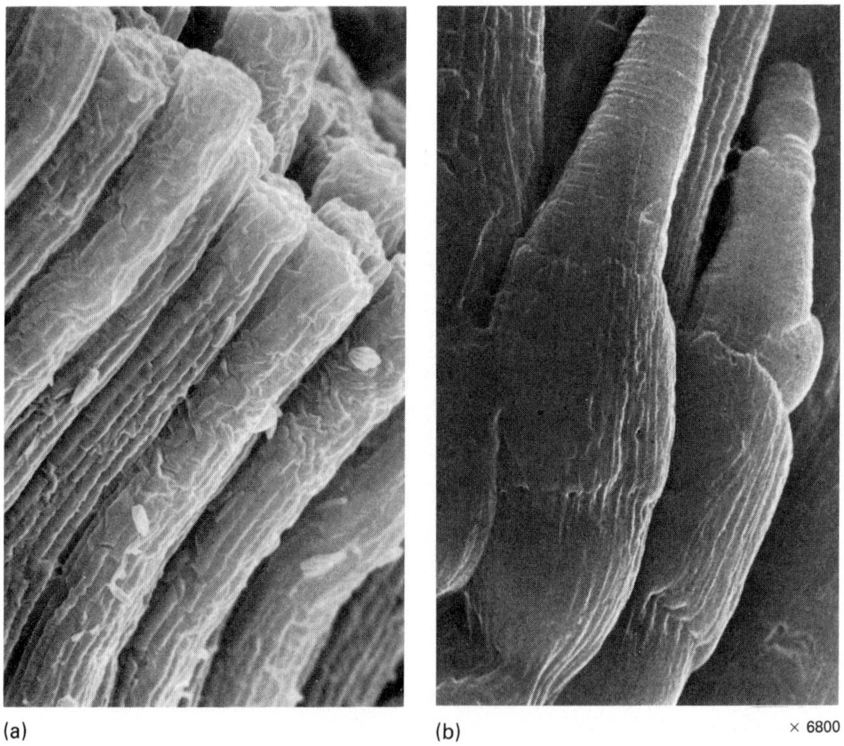

Figure 20.13
Scanning electron micrographs of (a) rods and (b) cones from amphibian retinas. The difference in shape is obvious; the difference in function is that the rods are more effective in dim light but are not sharp image formers, and the cones are used in color vision and in accurate focusing on an image.

retina is composed of layers of slender cells, the underneath layer consisting of photoreceptors, the **rods** and **cones,** and a complex of interacting, processing neurons (Figure 20.13). These sensory neurons send axons to the brain via the optic nerve, a bundle of axons.

There are about 125 million rods and 7 million cones in each human eye. The rod cells, which are not color-sensitive, function mainly in dim light and in peripheral vision. The cone cells are concentrated in the center of the retina directly behind the lens, especially in a small area called the *fovea,* where

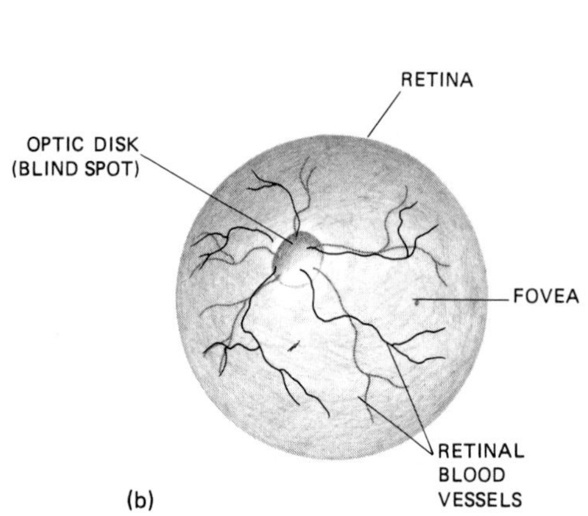

Figure 20.14
Finding your blind spot. (a) Hold the page directly in front of your right eye, keeping your left eye closed. While staring steadily at the target, move the page closer. When the page is about 15 to 20 centimeters (6–8 in.) from your eye, the X should disappear. This happens because the image of the X will at that point fall on the part of the retina known as the optic disk, or blind spot, where the optic nerve and blood vessels connect to the eyeball (b); the optic disk contains no rods or cones and is therefore insensitive to light. The presence of the blind spot is not annoying, because the two eyes are usually working in cooperation, and any object projected on the blind spot of one eye is projected on a sensitive spot of the other, so the object is constantly visible. Also, the eye usually focuses light on the fovea, the area of keenest vision, rather than on the blind spot.

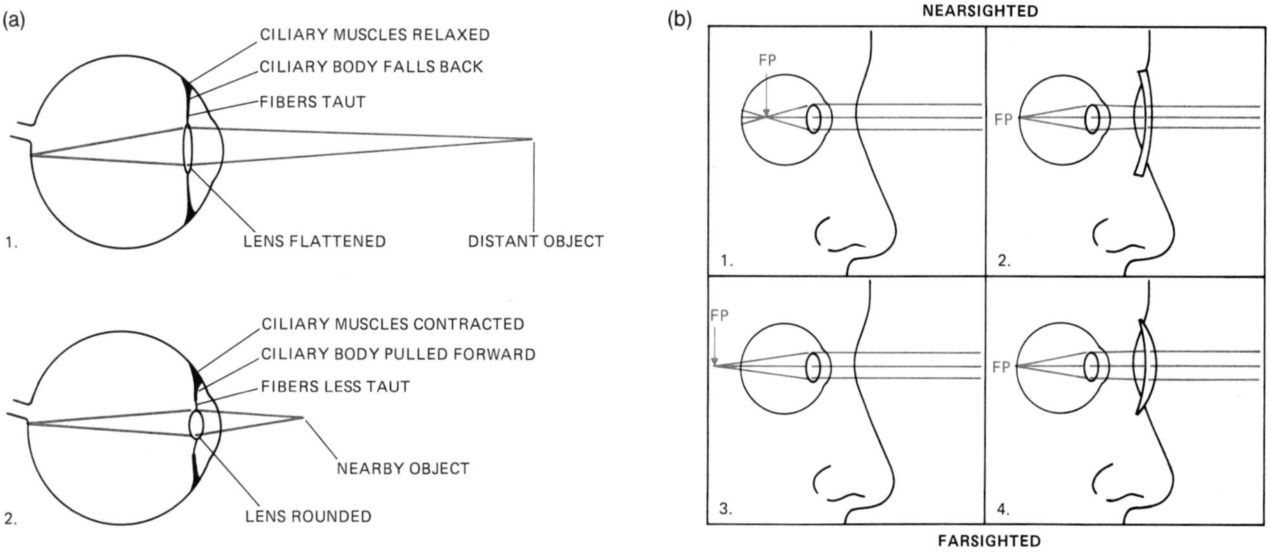

Figure 20.15
Making a sharp image on the retina. (a) How the eye *accommodates,* or focuses on near and distant objects. When the ciliary muscles relax, they fall back, tightening the fibers that suspend the lens. The tightened fibers pull the lens into a flatter shape, making distant objects come into focus on the retina. As shown in (1), a relaxed eye automatically focuses on distant things. Effort is needed, even if it is thoughtless effort, to focus on near objects (2). The ciliary muscles contract, pulling the whole ciliary body forward and reducing the size of the opening. This lessens the tension on the lens suspension fibers and allows the lens to assume a rounder shape, making near objects come into focus on the retina. Looking at a distant scene is restful to the eyes of a person doing long-term close work because the ciliary muscles are relaxed. (b) Correcting vision in eyes that are myopic (nearsighted), or hyperopic or presbyopic (farsighted). In a myopic eye the focal point (FP) is in front of the retina, but placing a reducing lens in front of the eye will make the focal point coincide with the retina, as shown in (1) and (2). If the focal point is behind the retina, placing an enlarging lens in front of the eye makes the correction, as shown in (3) and (4).

image formation is sharpest and color vision is most acute. The fovea contains only cones, and animals with the most cones in the fovea have the sharpest vision. (The human fovea has about 200,000 cones but some birds have up to eight times that number.) Humans can usually see most sharply by looking directly at an object so that the image falls on the fovea, but they can see better in dim light by *not* looking directly at an object, so that the image falls on the periphery of the retina, where the highly sensitive rod cells are located (Figures 20.14 and 20.15).

The biochemistry and biophysics of vision are incompletely known. We know that each rod contains a light-sensitive pigment called *rhodopsin,* which consists of a colorless protein, opsin, and a colored molecule, retinal (or retinene), a derivative of vitamin A. One photon—that is, one unit of light energy—is enough to affect one molecule of rhodopsin, and in humans visual stimulation can be achieved by as few as 5 to 10 molecules of affected rhodopsin. What is called the *primary light effect* occurs when light is absorbed by rhodopsin, which then separates into opsin and retinal. This reaction somehow causes the retina to respond, and information is ultimately relayed back to the brain. Afterward, the retinal is enzymatically recombined with opsin to generate rhodopsin again. For decades before these reactions were discovered, a deficiency in vitamin A was known to cause the reduced sensitivity to light known as night blindness.

Night vision is almost totally rod vision, since the color-sensitive cones require 50 to 100 times more stimulation than rods do. The human eye can adjust to night or day vision, but nocturnal animals, such as bats and owls, have only rods, and daytime

(a)

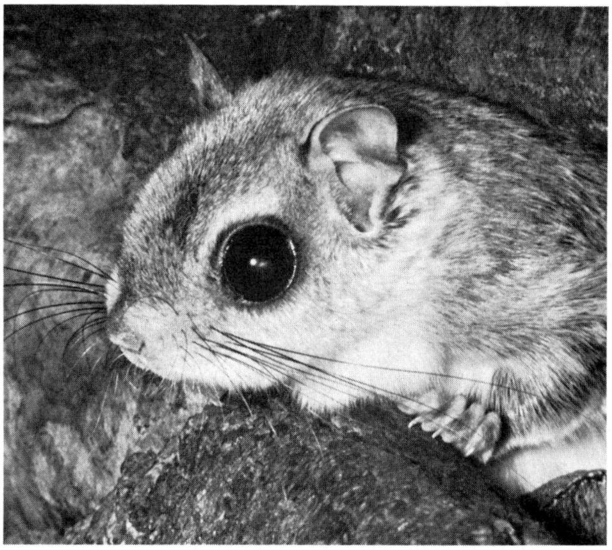

(b)

Figure 20.16
Extremes in the development of mammalian eyes. (a) A mole rat lives mostly underground in the dark and has tiny, almost useless eyes. (b) A flying squirrel is active mostly at night and must make the most of whatever light is available. Its eyes are relatively large, making their light-gathering power greater than that of human eyes. No animal can see in the dark, but some can see better in dim light than others can.

animals, such as some birds, have only cones, and these specialized animals lack the visual adaptability of humans (Figure 20.16).

The mechanism of color vision is not well understood. According to one theory, colors are perceived when red-sensitive, green-sensitive, or violet-sensitive cells (cones) in the retina are stimulated. The final color is determined by the combining of the different levels of excitation of each cell type (each type of cone). With the exception of primates and a few other species, most mammals are thought to be color blind. (Can an animal with cones be color blind, as many seem to be, or do we simply not know how to test animals accurately for color perception?) (See the full-color essay following page 414.)

The Touch Receptors

By "touch" we refer to different physical stimuli (such as heat, pressure, and pain) and different sense receptors. For example, there are microscopic receptors for direct touch in the skin, called *Meissner's corpuscles.* Touch stimuli are also received by hairs, which affect the skin when they are bent. (A bug crawling along a hairy arm will be felt even if its feet never touch the skin.) Firm pressure affects other, large receptors called *Pacinian corpuscles.* The skin receptors for heat and cold are not known, but were previously thought to be the Ruffini endings and the end bulbs of Krause, respectively.

Although the nerve impulse is identical for each type of receptor, the brain interprets the sensation according to affected nerve endings in the specific receptor. If a sufficiently intense stimulus is applied, more than one receptor may be stimulated, and the brain may be confused. When that happens, the person may be stimulated by intense cold but interpret it wrongly as heat. For example, one may get a sensation of cold by putting a hand in very hot water. Deep pressure is less likely than a creasing of the skin to stimulate pain-sensitive nerve endings; a tiny pinch hurts more than a big one. Nerve endings in muscles and joints give information on the position of various body parts. Without looking, a pianist knows how far to reach to find an exact note, and a tennis player watches the moving ball, not his own arms. There are more than half a million touch receptors on the surface of the human body, and most of them are on the lips, tongue, and fingertips.

That indefinable phenomenon known as pain is initiated by the action of free nerve endings that are present in most parts of the body, although intestines and brain tissue have no pain receptors. We do not know whether pain receptors respond directly to injured tissue or indirectly to some substance released by damaged cells. The perception of pain is to some extent a matter of attention. Who has not discovered an already dry cut without having any knowledge of when the wounding happened? There are about three million pain receptors distributed over the surface of the body. Such a large number is probably required to offer maximum warning and protection against harmful injury. Recently discovered compounds in the human brain called **endorphins** block the sensation of pain; they mimic the action of opiates.

Taste and Smell

Taste and smell, which are both chemically activated senses, are similar in their action. In fact, much of what we normally call taste, or *gustation,* is actually a function of our sense of smell, or *olfaction.* In order for a substance to be smelled, it must first be dissolved in the surface liquid on the membrane of the olfactory area. Thus taste and smell complement one another. A person whose nasal passages are blocked by a cold cannot "taste" food. But despite some similarities, taste and smell are separate sensations, and they will be so treated here.

The specialized receptor cells for taste in humans are located in the **taste buds** on the top and underside of the tongue, and to a lesser degree on the larynx and pharynx (Figure 20.17). Taste buds are stimulated by four basic taste sensations—sweet, sour, salty, and bitter. Although the areas of response to these tastes are located on specific parts of the tongue (Figure 20.18), there is no conclusive evidence that there are different types of taste buds for each taste sensation. In fact, the tastes we perceive are probably the result of a combination of different intensities of more than one basic sensation. Taste buds have a life of only about 10 days, and they are replaced with decreasing frequency as we get older, usually hampering our sense of taste somewhat.

Odors are more difficult to classify than tastes, but human sensitivity to odors is greater than it is to tastes. There are millions of receptor cells in the nasal passages. Even so, human olfactory sensitivity is less than that of most animals and tremendously less than that of insects. (See Chapter 29 for a more comprehensive discussion of chemoreceptors, especially smell, in animals.)

Human smell receptors are associated with the mucus-secreting membrane lining the upper nasal cavity. Smell receptors in humans are slender cells with thin filaments attached. The basal ends of these filaments are connected to the cranial nerve that leads to the olfactory lobes in the brain.

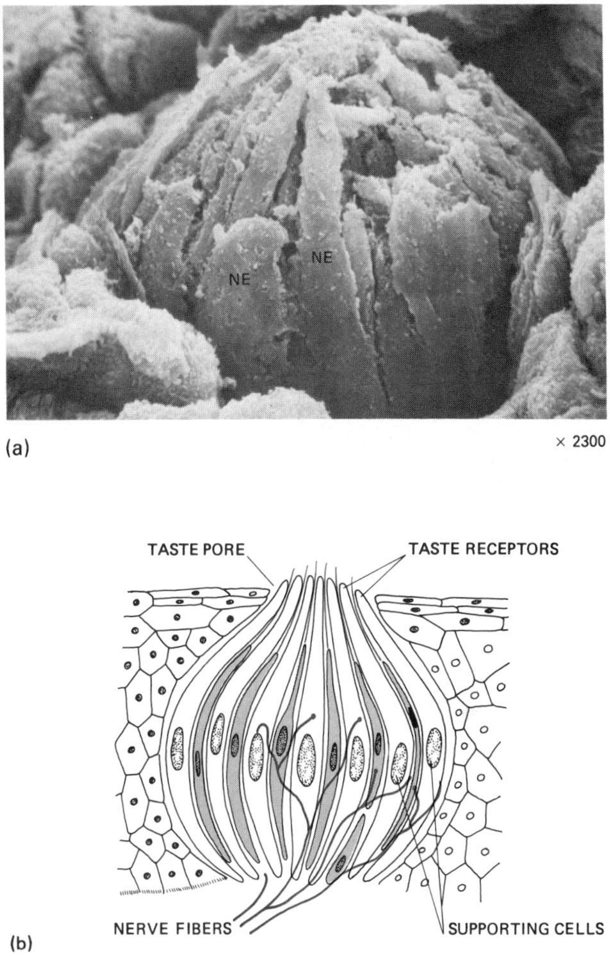

Figure 20.17
Taste buds on a tongue. (a) Scanning electron micrograph of a taste bud. (Compare with the drawing in b.) (b) Diagrammatic section through a taste bud, showing the arrangement of the nerve endings.

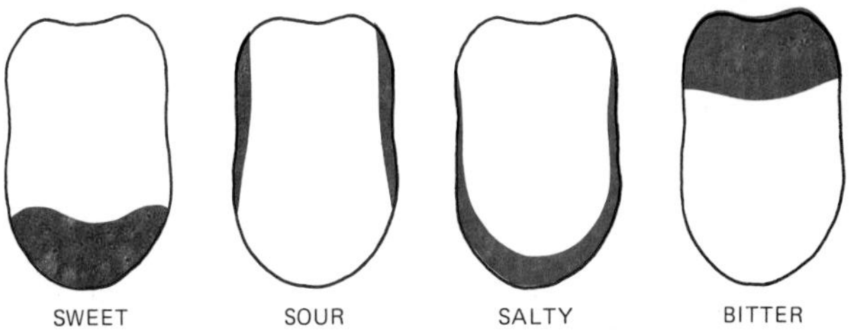

Figure 20.18
Distribution of various kinds of taste buds on a tongue. The number of "flavors" that can be detected strictly by taste is limited, but these, in combination with thousands of perceptible odors, give the impression of many tastes.

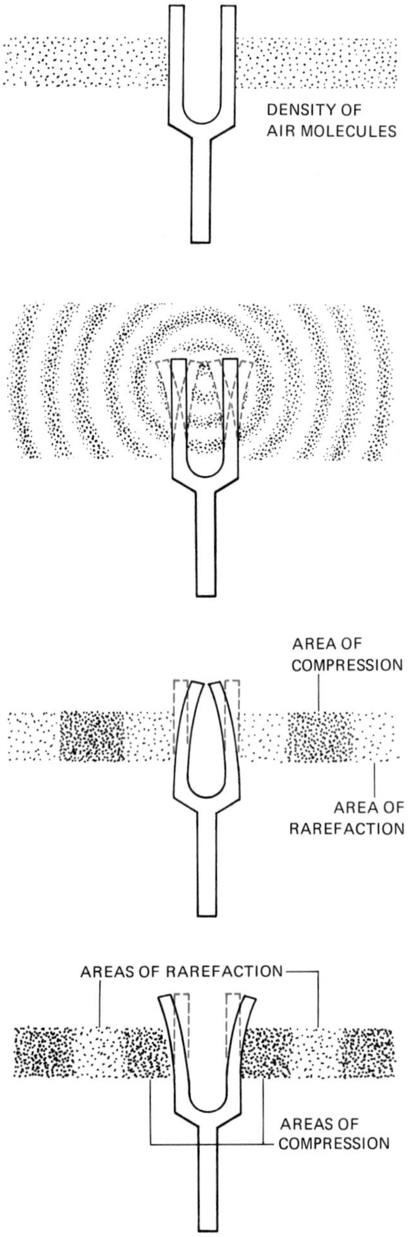

Figure 20.19
Sound generation by a tuning fork. When the prongs of a fork vibrate, they rock back and forth, sending into the surrounding air alternating compressions and rarefactions that spread out from the source like ripples in a pool. When a compression reaches an animal's eardrum, it pushes against the eardrum. When a rarefaction follows, it allows the eardrum to spring back. A tuning fork gives out a regular succession of waves producing a "pure tone" of a set frequency (vibrations per second), but most tones are mixtures of frequencies, whether they are generated by a vibrating string, a drumhead, explosions in a gasoline motor, animal vocal cords, resonant air columns in organ pipes, or the splashing of a stream. Loudness, or intensity, is a matter of the energy put out by the sound generator.

Of all the senses, the least understood is smell. One theory suggests that many different odors can be detected because of the many possible interactions between the molecules of the odor substances and the several kinds of receptor cells. One way to explain such odor variations: If odor molecules have specific chemical shapes that can fit into receptor sites in many combinations, the better the fit, the greater the firing rate of the afferent nerve fiber will be. Odor discrimination then results from the simultaneous but varying stimulation of receptor cells.

Hearing and Balance

Two such seemingly unrelated stimuli as sound and positional change are considered together because they are received in the same organ: the inner ear. Physically, sound is the alternating compression and rarefaction (decompression) of the medium (usually air) through which the sound is passing. In air that is disturbed by sound, waves of compression, in which air molecules are pushed together, are followed by waves of rarefaction, in which the air molecules are farther apart (Figure 20.19). Without a carrying medium such as air, there can be no sound. The waves in the air affect the receptor cells in the ear. The physical changes in the receptor cells are somehow changed to electrical impulses in the nervous system. The impulses are transmitted to the brain, where they are interpreted as sound. Psychologically, sound is the interpretation of nerve impulses from the auditory neurons.

Human ears are generally responsive to frequencies from about 50 cycles per second to about 20,000, but some people can hear from 16 to 30,000, especially during their early years. Hearing ability declines steadily from early childhood, because tissues in the inner ear lose their elasticity. In addition, vibrating hair cells begin to degenerate, and harmful calcium deposits may form (Figures 20.20, 20.21).

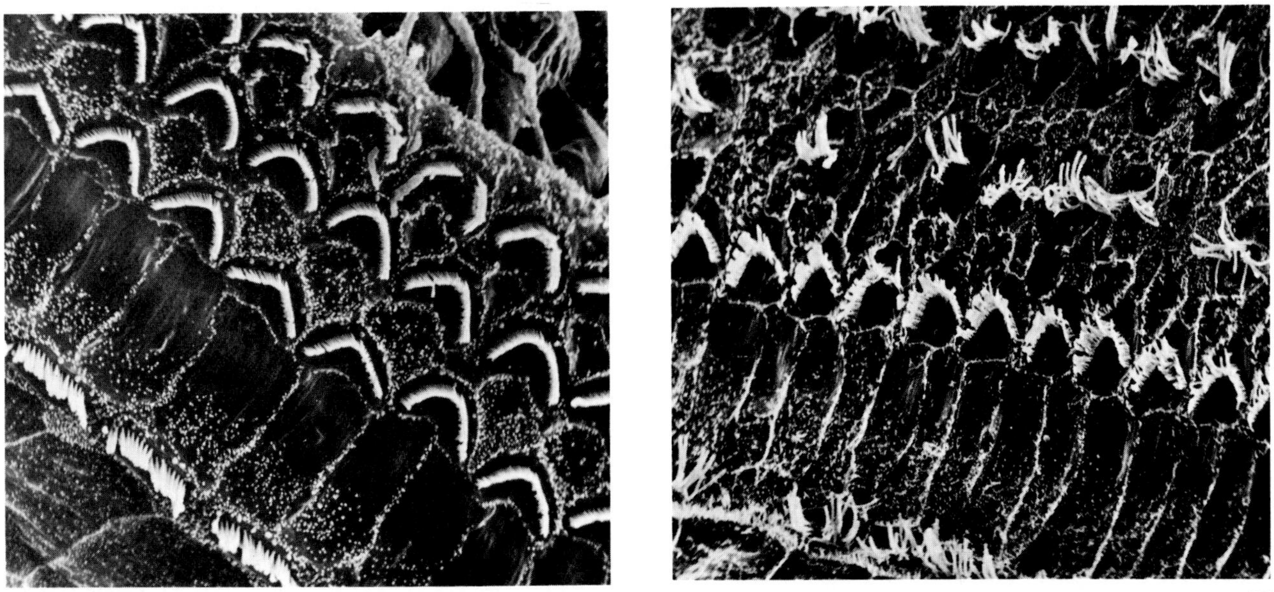

Figure 20.20
Injury to the inner ear by intense noise. (a) A scanning electron micrograph of a normal guinea pig organ of Corti, showing a regular pattern of hair cells, and (b) an injured organ of Corti after exposure to 24 hours of loud rock music (about 120 decibels). The injured ear shows most of the hair cells irreversibly shriveled and destroyed. Once hair cells in the organ of Corti are destroyed, they can no longer transduce sound, and continued exposure to dangerously loud noises will inevitably produce a permanent hearing loss. Current estimates are that noise is twice as bad now as it was in the 1960s, and it is expected to continue to double every 10 years. Studies have shown that factory workers have twice as much hearing loss as white-collar office workers. Approximately 10 million Americans use hearing aids, and many of these individuals may have suffered hearing impairment through prolonged exposure to sounds that may not immediately be thought of as excessively loud.

Very few naturally vibrating bodies produce simple vibrations or a "pure" tone such as that produced by a vibrating tuning fork. Natural objects produce instead combinations of frequencies. When the combinations are subjectively interpreted by our nervous system, they give various sounds their "quality," or timbre.

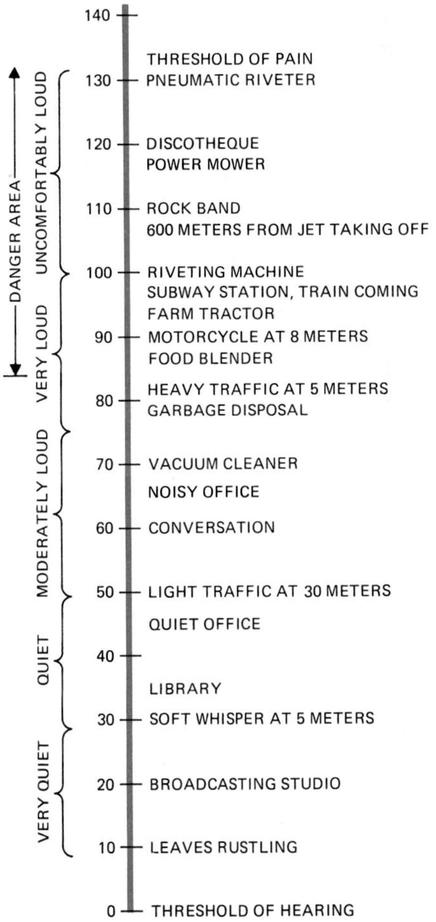

Figure 20.21
Decibel scale and common noise sources.

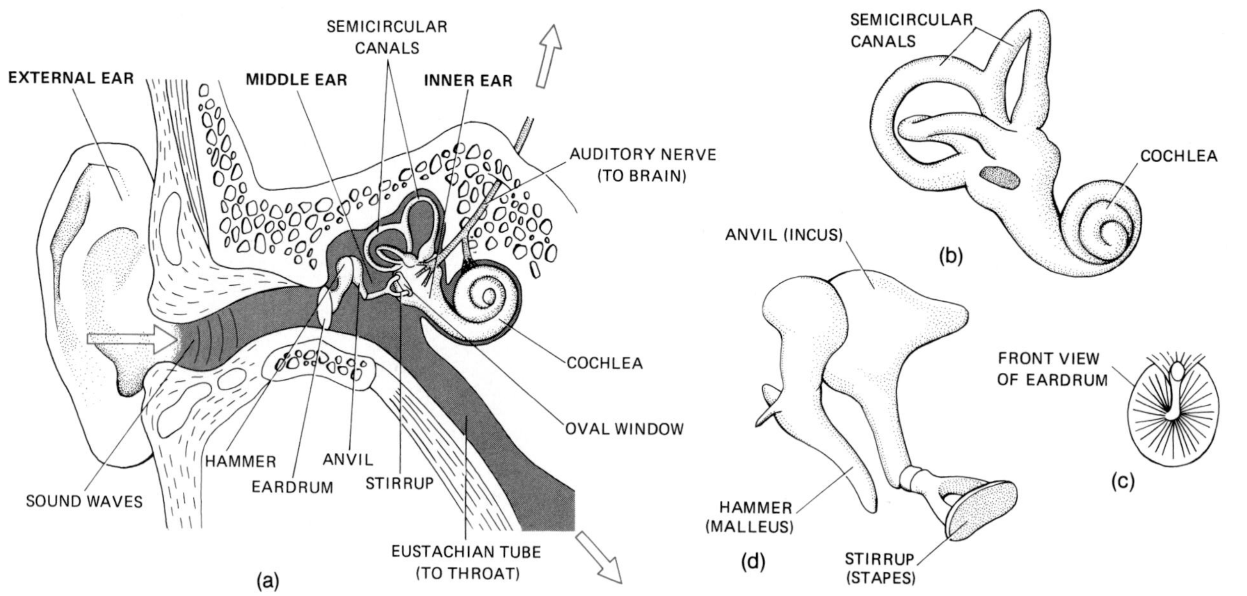

Figure 20.22
(a) Anatomy of the human external, middle, and inner ear. Sounds transmitted through air enter the auditory canal and set the eardrum to vibrating. The vibration is then transmitted to the auditory bones, the hammer, the anvil, and the stirrup. The bones, acting as levers, effectively amplify the vibrations, so that the fluid in the cochlea becomes agitated, causing selected hairs to be moved. If the hairs are bent, nerves are stimulated, and the message is sent to the brain for interpretation. Not all sound is heard through air. One's own voice is heard largely by bone conduction, and that is why most people are astonished when they first hear themselves speaking from a recording. (b) Enlarged view of the semicircular canals, the organ of balance, and the cochlea, which contains the nerve endings sensitive to sound. (c) The eardrum as seen from the inside of the ear. (d) The auditory bones, which transmit vibrations from the eardrum to the cochlea.

There are four basic parts of the auditory receptor (Figure 20.22): (1) A thin membrane, the tympanum, or eardrum, vibrates in response to air vibrations in the same way that a drum vibrates in response to the beat of a drumstick. (2) Three tiny bones, the hammer, the anvil, and the stirrup, transmit vibrations from the eardrum to (3) the oval window, a membrane-covered opening in the bone. (4) The inner ear, or labyrinth, contains a coiled tube called the cochlea (Latin, "snail shell"). These parts of the ear work together in the following way:

1. External sound causes the eardrum to vibrate.
2. The eardrum moves the hammer.
3. The hammer moves the anvil.
4. The anvil moves the stirrup.
5. The stirrup produces vibration in the oval window.
6. Vibrations in the oval window convert sound waves in the air into the vibrations within the fluid of the inner ear.

In the cochlea is the *organ of Corti,* which consists of membranes and some 30,000 sensory cells. Vibrations in the liquid in the labyrinth displace the receptor cells, and the displacement causes the cells to discharge nerve impulses. Sounds of different frequencies stimulate different cells, and the brain interprets the stimuli in terms of perceived pitch. Differences in loudness are perceived as a result of differences in the frequency of impulses along a neuron, with more frequent impulses (as many as 1000 per second) indicating louder sound.

It is not totally accurate to refer to the ear as the organ of balance, since the sense of balance is only partly seated there. Our sense of equilibrium is aided substantially by other senses, especially vision and muscle receptors, which constantly help us to locate our positions. (Try standing on your toes with your eyes closed. Without your eyes to guide your body, you will invariably begin to fall forward. Now stand on your toes with your eyes open and note the difference.) Nevertheless, the inner ear does contain specific parts that help the body to cope with changes in position and acceleration (Figure 20.23). The main receptors for equilibrium are

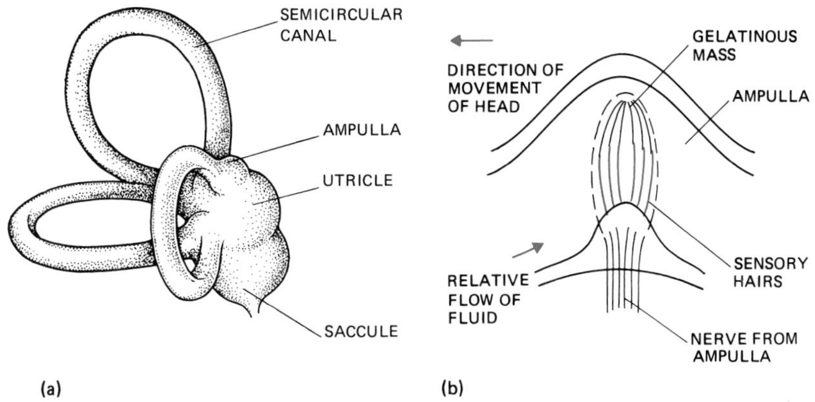

Figure 20.23
(a) The organ of equilibrium: the *semicircular canals* of the inner ear. They properly should be called the "circular canals" because they allow the internal fluid to move around completely. One is vertical and oriented front-to-back, one is vertical and oriented side-to-side, and one is horizontal. Any movement of the head can cause relative movement of the canal fluid over hairs in the bulbous swellings (*ampullae*). When the hairs are bent by the flow of the fluid over them, nerve impulses are set up and transmitted to the brain. (b) The passage of fluid over the hairs in an ampulla.

the utricle, the saccule, and the three fluid-filled semicircular canals. The two basal reservoirs of the canals, the *utricle* and the *saccule,* are filled with a liquid called endolymph. They also contain hair cells that are bent by the weight of attached crystals as the position of the head changes. The bending of the hair cells stimulates the vestibular nerve, which ultimately relays the message to the cerebellum, which in turn directs the necessary muscular adjustment. Receptors in the utricles and saccules regulate equilibrium in responses to changes in the *position* of the head (posture), whereas responses to *movements* of the head are initiated in the ampullae, swellings that join the ear canals with the utricle.

The ampullae are also filled with endolymph and contain cilia. The plane of each of the **three semicircular ear canals** in each ear is at right angles to the other two, so that at least one canal will be affected by every head movement. When the head moves, the fluid in the canals lags because of inertia. The relative movement of the endolymph stimulates the hairs in the ampullae, and proper nerve impulses are initiated to compensate for the head and eye movement. You may become dizzy or motion-sick when your head and eyes are rotated in one direction and the lagging endolymph rotates in the same direction but slower, and there is no opportunity for self-regulating countermovements.

CONSCIOUSNESS AND BEHAVIOR

No other animal is so much a creation of its brain as is the human animal. In fact, one of the greatest differences between humans and other animals is the exquisite complexity of the human brain—no other animal can think, learn, or communicate as effectively as we can. Of course, other animals *can* learn and remember, sleep and dream, and be mentally affected by drugs and other stimuli, but in the following sections we will emphasize these functions as they relate to the *human* brain, remembering that when we speak of the nervous system, and particularly the brain, we are stressing differences of *degree.* The human brain is the most complex of all.

Learning and Memory

Little is known about the neuronal processes of learning and memory. (See Chapter 29 for a discussion of learning patterns in animals.) Specific areas of the brain have been located for different types of learning, however, and it now seems certain, contrary to what was once thought, that no single area of the brain controls learning. In fact, long-term studies of children with learning disabilities due to brain damage have shown that alternate areas of the brain can be trained to "relearn" what had once been learned by the damaged portion. Apparently the brain can make such learning adjustments as long as there is a memory trace of what had been learned in the first place.

One theory of learning suggests that when one nerve circuit is used over and over again, synaptic connections are improved, and "preferred pathways" are created. Learning may also be a matter of changing preexisting pathways, thus producing new pathways of behavior. (Learning is often defined as a change in behavior based on experience.) Other theories of learning suggest that frequently used neurons grow in size and become more efficient, and that when the activity of neurons is increased there is also an increase in the number of glial cells surrounding the neurons. (The number of neurons does not increase after early childhood, but glial cells continue to reproduce throughout a person's lifetime.) While none of these theories can be proved, none can be dismissed either. More and more scientists believe that learning involves a variety of mechanisms, not just one.

In a sense, memory is related to learning because in order to remember something, we have to "learn" it first. Apparently, "memory traces" are formed during learning and are imprinted on the brain when neurons record and store information. Short-term memory helps us remember recent events that have no permanent importance (the score of yesterday's football game or a telephone number that can be forgotten as soon as the call is made), and long-term memory functions to help us remember information that is somehow important enough to last a lifetime. Scientists are not certain whether short-term memory and long-term memory use the same basic mechanism, but it has become clear that memory is related more to electrochemical activity in the brain than electrical activity alone, as was once thought. Recent experiments with rats have shown that the flow of the brain hormone norepinephrine accompanies the process of laying down a memory trace. If the hormone is blocked during the process, there is no memory of recently learned events. Also, it appears that memory is achieved only when chemical changes in the brain are given more than a few seconds to occur. Some researchers feel that such chemical changes represent the formation of "neuronal loops," closed circuits of neurons and synapses that register short-term memory.

Synapses may be related to memory, but no definite evidence is available. It *has* been shown that the hippocampal area of the brain is crucial for recent memory storage, and if the region called the hippocampus is destroyed, there is no retention of recent events. Long-term memory can be wiped out if the medial portion of the temporal lobes of each cortex are removed, and mild electrical stimulation of the temporal cortex can call back memories that were long forgotten. The temporal lobes probably function as a directory that permits the retrieval of stored memories, but it is not known exactly where memory is stored in the brain. In fact, rather than being stored in any one place, memory information is probably contained in large groups of brain cells that work together somehow to store and retrieve memories.

It has been shown that memory traces can be erased by chemicals that are known to inhibit protein synthesis, and some researchers have speculated that RNA, which codes for specific proteins, is also involved in coding for memory. The RNA theory arose when it was observed that RNA increases in proportion to brain activity, and decreases when brain activity slows down. Unfortunately, upsurges of RNA also occur during most normal activity of neurons, and other organs and glands besides the brain (such as the kidneys, thyroid, and pancreas) also show increases in RNA content whenever they are stimulated by their appropriate hormones.

Obviously, additional research is needed before we can fully explain the mechanism of memory. In the meantime, we can recall the explanation of J. M. Barrie, the Scottish writer of *Peter Pan,* who said that "God gave us memory so we could have roses in December."

Sleep and Dreams

The brain is always active; in fact, the human brain is more active in some areas when the body is asleep than when it is awake, and the brain requires a greater blood supply during sleep. Whether a person is asleep or awake, brain activity may be recorded on an instrument called an electroencephalograph. Both the machine and the record it produces, the electroencephalogram, are referred to by the initials **EEG.** The EEG shows the changing levels of consciousness in the brain, and it is frequently used to detect brain damage by locating areas of altered wave patterns. The four distinct wave patterns in the brain are known as alpha, beta, delta, and theta waves. Alpha waves, evident in relaxed adults whose eyes are closed, usually indicate a state of well-being. Beta waves are typical of an alert, stimulated brain. Delta waves are seen during deep sleep, in damaged brains, and in infants. Theta waves are usually found in children and in adults who are under stress.

It is generally agreed that sleep is one of the more important activities of human beings (lack of sleep will cause death faster than lack of food), but the exact benefits of sleep are not known, and the causes of sleep are equally unknown. It is particularly ironic that so little is known about sleep, because if you are 20 years old, you have already spent about eight years of your life asleep. By the time you are 60, you will have slept about 20 years.

The EEG has made possible the detection of at least four separate stages of sleep, which are described below.

1. During *Stage 1 Sleep* the rhythm of alpha waves slows down, and the individual experiences a floating sensation. This stage is usually not classified as true sleep, and individuals awakened from Stage 1 are quick to agree, insisting that they were merely "resting their eyes."

2. *Stage 2 Sleep* is indicated by the appearance on the EEG of short bursts of waves known as "sleep spindles." Sleep is not deep.

3. Delta waves appear during *Stage 3 Sleep.* Intermediate sleep is characterized by steady breathing, slow pulse rate (about 60), and decline in temperature and blood pressure.

4. *Stage 4 Sleep* is the deepest stage, also known as oblivious sleep. It usually begins about an hour after falling asleep. Although the EEG indicates that the brain acknowledges outside noises and other external stimuli, the sleeper is not awakened by such disturbances.

Sleep proceeds in cycles 80 to 120 minutes long. An important condition known as **REM** (rapid eye movement) **sleep** takes place during reentry into a new Stage 1 phase. It is during the REM period that dreams occur, and the brain is then almost completely separated from the outside environment. Breathing and heart rate may increase or decrease, testosterone secretion increases and penile erections occur, and twitching body movements and perspiration are common. (Such physiological changes during REM sleep do not affect males only. The clitoris also becomes erect during the REM period, and estrogen secretion probably increases as well.)

The REM period is thought to be essential for a general relaxation of normal buildups of stress and tension, and it is probably necessary for brain maturation. Dreams may serve to release tensions or fears that build up during the day. This theory is given credence by studies that show that seven out of eight adult dreams are somewhat unpleasant, and about 40 percent of the dreams of children are actually nightmares. Another theory suggests that dreams are an attempt by the brain to bring coherence to the accelerated firings of neurons in the brainstem during REM sleep. According to this theory, dreams occur when the arbitrary firings of the neurons are translated into images as close to our actual experiences as possible. The brain is attempting to take random impulses and direct them into themes with a "plot." The more bizarre dreams may be caused when the brain is unable to connect the neuron activity with any logical experiences or memories.

Babies, whose central nervous system is not fully developed, spend more than half of their sleeping hours in the REM state, but adults spend only about 15 percent of their total sleep in the REM condition. Growth hormone is secreted by children during deep sleep, so apparently sleep is essential for the normal growth of young children. It has been found that mentally retarded patients have less REM sleep than normal people, and in fact, the more severe the retardation, the shorter is the period of REM sleep.

Sleep is a highly individual process. No two people seem to require the same amounts of sleep (generally, we need less as we get older), and no two cycles proceed according to the same pattern. But even though some people do not remember their dreams, everyone *does* dream every time a full cycle progresses into the REM stage. We almost always awaken directly from REM sleep, and almost all sleepers awakened during their REM period will recall having been in a dream state.

Most people will experience extreme psychological discomfort after a few days of sleep (and dream) deprivation, although occasional rare individuals can live for years with only two or three hours of sleep per day. Without sufficient sleep, the ATP reserve in the body declines precariously, adrenal stress hormones are secreted into the blood steadily, and mental and physical levels of performance falter markedly. Large amounts of a chemical similar to LSD appear in the blood and may be the cause of hallucinations (such as those that happen to sleepy drivers) and psychotic behavior. Sleep is the only relief.

Drugs and the Nervous System

Throughout this chapter and the previous one we have seen examples of how important chemical reactions are for the proper functioning of the nervous system. Chemicals may also *interfere* with the nor-

(a) (b)

Figure 20.24
The effect of a drug (caffeine) on the web-building skill of a spider, *Araneus diadematus.* (a) An untreated spider can build a typical orb-shaped web with regular spokes and neat connecting lines. (b) A spider that has been fed sugar water containing caffeine tries to build a web but fails to keep the standard plan and makes a mess.

mal workings of nerve impulses (Figure 20.24). In the last chapter we saw how a synapse might be "short-circuited" by the action of chemical substances. Neuromuscular functions may be altered at the synaptic level by different chemicals, several of which are described here.

Curare Curare, a potent drug still used in parts of South America to poison arrowheads, functions by binding to the nerve receptor site normally occupied by acetylcholine. But unlike acetylcholine, curare does not allow a change in the permeability of the cell membrane, and it is not destroyed by acetylcholinesterase. Consequently, curare blocks the normal action of acetylcholine (which is still released) at the junction of nerve and muscle. The muscles, including the breathing muscles, are unable to contract, and death from asphyxiation follows. (Poisonous mushrooms have a similar effect on neuromuscular transmissions.) A minute dose of curare is sometimes used as a muscle relaxant in controlled surgical situations.

Pesticides and "nerve gases" Some organic phosphate pesticides (such as malathion) and "nerve gases" inhibit the action of acetylcholinesterase. When that enzyme is inhibited, acetylcholine is not destroyed, and muscle cells remain in a state of depolarization. As with curare, no new nerve impulses can be transmitted properly, and the muscles are paralyzed. Asphyxiation results.

Botulinus toxin Botulinus toxin, produced by the bacterium *Clostridium botulinum,* is responsible for a type of food poisoning known as botulism. Botulinus toxin blocks the release of acetylcholine from presynaptic junctions. Naturally, if acetylcholine is not present, the muscle cells cannot be stimulated by nerve impulses, and the typical result is death by asphyxiation. Botulism may be on the rise with an increase in home canning in this country. Unfortunately, botulinus toxin is an extremely powerful poison, and it has been said that about one-half pound of it would be enough to kill the human population of the entire world.

Many other alterations of the nervous system mechanism may be caused by drugs. Some of these drugs, and the effect they have on the nervous system, are outlined in Table 20.2.

SUMMARY

1. The *central nervous system* consists of the spinal cord and the brain.
2. The three main parts of the brain are the brainstem, cerebellum and cerebrum. The *brainstem,* the stalk of the brain, relays messages between the spinal cord and the brain. The *cerebellum* is the coordinating center for muscular movement. The *cerebrum* is popularly considered the region where thinking is done, and its control functions are numerous.
3. The *peripheral nervous system* enables the brain and spinal cord to communicate with the entire body. It is further divided into the *somatic* nervous system and the *autonomic* nervous system. The so-

TABLE 20.2
The Effects of Drugs on the Nervous System

DRUG	EFFECTS	MECHANISM OF ACTION
Stimulants		
Caffeine (coffee, tea, cola drinks)	Mild reaction: acceleration of heart rate, dilation of pupils, increase in blood sugar.	Stimulates sympathetic nervous system, facilitates synaptic transmission.
Nicotine (tobacco)	Medium reaction: acceleration of heart rate, dilation of pupils, increase in blood sugar.	Stimulates sympathetic nervous system, facilitates synaptic transmission.
Amphetamines (Dexedrine, Methedrine, or "speed," Benzedrine)	Powerful reaction: acceleration of heart rate, dilation of pupils, increase in blood sugar, reduced sense of fatigue. Depressant effect on appetite.	Stimulate sympathetic nervous system.
Cocaine	Temporary sense of well-being and alertness.	Stimulates central nervous system, inhibits uptake of norepinephrine.
Depressants		
Ethyl alcohol	Small amounts have a stimulant, then a sedative effect as the lower brain centers are depressed. Loss of dexterity, insensitivity to touch. Distorted vision, interference with hearing, difficulty in speaking. Loss of coordination and balance. Unconsciousness, coma.	Reduction of neuron function in the brain. Motor and sensory regions of cortex inhibited. Depression of visual, auditory, speech centers of cortex. Cerebellum inhibited. Depression of reticular formation.
Barbiturates (Seconal, Nembutal, Amytal)	Promote sleep; high doses induce respiratory failure. Barbiturates and alcohol combine to produce extreme central nervous system depression and may cause death.	Depression of reticular formation. Depression of medulla in large doses.
Tranquilizers (meprobamate: Miltown, Equanil; chlorpromazine: Thorazine; chlordiazepoxide: Librium; diazepam: Valium)	Reduce anxiety and tensions.	Block receptors of adrenalin and acetylcholine, depress reticular-formation activity.
Opiates (opium, morphine, heroin, codeine, methadone)	Reduce pain, induce muscle relaxation, lethargy. Highly addictive.	Depress thalamus.
Anesthetics (ether, chloroform, benzene, toluene, carbon tetrachloride)	Induce unconsciousness. Volatile hydrocarbons produce effects similar to alcohol intoxication.	Depress central nervous system. Block transmission of electrical impulse that triggers contraction of ventricles.
Hallucinogens		
Mescaline, psilocybin, LSD, dimethoxy-methylamphetamine	Distort visual and auditory perceptions. Enhancement of emotional responses. Hallucinations.	Mimic molecular structure of serotonin, a transmitter substance in parts of the brain. Duplicate effects of nervous system activity.
Marijuana, hashish	Produce a sense of well-being.	Unknown.

Adapted from John W. Kimball, *Biology,* Fourth Edition (Reading, Mass.: Addison-Wesley, 1978), pp. 498–502.

matic system involves "voluntary" movements of the skeletal muscles, and the autonomic system controls such "involuntary" activities as breathing and glandular secretion.

4. The autonomic nervous system is composed of the *parasympathetic* system and the *sympathetic* system. The first is mainly inhibitory, and the second is mainly stimulatory.

5. Structures that are capable of receiving stimuli are *receptors.* All receptors are capable of changing energy into nerve impulses, a process called *transduction.*

6. Practically all animals respond to light in some way, and light receptors vary from the image-forming, color-sensitive *camera-type eyes* of vertebrates to the simple photoreceptors of one-celled animals. Insects and other arthropods have *compound eyes.*

7. The retina is composed of layers of slender cells, one layer consisting of photoreceptors (*rods* and *cones*) and a complex of interacting processing neurons.

8. Direct touch receptors are *Meissner's corpuscles;* firm pressure affects receptors called *Pacinian corpuscles;* the skin receptors for heat and cold are not known.

9. Much of what we call taste, or *gustation,* is actually a function of our sense of smell, or *olfaction.* The specialized receptor cells for taste in humans are located in the *taste buds* on the tongue. The tastes we perceive are probably the result of a combination of different intensities of more than one basic sensation.

10. Of all the senses, the least understood is smell. One theory suggests that many different odors can be detected because of the many possible interactions between the molecules of the odor substances and the several kinds of receptor cells.

11. Both hearing and balance have receptors in the inner ear.

12. Little is known about learning and memory. Both processes probably involve electrochemical mechanisms, and neither process is located in only one part of the brain.

13. Sleep can be divided into four distinct stages and a transition stage, *REM sleep,* when dreams occur. Neither the exact benefits nor the causes of sleep are known.

ASK YOURSELF

1. What are the main parts of the human central nervous system?
2. What body activities are controlled mainly by the cerebrum?
3. What kinds of bodily activity are dependent upon the cerebellum?
4. What part of the brain controls such involuntary actions as breathing and vomiting?
5. What is the function of the thalamus in the human brain?
6. Contrast the actions of the sympathetic and parasympathetic nervous systems.
7. Why are vertebrate eyes called "camera type" eyes?
8. What is the "blind spot" in a human retina?
9. What is the advantage of looking directly at an object if you want to see it clearly?
10. Why is it possible to detect a dimly lit object better by *not* looking directly at it?
11. If you are punched in the eye, even in the dark, why do you "see stars"?
12. Why is Vitamin A important to vision?
13. Why does a "cold in the head" make food less tasty?
14. Using three flexible strips (pipe cleaners, flexible straws, or pieces of soft wire), make a model of the semicircular canals, conforming to the spatial relations of the actual canals in the middle ear.
15. What is the function of the eardrum?
16. In what part of the ear are mechanical movements transduced into nerve impulses?
17. How is REM sleep detected by someone observing a sleeper?

HOW ANIMALS MOVE

Animals move. This may seem too obvious to mention, but it acquires significance when we remember that plants *do not* move about. All animals use the same basic principle to move. Whether they creep, run, swim, jump, or fly, they move forward by pushing backward against the ground, water, or air, or they move backward by pushing forward. In so doing, the animal must change its shape—the moving animal must have moving parts.

HOW BIRDS FLY

Animal movement has always fascinated humans, but the flight of birds has had a special attraction. Some magnificent attempts to fly have been made by humans, but none has equaled the splendor of a bird in flight—or its economy. The 767 and the Concorde can fly, and so can bats and many insects, but birds are unique because they have feathers.

Birds are built to fly. Their thin, hollow bones and powerful breast muscles provide light weight and high power, two essential requirements for any flying machine. The refinements are remarkable. Teeth and heavy jawbones have been replaced by the gizzard, which does the work of teeth grinding food. Significantly, the gizzard is located near the bird's center of gravity, where it places little demand on the flying apparatus. The special structure of a bird extends to the inner organs and the entire mechanism that regulates homeostasis. A bird's heart is large and strong, to cope with the strain of flying, and the blood has a high concentration of energy-producing sugar. The respiratory system of a bird is even more specialized, and in proportion to body weight it is about four times more extensive than a human respiratory system. Besides lungs, birds have air sacs that extend throughout the body. In addition to improving the breathing system, the air sacs form an effective cooling mechanism. But most of all, it is the distinctive feathers of a bird that allow it to soar, glide, race, and hover.

A bird flies by applying power on the downstroke of its wings. As the wings pull down, the edge of each wing feather curls upward to produce a propellerlike action that lifts the bird up and pushes it forward. During the powerful downstroke the feathers overlap and present the maximum thrusting surface against the air. The downstroke ends on a level with the bird's bill. As the upstroke begins, the primary feathers at the tips of the wings twist and separate to permit air to pass through. This action makes it easier for the bird to lift its wings. Some forward propulsion and lift continue to occur as the wings move up and back against the air, but most of the driving force is produced as the downstroke begins at the start of a new cycle.

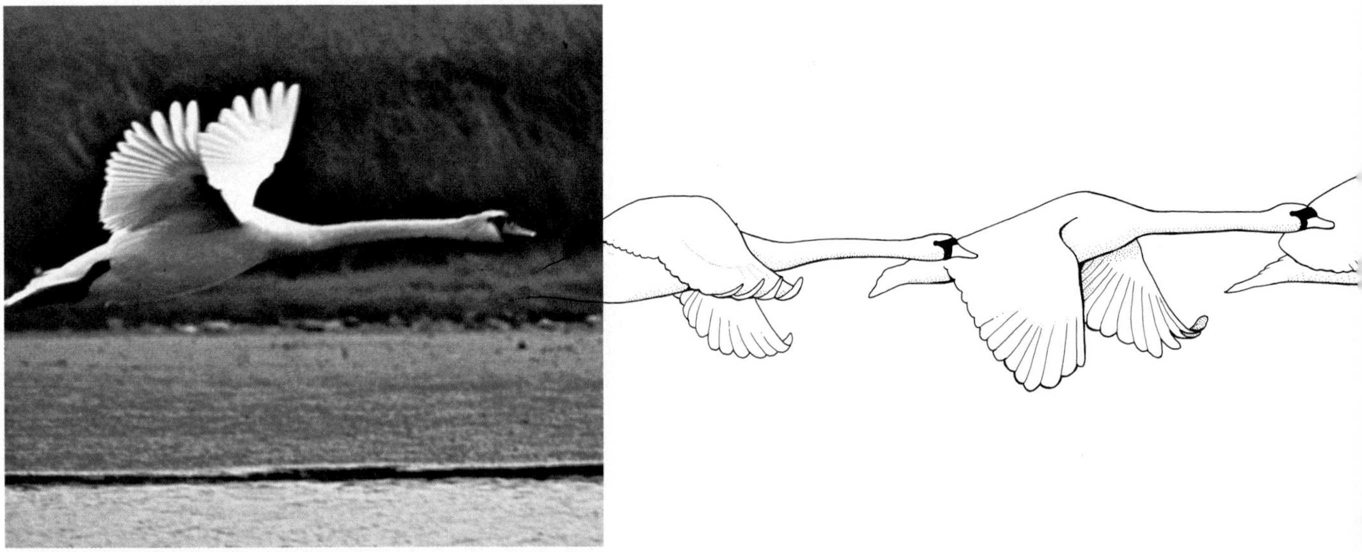

A hummingbird hovers with the aid of a special shoulder joint that permits the wing to swivel at the shoulder (many insects have such swivel joints). The forward stroke looks much like the conventional downstroke of other birds, except that because the hummingbird's body is held almost vertically, the wing stroke is parallel to the ground instead of downward. This provides lift but no forward propulsion. As the wing moves into the backstroke, it swivels almost 180 degrees at the shoulder. The edge of the wing is turned backward in a "mirror image" of the forward stroke, so that once again the hummingbird achieves lift without forward propulsion. By using its swivel action and a wing-beat rate of 50 to 70 beats per second, the hummingbird is able to hover like a helicopter while it sips nectar from a flower.

Adapted from an idea in *The Birds,* a Time-Life Book.

FEATHERS AND FRAMES: FORM FOLLOWS FUNCTION

Not only are feathers beautifully designed for flying; they also provide low weight and insulation against the weather. In fact, their properties of insulation are so effective that birds can usually live in climates that are too cold for other animals. The gross structure of the large contour feathers provides the bird with its overall sleekness. These feathers make streamlined flight possible. But the most impressive engineering is shown in the finer parts of a flight feather. The shaft is strong but lightweight, with a flexible tip so the bird can twist the feather at the end of a downstroke. Extending diagonally from the shaft are the parallel barbs of the feather, each made up of many smaller barbules that overlap and are hooked together by even smaller barbs. The result is an interwoven pattern that produces the strongest of all wings (including airplane wings), while still retaining its lightness. But as strong as feathers are, they are not totally resistant to wear. Most birds have oil glands that help offset the abuse that feathers take. These birds smear the oil on the feathers with their beaks—one form of preening. Other birds preen their feathers with a fine powder derived from the feathers themselves. Feathers are replaced about once a year, usually in late summer, after the nesting season. Some birds also acquire new, brightly colored feathers *before* nesting season, in time for their ritual courtship displays. Typically, most birds molt a little at a time so that their flying is not curtailed, but many water birds shed all their feathers at once, and they cannot fly until the new feathers grow in.

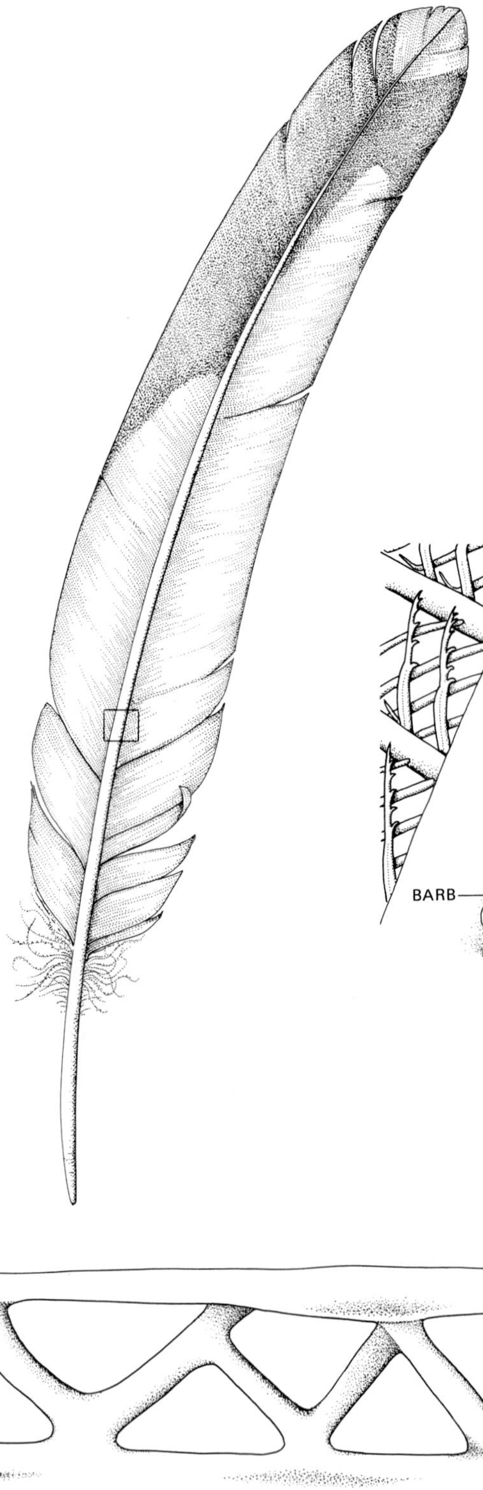

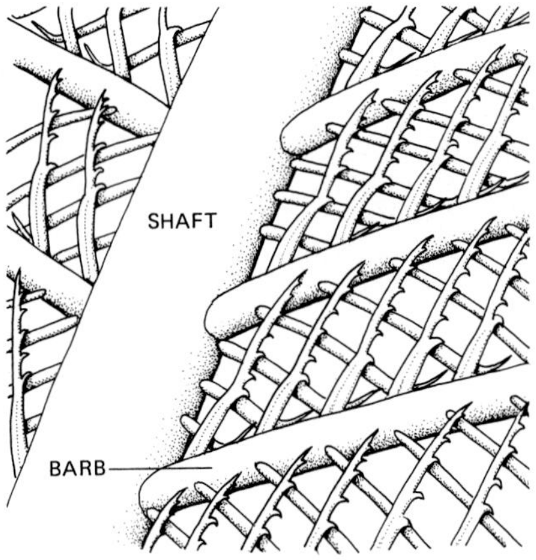

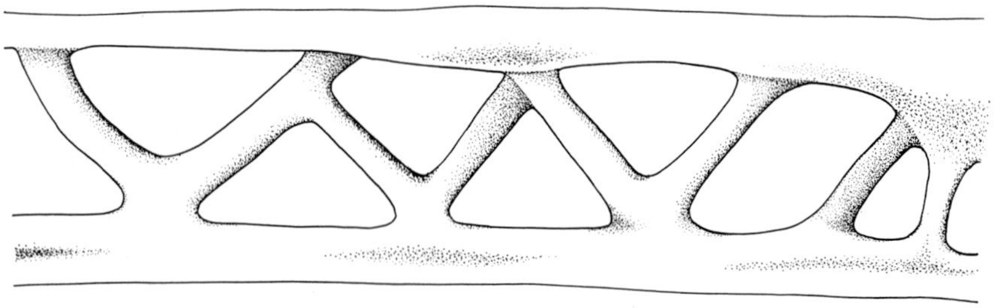

A long section of a bird's wing bone.

Besides being strong and light, some of the hollow bones of birds reveal an internal trusslike structure that makes them even stronger. The pattern of these struts has been duplicated in airplane wings and in steel beams used in the construction of buildings. As strong as such bones are, the skeletons of some birds actually weigh less than their feathers. ×10.

HOW SNAKES SLITHER

Snakes have evolved from animals that lost their limbs when they began to burrow. In the course of adaptive evolution, the entire body of a snake became an organ of locomotion. Instead of legs, snakes use the backbone and muscles to move. They have 100 to 400 vertebrae; most mammals have about 16. Snakes cannot move forward on totally smooth surfaces because there is nothing for them to push against. Also, in order to move forward, snakes must have room to change the curvature of their bodies, allowing their muscles to shorten on one side of the body while antagonistic muscles lengthen on the opposite side. For this reason, most snakes can move easily through a wavy glass tube, but they cannot move through a circular tube or a narrow straight one. There are exceptions and variations, as shown in the drawings on the next page, which illustrate the four basic kinds of snake movement. Most snakes are not limited to just one movement. In fact, they can usually use more than one kind of movement at the same time.

The most familiar method of snake movement is *lateral undulation.* The snake moves by pressing its curved body against stones, roots, or any other upraised surfaces along the ground. It pushes off from these contact points, each portion of its body following the portion ahead of it to trace a single path. A minimum of two contact sites on alternate sides of the body must be maintained simultaneously in order for a snake to move forward. The efficiency of lateral undulation decreases as the number of contact points increases, because the snake expends more energy for the forward motion it produces. The lateral undulation of a snake is basically the same movement used by a swimming fish.

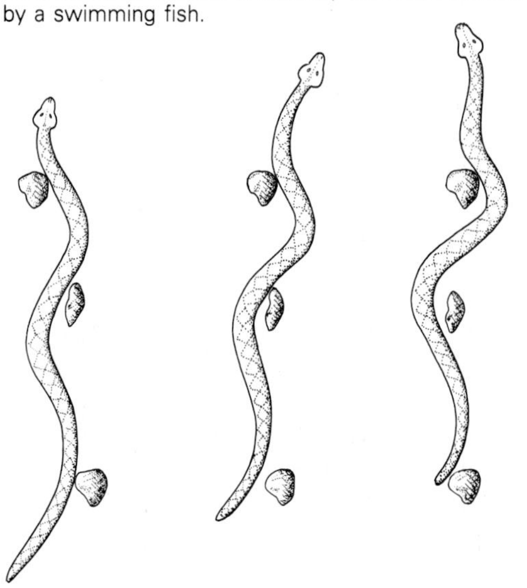

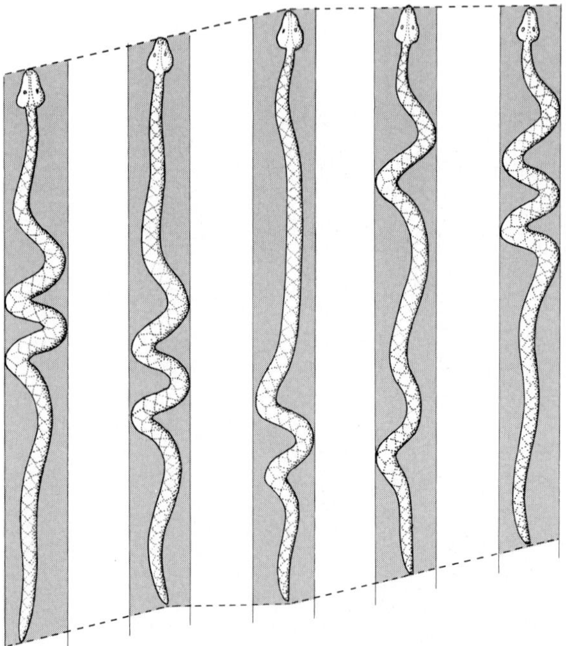

If a snake is enclosed within a narrow straight space, it cannot use lateral undulation and resorts to a *concertina movement* instead. In this method the midbody is folded into several S loops, which are pressed against the sides of the channel. As the snake moves its head forward, it also pushes against the channel to move the upper half of its body forward. Then it forms additional loops behind its head and pulls up its tail while pressing the forward loops against the channel walls.

Large snakes, such as pythons and boa constrictors, break the rule of not being able to travel in a straight line. These loose-skinned snakes have a powerful muscle that runs the length of their bodies. The muscle is attached at the sides to scutes, wide abdominal scales that look somewhat like tractor treads. As the series of scutes are held to the ground, the snake moves the muscle between them. By constantly removing the forward scutes from the ground and pressing down with the rear ones, the snake is able to propel itself forward. This *rectilinear movement* allows the snake to advance toward its prey in a straight line or to move across flat surfaces. Such movement is slow, and when the snake wants to move quickly it uses lateral undulation.

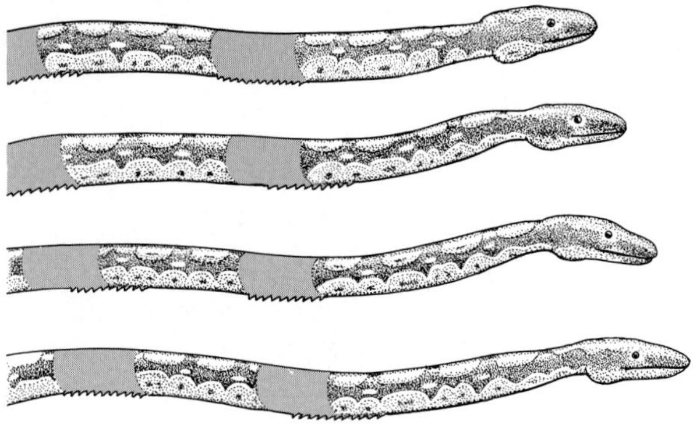

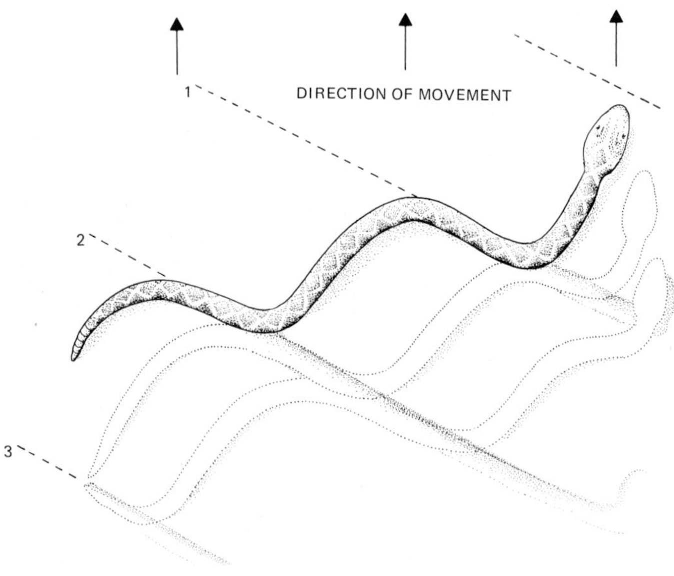

When a snake needs maximum speed it uses the *sidewinding* technique. A sidewinder like a desert viper literally moves sideways. Only three points of the snake's body touch the ground, at the neck, midbody, and tail. The head points in the direction of travel, and the rest of the body is inclined at about 60 degrees to the direction of travel. With the body in this wide-open position, the snake finds more contact points than it would if traveling in a straighter line, and progress is swift. The snake moves by forming two loops with its body and lifting its head off the ground. The neck was in one track (1), a point at the center of the body was in a second track (2), and the tail was in the third track (3). If the head moves toward a new track, the body will move into track 1 and the tail will shift into track 2. Between the tracks, the looped body forms bridges that do not touch the ground. These arches undulate back and forth as the snake progresses, but they remain off the ground, leaving only three straight tracks, with a little hook at the end where the head once rested.

Sometimes snakes do more than slither. Some poisonous snakes can rise up and strike forward, and the African desert viper, *Bitis caudalis,* can even jump to avoid hot ground surfaces. But in the sequence shown here, this tree snake seems to have outdone itself.

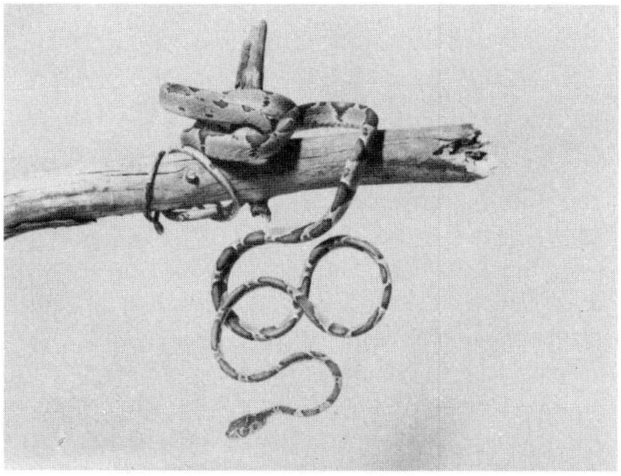

Preparing to move to a higher branch, the snake loosens its coils enough to extend its body, but it retains a firm grip with its tail.

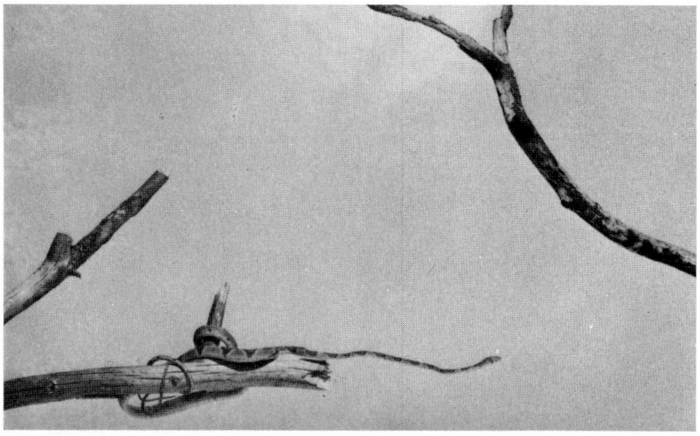

Tree-dwelling snakes have the ability to stiffen their entire bodies, as this one is doing as it stretches toward a higher branch.

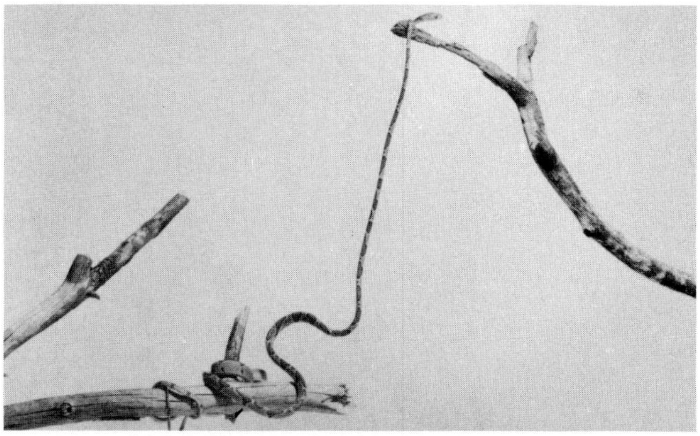

Using its coils as an anchor, the snake is able to raise its head and reach the new branch.

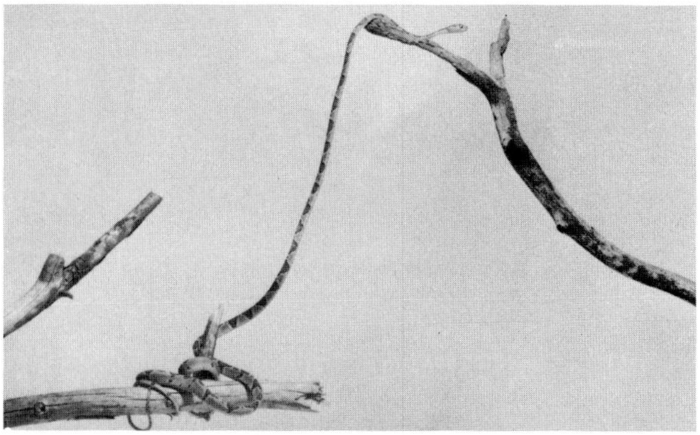

Moving forward, the snake begins to pull up its body as it uncoils from the branch below.

HOW ANIMALS RUN

The fastest land animals have the greatest combination of length of stride and rate of stride. But the real clue to speed is the ability to move as many joints as possible in the same direction at the same time. The cheetah, the acknowledged champion sprinter, has long legs in proportion to its body, a supple spine that speeds up the flexing motion of the body to go along with the legs, long and flexible feet, and a very small collarbone that is situated well forward on the body to allow a free swivel movement of the shoulder. Hooved animals, such as horses, antelopes, and gazelles, have no collarbone at all, and thus their shoulder flexibility is increased even further. Their speed is increased by their single-digit feet, which correspond to standing on tiptoe. Animals such as cats and dogs walk on their "fingers." Slower-moving animals, such as bears and human beings, keep the soles of their feet in contact with the ground when they walk. The tiptoe posture of hooved animals helps

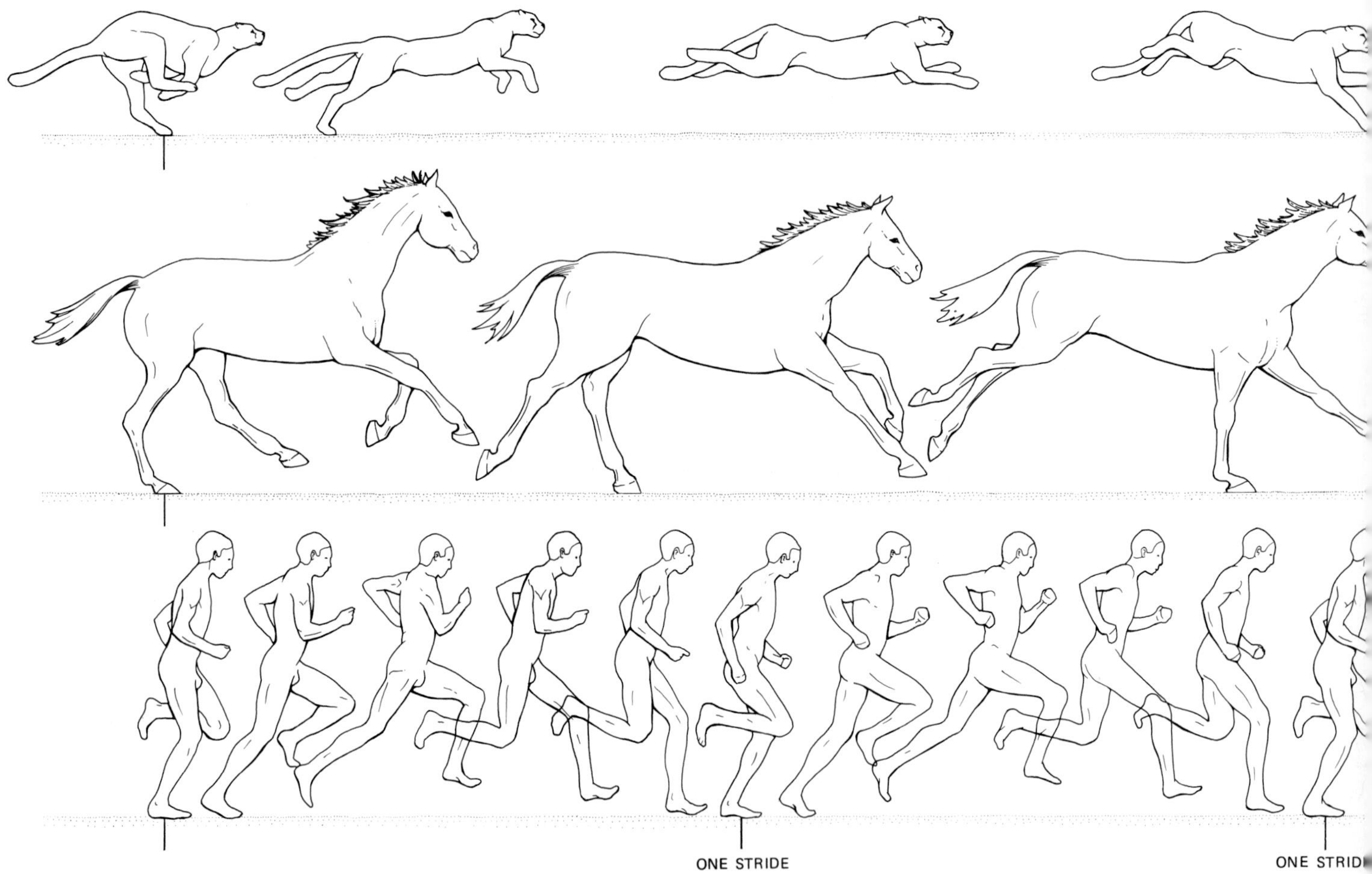

The strides of a cheetah, a horse, and a man are shown in these sequential drawings. The cheetah and the horse are in full gallop, their most powerful stride, and the man is running at top sprinting speed. Both the horse and the cheetah have strides of about 7 meters (23 ft), but the cheetah completes 3.5 strides per second, whereas the horse completes only 2.5. This ratio coincides exactly with the top speeds of the animals: The cheetah, a sprinter, can run short distances at about 112 km/hr (70 mph), and the horse may be able to sprint as fast as 80 km/hr (50 mph) for a short period. A horse can maintain a steady pace of about 24 km/hr (15 mph) for distances even greater than 50 kilometers (30 mi). An Olympic sprinter can run as fast as 36 km/hr (22 mph) for the 100-meter dash. Both the cheetah and the horse depend on their powerful shoulders and legs for their speed, but the cheetah is also aided by a supple spine that extends the length of its stride and complements the thrusting movement of its legs. The human spine, like the horse's, is too rigid to aid substantially in running, and both the increase the length of the leg substantially, and elastic ligaments in the feet of horses and other hooved animals give an extra springiness and lift. All these factors contribute greatly to an animal's speed. Equally important, especially over long distances, is the ability of the heart and lungs to keep the muscles well supplied with oxygen. Without energy-producing fuel, even the most streamlined system of levers (bones and muscles) would be useless.

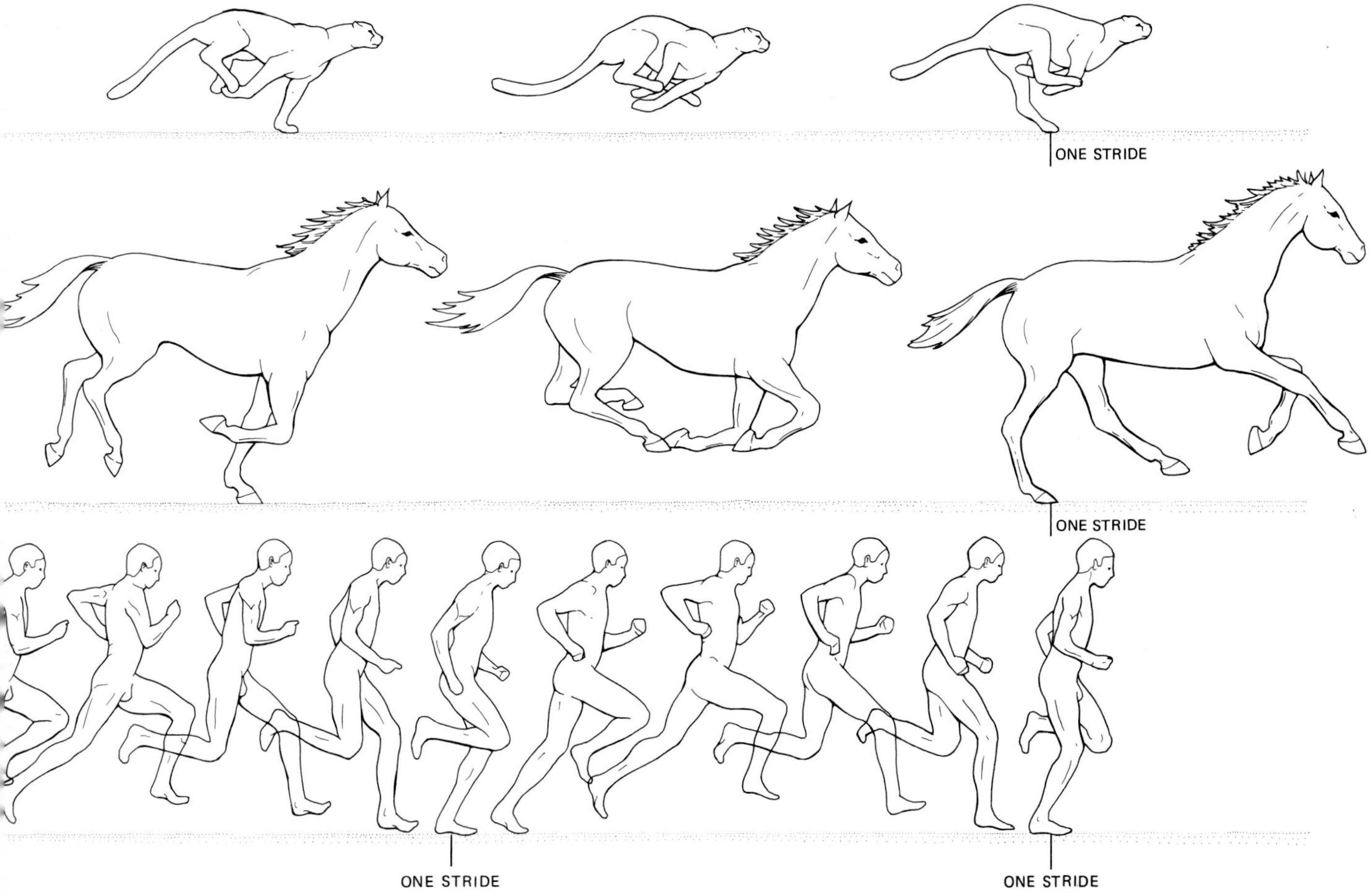

horse and the human sprinter rely mostly on the backward push of their legs on the ground to produce forward momentum. Humans are also assisted by the balancing action of the swinging arms. All three runners take all feet off the ground at some time during the sprint, and the cheetah is airborne almost half the time. The horse and the man leave the ground completely for about one quarter of their stride, the horse's unsupported leap being about four times the length of its body.

The European common hare (*Lepus europaeus*) uses its outstanding speed to run, hop, and zigzag away from predators. An adult hare is larger than a rabbit and can accelerate to 72 km/hr (45 mph) by using its extra-long hind legs. But some streamlined hunting birds, such as the falcon, are the fastest moving animals of all. They can fly at about 160 km/hr (100 mph) and can dive accurately at speeds up to 290 km/hr (180 mph). Rabbits, antelopes, and other animals can leap forward with great bounds, but the champion jumper is the flea. A flea can leap more than 100 times its own length. To equal the flea's leap, a person would have to long-jump the length of two football fields.

When the Australian crested dragon wants to move faster, it switches from four legs to two. Consequently, it is called the "bicycle lizard." The long tail remains on the ground for balance.

21
The Skeletal and Muscular Systems

SOME KEY POINTS

1. Invertebrates usually have tough external coverings, and vertebrates have internal skeletons.
2. Besides helping to support the body, bones store reserves of calcium and phosphorus, and produce red blood cells.
3. Bones may be spongy or compact, depending on their location and function.
4. The body is able to move because of several types of joints, and because of the coordinated lever action of bones and muscles.
5. The types of muscle are skeletal, smooth, and cardiac.
6. Muscle contraction occurs through an electrochemical process known as excitation-contraction coupling. The functional unit of muscle contraction is the sarcomere.

IN GENERAL, AS ANIMALS BECAME MORE COMplex they also became larger. And the larger they became, the more important it was for them to develop a reliable means of supporting their bodies, protecting their internal organs, and aiding movement. The skeleton provided support and protection, and together with the muscles, it permitted the body to move. Most animals have some firm supporting structure. Even protozoa may secrete shell-like tests (Latin, "pot") or cover themselves with grains of sand. Sponges make stiff internal fibers of geometrical elegance. Invertebrates commonly have external coverings, such as the shells of molluscs and the tough outer skeletons (*exoskeletons*) of crustaceans and insects. Bony, jointed internal skeletons (*endoskeletons*) were not produced until fishes appeared on earth, but from then on, all the vertebrates have had bones, except the sharks and their like, which have skeletons made of relatively elastic cartilage.

Animals with exoskeletons, like crayfishes, cannot grow any larger than their outer covering. They must shed their old skeletons and grow larger ones during the periods of molting. In contrast, animals with endoskeletons have fewer growth restrictions, as demonstrated by elephants, whales, and dinosaurs.

BONES DO MORE THAN SUPPORT THE BODY

The vertebrate skeleton is contained within the body and is composed of bone and cartilage (connective tissue) surrounded by soft tissue. The main function of the vertebrate skeleton is to provide support and protection for the internal organs, but bones are also the storehouse and main supply of reserve calcium and phosphorus, and they aid movement by providing a point of attachment for muscles. In the higher vertebrates (including humans), the bone marrow manufactures red blood cells and some white blood cells. Bone is not dry, brittle, or dead. It is a living, changing, productive tissue.

The Structure of Bone

Bone contains only about 20 percent water, as compared with other tissues of the body, which may have as much as 90 percent water. The solid portion of bone is composed mainly of inorganic minerals, especially calcium phosphate. The rest is organic material made up of bone cells (osteocytes) and bone collagen, a fibrous protein. The collagen and minerals are held together by a cementlike substance in a way that resembles the meshing of crystallized concrete (Figure 21.1). The strength of bone has been compared to cast iron, even though bone is much lighter and more flexible.

Besides having strength and flexibility, bone is alive and dynamic. The balance of calcium and other minerals in the bones, blood, and muscles depends largely on the ability of the minerals in the bones to move back and forth between the bones and other parts of the body. For example, most of the body's calcium is found in the bones, but calcium is vital for such functions as muscle contraction, the beating of the heart, and blood clotting. When the calcium supply in the body drops below the required level, calcium from the bones is deposited in the blood for transport to the site of the shortage. When the body is deprived of food for long periods, the blood withdraws large quantities of minerals from the bones. A pregnant woman uses some of the reserve supply of bone minerals to provide the fetus with the minerals it needs to build its skeleton.

The skeleton of the human embryo is almost entirely made up of cartilage for about two months, at which time it begins to harden. Cells of the flexible cartilage are replaced by bone cells, which remove calcium phosphate and other minerals from the blood and deposit them to form the bones. This process is *ossification*. We finally reach our adult height when all the bones have replaced the cartilage in the skeleton, and the bones no longer lengthen. This stage occurs earlier in women than in men, accounting for the relative tallness of men. Some cartilage remains in the adult body, in the external ear and the tip of the nose, for example.

Usually our bones begin to deteriorate by the time we are 40. Until then our bodies have contin-

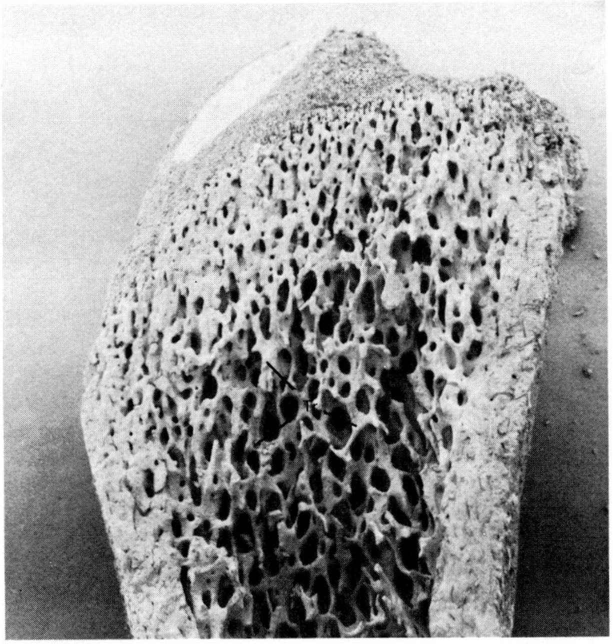

× 20

Figure 21.1
A section through a typical long bone reveals a meshwork of trabeculae (Tr), which provides a strong and flexible structure.

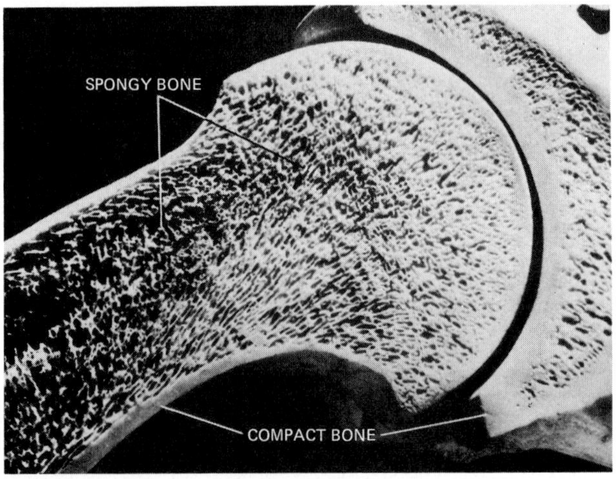

Figure 21.2
A section through a human shoulder joint. The shell-like, solid outer part of the bones contrasts with the porous inner part. Large bones commonly have such separation into *compact* and *spongy* regions. The streaked appearance of the spongy bone is due to the strands of hard material that are built in the direction of greatest pressure, achieving lightness without loss of strength.

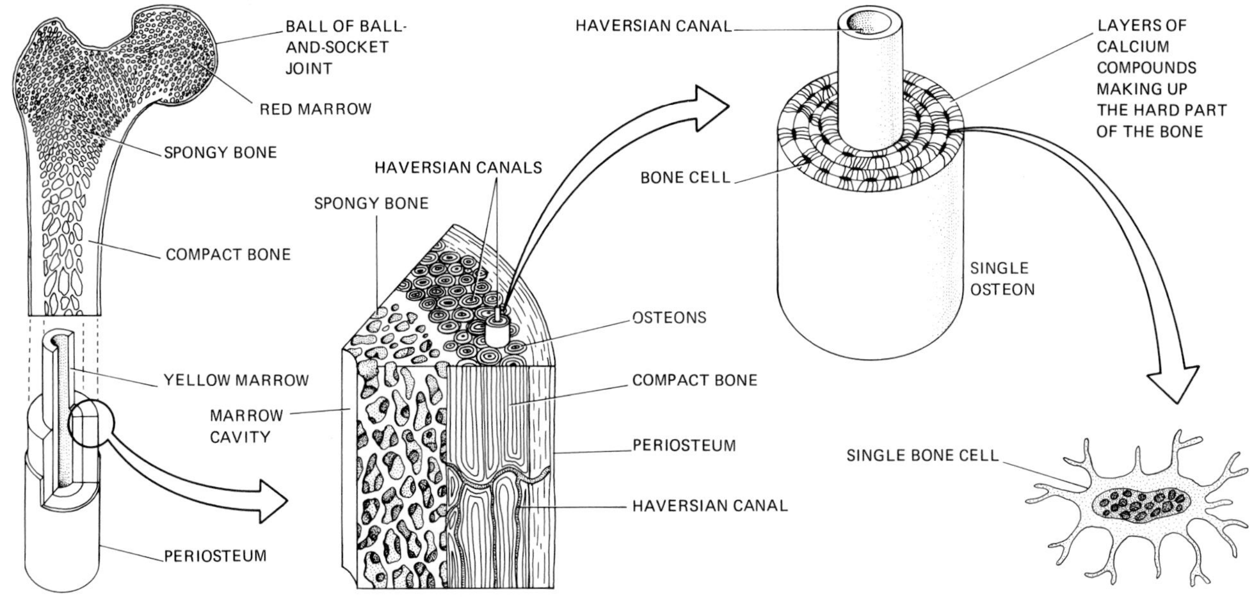

ued to build bone, even after the full skeletal height is reached, but once the deterioration starts, we lose bone cells. As intervertebral disks soften and weaken and the back begins to arch, the characteristic "humpback" of old age may result. Women are affected more than men, but sooner or later most of us are destined to shrink and creak.

Figure 21.3
Gross microscopic anatomy of a bone. An enlarged view of a portion of the compact bone shows the covering layer, the *periosteum,* and a number of cylindrical subunits, the *osteons,* penetrated by channels. Blood vessels pass through *Haversian canals.* Living cells are dispersed through the hard part of the bone (calcium carbonate and calcium phosphate).

The Classification of Bone

Bones may be classified according to their external shape as long, short, flat, or irregular, or according to their internal composition, as follows:

1. *Spongy (cancellous) bone* consists of an open, interlaced pattern of bony tissue designed to support maximum stresses and shifts in weight distribution (Figure 21.2). Spongy bone is found inside many bones, including ribs and the ends of long bones. The thigh bone (femur) is the longest, thickest, and strongest bone in the body; its latticed interior is so adaptable that it can shift its stress lines along the ridges if body stresses change. Because bones grow stongest where the stress is greatest, broken bones heal faster when some limited stress is placed on them during the healing process.

2. *Compact bone* is very dense, and contains concentric rings of calcified bone known as *osteons* (Figure 21.3). Within the osteons are cavities, in the center of which are *Haversian canals,* elongated tubes that carry blood vessels throughout the bone (Figure 21.4). The shafts of long bones are composed of compact bone, as in the midregion of the femur.

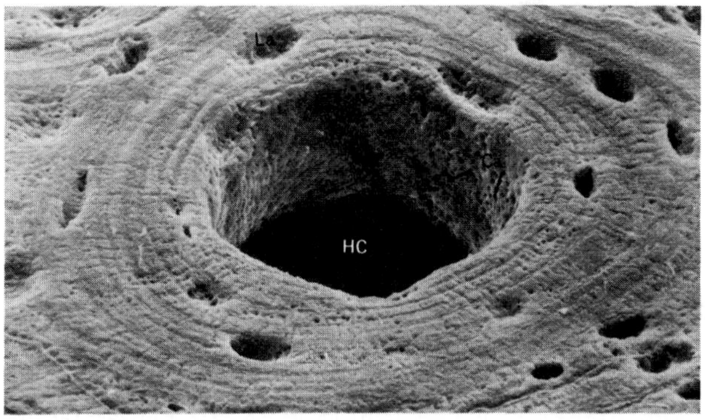

× 500

Figure 21.4
A scanning electron micrograph of a Haversian canal (HC), which channels blood vessels through a bone. Bone cells are housed within small depressions called *lacunae* (La), and tiny openings in the Haversian canals called *canaliculi* (Ca) provide passageways for the exchange of nutrients and gases. Obviously, a bone is more than just a support.

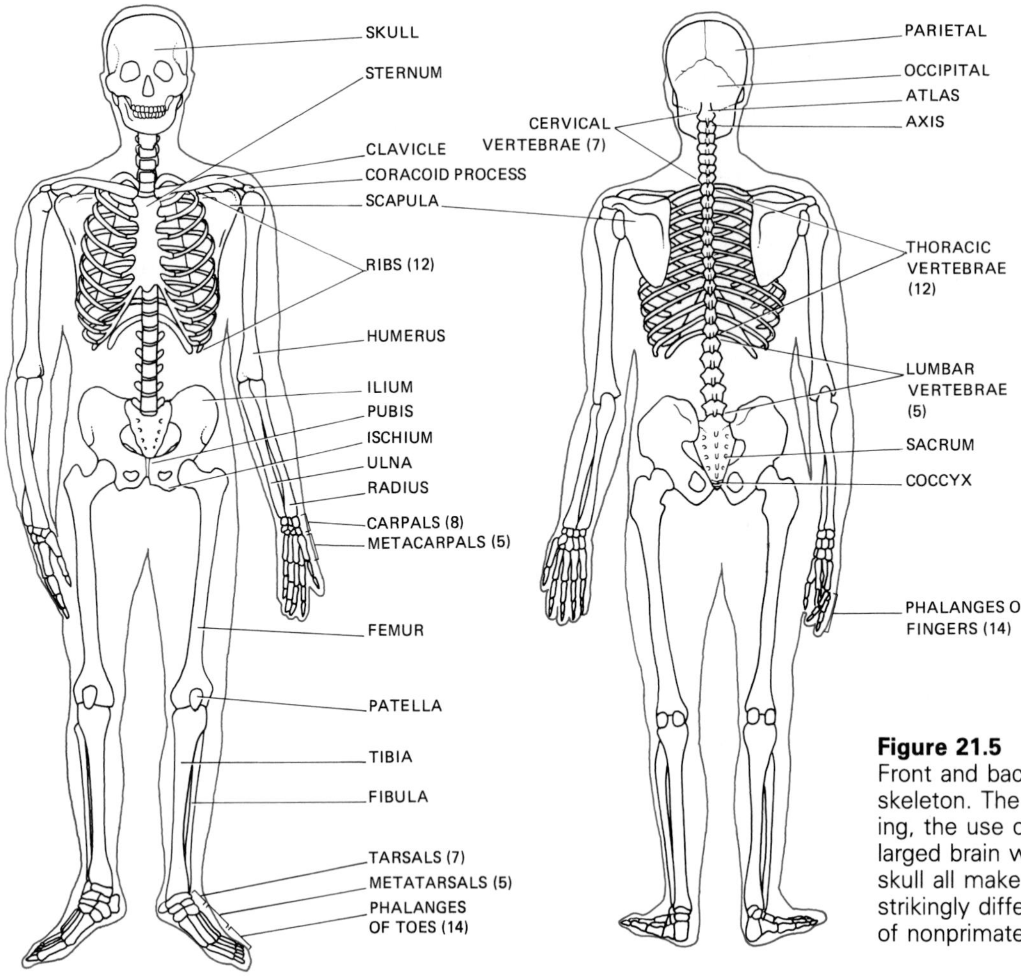

Figure 21.5
Front and back views of a human skeleton. The upright habit of walking, the use of hands, and the enlarged brain with accompanying large skull all make the human skeleton strikingly different from the skeletons of nonprimates.

THE VERTEBRATE SKELETON IS ESSENTIAL FOR SUPPORT, MOVEMENT, AND PROTECTION

The growth of a vertebrate skeleton depends on the proper balance of minerals, hormones, and vitamins. As we saw in Chapter 18, too much or too little growth hormone from the pituitary gland can produce giants or midgets, and dietary deficiencies can hamper proper skeletal development or even damage mature bones. Some bones normally become fused together as children grow, so that of the original 270 or so bones in the infant human body, only about 206 are present in the adult human skeleton (Figure 21.5). Of our 206 bones, 106 are found in the hands, wrists, feet, and ankles, where an intricate combination of bones allows the most varied movement.

The Organization of the Vertebrate Skeleton

The vertebrate skeleton, of which ours is not a typical example, consists of a central spinal column, which is a flexible stack of a varying number of bones called *vertebrae;* a skull, which protects the brain; two limb girdles to support the forelimbs and hindlimbs; and the limbs themselves (see Figure 21.5). The skull, vertebrae, and rib cage make up the *axial skeleton,* and the forelimbs, hindlimbs, shoulders, and pelvis form the *appendicular skeleton.* In terms of evolution, the axial skeleton was the earlier one.

In humans, the skull is composed of 28 bones, and only the jaw is freely movable. The head is about 12 percent of our total body weight in infancy, but only about 2 percent when we are fully grown. The spinal column contains 7 cervical (neck) vertebrae, which permit a broad range of lateral and vertical head movement; 12 thoracic vertebrae, which are attached to the ribs; 5 lumbar vertebrae in the small of the back, which support most of the body's weight; the sacrum, whose five small fused bones connect the pelvic girdle to the backbone; and a tail bone, or coccyx, of 4 small fused bones. (If a backache occurs, it will most likely happen in the lumbar or sacral areas.)

The rib cage has 12 pairs of flexible ribs, 10 of which are united by elastic cartilage in front to the breastbone, or sternum, to allow for expansion during breathing. Besides supporting the trunk, the backbone protects the spinal cord, which passes lengthwise through holes in the vertebrae. There are also openings between vertebrae through which pass the spinal nerves. Pads of elastic cartilage called *intervertebral disks,* located between the vertebrae, not only absorb shocks but allow the vertebrae to move smoothly against each other. The disks are filled with a jellylike substance that may leak out of a ruptured disk and cause great pain by irritating nearby nerves, producing muscle spasms that serve to protect the spinal cord by restricting movement. Severe physical injuries may actually crush the disk altogether, and major surgery may be required to repair the damage.

Backaches probably cause as much trouble for the average person as any other type of chronic

SUNLIGHT AND SKELETON BUILDING

When children or any young vertebrates (except fishes) are first developing and hardening their bones, they need a supply of the hormone *calciferol.* Without it they become literally rickety. They suffer pain and lassitude, their long bones are weak and ill formed, their heads are too large, and their rib cages are misshaped. The malady is acute in northern latitudes, where the winter sun is weak, and it is especially bad in smoggy cities, where children are kept indoors away from sunlight. Calciferol was for many years called Vitamin D and is still so labeled on the milk sold in the United States and most of Europe, but it is in fact a steroid that influences the absorption of calcium from the gut and its deposition in bones. Calciferol is therefore more properly called a hormone than a vitamin. Land-dwelling animals can synthesize the precursor of calciferol as a form of cholesterol, but they lack the enzyme necessary for the small but essential molecular change from the precursor to the active molecule of calciferol. That change, however, can occur if the cholesterol is irradiated with ultraviolet light, such as is present in sunlight. Children and baby animals exposed to enough sun do not develop rickets. The disease is unknown in regions where light is plentiful and growing infants are normally kept outdoors, nor does rickets develop even in such northern dwellers as Eskimos, who obtain their calciferol from the fish they eat. (Since fish are underwater animals, they receive no ultraviolet light, but they have the ability to make their own calciferol.)

Proper bone development can be guaranteed (other factors being adequate) if an infant is supplied with calciferol in its diet or is kept in enough sunlight to allow its cholesterol in the skin to be changed to calciferol, which can be absorbed. Although some calciferol is essential to the health of vertebrates, too much (like too much of any hormone) can be detrimental, and artificial intake of large amounts is dangerous.

Even now, a common sight in the high Andean villages of Ecuador is a tiny Indian mother carrying a large child in a sack on her back. The round, black eyes peering cheerfully out of the sack are frequently those of a two-year-old who in other parts of the world would be running about on its own legs. If you ask about the practice, you will learn that if the children are put down to walk before their bones are sufficiently developed, their legs will go crooked. Although Ecuador is on the earth's equator (hence the name of the country) and ultraviolet light is plentiful, the high altitude makes the air very cold. As a consequence, people cover themselves with clothes, and presumably they know from experience that rickets is a threat.

During the eighteenth and nineteenth centuries, when northern Europe was living through the Industrial Revolution, coal smoke so beclouded the skies of cities during the long dark days of winter that many of the factory workers, crowded by cold and poverty into slums, were permanently disabled simply because they did not know enough to get outside. Here is one of the saddest effects of human ignorance of basic biological information.

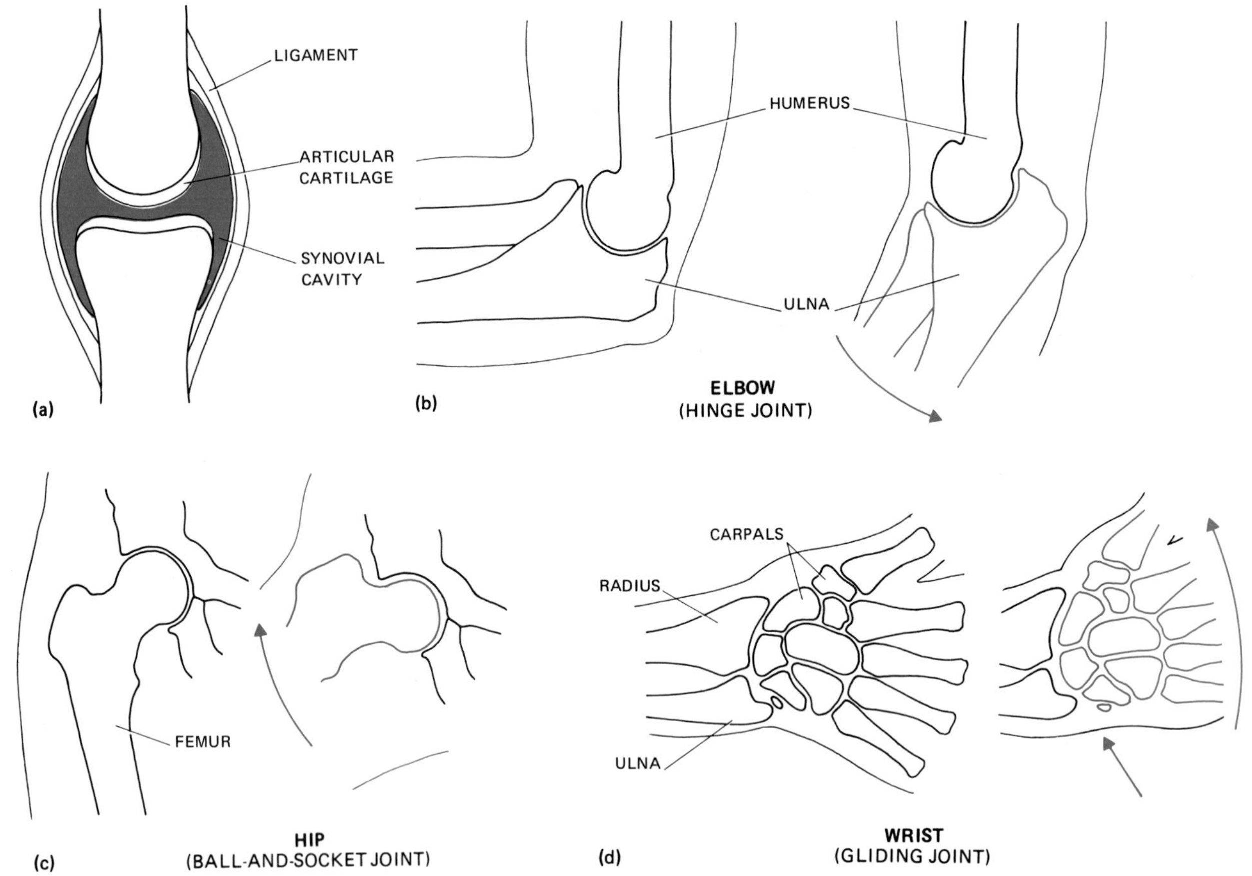

Figure 21.6
Joints. (a) Diagram of a long section through a joint. The two bones that move against one another are held together by tough, slightly elastic ligaments, which cover the joint and enclose the fluid in the synovial cavity. Where the bones approach each other, they are covered by marvelously slick articular cartilage, which offers minimal friction between the moving parts. (b) An elbow is an example of a hinge joint. Movement is restricted to one plane. (c) In the hip, a ball-and-socket joint, the thigh bone (femur) can work against the pelvic girdle freely up and down, sideways, and with a rotating motion. The shoulder joint is similar. (d) Wrists (and ankles) have gliding joints. Gliding joints do not have as great an angular range as hinge or ball-and-socket joints, but they do have freedom of direction, and they allow a variety of diverse bendings and turnings.

physical problem. The most important function of the backbone is to protect the spinal cord, and it performs that function admirably. "Backache" trouble does not originate with the cord, however. It starts with any of the 33 vertebrae and their supporting structures.

Problems with the kidneys, prostate gland, or liver can cause back pain. Even emotional problems may manifest themselves as backaches by producing muscle tension that leads to dull back pain. Overweight people may have backaches when their abdominal muscles weaken so much that the back muscles must pull in the opposite direction to hold the body erect. This is the same problem that many pregnant women have who suffer increasingly severe backaches as the pregnancy progresses.

Joints

To be effective, bones must not only be stiff enough to support muscles and internal organs; they must permit movement. This they do by means of the **joints,** or articulations. A joint is the place where two separate bones meet, or where a cartilage joins a bone. The structure of a joint depends on its function. Some joints are immovable, as in the skull;

some, like the hinged joints of the knees, bend essentially in one plane; some allow movement in several planes, such as the gliding joints of wrists; and some are free in all planes, such as the ball-and-socket joints of the shoulders and hips (Figure 21.6). Joints are held together by tough *ligaments* (there are about a thousand ligaments in the human body) and covered by special membranes, the synovial membranes. The moving parts of joints, where contact between two bones is made, are bathed in a viscous synovial ("egg white") fluid, which, along with the texture of the bone covering, makes a joint as nearly frictionless as any moving parts known to humans.

THE INTEGRATION OF BONES AND MUSCLES

Joints make a skeleton potentially movable, but the bones cannot move by themselves. But muscle cells, because they have the ability to contract, specialize in movement.* Bones (and exoskeletons, too) move as levers, with one part rotating about or being hinged to another. A muscle, capable of contraction, is attached to one bone at one end and across a joint to the next one at its other end. When the muscle contracts, the two bones are pulled toward one another, as when an elbow bends, or around on one another, as when the neck twists.

During contraction, a muscle grows shorter by about 15 percent of its resting length, as well as thicker. A contracted muscle is brought back to its original condition by the contraction of its opposing muscle, and it can be stretched to 120 percent of its resting length. A muscle does no work while being stretched. Because muscles work by contracting, not by expanding, they must operate in pairs, with one altering the relative position of two bones in one way, and its opposite muscle pulling them back in the other direction.

The Lever Action of Bones and Muscles

Usually, a muscle is attached to a bone by a **tendon** (Greek, "to stretch"), connective tissue composed of collagen. When a muscle contracts, one of the bones attached to it remains stationary while the other bone moves along with the contraction. The junction point where the muscle is attached to the stationary bone is the point of *origin*, and the point of attachment of the muscle to the bone that moves is the *insertion*.

* The human body contains about 600 muscles, which weigh about 18 kg (40 lb) in an average woman and 30 kg (65 lb) in an average man.

A contracting muscle bends a joint in a movement called *flexion*. Another complementary movement, *extension*, straightens a joint (Figure 21.7). The biceps muscle contracts to bend your elbow, and the *antagonistic* muscle, the triceps, contracts to straighten your elbow (see Figure 21.7). Most muscles or groups of muscles have an antagonistic mus-

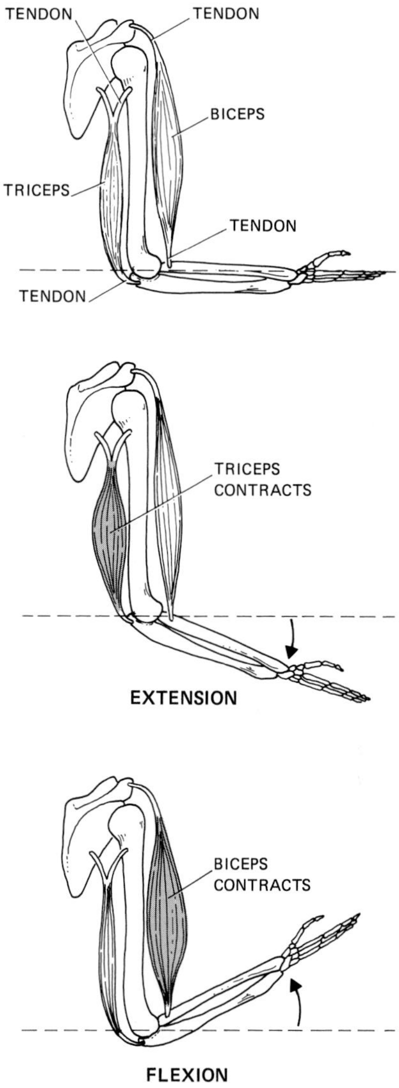

Figure 21.7
A pair of muscles working in extension and flexion. The forearm is extended when the triceps muscle contracts, and the biceps relaxes and is passively pulled longer. The forearm is flexed when the contraction and relaxation are reversed. Muscles do not extend themselves. They can only contract and then be pulled back to the elongated condition by the contraction of some antagonistic muscle.

cle or group to permit movement in opposite directions, or to allow a limb to be rotated. Basically, bones operate as a system of levers.

Although opposing muscle pairs usually work antagonistically, they are not necessarily equal. The biceps muscle of the upper arm is used to bend the elbow more than the opposing triceps is used to extend it. The boy who wants to show off his muscles exhibits his biceps, never his triceps. Jaw muscles are even more strikingly unbalanced. The closing muscles of a jaw are short, heavy, and powerful, but the opening muscles are slight and weak. Anyone can hold an alligator's mouth shut, but trying to hold one open is no simple matter (and is not wise).

The working muscle is not always close to the moving bone. The muscle used for standing on tiptoe, which works by pulling up the heel bone, is not in the ankle but behind the tibia, and it is called the calf muscle, or gastrocnemius. Hands are almost devoid of muscles except for thumb muscles and extensors of the little finger. Piano players do not get tired hands, no matter how strenuously they play. Their forearms, however, can become fatigued because that is where the finger-working muscles are located.

SKELETAL MUSCLE

Skeletal muscle (also called striated or voluntary muscle) is considered a "voluntary" type of muscle because it is under the control of the somatic ner-

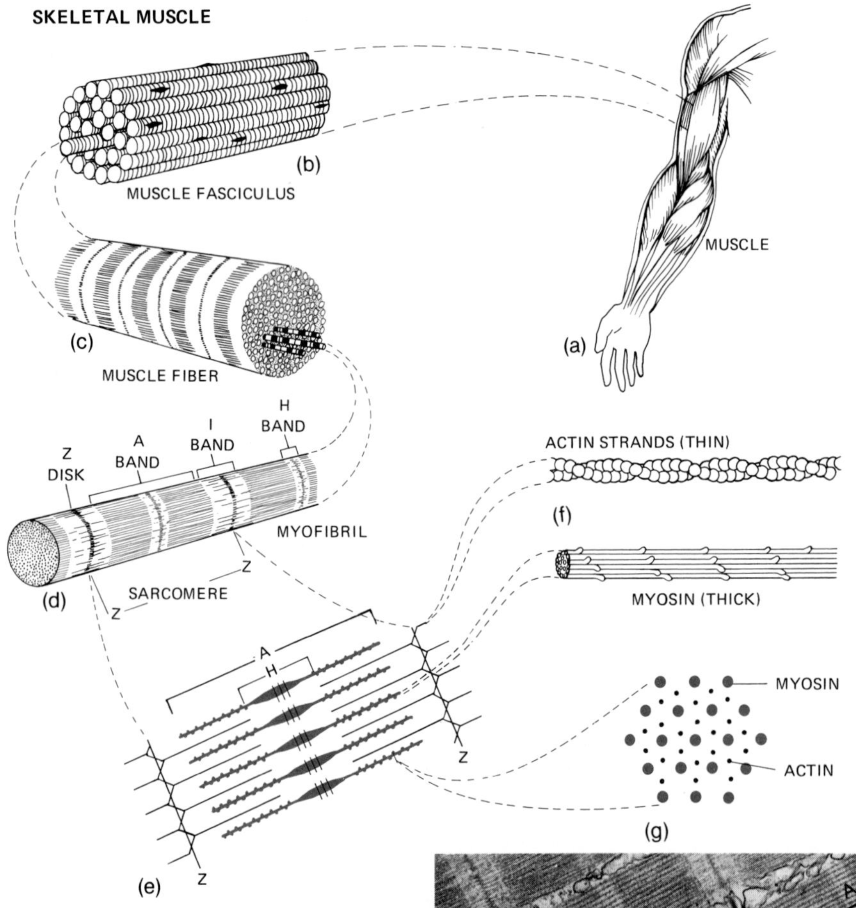

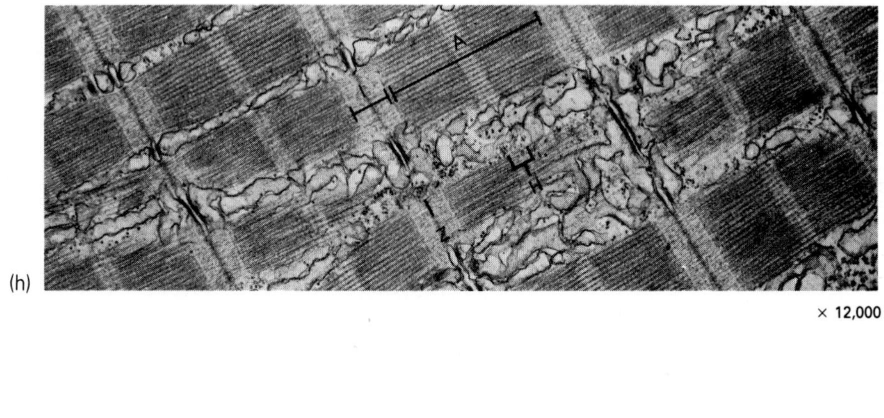

Figure 21.8
The anatomy of muscle, from the gross structure of an arm (a) to the molecules of muscle protein. A whole muscle is made of bundles (b), which in turn are made of muscle fibers (c). Each fiber can be frayed into fibrils, which show the banding that makes striated muscle striated (d). The striations are an expression of the position of different compounds distributed in the fibril (e). The two main muscle proteins are *actin* and *myosin,* and their overlapping or failure to overlap gives the muscle its banded appearance (f). (g) A cross-section of a myofibril. The thicker protein filaments are myosin and the thinner ones are actin. The electron micrograph (h) corresponds to the drawing (d), and shows the bands and disks illustrated in the drawing.

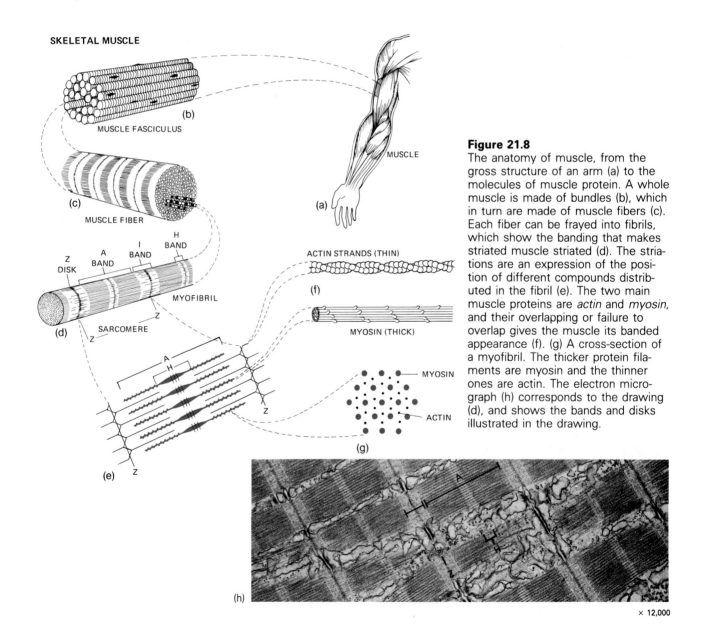

SPRAINS, STRAINS, DISLOCATIONS, AND FRACTURES

Practically all of us have had a strained ankle or a strained wrist. It hurts for a while but usually returns to normal after a few days, even without special treamtent. But what is a sprain, and what is a strain? A **sprain** is an injury to the ligaments around a joint, usually resulting in some tearing, and making movement of the joint painful. In a slight sprain,the ligaments may heal within a few days, but severe sprains (which are usually referred to as "torn ligaments") may take weeks or even months to heal. In some cases the ligament must be repaired surgically. A **strain** is an overstretching of a muscle or its tendon, and it may range from a simple "charley horse" to a violent wrenching that requires an operation to repair the torn tissue.

A **dislocation** is such a severe strain of the muscles, tendons, and ligaments that a bone is pulled out of its socket and fails to return to its proper position. The bone must be repositioned, usually by slipping it back into place by hand, a process called "reducing." If a dislocation can be reduced by simple manipulation, the aftereffects may be slight, but if surgery is required, the damage can take months to heal. Hip and shoulder joints are commonly dislocated, but any movable joint, such as the knuckles, knees, and elbows, may be "thrown out of joint."

A **fracture** is a broken bone. Broken bones that are reset soon after an injury have an excellent chance of healing perfectly because the living tissue and adequate blood supply at the fracture site stimulate a natural repositioning. It follows that the supple, healthy bones of children mend faster and better than the more brittle bones of older people. In either case, the fracture is almost always encased in a plaster cast until the bone sets permanently. In elderly people, bones contain relatively more calcified and less organic material, and consequently old bones break more easily. A fall that a child hardly notices can be serious in an older person. "To fall and break a hip," a common disaster among the elderly, could frequently be better stated as "to break a hip and fall," because the fragile old bones may crack merely under the strain of walking, making the legs give way. The hip is often broken before the body even hits the ground.

There are many different types of fractures. The simplest fracture is the *greenstick* (or incomplete) fracture, suffered mainly by children, whose elastic bones may bend as well as break. Consisting only of slight, longitudinal cracks, greenstick fractures are difficult to detect even in X-ray pictures, but they heal readily. Technically, a *simple fracture* is one in which the skin is not broken, whereas in a *compound fracture* broken ends of bone push through the skin. The most serious fractures require not only surgery but physical therapy afterward.

vous system, and it contracts when we consciously want it to. It is attached to and causes the movement of the skeleton. It is also responsible for such voluntary actions as the elimination of feces and urine. The fibers of skeletal muscle, composed of multinucleate cells, extend along the full length of the muscle.

The Cellular Structure of Skeletal Muscle

The organ called a **muscle** (from the Latin, "little mouse," because the movement of muscles under the skin resembles a running mouse) is actually an aggregate of muscle fibers (Figure 21.8a, b), just as a nerve is a bundle of nerve fibers. The cellular unit of muscles is therefore called the *muscle fiber* (Figure 21.8c). The adult human body probably has about six billion fibers in its 600 or so muscles. Ordinarily, one muscle is larger than another because the larger muscle contains more bundles of fibers.

Within the muscle fiber are smaller units called **myofibrils.** The long, thin myofibrils are made up of a fairly large protein, *myosin,* and a smaller protein, *actin.* The arrangement of these large and small proteins produces the alternating light and dark bands characteristic of skeletal ("striated") muscle (Figure 21.8c). The light bands, called I bands, contain actin; the darker bands, called A bands, contain myosin. Cutting across each I band, like a dime in a stack of pennies, is a disk called the Z disk. Within the A band there is a paler H band that also contains a disk, the M line. A section of a myofibril from one Z

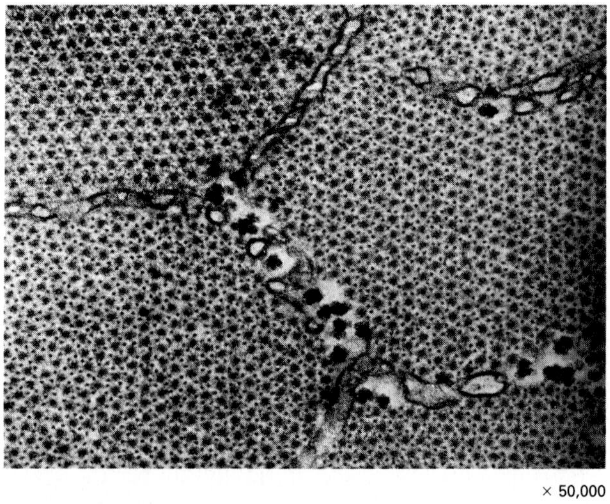

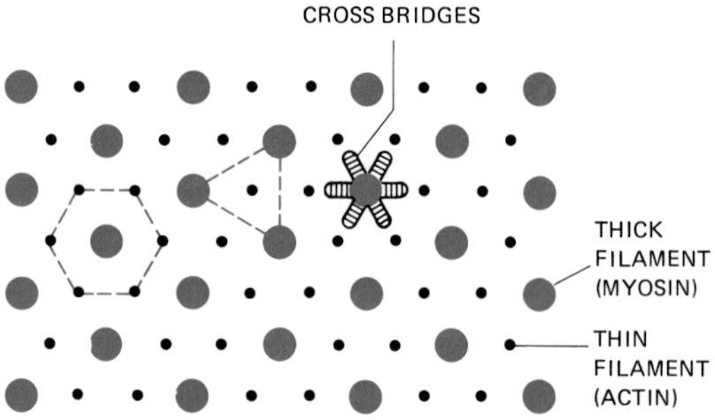

Figure 21.9
Electron micrograph of a thin cross section of muscle fibrils, showing the ends of the protein filaments. The larger spots are myosin and the smaller ones are actin. The diagram shows the spatial arrangement of the actin and myosin filaments, with cross bridges.

disk to the next, that is, one whole A band and two halves of the two adjoining I bands, constitutes a **sarcomere,** the functional unit of muscle contraction (Figure 21.8d).

The Molecular Mechanism of Muscle Contraction

The thin actin filaments and the thick myosin filaments are connected by molecular bridges, usually called cross bridges (Figure 21.9). During contrac-

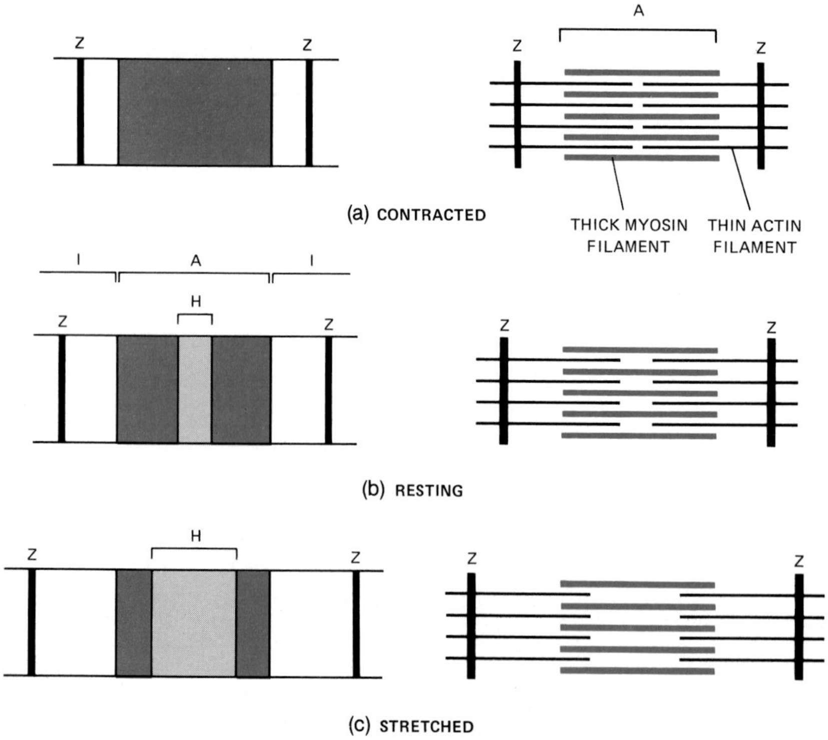

Figure 21.10
How the muscle protein filaments slip past each other as a muscle contracts and stretches. The whole action is known as the sliding-filament model of muscle contraction. The thin actin filaments and the thicker myosin filaments glide in and out. The ratchetlike cross bridges that reach from myosin to actin are not shown. (a) When a muscle contracts, the Z disks move close together, and the actin and myosin filaments overlap fully. (b) A muscle fiber in its resting state shows the Z disks farther apart, and the actin and myosin filaments overlapped less. In (c), which shows a stretched muscle, the two Z disks marking the two ends of a single sarcomere are far apart, and the actin and myosin filaments overlap only for a short portion of their tips. Note that the width of the A band does not change, no matter what state the muscle is in. The width of the H band does change, depending on how far the thin actin filaments slide into the A band. The H band disappears altogether during contraction.

tion the cross bridges slide back and forth across the filaments in a ratchetlike movement. The **sliding-filament theory** of muscle contraction states that muscles are shortened when the actin and myosin filaments slide past each other without changing their lengths. The I bands shorten during contraction, but the A bands do not (Figure 21.10). Striated muscle contractions occur so rapidly that each cross bridge may attach and release the active sites in the filaments as many as 100 times per second.

The action potential of muscle cells is similar to the action of nerve cells, described in Chapter 19. The electrochemical process of muscle contraction, known as **excitation-contraction coupling,** is initiated by an electrical stimulus in the membrane of the muscle cell (Figure 21.11). Like the activity of nerve cells, muscular activity is mediated by the action of ions. Two regulator proteins, troponin and tropomyosin, act as natural inhibitors of muscle contraction by preventing actin from forming cross bridges with myosin in a resting muscle, but when calcium ions are introduced into the muscle fibers, the muscle contracts immediately. This happens because the inhibitory effects of troponin and tropomyosin are offset by calcium ions, which bind to troponin and permit a restoration of activity in the cross bridge. The source of these calcium ions is the *sarcoplasmic reticulum,* a specialized form of the endoplasmic reticulum found in other cells (Figure 21.12).

Shortly after death, rigor mortis ("death stiffness") sets in and causes the corpse to become stiff. This phenomenon is caused by the loss of ATP in

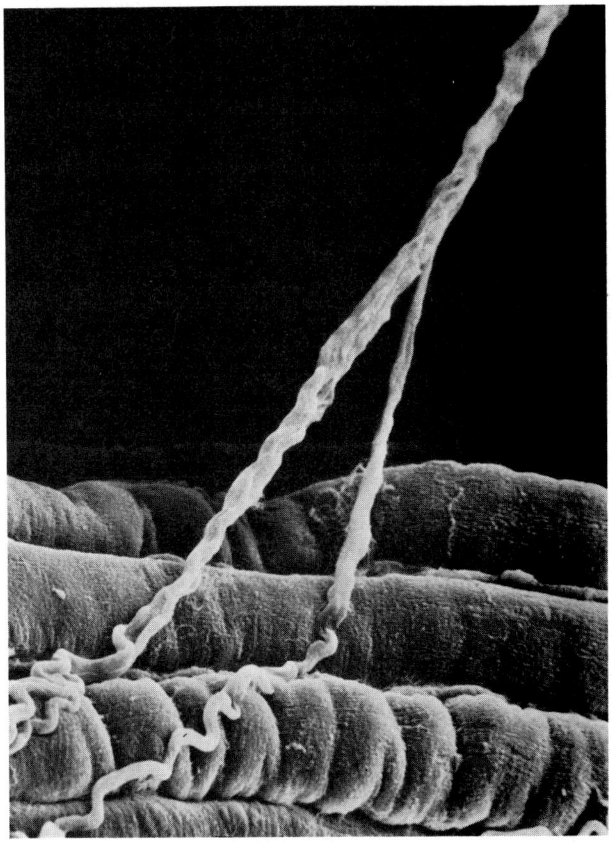
× 18,000

Figure 21.11
Scanning electron micrograph of nerve endings on muscle fibers. The job of a muscle is to contract, but contraction cannot be useful unless it is controlled, and the control comes by way of the nerves.

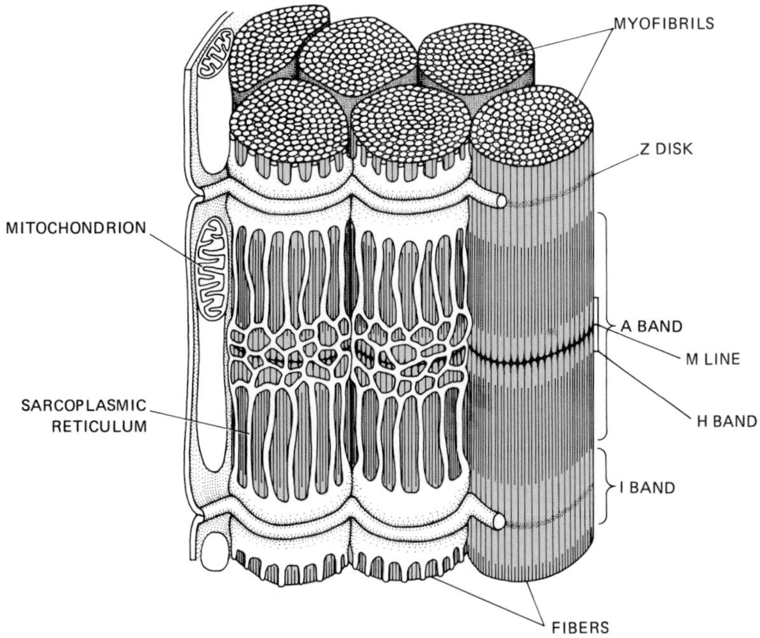

Figure 21.12
Vertebrate muscle from a frog. A three-dimensional drawing of a bundle of muscle fibers, reconstructed from electron micrographs of cross and long sections of muscle. The fibers run vertically. The right-hand fiber is represented as having had its sarcoplasmic reticulum peeled off to show the bands and the Z disk, a disk separating adjacent sarcomeres.

the dead muscle cells. Without ATP, actin-myosin cross bridges continue to be formed, but the resultant bonds cannot be broken. The dead muscles, therefore, are held in a locked, rigid position. Rigor mortis disappears as the dead body decomposes.

The Nervous Control of Muscle Contraction

As we saw in the previous chapter, afferent nerves carry messages from sensory receptors in the muscles to the brain and spinal cord, and efferent nerves from the brain and spinal cord provide motor impulses that lead to muscle contraction. Skeletal muscle fibers are packed together into bundles that average about 150 fibers, and each bundle is controlled by a single motor nerve fiber. Such a group of skeletal muscle fibers and the nerve fiber that stimulates them is called a **motor unit** (see Figure 21.11). The motor neuron makes contact with the appropriate muscle fiber through a synapse, a chemical bridge described in detail in Chapter 19. The synapse fulfills the possibilities for electrical stimulation of both the nerve fibers and the muscle fibers, and it makes body movement possible.

The nervous system transmits impulses to skeletal muscles continually, maintaining a constant state of weak muscle contraction. This slight tension, called *muscle tone,* serves to keep the muscles ready to react to the full stimulus of a contraction. Muscle tone also helps to keep the muscles from getting flabby through disuse. As discussed in Chapter 18, in a stressful situation secretions of adrenalin from the adrenal glands will prepare the muscles for extremely quick action.

Energy for Contraction

Energy release is the most obvious feature of muscle contraction. Aside from the mechanical work done, there is considerable heat generated, which indicates still further energy release. If energy is being released during muscle contraction, it must come from some biological source. As cellular energy usually does, it comes from ATP, which yields some of its energy as it is broken down to ADP and phosphate. The immediate reaction, substantiated by firm experimentation, involves an actin-myosin-ATP association, with release of phosphate and energy, which is used in the contractile process. Some of the ATP comes from the usual aerobic oxidation of foods, principally sugars and fats, via the Krebs cycle. Some comes from a special ATP-generating process involving a high-energy compound, *creatine phosphate,* which can form ATP from ADP by having its own phosphate transferred.

During times of heavy muscular activity, much ATP comes from anaerobic respiration, or *glycolysis,* which means "the loosening of sugar" (Chapter 5). When anaerobic respiration is the source of ATP, the incompletely oxidized result is lactic acid. Since lactic acid is toxic, it cannot be allowed to accumulate indefinitely in muscles. It must be removed by the circulating blood and then carried to the liver. Once lactic acid reaches the liver, it is converted back to glycogen. During strenuous exercise, lactic acid is generated faster than it can be oxidized by oxygen brought to the respiring tissues. The result is that muscles "feel tired" until adequate oxygen can be restored. Because of this temporary shortage of oxygen, heavy breathing continues even after exercise while the *oxygen debt* is being made up; that is, until the excess lactic acid is completely removed, and the ATP and creatine phosphate store is replenished.

TO SUM UP MUSCLE CONTRACTION

1. Muscle fibers contain smaller myofibrils. Myofibrils are made up of the proteins *myosin* and *actin,* which alternate to form A bands (myosin) and I bands (actin). One whole A band and the two halves of the adjoining I bands make a *sarcomere,* the functional unit of muscle contraction.
2. The actin and myosin filaments, connected by molecular *cross bridges,* slide past one another during muscle contraction, as described by the *sliding-filament theory.*
3. The electrochemical process of muscle contraction, *excitation-contraction coupling,* is initiated by an electrical stimulus in the membrane of the muscle cell.
4. Two regulator proteins, *troponin* and *tropomyosin,* prevent actin from combining with myosin in a resting muscle. Contraction occurs immediately when calcium ions from the sarcoplasmic reticulum are introduced into the muscle fibers.
5. Efferent nerves from the brain and spinal cord provide impulses that lead to muscle contraction.
6. The source of the energy released during muscle contraction is ATP.

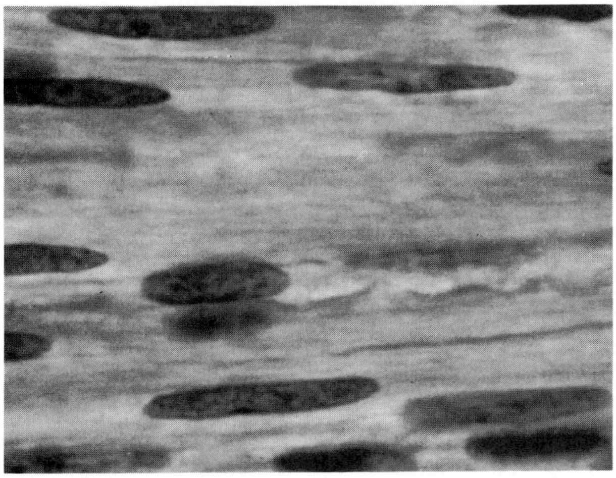

× 1100

Figure 21.13
Photomicrograph of a section of smooth muscle from the intestine of a snake. Smooth muscle is so called because it lacks the banding of striated muscle. It is slow and steady in action and as such is well suited to the organs in which it occurs. Intestines, for example, have smooth muscle in the walls that are effective in forcing food along.

SMOOTH MUSCLE

Smooth muscle is found mostly in the internal organs of the body (but not the heart), and for that reason it is also known as visceral muscle (Latin, *viscera,* "body organs"). Smooth muscle cells are long, tapered cells with one nucleus in each cell (Figure 21.13). Smooth muscle contains no striations as in skeletal muscle, and it has only a few myofibrils; it does contain actin and myosin filaments, however, and therefore the basic elements of muscle contraction may be accomplished.

One of the dominant features of smooth muscle is its slow and rhythmic action, which makes it suitable for the contractile control of the walls of the intestinal tract, uterus, blood vessels, respiratory passages, and urinary and genital ducts. Smooth muscle not only helps to move the contents of tubes such as the intestines but also regulates the diameter of blood vessels, for example, by slow and steady contractions that do not upset the normal circulation of the blood or the systematic beating of the heart.

× 2000

Figure 21.14
Cardiac muscle. Besides striated and smooth muscle, animals have a third type, which is especially suited to the repeated contractions a beating heart must provide. Cardiac muscle does have striations, much like those of ordinary striated muscle, but the arrangement of the sarcoplasmic reticulum is special. Like smooth muscle but unlike the usual striated muscle, cardiac muscle is not under direct voluntary control.

CARDIAC MUSCLE

Cardiac muscle is found only in the heart. Meant to resist prolonged wear, it is built for endurance rather than speed. Contractions of cardiac muscle last longer than those of smooth muscle. This type of action is necessary to give the heart enough time to pump blood from its chambers. Because of its strenuous and continuous pumping activity, cardiac muscle must have a constant supply of oxygen, and oxygen deprivation for more than 30 seconds or so may cause the cells to stop contracting, with resultant heart failure.

Cardiac muscle is striated like skeletal muscle and contains the same type of myofibrils and actin and myosin filaments (Figure 21.14). Until recently it was thought that cardiac muscle was composed of one mass of branching fibers, called a *syncytium*, but observations with the electron microscope now suggest an organization of closely packed, separate cells, each with its own nucleus.

Heart muscle can contract independent of any connection to the living animal. This can be illustrated when a frog's heart is surgically removed and then cut into little pieces; each piece will go on beating for as long as half an hour.

SUMMARY

1. Invertebrates commonly have external coverings, such as the tough *exoskeletons* of crustaceans and insects. Bony, internal skeletons (*endoskeletons*) were not produced until fishes appeared on earth, but after that all vertebrates, except sharks and their like, have had bones.
2. The main function of the vertebrate skeleton is to support and protect the body, but bones are also the main supply of reserve calcium and phosphorus, and the marrow produces blood cells.
3. Bones may be classified according to their internal composition as either *spongy* or *compact*.
4. The human skeleton is made up of an *axial* skeleton (skull, vertebrae, ribs) and an *appendicular* skeleton (arms, legs, and the structures that attach them to the axial skeleton).
5. The point where two separate bones meet or where a cartilage joins a bone is a *joint*, or *articulation*. The structure of a joint depends on its function. Joints are held together by *ligaments* and covered by *synovial membranes*.
6. A muscle is usually attached to a bone by a *tendon*. The junction point where the muscle is attached to the stationary bone is the point of *origin*, and the point of attachment of the muscle to the bone that moves is the *insertion*. Most muscles or groups of muscles have an *antagonistic* muscle or group to permit movement in opposite directions, or to allow a limb to be rotated. Basically, bones operate as a system of *levers*.
7. Muscle fibers are made up of *myofibrils*, which in turn are made up of the proteins *myosin* and *actin*. Within the myofibrils is a *sarcomere*, the functional unit of muscle contraction.
8. The *sliding-filament theory* of muscle contraction states that muscles are shortened when the thin actin and thick myosin filaments slide past each other without changing their lengths. The electrochemical process of muscle contraction, known as *excitation-contraction coupling*, is initiated by an electrical stimulus in the membrane of the muscle cell.
9. A group of skeletal muscle fibers activated by the same nerve fiber is called a *motor unit*.
10. The biological source of energy release during muscle contraction is ATP. The immediate reaction involves an actin-myosin-ATP association, with release of phosphate and energy.
11. The three types of muscles are *skeletal* (also called striated or voluntary muscle), *smooth* (found mostly in the internal organs of the body), and *cardiac* (found only in the heart).

ASK YOURSELF

1. How can an animal with its skeleton on the outside increase in size?
2. What do bones do besides support a body and provide attachment places for muscles?
3. What are bones made of?
4. Why do old people have fewer bones than children?
5. Describe four kinds of skeletal joints.
6. Explain how muscles in opposing pairs can bend or straighten a joint.
7. What are the structural, distributional, and functional differences between skeletal (striated) and involuntary (nonstriated) muscle?
8. How does a nerve impulse cause a muscle to contract?
9. How is calcium involved in muscle contraction?
10. What chemical events cause muscle fatigue?
11. Why do people breathe hard for a while after hard exercise?
12. Describe the "sliding-filament" theory of muscle contraction.

22
Nutrition and the Digestive System

SOME KEY POINTS

1. Digestion is the process of breaking down large food molecules into smaller ones.
2. Absorption is the passing of the smaller food molecules through the walls of the small intestine into the bloodstream.
3. The essential components of the human diet are carbohydrates, fats, proteins, vitamins, and minerals.
4. The dual processes of human digestion and absorption take place in the alimentary canal, starting with the mouth and ending at the anus.
5. Carbohydrates, fats, and proteins are broken down at specific locations throughout the compartments of the digestive system.
6. Digestion is completed in the small intestine, and most of the digested food molecules are absorbed from there into the bloodstream.
7. Enzymes and bile from the pancreas and liver aid in the digestive process.

THINK OF THE DIGESTIVE SYSTEM AS A TUBE within a tube. The inside tube is the digestive tract, including the stomach and intestines, and the outer tube is the body itself. Food that is in the stomach is still not inside the body proper, just as a raisin lying in the middle of the hole in a doughnut is not *inside* the doughnut. Food enters the digestive tract and undergoes **digestion,** the process of breaking down large food molecules into small food molecules. But the small molecules are useless unless they can get into the individual cells of the body. This is accomplished when the small food molecules pass through the cell membranes of the intestines into the bloodstream. This process is called **absorption.**

The twofold process of digestion and absorption has the same results in all animals: (1) it reduces large molecules of food to small molecules that can pass through cell membranes; (2) the small molecules can be used as a source of energy; and (3) the small molecules, such as amino acids, can be used by cells to rebuild, repair, and reproduce themselves. Besides being able to rebuild, repair, and reproduce its cells, an organism also needs a constant source of energy. Digestion and absorption make available the usable fuel that helps keep an animal alive. In this and the following chapters we will see how the body obtains and assimilates nutrients, how waste products are removed, and how oxygen and carbon dioxide are transported into and out of the overall body system.

WHAT ARE THE ESSENTIAL NUTRIENTS?

During the course of evolutionary adaptation to their environments, human beings have lost the ability to synthesize many of their cellular constituents. To remain alive, therefore, we must include certain essential components in our diets. These essential compounds are carbohydrates, lipids (fats), and proteins, as well as smaller molecules such as vitamins, which are essential to the proper functioning of enzyme systems, and minerals.

Carbohydrates

Glucose is the body's main energy source. It provides about two-thirds of the body's daily calorie requirement, and it is needed for cellular respiration. Glucose is produced from other carbohydrates, such as sugars and starches. The conversion of starches and some sugars takes place in the mouth and small intestine. Glucose may be used quickly for the synthesis of ATP, or it may be converted into glycogen and stored in the liver, where it is readily available when cells require energy (see p. 30).

Other sugars that can be easily used by the body are fructose (found in most fruits), maltose (in malt), and lactose (in milk). Some adults lose their ability to digest lactose.

Fats

Fats are an important part of every animal cell. They are essential structural components of the plasma membrane, the endoplasmic reticulum, mitochondria, and myelin sheaths around nerves. Animal

HUNGER AND STARVATION

Hunger may be nothing more than the desire for food, simple "appetite." One's appetite may be stimulated by the sight or smell of food, or even merely by reading about it. A person who feels hungry for dessert after an adequate meal clearly has no physiological need to eat. When the need for food exceeds intake, however, a different hunger, a real and physiological hunger sets in. Hunger pangs, one sign of hunger, are caused by the rubbing of the walls of an empty stomach, and they can be temporarily relieved by a drink of water. But when grim starvation occurs, not much water is needed to maintain the water balance in a body, because little is lost in excretion. A starving person can subsist on a quarter of a liter of water a day, as compared with 20 liters a day for a well-fed diabetic.

If the lack of food continues, body reserves must be used because there are unceasing demands, not merely for energy, but for energy that can be supplied only by specific compounds. To keep up body heat and to provide for movement, a number of kinds of molecules can be used, and they can be readily derived from carbohydrates, fats, or proteins; but a living brain demands a constant glucose supply except in really severe starvation, when some alternatives do become acceptable.

Continuing hunger results in shakiness of muscles and loss of mental alertness. A sometimes popular notion that fasting sharpens the mind is based in part on the observation that overeating causes lassitude, as indeed it does, and in part on a self-delusion similar to that which accompanies partial drunkenness. The half-starved or the half-drunk mind may feel acute because its most effective critical abilities are dulled; it is really more stupid than normal, so stupid, indeed, that it cannot recognize its own stupidity. In long-term starvation, the sufferer first uses reserves of fat, but when those are exhausted, muscle proteins are drawn on. Meanwhile, the starving person thinks constantly of food, finding reminders of it where a nourished person would never see them. Wastepapers are seen as having once wrapped food, and birds fly-

HOW ANIMALS EAT

Death comes swiftly in the natural world. To eat means to kill, and some animals may actually resort to cannibalism. Lions have been known to eat their cubs rather than begin a hunt, and in the frenzy of a group kill, hyenas may wound and then eat members of their own pack. Animals that hunt in packs usually spend much of their time trying to identify a weak or isolated member of the group they are stalking. Once the selection is made, the predators will chase the chosen victim until it is caught and killed, unless it manages to get away. If the prey has been chosen well, its handicap will be too much for it to overcome, and the hunters will eventually have their feast.

Cheetahs are so fast that few animals have a chance to escape unless the cheetahs are seen early, or unless inexperienced cubs such as these mishandle the chase. These young cheetahs stay close to their mother as they pursue an antelope and trap it as the mother watches. The kill is finally made by the mother, and only then do the apprentice hunters begin their meal.

Hyenas are not scavengers, but ferocious hunters that use the combined skills of the pack to subdue large animals like this African wildebeest. Once the prey has been wounded enough for them to bring it down, the hyenas begin to eat it alive. The hyenas use their powerful jaws to crush the bones and hide of the wildebeest. Only the skull and horns will be left. Lions are not nearly so thorough, and they are apt to abandon an animal before it is completely eaten. The scene on the next two pages usually follows as the vultures move in.

There is probably no more efficient hunter than a swooping killer bird. The types of prey and the devices for killing are almost as diverse as the birds themselves, but one factor is constant: When a killer bird strikes, it strikes with overpowering speed.

There are about 130 species of owls. About one-third of them are night hunters, aided by an acute sense of vision and a radarlike system of hearing. This barn owl probably did not need more than the light of the stars to make its catch, but it could have done just as well in total darkness if the mouse had made even the slightest sound.

Predatory birds like falcons, hawks, and eagles are some of nature's best-designed hunters. The peregrine falcon is streamlined for pursuit, diving at 290 kilometers per hour (180 mph), the fastest speed attained by any animal. This male osprey, a large fish-eating hawk, feasts on the head of a sea trout he has just caught in the shallows of Florida Bay. After the osprey has devoured the fish's head, he will carry the remainder of the carcass back to the nest in his powerful talons. His mate will tear the fish into little pieces for herself and the young ospreys.

Anhingas are also known in the southern United States as snakebirds or darters. The anhinga spears a fish with its pointed beak and then deftly tosses the fish into the air and catches it in the mouth. The fish is always caught head first so the scales and fins will not get stuck in the anhinga's throat.

LIFE Nature Library/Animal Behavior
Published by Time-Life Books Inc.

An eastern garter snake captures a toad and slowly maneuvers it into its mouth. The toad will be settled in the snake's gut, where powerful digestive juices will dissolve the limbs of the toad after three or four days. Most snakes have flexible skins and elastic jaws that allow them to swallow large objects. A snake's jaw can be opened relatively wider than a human jaw because the hinge where a snake's jawbone joins the skull is free to drop down, making the mouth enormous. A person with a snake's jaw and throat could swallow a football.

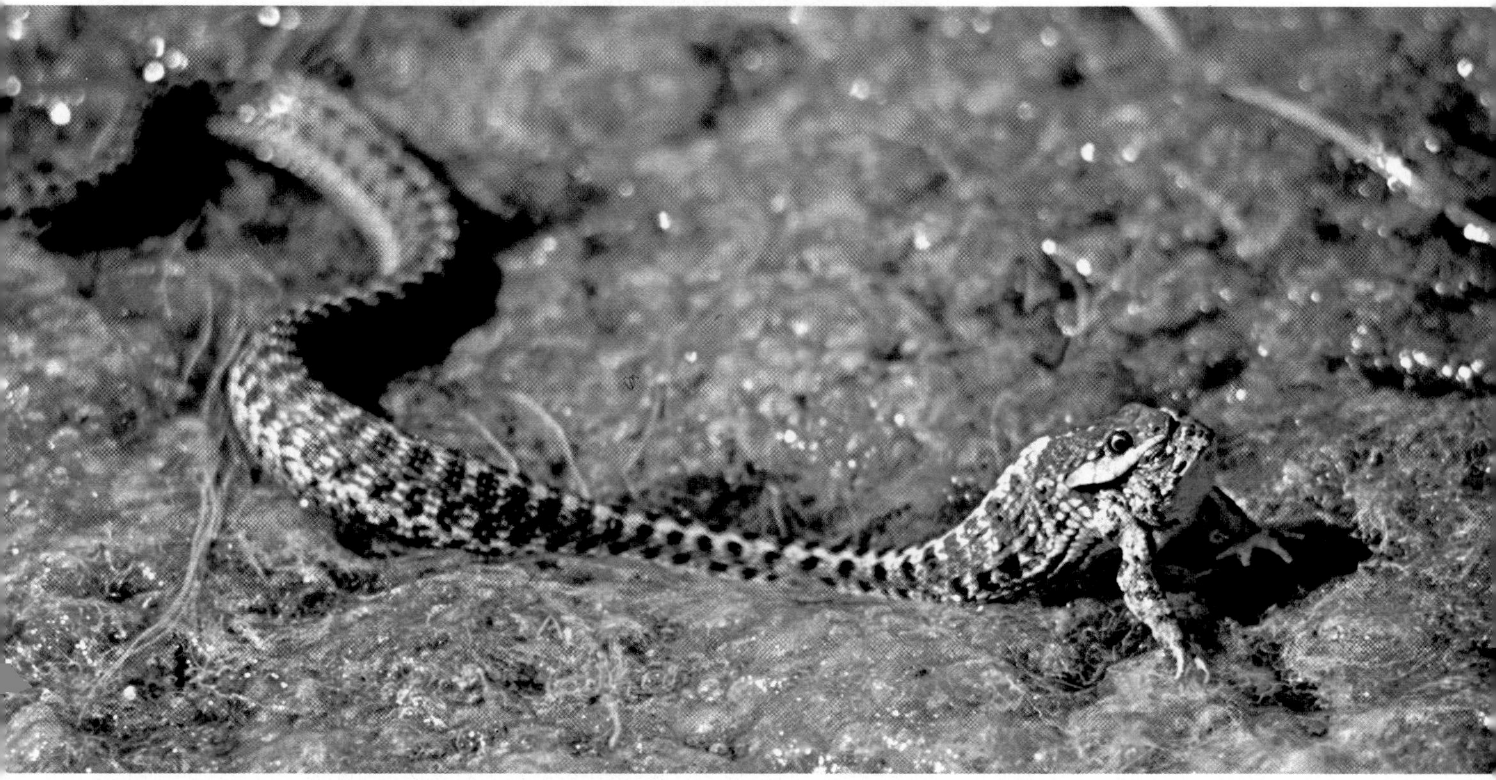

TONGUES. . .

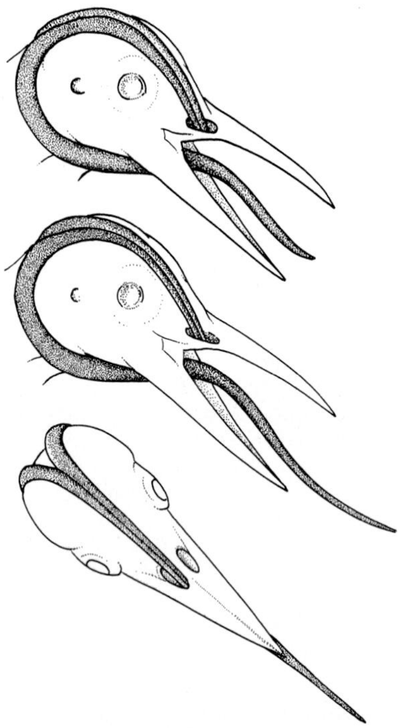

Woodpeckers may have perfected their hunting and eating skills as far back as 50 million years ago. In addition to a jackhammer beak and a specialized tail and claws that hold it securely in a vertical position against the tree trunk, the woodpecker has a flexible and retractable tongue that dips deep into bored holes to dislodge grubs and other food. The barbed tongue continues under the woodpecker's jaw, up around the back of the head, and is attached to the right nostril, leaving the left nostril clear for breathing. Because the tongue is held by elastic tissue, it can be retracted or protruded at will.

An anteater's tongue is usually about 60 centimeters (2 ft) long. In some species, like the giant anteater shown above, the muscles of the tongue are anchored on the rib cage, providing the strength and elasticity the tongue needs to snap at prey and pull back into the mouth at flashing speed. In the scaly anteater, the long muscles attach even farther back, on the pelvis. The anteater breaks into termite mounds and ant hills with its claws and then uses its sticky tongue to reach into crevices to catch the insects.

Frogs, toads, and many reptiles catch insects by flicking out a sticky tongue faster than the insect can act. Large insects that cannot be thrown neatly down a frog's gullet into its stomach are lined up lengthwise by the frog's forelegs and swallowed. A frog's pointed tongue is coated with viscous saliva. The tongue is usually held flat against the floor of the mouth, but it is unrolled and thrust outward when the frog sees a moving insect nearby. If the insect remains motionless, the frog will not strike.

Like frogs, chameleons also possess a long, sticky tongue that catches prey. But the chameleon's tongue is longer than a frog's, and the thrusting mechanism is more complicated. When not in use, the chameleon's tongue is folded up, accordion-style, at the back of the mouth. When an insect comes within range, the chameleon contracts one set of muscles that unfurl the tongue slightly, and then relaxes another set of muscles that run the full length of the tongue. This second action shoots the tongue forward at least the length of the chameleon's body, and the insect is trapped by the sticky bulb on the tip of the tongue.

The tongue of a honeybee is hollow, like a drinking straw. The bee sips nectar from a flower through the tube in its tongue and stores the liquid in its stomach. (Actually, enzymes begin converting the nectar to honey even before the bee returns to the hive.) Once back in the hive, the bee deposits the nectar in a special reservoir, where it is transformed into honey.

This African okapi is able to wrap its tongue around a leafy branch. It eats the leaves by pulling its tongue along the branch and stripping the leaves into its mouth.

. . . AND TEETH

A tiger's sharp teeth are typical of those of a carnivore that tears at its food. Long, pointed incisors at the front and strong canines are used for seizing and tearing, and the molars and premolars have been modified to shear like scissors.

Herbivores such as these hippopotamuses characteristically have flat molars and premolars that are suitable for grinding plants and for thorough chewing..

Rodents have long, curved front incisors that are used for gnawing; they have no canines at all. Because of their constant use, the incisors continue to grow, compensating for wear. The teeth remain sharp because only the front surfaces are coated with enamel, allowing the back side to wear down more quickly and creating a beveled edge.

The teeth of sharks are notorious. Not only are they long and razor-sharp, but they are lined up in several rows like soldiers, the young ones waiting to replace the old ones when they wear down or fall out. The new teeth are always larger than the ones they replace, keeping pace with the shark's overall growth. In ten years a tiger shark may use up about 24,000 teeth.

cells can usually synthesize their lipid constituents, but there are some limitations. Rats, for example, are unable to synthesize certain fatty acids. Young rats on a fat-free diet do not develop normally. They stop growing and do not mature sexually.

When fats are digested, they are broken down into their component parts, one of which is fatty acids. Fatty acids are long, thin molecules primarily consisting of hydrogen and carbon. If all the carbons have the maximum possible number of hydrogen atoms attached to them, they are called **saturated** fatty acids (see p. 32). If some of the carbons contain fewer hydrogen atoms, they are spoken of as "unsaturated" fatty acids. The human body can synthesize certain fatty acids, but others, labeled **essential** fatty acids, must be obtained from food. Most of the essential fatty acids are of the unsaturated type (see p. 32). Fatty acids, as well as the various kinds of fats they make up, are important to the body in a number of ways. They may serve as energy sources, but they may also play very important roles in the structure of cell membranes.

Cholesterol is a complex lipid with an entirely different molecular structure from fatty acids. It is an important constituent of all healthy living cells and an important component of any proper diet. However, too much cholesterol in the diet may contribute to circulatory diseases, such as the obstruction of arteries by fatty deposits (atherosclerosis). Animal fats and butter contain large amounts of this

---

ing are perceived as edible meat.

If some but not enough food is available, muscles waste and body fluids accumulate in the abdomen, causing skeletal spindliness of limbs and a puffy belly. In absolute starvation, the abdominal swelling does not occur. It must be emphasized that both quality and quantity of food affect the body reactions during times of insufficiency. Simple lack of calories has its own problems of inadequate energy and brain glucose, but a complication occurs when calorie intake by way of carbohydrate is sufficient, but protein content is too low. Then comes a disease of protein deficiency, *kwashiorkor,* with abdominal edema, splotchy colored skin, loss of hair pigment, and general weakness.

The effects of insufficient food and of unbalanced food are of special concern in childhood. Children who have too few calories do not grow to full size, and even if they get abundant nourishment in later life, they never overcome the initial stunting. The brain, too, is adversely affected both by too few calories and by too little protein. Studies have shown that the number of brain cells in an undernourished child or young animal is smaller than normal. Surprisingly, however, a smaller quantity of brain cells does not necessarily mean lowered intelligence. A famous investigation of Dutch army inductees showed that young men who had been born in the midst of the World War II famine had overcome their prenatal and immediately postnatal malnutrition so completely that there was no observable difference in intelligence between them and their better-fed siblings. The starving children and their mothers had lived for about half a year while sharing fewer than 500 kilocalories a day, but having survived to manhood, the young men were normally bright.

Assuming a varied diet that includes carbohydrates, proteins, fats, vitamins, and minerals, an average adult can use 3000 to 4000 kilocalories a day. A hardworking lumberjack cutting timber in the freezing woods may need twice that, and an inactive person can survive, though miserably, on half that.

complex lipid. It is believed that cholesterol and other lipid deposits on the walls of the arteries reduce the space available for blood flow. In its extreme forms this blockage can cause very serious circulatory defects, heart disease, and death. Unfortunately, the precise relationship of cholesterol in the diet to heart disease is not completely clear. Factors other than cholesterol intake may play an important part.

Proteins

The word "protein" comes from the Greek word that means "primary" or "holding first place." It is an apt name for such an important nutrient, since proteins, specifically *amino acids,* are required by all animals (see p. 34). Without proteins our bodies would waste away and die. No new cells would form, and the protein *hemoglobin* would not be available to carry oxygen through the blood.

The human body contains 20 different amino acids, which can be joined in practically unlimited combinations in protein molecules. The structure and amino acid composition of proteins vary from plant to animal and from animal to animal, even from human to human. For example, differences in protein complexes make it unlikely that one person can accept a skin graft from another. Because of structural differences, proteins must be broken down by the digestive system and rebuilt to meet the specifications of each organism. When you eat beef, you are not absorbing *beef* proteins. Rather, you digest the beef, breaking down the proteins to yield amino acids (the structural units of proteins), which are then absorbed. When your body synthesizes new protein, the amino acids are rearranged according to the genetic message of human DNA to make *human* proteins—more specifically, *your* proteins. In a similar way, a builder can construct many different types of houses with the same building materials.

Proteins are digested in the stomach and small intestine. The digested protein will eventually become skin, hair, fingernails, enzymes, hormones, and in general the tissues of the body. Those amino acids that the human body cannot synthesize and that must therefore be included in the diet are called the *essential,* or *required,* amino acids. Those that can be synthesized are equally necessary for survival, but they are not included in the technical term "essential amino acids." Humans require eight different essential amino acids in their diet (threonine, methionine, valine, leucine, isoleucine, lysine, phenylalanine, tryptophan). Children need two others (arginine, histidine) during their developing years. There are large amounts of protein in meat, eggs, fish, seeds, and dairy products.

Vitamins

Vitamins play such an important role in supporting the work of a number of enzymes, especially those involved in cellular respiration, that vitamin deficiencies cause many different diseases (Table 22.1). Because the body cannot synthesize vitamins from other food sources, foods containing the necessary vitamins must be included in the diet constantly. To be effective, vitamins must not be altered during the digestive process.

Vitamins are usually classified as either water-soluble or fat-soluble. They are ordinarily required in minute amounts, but their importance is enormous.

TABLE 22.1
Vitamins

DESIGNATION LETTER AND NAME	MAJOR SOURCES	FUNCTION	DAILY REQUIREMENT	EFFECTS OF DEFICIENCY
A (carotene)	Egg yolk, green or yellow vegetables, fruits, liver, butter	Formation of visual pigments, maintenance of normal epithelial structure	5000 I.U.*	Night blindness, skin lesions
B-Complex Vitamins (B_1 through Biotin)				
B_1 (thiamine)	Brain, liver, kidney, heart, whole grains, yeast	Formation of enzyme involved in Krebs cycle	1.5 mg	Stoppage of carbohydrate metabolism at pyruvate; beriberi, neuritis, heart failure, mental disturbance
B_2 (riboflavin)	Milk, eggs, liver, whole-grain cereals, spinach	Cytochromes in oxidative phosphorylation (hydrogen transport)	1.5–2.0 mg	Photophobia, fissuring of skin
B_6 (pyridoxine)	Whole grains, liver, milk	Coenzyme for amino-acid and fatty-acid metabolism	1–2 mg	Dermatitis, nervous disorders
B_{12} (cyanocobalamin)	Liver, kidney, brain; synthesis by bacteria in gut	Nucleoprotein synthesis (RNA); prevents pernicious anemia	2–5 mg	Pernicious anemia, malformed red blood cells
Folic acid	Meats, vegetables, eggs	Nucleoprotein synthesis; formation of red blood cells	0.5 mg or less	Failure of red blood cells to mature, anemia
Niacin	Whole grains, liver, chicken, yeast	Coenzyme in hydrogen transport (NAD, NADP)	17–20 mg	Pellagra, skin lesions, digestive disturbances, dementia
Pantothenic acid	Yeast, liver, eggs	Forms part of coenzyme A (CoA)	8.5–10 mg	Neuromotor disorders, cardiovascular disorders, gastrointestinal distress
Biotin	Egg white; synthesis by intestinal bacteria	Concerned with protein synthesis, CO_2 fixation, and nitrogen metabolism	150–300 mg	Scaly dermatitis, muscle pains, weakness
C (ascorbic acid)	Citrus fruits, tomatoes, potatoes, butter	Vital to collagen synthesis and intracellular substance	75 mg	Scurvy, failure to form connective tissue fibers
D_3 (calciferol)	Fish oils, liver, milk, sunlight	Increases calcium absorption from gut, important in bone and tooth formation	400 I.U.	Rickets (defective bone formation)
E (tocopherol)	Green leafy vegetables, wheat germ oil	Maintains resistance of red blood cells to hemolysis (bursting of red blood cells)	Not known	Increased fragility of red blood cells
K (naphthoquinone)	Liver, leafy vegetables; synthesis by intestinal bacteria	Aids in prothrombin synthesis by liver	Not known	Failure of blood to clot
Inositol	Fruits, nuts, vegetables	Aids in metabolism, prevents fatty liver	Not known	Fatty liver

Adapted from James E. Crouch and J. Robert McClintic, *Human Anatomy and Physiology,* Second Edition (New York: John Wiley, 1976), pp. 656–657. Used with permission.
* International units.

TABLE 22.2
Major Minerals

MINERAL	MAJOR SOURCES	FUNCTION	DAILY REQUIREMENT	EFFECTS OF EXCESS	EFFECTS OF DEFICIENCY
Calcium	Dairy products, eggs, fish, soybeans	Bone structure, blood clotting, muscle contraction, excitability, synapses	About 1 gm	None	Tetany of muscles, loss of bone minerals
Chlorine	All foods, table salt	Acid-base balance, osmotic equilibria	2–3 gm	Edema	Alkalosis, muscle cramps
Cobalt	Meats	Necessary for formation of red blood cells	Not known	None	Pernicious anemia
Copper	Liver, meats	Necessary for hemoglobin formation	2 mg in adults	Toxic	Anemia (insufficient hemoglobin in red cells)
Fluorine	Fluoridated water, dentifrices, milk	Hardens bones and teeth, suppresses bacterial action in mouth	0.7 part/million in water is optimum	Mottling of teeth	Tendency to dental caries
Iodine	Iodized table salt, fish	Synthesis of thyroid hormone	0.1–0.2 mg in adults	None	Goiter, cretinism
Iron	Liver, eggs, red meat, beans, nuts, raisins	Oxygen and electron transport	16 mg in adults	May be toxic	Anemia
Magnesium	Green vegetables, milk, meats	Bone structure, cofactor with enzymes, regulation of nerve and muscle action	About 13 mg	None	Tetany
Manganese	Bananas, bran, beans, leafy vegetables, whole grains	Formation of hemoglobin, activation of enzymes	Not known	Muscular weakness, nervous system disturbance	Subnormal tissue respiration
Phosphorus	Dairy products, meats, beans, grains	Bone structure, intermediary metabolism, buffers, membranes, phosphate bonds essential for energy (ATP), nucleic acids	About 1.5 gm	None	Unknown; related to rickets; loss of bone mineral
Potassium	All foods, especially meats, vegetables, milk	Buffering, muscle and nerve action	1–2 gm	Heart block	Changes in electrocardiogram, alteration in muscle contraction
Sodium	Most foods, table salt	Ionic equilibrium, osmotic gradients, excitability in all cells	About 6 gm in adults	Edema, hypertension	Dehydration, muscle cramps, kidney shutdown
Sulfur	All protein-containing foods	Structural, as amino acids are made into proteins	Not known	Unknown	Unknown
Zinc	Meats, eggs, legumes, milk, green vegetables	Part of some enzymes	Not known	Unknown	Unknown

Adapted from James E. Crouch and J. Robert McClintic, *Human Anatomy and Physiology,* Second Edition (New York: John Wiley, 1976), p. 658. Used with permission.

Minerals

At least 14 inorganic substances are essential for the maintenance of a healthy body, and their benefits range from the development of good bone structure to the activation of enzymes. Table 22.2 outlines the essential inorganic substances (*minerals*) and relevant facts about them.

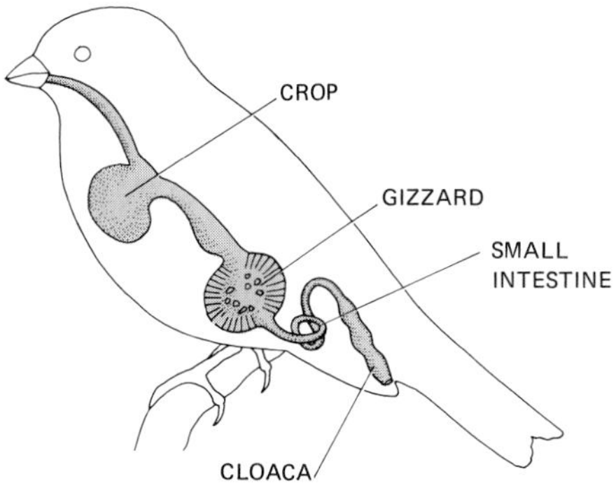

Figure 22.1
The digestive system of a bird. Food, swallowed whole, can be stockpiled in a swelling at the lower end of the esophagus, the *crop*. From the crop, it is passed into the *gizzard*, morphologically the stomach but with heavy muscular walls that rub the food with the gravel that the bird takes in with the food. The abrasive action in the gizzard takes the place of chewing in toothed animals. Once ground up in the gizzard, food is passed to the small intestine, where digestion and absorption occur. Gravel and undigested wastes are passed out through the *cloaca* (Latin, "sewer"), an enlargement at the terminus of the alimentary canal through which move feces, waste from the kidneys, and eggs or sperm.

THE DIGESTIVE PROCESS REDUCES LARGE, INSOLUBLE MOLECULES TO SMALL, SOLUBLE ONES

In its simplest terms, digestion is the process of breaking down large, complex, insoluble molecules of food into small, soluble molecules that can be absorbed and used by cells.

The digestive process, assisted by enzymes, takes place in the digestive tract, or **alimentary canal.** In humans, this muscular tube starts at the mouth and ends at the anus,and it is usually about seven to eight meters (23–26 ft) long. The major compartments of the alimentary canal are the oral (buccal) cavity (commonly but not quite correctly called the mouth), the esophagus (a slender tube connecting the throat with the stomach), the stomach, and the intestines. The glands associated with the digestive tube contribute mucus for lubrication, enzymes for chemical breakdown, agents for emulsification of fats, and ions for the control of acid-base relationships in the local environment.

ADAPTATIONS IN DIGESTIVE SYSTEMS

The process of reducing food particles in size and complexity is fundamentally the same in all organisms, but of course some differences occur. Some animals, such as earthworms and birds, who eat almost continuously, have a separate storage segment, the *crop*, from which food is released at controlled intervals (Figure 22.1). Another adaptation, the *gizzard*, is a muscular sac containing sand or small stones, which the toothless bird or worm has swallowed to help crush and grind food that would be broken down by teeth in other animal groups. Without such stones in the gizzard an animal such as a chicken starves to death, unable to begin the digestive process. Conscientious bird feeders supply sand as well as grain for wild birds when the ground is covered with snow.

Organisms like tapeworms, parasites that live in vertebrate intestines, do not have digestive systems (Figure 22.2). Tapeworms simply absorb the predigested food of their hosts and therefore do not need any digestive mechanisms of their own. The cow and some other grass-eating mammals have a large organ at the base of the esophagus called the *rumen*. The rumen contains protozoa and bacteria that break down cellulose, one of the constituents of cell walls, and that synthesize the necessary vitamins. Food is also stored and fermented in the rumen, and it can be regurgitated and rechewed several times before it passes on to the stomach.

THE HUMAN DIGESTIVE SYSTEM IS COMPARTMENTALIZED

Human beings are **omnivores;** that is, they can and do eat practically any type of food. Because of this varied intake of foods, the overall human digestive

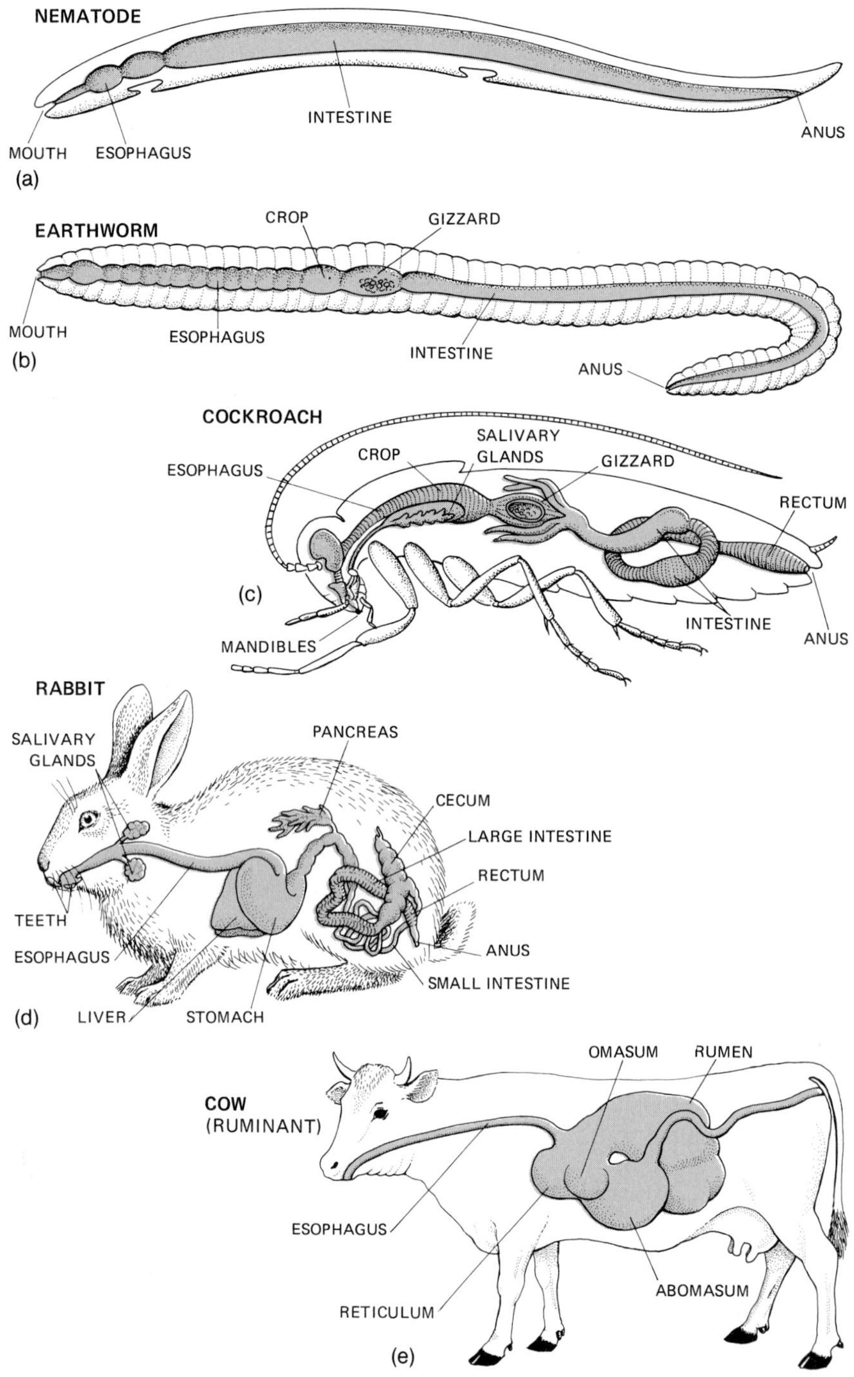

Figure 22.2
Comparison of the digestive tracts of several invertebrate animals and two mammals. Some of the simpler animals, such as jellyfishes, do not have any special structures that make a definite digestive tract. The main features of a digestive tract include a mouth for ingestion, in most animals a storage pouch, glands for the production of digestive enzymes, a more or less convoluted tube from which nutrients can be absorbed, and an anus through which wastes can be expelled. There is usually some means of reducing the size of ingested particles: a filter, a rasp, chewing teeth, or an internal grinder.

Some animals (the ruminants, such as cattle) have a double system, in which food is swallowed with minimal chewing and passed into the first two enlargements of the alimentary tract: the rumen and the reticulum. After it has been partially broken down by bacterial action, it is regurgitated and chewed further. The food is swallowed again and sent to two more containers, the omasum and the abomasum, where digestive enzymes act on it before it is passed to the small intestine for absorption of the nutrients. Most of the digestive action in a ruminant is accomplished by bacteria.

system (Figure 22.3) is less specialized than that of some other animals, whose systems are designed for only meat (carnivores) or only vegetation (herbivores), for instance. However, much specialization occurs within the different parts of the human digestive system. Probably the most impressive aspect of the system is its sequential coordination, which involves a network of controls, regulated by hormones and the central nervous system. These controls produce the required activities at the appropriate time and under the proper conditions. Most of the functions of the digestive system are autoregulatory; that is, they take place without our conscious control.

The human digestive system is compartmentalized, and each part is adapted to a specific function (Figure 22.4). The esophagus (Greek, "I shall carry" and "to eat") passes the food from the mouth to the

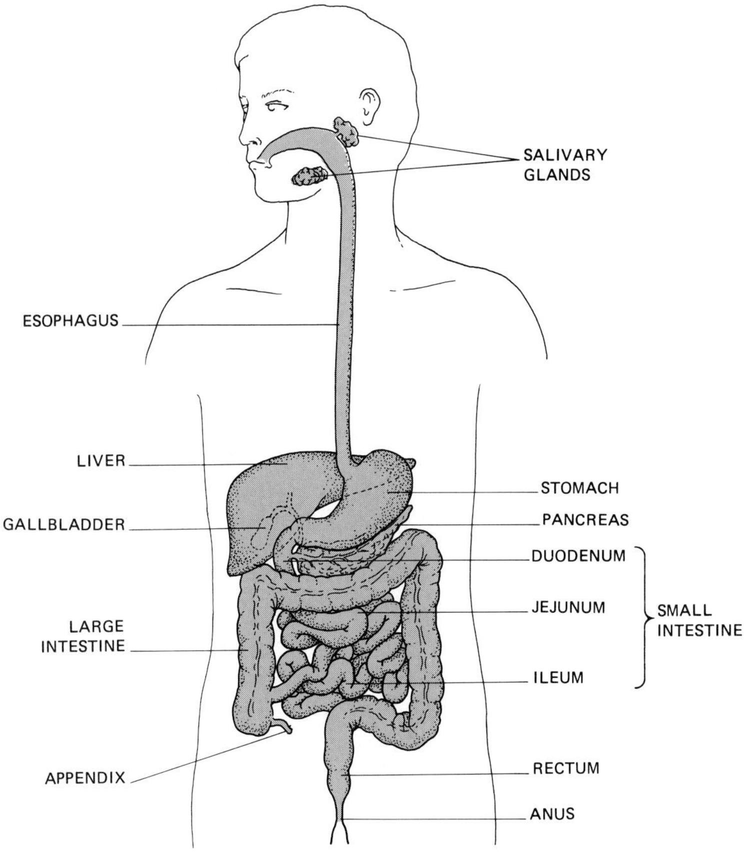

Figure 22.3
The human digestive tract. The drawing shows some parts, notably the small intestine, out of realistic proportion for the sake of clarity.

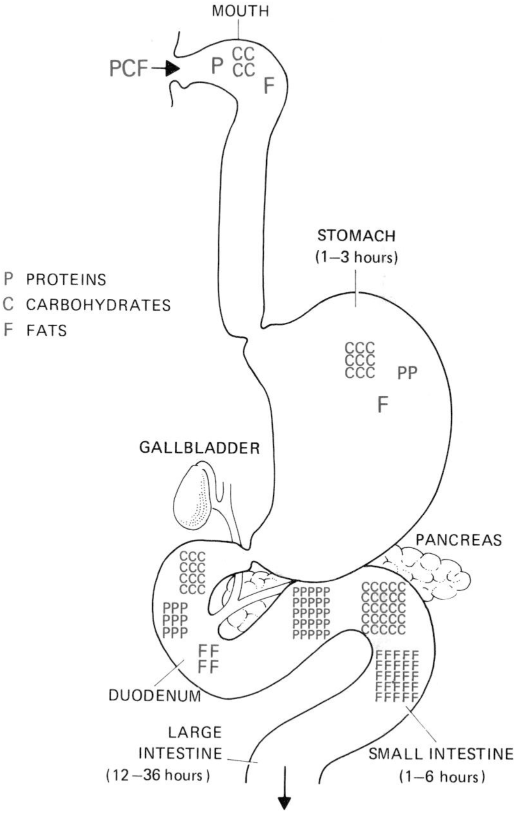

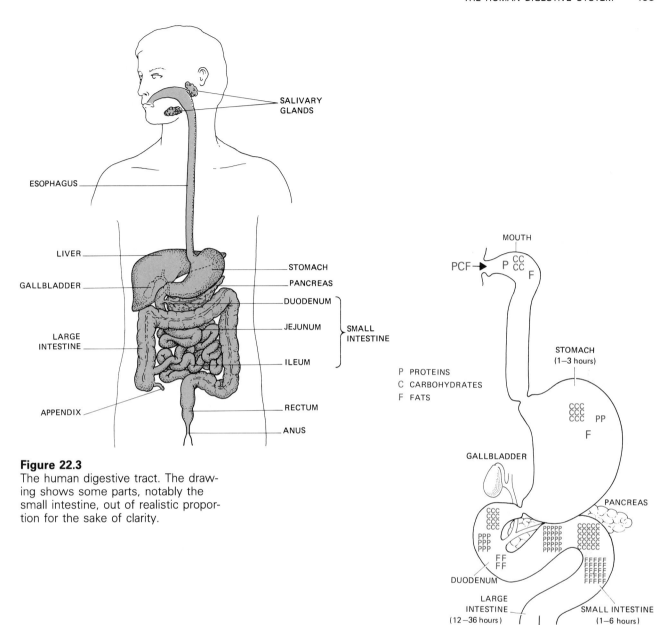

Figure 22.4
Compartmentalized digestion in the human body. Each compartment of the digestive system has a specific function in the overall process of digestion. Digestion begins in the mouth and ends in the small intestine, and carbohydrates, fats, and proteins are broken down at specific locations throughout the system.

The letter symbols PCF, shown entering the mouth in the diagram, represent undigested proteins, carbohydrates, and fats, respectively. The letter symbols are shown broken up into smaller and smaller sizes as proteins, carbohydrates, and fats are digested throughout the digestive system. For instance, the digestion of carbohydrates (C) begins in the mouth, and so carbohydrates are represented there with four medium-sized Cs, and the Cs keep getting smaller as digestion continues in the stomach, duodenum, and small intestine; but the digestion of fats (F) does not begin until they reach the duodenum, so the letter symbol F is still whole in the stomach. In the small intestine, the digestion of proteins, carbohydrates, and fats is completed.

In the *mouth,* some starches and sugars are broken down into simple sugars by the enzymatic action of ptyalin in saliva, but most food molecules are still too large to be absorbed into the bloodstream. In the *stomach,* the digestion of carbohydrates continues, and proteins are partially digested. Fats remain relatively unaffected. Some small molecules and inorganic salts may be absorbed from the stomach. In the *duodenum,* fats are emulsified by bile from the gallbladder, and secretions supplied by the pancreas break down carbohydrates, fats, and proteins. In the rest of the *small intestine,* the breakdown of carbohydrates, fats, and proteins continues, and most of the absorption takes place.

stomach; the stomach stores food and breaks it down with acid and some enzymes; the small intestine is the site of most of the chemical digestion of large molecules into smaller ones, and it is the place where most of the nutrients pass into the bloodstream; the large intestine carries undigested food, removes water from it, and releases solid waste products through the anus.

If we follow a mouthful of food through the digestive tract, we can observe the digestive process in greater detail.

Digestion in the Oral Cavity

In the mouth, where it all begins, the food is masticated (chewed) by the ripping, crushing, and grinding action of the teeth (Figure 22.5). As much pressure as 500 kilograms per square centimeter can be exerted by the molars, considerably greater than the pressure under the tires of a large Cadillac. At the same time, the food is moistened by saliva, which intensifies its taste and eases its passage down the delicate tissue of the esophagus. Saliva, secreted by the salivary glands, contains an enzyme, amylase, that converts large molecules of starch into small molecules of malt sugar. *This conversion from large molecules to small ones will continue throughout the digestive process.* The specific amylase in saliva, known as ptyalin, is the first of the many digestive enzymes at work in the digestive tract.

Now the moistened, softened parcel of food, known as a bolus, is ready to be swallowed. The first stage of swallowing is voluntary, and it is the final voluntary digestive movement until the waste products are expelled during defecation. The tongue serves as a digestive organ at this time by helping to push the bolus into the pharynx, or throat. The esophagus, a muscular tube about 25 centimeters (10 in.) long connects the pharynx to the stomach. When swallowing begins, a valve called the epiglottis presses down and prevents the food from entering the trachea (windpipe) or the larynx (voice box). If a tiny bit of food slips into the wrong channel, you will cough and choke until the passage is cleared, or you will die of suffocation.

The complete first stage of swallowing consists of four simultaneous movements (Figure 22.6): (1) The soft palate is raised to prevent food from entering the nasal cavity. (2) The muscles at the rear of the mouth form a thin slit, which keeps large objects from passing into the pharynx. (3) The vocal cords tighten, and the epiglottis swings back over the opening of the larynx. (4) The main muscle of the pharynx contracts and initiates a muscular wave called **peristalsis,** which pushes the food through the esophagus and into the stomach in 4 to 10 seconds.

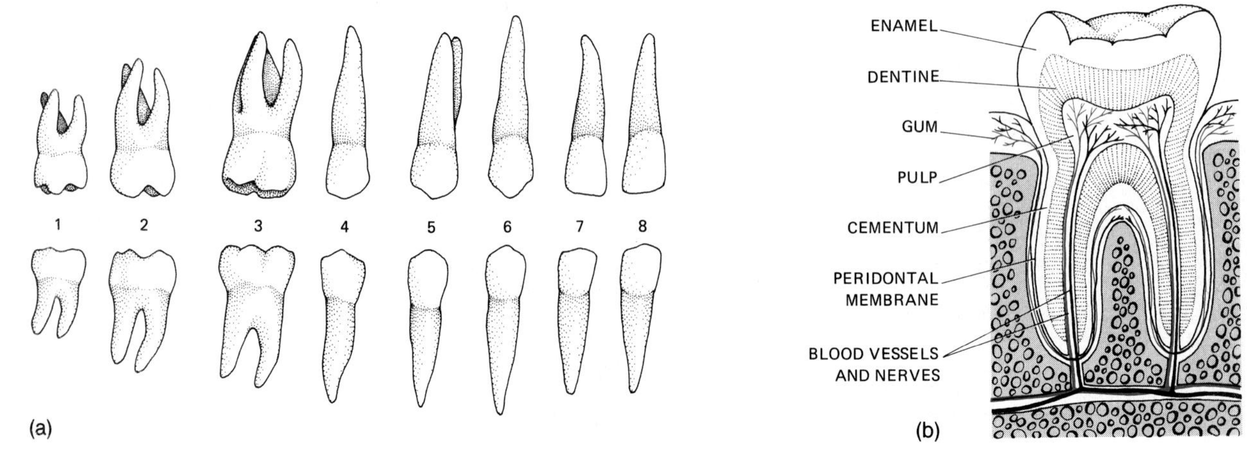

Figure 22.5
Human teeth. (a) Half a set of teeth: upper and lower from the right side. The short, prong-rooted molars (1, 2, 3) and premolars (4 and 5) are for grinding; the pointed canines (6) are tearing teeth; the chisel-edged incisors (7 and 8) are for nibbling and gnawing. After a few years' use, a set of teeth becomes so individual and distinctive, with its cracks, losses, shifts of position, and dental fillings, that it can be used for identification after other special peculiarities, such as fingerprints, are gone. Teeth are the hardest parts of the body. When hominid fossils are found, the teeth are usually virtually intact. (b) Diagram of a long section of a tooth. The hard, brittle outer covering is the enamel. Inward from the bony *dentine* is a soft *pulp,* containing nerves and blood vessels. The vulnerable parts of a tooth are the enamel and the peridental membrane, which covers the tooth below the gum line.

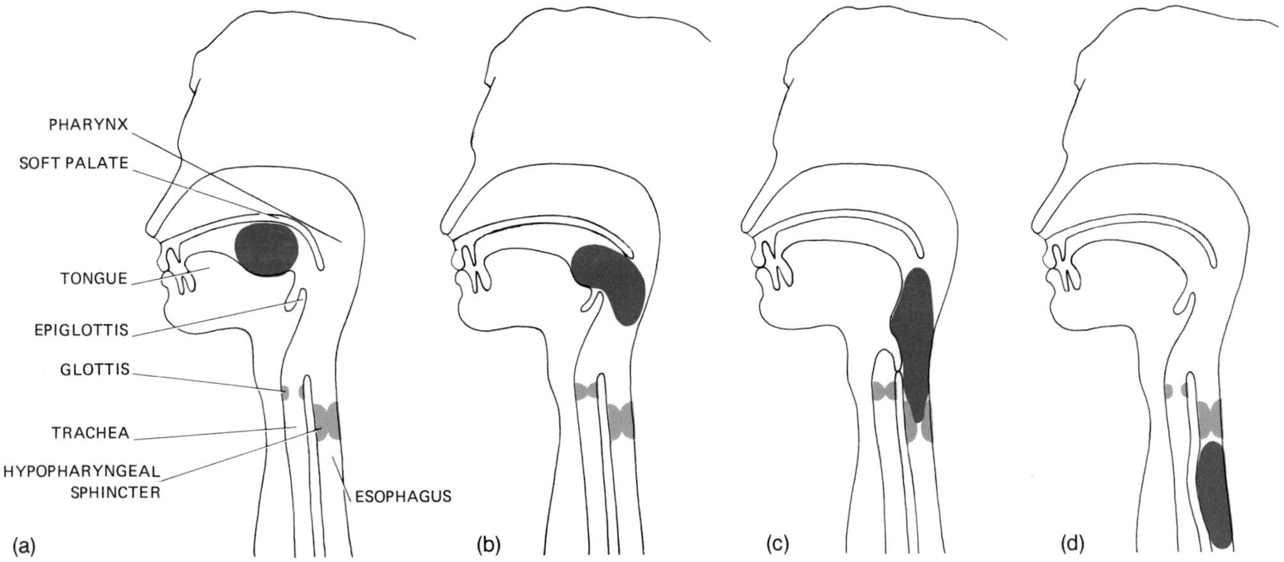

Figure 22.6
Action of the epiglottis in swallowing. (a) Air breathed in through the nose and pulled down through the trachea must move across the passageway through which food goes from mouth to stomach, and yet choking (erroneous entry of food into the trachea) is relatively rare. As food starts to go by the trachael opening, called the *glottis* (b), a flap of tissue, the *epiglottis*, folds down over the opening (c), allowing the food to slip past and into the esophagus (d). Once past the pharynx, a mass of food (a *bolus*) is beyond voluntary recall. From there on through the alimentary tract, it is propelled by peristalsis, the reflex action of waves of muscular contractions and relaxations. (Astronauts in a gravity-zero situation have no trouble swallowing.)

The muscular action of peristalsis continues to assist the passage of food through the rest of the digestive system.

Digestion in the Stomach

The stomach has been alerted that food is on the way, and secretions of hydrochloric acid and the stomach product pepsinogen have already begun in some of the 35 million glands lining the stomach. The pepsinogen is changed, by enzymatic action in the presence of hydrochloric acid, to an active protein-splitting enzyme: pepsin. Pepsin, helped by the acid, breaks large protein molecules into smaller molecules of peptones, proteoses, and amino acids. The acid condition is necessary for both the conversion of pepsinogen to pepsin and the digestion of proteins in the stomach. Although the stomach wall itself is largely made of protein, it is normally not digested because it is lined with a resistant, alkaline coat of mucus. Abnormally, a break in the mucous lining is thought to result in a stomach sore, or *gastric ulcer* (Figure 22.7). Actually, ulcers of the initial

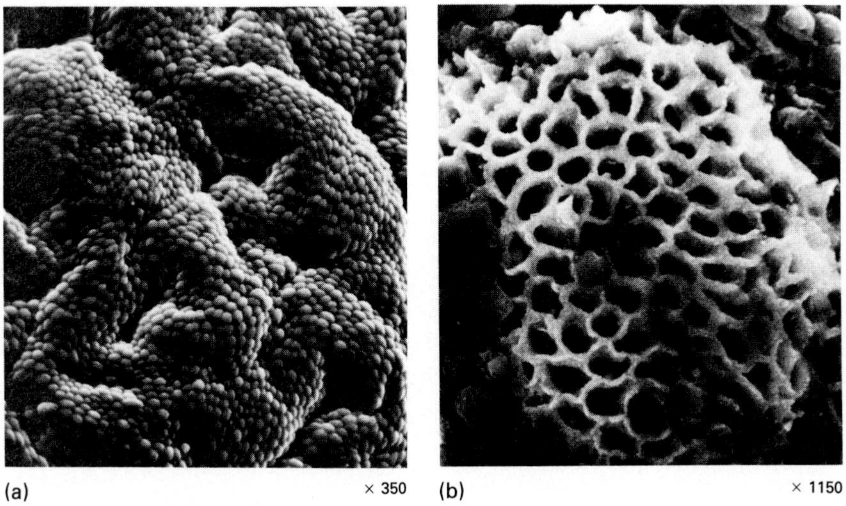

Figure 22.7
Normal and damaged stomach linings, as seen in scanning electronmicrographs. (a) A normal stomach lining, raised into ridges and furrows and pebbled with convex cell surfaces, has a continuous mucous coating that protects the living tissue from the digestive action of the stomach contents. (b) If the protection is destroyed, some of the cells can be killed, and the result is an open sore or ulcer.

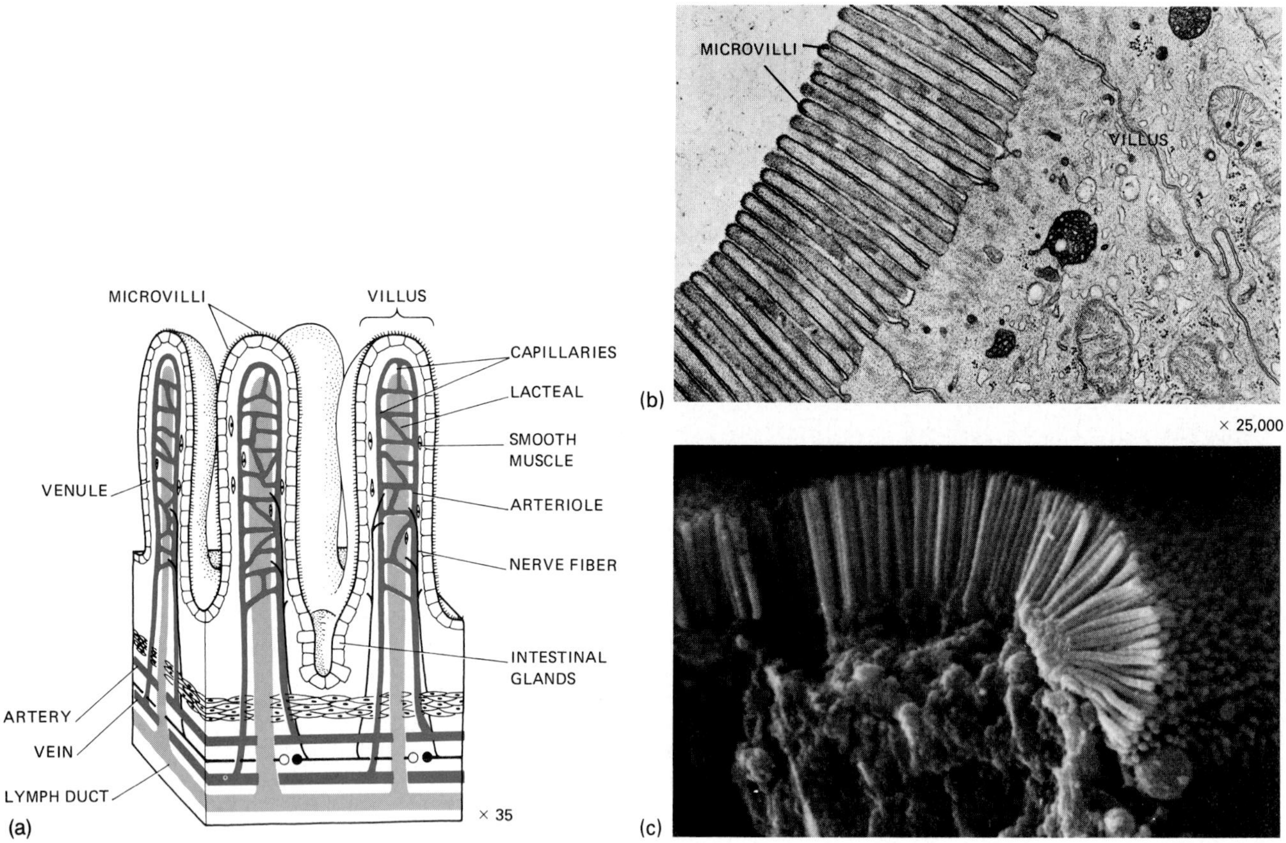

portion of the small intestine, the duodenum, are ten times as common as stomach ulcers. Acidified stomach contents are usually passed on to the small intestine where they are made alkaline, but if some stomach acid is inadvertently pushed upward to the mouth, the result is "heartburn."

The secretion of gastric juices in the stomach occurs in three phases:

1. The *cephalic* ("head") phase. When food is seen, smelled, or tasted, the stomach is stimulated by vagus nerves from the brain, and a small amount of juice is secreted.

2. The *gastric* ("belly") phase. Gastric juice is secreted when incoming food stretches the stomach. Further secretions occur when the hormone gastrin enters the stomach through the bloodstream and causes the production of large amounts of gastric juice.

3. The *intestinal* phase. Additional small amounts of gastric juice are secreted when the partially digested food enters the opening section of the small intestine.

Besides producing the gastric juices, the stomach stores large quantities of food, using a churning action to mix food into a soupy mixture called chyme (KIME). It then slowly empties the partially digested food into the small intestine. In fact, the most important function of the stomach is to regulate the flow of chyme into the small intestine. The stomach is well suited for storage because its muscles have little tone, allowing for enlargement up to four liters as necessary.

Food enters the stomach by way of the cardiac sphincter (so named because it is located near the heart), an opening regulated by a ringlike muscle (sphincter). When the food is ready, it leaves the stomach through the pylorus (Greek, "gatekeeper").

Digestion in the Small Intestine

After spending approximately one to three hours in the stomach (liquids pass through almost immediately), the food moves into the small intestine, where further contractions continue to mix the

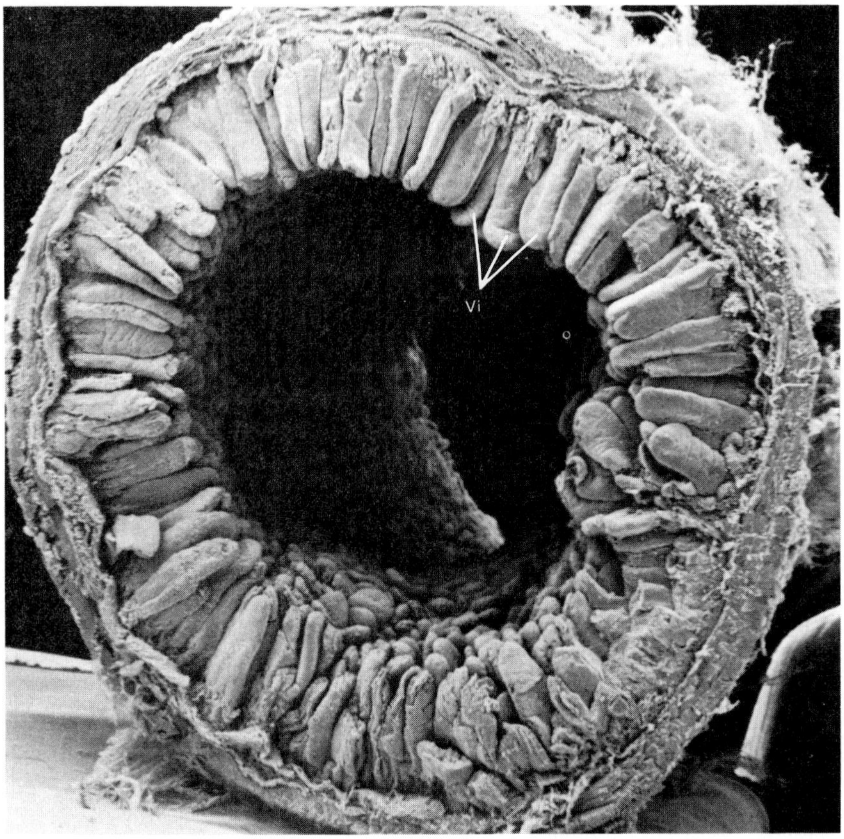

Figure 22.8
Lining of the small intestine. (a) Schematic drawing of a section through a small portion of the inner lining of a small intestine. Four villi are shown projecting into the central cavity of the intestine. The villi, which increase the absorptive surface, are capable of slow, limited movement, and they are provided with abundant blood vessels. The surface epithelial cells, which are subject to constant abrasion, are renewed by growth of new cells. Each villus is covered with smaller projections over the surface, the microvilli, which are just visible as a fuzzy edge to the villus viewed under an ordinary microscope. Thin sections in an electron micrograph (b) and a scanning electron micrograph (c) show a small portion of an epithelial cell with the microvilli (Mv) well defined. (d) A low-magnification scanning electron micrograph of a sliced portion of the tubular small intestine clearly shows how numerous villi (Vi) line the intestinal walls.

soupy liquid. It takes anywhere from one to six hours for the food to move through the six-meter-long (20-ft) small intestine, where the digestion of all three basic types of food is completed. It is there also that practically all of the digested molecules of food are absorbed into the bloodstream. Alcohol may be absorbed through the stomach walls, but the rest of the food mixture moves into the intestine.

The absorbing surface of the small intestine is increased by millions of fingerlike protrusions called **villi** (Latin, "shaggy hairs"), which give the appearance of the tufts of a Turkish towel or the velvety pile of a rug (Figure 22.8). The villi aid digestion further by adding their constant waving motion to the peristaltic movement already occurring throughout the small intestine. The thousands of cells making up each villus are the units through which absorption from the intestine takes place. The absorptive surface of each of these cells contains thousands of *microvilli,* which further increase the surface area for absorption (see Figure 22.8). An entire small intestine exposes some 250 square meters (300 yd^2) of surface area.

Enzymes for Digestion in the Small Intestine

Increased surface area and waves of motion are important to digestion in the small intestine, but even more important is the action of the pancreatic enzymes and bile secreted into the small intestine by the pancreas and liver through a common duct. Bile salts break down fat droplets into particles small enough for the enzymes to attack. The physical breakdown of fats is called *emulsification,* the same process by which soap breaks down grease. The liver secretes bile, which is collected, stored, and modified in the gallbladder.

These secretions of pancreatic juices and bile, together with intestinal enzymes, help to change the environment in the duodenum and entire small intestine from acidic to alkaline. The action of pepsin is stopped by the new alkaline environment that allows intestinal enzymes to function properly. Enzymes furnished by the pancreas and intestine break down carbohydrates, fats, and proteins. These enzymes are the *carbohydrases,* the *lipases,* and *proteases.* Note that enzymes almost always end with the "ase" suffix; the prefixes, such as "prote"

TABLE 22.3
The Major Digestive Glands and Enzymes

PLACE OF DIGESTION	SOURCE	SECRETION	ENZYME	DIGESTIVE FUNCTION
Mouth	Salivary glands	Saliva	Ptyalin	Begins carbohydrate digestion; breaks down starch into maltose
	Mucous glands	Mucus		Lubricates
Esophagus	Mucous glands	Mucus		Lubricates
Stomach	Gastric glands	Gastric fluid	Lipase	Converts fats to fatty acids and glycerol
	Gastric glands	Gastric fluid	Pepsin	Converts proteins to polypeptides
	Gastric mucosa	Hydrochloric acid		Activates pepsin; dissolves minerals; kills bacteria
	Mucous glands	Mucus		Lubricates
Small Intestine	Liver	Bile		Emulsifies fats; activates lipase
	Pancreas	Pancreatic fluid	Amylase	Converts starch to maltose and other disaccharides
			Lipase	Converts fats to fatty acids and glycerol
			Nuclease	Converts nucleic acids to mononucleotides
			Trypsin	Converts proteins to peptides and amino acids
	Intestinal glands	Intestinal fluid	Enterokinase	Converts trypsinogen to trypsin
			Lactase	Converts lactose to glucose and galactose
			Maltase	Converts maltose to glucose
			Sucrase	Converts sucrose to glucose and fructose
	Mucous glands	Mucus		Lubricates
Large Intestine	Mucous glands	Mucus		Lubricates

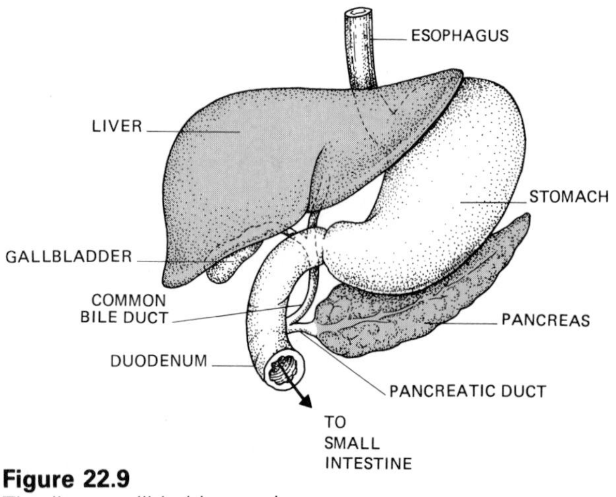

Figure 22.9
The liver, gallbladder, and pancreas secrete their products into ducts that merge as they enter the duodenum.

in protease, usually indicate quite clearly what substance is being acted upon (Table 22.3).

Digested carbohydrates and proteins are eventually transported to the liver, and digested fats are carried into the lymphatic channels (Chapter 24). Ultimately, almost all the digested food molecules will be passed into the bloodstream, to be circulated throughout the body as energy-producing and tissue-building nourishment.

The Liver as a Digestive Organ

The **liver** is the largest glandular organ in the human body, weighing about 1.5 kilograms (3 lb) in an adult (Figure 22.9; see also Figure 22.3). It is an extremely versatile organ, performing more than 500 separate functions, but this discussion will focus on

TABLE 22.4
The Major Digestive Hormones

HORMONE	SOURCE	STIMULUS	FUNCTION
Gastrin	Stomach	Food entering stomach, vagus nerve, protein	Stimulates secretion of gastric juices
Secretin	Small intestine	Acid and peptides in duodenum	Stimulates pancreatic juices containing bicarbonate; stimulates liver to produce bile
Pancreozymin (cholecystokinin)	Small intestine	Peptides and fat in duodenum	Stimulates gallbladder to release bile; stimulates pancreas to release enzymes

the liver as a digestive and metabolic organ. In the next chapter we will look at its functions as an excretory organ.

The liver is a chemical factory that converts the nutrients from food into usable substances and stores them until they are needed to nourish the tissues of the body. For example, the liver stores vitamins, minerals, fat, and iron, and converts glucose into glycogen and stores it as a reserve carbohydrate. The liver's main function as a digestive organ, however, is to secrete *bile,* an alkaline liquid that is stored in the gallbladder until it is needed in the duodenum during digestion. While stored in the gallbladder, bile becomes concentrated and easily able to perform its function of emulsification (breaking down globs of fat into small globules) and aiding in the absorption of fatty acids, cholesterol, fat-soluble vitamins (especially vitamin K), and other lipids from the small intestine. A person without sufficient bile in the small intestine is not able to metabolize lipids properly, and is likely to develop a serious lipid deficiency.

Besides its other functions, bile activates the pancreatic enzyme called *lipase,* which breaks down fat into fatty acids and glycerol. Being alkaline, bile also helps to neutralize the acidity of food as it leaves the stomach.

The liver is even more active in metabolism than in digestion, and its function is vital to the metabolism of proteins, carbohydrates, and fats. Some important metabolic functions of the liver in each food category are shown in Table 22.5.

TABLE 22.5
Some Metabolic Functions of the Liver

TYPE OF FOOD	FUNCTION OF THE LIVER
Protein	Removes amino acids from organic compounds
	Forms urea to remove ammonia from body fluids
	Forms plasma proteins
	Synthesizes certain amino acids, including nonessential amino acids
Carbohydrate	Stores glycogen
	Converts galactose and fructose to glucose
	Generates glucose from amino acids when necessary, to maintain a normal blood glucose level
	Forms important chemical compounds from some products of carbohydrate metabolism
Fat	Oxidizes fatty acids and forms acetoacetic acid
	Forms lipoproteins
	Forms cholesterol and phospholipids
	Converts carbohydrates and proteins to fat

The Pancreas as a Digestive Organ

The **pancreas** (see Figure 22.9) secretes three main digestive enzymes, which are capable of acting upon the three types of foods: lipase (fats), amylase (carbohydrates), and trypsin (protein). As mentioned in connection with the liver, *lipase* is most effective when it is activated by bile. Lipase splits large-molecule fats into smaller, absorbable glycerol and fatty acids. *Amylase* converts starch, glycogen, and most other carbohydrates into disaccharides, especially the sugar maltose. Pancreatic amylase is more powerful than ptyalin, which is found in saliva, and it acts upon uncooked starches as well as cooked ones. *Trypsin* continues the digestion of proteins that was begun in the stomach. It is inactive until it is converted into an active enzyme by enterokinase, an intestinal enzyme. Trypsin acts upon proteins that were only partially digested in the stomach and converts them into small peptides and amino acids. A fourth pancreatic enzyme, *nuclease,* breaks down nucleic acids into mononucleotides.

Action of the Large Intestine

Food passes from the small intestine into the large intestine in a watery condition. By now, digestion is complete, and the large intestine functions to remove water and salts from the liquid matter. Removal of water converts liquid wastes into *feces,* a semisolid mixture that is stored in the large intestine until ready to be eliminated through the anus during defecation. The final products of digestion remain in the large intestine from 12 to 36 hours.

Few bacteria are present in the stomach, but many bacteria inhabit the large intestine. These bacteria are harmless as long as they remain in the large intestine, and in fact they are useful in synthesizing vitamins K and B_{12}. Besides containing bacteria, the feces are composed of water, undigested food, such as cellulose (from cabbage or celery, for example), and other waste products from the blood or intestinal wall. The brownish color of the feces is caused by the presence of degradation products of red blood cells in the form of bile pigments (bilirubin).

The elimination of most of the body waste products is conducted not by the digestive tract but by the kidneys and lungs. The next two chapters will examine these organs, as well as another extremely important organ, the heart.

TO SUM UP DIGESTION AND ABSORPTION

1. Digestion starts in the mouth, with the chewing action of teeth, and with enzymes that begin to break down carbohydrates.
2. Muscular contractions help the moistened, chewed food travel to the stomach, where the breakdown of proteins begins and the breakdown of carbohydrates continues. Muscular contractions that push the food along continue throughout the digestive system.
3. Gastric juices are secreted in the stomach, and they help digest the churning food.
4. The partially digested food from the stomach enters the duodenum, the anterior end of the small intestine. The breakdown of fats begins in the duodenum.
5. Final digestion takes place in the rest of the small intestine, where villi and microvilli increase the surface area of the intestine and enhance the absorption of digested food molecules into the bloodstream, where they will circulate throughout the body.
6. After digestion and absorption are complete in the small intestine, the watery mixture that passes into the large intestine is converted into the feces, which is eliminated through the anus during defecation.

SUMMARY

1. Food cannot be used until it has been reduced in size enough to enter the cells of the body, not just the digestive tract. Large, insoluble food molecules are broken down into small, soluble molecules through the process of *digestion.* The small molecules produced by digestion are taken into the bloodstream across the cell membranes of the intestines through the process of *absorption.*
2. Food provides energy and the material to build and repair cells. Humans must include in their diets certain essential components, such as *carbohydrates, fats,* and *proteins.*
3. *Vitamins* play an important role in supporting the work of enzymes, especially those involved in cellular respiration. At least 14 *minerals* are also essential to the human body.
4. The digestive process takes place in the digestive tract, known as the *alimentary canal.*
5. Humans are *omnivores,* and their digestive system is designed to digest and assimilate almost any type of food.

6. Digestion begins in the mouth, where enzymes in saliva convert large molecules of starch into small molecules of sugar. The act of swallowing is the final voluntary digestive act until defecation. *Peristalsis* helps to propel food along the digestive tract.

7. *Gastric juices* in the stomach act on protein to convert it into smaller molecules of amino acids. Besides mixing the food and producing gastric juices, the stomach stores *chyme* and regulates its flow into the small intestine.

8. Enzymes furnished by the pancreas and the intestine and *bile* secreted by the liver break down carbohydrates, fats, and proteins in the small intestine. Practically all the digested molecules of food are absorbed into the bloodstream from the small intestine.

9. By the time food reaches the large intestine, digestion is complete. The large intestine functions to remove water and salts from undigested food and to produce feces.

ASK YOURSELF

1. Why must most foods be digested before they can be used by an animal?
2. If you eat pure glucose, do you need to digest it? Explain.
3. Why are fats necessary in a balanced diet?
4. Trace the path of a fat from the time it is eaten till it is distributed to body tissues.
5. What is meant by the phrase "essential amino acids"?
6. Where in a human alimentary tract are proteins digested?
7. What is the function of the epiglottis?
8. What makes food move along an alimentary tract?
9. What is the usefulness of intestinal villi?
10. What is the function of bile in the digestion of fats?
11. What digestive enzymes are produced by the pancreas?
12. What is the function of the large intestine?
13. What is the source of the brown color of feces?

23
The Excretory System and Temperature Regulation

SOME KEY POINTS

1. All animals need to regulate the amount of water within their systems.
2. Invertebrates have varying mechanisms for the regulation of water balance and the elimination of wastes. In vertebrates, the main regulator of water balance and waste elimination is the kidney system.
3. Kidneys are filters that cleanse the blood and dispose of wastes through the formation and elimination of urine.
4. The liver is involved in many regulatory functions, including excretion and detoxification.
5. The skin has many functions, including protection, waste disposal, and temperature regulation.
6. Some animals (such as amphibia) are unable to stabilize their body temperatures as outside temperatures change, while other animals (such as human beings) can maintain fairly constant body temperatures as external temperatures change.

LIFE ON EARTH NEEDS A VARIETY OF QUANTITIES and concentrations of water in order to survive. At one extreme are the freshwater fishes, which must keep body fluids sufficiently concentrated in a highly dilute environment. At the other extreme are desert animals, which have practically no liquid water available and must obtain and conserve every molecule of water they can. Most terrestrial mammals are in between. They must conserve water, but at the same time they must remove excess salts and body wastes, especially nitrogenous ones. About two-thirds of the body weight of animals is water, and varying amounts of water are constantly entering and leaving the cells (through metabolic activity) and the body. The balance of water must be maintained, and only slight variations are tolerable.

Humans lose water by evaporation from moist lung surfaces and through the skin. While some water, salts, and organic substances are given off in sweat, the prime regulator of water balance and waste elimination is the kidney system.

SALT AND WATER BALANCE IN ANIMALS

Most animals that live in water environments have specialized methods of maintaining salt and water balance. But because the conditions of the open sea do not change much, oceanic invertebrates have little need for complex internal regulators. The bodies of these animals have a lower salt concentration than the seawater that bathes them, and their permeable body surfaces allow their interior fluids to conform to any changing concentrations of seawater (Figure 23.1). The lack of internal regulatory processes in such animals can be demonstrated by placing a specimen in fresh water, where salts are rapidly lost and water is taken up. If oceanic invertebrates are forced to remain in fresh water, they will soon die.

Animals living in fresh water have a great need to prevent salt loss and to limit water uptake. In the freshwater environment, simple physical diffusion will favor the loss of salts from the animal's body, and osmosis will favor the uptake of water. Internal regulatory mechanisms counter these tendencies and permit animals to live in a low-salt environment of fresh water, where they maintain a salt concentration higher than that of the environment.

Freshwater invertebrates keep their salt concentrations higher than that in the water by pumping water molecules out, by absorbing salt, and by actively transporting sodium and chloride from the environment into the body. If it is not possible to limit the water uptake, the active excretion of water will maintain the balance necessary to sustain life.

Terrestrial animals have opposite problems. If they are to get rid of body wastes and excrete excess salts, they must lose water by evaporation or through urination. For the terrestrial animal, water is always a scarce commodity. Terrestrial animals may acquire water in three ways: by eating food with water in it; by releasing water from food by oxidation; and by drinking water. Some terrestrial animals will use all three methods, but others will rely almost entirely on one form only.

Some animals, such as marine birds, turtles, and lizards, are unable to produce highly concentrated urine the way mammals can, and these animals excrete salt through special salt glands located near the eyes. As part of their overall adaptation to flight, birds have no bladder for the storage of urine. Their kidneys excrete a semisolid uric acid as part of the intestinal waste.

TABLE 23.1
Water Content of Animal Bodies

ORGANISM	PERCENTAGE OF BODY COMPOSED OF WATER
Pea weevil	48%
Human being	65%
Kangaroo rat	65%
Herring	67%
Chicken	74%
Frog	78%
Lobster	79%
Earthworm	80%
Jellyfish	95%

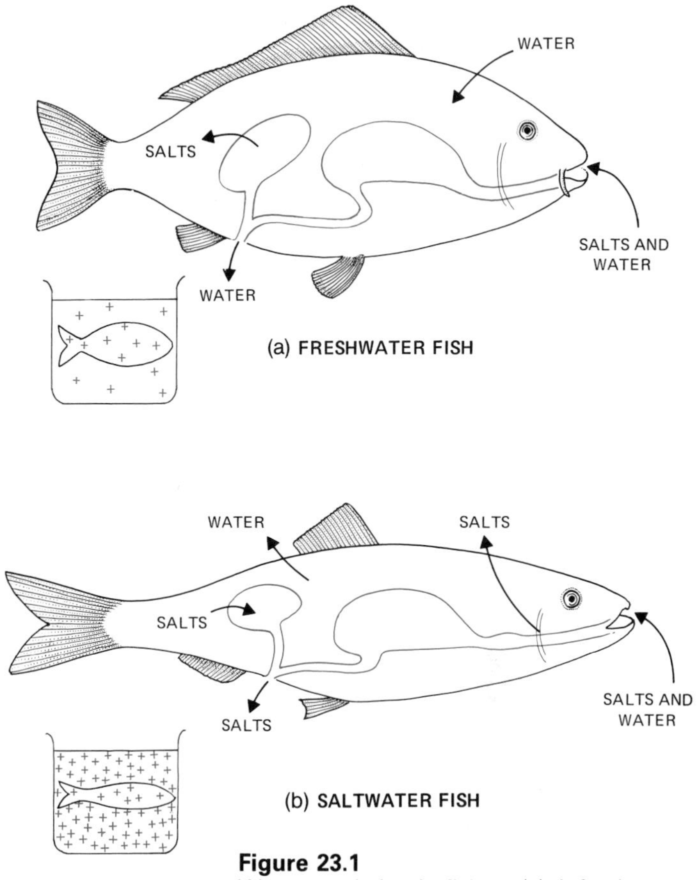

Figure 23.1
Water regulation in fishes. (a) A freshwater fish must maintain a salt concentration in its body fluids higher than that in the surrounding water. If it loses too much salt or takes in too much water, it dies. The water that does come in through the skin and gills is passed out in plentiful urine, and what little salt does come in is filtered out and retained. (b) A saltwater fish has a different requirement. It must maintain a salt concentration in its body fluids lower than that in seawater. It drinks a lot, urinates little, and excretes concentrated salt solution from glands in the gills. It also loses water through the skin.

INVERTEBRATE EXCRETORY MECHANISMS

Excretory mechanisms vary greatly among invertebrates. Many protozoa and freshwater sponges have simple intracellular organelles called **contractile vacuoles,** which are essentially reservoirs of excess water. Because freshwater protozoa contain

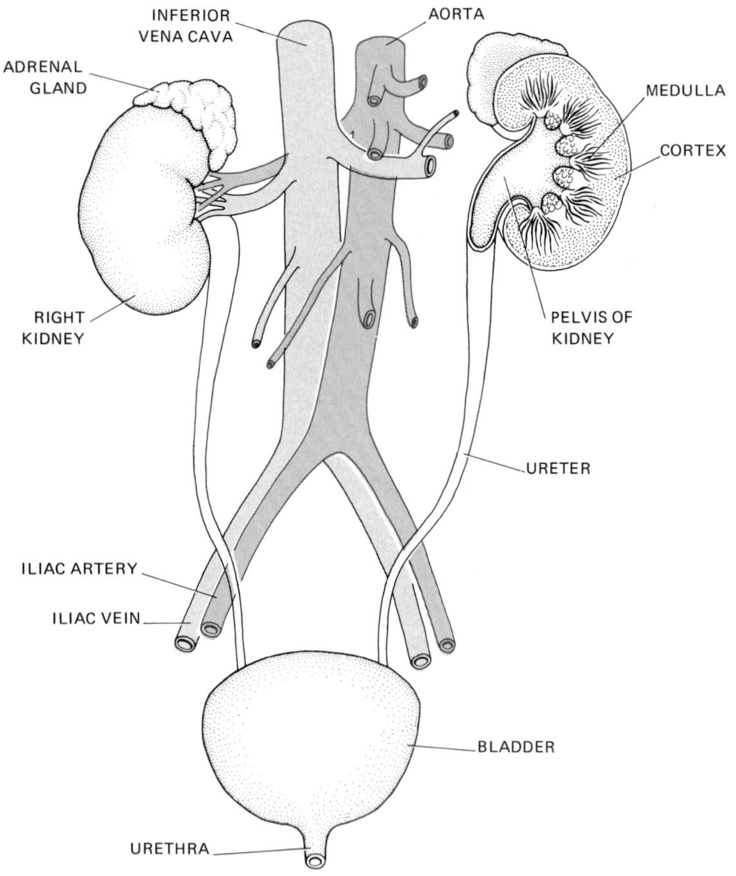

Figure 23.2
The human urinary system. The kidneys are the main blood filters. Blood from the aorta passes through the kidneys, where potentially dangerous waste materials are removed, and returns to the main bloodstream by way of veins to the inferior vena cava. Water, salts, urea, pigments, and other wastes are constantly accumulated in the bladder, which is emptied periodically.

more salt than the water they live in, they absorb water through osmosis. This excess water is accumulated in the expanding contractile vacuole until it is finally emptied back into the outside water. Protozoan waste products are actually eliminated directly from cell membranes into the outside water, so the contractile vacuole is more accurately described as an organ of water balance, not excretion.

Many invertebrates have excretory organs called **nephridia.** Mechanical differences exist from one organism to another, but a nephridium is basically a tubular system that carries wastes to the outside. Flatworms have an arrangement of branched tubes called a **flame cell system,** which removes liquid wastes from intercellular spaces. The wastes are swept into excretory ducts by flagella (which reminded early observers of a flickering flame), and the liquids are finally excreted through excretory pores on the surface of the body. In this way, the flame cell system maintains water balance at the same time that it excretes wastes. Segmented worms have a system of so-called true nephridia, which adds a network of blood vessels and a collecting bladder to the system. The process of urine formation is basically the same in both types of nephridia.

The excretory systems of insects and spiders are called **Malpighian tubes,** an arrangement well suited for land animals in dry environments. As salts and wastes are secreted into the tubes, an osmotic gradient is created that allows water to flow into them. A urinelike fluid is then drawn into the intestine, where rectal glands reabsorb water and most of the potassium in the salts. The remaining waste, usually uric acid, is then excreted through the rectum. Such a system was probably the forerunner of the excretory systems of land-dwelling vertebrates.

THE VERTEBRATE KIDNEYS AND EXCRETORY SYSTEM

Human adults take in about 2500 milliliters (2½ qt) of water each day, most of it through foods and liquids they eat and drink.* Under normal conditions,

* Only about 47 percent of a person's daily water intake comes from drinking. As much as 39 percent is supplied by so-called solid food (meats are 50–70 percent water, most vegetables contain more than 90 percent water, and even "dry" bread is about 35 percent water), and the body's own metabolism produces about 14 percent of the daily water requirement.

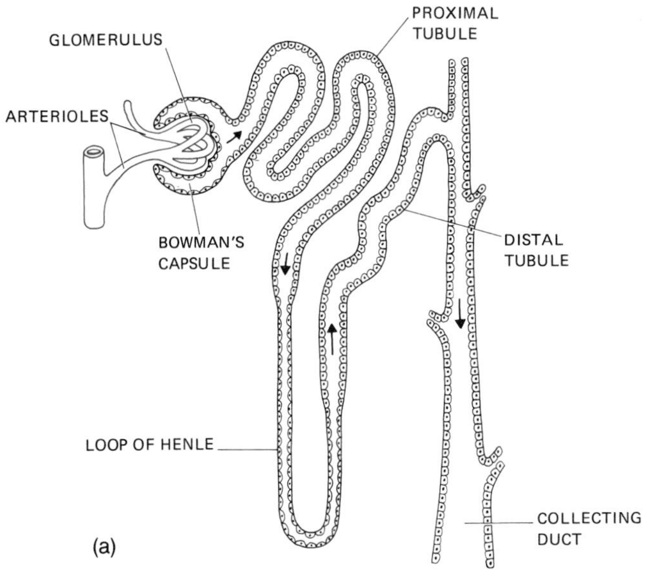

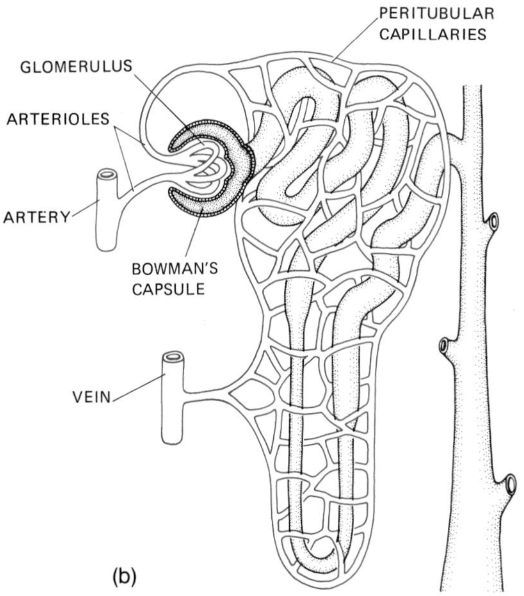

TABLE 23.2
Average Normal Routes of Water Gain and Loss in Adults

	MILLILITERS PER DAY
Intake:	
In liquids (drunk)	1200
In food (eaten)	1000
Metabolically produced	350
Total	2550
Output:	
Evaporation (skin and lungs)	900
Sweat	50
Feces	100
Urine	1500
Total	2550

From Arthur J. Vander, James H. Sherman, and Dorothy S. Luciano, *Human Physiology: The Mechanisms of Body Function,* Third Edition, p. 367. Copyright 1980 by McGraw-Hill Book Company, New York. Used with permission of McGraw-Hill Book Company.

the same amount of water is given off daily, about a third of it evaporated from the skin and exhaled from the lungs, and more than half excreted in urine (Table 23.2). The remarkable organs that regulate this constant water balance are the **kidneys,** a pair of bean-shaped organs, each one about the size of a large bar of soap. The kidneys lie behind the lining of the abdominal cavity, one kidney on each side of the spinal column, at about waist level (Figure 23.2). Each kidney contains more than one million **nephrons,** the functional units of excretion. Each nephron is an independent urine-making machine.

Each nephron contains a long, thin excretory tubule that originates as a cupped sac called **Bowman's capsule** (Figure 23.3). The excretory tu-

(c)

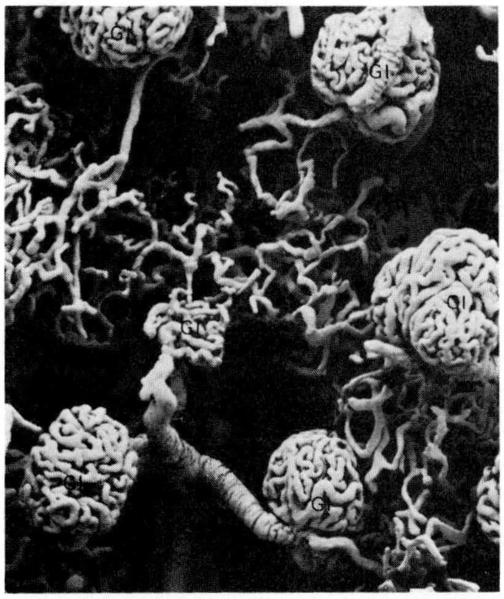

× 150

Figure 23.3
Microscopic anatomy of the urine-collecting unit. (a) A *nephron* in outline, with blood supply removed. Fluid enters the central canal of the nephron in *Bowman's capsule,* passes through the proximal tubule, the loop of Henle, and the distal tubule to the collecting duct, by which time it has been changed into urine. (b) A nephron with its accompanying blood supply. An artery sends a twisted ball of capillaries into the cup of Bowman's capsule. From the ball, the *glomerulus,* blood fluids are forced under pressure into Bowman's capsule: water, sugars, salts, and various wastes. From Bowman's capsule there passes a vein that breaks up into capillaries surrounding the tubules, and these peritubular capillaries reunite after sending blood past the tubules. After having been filtered through the nephron, the purified blood leaves by way of a vein to rejoin the general circulation. (c) A scanning electron micrograph of glomeruli (Gl), and arterioles entering and leaving the glomeruli.

ARTIFICIAL KIDNEYS

If the kidneys fail, toxic wastes build up in the body until cells and organs begin to deteriorate and eventually die. Many tens of thousands of Americans suffer from kidney failure, but more than 16,000 of them are leading relatively normal lives because of the successful use of the artificial kidney, or *dialysis* (diffusion).

The principle of diffusion is basic to the working mechanism of the artificial kidney. Very simply, the artificial kidney is a mechanical device that pumps 5 to 6 liters of blood from the body through a membrane of cellophane. The blood is rinsed by a briny solution, and waste products, such as urea, uric acid, excess water, creatinine, sodium, and potassium, diffuse through the pores of the cellophane by osmotic pressure. The cleansed blood is then rerouted back into the body.

The procedure is as follows: One cellulose tube is permanently linked to an artery (usually in the arm), and another is attached to a nearby vein. The two tubes are connected by a rubber shunt when not in use. Several shunt variations are also used, including a bovine graft (part of a cow vein) that links the artery and vein when dialysis is not taking place. During dialysis the tubes are hooked up to the machine. Blood is pumped from the artery through an oxygenated salt solution similar in ionic concentration to body plasma. Since the concentration of wastes is higher than the normal concentration of the plasmalike fluid, the wastes will automatically diffuse through the permeable membrane of the tubes into this rinsing fluid. The membrane is porous to all blood substances except proteins and red blood cells. The wastes are eliminated from the body, and the purified blood is free to flow back into the body. Dialysis is sometimes also used to add nutrients to the blood. For instance, large amounts of glucose may be added to the salt solution, so that the glucose may be diffused into the blood at the same time that wastes are being removed.

It takes about six hours and 20 passes through the bathing fluid to complete a full cycle of dialysis, and most patients receive treatment two or three times a week. Unfortunately, even this incredibly successful machine (it can remove wastes from the blood 30 times faster than a natural kidney) provides only partial relief for kidney-failure victims. All patients remain uremic (Greek, "urine in the blood") to some degree, and some victims are unable to receive enough cleaning of the blood to make the treatments worthwhile. In such cases kidney transplants are sometimes attempted, but so far the usual problems of tissue rejection have minimized the success of the transplant operations.

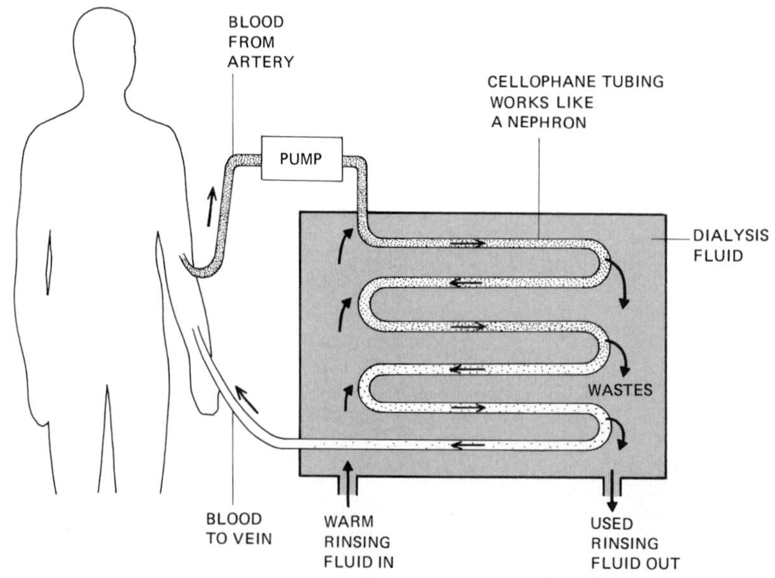

bules are coiled and winding, and if they were all straightened out, they would be about 80 kilometers (50 mi) long in an adult. Each of the million or so excretory tubules leads into a large collecting duct that eventually drains into the central cavity of the kidney. This cavity is connected to the ureter, a long tube that transports urine from the kidney to the urinary bladder, where the urine is held until it is excreted through the urethra to the outside.

Kidneys are filters. Each day they filter about 1900 liters of recirculated blood. It is the kidneys, not the intestines as one might think, that are the body's main regulators of waste disposal. Not only do the kidneys regulate water balance; they also clean and filter the blood and control its chemical concentration.

How do the blood vessels and the nephrons work together to filter the blood and produce urine? Blood enters each kidney through a renal artery, which branches into smaller vessels called arterioles. Each arteriole enters a Bowman's capsule and subdivides even further to form a tiny ball of capillaries, a clump of small blood vessels known as the **glomerulus** (see Figure 23.3). The blood in the glomerulus is separated from the Bowman's capsule by a thin layer of tissue. Some of the fluid portion of the blood passes from the glomerulus into Bowman's capsule and along the excretory tubule. The blood vessels that leave the glomerulus are intimately associated with the excretory tubule. As we will see in the following sections, the blood returns to the body circulation through a renal vein after it has passed through the network of blood vessels leading from the glomerulus. The intertwining of these blood vessels with the excretory tubule facilitates the crucial exchange of materials to and from the blood.

THE MECHANISM OF URINE COLLECTION

When amino acids from proteins are oxidized, most of the final breakdown products are in the form of carbon dioxide (exhaled from the lungs) and water. But the nitrogen from amino acids is removed and made into urea, which must be excreted in urine. If nitrogen wastes accumulate in the tissues, they cause poisoning, starvation, and eventual suffocation of the tissues. Another crucial process accomplished by the excretion of urine is the maintenance of the body's salt content by extraction of excess salt from the blood.

Glomerular Filtration, Tubular Secretion, and Tubular Reabsorption

In the formation of urine a series of events occurs, the end results of which are the elimination of wastes and the conservation of necessary water, salts, and other useful compounds. Three processes take place during the formation of urine from the blood (Figure 23.4): (1) When blood circulates from the arteries into the glomerulus of the kidneys, it is under high pressure. This pressure forces some of the plasma fluid into Bowman's capsule, but the blood cells and large proteins remain within the glomerulus, unable to filter through. This process is known as **glomerular filtration.** (2) Certain substances are secreted from the blood in the peritubular capillaries *into* the nephric fluid as it passes through the nephric tubules. In this way, such waste products as hydrogen and potassium ions,

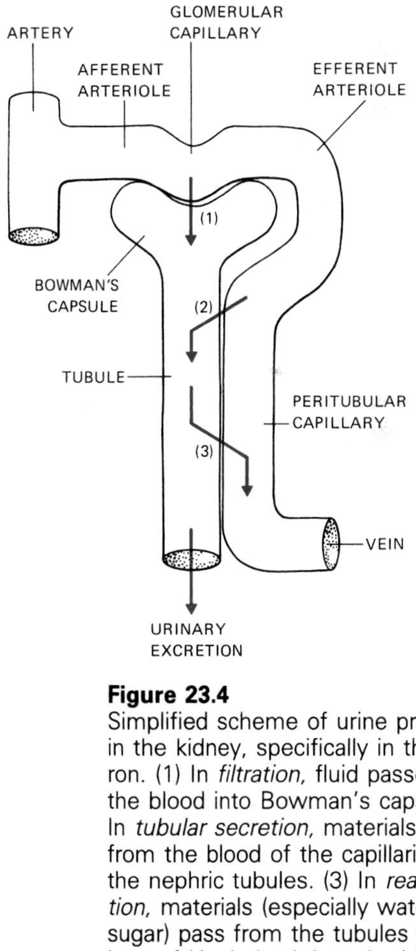

Figure 23.4
Simplified scheme of urine production in the kidney, specifically in the nephron. (1) In *filtration,* fluid passes from the blood into Bowman's capsule. (2) In *tubular secretion,* materials pass from the blood of the capillaries into the nephric tubules. (3) In *reabsorption,* materials (especially water and sugar) pass from the tubules and the loop of Henle back into the blood of the capillaries. The remaining fluid in the collecting tubule is urine. Tubular secretion and reabsorption are carried on simultaneously.

drugs, and foreign organic materials may be excreted. This process is **tubular secretion.** (3) As the fluid moves through the nephron, useful substances, such as water, some salts, glucose, and amino acids, that were initially lost from the blood during filtration are returned to the blood by active transport. This process is called **tubular reabsorption.**

Urine and Urination

While the kidney is forming urine, it is regulating the concentration of substances in the extracellular fluid by removing wastes and conserving substances that are present in normal or subnormal quantities. About 190 liters of water filter through the glomeruli each day, but only about 1½ liters are excreted in urine. The difference, about 99 percent, is reabsorbed. Urine is composed mostly of chloride, sodium, potassium, urea, creatinine, and uric acid.

The volume and concentration of urine are regulated by two hormones and an enzyme. The hormones are *ADH* (antidiuretic hormone), or vasopressin, and *aldosterone;* the enzyme is *renin* (not rennin, another enzyme). Renin is important for the regulation of sodium balance, and apparently it also plays a role in water balance by stimulating the thirst reflex in the brain. Thirst is also indicated by other signals, such as a dryness of the mouth. ADH controls the permeability of the walls of the collecting ducts (see Figure 23.3a). When dehydration occurs, more ADH is released and more water is withdrawn from the urine. The opposite effect is achieved during overhydration. Aldosterone, an adrenal cortical hormone, stimulates the reabsorption of sodium from the kidney and thereby reduces the amount of sodium in the urine.

Urine from the kidneys flows through the ureters to the *urinary bladder,* a thick, muscular sac with a capacity of about one liter (see Figure 23.2). However, the bladder rarely fills to its full capacity, as we shall see. Every 10 or 20 seconds, peristaltic waves force small amounts of urine from the renal pelvis into the bladder. The bladder becomes distended as it fills with urine, and when an adult accumulates about 300 milliliters (9 fluid ounces), the sensory endings of the pelvic nerve in the bladder wall are stimulated, and there is an urge to urinate. In an adult, this urge can be counteracted by consciously controlling the external urethral sphincter, but an infant lacks the development of the higher brain centers and cannot control the sphincter. Urination in an infant is a spinal reflex action initiated by the distention of the bladder. In an adult, those same impulses generated by stretch receptors in the bladder are mediated by control centers in the cerebrum.

An adult normally eliminates about 1½ liters of urine a day, but several conditions may increase the output considerably. Increased stress or fear may cause increased urine production. Sometimes the production of urine may not actually increase during stress or excitement, but the urge to urinate may occur anyway. Pregnant women know that as the fetus develops it puts more and more pressure on the woman's bladder, causing frequent urinations. Cold weather also increases the urge to urinate. This happens because the blood is moved away from the skin and into the internal organs in an effort to conserve heat. The extra blood entering the organs subsequently causes the kidneys to filter more blood and produce more urine. Certain foods, such as mustard, pepper, tea, coffee, and alcohol, increase urine production. In contrast, nicotine decreases the production of urine. Fortunately, urine production decreases while the body is asleep, so that we are usually not disturbed during the night.

TO SUM UP THE KIDNEYS AND EXCRETION

1. The kidneys contain nephrons, the functional units that actually produce urine.
2. The excretion of urine removes potentially harmful nitrogen wastes, and also helps keep a healthy balance of salt in the body.
3. When arterial blood enters the glomerulus, some plasma fluid is forced into Bowman's capsule, but blood cells and large proteins are retained in the bloodstream; this process is *glomerular filtration.*
4. As fluid passes through the nephric tubules, waste products from the blood in the peritubular capillaries are excreted into the tubules; this process is *tubular secretion.*
5. As the fluid moves through the nephron, useful substances are returned to the body via the peritubular capillaries; this process is *tubular reabsorption.*
6. The rate of urine production is regulated by blood pressure, temperature, and hormones.
7. Urine from the kidneys flows into the urinary bladder through the ureters, and eventually leaves the body through the urethra.

THE LIVER AS AN EXCRETORY ORGAN

One of the liver's many jobs is to serve as an excretory organ, and it accomplishes this task in several ways. After amino acids have been absorbed by the blood, they are received by the liver. In the liver, most of the ammonia from the amino acids is converted to *urea*, which is eventually excreted in the urine. If the liver is seriously damaged or diseased, ammonia accumulates in the blood, and may cause brain damage.

The liver also becomes involved with excretion by modifying drugs and other waste products before they can be excreted in bile or urine. Some typical substances altered by the liver prior to excretion are penicillin; erythromycin; estrogen, cortisol, and other steroid hormones; and calcium, which is removed from the blood into the bile and then excreted in the feces. Finally, specialized phagocytes in the liver called *Kupffer cells* remove old blood cells and toxic materials, which are excreted in the feces.

THE SKIN

Like the liver, the skin has many functions, including protection, excretion, and the regulation of body temperature. Skin covers the entire body, and in fact, it is the body's largest organ, occupying almost two square meters (18 ft^2) of surface area (Figure 23.5). It meets and is continuous with the inner lining of the body at external openings: the mouth, nostrils, and ears; and the anal, urinary, and genital openings. A 70-kilogram (155 lb) man has about 11 kilograms (25 lb) of skin.

Skin protects against injury and bacterial infection; it keeps in warmth during cold times (especially in furry animals) and dissipates heat during hot times by losing water through evaporation. Through perspiration, a small amount of waste materials such as urea are excreted through the skin. Up to one gram of waste nitrogen may be eliminated through the skin every hour. Skin helps the kidneys regulate water balance; it protects against injurious radiation by absorbing ultraviolet light and uses those same ultraviolet rays for conversion of cholesterol, secreted by the skin, into essential vitamin D; it produces hair, fingernails, and toenails (or those variants called hooves) and horns in animals that have them; it produces sweat and such variants thereof as milk and ear wax; and it yields odors that are believed to be important, even in people, as sex attractants. It is well known that the ridges on fingertips (fingerprints) can be used to identify individuals, but it is not well known that skin from any other part of a human body could be used just as well.

Skin is flexible; it can be folded and straightened indefinitely without cracking or wearing out. It is adaptable; it can be as thin as 0.07 millimeter over the inner side of an arm, or up to several millimeters on soles of feet. Even unborn babies have thickened skin on their soles and palms. Skin is sensitive; from

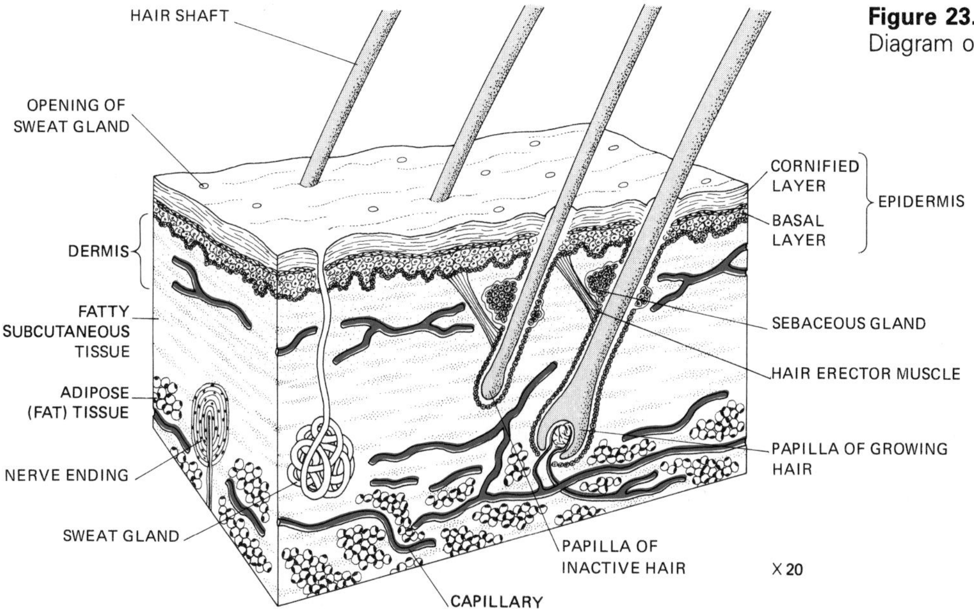

Figure 23.5
Diagram of a block of human skin.

its nerve endings come impulses that inform the individual of pressures, temperatures, textures, or injury. Skin is complex in its origin; the outer layer is derived from the embryonic ectoderm, the inner layers from mesoderm.

Structurally, skin is one of the more complex organs in the body. The outer layer, the **epidermis,** has layers of flattened, cornified (horny) cells, which are continually being worn away and replaced by new ones from the basal layer, where protein synthesis is especially active. The second layer, the **dermis,** contains an array of tissues: hair follicles that push out a continuous filament of hair shaft (up to several meters if not cut or worn off); sweat glands, simple coiled ducts that give off water and salts to the outside; sebaceous glands, like microscopic bunches of grapes, secreting the oily substance that keeps skin soft or makes fur and feathers waterproof; muscles that move the hair shafts; nerves; blood vessels; and an insulating layer of fatty tissue.

Skin is reactive. It can darken in response to ultraviolet light by the creation of melanin from the amino acid tyrosine; or in response to hormonal changes, the melanin can be intensified in certain areas, as in some women the face is darkened by the "mask of pregnancy." It can release sweat when the skin is warm or when fright stimulates a "cold sweat." Cold or excitement can also make the hairs stand on end, or it can make the muscles that erect

DESERT RATS, CAMELS, AND "WARM-BLOODED" FISHES

Most desert animals need water as regularly as we do (their bodies contain about the same percentage of water as ours—65% of body weight), and they usually get it from the water in vegetation. Cactus plants, the most likely source of liquid refreshment, contain about 90 percent water. But the desert rat, usually called the kangaroo rat, eats dry seeds and dry plants that contain very little water. And yet the kangaroo rat has adapted successfully to desert life. Where does its water come from? Mostly from oxidation of food and the tiny amounts of moisture in the air, hardly enough to keep the little rodent supplied with enough water unless its water loss is kept low. The key to the balance between low water intake and low water loss is that the kangaroo rat is a nocturnal animal. It remains in its relatively cool and humid burrow all day and avoids the dangerous heat of the sun, which would cause water to evaporate from the body. Besides, the kangaroo rat's kidneys are so efficient that they can excrete urine as concentrated salt water. Under experimental conditions kangaroo rats drank salt water and successfully excreted both the excess urea and the salt it contained. So far as is known, no other mammal can drink large quantities of salt water and survive. When most mammals (including humans) drink salt water, they actually increase their thirst, because the body becomes dehydrated in the process of getting rid of the excess salt.

Another famous desert animal is the camel. So legendary is the camel's ability to conserve water, in fact, that as many myths as facts about the camel exist. In the first place, the camel does *not* store water in its hump. Nor does it possess an extra stomach for that purpose. Instead, the camel can survive, in the somewhat milder winter months at least, on water-filled plants, and it rarely needs to drink water directly. Apparently the camel does not "store" water at all but merely consumes enough to retain a healthy water-to-body-weight ratio. A thirsty camel will lose weight, which is replaced almost instantly when it drinks water. Camels can drink almost 103 liters (27 gallons) of water in ten minutes. Like the kangaroo rat, its urine contains huge quantities of urea and salt and very little water, and camels may also be able to drink water containing large amounts of salt.

The camel's nose, not its hump, provides a unique water-saving device. The lining of a camthe hair shafts draw so tight as to pucker the surface, making "goose pimples."

REGULATION OF BODY TEMPERATURE

Since the designations "cold-blooded" and "warm-blooded" are too general and misleading to be suitable for scientific use, we will use instead the terms **poikilothermic** (Greek, "varied heat") and **homeothermic** (Greek, "same heat"). Animals such as reptiles, fishes (there are exceptions; see the essay on p. 470), and amphibians are poikilothermic; that is, they are unable to stabilize their body temperatures, which change along with the surrounding temperatures. Birds and mammals, however, are homeothermic; that is, they are able to maintain fairly constant body temperatures independent of external temperatures.

Birds and mammals vary their physical activities greatly, and yet their body temperatures usually stay within a range of 2°C. Mammals have body temperatures of 36°C to 38°C (37°C is normal for humans), and birds range between 40°C and 42°C. Homeothermic animals produce heat through metabolic activity, such as oxidation of food, cellular metabolism, and muscular contraction. Shivering is merely an increase of muscular activity that increases heat production through muscular contraction. Cooling is accomplished mostly by evapora-

el's nose consists of rolled-up tissue that creates a large surface area. At night, the nasal tissue absorbs water from the air the camel exhales. The camel is able to recapture and conserve water from its own breath, and in so doing, it saves about 70 percent of the water it would lose if it exhaled a warm, moist breath the way people do.

Besides all these survival abilities, the camel's body temperature fluctuates with shifts of outside temperature, and in this way perspiration is reduced. The camel's temperature may go as high as 40°C (105°F) during the day and as low as 34°C (93°F) at night. Finally, this remarkably adaptable animal has a twofold insulation system. Its hair insulates against the heat, and its fatty hump concentrates the body fat in one place instead of distributing it all over the body. Thus, a lack of insulative material between the hair and skin permits heat to flow outward, and a cooling effect results.

One more example of unusual methods of temperature regulation concerns the "warm-blooded" or homeothermic fishes, such as the bluefin tuna. This time, the circulatory system, not the kidney, is the crucial mechanism. The *rete mirabile* ("miracle net") is a tissue containing a rich network of closely packed veins and arteries that not only permits the transport of oxygen through the blood but also creates a countercurrent heat exchange system that warms the tissues and increases the power of muscles. This is accomplished by a circulatory bypass that prevents body heat from escaping through the gills as usual, directing it instead back to the muscle.

As body temperature increases within an acceptable range, the muscle power also increases, because warmer muscles can contract more rapidly than cooler ones. The bluefin tuna and the mackerel shark, two fishes equipped with a rete mirabile system, are predators, and their increased muscle power enables them to swim faster, thereby becoming more effective predators. Interestingly, the bluefin tuna (and the other tunas with a rete mirabile) and the mackerel shark are not related. They have independently evolved similar means of adapting to an environment where controlled body temperature makes for successful living. This is a vivid example of parallel, or convergent, evolution, which we will examine in some detail in Chapter 26.

tion of water through the skin (perspiration) and lungs. (Desert animals have elaborate cooling mechanisms. See the essay on p. 470.)

Adaptations to Cold

Homeothermic animals must adapt to temperatures ranging from about −60°C to 60°C. Although most homeotherms have body temperatures close to 38°C, they still manage to adapt to these extreme temperature ranges extending from the Arctic zone to the tropic zone (Figure 23.6). For instance, an Eskimo's body temperature is about 37°C, and yet the external environment may be 50 to 100 degrees cooler. What are the main methods of adapting to the cold?

Homeothermic animals can cope with extreme cold by adapting to it, escaping from it, or enduring it. Besides increasing muscular activity and generating body heat through the metabolism of food, some animals retain body heat by using some form of insulation (fur or clothes, for example). An Arctic fox with winter fur can live comfortably in external temperatures more than 85°C colder than its own body temperature. In contrast, tropical animals would have difficulty adjusting to temperatures 15 or 20 degrees lower than their body temperature. In such conditions, the tropical animals would have to increase their metabolic rate to produce heat. (A naked human has little natural insulation, and temperatures lower than 20°C are uncomfortably cold.)

Small animals, such as mice, have a relatively larger body area exposed to heat loss than larger animals do, and they cannot rely on insulative heat alone. Instead, such animals as Arctic hares and mice live under the snow, where the temperature usually drops no lower than −5°C. Birds such as herons that spend a great deal of time wading through shallow water looking for fishes and other small aquatic animals to eat often stand on only one leg (Figure 23.7). This is because the birds' legs are not insulated, and body heat is lost through the bony legs. By standing on one leg, the birds conserve some body heat.

Many birds and other homeothermic animals counter the cold of winter by removing themselves from it altogether. These animals migrate to warmer climates, an activity discussed further in Chapter 29.

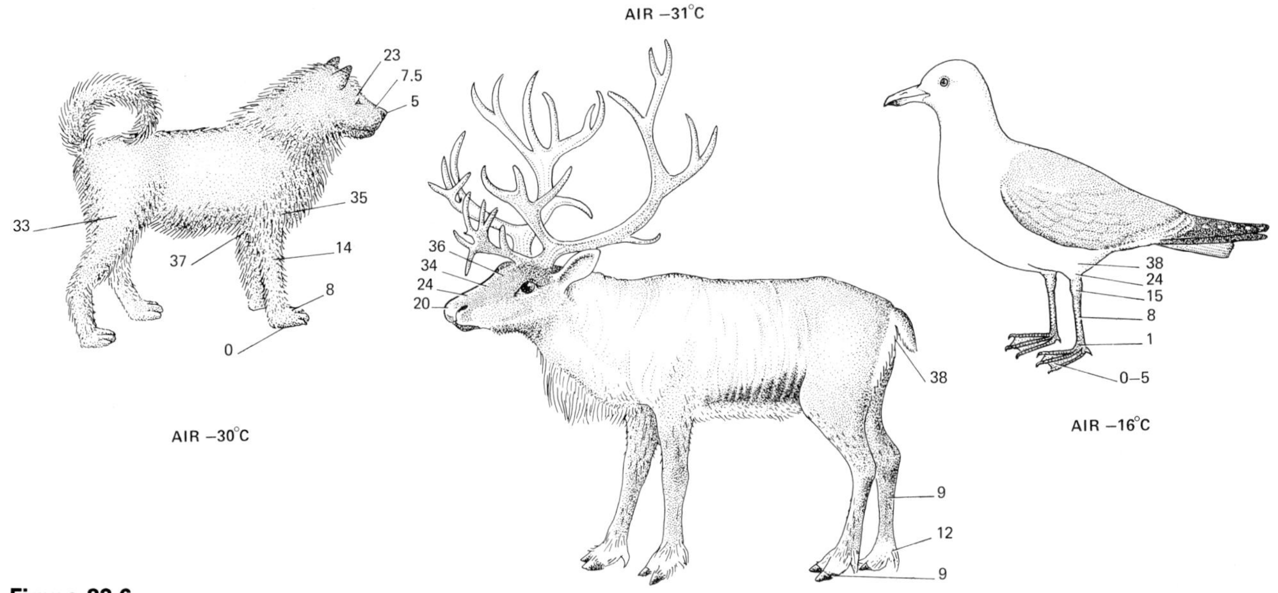

Figure 23.6
Even the best-insulated Arctic animals have unprotected areas that must be left unencumbered by heavy fur or feathers if they are to function properly. Feet and noses, for instance, are usually naked to the cold. A *countercurrent heat exchange* between warm arterial blood to the foot and cold venous blood from the foot permits heat to pass directly from the artery to the vein. This exchange rechannels the heat back to the body, away from the extremities. As a result, the extremities remain colder than the body, usually close to the temperature of the outside environment. (This *rete mirabile* or "miracle net" system of heat retention is also discussed in the essay on p. 471.) If the webbed feet of Arctic gulls were not as cold as the water in which the gulls swim, the feet would lose more heat than the body could supply. Furthermore, if a gull with warm feet stood still on the ice, its feet would melt the ice. Before the gull could move again, the ice might refreeze and trap the bird. Having cold feet has its advantages. (All numerals refer to degrees Centigrade.)

Also discussed in that chapter is hibernation, a method used by some animals to endure the seasonally cold weather without actually leaving their home sites.

Adaptations to Heat

Desert animals have evolved ways to live successfully in their extremely hot and dry environments (see also the essay on p. 470). Camels are the best-known examples, but many other animals are able to maintain homeostasis in hot climates. The main consideration for desert animals is to reduce water loss by evaporation. Many small desert animals are nocturnal; they avoid the sun by living in burrows during the day. Where drinking water is not available, animals derive enough water from solid foods. Desert animals are also able to conserve water by excreting concentrated urine and dry feces.

Animals too large to live in burrows keep cool in several ways. Light-colored fur reflects the sun's rays and keeps heat from penetrating the body. Hairless portions of the body, usually unexposed to the sun, contain capillary networks that permit heat loss, and animals with horns or large ears (such as gazelles and African elephants) use these parts of the body as cooling centers through the process of radiation (Figure 23.8).

Many animals, including humans, are effectively cooled by the evaporation of perspiration, but dogs hardly sweat at all. Instead, they lose heat by panting, an increased rate of breathing that facilitates water evaporation from the tongue and upper respiratory tract. The accompanying excessive loss of carbon dioxide does not cause unconsciousness in a dog as it would in a human.

Figure 23.7
Some wading birds, such as this American avocet, *Recurvirostra americana,* conserve body heat by standing on one leg, thus limiting the amount of heat that is lost through their uninsulated legs.

Figure 23.8
A contrast in cooling systems. (a) An elephant, *Loxodonta africana,* has ears so large that one of them can hide its whole head. The thin flaps, richly supplied with blood vessels, act as radiators that dissipate heat. The ears are one of an elephant's most useful cooling devices. (b) At the other extreme of ear size, the European pigmy shrew, *Sorex minutus,* has minute external ears. Such a small animal has so much surface in proportion to its bulk that it needs to conserve all the heat it can.

(a) (b)

SUMMARY

1. Seawater animals that lack internal regulatory processes to maintain a salt and water balance in their bodies soon die in fresh water. In contrast, freshwater invertebrates can maintain the concentration of their blood and their body fluids even in diluted seawater.
2. Terrestrial animals must acquire large amounts of water, and they must also eliminate salts and other body wastes through the lungs and skin, and through excretion of urine and feces.
3. Excretory mechanisms vary greatly among invertebrates. Many protozoa and freshwater sponges have simple intracellular organelles called *contractile vacuoles.* Many invertebrates have excretory organs called *nephridia;* each of which is basically a tubular system that carries wastes to the outside. Flatworms have a dual arrangement of branched tubes called a *flame cell system,* which maintains water balance at the same time that it excretes wastes. The excretory systems of insects and spiders are called *Malpighian tubes.* A similiar system was probably the forerunner of the excretory systems of land-dwelling vertebrates.
4. Humans lose water by evaporation from moist lung surfaces and from the skin. Water, salts, and organic substances are given off in sweat, but the prime regulator of water balance and waste excretion is the *kidney system.*
5. Urine formation essentially consists of forcing into nephric tubules most of the blood except the proteins and blood cells (*glomerular filtration*), and then retrieving from the tubules everything except some water, salts, urea, and other wastes (*tubular reabsorption*).
6. While the kidney is forming urine, it is regulating the concentration of substances in the extracellular fluid by removing wastes and conserving substances that are present in normal and subnormal quantities. Urine is formed from the capillaries of the *glomerulus.*
7. The vertebrate *liver* serves as an excretory organ, mainly by forming urea and detoxifying some materials prior to excretion.
8. Skin has many functions, including excretion, the regulation of body temperature, protection, and the reception of sensory impulses.
9. *Poikilothermic* animals have little or no ability to stabilize their body temperatures, which change along with the surrounding temperatures. *Homeothermic* animals are able to maintain fairly constant body temperatures independent of external temperatures.

ASK YOURSELF

1. What would happen to a freshwater worm that got washed into the sea? Explain in terms of water-salt balance.
2. Besides drinking, how do animals obtain water?
3. What are the general functions of kidneys?
4. What are the differences between glomerular filtration, tubular secretion, and tubular reabsorption in kidney function?
5. In what organ in a mammalian body is urea mainly produced?
6. What kind of food is the source of urea?
7. What are the functions of human skin?
8. How are "goose pimples" formed by human skin?
9. What are the ways in which a warm-blooded animal can keep heat from escaping from its body?
10. How can a warm-blooded animal avoid becoming overheated?

24
The Respiratory and Circulatory Systems

SOME KEY POINTS

1. Not all animals breathe as human beings do, but all animals must have a mechanism for obtaining oxygen and eliminating carbon dioxide.
2. Warm-blooded animals need large amounts of oxygen. Their respiratory systems consist of lungs, passages from the outside to the lungs, and a muscular mechanism to move air into and out of the lungs.
3. A negative feedback system controls breathing rate. A high content of carbon dioxide in the blood increases the rate, and a low content decreases it.
4. Oxygen is carried in the blood by hemoglobin, the respiratory pigment in red blood cells.
5. Blood has several functions in the human body, including the transportation of dissolved foods, carbon dioxide, oxygen, and hormones.
6. Blood is composed of plasma (water, dissolved solids, and dissolved gases) and formed elements (red blood cells, white blood cells, and platelets).
7. The heart is a double pump. It pumps oxygen-poor blood from the body to the lungs, and pumps oxygen-rich blood from the lungs throughout the body.
8. Blood vessels are arteries, capillaries, and veins.
9. The lymphatic system and immune system help keep the body free of infection and foreign substances.

MOST ANIMALS CAN LIVE FOR DAYS WITHOUT food or water, but it takes only a few minutes without oxygen for the muscles, the heart, and the brain to fail. The intake of oxygen is called **respiration,** an unfortunate term because it has two related but different biological meanings: (1) Respiration at a cellular level means the oxidation of food with the consequent release of energy (Chapter 5). (2) Respiration at the gross anatomical level means *breathing* (air containing oxygen enters the lungs, and air containing waste gases is expelled from the lungs), and it is the breathing, or respiratory, system that will be discussed here. This chapter will also deal with the circulatory system, which is related to the respiratory system because in many animals oxygen is carried (circulated) by the blood to all parts of the body.

THE EVOLUTION OF GAS EXCHANGE IN ANIMALS

Although most animals do not breathe as humans do, they must have some sort of mechanism for obtaining oxygen and eliminating carbon dioxide. All animals exchange gases across a respiratory surface. In some animals this may be the entire body surface. Other animals accomplish gas exchange by using gills, lungs, or tracheae. These methods will be considered in the following sections.

Respiration by Surface Diffusion

Many small organisms, such as protozoa, earthworms, and small crustaceans, respire by the direct diffusion of gases through the surfaces of their bodies. Because of this direct diffusion, no other special

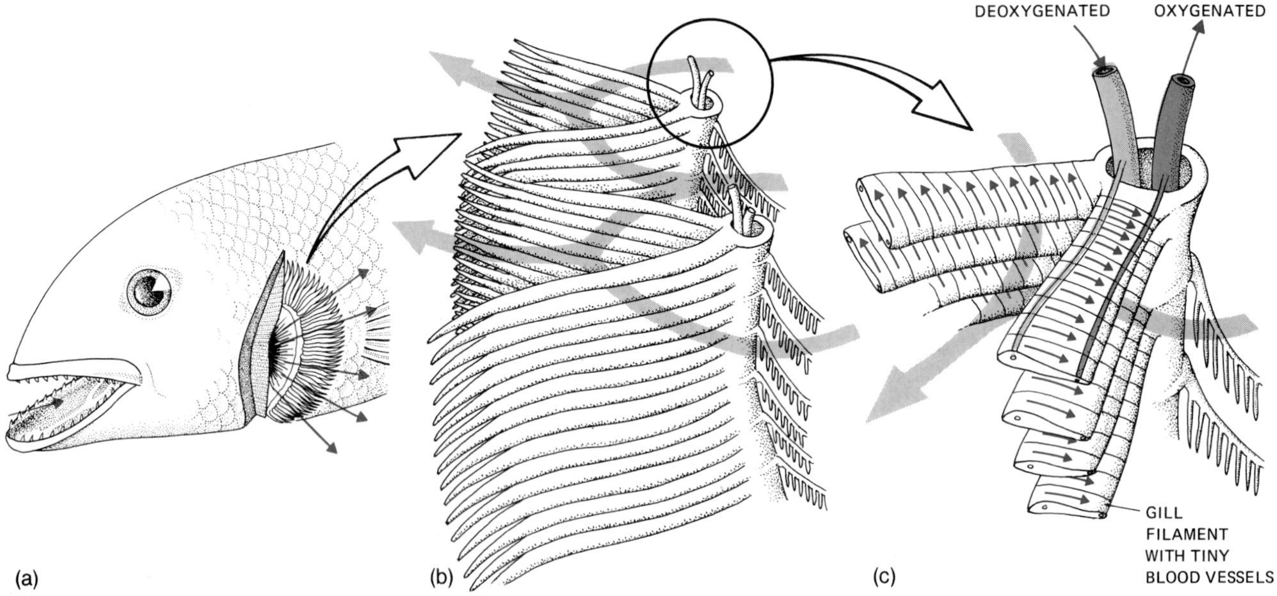

Figure 24.1
The mechanism by which a fish obtains oxygen from water. (a) The fish gulps water into its mouth, and the oxygenated water passes over the gills and then out the gill slits in the sides of the pharynx. (b) The gills are finely divided, offering abundant surface for gas exchange, and they are provided with an array of blood vessels close to the surface. (c) The blood entering the gills from the fish's body is lower in oxygen and higher in carbon dioxide than the water flowing past. Since each gas tends to diffuse to a region of lower concentration, oxygen moves from the water into the blood, and carbon dioxide moves from the blood into the water. (d) The exchange is helped by *countercurrent flow*. The blood and water are moving in opposite directions. Such an arrangement results in an efficient system because at any place along the countercurrent exchanger the difference in concentration of a diffusing substance is maximized. If blood and water flowed along together in the same direction, even if there were concentration differences, the relative difference in concentrations in blood and water would be less, and the rate of exchange would be lower.

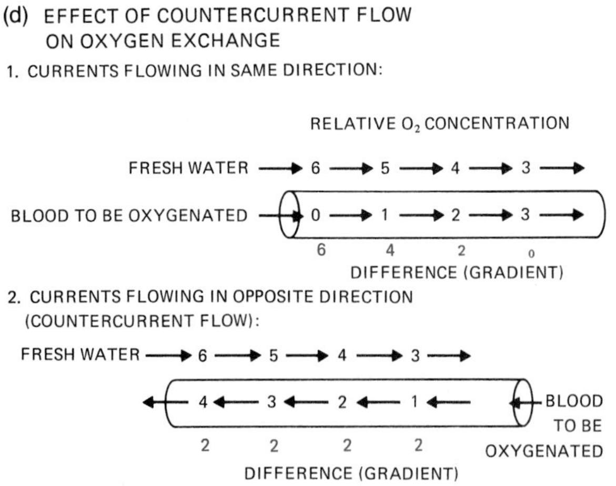

respiratory apparatus is necessary. Animals smaller than one millimeter in diameter are most successful with surface diffusion. However, other larger aquatic animals, such as eels and frogs, supplement lung or gill respiration with gaseous diffusion through the skin, especially when they are totally immersed in water.

Gills

Used by water-dwelling animals (except mammals), especially fishes, gills are the most efficient means of respiration in water. But efficient as gills are, fishes still expend about 20 percent of their total energy extracting oxygen from water. In contrast, mammals use only about two percent of their energy for breathing. There are only small amounts of dissolved oxygen contained in water. In order for a fish to supply sufficient oxygen to its blood, it must constantly take in water through the mouth. The water is forced over the gills, and it passes out of the body through the gill slits. Fish **gills** are soft filaments filled with blood vessels. Gills allow blood to flow in a direction opposite to the flow of water, thus establishing an efficient *countercurrent flow* (Figure 24.1), which is aided by a gill pump and the swimming motions of the fish. A gill may be defined as an enlarged, outward-turning respiratory surface, whereas a lung is a cavity formed by the turning *in* of the respiratory surface.

Figure 24.2
The respiratory system of a bird. (a) Diagram of the anatomy of a bird's trachea, lungs, and air sacs. The air sacs, which are directly connected with the lungs, branch throughout many parts of the bird's body, including the hollow bones. The requirements of flight, including a need for lightness and a high rate of muscular activity, plus the demands of a relatively high temperature, make it necessary for a bird to have an efficient method of delivering oxygen to the tissues and removing carbon dioxide. (b) A scanning electron micrograph of the interior of a bird's lung. The network of air tubes ensures continuous passage of air close to the capillaries. In a bird's lungs, the air flows *through* the tubes. This is a more efficient system than in mammalian lungs, which simply pump air in and out.

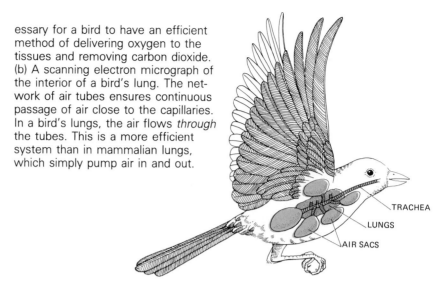

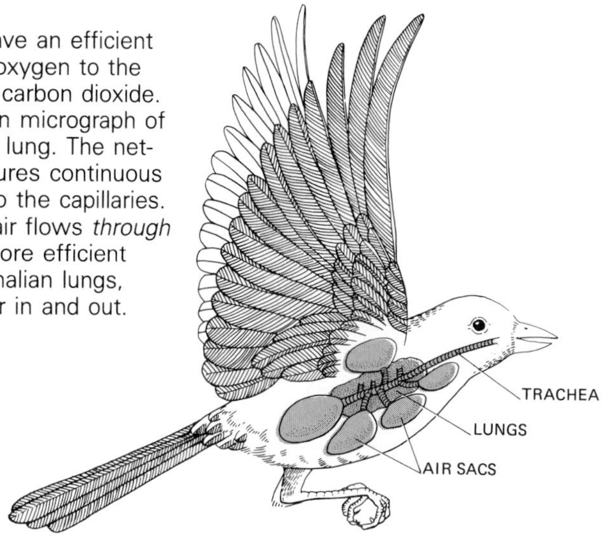

(a)

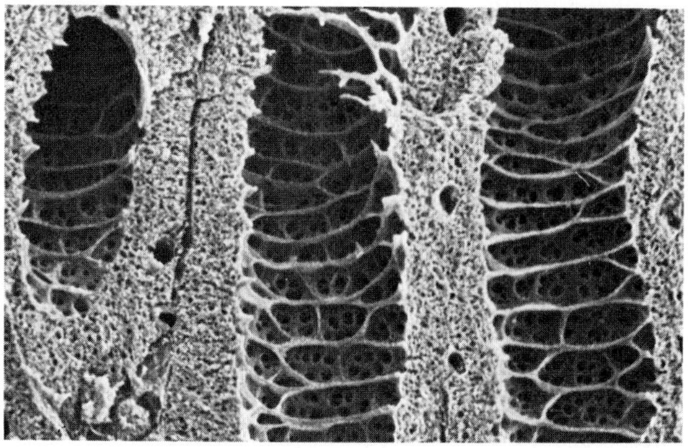

(b) × 60

Lungs

Land-dwelling animals developed **lungs,** which are moist, saclike organs containing a fine network of blood vessels. The first lungs probably appeared in animals like lungfishes, and amphibians still retain simple lungs. In the more highly developed organisms, the internal structure of the lung is more complex, having evolved from a simple sac to the highly folded and lobulated mammalian lung, which contains millions of tiny air sacs called *alveoli.* Each alveolus is surrounded by a network of blood vessels.

The lung adequately serves the respiratory needs of terrestrial vertebrates, but it is an inefficient mechanism nevertheless, especially in comparison with gills. Water flows directly and efficiently past a gill, but air must enter *and* leave the lung through the same opening. Without a free one-way flow of air, problems of lung ventilation occur, and they are further complicated by the fact that the lungs are not in direct contact with the outside air sources. Instead, a series of "dead-air spaces" connect the lungs to the outside, thus setting up an area of unventilated and uncirculated air in the respiratory tubes that never reaches the lungs. Only about 70 percent of the human lung is filled with fresh air after each inhalation. Bird lungs are more efficient, as shown in Figure 24.2.

Tracheae

Insects and some other land-dwelling arthropods have a system of air tubes called **tracheae** (not to be confused with the human trachea, or windpipe), which branch out from surface openings called spiracles to all parts of the body (Figure 24.3). The spiracles are made operative by the movement of the body. The internal ends of the tracheae are air capillaries that carry oxygen directly to the body cells. Carbon dioxide is transported from the body cells through the tracheae to the outside.

RESPIRATION IN HUMAN BEINGS

Homeothermic animals, with their need for large amounts of oxygen, evolved a system of air breathing. One of the advantages air-breathers have over aquatic animals is that oxygen diffuses out of the air about 300,000 times faster than it diffuses out of water. In this section we will consider the human breathing mechanism and the way in which gases are exchanged in the lungs prior to being transported by the blood.

The Organization of the Respiratory System

The respiratory system in mammals is made up of the lungs, the several passageways leading from the outside to the lungs, and the muscular mechanism that moves air into and out of the lungs (Figure 24.4). Before the air we breathe reaches the lungs, it must enter the nose or mouth and pass through the trachea (windpipe). The trachea branches into tubes called bronchi, which enter the lungs and divide into smaller and smaller tubules, finally terminating in air sacs called **alveoli,** the sites of gas exchange between the lungs and the bloodstream (Figure 24.5). Each lung contains about 300 million alveoli, each surrounded by many blood capillaries. Gas exchange is facilitated by the thin membrane walls (only one cell thick) that separate the alveoli from the blood vessels and also by the enormous surface area of the alveoli, which amounts to about 70 square meters. Unlike many smaller animals, humans receive oxygen indirectly, through the blood.

Throughout the upper respiratory tract the air is filtered, warmed, and moistened. (Air breathed through the mouth receives fewer of these benefits than air taken in through the nose.) Epithelial glands in the passages secrete mucus (about a liter

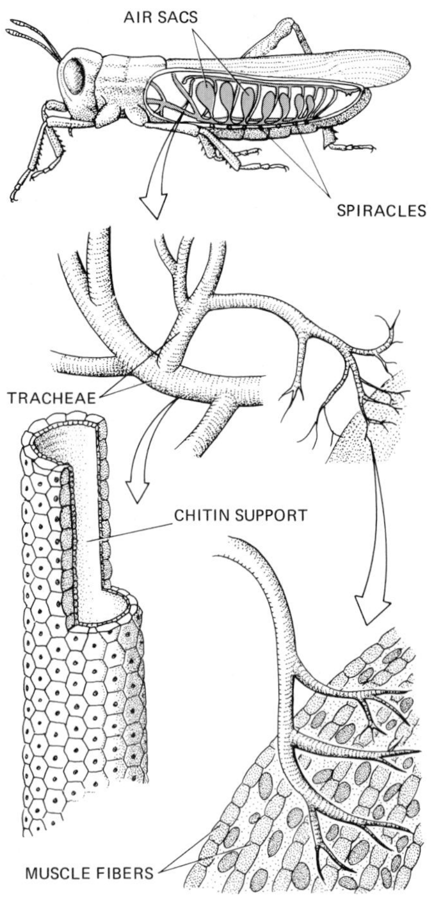

Figure 24.3
Breathing mechanism in an insect. Air goes in and out through small openings along the sides of an insect's abdomen. These *spiracles* lead into air tubes, the *tracheae,* which branch and rebranch until fine endings reach the tissues. Pulsations of the insect's body create alternating pressure and suction. Insects with aquatic larval stages have various gill-like structures that provide gas exchange, but an adult insect can be drowned only if its abdomen, not its head, is kept under water.

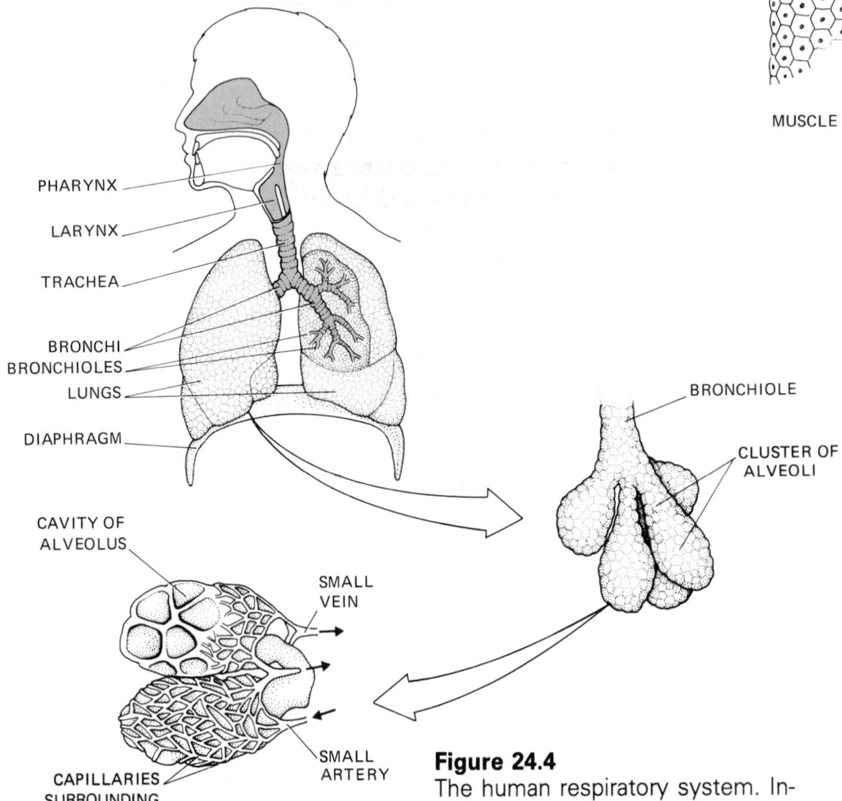

Figure 24.4
The human respiratory system. Inhaled and exhaled air passes through the pharynx (the swallowing apparatus), then the larynx (the Adam's apple, or voice box, containing the vocal cords), the tubular trachea (the windpipe), the two bronchi, which branch into ever smaller bronchioles, and finally into the alveoli (the terminal air sacs) of the lungs. The thin walls of the alveoli are richly supplied with capillaries and are kept moist. Respiratory gases are exchanged through the walls of the alveoli and the capillaries.

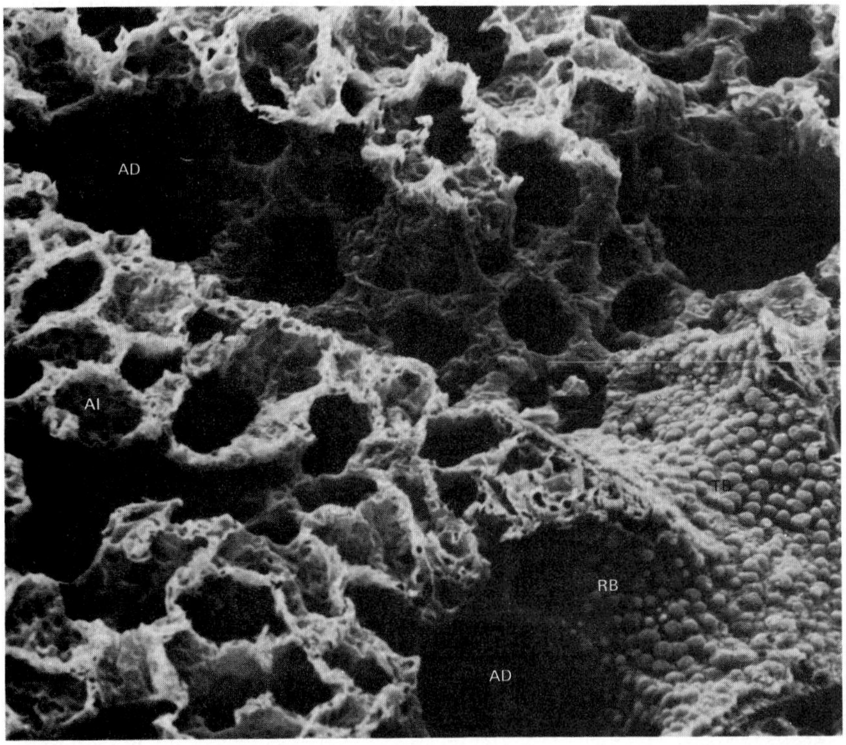

Figure 24.5
Scanning electron micrograph of the inside of a human lung. The darker holes show alveoli (Al) and alveolar ducts (AD), which lead into deeper alveoli. Note the continuity between terminal bronchioles (TB), respiratory bronchioles (RB), and alveolar ducts (AD). This picture gives an idea of the convolutions that add to the amount of surface available for gas exchange.

each day), which traps foreign particles that are inhaled or swallowed, so that they are eventually eliminated in the feces. The mucus is kept moving, and a new film of mucus is secreted about every 20 minutes. An important role in this purifying mechanism is played by the cilia that sweep the mucus along the tract (Figure 24.6). Air pollutants, such as cigarette smoke, can paralyze these cilia and cause the mucus to clog, a reaction frequently resulting in "smoker's cough." Phagocytes, cells that engulf foreign substances, are also neutralized by air pollutants. When cilia in the nose are partially paralyzed by cold air, there is an overproduction of mucus, which is not swept back into the throat as usual. When this happens, the mucus drips foward and causes a runny nose.

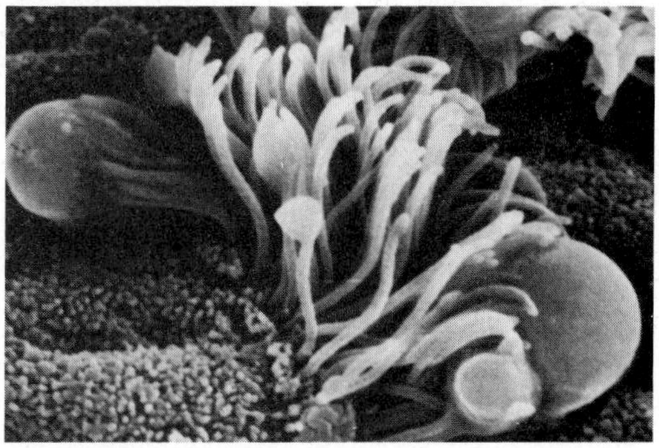

Figure 24.6
Scanning electron micrograph of the interior of a hamster trachea. The upstanding objects are cilia, which wave continuously, carrying mucus and inhaled solid particles, such as dust, upward from the direction of the lungs. This cleaning operation keeps the breathing passages clear. Most of the lining of mammalian respiratory surfaces is provided with cilia, including the bronchi, tracheae, and nasal passages.

The Mechanics of Breathing

The lungs, which are composed mainly of the elastic tissue of the alveoli, contain very little muscle. They are covered by a layer of epithelium called pleura and are held within the bony chest cavity. The expansion and contraction of the lungs is accomplished by the **diaphragm,** a muscular partition between the thoracic and abdominal cavities, and by abdominal muscles and muscles between ribs.

The intake of air, **inspiration,** occurs when air rushes into the lungs to equalize a reduction of air pressure in the thoracic cavity (Figure 24.7). The mechanism operates as follows:

1. Rib muscles contract and elevate the ribs.
2. The muscles of the diaphragm contract, lowering the diaphragm and increasing the volume of the chest cavity.
3. Abdominal muscles relax to compensate for the compression of abdominal organs.
4. The increased size of the chest cavity causes the pressure in this cavity to drop below the atmospheric pressure, and air rushes through the respiratory passages into the lungs, equalizing the pressure.

Expiration, or the expulsion of air from the lungs, occurs in the following way:

1. Muscles of the ribs and diaphragm relax, allowing the thoracic cavity to return to its original, smaller size.
2. Abdominal muscles contract, pushing the abdominal organs against the diaphragm.
3. The elastic lungs contract as the air is expelled.

An ordinary breath will move about 500 milliliters (1 pint) of air in and out of the lungs, but a deep breath may increase the lung capacity to about 4000 milliliters (4 liters). Trained athletes may have a lung capacity of more than six liters.

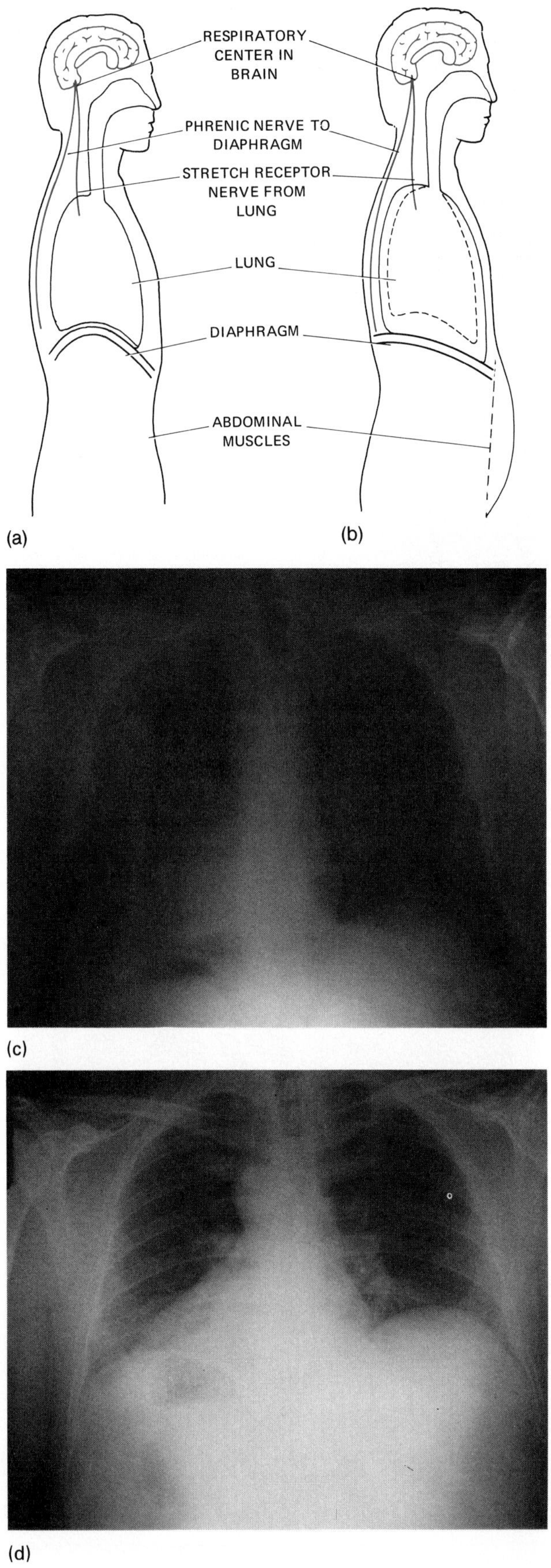

Figure 24.7
The mechanism of breathing. When the respiratory center in the medulla sends an impulse to the diaphragm, the diaphragm muscles contract. This contraction, shown in (b), flattens the diaphragm, pushes out the abdominal muscles, and enlarges the chest cavity. The resulting reduced pressure in the lungs makes air rush into the lungs. Diaphragm contraction is aided by the contraction of the *intercostal muscles* between the ribs. When these muscles contract, the ribs are raised in a hingelike action, further expanding the rib cage. When the lungs are filled with air, stretch receptors in the lungs send impulses to the respiratory center, which sends back impulses causing the diaphragm to relax. Pressure of the abdominal wall against the internal organs makes the diaphragm rise into a dome shape as shown in (a), expelling air from the lungs. X-rays of the chest show the changes during exaggerated inhalation (c) and exhalation (d).

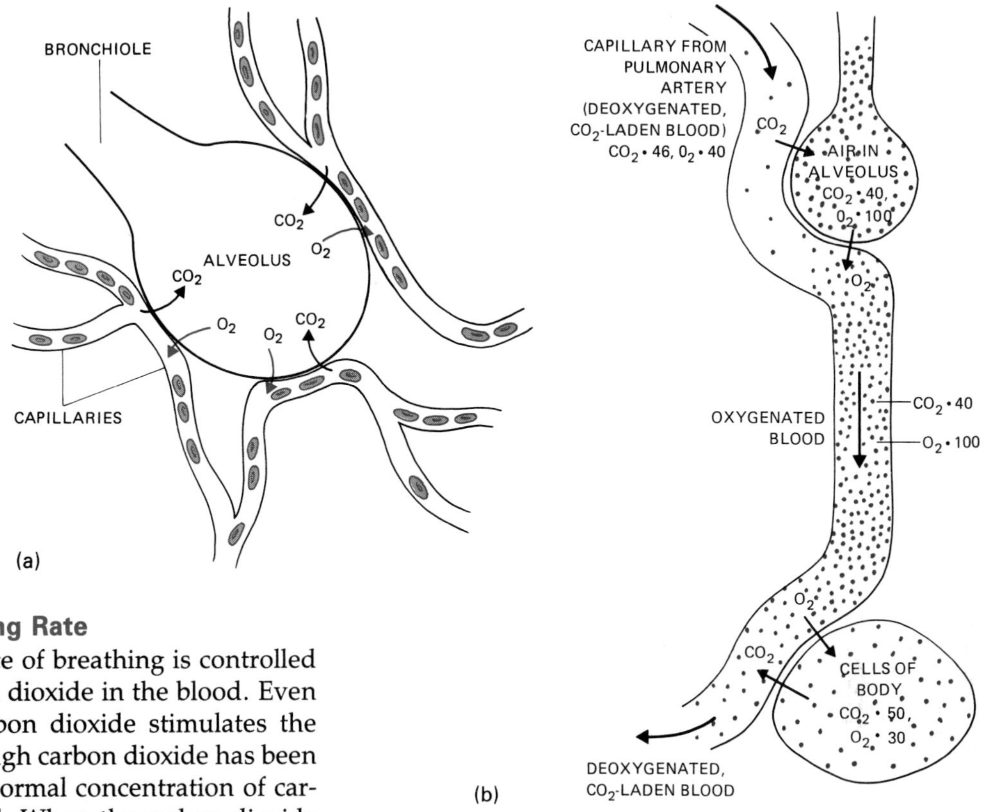

Figure 24.8
Scheme of gas exchange in lungs and tissues. (a) Waste carbon dioxide from blood in lung capillaries enters the alveoli, and fresh oxygen passes from the alveoli into lung capillaries. (b) When blood passes through the lung capillaries and comes close to the alveoli of the lungs, the concentration of oxygen in the alveoli is high and that of carbon dioxide is low. The opposite is true of the concentrations in the blood. Since diffusion favors a movement from a region of high concentration to a region of low concentration, the two respiratory gases are exchanged. The blood passing through the lungs loses carbon dioxide and gains oxygen while the lungs lose oxygen and gain carbon dioxide. The oxygenated blood is distributed throughout the body, and when it reaches capillaries in tissues, the situation is reversed: The high concentration of oxygen in the blood favors diffusion into the respiring cells, where the oxygen is low. Carbon dioxide moves in the opposite direction. Thus living cells are constantly supplied with oxygen from the lungs, and the waste carbon dioxide is eliminated.

The Control of Breathing Rate

The regulation of the rate of breathing is controlled by the amount of carbon dioxide in the blood. Even a slight increase in carbon dioxide stimulates the breathing rate until enough carbon dioxide has been released to restore the normal concentration of carbon dioxide in the blood. When the carbon dioxide level of the blood rises, the respiratory center in the brain sends out nerve impulses that increase the rate of operation of the rib muscles and diaphragm, resulting in faster breathing.*

Thus a high carbon dioxide content of the blood causes an *increase* in breathing, and a low carbon dioxide content causes a *decrease,* a classic example of negative feedback control. Note that it is mainly the carbon dioxide concentration, not the oxygen concentration, that regulates the breathing rate. A person breathing in an atmosphere low in oxygen may not feel discomfort as long as the carbon dioxide proportion does not rise. However, although the main regulator of the breathing rate is the level of carbon dioxide in the blood, the breathing rate is also increased somewhat by a drop in the level of oxygen in the blood.

Gas Exchange in the Lungs

The blood coming to the lungs is low in oxygen. Through the process of diffusion, oxygen from the air in the alveoli will move into the blood, and carbon dioxide and water in the blood will move into the alveoli (Figure 24.8). Oxygen and carbon dioxide

---

* Small mammals have a faster breathing rate than large mammals do, but large mammals usually live longer than small mammals do. As a result, all mammals, whatever their size, take about the same number of breaths in a lifetime.

move in different directions, but both move from areas of relative abundance (high concentration) to areas of less abundance (lower concentration). Both gases move freely through the walls of the capillaries and the alveoli.

THE GAS-CARRYING ABILITY OF BLOOD

On its own, the water of blood can carry only about 1 percent of an adult's oxygen requirement. But **hemoglobin,** the respiratory pigment in the red blood cells, has such an affinity for oxygen that it binds about 98 percent of the oxygen available in the lungs. In this way, hemoglobin acts as a carrier and makes possible the effective transport of oxygen throughout the body. Hemoglobin is a protein; each molecule contains five percent *heme,* an iron-containing compound, and 95 percent *globin,* a protein. Although the heme makes up only a small portion of this molecule, it is absolutely essential to the function of hemoglobin as an oxygen-carrying pigment. It is the heme that gives hemoglobin its red color.

The Transport of Oxygen in Blood

Oxygen is abundant in the lung alveoli, and it combines readily with hemoglobin to form oxyhemoglobin. The oxyhemoglobin is then transported in the red blood cells to the tissues, where oxygen is less abundant. Because oxygen is constantly being used up by cellular oxidation, functioning tissues have less oxygen than the lungs. Because of the low concentration of oxygen in the tissues, the oxygen in oxyhemoglobin is free to diffuse into the cells of the tissues. Varying amounts of oxygen are released, depending on cellular needs.

Unfortunately, some toxic agents bind to hemoglobin even more readily than oxygen does. Air pollutants such as insecticides and sulfur dioxide bind to hemoglobin and prevent its effective carrying of oxygen. Such poisons cause a numbness or dizziness that is characteristic of a lack of oxygen. Probably the best-known hemoglobin poison is carbon monoxide, found in automobile exhaust fumes. Carbon monoxide binds to hemoglobin about 210 times faster than oxygen and forms a stable compound. Carbon monoxide at concentrations of 0.1 or 0.2 percent in the air is dangerous, and increased amounts may induce death by blocking the uptake of oxygen by hemoglobin. When this happens, tissues die of asphyxiation.

One of the most serious lung problems is *emphysema,* a condition that develops when alveoli lose their elasticity, usually because of extreme air pollutants that clog breathing passages and destroy sensitive tissues. When this happens, the alveoli do not retain enough elasticity to expand and collapse rhythmically. Waste carbon dioxide cannot be expelled properly, and it is trapped within the alveoli. At the same time, oxygen is not distributed freely to the blood. Fortunately, a normal adult has about eight times as many alveoli as needed for routine activities, and acute cases of emphysema are not so prevalent as they might be otherwise.

The Transport of Carbon Dioxide in Blood

Carbon dioxide is more soluble in water than oxygen is, and it diffuses easily through capillary walls from the body tissues. Once in the blood, carbon dioxide is transported in three ways:

1. About 67 percent of the carbon dioxide reacts with water in the blood to form carbonic acid. In the lungs, where carbon dioxide is less abundant than in the blood, this reaction will operate in the reverse direction, and carbon dioxide will be lost from the blood.
2. About 25 percent of the carbon dioxide reacts with hemoglobin and is carried by the hemoglobin from the tissues to the lungs. When hemoglobin carrying carbon dioxide arrives in the lungs, the carbon dioxide is exchanged for oxygen.
3. The remaining 8 percent of carbon dioxide is dissolved directly in the blood as molecular carbon dioxide (CO_2).

BLOOD SERVES MANY FUNCTIONS IN THE HUMAN BODY

Although animals have evolved various ways to transport food, gaseous waste products, and regulatory substances throughout their bodies, the most common method uses a circulating fluid. **Blood** is the term used for circulating fluids in general, even though some blood, as in invertebrate animals, may be blue or yellow and may perform functions somewhat different from those of blood in humans. Blood in the human body serves a number of functions. It carries dissolved foods, transports carbon dioxide and oxygen, distributes chemical regulators

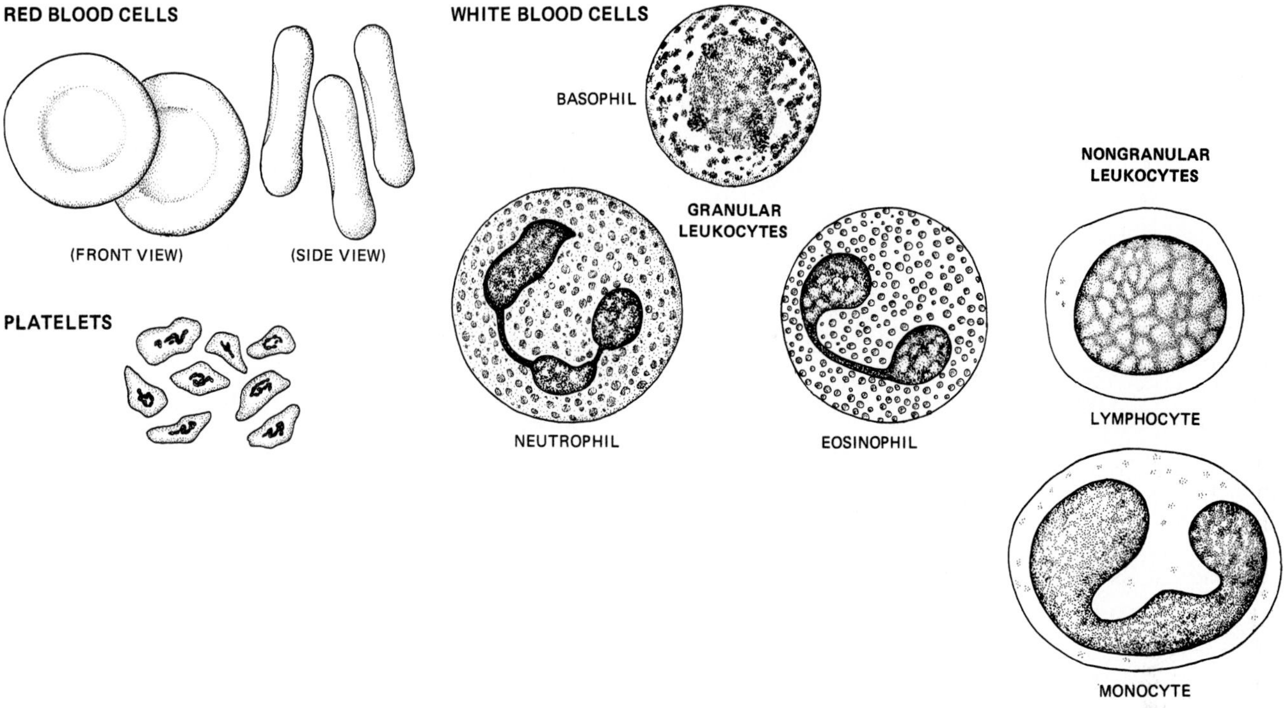

Figure 24.9
Cells and platelets in human blood. Five different *leukocytes,* or "white blood cells," are recognized on the basis of their granular contents and their staining reactions. All leukocytes are nucleated, and all act as defenses against infections and other foreign materials. They are not actually white, but they are unpigmented, in contrast to the *erythrocytes,* or red blood cells, which are red with *hemoglobin* and lack nuclei. *Platelets* are particles, not cells, but they are visible with a light microscope. Platelets function in blood clotting.

(hormones), acts as a defense against infection, helps maintain an even temperature, and removes waste products.

Composition of Blood

The liquid portion of mammalian blood is called **plasma** (about 55%), and the solid portion is known as **formed elements** (about 45%), mostly blood cells suspended in the plasma (Figure 24.9). The following tabulation summarizes the composition of mammalian blood:

PLASMA

1. *Water,* 90 percent.
2. *Dissolved solids,* consisting of the plasma proteins (albumin, globulins, fibrinogen), glucose, amino acids, electrolytes (compounds that separate into ions when dissolved in water), various enzymes, antibodies, hormones, metabolic wastes, and traces of many other organic and inorganic materials.
3. *Dissolved gases,* especially oxygen, carbon dioxide, and nitrogen.

FORMED ELEMENTS (solid components)

1. *Red blood cells* (erythrocytes) transport oxygen to body cells and remove carbon dioxide.
2. *White blood cells* (leukocytes) serve as scavengers and immunizing agents by destroying bacteria at infection sites.
3. *Platelets* (thrombocytes) help blood to coagulate.

Each milliliter of human blood contains about a billion **erythrocytes** (red blood cells), and there are about 25 trillion erythrocytes in the human body; 40 to 50 red blood cells in a row would reach across the period at the end of this sentence. Erythrocytes make up about half the volume of human blood. The erythrocyte is a disk, slightly indented on both

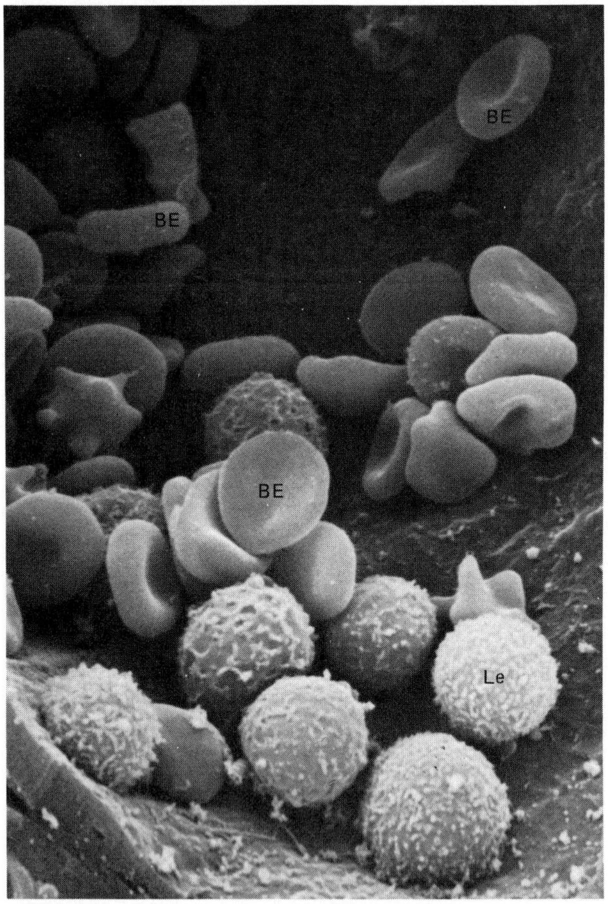

(a) × 2500

(b) × 2000

Figure 24.10
(a) A scanning electron micrograph of human blood cells, showing biconcave (concave on both sides) erythrocytes (BE) and leukocytes (Le).
(b) Scanning electron micrograph of human erythrocytes from a patient suffering from sickle-cell anemia. Instead of being normal biconcave disks, these cells are variously distorted. Some are stretched out so long and thin that they bear some resemblance to a sickle blade—hence the name of the disease.

sides, providing a larger surface for gas diffusion than a flat disk or a sphere (Figure 24.10).

An erythrocyte exists for about 120 days in the bloodstream before it finally fragments. It is then engulfed by scavenger cells called macrophages in the liver, bone marrow, or spleen. Every *second* about 7 to 10 million erythrocytes are destroyed, and the same number are created. (Decreased production or increased destruction of red blood cells produces a condition known as *anemia.*) The iron from hemoglobin is retained and used again. The rest of the heme is converted to bilirubin, a bile pigment, which is excreted in the feces. If the liver is faulty, bilirubin may accumulate in abnormally high amounts and cause the skin to turn yellow—a condition known as *jaundice.* During its brief life span an erythrocyte may travel 1100 kilometers (700 mi) through the bloodstream (about 75,000 round trips from the heart to other parts of the body and back to the heart), bending to squeeze through the tiny capillaries.

The term **leukocyte** (or white blood cell) is used to cover a number of slightly different cell types that circulate in the blood. Leukocytes are slightly larger than erythrocytes (see Figure 24.10a). Unlike erythrocytes, leukocytes have nuclei, and they are capable of independent ameboid movement to pass through blood vessel walls into tissue spaces. In adults, there are about 700 erythrocytes for every leukocyte.

The concentration of leukocytes in adults may increase as a result of an infection from 8000 to 25,000 per cubic millimeter, about the same number that newborn infants have. Some leukocytes are actively engaged in engulfing and digesting foreign material, including bacteria. Specialized leukocytes called lymphocytes play an important role in producing antibodies to establish an immune response to foreign substances (see Figure 24.20).

Leukemia is a disease characterized by uncontrolled leukocyte production. The white blood cells tend to use the oxygen and nutrients that normally go to other cells, causing the death of otherwise healthy cells. Because almost all the new white blood cells are immature and incapable of normal function, victims of leukemia have little resistance to infection. In most cases, the combination of starving cells and lack of an immune system is enough to cause death. The cause of leukemia is unknown, although a viral infection is considered likely.

Platelets (thrombocytes) are minute disks that serve as starters in the process of blood clotting, to be described below.

How Blood Clots

The fluidity of blood is delicately balanced. If blood was not liquid, it could not function, and yet if it could not solidify, any small wound would allow the blood to escape unhindered, as it does in victims of "bleeder's disease," or hemophilia.

When a tissue is damaged in a normal person and blood escapes from a blood vessel, the following sequence takes place:

1. Platelets disintegrate and release the enzyme thromboplastin.
2. Thromboplastin combines with calcium to convert the inactive protein prothrombin into the active enzyme thrombin.
3. Thrombin acts as a catalyst to convert the plasma protein fibrinogen into the insoluble, stringy protein fibrin.
4. The fibrin threads entangle the blood cells and create a clot (Figure 24.11).

This whole process may be summarized as follows:

1. $\text{prothrombin} \xrightarrow[\text{calcium}]{\text{thromboplastin} + } \text{thrombin}$

2. $\text{fibrinogen} \xrightarrow{\text{thrombin}} \text{fibrin}$

Why doesn't blood clot in the blood vessels? Because the enzyme thrombin is not present in the circulating blood. Besides thrombin and the other coagulation (clotting) agents mentioned above, as many as 35 compounds may be required for blood coagulation, a necessary precaution to prevent clotting when no bleeding has occurred. Mosquitoes and other blood-sucking animals are able to counteract the clotting action of thrombin by having an anticoagulant present in their saliva. Without this mechanism the victim's blood would clot in the delicate mouth parts of the mosquito. On the other hand, the entry of too much anticoagulant into an animal's bloodstream can be fatal. A recent plague of bloodsucking bats (*Desmodus rotundus*) in Central and South America threatened to ruin the cattle industry until a special anticoagulant was injected into the rumens of the cattle. The blood of the specific vampire bat was unable to clot after a full meal of the blood of the treated cattle. As a result, the small capillary breaks that normally occur in a bat's wings during flight did not seal as usual. The bats died of internal bleeding or were weakened to the point that they were unable to hunt for food, and they starved to death. The anticoagulant apparently had no adverse effect either on the cattle or on the consumers who eventually ate the beef.

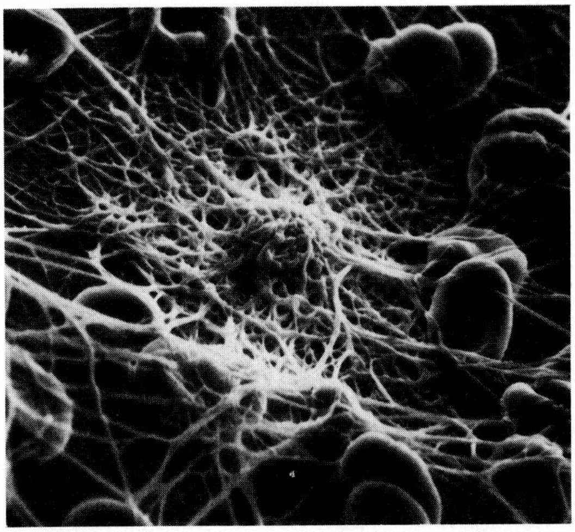
× 3000

Figure 24.11
A scanning electron micrograph of a portion of a blood clot. The tangled threads are *fibrin*, which binds the clot into an insoluble mass, and the circles are blood cells, enmeshed in the fibrin filaments.

THE HEART IS A DOUBLE PUMP

In vertebrate animals, including humans, blood flows in a *closed system*, remaining essentially within the carrying vessels. Most animals, however, have *open systems*, in which blood oozes generally throughout the body in relatively large, loosely jointed cavities, and it is kept moving partly by a rather simple heart and partly by body movement

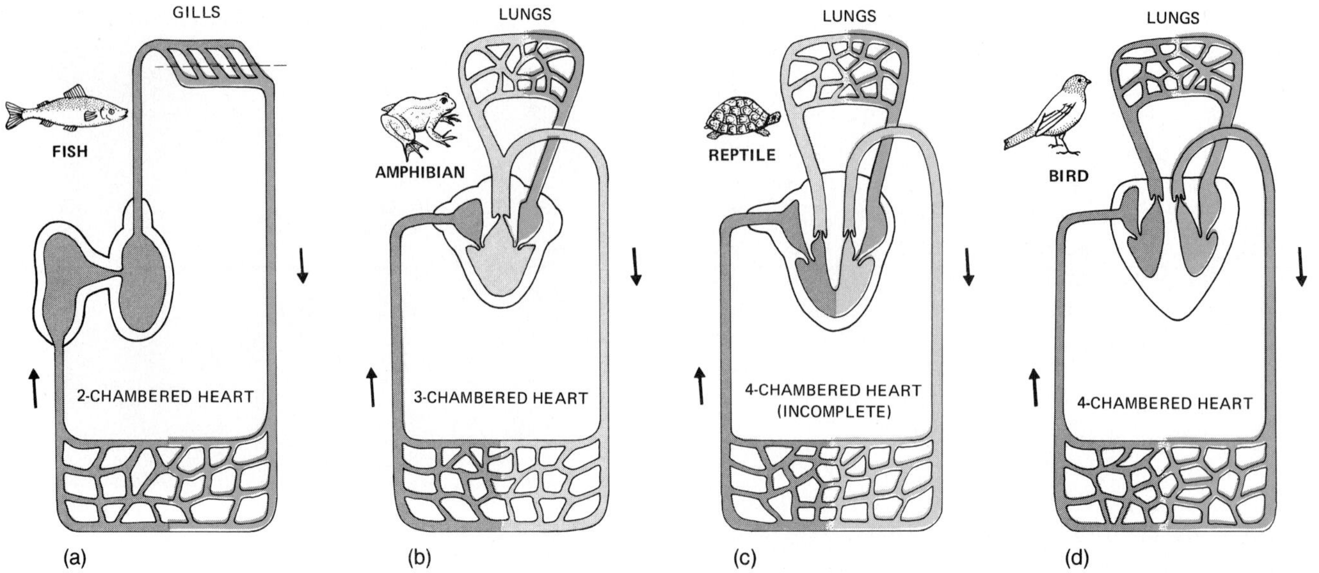

Figure 24.12
Comparison of circulatory systems. (a) The two-chambered fish heart is a single pump, forcing deoxygenated blood (gray) from body tissues out through the gills, from which oxygenated blood (brown) goes to the body tissues. (b) In a frog (amphibian), deoxygenated blood (darker gray) enters the three-chambered heart, where it is mixed (lighter gray) with blood from the lungs (brown). A mixture of oxygenated and deoxygenated blood (lighter gray) is sent out over the body. (c) A turtle (reptile) has two incompletely separated ventricles. Deoxygenated blood from the body (darker gray) is partially mixed with oxygenated blood from the lungs (darker brown) before being sent to the lungs (lighter gray). Blood sent from the left ventricle to the body is a mixture of oxygenated and partially deoxygenated blood (lighter brown). (d) Birds and mammals separate deoxygenated blood (gray) from oxygenated blood (brown). The right side of the heart is only an accessory pump that sends blood to the lungs to be reoxygenated.

(Figure 24.12). Such circulation is typical of crustaceans and insects. In the mammalian circulatory system, which will be treated here, blood is pumped from a muscular heart out through elastic arteries, then through microscopically fine vessels, the capillaries, and finally through veins back again to the heart.

The Quantity of Blood Circulated

A heart begins beating in a human embryo within a few weeks, and it must continue through the life of the individual at a rate of about 70 times a minute, 100,000 times a day, or about 2½ billion times during a 70-year lifetime.* The heart rarely misses a beat, but if it skips four or five in a row, the body will probably fall down dead.

The human body contains about 5½ liters (12 pints) of blood, and the heart takes slightly more than a minute to pump a complete cycle of blood through the body. In times of strenuous exercise the heart can quintuple this output. And yet, the heart is a relatively small organ, no larger than your fist, and it weighs only about 300 grams (11 oz).

The Mechanics of Heart Action

The interior of the human heart is separated into halves by a thick vertical wall called a septum. The left section is somewhat larger than the right section, and each section acts as an individual pump. Each half of the heart consists of an **atrium** (formerly called an auricle) on top, and a larger, stronger, thicker-walled **ventricle** beneath it (Figure 24.13a). The atria serve as reservoirs, and both contract at approximately the same time to force blood into the ventricles. Atrioventricular (A-V) valves permit

* The specialized muscle tissue that performs this extraordinary feat is contained only in the wall of the heart. The only comparable muscular competence occurs in a woman's uterus in preparation for childbirth and during actual labor and delivery, but the uterus does not have to perform at this level for 70 years or so. The heart does.

blood to flow from the atria to the ventricles, but prevent the blood from seeping back into the atria.

The heart is a double pump. It takes oxygen-poor blood returning from the body and pumps it into the lungs (this is the first pump), where waste carbon dioxide is exchanged for fresh oxygen. The oxygen-rich blood re-enters the heart from the lungs, and the heart pumps it throughout the body (this is the second pump). Figure 24.13(b) shows the complete circulation of blood through the heart, which can be broken down into sequential steps as follows:

1. Oxygen-depleted blood from the body flows through the veins into the right atrium at about the same time that oxygen-rich blood from the lungs flows into the left atrium through the arteries. (See the drawings on p. 488.) The blood that enters the right atrium (light brown in the drawing on the next page) is low in oxygen and high in carbon dioxide

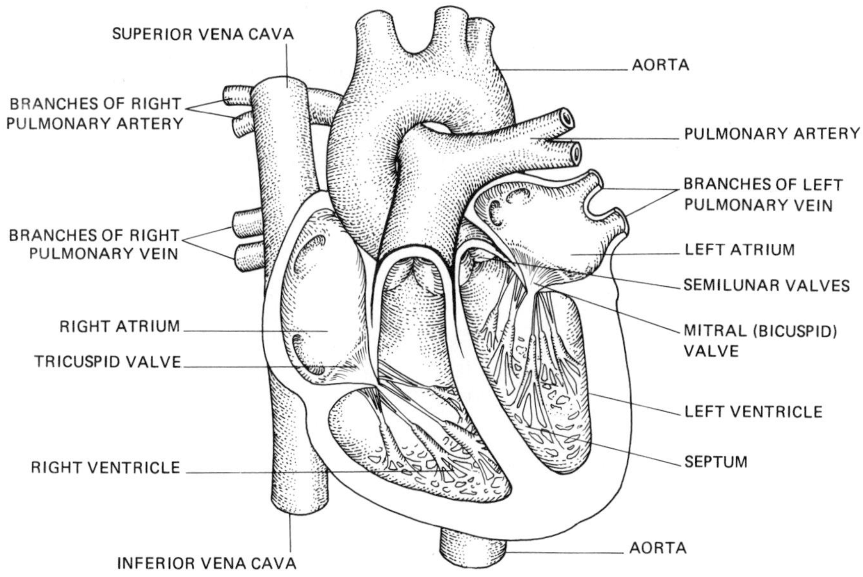

(a)

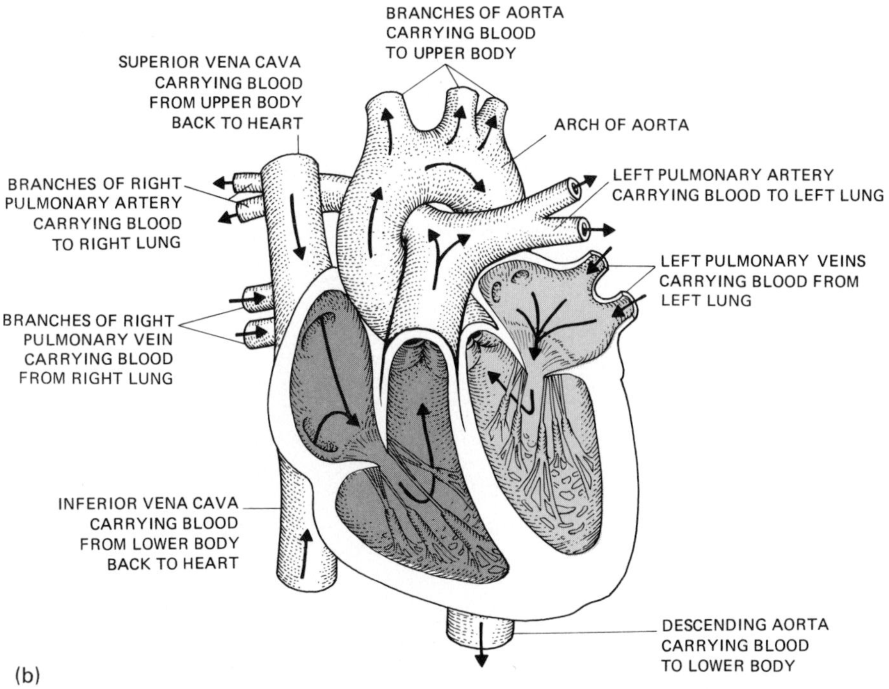

(b)

Figure 24.13
(a) Dissection of a human heart, as seen from the front, with the ventral part of both atria and both ventricles removed. (b) The path of blood through the heart and lungs.

because it has just returned from supplying oxygen to the body tissues. The blood entering the left atrium (dark brown) is rich in oxygen because it has just passed through the lungs, where it has picked up a fresh supply of oxygen and released its waste carbon dioxide.

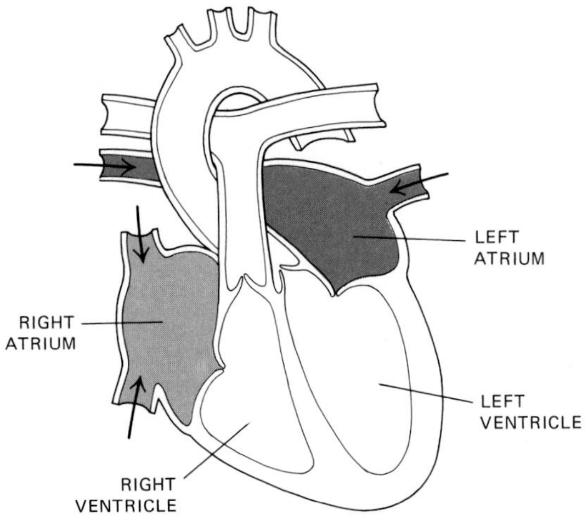

1. Blood enters the atria.

2. The heart's natural pacemaker (see Figure 24.14) fires an electrical impulse and causes both atria to contract rather weakly. Blood is squeezed through the one-way A-V valves into the relaxed ventricles.

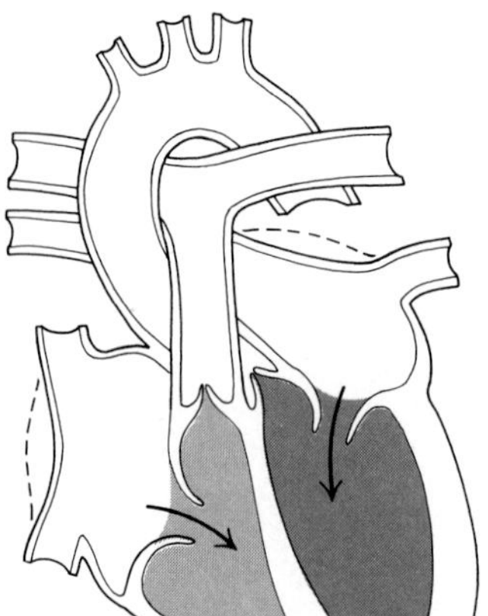

2. Blood is forced into the ventricles.

3. The ventricles, gorged with blood, hesitate for an instant.

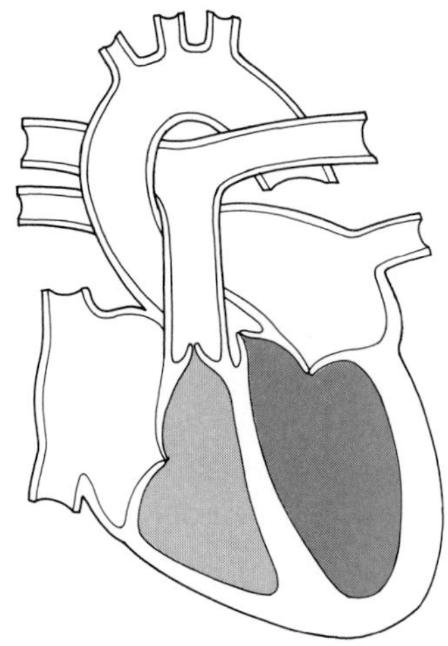

3. The ventricles relax momentarily.

4. On signal from another electrical impulse, the ventricles contract, creating a pressure that closes the A-V valves between the atria and the ventricles while opening the semilunar (S-L) valves leading out of the ventricles. The right ventricle forces oxygen-poor blood out through the pulmonary arteries to the lungs. The left ventricle forces oxygen-rich blood through the largest artery in the body, the aorta. The aorta branches into the arteries that carry oxygen-rich blood to all parts of the body. By this time, the atria are ready to be refilled for another cycle.

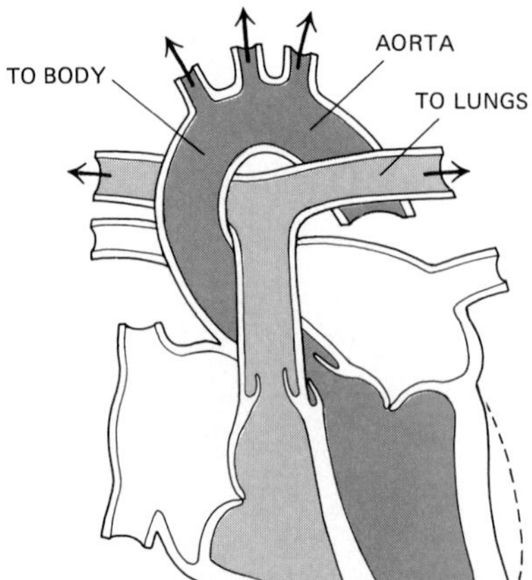

4. The ventricles contract, sending blood to the body and lungs.

The left and right ventricles pump almost simultaneously, so that equal amounts of blood enter and leave the heart. The pumping pressure of the thicker-walled left ventricle is greater, however, because it must supply blood to all parts of the body, whereas the right ventricle supplies only the lungs.

A complete heart cycle, or beat, consists of a ventricular *contraction,* or **systole** (SIS-toe-lee), and a ventricular *relaxation,* or **diastole** (die-ASS-toe-lee), when the ventricle receives blood from the contracting atrium. The only rest the heart gets is about half a second during the brief period between ventricular contractions; also, the heartbeat usually slows down during sleep because most of the capillaries are inactive then. The characteristic "lub, dup" heart sounds are caused not by the contractions of the heart, but by the sudden closing of the heart valves. When a valve leaks, the condition is known as a heart "murmur," and the backflow of blood may be heard as a whooshing sound.

The lower the metabolic rate of an animal, the slower the rate of its heartbeat. Usually, the smaller the animal, the higher will be its heart rate. An average human heart rate is about 72 to 84 beats per minute, for example, but other animals have the following average rates per minute: shrew (the smallest mammal), 800; robin, 600; mouse, 400; rabbit, 200; cat, 125; fish, 60; elephant, 30.*

The Control of Heart Muscle Activity

The origin of a heartbeat is a specialized section of muscle tissue, called the *sinoatrial* (S-A) *node,* near the entrance to the right atrium. The S-A node is known as the main **pacemaker** of the heart (not to be confused with an *artificial* pacemaker), and its original contraction reaches across both atria to the *atrioventricular* (A-V) *node* and spreads along the walls of both ventricles to specialized fibers in the tips of the ventricles (Figure 24.14). Now the simultaneous contractions may begin at the tips of the ventricles and squeeze the blood out through the pulmonary arteries and aorta.

The strength of heart contractions can be changed as body needs vary. More blood is pumped

* Animals with slow pulse (or heart) rates usually live longer than animals with rapid heartbeats. A rabbit, for instance, lives only about five or six years, and an elephant lives about 60 years. Human beings seem to be an exception. We have a faster pulse rate than elephants, and yet we generally live about 10 years longer. Most animals, no matter what size, average about one billion heartbeats per lifetime, but humans, the exception, average about 2½ billion beats per 70-year lifetime.

Figure 24.14
How heartbeat is controlled. When the atria are filled with blood, an impulse is generated in the sinoatrial (S-A) node. As the impulse spreads throughout both atria, their muscles contract. The impulse from the S-A node also stimulates the atrioventricular (A-V) node. This second node, upon sending out its own impulse over the ventricle, causes contraction of the ventricular muscles. The two nodes serve as part of the "pacemaker" system, keeping the heart beating rhythmically. The numbers on the diagram give the time, in fractions of a second, required for the impulses from the S-A node to pass through the heart. The electrocardiogram shows the changes in electrical potential as the impulse progresses. The peak occurs about 0.16 second after the initiation of the impulse, when the ventricles are starting their maximum contraction.

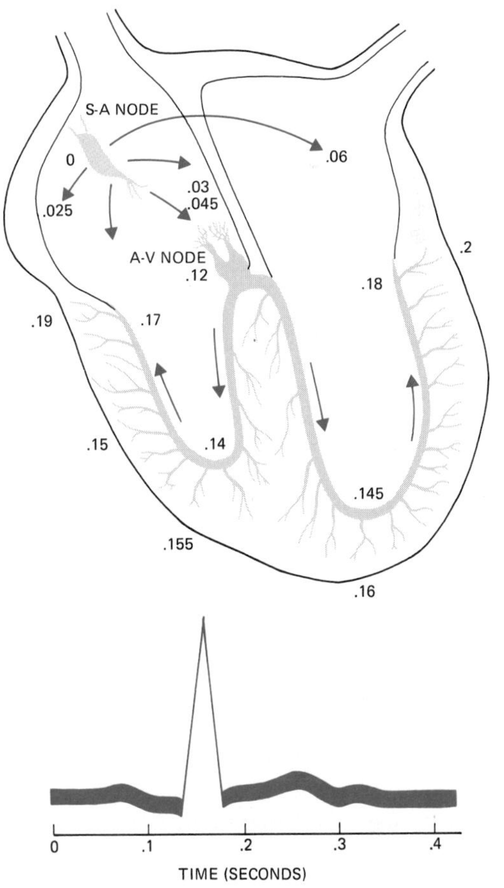

THE UNHEALTHY HEART

"Heart disease" may be defined as any disease that affects the heart, but it can also be a disease of the blood vessels. A more appropriate term is "cardiovascular diseases," which includes both heart and blood vessel disorders. About 29 million Americans have cardiovascular diseases, which are responsible for more deaths than all other causes of death combined. Although it may be decreasing, the American death rate from cardiovascular diseases is still one of the highest in the world. Among the disorders of the cardiovascular system are atherosclerosis, coronary artery disease, stroke, and high blood pressure.

Atherosclerosis is characterized by deposits of fat, fibrin (a clotting material), cellular debris, and calcium on the inside of arteries. These built-up materials stick to the inner walls of the arteries, narrowing the space for passage of blood and reducing the elastic stretchiness. The condition is dangerous because it cuts down blood flow, and if the artery is closed off entirely, no blood can flow. If the closed artery is in the heart, a *heart attack* occurs. If the artery takes blood to the brain, a *stroke* occurs.

Although no one really knows the basic cause of atherosclerosis, several factors are known to increase its progress. Among them are cigarette smoking and the amount of animal fat and cholesterol in the diet. Other factors that may have an effect are hypertension, diabetes, age, stress, heredity, and the male sex hormones.

There are three ways in which atherosclerosis can cause a heart attack: (1) It can completely clog a coronary artery. (2) It can provide a rough surface where a blood clot (*thrombus*) can form and grow to the point that it closes off the artery and causes a heart attack called a *coronary thrombosis.* (3) It can partially block blood to the heart muscle and cause the heart to stop beating rhythmically.

The usual warning signs of a heart attack (they are not always the same) are (1) a heavy, squeezing pain in the center of the chest; (2) pain that may spread into the shoulder, arm, neck, or jaw; (3) sweating; and (4) nausea, vomiting, and shortness of breath. Sometimes these symptoms go away and come back later. Quick medical attention is imperative.

Angina pectoris (Latin, "chest pain") occurs when not enough blood gets to the heart muscle. Sometimes exercising or stress will cause angina. Stopping the stress may relieve the pain, drugs may be prescribed, or surgery may be necessary to replace the damaged artery. Angina does not always lead to heart attack. Sometimes alternative circulation develops, more blood reaches the heart muscle, and the pain decreases. Angina may even disappear altogether if the heart muscle is receiving enough blood.

A **stroke** occurs when something cuts off the brain's blood supply. There are several ways a stroke may happen. (1) Atherosclerosis in the arteries of the brain or neck may block the flow

during exercise, for instance. Likewise, the *rate* and *rhythm* of cardiac muscle activity can be altered by several factors, including blood temperature, concentration of ions, and the chemical environment. But the main factor is nervous activity, and the control center is located in the medulla of the brain.

When the control center in the medulla stimulates the *cardiac* (sympathetic) nerves, norepinephrine is liberated at the nerve endings, and the heart rate increases. In contrast to these cardiac accelerator nerves, the *vagus* (parasympathetic) nerves are *inhibitory* nerves. When the vagus nerves are stimulated, acetylcholine is liberated at the nerve endings, and the heart rate decreases. This slowdown is caused by the hyperpolarization of the nerve membranes. (Hyperpolarization occurs when the cell interior becomes even more negative than the outside when it is at rest. This condition makes it difficult for the nerve cell to generate an action potential.) The accelerator and inhibitory nerves are continually active, both operating in a feedback system. Both sets of counterbalancing nerves terminate in the S-A node, directing the activity of the pacemaker.

of blood. (2) A blood clot (thrombus) may form in the atherosclerotic vessel, closing off the artery and causing a *cerebral thrombosis.* (3) A traveling blood clot (*embolus*) can become wedged in a small artery of the brain or neck; this kind of stroke is an *embolism.* (4) A weak spot in a blood vessel may break; this is a *cerebral hemorrhage.* (When the weak spot bulges, it is called a cerebral *aneurysm.*) (5) In rare cases, a brain tumor may press on a blood vessel and shut off the blood supply.

Because the brain controls the body's movements, any part of the body can be affected by a stroke. The damage may be temporary or permanent. If a brain artery is blocked in the area that controls speaking, speech will be affected. Muscles or vision may be affected. Even memory can be affected. "Little strokes," which may be warning signs of an impending major stroke, should be treated by a physician.

High blood pressure (*hypertension*) is blood pressure that is too high all the time, instead of going up or down in a normal way. Hypertension may be an inherited problem. Usually there are no external signs that blood pressure is high, and only regular medical examinations can detect the condition.

During hypertension, the artery walls are hard and thick, and the important stretchiness of arteries is reduced. Once the stretch is gone, the heart must work harder to pump enough blood, and if hypertension persists, the heart may become enlarged, a condition called *hypertensive heart disease.* High blood pressure can also cause a stroke if the extra force of the blood breaks an artery in the brain and a cerebral hemorrhage occurs.

More than one million Americans die annually from diseases related to hypertension. About 25 million Americans have high blood pressure and almost that many are borderline cases. Most forms of treatment involve the reduction of sodium intake (sodium builds up inside blood vessels, hence the need to stop using table salt), but no matter what drug therapy is prescribed, the purpose is to dilate the arteries so that blood pressure is lowered.

An **artificial pacemaker** is an electronic device that takes over the job of the heart's natural pacemaker. Before the heart muscle will contract, it must receive an electrical impulse, which normally comes from the heart's natural pacemaker, a center of specialized muscle tissue. As long as these impulses are sent regularly, the heart pumps at a steady pace. But if something interferes with the electrical impulses, the natural pacemaker cannot do its job. The heart then pumps too quickly, too slowly, or irregularly. The result is that not enough blood—carrying oxygen and nourishment—gets to the body cells. The answer to this problem is an artificial pacemaker, which is placed in the body (in a rather simple operation) to make the heart beat normally. If checked regularly by a physician, an artificial pacemaker can do its job indefinitely.

THE CIRCULATORY SYSTEM HELPS TO MAINTAIN HOMEOSTASIS

In order to perform its many functions properly, blood must circulate continuously throughout the body. The center of the circulatory system is the heart, but it is the circulatory vessels—arteries, capillaries, and veins—that help maintain cellular homeostasis by supplying cells with the substances they need for metabolism and regulation, and by carrying off their waste products for excretion in the kidneys and lungs (Figure 24.15). A "second circulatory" system is the lymphatic system, which returns to the blood excess extracellular fluid, proteins, and other substances that are lost from the blood (see p. 494).

Arteries, Capillaries, and Veins

Blood moves through a system of three types of vessels: *Arteries* carry blood *away* from the heart; *capillaries* are small, thin-walled vessels that permit exchange of nutrients, oxygen, and carbon dioxide between the blood and tissues; *veins* carry blood *toward* the heart (Figure 24.16).

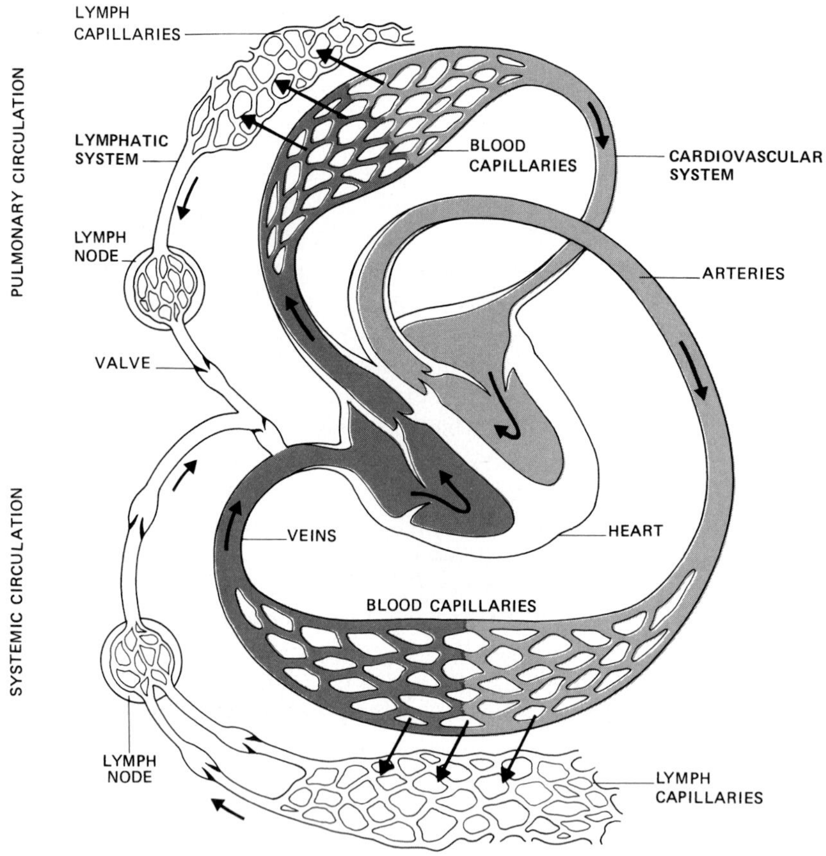

Figure 24.15
The human blood and lymphatic systems. In this schematic diagram, the heart is drawn disproportionately large in order to show its internal structure, and the major branches of arteries and veins are indicated merely as networks of capillaries. Blood vessels are in color, and the lymphatic system is in white. Oxygenated blood is shown as dark brown, and deoxygenated blood is light brown. The blood moves in a closed system, with at least the formed elements—cells and platelets—remaining mostly in the blood vessels. The lymphatic system, in contrast, is not a closed system. The plasma fluid from the blood oozes out from capillaries into the tissues and is worked into the lymphatic system; no circuit similar to that in the blood system is apparent. The direction of lymph flow is not determined by a lymphatic heart; it is aided by one-way valves, the general movement of the body, and the directional flow of blood. (See also Figure 24.18.)

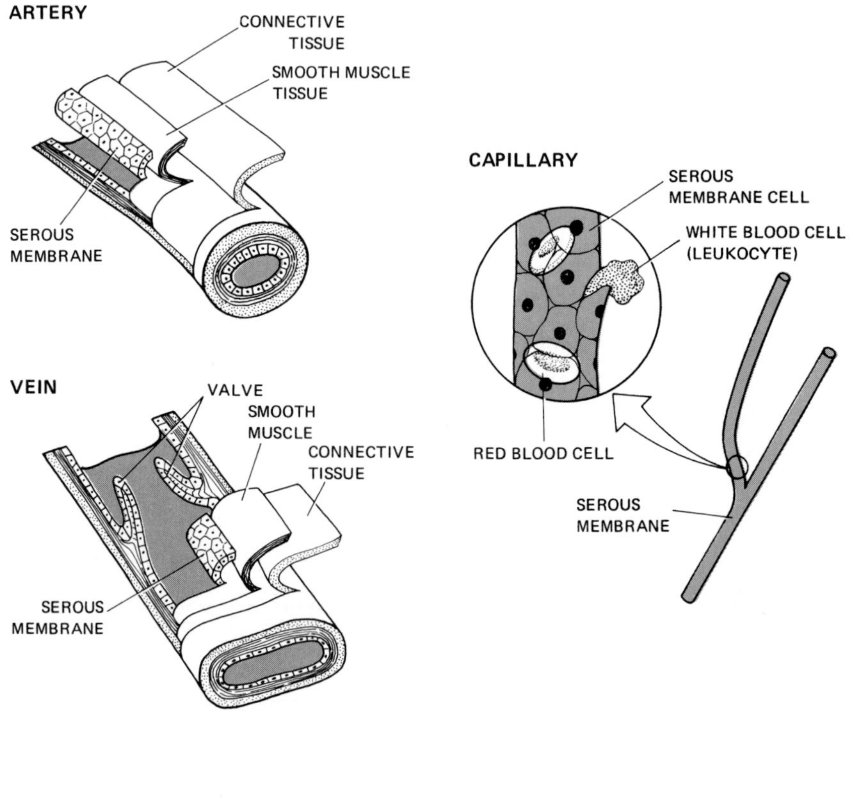

Figure 24.16
Varieties of mammalian blood vessels. *Arteries* are rather firm-walled but elastic, with an internal lining (the serous membrane), a coating of smooth muscle, and a sheath of connective tissue. The walls expand slightly with each pulse of pressure from a heartbeat; they can also be either relaxed or constricted by nervous control. The *veins* are generally softer and more flexible than arteries. They have the same kinds of tissues as the arteries, but they lack the stiffness, and they collapse if blood pressure is not maintained. The *capillaries* are microscopically fine vessels, with walls mostly only one or a few cells thick. They allow passage of water and dissolved materials, and even some blood cells (especially leukocytes). The capillaries are distributed everywhere except in the dead outer layers of skin and in such special places as the lenses of the eyes.

BLOOD PRESSURE

Arterial blood pressure depends on the volume of blood in the arteries and the elasticity of the arterial walls, as well as on the rate and force of heart contractions. If the arteries are supple, they can be stretched by large volumes of blood without an appreciable rise in blood pressure. Conversely, blood pressure will rise easily in arteries that have become hard, internally coated, or inelastic. When blood is ejected into the arteries by the ventricles during systole, an equal amount of blood is not simultaneously released out of the arteries. In fact, only about one-third of the blood leaves the arteries during systole, and the excess volume raises the arterial pressure. After systole, when the ventricular contraction is over, the arterial walls return to their unstretched condition as blood continues to leave the arteries. Pressure slowly decreases, but before all the blood has left the artery, the next ventricular contraction occurs, and the pressure begins to build up again. Because of this consistent rhythm, the arterial pressure never reaches zero, and there is always enough pressure left to keep the blood flowing.

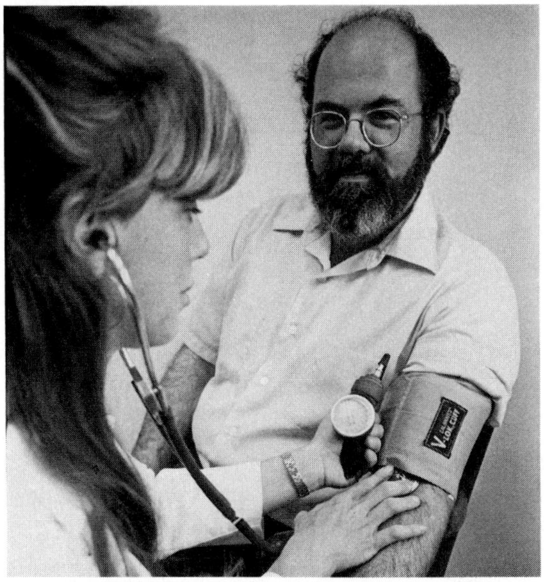

The determination of blood pressure. With a sphygmomanometer *(Greek, "thin pulse measurer"), the maximal (systolic) and minimal (diastolic) pressures in arteries can be determined in a safe and noninvasive manner.*

Blood pressure levels are expressed by two numbers, both representing the height (in millimeters) of a column of mercury. The high number is the **systolic** pressure (when the heart contracts), and the low number represents the interval between heartbeats (the diastolic pressure). The lower figure is critical, because the higher the **diastolic** pressure is, the less rest a heart will get. An overworked heart simply does not last as long as a well-rested one. A normal adult blood pressure is 140/90 or *less.* Blood pressure is considered high, or hypertensive, in an adult when the systolic reading exceeds 160 and the diastolic reading is higher than 95. The *pulse pressure* is the difference between systolic and diastolic pressure. In the example above of "normal" blood pressure, 140/90, the pulse pressure equals 50. (This is not the same as the pulse *rate,* or the number of beats per minute, which usually ranges from 72 to 84 beats per minute in normal adult humans. The pulse rate can easily be felt at the inside of the wrist.)

The instrument used for measuring blood pressure is the easily recognized but not easily pronounced *sphygmomanometer* (sfig-moh-muh-NOM-ih-ter). The sphygmomanometer consists of an inflatable cuff, a rubber bulb, and a column of mercury. The hollow cuff is wrapped around the upper arm and inflated to a pressure above the systolic pressure (the tightened cuff closes the arteries in the upper arm and prevents blood from flowing to the lower arm). A stethoscope is placed over the artery just below the cuff, and the pressure in the cuff is slowly released until the arterial pressure is greater than the pressure of the cuff. At that point, a recognizable sound can be heard through the stethoscope. This sound signals the high-velocity release of blood, and the figure on the mercury column at which it occurs represents the *systolic* blood pressure. When the cuff pressure is lowered further, a louder sound is heard, and then the sound gradually becomes softer. When the sound stops altogether, the *diastolic* blood pressure is noted on the column. The absence of sound indicates a free-flowing gush of blood through the open artery.

Arteries Of all the types of blood vessels, the **arteries** have the greatest pressure, and the arterial walls, which are appropriately strong and elastic, adjust to the great pressure of the contraction of the ventricles during systole. Blood travels through the aorta at a speed of about 40 cm/sec (0.9 mph). In people who suffer from hardening of the arteries (*arteriosclerosis*), the arteries are too rigid to expand properly, and the systolic blood pressure rises higher than the normal range. The arteries must also be able to squeeze down during diastole, the period of lowest arterial pressure. Arteries branch into smaller arterioles, whose walls contain smooth muscle like those of the arteries. By contractions of this smooth muscle, the arterioles control the varying flow of blood to organs as they require it.

× 750

Figure 24.17
A microscopic view of blood flowing through capillaries. The smallest capillaries have such a narrow passageway that the red blood cells move through in single file, allowing time for the necessary gas, nutrient, and waste exchanges to take place between blood and functioning cells.

Capillaries Arterioles enter the body tissues and branch out further to form **capillaries,** the link between arteries and veins. Capillaries are the smallest and most numerous blood vessels. If all the capillaries in an adult human body were connected, they would stretch out about 96,000 kilometers (60,000 mi). Such an abundance of capillaries makes available an enormous surface area for the easy exchange of substances between the blood and nearby cells. Capillaries are so narrow (less than 0.01 mm in diameter) that red blood cells must squeeze through in single file (Figure 24.17). Because the capillaries are so narrow, a high-fat diet may present circulatory problems. After a high-fat meal, fatty globules are present in the blood. These globules may adhere to red blood cells, making it difficult for the heart to push the blood through the capillaries.

The thin walls of the capillaries are *selectively permeable;* that is, they allow some dissolved substances to filter through and hold back others. Such substances as gases, waste products, salts, sugars, and amino acids pass freely through the capillary walls, but the large proteins present in the fluid part of the blood pass through only with difficulty. Likewise, the red blood cells cannot pass through the walls of the capillaries.

Veins As capillaries leave the cells and tissues of the body, they join together to form **veins.** Venous walls are thinner and less elastic than arterial walls, and their diameter is greater. Blood pressure in the veins is low, and the venous blood is helped along by the contraction of surrounding muscles and is prevented from flowing backward by one-way valves. These valves are especially important in the legs, where gravity cannot assist the return of blood to the heart. If a person stands still for a long time, the leg muscles are not able to push the venous blood up. In such cases, the venous blood returns more slowly to the heart and tends to accumulate in the veins. Reduced return of blood to the heart may result in poor blood flow to the rest of the body, including the brain. This could cause fainting. Another problem is that the veins in the feet and legs may be stretched somewhat as tissue water accumulates there. If the one-way valves in these veins weaken and the veins become permanently dilated, *varicose veins* result.

The Lymphatic System

The **lymphatic system** is sometimes called the "second circulatory" system (Figure 24.18). However, it is quite different from the system that transports blood around the body. The lymphatic system is not a closed, circular system, nor does it have a pump. It is made up of a network of thin-walled vessels that carry a clear fluid called **lymph.** The system begins with very small vessels, lymph capillaries, in contact with the tissue. These vessels join together to make larger ducts, which pass through specialized structures called lymph nodes and continue on to drain their contents into a vein in the neck.

The composition of lymph is similar to that of blood. Lymph contains water, some plasma proteins, electrolytes, and white blood cells (lymphocytes), but it lacks red blood cells and most of the blood proteins. Lymph is derived from the fluid portion of the blood that passes from the arterial ends of capillaries out into the spaces around cells. The fluid lymph surrounds and bathes the cells of the body.

Some of the fluid derived from the blood returns to the blood in the venous end of the capillaries. What remains may enter the lymphatic system and move slowly from remote regions of the body toward the lymph nodes in the groin, arms, armpits, and neck. The lymph nodes are specialized tissues that act as sieves, removing cellular debris, old cells, and foreign particles from the lymph fluid before the fluid rejoins the venous blood. When infection or injury occurs, lymph nodes may enlarge as they collect dead tissue cells or many lymphocytes. This produces the familiar "swollen glands" that result from some diseases and are evident frequently in the neck (Figure 24.19).

Figure 24.18
Distribution of the lymphatic system in a human body. Although lymphatic vessels are distributed throughout practically all living tissue, they are concentrated in some regions, especially the neck, upper chest, armpits, and lower abdomen. If an infection occurs, lymph *nodes* may become so enlarged that they can be felt as lumps in the neck or groin. Adenoids, tonsils, the thymus gland, and the spleen are composed of lymphoid tissue. (See also Figure 24.15.)

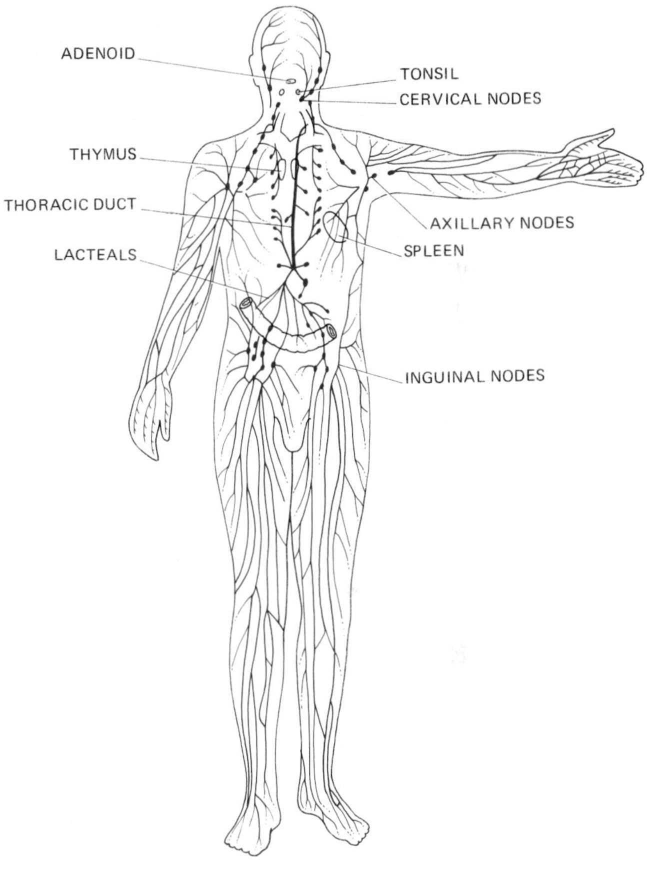

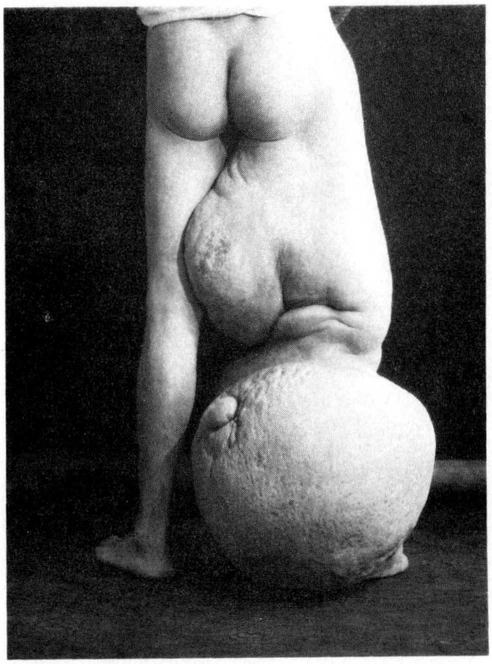

Figure 24.19
The effect of stoppage of lymph flow. A tropical roundworm, filaria, can be injected into the human bloodstream by mosquitoes. The parasite burrows into tissues and obstructs the return of lymph to the lymphatic ducts. The result is a swelling of the affected part, an affliction known as *elephantiasis.* An unusually severe case is shown here. The damage is not locally painful, but continued interference with the movement of lymph can be fatal. No cure is known.

Basically, the lymphatic system performs three major functions:

1. It returns to the blood excess fluid and proteins from the spaces around cells.
2. The lymph nodes filter and destroy bacteria before returning lymph fluid to the blood.
3. The lymphatic system plays a major role in the transport of fat materials from the tissue surrounding the small intestine to the blood. This action is accomplished by a portion of the lymphatic system that leads from the small intestine to an opening in the inferior vena cava.

IMMUNOLOGY

Several observations on the reactions of people and animals to disease led to the discovery of the *immune systems* and hence to the whole science of **immunology.** One observation was that even during the

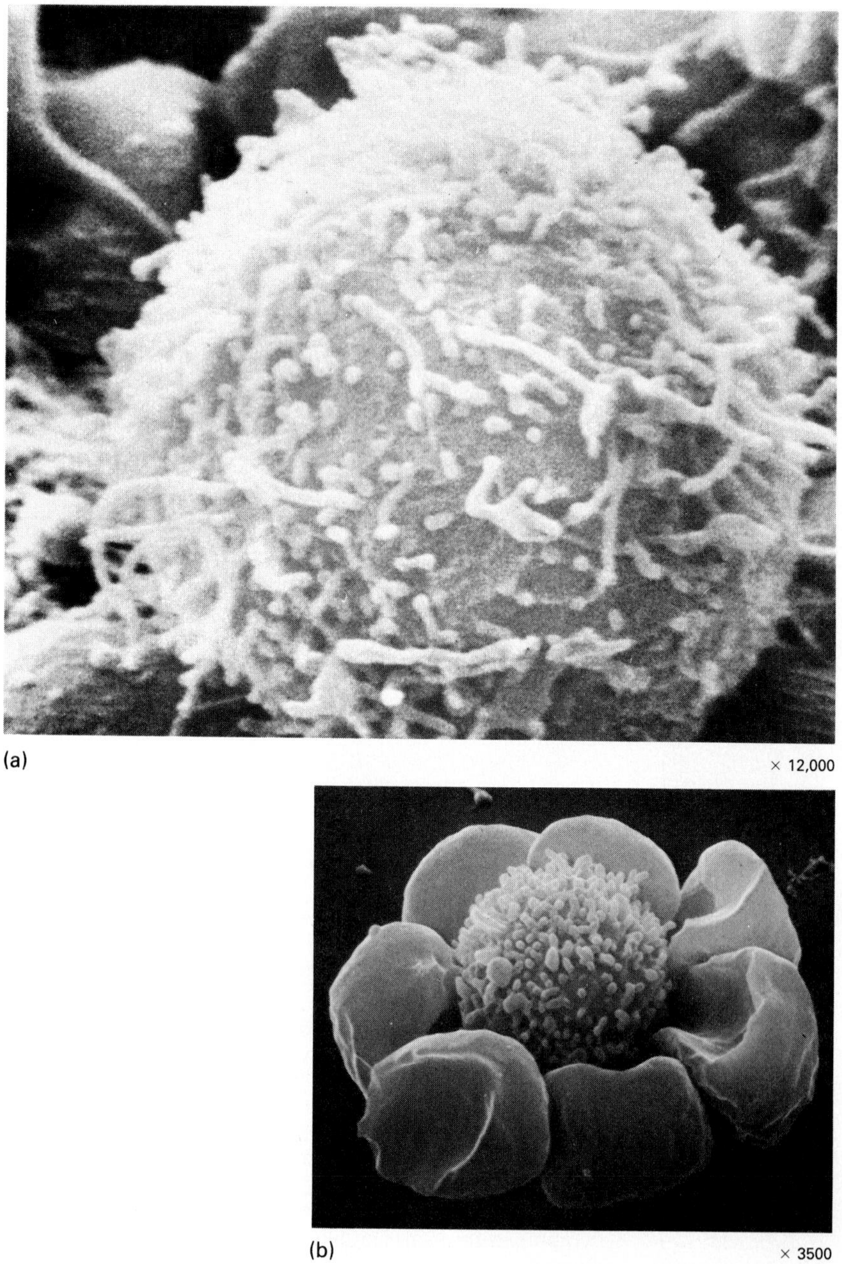

(a) × 12,000

(b) × 3500

Figure 24.20
Scanning electron micrographs of lymphocytes. (a) Unlike red blood cells, lymphocytes have nuclei and are capable of undergoing mitosis, increasing enormously in number when an infection occurs. (b) The photograph of this flowerlike cluster has in the center a mouse spleen lymphocyte containing proteins (immunoglobulins) specific for sheep red blood cells. When such a cell comes into contact with sheep red cells (the surrounding "petals" in this photograph), the two kinds of cells are firmly bound. Once a lymphocyte has developed such a specific ability, it can transmit that ability to the cells resulting from its divisions. All the lymphocytes descended from such a cell retain its immune reaction; they are said to have an immunological memory.

IMMUNITY FROM NAZI SLAVERY

During World War II, when the Germans were killing or enslaving Poles and Jews as racially inferior people, an immunological trick was employed to save lives in a political way. E. Lazowski, now at the Illinois Children's Hospital in Chicago, and S. Matulewicz, now at the Université Nationale in Zaire, knew that an infection by a relatively harmless bacterium, *Proteus* OX-19, could cause a blood reaction resembling that of epidemic typhus, the dreaded louse-borne "disease of human misery." The Weil-Felix reaction had been known since an earlier epidemic in Poland in 1916, and it was routinely accepted as a diagnostic test for true typhus. The doctors reasoned that a blood test of a healthy person who had been injected with a dead suspension of *Proteus* OX-19 would indicate typhus.

For more than 25 years there had been no epidemic typhus in Germany, and consequently the Germans had lost much of their immunity to the disease. Knowing that, German doctors had no desire either to import the disease into their country or to go themselves into places where they could come into contact with it. Thus any Pole who showed a positive Weil-Felix reaction was presumed to be carrying typhus and was not wanted in Germany.

A Polish slave laborer, allowed a 14-day home leave, volunteered to be the first test case of an artificially induced immunological experiment. Although the trial carried risks, some known and some unknown, both to the Polish doctors and the worker, he preferred those risks to a return to slavery. He took the *Proteus* injection, officially disguised as "protein stimulation therapy." His blood sample, sent to the German laboratories, prompted a telegraphic reply: Weil-Felix positive. The local German officers, interpreting the telegram to mean typhus, excluded both the worker and his family from Germany in order to keep out typhus-carrying lice.

With the technique established, Lazowski and Matulewicz began injecting many Poles with *Proteus* OX-19, thus producing false typhus-positive blood tests, until a dozen villages were officially declared an epidemic area, relatively exempt from German interference. Suspecting that Polish doctors might send mislabeled blood samples, the German doctors tried to take samples only in their own laboratories, but of course their results were also positive because the reaction was a true biological one, even though it had been artificially induced. As Lazowski and Matulewicz reported in 1977, "Our private immunological war gave us a deep satisfaction because we knew that we had saved the lives of people who would otherwise have been killed."

worst of epidemics—for instance, during the plague—not everyone caught the prevalent disease. Some individuals were able to mingle freely with the dying and walk away untouched. Some people are immune to certain diseases. Another observation is that a person who has once had a disease, say mumps, is highly unlikely to suffer from the same disease again. One has, so to speak, earned one's immunity. Medical doctors thus learned empirically that they could confer immunity artificially by administering a weakened form of a disease-producer. Edward Jenner (1749–1823) is frequently credited with originating the practice of giving a vaccination (Latin, "vaca," cow) with cowpox to keep a possible victim from getting deadly smallpox. In fact, various kinds of vaccinations had been in use in many places for years even before Jenner's discovery. Louis Pasteur later applied the same principle when he inoculated sheep against anthrax, and today such immunizing methods as treatment with the Salk vaccine against poliomyelitis are still based on the same idea.

The essential action in the development of an animal's immunity to any specific outside influence is the synthesis of a specific protein, known as an **antibody.** After the entrance of a foreign material, whether it is a bacterium, a virus, or an inhaled or swallowed substance, an animal's body may start the production of an antibody. The foreign material, known as an **antigen,** somehow stimulates the synthesis of a particular kind of molecule, the antibody, whose exact molecular shape is such as to match the molecular shape of the antigen. The antibody can tie up the antigen firmly enough to render it ineffective by coagulating it in a way roughly equivalent to what happens to the white of an egg when it is boiled.

If an invading bacterium is the antigen, and the antibody can be produced fast enough to keep ahead of the growth of the invader, the animal will destroy the bacterium. This would be a case of naturally developed immunity. In human medicine, a person who has been exposed to a possibly contagious infection may, as a precaution, receive an injection of a mild, perhaps weakened or heat-killed pathogen (a disease-causing agent, often a bacterium). The hope is that the body system will produce enough antibody of the correct type to avoid disease. Thus *active immunity* may be achieved. However, if a disease is found only after it has progressed to an uncomfortable or dangerous stage, the sufferer may be given a dose of actual antibody. Such an antibody is often prepared by injecting the causative bacteria into an animal such as a horse, and allowing the horse to manufacture the proper antibody. Serum containing the antibody is taken from the horse and administered to the human patient, to provide *passive immunity*.

The same system that protects an animal from infection can be annoying or even deadly. Developing an immunity against typhoid fever is desirable, but inadvertently developing an immunity or sensitization to wheat flour makes eating difficult. A person with a sensitivity to what most people consider ordinary food is **allergic,** and allergic reactions are expressions of the same kind of antigen-antibody activity that helps an animal body defend itself against bacteria and viruses. (See the essay on this page.)

Blood transfusions must be strictly monitored because of the antigen-antibody system. A basic principle of blood transfusion is that a transfusion must not be made if the cells of the donor's blood will be *agglutinated* (clumped together) by the recipient's plasma (see p. 257).

Now that organ transplants are mechanically feasible, one of the critical details that must be attended to is the possibility of an immune reaction that may cause the recipient to reject the new organ from a different person with a different set of proteins. In a related way, some women chronically have miscarriages because they lack a factor that prevents the immunological rejection of the embryo in the mother's uterus. Because the embryo has proteins that resemble its father's as well as its mother's proteins, the mother's body may react to the embryo as though it were a "foreign" antigen unless the mother possesses the proper antibody. Obviously, most women have an immunological mechanism that makes chronic miscarriages unusual.

ALLERGY: WHAT IS IT?

Most people have some kind of allergy, but few understand what an allergy is. We blow our noses, wipe our eyes, and take "anti-allergy" pills, but even if we deduce that an "antihistamine" is a drug that opposes a "histamine," who knows what a histamine is? To start with, hay fever is not a fever and is not caused by hay. And antihistamines do not attack the allergy; they fight against the work your own body is trying to do to defend itself against the foreign substance that starts the allergic reaction.

What we call hay fever is generally a reaction to pollen in the air, a seasonal phenomenon that is usually caused by tree pollen in the spring, grass pollen in summer, and weed pollen in the late summer and fall. Ragweed does the most damage because it produces huge amounts of pollen. A single ragweed plant produces more than 10 billion pollen grains. (Such a prodigious output is necessary to offset the hazards encountered by plants that depend on the wind to carry pollen to the female flower. The output can be compared to the large amounts of sperm released at each ejaculation.)

An **allergy** is an abnormal body reaction, and the agent that causes the unusual reaction (coughing, sneezing, headache, hives, etc.) is an *allergen* or *antigen.* The most common allergen is dust, and other well-known allergens are pollen, food, and chemicals. When you are allergic to something (chocolate, for instance), your body reacts to it as though it were a foreign substance. Your body is triggering a typical immune reaction, the same sort of procedure that produces beneficial immunities against infections or that causes troublesome rejections of organ transplants.

Once the allergen is recognized as a foreign substance, your body produces specific chemicals that stimulate the formation of an *antibody.* One of these chemicals is *histamine,* which causes hypersensitivity of some body tissues. So the purpose of an antihistamine is to try to eliminate the annoying effects produced by histamine, and not by the allergen.

The first time an allergen enters the body, there is an incubation period during which antibodies are being formed, and no allergic symptoms occur. Subsequently, the allergen will trigger specific allergic reactions. These reactions are usually out of proportion to the actual need, but the body is effectively defending itself, and it has no way of knowing that the allergen is not indeed a serious threat. Ordinarily, two specific occurrences accompany allergic reactions: Certain body cells release chemicals, and smooth muscles contract. We do not know how such chemical substances as histamine, serotonin, and acetylcholine are released from cells, but we have learned that the location of the meeting between allergen and antibody determines the severity of the allergic reaction. If an antibody encounters an allergen in the bloodstream, the allergen is neutralized without any side effects. However, if antibody meets allergen in or on a cell, neutralization occurs, but the cell is disturbed also. It is this disturbance and release of toxic chemicals that produces the specific allergic reaction.

Histamine is the best understood of the chemical substances released from cells. It dilates blood vessels and causes the contraction of smooth muscle, and it is probably involved in most allergic reactions to some degree. The most troublesome products of histamine are *wheals,* localized swellings that vary according to their location. Skin hives are external wheals. Wheals in the mucous membranes of the nose cause stuffiness and are usually accompanied by a runny nose. Such respiratory problems as bronchial asthma are caused by wheals in the air passages, and they can interfere with breathing.

Hundreds of other types of allergies exist, and thousands of allergens have been identified. Some of the most common offenders are chemicals such as lacquers, animal danders (flaking skin) and hair, eggs, milk, wine, wheat flour, legumes, chocolate, tobacco smoke, coffee, nuts, citrus fruits, wool, and shellfish.

If you are allergic to strawberries, will your children inherit the allergy? Maybe, but it is more likely that they will inherit the susceptibility to allergy rather than an allergy to a specific substance.

Many people are allergic to pollen or cat hair. But recently a more dramatic allergy has come under suspicion. Some women may be infertile because they are allergic to their husband's sperm. These women may create antibodies that prevent sperm from entering the outer covering of the egg.

SUMMARY

1. *Respiration* at a cellular level means the oxidation of food with the consequent release of energy. *Respiration* at the gross anatomical level means *breathing* (air containing oxygen enters the lungs, and air containing waste gases is expelled from the lungs).

2. Respiration is relatively simple for organisms like single-celled protozoa. They obtain oxygen by direct diffusion through their bodies, and they lose carbon dioxide wastes by diffusion also. But larger, more complex animals have a body surface area that is smaller in relation to their volume and less permeable, and for these animals more complex systems of gas exchange evolved.

3. *Gills* are an efficient means of respiration in water. They are filaments that allow the blood to flow in

a direction opposite to the water flow, thus establishing an efficient *countercurrent flow.*

4. The more developed the organism, the more complex is the internal structure of the *lung.* Lungs have evolved from single sacs to highly folded and lobulated mammalian lungs, in which millions of tiny air sacs called *alveoli* appear, each surrounded by a network of blood vessels.

5. Insects and some other land-dwelling arthropods have a system of tubes called *tracheae,* which branch out from surface openings called *spiracles* to all parts of the body.

6. The respiratory system in mammals is made up of the lungs, the several passageways leading from the outside to the lungs, and the muscular mechanism (*diaphragm* and abdominal and rib muscles) that moves air into and out of the lungs.

7. The intake of air, *inspiration,* occurs when air rushes into the lungs to equalize a reduction of air pressure in the thoracic cavity. *Expiration* is the expulsion of air from the lungs.

8. The *respiratory center* in the medulla of the brain receives and coordinates information about the changing needs of the body for oxygen. A high carbon dioxide content of the blood causes an *increase* in breathing rate, and a low carbon dioxide content causes a *decrease.*

9. Through the process of diffusion, the blood that enters the lungs receives oxygen from the alveoli, and carbon dioxide and water diffuse from the blood into the alveoli and subsequently to the air.

10. *Hemoglobin,* the *respiratory pigment* in the red blood cells, has such an affinity for oxygen that it binds about 98 percent of the oxygen available in the lungs. When oxygen combines with hemoglobin, the resultant compound is *oxyhemoglobin.* Some toxic agents, such as carbon monoxide, bind to hemoglobin even more readily than oxygen does.

11. Carbon dioxide is transported by the blood in three ways: (1) About 67 percent of the carbon dioxide combines with water to form carbonic acid; (2) about 25 percent of the carbon dioxide reacts with hemoglobin and eventually is released in the lungs in exchange for oxygen; and (3) the remaining 8 percent is dissolved in the blood as molecular carbon dioxide.

12. *Blood* is the term used for circulating fluids in general. Blood in the human body carries dissolved foods, transports carbon dioxide and oxygen, distributes hormones, defends against infection, helps maintain an even temperature, removes waste products, and performs several other functions.

13. The liquid portion of mammalian blood is *plasma* (water, dissolved solids including proteins, and dissolved gases), and the solid portion is *formed elements* (red blood cells, white blood cells, and platelets).

14. Blood *clots* at the site of a wound when platelets initiate the release of a series of enzymes and proteins. The ultimate product is a stringy protein, *fibrin,* which entangles the blood cells and forms a clot.

15. The *heart* is a double pump, each half consisting of an *atrium* and a *ventricle.* With the help of a series of one-way valves, blood enters the right atrium, passes to the right ventricle, then flows out through the pulmonary arteries to the capillaries of the lungs. It returns to the left atrium and then passes through the left ventricle and out again through the aorta. A complete heart cycle, or *beat,* consists of the ventricular contraction, *systole,* and the ventricular relaxation, *diastole,* when the ventricle receives blood from the contracting atrium.

16. The *strength* of heart contractions can be changed as body needs vary, and the *rate* and *rhythm* of cardiac muscle activity can be altered by several factors, including blood temperature, ionic concentration, and chemical activity. But the main factor is *nervous activity,* and the control center is located in the medulla. The heartbeat originates in a specialized section of muscle tissue called the sinoatrial node, or *pacemaker* of the heart. Accelerator and inhibitory nerves act in a feedback system.

17. Blood moves through a system of three types of vessels: *Arteries* carry blood away from the heart; *capillaries* are small, thin-walled vessels that permit diffusion between the blood and tissues; *veins* carry blood toward the heart.

18. The *lymphatic system* returns excess fluid and proteins from the spaces between cells to the blood, and it filters and destroys harmful material.

19. When foreign substances (*antigens*) get into the bloodstream, they combine with specific *antibodies* and an inactive substance is formed, resulting in *immunity.* An *allergy* may result when an antigen that does not normally evoke antibodies does so, causing discomforting physiological symptoms.

ASK YOURSELF

1. Distinguish between respiration as a cellular activity and respiration by lung expansion.
2. How does a fish obtain oxygen?
3. Describe an example of "countercurrent exchange."
4. Could you drown a grasshopper by holding its head under water? Explain.
5. What muscles are used for inspiration of air by a mammal? For forceful expulsion?
6. Which causes greater discomfort to a person: excess carbon dioxide in the air or reduced oxygen?
7. What part of the nervous system regulates breathing in humans?
8. Why is carbon monoxide poisonous?
9. How is carbon dioxide transported in the bloodstream?
10. What are the structural and functional differences between red and white blood cells?
11. What steps are followed when a blood clot is formed?
12. Describe the natural pacemaker in a human heart.
13. What human artery carries deoxygenated blood?
14. What causes the sounds made by a normal beating heart?
15. What causes a "heart murmur"?
16. Why do soldiers standing at attention during inspection frequently faint?
17. What are varicose veins?
18. What causes "swollen glands" in the neck, armpits, or groin during a bacterial infection?
19. Distinguish between an antigen and an antibody.
20. What is the difference between active and passive immunity?

PART SEVEN
EVOLUTION, ECOLOGY, AND ANIMAL BEHAVIOR

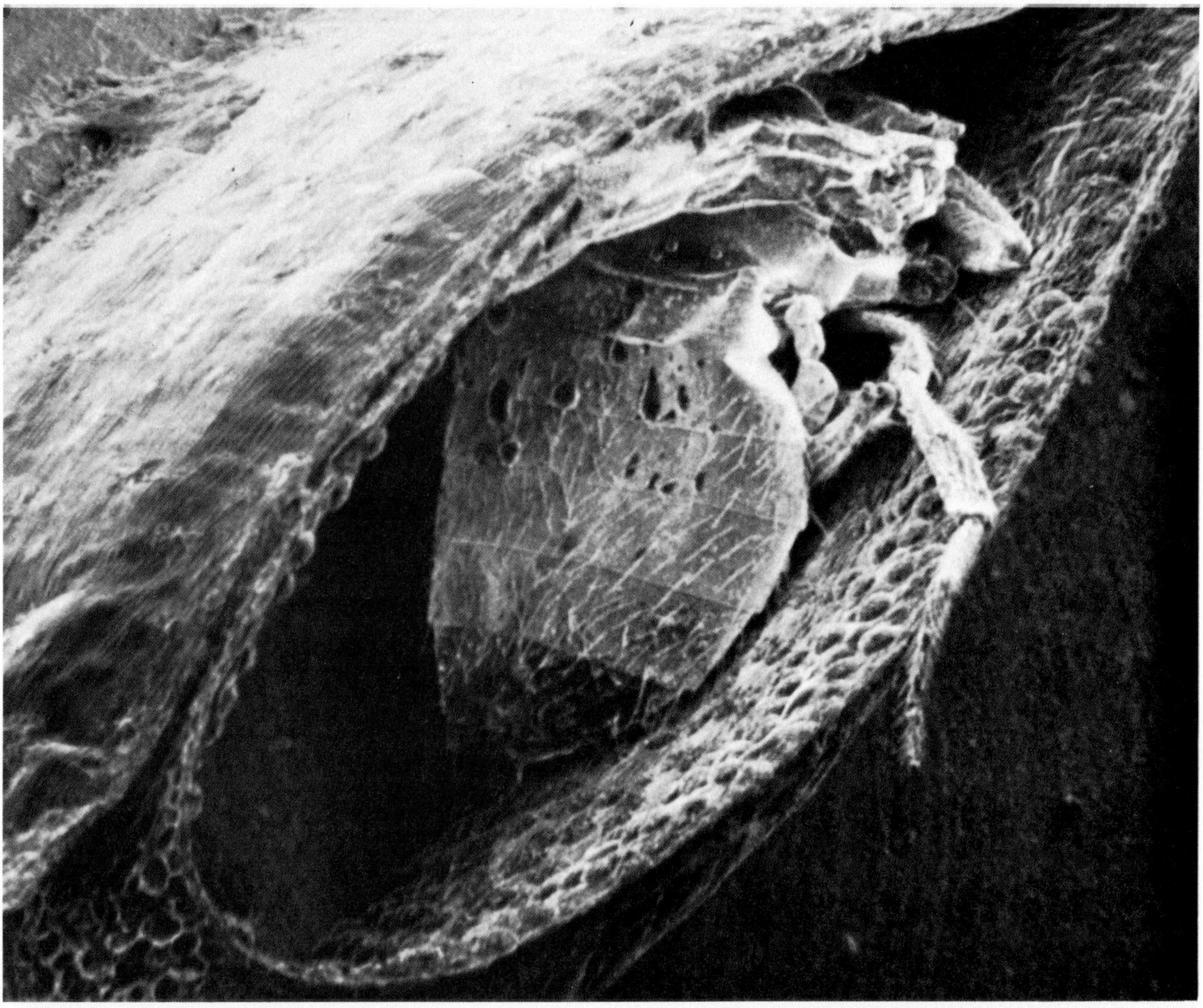

Venus fly-trap leaf, cut longitudinally, which has partially eaten a fruit fly, *Drosophila melanogaster.* × 100

25
Darwinian Evolution

SOME KEY POINTS

1. Darwin's idea that species can become modified was the basis of the theory of evolution.
2. An organism has the best chance for survival when it inherits those traits that make it best suited to its environment.
3. Individuals do not evolve; *populations* evolve.
4. If Darwin had been aware of Mendel's genetic research, he might have been able to explain the mechanism of evolution in terms of genetic mutations.
5. Comparative anatomy, the embryological development of animals, the geographical distribution of organisms, fossil organisms, and the overall idea of the vast expanse of geological time helped Darwin formulate his ideas about evolution.

EVOLUTION MEANS CHANGE OR LITERALLY, "AN unrolling." In the biological sense, evolution specifically means *genetic* change. Even though genetic changes are usually detrimental in the environment in which they arise, once in a while a genetic change bestows an advantage. Such a change is the basis of evolution and the means by which new species may arise.

The concept of evolution pervades all biology and underlies all its major generalizations. We can start with the recognition of a unity in all life on earth. Practically all organisms build the same amino acids into their proteins, use the same enzymes to carry on their energy-exchanging activities, construct their membrane systems in similar ways, and regulate their productivity by means of similar molecules, such as cyclic AMP in human livers and in slime molds. The more we learn about the chemistry and activities of organisms, the more striking the similarities become and the less important are the easily observed differences. Because of this underlying unity, biologists believe further that the present differences among organisms, bewildering in their diversity, have come about by the genetic modification of earlier, simpler living things. This belief makes possible the integration of isolated facts and gives us confidence that general conclusions about evolution are valid.

Consider a single subcellular particle, a ribosome. If ribosomes are found to look and act in a special way in bacteria, that bit of information may be interesting to a microbiologist, but by itself it means relatively little. However, if ribosomes are found to look and act similarly in corn roots, human heart muscle, and tomato worm eggs, then it becomes more reasonable to think that ribosomes are particles of general, fundamental importance to living things. Without an evolutionary connection between such apparently unlike organisms, a few bits of information would remain interesting to a few specialists, but they would be lacking in the broad significance that can make them interesting to every thinking human. As it is, we can look on all the earth's inhabitants as members of an enormously varied but ultimately related superfamily. As any satisfying scientific concept should do, the concept of evolution simplifies and unifies knowledge.

THE IDEA OF EVOLUTION

The earliest students of organisms were primarily interested in the human uses of plants and animals, and they made schemes of classification purely for their own convenience. When they were making such schemes, similarities between kinds of organisms did become apparent, and those who looked closely noticed that the several sorts of birds known as hawks had features in common, or that the many varieties of beans, for all their little differences, were still beans. Such observations prompted speculation about the origin of species.

The concept of evolution was stated in its modern form by Charles Darwin in 1859, but Darwin himself was never able to explain exactly how evolution happens because he lacked an understanding of genetics. But what were the clues that helped Charles Darwin to convince most of the world that evolution *did* (and does) happen?

Geological Sources

Much of the basis for the growing interest in evolution came from geology. James Hutton, generally regarded as the father of modern geology, stated in 1785 that the landscape of the earth was created by a continuing process of slow, steady, and natural change, not by sudden catastrophes, as propounded by other scientists of that time.

The findings of geologists implied a long history of the earth. In order to harmonize these new ideas with biblical teachings, Baron Georges Cuvier and others proposed what came to be known as the "great compromise." According to Cuvier, the period from the Flood of Noah to modern times was 6000 years—the age of the earth and of humans. Prior to the Flood, Cuvier proposed, was the supernatural time of "geology," when violent catastrophes occurred and all sorts of strange animals existed.

All was calmly accepted until 1830, when Charles Lyell published *Principles of Geology*, which supported Hutton's idea that the so-called geological period was not a supernatural time. Lyell's book renewed worldwide interest in Hutton's notion that there was "no sign of a beginning or of an end." Lyell and Hutton both realized the dramatic and sudden effects of floods, earthquakes, volcanoes, and other geological "catastrophes," but in terms of geological change they could see only an endless cycle of uniform processes. If Lyell and Hutton were correct, the earth had to be much older than the few thousand years claimed in the belief of their day. This was one of the critical issues that Darwin had to confront. He also had to confront the issue he never really solved: What *causes* evolution?

Lamarckian Evolution

One of the first scientists to attempt an explanation for the *causes* of evolution was Jean Baptiste Lamarck, who in the early 1800s proposed the following ideas:

1. As environmental conditions change, plants and animals develop new traits to cope with the change. If vegetation becomes scarce, for example, giraffes will be forced to stretch their necks higher and higher to reach tree leaves.

2. *The theory of use and disuse.* The more an individual uses a part of its body, the more developed that part becomes. Conversely, as an individual neglects to use a body part, that part disintegrates and will eventually disappear. For instance, athletes develop their muscles by constant exercise (*use*), whereas snakes used their legs less and less until those legs finally disappeared (*disuse*). (We now know that the theory of use and disuse is correct for an individual organism during its lifetime, but not from generation to generation.)

3. *The theory of the inheritance of acquired characteristics.* Individuals who have acquired or developed a characteristic as a result of use or disuse will pass that characteristic on to their progeny. In other words, the muscular athlete will foster offspring whose muscles are well developed. (There is no evidence to support this theory.)

Lamarck believed that animals strive to adapt to environmental changes, and in so doing, they acquire new characteristics; for example, the giraffe acquires a long neck by stretching for leaves. According to Lamarck, such an environmentally derived feature will be inherited by offspring. Lamarck's explanation seemed reasonable, but the theory of the inheritance of acquired characteristics did not stand up to experimental tests.

Figure 25.1
Charles Darwin (1809–1882), of whom science historian Erik Nordenskjöld said: "If we measure him by his influence on the general cultural development of humanity, then the proximity of his grave to Newton's is fully justified."

Numerous experiments over the last 200 years indicate that changes in body parts—shape, color, etc.—acquired by parents are *not* transmitted to their offspring. Only changes (mutations) in the DNA of the sex cells of the parents can be passed on to offspring. If the DNA of the sex cells is not affected, as it is not in the muscular athlete or the stretching giraffe, future generations will not be affected. If you change the shape of your nose through plastic surgery, you are not guaranteeing an improved nose for your future offspring. Obviously, plastic surgery does not affect your sex cells, and only changes in your sex cells can affect your future offspring. Today biologists are able to explain that the thing that can be inherited is a *genotype,* not a *phenotype* as Lamarck unknowingly suggested.

It is unfortunate that such an original, creative, and skillful scientist as Lamarck has been belittled because a small portion of his work turned out to be incorrect. Lamarck was wrong, but he laid the groundwork for Charles Darwin.

CHARLES DARWIN AND THE VOYAGE OF THE BEAGLE

In 1831, the 22-year-old Charles Darwin (Figure 25.1) left England on a five-year voyage around the world aboard H.M.S. *Beagle,* a brig designed specifically for scientific research. The ship would chart the coastline of South America, make longitudinal and astronomical measurements, and collect samples of local wildlife (Figure 25.2).

Figure 25.2
The route of Darwin's trip around the world on H.M.S. *Beagle.*

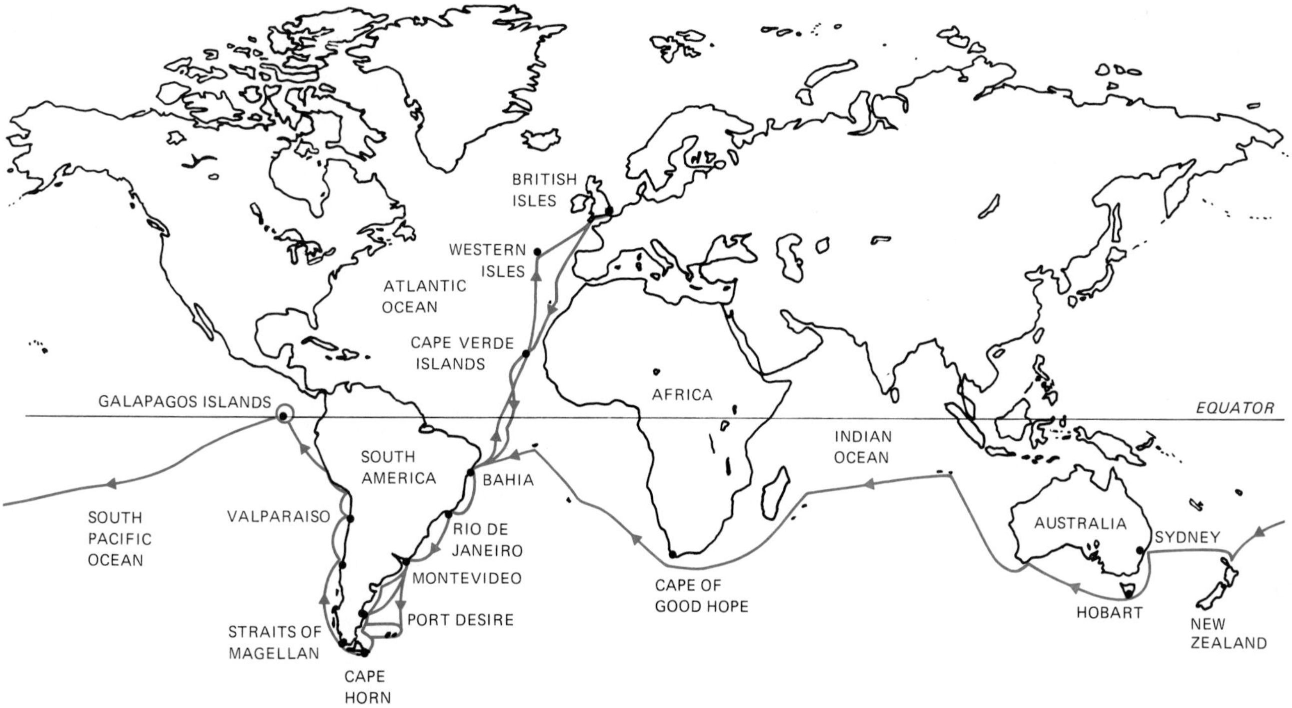

Young Charles had shown little aptitude for medicine, the law, or the clergy, but he had developed an avid interest in nature while studying theology at Cambridge for three years. Darwin's profession was finally chosen for him when Captain Robert Fitz Roy invited him to become the *Beagle's* naturalist. Fitz Roy made it clear to Darwin that as part of his duties he would be expected to help refute the radical new ideas of geology by gathering as many specimens as he could to celebrate the wonders of nature. (Captain Fitz Roy realized the irony of that situation when Darwin published his theory of evolution 28 years later. Another irony was that Darwin's former botany professor gave him a book to take along on the voyage, warning him not to take it seriously. The book was *Principles of Geology,* by Charles Lyell.)

Darwin did indeed find a glorious array of natural wonders and he meticulously accumulated a mass of information about them. But along with the many examples of nature, he also began to ask questions about the overwhelming diversity of living things. This was the question that began to nag at Darwin: What was the reason for so many different kinds of plants and animals? He had read the "heretical" book by Lyell, and he readily accepted the idea that the earth's landscape was gradually being altered through vast expanses of time. Was it possible that plants and animals had been changed as well?

The voyage of the *Beagle* ended in 1836, and Darwin returned home with his head (and his notebooks) full of questions about the life he had seen. He began a systematic analysis of his collections. Darwin did not start his investigations with the intent of proving anything, certainly not to formulate a theory of evolution. Instead, he merely applied a lively curiosity to the natural world; having observed thousands of plants, animals, and the earth, he sought an explanation that would integrate what he had learned. He wanted to know why and how, as well as what.

Darwin's observations during his trip aboard the *Beagle* were the basis for his later theories of evolution. Three items in particular seem to have impressed him:

1. Fossils found in South America appeared related to (but not the same as) the living animals of South America; in fact, Darwin realized later that these fossils were more like the living animals in the same country than like fossils of the same age in other countries.

2. Animals living in different climates of South America appeared related to (but not the same as) each other; here again, the similarities were greater among South American animals in hot and cool climates than between South American animals in a hot climate and animals of another country with a comparably hot climate.

3. Animals on islands appeared to be related to (but not the same as) animals of the closest mainland; in addition, some island animals seemed to show a relationship to other species on the same island. These observations eventually suggested to Darwin the idea of the *modification of species,* an idea that was the basis of the theory of evolution.

The Influence of Malthus

Shortly after Darwin began what he called his "systematic enquiry," he read *An Essay on the Principle of Population,* which had been written in 1798 by Thomas Malthus, a political economist. Malthus had written that organisms possess the theoretical ability to breed at a geometrical rate, thus eventually running out of food and space. This idea had been the cause of lively, often bitter debate ever since Malthus had first proposed it, but theoretically it is unassailable.

If a single bacterium could divide and its offspring could divide every half hour without any hindrance, within three days the earth could not hold them all. This much is simple arithmetic. However, the anti-Malthusians pointed out that in reality such unlimited growth does not occur. Besides, observations of animals in the wild, as well as experiments with caged ones, have shown that many animals have a number of self-regulating devices, including failure to mate and infant abandonment. More recently, neo-Malthusians have looked at the human species, noted the exponential growth of the world population, and point out that here at least is a species that seems to be determined to crowd itself out of home and food. Regardless of whether or not Malthus was completely realistic, the reproductive *potential* is in fact present in the form of innumerable gametes.

Darwin was aware of the phenomenon of variation within species. It was the consideration of the potential for variation that made its impression on Darwin, and he noted in his autobiography: "In October 1838, that is fifteen months after I had begun my systematic enquiry, I happened to read for amusement Malthus on Population, and being well prepared to appreciate the struggle for existence which everywhere goes on, from long-continued observation in the habits of animals and plants, it at once struck me that under these circumstances favourable variations would tend to be preserved and unfavourable ones to be destroyed. The result of this would be the formation of a new species. Here, then, I had at last got a theory by which to work."

NATURAL SELECTION: THE KEY IS VARIATION AND SURVIVAL

Charles Darwin was a methodical man. Twenty-two years after the voyage of the *Beagle,* he was still working on his definitive study. Darwin, in fact, almost waited too long. In 1858, Alfred Russel Wallace also formulated a theory of evolution, based on his studies in Brazil and the East Indies (Figure 25.3). Wallace's interpretation was remarkably similar to Darwin's (both had been influenced by Malthus), and when Wallace sent the manuscript of his essay to Darwin for his opinion, Darwin was astounded. Although Darwin's first instinct was to give Wallace full credit for the theory, the two men agreed to present their papers in the same issue of the *Journal of the Linnean Society.*

The next year, 1859, Darwin finally finished his book, *On the Origin of Species by Means of Natural Selection, or the Preservation of Favoured Races in the Struggle for Life;* the popular title is *The Origin of Species.* The book was so well thought out, so precisely documented and filled with examples, that it was almost futile to dispute the evidence. Darwin's theory contained the following points:

1. Organisms tend to produce more offspring than the environment can support.
2. Although organisms tend to reproduce at a geometrical rate, the overall population does not increase at that same high rate. In fact, populations tend to remain fairly constant, probably because many individuals do not live long enough to reproduce.
3. Because so many individuals are introduced into a limited environment, they must compete for the available resources.
4. Individuals within any species vary considerably.
5. Individuals who inherit beneficial traits are better adapted to survive.
6. The organisms that survive transmit their favorable variations to their offspring.

Figure 25.3
Alfred Russel Wallace (1823–1913), the English explorer, naturalist, author, social reformer, and coauthor with Darwin of the principle of organic evolution. This modest, intrepid traveler made unbelievable journeys, frequently with no company but savages, collecting, describing, and observing, until he was forced to the conclusion that species are changeable.

Darwin suggested that new species may descend from the old ones. This theory contradicted the belief that once a species is created, it is incapable of change. The uproar was instantaneous, but there was no available evidence with which to mount a real counterattack. Darwin had shown that "descent with modification" had probably occurred, and no one would ever be able to think of living things in quite the same way again.

A crucial part of Darwin's theory stated that the best chance for survival came when offspring inherited those traits that best suited them to their environment. In a way, then, the environment was "selecting" those individuals who would survive. Darwin called the entire process **natural selection;** it is sometimes referred to as *the survival of the fittest.*

What Role Does the Environment Play?

There is no evolutionary rule that says evolution must proceed from the simple to the complex. If environmental conditions are such that a less complex mechanism is advantageous, then the simpler

Figure 25.4
A polar bear (*Thalarctos maritimus*), an example of an animal well suited to its environment. Its size, its feeding habits, its insulation, its metabolism, and its coloration all make the polar bear able to live in frozen regions that would kill a less well-adapted animal.

organisms will flourish. The main point is that *organisms that are able to adapt to changing conditions will survive and produce offspring.*

In Darwin's own words, "The more diversified the descendants of a species became in structure, constitution and habits, the better they would be able to seize on widely diversified places in nature." A plant that was able to get by with little water (a cactus, for example) would survive in the desert. Animals that were well insulated against the cold and had protective coloring besides (polar bears, for example) would survive in the Arctic (Figure 25.4). The Arctic environment did not create the polar bear. Bears already existed when the environment where they lived became colder and colder, or bears moved northward for food when their resources dwindled or competition increased. When the environment changed, some ancestor of the polar bear had the proper genetic makeup for survival. If the polar bear had been unable to exist in a frigid climate, it would be extinct. Organisms that can adapt survive. Otherwise they die.

An important point should be reinforced here: *Individuals do not evolve; populations evolve.* The eminent geneticist G. Ledyard Stebbins has summarized this issue admirably: "Individual animals, plants, or microorganisms develop, reproduce, and die; but they do not evolve. Evolution takes place in populations, and can be understood only by studying populations in their entirety, not just the individuals of which they are composed. Individuals retain the same genetic constitution throughout their lives; the genetic constitution of populations varies to a greater or lesser degree from one generation to another."

Recently, humans have created entirely new strains of some insects by using DDT as a pesticide and thereby introducing an environmental change. Theoretically, DDT is an effective killer, but some insects within a species were naturally resistant to DDT, and these mutants survived the chemical onslaught that might have made their species extinct. These "superbugs," temporarily in the minority, multiplied rapidly because of the dwindling competition, and they passed on their resistance to their offspring. Those insects that were resistant to DDT became more common than the sensitive ones. (DDT acted as the agent for selection; it did not create resistance in the insects.) Naturally, DDT then became less useful as an insecticide.

DARWIN'S THEORY UPDATED

Darwin was not aware of the importance of Gregor Mendel, even though their theories were presented only seven years apart. But remember, *most* people had never heard of Mendel until the twentieth century. As a result, Darwin was unable to explain natural selection in terms of genetics, and therefore he was incapable of answering one of the critical questions asked about his theory: What made evolution work?

If Darwin had known about chromosomes, genes, and DNA, his theory might have been stated more like this:

1. Organisms can produce more offspring than the environment can support.

2. Although organisms tend to reproduce at a geometrical rate, the overall population does not increase at that same high rate. In fact, populations tend to remain fairly constant, because many individuals do not live long enough to reproduce, and because many animals have a number of self-regulating devices, including failure to mate, infant abandonment or infant destruction, and reduced fertility and litter sizes.

3. Because so many individuals are introduced into a limited environment, they must compete for the available resources. Such a competitive struggle for existence is not limited to physical combat but includes such competitive activities as the selection of mates and obtaining nesting sites and food.

THE COCKROACH: EVOLUTIONARY PERFECTIONIST

Nobody likes a cockroach, except perhaps biologists, who recognize a marvel of evolutionary perfection when they see one. The cockroach, along with the shark, the opossum, the horseshoe crab, and several other organisms, is a truly prehistoric being, which no longer evolves in any major way because it is already so well suited to its environment.

Cockroaches flourished during the Carboniferous Period, about 300 million years ago, and fossil remains of cockroaches indicate that they have changed very little since then. How have cockroaches managed to survive for so long, and in what ways are they so perfectly evolved? Perhaps the most obvious answer to these questions is that the design of the cockroach is generalized, simple, and functional.

But besides their functional design, cockroaches are not without intelligence. In laboratory experiments roaches have demonstrated an ability to learn, and it is probably that ability that allows them to use fully their functional assets.

A cockroach either sees or otherwise senses a source of trouble and moves away from it in about 2/100 of a second. (Male sprinters competing in the 100-meter dash at the 1976 Olympic Games reacted to the starter's gun in about 12/100 of a second, about six times slower than the cockroach's reaction time.) The central nervous system of the cockroach, though relatively simple, is adapted for quick decision-making and movement. Two feelers, or *cerci,* at the base of its abdomen pick up deflections of air and relay a message to the nervous system. If the stimulus is sufficiently strong, the cockroach uses its six powerful legs to move quickly away from the intrusion.

Cockroaches do most of their prowling in the dark, when the danger of being discovered is minimized. When a cockroach does leave the safety of its nest, it is usually "searching" for a meal or a mate. When hunger is the stimulus, the cockroach uses its antennae to sense the presence of food, *any* food. Another reason the cockroach has survived so long is that is will eat just about anything, including soap and glue. In fact, the binding of the book you are holding has been put together with types of glue, cardboard, and other cover materials that are intentionally made unattractive to mice and insects. Some older books make delightful homes for roaches, not to mention bookworms. Because they will eat anything, cockroaches can also live anywhere—another advantageous trait for survival.

The body of a cockroach is not only sturdy, it is flat enough to squeeze through minute openings to escape danger. This remarkable body can live without air for several hours and can withstand almost 100 times as much radiation as a human body can. It is no wonder that no insecticides are totally effective, and even the most successful repellent doesn't always work. Unfortunately, the cockroach's evolutionary process has not ceased altogether. An occasional mutant cockroach will survive even the most potent poison, and the unaffected strain will continue to live and reproduce, rapidly producing offspring that will inherit their parents' immunity.

Obviously, poison will not be the factor that finally eliminates the cockroach. If human beings ever eliminate the cockroach, it will probably be through a mechanism that prevents the birth of live and healthy offspring. Until then, the battle between humans and cockroaches continues, a draw at best.

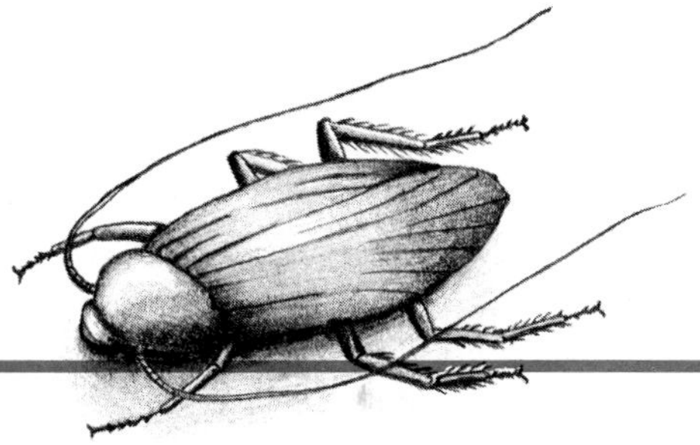

4. Individuals within any species vary considerably because genetic changes occasionally occur.

5. Ordinarily, variations caused by gene mutations, chromosome mutations, or gene recombination are either harmful or useless. In the course of time, however, there occur some mutations that give their possessors an improved chance of survival in their environment.

6. In a changing environment, those organisms with favorable genetic variations survive. The sur-

viving organisms then transmit their DNA to their offspring, and over long periods of time entirely new species may evolve. Organisms that have successful genetic variations not only live longer but produce more offspring, who in turn inherit the favorable variation.

THE EVIDENCE FOR EVOLUTION

The facts that Darwin collected can be grouped into a small number of categories, each of which we will discuss briefly to introduce the kind of information that has convinced biologists of the validity of evolutionary theory. These categories are (1) comparative anatomy, especially of the vertebrate animals, (2) the embryological development of animals, (3) the variability of plants and animals under the special care of humans, (4) the distribution of organisms on earth, and (5) the structure and distribution of fossil organisms.

Comparative Anatomy

Why, Darwin asked, is there such similarity in the structure of the forelimbs of such diverse animals as humans with hands, horses with hooves, bats and birds with wings, and seals with flippers? He compared the bones of those organs and found many major similarities in all instances. It seemed reasonable to Darwin that such similar forelimbs have similar origins, even though they may be used for different purposes now. When different species have similar body parts derived from the same common ancestor, those parts are called *homologous* (Figure 25.5). If different organisms have parts that are used for similar purposes but are not similarly constructed or derived from the same common ances-

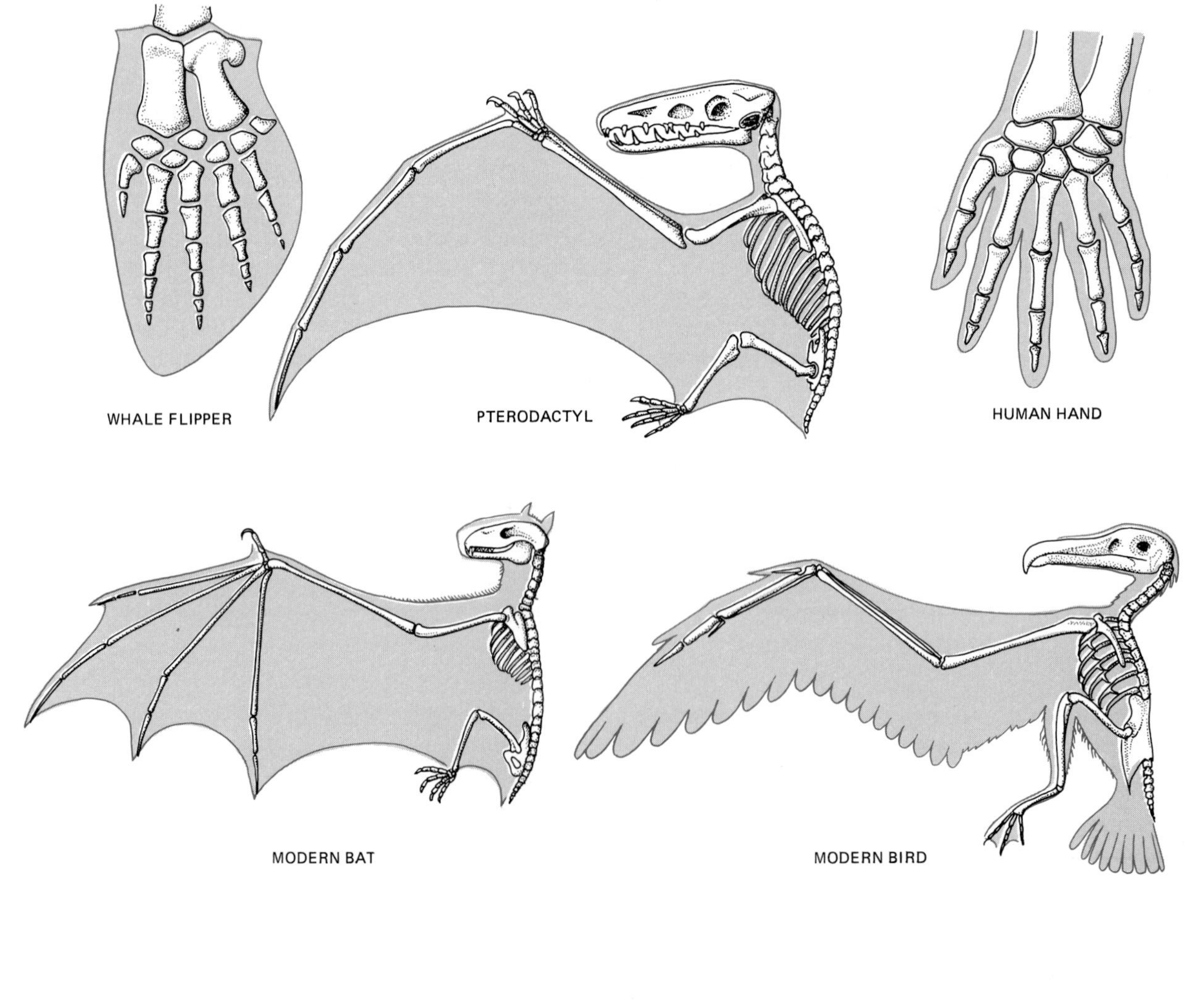

Figure 25.5
Homologous organs have structural and developmental similarity but may have functional diversity. Compare the bony structure of a human hand with that of a whale's flipper, or the wings of an extinct pterodactyl, a modern bat, and a modern bird. All are homologous. All have fundamentally the same plan of skeletal support, and all arise in similar fashion during embryonic life.

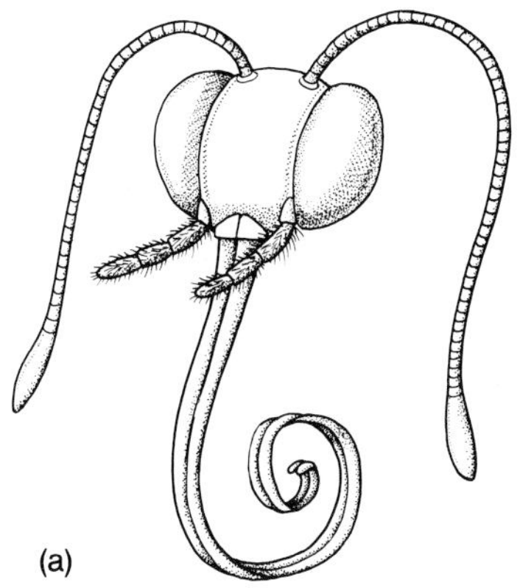
(a)

(b)

Figure 25.6
Analogous organs have functional similarity but structural and developmental differences. (a) The proboscis of a butterfly and (b) that of an elephant look somewhat alike in spite of the difference in size, and they are used in essentially similar ways, at least in part; but in their embryological development and in their fundamental structure they are not in any way comparable.

tor, those parts are called *analogous* (Figure 25.6). Homologous parts have been derived from the same common ancestor, whereas analogous parts have not.

Considering further, Darwin asked why animals have structures they do not use. He noted the presence in humans of ear, tail, and skin muscles, of an appendix at the junction of the large and small intestines, and of a rudimentary "third eyelid," or nictitating membrane (Figure 25.7). None of these structures is used by humans, but all are functional in other vertebrate animals. A number of such structures are known, and they are called **vestigial** (Latin, "footprint," "trace") because they are believed to be lingering remnants of once useful parts. (In rabbits, for instance, the appendix serves as a temporary storage bin for the cellulose-rich diet of the rabbit. It thus allows extra time for the appropriate enzymes to break down the cellulose.) If evolution has occurred, vestigial organs are understandable as structures that have almost but not quite disappeared. On the other hand, if evolution has not occurred, then vestigial structures seem beyond understanding.

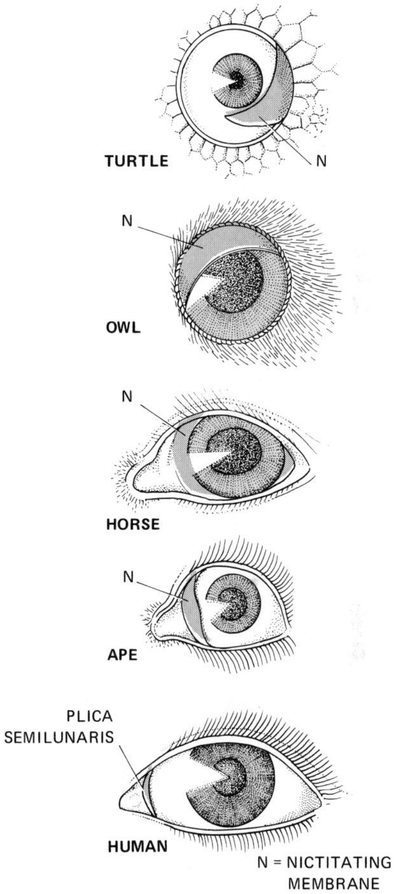

Figure 25.7
An organ that is functional in some species and vestigial in others: the nictitating membrane, or "third eyelid." Even though the membrane is consistently present in all vertebrates above fishes, it is frequently nothing more than a tiny fold of muscle. In the eyes of birds, however, it is a readily workable, translucent membrane that can be flicked over the eyeball, cleaning it without requiring the bird to close its opaque "real" eyelids. In the human eye, it is barely visible and cannot be moved.

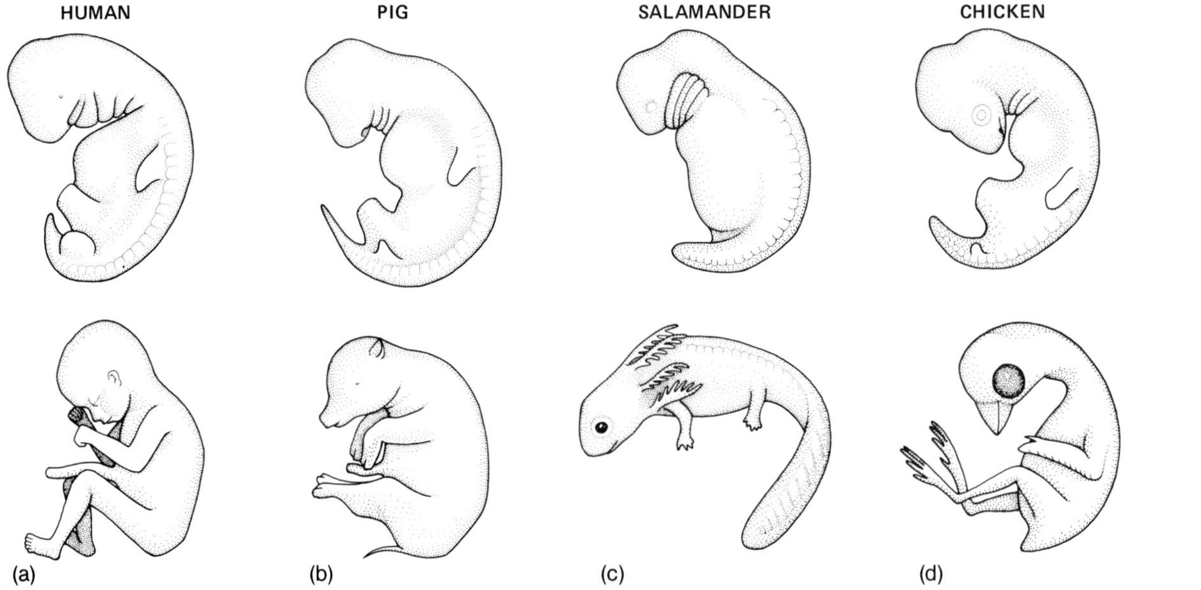

Figure 25.8
Similarity among different species of animals during early embryonic stages. Embryos of (a) a human, (b) a pig, (c) a salamander, and (d) a chicken are very much alike at the stage when the head is beginning to develop, the gill arches are pronounced, the limbs have budded, and the tail is conspicuous (top). Only after specific peculiarities appear do the animals become readily recognizable (bottom).

Figure 25.9
Varieties of a single species (*Brassica oleracea*). The original wild species, first cultivated in the eastern Mediterranean region, is something like modern collards, of which kale is a crinkly variety. Wherever humans cultivated different varieties, the primitive types of cabbagelike plants were selected for different features and were adapted to different climates. Some varieties became hard-headed, as in modern cabbages; some grew swollen stems, as in kohlrabi; some made masses of flower buds, as in broccoli and cauliflower; and some made clusters of leaf buds, as in Brussels sprouts.

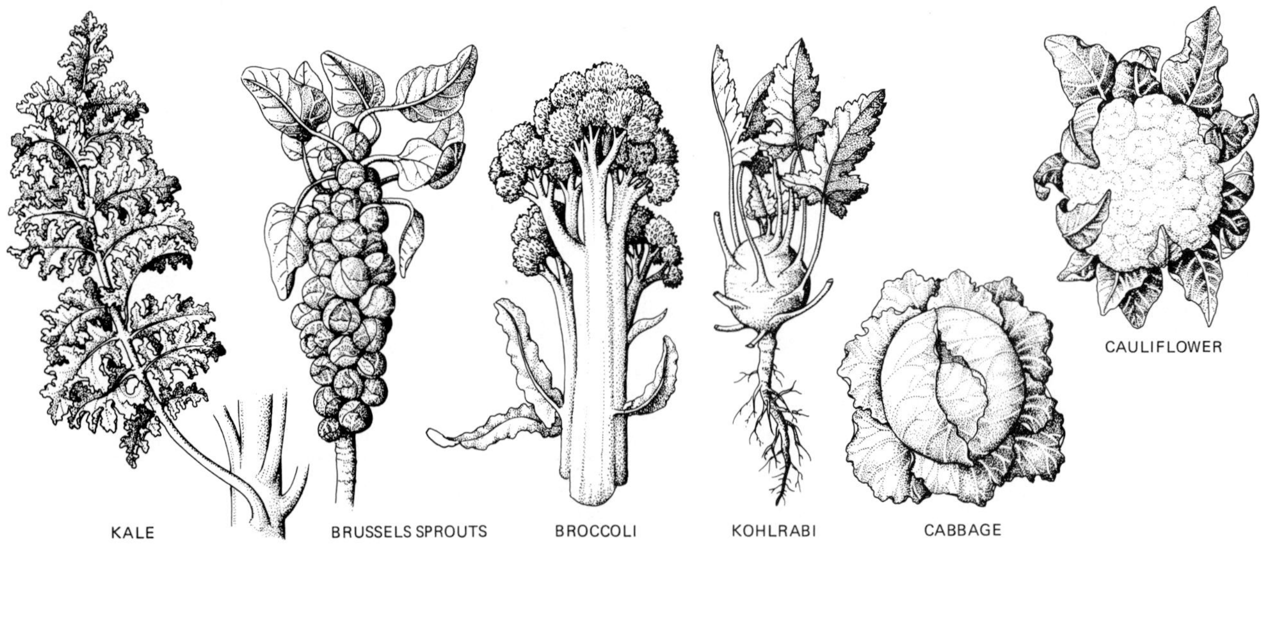

Embryological Development of Animals

Darwin also considered the embryological development of animals (Figure 25.8). Very young embryos of fishes, birds, and mammals are strikingly similar. Humans have a tail during the early stages of embryonic growth. Indeed, for a while a human embryo has a tail that is longer than the legs, and an occasional human infant is born with a tail, which is quickly and quietly snipped off by the attending doctor. All mammalian embryos have gill slits for a while, but they close eventually, leaving only slight evidence of their presence. In fishes, gill slits remain open and functional throughout life. If mammals evolved from a fishlike ancestor, then the possession of gill slits in the mammal embryo is understandable as a remnant from earlier times.

Variations among Domestic Species

We do not usually see evolution happening. It is too slow to be readily seen in one or even a number of human generations, and it is consequently not obvious on casual observation. There are instances, however, that show that evolutionary change can occur within relatively short times, and these are notable in the plants and animals that have come under the influence of human cultivation. Consider, for example, the tough, weedy plant, *Brassica oleracea,* which has been subjected to artificial selection as a food plant for only a few centuries. It has given rise to a surprising diversity of garden vegetables: ordinary cabbage, red cabbage, Savoy cabbage, kale, kohlrabi, Brussels sprouts, cauliflower, and broccoli (Figure 25.9). These are all varieties of the same species, developed rapidly under the intense selection pressure of cultivation. Similarly, domesticated horses, sheep, dogs, cattle, and camels now exist in forms unknown in any wild state, having been artificially guided in their evolution by humans (Figure 25.10). Darwin called special attention to the many races of pigeons bred by pigeon fanciers, pointing out that evolution can indeed be observed within limited time.

The Geographical Distribution of Organisms

During his cruise on the *Beagle,* Darwin's curiosity was aroused by the distribution of species on the earth. He was particularly interested in island flora and fauna because these are usually special. The most famous of "Darwin's islands" are those in the Galapagos archipelago off the west coast of Ecuador in South America. There he studied many life forms but most particularly the tortoises and birds. He saw that the finches varied somewhat from island to island, but they were more like the finches of the

Figure 25.10
Variation within an animal species: dogs (*Canis familiaris*). Dogs have been domesticated and bred for special traits for such a long time that the original wild dog is not known.

Figure 25.11
Divergence among species following geographical isolation and specialization of feeding habits. These finches from several of the Galapagos Islands are in many ways alike, but each has its own separate home island and its own kind of chosen food. They have been separated from one another long enough to have become distinct species but not long enough to lose their status as finches. See also the essay on Darwin's finches on the next page.

South American mainland than they were like European finches (Figure 25.11). A reasonable explanation is that the island finches had come from the mainland, but they had been separated from the ancestral types long enough to have changed in form and habit. Other examples are available. Many caves harbor blind fishes and other blind animals, but such animals have useless eye sockets or remnants of eyestalks. The notable fact is that blind fishes in American caves are similar to the normal American fishes outside, and the blind fishes in European caves are similar to the normal European fishes outside. Once more, if we postulate an evolutionary development, the facts make sense; if we do not, they are baffling.

Fossil Organisms

Finally, the earth itself offered evidence. There were seashells heaped in inland valleys, and layer upon layer of rocks containing fossilized remains of long-dead animals. Such fossils indicate that the world is not as everlastingly changeless as it may seem to an observer who has only one lifetime in which to study. The study of fossils helps establish some generalities. For example, by and large the deeper the fossils are buried in rock, the simpler they are. In addition, many fossils are recognizably similar to animals and plants now living, but they are not identical. Fossils in the upper layers of rocks are more like modern living species than are fossils in deeper layers. Such facts puzzled the scientists of

Darwin's time, as they had puzzled people for centuries.

There are, it is true, problems with the study of fossils, or the science of *paleontology*. Sometimes sections of the earth have been flipped over like a pancake, leaving the deeper layers with their more ancient fossils uppermost, and burying the more complex fossils deep in the earth. Then there is the perpetual difficulty of incomplete preservation. Few of the organisms that lived became fossilized, and those that were fossilized were usually imperfectly preserved, so that a complete history of organic evolution is lost forever. One may try to imagine how many of today's billions of humans will still be physically in existence as fossils 10 million years from now. Reading the fossil record is frequently compared to reading a book from which all but a few lines have been lost.

Darwin's information on paleontology, compared with the amount available today, was limited, but he knew enough to perceive the important principles. For one thing, geologists of the nineteenth century were showing that the earth was much older than anyone up to that time had suspected

DARWIN'S FINCHES

As part of his tour aboard the *Beagle,* Charles Darwin visited the Galapagos Islands, about 1000 kilometers (600 mi) off the west coast of Ecuador. These islands probably originated as active volcanoes more than one million years ago. It is believed that the Galapagos were never attached to the mainland, and if so, the descendants of the animals that Darwin saw would have had to come by air or sea.

Because of the isolation of the islands, the assortment of wildlife was limited to a few kinds of birds, mammals, and reptiles, but there were 13 different species of finches. Darwin was not struck by the wide diversity of finches until he returned home to England, where he could leisurely examine his notebooks and specimens. He thought it unlikely that so many different species of the finch could have been created originally, especially since there were not very many other kinds of birds on the islands. It was more reasonable to assume that the 13 species had descended from a single species, and that they were modified through time to adapt to different environments.

Darwin separated the types of finches into ground finches, tree finches, and a single species of warbler finches. The main difference in the birds was their beaks. Otherwise the birds resembled one another quite closely. Darwin saw that the beak shape was related to the bird's diet—vegetation, insects, seeds, cactus, or some combination of these. Not only were the beaks adapted to the kind of food the bird ate, but the beaks of the seed eaters varied in size, depending on the size of the preferred seeds.

The woodpecker finch has become the most famous of the group. Although its long beak resembles that of a woodpecker, the bird lacks the long tongue of a woodpecker, and therefore it cannot dig out insects from their hiding places within crevices. But the woodpecker finch carries a small twig or cactus spine in its beak, and it uses this tool to drive insects out into the open. If this bird lived on the South American mainland, it would have to compete with woodpeckers, and since it is less efficient than the woodpecker, it would probably lose out in an ongoing struggle for food. However, the woodpecker finch on the Galapagos Islands is unrivaled in its special environmental niche.

The study of the Galapagos finches suggested to Darwin how species may descend from previous ones by way of geographical isolation. The relatively recent evolution of the finches gave Darwin the opportunity to see an unusually clear example of adaptation, and his experience with the finches may very well have been the most important factor in his formulation of the theory of natural selection.

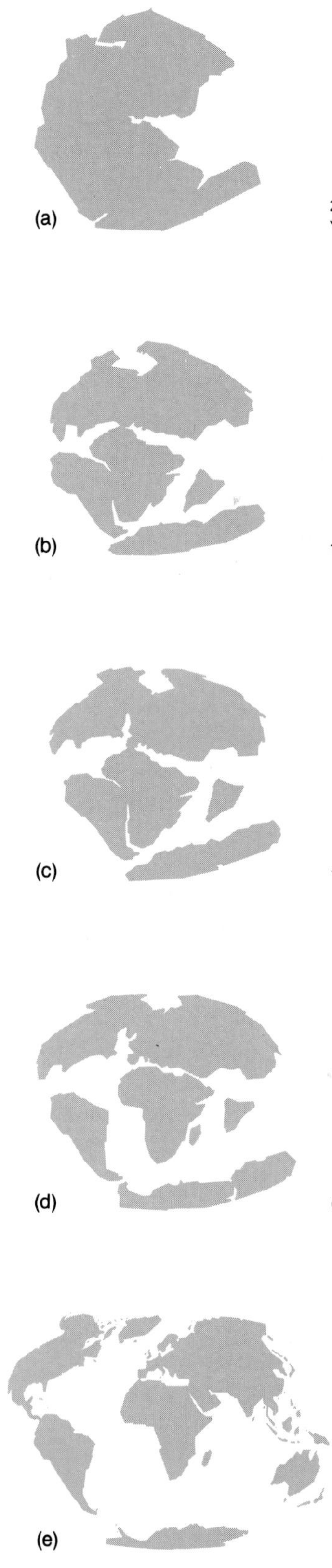

(a) 200 MILLION YEARS AGO

(b) 180 MILLION

(c) 135 MILLION

(d) 65 MILLION

(e) PRESENT

Figure 25.12
Pangaea and the breakup of the continents. Many geologists think that some 200 million years ago, the part of the earth that was above sea level was united in one great land mass, called Pangaea. At that time, all living things had access to each other by land, so that strict isolation was unlikely. When portions of Pangaea were separated from the main body, however, some populations were separated from the ancestral ones and thus had a chance to evolve in different ways. The series of pictures shows the progressive changes as they may have occurred at different times. (a) Pangaea, the original single great mass. (b) The separation of continents, 180 million years ago. (c) Further separation of continents, 135 million years ago. (d) By 65 million years ago, the face of the earth was approaching its present configuration. (e) The continents as they stand now.

(Figure 25.12). If, as some historical calculations seemed to show, the earth was only some 6000 years old, and evolution was as slow as observations indicated, there simply was not time enough for plants and animals to have undergone drastic changes. By using calculations based on physical evidence, however, geologists began pushing the date of the origin of the earth further and further back. By Darwin's time, estimates had changed from thousands of years to millions, and it seemed likely that there might indeed have been time enough for evolution to work. Today's even more accurate dating techniques allow for billions instead of millions of years, so that we know there has been abundant time for evolution, slow as it is, to work its changes.

EXPERIMENTAL EVOLUTION

Darwin's hypothesis did not go unchallenged. People were reluctant enough to concede that the earth is not the center of the solar system, or that the solar system is not the center of the universe, but they did slowly come to accept those ideas. Darwinism, however, was the worst "heresy" of all. It was bad enough that the earth is not the center of the universe, but the idea that humans developed from some subhuman creature was hopelessly repugnant to many. Even today, well over a century since the publication of *The Origin of Species,* strong objections are expressed against the concept of organic evolution, in spite of the mass of evidence that biologists find convincing.

Two challenges have been made to proponents of evolution: Can scientists create a recognizable old species, and can they make a brand-new one? The answer to both challenges is yes, they can and have done so.

In 1930, Arne Müntzing crossed two species of weedy little mints, using pollen and ovules that

Figure 25.13
A natural Linnean species, *Galeopsis tetrahit,* resynthesized by a plant breeder. Two species of hemp nettles, *G. pubescens* and *G. speciosa,* when hybridized, yielded a plant that was indistinguishable from *G. tetrahit* in all its structural details. Further, it was completely fertile with *G. tetrahit.* The chromosomes of the artificially produced *G. tetrahit* matched those of the wild species. If scientists can recreate a species in the few years that geneticists have been working, it is easy to see how new species could have arisen again and again in nature over the thousands of centuries during which organisms have had a chance to evolve.

were abnormal in their chromosome constituents, and the result was a plant well known as the hemp nettle, *Galeopsis tetrahit*. The new, artificially produced plant was indistinguishable from the wild type of hemp nettle, it had the same chromosome configuration as the wild species, and it could fertilize or be fertilized by the wild species. In this instance, an old, successful species was recreated by direct human interference (Figure 25.13). Since then, many species of established plants have been similarly duplicated.

In the 1940s, G. Ledyard Stebbins produced a tetraploid grass from a diploid South African species, *Ehrharta erecta*. (A *tetraploid* plant has four sets of chromosomes, twice as many as a normal diploid.) The new tetraploid grass was a new species in that it was not interfertile with the plants from which it was derived, and it had different growth requirements. Stebbins planted the tetraploid in a number of places in California, and he reported 28 years later (in 1971) that it was surviving. In Stebbins' words, "The tetraploid did not occur anywhere until I produced and established it." We can not only observe evolution; we can make it happen.

SUMMARY

1. In the biological sense, *evolution* means *genetic* change. It is the mechanism that provides for a diversification of all living things. Biologists believe that the present differences among organisms have come about by the genetic modification of earlier living things.

2. *James Hutton's* theory of slow and steady geological change provided a basis for Charles Darwin's new theories of evolution.

3. An early scientist who attempted an explanation for the causes of evolution was *Jean Baptiste Lamarck,* who proposed the *theory of use and disuse* and the *theory of the inheritance of acquired characteristics.* Biologists have found that only the former theory is basically correct.

4. The population theories of *Thomas Malthus* provided a concept of a continuing struggle for existence that supplied a basis for Darwin's early speculations about evolution.

5. In 1859 *Charles Darwin* proposed his theory of *natural selection* in his book *The Origin of Species.* Two basic assumptions of the theory were that new species may arise from former ones, and that organ-

isms that are best suited to adapt to environmental changes will survive.

6. The facts that Darwin collected to support his theory can be grouped into the following categories: *comparative anatomy, embryological development of animals, variability of plants and animals under the special care of humans, distribution of organisms on the earth,* and *structure and distribution of fossil organisms.*

7. Strong objections were expressed against Darwin's concept of organic evolution, and criticism is still voiced today, in spite of the evidence that biologists find convincing.

ASK YOURSELF

1. How does the idea that "similarity indicates relationship" apply to biological thinking?
2. Give an example of Lamarck's theory of "use and disuse."
3. Why does no biologist now accept the theory of "inheritance of acquired characters"?
4. How did Malthus contribute to Darwin's ideas on evolution?
5. How did a consideration of homologous parts of organisms contribute to the theory of evolution?
6. Give examples of some plants or animals that have shown striking variation under the influence of human cultivation.
7. Why are the Galapagos Islands frequently mentioned in connection with the history of evolutionary theory?
8. What general conclusions can be drawn from observations on the form and distribution of fossils?
9. Have biologists been able to bring into being a new species, that is, one not known from nature? Give an example.

26
The Mechanisms of Evolution

SOME KEY POINTS

1. Populations can change when changes occur in DNA or when environmental forces encourage variation.
2. If organisms do not adapt to a changing environment, they die. If populations do not evolve to keep them suited to a changing environment, they become extinct.
3. For sexual beings, a "species" can be defined as a group of organisms capable of producing fertile offspring.
4. The gene pool is the total genetic material of all individuals in a species (sexually interbreeding population).
5. New species can develop, either in one generation or gradually over many generations, by changes in chromosome number, geographical isolation, or reproductive isolation.
6. Two species can evolve and become so dependent on each other that one cannot survive without the other. Such a process is called coevolution.

ONCE BIOLOGISTS HAD THE LOGIC OF THE IDEA of evolution brought to their attention, they began searching for the specific ways in which it could function. When *The Origin of Species* appeared in 1859, nothing was known of chromosomes or their action, or of any modern genetic principles, or of the biochemistry of DNA, RNA, or proteins, or of the physiological effects of environmental forces. Only after an enormous amount of information was accumulated could an explanation of the mechanisms of evolution be developed. Much remains to be learned, of course, but the essential methods of evolution are reasonably well understood.

Darwin could see that variation within a species is common. Differences among individual people are accepted and used as everyday means of recognition. Pet dogs and cats are almost as easily recognized as individuals. We do not usually pay close attention to nonhuman organisms, but it is still true that every pigeon, every starling, every plant of crabgrass is unique. Although Darwin knew that variation occurs, he did not know the source of the variation as we do now.

VARIATIONS CAN BE INDUCED GENETICALLY OR ENVIRONMENTALLY

The differences among individuals are due in part to the organization of an organism's DNA (this is the part that is important in evolution) and in part to the environmental forces that alter individual development (this is of no direct importance in evolution). Each of these factors should be studied separately, but they work together simultaneously in a developing organism.

First, the special combination of nucleotide bases that makes up an individual's genetic potential is established when a sperm nucleus meets an egg nucleus. From the moment of fertilization, the biological capabilities of an individual are fixed, being determined by the molecular arrangement of the maternal and paternal DNA.

Second, the way in which the special DNA can express itself is influenced by the conditions surrounding the organism as it develops. For example, if a homozygous human zygote is genetically endowed with the capacity for growing straight, black hair, then that is what the resulting person will have, and he or she can pass only that capacity on to the next generation. However, variation in the appearance of the hair can be caused by alteration in the environment. Severe malnutrition or artificial bleaching can lighten the color, but the DNA in an individual's gonads remains unchanged.

What an organism *can* do is determined by its genes; what it *does* do is determined by its genetic ability operating in conjunction with the environmental forces that influence the expression of that genetic ability.

Mutation and Recombinants as the Basis of Variation

Genetic variation is essential to evolutionary change. The next question is: What are the sources of genetic variation? The answer is: (1) **mutation,** or change in nucleotide sequences in DNA, and (2) **genetic recombinations** of old nucleotide sequences. Some mutations are caused by such environmental factors as radiation or by chemicals, *mutagens,* in cells, but most mutations seem to be the result of mistakes in the copying of DNA sequences when replication occurs. These rare mutations have been called the raw material of evolution.

Even without new mutations, new visibly expressed phenotypes can be made by recombination. In breeding programs designed to produce new crop varieties, the actual number of new mutations must be relatively low, yet new varieties do appear after crosses between old varieties, and they appear quickly and in large numbers. When one mutant gene, already present in the gene pool, meets a second gene, they may both be expressed phenotypically, and when they are, you have essentially a new variety of organism.

Without knowing anything about mutations, chromosome recombinations, or other genetic events, Darwin could see that the number of gametes and the numbers of possible offspring were great enough to allow for variation and consequent selective advantages. The mutations that confer some survival advantage on their possessors are the mutations that are most likely to be passed on to the next generation, even though the advantage may be slight. Change can pile on change, generation after generation, as long as each change helps an organism to survive and to reproduce. New mutations, plus new combinations of old and new genes, make new varieties, adding to the total variability of a species.

Great Changes or Small?

Most mutations cause slight changes, and therefore evolution has usually progressed in small moves. Occasionally, however, a single mutation can be responsible for a change of considerable magnitude. There are known examples in which legs of animals are severely shortened, as in the famous Ancon sheep (Figure 26.1), or tails are eliminated, as in Manx cats. These examples confer no advantage, except to owners who like short-legged sheep or tailless cats, but it is conceivable that loss of a long, heavy tail by a flying reptile could be an enormous advantage, enabling the tailless one to outstrip its contemporaries in seeking food, escaping from hungry hunters, or overtaking a mate (Figure 26.2).

ADAPTATION IS CRUCIAL IN A CHANGING ENVIRONMENT

In this unstable world, lands and seas and climates change. When an organism finds itself in a changed environment, it too must change or die. Whether its old home has altered or whether it was swept to a new one, it has only those two choices. In its old environment, the organism must have been well suited to its surroundings, or it would not have survived there; in the changed circumstances, it will probably be less well suited. The changing of an organism that brings the organism into better harmony with its environment is called **adaptation.**

Adaptation must not be thought of as a striving on the part of an organism, certainly not a conscious striving. Rather, it is a result of the combined effects of variation, necessarily already present or available as mutation, and the selecting power of the environment. For example, plants in a population have differing capacities for producing cutin, a waterproof wax (see Figure 1.19). Some individuals are heavily

Figure 26.1
A mutation in sheep. The Ancon breed appeared spontaneously in New England. The owner, Seth Wright, bred it, but about 1870 the breed died out, only to occur a second time in Norway about 1920. A normal sheep is in the middle.

covered with a protective layer, and others are only thinly covered. If the climate becomes drier, as it did in the Sahara Desert, plants with a thicker cutin layer will not dry as fast as those with a thin cutin layer and may live to set a crop of seed. They have been "selected." Succeeding generations will also show variability, and steadily and inexorably those plants with the best protection against drying will be the only ones to live and reproduce. In this instance, only one feature, cuticular covering, has been pointed out, but in reality a plant would have to possess a whole mosaic of features that work together. It is important to remember that the species, not the individual, evolves.

Of the more than two million known species now living, plus many that are still unknown, few if any have been in existence for even 15 million years. In terms of the history of life, adaptation works part of the time, or we would not be here, but it has failed more often than it has succeeded. It is sufficient that some few species have changed fast enough and in the proper directions to keep life going.

Adaptation is a never-ending process. As organisms change to become suited to a changed environment, they then cause still further environmental changes, at which point they must change again. That changes the environment again, and the next round of changes has begun.

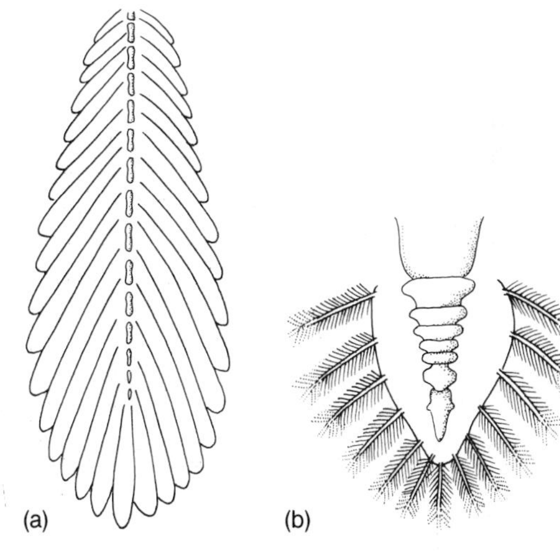

Figure 26.2
Tails of primitive and modern birds. (a) Archaeopteryx had a long tail with a number of separate slender bones and a row of feathers down each side. (b) Modern birds have short tails, with a few bones fused into a solid stub, and with the feathers produced so close together that they make a tuft.

If, as is generally thought, the earth's original atmosphere contained little or no free oxygen, then early organisms, like most present-day anaerobic bacteria, found oxygen toxic. But as photosynthetic organisms increased, they increased the proportion of oxygen in the air, and most nonphotosynthetic organisms had to evolve a respiratory chemistry that would allow them to live in an oxygen-rich atmosphere. Meanwhile, green plants took advantage of available carbon dioxide and water, and they poured ever more oxygen into the air. Even now, too high a concentration of oxygen can be dangerous. We are adapted to a concentration of about 20 percent, and premature babies kept in high concentrations of oxygen to help them breathe may be blinded because of a disturbed oxygen supply to their eyes.

Industrial Melanism Any organism can be thought of as a collection of adaptations, but some special ones are so dramatic or so instructive that they are considered classic examples. One striking instance is the development of what is called **industrial melanism,** a darkening of an organism's color associated with the black deposits of soot on the landscape around coal-burning factories. Before about 1850, the peppered moth of England, *Biston betularia,* was light-colored, speckled with black. However, one practically black one was collected in 1848 (Figure 26.3). Over the next 50 years, the proportion of the dark form, called *melanic,* increased until melanics came to make up almost the entire population in those regions where heavy smoke darkened the tree trunks and killed the pale lichens that grew on them. In the cleaner areas of the south, the light forms continued to predominate. By releasing moths of both forms in industrial and nonindustrial counties, watching birds feeding, and counting how many of each form could be recaptured, H. B. D. Kettlewell of Oxford University showed that birds could easily find light-colored moths against a dark background. Conversely, dark-colored moths against

THE EVOLUTION OF ANIMAL BEHAVIOR

The scientific study of behavioral evolution has developed more slowly than that of structural, cellular, or biochemical evolution. The reasons are several. First, behavior is harder to reduce to quantitative data, and it is generally thought to be harder to deal with experimentally. Further, recognizable evolutionary trends are not usually apparent. Also, firm proof is lacking that there is a gene for any particular aspect of complex social behavior. Finally, the species with the most complex behavior, and the species most interesting to us, is the one on which behavioral experimentation is most severely restricted—the human species. The topic of behavioral evolution has been befogged by many books and articles based on speculation or intuitive thinking, or written by people who had some private cause to defend, but without convincing facts.

Nevertheless, in the last 25 years, studies on behavior with an evolutionary emphasis have become increasingly common, and by now a fair body of knowledge has accumulated. Some of the most informative studies have been made on insect societies, especially on ants, bees, wasps, and termites. Different species of wasps have different social patterns. Those presumed to be the most primitive are the solitary wasps, which raise their young without any formal social organization. Other species have small, rather loosely organized nests in which a number of individuals live together but show little or no specialization of duties. More advanced species have colonies in which certain individuals have definite roles in the life of the colony. The most advanced species are the vespid wasps, with their rigid system of castes, in which queens, workers, and other types live rigorously ordered lives. In wasps, at least, there seems to be an evolutionary sequence of behavioral steps.

In other animal groups, by contrast, no patterns are discernible. Social habits in carnivores have evolved in such various ways that no gen-a dark background were difficult to find. He concluded that there was a strong selective pressure against light-colored forms in industrial areas and against dark-colored forms (melanics) in the cleaner regions. With stricter air-pollution standards, the selective pressure would once again favor the lighter moths.

Cryptic Coloration An adaptation that helps camouflage an animal by providing it with a color that makes it inconspicuous is called **cryptic coloration**. The world abounds in cryptically colored creatures. Striped tigers are hard to see among the striped shadows of the grass in which they customarily hunt. Some animals, notably fishes and reptiles, can change color in a few minutes to match backgrounds. Some chameleons are so adept at switching from green to brown and back again that they are regarded by humans as the standard of inconstancy. Flounders can become practically invisible against a variety of backgrounds, even making a

eral statements can be made about them. For example, African hunting dogs are highly organized, but raccoons and wildcats are solitary. Some carnivores have developed the habit of defending a home territory, and some have not. In some (hyenas) the females are dominant, and in others (wolves) it is the males. Generalizations that might be useful in deciding evolutionary trends cannot be made from such diverse types.

In 1975, Edward O. Wilson of Harvard University published *Sociobiology: The New Synthesis,* in which he attempted to integrate ideas on genetics, evolution, and animal behavior, including human social behavior. The work was praised by some biologists as an important contribution to evolutionary thought, and it was damned by others as having reactionary political meaning and making unwarranted conjectures. In time, new information and new insights will resolve the disagreements, and Wilson will be either vindicated or forgotten.

Figure 26.3
Light and dark (melanic) forms of the British peppered moth, *Biston betularia.* It should be noted that both kinds of moths appear in both pictures.

reasonably good attempt at duplicating a black-and-white checkerboard (Figure 26.4).

Mimicry Another striking adaptation is **mimicry,** in which one species develops a resemblance to something different from its near relatives. (See the color portfolio on the next page.) The object copied may even be in a different kingdom. Among insects, famous examples are the stick insects, which look like dead twigs, and the green katydid, which, when resting, looks remarkably like a leaf, complete with midrib and leafstalk (Figure 26.5). In plants, the orchid, *Ophrys*, looks and especially smells so much like a female bee that pollinating males try to mate with the flower (see Figure 15.29). More commonly, a harmless insect may resemble a poisonous or at least distasteful species. Insect-eating birds, having once tasted a repulsive type, will not attack a harmless mimic (Figure 26.6).

(a)

(b)

(c)

(d)

Figure 26.4
Cryptic coloration. (a) A motionless white-tailed deer, *Odocoileus virginianus*, is difficult to see among thick vegetation. (b) A leopard, *Panthera pardus*, can move slowly through sun-dappled brush and still remain camouflaged. (c) A snowshoe hare, *Lepus*, changes color with the seasons, becoming almost invisible to predators and prey alike. (d) Flounders and other flat, bottom-dwelling fish can change their pigment patterns. They can expand or contract their dark pigment cells within a few minutes.

HOW ANIMALS AVOID PREDATORS

Animals have ways to keep from being eaten. And if an individual animal stays alive, so may its species. Some animals, like leopards, kill before *they* are killed. Others, like gazelles, run and dodge to elude their enemies. And others, like elephants or rhinoceroses, are just too big or too formidable for most other predators to bring down. But in small and basically defenseless animals, more elaborate strategies have evolved. Probably the most common defense is to remain motionless. Many insects, for example, try to avoid being swallowed by looking as inconspicuous as possible. Some insects blend in so well with their surroundings that they are virtually invisible. But if they are discovered by a predator anyway, some insects may use a startling bluff, looking like a super-predator that scares away the intruder. In many cases, there is yet a third phase to the defense of the prey, and warning coloration or scare tactics may be accompanied by a noxious taste or a chemical repellent. If a noxious animal is killed anyway, it still benefits the species because the predator usually does not attack similar prey for a long time, if at all. Of course, no animal consciously knows what will deter a predator. Rather, an accidental change in appearance or behavior—a mutation—may prove to be favorable in the process of natural selection. The predator is the selector. If it is put off by the defense of its prey, then that prey will be "selected" to survive. If not, the species may become extinct.

Although this Australian leaf-tailed gecko is about 20 centimeters (8 in.) long, it can still remain safely camouflaged all day by resting motionless on a background that resembles its own coloring. In addition, the leaf-shaped tail helps to distract a predator away from the more vulnerable body. In the cover of night the lizard opens its eyes and creeps away in search of its own prey.

These insects survive by pretending to be something they are not. The specific device may help the insect blend in with its natural surroundings, or it may serve to bluff a predator into hunting for less "ferocious" prey, but one goal remains constant: Stay alive long enough to reproduce.

A larva of a swallowtail butterfly lies bent and motionless on a leaf, mimicking a bird dropping. This disguise may seem farfetched, but apparently it works. Some small tropical frogs also mimic bird droppings. They tuck their legs under their bodies and remain motionless on leaves all day. At night they are free to move about.

Looking remarkably like a battered leaf, this grasshopper hangs upside down from a twig, with its pointed head and thin antennae resembling a stem. Once the grasshopper moves, the adaptive concealment loses its effectiveness, but even if the grasshopper is threatened on the ground, it may be able to fool a predator into mistaking it for a fallen leaf as long as it remains absolutely still.

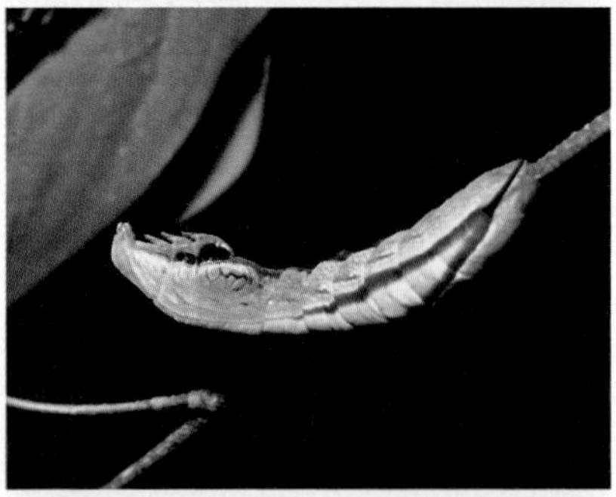

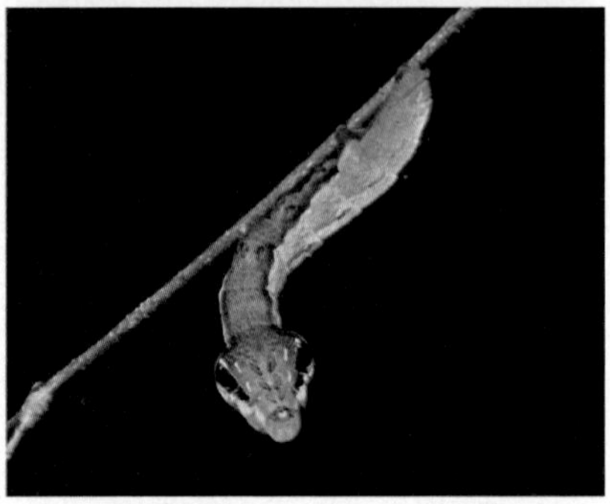

The caterpillar of the *Leucorhampa* hawkmoth is sometimes called a snake caterpillar, with good reason. When disturbed, it moves away from its usual flat position on a twig and changes the shape of its rear end to look like a snake's head. The "eyes," hidden before, now appear menacingly. It is typical for an animal to present "eyespots" on parts of its body other than its head. A sphinx moth, for instance, has hidden eyespots on the expendable edges of its hind wings, and a hornworm tucks its head safely under its body when it displays its phony head and eyes farther down on its body, where damage may not be serious if a predator strikes.

Insects are especially effective at using the evolutionary benefits of coloration because of their rapid reproductive cycles, but other animals, such as some frogs, also use color to warn or startle predators.

The "painted frog" of Panama uses its unusually dramatic colors to remind predators that its multicolored skin is highly poisonous.

Some animals have defense postures designed to avoid conflict completely, and they resort to chemical warfare only when other techniques do not work. This South American frog (*Physalaemus natteri*) tries to evade a predator by flashing two "eye spots" on its rear end. Perhaps the predator will think these monstrous "eyes" belong to one of its own predators, and it will either go away or remain startled long enough for the frog to escape.

These dashing little frogs live in Central and South America. They are popularly known as "arrow poison frogs" because glands in their skin secrete venom used by local Indians on arrowheads to paralyze prey.

Some animals can avoid conflict altogether if they detect a predator soon enough. A deer that spots a predator raises its tail to expose a white underside. Presumably this signal warns other deer that a predator is near, but it may also notify the predator that the deer has seen it early enough to escape. The signal effectively says: "I have seen you, and I can get away if you try to chase me." In most cases the predator does not bother to attack. Generally an animal will not expend energy in a fruitless pursuit.

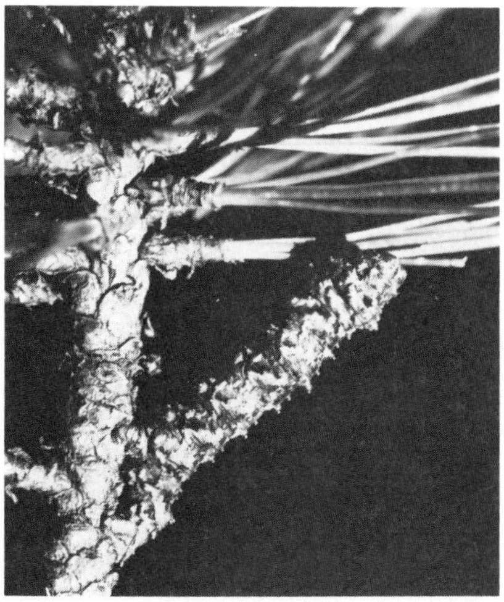
(a)

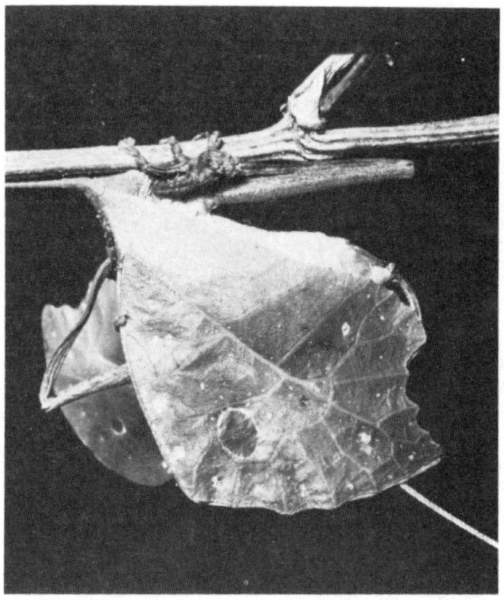
(b)

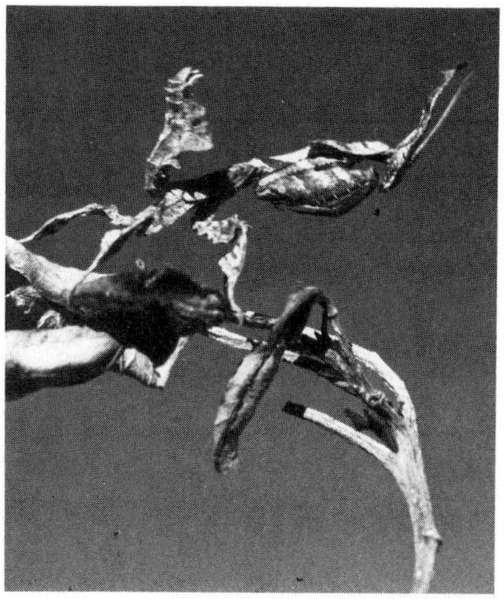
(c)

Figure 26.5
Mimicry in insects. (a) A moth caterpillar not only has the form and color of a pine twig, but holds itself in a proper position at an angle and remains motionless. (b) A katydid has wings so leaflike that they even have veins and torn edges. Until it moves, such an insect is not likely to be noticed by a hunter. (c) A devil's mantis looks like a half-eaten leaf, even to the extent of being the same color as the actual leaf.

(a)

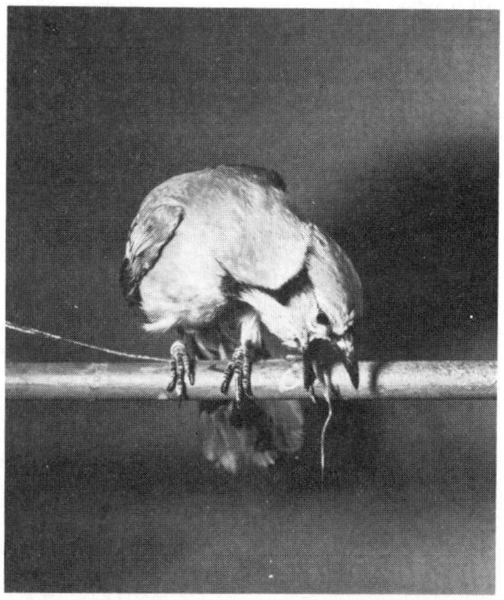
(b)

Figure 26.6
A blue jay will eat a monarch butterfly if it has never tried one before (a), but the butterfly contains an emetic that causes the jay to vomit (b). Having learned that monarchs are nasty, the jay will not attack a viceroy butterfly if offered one. The viceroy is harmless (c), but looks so much like a monarch (d) that the similarity is enough to repel the jay. Both the monarch and the viceroy are protected, one by its own toxin and the other by its similarity to the noxious species.

(c)

(d)

Figure 26.7
Adaptive coloration in the wood duck, *Aix sponsa.* The male is splashed with color, and during courtship he erects his crest and shakes his body to show off his decorative feathers. He is easy for the female to recognize. The female's brown- and white-flecked plumage is useful in making her difficult to see when she is nesting and raising vulnerable ducklings.

Warning coloration and adaptive coloration In contrast to cryptic coloration, **warning coloration** is conspicuous. A strikingly colored insect may be so noxious that preying birds learn to let it alone. It seems that each individual bird must learn for itself which bright-hued insects are not good, and therefore some of the insects are sacrificed.

Adaptive coloration in birds is a compromise between being easy to find by a predator (that is bad) and easy to recognize by a possible mate (that is good). In birds that court, the female normally accepts only males of her own species. Since males are less discriminating, and since interspecific hybrids are usually biological failures, it is important to a female to make certain that she chooses correctly, and strong coloration simplifies the choice. Female birds tend to be rather drab. Selection pressure seems to favor inconspicuous colors in females (thereby improving their chances of escaping predation), but conspicuous colors in males. The males thus have to take their chances with predators if they are to be successful breeders. The system functions; male birds are indeed generally the showy sex (Figure 26.7).

TO SUM UP ADAPTATION

1. The changing of an organism that brings it into better harmony with its environment is called adaptation.
2. Adaptation is a result of variation, usually accomplished through mutations and genetic recombination of parental traits, plus the natural selective power of the environment.
3. One example of a specific adaptation is industrial melanism, which is a selection for a darker-colored organism that is best camouflaged when industrial factors such as pollution darken the environment.
4. Cryptic coloration is natural camouflage (striped tigers in tall grass) or the ability to change body color quickly as the immediate environment changes (chameleons).
5. Mimicry is the process by which one species develops a resemblance to another species, or even an inanimate object. Such resemblance helps it survive in its environment.
6. Warning coloration is a distinctive body coloration that warns predators that the organism is poisonous or otherwise unattractive to eat.
7. Adaptive coloration helps some animals to remain inconspicuous, but also helps the males to be conspicuously colored so that mating and reproduction are facilitated.

"SPECIES" IS DIFFICULT TO DEFINE

Until Darwinism disturbed their serenity, biologists felt comfortable about the meaning of a species. It was a group of individuals that resembled one another so closely that some careful student of the group could confidently assign them a name. There was no concern for mutability, little for variability, and practically none for hybridization. The idea of evolution changed all that. Even though the very book that brought about the change had the word "species" in its title, exactly what a species is became a slippery concept, changing and eluding those who sought a simple, firm definition.

If a stand of flowers with red petals lives on one side of a mountain, and an almost identical stand, except that the flowers have pink petals, lives on the other side, are they the same species? One first tries to determine whether the two stands are interfertile. If they are, they may well be the same "species," that is, very closely related. Thus *taxonomy*, or the study of relationships as expressed by classification, changed from an observational to an experimental discipline. No longer satisfied with merely looking at their specimens, taxonomists tested them for interfertility and transferred them to different habitats in an effort to discover which characteristics could be affected by environment and how much they could be changed. In one such experiment, subspecies of *Potentilla glandulosa* were planted at various elevations in the California mountains. The results show that the actual development of the organisms was greatly influenced by the environment (Figure 26.8).

The species is still the basic grouping of organisms for students of classification and of evolution, but there is no definition that meets the requirements of all the investigators. Many zoologists, who deal with sexually reproducing animals, define a species as a group of organisms capable of producing fertile offspring. That satisfactory working definition for most purposes is without meaning, however, when it is applied to an essentially asexual population of algae or fungi. One extreme group of taxonomists denies that there is any such thing as a species, claiming that the world is populated only by individuals, some of which more or less resemble others, and that a species exists only in the imaginations of classifiers. Meanwhile, for practical purposes, we think and talk and write about species, hoping to learn as much as we can about the earth's living things. The old aphorism concerning life itself comes back: It is easier to study something than to define it.

DEMES ARE GROUPS OF INDIVIDUALS ISOLATED FROM THEIR SPECIES

In nature, free passage of genetic material within a species may be restricted even when there is the biological possibility of such passage. Many animals range over a small territory and limit their mating to members of their immediate group. A group of individuals partially or entirely isolated from the other members of the species is a **deme.** City rats, for example, tend to form colonies, sometimes confined within a single vacant lot or even in a well-defined portion of a garbage dump. Mating, feeding, nesting, and raising young are carried on by the group as a tight unit. Members do not leave unless they are driven out by hunger or crowding, and strangers are not welcome. Several demes of mice may inhabit one barn, the deme in the cow stalls having little or no contact with the deme in the hayloft. On occasional chance meetings on neutral ground, the mice of one deme may ignore those of the other.

Still, demes have considerable fluidity. The areas they occupy are usually small and may be subject to change. When a vacant lot is cleared or built upon, the local rat population is scattered. If they are aggressive enough, the rats may try to work their way into other demes, carrying their genes

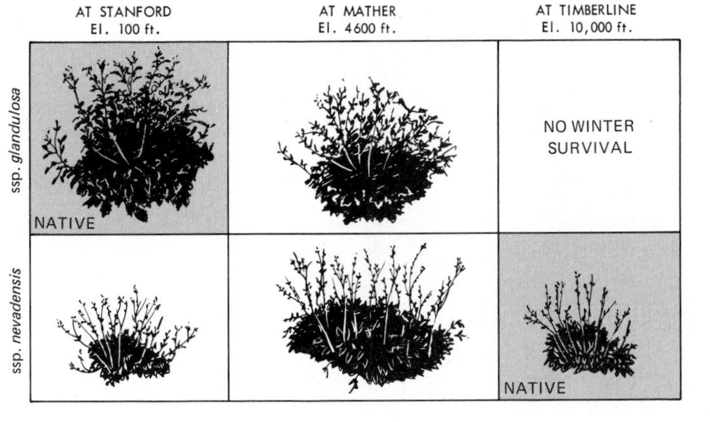

Figure 26.8
The effects of environment on development. *Potentilla glandulosa* has several subspecies in California, including *Potentilla glandulosa glandulosa,* and *Potentilla glandulosa nevadensis.* When grown at different elevations, each subspecies changes. *P. g. glandulosa,* native at low elevations, shows reduced growth at 1400 meters (4600 ft) and fails at 3000 meters (10,000 ft). *P. g. nevadensis,* native at timberline (3000 meters), grows well at 1400 meters but does no better at low elevations than it does in the high mountains.

with them. In nature, environmental instability is common. A deme of fishes in one pond may lose a few individuals to an adjacent pond during a flash flood. Two demes of strictly forest-inhabiting asters in separate patches of woods may be separated by fields, in which they cannot grow. But occasional seeds may blow across the fields that form barriers, or the fields may be abandoned to grow up into brush and then into forest, at which time the demes can intermingle. Therefore they do not diverge enough genetically to suffer reproductive isolation, and they remain within the species.

GENE POOLS

The **gene pool** is the total genetic material of all individuals in an interbreeding population (species). In a large breeding group, there may be thousands of rarely expressed recessive characteristics. Most of them are probably detrimental, some apparently neutral, and some beneficial. There is always the possibility that a new combination of already existing features may confer an advantage on the species.

In developing new varieties of plants, for example, breeders take advantage of the alleles that exist and try to bring them into combinations that give desired qualities (Figure 26.9). One variety of strawberries, for example, may have fine flavor but is susceptible to disease; a second may be disease resistant with little else to recommend it; a third has firm lasting qualities so that the fruit can be shipped.

As long as all the necessary alleles are present in the general pool, they can be brought together. Of course, some of the trials will produce berries that spoil easily, taste bad, and are prone to easy infection. But after enough crosses are made, the breeder should obtain the best combination of taste, durability, and resistance. Without a diversified gene pool to draw on, such results would be difficult if not impossible. It is therefore wise to maintain stocks of seeds and domestic animals, even though some strains may seem unattractive, because no one knows what recessive alleles are being perpetuated in the gene pools. Nor can anyone foresee what genetic reserves will be needed in the future. Who would have guessed a hundred years ago that durability during transport would be a necessary quality in a commercial strawberry? Is there in the gene pool of beans an allele that will confer drought resistance, so that submarginal land could be used to help feed hungry people?

THE HARDY-WEINBERG PRINCIPLE IS EFFECTIVE ONLY IF GENETIC STABILITY IS ASSUMED

If one is concerned with the genetics of individuals or at most of small breeding groups in controlled matings, the genotype of an individual can be found, provided that the phenotypes of ancestors or offspring are known. Dominance, recessiveness, and phenotypic ratios can be observed. Then, by the application of Mendelian principles, genotypes can be determined. With wild populations, however, controlled matings are seldom possible, and

(a) (b) × 0.75

Figure 26.9
The results of selective breeding in plants. (a) Native wild strawberry. (b) Some varieties of cultivated strawberries have been bred for special qualities until they are scarcely recognizable as the original wild species.

with human beings, sufficiently detailed pedigrees are unobtainable. Consequently, when studying whole populations, we must use different methods.

In 1908, the English mathematician G. H. Hardy and the German physician Wilhelm Weinberg independently developed a formula for analyzing the frequency of alleles in large populations. By using the formula, now called the **Hardy-Weinberg principle,** we can calculate the approximate frequencies of different alleles in a population.

The Hardy-Weinberg formula is usually expressed as

$$p^2 + 2pq + q^2 = 1,$$

where p is the frequency of a specific dominant allele, and q is the frequency of the corresponding recessive allele. In words, this formula states that "the frequency of the homozygous dominant genotype (p^2) plus the frequency of the heterozygous genotype ($2pq$) plus the frequency of the homozygous recessive genotype (q^2) equals all the genes in the population (1)." The numeral 1 in this sense means 100 percent of the genes in the population.

If we can find the frequency of the expressed recessives, we can find the frequency of the recessive allele and of the dominant allele and the number of heterozygous individuals in the population. The heterozygotes are those in which the recessive allele is present but not expressed; they are still "carriers" of the allele.

Allele frequencies can be determined for such human features as eye color, ear shape, blood groups, physical deformations, and predispositions to disease. For example, the hereditary recessive disease cystic fibrosis occurs among Caucasians in the United States in 26 out of every 100,000 individuals. Knowing that, we can calculate how many carriers there are in that population, using the Hardy-Weinberg equation. Twenty-six out of 100,000 is about one out of 3846. We can call the recessive allele for cystic fibrosis *a*, and the normal allele *A*. The homozygous recessive, *aa*, is $\frac{1}{3846}$, and the frequency of *a* in the population is the square root of $\frac{1}{3846}$, which is close to $\frac{1}{62}$. If one out of every 62 alleles is *a*, then the rest, or $\frac{61}{62}$ are *A*. The number of carriers is $2Aa$ (or $2pq$ in the general equation). This gives us $2 \times \frac{61}{62} \times \frac{1}{62}$, or $\frac{122}{3844}$, which is $\frac{1}{31}$. Thus we know that one Caucasian out of every 31 in the United States is a carrier. If we use the whole equation, we get

$$(61/62 \times 61/62)AA + (2 \times 61/62 \times 1/62)Aa + (1/62 \times 1/62)aa = 1 \text{ (that is, 100\%)}.$$

$$(3721/3844)AA + (122/3844)Aa + (1/3844)aa = 100\%.$$

In Japan, by way of contrast, there is only one case of cystic fibrosis for every 100,000 Japanese individuals. Using the Hardy-Weinberg formula, we can calculate the incidence of carriers in Japan as follows: If the frequency of *aa* is one out of 100,000, then the frequency of *a* is the square root of 1/100,000, or approximately 1/317. If one out of every 317 alleles is *a*, then 316 out of 317 are *A*. The heterozygous carriers are $2Aa$, or

$$2 \times 316/317 \times 1/317 = 632/100{,}000, \text{ or about } 1/158.$$

Therefore, one Japanese person out of 158 is a carrier. There are five times as many carriers per 100,000 for cystic fibrosis in United States Caucasians as there are Japanese carriers.

Another recessive characteristic is albinism (Figure 26.10), which in humans occurs about once

Figure 26.10
An albino deer. Most, perhaps all, animals occasionally produce albino individuals. Albino rats, mice, and rabbits are well known, but nonpigmented forms also occur in robins, blackbirds, squirrels, fishes, snakes, and even in humans and plants.

in every 20,000 births. The Hardy-Weinberg formula shows that although the number of observed albinos is low, 1.4 percent of the population are carriers, or about one in every 71 individuals. As we have seen, most recessive alleles in a population are present in *carriers*. The high number of recessive alleles in populations is frequently surprising to people who are not familiar with the statistical laws of chance.

We must make clear that the Hardy-Weinberg principle assumes the following conditions:

1. Natural selection does not occur.
2. The population does not gain or lose individuals through immigration or emigration.
3. No mutations occur.
4. Mating is random.
5. The population is large enough to be unaffected by random changes in the frequencies of genes.

Under actual conditions, such genetic stability is probably nonexistent, so why should we be concerned with the Hardy-Weinberg principle at all? Actually, the principle has practical use. For instance, it is employed by medical advisors in counseling prospective parents about the probability of certain gene frequencies in their future offspring, and anthropologists can use it in studying differences among human groups. One may also use it to gain valuable information about changes in populations, because when alleles do *not* follow the Hardy-Weinberg principle we know that something in the environment is selecting for a particular allele.

GENETIC DRIFT IS THE RESULT OF CHANCE

Most breeding groups include large numbers, sometimes many thousands of individuals. In a large population, a rare mutation has a poor chance of being expressed. Not all breeding populations, however, are large. Small groups may be isolated by floods, landslides, or emigration, and a colony of perhaps as few as a hundred individuals may be restricted to interbreeding (Figure 26.11). With such a low number, a rare recessive allele may exist in only one organism, and that organism may die before passing the allele on. For that population, then, the allele is permanently lost. In contrast, if there happen by chance to be two individuals with the allele, and it is expressed as a homozygous recessive, the once-rare allele may become more common in the isolated community than it was in the larger population from which it came. **Genetic drift** is a chance variation in the frequencies of alleles in a population. In a large population, frequency changes are less likely, but genetic drift can be significant in a population of only a few hundred individuals.

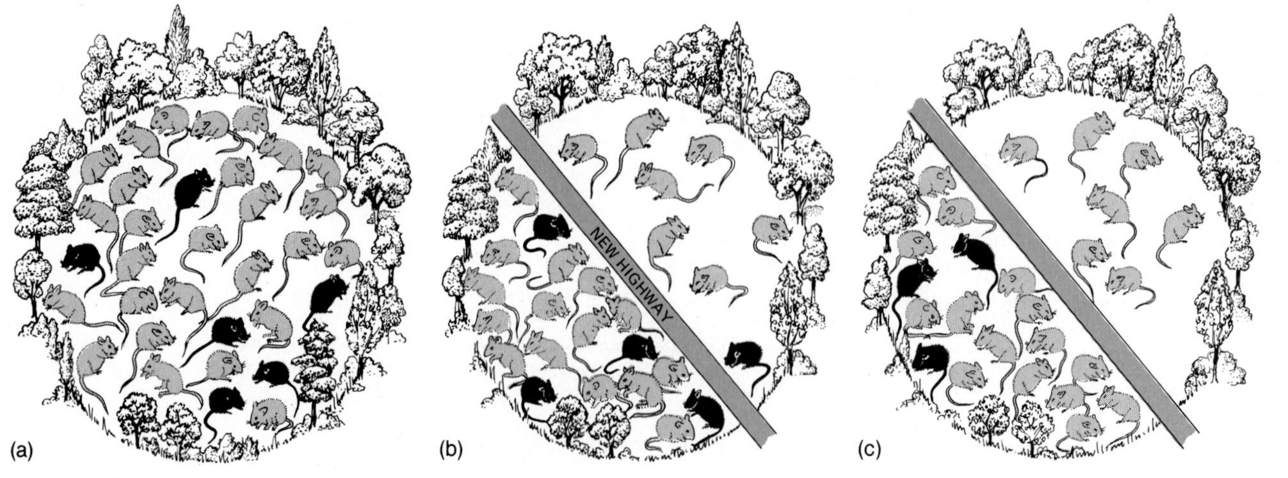

Figure 26.11
Genetic drift. (a) In a small woodlot, a population of mice consists of mostly gray individuals, but there are a few black ones. (b) A new highway through the woodlot effectively cuts the mouse population into two separate groups. One group still has some black mice, but the other group happens to have only one individual carrying any black alleles. (c) Before it has a chance to reproduce, the lone black mouse in the isolated group is killed, leaving the little population all gray. When a population is small, a relatively minor chance event can alter the genetic makeup of the population.

SPECIATION IS THE PRODUCTION OF NEW SPECIES FROM OLD ONES

The development of new species, or **speciation,** can occur in one generation, or it can be a slow process extending over many generations. In either method, the result is a new population of organisms that cannot produce fertile offspring when mating with unchanged descendants of the old ancestral line. If a population is maintaining itself and continuing to build up a load of new mutations, it may be expected that the species will become perhaps more variable. But as long as breeding continues randomly and the environment remains stable, there will be no evolution in any particular direction. In biological history, however, populations have not remained static. We shall now see some of the ways in which new species are derived from old ones.

Polyploidy

An abrupt appearance of a new species can be brought about by a change in chromosome number. Plants have used chromosome changes in the formation of new species much more commonly than animals. One way for chromosome alteration to bring about speciation is by the phenomenon of **polyploidy,** the replication of entire chromosome sets. Although plants usually form haploid gametes, in which there is one set of chromosomes, plants will occasionally produce an unreduced gamete with two sets, that is, a diploid gamete. Such gametes are rare, and the chance that one diploid gamete will fertilize a second diploid gamete is even rarer. With the huge number of gametes that are actually produced, however, mating between pairs of diploid gametes does happen. The resulting zygote, which has four chromosome sets, is a *tetraploid.* Tetraploid plants are frequently larger and more vigorous than diploids, and they are not interfertile with diploids. A new tetraploid is functionally a new species. Tetraploids can be artificially produced by treating plants with a toxic alkaloid, colchicine, derived from the autumn crocus, *Colchicum autumnale* (Figure 26.12). Tetraploid plants may be marketed as "tetra snaps" (tetraploid snapdragons) or "tetra tomatoes."

When a new species arises as the result of tetraploidy, or polyploidy of any kind, no new genetic material has been added. There are no mutations, no new nucleotide sequences; there is only a new combination of pre-existing genes. It is like making a new building from an old one, merely rearranging the components.

× 0.5

× 0.5

Figure 26.12
The effect of doubling the chromosome number. The flowers above are standard diploid snapdragons (*Antirrhinum majus*). Those below are tetraploid as a result of crossing colchicine-induced diploid egg cells and diploid pollen. The tetraploid form not only is visibly larger but has larger cells.

Adaptive Radiation

The common way for speciation to occur is by slow steps, with geographical isolation providing the necessary conditions. An occasional bird, insect, or seed may be blown or washed from its place of origin to new shores, perhaps far away, where it is isolated from the rest of the population, and where new environmental forces act on it. If a portion of a population is cut off from the main body of that population, it will most likely carry with it only a fraction of the gene pool of that population. Being thus separated, with a slightly different gene pool to draw on, and subject to somewhat different selection pressures, the isolated population can be on the way to evolving into a new species.

The results of such isolation are observed most clearly in islands far from a mainland, and indeed it was the peculiarity of island plants and animals that so stirred Darwin's curiosity. But organisms need not leave home to become isolated. Geological events may be responsible. Mountains are pushed up, rivers change courses, seas rise and fall, ice sheets advance and melt, hills erode, mountainsides slump in avalanches—all these events change the continuity of land areas and of lakes and seas. Such changes leave some populations in places where they are cut off from larger populations, or conversely, populations may be thrown in with other populations from which they were previously separated. Events like these have been happening for billions of years. Countless mutations have occurred, and enormous numbers of plants and animals have been subjected to selection forces.

Eventually, variation and selection contribute to the ever-widening separation of two physically isolated populations. When they reach a state of reproductive incompatibility, a new species has evolved. The slow process of speciation by isolation is known as **adaptive radiation,** because organisms *radiate* out from one place and *adapt* to new environments.

Reproductive Isolation

The immediate causes for loss of interfertility between emerging species are many and various. Some simply prevent mating. An emerging species may have taken up a new habitat, where it does not meet the other "species." Or it may mature its gametes at a later season, so that fertilization is thrown out of timing. Since mating behavior in animals is frequently quite specific, the diverging populations may give different courtship signals and may not recognize each other as possible mates. Or structural changes in animal bodies may make copulation physically impossible. In any case, gametes may be incapable of fusion, or the female reproductive tract may not allow survival of sperm cells. Even successful fertilization of an egg does not ensure successful reproduction, because if the genes of the gametes are not functionally compatible, a number of disasters can follow. The embryo may abort, or the young animal may be (and frequently is) ill suited to the environment. Finally, even if the hybrid survives to maturity, it may not produce viable gametes. The mule is the hybrid offspring of mating a female horse with a male donkey. The mule is infertile, however, and it cannot reproduce. Because mules are infertile, they do not constitute a species.

Sympatric and Allopatric Species

Although any one of the methods discussed in the paragraph above is enough to prevent the production of fertile offspring, usually several isolating mechanisms act at once. In nature, species that appear very similar in shape, behavior, color, and so on, and occupy the same geographical region commonly have a number of differences that prevent interbreeding. Some such differences are differences in specific habitat, timing of sexual maturity, or incompatibility of physical shape.

Some species of violets live together in the northeastern United States. Such species are called **sympatric** (Latin, "together in the country"). Other species of violets that are similar but inhabit the prairies or the west coast are isolated both geographically and reproductively and are therefore called **allopatric** ("another country"). Even a species that appears uniform over a wide geographical range may have gradational changes in its gene pool. Consequently, in spite of the possibility for continuous gene flow through the whole population, the individuals in one locality can be so distant from individuals at the other end of the range that they are effectively isolated genetically. In such an instance, should we call the whole population a single species, or if not, where should we draw the lines? One possible answer is that humans should not attempt to draw lines when nature has not done so.

CONVERGENT EVOLUTION: CHANGES TOWARD SIMILARITY

Much evolutionary discussion centers on change leading to differences, and indeed differences make the biological world interesting. Sometimes, however, changes can occur in such a way that organisms that are not closely related come to have cer-

Figure 26.13
An example of convergent evolution. (a) In the western hemisphere, most thick, spiny, succulent plants are cacti. (b) In the eastern hemisphere, plants that are vegetatively similar are spurges. A cactus and a spurge may look so much alike that telling them apart when they are not blooming is difficult. (c) The cactus flower, with its numerous petals and stamens, is unmistakable. (d) The flower of the spurge (*Euphorbia mammilaris*) is obviously different from the flower of a cactus.

tain similar characteristics. The process of changing toward similarity instead of difference is **convergent evolution.** Birds and bats have wings that are at least superficially similar, but birds have evolved from reptiles, and bats from a flightless mammalian ancestor. Probably the most striking example of convergent evolution is that of the Old World spurges and the New World cacti. Both, at least in some species, live in dry habitats, where a thick stem, reduced leaves, and a thick, waterproof covering of wax are of great survival value. Except when they are flowering, some species of spurges are so cactuslike that one can scarcely distinguish between the two (Figure 26.13). They are not even in the same family. Dozens of such examples are known, all showing the shaping force of the environment in directing the adaptive changes of which organisms are capable.

COEVOLUTION

The phrase "survival of the fittest" may indicate that in competition with members of its own species the powerful and pitiless are winners. That may be so sometimes. Surely a weak, timid lion is in trouble, and a diffident rat is a "social outcast." Nevertheless, genetic variations are usually tested against the total environment, which includes both living and nonliving factors. Any number of behavioral or structural features may confer an advantage. A rabbit that escapes the notice of a hunter by freezing into immobility survives as a result of a behavioral pattern. A green lizard on a green leaf escapes, too, because its color makes it inconspicuous. What an individual has or does is obviously important. What individuals do in groups, however, is also important, and evolution is not all competition.

Cooperation is frequently conspicuous, not only within species, as in wolf packs and herds of buffalo, but between species. When two species undergo change in such a way that both can operate more effectively as a team, the process is called **coevolution.** Some elaborate examples of coevolution are found in flower-animal relations. Bat-pollinated flowers have odors attractive to bats and suitable physical structures that allow bats to get at the pollen. They even produce high-protein pollen, and the pollinating bats have specially adapted tongues (Figure 26.14). Similar coevolutionary trends can be seen in flowers pollinated by butterflies, with each partner adapting to the other.

Recently, Stanley A. Temple at the University of Wisconsin suggested a fascinating example of coevolution. According to Temple, it is not a coincidence that the seeds of a large tree (*Calvaria major*) that still exists on the island of Mauritius in the Indian Ocean ceased to germinate at the same time that the dodo bird became extinct at the end of the seventeenth century (Figure 26.15). Apparently the dodo bird fed on the fruit of the Calvaria tree. The large pit inside the fruit was ground into small bits by the dodo's gizzard. (Most birds swallow small stones, which are stored in the gizzard and are used to grind food.) In the course of coevolving with the dodo, the Calvaria's pit became harder and harder, protecting the seed within. Finally a point was reached where the seed could not be released naturally through the tough outer shell of the pit. And that is where the dodo comes in as a partner in coevolution. The dodo, keeping up with the hardening process of the pit, was still able to grind the pit in its gizzard and liberate the seed. As soon as the dodo was gone (probably because European settlers inadvertently hunted it into extinction), the Calvaria seeds were unable to germinate.

Although the Calvaria trees on Mauritius are slowly disappearing, those that are still there keep producing viable seeds within the pits of their fruit, but the seeds remain locked inside. Temple has fed some of these pits to turkeys, whose gizzards supposedly resemble those of the dodo birds. The 10 seeds that ultimately were freed by the grinding gizzards were planted, and three actually germinated.

There are several cases on record where animals became extinct when plants disappeared, but if Temple's hypothesis is correct, this will be the first time that a plant has been known to disappear because an animal became extinct.

A similar example of evolution was observed by British scientist Michael Goulding. This time, the

Figure 26.14
An example of coevolution. Bats and bat-pollinated flowers have in a number of instances developed structures that are mutually useful. The flowers produce sugar-rich nectar and protein-rich pollen, both of which nourish the bats. In return, the flowers are fertilized and can set viable seed. On their side, bats have developed highly specialized tongues, which they can stick out, as shown in the photograph, to reach deep into a flower. By now, both bat and flower have become so completely dependent on each other that neither could survive without the other.

Figure 26.15
The dodo bird that was thought to be responsible for the germination of Calvaria seeds in its day. The phrase "dead as a dodo" refers to the fact that these flightless birds have been extinct for more than two centuries.

partners in coevolution are Amazonian trees that grow in the floodplain forests of South America, and large fish whose teeth and jaws are strong enough to crack open the fruits and nuts dropped into the water by the trees. The fruits and nuts are the main food source for the fish, and the seeds of the trees are dispersed when the fish defecate live seeds. It is thought to be the first recognized example of coevolution between plants and fish.

Coevolution provides an excellent example of how the course of evolution can be affected by the sometimes unlikely cooperation of living organisms. It also demonstrates the basic principle of ecology, the interaction between living things and their environment. Figure 26.16 illustrates four more examples of evolution and ecology at work.

INORGANIC EVOLUTION

Evolution has been considered so far in this book in the sense that Darwin used the term, that is, with regard to progressive changes in living things. But soon after Darwin, the idea of evolution pervaded

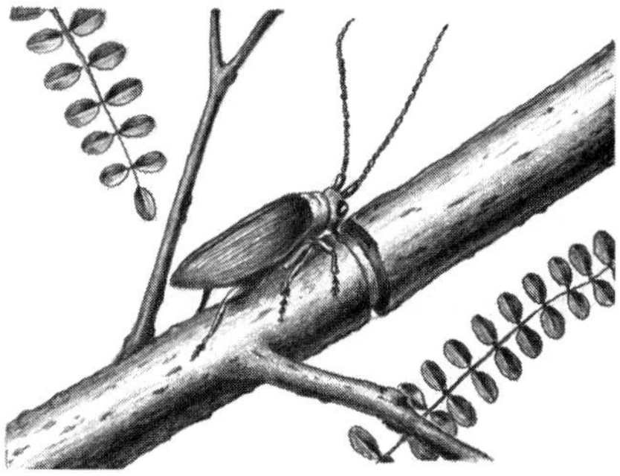

(a)

(b) × 4

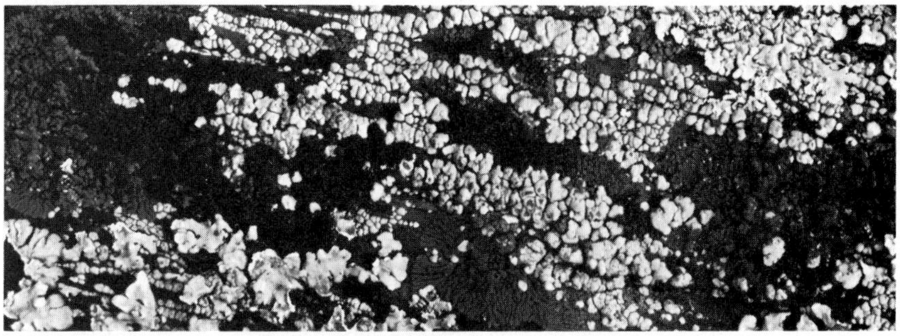

(c) × 1

(d) × 5

Figure 26.16
Partners in coevolution. (a) When a female mimosa girdler beetle finds a mimosa tree, she cuts a slit in the end of a branch and lays her eggs in it. She then gnaws a groove around the branch below the eggs. This kills the branch, which later falls to the forest floor and scatters the eggs. The tree benefits from the pruning, which usually doubles its life to 40 years or so. (b) A hermit crab is protected by the stinging tentacles of a sea anemone. In return for protection against predators, the crab brings food to its home and shares it with the anemone. (c) Lichens seem to be individual organisms, but are in fact associations of algae and fungi. The algae furnish food for the fungi, and the fungi—as some biologists think—provide water and protection for the algae. (d) A colony of insects called aphids suck plant sap for nourishment. Ants "milk" the aphids for the partially digested sap, and in return receive protection during winter in the ants' nests. Aphids ride on the ants' backs from plant to plant.

disciplines other than biology, with the result that astronomers began to think of cosmic evolution and geologists of earthly evolution. You can see a kind of evolution even in mechanical objects. Modern automobile fenders, for example, scarcely seem related to the flat strips of iron that were used over the wheels of cars of the 1890s, but if one looks at intervening models, one can see a steady, progressive series of changes leading up to the current styles. If automobiles were alive, one could write a convincing article on the evolution of fenders.

Up to now we have emphasized the influence of the environment as a shaping force in the evolution of plants and animals, but have been less concerned with the actions of plants and animals as shaping forces in the changing of environments. If evolution is progressive change in general, then an environment can be said to evolve as surely as its living inhabitants do. Next we will see how plants have altered the oxygen content of the air, how plants and animals make soil from rocks, and how successive waves of invading organisms can alter the aspect of a region. In succeeding chapters we will be concerned with interactions among organisms and interactions between organisms and their environments.

SUMMARY

1. The differences among individuals are due in part to the organization of an organism's DNA (this is the part that is important in evolution) and in part to the environmental forces that alter individual development (this is of no direct importance in evolution).

2. *Mutations* are changes in nucleotide sequences in DNA. Mutations, plus genetic recombinations of old nucleotide sequences are the raw material of evolution. New mutations, plus new combinations of old and new genes, make new varieties, adding to the total variability of a species.

3. The changing of an organism that brings the organism into better harmony with its environment is called *adaptation.* A striking example of adaptation is *industrial melanism. Cryptic coloration, mimicry, warning coloration,* and *adaptive coloration* are several types of adaptation that may occur in plants or animals.

4. The *species* is still the basic grouping of organisms, but there is no definition that meets the requirements of all the investigators.

5. A group of individuals partially or entirely isolated from the other members of the species is a *deme.*

6. The entire genetic stock of a population is a *gene pool.* In a large breeding group, there may be thousands of rarely expressed recessive characteristics. Most of them are probably detrimental, but some are apparently neutral, and some are beneficial.

7. The *Hardy-Weinberg formula,* in conjunction with the Mendelian principle of segregation, can be used in calculating the frequency of an allele in a large population. Genetic stability is assumed as a condition for the effectiveness of the Hardy-Weinberg formula.

8. *Genetic drift,* a chance variation in the frequencies of alleles in a population, is especially likely to occur in small populations.

9. The development of a new species, *speciation,* can occur in one generation, or it can be a slow process extending over many generations. In either method, the result is a new population of organisms that cannot produce fertile offspring when mating with unchanged descendants of the old ancestral line.

10. One way for chromosome alteration to bring about speciation is by the phenomenon of *polyploidy,* the replication of entire chromosome sets. When a new species arises as the result of polyploidy, no new genetic material has been added. There are no mutations, no new nucleotide sequences; there is only a new combination of pre-existing genes.

11. The slow process of speciation by isolation and increasing divergence is known as *adaptive radiation,* because organisms *radiate* out from one place and *adapt* to new environments.

12. *Reproductive isolation* occurs when the union of gametes is not accomplished. Although any one of many methods acting alone is enough to prevent the production of fertile offspring, usually several isolating mechanisms act at once.

13. Sometimes changes can occur so that organisms that are not closely related come to have certain similar characteristics. The process of changing toward similarity instead of difference is *convergent evolution.*

14. When two species undergo change in such a way that both can operate more efficiently as a team, the process is called *coevolution.*

ASK YOURSELF

1. Distinguish between mutation and genetic recombination.
2. What are some ways in which an individual animal can adapt to an environmental change?
3. Give some examples of cryptic coloration used by animals.
4. How can biological mimicry be useful to an animal?
5. How do zoologists usually define "species"?
6. Why is it difficult to make a definition of *species* satisfactory to all biologists?
7. What is a deme?
8. How can the Hardy-Weinberg equation be used by genetic counselors?
9. Under what circumstances is genetic drift likely to occur?
10. How can a polyploid individual come into being?
11. What is meant by the phrase "adaptive radiation"?
12. Give an example of convergent evolution.
13. In a case of coevolution of a flower and an insect, who benefits?

27
The Ecology of Populations

SOME KEY POINTS

1. Ecology is concerned with populations, communities, ecosystems, and the biosphere.
2. Human beings have an increasing effect upon the environment.
3. The methods of studying ecology include field studies, laboratory experiments, and mathematical models. The growth of populations is being studied actively.
4. The dwelling place of an organism is its *habitat,* and the specialized method of living within a habitat is an organism's *niche.*
5. Population growth is affected by the actions of living organisms, such as predation and behavior, and by nonliving factors such as water and temperature.

THE PRESENT AIMS OF ECOLOGY ARE TO LEARN what organisms inhabit particular places and how many individuals are in each place, to identify and quantify the forces that determine distribution and abundance of organisms, and to find out what places organisms can inhabit.

In a broader sense, **ecology** is the study of the interactions of organisms with each other and with their environment. Viewed in this way, however, ecology is so inclusive that practically all biology is ecology, and, as a scientific discipline, it becomes diffuse and unwieldy. Ecology does indeed intersect several other biological subfields, especially genetics, plant and animal physiology, mathematics, evolution, animal behavior, and taxonomy, and an ecologist needs to be familiar with those subjects.

ECOLOGY AND THE LEVELS OF ORGANIZATION

We have said before that we can recognize a number of levels of organization of matter and the living world (Chapter 1), and that each level has features not shown by the next lower level. We started this book at the simplest level: the elementary particles of matter. From there, we worked our way through more and more complex levels, until finally we arrive at the most complex of all. The age, amount, and certainty of our knowledge decreases as we go up the scale. More is known about molecules and cells than about whole organisms, and more is known about organisms than about communities of organisms. Ecology, like psychology, is a young science in comparison with chemistry and other long-established fields of study.

Where in the hierarchy of organizational levels does ecology belong? In the following list, the levels with which ecology is concerned are printed in italics.

Subatomic particles
Atoms
Molecules
Cell organelles
Cells
Tissues
Organs and organ systems
Organisms
Populations of organisms
Communities of populations
Ecosystems
The biosphere

As understood by ecologists, a **population** is a group of individuals of one species, usually occupying a defined space. A population rarely, if ever, occurs in nature as a single species isolated from other species. A **community** is an interacting mixture of populations, perhaps as few as two species in an artificial, experimental situation, or as many as hundreds in a complex, natural one. An **ecosystem** includes more than the organisms of a given community, as it involves not only the living inhabitants of a space, but also such nonliving factors as water, temperature, light, and inorganic nutrients. Finally, the **biosphere** is that part of the earth that consists of living beings, a counterpart of the air, which makes the *atmosphere,* and the solid earth, which makes the *lithosphere.* Populations, communities, ecosystems, and the biosphere are the domain of ecology.

Figure 27.1
A shanty town in southern Africa, where lack of sanitation, adequate food and space, all the results of poverty, cause human unhappiness.

ECOLOGY IS RELATED TO HUMAN AFFAIRS

In our aim of understanding the distribution and abundance of organisms, we touch the life of human beings in many ways. There is the prime need to nourish mankind. Producing enough food by agriculture, animal husbandry, fishing, and perhaps some day by artificial methods is a matter of regulating the distribution and abundance of organisms, and is therefore of ecological importance. Similarly, the environmental changes that come under the general heading of pollution are of interest to ecologists because they affect the distribution and abundance of humanity itself, as well as of the plants and animals with whom we share the earth. Indeed, in popular belief, ecology is—inaccurately—thought of as being primarily concerned with the pollution of earth, air, and water.

Then there is the matter of human population. We need to know what has led to the present overabundance of this unique species, an overabundance that has led to our present polluted world with its small pockets of affluence and its greater pockets of poverty, crowding, and personal misery (Figure 27.1). We do not know where we will finally go, but the intent of ecologists is to understand the principles of population dynamics. If they can do that, they might, in collaboration with sociologists and psychologists, make reasonable prophecies and perhaps offer sensible and acceptable suggestions for avoiding greater disasters than we have already experienced.

(a)

(b)

Figure 27.2
(a) A dilapidated slum, where wrecked buildings and scattered refuse make a scene of squalor. (b) A field of wild flowers, with a mountain backdrop, gives an esthetic sense of cleanliness and open freedom.

Finally, ecological thinking touches on a nonscientific portion of human life, and that is our perception of the earth and our feelings about it. This aspect of ecology leads into esthetics, or the elusive quality of beauty. It is the distribution, diversity, and abundance of organisms that make the world either beautiful, various, and new or dull, monotonous, and sordid. Contrast, for instance, two images: One is that of a vacant city lot, supporting a few limp pigweeds and a couple of dozen furtive rats. The other is that of a mountain meadow deep in grass, with flecks of color supplied by buttercups, swallowtails, tanagers, and iridescent tiger beetles. (Figure 27.2). That is where it is the business of ecology to understand, to predict, and to control, for the benefit of humanity and of the nonhuman earth-dwellers as well. Ecological forces affect all of us all the time, whether we recognize them and think about them or not.

THE METHODS OF ECOLOGY ARE VARIED

The earliest ecological data were based on observations of nature, or what biologists call "in the field." Kerner von Marilaun, for example, watched the changes in the numbers and kinds of plants in the Hungarian marshes along the Danube River in the early 1860s, before the word "ecology" was invented. One of the longest-lasting and most detailed sets of data comes not from intentional biological investigation, but from the pelt records of the fur traders of the Hudson Bay Company, from the seventeenth century to the present.

One difficulty with field observation is that conditions are not readily controllable, with the result that it is only possible to separate cause and effect from coincidence by application of rigorous statistical methods to carefully collected field data. Nature is so complex that observers can overlook important factors and not be aware of the oversight.

Recognizing the difficulties inherent in field studies, ecologists turned to experimental copies of nature. Species of organisms are intentionally introduced into new places, and their survival or extinction is monitored. Or selected areas are cleared of organisms to allow studies of repopulation. Or areas are fenced or otherwise isolated to keep chosen plants or animals in or out. Laboratory experiments include such controllable units as greenhouses, terraria, aquaria, and microbiological culture vessels. In these confined areas, environmental factors (temperature, water, etc.) can be regulated, and accurate counts can be made of numbers and kinds of organisms.

Before high-speed computers were available, the statistical calculations of population data were slow, laborious, and limited. Now, ecologists develop equations, originally based on observations of natural phenomena, that can be used to make long-term projections of population changes. Provided that the initial observations and assumptions were correct, such *mathematical models* can predict the numbers of individuals in future generations and the frequency of genes in those generations. In using such models, ecologists and geneticists are working with common interests.

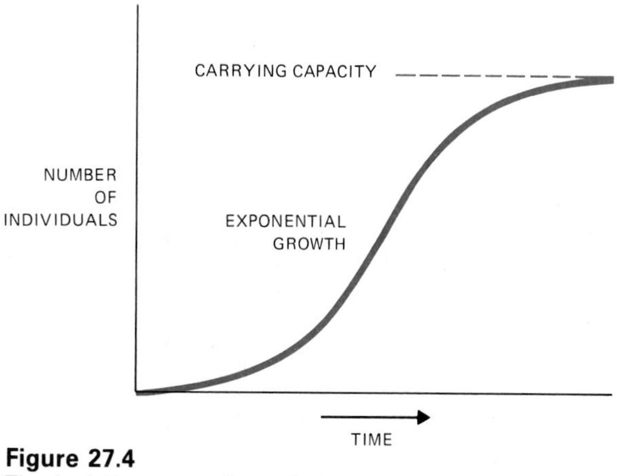

Figure 27.4
The logistic curve of population growth. In a real population, some factors limit growth, so that the curve comes to have a less steep slope and finally flattens out. When there are as many individuals present as the environment can support, there is no further growth: The *carrying capacity* of the environment has been reached, and the population has a *zero growth rate*.

Recording and Estimating Population Changes

The most accurate way to find out how many individuals make up a population is to count them. Counts can be made if the population is restricted enough, as it can be with caged animals or microbial cultures. It is possible to start with a pair of mice or insects, or a single bacterium or protozoan, and keep a record of how reproduction proceeds. Less precise, but still useful data can be had from human census records. Official census figures are available, however, for only a few centuries at best, and they are less than complete. Even the most recent government census information is disputed. Nevertheless, censuses are useful bases for estimates and provide students of populations with a working start.

Obviously, if the number of individuals in a population being born and immigrating is greater than the number dying and emigrating, the population is growing. If the reverse is true, the population is shrinking, and, if shrinkage continues, the population will eventually become extinct. To take a simple example, let us assume that we start with one bacterial cell and provide it with adequate warmth, nutrients, and water such that it can divide in half an hour. The population will then make two cells in 30 minutes, four in 60 minutes, and so on. Such growth can be shown graphically, and because it expresses an exponential increase, it is called an **exponential growth curve** (Figure 27.3). Whether the increase is a doubling every 30 minutes, or a one percent rise every century, the final shape of the curve is the same: The population will start to approach infinity.

Unlimited growth is impossible, however. A limit will be imposed by dwindling food supplies, accumulation of toxic wastes, simple lack of space, or a number of other factors. What actually happens is that a small population may get started and grow exponentially until one or more of the limiting forces act to slow down the growth rate. Then the slope of the curve will become less and less steep, until it flattens out, indicating that growth has stopped. Then the population has stabilized; a **zero growth rate** has been achieved (Figure 27.4). Such an S-shaped curve is a **logistic curve** of population growth and is typical of many observed populations in nature.

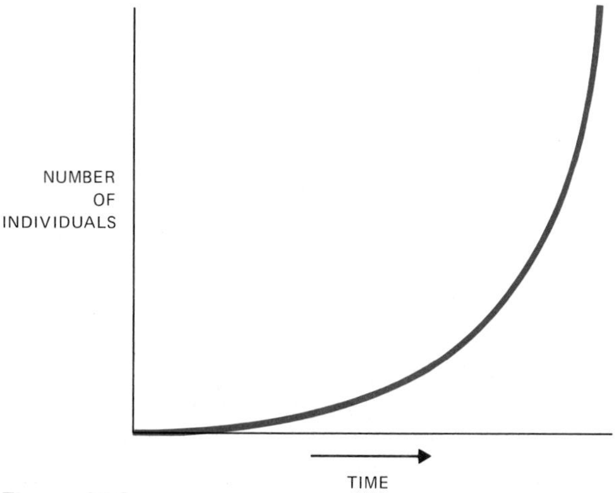

Figure 27.3
The exponential curve of population growth. As long as a population is growing without limit, the absolute number of individuals will keep increasing. The slope of the curve gets steeper and steeper until it is almost vertical, which means that the population is growing too fast for the rate to be measured.

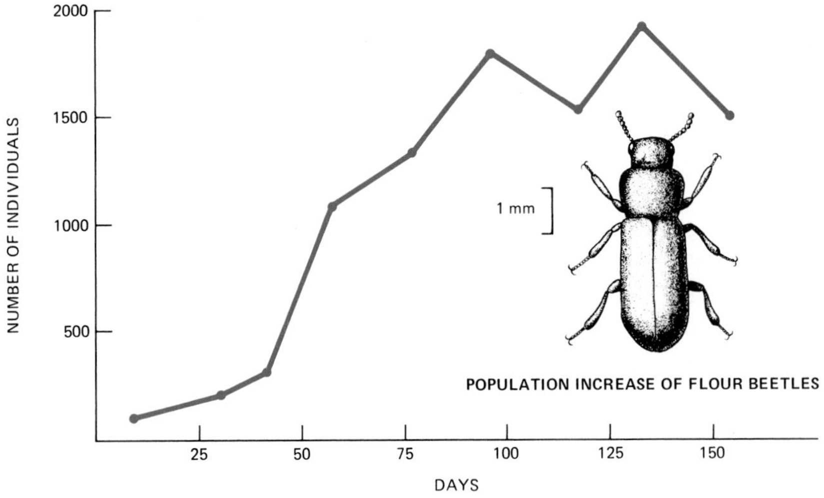

Figure 27.5
A realistic representation of the growth of a population of flour beetles (*Tribolium confusum*). The general shape of the curve is similar to that in Figure 27.4, with a period of initial slow growth, followed by fast growth, and finally by a period of relative stability. Along the way minor disturbances cause fluctuations, shown as peaks and valleys.

Notice that the curve in Figure 27.3 is the same as the part of Figure 27.4 that represents the early stages of population growth. The flattened part of the curve in Figure 27.4 shows that the population has reached an equilibrium. This means that the region where the organisms live contains as many individuals of that species as can live there, the number being determined by many factors, including the availability of food, water, and light and the presence of parasites or competing species. The number of individuals that remains relatively stable indefinitely is the **carrying capacity** of that environment and is set by both the living *(biotic)* and the nonliving *(abiotic)* factors of the environment. If the number falls below the carrying capacity, the number will tend to rise again; if it rises above, it will be forced to fall back.

A smooth S-curve is at best an approximation, because real populations do not change numbers at constant rates, nor do they remain steady once they have stabilized. Instead, they grow irregularly, and then tend to fluctuate, sometimes dropping temporarily below the average and sometimes rising above it (Figure 27.5).

HUMAN POPULATION: THE UNITED STATES

As with any growing population, the rate of growth of the population of the United States is determined by four regulators: immigration, emigration, the birthrate, and the death rate. Reasonably good records of all such data are available for nearly two centuries; they not only give a history of what the population has been but also provide a basis for guessing what it may be in the future. Figure 27.6 shows that the number of people in the United States has been increasing steadily. The increase in absolute numbers is shown by the greater steepness of the slope of the curve. This increase has happened in spite of wars and depressions and fluctuations in birthrates and immigration.

The birthrate at the beginning of the nineteenth century was over 50 births per 1000 inhabitants per year, and it has been generally declining ever since. It dropped to below 20 during the depression years of the 1930s, rose to over 25 after World War II (the "baby boom" of the late 1940s and 1950s), and has been falling again since the early 1960s. It dropped to about 16 in the late 1970s. Immigration has varied even more widely. During stress times, as during wars and depressions, immigrants were fewer, with only a few thousand people a year entering the country. But during the early decades of this century, the United States was admitting a million a year. Emigration has never had much impact; not many people want to leave. The death rate, with minor shifts, has remained close to 10 per 1000 per year for half a century.

In spite of the variations in factors that typically determine populations, the population of the United States continues to follow the pattern of any population that is increasing. We do not, of course, know what is coming next. But a comparison of Figure 27.6 with Figure 27.4 shows one point plainly: The population curve for the United States from 1800 to the present is a copy of the first part of the population curve for any theoretical population. We know that we cannot have a population approaching infinity. The question is, when will the curve begin to flatten?

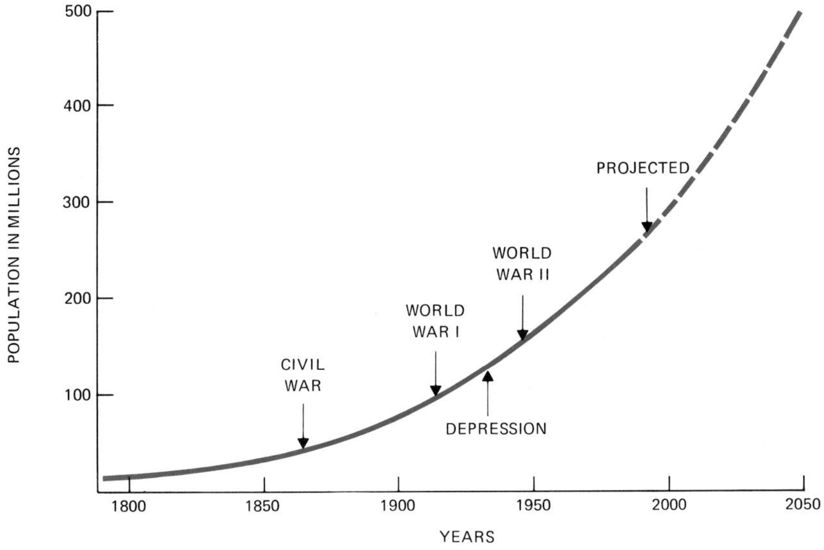

Figure 27.6
Growth of the population of the United States of America from 1790 to 1980, with an estimate of continued growth to 2050. (Based on figures from the U.S. Bureau of the Census.)

HUMAN POPULATION: THE WORLD

We turn now from consideration of the population of a single country to that of the whole world. The questions for us as members of the human species to consider are: What is happening now? What is likely to happen next? What can we do to influence future developments?

In order to learn what is happening now, we must first look farther back into the past. Ancient history, in the minds of an unfortunately large proportion of the present citizenry, is something that happened 25 years ago. But unless we compare the current population to human populations of earlier times, centuries ago, we cannot see where we stand now and consequently cannot even guess how many of us there will be a century from now. Figure 27.7 shows the number of people on earth for the past two thousand years.

Although no accurate numbers can be obtained for early years, reasonable estimates indicate that the number of people living by about 6000 B.C. may have been around five million. At the beginning of the Christian era, that number may have grown to between 100 and 300 million, a twenty- to sixtyfold increase in some 6000 years. Much of the increase

Figure 27.7
World population from 1 A.D. to 1981, with three possibilities toward the year 2000.

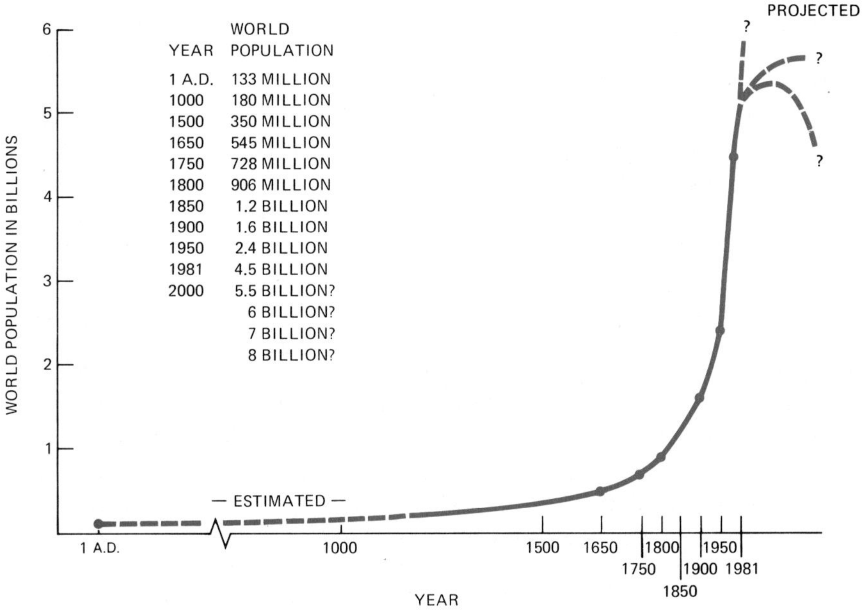

Figure 27.8
The population of the world. (a) A map of the world, showing the actual relative sizes of countries. (b) A world population map, showing the sizes of countries in proportion to their populations. Note especially the disproportionately large sizes of Japan, India, and China. In 1982 China's population exceeded one billion, almost one-quarter of the world's population.

was due to agricultural improvements. Change came slowly, and by 1650, there were still only some 500 million people living.

Then with the industrial revolution, the *rate of increase* began to accelerate. It took about 200 years between 1650 and 1850 for the world population to double. (From here on, official censuses give figures that are more dependable than the estimates of earlier times.) After 1850, it took only 80 years to double again, and only 45 years for the next doubling. It is expected to double once more by the year 2015.* It probably took something over a million years after the appearance of the first humans for the earth to have a billion people, but it took 80 years to get its second billion, and only 30 for the third. The 1981 estimate is about 4.5 billion (Figure 27.8).

What will happen next? Although guaranteed prophecies are impossible, reasonable projections

* Source: Population Reference Bureau, Washington, D.C.

can be made, based on past experience. By 1990, according to low estimates, we may have a world population of 5.5 billion; by high estimates, 7 billion. In the past, even high estimates have proved to be too low when actual times and figures were obtained. Unless we suffer some unforeseen major catastrophe, our population will surely continue to increase for a while. But the earth's space and resources are finite, and we cannot continue to increase our numbers at the present rate indefinitely. When the human species becomes so numerous that it has reached the carrying capacity of the environment, we must stop the increase. Another alternative, a scary one, is a precipitious drop, possibly even to extinction. After all, more species have perished than have survived, and we know of enough examples to make us wary. A century ago, passenger pigeons were so numerous in the American Midwest that no one could estimate their numbers, and they were shot by the trainload for shipment to city markets. By 1910, one carrier pigeon was known to be alive. It is now stuffed and sits in a museum.

Finally, what can be done? The answer is easier to state than to implement: cut down the reproductive rate. If we do not, some external force will do it for us. Some of the obvious possible forces that could reduce the population are mass starvation and uncontrollable disease. A major nuclear war could eliminate most of humanity quickly. Less obvious are "social diseases," such as have been seen in experimental animals. Adults fail to mate, litter size is reduced, babies are abandoned, cannibalism arises, and apparently healthy animals simply lie down and die. Until such catastrophes occur, however, we will keep on for a while, adding some 75 million bodies a year to the growing burden. As that goes on, we have less space, less food, less privacy, and more dirt (Figure 27.9).

We cannot continue breeding as we have been for the last two centuries. Most people are not aware of this, particularly at the time when they are considering starting a family, when they have not lived long enough to see for themselves what is happening. Or they may come from families or cultures with a long tradition that puts high value on large families. But the possibilities of mass suffering are so apparent to those who know the facts that a number of governmental and unofficial projects are in existence to inform people of the dangers. The sad part of the story is that those most in need of information are the hardest to reach. They live in places where birthrates remain high, death rates have been reduced by sanitation and medical care, and poverty is serious. Besides, many of these same people who would profit by reduction of the birthrate resent intrusion in their lives and suspect that the outsiders who are trying to influence them are acting for selfish reasons.

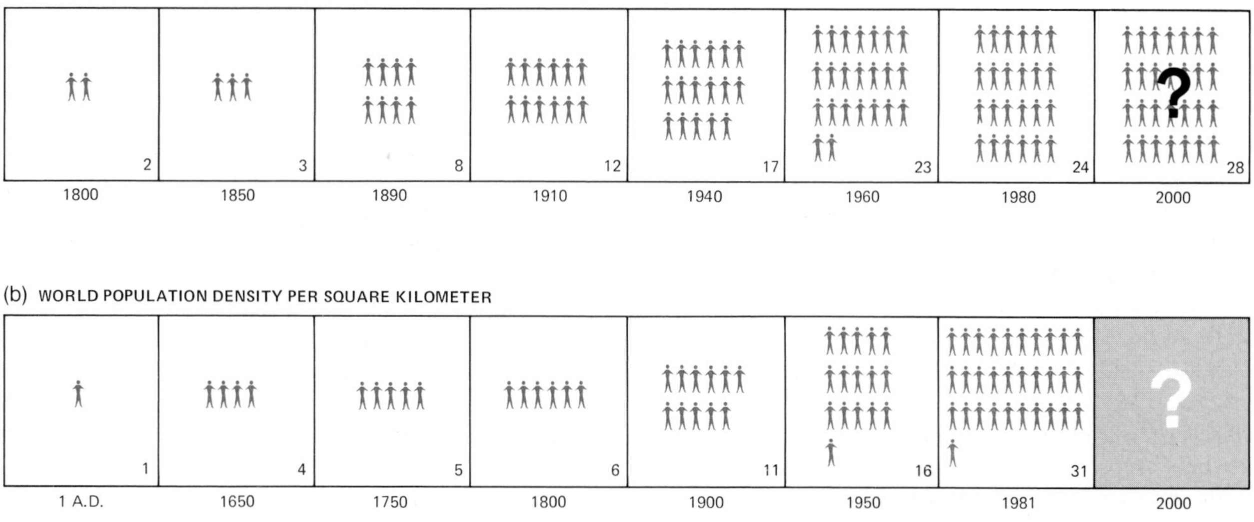

Figure 27.9
(a) Population density per square kilometer in the United States from 1800 to 1980, with a projected figure for the year 2000, based on estimates of the Population Reference Bureau.
(b) Population density per square kilometer of the world from the year 1 A.D. to 1981.

(a) × 2

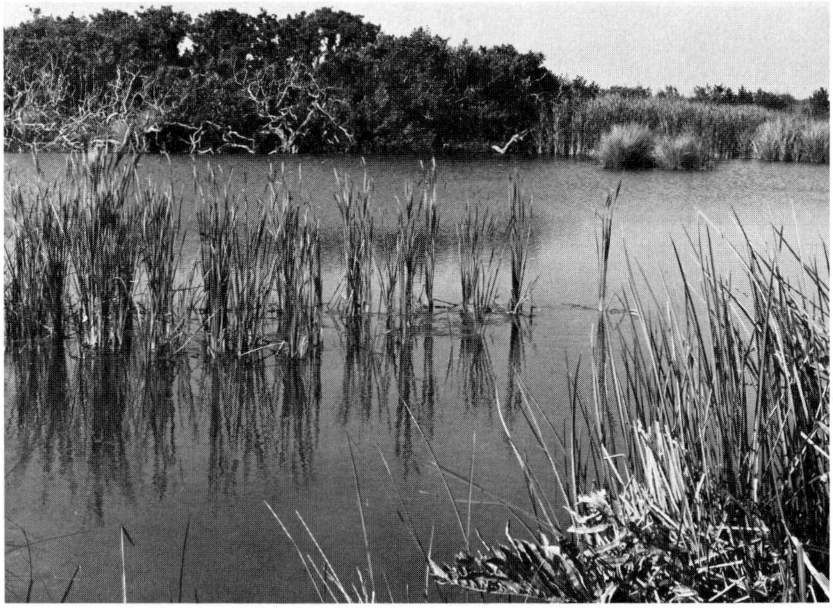

(b)

Figure 27.10
Examples of habitats. (a) Small mushrooms, *Marasmius*, growing on the restricted space of a piece of decaying wood. (b) Here, on the edge of a pond, the light is bright, the soil is moist and heavily organic, and the plants are low-growing, mostly sedges and cattails. Animals may include redwing blackbirds, an occasional marsh hawk, garter snakes, land snails, and many insects.

HABITATS AND NICHES

Any organsim obviously has to have a place to live. The actual dwelling place is a **habitat.** Some familiar examples of habitats are marshes, grassy prairies, tidal estuaries, ponds, and—in the past few years—the edges of highways. A habitat may be as restricted as a single rotting log with only a few species living in it, or as extensive as a woodland containing hundreds of species and millions of individuals (Figure 27.10).

The example of a single rotting log can be called a *microhabitat*. Even such a restricted space as the log may have within it still smaller microhabitats, each differing from the others in factors that are commonly thought of as climatic. Climate is usually treated in a broad way to include a fairly large area, at least on the order of square kilometers. Wind, rainfall, and temperature, recorded in one weather station or a few scattered ones, are reported as being typical of a whole region. But as far as plants and animals are concerned, countless small places vary from the "typical." Local differences in elevation, direction of slope of the ground, shading by hills or buildings or trees, or nearness to water can make considerable differences in *microclimates*. Consider, for example, two microclimates, with their microhabitats, under a large tree. The rainfall at the periphery of the longest branches may be twice that close to the trunk only six meters (20 feet) away. Ground-dwelling insects are sensitive to differences smaller than that, and as a result species distribution of insects is so strongly affected that wholly different faunas may inhabit the two microhabitats. Small plants and many invertebrate animals have narrow, specific requirements, and they live only in places that satisfy those requirements (Figure 27.11).

In a habitat, each species has its functional place, that is, its special ways of using the habitat. This working position is its **niche.** The niche of a species in a particular habitat is the sum total of the ways in which the population gathers resources. An oak in a forest may provide a physical dwelling place—a habitat—for flying squirrels, gall wasps, bark beetles, mushrooms, blue jays, yeasts, and dozens of other creatures, but each species has its own niche, or way of finding food, space, and shelter. The various and sometimes highly specific needs of organisms are so numerous and so complex that no complete catalogue has ever been made. Most habitats are usually shared by many species, but rarely do two or more species in a given habitat have identical niches (Figure 27.12). Even if they seem to do so, there may be subtle distinctions that escape detection.

Figure 27.11
Examples of microhabitats. In this stream through woods, there are hundreds of small variations in exposure to light, wind, and water. A single rock may have areas that are different enough to encourage distinctly different populations. Part may be permanently submerged, part wet only during rain or unusually heavy water flow, part (on the south side) hot, and part (on the north side) cool. A spot under a dense bush is cooler, dimmer, and moister than a spot under a break in the overhead trees.

(a)

(b)

(c)

Figure 27.12
Animals often share the same habitat. (a) A bumblebee moth and a skipper both drink nectar from a thistle inflorescence. (b) A giraffe, a zebra, and a roan antelope all share the same water hole, but the three animals prefer different type of leaves. (c) A great blue heron and an American alligator seem not to notice each other as they stalk their prey in the Florida Everglades.

ABIOTIC FACTORS AFFECTING DISTRIBUTION AND ABUNDANCE

Inasmuch as every organism lives in a physical world, the abiotic factors in the environment necessarily are critical in determining what lives where. An effect may be immediate and direct, as in the lack of boron in some Florida soils. Or it may be indirect, as when temperature limits the distribution of a species by limiting its food supply. We include here only those factors that are well known and susceptible to experimentation.

Water

Water is essential to life. Its abundance or scarcity is consequently of primary importance in the distribution and abundance of organisms. Where rain is plentiful, plants and animals are plentiful too (Figure 27.13), as in the Olympic peninsula of Washington State or the North Carolina Smoky Mountains. Where water is intermittently lacking, growth is specialized and restricted, as in the short grass prairies of Nebraska. Constant dearth of water makes a desert.

Water must not only be present, it must be *available*. Frozen water is not available to plants, and neither is water that is diluted by too much dissolved salt. Freshwater organisms can rarely survive in the sea, and desert-dwellers have special ways of making the most of what little water they can get, either by conserving it carefully or by mechanisms for extracting it with high effectiveness.

Like other environmental factors, however, moisture does not work alone, but in connection with other factors, especially temperature. Plants in a dry, *cool* region are different from those in a dry, *warm* one.

Water supply is not entirely dependent upon rainfall. A south-facing slope is warmer and dries faster than a north-facing one (in the northern hemisphere, that is), and the difference is evident in tree distribution. In Idaho, grand fir and western red cedar are commoner on north slopes, ponderosa pine and Douglas fir on south slopes.

The nature of soil also affects water loss. Two species of salamanders in the Blue Ridge Mountains are localized, depending upon soil types. One, the red-backed salamander (*Plethodon cinereus*), inhabits areas where the soil has good moisture-holding ability; the second, the ravine salamander (*Plethodon richmondi*), inhabits slopes whose soil has less moisture-holding ability (Figure 27.14).

Temperature

Temperature can work directly on some organisms. Cold insects and reptiles simply cannot move; plants cease photosynthesis when they are too hot. In many instances, though, temperature exerts its effects indirectly. The royal tern is a bird of warm seas, restricted in its range by the food it can find. On the Pacific Ocean side of the Americas, warm water reaches from Lower California to Ecuador for 30 degrees of latitude, with the north limit imposed by the cold California current and the south by the Peru current. On the Atlantic side, the water is warm for about 70 degrees of latitude, from South Carolina to Central Argentina. The broader Atlantic reaches of warm water support the tern's fish re-

Figure 27.13
A mule deer in a lush meadow in the Olympic Mountains of Washington. Warm, moisture-laden air from the Pacific Ocean brings abundant rain to the region, ensuring ample water and mild temperatures. The result is a rich growth of organisms.

Figure 27.14
Distribution of species depending on moisture. The red-backed salamander (left) lives in moister soil in wooded areas. The ravine salamander (right) lives on drier slopes whose soil has less water-holding ability.

quirements and consequently allow the bird a greater range. It could itself tolerate some cooler climate, but it must eat (Figure 27.15).

The upper temperature limits of some biological activities are, from a human view, low enough to be called "cold." Peaches cannot fruit without a freezing period. Peaches therefore do not grow in frost-free countries. Some barnacles do not produce gametes until the temperature of autumn air drops at least to 10°C (50°F). Such barnacles could not reproduce in the tropics.

Temperature also interacts with other factors. The microscopic fungus *Phoma* grows better in seawater than in freshwater in the tropics (Panama), but it cannot survive in full seawater salinity in cool regions (Ireland). Here the balance is between temperature and salinity; temperature alone cannot make such a difference, as laboratory cultures show.

Light

The immediate and obvious effect of light on the distribution and abundance of organisms is its ability to provide energy for photosynthesis. When light (and of course the other photosynthetic requisites) is present, plants grow. Green plants do not grow in caves or deep in the sea. But the energy trapped by photosynthesis can be dispersed, so that even deep-dwelling fish far from green plants can still make use of it. The energy of such organisms is several steps away from its source, but in general the abundance of life forms is closely connected with the availability of plant-produced food, which is light-dependent.

Light does more than supply energy. As one moves north from the equator, where days and nights are always of equal length, the nights become shorter and shorter in summer and longer in winter. The differences in day length not only make temper-

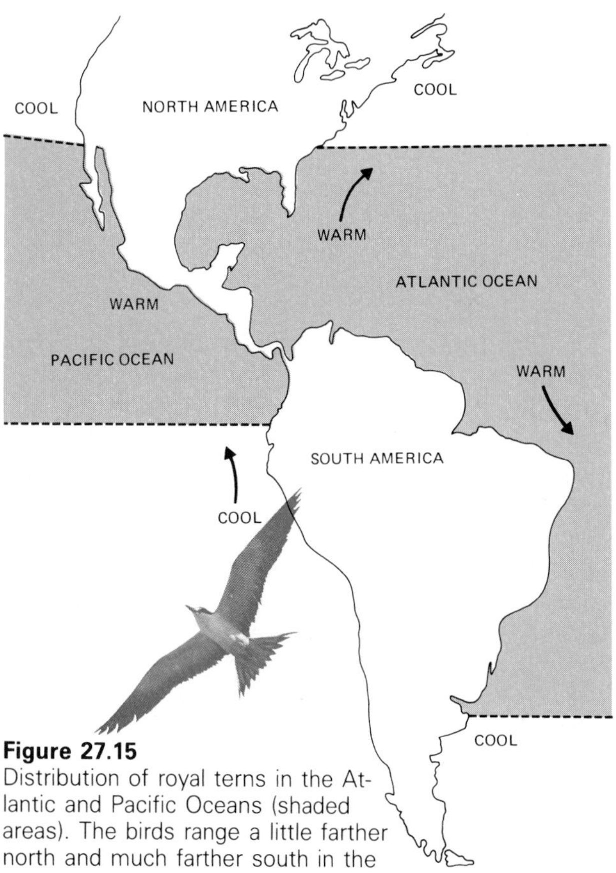

Figure 27.15
Distribution of royal terns in the Atlantic and Pacific Oceans (shaded areas). The birds range a little farther north and much farther south in the Atlantic Ocean than they do in the Pacific because the warmer band of Atlantic water produces the fish needed by terns.

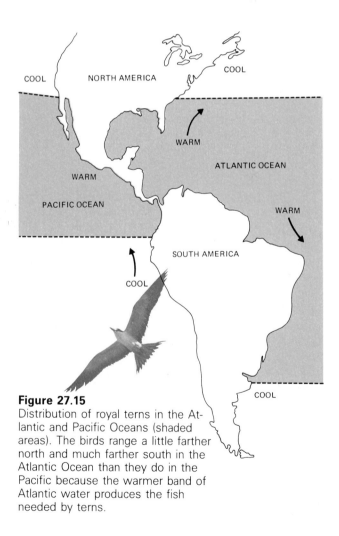

ature differences, but affect the breeding and migration habits of animals and the growth habits of plants (Chapters 29 and 17). As a result, strictly long-day plants such as winter barley do not live in the tropics because they cannot reproduce there. Animals are similarly restricted to places where their reproductive behavior is suited to the day length.

Mineral Nutrients

Except for carbon, hydrogen, and oxygen from air and water, all the elements needed by plants and animals come from the soil (Chapter 17). Even the salts of the sea come from soil. Thus soil as the source of what are known as the **mineral nutrients** helps determine distribution and abundance of organisms.

It has already been emphasized that any working system, including the living world, has to have an energy source, an organization, and a source of materials. Living things have their organization, governed by their DNA. The energy comes from the sun, which warms the earth and supplies light for photosynthesis. As far as energy goes, the biological world is an *open system,* that is, it receives a virtually endless input of energy, which is used for growth and other activities and is continuously reradiated into space as heat. The materials of which organisms are built, however, are limited to what is present. The material world is a *closed system,* which means that the elements required for life must be used over and over, or to use a current popular word, recycled.

Plants take an element, iron, for example, from the soil for their own use. Animals then get their iron by eating plants. They excrete iron, or they die or are eaten by other animals, and eventually their iron is returned by the action of bacteria and fungi once more to the soil. Or a plant may incorporate the oxygen of atmospheric carbon dioxide into sugar. If the sugar is eaten by an animal and oxidized during respiration, its oxygen goes back into the air, again as carbon dioxide. In an opposite direction, animals use for the oxidation of their food the same oxygen that plants give off in photosynthesis.

In a somewhat similar way, all biologically useful elements pass from soil, air, or water through organism after organism and return from time to time to soil, air, or water. Without this endless recycling, essential elements would be bound in unusable forms, and life as we know it could not continue. The ability of biological activity to make major changes in the earth's materials is shown by the present concentration of oxygen in the atmosphere. There is good reason to think that the original atmosphere had little or no free oxygen, but as a result of photosynthesis it is now about one-fifth oxygen. If it were not for the recycling ability of organisms, the various elements would become so tied up in organic compounds that further growth would be impossible.

All the elements used by plants and animals are used over and over, and "cycles" have been described in detail for carbon, oxygen, hydrogen, sulfur, nitrogen, phosphorus, iron, potassium, calcium, magnesium, and the trace elements. Here, instead of treating a number of possible elements, we have chosen nitrogen as one example and will show some of the ways in which it moves from air, where it makes up about four-fifths of the atmosphere, to plants, to animals, and back to air.

As nitrogen moves into and out of living things, it provides one of the most complex cycles known (Figure 27.16). As a gas, nitrogen (N_2) is not usable by most organisms. It can, however, be "fixed," or combined with hydrogen to make ammonia (NH_3) or with oxygen to make oxides of nitrogen (NO_2, NO_3). Atmospheric nitrogen can be fixed by commercial processes to make fertilizer, or nitrogen compounds can be formed by lightning, fires, and internal combustion engines and brought to earth by snow and rain. Some few microorganisms can fix nitrogen directly from the air. The best known of these are some blue-green algae and bacteria. Blue-green algae in flooded rice fields can fix nitrogen to a usable form so effectively that the addition of artificial fertilizer is not necessary.

The main bacterial nitrogen-fixers are the so-called nodule-forming genera that live in the roots of seed plants. Such bacteria bore into root hairs, grow colonies in the cells of the host plant, and stimulate the host to grow lumpy knots of tissue, the nodules (Figure 27.17). There the bacteria use free nitrogen to make amino acids, and the host plant profits by the availability of a usable form of nitrogen. The best-known nodule-producing plants are such legumes as beans and clover, but others—alder trees, for example—have similar nitrogen-fixing symbionts, and there are probably many species that serve as host plants but have not been discovered. People, even biologists, pay too little attention to what goes on under the surface of the ground.

Once nitrogen has been fixed as ammonia or nitrate (—NO_3), it can be used by plants (Chapter 17). Plants then furnish the nitrogenous compounds that animals need. Animals in turn excrete nitrogen

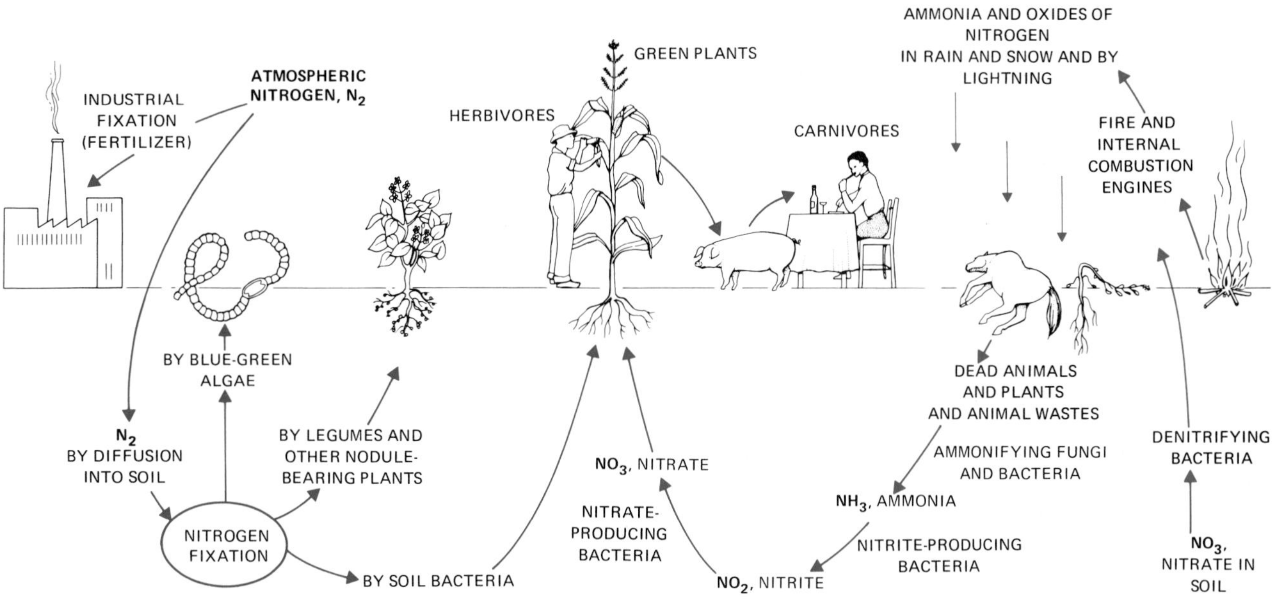

Figure 27.16
The flow of nitrogen in earth and organisms. Except in special instances, nitrogen is on the move from atmosphere to plant to animal to soil in an endless string of possibilities. Although nitrogen is the most voluminous component of the atmosphere, nitrogen-rich foods are the scarcest and most expensive of human nutrients.

(as urea, for example), or are eaten by other animals, or die. Animal excreta and dead animals are broken down by bacteria and fungi, releasing the nitrogen from proteins and other nitrogen-containing compounds as ammonia, which dissolves readily in soil water. In the soil, ammonia can be oxidized by relays of bacteria first to nitrite ($—NO_2$) and then to nitrate ($—NO_3$). Nitrate, being the ion that most plants can use as a nitrogen source, can once again be made into plant material, and the cycle goes on.

Some bacteria reverse the fixing process. Especially in boggy, poorly aerated soils, these *denitrifying bacteria* reduce nitrate to elemental nitrogen, which diffuses into the air.

The essential mineral elements are so distributed over the earth that most soils provide adequate nutrition for plant growth. Some regions, however, are strangely deficient in one element or another, and most soils can be made more productive if they are fertilized by the addition of phosphorus, potassium, and nitrogen. When trace elements are lacking, minute amounts of the limiting element are enough to cure deficiency diseases. In Florida, it is enough merely to drive a zinc-coated nail into an orange tree to relieve the symptoms of a disease fittingly call "little leaf." In Australia, millions of acres of rangeland lack molybdenum, but the trouble can be remedied by adding as little as two ounces of molybdenum per acre once every ten years. As recently as 100 years ago, no one knew that plant distribution could be affected by such subtle factors (Chapter 17).

× 3

Figure 27.17
Nodules on the roots of a soybean plant. The lumps are formed by the plant after entrance of the symbiotic bacteria. If the nodules are left in the ground after the plant is harvested, the soil can have its usable nitrogen content increased.

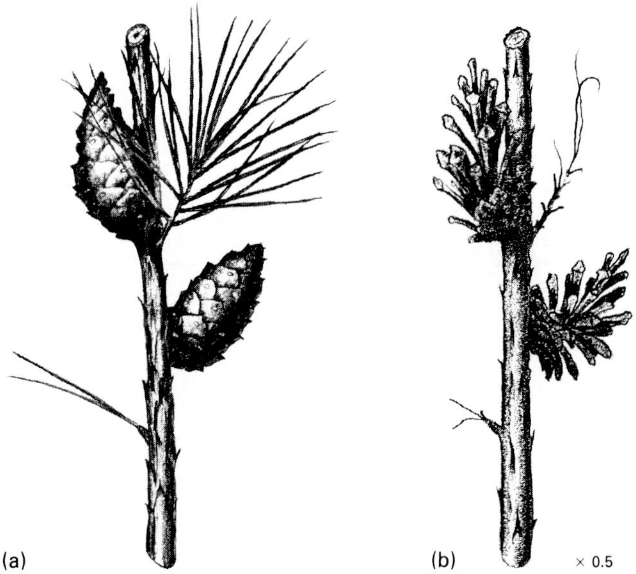

Figure 27.18
(a) The cone scales of a jack pine (*Pinus banksiana*) stay shut tight, enclosing the seeds so effectively that they cannot germinate. (b) Only after being heated by a forest fire do the scales spread open and let the seeds fall out. Then they can germinate and keep the species going. Fire is a critical factor in maintaining this kind of pine forest.

Fire

Forest and prairie fires can be destructive and dramatically terrifying, and public campaigns against fires have made Smokey the Bear a familiar symbol. But not all burning is bad. Indeed, some ecosystems maintain a measure of stability only if they experience an occasional fire; they have persisted for centuries because lightning can start fires often enough to cause certain effects. The cones of the jack pine of Canada and the northern United States do not open to release their seeds until they have been heated (Figure 27.18). Fire thus makes possible the start of new seedlings. The West Coast redwoods profit by periodical burning, because fire cleans off accumulated debris on the forest floor, providing a better environment for seed germination. The large trees have such thick, insulating bark that they are not hurt by small fires. If too long a period of time passes between fires, however, there may be such a buildup of litter on the forest floor that when fire does come it is too hot even for large trees to endure.

Grass fires and brushfires can also have a beneficial effect. The tender growing tips of plants are killed, but the tougher or better-protected older parts are unharmed. The result is increased branching, due to the removal of apical dominance. Many grazing lands are fired intentionally to clear the old dead scurf and to encourage the growth of new shoots (Figure 27.19). Indiscriminate burning can of course do more harm than good, but properly controlled fire can be ecologically helpful.

Figure 27.19
Intentional burning of grassland. Burned-over grasses provide more food for grazing animals than unburned ones.

Figure 27.20
Birds are effective distributors of plants and animals. They eat seeds which they pass unharmed through their digestive systems, and carry other seeds as well as spores, microorganisms, and insects on their feet and feathers.

BIOTIC FACTORS AFFECTING DISTRIBUTION AND ABUNDANCE

If a species has established itself in a habitat, some pioneer individuals obviously must have arrived there, stayed there, and reproduced. Several forces, acting together, determine whether a species can or will immigrate into a new territory, and, having arrived, whether it can or will remain. Once a species has come to inhabit a place, then other factors determine whether or not it can grow into a stable population. The following are the major factors in the establishment, maintenance, and regulation of populations.

Dispersal

Organisms have many ways of moving about and getting into new places. Microscopic species and seeds blow long distances in the wind, wash about in water, and are carried on or in the bodies of birds, beasts, and insects (Figure 27.20). Ships, railroad cars, and other human artifacts distribute plants and animals wherever they go. Such dispersal may be intentional (pigs, goats, cats, starlings) or unintentional (rats, cockroaches, pestiferous weeds). Broad-leaved plantain, one of the commonest weeds in America, followed European settlers so dependably that some Indians called it "white man's foot" (Figure 27.21).

Plants do not actively emigrate, but animals do. Birds can fly across oceans, even from east to west against prevailing winds, although more succeed in going the other way. Many animals roam about in an exploratory manner, especially when they are

× 0.5

Figure 27.21
Example of a species, accidentally introduced, that successfully established itself in a new region: broad-leaved plantain, brought by chance from Europe by colonists. It was well established by the early eighteenth century and now lives in countless numbers in lawns, parks, fields, and roadsides practically all over temperate America.

Figure 27.22
A sand grouse, *Syrrhaptes paradoxus,* a dweller in generally semidesert habitats, occasionally swarms over Europe, probably driven from its native territory by hunger.

young, and may thus take themselves to new places. During difficult seasons, mass emigrations occur, when large numbers of birds, mammals, or insects, faced with drought or starvation, move out of their original terrain. Lemmings are famous for their explosive migratory movements. Less well known, but spectacular in their own way, are the sand grouse from the Kazakh Republic in south central Russia, which have from time to time spread all over Europe in astonishing numbers and then died out (Figure 27.22). It baffles biologists that such flights from famine do not usually result in the establishment of new colonies. They do, however, indicate one way in which species can be dispersed.

Figure 27.23
Birds whose choice of habitat is determined by behavior. The European tree pipit (*Anthus trivialis*) picks a nesting site that provides it with a high landing place after a song flight.

Behavior

An animal species may accept or abandon a new region for some behavioral reason. Animal behavior in general will be considered in Chapters 29 and 30, but it deserves mention here in connection with the establishment and maintenance of new animal colonies, emphasizing the pervasiveness of ecology into other biological subfields. A couple of examples will show that a place may be habitable for food and shelter but not acceptable because an animal does not "like it." Prairie deer mice, which are accustomed to open fields, tend to avoid woods, even though if penned in a woody area they may survive well enough. The little British bird, the tree pipit, prefers wooded places, apparently for the sole reason that it uses a high perch as a landing platform following a song in flight (Figure 27.23). A lone telephone pole in the midst of a field will do; but without such a perch, it heads for the trees.

Even insects are known to show behavior that limits their distribution. A mosquito, *Aedes culicifacies,* will not lay eggs unless she has access to an open water surface. Blades of grass, or even glass rods sticking up from the water, will keep a female from approaching to lay eggs.

Competition

Competition is inevitable when two individuals come into conflict for possession of an environmental resource. The need may be for food, water, light, space, or shelter, and the competition may be between members of the same species or of different species.

If competition is among members of one species, the short-term result may be that the number of individuals inhabiting a place is affected. In the long run, competition may result in genetic changes in the population, but that requires many generations and great lengths of time. Genetic or evolutionary changes may require hundreds or even millions of years (Chapter 26). Ecological time, however, covers relatively few years, and the immediate results of competition among members of a species are in regulation of population density rather than in genetic changes. In animals, the factors most commonly involved in such intraspecific competition are food and space for rearing young. In plants, competition is for soil nutrients, water, and light.

In a famous set of experiments on interspecific competition, the Russian biologist G. F. Gause established cultures of protozoa, *Paramecium aurelia* and *P. caudatum.* He grew them first separately, and they lived. Then he grew them in mixed culture so that they were in competition with each other. The result was that in the mixed culture *P. aurelia* survived but *P. caudatum* died out (Figure 27.24). From such experiments came what is known as Gause's principle of *competitive exclusion,* which is that no two species can continue to occupy the same niche at the same time; that is, if they are in direct competition, one will live and the other will die out.

Some known examples seem to confirm the principle of competitive exclusion. Ranges of some closely related species of animals overlap little or not at all. Absence of overlap can be seen in the several species of pocket gophers (rodents) of the Rocky Mountains and in the salamanders (amphibians) of the Blue Ridge Mountains. When such nonoverlapping ranges are found, we suspect that there is interspecific competition.

British barnacles give an instructive illustration. One genus, *Balanus,* lives on rocks covered by water at high tide and exposed at low tide. A second genus, *Chthamalus* (pronounced "thamalus"), is actually able to inhabit the same space if given the chance, but thrives better than *Balanus* does at the higher rock levels between the tide levels. When both genera occur together, *Balanus,* the more effective competitor at lower levels, crowds *Chthamalus*

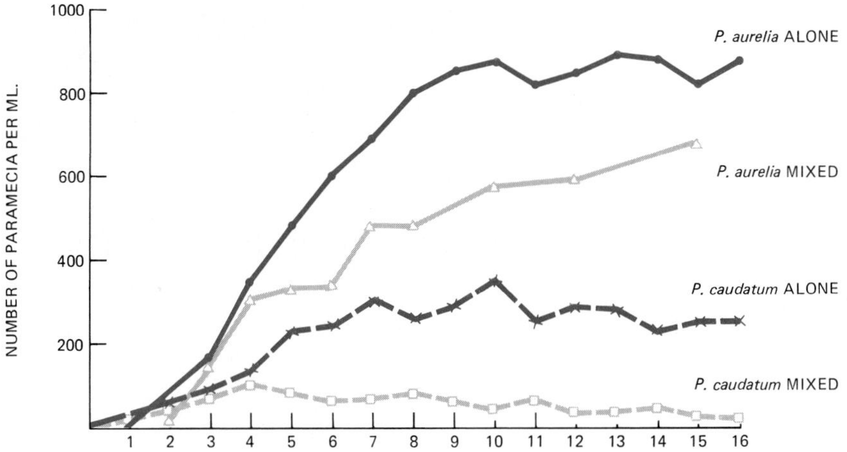

Figure 27.24
Growth curves for two species of *Paramecium,* alone and in mixed culture. *P. aurelia* and *P. caudatum* have similar requirements, but *P. caudatum* cannot compete successfully with *P. aurelia* and dies out.

into the upper levels. Both survive, but only at the levels where each has an advantage (Figure 27.25).

It may seem reasonable to assume that these spaces are exclusive because of competition. But other examples can be found in which apparently similar species do occur together. Two closely related birds are a cormorant (*Phalacrocorax carbo*) and a shag (*P. aristotelis*). Both nest on sea cliffs and feed on fish from the ocean, and it might seem on casual inspection that they are in direct competition. Yet they occupy the same range. Close observation of the behavior of the two species, however, shows that they do in fact have slightly different preferences for nesting sites, and they feed in different places, the cormorant closer to shore and the shag farther out at sea (Figure 27.26).

Is the competitive exclusion principle universally applicable? Biologists do not agree, because some examples in nature seem to support an answer of "yes," and others support "no." In considering cormorants and shags, we see that what at first glance looked like competition proved not to be. This example tells us that if we could know all the requirements of each species in complete detail, we might be able to answer our question confidently. Meanwhile, research on the problem continues.

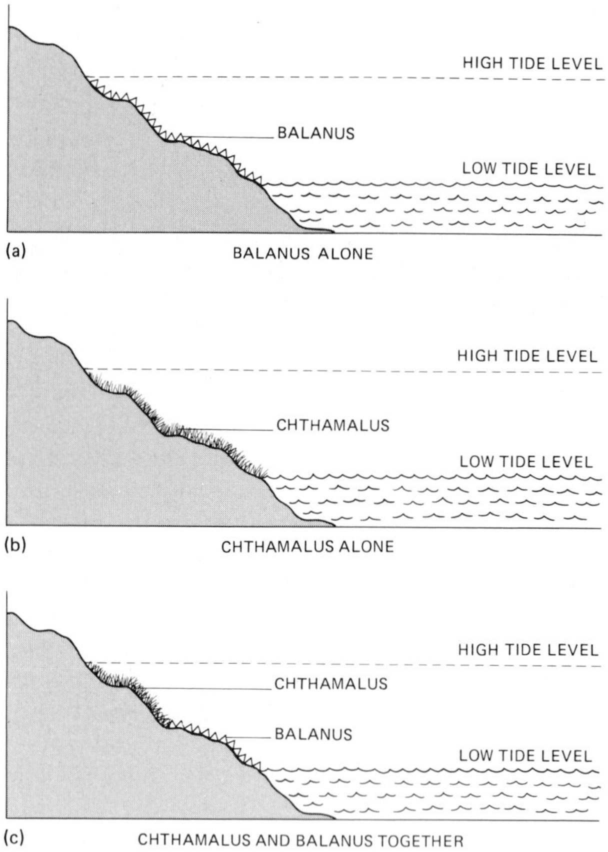

Figure 27.25
Competition between two species of barnacles on the shore between tide levels. (a, b) Both species are capable of survival through the intertidal range, as they show when one or the other is artificially removed. But when both species live together on the same shore (c), *Balanus* occupies the lower levels and *Chthamalus* predominates on the upper reaches.

Figure 27.26
A cormorant (a) and two shags (b). These are two similar species that seem to compete but in fact occupy different parts of their common environment and thus coexist.

(a) × 0.2

(b) × 0.1

(a)

(b)

(c) × 2

(d)

Figure 27.27
Predation in animals. (a) Snakes are usually effective predators, but this emerald tree boa, *Corallus caninus*, is especially effective because of its natural camouflage and such dazzling speed that it is able to catch a bird. (b) This time, a bird seizes a reptile. The desert roadrunner, *Geococcyx californianus*, is adept at catching rather large prey such as this collared lizard, *Crotaphytus collaris*. (c) A diving beetle, *Dytiscus marginalis*, is strong enough to capture a fish larger than itself. (d) A young baboon, *Papio ursinus*, practices its own form of predation: It picks up an ostrich egg when the mother ostrich is not watching and carries it off for a juicy meal.

Predation

In popular usage, predation is hunting. To an ecologist, predation has a broader meaning than "cat catches mouse" and includes any consumption of one organism by another. Thus a robin is a predator of earthworms, a gypsy moth larva is a predator of oak trees, and a malaria parasite is a predator of mosquitoes and vertebrates (Figure 27.27). Predation is a concern of ecologists because it affects the abundance and distribution of organisms.

In theory, if an edible species (the prey) is abundant, a species that eats it (the predator) will increase in numbers until the prey is diminished in number. When that happens, the predator will begin to diminish too, and being deprived of food, will become so scarce that the remaining prey will

INNOVATIVE PREDATORY BEHAVIOR IN ANIMALS

The predatory behavior of animals is often bizarre, at least to humans. The scorpion, for instance, feeds on live bees in a surgical and spectacular fashion. The scorpion grabs its prey in its forward pincers, holds it steady, and paralyzes it with a slashing sting from its upcurled tail. The poison is already dripping on approach. Sometimes, if the bee is held in a particularly vulnerable position, the scorpion will merely stab it to death, conserving its venom. Now the most interesting sequence occurs. After the bee is paralyzed or freshly killed, the scorpion bites a tiny hole in the bee and injects digestive enzymes that liquefy everything but the shell. The scorpion then sucks out its meal, and the hollow shell of the bee remains intact, like an empty, decorated Easter egg, lifeless but still beautiful.

Another ingenious predator is *Mastophora,* the bolas spider. Its method of capturing its prey is so specialized that it might seem destined for certain failure. But the unusual prey-catching technique of the bolas spider works. Instead of spinning a conventional web, the American bolas spider spins a thread weighted at the end with an extra blob of silk. When an insect enters the striking area, the spider swings its line like a gaucho throwing his bola (hence the name bolas) and grabs its prey (see drawing). The unorthodox method seems to be effective. The bolas spider has relatives in Australia and Africa. The Australian cousin is called the angler spider because of the way it casts its single line like a fisher at its prospective dinner.

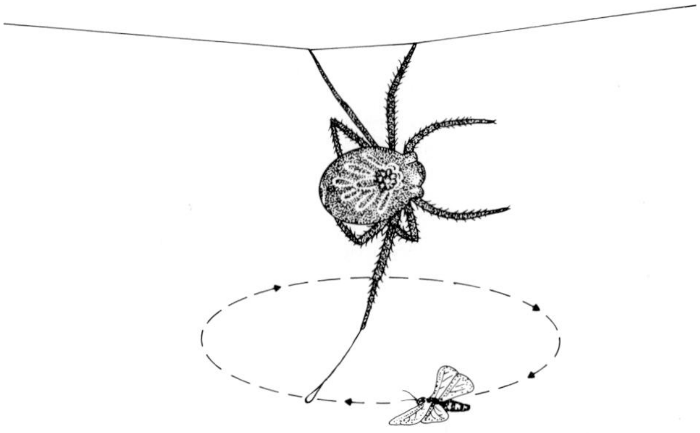

A BOLAS SPIDER TRAPPING ITS PREY

The anatomy of the praying mantis is perfectly designed for killing other insects. Furthermore, it is one of the few higher organisms that will attack, kill, and eat members of its own species. It is not really unusual among insects for a female to kill and eat her mate, but the female praying mantis consumes her mate from the head down, *during* copulation. The male is as persistent as his counterpart. With head already eaten, what is left of him continues the strange intercourse until finally his sex organs are gone, too.

In evolutionary terms, the seemingly inexplicable body control of the male praying mantis can be understood. In the process of natural selection, male mantises have been "favored" if their nervous systems make it possible for the lower abdomen to function without neural instructions from the head. Copulation, and therefore the survival of the species, can take place in spite of the cannibalistic female, which has been known to devour as many as seven willing mates in two days.

begin to increase once more. That will lead to a second increase of the predator, and a new round will have begun. This cycling of prey and predator can be expressed mathematically and shown graphically as a series of peaks on two curves, with the predator always lagging behind (Figure 27.28).

This scheme assumes a simple one predator–one prey system, something that rarely if ever actually exists. The nearest approach to this idealized situation is the interaction of the Canadian lynx and the snowshoe hare. Fortunately, records of lynx and hare furs have been kept since colonial days by fur traders, and from those data, an actual predator-prey relationship can be constructed (Figure 27.29). Besides being rare, this example raises problems. It does not take into account that fur trappers are predators on both lynxes and hares, nor that hare populations show cyclic fluctuations even in places where there are no lynxes. In nature, regular cycles are altered by such variables as multiple prey-predator relationships, climatic changes, and other, undetermined factors that affect the food supply of the

Figure 27.28
A population curve for a hypothetical predator-prey relationship based on a mathematical model, showing the fluctuations in numbers with the passing of time. The increase or decrease of the predator follows the increase or decrease of the prey.

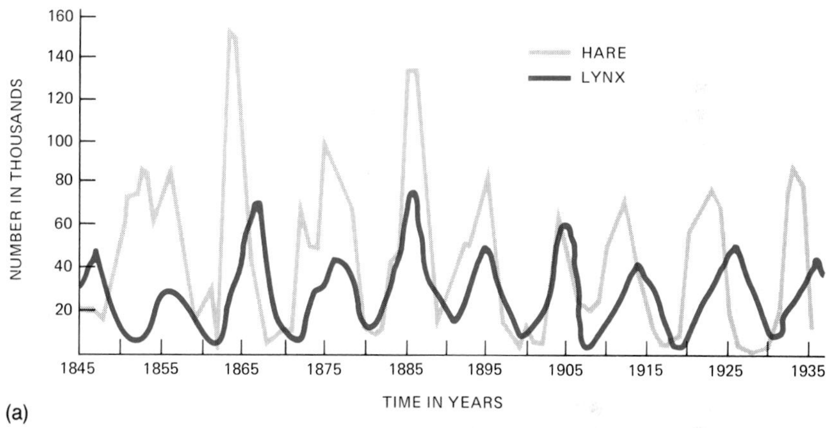

(a)

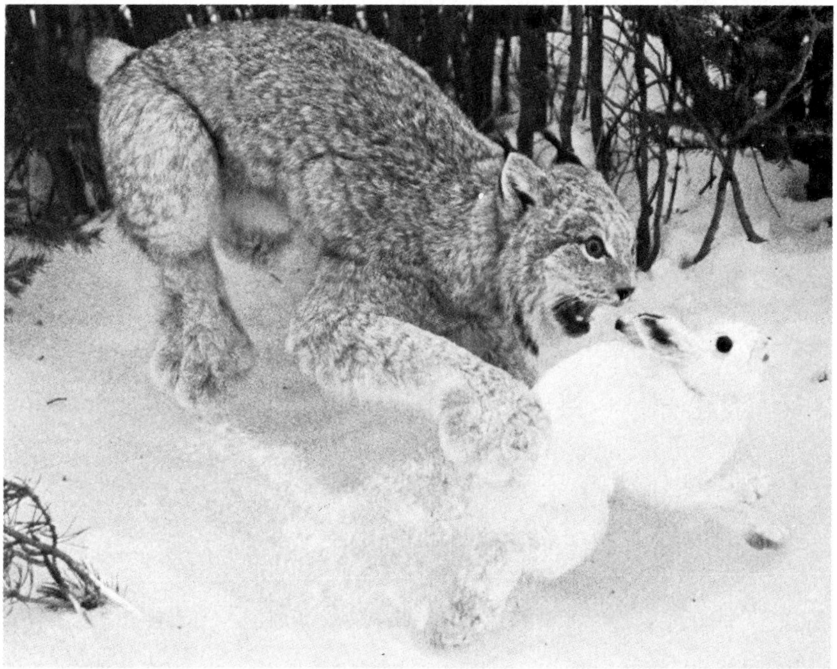

(b)

Figure 27.29
One of the rare actual long-term records of a predator-prey relationship: the lynx and the hare. (a) A rise in the number of hares is generally followed closely by a rise in the number of lynxes. When the lynxes get numerous, the hares start to decline and continue to do so until the lynx population drops low enough (starved by lack of hares) to let the hares start increasing again. (b) A Canadian lynx about to devour a snowshoe hare.

Figure 27.30
The North Island saddleback is a poor flier that feeds by hopping on the forest floor in search of insects and fallen fruits. With the introduction of new predators, it would have been preyed upon into extinction, but the New Zealand Wildlife Service has given it a refuge on an island off the coast of the mainland. Some natural sanctuaries also exist, where animals such as these are relatively safe from their predators.

prey. Nevertheless, a start is made toward studying the interactions between two species when one hunts and the other is hunted.

Other schemes than cyclic fluctuations are possible, especially when several species are involved. A predator population level can remain fairly stable if it preys on several species, catching first one and then another. It is also possible for the predator to drive the prey into extinction unless some of the prey find a safe hiding place, a *refugium,* like a medieval sanctuary, where a few may survive (Figure 27.30).

Parasitism

Parasitism is a special case of predation. In everyday usage, a parasite is small in relation to its "prey," usually called a *host.* Ordinarily one would not refer to a tiger as a parasite with a deer as its host, nor a virus as a predator with a man as its prey, but the biological relationships are similar.

One might expect therefore that the same theoretical treatment would apply to both parasite-host and hunter-and-hunted populations. But a predator kills its prey and lives on it, while a parasite may or may not kill its host. A virulent parasite, in killing its host, perishes along with it, but a "smart" parasite—that is, one that has achieved some balance with its host—does *not* kill the host and thus survives along with it.

Parasites can have a powerful effect on the host population. Chestnut trees were a main part of the eastern American forests at the beginning of this century, but they have been brought essentially to extinction by an accidentally introduced blight fungus. Seeing how effectively a parasite can destroy a host, people conceived the idea of introducing parasites intentionally in order to control such pests as insects and unwanted weedy plants. Thus the practice of *biological control* was born.

Some efforts at biological control have been successful. After prickly pear cacti were brought into Australia, they spread wildly and infested some 60 million acres of grazing land. A natural parasite was intentionally freed in the cactus country. This parasite was a moth whose larvae eat prickly pears so voraciously that they effectively keep the noxious plants under control (Figure 27.31). Another rangeland pest was a European weed known in California as Klamath weed (Figure 27.32). It was damaging to cattle that ate it, and it crowded out the forage grasses; after the importation of several species of parasitic beetles, the Klamath weed population was brought within acceptable limits.

When parasites such as those attacking Klamath weed are deliberately brought to a new place,

Figure 27.31
Biological control of an unwanted plant. (a) Prickly pears in Australia, growing so densely as to make the land unfit for cattle. (b) The same place three years after the introduction of the parasitic moth *Cactoblastis cactorum*. The prickly pears have been practically eliminated.

the introducers must first make sure that the intended victim is the *only* victim, or the parasite may infest and kill desirable plants as well. We would be dismayed to learn too late that our supposed ally against a pest also destroyed a valuable crop plant.

Mutualism

Not everything in nature is competitive. Cooperation within species, among different species, and even within cells is frequently effective in helping organisms survive.

Symbiosis, meaning "living together," is a general term that covers all aspects of biological contacts among species except those in which one member profits and the other suffers. When both members in a cooperative effort profit by the relationship, it is **"mutualism."**

Some examples of mutualism are obvious. Some fishes and invertebrates, especially shrimps, make such a regular practice of picking parasites and bits of dead tissue from other marine-dwellers that regular cleaning stations are established, where fish come to be cared for. A fish being cleaned stays still while the cleaner goes over its body, removing parasites; when the cleaner approaches the fish's head, the "client" opens its gill covers or even its mouth to let the cleaner get at the difficult spots

Figure 27.32
A branch of Klamath weed and one of the beetles imported to eat it. This beetle (arrows) eats leaves; others bore into the roots.

Figure 27.33
An example of cleaning symbiosis. A Caribbean grasby, *Petrometopon cruentatus*, lets its mouth gape open and keeps it open while the sharknose goby, *Gobiosoma evelynae*, provides a sort of dental hygiene in return for a meal. The fish being cleaned, even one as ferocious as a barracuda, usually does not eat the cleaner, even when the smaller fish swims inside the mouth of the would-be predator.

(Figure 27.33). The cleaned fish is relieved of pests and the cleaner gets lunch delivered free, in what is plainly a mutual gain.

Other symbiotic relations prove to be mutualistic only on close examination. An instructive instance is the ant-acacia symbiosis. Bull-horn acacias are small, prickly trees of Central America, whose thorns resemble a pair of cattle horns. In some species, the thorns are penetrated while soft and young by fierce stinging ants (Figure 27.34). The ants depend upon a continuous supply of young acacia tissue for food, and in return they keep at bay both herbivores who are potential predators (they sting like fury) and encroaching plants (they destroy them). When artificially deprived of their protective ants, these acacias are stunted or may die. What seemed to be parasitism turns out to be mutualism.

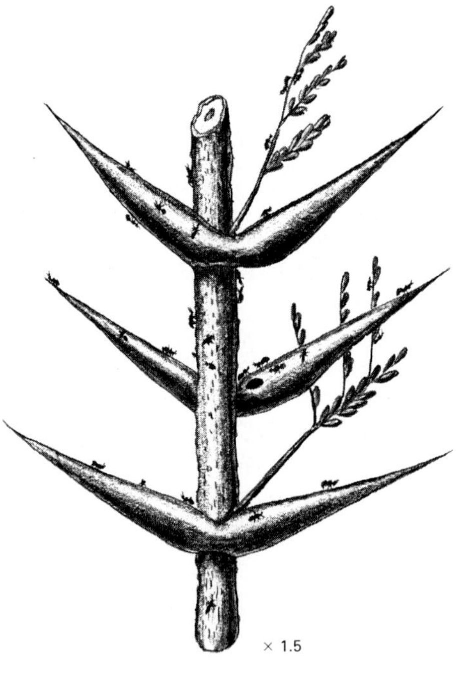

Figure 27.34
An ant–plant example of mutualism. This bull-horn acacia has hard, sharp, hollow thorns inhabited by colonies of stinging ants. The ants bore an entrance hole in each young thorn (one hole can be seen). Ants come out in force to bite any animal unwary enough to touch the plant.

BIOTIC FACTORS USUALLY DEPEND ON DENSITY; CLIMATIC FACTORS DO NOT

Some of the ecological factors described in this chapter produce their effects regardless of the number or concentration of the affected organisms. These are known as **density-independent factors.** Climatic factors are more or less density-independent. A cold region is as cold in a densely inhabited place as in a sparsely inhabited one.

Reactions among organisms are, in contrast, determined by the numbers and concentrations of organisms. Where there are few or scattered organisms, competition is slight; it increases with the increasing density of a population. A predator or a parasite is more likely to meet a victim if there is a concentration of potential victims. Competition, predation, and parasitism are thus examples of **density-dependent factors.**

Separating the two kinds of factors is not always clear. If winter drives animals into underground shelters and the number of acceptable places is limited, then cold, which is usually a density-independent factor, becomes density-dependent.

SUMMARY

1. *Ecology* is the study of the *distribution* and *abundance* of organisms and their interactions with each other and with their environment.
2. Ecology is concerned with *populations, communities, ecosystems,* and the *biosphere.*
3. Many of the findings and principles of ecology are applicable to human affairs.
4. Ecological methods include field studies, laboratory experiments, and mathematical models.
5. The growth of a theoretically unlimited population can be shown graphically by an *exponential curve.*
6. The growth of a real population, always limited by the *carrying capacity* of the environment, can be shown graphically by a *logistic curve.*
7. The population of the United States, and of the whole world, has shown a steady growth since its beginning. Its growth curve resembles that of any young population and is at present in an exponential growth phase.
8. An organism's actual dwelling place is its *habitat.* The organism's special working position within the habitat is its *niche.*
9. The factors affecting population growth are living (*biotic*) and nonliving (*abiotic*). The abiotic factors are *water, temperature, light, minerals,* and *fire.* The biotic factors are *dispersal, behavior, competition, predation, parasitism,* and *mutualism.*
10. In general, biotic factors are *density-dependent,* and climatic factors are *density-independent.*

ASK YOURSELF

1. Criticize the statement "ecology is the study of pollution."
2. Why is exponential growth of a population inevitably limited?
3. What is the "carrying capacity" of an environment?
4. What data must be available for the determination of the growth rate of a population?
5. What are some ways in which the human population of the earth might be reduced?
6. Differentiate between a "habitat" and a "niche."
7. What is a microclimate? Give examples.
8. What abiotic factors influence the numbers and distribution of organisms?
9. Explain why mineral elements used by organisms must be recycled but energy is not.
10. In what ways can atmospheric nitrogen be brought into a form usable by organisms?
11. Explain how fire can help stabilize an ecosystem.
12. What is meant by the phrase "competitive exclusion."
13. What is the difference between parasitism and mutualism?

28
The Ecology of Communities

SOME KEY POINTS

1. Communities are made up of a number of different populations. Some important considerations about communities are how energy is used, how populations are distributed, how the earth's major communities are distributed, and how communities change.
2. A total environment, including plants, animals, and climate, is an ecosystem.
3. The capture of energy from the sun and its conversion into usable food make up the primary productivity of a system.
4. Food production for human beings is limited by energy, mineral nutrients, and space.
5. In a reasonably stable ecosystem, the input and output of energy are equal.
6. In a food web, organisms are on different levels depending on what is being eaten by whom.
7. Biomes are major vegetational units of the earth; they include oceans, deserts, forests, and grasslands.
8. Changes in an ecological system produce biological successions. Changes continue until a somewhat stable climax community is developed.
9. Human beings have changed the ecological balance more than any other species has. Some human alterations have produced negative results.

A NUMBER OF INDIVIDUALS OF THE SAME species in a space makes a population; a number of populations in a space makes a **community.** Just as a population is more complex than an individual and is studied in special ways appropriate to it, so a community is more complex than a population and, in turn, requires its own special methods of study.

A community may be considered in several ways: (1) the way in which energy enters the community and is passed through its members; (2) the way in which the populations are distributed in space; (3) the geographical distribution of the earth's major communities; and (4) the way in which populations change with the passage of time. In this chapter we will consider each of these ways of studying communities.

The makeup of communities may be studied simply in order to understand the workings of the biological world, that is, for theoretical or academic reasons. Practically, however, a knowledge of communities can be useful when that information can be applied to agricultural problems or when one is generating *impact statements.* In most developed countries and especially the United States, such major new developments as shopping centers, dams, housing developments, and industrial complexes cannot be legally started until an ecological "impact statement" has been prepared by a governmental or independent agency. Such a statement should include an inventory of existing conditions (soils, plants, animals, water, drainage, minerals, etc.) in a territory under consideration and areas outside it that might be affected.

There is also an esthetic side to the studies of natural communities. If the day comes when the inhabitable parts of the earth that are not paved are covered with tanks of high-protein alga cultures, some odd people may wish to know what used to live in the Pennsylvania mountains, the Louisiana swamps, or the Oregon forests. Perhaps only the artists will care, but it will be the scientists who can tell them.

ECOSYSTEMS INCLUDE BOTH LIVING AND NONLIVING FEATURES

In describing populations and communities, we were concerned primarily with organisms (Chapter 27). Organisms, however, live in a physical world, and we must take into account more than just organisms when we study ecological relations. A total environment, including all the plants, animals, microbes, soil, and climatic factors, is an **ecosystem.** An ecosystem may be as small as a pond (Figure 28.1), or as large as a woodland covering hundreds of square kilometers.

No ecosystem—unless the whole earth is counted as one great ecosystem—is neatly bounded. Each one grades into the next, sometimes gently and slowly, sometimes with sharp demarcation. Along the boundary between two distinct ecosystems, as between a woodland and a meadow, there is a transition zone, an *ecotone.* An ecotone has some, but not all, of the features of both adjoining ecosystems, and may contain plants and animals peculiar to itself, or concentrated there. Asters and wild strawberries, for example, frequently grow profusely at the edge of a forest.

PRIMARY PRODUCTIVITY RELIES ON ENERGY FROM THE SUN

Any functional system requires energy to drive it, and the biosphere and all the ecosystems that it comprises are no exception. The energy source is the sun, and sunlight energy is fed into the system by way of photosynthetic organisms. The capture of radiant energy and its conversion into usable food constitute the **primary productivity** of a system.

The rate at which photosynthesis proceeds is regulated by several factors, all working together:

1. *Light,* being the source of energy, is of prime importance in determining the photosynthetic rate. The intensity of illumination is affected by latitude, by the amount of particulate matter in the air (smoke, smog, tiny biologically produced droplets or water mist), by the direction of slope of a surface (a south-facing slope in the northern hemisphere gets more direct sunlight than a north-facing one), and by the shadowing of taller plants which vary

Figure 28.1
An ecosystem. This small pond is a well-defined place that harbors a number of species, most of which live their entire lives here.

from species to species and from season to season. For aquatic plants, light is diminished by depth, since water itself has some absorbing power, by turbidity, and by the density of the plants themselves.

2. *Carbon dioxide* concentration is usually fairly constant because winds keep the atmosphere well mixed, and water can dissolve as much carbon dioxide as the atmosphere provides. However, plants could use a higher concentration than they actually have if they could get it. The amount of carbon dioxide is usually the limiting factor in photosynthesis in daylight, although in deserts it may be water that is limiting.

3. *Temperature* affects energy input by photosynthesis. Except for light activation of chlorophyll, photosynthesis is a chemical process, and the rate of chemical processes depends upon temperature. Naturally, a water-dependent activity such as photosynthesis cannot proceed if the temperature is low enough to freeze water. There are upper temperature limits as well, these being determined by the species of plant.

4. *Water*, as one of the raw materials of photosynthesis, is always needed. It must be present in sufficient quantity not only to allow photosynthetic buildup of food, but to fill all the other water requirements of plants.

5. *Inorganic nutrients* must be present in adequate supply. Some, such as nitrogen, potassium, phosphorus, and sulfur, are needed in relatively large amounts, and some in quite small amounts. In the sea, iron, phosphorus, and potassium are usually in short supply. These elements are never really abundant in the sea, and the small amounts present are taken up by the marine autotrophs. If these autotrophs die and sink before they are eaten by some animal, they carry the scarce elements with them to the bottom. The elements are then out of circulation, but sometimes special water movements stir them up. The most famous of the ocean currents that brings about this *upwelling*, the Humboldt current along the west coast of South America, brings the deep, cold water, rich in mineral nutrients, to the illuminated surface. There productivity is so intense that the Peruvian coast is one of the richest biological areas in the world, enabling thousands of tons of fish and fish-eating birds to thrive (Figure 28.2).

Measurement of Productivity

When productivity of land plants is measured, one common procedure is to weigh a measured area of leaf at the beginning of a day and compare that with the weight of a similar sample after a given time in light. From such data, making corrections for respi-

Figure 28.2
The biological effect of upwelling in the ocean. In this view of the Peruvian coast, the flocks of sea birds feed on millions of fish in the surface waters.

Figure 28.3
A champion energy trap: a field of sugarcane. An area like this can take from the sun as much as 25,000 kilocalories per square meter per year.

ration and transport of sugars away from the site of photosynthesis, one can estimate the total productivity of an entire ecosystem. In aquatic systems, a sample of water is illuminated, and the rate of oxygen production is measured, while a similar sample is kept in the dark for measurement of oxygen consumption. The actual productivity then is about equal to the measured oxygen output of the illuminated sample plus that which, it is assumed, is simultaneously being used in respiration. The amount of chlorophyll in such a sample can be determined by measuring the amount of light at specific wavelengths absorbed. Once these data are available, other samples can be studied merely by measuring light absorption, a simple matter with a spectrophotometer. Chlorophyll determinations are widely used as indicators of the productivity of waters.

From data on productivity, it is possible to estimate the energy-trapping abilities of various ecosystems. The usual way of expressing the figures is in kilocalories per square meter per year (kcal/m^2/yr). As would be expected, deserts are the lowest in production, with a rough average of about 500 kcal/m^2/yr. Productivity in mountain forests, arid grasslands, and the seas along continental shelves is variable but somewhat higher, ranging between 500 and 3000 kcal/m^2/yr. Moist grasslands or forests and ordinary croplands may trap between 3000 and 10,000 kcal/m^2/yr, but oceans beyond the continental shelves are relatively poor, at around 1000. The truly heavy producers are special marine ecosystems, such as brackish estuaries and coral reefs, or intensively cultivated croplands, whose totals may reach 10,000 to 20,000 kcal/m^2/yr, with exceptional places achieving up to 25,000. The most productive areas in the world are sugarcane fields (Figure 28.3), which have everything helping: dense plant cover, abundant water, full sun, adequate mineral supply, and a photosynthetic system that utilizes 4-carbon acids (C-4 plants, described in Chapter 4). Theoretical values up to 50,000 kcal/m^2/yr are possible, but not on a large scale or for protracted times.

If we know the productivity figures for different kinds of ecosystems and the areas covered by each of them, we can calculate about how much energy conversion occurs throughout the world in a year. The oceans cover about 360 million square kilometers (140 million square miles). With an average productivity of about 1000 kcal/m^2/yr, the total marine

TABLE 28.1
Estimated Gross Primary Production (annual basis) of the Biosphere and Its Distribution among Major Ecosystems

ECOSYSTEM	AREA, MILLIONS OF KM^2	GROSS PRIMARY PRODUCTIVITY, KCAL/M^2/YR	TOTAL GROSS PRODUCTION, 10^{16} KCAL/YR
Marine			
Open ocean	326.0	1,000	32.6
Coastal zones	34.0	2,000	6.8
Upwelling zones	0.4	6,000	0.2
Estuaries and reefs	2.0	20,000	4.0
Subtotal	362.4	—	43.6
Terrestrial			
Deserts and tundras	40.0	200	0.8
Grasslands and pastures	42.0	2,500	10.5
Dry forests	9.4	2,500	2.4
Northern coniferous forests	10.0	3,000	3.0
Cultivated lands with little or no energy subsidy	10.0	3,000	3.0
Moist temperate forests	4.9	8,000	3.9
Fuel-subsidized (mechanized) agriculture	4.0	12,000	4.8
Wet tropical and subtropical (broad-leaved evergreen) forests	14.7	20,000	29.0
Subtotal	135.0	—	57.0
Total for biosphere (round figures; not including ice caps)	500.0	2,000	100.0

From E. P. Odum, *Fundamentals of Ecology,* Third Edition (Philadelphia: W. B. Saunders, 1971), p. 51. Used with permission.

productivity is about 0.4 billion billion kcal/yr. Terrestrial productivity is estimated at about 0.6 billion billion kcal/yr (Table 28.1).

Limits to Productivity

The human population has a special interest in primary productivity because that is the source of all food. With an ever-increasing number of people to be fed, it is well to examine the reasons why we do not have an unlimited amount of food. The limitations can be in (1) energy, (2) materials, or (3) space.

The primary energy source for photosynthesis, sunlight, is present in abundance. Ordinarily, plants convert about one percent of the available sunlight energy into carbohydrate. The rest may be reflected, it may not reach a chloroplast, or it may be transmitted or dissipated as heat (see Figure 28.4). So far, efforts to improve the productivity of plants in natural light have been such as to require additional energy input from other sources. Closer planting to make fuller ground cover, introduction of genetically improved strains of plants, and more intensive chemical fertilizing of soil all require materials and machinery that use energy, usually petroleum. Nevertheless, one possible factor that, with better knowledge and skill, could be used more effectively is the energy of the sun.

Photosynthesis does not run on just energy, carbon dioxide, and water. It requires as well the mineral elements that necessarily go into the building of protoplasm. In nature, those elements are more or less uniformly distributed, but "more or less" is not good enough for intensive productivity. Soils are sometimes deficient in certain elements, and the sea is notably low in iron, phosphorus, and potassium. Efforts can be made to provide artificial upwelling comparable to the natural upwelling of nutrients in the Peruvian Humboldt current, but that, too, takes energy. It also takes energy to process and distribute the nitrates, phosphates, and potassium in chemical fertilizers required by high-intensity agriculture. Answers to the problems of material supply are complex and still not available.

Space for growing things is limited. Until the present century, whenever land became unfit for cropping as a result of mineral depletion or erosion, new land was opened. But most of the earth's usable land is already under cultivation and new land is rapidly running out. Even now, much inferior land is being cultivated, and the more it is used, the poorer it becomes. Some optimists have looked to such unused tropical expanses as the Amazon basin. However, not only are these areas limited in extent, but tropical soils deteriorate more rapidly than temperate soils do. Opinions and predictions vary, but a general belief is growing that with the ever-increasing demands of human population, agriculture will be harder and harder pressed to produce food for all the people. One hopeful group used to think that the vast expanses of the oceans could be made to provide practically unlimited amounts of food, but with more technically effective harvesting methods and more intensive fishing, the waters of the earth have been yielding less, not more.

Much hope was generated by the Green Revolution, which produced new high-yielding varieties of grains, especially rice. It also produced new combinations of amino acids, which promised to overcome some of the nutritional deficiencies of improperly fed people. But the Green Revolution, too, has drawbacks. It demands a high-energy input in tillage and fertilizer, which the less-developed countries do not have and cannot afford. And it depends on *monoculture,* the use of a single crop, with the built-in danger of any monoculture: susceptibility to unexpected disease. The 1970 corn crop in the United States, for example, was severely reduced by a fungus invasion, and the classic example is the Irish potato, whose destruction in Ireland in the late 1840s produced effects far beyond the boundaries of the country (Chapter 9).

With the birthrate continuing high, especially in less-developed countries, and the death rate dropping, the population increases and increases. Although food production is increasing too, it is lagging behind—a frightening realization for those who study the facts. It behooves the students of today, who will have to deal with tomorrow's already visible problems, to acquaint themselves with the reality of too many people and not enough food. We are too definitely a part of the web of life to escape.

ENERGY FLOW IN ECOSYSTEMS

Energy input for the biosphere by photosynthesis was described in Chapter 4. The concern up to now has been with the mechanisms of energy trapping and with the buildup of food in plants. But the biosphere contains many nonphotosynthetic organisms: fungi, bacteria, and animals. These, in ecological terms, are the *consumers,* as contrasted with the photosynthetic *producers.*

Primary Productivity

In an ecosystem, the main, and for practical purposes, the only energy input is sunlight. Most of the sun's energy—as much as 98 percent of the total—is transformed into heat, and this portion therefore does not contribute to an ecosystem's primary productivity (Figure 28.4). The heat energy is eventually lost by radiation back into space, but it serves more than one function. It maintains the temperature of soil and water; it causes differences in air temperature, which not only result in air currents

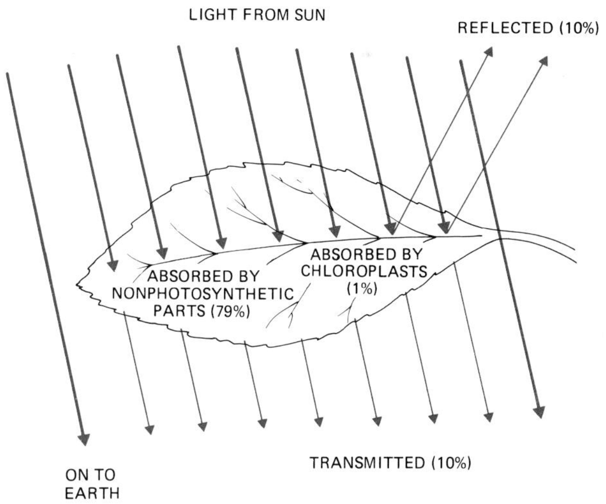

Figure 28.4
Little of the energy from the sun is trapped in photosynthetic productivity. Most is reflected back into space, is absorbed by nongreen parts of plants, passes right through the leaves, or fails to strike any plant part. Only 1 or 2 percent of the light energy is used by chloroplasts.

but affect the evaporation and precipitation of water; and it keeps the green plants warm enough to be functional.

Energy Transfers

The energy trapped by photosynthesis has several possible fates. Some of it is used by the same plants that trapped it, as an energy source for the non-green parts and for all living parts of plants during darkness. The rest of the energy is passed on when the green plants are eaten by herbivores (insects or other plant-eating animals) or when the plants die and are decomposed by fungi, bacteria, or *detritus-eaters.* (Detritus means particles of plant and animal material.) Eventually all the energy that came into the system as light is dissipated from the system as heat, but meanwhile, as it flows through the system, the energy is used to maintain the life of all the individuals within it.

Energy Balance

A reasonably stable ecosystem achieves a state of **dynamic equilibrium,** which means that over a period of time the energy input will equal the outflow. Such a system has been compared to a water tank with a tap for inflow and an overflow drain. The rates of inflow and outflow are not constant, one sometimes exceeding the other, but in general the two rates keep an average level in the tank. In an ecosystem, input is faster than output during a summer day, but during darkness or in cold weather, when photosynthesis slows down or stops while respiration continues, output is faster than input. As long as the dynamic equilibrium is maintained, the ecosystem remains stable.

If equilibrium is disturbed, the ecosystem changes. The energy input may exceed the outflow if a new species of more-efficient autotrophs invades the system or a plant-eating insect species is eradicated. In such situations the total amount of organic material will increase and will continue to increase until a new equilibrium is reached. When that happens the system will be different from what it was. Another possibility is that primary productivity may drop, so that effective input is lower than respiratory outflow. A swamp may be drained, a fungus may almost wipe out the major producer, or a whole climate may be altered. As with cash flow, so with energy flow: Outgo cannot continue indefinitely to exceed income. In an ecosystem, the respiratory release of energy can continue only at the expense of stored food, and when that is reduced, the consumers will be reduced, and the reduction of consumers will continue until a new equilibrium is achieved. The new equilibrium will be at a new level, and the ecosystem will have changed.

The rate of energy flow is an indication of the activity of an ecosystem. A highly productive system, such as a warm tidal marsh, with a primary productivity of 10,000 kcal/m^2/yr, will support a large number of active consumers. At the same time, it will have a large amount of organic material, the **biomass.** In contrast, a desert, with a primary productivity of less than 500 kcal/m^2/yr, will support few consumers, and will have a small biomass. In the real world, however, the relation between energy flow and biomass is not necessarily proportional. A northern forest, for example, may have a large biomass, but relatively low productivity because of the low temperature.

Trophic Levels

The primary producers make up only a part of the biomass of an ecosystem. They may be eaten by insects or grazing animals or seed-eating birds, which in turn may be eaten by carnivorous animals. Carnivorous animals themselves may then be eaten by still others. The distance of a consumer from the primary producer, in terms of the number of passages of food through organisms, determines the **trophic level** the consumer occupies. A mouse that eats seeds is getting its energy directly from photosynthetic plants. It is said to be a *primary consumer,* with the plants as producers. A snake that eats mice is a *secondary consumer,* and a hawk that eats snakes is a *tertiary consumer.* If a fox should eat a hawk, the fox would be a *fourth-level consumer.* These levels are not fixed, as can be seen on examination of human feeding habits. When people eat cereals, they are primary consumers, but when they eat beef, which fed on grass, they are secondary consumers. Or a human may eat a fish, which lived on smaller fish, which fed on minute crustaceans, which fed on algae, and in those circumstances the human is a fourth-level consumer.

Feeding patterns in nature are further complicated by a variety of possibilities. A plant may die and be decomposed by fungi and bacteria. A seed-eating bird may meet the same fate, or a parasitic fungus may destroy a plant-eating insect. The result is an intricate series of possible events. A **food web** is the sequence of plant-herbivore-carnivore feeding in which energy and materials pass through a number of organisms before the energy is finally lost to the system as heat (Figure 28.5).

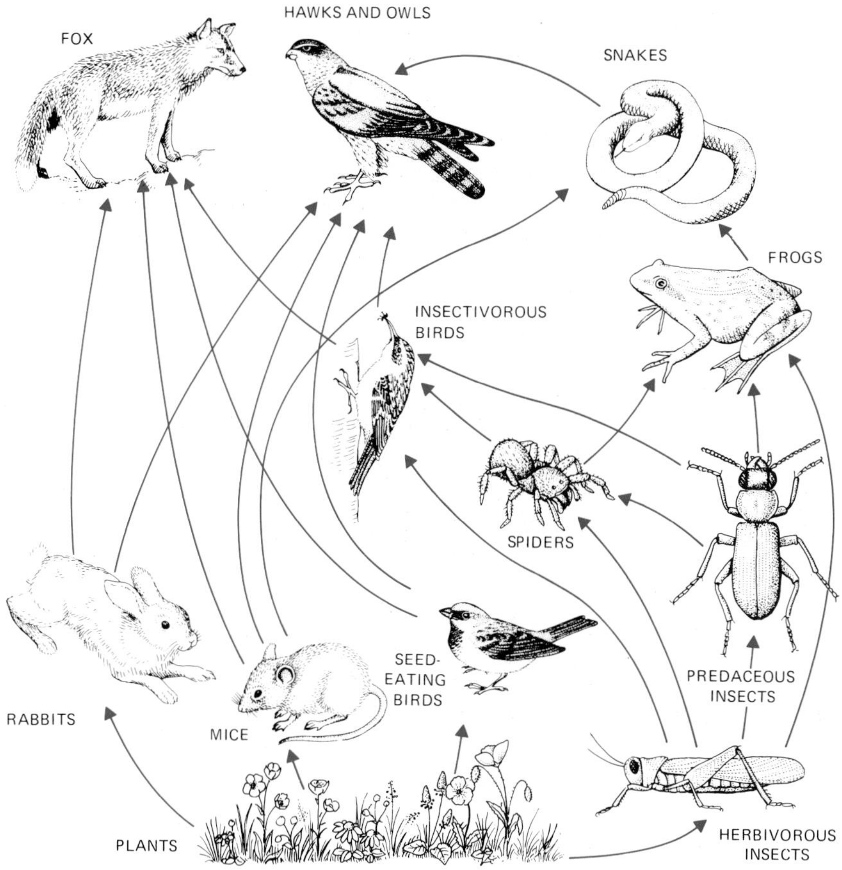

Figure 28.5
A simplified scheme of a *food*—that is, energy—*web.* Photosynthetic plants build up reserves adequate for their own needs, with enough extra to feed the herbivores, exemplified here by rabbits, mice, seed-eating birds, and a host of plant-eating insects. (The arrows indicate what is eaten by whom.) Intermediate feeders prey on some of the herbivores, with spiders, various amphibians, and insect-eating insects and birds living one step away from the primary producers. Large carnivores, such as birds of prey and some mammals (a fox in this array), eat the smaller carnivores, in addition to some of the herbivores. In nature, the transfer of energy is usually even more complex than this drawing suggests.

Thermodynamics of an Ecosystem

The only closed ecosystems are artificial ones, such as a balanced aquarium, in which the materials are repeatedly recycled by plants and animals and the only input is light energy. In nature, ecosystems are open to changes, with inflow and outflow of water, inwash and outwash of soil, wind blowing, immigration and emigration of organisms, and the carrying in and out of materials by animals. Still, for purposes of investigation and analysis, it is acceptable to treat a single ecosystem, such as a pond or a woodlot, as a unit for study.

Transfer of energy in ecosystems, as in single organisms, conforms to the laws of thermodynamics. Recall that these laws state (1) that energy is neither created nor destroyed, and (2) that with any energy conversion, some of the energy is lost to the system in the form of heat. In an ecosystem, all the entering energy in the form of sunlight can be accounted for either as chemical energy in food or as heat that simply warms the place, and all the energy is eventually dissipated from the ecosystem back into space. That is in accordance with the first law.

The second law operates during every conversion of energy that occurs within the system, including the original photosynthetic buildup of food, the respiratory activity of the green plants themselves, and the movements and metabolism of every consumer at every trophic level. The energy loss has practical consequences. Some sunlight energy is lost to the system as heat even as it is being partially converted to chemical energy in food; more is lost in the general life processes of the green plant; still more is lost with every physical and biological action of any herbivore that eats the plant, and of any carnivore that eats the herbivore. At every trophic level, less and less of the energy originally stored photosynthetically is available as food.

Energy Pyramids

If a food web in an ecosystem is known, the amounts of energy required for the lives of the various inhabitants can be calculated. The resulting figures give an indication of what amounts are available at different trophic levels, and they can be used to construct **energy pyramids.** The base of a pyra-

PLANTS THAT EAT ANIMALS

Some habitats offer special challenges to survival, perhaps by having too much or too little of some factor. One such habitat is a bog in which the essential element nitrogen is in short supply. Where such a shortage occurs, some organisms find means of survival anyway.

Plants, which are the primary food-producers, are typically eaten by animals, but in places where nitrogen is limited, several genera and dozens of species of plants have evolved methods of reversing the old order. They can capture and digest their own meat. They manage this feat with traps like those used by human trappers: gums, pits, and springs.

The simplest traps are those of sticky plants. The glistening hairs on sundew leaves are tipped with adhesive droplets, to which wandering insects are effectively glued (see photos). When an insect is caught, nearby hairs bend in to meet it, secrete protein-digesting enzymes, and absorb nourishment for the plant. The shining leaves of butterworts can roll slowly up, enclosing insects captured by sticky hairs.

A young grasshopper caught on a sundew plant's tentacles is doomed to a slow death.

A small insect is trapped on the sticky leaf of a sundew plant. The leaf rolls up, and the digestive glands of the plant remove the fluids from the insect's body. When the insect is drained, the leaf opens again.

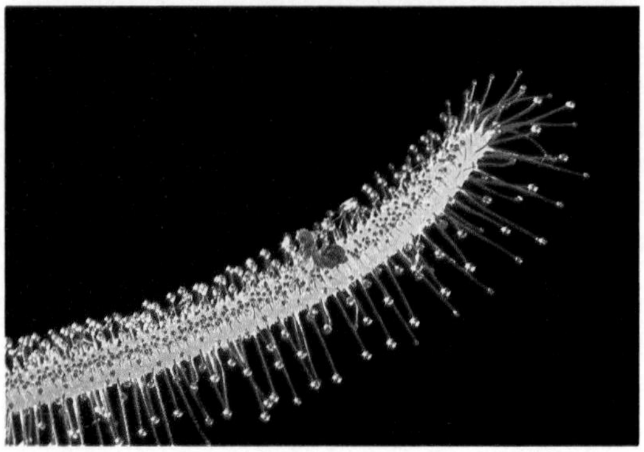

An unsuspecting insect triggers the closing mechanism of a Venus flytrap, and is trapped as the toothed edges snap shut. The complete digestion of an insect may take 10 to 35 days.

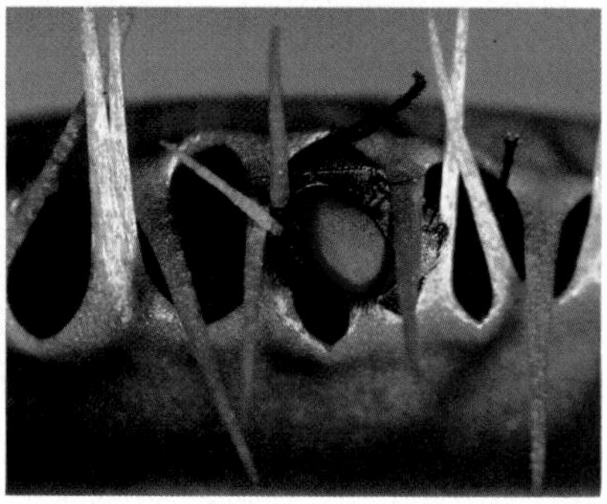

Bladderworts have underwater traps that suck in tiny swimming animals by allowing a rush of water into the bladders when a trigger is tripped.

The pitcher plants have leaves modified into cups that catch rain, and they are provided with downward pointing hairs that make it easy for an insect to go down but hard to climb up. Down in the leaf cups, insects drown and are digested. The pitcher plants in the eastern United States are *Sarracenias;* those in the west are *Darlingtonias.* The most spectacular cups are those of the oriental *Nepenthes* (meaning "without care"), named for the wine cup that Homer tells us Helen used when she spiked wine to stupefy men. In addition to the insect trap, one species, *Nepenthes bicalcarata,* has at the cup opening a pair of downward-pointing spines. If a tarsier (a small, primitive primate) tries to rob the plant of its prey, the spines prick it in the back of its neck. Leaves of pitcher plants usually have a number of dead insects floating in the contained water, but some species of ants, like people living in constant danger on the side of an active volcano, regularly inhabit the *Nepenthes* stems, hollowing out homes there.

The most astonishing of the insect-catchers is the Venus flytrap, native to the coastal region of North and South Carolina (see photos). This famous plant has hinged leaves with toothed edges and sensitive hairs on the upper surfaces. When an insect touches a hair twice (only once is not enough), the blade neatly hinges shut like the jaws of a dredge. It takes about a second to cage a victim. The mechanism is incompletely understood, but it involves changes in turgor of the hinge cells. Some people who are in flytrap country for the first time walk with careful steps and wary eyes until they actually see the tiny plants, which are capable of holding ants but nothing much larger.

In all this, the most impressive fact is the multiplicity of ways in which plants have solved a single problem. The human species, for all its intelligence, has done no better in providing itself with protein.

mid represents the energy in primary producers, and the upper levels represent the proportion of energy available in successive trophic levels.

The flow of energy in a large freshwater spring in Silver Springs, Florida, can serve as an example (Figure 28.6). The main primary producers are freshwater eelgrass and algae; the saprobes, which use dead matter, are fungi, bacteria, and detritus-eaters; the herbivores are insects, snails, and turtles; the third trophic level contains such carnivores as insects and small fishes; and the top carnivores are larger fishes such as basses. Figure 28.6 shows the small amount of energy at the top of the pyramid, as compared with the amount captured by the primary producers. Note, too, the rapid diminution of energy retained at succeeding trophic levels.

The herbivores get for their own use only about 10 percent of the energy available from the green plants. The carnivores that eat herbivores retain only about 10 percent of that energy. The top carnivores similarly retain 10 percent of the energy of what they eat. Thus the top carnivores, three levels removed from the primary producers, get only one-thousandth of what was originally available (one-tenth of one-tenth of one-tenth).

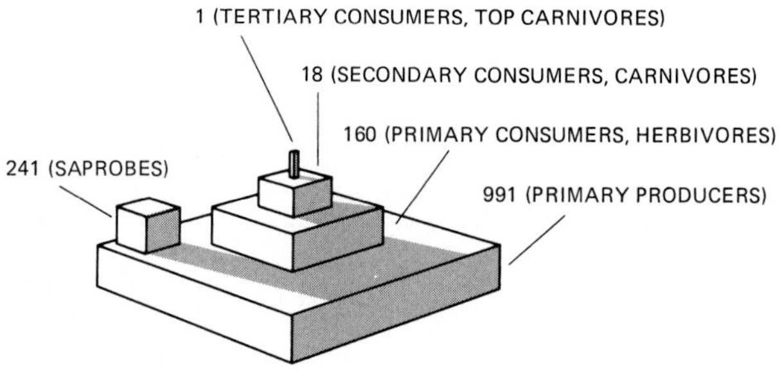

Figure 28.6
An energy pyramid, based on productivity and consumption in an aquatic ecosystem. The figures, in cal/m^2, show how little energy (1 cal) is harvested by the main carnivores in comparison with the amount (nearly 1000 cal) produced by the green plants.

Human Beings as Energy Consumers

The energy loss during passage to higher trophic levels is important in the consideration of human populations and human food supply. For example, if we eat beef that was fed with alfalfa fodder, we obtain only 1 kcal from 143 kcal of beef (Figure 28.7). The animal required 1795 kcal to produce those 143 kcal. Much of the energy that beef cattle obtain from plants is spent keeping warm and walking. (Pigs are more efficient than cattle in producing protein, and chickens are still more efficient.) A greater human use of the energy from primary production can be made if the human being is the herbivore, occupying the second trophic level rather than the third. Most Americans are meat-eaters from habit and personal choice, but the second law of thermodynamics suggests, at least for ecological efficiency, that we should be mostly vegetarians.

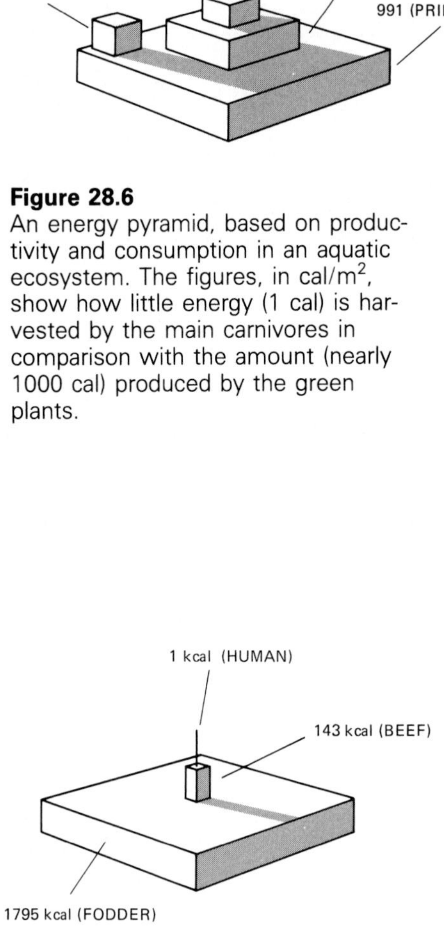

Figure 28.7
A pyramid illustrating the use of beef, fed on alfalfa, for human consumption. Plants must produce enough to store 1795 kcal to provide, by way of the intermediate beef, a single kcal for humans.

A number of schemes have been devised in an effort to shorten the food chain for human food. Some of these schemes are the cultivation of algae for food, the breeding of crop plants that are more nearly completely usable (we eat less than 30 percent of a wheat plant), the harvesting of yeasts grown on carbohydrates from hydrolyzed cellulose (of which millions of tons are wasted annually), and even the growing of bacteria on petroleum for their protein. By moving only one step farther from the primary producers, we might cultivate shellfish, such as oysters and mussels, in prepared beds or turn our attention to grazing animals that are now not generally eaten, such as some species of antelope. A number of other novel methods of producing more food and producing it more efficiently have been suggested, but so far they are prohibitively expensive, or they present technical prob-

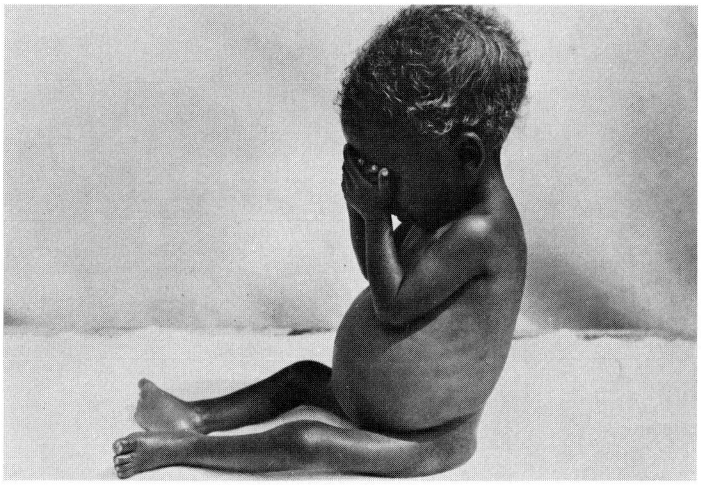

Figure 28.8
Kwashiorkor. This child in Kenya in a time of famine was suffering from extreme protein deficiency, and the results are striking. Many children in regions where the main nourishment is carbohydrate show milder but still weakening symptoms of kwashiorkor.

lems, or they yield such unaccustomed products that people would rather starve than eat them.

The problem of feeding the people of the world becomes increasingly enormous. As long ago as 1967, an estimated 1.5 billion people were undernourished (simply had insufficient quantity of food) or malnourished (had food lacking in such essential components as protein and vitamins); and the numbers grow. Inadequate intake of calories results in *marasmus,* a disease in which children have spindly legs, wrinkled, old-looking faces, sluggish muscles, and no resistance to respiratory infections. Deficiency of protein, regardless of how much food is eaten, produces *kwashiorkor* (Figure 28.8). This malady, estimated to afflict as many as 25 percent of African children, is from a West African word meaning "sickness a child has when another child is born." The shrunken muscles, pot belly, skin sores, bleached and falling hair, and intestinal upsets begin when a child is weaned from its mother on the arrival of another child or is put on an artificial diet because nursing is becoming unfashionable. It can be cured, sometimes with dramatic speed, by the inclusion in the child's diet of sufficient protein containing the proper balance of amino acids.

In the United States, malnutrition is mainly a matter of diet preferences, but it can and does occur, especially because of inadequate intake of vitamins. One result of the recognition of American malnutrition has been the spread of the nonsense phrase "empty calories." A calorie is a calorie, whether it is included in a sugar lump or a spoonful of the best-balanced food in the world. True, one could obtain the daily 2500 Calories by eating pure starch and would soon die of starvation because protein and vitamins are lacking. But one could also starve on a strict diet of vitamin pills and protein concentrate. For proper nutrition, both energy *and* specific kinds of molecules are essential.

TO SUM UP ENERGY FLOW

1. In ecological terms, nonphotosynthetic organisms are *consumers,* and organisms capable of photosynthesis are *producers.*
2. The main energy input in an ecosystem is sunlight, and even the "waste" heat that does not contribute to an ecosystem's primary productivity serves many useful functions.
3. An ecosystem achieves *dynamic equilibrium* when its energy input equals the outflow.
4. The activity of an ecosystem is usually in direct proportion to the rate of energy flow.
5. An animal's *trophic level* in a community is determined by what it eats. An herbivore is a *primary* consumer, an animal that eats herbivores is a *secondary* consumer, and a carnivore that eats other carnivores is a *tertiary* consumer.
6. A *food web* is a sequence of plant-herbivore-carnivore feeding in which energy and materials pass through a number of organisms before the energy is finally lost as heat.
7. Transfer of energy in ecosystems conforms to the laws of thermodynamics: energy is neither created nor destroyed, and some energy is lost to the system as heat.
8. The amounts of energy available at different trophic levels can be shown by constructing *energy pyramids.*

Figure 28.9
Horizontal layers in a forest community. The large trees form the canopy and receive most of the available light. The small trees and shrubs form an understory. Light is diminished as it filters through the layers of leaves, until as little as 1 percent may reach the lowest levels.

(a) × 0.2

(b) × 0.2

Figure 28.10
Plants of the forest floor. (a) Pink lady's slipper (*Cypripedium acaule*) takes advantage of early spring to make food while trees are leafless. (b) Cinnamon fern (*Osmunda cinnamomea*) thrives on dim light.

THE DISTRIBUTION OF ORGANISMS IN SPACE: COMMUNITY STRUCTURE IN A TEMPERATE DECIDUOUS FOREST

Up to now we have been concerned with a number of physical and biological factors that affect the growth and general distribution of organisms and populations. Next we look at the spatial distribution of some selected organisms in one particular kind of community, a temperate deciduous forest.

On land, organisms are closely tied to the soil, plants almost exclusively so. For this reason, the height that a plant achieves may be important to its well-being, for if it grows tall it avoids being overshadowed. In recognition of this fact, we can describe the way in which plants arrange themselves in a community with respect to height. The tallest trees make a high covering, the **canopy,** which gets the most sunlight. Under the canopy, partly shaded, is an **understory** of smaller trees or large shrubs. Finally, close to the forest floor is a **carpet** of herbs, grasses, and small shrubs. These layers are not neatly arranged, but they do in general follow a recognizable pattern (Figure 28.9).

Light, being the energy source for any community, is a controlling factor in the arrangement of the various species. The tallest trees forming the canopy do indeed receive full sunlight, but they must pay for the advantage by having to produce a massive supporting structure and by being more likely to suffer storm damage. The understory is less massive and less dense, and it is illuminated by such light as filters through the leaves above. In a representative forest, if (for example) 5000 footcandles of light are available at the top of the canopy, the understory can catch only about one-tenth of that, or 500 footcandles. The lowest plants get only about one-tenth of that, as low as 50 footcandles.

The plants on the forest floor must have physiological adaptations to low light intensities, or be restricted to patches of light provided by occasional breaks in the canopy, or do their main photosynthetic work early in the growing season before the leaves appear on the trees above (Figure 28.10).

Animals, too, occupy special levels (Figure 28.11). Some (earthworms, springtails, some beetles) live mostly below the soil surface. Many insects and ground-feeding and ground-nesting birds such as oven birds stay on or close to the ground. Warblers of various species have height preferences, some feeding in low levels, others in upper levels. Many owls stay in low branches; crows choose higher ones.

Organisms are distributed horizontally as well as vertically. Complete regularity of surface distribution is rare, and when it does occur it is almost always the result of human regulation, as in an orchard. Animals that establish territories for feeding and nesting may be fairly evenly spaced from one another. Desert shrubs, living where water is scarce, are also sometimes distributed with some regularity. Otherwise, organisms tend to be scattered either randomly and individually, or in clumps. The clumps themselves may be randomly distributed, or they may be regulated by such environmental factors as slope, soil types, water supply, or the presence of other organisms.

(a)

(b)

(c)

Figure 28.11
Animals that inhabit chosen elevations. (a) Owls prefer low perches from which they can swoop on prey. (b) Rattlesnakes remain on the ground. (c) Bobwhite quail feed on the ground and walk frequently, but fly well.

GEOGRAPHICAL DISTRIBUTION OF THE EARTH'S MAJOR COMMUNITIES: THE BIOMES

The great biological units on earth are called **biomes.** The kinds of plants in a biome are determined by the climate, latitude, and geological history, and to some extent by the animals of a region. The kinds of animals are largely determined by all the above, but especially by the plant life. The main biomes are the aquatic ones, forests, grasslands, deserts, and tundra, and these categories may be subdivided according to special regions. Aquatic systems may occur as open ocean, shores and estuaries, or swamps and marshes; forests may be temperate deciduous, temperate coniferous, northern coniferous, tropical and seasonal, or tropical rain forests. Similarly, the other large vegetational types can also be subdivided.

When such broad areas as biomes are described, much detail must be omitted. No biome has a continuous, homogeneous vegetation cover. Thousands of local variations in climate, soil, elevation, and physical disturbances cause changes in the types of plants and animals. When a biome is delimited and described, the weaknesses of generalities become apparent to anyone who travels through a region.

Another difficulty with treating biomes as discrete entities is that they do not end abruptly. Where one meets another, there are usually transition zones or ecotones which are likely to contain special plants and animals. Despite these reservations, we can gain some idea of the distribution of organisms on earth and the forces that determine their survival.

Here we shall sample only eight of the major biomes, those that cover most of the earth's surface, describing the kinds of plants and animals found and special climatic features.

The Ocean Biome

Traditionally, biomes are thought of as terrestrial, but the oceans can also be considered one great biome (Figure 28.12). Their very immensity demands attention; they cover some 70 percent of the earth's surface. Despite the occasional great depth—down to 11,000 meters (36,000 ft) in the Marianas Trench—animals live in all parts. The endless mixing of the waters has resulted in a remarkably constant proportion of various salts. Where large rivers enter the sea, the salt water is diluted by fresh water, but even then it is the total concentration that is changed, not the relative amounts of the various salts. (In salty inland seas, not connected with the ocean, departures from the standard seasalt proportions can be found.) For this reason , we may consider the oceans a single biome, at the same time recognizing that the organisms in them are so numerous and so varied that we can give them only a superficial description here.

Figure 28.12
One aspect of the ocean biome: a wave breaking on a rocky coast.

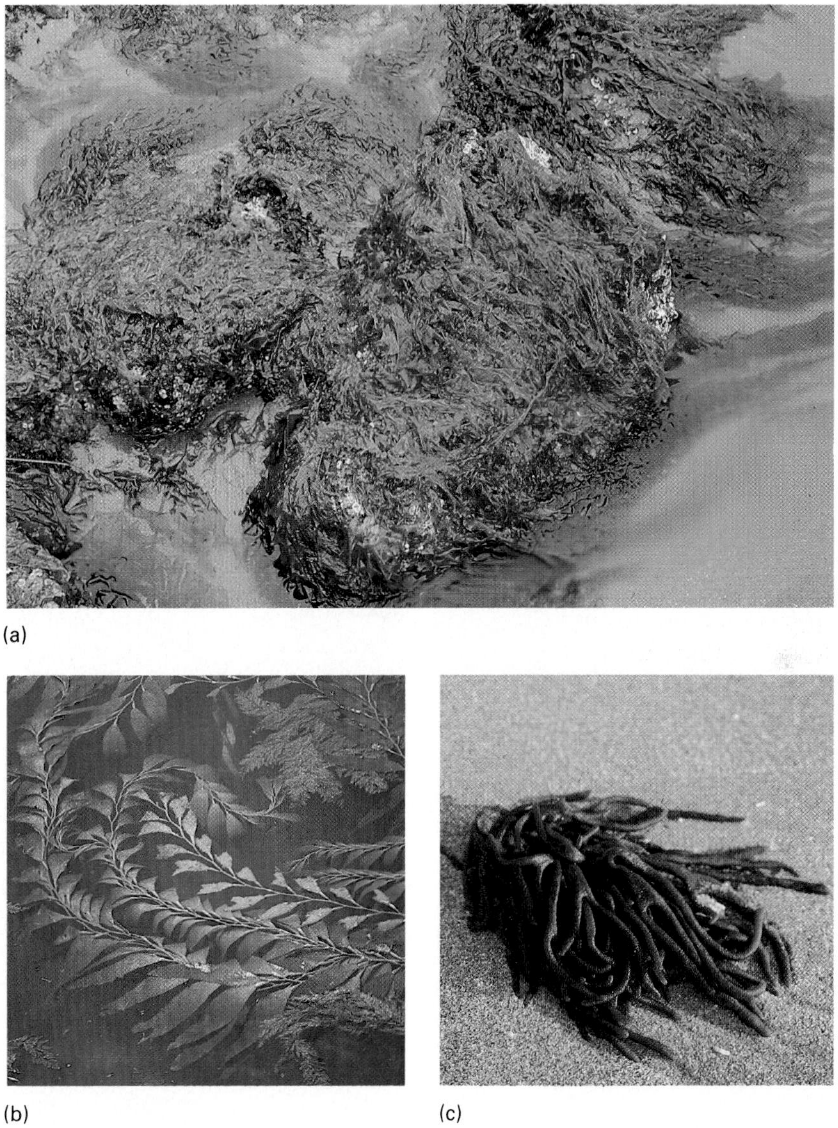

Figure 28.13
Seaweed at the ocean's edge. (a) Sea lettuce. (b) Giant kelp (*Macrocystis pyriteva*) in California. (c) A green seaweed (*Codium*) stranded on a beach.

The oceans present the sea-dwellers with some variables that other dwellers never meet. The differences in pressure, from the surface to the deeps, are greater than anything terrestrial organisms ever have to endure. Some invertebrates in the deepest parts of the **benthic zone,** or *benthos,* sustain pressures as great as 1000 atmospheres (about 1000 kg/cm^2 or 15,000 $lb/in.^2$). Animals from the deeps do not survive when brought to the surface.

In some ways the sea is a less variable place to live than the land. The alkalinity of seawater is buffered at a fairly constant pH 8.2, and the temperature seldom goes above 27°C (81°F) or below 1°C (34°F) (seawater with 3.5 percent salinity freezes at about −1.9°C). Compare that with a high of 58°C (136°F) in the Libyan desert and a low of −68°C (−90°F) in Siberia.

Along the shores, however, some special variations occur. Inhabitants of the **littoral zone** along the edges of the oceans are affected by tides that rise and fall about twice a day. The tides change the water level from a few centimeters to as much as 16 meters (50 ft), thus inundating or exposing some organisms repeatedly. In addition, outflow of fresh water from land causes fluctuations in total salinity in these coastal areas.

Plants in the sea are limited to the **euphotic zone** (Greek, "good light"), the upper levels into which light can penetrate (Figure 28.13). The larger plants are mostly seaweeds along shores (but see p. 176 for deviation from this in the Sargasso Sea). However, microscopic algae, the *phytoplankton* (Greek, "plant wanderers") are distributed across the entire surface, and they are especially concen-

Figure 28.14
Representative animals of the sea. (a) A fiddler crab, inhabitant of marshy estuaries. (b) A sea anemone in a tide pool. (c) Sea lions basking on a beach. (d) A starfish attacking a mussel. (e) An octopus.

trated in colder waters and in upwellings such as the Humboldt current. Animals in oceans are as different from land animals as the two sets of living conditions would lead one to expect (Figure 28.14). Whereas insects are the common terrestrial invertebrates, crustaceans are the marine ones, and whereas mammals and birds are the main terrestrial vertebrates, fishes are the marine ones. Many of the familiar commercial fishes are taken from the euphotic zone: herring, mackerel, tuna, and sardines. In deeper water, the fishes are adapted to cold, darkness, and pressure, and they have evolved in ways that appear strange to people who are not accustomed to them. They may have long thin tails, huge eyes or practically none, glowing spots, or gaping mouths, and the names given them reflect the feelings of the fishers: gulpers, swallowers, hatchetfish, and viperfish. Some fish, echinoderms, and sponges live even in the deepest parts of the sea that have been dredged, where it is perpetually dark and cold. In view of recent geological evidence of the activity of ocean bottoms, it is possible that some of the benthic species have been there longer than the sea floor itself.

Tundra

Stretching all across northern America, Europe, and Asia just below the Arctic region is a flat, treeless expanse of territory, marked by long, cold winters and short, cool summers; this is the **tundra** (Russian, "bare hill") (Figure 28.15). The main plants are short grasses, sedges, and lichens, with a mixture of dwarf woody willows. The surface thaws down to a depth of a few centimeters each summer, enough to allow the plants a meager living during only two months of growing weather. Below that, hundreds of meters deep in some places, is the **permafrost,** the permanently frozen ground. In spite of the sparse vegetation, animals abound, especially insects. Mosquitoes are traditionally the scourge of the Arctic. A number of migratory birds nest in the tundra (curlews, for example); and Arctic hares, foxes, bears, lemmings, musk-oxen, reindeer, and caribou can exist on the available food (Figure 28.16). Isolated high-altitude plateaus on more southerly mountains have such tundralike plants and climates that they are frequently called tundra. The southern "tundras," however, have longer days than their Arctic counterparts, and there is no permafrost, although the vegetation in both places is similar.

Figure 28.15
An expanse of tundra in the vicinity of Hudson Bay, with lichen-covered rocks and sparse, low-growing grasses and shrubs.

(a)

(b)

(c)

Figure 28.16
Organisms of the tundra. (a) A caribou bull in late August. (b) These new plants coming through the snow demonstrate the rapid life cycle during the short growing season in the tundra. (c) *Bryanthus* sheltered by rocks.

Figure 28.17
A pond in the taiga region. The dense stand of spruces shelters a variety of fur-bearing animals.

(a) (b) (c)

Figure 28.18
Organisms of the taiga. (a) Larch (*Larix*), a deciduous conifer. (b) A marten in a spruce tree. (c) A moose wades through one of the cold ponds typical of a taiga region.

Northern Coniferous Forests

South of the tundra and reaching across northern Siberia and northern Canada, the northern coniferous forests, also called **taiga** (TIE-gah), are enormous stretches of spruce, pine, fir, and larch (Figure 28.17). Except for the larch, these trees are evergreens, and broad-leaved plants are scarce. The climate is cold, the growing season short, and the land poorly drained. The result is a boggy terrain providing ample breeding ground for mosquitoes. Productivity is relatively high for such a cool region, and trees can grow well enough to provide the bulk of the world's timber. There is also food for such large animals as moose, which can wade through the wet, mossy bogs called *muskeg,* and for some of the world's most valuable fur-bearing animals: marten, mink, beaver, and fisher (Figure 28.18). Other taiga-dwellers are squirrels, lynx, snowshoe rabbits, and such birds as ptarmigans and pine siskins.

Temperate Deciduous Forests

The striking feature of temperate deciduous forests is the change from winter's leafless, dead-looking aspect to summer's abundant green (Figure 28.19). Such forests grow where winters are cold (but not with Arctic cold) and summers hot (but not with tropical heat) and where rainfall is ample (90–200 cm/yr) and distributed fairly evenly throughout the year. In contrast to taiga and tundra, which are circumpolar, the temperate deciduous forests are discontinuous. There is one large area in eastern North America, one in western Europe, and one in eastern Asia, and there are some small ones on islands.

These forests are not of uniform biotic composition. Oak trees are indeed dominant in many of the regions, and members of the oak family (including beeches) are widespread. Even in the United States, however, differences can be found between the New England region with its maples, the north central states with their lindens, the east central states with their tuliptrees and gums, and the mixtures in the Appalachian Mountains. Scattered through the whole eastern United States are stands of conifers, mixed with the deciduous trees.

In the United States, forest animals include such mammals as deer, elk (recently reintroduced),

Figure 28.19
Interior of a temperate deciduous forest such as once covered most of the eastern United States. There is a mixture of oaks, hickories, maples, tulip trees and various gum trees. The understory of small trees (dogwoods, hawthorns, sassafras) is vigorous, but few plants live on the forest floor.

(a)

(b)

Figure 28.20
Representative animals of the temperate deciduous forest. (a) Elk, (b) flying squirrel, (c) cottontail rabbit, and (d) raccoon.

(c)

(d)

opossums, raccoons, foxes, weasels, and squirrels (Figure 28.20). Birds include hawks, owls, turkeys, turkey vultures, and many warblers. Reptiles are mostly snakes and turtles. Thousands of species of insects and other invertebrates live in the varied habitats that a forest provides.

It is notable that the extent of the temperate deciduous forests coincides with some of the most intense human activities. These areas are the United States east coast, the European west coast, and China. This means that these forests, considered as a biome, have been changed so radically from their original condition as to be, in many places, unrecognizable (see Figure 28.19).

Temperate Grasslands

Since the areas now or formerly occupied by temperate grasslands furnish most of the world's grain and cattle, they are of critical importance to human economy. They occupy a large part of the inhabited earth: portions of the American middle west, eastern Europe and all across Asia north of the Himalayan Mountains, the southeastern coast of South America, great stretches of eastern Africa, and parts of Australia (Figure 28.21). Grass is the dominant vegetation in places where there is too little rainfall to allow forest development, that is, from 25 to 75 centimeters (10–30 in.) a year. Some grasslands are actually moist enough to support trees, but trees cannot establish themselves. The sod may be so

Figure 28.21
Bison grazing on the American short-grass prairie.

thick with roots that tree seedlings cannot get started, or recurrent fires, which do not kill grass, kill any tree seedlings that start. In such places, the dominance of grass is determined by fire, not by climate (see page 554). It is now thought that the eastern prairie grasslands of the United States were maintained by fire. We say "were" instead of "are" because such grasslands are gone except for a few "museum specimens" preserved for study and curiosity. Cattle, wheat, corn, and construction have taken the rest.

In the American grasslands, the eastern part was "high-grass" prairie, and the western part "short-grass" prairie. The high-grass prairie is now the corn belt, but before it was converted to agriculture, there was a stand of grass, two meters (6 ft) tall at the height of the season. The fertility of the soil in the high-grass prairie has been one of the main contributors to the country's wealth. The short-grass prairie to the west, which is now wheat and cattle country, gradually merges into bunch-grass territory, drier and with sparser plant cover. Where cattle are unprofitable, sheep can live, but moving westward from the grasslands, one meets ever less rainfall. The plants are thinner and less productive, and the country slowly becomes desertlike. From the beginning of history, grasslands have been the sources of great wealth, but they have also been the scenes of some of the world's most spectacular economic disasters (Figure 28.22).

Figure 28.22
Grassland. A rare photograph of a Wyoming prairie in 1870. The thick stand of grass in this region is now practically gone, eaten by cattle. It has been replaced mainly by sagebrush.

Figure 28.23
A desert. Most of the plants in this region of Death Valley in California are succulent or spiny; they are so widely spaced that much of the surface area is bare.

(a) × 0.5

(b) × 1

(c) × 0.5

(d) × 0.2

(e) × 1.5

Figure 28.24
Representative desert organisms. (a) Prickly pears (*Opuntia*) in blossom. (b) Phlox (*Phlox gracilis*). (c) A desert lizard. (d) A striped racer (*Masticophis lateralis*) of the American southwestern deserts. (e) A scorpion.

Deserts

In an area receiving less than about 25 centimeters (10 in.) of rainfall a year, the plant cover is so sparse that the region is called a **desert** (Figure 28.23). The great deserts of the world are the Asian Gobi Desert; the Australian desert, which takes up almost half the continent; and the Sahara, whose 8.3 million square kilometers (3.2 million mi^2) rivals in size the entire United States (9.3 million km^2, or 3.6 million mi^2). The Sonoran Desert and Death Valley in the United States, well known to us because they are near, are relatively small. One of the driest places in the world is the Atacama Desert in Chile: The first rain ever known there fell in 1971.

Deserts are not usually completely waterless, nor are they barren. Most have some plants, those adapted to survival in dry places, whether hot or cool. Some store water in juicy stems (cacti), some with long roots shed their leaves during the worst seasons (creosote bushes), and some are quick-growing annuals that can germinate, grow, and produce flowers and seeds during a short rainy season (poppies). Desert plants help themselves further by spacing themselves far enough apart to avoid competition for the scant moisture there is and by using the intense heat and light efficiently (see C-4 photosynthesis in Chapter 4).

Desert animals, too, have special adaptations for survival under stress (Figure 28.24). Some, like the kangaroo rat or its eastern hemisphere counterpart, the jerboa, live on dry food but conserve their metabolic water (see the essay on p. 470). Some, like the horned lizards ("horned toads"), go underground where there is some moisture, coming out to feed only in the cool of night. Others (desert rats) eat only juicy plants, and still others (camels) must drink but can wait a long time between drinks.

Tropical Grasslands

Tropical grasslands occur where cold weather does not impose limits on vegetation, but lack of water does. Some occur in east central South America and in Australia, but the famous ones are in Africa south of the Sahara, where the grasslands are dotted with scrubby trees, making parklike **savannas** (Figure 28.25). Inhabited by the greatest concentration of large mammals on earth, the savannas are known for the aggregations of antelopes, lions, rhinoceroses, and elephants that make the east African plains a sort of Noah's Ark in the imaginations of humans the world over. The tropical grasslands have a continuously warm climate, rainfall of 100 to 150 centimeters a year, which comes seasonally with alternating wet and dry seasons, and periodic fires. This combination of features has produced a unique biome. The number of species is small, in contrast to the diversity in the rain forests, because all the grasslands species must be adapted to long, hot, dry periods. Drought-enduring grasses dominate the vegetation, but some trees, rather widely separated, survive, especially the picturesque flat-topped acacias and the legendary baobab trees. Huge groups of consumers (grazers, carnivores, and birds) can be

Figure 28.25
Tropical grasslands. Rain is usually seasonal, so that the turf grows luxuriantly during wet months and dries down to dormancy the rest of the year.

Figure 28.26
Uganda kob are representative grazers of the African grasslands.

supported because the primary productivity is high (Figure 28.26). The African grasslands are being watched with unequaled interest by ecologists, hunters, demographers, and plain "nature lovers" because they are being diminished. In many places they have already been destroyed by the growing demands of people for living space.

Tropical Rain Forests

In those few places where the seasons are always warm and wet, tropical rain forests grow: in parts of Central America, the Congo lowland in Africa, the west coast of India, the west coast of Indochina, some of the South Pacific islands, and the greatest of all in the Amazon basin (Figure 28.27).

(a)

(b)

Figure 28.27
(a) A tropical rain forest as seen from the outside. (b) A tropical rain forest as it appears to an observer under the canopy of the trees.

(a) × 0.1

(c)

(b) × 0.02

Figure 28.28
Common rain forest plants. (a) Epiphytes, (b) tree ferns, (c) vines and air roots.

A tropical rain forest is not a jungle. A jungle is full of tangled vegetation and may follow the destruction of a rain forest, but a virgin rain forest is open along the ground. In a true rain forest you can walk freely between the trunks of trees such as grow nowhere else: tall, straight, and unbranched for perhaps 50 meters (165 ft), but with fluted, spiny, or warted trunks that flare out at the base into serpentine roots like the flying buttresses of medieval cathedrals. The great crevices between such root buttresses, which are like rooms, harbor all manner of animals, especially colonies of bats. The forest floor is dark, moist, and littered with rotting fruits, leaves, and flowers (Figure 28.28). Even where an occasional shaft of sunlight pierces the dense canopy, water drips. The frequent rains do not come straight to the earth. They are caught by the upper leaves, which continue to let water fall for a while, so that it "rains" on the ground after the rain above has ceased. Massive, distorted vines climb to the tops of trees, and curtains of thin roots drop from such heights that you cannot see where they come from. Plants grow on plants. These are the *epiphytes*—orchids and bromeliads, and ferns, stuck to trunks and branches, all vying for the little light that filters down.

(a)

(b)

(c)

(d)

Figure 28.29
Rain forest animals. (a) A "jungle cat" in a tree. (b) A scarlet macaw. (c) Monkeys are at home swinging through branches in rain forests. (d) A striped bark beetle (*Chrysochoa ocellata*).

A rain forest may be dark, but it is never still or quiet (Figure 28.29). Besides the dripping of water, there are always dead branches snapping and falling, flocks of parrots screaming through the leaves, insects buzzing and stridulating, or the misnamed howler monkeys roaring like lions, so high in the branches that you need binoculars to see them. An occasional morpho butterfly wavers silently past, flashing out a glint of electric blue when it passes a spot of sunlight.

Here evolution has figuratively "run wild." A Pennsylvanian who knows a couple of dozen common trees can recognize most of the trees around his home, but a Brazilian who knows that many knows nothing. You may see a tree, spectacularly massed with golden flowers, and not find another like it in the next hour of walking. Unlike the temperate forests, where black oaks grow by the millions in county after county, the rain forest is made up of hundreds of species, mixed and scattered in baffling profusion. If you have any enthusiasm at all for exotic nature, get to a rain forest before progress clears them out.

POPULATION CHANGES OVER TIME: BIOLOGICAL SUCCESSION

Some communities are relatively stable, at least to human observers with short life spans and short memories. Actually no community is permanent, as even the most durable ones we know of (the California redwood forests, for instance) are only a few thousand years old. Change is a common feature of all living things.

Communities change as the species they comprise change, and the changes can be fast enough in some places to be watched and recorded. The changes that occur when a series of populations comes into a region make a **biological succession.** It is fastest and most easily monitored in a place that has been disturbed, perhaps by a climatic change (warming of a region followed by melting of ice), a geological action (eruption of a volcano or a landslide), or a large-scale human action (clearing of a forest, cultivation for agriculture, and subsequent abandonment).

Primary Succession

A **primary succession** occurs when an area, potentially habitable, is opened for colonization, but is essentially devoid of life and has little or no soil. Such a situation developed when the last ice sheet melted back from the northern United States about 10,000 years ago, and a similar situation is developing now in Great Glacier Bay in Alaska (Figure 28.30). There, biologists are watching the advance of new communities as a glacier has been melting back noticeably for the last 150 to 200 years. A **secondary succession** occurs when some event opens an area that was previously inhabited and is already covered with soil. A secondary succession follows a severe fire or the abandonment of a cultivated field.

An area available for colonization by a primary succession may be dry and rocky, or it may be a water-filled depression. In either instance, the first invaders, called **pioneers,** will be organisms especially adapted to the special conditions prevailing. The pioneers act on the environment and then die. By their action they change the environment so as to make it uninhabitable for themselves but favorable to some different organisms. These later arrivals act similarly, preparing the way for still other types of organisms. Such events are repeated, each time with some variation, until further changes are slight. At that time the succession is finished, and an ecological **climax** is achieved.

As an example of succession on a dry area, we can follow the events in a temperate region that has been scoured by a moving glacier. After the glacier has melted, a soilless expanse of bare rock is exposed. The pioneers in such places are lichens, as they are the only organisms able to stand the temperature extremes and severe drying they must endure. Besides, they attach themselves successfully

Figure 28.30
Great Glacier Bay in Alaska is a place where ecologists are watching biological succession following the retreat of a glacier.

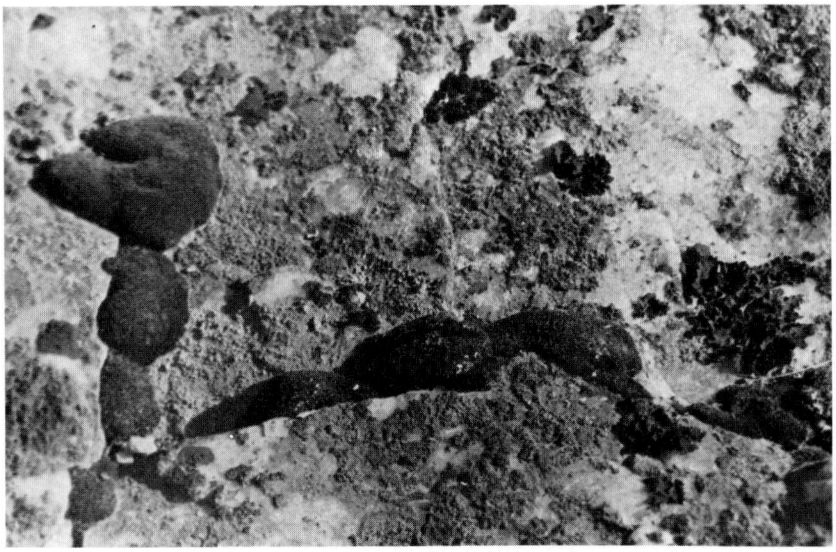

Figure 28.31
A close-up view of a primary succession on a rocky area. The rough, scattered masses are the pioneer lichens. The dark blobs are rock mosses filling the cracks.

to smooth, flat surfaces. As lichens grow, they send minute extensions of themselves into microscopic cracks and give off dissolving acids that break loose small rock particles. This is the beginning of soil. The bodies of dead lichens mix with the broken rock particles, adding organic matter to the soil that is being built up. After hundreds or thousands of years, the lichens, helped by rain, freezing, and thawing, can fill larger and larger cracks in the rocks. When there is enough soil to support new invaders, the lichens are displaced. By their own activities they have changed the environment so that they are no longer suited to it. The newcomers are likely to be mosses, and they carry on where the lichens stopped, creating more soil (Figure 28.31). The mosses crumble more rock and add their small corpses to the increasing soil supply, and they keep doing that until they have made the place suitable for still another invader.

The next arrivals are likely to be grasses, but they must be drought-tolerant because there is little water-holding ability in the still scanty soil. But they survive and carry the process further, essentially repeating the earlier story. Grasses in their turn are supplanted by drought-tolerant shrubs such as blueberries. By this time, soil is abundant enough to see, and it covers expanses of rock. The grasses and shrubs, being better producers than the earlier lichens and mosses, speed up the conversion of a bare spot to a plant-covered one.

When soil builds up and holds water increasingly well, small trees can begin to grow if they can tolerate the still somewhat difficult conditions. Oaks are efficient invaders at such a time, as are some birch trees (Figure 28.32). But they in their turn provide only one chapter in the story, as they continue to make water-holding soil.

The roots of plants, the activities of soil animals, the weathering of rock, and the addition of decayed organic material can, after centuries, produce a soil that is deep enough and holds water well enough to support a forest of moisture-loving species. In the somewhat less humid places, these species are oaks and hickories; in moister areas, they are beeches and maples.

When a mature forest has been established, further change is slow, in contrast to the rate of change during a succession. During the shrub stage of a succession, pine trees may come in. When the pines grow up, they shade their own seedlings, which do not tolerate shade as well as maple seedlings do. Invading maples consequently can displace pines, as they can reproduce in their own shade. Since maple trees do not make the environment unsuitable for themselves, they can form a community that is self-perpetuating. When a community has become relatively stable, it is said to have reached a climax.

Succession need not start in a dry place. It may start in a pond, in which progress is from a watery place toward a drier one (Figure 28.33). There the series is from aquatic plants, to marshy types, to a forest. The same principle applies in both wet and dry places. Each invader prepares the way for the next, until the final climax community can perpetuate itself.

Figure 28.32
A dry area in a stage of development later than that shown in Figure 28.31. The bare rock is covered with small plants, and around the edge of that patch of small plants the soil has become deep enough to support low bushes.

Figure 28.33
Succession in a boggy place. An old lake bed in the Algonquin Forest of Ontario, now filled with soil and supporting low-growing plants, is surrounded by a ring of encroaching bushes, which in turn are encircled by large spruce trees.

A Climax Community

An ecological climax can be recognized by comparing the rate of photosynthetic productivity with respiratory energy loss. This can be done by comparing the buildup of organic matter with its breakdown. When photosynthesis is producing more material than is being used in respiration, then the quantity of organic matter is increasing, and the community is unstable. When productivity equals respiration, or when species are able to perpetuate their own kind, the community is at climax. Note especially that the species that succeed in keeping a permanent place in a community are the ones that do not make the environment unfit for their own kind.

Secondary Succession

Primary successions are usually so slow that changes during one observer's lifetime are seldom visible. A secondary succession, in contrast, can be seen to progress in a few years. A **secondary succession** is one that follows destruction of an established community by cultivation or fire. Soil is already present, pulverized and containing some organic matter; and seeds, spores, eggs, and bits of living plant material are already there or do not have far to go to get there.

After a field has been left untilled, the first invaders are annual weeds, especially ragweeds and crabgrass. They are replaced by perennial weeds,

Figure 28.34
Secondary succession. This field, once cultivated, is covered with goldenrod and blackberries. To the right of center is a young red cedar. The small trees on the extreme right are sassafras, moving in toward the middle of the field.

whose underground rootstocks persist for several years. Perennials, particularly the goldenrods, are notably durable; they can be seen in old fields everywhere. After 10 to 15 years of weediness, shrubby plants begin to grow (Figure 28.34). These plants are the blackberries, poison ivy, sumach bushes, and black cherry seedlings, followed in many places by red cedar. An advancing tide of red cedars can be seen in many an old field. In the eastern United States, where this kind of succession has been most closely followed, the cedars are slowly replaced by red maples, oaks, and the other trees that are part of the deciduous flora. It takes from 50 years to a century for the open field to begin to look once more like a typical deciduous forest community, a second century for it to achieve a measure of maturity, and as much as two more centuries for it to return to the impressive condition of its virgin state.

In all successions, the climax will depend on many forces, especially climatic ones. Only in a well-watered land can succession continue on to a forest. If water is in short supply, a climax can be reached at some stage short of a forest, and an area may be stabilized as grassland or scrub forest.

The idea of ecological succession is only about a century old. Since development of a climax community seems to take perhaps many centuries, details of how primary succession proceeds must be inferred from observed stages. Consequently, ecologists do not agree on how communities have developed. Secondary successions, in contrast to successions from an uninhabited area, are fast enough to be well documented by photographs and personal observation by individual biologists.

The study of successions raises questions beyond the vegetational and faunal changes. One may ask whether human beings are so altering their own habitat as to make it impossible for the human species to survive in it.

TO SUM UP BIOLOGICAL SUCCESSION

1. The changes that occur when a series of populations comes into a region make a *biological succession*.
2. A *primary succession* occurs when an uninhabited area that has little or no soil is opened for colonization. A *secondary* succession takes place on previously inhabited land that has been disturbed by fire or cultivation followed by abandonment.
3. A *climax community* is one that is relatively stable in time, and in which the rate of primary productivity generally equals the rate of respiration of the system.

HUMAN ECOLOGY

We have been concerned so far largely with nonhuman populations and environments. Now we turn to some problems that are specifically human problems. Aside from those of population growth, which were treated in Chapter 27, there are problems resulting from other human activities.

We have come in recent years to regard the output from our own species as pollution. Pollution (which originally meant merely "that which has not been washed") now means "something changed or added by *people*." We do not regard as pollution the deep deposits of bird droppings on the Peruvian guano islands, or the aerosols (tiny resinous droplets) from evergreen forests, or the outpourings of hot water from underground springs. We are not responsible for them and call them natural, as if the products of human action were unnatural.

Looked at with an impersonal view, human-produced pollution is as natural as any earthly output. But there are two important differences between what *we* do and what all other species do. First, in the magnitude of our effects, we have no near rivals. Second, we are the only organisms capable of recognizing that we are making our environment unfit for our own survival, and we are presumably smart enough to make constructive changes before it is too late.

Pollution

Most pollution is like a weed: not intrinsically "bad," but out of place. Humans do not make mercury, sulfur, or heat; but they do move mercury, sulfur, and heat from a place where it does no biological harm to some other place, where it can and frequently does do harm. Shifts and concentrations of materials and energy can result in pollution, but modern industry has introduced still another threat in the form of synthetic chemicals. These chemicals are dangerous partly because the organisms of the earth have never evolved enzymes capable of digesting them and partly because organisms have no defenses against them. The sawdust and slabs from a sawmill are a temporary nuisance, which can be reduced within a few years to reusable water and carbon dioxide; but the polychlorinated biphenyls (PCBs) from plastics, poured into flowing streams, remain to do their biological damage for indefinite periods of time (Figure 28.35). Many products of modern chemical syntheses are neither attackable by enzymes—that is, they are not **biodegradable**—nor readily oxidized or otherwise destroyed by such agents as air and water. This fact has been most fully documented for the chlorinated hydrocarbon pesticides such as dichloro-diphenyl-trichloro-ethane (DDT) and its chemical relatives, aldrin and dieldrin.

The easiest of the pollutants to control is the waste from human biological activities. Techniques for reuse of sewage-laden water are available. Theoretically, water can be used over and over indefinitely. After all, rainwater is distilled and recondensed, and it is biologically and esthetically quite acceptable. The organisms of this earth have been using recycled water for millions of years without visible harm. The difficulty is cost. The mere removal of solids from sewage, or the *primary treatment,* is relatively inexpensive. *Secondary treatment,* by which biological materials are mostly removed, costs two to three times as much. *Tertiary treatment,* depending on whether chemicals and nutrients are simply reduced or are cleaned out enough to make the water drinkable, costs five to ten times as much as primary treatment. At present, the processes of filtration and chlorination can make water clear and free from living bacteria, but they have no effect on dissolved chemicals.

Figure 28.35
Pollution of a stream, showing foam from excess detergent.

Figure 28.36
The effects of atmospheric pollution. (a) A damaged spruce seedling. (b) Normal (right) and air-pollution damaged (left) clover. (c) Output of smoke and fumes from a chemical factory.

Air pollution Most of the serious air pollutants result from industrial activities and from heating and transport (Figure 28.36). Sulfur dioxide comes from burning sulfur-containing coals and oils, and is troublesome because it can combine with water to form sulfuric acid. If it is inhaled, it makes sulfuric acid in the lining of lungs. If it combines with atmospheric water, it is brought to earth in the form of acid rain. Rain in unpolluted air is already slightly acid, pH 5.7, because it picks up carbon dioxide from the air, making carbonic acid. But in regions of heavy sulfur dioxide concentrations, rain may be a hundred times as acid, pH 3.7 or even lower. Acid rain is a fairly recent arrival, as shown by the sudden deterioration of acid-soluble stonework in buildings and by the destruction of fish in the lakes of the Adirondack Mountains (Figure 28.37). Oxides of nitrogen, especially from automobile exhausts, combine with water to make nitrous acid. Carbon monoxide, also from exhausts, combines preferentially with the hemoglobin of blood, thereby excluding oxygen. Lead from antiknock additives as well as from other sources is dangerous because it is, unlike many poisons, *cumulative,* that is, it is neither degraded in the body nor excreted, so that it continues to build up in concentration as more is taken in.

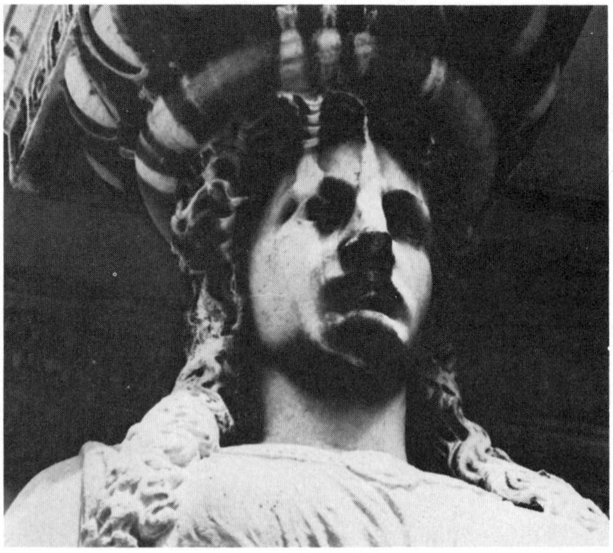

Figure 28.37
This sculpture was created from sandstone about 280 years ago and kept its detail for two centuries as this photograph taken in 1908 shows. At present, as the result of the effects of acid rain, it has been reduced to an almost formless lump.

Particulate pollutants such as carbon—coal dust and soot—and mineral dusts are mechanical irritants. Microscopic particles, especially if they are sharp and hard, work their way into living cells of such soft tissues as the linings of the lungs and digestive tract. Once inside a cell, foreign objects can either kill it directly or stimulate it to grow abnormally. (See the essay "Cancer from the Air" on p. 600.)

When solid particles and minute water droplets occur together, the combined smoke and fog make **smog,** a word that did not appear in English until nearly the middle of the present century. Smog by itself is nasty and smelly enough, but when it contains petroleum products as well and is acted upon by the ultraviolet rays of sunlight, the result is the formation of a number of new compounds. The mixture causes watering of the eyes, burning and itching of the throat and lungs, and vague discomfort. People with serious respiratory difficulties, such as asthma and emphysema, are especially advised to stay indoors and refrain from smoking when smog outdoors is heavy. It is ironic that this *photochemical effect* is caused by the same energy from the sun that provides the energy for photosynthesis.

Smog is most troublesome in places subject to atmospheric **inversions.** An inversion is an atmospheric condition in which a layer of air near the ground is cooled at night and becomes trapped by a layer of warmer air above, the reverse of what normally happens. If an industrial site puts out pollutants during a period of inversion, the unwanted matter is kept in place near the ground instead of being blown away and diluted. The result may be a heavy and persistent smog. The classic smog area is Los Angeles, where everything combines to make trouble: a heavy concentration of smoke-emitters surrounded by a ring of hills. An inversion there acts like a great blanket, effectively holding polluted air in place (Figure 28.38).

Radiation pollution Radiation pollution from human activities has been a completely new problem since the end of World War II. Certainly there was radiation before we made and released radioactive materials. The cosmic radiation from the sun, X-rays, and the radiation of radioactive isotopes of elements in the earth were here, and they remain the principal source of radiation bombardment of organisms. Still, by-products from atom bomb explosions have been released in the air and are a cause for concern. These by-products, called **fallout,** descend like snow after an atomic blast. Because they can release enough energy to disrupt molecules, they are known as *ionizing*. Some radioactive isotopes emit *alpha* or *beta* particles, both of which have such poor

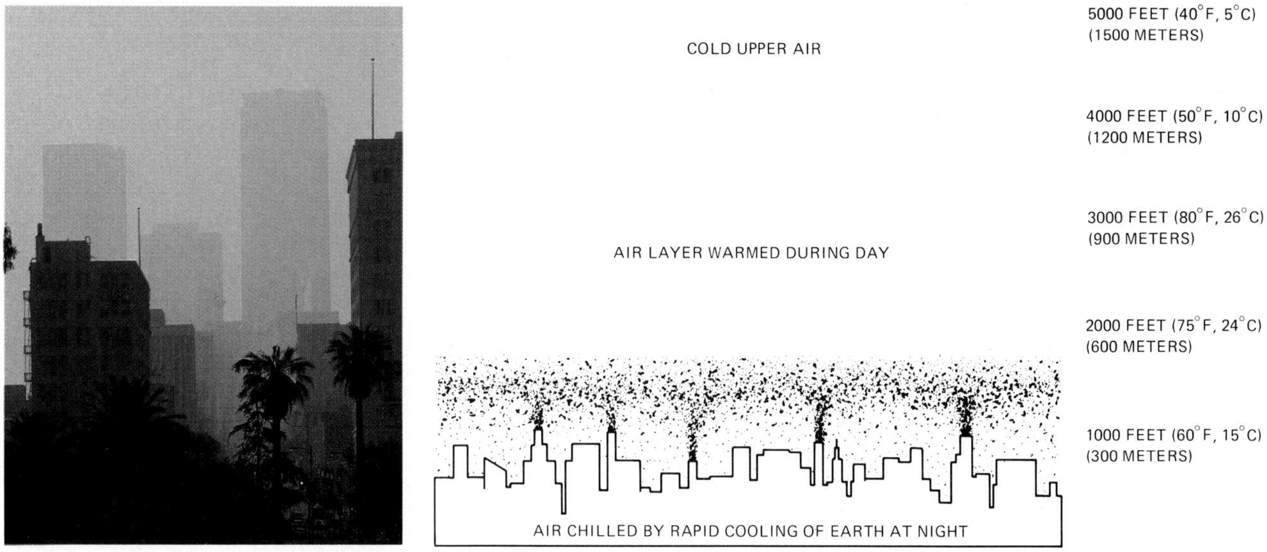

(a) (b)

Figure 28.38
The effect of an inversion on the smog-holding ability of a region. (a) Los Angeles. (b) A layer of warm air, in this instance between 2000 and 4000 feet, is sandwiched between the cold upper air and the cool air close to the ground. The lowest layer with its load of smoke and fog is held down by the warmer air above, and the pollution is not dissipated by being blown away.

CANCER FROM THE AIR

Asbestos is one of the best-known cancer-producing agents. Made of fine fibers, it can float as dust in air and can be inhaled. Once asbestos is drawn into the lungs, the sharp pointed particles, like microscopic needles, can penetrate cells, damaging interior lung surfaces. The fibers can also work their way completely through lung tissue and come to rest in the lining of the chest cavity, the mesothelium. In either place, asbestos can stimulate abnormal growth: lung cancer or cancer of the mesothelium (mesothelioma). Indeed, mesothelioma is not known to be caused by any agent except microscopic fibers. Even if asbestos does not produce cancer, it can cause asbestosis, a dangerous, even fatal scarring of lung tissue.

Cilia in the air passages can sweep some asbestos fibers back up from the lungs to the pharynx. When swallowed, they can cause cancers of the esophagus, stomach, large intestine, and rectum.

Asbestos is a special form of the mineral serpentine, frequently found as crocidolite in Africa or as chrysotile in the United States. Since asbestos occurs as soft fibers, it can be matted into feltlike sheets, woven into cloth, or added to such building materials as wallboard and pipes. Like sand, it is a silicate and does not burn. That quality has been known for a long time. Marco Polo told his unbelieving thirteenth-century listeners about an oriental cloth that was not consumed by fire, and Benjamin Franklin used to amuse his guests by tossing a sheet of "paper" into flames and watching it lie there undamaged. (He had made it of Pennsylvania asbestos.) Because of the flame resistance and insulating properties of asbestos, millions of tons have been used in the building industry.

Since the early 1900s the public health menace of asbestos has been increasingly recognized. In Africa, Europe, and America, workers in asbestos-processing plants have been found to suffer from lung cancer eight to ten times more frequently than the rest of the population; in some places as many as 40 percent of the people in constant contact with asbestos were being killed by various diseases due to asbestos fibers. Even families of workers were stricken when they inhaled dust brought home on fiber-laden clothes. Men who sprayed asbestos onto walls to fireproof them released tons of dust into the air to be breathed in by everyone.

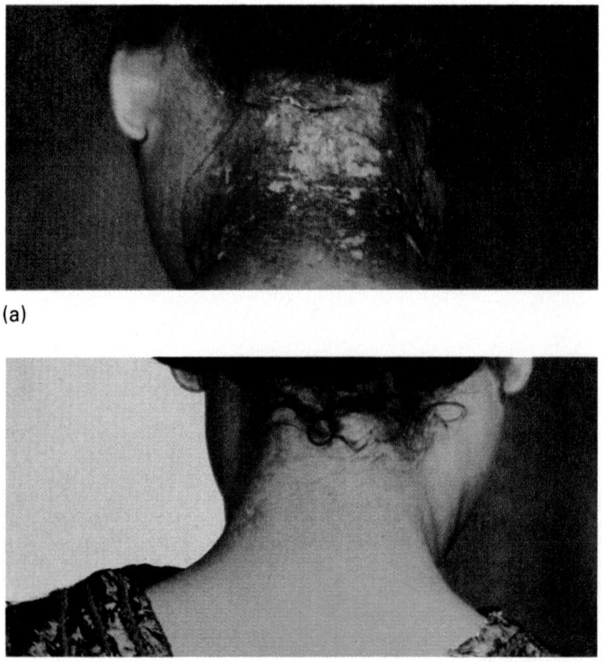

(a)

(b)

Figure 28.39
(a) Burn on the neck of a Rongelap Islander a month after she was exposed to radiation from fallout following an atomic bomb test. (b) After a year, the burn was healed.

penetrating power that simply wearing clothes during fallout is good protection. Other isotopes, such as carbon-14, emit *gamma* rays, which have great penetrating ability and are therefore capable of causing ionizations and potential damage deep within living tissues.

The durability of activity is as important as the kind of radiation. Some isotopes, say nitrogen-16, have a half-life of only 8 seconds. Others have long half-lives: carbon-14 at 5730 years, or thorium-232 at ten billion years. Strontium-90, with a half-life of 28 years, has received particular attention because it is

Industrial use of asbestos is now being controlled, and substitutes are being sought, but problems remain. One is that asbestos-bearing serpentine rocks are common in many parts of the world, and when they are used for highway paving, they can release deadly fibrous dust. In Montgomery County, Maryland, the concentration of airborne asbestos (originating in a local quarry and spread on roads) was found in 1977 to be a thousand times higher than in regions where serpentine is not used for paving. One might ask if the incidence of lung cancer is unusually high in that area. It is not, at least at present, because the quarrying and dispersal of asbestos-containing rock have not been carried on there for the 20 to 30 years it takes for the damage to become apparent. Now, however, roads, playgrounds, and parking lots bearing loose serpentine will have to be sealed to keep down dust.

A second problem appears when other tiny fibers are used. The hope that glass fiber could be used commercially instead of asbestos is weakened by the discovery that in the village of Karain in Turkey, 24 of the 55 people who died between 1970 and 1974 died of mesothelioma. (Strangely enough, Karain means "stomachache" in Turkish.) The only possible cause found in Karain so far is a fibrous mineral called zeolite, which is present in the drinking water. Zeolite is not asbestos, but it does have fibers of comparable size. If nonasbestos zeolite particles, about 3 micrometers thick and 15 micrometers long, can cause cancer, then perhaps fiberglass is also dangerous, and one more potential asbestos substitute becomes suspect.

Still another difficulty appears. Asbestos and cancer are clearly linked. Tobacco smoking and cancer are also linked. The insidious thing about asbestos, bad enough by itself, is that when it is coupled with tobacco smoke, the results are not simply added. The phenomenon of *synergism* emerges. Asbestos and tobacco smoke together are much more dangerous than either is alone. According to Irving Selikoff of the Mount Sinai School of Medicine in New York, "Asbestos workers who smoke have approximately eight times the lung cancer risk of other smokers and 90 times the risk of individuals who neither smoke nor work with asbestos." With respect to the children who live near the Maryland serpentine quarry, Selikoff has pointed out that they "have good reason never to smoke cigarettes."

readily taken into animal bones, and it may be passed through mother's milk to nursing babies. At present, even after massive statistical analysis of the Japanese survivors of the first atomic bombs and of victims of accidents, safe standards of acceptance for ionizing radiation have not been agreed upon. Lethal doses depend on the species under consideration, the usual measurement being a *rem* (roentgen-equivalent-man), a unit of delivery of energy. Some microorganisms, such as spore-forming bacteria and fungi, can survive doses larger than a million rem, but a dose of about 500 rem is considered lethal for a human being. In spite of differences of opinion as to the degree of danger from radioactivity, there is no disagreement that ionizing radiation is potentially destructive to organisms. In sufficient doses it can kill, and in lesser doses it can cause cause burns, cancer, and genetic change (Figure 28.39). Scientific study can only provide the factual information; the decisions on nuclear testing and on nuclear power production must be social and political.

Noise pollution The industrial world is a noisy world, filled with the roar and clangor of machines, motors, sirens, planes, and amusement devices. Only in recent years has evidence been accumulating that high-intensity noise can be destructive. Aside from the obvious possible damage to ears, noise damage has been implicated in atherosclerosis, coronary heart disease, and increased blood pressure. David Lipscomb (University of Tennessee) demonstrated the destruction of cochleal cells in the ears of guinea pigs that had been subjected to 24 hours of loud rock music (see Figure 20.20). Noise-reduction programs are increasing, especially near airports, in the noisier factories, and in road-building activities.

Thermal pollution Another kind of nonmaterial pollution occurs when streams are used to cool large industrial activities. As a result, the streams become heated. The resulting **thermal pollution** can cause changes in the plant and animal life in an affected stream, since most species have temperature ranges that are fairly definite. Some species are adapted to a wide range of temperatures. Others can exist only within a narrow range with sharp limits. Thermal pollution will not affect some organisms, unless of course temperatures become excessive, but others may be eradicated by a change of only two or three degrees. The principle followed by many ecologists with regard to such environmental alterations is that any change is likely to bring about unforeseen and unforeseeable effects on organisms. For instance, if a nuclear power plant is planned with a river as a cooling system, everyone involved should try to know in advance how much heating of water will follow, what organisms inhabit the water, and what is likely to happen to them, basing predictions on models and on past experience. Recognition of thermal pollution, like that of noise pollution, is so recent that relatively little firm information is available, and the ecologically sensible guide is to proceed with caution. It is easier to prevent mistakes than to correct them.

SUMMARY

1. A *community* is an aggregation of populations living together in a space.
2. An *ecosystem* is a community of populations, plus the physical and climatic aspects of the environment. An *ecotone* is a transition zone between ecosystems.
3. The photosynthetic activity in an ecosystem is its *primary productivity*. Its rate is determined by light, carbon dioxide concentration, temperature, the presence of water, and the availability of mineral nutrients.
4. The primary productivity of a region is measured in kilocalories per square meter per year. Deserts have the lowest productivity on earth, with 500 or fewer $kcal/m^2/yr$. Some unusually productive areas, such as marine estuaries and intensively cultivated croplands, may yield up to 20,000 $kcal/m^2/yr$.
5. Production of food for human use is limited by available energy, mineral nutrients, and space.
6. Sunlight provides energy for photosynthesis and warmth. About 1 or 2 percent of the sun's energy striking a plant is used in photosynthesis. The energy fed into a system comes via photosynthesis; the outflow is a result of respiration, which liberates energy as heat. An ecosystem is stabilized in *dynamic equilibrium* when the photosynthetic input balances the energy loss by respiration of all the members of the system.
7. What an animal eats determines its *trophic level* in a community. An herbivore is a *primary* consumer; a carnivore that eats herbivores is a *secondary* consumer; a carnivore that eats carnivores is a *tertiary* consumer.
8. The complex producing and feeding activities in a community are designated as a *food web*. When a consumer feeds, it retains for its own energy only about one-tenth of the energy contained in the next lower level. Humans eat at several trophic levels, ranging from first to fourth.
9. *Forest communities* have a layered structure with a high canopy, a middle understory, and a low carpet of vegetation.
10. The largest vegetational units are *biomes*, distributed on the earth according to climate and topography.
11. The major biomes are the oceans, tundra, northern coniferous forests (or *taiga*), temperate deciduous forests, temperate grasslands, deserts, tropical grasslands, and tropical rain forests. Many subdivisions of these categories occur.
12. When a region is changed by fire, biological activity (including human activity), or climate, the organisms change. Some species replace others in a *biological succession*.
13. A *primary succession* occurs when organisms invade an uninhabited region, either dry and rocky, or aquatic. A *secondary succession* occurs on disturbed but previously inhabited land.
14. Successive invaders during a biological succession render the environment unfit for themselves, but prepare it for the next invaders. The succession continues until a *climax community* is developed. A climax community is self-perpetuating and has a balance of primary productivity versus respiratory energy release.
15. Human beings have altered the earth's environment more than any other species has. Some of the results of human activity are pollution of water with heat, synthetic chemicals, and metals, and of air with noise, particles forming smog, and chemicals. Radiation pollution of earth, air, and water is a recent problem intensified by artificial production of radioactive elements.

ASK YOURSELF

1. What is an ecosystem?
2. Give an example of a synergistic effect.
3. Why are some parts of oceans more productive than others?
4. Why is a field of sugarcane exceptionally productive?
5. About what percent of sunlight energy falling on a plant is converted to the chemical energy of food?
6. What are the inherent dangers of depending heavily on the Green Revolution for the world's food supply?
7. About what percent of the energy taken in by a beetle eating a plant does the beetle keep for its own build-up?
8. What becomes of the energy that a predator gets by eating but fails to keep for itself?
9. What is the difference between the causes of marasmus and kwashiorkor?
10. In what ways do oceans provide a more stabilized environment than does land?
11. What climatic factors determine the establishment of a temperate deciduous forest?
12. Why are grasslands important to human economy?
13. What is the difference between a jungle and a tropical rain forest?
14. What is the difference between a primary and a secondary ecological succession?
15. What is the evidence that the human species is acting as an agent in ecological succession?
16. Can sewage water be made safe for human consumption?
17. How can an atmospheric inversion be dangerous to human health?
18. What are the sources of ionizing radiation on earth?
19. Why is radiation dangerous to living things?

29
The Patterns of Animal Behavior

SOME KEY POINTS

1. Animal behavior that promotes the survival of the individual is naturally selected.
2. An animal demonstrates learning when it is able to modify its behavior based on past experience. Instinct is less flexible than acquired learning, but it is usually reliable enough to encourage survival.
3. Animals have built-in "biological clocks" that help them regulate their lives according to seasonal variations or other time changes.
4. Migration and homing (which can be considered as short-range migration) are typical animal activities that require a complex sense of time and direction.
5. Animals perceive the world around them with the aid of acute senses, and many are able to communicate those perceptions to other members of their species.

ANIMAL BEHAVIOR, LIKE EVERY OTHER BIOLOGICAL activity, has developed through the action of natural selection. During the course of evolution, behavior from courtship to feeding to camouflage has been naturally selected when it has promoted the survival of the individual. John Alcock (Arizona State University) describes the complexity of the study of animal behavior when he says that "the study of behavior is extremely broad-ranging, involving such diverse phenomena as the action of a gene, the structure of a nervous sytem, the relations of an animal to its environment, and evolutionary events lasting millions of years."

ETHOLOGY: THE STUDY OF ANIMAL BEHAVIOR

In 1973, the Nobel prize in Physiology or Medicine was awarded to Konrad Lorenz, Nikolaas Tinbergen, and Karl von Frisch, who had devoted about 50 years apiece to field studies of insects, birds, and animals of many sorts. Lorenz, Tinbergen, and von Frisch made the study of **ethology**, or animal behavior, feasible by demonstrating methods and techniques that would soon become standard.

When Charles Darwin published *The Origin of Species* in 1859, he presented the notion that animals behave the way they do because of a continuing process of evolution. Living things, Darwin said, must be able to adapt to changing conditions. They are not created with fixed responses that were indelibly set eons ago. Darwin's more specialized publication, in 1872, of *Expression of the Emotions of Man and Animals* was the first scientific exploration of the causes of behavior patterns of plants and animals.

Pioneers of animal behavior followed Darwin's revolutionary groundwork. Jean Henri Fabre in France observed insects in their natural habitats. Others were Sir John W. Lubbock and C. Lloyd Morgan in England and Jacques Loeb and Herbert S. Jennings in the United States. Sir Charles Sherrington and Edgar D. Adrian shared a 1932 Nobel prize for their work with nerve and skeletal muscles. And of course, Ivan Pavlov in Russia discovered the "conditioned reflex" during his experiments with dogs and won a Nobel prize for his work. Ethology continued to develop into a modern science with the complementary ideas of psychology, natural history, genetics, and evolution.

THE EVOLUTION OF ANIMAL BEHAVIOR

The study of animal behavior has revealed that as an animal's brain becomes more complex, its behavior patterns are less rigid. Insects seem to be programmed with an unshakable set of instructions (honeybees are genetically programmed to such an extent that they are essentially robots), whereas "higher organisms" need not rely totally on inflexible behavior patterns.

A typical example of rigid insect behavior concerns dead ants and the specific odor they emit. Worker ants, recognizing the death odor, will cart the corpse to a garbage site away from the nest. However, if a live ant is experimentally dabbed with the substance that releases the "dead ant" odor, the ant will be promptly carried off to the garbage dump by the workers, even though the "dead" ant shows definite signs of life.

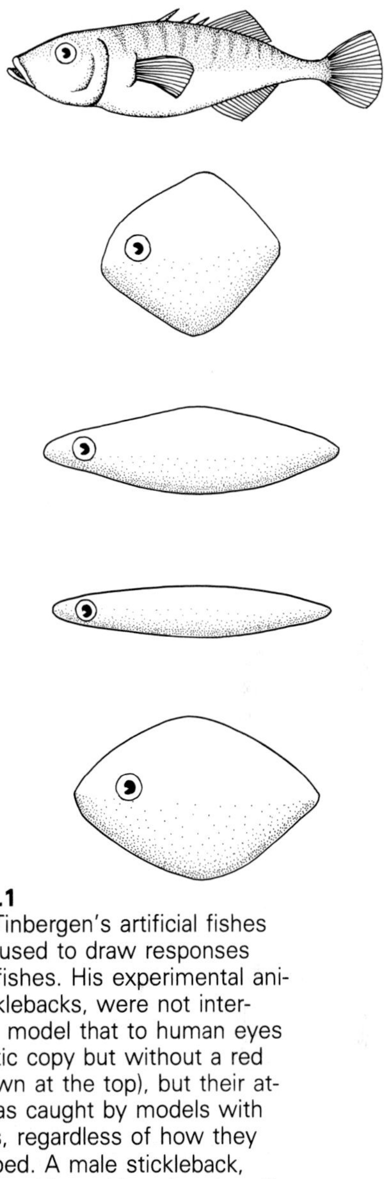

Figure 29.1
Some of Tinbergen's artificial fishes which he used to draw responses from live fishes. His experimental animals, sticklebacks, were not interested in a model that to human eyes is a realistic copy but without a red belly (shown at the top), but their attention was caught by models with red bellies, regardless of how they were shaped. A male stickleback, seeing something with red on it, will either attack it as if it were an intruding male rival, or court it as if it were an acceptable female, depending on the time of year. A specific detail, whether of structure, color, or movement, that elicits a response is called a "releaser."

The Basic Conditions of Animal Behavior

Complex animals are able to rely on their own experience and specific needs when responding to stimuli. But even though the response of a complex animal may be more flexible than that of a simpler animal, all organisms possess the property of **irritability,** the basic ability to respond to stimuli. All animals must be able to sense and receive stimuli, conduct the information by way of a nerve impulse, and use an effector to initiate the proper response. In complex animals like humans, the once-simple sensory cells have evolved into sense organs, nerve cells have been elaborated into the brain, and effector cells have become glands and muscles. But the fundamental components of behavior remain the same for all living things.

In Chapter 19 we observed that complex behavior becomes possible when synaptic relationships and reflex arcs develop to the point where they provide more intricate pathways for a nerve impulse. As the nervous system itself becomes more complex, it is ultimately the greater capacity of the brain that makes it possible for an animal to analyze and interpret stimuli in a selective way. When the response to any given stimulus becomes flexible, the organism is freed from **stereotyped behavior,** or *fixed-action patterns* (Figures 29.1 and 29.2). Individu-

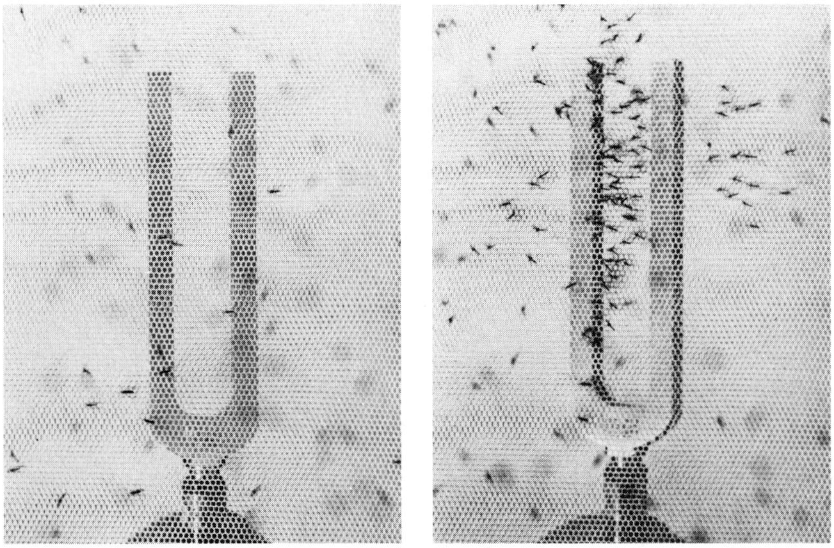

Figure 29.2
Sound as a sex attractant in mosquitoes. The silent tuning fork at left draws few male mosquitoes to the screening, but the humming fork at right has attracted a swarm. Males are attracted to the sound frequency (or a close approximation) emitted by the vibrating wings of female mosquitoes. The mating attractant is a tone of a particular pitch; the feeding attractant is temperature. Mosquitoes attack human ankles with greater frequency than they go for hands because the ankles are thin-skinned and emit more detectable warmth than hands do.

als who can adjust to a changing environment are better suited to find food and to avoid predators or unpleasant conditions. And, of course, the improved senses are passed on to the offspring, once more demonstrating the effectiveness of the natural process of selection.

INSTINCT AND LEARNING

Controversy accompanies any attempt to separate the concepts of instinct and learning. To further complicate the issue, there are no simple, clear-cut categories of learning, although we must attempt to separate certain basic kinds of learning patterns if we are to proceed at all. We will consider **learning** to be the modification of behavior as a result of changes in individual experience (Figure 29.3). Generally, such behavioral changes increase the organism's chances for survival, and hence they are advantageous to the species.

Instinct

The major strength of **instinctive,** or inherited, behavior is that it is predictably correct and safe. The very nature of instinctive behavior makes it reliable, since it has survived the natural selection process and has proved to be advantageous to the organism. The reliability of instinctive behavior patterns is especially important when the organism has to make the correct choice the first time the action occurs. If an animal being pursued by a predator reacts improperly, it may not have the opportunity to react again.

Usually an animal has developed its behavior patterns by the time it needs to use them. Many animals have fully developed behavior patterns at birth because those patterns were genetically maturing during the embryo stage. Other behavior patterns, such as the ability to mate, usually develop along with the animal's growth and the maturation of the specific organs involved. They may also require the experience of watching and imitating adults.

Although instinctive behavior may develop genetically, its application to a real situation may have to be learned. Chimpanzees in zoos may use thin sticks to poke at holes, with no obvious purpose. If the same chimpanzee had been born in the wild and remained there, it would have learned to apply its innate tool-using ability to probe for termites in holes in logs and termite nests (Figure 29.4). The tendency to perform behavior may appear spontaneously, but practice may perfect its execution.

Habituation

It is generally accepted that habituation is the least complex kind of learning. **Habituation** takes place when a stimulus no longer provokes a behavioral response. In other words, the animal that has learned that a stimulus is meaningless will ignore it.

(a) (b) (c)

Figure 29.3
Learning what to fear and what not to fear. (a) When objects (a goose and a hawk in this case) pass over an inexperienced young goose, it drops down and stays still. (b) If geese pass over a gosling again and again without hurting it, the gosling ceases to crouch. (c) If an unfamiliar object appears, however, the gosling may assume the old crouching stance again.

By ignoring meaningless stimuli, an animal is better able to devote its full attention to stimuli that do have biological significance (see Figure 29.3).

Conditioning

Another type of learning is **conditioning,** in which the animal is rewarded or punished according to its reaction to a given stimulus. Included in the general area of conditioned learning is the *trial-and-error* method, adopted by animals adjusting to unfamiliar stimuli in their environments. As with many types of learning, successes and failures, rewards and

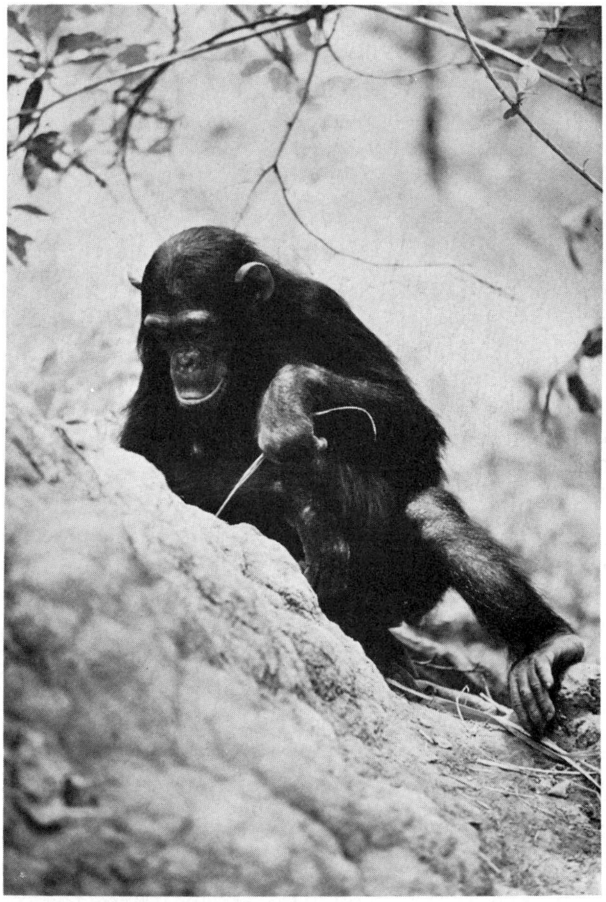

Figure 29.4
Some animals take advantage of natural objects to help them obtain food. Such actions can be called "tool using." In this photo a chimpanzee gets termites out of their nest by poking a stem into the entrance hole.

punishments are involved with conditioning and trial-and-error learning (Figure 29.5). Early, in a learning task, 100 percent reinforcement gives optimal results, but once an animal is trained, it performs better and "forgets" less often if it is rewarded only occasionally.

Probably the most famous case of experimental conditioning was developed by the Russian physiologist Ivan Pavlov in the early 1900s. Using dogs in his experiments, Pavlov developed the idea of the conditioned reflex when he rang a bell just before he fed an experimental dog. The dog did not at first salivate when the bell was rung, but when the dog was actually fed, it salivated at the sight of the meat. Soon Pavlov was able to cause the dog to salivate at the sound of the bell, in anticipation of the meat to follow. This type of conditioning is called *classical conditioning*.

Another form of conditioning is *operant conditioning*, popularized in the 1930s by the American psychologist B. F. Skinner. The usual demonstration of operant conditioning involves placing an animal in a cage and providing a lever that, when pressed by the animal (typically a rat), will make food available through a small opening in the cage. As soon as the rat learns to make the association between the depression of the lever and the reward, it depresses the bar often. The difference between classical conditioning and operant conditioning is that the former depends on a *reflexive* response to a stimulus whereas the latter uses the promise of a positive reinforcement to trigger a voluntary action, such as pressing a lever.

(a)

(b)

(c)

(d)

(e)

Figure 29.5
An animal learns by punishment and reward. (a) A toad, accustomed to eating insects, is presented with a stingless robber fly. (b) The toad eats the fly. (c) A bumblebee, resembling the robber fly but equipped with a wicked sting, is dangled on a string before the toad. (d) As toads do, this one snaps at a moving insect, only to receive what is to a toad a serious dose of venom from the bumblebee. (e) The toad now backs off from a robber fly, having learned that anything resembling a bumblebee is best let alone.

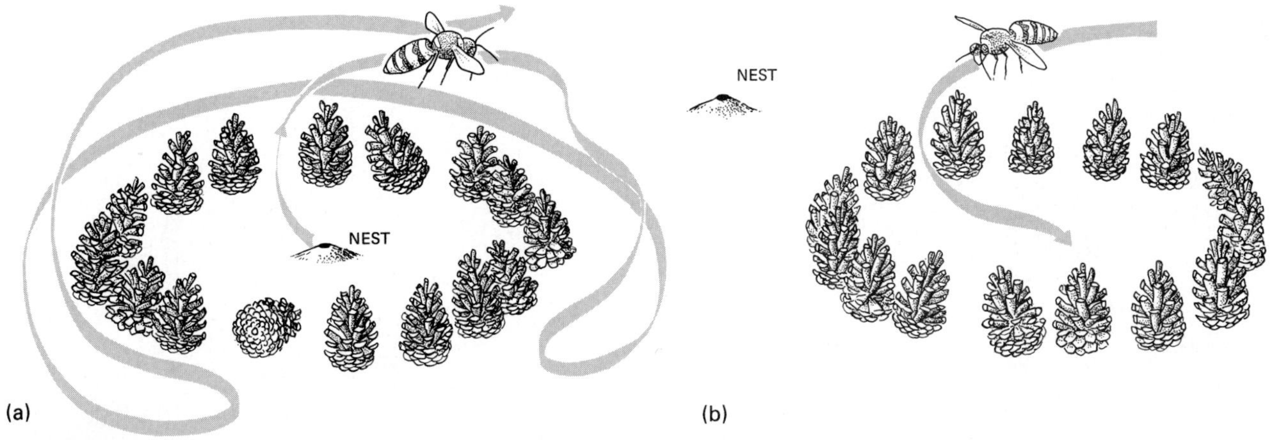

Figure 29.6
Learning in a wasp. A female digger wasp, *Philanthus triangulum,* comes and goes from her sandy nest, finding her way by looking at landmarks. (a) Once, on emerging from her nest, she found that a ring of pinecones had been placed around it (by the experimenter, Niko Tinbergen). She made an exploratory flight around the new arrangement and then took off to forage. While she was gone, Tinbergen moved the cones away from the hole. (b) When the wasp came back, she flew to the center of the ring to seek her nest. She had apparently "memorized" the landscape.

Latent Learning

Animals do not need to be prodded to learn. **Latent learning** takes place when an animal actively seeks to find out more about its environment. Such behavior may simply be called curiosity, but whatever the label, this exploratory learning needs no particular reinforcement or reward. Biologically, latent learning is important for an animal to prepare it for important changes in its environment. A rat that masters the idea of a maze or a digger wasp that can find its nest after wandering great distances in search of prey has an advantage over an animal that has not learned to explore and retrace its path, or one that needs a reward to learn anything (Figure 29.6).

Imprinting

Rarely has a term been as completely associated with one person as "imprinting" is with the Austrian ethologist Konrad Lorenz. In the 1930s, Lorenz originated the term **imprinting** to show how newly hatched goslings permanently regard as their mother the first creature they can follow (Figure 29.7). Imprinting takes place during the first few hours of a bird's life, and it is no longer possible after that crucial early period has passed. Current research indicates that imprinting is not necessarily permanent.

The sort of imprinting shown in Figure 29.7 is obviously artificially imposed upon animals. In a natural setting, the imprinting instinct will work

Figure 29.7
This photograph of Nobel prize winner Konrad Lorenz and his "children" has achieved modest fame as an example of what can be accomplished by *imprinting* of young animals. If newly hatched birds are presented with some object as soon as they can perceive anything, they will accept that object as a "parent," whether it actually is a parent, a mechanical toy, or Konrad Lorenz. Since the parent is in fact the first thing a young bird is most likely to meet, the system is effective in nature.

Figure 29.8
Intelligent behavior, or insight learning, in a chimpanzee. Meeting a situation he has never seen before, this hungry ape managed to get the inaccessible bananas by stacking boxes. Such action implies a certain amount of imagination.

perfectly well, with no sexual or other complications, because the first moving object the young animal sees is usually its mother, and it will follow her in a normal way.

Insight Learning

When an animal is able to associate apparently unrelated experiences, it is demonstrating **insight.** Chimpanzees, of all the animals except humans, seem to have developed the highest degree of insight learning, although it can be observed in many of the higher animals. Animals capable of insight learning seem to practice a sort of mental trial-and-error process, analyzing the possibilities for the solution of a problem before actually setting out to tackle it. We have already mentioned that chimpanzees have a tool-using ability, which they can apply to locate and capture termites, for instance. But besides being able to attach sticks together to reach an object, or to pile boxes one on top of the other for the same purpose (Figure 29.8), two chimps may even cooperate, with minimal nonverbal communication, to accomplish other goals.

THE RHYTHMS OF ANIMAL BEHAVIOR

Natural cyclical events, such as the rising and setting of the sun and moon, seasons, tides, and light and dark, obviously affect the behavior of animals and, for that matter, of plants. The overall behavior of most organisms is certainly related to the great cycles and rhythms of nature, and it is instructive to investigate animal rhythms and the ways in which they depend on both genetically and environmentally imposed rhythms.

Biological Clocks

It has been known for hundreds of years that some plants have predictable, cyclical behavior. Sunflowers turn toward the rising and then toward the setting sun. More recently it has been found that bean leaves which rise high in the morning and droop at night, can continue to rise and droop at the appropriate times for several days even after they have been kept in continuous darkness. Beans plants (and many other plants) must have some internally regulated timing mechanism, or "biological clock." We now know that animals have such time-measuring "biological clocks," too, and we can observe definite cycles of behavior at specific times of the day or year, even if the visible environment is kept constant. One type of biological clock distinguishes the changing length of the days throughout the seasons. Many organisms respond to seasonal variations in day length by changing their metabolism or behavior, and such a response is called **photoperiodism** (see also Chapter 17). Studies have shown that the photoperiodic mechanism is located in the brain, and, at least in insects, the eye is not involved. How does a biological clock work? Although we do not have a detailed understanding of the mechanism, it is likely that a biological clock is run by hormones, by oscillating responses to light and dark, or by a combination of these mechanisms.

The term **circadian rhythm** has been applied to daily cycles of behavior (from the Latin, *circa,* "about," and *dies,* "day"). *Circannual* rhythms are cyclical behavior patterns on an annual basis. Whether circadian or circannual, the rhythms of a biological clock are genetically programmed, and they need not be learned. Biological clocks are usually complemented by a reaction to changes in heat or light. Environmental clues are valuable in providing a check on programmed rhythms because those rhythms must be reset occasionally when they deviate slightly from a 24-hour or 365-day pattern. Actually, no rhythm of a biological clock corresponds exactly to the length of a day or a year.

Migration

Some animals periodically leave their homes in favor of other locations for purposes of feeding or mating. This process is called **migration.** Ordinarily, migratory expeditions take place in the fall, and the animals return to their original homes in the spring. Even when environmental changes are not apparent, a biological clock provides the signal for action. Instead of waiting for an actual event like winter to occur, the animal is able to anticipate such an extreme change and prepare for it.

In a series of classic experiments, E. G. F. Sauer of the University of Freiburg in Germany investigated the annual migration of warblers to Africa. Each year, starting in August, the birds left their nests in Germany and continued their migrations until the end of October. The next spring the warblers returned to the exact sites they had left in the autumn, and even young warblers, who had never made the flight before, successfully navigated to Africa and back. Such competence is especially notable since warblers migrate singly, without the support of a familiar and experienced group. Warblers usually migrate at night, navigating by the stars, and they use a built-in time sense that coordinates with the seasonal position of the stars.

HOMING

One of the most fascinating aspects of animal orientation is **homing.** By homing, animals that leave their homes in search of food, for instance, are able to return without wandering aimlessly. Homing pigeons do not need landmarks to find their way; pigeons carried from their coops in covered containers are able to fly back without any trouble. It is not uncommon for pigeons to fly home in an almost direct route at about 80 km/hr (48 mph), covering almost 1000 kilometers (600 mi) a day. Homing pigeons and other airborne navigators use the sun or stars as a compass, but they probably also orient themselves according to the earth's magnetic field, and use odor clues and the phases of the moon as well.

A sense of time is important to animals that must find their way over long distances, and animals that use the sun or stars to navigate have a time sense that makes allowances for the shifting positions of the celestial bodies. If an animal can make such a compensation, it presumably can determine its direction and distance away from its home base.

Honeybees and some other insects also have this ability to correlate time, distance, and direction according to the changing position of the sun. In fact, honeybees can navigate when the sun is obscured, using only patches of blue sky. Scientists have found that bees can detect polarized light.* By using the fixed relationship between the pattern of polarized light in the sky and the position of the sun, the bees can make the necessary adjustment to locate the sun. All skylight is polarized, but the honeybees make use of only the polarized light in the ultraviolet end of the spectrum, to which they are most sensitive. Bees may also have the same sort of built-in magnetic compass that pigeons do, and it seems that they may also be guided by other backup systems, including smell. Recent evidence suggests that some birds may also use polarized light for navigation when the sun is hidden by clouds or mountains.

* We speak of light from the sun as being unpolarized, and waves of unpolarized light normally vibrate freely in all directions. But when unpolarized light from the sun is scattered by water droplets in the earth's atmosphere, the horizontal light waves are filtered out, and the light becomes "polarized." (Polarized light waves vibrate only up and down in a vertical plane.) The vibration direction of the polarized light that reaches the earth from a patch of sky is at right angles to the direction of the sun from that piece of sky. Bees seem to "know" this fact and adjust for the angle between the light and the sun.

Figure 29.9
Map of the western hemisphere, showing four of the most spectacular bird migration routes. Some birds travel hundreds or thousands of miles each year, going to new feeding grounds after the nesting season is over.

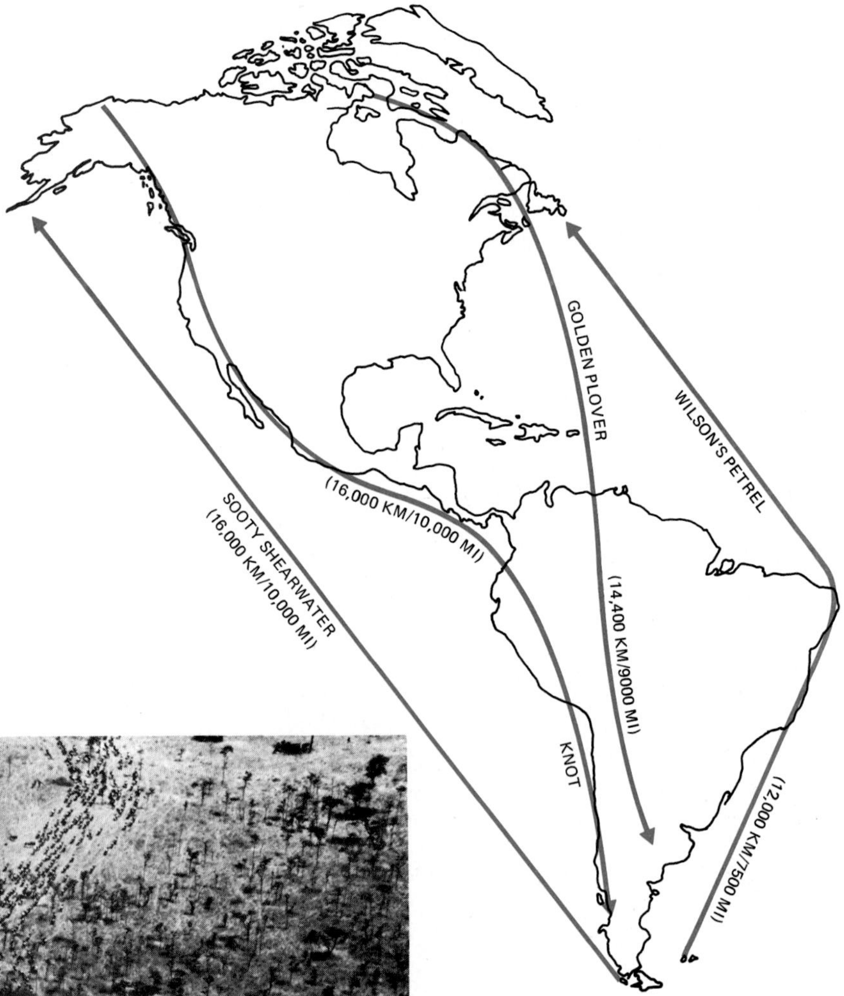

(a)

(b)

(c)

Figure 29.10
Animal migrations. (a) Wildebeests move in great numbers over kilometers of African plains, seeking food as the seasons change. (b) Monarch butterflies fly south in late summer and fall, moving by the millions from the United States and Canada to spend the winter in Mexico. Even while on their way, they can accumulate in clusters, as those shown here have done. (c) The end of a migration. The nesting ground of the Adélie penguins at Cape Crozier in Antarctica is crowded with hundreds of thousands of birds who go there annually to nest and rear their young. The adults feed in the open water, visible in the distance, then climb onto the barren, stony ground, which they prefer for nesting, and feed the chicks. The main feeding areas are many kilometers from the breeding ground, but the penguins undertake mass movement so as to make the best of a difficult life, eating where food is available, and founding a rookery where reproduction is safe.

Birds have an impressive ability to migrate over long distances (Figure 29.9). A long-winged shearwater taken from its nest in Scotland returned from 4800 kilometers (2900 mi) across the Atlantic in 12½ days. Migrating birds are fueled by their own fat, which they build up in preparation for their flights. A typical migrating bird may store as much as 20 percent of its body weight in fat, and ruby-throated hummingbirds, which fly nonstop 800 kilometers (480 mi) across the Gulf of Mexico twice each year, store almost 50 percent of their weight as fat for fuel.

Most animals do not migrate each year. (Examples of nonmigratory animals are the Cape buffalo of southern Africa, wild turkeys, white-tailed deer, and wolves.) If the migratory impulse is primarily a search for better feeding grounds, then an animal apparently does not need to move from its home ground when food remains plentiful. But migration seems to be an inherited activity, and many organisms, including insects, fish, birds, and mammals, migrate when their environments become precarious. Whales may migrate thousands of kilometers during the summer in search of plankton, and flimsy monarch butterflies travel from Canada to southern California and Mexico every fall (Figure 29.10). Antelopes and wildebeests migrate across the Serengeti Plain in Africa when the rains renew their favorite grazing land upcountry. Emperor penguins, instead of seeking a warmer climate during the winter, look for a cold and solitary spot, where they can mate and raise new chicks away from predator seals and sea leopards. Adélie penguins use the sun as a compass, and 300,000 or more of the birds may assemble at the same rookery every year.

Hibernation

Some animals cope with environmental problems by **hibernating** (Latin, "winter") instead of migrating. Although animals usually hibernate during the winter to avoid the extreme cold and lack of food, some animals begin their hibernation periods as early as September 1. Hibernating animals use stored body fat at a very slow rate and need little oxygen because of their lowered body metabolism and temperature. Many endocrine glands stop working altogether. A ground squirrel, for instance, usually breathes between 100 and 200 times a minute. During hibernation the squirrel breathes only once or twice a minute, and its heart beats about five times a minute (Figure 29.11). Hibernating animals do not usually remain asleep without interruption all winter, and some animals may actually leave their dens for short periods.

During periods of severe drought or heat, some animals, like the African lungfish, curl up in a ball in

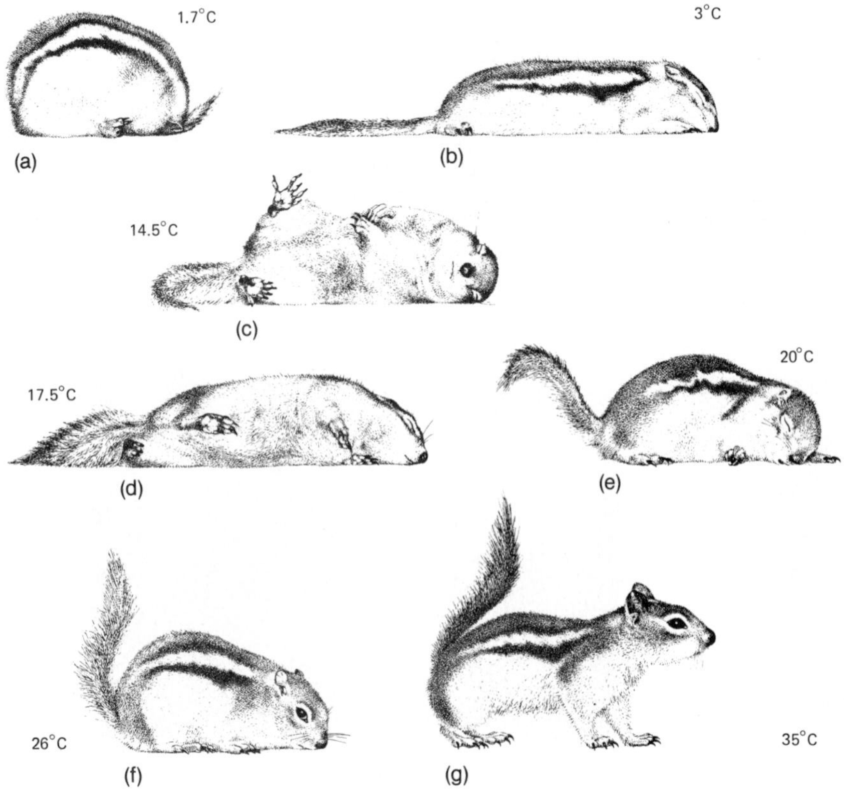

Figure 29.11
Awakening from hibernation. A ground squirrel passes from a state of lethargy, with most of its metabolic processes slowed almost to a zero rate, to an active condition as its temperature rises. (a) At 1.7°C, the squirrel is rolled into an inert ball, insensitive to most stimuli, barely breathing, and with minimal circulation. (b) At 3°C, it can move enough to stretch. (c) At 14.5°C, it can roll slowly over, but it is lethargic and unresponsive. (d) At 17.5°C, it moves a little faster and more frequently, but it is far from being awake. (e) At 20°C it is no longer so completely relaxed. (f) At 26°C, it opens sleepy eyes and is alert enough to hold its tail in the characteristic erect position. (g) Finally, with the arrival of full warmth at 35°C, it achieves full alertness: eyes wide, muscles tense, standing up and ready to go.

mucus-lined mud holes, slow down their body functions, and *estivate,* the summertime equivalent of hibernating.

ANIMAL PERCEPTION AND COMMUNICATION

Of all the animals, only humans can speak, but most animals communicate. Lacking the ability to use words, animals have developed their own methods of communication, aided by senses such as taste and smell, touch, hearing, and vision. In a murky underwater environment, where vision would be of little use, fish have well-developed chemical-sensing abilities. Bats, even more surrounded by darkness than fish are, use a sensitive and highly developed "radar" system to find their way. Birds can maintain a precise visual sense through vast expanses of air, where smells and sounds might be hopelessly jumbled. Whatever the medium, each organism has evolved its own best sensory method for survival and has applied that method toward communication within the species.

Chemical Perception and Communication

The ability to respond to chemicals is probably one of the earliest senses developed by living organisms. Insects are especially reliant on chemical sensitivity (Figure 29.12), but most of the higher organisms, including humans, use a chemical sense to one degree or another.

The champion odor detector is probably the salmon. Shortly after they are hatched, salmon swim downstream to the open sea, where they grow to maturity. Two, three, or even five years after a salmon has left its birthplace, it will unerringly retrace its path on a strenuous journey upstream to spawn in its original birth-stream. It is believed that the salmon's incredible homing behavior is due to its ability to imprint the specific odor of the water where it was born, to remember it, and to isolate it from all the other odors in the water between the ocean and the original stream upriver.

Foraging worker ants leave a scent trail for other ants when they have found food. The ants touch their extended stingers to the ground and secrete a substance that is usually detected only by members of the same species. The more satisfactory the food discovery, the more the ants will release their food scents onto the trail; in this way the trail becomes quantitatively stronger if the food is worth advertising. If another animal accidentally crosses the scent trail and disconnects it, the ants will be-

(a) × 4

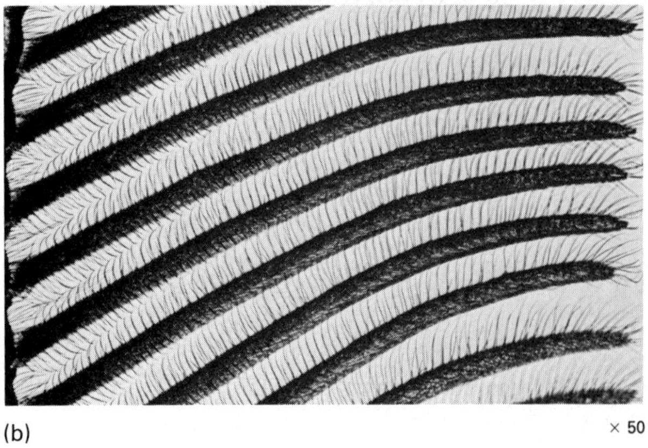

(b) × 50

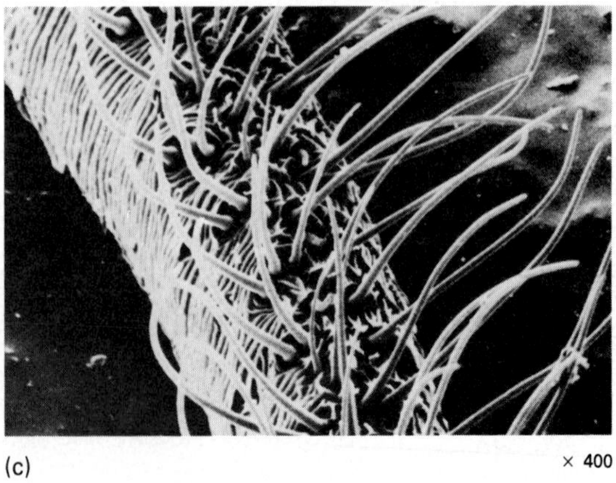

(c) × 400

Figure 29.12
The sensitive smelling apparatus of a moth, *Bombyx mori.* (a) A male moth with feathery antennae, exposing to the air a great surface that is capable of catching stray molecules. (b) A portion of an antenna, showing the branches and slender sensory hairs to which such substances as pheromones from a female moth can stick. (c) A scanning electron micrograph of a male moth's sensory hairs.

come disoriented until they make a new connection with an ant on the other side of the break. (Army ants will be discussed in the next chapter.)

The scent trail of the ant is formed by a secretion that is effective in communicating information by odor to members of its *own* species. Such a species-specific chemical substance is called a **pheromone** (Greek, *phero* "carry," and *hormon,* "arouse"). Pheromones have been referred to as social hormones. Water is the ideal medium for pheromones, and many species of fish communicate through odors. When one fish of a school is injured, for instance, it exudes an odor that serves to scatter the rest of the school away from the source of injury. Much interest has been shown in pheromones recently, especially because synthetic pheromones may eventually be successful in attracting and controlling insects that ruin agricultural crops. Certain species of insects seem to combine several pheromones to form more complicated messages. Ethologists are finding more and more evidence that pheromones usually work in combinations, sometimes with other stimuli.

Sound Perception and Communication

If humans could hear all the animal sounds in nature, they would be overwhelmed by the intense noise level. In fact, any given animal is unaware of much of what is going on around it. What is important is that the animal be able to perceive whatever it needs for its survival. Too much awareness of "reality" will merely reduce an animal's capacity to respond to the stimuli that are really important in its own natural world. Of course, this applies to all the senses, not just sound.

When we consider the subject of animal sounds, we usually think of dolphins or bats because of the sophisticated "sonar" and "radar" systems they have developed. However, rodents such as rats and squirrels also use ultrasonic frequencies to communicate with other members of their species, and moths have developed such sensitive hearing abilities that they can sometimes elude even a predator bat.

On a summer evening in 1956, Kenneth Roeder of Tufts University was giving an outdoor party at which one of his guests ran his finger around the damp edge of a glass, causing a whistling sound. Roeder noticed that at that moment the several moths flapping around a nearby light suddenly fell to the ground. The moths remained motionless for about a minute and then flew away, obviously alive and apparently unharmed. Roeder inferred a connection between the whistling sound of the glass

Figure 29.13
Specialized echo receptors. A bat's ears are constructed so as to take maximum advantage of sounds that the bat originally emitted and that are reflected off some object, whether it is an obstacle in the bat's path or an edible flying insect. The large, forward-pointing, concave ears funnel the faint echoes into the auditory canal.

and the swoon of the moths, and he and Asher Treat of City College of New York began an investigation.

Perhaps the findings of Roeder and Treat will be more significant if we first examine a related portion of the story: the precise auditory mechanism of the bat, a mechanism resembling radar. Bats are nocturnal, and they can navigate successfully and locate and capture prey in complete darkness. Donald R. Griffin and others have demonstrated that bats emit high-frequency sounds that are reflected back to them when the sound waves encounter objects of almost any size. (Bats can even locate tiny fruit flies in complete darkness.) This process of locating objects by the reflection of sound waves is called **echolocation** (Figure 29.13). By using echolocation, the bat can fly in a dark room filled with obstacles as unobtrusive as thin wires without bumping into anything. Besides being able to avoid obstacles, the bat is not distracted by the usual low-frequency sounds in the environment. However, if its ears are plugged, if it is made mute, or if high-frequency sounds are introduced, the bat will crash into obstacles and will stop flying until the interfering sounds are removed or its hearing and voice are

THE EARLY BIRD IS EARLIER THAN WE THOUGHT

On the basis of current research, it seems likely that chicks and their mothers use sounds to communicate with each other while the chick is still in the egg. What is more, the unhatched chick and its siblings, also in their shells, communicate with each other before the eggs are hatched. Think back to the concept of imprinting, and imagine how much easier it would be accomplished if mother and child were indeed able to establish a bond through prenatal communication. Even if we think of imprinting with Konrad Lorenz playing the role of mother goose, we can still accept the idea of prenatal communication since we know that Lorenz imitated bird sounds during imprinting and afterward. He may have imprinted his birds so successfully because he supplemented other factors of imprinting with appropriate bird sounds.

An embryo may be primed for the call of its mother by hearing its siblings. If an embryo has been denied any prenatal communication with its siblings, it will take longer than other birds to respond to the mother's calls. Apparently the chick is initially primed by hearing its own voice within the shell, and experimental embryos whose vocal cords have been cut do not recognize the sounds of their species until two days after hatching. The removal of the priming agent retarded but did not destroy the chick's receptivity.

One of the most interesting discoveries of embryonic behavior has been the phenomenon of synchronous hatching, which has been observed in bobwhite quails. Approximately 10 to 15 hours before hatching takes place, distinct clicking communications among embryos can be heard. It seems that the clicking sounds serve to synchronize hatching by slowing down the hatching impulse in the more developed embryos while speeding it up in embryos whose development has lagged.

Human prenatal communication cannot be tested precisely, although most physicians agree that the psychological state of a pregnant woman will affect the behavior of the fetus.

restored. (Few humans can hear sound-wave frequencies higher than 20,000 cycles per second. Bats produce and hear sounds higher than 100,000 cycles per second.)

The echolocation system of a bat is so sensitive that it can detect differences in sound patterns of objects. Besides using echolocation for navigation, some bats can distinguish edible prey by echolocation. They can make the distinction successfully, even though the bat's outgoing signal is 2000 times stronger than the returning signal. Bats locate moths in their usual way, and further, bats can identify moths by the special alternating sound pattern produced by the moths' up-and-down wing motion. If bats are such impressive navigators and hunters, do they ever miss? Indeed they do, and the experiments of Roeder and Treat provide some interesting information, not only about sound perception and communication, but about evolutionary adaptation.

If bats are interested in capturing moths, moths are equally interested in avoiding bats. Roeder and Treat discovered that Noctuids and other moths have simple ears near the thorax that allow them to detect the oncoming beeps of bats. When in danger, moths may collapse their wings and fall (as they did at Roeder's party), go into a "power dive," or begin a series of twists and turns like an air-show stunt pilot. Each tactic is an evasive one, designed to avoid the bat (Figure 29.14). It has also been suggested that the Arctiid moth may be capable of producing its own high-frequency sound to further distract predators.

The members of the cetacean order—dolphins, whales, and porpoises—possess an underwater "sonar" echolocation system equivalent to the bat's "radar." When naval engineers T. G. Lang and H. A. P. Smith in Pasadena, California, conducted dolphin experiments in 1965, they noted discernible clicks, grunts, and squeaks, and six different kinds of whistles when the dolphins communicated with each other. Lang and Smith assigned meanings to some of the whistles and sequences of sounds, and other scientists have trained dolphins to react to simulations of dolphin sounds, but the language of the dolphin remains basically undeciphered.

Visual Perception and Communication

Many animals use visual displays as part of courtship rituals, and although such sexually motivated displays are not the only expression of visual communication in animals, they are certainly among the most interesting and appealing phenomena in nature.

The three-spined stickleback is a freshwater fish about the size of a goldfish, but unlike the goldfish, the stickleback has a drab gray-green color. During the mating season, however, the male stickleback abandons his protective dullness for colors befitting a springtime suitor. Figure 29.15 illustrates the mating behavior of the stickleback, which progresses in distinct stages from the establishment of a territory to courtship, mating, and caring for the newborn.

Mallard ducks provide another example of visual communication as part of mating behavior. One of the 10 possible patterns is seen in the male duck as part of a full "ritualized" courtship sequence (Fig-

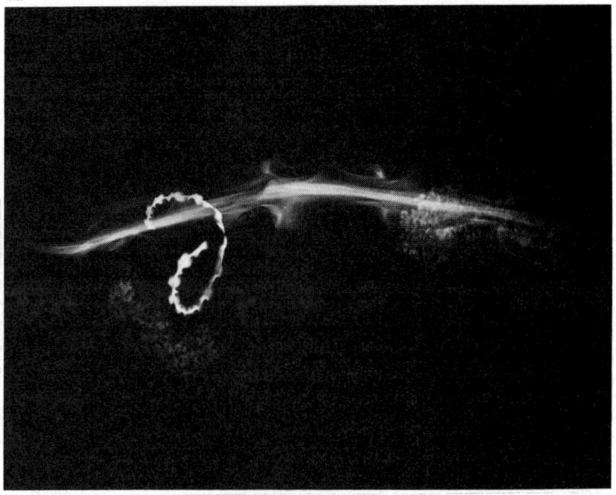

Figure 29.14
Record of a sound-guided kill. Photographed against a dark background, the long-exposure picture of a bat catching a moth shows the bat coming in from the right (the horizontal streak). The moth (the irregular central loop) tried to evade the hunter, but its flight is ended at the point where the two streaks meet. The bat could pick up the echoes of its own squeaking from the moth's wings and body, and it could do so with such accuracy that it put its prey directly on target, despite the moth's wavering flight.

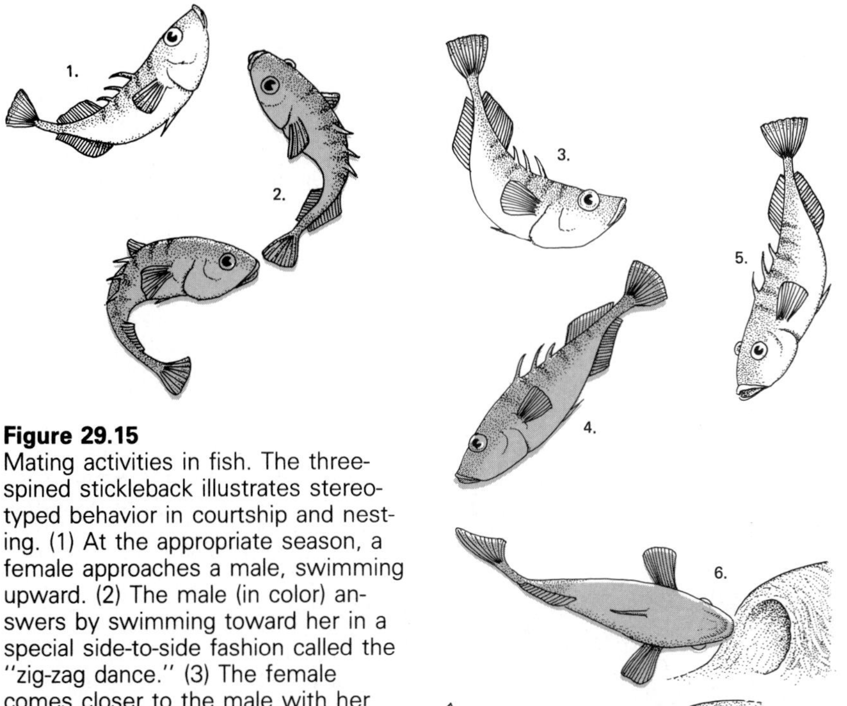

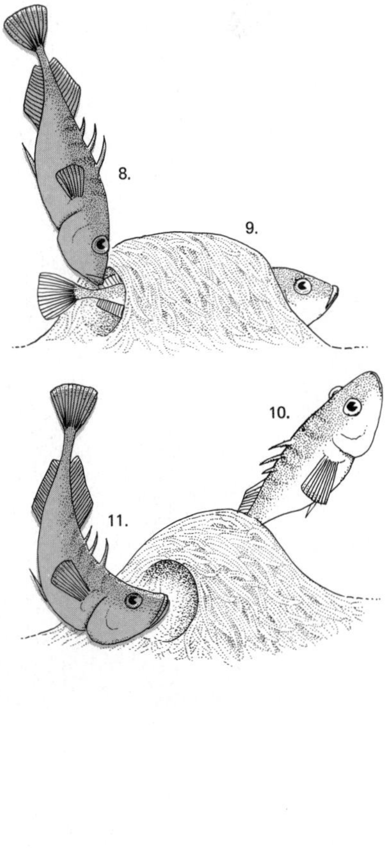

Figure 29.15
Mating activities in fish. The three-spined stickleback illustrates stereotyped behavior in courtship and nesting. (1) At the appropriate season, a female approaches a male, swimming upward. (2) The male (in color) answers by swimming toward her in a special side-to-side fashion called the "zig-zag dance." (3) The female comes closer to the male with her head turned upward. (4) The male leads the way to the nest. (5) The female follows him. (6) The male stops near the nest. (7) The female swims into it. (8) The male stands on his head and shakes with a motion known as the "tremble-thrust." (9) The female lays her eggs and (10) swims away. (11) The male enters the nest and deposits his sperm on the eggs.

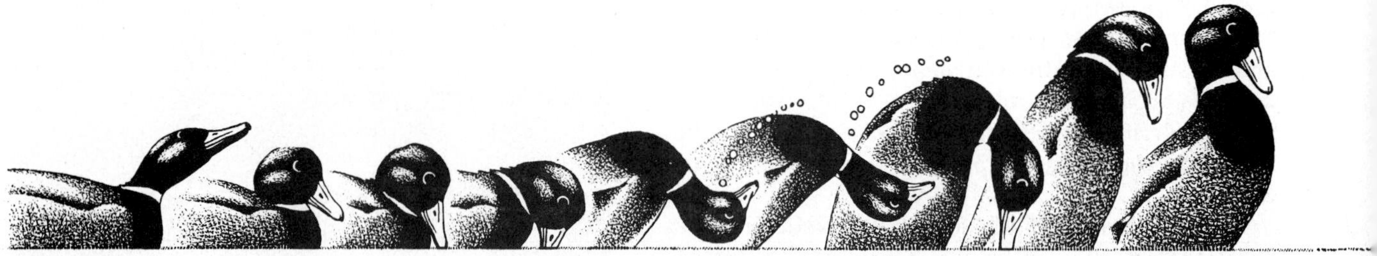

Figure 29.16
Mating display in a male mallard duck. Female ducks, dull of color, do the choosing in mating, but they normally accept only those males whose colors and movements are species-specific. A male mallard proves that he is a male mallard by having typical feather markings (green head, etc.) and by going through a set of exercises peculiar to mallards. In this series of pictures, a male mallard is shown in a part of the duck ballet that includes, among other actions, dipping his bill in the water, flipping a few drops, and raising his body high out of the water. Tinbergen calls this manuever the "grunt-whistle" display. A female mallard, recognizing the signs, may accept the male duck and allow him to copulate.

ure 29.16). The term **ritualization** was first used by Sir Julian Huxley in 1914 to describe display movements that have been modified through evolution to become stereotyped communication signals. Ritualized displays like those of the mallard duck communicate such specific messages that only members of the same species will respond.

Not everyone has seen a three-spined stickleback or a mallard duck, let alone their mating rituals. But the chances are excellent that we have all seen fireflies flashing in the dark. That, too, is a mating call. The mating signals of the firefly are simple and economical compared with the complex and ritualized displays of the mallard duck. And yet, each species of firefly produces a species-specific signal through differences in intensity, frequency, and duration of the flash.

Because fireflies are probably distasteful to predators, they can practice an extremely conspicuous form of communication in a situation where it would normally be disadvantageous to reveal their nocturnal positions. As males fly over the grass, they emit regular flashes that advertise their positions to available females hidden in the grass below. A female emerges partially and flashes a response to a male. This sequence of calls and responses is repeated several times until the male finally approaches the female, and mating takes place. (For more information about biological luminescence in general, review the essay on p. 42.)

Recently, James E. Lloyd of the University of Florida observed that in at least 12 species of fireflies the female can mimic the courtship response of other species. Once the foreign male is lured close enough by the female's flashing, she eats him.

Visual perception is important to many animals, including most birds, who depend almost entirely on their extraordinary vision to locate their prey. (See the full-color portfolio, What Animals See, following p. 414.) Insects use several perceptive

Figure 29.17
An attempt to compare the vision of a bee with that of a human being. The flower on the left was photographed in the usual way to make it look as nearly as possible like the flowers we see. The same flower shown on the right was photographed so as to accentuate the parts that reflect ultraviolet light. Since it can be experimentally demonstrated that bees are sensitive to ultraviolet (and human eyes are not), we can infer that the bees "see" the flower as it is on the right.

senses, but one interesting example of the specialized visual perception of bees is shown in Figure 29.17. Apparently, bees are able to recognize the part of a flower that contains nectar (and pollen) because that part reflects ultraviolet light, which bees can see.

Perception by Electricity and Nonvisible Radiant Energy

Adaptation to the environment is a crucial factor for survival, and some animals, like the African fish *Gymnarchus niloticus,* have evolved highly specialized sensory systems to cope with their environments. The poorly sighted *Gymnarchus* perceives by using electricity. Weak electrical discharges are generated from highly modified muscle tissue in the fish's tail, and changes in the environment are detected when the resultant electrical field from the tail to the sensory organs in the head is disturbed. Unlike electric eels and other powerful electric fish, *Gymnarchus* and certain other fishes, such as the Gymnotids and skates, use their electrical sense to locate objects and prey rather than to shock predators or paralyze prey.

Some animals have the ability to locate objects through an awareness of radiant energy in the form of heat. Rattlesnakes and other pit vipers "see" their camouflaged victims by using 150,000 nerve cells in heat-sensitive organs on both sides of the reptile's head. A female ichneumon wasp is able to locate a beetle larva several centimeters inside a piece of wood, where the larva has burrowed. After searching the surface of the wood with her antennae, the wasp drills a hole with her ovipositor through the wood into the larva. She then deposits an egg directly into the paralyzed larva. When the egg hatches, the newborn wasp will use the beetle larva as its first food source. It is thought that the wasp detects the heat of the larva, even though the larva may have penetrated deep within a tree.

SUMMARY

1. Animal behavior, like every other biological activity, is a product of evolution by natural selection, and it has been refined by selection so as to promote the survival of the species.
2. Although the response of a complex animal may be more flexible than that of a simpler animal, all organisms possess the property of *irritability,* the ability to respond to stimuli.
3. *Learning* may be considered as the modification of behavior resulting from changes in individual experience. Some types of learning are *instinct, habituation, conditioning, latent learning, imprinting,* and *insight learning.*
4. When an organism exhibits a metabolic or behavioral change in response to seasonal variations in day length, the phenomenon is called *photoperiodism.* Daily cycles of behavior are known as *circadian rhythms.*
5. *Migration* occurs when animals periodically leave their homes in favor of other locations for purposes of feeding and mating. Migratory expeditions usually take place in the fall and spring. Some animals *hibernate* during the winter, using stored body fat at a slow rate and consuming very little oxygen because of a lowered body metabolism. *Estivation* is the summertime equivalent of hibernation.
6. Animals have developed their own special methods of *perception* and *communication.* In order to communicate, some animals (like social insects) use *pheromones,* species-specific chemical substances referred to as social hormones. Other animals (bats and dolphins, for example) use *echolocation,* a process of locating objects by the reflection of sound waves. Sticklebacks and mallard ducks use visual communication, including stereotyped, instinctive communication signals called *ritualization.* Perception by *electricty* and *nonvisible radiant energy* is possible for some animals.

ASK YOURSELF

1. What is stereotyped behavior in an animal?
2. What does "habituation" mean in animal behavior?
3. What is the difference between a conditioned reflex and operant conditioning?
4. Give an example of behaviorial imprinting.
5. What clues do migrating animals use to help them find their ways across the face of the earth?
6. What physiological changes does an animal undergo when it goes into hibernation?
7. What is a pheromone?
8. How can a bat fly through a complex environment in the dark?
9. How do fishes use perception of electricity to help them survive?

30
Animal Societies

SOME KEY POINTS

1. In animal societies the single member may lack individuality for the overall good of the group.
2. Although animals will fight, they will usually try to avoid conflict if possible.
3. Courtship and mating employ definite behavior patterns that are recognizable to both partners within the same species.
4. Social insects such as honeybees and army ants are genetically programmed to perform specialized functions that exclude individual choice. In contrast, vertebrate societies allow for some individual intelligence and flexibility that favor their chances to cope with a changing environment.

ORGANISMS LIVING TOGETHER IN ORGANIZED groups are said to live in **societies.** In a human society individuals usually consider themselves more important than the group, and relationships between members of the society are based on personal recognition. In many nonhuman societies, the group is more important than the individual. In a society of army ants, for example, as far as we know, there are no individual relationships between members. In evolving a society that is like one giant organism, army ants have lost their individuality. Mammals, in contrast, have tended to evolve complex societies in which flexibility has been favored.

During the evolution of social groups, animals have evolved special ways of behaving that help groups to survive. In this chapter we will describe some of them.

TERRITORIALITY AND AGGRESSION

A **territory** may be defined as a specific area guarded against intrusion by other members of the same species, although in certain cases it may be guarded against other species.

Establishing a Territory

Territories are usually established and defended for many reasons, although some reasons may appear to be more dominant than others. A hummingbird, for instance, seeks a territory mainly for food, colonies of bats use their territories primarily for shelter, and the Uganda kob (an African antelope) selects a territory for mating.

Generally it is males who establish and defend a territory, and this is especially true during the

Figure 30.1
Defense posture. When the musk-oxen of northern Canada are threatened, they form a circle, with the bulls facing outward and the cows and calves protected in the middle. None of their natural enemies are able to harm them when they are in such a position.

mating season. Males tend to defend the center of their territories most vigorously (Figure 30.1), and as males encounter intruders toward the edge of their territory they are just as likely to flee as to fight. The point where such ambivalent behavior may be observed marks the **boundary** of the territory.

Many animals declare their territories by placing scents in conspicuous locations around the area. A male dog urinates on objects that define his concept of the boundaries of his territory. Antelopes gently deposit secretions from glands at the corners of their eyes on branches and other foliage. Male rabbits use chin and anal glands to mark territories directly or use their feces as territorial markings. Birds indicate their territories by singing, gibbons by howling, and crickets by chirping. Inherent in any establishment of territory is a declaration of living rights (Figure 30.2). If a conflict arises anyway, especially among members of the same species, the animals will use familiar gestures of threat or appeasement. In either case, they avoid actual combat

(a)

(b)

Figure 30.2
(a) King penguins on their breeding ground declare separate spaces just beyond pecking range. When conditions become crowded, the penguins defend their territories by flailing their flippers and jabbing intruders with their sharp beaks. Other animals, too, maintain spaces that are comfortable for them, such as the gulls perched on a roof at recognized distances from each other in (b). Indeed, people usually establish temporary "territories" in public places, and intrusion on such territory is considered aggression.

Figure 30.3
Fighting. Most encounters between animals of the same species stop short of physical combat. Posturing and bluffing are usually enough to send one potential combatant away, but many animals do fight, especially males fighting for territory or access to females, and in some instances injury is serious. A tag-eared, scar-faced, limping tomcat is a familiar example of a fighting male, but even small birds can on occasion tear into each other so violently that one is killed.

whenever possible, but fights sometimes do occur (Figure 30.3).

Threat and Appeasement

Threat gestures are used to intimidate and drive off an intruder without actually fighting. One of the most studied threat displays is that of the herring gull, which will stand as upright as it can (most animals make themselves look as large as possible when they are threatening), point its head slightly downward, and display its wings outward without actually spreading them (Figure 30.4). In its **appeasement gesture,** however, the gull assumes a submissive position directly opposite to the threat position, lowering its body and shortening its neck, holding its wings in close, and pointing its bill upward.

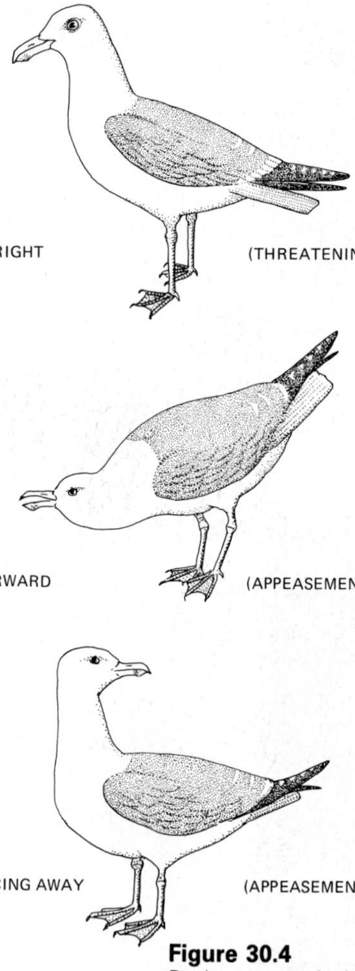

Figure 30.4
Body posture in herring gulls, which is understood by other gulls. The upright position is a threatening position. The head-ducking-forward position signifies appeasement, as does the gesture of facing away. George Schaller, a student of gorilla behavior, found that he could approach wild gorillas safely if he stopped walking forward and turned his head away at any sign of apprehension on the part of the gorilla.

Courtship and Mating

Definite signals are used within a species to attract a sexual partner, to deter aggressiveness or threatening behavior, and to synchronize male and female actions so that copulation may take place (Figure 30.5). Conspicuous displays are important to bring the male and female together, but they are also a crucial factor in avoiding mating between members of different species, since such interspecific mating may produce offspring that are infertile or die soon after birth. Just as songs by male birds both threaten males and attract females, so threat and appeasement gestures serve the dual purpose of avoiding fighting and initiating courtship and mating.

Courtship often possesses the conflicting elements of approach and attack behavior because animals normally avoid close contact with other animals, even members of the same species. Therefore the body contact initiated during courtship and mating arouses ambivalent motivations of escape behavior and mating behavior at the same time (Figure 30.6). If the proper courtship displays have been sig-

Figure 30.5
When prairie dogs of the same social group meet, they "kiss" and groom each other. If members of a different group meet, they are likely to make threatening gestures or even fight.

(a)

(b)

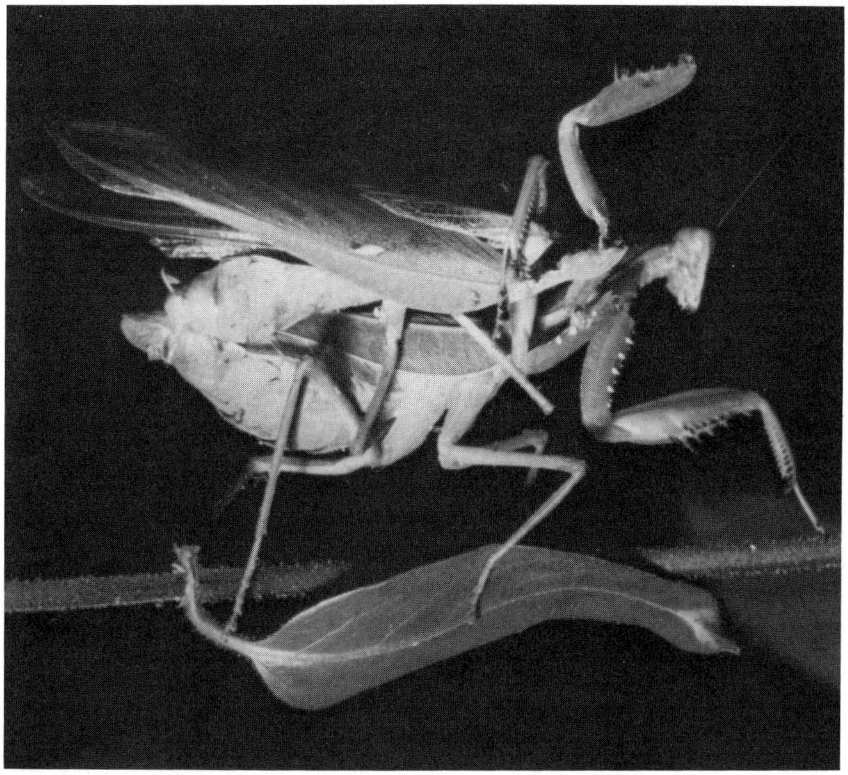

(c)

Figure 30.6
Courtship and mating in animals takes many forms. (a) A male bighorn ram, *Ovis canadensis,* races after a female who has attracted him with her scent trail. Her avoidance behavior is part of their ritualized courtship, which normally concludes with mating in the autumn and the birth of a lamb in the spring. (b) A pair of wandering albatrosses, *Diomedia exulans,* have a loud pre-mating ritual that includes hissing and rattling. At the climax of their courtship display, the male and female bird face each other, spread their wings, and screech to the sky that they are ready for mating. (c) One of the more unusual mating procedures occurs when the female mantis actually devours her mate *while* they are copulating. Although the female does not always eat the male during mating, it happens often enough so that he is a wary suitor. In this photograph, the male's head and upper body have already been eaten by the female, but he will continue to copulate until his midsection is gone.

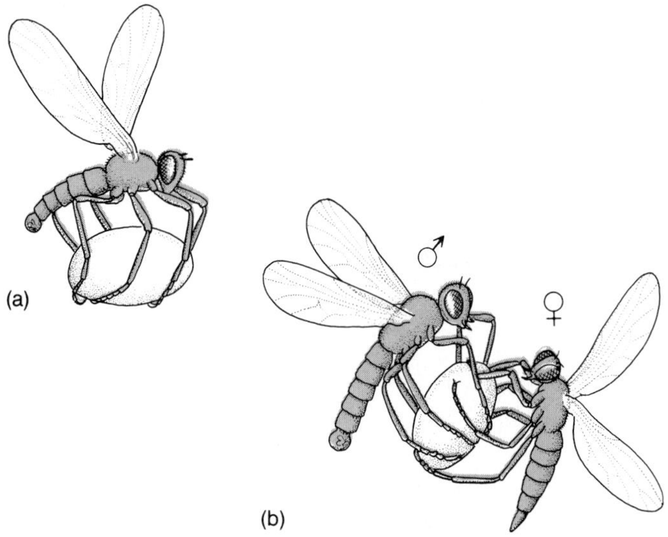

Figure 30.7
Symbolic courtship. (a) Some species of balloon flies practice what appears at first sight to be a meaningless ritual, as a male brings a silken balloon he has made to a female. (b) After she accepts the present, they mate. Such action becomes understandable when one knows the activities of related species. In some species, a male brings the female a small captured insect for a food offering, possibly to keep her too busy to eat *him*. In other species, he attaches bits of silken fiber to his gift, making it more distinctive. In still other species, he encases the food in a full covering of silk. The final step is to make the wrapping so elaborate that the contents are omitted, and an empty balloon becomes as acceptable as a real present.

naled, the mating instinct will be stronger than the urge to escape, and mating will take place. But until both partners have demonstrated that no real danger exists, the presence of a potential mate intruding on a territory will arouse conflict possibilities as well as sexual ones.

During courtship among birds it is not unusual for the male to offer food to the female. This action is a likely evolution from the food-begging of young birds, a posture repeated by the female bird as an appeasement gesture toward the aggressive male. Some male insects also include a food offering in their courtship displays, even though such an offering is often merely symbolic (Figure 30.7). Mouth-to-mouth courtship feeding by bullfinches is almost certainly derived from similar gestures used by the parent bullfinches when they feed their chicks, and mouth-to-mouth courting among chimpanzees is also similar to one method of feeding the young. It is speculated that kissing in humans may have originated from the primate courtship practice of touching the lips and making smacking and sucking sounds.

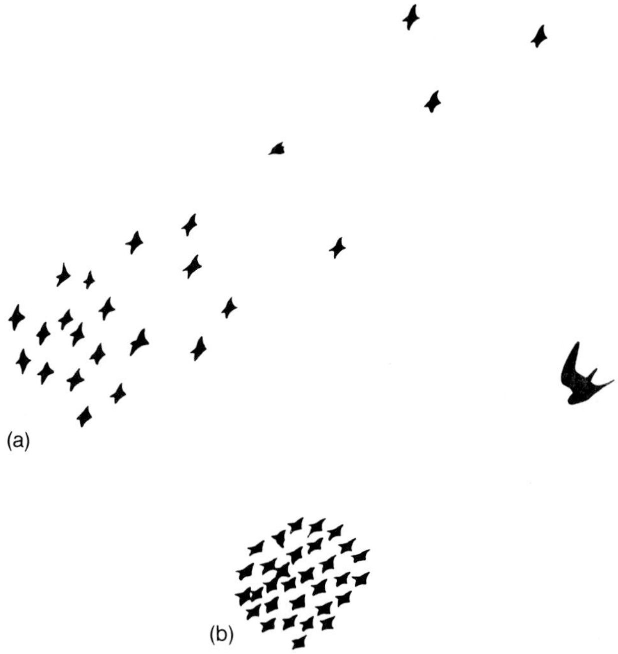

Figure 30.8
Alarm response in starlings. When threatened by a hawk, a loose flight of birds (a) draws into a more compact flock (b). It may be more difficult for a hawk to single out a victim from among a dense flock, or diving through such a flock may be dangerous for the hawk, which can knock down one starling easily enough but may injure itself by hurtling into several. Whatever the reason, birds (and some fishes) bunch up in tight groups when they are threatened.

COOPERATION AND COMMUNICATION

Animals in a society exhibit a high degree of cooperation, made possible by a species-specific system of communication. In the simplest case, the societal relationship of a mother and her child constitutes the primary unit of a social organization, but in a "true" society we look for a group of individuals working together. One aspect of this system of mutual benefit is **altruistic behavior.** In one type of altruistic behavior, an individual sacrifices its own reproductive opportunity to help a genetically related

Figure 30.9
Wild dogs cooperate to hunt and kill a wildebeest that has been separated from its herd. Unlike wild cats such as lions and tigers, wild dogs cannot make a clean kill on their own, and the cooperation of a pack is always necessary to pull down a large animal.

individual raise its offspring. Year-old scrub jays, for instance, sometimes give up their first reproductive opportunities in assisting their mother with the care of the newest brood.

Animals do not make a conscious decision to be altruistic or to concentrate on cooperative ventures that will include their genotype in the overall gene pool, but natural selection has favored those organisms that are programmed to propagate their genotype. An altruistic sibling might best perpetuate his or her genes through the successful reproduction of the brothers' and sisters' genes.

Probably the most obvious advantage of groups is massed defense against predators (Figure 30.8; see also Figure 30.1), as predators may be threatened or confused by large groups. Other advantages can be seen in predatory animals such as lions and hyenas, which hunt in groups (Figure 30.9). During the winter, bees increase the heat within their hives by constantly buzzing and moving their bodies to generate heat, sometimes producing an internal temperature 15 to 20 degrees higher than outside the hive. Insect societies provide countless examples of successful group living, and we will look at some of them now.

SOCIAL INSECTS

Most insects are among the most solitary of animals, but honeybees and army ants, for example, clearly are social insects.

Honeybees

The social organization of the honeybee is highly developed (Table 30.1). The honeybee society con-

TABLE 30.1
Division of Labor in Honeybee Workers

1. *First three days.* The newly emerged worker leaves its cell, begins to clean cells.
2. After cells are cleaned, the queen deposits a new egg in each.
3. *Fourth day.* Worker collects honey and pollen from stores. Begins to feed the older larvae.
4. *Seventh day.* Begins feeding younger larvae, which receive an easily digested "milk" secreted by glands in the worker's head. Begins making short reconnaissance flights, without gathering nectar or pollen.
5. *Tenth day.* Stops working with brood and takes on various household duties. Takes nectar from incoming foragers, deposits it in cells or feeds it to other bees, packs the pollen brought by foragers into the pollen cells, builds new cells using wax secreted by its wax gland, carries away dead bees and rubbish.
6. *Twentieth day.* Becomes a guard, inspects every bee arriving at the hive entrance. Twenty to thirty guards stand watch simultaneously, driving off intruders.
7. After a short period as guards, the workers become foragers, flying into the country and collecting nectar and pollen. They remain as foragers until they die about three weeks later.
8. Some foragers are scouts, looking for new food sources, and they communicate the position and quality of a food source by means of their waggle dance.

Adapted from N. Tinbergen, *Social Behavior in Animals,* Second Edition (London: Methuen; New York: John Wiley, 1965), pp. 101–102, from data originally formulated by Martin Lindauer.

Figure 30.10
A beehive is a busy but well-regulated entity. Its seeming confusion is illusory. Many activities are going on, some all at once and others changing with the seasons, but each activity is precisely determined. Some of the main functions of a hive are shown here. (a) Most of the workers (sterile females) are engaged in bringing nectar and pollen from outside, furnishing the food for the colony and the material for building the waxy comb in which the young are raised. (b) Some "engineers" make wax from the pollen and construct the brood cells, objects that have long been a subject of admiration for their geometrical precision. (c) When workers, making up the great bulk of the population, emerge from their pupae and leave their cells, nurses prepare the empty cells for the next brood. (d) The duty of some workers is to ventilate the hive. They stand at the entrance and fan their wings to maintain the best combination of temperature and humidity for the health of the bees and the concentration of the honey. The evaporation of water from dilute nectar is like the boiling down of maple sap to make syrup. (e) Other workers defend the hive against accidental or threatening intruders, as shown by the attack against a wandering caterpillar. (f) One worker, having discovered a food source, is climbing the walls in a pattern that is understood by the other workers. They will take off to find the food source. (g) During cold weather, masses of bees clump together and vibrate their wings in an action something like shivering. They can generate enough heat to keep them from cold damage. (h) The males (drones) do nothing but wait for a chance to mate with a queen. Their lack of productivity is so well known that any lazy parasite on society is called a drone. The single queen in a hive is just an egg-layer. She is fed and cared for by the workers, and in return she produces the thousands of eggs needed to keep the hive active. She can determine whether the eggs will be fertilized and yield diploid females, or will not be fertilized and yield haploid males. Whether the female will develop into a sterile worker or a fertile queen is determined by the nourishment provided to the growing larva. If queens are to be produced, special brood cells are constructed, and the larvae in them are fed with the queen-making diet. (i) When several queens emerge at once, they fight it out for supremacy until only one is left. She will leave the hive, fly high into the air, and there mate many times with accompanying drones. She stores enough sperm to last her lifetime. The whole intricate system works because each individual can react in response to what it sees, smells, tastes, and (probably) hears.

sists of one queen, sterile female workers, and male drones, totaling between 20,000 and 80,000 in each hive (Figure 30.10). Why do so many honeybees live together, and how is their society regulated to avoid conflict and chaos? It is clear that the overall benefit of the hive supersedes a bee's individual needs, and even the apparently rigid division of labor shown in Table 30.1 can be altered by means of the bees' remarkable communications system to cope with environmental changes.

The dominant member of the honeybee society, the single female queen, does not take part in the routine activities of the hive. Her main function is to produce eggs, and she lays more than 1000 eggs each day of her life, which may last 5 to 10 years. The queen controls the population of the hive through the secretion from her mandibular glands of a pheromone popularly known as "queen substance." Actually, the queen secretes several pheromones, but it is a single compound, 9-ketodecanoic acid, that prevents the production of other queens and also inhibits the ovarian development of workers, so that no new eggs will be introduced into the hive. Thus the queen eliminates any rivals. The queen substance is licked off the queen by the workers (Figure 30.11). They quickly pass it on to all the other workers in a process of regurgitation (vomiting). The exchange of food and pheromones through regurgitation is essential for communication within insect societies. If the queen is removed from the hive, or if her pheromone production is otherwise decreased or halted, the workers will sense the loss in a matter of minutes and will become agitated.

Figure 30.11
A large queen bee (center) being groomed by workers.

Figure 30.12
A spring swarm of honeybees. Scouts return to the swarm and report the location of likely new homes for the colony.

Royal cells are not built as long as the mother queen is secreting "queen substance." But if the queen's production of 9-ketodecanoic acid decreases, as it usually does at the beginning of the spring reproductive season, the workers begin to construct large royal cells in preparation for a new queen. The queen lays one fertilized egg in each royal cell, and the workers feed a "royal jelly" to the larvae, enhancing the development of the larvae into queens. New queens, when one is needed, will develop into adults only 16 days after the eggs have been laid, compared with 21 days for a worker and 24 days for a drone. The old queen's decreased pheromone production signals another change in the behavior of the workers, who feed and groom her less and eventually expel her from the hive, which by now is overcrowded because of the new broods that have been produced during the peak spring nectar season. Soon afterward, the old queen and about half the workers will swarm to the site of a new hive, where they will form a new colony (Figure 30.12).

The old hive does not continue without a queen for long. Before leaving the hive, the queen also lays an unfertilized egg in each of the newly constructed drone cells. These unfertilized eggs develop into haploid male drones, which leave the hive and await the emergence from the hive of a new virgin queen. The young queen then disperses sufficient quantities of 9-ketodecanoic acid to attract drones waiting in a swarm downwind. The queen makes several nuptial flights a day, perhaps 15 flights in a week, until she has mated with several drones and accumulated enough sperm to fulfill the needs of a lifetime's production of fertilized eggs. All but a few of these eggs will eventually become sterile female workers. During mating the genitalia of the drone explode as he deposits his sperm in the queen's genital chamber, and he dies almost instantly. Drones that do not mate are driven out of the hive by the workers in the fall, when the food supply becomes low. After the new queen has completed her mating flights she returns to the hive to join workers in a new swarm, or she returns to the hive, kills the other emerging virgin queens, and begins a new colony within the hive.

Most species of bees are solitary. What are the advantages of living in a society as complex as that of the honeybee? The greatest advantage is probably the increased efficiency of foraging, especially because flowers are a very spotty food source, and many honeybees working individually to acquire nectar would not be nearly as efficient as a highly communicative and cooperative society.

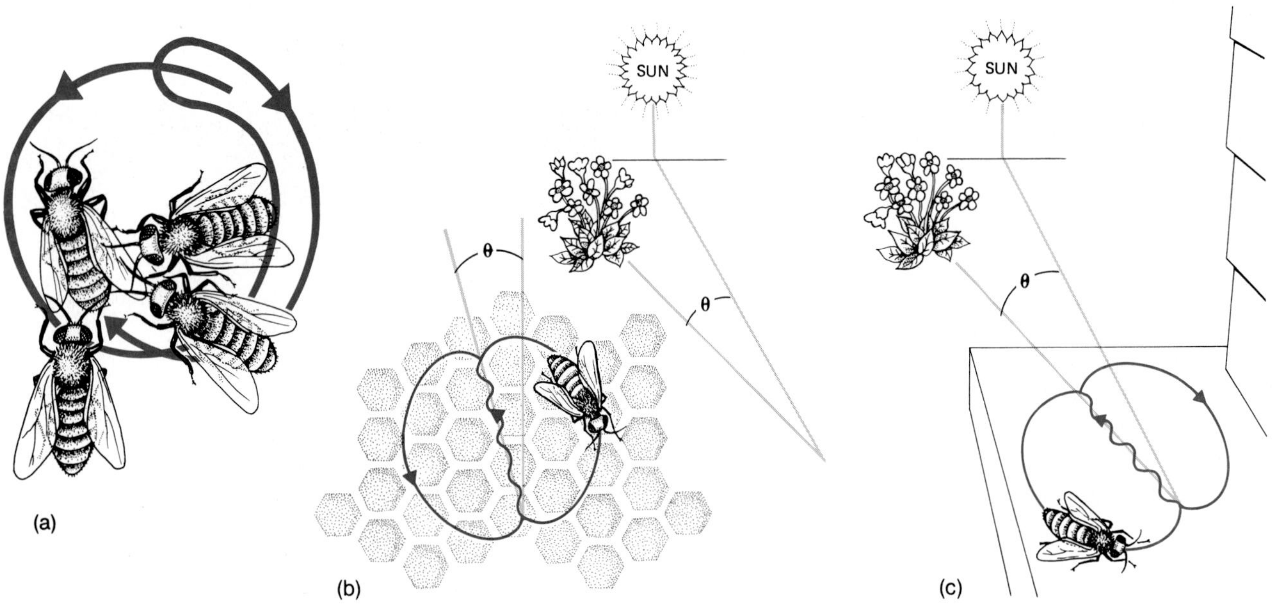

Figure 30.13
The round dance and the waggle dance. (a) When the food source is less than 75 meters (250 ft) from the hive, the scout bee will perform the *round dance,* a movement of alternating clockwise and counterclockwise circles. The richer the food supply, the more vigorous will be the dance. The direction of the nectar is not indicated, but the worker bees detect the scent of the flower on the scout bee, and a drop of nectar from the flower is also regurgitated by the scout and sampled by the worker bees. Besides these clues, the scout has deposited its own scent on the flower, and this odor is easily detected by the searching workers.
(b) A scout bee whose food source is not in the immediate area of the hive will perform the *waggle dance,* a figure eight with a straight line between the loops. Usually the scout bee performs the waggle dance on the vertical surface of the dark hive. In this case, the straight run of the bee between loops will be translated into an angle related to the imaginary vertical line of gravity. The scout bee has transposed this same angle from its own orientation method of relating the nectar source to the position of the sun. As shown in the figure, both angles are the same. When the worker bees leave the hive they retranslate the gravity-related angle they have seen into the same angle in relation to the sun. A similar ability to transpose angles of orientation from the sun to gravity and vice versa is not uncommon among such insects as the ant and the dung beetle. If the scout bee performs the straight-run waggle upward, the worker bees will fly toward the sun; a downward waggle tells the bees to fly away from the sun. Distance is indicated by the number of figure eights—the more performed, the closer the food source and vice versa. Humming sounds made by the wings of the scout bee during the waggle dance probably provide redundant information about the food source.
(c) Occasionally a scout bee will perform its waggle dance on a horizontal surface outside the hive. In this case, the straight position of the figure eight of the dance will point directly to the food source.

The language of the bees The foraging worker honeybee leaves the hive in search of food. When she finds it, she returns to the hive and achieves a remarkable feat: She not only communicates to the other bees that food has been located, but informs them of its direction and distance from the hive. Therefore, without actually being led to the food, the other bees know exactly how to find it. Such precise information is relayed by either of two kinds of dance: the **round dance** or the **waggle dance** (Figure 30.13). Bees that have been raised outside the hive will respond accurately to the dances the first time they perceive them, indicating instinctive behavior.

Besides indicating the distance, direction, and quality of a food source, bees can also relay information about a new hive location when the present hive becomes overcrowded. Scout bees use the waggle dance to communicate with the waiting queen and other bees, and these worker bees may investigate many potential sites. The quality of the sites is translated through the enthusiasm of the dance, until one site is finally chosen after a few days. Then about half the inhabitants of the hive, as many as 40,000, will swarm to the new location.

Karl von Frisch deciphered what he called "the language of the bees" during a series of experiments that culminated in the 1950s. Recently, several biologists have presented arguments opposing von Frisch's interpretation of the language of the bees. They propose that the factors of communication are odor and sound, but further experiments tend to confirm the original findings of von Frisch, with the possibility that bees use pheromones, built-in magnetic compasses, and even smell to communicate the location of food sources in occasional instances when they do not use the dances.

Army Ants

Army ants, usually found in tropical areas such as South America and Central America, are organized into nomadic colonies of 100,000 to 500,000 or more. They have the most complex animal society operating outside a fixed homesite. A colony of army ants consists of a single queen, a brood of young ants in varying stages of development, and sterile female workers. (Winged males develop during the mating season.) As with the honeybee, a division of labor exists among worker ants, but the division in the ants is formulated according to the worker's size instead of its age. The smallest workers feed the larvae; the medium-sized workers, making up the majority of the colony, spend most of their time raiding for food; and the largest workers, known as soldiers, are armed with large heads and powerful mandibles that make effective weapons against predators (Figure 30.14).

Army ants are raiders, typically living off other arthropods encountered on their systematic marches, which may advance at a rate of 35 meters (115 ft) an hour in an unbroken column 300 meters (1000 ft) long. Such an organized procession should be an unlikely activity for the near-blind ants, but they are guided by their own odor trail of pheromones, tipped onto the ground from the stingers of the lead ants and reinforced by the ants that follow (Figure 30.15). This *nomadic phase* of the ants' migra-

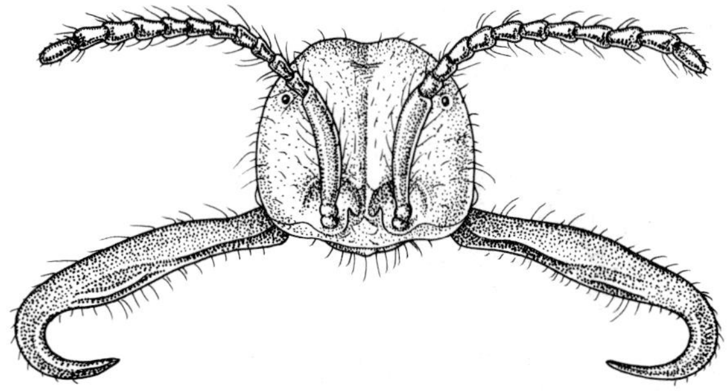

Figure 30.14 × 50
Soldier ants of the genus *Eciton* have large heads and powerful, sharp jaws that make formidable weapons.

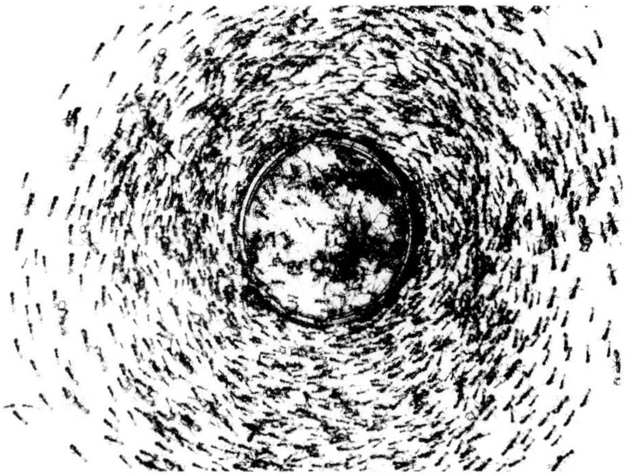

Figure 30.15
A whirlpool of army ants. When on the march, these ants follow a scent trail laid down by the leaders. In the experimental situation shown in the photograph, a circular trail was placed and ants started along it. If permitted to do so, they would probably keep going around until they starved, because they do not use their eyes or any individual initiative.

Figure 30.16
Army ants on the move. The mass advances like a living carpet, gliding along and conforming to every bump and depression on the surface of the ground, eating every unwary insect or even much larger animal they touch. Fortunately, they move at an ant's pace, and a person can walk slowly from them, but the immensity of their numbers and their reputation for destructiveness make observers watch their step.

tory activity lasts 17 days, during which they systematically move forward at dawn to conduct predatory raids, biting, stinging, and dismembering their prey, usually other insects (Figure 30.16). The raider workers doggedly lead the way, followed by soldiers and larva-carrying workers. The huge queen, almost completely hidden by her retinue of tumbling workers, brings up the rear.

Unlike most species of ants, army ants do not build permanent underground nests. Each night after their raids are completed, army ants form temporary nests or "bivouacs," by clustering together with their legs hooked to one another to form bristling, dangling cylinders that surround and protect the queen and brood. The nomadic stage ends abruptly, and the colony then halts its migration, beginning a *statary period* of 19 or 20 days. During the statary period infrequent raids take place, and very little prey is returned to the nest. The nomadic and statary phases coincide exactly with the queen ant's reproductive cycle.

The statary phase begins when the larvae from the last clutch of eggs have progressed to the pupal stage. After a week, the queen produces 20,000 to 30,000 new eggs, and the daily raids begin to increase. On the twentieth day a new generation of workers is born, and daily foraging raids resume at full speed. From these observations it can be concluded that the cyclical behavior pattern of the army ant is regulated not by external stimuli, such as the need for food, but rather by interactions between the adults and brood.

Like the honeybees, male army ants serve only a reproductive function. Winged males are produced during the summer to prepare for their nuptial flights, which take place away from the nest. After mating, the males do not return to the nest, and without social contact with the colony they soon die.

SOCIAL ORGANIZATION IN VERTEBRATES

Many vertebrates are social animals. Primates such as African baboons and Japanese macaques provide vivid examples of stratified vertebrate societies. As with other socially oriented animals, the behavior of baboons and macaques results from a combination of social and biological forces.

African Baboons

Baboon troops, usually containing from 10 to 200 members, are held together by a strict social order that permits individual relationships to function within the cohesive framework of the group. The troop functions and moves as a unit. The dominant males travel in the center of the troop, surrounded by females and juveniles (Figure 30.17). At the rim of the troop are the lower-ranking adult males, alert to join forces against predators that may attack from any direction. So well organized is the baboon defense system that perhaps only the lion is a real threat, and this ability to defend against predators is one of the major reasons for the evolutionary success of baboons. Baboons that leave the troop because of injury or for any other reason have little chance of surviving on their own.

Within each baboon troop there is a firm dominance hierarchy ruled over by one or more high-ranking males. Dominant males protect lower-ranking members of the troop and exercise a general role of establishing and maintaining peace and order within the troop. Baboons usually avoid fights within the species, and low-ranking members are always willing to discontinue their disputes when the dominant male intervenes. One of the benefits of a dominant male's position is the sexual receptivity of the females in heat (estrus), who approach a dominant male and make the formal gesture of presenting her swollen genitalia. When a female in estrus is accepted by the male, the two form a consort and remain together throughout the estrous period, grooming each other and copulating frequently (Figure 30.18). Usually the pair separates when the estrous period ends, but if the female has become pregnant, she may continue to follow the male closely.

Infant baboons are extremely attractive to all members of the troop. They will be fondled and groomed incessantly by males and females alike, even though they constantly cling to their mothers for one or two months. Apparently this early mother-child bond is a durable one; juvenile baboons retain their maternal attachments throughout their early years. As the infant develops, it spends more and more time with other young baboons in play groups, where it seems to practice adult gestures and begins to learn the social mannerisms of the adults. The importance of imitative play groups can be seen in experiments with male juveniles raised apart from the troop, who are unable to

Figure 30.17
Baboons moving in a group. Dominant adult males, large and heavily maned, protect females and small infants in the center of the group. Young juveniles are shown below the center, and older juveniles are shown above. Other adult males, also large and heavily maned, guard the front, rear, and sides of the troop. Adult females in heat, indicated by their dark bottoms, are accompanied by adult males.

Figure 30.18
Adult female baboons often groom a dominant male.

Figure 30.19
Behavior by imitation. The usefulness of having something to imitate is shown by baboons in successful and frustrated attempts to mate. (a) A male mates with a female, using her legs as a platform in what is a normal procedure for baboons. (b) The clumsy efforts (and failure) of a male in the presence of a receptive female. Having been raised in isolation and not having had the advantage of seeing others copulate, he has only an imperfect idea of how to behave.

mount inexperienced females properly because they have never watched adult baboons copulating (Figure 30.19).

A complex communication system exists in most primate societies. African baboons have a repertoire of vocal expressions ranging from a grunt to a shrill bark of warning. Friendly gestures may include lip-smacking and genital presentation. Scratching, shoulder-shrugging, and other bodily gestures are used to relate feelings of fear, escape, or uncertainty. Slapping the ground with the hands indicates threat or attack. An adult male will gently bite the nape of a female's neck to express sexual responsiveness. Besides tactile, visual, and auditory signals, baboons also communicate by odors.

Japanese Macaques

Japanese macaque monkeys usually live in troops containing 50 to 150 members. A typically male-dominated social hierarchy exists, and a male's rank is in direct proportion to the status of his mother. Each macaque participates in specific activities according to its sex, age, and rank, and it appears that these specialized activities are a major factor in the cohesiveness of the social system. As with baboons, a dominant male macaque has the highest-ranking position as leader and defender of the troop.

An example of the kind of behavioral flexibility that does not exist in the rigid societies of insects has been observed in a confined macaque troop. One day after a snowstorm, a low-ranking male stopped eating a snowball and began to roll it in the snow until it was the size of a large beach ball. It is now common for this male and other macaques to roll large snowballs after every snowfall. Juvenile monkeys seem to delight in playing with the snowballs, and adults are content just to sit on them (Figure 30.20). Rather than having a definite and prescribed function, the snowballs seem merely to represent manifestations of the curiosity, adaptiveness, and skills of the monkeys. Such an ability to adapt their behavior to changes in their environment has certainly favored their survival.

Figure 30.20
Japanese macaques playing in the snow.

SUMMARY

1. Organisms living together in organized groups are said to live in *societies.*

2. A *territory* is a specific area guarded against intrusion by other members of the same species. Territories are most often established for purposes of food, shelter, or mating. Animals may declare their territorial claims in several ways, including scent markings and sounds; inherent in each method is a declaration of living rights.

3. *Threat gestures* are used to intimidate and drive off an intruder without actually fighting, and *appeasement gestures* indicate a submissive position directly opposite to threat gestures.

4. Special species-specific signals are used to attract a sexual partner, to deter aggressiveness or threatening behavior, and to synchronize male and female actions so that copulation may take place. Sometimes courtship includes conflict behavior because animals normally try to avoid close contact with other animals.

5. Animals in a society exhibit a high degree of cooperation, made possible by a species-specific system of communication. One aspect of this system is *altruistic behavior.*

6. Most insects are solitary, but honeybees and army ants have rigid social organizations.

7. Besides indicating the distance, direction, and quality of a food source, honeybee scouts can also relay other information to members of the hive by using the *round dance* or the *waggle dance,* part of the "language of the bees."

8. Some primates, such as African baboons and Japanese macaques, live in stratified societies. As with other socially oriented animals, the behavior of baboons and macaques results from a combination of social and biological forces.

ASK YOURSELF

1. How may an animal let other animals know where its territorial boundaries are?
2. What useful purposes may be served by courtship displays?
3. Give several examples of cooperative behavior within a species.
4. How is a new queen in a bee colony produced?
5. If a honeybee finds a food supply, how does it let the rest of the colony know about it?
6. What regulates the movement or lack of movement of a colony of army ants?
7. Why are sociologists and anthropologists interested in the activities of baboon troops?
8. What evidence is there that primates are less limited in their behavior than insects?

31
The Origin and Development of Life on Earth

SOME KEY POINTS

1. The universe originated when densely packed gases exploded and produced the raw materials for life. The earth's primitive atmosphere included compounds that contained hydrogen, nitrogen, oxygen, and carbon.
2. From the basic elements, complex molecules are thought to have been built spontaneously over long periods of time.
3. The first cells were probably something like present-day bacteria, simple cells without nuclei. Later, more complex cells with nuclei evolved.
4. Photosynthesis made it possible for life to evolve on land.

FOR UNCOUNTED YEARS, THE HUMAN MIND HAS puzzled over the problem of where life came from and how the earth came to be. Every social group, no matter how simple or isolated, has a story of creation. The expressions range from the general and poetic, as in the Genesis account ("In the beginning God created heaven and earth") to the elaborate and fantastic, complete with monsters, superheroes, imaginary beasts of outlandish anatomy, hundreds of gods and goddesses, storms, earthly upheavals, floods, spirits, and magic. We can never know how life began; we were not there. But with experimentation we can begin to understand how it *might* have happened.

WAYS OF EXPLAINING THE ORIGIN OF LIFE

A number of explanations of the origin of life have been developed, all of them philosophically interesting, but each presenting its own problems. An ancient and widely followed belief is that of the Hebrew histories: God created life. Since such a belief is beyond experimental attack, it is outside the defined limits of biology as a science. *Special creation* is a matter of faith and cannot be proved or disproved. A slightly less mystical idea is contained in *vitalism,* according to which life has a force and quality of its own, something different and apart from the forces and mechanisms of the inanimate world. Like special creation, vitalism is not subject to experimentation, and consequently it will not be considered further in this text. A modification of vitalism is *panspermia,* the idea that some kind of "seeds of life" are universally distributed everywhere, needing only suitable conditions for development. This idea approaches the notion of spontaneous generation, to be treated in some detail below.

The possiblity that life came to the earth from outer space has been considered repeatedly over the years. Several bits of evidence make it impossible to dismiss extraterrestrial life out of hand. Meteorites, investigated with precautions to exclude earthly contamination, have been shown to contain particles that resemble microscopic fossils and organic compounds. In addition, radio telescope data and spectroscopy of other planets and of space itself reveal the presence of a wide variety of organic compounds. The ingredients necessary for the making of cells are out there. One difficulty with the idea of extraterrestrial life is that it explains nothing and pushes the problem away in time and space, making practical work impossible.

The most promising efforts to find a plausible explanation for the origin of life came from the hypothesis that if the necessary energy, required chemical elements, and suitable physical conditions once permitted the synthesis of living stuff, we might be able to repeat the original event. Although no living cells have been manufactured with present techniques, some suggestive results have been obtained, and these will be the subject of much of this chapter.

Spontaneous Generation

For centuries it was believed that plants and animals could produce themselves without parents. Mold seems just to appear on bread; dead carcasses left exposed soon become crawling heaps of maggots; an old bottle of wine comes to contain a teeming mass of worms. One convincing experiment was made by a seventeenth-century Italian, Francesco Redi, who showed that meat exposed to air came to contain maggots only if flies were allowed to reach it. Meat kept in covered jars remained maggot-free. But the idea of **spontaneous generation** died hard.

No one understood how microorganisms got into things until Louis Pasteur (Figure 31.1) in 1864 devised a simple means of overcoming all objections to earlier experiments. His flasks, filled with sterilized but easily fermentable substances, such as wine or urine, had openings to the air, but the necks were bent so as to trap foreign particles and prevent them from entering the flasks (Figure 31.2). His

Figure 31.1
Louis Pasteur (1822–1895), one of the great scientists of all time. A man of enormous patience, compassion, imagination, and insight, Pasteur (pronounced Pa-STIR) has to his credit a list of accomplishments that no other scientist can equal: contributions to knowledge of molecular symmetry, fermentation, "spontaneous generation," the germ theory of disease, sterile surgery, and the use of weakened pathogens in the treatment of disease.

Figure 31.2
Pasteur's demonstration that organisms of decay come only from preexisting organisms of decay. This simple but ingenious procedure ultimately (but not immediately) stilled the voices of the proponents of spontaneous generation. It answered the objections to previous experiments, especially the objection that free air is necessary for spontaneous generation, and was so straightforward and convincing that it earned the highest praise scientists can give an experiment: It was elegant.

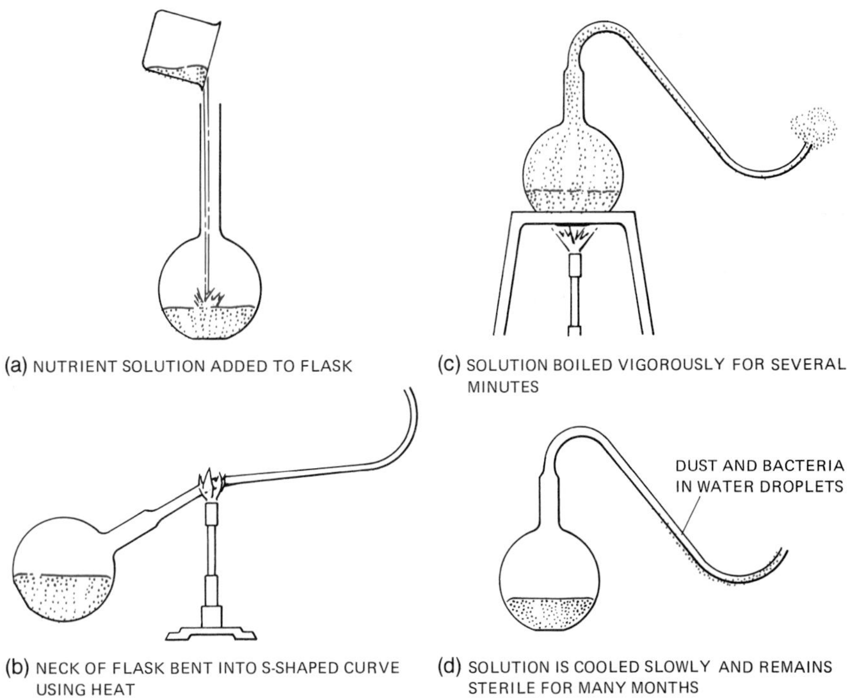
(a) NUTRIENT SOLUTION ADDED TO FLASK
(b) NECK OF FLASK BENT INTO S-SHAPED CURVE USING HEAT
(c) SOLUTION BOILED VIGOROUSLY FOR SEVERAL MINUTES
DUST AND BACTERIA IN WATER DROPLETS
(d) SOLUTION IS COOLED SLOWLY AND REMAINS STERILE FOR MANY MONTHS

Figure 31.3
How the earth may have appeared during its formative period, with gases and vapor billowing from cracks in the ground. The picture is from a contemporary scene: the Nyamlagira crater in the Congo region of Africa.

preparations remained clean indefinitely, showing that even bacteria are derived only from parent bacteria. Thus matters were settled as far as the life of cells goes, but the question of where the original cells came from remained as difficult as ever.

THE ORIGIN OF THE UNIVERSE AND THE BEGINNING OF LIFE

Wonder about the beginnings of life continued, but early in the twentieth century help was forthcoming from chemists, geologists, and paleontologists, who tried to look back in time and determine how the universe was born and how life began. Most scientists agree on what happened, even *when* it happened, but the question still remains: What *made* it happen?

The Origin of the Universe

The universe was born in a flash of light and energy 15 to 20 billion years ago.* Such a relatively precise date is possible because astronomers have been able to measure the speed at which galaxies (clusters of stars) are moving away from one another toward infinite space. If we go backward in time and reverse the direction of the moving galaxies, we can imagine that they were once closer together. Inevitably, at an even earlier time, the galaxies were intermingled into one huge mass. The temperature and pressure must have been enormous. When the pressure and temperature became too great to sustain the dense mass any longer, the entire mixture flew apart in a cosmic explosion of unimaginable power. At that moment, the universe was born. Astronomers have been able to detect, even now, traces of the radiant afterglow of that "big bang."

The Primordial Atmosphere and the Origin of the Earth

Out of that stupendous explosion came the primordial gases, the first "atmosphere," probably consisting mainly of helium and hydrogen. Clouds of primordial gas continued to swirl, and the atoms within became compressed. The temperature rose along with the increasing pressure, until finally, through billions of years, the hydrogen within the compressed ball of gas exploded in a series of nuclear reactions. (All the elements originated from the nuclear reactions that ultimately converted hydrogen to all the more complex elements of the universe.) When a cloud reached its maximum state of compression, it became a star. Our sun is such a star, and it was probably formed along with the rest of our solar system five or six billion years ago.

The earth originated about 4.6 billion years ago as one of the cooling satellites that somehow formed around the central body of the sun, which even now continues to emit radiant energy. While the earth

* It is probably impossible for most of us to imagine the reality of a number as large as 20 billion, but perhaps we can at least understand something of the enormousness of 1 billion. If you had a billion dollars and you spent a dollar a minute ($1440 a day), you would get rid of only a little more than half a million dollars a year. It would take about 2000 years to spend the entire billion. If you decided to invest your billion dollars at 10 percent interest, you would have to spend $190 a minute ($273,600 a day) just to spend the first year's *interest*. A billion years is a long time.

To try to understand the magnitude of *20* billion, imagine this: If you bought a $50,000 sports car *every day*, you would have to live more than 1000 years to be able to spend 20 billion dollars.

was still in a somewhat gaseous state, the lighter elements, such as hydrogen, remained near the outer surface, and the heavier elements, such as nickel and iron, settled toward the center. The primitive earth, first a seething mass of volcanic activity, gradually cooled enough to form an outer rim of rock, but at its center the earth remained a core of molten rock and metal.

Along with the eruption of molten rock, gases were released into the earth's atmosphere: nitrogen, ammonia, hydrogen, water vapor, methane, and hydrogen sulfide.* Some of the lighter gases, such as helium, had already escaped into space before the gases had become dense enough to succumb to the earth's gravity. As more and more gas escaped from the earth's core, the accumulation of vapor condensed, and the rain began (Figure 31.3). The oceans were forming, and the raw materials of life were there. But one component of the modern atmosphere was still lacking: free oxygen.

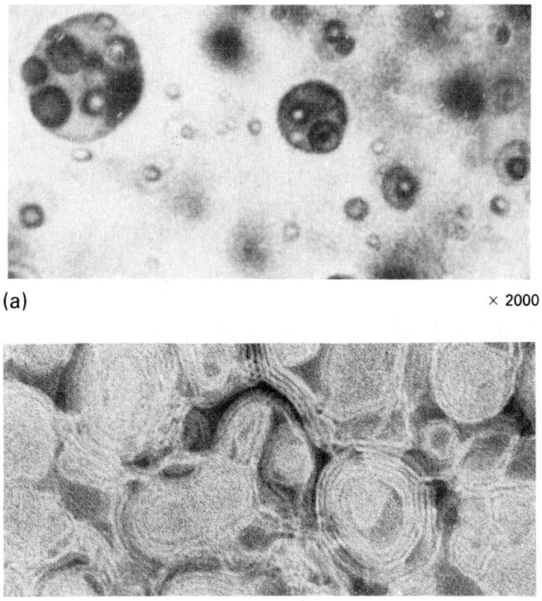

(a) × 2000

(b) × 3200

Figure 31.4
Particles produced artificially from organic but nonliving materials. These bits, seen at high magnification, appear variously as microscopic droplets, balls within balls (a), or polygonal masses surrounded by membranelike layers (b). Objects resembling cells have been made by many ingenious techniques, but no indefinitely self-reproducing entities have so far been achieved.

The Spontaneous Production of Organic Molecules

Imagine a moderately warm land, lashed by lightning, engulfed in an atmosphere without oxygen, and well supplied with water. The image of such a world has prompted many biologists to think of the ways life may have started. In the early 1920s, two scientists in particular, the Englishman J. B. S. Haldane and the Russian Alexander I. Oparin, independently speculated on how life might have arisen from nonliving matter. Either life was created instantaneously by a divine being, or it arose *gradually* from materials that were once lifeless. The latter alternative was the only scientifically acceptable one. The important factor was *time.* Given enough time, there could be a gradual progression from the inorganic to the organic to the biological. Because of the work of men like Hutton, Lyell, and Darwin, scientists were finally able to imagine the earth, and even the universe, in terms of vast expanses of time. Nobel laureate George Wald has said, "What we regard as impossible on the basis of human experience is meaningless here. Given so much time, the 'impossible' becomes possible, the possible probable, and the probable virtually certain."

Haldane in England and Oparin in Russia discussed the chemical reactions that might occur where water, energy (in the form of heat, lightning, and ultraviolet light†), and a hydrogen-dominated atmosphere of methane (containing carbon), ammonia (containing nitrogen), and hydrogen were available. Haldane suggested that these factors would produce simple organic molecules, and he coined the phrase "hot dilute soup" (sometimes called the "primordial soup") to describe the teeming mixture of chemicals in the ancient seas, where life probably began. Oparin envisioned the production of organic molecules and went further to suggest that these primitive organic molecules would form into aggregates, or droplets, which he called **coacervates** (koh-ASS-uhr-vates) (Figure 31.4). Some of the droplets, he argued, would be more stable than others; that is, they would "survive" longer, have a chance

* These simple molecules of the earth's primitive atmosphere were the source of the elements of the present atmosphere. For instance, oxygen was present in the form of water vapor (H_2O), carbon was in the form of methane (CH_4), nitrogen was in the form of ammonia (NH_3), and obviously, hydrogen was abundant, occurring as a separate gas (H_2), in hydrogen sulfide (H_2S), and as a component of all the molecules mentioned above.

† It is now believed that radioactivity was also an important energy source in the primitive earth.

LIFE BEYOND THE EARTH: EXOBIOLOGY

Since we know what we do of the universe and of the processes of evolution, it is reasonable to think that there is some kind of life on planets of other solar systems. The earth is the proper size and contains the proper ingredients for the life that has developed here. A planet that is too massive would have such a gravitational pull that mobility would be improbable; one that is too small would not have enough pull to retain the necessary gases. A habitable planet must be neither too hot nor too cold, and it must have an atmosphere that can be used. None of the other planets in our own solar system seems to offer a hospitable environment, but there are some 10^{20} (100 million million million, or 100 billion billion) stars known to astronomers, each a sun in its own right, and each possibly with its planets. If we assume that only one star out of every billion carries a suitable planet where some kind of life might have evolved, that still leaves 10^{11} (100 billion) possible planets. Some astronomers have estimated that the possibilities for life on other planets may be as high as 10^{18}, or 1 billion billion.

Some kind of life, then, appears to be most probable, but we can wonder if those other evolutionary processes led to anything like the human species. The answer is "Probably not." Variability is the essence of evolution, and it is therefore highly unlikely that any two evolutionary streams would have led to identical or even similar types. Consider such recent islands as the Galapagos, with their scant million-year history, on which life has already diverged from that on the nearest mainland. It is reasonable, however, to expect the evolution of intelligence, that is, the ability to understand and perhaps to manipulate the environment, even to make efforts at communicating with us. Such thoughts have prompted the suggestion that we should try sending radio messages out into space and keeping radio watch for incoming messages. Indeed, such experimentation is proceeding now. The little green men that cartoonists like to draw may be smarter than we are, but if they could visit the earth, they would find some organisms here that might startle them. If you were a perceptive being from Beta Cygni, what would you think of such earth-born creatures as an octopus, a Venus flytrap, a praying mantis, and a vampire bat?

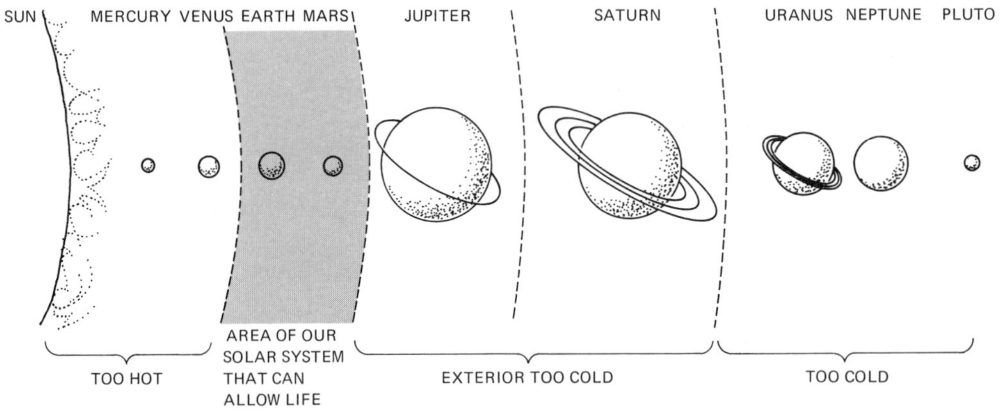

to enlarge, and perhaps to divide. Such droplets, though not alive, could be thought of as prebiotic, and they might eventually lead to the formation of real biological cells.

Charles Darwin, more than half a century before, had suggested how life may have originated "in some warm little pond, with all sorts of ammonia and phosphoric salts, light, heat, electricity, etc., present." In 1947, the British scientist J. D. Bernal proposed that primordial organic compounds in the ocean were too dilute to produce the more complex molecules needed to initiate life. He suggested that the clay in lagoons and shallow pools near the oceans would be a perfect medium for the condensation and protection of organic compounds.

Through millions of years of random combinations, molecules were formed that made possible the replication of entire organisms: amino acids and proteins, which provide the structural and functional compounds of cells, and nucleic acids, the only molecules capable of self-replication.

The Oparin-Haldane description of the earliest life forms has its critics. Other forceful but less popular suggestions are that prebiotic particles were first formed not in water but in the atmosphere, and that photosynthesis predated all other biochemical events.

Theoretical suggestions, such as those of Haldane and Oparin, led to actual laboratory attempts to recreate a primitive climate. In 1953, Stanley L. Miller, then a graduate student at the University of Chicago, experimented with a closed vessel containing a mixture of gases that were supposed to have been present in prebiotic times (Figure 31.5). Miller introduced electrical sparks into the vessel as a substitute for lightning. After about a week of sparking, the vessel contained a number of organic molecules suspended in the water, including a variety of simple amino acids, from which proteins are made. This experiment showed that if ammonia, methane, and water are supplied with energy in a *reducing* (hydrogen-rich) atmosphere, there will be produced many of those compounds that are characteristic of modern cells. Subsequent experiments have produced even more complex compounds, including nucleotides, the precursors of nucleic acids.

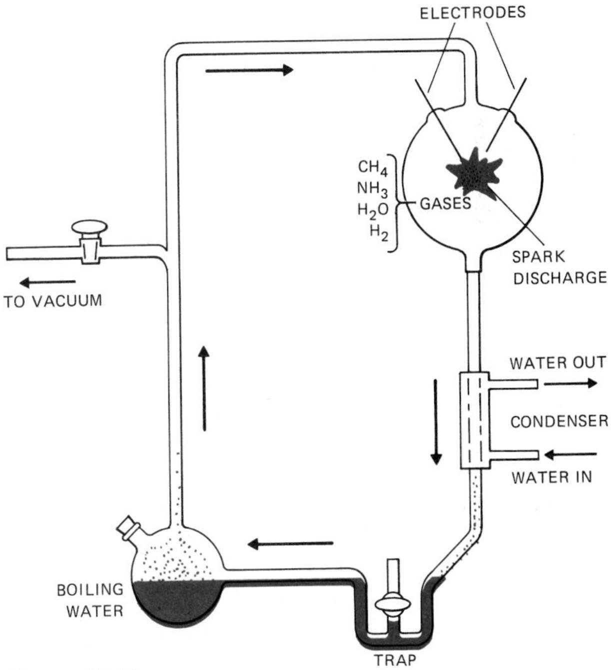

Figure 31.5
The production of amino acids from methane, ammonia, hydrogen, and water, energized by electrical sparks. The diagram shows the apparatus Stanley L. Miller used to simulate a primordial atmosphere and to obtain compounds that are generally regarded as biological from nonbiological ones.

The Evolution of Cells

The first cells appeared in the form of bacteria in the primordial soup about 3.5 billion years ago. On a very small scale, photosynthesis probably started about 3 billion years ago, but at that time only in simple bacteria and blue-green algae. Cells were successful because they were efficient. Even the simplest cells speeded up the evolutionary process, reproducing more rapidly than free-floating strands of DNA. They provided the basis for the varied forms of all the multicellular life to come.

There are two ways in which the first cells may have arisen. According to one view, something resembling a membrane came first to give a cellular covering of cytoplasm. A contrasting view holds that the genetic material came first—the "naked gene"—and was later surrounded by cytoplasm. Since it is difficult to imagine how either way alone could lead to a complete cell, it is reasonable to think that both worked at the same time. These first cells must have been **prokaryotic.** They were cells without nuclei, such as bacteria, but still containing the strand of nucleic acid (probably RNA) that makes replication possible. Prokaryotes continue to flourish today.

Then, perhaps a billion years after the appearance of the prokaryotic cells, the first cell nuclei appeared. These more complex cells, called **eukaryotic,** had not only nuclei but all the usual components and compartments of the familiar cells of higher organisms. Once a cell appeared that we could now call a "plant cell," the stage was set for the revolution that brought some life out of the water and onto the land.

Photosynthesis: The Foundation for a World of Oxygen

Before the development of photosynthesis, life depended on fermentation (see Chapter 5) and the molecules that were being used to yield energy were not being replaced. It was a matter of time before life might die out and have to start all over again. But before the organic energy sources were used up, some cells developed chlorophyll, and photosynthesis evolved. (The waste product of fermentation, carbon dioxide, was essential to the process of photosynthesis.) When cells became able to convert the

MICROORGANISMS WITH AN ANCIENT HISTORY: ARCHAEBACTERIA

Bacteria, or something like bacteria, have been living on this earth longer than any other surviving organisms. They have left their remains as fossils, the oldest now known having been located in African and Austrialian rocks about 3.5 billion years old, and there must be still older ones.

Until recently, all bacteria were commonly thought to be similar enough to be included in one biological phylum, certainly within one kingdom. But new techniques, especially of nucleotide sequence analysis of nucleic acids, show that bacteria are separable into two sorts: the usual ones, known as eubacteria, and a special group that have been named the **archaebacteria.**

Archaebacteria are like common, familiar bacteria in size and structure, but they live in places that seem odd: oxygen-free swamps (the methane-producers), the intestines of animals, extremely salty seas, scalding hot acid sulfur springs, and smoldering coal heaps. Their chemical makeup is not like that of the eubacteria. Eubacteria have cell walls containing a peculiar bacterial compound, muramic acid; archaebacteria do not. Eubacteria are sensitive to certain antibiotics such as chloramphenicol; archaebacteria are not. In the two groups, the RNAs contain different specific nucleotides, and the fatty components are consistently different in their construction.

The regularity with which these differences appear has led to the suggestion that the archaebacteria are so distinct from all other life forms that we should classify them in a kingdom of their own, as we have done for the fungi, plants, and animals. The supposition is that the archaebacteria diverged from the line of eubacterial evolution so long ago that each of the two kinds of bacteria has evolved its own way of living. That means that the archaebacteria have an impressively long independent life.

Another special archaebacterial feature is the possession of a ribosomal protein that is like that of the eukaryotic cells of higher plants and animals. If similarity indicates relationships, it could be that some most ancient ancestor contributed to the evolution of archaebacteria, eubacteria, and all the other organisms on earth, and that chemical traces of that old ancestor are still present in decaying swamps as well as in the bodies of living people.

energy from the sun into chemical energy for their own use, they became independent, and as long as the sun shone on the land, they could evolve there as well as or even better than in the ocean.

Before photosynthesis, the atmosphere was mostly nitrogen and some other gases, including carbon dioxide. Now oxygen began to be abundant, and a protective layer of oxygen (in the form of ozone, O_3) shielded the earth from the harmful ultraviolet rays from the sun. Organisms did not *have* to leave the water, but for the first time, there was an option. Before photosynthesis, underwater organisms were protected from the harmful atmospheric radiation; now organisms were free to emerge and even migrate toward the dry land. Of course, organisms had to "learn" to adapt to free oxygen. That obstacle was overcome by the evolution of cellular respiration. The chemical chain of evolutionary events was ever more efficient, from fermentation to photosynthesis to cellular respiration.

WHY DOESN'T SPONTANEOUS GENERATION CONTINUE TODAY?

If living organisms *did* once develop from nonliving compounds, why hasn't the process continued? One answer to the question is that conditions have changed. The atmosphere is different from what it was, both in its components (which could be used in making amino acids) and in the radiation it allowed to reach the surface of the earth. Now the atmosphere has little free ammonia and a lot of oxygen, and that mix is not conducive to the synthesis of protein precursors. Besides, when the earliest organic molecules were formed and were followed by primitive cells there was nothing to destroy them. If an occasional macromolecule now happens to be made spontaneously, any one of countless microorganisms is likely to be nearby, and the new molecule is digested or engulfed before it can do anything further.

In view of the variety of experimental conditions under which scientists have produced biological molecules, it seems likely that in the prebiotic world something approaching living entities may have been elaborated many times in many places and with many results. Whatever happened, we can be reasonably certain that if prebiotic bits accumulated, for a while they were safe from being eaten because there was nothing to eat them; they would not decay because there were no enzymes to act on them; and they would not be readily oxidized because there was little free oxygen. We have little hope of finding actual remains of the earliest prebiotic and biotic structures, and we may never know where or when life started on earth.

GEOLOGICAL TIME

For some time, scientists have agreed that the universe began with a "big bang" about 15 billion years ago, and in the last few years estimates have gone back as far as 20 billion years. The earth was probably born about 4.6 billion years ago, and the first bacteria may have originated about a billion years later. Multicellular organisms most likely inhabited the earth about a billion years ago (Table 31.1). Once organized creatures became established, they continued in uninterrupted progress until the present array evolved.

The Geological Eras

The major divisions of geological time are called **eras,** and they are named for convenience simply by the estimated ages of fossils that were deposited during each time. A number of methods have been used to estimate the lengths of time involved, but the most recent and most accurate is one that uses the decay of radioactive elements, such as uranium (U^{238}), potassium (K^{40}), and, for some recent samples, carbon (C^{14}) (see Chapter 1). If the amount of

THE ROLE OF PHOTOSYNTHESIS IN THE HABITATION OF THE LAND

Photosynthesis is such a fundamental and important process that it may be natural to think that it has existed since the earth originated about 4.6 billion years ago. In fact, photosynthesis probably started about three billion years ago, and then only with simple bacteria and blue-green algae.

Life began in the water at least 3.5 billion years ago; three billion years later it managed to emerge onto the land. Without photosynthesis, however, life could not have evolved even long enough for it to begin the conquest of dry land.

In a sense, some cells "invented" photosynthesis and thereby opened a new energy source for exploitation. Fermentation (energy release without oxygen intake) came first, and although it produced the minimum energy necessary for life, it was a limited and wasteful process. According to present opinion, the atmosphere of the earth was essentially devoid of free oxygen before the onset of photosynthesis. Fermenting cells did not use free oxygen, but the carbohydrates that were used to produce energy were being consumed without being replaced. However, before fermentation ran out of carbohydrate energy sources, photosynthesis evolved, first in simple organisms and then in higher green plants. Of course, those cells that developed chlorophyll and were able to convert sunlight into chemical energy gained a tremendous advantage over cells that remained dependent on existing sources of nourishment.

Once chlorophyll evolved, it used carbon dioxide and sunlight to produce ATP and oxygen through photosynthesis. As oxygen in the atmosphere increased, so did ozone (O_3), which could absorb much of the sun's lethal ultraviolet radiation. About 350 million years ago, life on the land and in the air became possible because of the protective shield of ozone and the abundance of free oxygen. (Actually, organisms had to adapt to the free oxygen, since it was a totally new factor. The great importance of free oxygen was that it made cellular respiration possible later on.)

Why did some sea life move to the land? Why not remain in the ocean without the evolutionary migration toward the shore and onto the land? Probably because the shore provided a new habitat where the bounty was great. Sunlight, land, water, oxygen, carbon dioxide, minerals, and virgin territory with minimal competition were plentiful.

TABLE 31.1
Events in the Origin and Development of Life on Earth*

EVENTS IN THE ORIGIN OF LIFE	YEARS BEFORE PRESENT
Cro-Magnon (modern *Homo sapiens*) appears	10,000–35,000
Neanderthal appears	35,000–100,000
Homo sapiens appears	200,000
Homo erectus appears	3 million
Australopithecus (ape-man) appears	5 million
First apes appear	15 million
First monkeys appear	35 million
First tree-dwelling mammals appear	60 million
Dinosaurs disappear	75 million
First mammals and birds appear	150–200 million
Dinosaurs appear	175–200 million
First reptiles appear	300 million
First amphibians appear	325 million
First fishes migrate to land	350 million
First fishes appear	400–500 million
First animals with external skeletons	600 million
Multicellular organisms appear	1 billion
Photosynthesis begins	3.3 billion
First cells (bacteria) appear	3.5 billion
Oldest rocks	3.8 billion
Origin of the earth	4.6 billion
Origin of the sun and our solar system	5–6 billion
Origin of the universe	15–20 billion

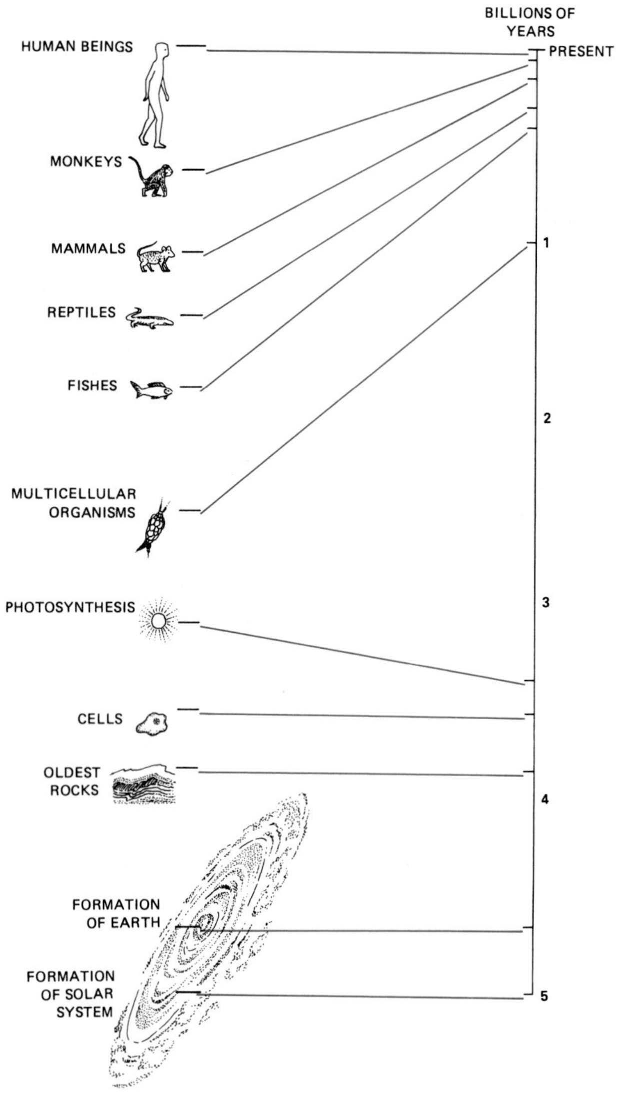

*Such a listing is necessarily approximate and temporary. Just a few years before this writing, the birth of the universe was marked at about 10 billion years ago; now the date is more firmly established at 15–20 billion years ago. Books go out of date even while they are still on press, as information about the earliest *Homo* fossils keeps moving the date further back in time. That seems to be the pattern: The more we find out, the further back in time we are pushed.

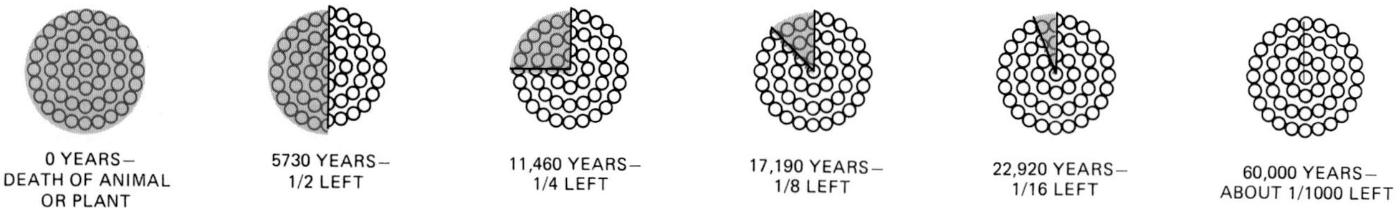

Figure 31.6
The principle used in radiocarbon dating. Knowing the amount of radioactive carbon in a modern biological sample and knowing that the half-life of C^{14} is 5730 years, one can estimate the age of an unknown sample. A sample 5730 years old would have only half as much C^{14} as a modern sample and a sample 11,460 years old would have half that much, or one-fourth of the original amount. The figures are based on the supposition that there is a definite amount of carbon dioxide containing C^{14} in the air and that the proportion has remained constant through the years (which can be said for "recent" history, or the last million years). When radioactive carbon dixoide is taken into a plant in photosynthesis, it may be locked in when the plant dies, to remain perhaps as a piece of wood or to be eaten by an animal and incorporated into the animal's tissues. Thus biological samples indicate, by the amount of C^{14} they exhibit, how long they have been dead.

a radioactive element originally present can be known, as well as the rate at which it decays, then a measurement of the amount remaining in a sample will tell how old the sample is (Figure 31.6).

Precambrian time lasted from the beginning of rock formation (3.8 billion years ago) up to about 600 million years ago, or about 3 billion years—longer than all the other eras combined. The next was the *Paleozoic* era (meaning the era of "old animals"), which lasted about 350 million years. The third, the *Mesozoic* era (meaning the era of "middle animals") extended for about 160 million years, up to the most recent, the *Cenozoic* era (meaning the era of "new animals"), in which we are living, and which dates from about 65 million years before the present (Table 31.2).

TABLE 31.2
Geological Time Scale and Important Biological Events

ERA	PERIOD	EPOCH	MILLIONS OF YEARS AGO	DURATION IN MILLIONS OF YEARS	PLANTS	ANIMALS
Cenozoic ("new animals")	Quaternary	Recent Pleistocene	2	2	Increase in herbaceous plants	Appearance of human beings
	Tertiary	Pliocene Miocene Oligocene Eocene Paleocene	65	63	Modern flowers	Birds, mammals; decline of reptiles
Mesozoic ("middle animals")	Cretaceous		145	80	Increase of flowering plants	Increase of mammals; modern birds; dinosaurs dying
	Jurassic		190	45	First flowers; abundant conifers	Primitive birds and mammals; dinosaurs abundant
	Triassic		225	35	Conifers dominant	First dinosaurs; first mammals
Paleozoic ("old animals")	Permian		280	55	Decline of nonseed plants	Increase of reptiles and insects
	Carboniferous		345	65	Coal formation; primitive conifers, many vascular nonseed plants	Early reptiles; increase of amphibians
	Devonian		395	50	Development of vascular plants: club mosses, ferns	First amphibians; sharks; numerous invertebrates
	Silurian		430	35	Algae; first vascular plants	Insects; increase of fishes
	Ordovician		500	70	First land plants; aquatic algae	First fishes
	Cambrian		570	70	Algae	Marine invertebrates
Precambrian Time			3800	2930	Blue-green algae, bacteria	Protozoa

SUMMARY

1. A number of explanations of the origin of life on earth have been developed, such as *special creation, vitalism, panspermia,* and the introduction of organic compounds from outer space. The theory of *spontaneous generation* proposed that life could appear spontaneously from nonliving materials, but this notion was attacked as long ago as the seventeenth century.
2. The universe probably originated about 15 to 20 billion years ago as part of a cosmic explosion explained by the "big bang" theory. The first "atmosphere" probably consisted mainly of hydrogen and helium.
3. The sun was probably formed about five or six billion years ago, and the earth is thought to have formed about 4.6 billion years ago. As the earth cooled, such gases as nitrogen, ammonia, hydrogen, water vapor, methane, and hydrogen sulfide escaped from the molten core. The oceans began to form but free oxygen was still lacking.
4. Given enough time, there could be a gradual progression from the inorganic to the organic to the biological. Scientists, such as J. B. S. Haldane and Alexander I. Oparin, discussed the chemical reactions that might occur in a reducing atmosphere where water, energy, ammonia, methane, and hydrogen were available. Haldane postulated the synthesis of numerous organic compounds, and Oparin went further to suggest that these primitive organic molecules would form prebiotic aggregates called *coacervates,* which might eventually lead to the formation of real biological cells.
5. Several scientists have succeeded in recreating a primitive climate. Stanley L. Miller demonstrated that simple amino acids could be produced if ammonia, methane, hydrogen, and water are supplied with electrical energy in a reducing atmosphere.
6. The first cells appeared in the form of bacteria about 3.5 billion years ago. *Prokaryotic* cells came first, followed by more complex, nucleated *eukaryotic* cells. According to one view of how the first cells originated, something resembling a membrane came first to give a cellular covering of cytoplasm. A contrasting view holds that the genetic material came first—the "naked gene"—and was later surrounded by cytoplasm. Both processes may have developed together.
7. Before the organic energy sources for fermentation were used up, some cells developed chlorophyll, and photosynthesis evolved. Then some organisms adapted to free oxygen, using the process of cellular respiration.
8. Organisms have not continued to evolve from nonliving compounds because the primordial conditions have changed.
9. The major divisions of geological time, from the oldest to the most recent, are the *Precambrian, Paleozoic, Mesozoic,* and *Cenozoic.*

ASK YOURSELF

1. What are the physical requirements for a planet—or any body in space—to permit the development of life?
2. Why do biologists think the first cells on earth were something like present-day bacteria?
3. Why do biologists generally not concern themselves with the ideas of "special creation" or "vitalism"?
4. What major component of the earth's present atmosphere is thought to have been absent from the original atmosphere?
5. What experimental evidence is there for the synthesis of organic compounds without the presence of living cells?
6. Why does living matter not arise from non-living matter at present?
7. How do scientists arrive at estimates of the age of the earth?
8. Using a ten-foot length of string to represent time since the formation of the earth, measure and mark the places on the string to indicate the time of the following events: (1) appearance of insects, (2) appearance of flowering plants, (3) appearance of the human species.

Metric Conversion Factors

ENGLISH TO METRIC	METRIC TO ENGLISH
Weight (Mass)	
Ounces × 28.3495 = grams	Grams × .03527 = ounces
Pounds × .4536 = kilograms	Kilograms × 2.2046 = pounds
Tons × 1.1023 = tonnes (1000 kg)	Tonnes × .9072 = tons
Length	
Inches × 25.4 = millimeters	Millimeters × .0394 = inches
Inches × 2.54 = centimeters	Centimeters × .3937 = inches
Feet × .3048 = meters	Meters × 3.2809 = feet
Yards × .9144 = meters	Meters × 1.0936 = yards
Miles × 1.6093 = kilometers	Kilometers × .621377 = miles
Area	
Square inches × 6.4515 = square centimeters	Square centimeters × .155 = square inches
Square feet × .0929 = square meters	Square meters × 10.7641 = square feet
Square yards × .8361 = square meters	Square meters × 1.196 = square yards
Square miles × 2.59 = square kilometers	Square kilometers × .3861 = square miles
Acres × .405 = hectares	Hectares × 2.471 = acres
Square miles × 259.07 = hectares	Hectares × .00386 = square miles
Measure	
Fluid ounces × 29.47 = milliliters	Milliliters × .0339 = fluid ounces
Quarts × .9433 = liters	Liters × 1.06 = quarts
Gallons × 3.774 = liters	Liters × .265 = gallons
Cubic inches × 16.4 = cubic centimeters	Cubic centimeters × 0.06 = cubic inches
Cubic feet × .02832 = cubic meters	Cubic meters × 35.3156 = cubic feet
Cubic yards × .7645 = cubic meters	Cubic meters × 1.308 = cubic yards
Pressure	
$lb/in.^2 \times .0703 = kg/cm^2$	$kg/cm^2 \times 14.2231 = lb/in.^2$

Temperature

When you know the Fahrenheit temperature:

$$°C = \frac{(°F - 32)}{1.8}$$

When you know the Celsius temperature:

$$°F = (°C \times 1.8) + 32$$

°C		°F
100	Water boils	212
90		194
80		176
70		158
60		140
50		122
40		104
37	Normal body temperature	98.6
30		86
20		68
10		50
0	Water freezes	32

Glossary

This glossary includes pronunciation, derivation, and definitions for technical terms, phrases, and abbreviations used in the text. Derivations from the French (Fr.), Italian (It.), Russian (Rus.), Latin (L.), and Greek (Gr.) are given as an aid to understanding, not as a scholastic exercise.

ABSCISIC ACID (ab-SIZE-ik) [L. cut off]. A plant growth-regulating substance that, among other things, regulates leaf fall and acts as an antagonist to auxin in such processes as cell elongation.

ACETYLCHOLINE (ASS-a-TEEL-KO-leen). A chemical substance, released by synaptic vesicles in the nerve endings, that crosses the synaptic gap and excites the next nerve fiber.

ACID [L. sour]. A compound that yields hydrogen ions (protons) when added to water; turns blue litmus paper red; reacts with bases to form salts. (*See* BASE)

ACROMEGALY (ak-roh-MEG-uh-lee) [Gr. extremity + great]. A condition in which the skeleton is thickened and the face, hands, and feet continue to grow after skeletal development is complete because of excess growth-hormone secretion.

ACROSOME (AK-roh-sohm) [Gr. extremity + body]. A cuplike body at the anterior tip of sperm, containing several enzymes that aid the sperm in penetrating an egg.

ACTIN (AK-tin). A protein that makes up the thin filament in muscle. Actin and myosin, a larger protein, are the ultimate structural units of contractility.

ACTION POTENTIAL. The temporary reversal of relative charge on the inner and outer surfaces of a nerve or muscle cell membrane, passing along the membrane as a wave of excitement.

ACTIVE TRANSPORT. The movement, by a process requiring expenditure of energy, of molecules into or out of a cell, across a cell membrane.

ADAPTATION. A change in an organism that brings it into better harmony with its environment.

ADAPTIVE RADIATION. The process of speciation as isolated populations adapt to new environments.

ADDISON'S DISEASE. A disease caused by underactivity of the adrenal cortex. Symptoms include anemia, increased blood potassium, decreased blood sodium, and bronzing of the skin.

ADENINE (ADD-uh-neen). A purine, a nitrogenous base that is a constituent of nucleic acids and of such compounds as ATP, ADP, AMP.

ADENOSINE MONOPHOSPHATE. (*See* CYCLIC-AMP)

ADENOSINE TRIPHOSPHATE (ATP) (ah-DEN-oh-seen try-FOSS-fate). An organic compound containing adenine, ribose, and three phosphate groups; stores energy in chemical bonds and thus serves as an energy source for chemical reactions in living cells.

ADRENAL (uh-DREEN-ul) [L. upon + kidney]. The paired glands that rest on the kidneys; each is composed of inner medulla and outer cortex. Adrenal medulla chiefly secretes adrenalin. Adrenal cortex chiefly secretes aldosterone and hydrocortisone.

ADRENALIN. (*See* EPINEPHRINE).

ADVENTITIOUS ROOTS (ad-ven-TISH-us) [L. not properly belonging to]. A root produced from an organ other than the radicle or another root.

AEROBIC (air-ROH-bik) [Gr. air + life]. Processes or organisms that require free atmospheric oxygen.

AFFERENT (AF-uh-rent) [L. to bring to]. Bringing substances or impulses toward the heart or central nervous system; denoting directional action of veins, nerves, etc.

AGGLUTINATION (uh-GLOOT-in-AY-shun) [L. to glue to a thing]. The clumping of cells or particles in a fluid. For example, clumping of blood cells occurs when incompatible blood types are mixed.

ALDOSTERONE (al-DOSS-tuh-rone). An adrenal cortical hormone, which stimulates the reabsorption of sodium by the kidneys, and thereby reduces the amount of sodium in the urine.

ALLELE (uh-LEEL) [Gr. of one another]. One form of a gene, responsible for one of two or more contrasting traits.

ALLELOPATHY (al-uh-LAW-puh-thee) [Gr. other + suffer]. The phenomenon of keeping plant competitors at bay by the secretion of toxic chemical compounds.

ALLERGY (AL-er-gee). A hypersensitivity to some substance in the environment.

ALLOPATRIC SPECIES (al-oh-PAT-rik) [Gr. other, different + L. country]. Similar species that are isolated both geographically and reproductively.

ALL-OR-NONE LAW. The concept that an event happens to its fullest possibility or not at all; usually used to refer to the firing or lack of firing of a nerve impulse.

ALPHA PARTICLES (ALPHA "RAYS"). Helium nuclei emitted from radioactive isotopes of elements.

ALTERNATION OF GENERATIONS. A reproductive cycle in which the haploid gametophyte, or sexual phase, is followed by the diploid sporophyte, or asexual phase. Spores, produced from the sporophyte, give rise to new gametophytes.

ALTRUISM (AL-troo-iz-um) [Fr. others]. Willingness of an individual to sacrifice something or all of itself for the benefit of the species genotype; in humans, unselfish idealism.

ALVEOLUS (al-VEE-uh-lus) [L. hollow]. A tiny air sac in a lung, surrounded by a network of blood vessels; the site of gas exchange between the air and the capillaries.

AMINO ACID. An organic compound containing an amino group, $—NH_2$, and a carboxyl group, —COOH; the major components of proteins are amino acids.

AMNION (AM-nee-on) [Gr. lamb]. One of the special extraembryonic membranes of reptiles, birds, and mammals; a fluid-filled sac in which the embryo floats.

AMYLASE (AM-uh-lace). An enzyme that hydrolyzes starch to sugar.

ANAEROBIC (an-uh-ROH-bik) [Gr. not + air + life]. Not requiring free atmospheric oxygen.

ANALOGOUS (an-AL-uh-gus) [Gr. conformable, proportionate]. Pertaining to parts of organisms used for similar functions but not similar in origin or in construction, such as wings of insects and birds. (*See* HOMOLOGOUS)

ANAPHASE (AN-uh-faze) [Gr. up, back, again + phase]. A stage in mitosis or meiosis in which the chromosomes move from the equator toward the poles of the spindle.

ANDROGENS (AN-druh-jenz) [Gr. man + to produce]. Sex hormones or other substances that have masculinizing effects, e.g., testosterone.

ANEMIA (uh-NEE-mee-uh) [Gr. no blood]. A condition resulting from a decreased production or increased destruction of erythrocytes in the human body; a lack of hemoglobin.

ANGIOSPERM (AN-jee-uh-sperm) [Gr. vessel + seed]. A flowering plant, whose ovules are enclosed within ovaries.

ANGSTROM UNIT. One ten-thousandth of a millimeter, or one ten-billionth of a meter.

ANTAGONISTIC. In anatomy, opposing muscle pairs that permit movement in opposite directions.

ANTERIOR. Toward the front or head end.

ANTHER (AN-ther) [Gr. flower]. The pollen-producing organ of a flower.

ANTHERIDIUM (an-thuh-RID-ee-um). An ovoid, multicellular, male sex organ that produces sperm; found in mosses and ferns.

ANTHOCYANINS (AN-tho-SIGH-a-nins) [Gr. flower + blue]. Sugar-based vacuolar pigments in plants, causing shades of reds, blues, and violets.

ANTHROPOMORPHISM (AN-thro-po-MORE-fiz-um) [Gr.man + form]. The practice of regarding animals in terms of human characteristics.

ANTIBIOTIC (an-tee-by-OTT-ik) [Gr. against + pertaining to life]. Substance formed by microorganisms that has the capacity to inhibit the growth of other microorganisms; used in the treatment of infectious diseases in humans and animals.

ANTIBODY [Gr. against + body]. A specific protein synthesized in response to the presence of a foreign material. (*See* ANTIGEN)

ANTICODON (an-tee-KOH-don) [L. against]. The sequence of three nucleotides in transfer RNA that pair complementarily with three nucleotide codons of messenger RNA. (*See* CODON.)

ANTIDIURETIC HORMONE (ADH) (an-tee-dy-yuh-RET-ik). A hormone, produced by the hypothalamus and released by the posterior lobe of the pituitary, that aids water reabsorption in the kidneys and stimulates the smooth muscles of the arteries.

ANTIGEN (AN-ti-jen) [Gr. against + to produce]. A foreign material that elicits the synthesis of a particular kind of molecule, the antibody, within an organism. (*See* ANTIBODY)

APHASIA (uh-FAY-zhuh) [Gr. without + speech]. The inability to understand or to produce written or spoken words, due to damage to the speech areas of the cerebrum.

APICAL BUD (AY-pi-kul). A mitotically active bud, at the tip of the stem of vascular plants, which adds new cells at the tip and thus adds to the height of the plant.

APPEASEMENT GESTURE. A submissive position directly opposite to a threatening posture.

APPENDICULAR SKELETON (ap-en-DIK-yu-ler). The structures that attach to the axial skeleton, such as arms and legs.

ARCHEGONIUM (ahr-ki-GO-nee-um) [Gr. the first of a race]. A multicellular female sex organ that produces eggs; found in mosses, ferns, and conifers.

ARTERIOSCLEROSIS (ahr-TEER-ee-oh-skloh-ROH-sis). [Gr.artery + hardness]. A disorder in which the arteries harden, with resultant higher-than-average blood pressure.

ARTERY. A vessel that carries blood away from the heart.

ARTICULATION (ahr-TIK-yu-LAY-shun) [L. joint]. Joint, the point where two separate bones meet or where a cartilage joins a bone; any place where two separate pieces of an organism are joined.

ASCUS (ASS-kus) [Gr. bag]. The spore-carrying cell common to the fungi in the Class Ascomycetes.

ASSOCIATION. The recombining of ions to form an uncharged molecule. (*See* DISSOCIATION)

ATHEROSCLEROSIS (ATH-uh-roh-skloh-ROH-sis). A deposition of fat, fibrin, cellular debris, and calcium on the inside of arteries.

ATOMIC NUMBER. The number that represents the number of protons in the nucleus of an atom of a chemical element.

ATOMIC WEIGHT. The average weight of an atom of an element relative to the weight of an atom of carbon 12.

ATP. (*See* ADENOSINE TRIPHOSPHATE)

ATRIUM (AY-tree-um) [Gr. hall]. A chamber of the heart, which receives blood from a vein and pumps it to the next chamber, the ventricle; formerly called an auricle.

AUTONOMIC NERVOUS SYSTEM (ott-oh-NOM-ik). A portion of the vertebrate nervous system composed of nerve fibers that synapse in ganglia outside the central nervous system. It controls such "involuntary" activities as breathing and glandular secretion.

AUTOSOME (OTT-oh-sohm) [Gr. self + body]. A chromosome not directly associated with sex determination.

AUTOTROPH (OTT-oh-trohf) [Gr. self + to nourish]. An organism capable of synthesizing its food from inorganic sources. (*See* HETEROTROPH)

AUXIN (OX-in) [L. increase]. A plant growth-regulating substance, indoleacetic acid (IAA).

AXIAL SKELETON (AX-ee-ul). The skull, vertebrae, and ribs.

AXIL [L. armpit]. The part of a stem of a vascular plant just above the point of attachment of a leaf.

AXON (AX-on). A part of a neuron, a long process extending from the cell body. It usually carries impulses away from the cell body.

BACTERIOPHAGE (bak-TEE-ree-uh-fayj) [L. little rod + Gr. to eat]. A virus that infects bacteria.

BASE. A compound that accepts hydrogen ions when added to water; turns red litmus paper blue; reacts with acids to form salts. (*See* ACID)

BASIDIUM (buh-SID-ee-um) [L. a little pedestal]. A specialized reproductive cell in fungi of the Class Basidiomycetes, in which meiosis occurs and which produces external spores.

BENTHIC ZONE (BEN-thik). The ocean deeps, below the region of light penetration.

BETA PARTICLES (BETA "RAYS") (BAY-tuh). Electrons emitted from radioactive isotopes of elements.

BILE. Organic alkaline substance produced in the liver and stored and modified by the gallbladder, which emulsifies the fats in the small intestine and helps neutralize the acidic chyme.

BILIRUBIN (bil-ee-ROO-bin) [L. bile + red]. A bile pigment, converted from hemoglobin without iron, that is excreted in feces.

BIODEGRADABLE. Capable of being destroyed by enzymes from organisms.

BIOFEEDBACK. The conscious control of "automatic" body functions, such as blood pressure and rate of heartbeat; any self-regulating biological system.

BIOLUMINESCENCE (by-oh-loo-mi-NESS-ens) [Gr. life + L. light]. The emission of light by living organisms or by enzyme systems prepared from living cells; the direct conversion of chemical energy to light energy.

BIOMASS. The total weight of organisms in an ecosystem.

BIOME (BY-ohm). A major division of the earth's surface, largely determined by climate but characterized by the plants and animals inhabiting it.

BIOSPHERE. The living portion of the earth.

BIRAMOUS (by-RAY-mus) [L. two + branch]. Having two branches, such as the swimmerets of crustaceans.

BLASTOCYST (BLAS-toh-sist) [Gr. bud, sprout + bag]. The entire mass of trophoblast and embryo.

BLASTOMERE (BLAS-toh-meer) [Gr. germ + part]. A cell of the blastula in an embryo.

BLASTOPORE (BLAS-toh-pohr) [Gr. germ + a passage]. A small opening at one side of the double-walled ball of an embryo during the gastrula stage. It is the region where invagination occurred.

BLASTULA (BLAS-chuh-luh) [Gr. germ]. A hollow ball resulting from cellular cleavages early in embryonic development.

BOHR EFFECT [Christian Bohr, 19th-century Scandinavian physiologist]. High carbon dioxide level in the blood causes more oxygen to be released from hemoglobin.

BOLUS (BOH-luss) [L. morsel]. A moistened, softened mass of food being passed through the alimentary canal.

BOOK LUNGS. Thin-layered folds of tissue for gas exchange in organisms such as spiders.

BOWMAN'S CAPSULE [Sir William Bowman, 19th-century British physician]. In a kidney, one of the microscopic cup-shaped structures that filter water and other substances from blood.

BRAINSTEM. The stalk of the brain, relaying messages between the spinal cord and the brain. It is divided into the medulla, pons, and midbrain.

BRONCHUS (BRONG-kus) [Gr. windpipe]. One of the two posterior branches of the trachea serving as a passage for air to and from the lungs.

BUFFER. A compound, with dissociated ions, that resists change in pH.

BUNDLE SHEATH. A cellular covering around the veins of leaves.

CALORIE [L. heat]. The amount of energy required to raise the temperature of a gram of water one degree Celsius, specifically from 14.5° to 15.5°.

CALVIN CYCLE. The process by which carbon dioxide is reduced to carbohydrates during photosynthesis.

CAMBIUM (KAM-bee-um) [L. change]. A plant tissue, usually under the bark of trees, whose cells by repeated division increase the girth of the stem, adding new wood and bark.

CANCER [Gr. crab]. A malignant tumor characterized by abnormal, uncontrolled growth.

CANINES (KAY-nines) [L. dog]. Long sharp teeth adapted to tearing flesh.

CAPILLARY (KAP-uh-layr-ee) [L. hair]. A small, thin-walled vessel that permits exchange of nutrients, oxygen, and carbon dioxide between the blood and tissues; the connecting vessels between arteries and veins.

CAPILLARY WATER. Water that is present in the soil in liquid form but held firmly enough so that it does not drain away to the water table below. It is an important source of water to growing plants.

CARBOHYDRATES. Organic compounds composed of carbon, hydrogen, and oxygen, in which the ratio of hydrogen to oxygen is 2:1, e.g., sugar, starch, cellulose.

CARBON MONOXIDE (CO). A common colorless gas, a product of internal combustion engines, dangerous to animals because it can displace oxygen in the red blood cells.

CARDIAC (KAR-dee-ak) [Gr. heart]. Pertaining to the heart.

CARPELLATE (KAR-puh-late) [Gr. fruit]. Ovulate; referring to the plant or plant part that bears the female gametophytes and eventually the embryos in flowering plants.

CATALYST (KAT-uh-list) [Gr. a throwing down]. A chemical that speeds up a chemical reaction without permanently entering into the reaction itself.

CELL MEMBRANE. The outermost cytoplasmic layer of the cell, consisting mainly of phospholipid and protein; about 100 Å thick; selectively permeable; can initiate and control the active transport of substances.

CELLULOSE (SELL-yu-lohs). A polymer of glucose; the main component of the cell wall in land plants.

CENTRAL NERVOUS SYSTEM. The brain and spinal cord.

CENTRIOLE (SEN-tree-ohl) [L. center]. A self-replicating, cylindrical organelle located just outside the nucleus of animal and some plant cells; occurs in pairs, acting as focal points at opposite ends of the cell during nuclear division.

CENTROMERE (SEN-troh-meer) [Gr. center + part]. The region of spindle attachment on a chromosome, which ensures that a chromosome will proceed toward a pole in cell division.

CEPHALOTHORAX (SEF-uh-loh-THO-raks) [Gr. head + thorax]. The fused head and thorax of the chelicerates and crustaceans, such as spiders, crabs, and shrimps.

CEREBELLUM (ser-uh-BEL-um) [L. brain]. A part of the brain; the coordinating center for muscular movement.

CEREBRUM (se-REE-brum) [L. brain]. A part of the brain divided into two hemispheres, with each hemisphere divided further into lobes. Each lobe contains special functional areas, including areas for speech, vision, movement, learning, and memory.

CERVIX (SER-viks). The "neck" of the uterus, which opens into the vaginal canal.

CHELICERAE (kih-LISS-uh-ree) [Gr. claw + horn]. The first pair of appendages on the cephalothorax; the pinching claws in horseshoe crabs and the poison-injecting claws in spiders.

CHEMIOSMOTIC HYPOTHESIS (KEM-ee-ahs-MOT-ik). The theory that energy in mitochondria and chloroplasts is made available because of a difference in hydrogen ion concentration on two sides of a membrane.

CHEMORECEPTOR (KEE-moh-ree-SEP-ter). Sensory cell or organ capable of perceiving chemical stimuli, such as those of smell or taste.

CHIASMA (kye-AZ-muh) [Gr. X]. Exchange of parts between two chromatids during prophase I of meiosis, when homologous chromosomes are paired.

CHITIN (KYE-tin) [Gr. tunic]. An insoluble, tough, horny polymer of acetyl glucosamine, forming the cell wall of some fungi and the exoskeleton of the arthropods.

CHLOROPHYLLS (KLOR-oh-fills) [Gr. green + leaf]. The green pigments in plants that absorb light at the outset of photosynthesis.

CHLOROPLAST (KLOR-oh-plast) [Gr. green + formed]. A plastid that contains chlorophyll; photosynthesis is initiated when light strikes a chloroplast.

CHOLESTEROL (koh-LESS-tuh-role). A lipid that is an essential part of the human diet. Besides acting as a usual source of energy and assisting tissue maintenance, it is also needed for the production of bile acids and sex hormones.

CHOLINESTERASE (koh-lin-ESS-tuh-race). An enzyme that attacks and neutralizes the neurotransmitter acetylcholine.

CHORION (KOHR-ee-on). An extraembryonic membrane that forms an outer membrane around the embryo; in mammals the chorion makes up most of the placenta.

CHORIONIC GONADOTROPIN (kohr-ee-ON-ik goh-nad-oh-TROH-pin). A female sex hormone produced in the embryonic membranes and the placenta, which keeps the corpus luteum from disintegrating and stimulates steroid secretion from this structure.

CHROMATID (KRO-muh-tid). One of the two daughter strands of a newly duplicated chromosome.

CHROMATIN [Gr. color]. The readily stainable nuclear material of the chromosomes, composed of DNA and proteins.

CHROMATOPHORE (kro-MAT-uh-fore) [Gr. color + to bear]. A granule containing pigments in bacterial cells or in skin of some animals.

CHROMOPLAST (KRO-moh-plast). A pigment containing plastid in a plant cell, usually red, orange, or yellow. Chloroplasts are green chromoplasts.

CHROMOSOME (KROHM-uh-sohm) [Gr. color + body]. The nucleoprotein structure that contains the hereditary units, the genes.

CHRYSALIS (KRISS-uh-liss) [Gr. gold, referring to the gold-colored pupa of a butterfly]. The pupal stage in butterflies, in which the young is enclosed in a firm case.

CHYME (KIME). A soupy slurry of partially digested food.

CILIA (SILL-ee-uh) [L. eyelid]. Short bristlelike locomotor organelles on the free surface of cells.

CIRCADIAN RHYTHM (sir-KAY-dee-un) [L. about + day]. Daily cycle of behavior.

CITRIC ACID CYCLE. (*See* KREBS CYCLE)

CLEAVAGE. The cell divisions, initiated after fertilization, that mark the beginning of embryonic development.

CLIMAX. In ecology, a community of organisms that has achieved stability and can resist changes by invading species; in coitus, an orgasm.

CLITELLUM (kly-TELL-um) [L. a pack saddle]. A band around the body of an earthworm, functioning during mating and providing a cocoon containing eggs.

CLITORIS (KLIT-uh-riss) [Gr. small hill]. Small erectile body just anterior to the vaginal opening; it is homologous to the male penis.

CLONE (KLOHN). A genetically uniform population of cells or organisms derived asexually from a single ancestor.

COACERVATES (koh-ASS-uhr-vates) [L. cluster together]. Prebiotic aggregates formed by organic compounds, which may eventually have led to the formation of biological cells.

COCHLEA (KOCK-lee-uh) [L. snail]. The coiled tube in the inner ear, containing the sensitive cells that make sounds perceptible.

COCOON [Fr. shell]. The pupal stage in moths. The developing young is covered by an envelope, often of silk.

CODON (KOH-don). A sequence of three nucleotides in messenger RNA that codes for a single amino acid.

COELOM (SEAL-um) [Gr. hollow]. The body cavity of an animal, between the body wall and the internal organs.

COENZYME (koh-EN-zime) [L. with + Gr. in + yeast]. A compound, such as a vitamin, that helps activate an enzyme.

COEVOLUTION (KOH-ev-oh-LU-shun). The evolution of two different species in ways that make them mutually dependent.

COITUS (KOH-i-tuss) [L. to come together]. The act of sexual intercourse.

COLEOPTILE (koh-lee-OP-tile) [Gr. sheath + feather]. The sheath that covers the first leaf in a grass seedling.

COLON (KOH-lun). The large intestine.

COMPANION CELL. A slender, nucleated cell in the phloem of vascular plants.

COMPLETE METAMORPHOSIS (met-uh-MORE-fuh-sis) [Gr. after, beyond, over + shaping]. In insects, the change from egg to larva to pupa to adult.

COMPOUND. A substance whose molecules are composed of more than one kind of atom; every molecule of one compound contains the same elements joined in the same proportions by weight.

COMPOUND EYE. An arthropod eye that is composed of many tubular units, ommatidia, each with its own lens and light-sensitive receptor cells.

CONDENSATION. A reaction in which small, simple molecules are combined into larger, more complex ones, with the elimination of one or more molecules of water. (*See* HYDROLYSIS)

CONDITIONING. A type of learning in which an animal is rewarded or punished according to its reaction to a given stimulus.

CONDUCTIVITY. The ability of neurons to conduct a current along a nerve cell. The other basic property of neurons is excitability. (*See* EXCITABILITY)

CONES. (1) Light-sensitive cells in the retina of the eye; used in color vision. (2) Spore-bearing structures in nonflowering plants, such as pine trees.

CONIDIUM (kuh-NID-ee-um) [Gr. dust]. An asexual fungus spore.

CONJUGATION (kon-juh-GAY-shun) [L. together + yoke]. Sexual union of similar cells as in some algae, protozoa, and bacteria.

CONTRACTILE VACUOLE (kon-TRAK-tul VAK-yu-ohl). An intracellular organelle of water balance, which acts as a reservoir of excess water; common in protozoa. The excess water is accumulated and then emptied to the outside.

CONVERGENT EVOLUTION (kon-VER-jent) [L. together + to incline]. The process of changing toward similarity so that organisms that are not closely related come to be alike.

CORPUS CALLOSUM (KOHR-puss kal-LOH-sum) [L. body + hard]. A tough bridge of nerve fibers that connects the two cerebral hemispheres and relays nerve impulses between them.

CORPUS LUTEUM (KOHR-puss LOO-tee-um) [L. body + yellowish]. A temporary endocrine gland in mammals that secretes the hormone progesterone; it is formed in an ovary after the discharge of an egg.

CORPUSCLE (KOHR-pus-el) [L. body]. Any formed structure; usually a blood cell.

CORTEX (KOHR-teks) [L. bark]. (1) The outer portion of an organ, as in the brain or kidneys. (2) The tissue in the outer bark of plants.

CORTISOL. (*See* HYDROCORTISONE)

COTYLEDON (kot-uh-LEE-dun) [Gr. cup-shaped hollow]. Seed leaf of the embryo of a plant.

COVALENT BOND (koh-VAY-lent). A chemical bond between two atoms, formed by the sharing of one or more electrons.

CREATINE PHOSPHATE (KREE-uh-teen). A phosphate compound that is a source of energy in the contraction of muscle.

CRETINISM (KREET-in-iz-um) [Fr. Christian]. A disorder, characterized by irregular development of bones and muscles and mental retardation, resulting from underactivity of the thyroid gland during infancy. Often confused with Down's syndrome (mongolism).

CRISTAE (KRIS-tee) [L. cock's comb]. The inner membranes of mitochondria folded and doubled on themselves, forming incomplete partitions.

CROP. A storage segment in the digestive system of earthworms and birds, from which food is released at controlled intervals.

CROSSOVER. A process during prophase I of meiosis, whereby nonsister chromatids exchange parts; also the genetic results of such an exchange.

CRYPTIC COLORATION (KRIP-tik) [Gr. hidden, concealed]. An adaptation that helps camouflage an animal by providing it with a color that makes it inconspicuous.

CUSHING'S DISEASE. A condition usually caused by a cortical adrenal tumor that overproduces glucocorticoids. A tendency toward diabetes is caused by increased blood sugar.

CUTICLE (KYOOT-i-kul) [L. skin]. A waxy layer coating the outer wall of epidermal cells in plants.

CYCLIC-AMP (CYCLIC 3′, 5′-ADENOSINE MONOPHOSPHATE). A chemical substance that plays a major role in mediating cellular activity in animals, bacteria, slime molds, etc.

CYCLIC PHOSPHORYLATION (SIK-lik FOSS-for-i-LAY-shun). A process in some plants during which electrons are recycled and phosphorus is transferred; the only product is ATP. (*See* PHOSPHORYLATION)

CYTOCHROME (SIGH-toe-krohm). Protein-plus-pigment molecule containing iron; acts as a carrier of electrons from the Krebs cycle during respiration in mitochondria and during photosynthesis in chloroplasts.

CYTOKININ (sigh-toe-KYE-nin) [Gr. cell + motion]. A plant growth-regulating substance, which stimulates cell divisions.

CYTOPLASM (SIGH-toe-plazm). Nonnuclear portion of a cell.

CYTOSINE (SIGH-toe-seen). A pyrimidine, a nitrogenous base that is a constituent of nucleic acids.

DEDIFFERENTIATION. Reversion of a specialized cell to an unspecialized condition.

DEGENERACY. The ability of more than one nucleotide triplet (codon) in DNA or mRNA to specify one amino acid.

DELETION. A chromosome change in which a segment of a chromosome, involving one or more genes, is lost.

DEME (DEEM). A group of individuals, partially or entirely isolated from the other members of the species.

DENATURATION (dee-nay-chu-RAY-shun). Alteration of three-dimensional structures and physical properties of a protein.

DENDRITE (DEN-dryte) [Gr. tree]. Short, threadlike process, extending out from the cell body of a neuron. Dendrites carry impulses toward the cell body.

DENDROCHRONOLOGY (DEN-droh-kruh-NOLL-uh-jee) [Gr. tree + time + study]. The technique of reconstructing past events, such as buildings or climatic changes, by the study of annual rings in wood.

DEOXYRIBONUCLEIC ACID (DNA) (dee-AHK-see-rye-boe-new-KLAY-ik). A double-stranded nucleic acid that is a constituent of chromosomes; contains genetic information coded in specific sequences of its nucleotides.

DETERMINATE CELLS. The cells of the early embryo that are so prearranged that they are destined to become some specific part of the future animal.

DETRITUS (deh-TRY-tuss) [L. rubbing away]. Microscopic particles in soil; in a biological sense, finely divided organic matter.

DIABETES MELLITUS (die-uh-BEE-teez MEL-it-us) [Gr. to pass through + honey]. An inherited metabolic disease caused by insufficient production of insulin. Cells are unable to use glucose, which is excreted in the urine.

DIALYSIS (die-AL-uh-sis). A technique of removing diffusible substances from a mixture of substances by allowing the diffusible components to pass through a selectively permeable membrane.

DIAPAUSE (DIE-uh-pawz) [Gr. pause]. A period of insect dormancy between the larval and mature stages.

DIAPHRAGM (DIE-uh-fram). A muscular partition between the thoracic and abdominal cavities serving as part of the muscular mechanism for the respiratory system; any separating partition.

DIASTOLE (die-ASS-toe-lee) [Gr. expansion]. The time of ventricular relaxation, when the ventricles receive blood from the contracting atria.

DICOTYLEDON (die-kot-uh-LEE-dun) [Gr. two + cotyledon]. A flowering plant whose seeds have two embryonic leaves, e.g., beans, walnuts, oaks. Frequently shortened to "dicot."

DICTYOSOME (DIK-tee-uh-sohm). The Golgi body in cells.

DIFFERENTIATION. The process of developmental change from an unspecialized cell to a specialized cell.

DIFFUSION. The movement of molecules from a region of greater concentration toward a region of lesser concentration as a result of their kinetic energy.

DIGESTION. The process of breaking down large, complex, insoluble molecules of food into smaller, less complex, soluble molecules.

DIHYBRID CROSS (die-HY-brid) [Gr. two + L. mongrel]. A cross between individuals differing in two inheritable traits, or in which only two such different traits are considered by an experimenter.

DIPLOID (DIP-loyd) [Gr. double]. A cell having two full sets of chromosomes in the nucleus.

DISLOCATION. Such a severe strain of the muscles, tendons, and ligaments that a bone is pulled out of its socket and fails to return to its proper position.

DISSOCIATION. The separation of a molecule into ions. (*See* ASSOCIATION)

DOMINANT. Pertaining to a gene that expresses itself to the exclusion of its recessive allele.

DORSAL (DOHR-sul). Pertaining to a position toward the back surface. (*See* VENTRAL)

DOWN'S SYNDROME (MONGOLISM, TRISOMY 21). A genetic disorder in which an extra chromosome is present. The afflicted person is mentally retarded.

DUCTUS ARTERIOSUS (DUCK-tus ahr-TEER-ee-OH-sis). The arterial duct that connects the pulmonary artery to the aorta in the fetal circulatory system of mammals.

DUODENUM (doo-oh-DEE-num). The portion of small intestine posterior to the stomach.

DUPLICATION [L. to double]. A chromosome change in which a segment of a chromosome is repeated.

DYNAMIC EQUILIBRIUM. In ecology, a condition that exists in a stable ecosystem when over a period of time the energy input equals the outflow.

ECDYSONE (EK-duh-sohn) [Gr. escape]. The insect molting hormone that favors the formation of the pupa.

ECG (ELECTROCARDIOGRAM) (uh-LEK-troh-KAR-dee-oh-gram) [Gr. electricity + heart + writing]. A record of the electrical activity of the heart.

ECHOLOCATION (EK-oh-loh-KAY-shun). A process of locating objects by the reflection of sound waves, used by such animals as bats and dolphins.

ECOLOGY (ee-KOL-uh-jee) [Gr. house + study]. The study of organisms in relation to all the forces that act upon them.

ECOSYSTEM (EE-koh-sis-tum). A space containing an interacting group of organisms, more or less self-contained, such as an enclosed field, a pond, or a small island.

ECOTONE (EE-kuh-tohn) [Gr. home + stretch]. A border region where two ecosystems meet, as at the edge of a forest.

ECTODERM (EK-toh-derm) [Gr. outer + skin]. The outer germ layer of an embryo, which gives rise to the skin, hair, fingernails, horns, hooves, brain with associated nervous tract, and eyes.

ECTOPIC PREGNANCY (ek-TOP-ik) [Gr. outside + place]. Abnormal pregnancy in which the zygote becomes implanted somewhere other than in the uterus.

EEG (ELECTROENCEPHALOGRAM) (uh-LEK-troh-en-SEF-uh-loh-gram) [Gr. electricity + brain + writing]. A record of the electrical activity of the brain.

EFFERENT (EF-uh-rent) [L. to bring out]. Carrying substances or impulses away from heart or central nervous system; denoting directional action of arteries, nerves, etc.

EJACULATION (ee-JAK-yu-LAY-shun) [L. to throw out]. The forcible expulsion of the semen during male orgasm.

ELECTROLYTE (e-LEK-tro-LITE). A substance that ionizes in solution and is then capable of carrying an electric current.

ELECTRON. A subatomic particle, the unit negative charge.

ELECTRON TRANSPORT SYSTEM. A group of enzymes closely associated with one another on the inner surface membrane of mitochondria. Electrons move from one of the enzymes to the next, resulting in alternatively reducing and oxidizing the enzyme and transferring energy by the synthesis of ATP from ADP and inorganic phosphate. (*See* CYTOCHROME)

ELEMENTS. Substances whose molecules consist of only one kind of atom.

EMBRYO SAC. The mature female gametophyte of the angiosperms, usually having eight nuclei.

EMULSIFICATION. The physical breakdown of fats.

ENDERGONIC (en-der-GAHN-ik) [Gr. within + work]. Activities or reactions that require energy. (*See* EXERGONIC)

ENDOCRINE GLAND (EN-doh-krin) [Gr. in + separate]. A ductless gland that secretes its product into the bloodstream without an emptying tube. (*See* EXOCRINE GLAND)

ENDODERM (EN-doh-derm) [Gr. inside + skin]. The inner germ layer of an animal embryo lining the archenteron. It gives rise to the lining of the digestive tract, liver, lungs, and pancreas.

ENDOMETRIUM (en-doh-MEE-tree-um) [Gr. inside + uterus]. The nutritious inner layer of the uterine wall.

ENDOPLASMIC RETICULUM (en-doe-PLAZ-mik reh-TIK-yu-lum) [Gr. within + L. net]. A labyrinthine complex of double membranes in the cytoplasm of a cell, acting as a system of internal channels through which various materials move.

ENDOSKELETON [Gr. within + a dried body]. Internal skeleton, which provides support and rigidity of form; typical of vertebrates.

ENDOSPERM. An amorphous mass of tissue, usually triploid, which functions as a nourishing tissue for growing embryos in the seeds of flowering plants.

ENDOTHERMIC [Gr. inside + heat]. Reactions that occur with absorption of heat or other energy; endergonic; the opposite of exothermic or exergonic.

ENERGY. The capacity to do work.

ENERGY LEVEL SHELLS. The electrons in an atom are distributed in several different energy levels. When an electron changes from one shell to another, it either releases or absorbs energy.

ENZYME (EN-zime). A protein that speeds up a chemical reaction without permanently entering into the reaction itself; an organic catalyst.

EPICOTYL (EP-i-kot-ul) [Gr. upon + cup]. Stem above the attachment of seed leaf.

EPIDERMIS (ep-i-DER-mis). An outer covering. In plants it is the outermost layer or layers of leaf and of young stems and roots. In animals it is usually many cell layers thick.

EPIDIDYMIS (ep-i-DID-uh-mis) [Gr. upon + testes]. A coiled tube, lying on the testis, that stores sperm cells formed in the testis.

EPIGENESIS (ep-i-JEN-uh-sis) [Gr. upon + to be born]. The theory that development proceeds from the fertilized egg, and that the specialized body parts emerge during embryonic growth.

EPIGLOTTIS (ep-i-GLOT-is) [Gr. upon + tongue]. A valve that presses down and prevents food from entering the trachea or larynx during swallowing.

EPINEPHRINE (ADRENALINE) (ep-i-NEFF-rin) [Gr. upon + kidney]. A neurohormone, secreted by the adrenal medulla, that produces a "fight or flight" condition, which permits the body to react quickly and strongly to emergencies.

EPIPHYTE (EP-i-fite) [Gr. upon + plant]. A plant that lives attached to another plant but does not parasitize it.

ERA. A major division of geological time. From the oldest to the most recent, the eras are the Precambrian, Paleozoic, Mesozoic, and Cenozoic.

ERYTHROBLAST (ih-RITH-roh-blast) [Gr. red germ]. Cell in bone marrow, which is nucleated until it develops into a red blood cell.

ERYTHROCYTE (ih-RITH-roh-site) [Gr. red + cell]. A red blood cell.

ESOPHAGUS (ih-SOFF-uh-guss) [Gr. I shall carry + to eat]. A slender tube connecting the throat with the stomach.

ESTROGEN (ESS-troh-jen). One of the female sex hormones, secreted by ovarian follicles, which stimulates the thickening of the smooth muscle and the glandular lining of the inner walls of the uterus. It also promotes the development of secondary sex characteristics.

ETHOLOGY (ee-THOL-uh-jee) [Gr. character, habit + study]. The study of animal behavior.

ETHYLENE (ETH-uh-leen). A plant growth-regulating substance, C_2H_4, which hastens the ripening of fruits.

EUKARYOTE (you-CARRY-ote) [Gr. true + kernel]. An organism or cell with a membrane-bound nucleus; possesses mitochondria, plastids, flagella, and other organelles. (*See* PROKARYOTE)

EUPHOTIC ZONE (yu-FOTT-ik) [Gr. good + light]. The upper layers of water in seas, where light penetrates.

EUTROPHIC (yu-TROH-fik) [Gr. good + food]. Pertaining to a body of water containing an abundance of organisms, especially microscopic plants.

EVOLUTION [L. an unrolling]. Progressive change.

EXCITABILITY. The capacity of neurons to respond to stimuli; the other basic property of neurons is conductivity. (*See* CONDUCTIVITY)

EXCITATION-CONTRACTION COUPLING. The chemical process of muscle contraction, initiated by an electrical stimulus in the membrane of the muscle cell.

EXERGONIC (ek-ser-GON-ik) [L. out + Gr. work]. Pertaining to activities or reactions that liberate energy. (*See* ENDERGONIC)

EXOBIOLOGY. The study of life beyond the earth.

EXOCRINE GLAND (EK-suh-krin) [Gr. out + separate]. A gland, such as a salivary or sweat gland, which exudes its secretion by a distinct duct or opening; contrasted with endocrine glands.

EXOSKELETON [L. outside + Gr. a dried body]. External covering, which provides protection, support, or prevention of dehydration; typical of invertebrates.

EXOTHERMIC (ek-soh-THER-mik) [L. outside + Gr. heat]. Pertaining to reactions that occur with liberation of heat or other form of energy; exergonic.

EXTENSION. A movement that straightens a joint. (*See* FLEXION)

FACILITATED TRANSPORT [L. easy]. The movement of molecules across a membrane by means of a postulated carrier that "escorts" a molecule across; no energy is required.

FALLOPIAN TUBE. (*See* OVIDUCT)

FALLOUT. Radioactive debris released by atomic explosions, emitting alpha, beta, and gamma rays for varying lengths of time, depending on the radioactive element involved.

FERMENTATION [L. leaven]. The partial decomposition of organic molecules in the absence of free oxygen. Such substances as alcohol and lactic acid may be produced.

FERTILIZATION [L. to bear, produce]. The fusion of a sperm and egg to form a zygote.

FETUS (FEE-tuss) [L. offspring]. The highly developed human embryo, from the third month of pregnancy until birth.

FISSION (FISH-un) [L. to cleave]. Asexual reproduction consisting of the division of a single cell and the separation of the two resultant cells, e.g., in protozoa, bacteria, some algae.

FLAGELLUM (fluh-JELL-um) [L. whip]. A long, whiplike locomotor organelle on the surface of a cell.

FLAME CELL. The excretory structure in flatworms, so called because of the wavering flagella that resemble a candle flame.

FLEXION (FLEK-shun). A movement that bends a joint. (*See* EXTENSION)

FLORIGEN (FLORE-uh-jen) [L. flower-maker]. A postulated hormone responsible for the induction of flower buds.

FLUORESCENCE (flure-ESS-ens). Emission of light by a compound that is energized by photons or electrons.

FOLLICLE (FOLL-i-kull) [L. small bag]. Center of egg production in mammalian ovary. Each follicle houses an ovum that is in the process of being matured and released.

FOLLICLE-STIMULATING HORMONE (FSH) [L. small bag]. A female gonadotropic hormone that causes one or more immature follicles to develop.

FOOD WEB. The passage of food in an ecosystem from the primary producers through all the various consumers.

FORAMEN OVALE (fo-RAY-men oh-VAH-lee) [L. oval hole]. (*See* OVAL WINDOW)

FOVEA (FOH-vee-uh) [L. small pit]. A spot at the center of the retina directly behind the lens where the concentration of the cones is greatest. This is where image formation is sharpest and color vision is most acute.

GAMETE (GAM-meet) [Gr. to marry]. A sexual reproductive cell; an egg or sperm.

GAMETOPHYTE GENERATION (guh-MEET-uh-fite). A gamete-producing haploid phase of plants having alternation of generations.

GAMMA RAYS. Electromagnetic rays emitted by radioactive isotopes of elements. Gamma rays have much greater penetrating power than alpha or beta rays.

GANGLION (GANG-glee-un) [Gr. knot]. A cluster of cell bodies of neurons located outside the central nervous system.

GASTRIN (GAS-trin) [Gr. stomach]. A hormone, secreted by certain mucosal cells of the stomach, that stimulates the production of hydrochloric acid when food enters the stomach.

GASTROVASCULAR CAVITY (gas-tro-VAS-kyu-lar) [Gr. stomach + L. a little vessel]. Saclike cavity in the body of coelenterates in which one opening serves for both intake of food and outflow of waste.

GASTRULATION (gas-truh-LAY-shun) [Gr. stomach]. The phase of embryonic development when there is infolding of the blastula to form the gastrula.

GENE (JEEN) [Gr. to produce]. A specific segment of DNA that controls a specific cellular function, either by coding for a polypeptide or by regulating the action of other genes; the foundation of inheritable traits.

GENE POOL. The total genetic material of all individuals in an interbreeding population (species).

GENETIC CODE (jen-ET-ik). The complete set of triplet nucleotide symbols that indicate specific amino acids.

GENETIC DRIFT. The chance genetic changes resulting from isolation of small groups of organisms.

GENOME (GEE-nome). One complete set of genes in an organism. A haploid organism has one genome; a diploid has two.

GENOTYPE (JEE-nuh-type) [Gr. to produce + type]. An individual's genetic makeup. (*See* PHENOTYPE)

GERMINATION [L. sprout]. The beginning of growth of a seed or spore.

GIBBERELLIN (jib-ber-ELL-in). A plant growth-regulating substance that affects growth in height and starch digestion of seeds.

GIGANTISM. A disorder of the pituitary gland, caused by oversecretion of growth hormone during the period of skeletal development.

GILL. The respiratory organ of aquatic and some terrestrial organisms. Gills extract oxygen and release carbon dioxide.

GILL SLITS. Openings in the region of the pharynx leading from the internal alimentary canal to the outside of the animal; characteristic of chordate animals.

GIZZARD (GIZ-erd) [L. cooked entrails of poultry]. A thick-walled muscular compartment, formed in the digestive system of birds and earthworms, where food is ground.

GLIA (GLEE-uh) [Gr. glue]. Nonconducting cells of the nervous system that protect and nourish the neurons; they may also function to store memory traces.

GLOMERULAR FILTRATION. Initial removal of salts, wastes, and water from blood into Bowman's capsule; the first step in urine production.

GLOMERULUS (gluh-MARE-yu-luss) [L. a ball]. A ball of capillaries intimately associated with the Bowman's capsule in the nephrons of the kidney.

GLUCAGON (GLOO-kuh-gon). A hormone, secreted by alpha cells in the pancreatic islets of Langerhans; increases blood glucose concentration by stimulating the liver to convert glycogen to glucose.

GLUCOCORTICOIDS (GLU-co-COR-ti-coids). A group of hormones concerned with glucose-glycogen balance in blood.

GLYCOLYSIS (gly-KOL-i-sis) [Gr. sweet + solution]. The initial anaerobic breakdown of glucose to an intermediate compound, occurring in the cytoplasm of a cell.

GLYCONEOGENESIS (GLY-koh-nee-oh-JEN-i-sis) [Gr. sweet + new + origin]. The process whereby amino acids or fatty acids are converted into glycogen.

GOITER (HYPOTHYROIDISM) (GOY-ter). A condition characterized by an enlarged thyroid caused by insufficient iodine in the diet.

GOLGI BODY (GOAL-jee) [Camillo Golgi, 19th-century Italian histologist]. A collection of membranes associated with endoplasmic reticulum; functions in concentrating and delivering materials to be secreted.

GONAD (GOAN-ad) [Gr. generate]. An organ that produces gametes, either ovaries or testes.

GONADOTROPIC HORMONE (GO-nad-o-TROW-pic). A hormone that stimulates growth of a gonad.

GONORRHEA (gon-uh-REE-uh) [Gr. seed, reproductive organ + flow]. A venereal disease caused by a bacterium, *Neisseria gonorrhoeae,* usually contracted through sexual intercourse.

GRANA (GRAY-nuh) [L. grain]. Stacks of flattened, photosynthetic, membrane-bound vesicles in a chloroplast.

GRAVES' DISEASE. (*See* HYPERTHYROIDISM)

GUANINE (GWAH-neen). A purine, a nitrogenous base that is a constituent of nucleic acids.

GUSTATION [L. taste]. The sense of taste.

GYMNOSPERM (JIM-nuh-sperm) [Gr. naked + seed]. A type of seed plant whose seeds are not enclosed in an ovary but are borne on the surface of sporophylls.

HABITAT (HAB-i-tat) [L. living place]. The local physical dwelling place of an organism, such as a bare rock, a moist, cold, evergreen forest, or a tropical estuary.

HABITUATION (ha-bit-yu-AY-shun). A simple type of learning in which an animal eventually ignores a stimulus.

HALF-LIFE. The amount of time required for half the atoms of a given radioactive sample to decay to a more stable form.

HAPLOID (HAP-loyd) [Gr. single]. Having only one complete set of chromosomes, that is, one genome in the nucleus of a cell.

HARDWOOD. The wood of an angiosperm tree.

HARDY-WEINBERG PRINCIPLE. Mathematical formula used in calculating the approximate frequencies of different alleles in a population after measuring the frequency of phenotypes in a representative sample of the whole population.

HAVERSIAN CANALS (huh-VER-zhun) [Clopton Havers, an English physician]. Elongated tubes that carry blood vessels through bone.

HEMIZYGOUS (hem-ee-ZY-gus) [Gr. half + yoke]. Having only one allele, e.g., a sex-linked gene in XY males.

HEMOGLOBIN (HEE-moh-gloh-bin) [Gr. blood + L. globe]. The iron-containing, respiratory pigment in the red blood cells, which transports oxygen and carbon dioxide throughout the body.

HEMOPHILIA (hee-moh-FILL-ee-uh) [Gr. blood + love]. The "bleeder's disease," in which blood does not clot normally.

HEPATIC PORTAL SYSTEM (huh-PAT-ik). A system of vessels in which the venous blood coming from the intestines passes through capillaries in the liver before moving on to the posterior vena cava.

HERPES VIRUS (HER-pees) [Gr. crawl]. A group of viruses causing lesions of lips, gums, and genitals.

HETEROGAMY (het-uh-RAH-guh-mee) [Gr. other + marriage]. A type of sexual reproduction resulting from the mating of two gametes that are of a different size or shape or both.

HETEROSPORY (het-uh-AHS-poh-ree) [Gr. other + seed]. Production of two kinds of spores: megaspores and microspores.

HETEROTROPH (HET-uh-ruh-trohf) [Gr. other + feeder]. An organism that must acquire its food because it cannot manufacture it from inorganic sources. (*See* AUTOTROPH)

HETEROZYGOUS (het-uh-ruh-ZY-gus) [Gr. other + yoke]. Possessing two different alleles at a given locus for a given characteristic on a pair of homologous chromosomes.

HIBERNATION [L. winter]. The dormant state of animals, with concomitantly lowered body metabolism.

HISTOLOGY (hiss-TOL-uh-jee). The study of biological tissue.

HISTONE (HISS-tone). A basic protein associated with DNA in chromosomes.

HOMEOSTASIS (hoe-mee-oh-STAY-sis) [Gr. unchanging + standing]. The maintenance of a measure of physiological stability in spite of environmental changes.

HOMEOTHERMIC (hoe-mee-oh-THER-mik) [Gr. unchanging + heat]. "Warm-blooded"; having a body temperature that remains fairly constant, independent of external temperatures.

HOMOLOGOUS (huh-MOLL-uh-gus) [Gr. agreeing, corresponding]. Pertaining to similar body parts in different species, arising from common construction and development, such as bird wings and whale flippers. (*See* ANALOGOUS)

HOMOLOGOUS CHROMOSOMES [Gr. same]. Chromosomes occurring in matching pairs from each of two parents. The members of such a pair are called homologs.

HOMOSPORY (hoe-MAHSS-poh-ree) [Gr. same + seed]. Production of only one kind of spore. (*See* HETEROSPORY)

HOMOTHALLIC (hoe-muh-THAL-ik) [Gr. same + a sprout]. Pertaining to an organism that is self-fertile. (*See* HETEROTHALLIC)

HOMOZYGOUS (hoe-muh-ZY-gus) [Gr. same + yoke]. Possessing two identical alleles at a given locus on homologous chromosomes.

HORMONE [Gr. to arouse]. A chemical compound secreted from an endocrine gland into the bloodstream and carried throughout the entire body. It affects a specific "target" organ or tissue, regulating and coordinating its activities.

HOST. An organism that supports another organism nutritionally. (*See* PARASITE)

HUMUS (HYU-mus). A hygroscopic, decay-resistant mix of protein and lignin in soil.

HYBRID (HY-brid) [L. mongrel]. A union of dissimilar biological entities, as in species hybrids, genetic hybrids, hybrid cells, or hybrid nucleic acids.

HYDROCORTISONE (HI-dro-COR-ti-sone). An adrenal hormone involved, among other activities, in immune responses and inflammation.

HYDROGEN BOND. A weak chemical bond formed by the attraction of a hydrogen atom bearing a slight positive charge to an oxygen or a nitrogen atom bearing a slight negative charge.

HYDROLYSIS (hy-DROL-uh-sis) [Gr. water + loosening]. The splitting apart of a compound into two molecules, with the addition of a water molecule.

HYDROPHILIC (hy-druh-FILL-ik) [Gr. water + loving]. Referring to organisms that live where water is abundant.

HYDROSERE (hy-druh-SEAR) [Gr. water + English "barren"]. A wet region cleared of organisms, open for invasion, such as a glacial lake.

HYMEN (HY-men). A fold of skin that partially blocks the vaginal entrance.

HYPERTENSION. High blood pressure.

HYPERTHYROIDISM. A condition caused by an oversecretion of TSH by the pituitary or by a thyroid tumor. Also called Graves' disease.

HYPHA (HY-fuh) [Gr. web]. A threadlike filament of a fungus.

HYPOGLYCEMIA (hy-poh-gly-SEEM-ee-uh) [Gr. under + sugar]. A condition of low blood sugar, caused by excessive secretions of insulin.

HYPOTHALAMUS (hy-poh-THAL-uh-mus) [Gr. under + inner chamber]. A portion of the brain, below the thalamus, that controls many physiological and endocrine activities.

HYPOTHYROIDISM. Underactivity of the thyroid gland. The resultant insufficiency of iodine usually causes a goiter, a swelling in the neck. (*See* GOITER)

IMAGO (ih-MAY-go) [L. image, likeness]. The adult stage of insects.

IMMUNOLOGY (im-yuh-NOL-uh-jee) [L. safe + Gr. study]. The science dealing with the process whereby organisms develop a chemical resistance, an antibody, to a foreign substance, its antigen, such as pollen, viruses, bacteria, or feathers.

IMPLANTATION [L. into + to set]. The sinking of the blastocyst into the endometrium of the mammalian uterus.

IMPRINTING. The rapid fixing of social preferences. For example, newly hatched goslings will form an attachment to the first creature they can follow as their mother.

INCISOR (in-SIZE-er). The front teeth of mammals, adapted to chiseling action.

INCOMPLETE METAMORPHOSIS (met-uh-MORE-fuh-sis) [Gr. beyond, over + shaping]. The gradual transition of some insects, by several stages of development, from the nymph to the adult. The young are small copies of adults.

INDEPENDENT ASSORTMENT, LAW OF. A generalization of Mendel, which states that when one pair of alleles segregates during sexual reproduction, its manner of segregation is not affected by the manner of segregation of a second pair of alleles.

INDETERMINATE CELL (in-duh-TERM-uh-nat). A cell of the early embryo from which no preordained part of the future animal will necessarily differentiate.

INDUCTION [L. to lead in]. The influence of one type of tissue on the developmental pattern of another type of tissue.

INFLORESCENCE (in-flo-RES-ens). A flower cluster.

INORGANIC. A chemical compound containing no carbon.

INSERTION. In anatomy, the point of attachment of a muscle to the bone that moves. (*See* ORIGIN)

INSIGHT LEARNING. A type of learning whereby an animal is able to associate apparently unrelated experiences.

INSTINCT. Fixed action or stereotyped behavior; a controversial and poorly defined term, which is not generally acceptable to students of behavior.

INSULIN (IN-suh-lin). A hormone, secreted by beta cells in the islets of Langerhans portion of the pancreas; facilitates glucose transport across cell membranes. Insulin enhances the conversion of glucose to glycogen.

INTERNEURON (in-ter-NYU-ron) [L. between + Gr. nerve]. A nerve cell that lies entirely within the brain or spinal cord. It carries impulses from sensory neurons to motor neurons.

INTERPHASE (INT-er-faze) [L. between + Gr. phase]. The period between cell divisions, during which DNA duplicates.

INTERSTITIAL CELL (in-ter-STISH-ul). A cell among the seminiferous tubules in testes that secretes the male sex hormones, especially testosterone.

INTRON. An interruption in DNA by stretches of nucleotide strands that do not code for any amino acids.

INVAGINATION (in-vaj-i-NAY-shun) [L. within + sheath]. The process of infolding, as in a blastula to form a double-layered cup.

INVERSION, CHROMOSOMAL. A type of chromosome change in which there is a reversal of a segment in a given chromosome.

ION (EYE-on). An electrically charged atom or group of atoms, the result of a gain or loss of electrons. Electrically charged groups of atoms are sometimes called complex ions.

IONIC BOND (eye-ON-ik). The chemical method of holding together two atoms by the transfer of one or more electrons from one atom to the other.

ISLETS OF LANGERHANS (EYE-lets; LAHNG-er-hanz) [Paul Langerhans, 19th-century German anatomist]. Cluster of pancreatic cells, containing two groups of cells: alpha and beta. The alpha cells secrete glucagon; the beta cells secrete insulin.

ISOGAMY (eye-SAHG-uh-mee) [Gr. equal + marriage]. Sexual reproduction resulting from the mating of two indistinguishable gametes.

ISOMER (EYE-soh-mer) [Gr. equal + part]. Compounds with identical components but different atomic arrangements in their molecules.

ISOTOPE (EYE-soh-tope) [Gr. equal + place]. An atom of an element with the same number of protons and electrons as other atoms of the same element but a different number of neutrons.

JAUNDICE [Fr. yellow]. A liver disease in which abnormally high accumulations of bile pigments cause the skin to turn yellow.

JUVENILE HORMONE. An insect hormone that prolongs the larval stage through several molts.

KARYOTYPE (CARRY-oh-type) [Gr. nucleus + type]. The size, shape, and appearance of a metaphase chromosome set of an individual or species.

KIDNEY. The organ that produces urine by filtration of water, salts, and nitrogenous waste from blood; the main water-regulating organ in land animals.

KILOCALORIE (KILL-oh-CAL-uh-ree) [Gr. thousand + L. heat]. One thousand calories; also written as kcal or Calorie.

KLINEFELTER'S SYNDROME (KLYNE-fel-terz). A type of genetic disorder in which there is an extra X chromosome, giving an XXY karyotype. The afflicted individual is sterile and has underdeveloped male genitals.

KREBS CYCLE (TRICARBOXYLIC ACID CYCLE; CITRIC ACID CYCLE) (KREBZ) [Sir Hans Krebs, a British biochemist]. The cyclic series of reactions in mitochondria by which certain organic acids yield carbon dioxide and hydrogen.

KWASHIORKOR (kwash-ee-OHR-ker). A disease caused by insufficient protein in the diet, resulting in poor muscle development and skin difficulties.

LACTATION [L. milk]. The production of milk in mammals.

LACTIC ACID (LAK-tic) [L. milk]. An organic acid, $C_3H_6O_3$, formed by incomplete oxidation of pyravic acid. (*See* FERMENTATION)

LAMELLA (la-MEL-la) [L. plate]. Any layer; applied to layers in bone, mollusc shells, leaves, and the pectic layers between plant cells.

LARVA [L. ghost]. The young developmental stage in some insects, unlike the adult in appearance; a grub, maggot, or caterpillar.

LARYNX (LAR-inks) [Gr. upper part of windpipe]. The voice box; a cartilagenous structure at the top of the trachea that contains the vocal chords.

LATENT LEARNING. An exploratory learning that takes place when an animal actively seeks to find out more about its environment; curiosity.

LATERAL BUD. A bud in the axil of a leaf. (*See* APICAL BUD)

LEAF PRIMORDIUM (pry-MORE-dee-um) [L. first + to begin to weave]. An incipient leaf, just below the apical growing point of a vascular plant.

LEARNING. In experimental psychology, the modification of behavior as a result of changes in individual experience.

LEUKEMIA (lyu-KEE-mee-uh) [Gr. white + blood]. A disease characterized by uncontrolled white blood cell production.

LEUKOCYTE (LYU-koh-site) [Gr. white + cell]. A white blood cell.

LIGAMENT. Tough sheet of fibrous tissue connecting two or more separate bones or cartilages.

LIGNIN (LIG-nin) [L. wood]. Various phenolic compounds which, together with cellulose, make up the bulk of wood.

LIPID (LIP-id). An organic compound insoluble in water but soluble in ethers and alcohols, e.g., fats, oils, steroids.

LITTORAL (LIT-uh-rul) [L. shore]. Referring to the region at the edge of a body of water, including both water and land, usually in oceans.

LIVER. The largest gland, which has many important functions, including among others the storage of foods, production of bile salts, conversion of stored glycogen to glucose, deamination of amino acids, and breakdown of hemoglobin.

LOOP OF HENLE (HEN-lee) [Friedrich Henle, 19th-century German anatomist]. A narrowly U-shaped portion of the nephron. It acts as a countercurrent multiplier.

LUCIFERIN (lyu-SIF-uh-rin) [L. light + to bear]. A substance, isolated from bioluminescent organisms, that emits light when acted upon by the enzyme luciferase.

LUTEINIZING HORMONE (LH) (LYU-tee-un-eye-zing) [L. yellowish]. One of the female gonadotropic hormones, which participates in egg maturation and in converting the old follicle, after ovulation, into the endocrine gland corpus luteum.

LYMPH (LIMF). Clear circulating body fluid, which contains water, some proteins, electrolytes, and white blood cells.

LYMPHATIC SYSTEM (lim-FAT-ik). The "second circulatory" system, made up of a network of thin-walled veins and capillaries that carry lymph fluid. It returns excess fluid and proteins from the spaces between cells to the blood, and it filters and destroys harmful material.

LYMPHOCYTE (lim-FO-site) [Gr. water + cell]. Any of several white blood cells involved in immunity.

LYSOGENIC (lye-soh-JEN-ik) [Gr. loosening]. The inactive state of a virus in a living bacterial cell.

LYSOSOME (LYE-soh-sohm) [Gr. loosening]. A cytoplasmic organelle rich in hydrolytic enzymes.

LYTIC (LIT-ik) [Gr. loosening]. Capable of causing dissolution; used to describe the stage in which an infecting phage destroys a bacterial host cell.

MACROPHAGE (MAK-roh-fayj) [Gr. large + to eat]. A scavenger cell, found in the liver, bone marrow, or spleen, that engulfs red blood cells.

MALPIGHIAN TUBES (mal-PIG-ee-un) [Marcello Malpighi, 17th-century Italian anatomist]. The excretory system of insects and spiders.

MANDIBLE (MAN-duh-bul) [L. chew]. Jawlike appendage of the mouth in arthropods; the jawbone in vertebrate animals.

MARASMUS (muh-RAZ-mus). A disease caused by insufficient quantity of food, as contrasted with inadequate quality. It results in general weakness and lack of muscle tone.

MARSUPIAL (mar-SUPE-ee-ul) [L. pouch]. A mammal that bears young in poorly developed condition and then keeps them in an abdominal pouch. Each embryonic baby attaches itself to a nipple and nurses until it achieves a measure of independence.

MECHANORECEPTOR (MEK-uh-noh-ree-SEP-ter). Sensory cell or organ capable of perceiving changing mechanical stimuli, such as those of touch or hearing.

MEDULLA (muh-DULL-uh) [L. middle]. (1) The inner portion of an organ. (2) The posteriormost part of the brain, which controls such vital functions as heart rate, breathing rate, and blood pressure.

MEDUSA (muh-DEW-suh) [L. myth., *Medusa,* a woman with snakes for hair]. A free-swimming, reproductive, umbrella-shaped form in the life cycle of some coelenterates; a jellyfish.

MEGAGAMETOPHYTE (meg-uh-guh-MEET-uh-fite) [Gr. large + marriage + plant]. Haploid egg-producing plant; a female gametophyte.

MEGASPORE (MEG-uh-spohr) [Gr. large + seed]. A large spore that germinates to form a female gametophyte.

MEGASPORE MOTHER CELL. A diploid cell destined to undergo meiosis, producing four haploid megaspores.

MEIOSIS (my-OH-sis) [Gr. diminution]. The two nuclear divisions that result in the reduction of chromosome number from diploid to haploid.

MEISSNER'S CORPUSCLE (MICE-nerz KOR-pus-ul) [Georg Meissner, 19th-century German histologist]. A specialized receptor in the skin that is sensitive to direct touch.

MELANIN (MEL-uh-nin) [Gr. black]. A pigment in skin, eyes, or hair, giving brown to black color.

MELATONIN (mel-uh-TOE-nin). A chemical formed by the pineal gland. It affects skin pigmentation and seems to react to changes of light through the eye.

MENARCHE (MEN-ar-kee) [Gr. month + beginning]. The first menstrual period of a human female, which occurs during puberty.

MENINGES (meh-NIN-jeez) [Gr. membrane]. The three protective membranes enveloping the brain and spinal cord.

MENOPAUSE (MEN-oh-pawz) [Gr. month + cessation]. Permanent termination of the menstrual period in a woman.

MENSTRUAL FLOW (MEN-strew-al) [L. monthly]. The periodic sloughing off of the uterine wall along with cells and blood if no fertilization follows ovulation.

MERISTEM (MERRY-stem) [Gr. divide]. Plant tissue capable of indefinite growth by cell division.

MESODERM (MEZ-oh-derm) [Gr. middle + skin]. The middle germ layer of an embryo, between the ectoderm and endoderm, giving rise to muscle, connective tissue, bone, circulatory system, lining of coelom, and urogenital system.

MESOPHYTIC (mez-oh-FIT-ik) [Gr. middle + plant]. Referring to plants that require moderate amounts of water.

MESSENGER RNA (mRNA). A single-stranded nucleic acid that is transcribed from a DNA template; contains the "genetic message" to be translated into proteins.

METAMORPHOSIS. (*See* COMPLETE METAMORPHOSIS *and* INCOMPLETE METAMORPHOSIS)

METAPHASE (MET-uh-faze) [Gr. after, beyond, over + to cause to appear]. A stage in nuclear division during which the shortened chromosomes are at the equatorial plane of the spindle.

MICROCLIMATE (MY-kroh-KLY-met). The special conditions of light, temperature, and moisture that occur in a narrowly restricted area, as on one bank of a stream, under a bush, in a small woodland opening.

MICROGAMETOPHYTE (MY-kroh-gah-MEET-uh-fite) [Gr. small + marriage + plant]. A haploid sperm-producing plant; a male gametophyte.

MICRON (MY-kron) [Gr. small]. A unit of measurement denoting one millionth of a meter; one micrometer; one thousandth of a millimeter.

MICRONUTRIENT [Gr. small]. One of the elements required in very small amounts by plants and animals; a trace element, such as zinc, boron, chlorine.

MICROPYLE (MY-kroh-pile) [Gr. small + gate]. An opening in the integuments of the ovules of seed plants through which a pollen tube can enter.

MICROSPORE [Gr. small + seed]. A spore that germinates to form a male gametophyte.

MICROTRABECULAE (MY-kro-truh-BEK-yuh-lee) [Gr. small + L. beam]. A mesh of filaments that suspends the organelles in a cell, holding them more or less firmly in place.

MICROTUBULE (my-kro-TOO-byul) [Gr. small + tubule]. A straight, thin cytoplasmic structure made of the protein tubulin; associated with movement and cytoplasmic structure.

MIGRATION. Seasonal movement of animals, usually in spring and fall, for feeding, mating, or escape from unfavorable weather.

MIMICRY (MIM-ik-ree) [Gr. to imitate]. The resemblance of an organism to some other animate or inanimate object, presumably increasing its chance of survival.

MITOCHONDRION (my-toe-KON-dree-on) [Gr. thread + granule]. A self-replicating cytoplasmic organelle, called the "powerhouse of the cell," the site of Krebs cycle activity and the cytochrome system. Its main work is the production of ATP for use in cellular respiration.

MITOSIS (my-TOE-sis) [Gr. thread + state, condition]. Nuclear division that results in the formation of two new daughter nuclei with identical chromosome complements.

MOLECULE (MOLL-ih-kyul) [L. little mass]. The smallest unit into which a substance can be divided while still retaining the chemical properties of that substance.

MOLTING [L. to change]. The shedding and replacement of an old exoskeleton in crustaceans and other animals or of feathers in birds.

MONGOLISM (MON-guh-liz-um). A type of genetic disorder in which an extra chromosome is present. The afflicted person is mentally retarded; also called Down's syndrome.

MONOCOTYLEDON (MON-uh-kot-uh-LEE-dun) [Gr. single + cotyledon]. A flowering plant with one seed leaf in the embryo. Frequently shortened to "monocot".

MONOHYBRID CROSS [Gr. one + mongrel]. A cross between individuals differing in only one inheritable trait, or in which only one trait is considered by an experimenter.

MOTOR UNIT. A group of skeletal muscle fibers activated by the same nerve fiber.

MULTIPLE ALLELES (uh-LEELZ). Genes that are present in a population in more than two forms, although only two can be present in a single diploid organism; for example, the human ABO blood group.

MULTIPLE GENES. Two or more different pairs of alleles at different loci on chromosomes, capable of adding quantitatively to such traits as size and color.

MUTAGEN (MEW-tuh-jen) [L. to change + Gr. to produce]. An agent that causes mutation.

MUTATION (mew-TAY-shun) [L. change]. An inheritable change in a gene; a change in nucleotide sequences in a DNA molecule.

MYCELIUM (my-SEE-lee-um) [Gr. fungus]. Fungus tissue, composed of hyphae.

MYCORRHIZA (my-koh-RYE-zuh). [Gr. fungus + root]. Fungus inhabiting the roots of higher plants, feeding off the host, and aiding the host in the intake of minerals.

MYELIN (MY-uh-lin) [Gr. marrow]. A fatty protein material,

wrapped around the axon of some nerve fibers, that aids in the conduction of electrical impulses.

MYOFIBRIL (my-oh-FYBE-ril) [Gr. muscle + L. small fiber]. A subdivision of a muscle fiber.

MYOGLOBIN (my-oh-GLOW-bin) [Gr. muscle + ball, globe]. An iron-containing protein that acts mainly in muscle cells, quickening release of oxygen during increased activity.

MYOSIN (MY-uh-sin) [Gr. muscle]. A 100-Å-thick protein that makes up the thick filaments of the muscle. Myosin and actin, a smaller protein, are the ultimate structural units of contractility.

MYXEDEMA (mix-uh-DEE-muh) [Gr. mucus + swelling]. A disorder, characterized by swollen facial features, dry skin, low basal metabolic rate, tiredness, and possible mental retardation, resulting from an underactive thyroid gland during adulthood.

NAD. (*See* NICOTINAMIDE ADENINE DINUCLEOTIDE)

NANOMETER (NAN-uh-mee-ter). One billionth of a meter; used for measuring wavelengths of light.

NATURAL SELECTION. The process of the environment "selecting" for those individuals with traits that best suit them to their habitat; they are better able to pass on those traits to their offspring; the survival of the fittest.

NEMATOCYST (NEM-uh-tuh-sist) [Gr. thread + bladder]. A stinging cell on a tentacle of a coelenterate, used for defense and capturing prey.

NEPHRIDIUM (nuh-FRID-ee-um) [Gr. kidney]. The excretory organ of the earthworm and the other annelids.

NEPHRON (NEFF-ron) [Gr. kidney]. The functional unit of excretion of the kidney, separating water, minerals, and nitrogenous waste from the blood, and making urine.

NERVE [L. sinew]. A bundle of fibers enclosed in a connective tissue sheath, through which stimuli are transmitted from the central nervous system to the peripheral nervous system or vice versa.

NERVE IMPULSE. An electrochemical impulse passing along a nerve fiber.

NEURON (NYU-ron) [Gr. nerve]. A nerve cell, the fundamental unit of the nervous system. It is made up of three parts: the cell body, the dendrites, and the axon.

NEUTRON (NU-tron). An elementary particle in an atomic nucleus, almost equal to a proton, but without charge.

NICHE (NITCH) [It. a hollow]. The physiological, behavioral, or nutritional specialty of an organism, permitting it to survive in a given habitat; determined by the physiological or behavioral actions of a given species.

NICOTINAMIDE ADENINE DINUCLEOTIDE (NAD). A coenzyme that functions as a hydrogen acceptor in cellular oxidation.

NODE [L. knot]. The part of a stem bearing a leaf; any differential structure along an otherwise uninterrupted line, e.g., a node of Ranvier in nerve fibers.

NONDISJUNCTION. Failure of homologous chromosomes to separate during meiosis.

NOTOCHORD (NOTE-uh-kord) [Gr. back + cord]. The rod-shaped body serving as internal skeleton in the embryos of all chordates and in the adults of some primitive ones; replaced by a vertebral column in adult vertebrates.

NUCLEIC ACIDS (new-KLAY-ik). Macromolecules that are polymers of nucleotides; DNA and RNAs.

NUCLEOLUS (new-KLEE-uh-lus). A non–membrane-delimited assembly center in the nucleus, responsible for partial construction of ribosomes.

NUCLEOTIDE (NEW-klee-uh-tide). A molecule consisting of ribose or deoxyribose, a nitrogenous base, and a phosphate group; one of the units from which nucleic acids are synthesized.

NUCLEUS (NEW-klee-us) [L. kernel]. In cell biology, the control center of the cell containing the necessary information to direct the metabolism, replication, and heredity of cells.

NYMPH [L. young woman]. A young insect that appears somewhat like an adult but with juvenile features.

OBLIGATE. Allowing no alternative, as in obligate anaerobe, obligate aerobe, or obligate parasite.

OLFACTION [L. to smell]. The sense of smell.

OMMATIDIUM (oh-muh-TID-ee-um) [Gr. small eye]. In arthropods, a single component of a compound eye, a slender tube with a light-sensitive receptor at the base.

OMNIVORE (OM-ni-vore) [L. all + eat]. An organism that can and does eat both plant and animal material.

OOGENESIS (oh-oh-JEN-eh-sis) [Gr. egg + generation]. The process of maturation of eggs.

OPERATOR GENE. A gene that does not contribute directly to polypeptide synthesis but acts as a switch to "turn on" structural genes.

OPERCULUM (oh-PER-kyu-lum) [L. cover, lid]. (1) A tightly fitting, horny plate that covers the opening in the shells of some gastropods. (2) A bone-supported flap that covers the gill slits in fishes. (3) The lidlike cover on the spore capsule of mosses.

ORGANELLE (or-guh-NELL) [Gr. bodily organ]. A specialized, subcellular structure, such as a mitochondrion, flagellum, lysosome, Golgi apparatus.

ORGANIC. Pertaining to a chemical compound, usually complex, containing carbon; in the current popular sense, derived from living organisms.

ORGIN. In anatomy, the point of attachment of a muscle to the bone that remains steady when the next bone moves. (*See* INSERTION)

OSMOSIS (OZ-MOE-sis) [Gr. impulsion]. The simple movement of a solvent through a selectively permeable membrane.

OSMOTIC CONFORMER (OZ-MOT-ik kon-FORM-er). Animal that is incapable of regulating the concentration of dissolved material in its body fluid.

OSSIFICATION (oss-uh-fi-KAY-shun) [L. bone]. A process of deposition of calcium phosphate and other minerals from the blood to form bones, replacing cartilage cells.

OVAL WINDOW. An opening between right and left atria in mammalian fetal hearts; also called foramen ovale.

OVARY (OH-vuh-ree) [L. egg]. An egg-producing organ.

OVIDUCT (OH-vuh-dukt) [L. egg]. A tube through which eggs pass from the ovary; in mammals, a Fallopian tube.

OVULATION (awv-yu-LAY-shun) [L. little egg]. The release of a matured egg from a follicle of an ovary.

OVULE (AWV-yule) [L. little egg]. In seed plants, the structure that first forms the megasporangium and later the seed.

OVUM (OH-vum) [L. egg]. A female reproductive cell; an egg.

OXIDATION (OX-si-DAY-shun). A chemical reaction in which electrons are removed from the substance being oxidized. (*See* REDUCTION)

OXYGEN DEBT. The accumulation of lactic acid during heavy muscular activity, subsequently relieved by deep breathing.

OXYTOCIN (OX-si-TOE-sin). A female sex hormone secreted by the pituitary gland; it stimulates uterine contractions.

PACINIAN CORPUSCLE (pah-SIN-ee-un) [Filippo Pacini, 19th-century Italian anatomist]. A specialized skin receptor sensitive to firm pressure.

PALEONTOLOGY (PAIL-ee-on-TOL-uh-jee) [Gr. ancient + science]. The study of fossils.

PALISADE PARENCHYMA (puh-RENG-ki-muh) [Gr. in + pour]. Tissue of chloroplast-bearing, photosynthetic, columnar cells under the upper epidermis of a leaf.

PANCREAS (PAN-kree-us). An elongated gland under the stomach, which secretes digestive enzymes into the small intestine. The islets of Langerhans in the pancreas secrete the hormones insulin and glucagon into the bloodstream.

PANSPERMIA (pan-SPER-mee-uh) [Gr. all + seed, semen]. The idea that "seeds of life" of some kind are universally distributed everywhere, needing only suitable conditions for development.

PARASITE. An organism that obtains its nourishment from another organism, usually by remaining in contact with its host. (*See* HOST)

PARASYMPATHETIC SYSTEM (PAR-uh-sim-pa-THET-ik) [Gr. beyond + with + feeling]. A part of the autonomic nervous system that helps to regulate body functions. The action is mainly inhibitory.

PARATHYROIDS [Gr. near, beyond + shieldlike]. Four tiny glands, embedded in the thyroid gland, that secrete parathormone (PTH), which is responsible for increasing the calcium level and for decreasing the phosphate level in blood.

PARKINSON'S DISEASE. A condition caused by damage to the thalamus; characterized by involuntary tremors.

PARTHENOGENESIS (par-then-oh-JEN-uh-sis) [Gr. virgin + production]. The development of an unfertilized egg, as in honeybees and wasps.

PASSIVE TRANSPORT. The movement of molecules across a membrane purely by diffusion.

PEDICELLARIA (PED-i-sell-AY-ree-uh). A pair of pincers that act as tiny pliers with crossed jaws, on the skin of echinoderms.

PEPSIN [Gr. digestion]. An enzyme, secreted by the stomach mucosa, that converts proteins to polypeptides.

PEPTIDE LINKAGE. A strong covalent bond between the carboxyl group of one amino acid and the amino group of another.

PERICYCLE (per-i-SY-kul) [Gr around + circle] In roots and stems, a layer of parenchymal cells between the phloem and the endodermis. It gives rise to lateral roots.

PERISTALSIS (per-i-STOLL-sis) [Gr. around + contraction]. A process that produces waves of muscular contractions and relaxations along the digestive tract. The alternating movements help to propel food along.

PERMAFROST (PER-muh-frost). The permanently frozen deeper layers of soil in the tundra regions of the northern hemisphere.

PERMEABILITY (per-mee-uh-BIL-i-tee) [L. through + to pass]. The ability to allow passage of substances, said of a membrane.

PETIOLE (PET-ee-ohl). A leaf stalk.

pH SCALE [p(otential of) H(ydrogen)]. A numerical scale from 0 to 14 that measures the concentration of hydrogen ions free in water. The smaller the number below 7, the more acidic is the solution; the higher the number above 7, the more basic the solution. Pure water is neutral, with a pH of 7.

PHAGE (FAYJ) [Gr. to eat]. A virus that infects bacteria; also called bacteriophage.

PHAGOCYTE (FAG-oh-site) [Gr. to eat + cell]. A blood cell that has the ability to engulf and digest foreign material.

PHAGOCYTOSIS (fag-oh-sy-TOE-sis) [Gr. to eat + hollow vessel + state, condition]. Engulfing of bacteria or other particles by a cell.

PHARYNX (FAR-inks) [Gr. throat]. The part of the digestive tract from which the gill slits develop; a muscular ring in an alimentary tract, used by many animals for swallowing.

PHENOTYPE (FEE-nuh-type) [Gr. to show + type]. The appearance or discernible characteristic of an individual, as determined by the genetic makeup and environmental influences. (*See* GENOTYPE)

PHEROMONE (FEH-roh-moan) [Gr. carry + arouse]. A compound, released into the environment by an individual, that is capable of causing a reaction in another individual, usually of the same species.

PHLOEM (FLOH-em) [Gr. bark]. The conducting tissue in vascular plants, usually in the bark, carrying dissolved nutrients.

PHOSPHOLIPIDS (fos-foh-LIP-idz). A group of fatty compounds that contain fatty acids, glycerol, and a negatively charged phosphate.

PHOSPHORYLATION (fos-for-i-LAY-shun) [Gr. light + to carry]. The addition of a phosphate group to an organic molecule.

PHOTON (FOE-tahn) [Gr. light]. A unit of energy in electromagnetic radiation.

PHOTOPERIODISM (fot-toh-PEER-ee-ud-iz-um). The ability of some organisms to respond metabolically to variations in day length.

PHOTORECEPTOR (foh-toh-ri-SEP-ter). Sensory cell or organ that is sensitive to light stimuli.

PHOTOSYNTHESIS [Gr. light + putting together]. The synthesis of carbohydrate from water and carbon dioxide, using the energy of light captured by chlorophyll.

PHYLLOTAXY (FILL-oh-tak-see) [Gr. leaf + arrangement]. The geometry of leaf arrangement.

PHYTOCHROME (FITE-oh-krohm) [Gr. plant + color]. A light-sensitive pigment involved in seed germination, flowering, morphogenesis, anthocyanin synthesis, etc.

PINEAL GLAND (PIN-ee-al) [L. relating to the pine]. A small round gland located in the midbrain; secretes melatonin, a light receptor in primitive vertebrates. In humans it may act as a "biological clock" that affects secretions of sex hormones.

PINOCYTOSIS (pin-uh-sy-TOE-sis) [Gr. to drink + cell + state, condition]. The engulfing of a droplet by a cell membrane.

PITUITARY (HYPOPHYSIS) (pi-TU-i-teh-ree) [L. secreting phlegm]. The so-called "master gland"; an endocrine gland that controls most of the other endocrine glands. The secretions of the anterior pituitary are controlled by hormones of the hypothalamus. Some of the regulatory hormones secreted by its two lobes are: TSH, STH, FSH, LH, ACTH, prolactin, vasopressin, and oxytocin.

PLACENTA (pluh-SEN-tuh) [L. flat cake]. In mammals, the tissues composed partly of uterine wall and partly of extraembryonic membranes, through which wastes and nutrients pass for a developing embryo. In plants, the place of attachment of ovules in a flower.

PLANKTON (PLANK-tun) [Gr. wanderers]. The microscopic, actively swimming plants and animals of lakes and seas.

PLANULA (PLAN-yu-luh) [L. flat]. Ciliated larva of coelenterates.

PLASMA (PLAZ-muh). The liquid portion of mammalian blood; composed of water, dissolved solids including proteins, and dissolved gases.

PLASTID (PLAS-tid) [Gr. formed + small]. A self-replicating cytoplasmic organelle in plants functioning in photosynthesis and/or nutrient storage; e.g., chloroplast.

PLATELETS (THROMBOCYTES). Minute disks that serve as starters in the process of blood clotting.

POIKILOTHERMIC (poy-KEE-luh-therm-ik) [Gr. varied + heat]. "Cold-blooded"; having a body temperature that varies with that of the environment.

POLAR BODY. In the maturation of animal eggs, the small, nonfunctional products of meiosis, in contrast with the large, functional egg cell.

POLAR MOLECULE. A molecule whose two ends bear different electrical charges.

POLIOMYELITIS (POH-lee-oh-my-uh-LY-tis). Infantile paralysis due to viral infection causing the destruction of the neurons leading to skeletal muscle.

POLLEN. The product of the anther in a flower, or of staminate cones in conifers. A pollen grain contains the sperm nuclei plus additional nonsexual nuclei.

POLLEN TUBE. A filament containing male sex nuclei, growing from a pollen grain and delivering the nuclei to the egg in the ovule.

POLLINATION. The movement of pollen from anther to stigma in a flowering plant, preceding fertilization.

POLYMER (POL-i-mer) [Gr. many + part]. A large molecule built of many smaller, identical units; as starch, a polymer of glucose.

POLYPEPTIDE (pol-i-PEP-tide). A chain of linked amino acids.

POLYPLOIDY (POL-i-ployd-ee) [Gr. many + folds]. The condition of more than two sets of chromosomes (genomes) in a single nucleus.

POLYSOME (POL-i-sohm). A string of five or more ribosomes linked together by a strand of mRNA during active protein synthesis; a polyribosome.

POSTERIOR. Situated back of or in the back part of an organ, structure, or organism.

PREFORMATION. A hypothesis of development of new organisms, which states that a fertilized egg contains a fully developed miniature animal that merely needs the nourishment of a female uterus for it to grow into a normal-sized baby.

PRIMARY TISSUE. Cells derived from the apical meristems in plants.

PRIMITIVE STREAK [L. first (in point of time)]. A strip of ectodermal thickening, developing on an embryo, marking the future longitudinal axis of the embryo.

PROBOSCIS (proh-BOSS-iss). Any elongated and usually tubular structure extending from the head of an animal; an elongated nose.

PROGESTERONE (proh-JES-tuh-rone) [L. before + to bear, carry]. One of the female sex hormones, produced by the corpus luteum. Together with estrogen, it regulates the menstrual cycle and maintains pregnancy.

PROGLOTTID (proh-GLOT-id) [L. before + the tongue]. A body segment of a tapeworm.

PROKARYOTE (pro-CARRY-ote) [L. before + Gr. nucleus]. A cell that has no membrane-delimited nucleus and no membrane-delimited organelles. Its chromosomes are "naked" DNA. (*See* EUKARYOTE)

PROLACTIN (proh-LAK-tin) [Gr. before + L. milk]. A female sex hormone, secreted by the pituitary gland, which allows mammary glands to secrete milk after childbirth.

PROPAGULE (PROP-uh-gyul) [L. plant cutting]. Any reproductive body of an organism, such as a fertilized egg, seed, spore, or living vegetative part.

PROPHASE [L. before + Gr. appearance]. The first named stage in nuclear division, during which chromosomes shorten and thicken, and spindle fibers form.

PROPRIOCEPTOR (proh-pree-oh-SEP-ter) [L. one's own]. Sensory cell or organ sensitive to changing movements and position of one's own body.

PROSTAGLANDIN (pros-tuh-GLAN-din). One of a group of fatty-acid hormones whose effects include contraction of uterine muscle, inhibition of progesterone secretion by the corpus luteum, and lowering of blood pressure.

PROSTATE GLAND (PROS-tate) [L. coming before]. A gland in male mammals that furnishes fluid for the transport of sperm.

PROTEIN [Gr. first]. Macromolecule composed of chains of amino acids linked by peptide bonds.

PROTON. An elementary particle of an atomic nucleus, with a positive electrical charge.

PROTOPLAST (PROH-toh-plast). The content of any living cell; also a plant cell whose external coats (walls and lamellae) have been digested away, leaving the plasma membrane naked.

PTYALIN (TIE-uh-lin). The amylase in saliva.

PUBERTY (PEW-burr-tee) [L. adult]. The period during which sexual maturation is achieved.

PULSE. The systolic pressure that can be felt in an artery near the surface of the body.

PUPA (PEW-puh) [L. doll]. The stage in the development of some insects between the larva and the adult.

PURINE (PURE-een). A nitrogenous base composed of a double ring of carbon and nitrogen atoms; a constituent of DNA, RNA, ATP, NAD, etc.; e.g., adenine, guanine.

PYRIMIDINE (py-RIM-uh-deen). A nitrogenous base composed of a single ring of carbon and nitrogen atoms; a constituent of DNA and/or RNA; e.g., thymine, cytosine, uracil.

PYRUVIC ACID (py-ROO-vik) [L. pear]. An organic acid, $C_3H_4O_3$, which is an intermediate metabolic product in cellular respiration.

QUANTASOME. A unit of structure, containing chlorophyll, on the surface of thylakoid membranes in chloroplasts; thought to be photosynthetic units.

RADICLE [L. root]. The morphological lower tip of the embryo in a seed. The first root grows from its tip.

RADIOIMMUNOASSAY (RIA) (RAY-dee-oh-im-myu-noh-ASS-ay). A technique that measures hormone levels in the blood.

RADULA (RAJ-uh-luh) [L. scraper]. A mouth part in some molluscs, used as a rasping organ.

RAY. In plant anatomy, a strip of cells streaking across the grain of wood from the pith toward the bark.

RECEPTOR. A structure that is capable of perceiving stimuli. Receptors are capable of changing energy into nerve impulses.

RECESSIVE (ree-SESS-iv). Pertaining to an allele that is masked phenotypically in the presence of its dominant allele in a heterozygote.

RECOMBINANT DNA (ree-KOM-buh-nent). A segment of DNA enzymatically removed from one chromosome, united with a strand of DNA from a second organism (frequently a virus), then inserted into a new host cell, sometimes into the host chromosome.

RED BLOOD CORPUSCLE (ERYTHROCYTE) (ih-RITH-ruh-site). One of the formed elements in the blood, responsible for the transport of oxygen and carbon dioxide.

REDUCTION. (1) Chemical: a reaction in which electrons are added to the substance being reduced. (*See* OXIDATION) (2) Biological: the change in chromosome number from diploid to haploid.

REFLEX. A simple behavior pattern consisting of an automatic response to an external stimulus.

REFLEX ARC [L. bent back]. A group of two or three neurons regulating a fairly simple body movement without conscious participation of the brain.

REGENERATION. Regrowth of lost or damaged tissue or part of an organism.

REGULATOR GENE. A gene that provides codes for synthesis of repressor proteins.

RELEASER. In animal behavior, a signal such as an odor, sound, or movement that evokes a specific, unlearned response.

REM. A unit of measurement of the energy from radioactive elements; an acronym for "roentgen equivalent man."

REM SLEEP (for RAPID EYE MOVEMENT). A transition stage of sleep when dreams occur.

RENIN (REE-nin). A hormone from the kidney, which is important in the regulation of sodium balance and also plays a role in water balance by stimulating the thirst reflex in the brain.

RENNIN (REN-in). An enzyme, secreted by mucosa of the stomach, that curdles milk.

REPOLARIZATION. A condition that results when the outside membrane of a nerve fiber returns to a positive charge and the inside of the membrane becomes relatively negative again.

REPRESSOR. The protein that can combine with and repress action of an associated operator gene.

RESPIRATION (res-pi-RAY-shun) [L. again + breathe]. (1) Breathing. (2) The cellular processes by means of which energy for biological use is released from food.

RESTING POTENTIAL. The small voltage (1/10 volt) that normally exists across the surface membrane of many cells, especially muscle and nerve. This voltage is due to an imbalance in the distribution of ions. It changes rapidly if the cell is touched or stimulated. (*See* ACTION POTENTIAL)

RETICULAR FORMATION (rih-TIK-yu-lar) [L. net]. A net of nerve cells in the thalamus that filters all incoming stimuli, discarding unimportant ones and sending others to the proper decoding centers in the brain.

RETINA (RET-in-uh) [L. net]. The light-sensitive layer of the eye. It contains the rods and cones and a layer of sensory neurons.

Rh FACTOR, also rhesus factor. A group of blood cell antigens, clinically important because they can cause destruction of blood cells in certain new born infants, or erythroblastosis.

RHODOPSIN (roh-DOP-sin) [Gr. rose + sight]. The light-sensitive pigment of rod cells in the eye, consisting of a colorless protein, opsin, and a colored carotenoid molecule, retinal (or retinene).

RIBONUCLEIC ACID (RNA) (rye-boe-new-KLAY-ik). A single-stranded nucleic acid containing the sugar ribose. It is transcribed from the DNA template; found in both nucleus and cytoplasm.

RIBOSOMAL RNA (rRNA) (rye-boh-SOHM-ul). The ribonucleic acid that is a constituent of ribosomes.

RIBOSOME (RYE-boh-sohm). A cytoplasmic structure composed of ribonucleoprotein; the site of protein synthesis.

RITUALIZATION. Display movements that have been modified through evolution to become stereotyped communication signals.

ROD. A photoreceptor in the retina of the eye, containing light-sensitive pigment (rhodopsin).

ROOT CAP. A thimble-shaped mass of loosely connected cells protecting the root tip.

ROUND DANCE. Part of the "language of the bees," which consists of a movement of alternating clockwise and counterclockwise circles to indicate the direction, distance, and quality of a food source, or to relay information about a new hive location.

RUMEN (ROO-men). An enlargement of the esophagus in cows and other grazing animals. Partial digestion of cellulose occurs there.

SARCOMERE (SAR-koh-meer) [Gr. flesh + part]. A single muscle unit extending from one Z line to the next.

SARCOPLASM (SAR-koh-plaz-um) [Gr. flesh + something molded]. Viscous cytoplasm of a muscle fiber.

SARCOPLASMIC RETICULUM (sar-koh-PLAZ-mik reh-TIK-yu-lum) [Gr. flesh + molded + L. a network]. A complex network of fine tubules and sinuses in the sarcoplasm of a muscle fiber.

SATURATED FAT. Fat with fatty acids whose carbon atoms have the maximum number of hydrogen atoms attached to them. All the carbon atoms are hydrogenated with no double bonds between adjacent carbons.

SAVANNA (seh-VAN-uh). Grassland with scattered trees.

SCROTUM (SCROH-tum). An external sac between the male thighs containing testes and parts of spermatic cords.

SECOND-MESSENGER CONCEPT. A proposed mechanism for the control of the endocrine system, stating that a hormone, the "first messenger," causes the conversion of ATP to cyclic AMP inside a target cell. The cyclic AMP, the "second messenger," then diffuses throughout the cell and causes the cell to respond with its specific endocrine function.

SECRETIN (sih-KREET-in). A hormone, secreted by the epithelial cells of the duodenum, that stimulates secretion of pancreatic enzymes. Secretin was the first hormone to be discovered (in 1903).

SEED. A mature ovule of a seed plant, containing an embryo.

SEGMENTATION [L. cut]. The growth pattern in many animals in which there is serial repetition of parts, notably in earthworms, abdomens of insects, and backbones of vertebrate animals.

SEGREGATION, LAW OF. One of the fundamental principles of Mendelian genetics, which states that contrasting characteristics in an individual separate when gametes are formed, each gamete receiving the gene for only one of each pair of alleles.

SEMEN (SEE-men) [L. to sow]. A secretion of the male reproductive organs consisting of sperm cells and fluids secreted by the seminal vesicles and prostate gland.

SEMICIRCULAR EAR CANALS. Structure in the inner ear, important in maintaining balance.

SEMINAL VESICLE (SEM-uh-nul VES-uh-kul). The portion of the male reproductive system that secretes nutritive fluids for sperm.

SEMINIFEROUS TUBULE (sem-uh-NIF-uh-rus) [L. sperm + bearing]. A coiled tube, contained inside testes, that produces the sperm cells.

SEPTUM. A partition, as between nostrils, between heart compartments, or between cells.

SERE (SEER). In ecology, the repopulation of a space that has been completely cleared of living organisms.

SEX-LINKED. Determined by a gene on a sex chromosome.

SIEVE CELL (SIV). A nonnucleated, elongated plant cell in the phloem, with perforations in its ends and sides through which materials, especially water and nutrients, can pass.

SIGN STIMULUS. In animal behavior, a major signal that initiates a given movement or reaction; releaser.

SINOATRIAL (S-A) NODE (sy-noh-AY-tree-ul). The specialized section of muscle tissue where the heartbeat originates.

SKELETAL MUSCLE (STRIATED MUSCLE, VOLUNTARY MUSCLE). A "voluntary" type of muscle under the control of the somatic nervous system.

SLIDING-FILAMENT THEORY. A theory stating that muscles are shortened when the thin actin and thick myosin filaments slide past each other.

SMALL INTESTINE. That part of the alimentary canal between the stomach and large intestine, the site of most digestion and absorption of nutrients.

SMOOTH MUSCLE (VISCERAL MUSCLE). Long, tapered cells, with one nucleus in the fiber of each cell, that line internal organs of the body (but not the heart). Smooth muscle contains no striations and moves slowly.

SOCIETY. Organisms living together in organized groups.

SOFTWOOD. The wood of a coniferous tree.

SOLUTE (SOL-yute). Substance dissolved in a medium.

SOLVENT. The medium in which substances are dissolved.

SOMA (SOH-muh). In general, a body. In neurobiology, that part of a nerve cell containing the nucleus.

SOMATIC NERVOUS SYSTEM (soh-MAT-ik). Nervous system composed of nerve fibers that lead from the brain and spinal cord to the skeletal muscles. It involves "voluntary" movements of the skeletal muscles.

SOMATOTROPIC HORMONE (STH) (soh-MAT-oh-TROH-pik) [Gr. body + turn, change]. A hormone, secreted by the anterior lobe of pituitary, that regulates the growth of the skeleton.

SPECIATION (spee-shee-AY-shun) [L. sort, kind]. The development of new species.

SPECIES (SPEE-sheez) [L. kind]. (1) A group of individuals capable of producing fertile offspring. (2) A taxonomic unit considered satisfactory to a specialist in biological classification.

SPERMATID (SPUR-muh-tid). One of four haploid cells formed by the meiosis of spermatocytes during spermatogenesis. A spermatid develops directly into a sperm cell.

SPERMATOCYTE (sper-MAT-o-site) [Gr. seed cell]. A diploid cell destined to undergo meiosis before the formation of spermatids.

SPERMATOGENESIS (sper-MAT-O-JEN-e-sis) [Gr. seed beginning]. The process of sperm production.

SPERMATOZOA (spur-mah-toh-ZOH-uh) [Gr. seed + animal]. The sperm cells.

SPICULE (SPIK-yule) [L. a spike]. A crystalline skeletal element, made of siliceous material, in the middle cell layers of sponges.

SPINAL CORD. A part of the central nervous system that passes through the protective vertebrae, from which 31 pairs of spinal nerves branch to each side of the body. It is the connecting link between the brain and most of the body.

SPINDLE. The aggregation of microtubules associated with chromosomes during nuclear division.

SPINNERET (spin-uh-RET). An appendage at the tip of the abdomen that is capable of extruding a silky filament, which is used for web-building by spiders.

SPIRACLE (SPYR-uh-kul) [L. to breathe]. A breathing hole along the sides of an insect abdomen, allowing for oxygen and carbon dioxide exchange.

SPLIT GENES. Genes with interrupting nucleotide strands called introns; the introns do not code for any amino acids. (*See* INTRON)

SPONTANEOUS GENERATION. A theory that life can arise from nonliving matter.

SPORANGIUM (spoh-RAN-jee-um) [Gr. seed vessel]. Any plant part that produces spores.

SPORE [Gr. seed]. A reproductive cell in plants; in general, any nonsexual reproductive cell.

SPOROPHYLL (SPOHR-uh-fil) [Gr. seed + leaf]. A spore-bearing leaf.

SPOROPHYTE GENERATION (SPOHR-uh-fite). The spore-producing, diploid phase of plants having alternation of generations.

STAMEN (STAY-men) [L. thread]. The pollen-producing organ in a flower, consisting of the pollen sac (anther) and usually a supporting filament.

STERILE (STAIR-ul). (1) Incapable of reproduction. (2) Devoid of life (culture media, microbiological technique). (3) Sustaining little or no growth (soils).

STERNUM (STIR-num) [Gr. breastbone]. The breastbone.

STEROID (STEER-oid) [Gr. solid + like]. A fat-soluble compound of four connected carbon rings, occurring in many hormones.

STOLON (STOH-lun) [L. shoot]. A horizontal stem that grows along the ground and forms adventitious roots and shoots, as in a strawberry plant.

STOMA (STOW-muh) [Gr. mouth]. A microscopic opening between epidermal cells on leaf and stem surfaces, surrounded by a pair of guard cells.

STRIATED MUSCLE (STRY-a-ted) [L. striped]. (*See* SKELETAL MUSCLE)

STROMA (STROH-muh) [L. bed covering]. Nongrana regions of the chloroplasts. The sites of the "dark reactions" of photosynthesis; any relatively unorganized part or tissue.

STRUCTURAL GENE. A gene that produces mRNA and is therefore instrumental in the production of polypeptides. It functions in the synthesis of enzymes.

SUBSTRATE. Molecule on which an enzyme works.

SUCCESSION. In ecology, a series of species, each replacing the former, until a self-perpetuating community is established.

SYMBIOSIS (sim-bee-OH-sis) [Gr. with + life]. The close association of more than one species of organisms living together in some special physical or nutritional relationship.

SYMPATHETIC SYSTEM [Gr. with + feeling]. A part of the autonomic nervous system that controls metabolic activities. Its action is mainly stimulatory.

SYMPATRIC SPECIES (sim-PAT-rik) [Gr. with, together + L. country]. Closely related species that occupy common geographical regions.

SYNAPSE (SIN-aps) [Gr. union]. The junction between the axon terminal of one neuron and the dendrite or cell body of the next neuron.

SYNAPSIS (si-NAP-sis) [Gr. union]. The pairing of homologous chromosomes during prophase I of meiosis.

SYNDROME (SIN-drohm). A group of symptoms and signs that, taken together, characterize a condition or disease.

SYNERGISM (SIN-er-jiz-um). The enhanced effect of two or more factors working together. The final result is greater than it would be if either factor worked alone.

SYNOVIAL MEMBRANE (si-NOH-vee-ul). Special membrane covering bones at joints.

SYPHILIS (SIF-i-lis). A venereal disease, caused by the bacterium *Treponema pallidum*, contracted usually by sexual intercourse. It causes paralysis, insanity, and death.

SYSTOLE (SIS-toe-lee) [Gr. drawing together]. The ventricular contraction of the heart, by which the blood is forced out of the ventricles.

TAIGA (TIE-gah) [Rus.]. The cold, marshy, evergreen forest regions of the northern hemisphere.

TAXIS (TACK-sis) [Gr. arrangement]. A movement, usually by microorganisms, toward or away from a stimulus such as light, heat, or gravity.

TAXON (TACKS-on) [Gr. arrangement]. A category of classification of organisms.

TAXONOMY (tack-SON-uh-mee) [Gr. order + law]. The study of classification, especially with the aim of arranging organisms in ways that show their evolutionary relationships.

TELOPHASE (TELL-uh-faze) [Gr. end + phase]. The final stage in nuclear division, in which two daughter nuclei are formed; usually followed by cytoplasmic division.

TENDON [Gr. to stretch]. A fibrous band that connects a muscle to a bone.

TERRITORY. In animal behavior, a specific area guarded against intrusion by other members of the same species.

TEST [L. jug]. Any hard protective shell; especially the shells of sea urchins and certain protozoa.

TEST CROSS. A cross between an experimental organism of unknown genotype and one with recessive phenotype to determine the genotype of the unknown.

TESTES (TESS-teez) [L. jug]. Male gonads that produce sperm cells.

TESTOSTERONE (tess-TAHSS-tuh-rohn). The primary androgenic hormone that affects the production of sperm, the development of sex organs, and the appearance of secondary sex characteristics.

TETRAD (TEH-trad) [Gr. four]. The complex of four chromatids during the pairing of homologous chromosomes during prophase I of meiosis; any quartet of objects, such as four microspores.

THALAMUS (THAL-uh-mus). A part of the brain, below the cerebrum, which acts as a monitor to all incoming and outgoing impulses of the central nervous system and modifies some of them.

THERMAL POLLUTION. A kind of nonmaterial pollution that occurs when streams are used to cool large industrial activities.

THERMODYNAMICS (THER-moh-dy-NAM-iks) [Gr. heat + movement]. The study of energy relations in biological, chemical, and physical actions.

THERMORECEPTOR. Sensory cell or organ that is sensitive to changing temperature.

THRESHOLD. A minimum stimulus required to instigate a nerve impulse.

THYLAKOID (THIGH-luh-koyd) [Gr. sack, pouch]. A disklike, photosynthetic, membrane-bound vesicle in a chloroplast.

THYMINE (THIGH-meen). A pyrimidine, a nitrogenous base that is a constituent of DNA.

THYMUS (THIGH-mus). A lymphoid gland, located under the breastbone, active during childhood. Its main function in humans may be the development of the immunological system.

THYROID. A gland in the neck that secretes the hormones thyroxine, which controls the rate of metabolism and growth, and thyrocalcitonin, which lowers the calcium level in the blood.

THYROID-STIMULATING HORMONE (TSH). An anterior pituitary product that stimulates the production of thyroxine.

THYROXINE (thigh-ROCK-seen). A thyroid hormone, consisting of an amino acid containing iodine, which controls the rate of body metabolism and growth.

TISSUE. A mass of cells all similarly differentiated structurally and functionally.

TOTIPOTENCY (tote-i-POE-ten-see) [L. all + powerful]. The ability of a cell (or nucleus) to develop (or regulate the development) into any kind of differentiated cell; having complete genetic ability.

TRACHEA (TRAY-kee-uh) [Gr. rough]. In vertebrates, the windpipe. In plants, the open-ended water-conducting cell in wood. In insects and some other land-dwelling arthropods, an air tube in the body.

TRACHEID (TRAY-kee-ud) [Gr. rough]. A thickened and elongated water-conducting cell with closed tapered ends, occurring mainly in softwoods.

TRANSCRIPTION [L. across + to write]. Process whereby messenger RNA is synthesized from the DNA template.

TRANSDUCTION [L. to lead across]. (1) The ability of all receptor cells to change received stimuli into nerve impulses. (2) The genetic change in a bacterium, induced by introduction of foreign DNA via a virus carrier.

TRANSFER RNA (tRNA). A form of ribonucleic acid that serves as an adapter molecule in the synthesis of proteins. Each tRNA can combine with a particular amino acid and place it in its proper position in a growing polypeptide.

TRANSFORMATION. Phenotypic change in a bacterium caused by the passage and incorporation of DNA directly from one type of bacterium to another.

TRANSLATION. The process by which a protein molecule is synthesized from amino acids according to the specific nucleotide sequence of the mRNA molecule.

TRANSLOCATION [L. through + place]. (1) The transfer of a piece of one chromosome to another part of the same chromosome or to a different chromosome. (2) Movement of material from one part of an organism to another.

TRICARBOXYLIC ACID CYCLE. (*See* KREBS CYCLE)

TROPHALLAXIS (troh-fuh-LAK-sis) [Gr. food + exchange]. The exchange of food and pheromones through regurgitation. In insect societies it is essential for communication.

TROPHIC LEVEL (TROH-fik). In ecology, the number of steps a consumer is removed from the primary producers. A vegetarian is a first-level consumer; a carnivore is higher.

TROPHOBLAST (TROH-fuh-blast) [Gr. food + sprout]. The growing mass of cells on one side of an embryo, which will give rise to the placenta and the extraembryonic membranes.

TROPISM (TROH-piz-um) [Gr. a turning]. A growth movement whose direction is determined by the direction from which the stimulus comes.

TUBULAR REABSORPTION. The process in the kidney that returns useful substances to the blood by active transport after those substances have been initially lost from the blood during glomerular filtration.

TUBULAR SECRETION. The process in the kidney by which waste products are secreted into the renal fluid as the fluid passes through the nephric tubules.

TUNDRA (TUN-druh) [Rus.]. The cold, treeless plains of the northern hemisphere, with permanently frozen subsoil.

TURNER'S SYNDROME. A genetic disorder in which there is only one sex chromosome, the XO condition. The afflicted individual is female but has abnormal body structure.

TYMPANUM (TIM-puh-num) [Gr. drum]. The eardrum.

UMBILICAL CORD (um-BIL-i-kul) [L. the navel]. A cord that carries wastes from the embryo to the placenta and carries oxygen and nutrients from the placenta to the embryo.

UNSATURATED FAT. Fat whose constituent fatty acids contain some carbon atoms joined by double bonds, instead of having two hydrogen atoms each.

URACIL (YUR-uh-sill). A pyrimidine, a nitrogenous base that is a constituent of RNA.

URETER (YUR-et-er) [Gr. to urinate]. A tube carrying urine from the kidney to the urinary bladder.

URETHRA (yu-REE-thruh) [Gr. to urinate]. The canal that conveys urine from the bladder to the outside of the body. It also transports sperm outside the male body.

UTERUS (YU-tuh-rus) [L. womb]. The womb; the pear-shaped organ located just beyond the vagina, which houses, nourishes, and protects the developing fetus within its muscular walls.

VACUOLE (VAK-you-ole) [L. empty]. A membrane-bound vesicle located in the cellular cytoplasm, containing mostly water with salts, proteins, crystals, and pigments.

VAGINA (vuh-JINE-uh). A muscle-lined tube that receives sperm from the penis during sexual intercourse and allows the fetus to pass down from the uterus during childbirth.

VASOPRESSIN (ADH) (vay-zoh-PRESS-in) [L. vessel + lower]. A hormone, produced by the hypothalamus and released by the posterior lobe of the pituitary, that aids water reabsorption in the kidneys and stimulates the smooth muscles of the arteries. (*See* ANTI-DIURETIC HORMONE)

VEIN. A vessel that carries blood toward the heart.

VENEREAL (vuh-NEER-ee-ul) [L. myth. *Venus,* goddess of love]. Relating to or resulting from sexual intercourse. Examples of venereal diseases are syphilis and gonorrhea.

VENTRAL. Pertaining to a position toward the belly surface of an organ, structure, or organism. (*See* DORSAL)

VENTRICLE (VEN-tri-kul) [L. stomach]. A cavity in an organ, such as the heart chamber, posterior to the atria, or the cavities in the brain.

VESTIGIAL (ves-TIJ-i-ul) [L. trace]. Existing in a condition reduced from that of an earlier time. In a morphological sense, a part that has been diminished through evolution from a once useful condition; e.g., human tail bones, scale leaves of cacti.

VILLI (VILL-eye) [L. shaggy hair]. Fingerlike projections from a free surface of a membrane, as on the lining of the small intestine or the chorionic membrane of the placenta.

VISCERAL MUSCLE (VISS-er-al). (*Sea* SMOOTH MUSCLE)

VITALISM [L. life]. An explanation of biological phenomena holding that life has a force and quality of its own, something different and apart from the forces and mechanisms of the inanimate world.

VITAMIN. An organic substance present in minute amounts in food, essential in supporting the work of enzymes, especially those involved in cellular respiration.

VIVIPAROUS (vi-VIP-uh-rus) [L. alive + to bring forth, produce]. Organisms that nourish the young within the uterus until they are born alive.

VOLUNTARY MUSCLE. (*See* SKELETAL MUSCLE)

WAGGLE DANCE. Part of the "language of the bees," which consists of a movement of a figure eight with a straight line between the loops to indicate the direction, distance, and quality of a food source, or to relay information about a new hive location.

WARNING COLORATION. Conspicuous coloration that helps to warn predators of poisonous or other objectionable features.

WHITE BLOOD CORPUSCLE (LEUKOCYTE). One of the formed elements of the blood; they serve as scavengers and immunizing agents.

XYLEM (ZYE-lem) [Gr. wood]. The wood tissue through which most of the water and minerals of a plant are conducted.

ZYGOSPORE (ZY-goh-spohr) [Gr. yoke + pair + seed]. A thick-walled spore developing from the sexual fusion cells in some algae and fungi.

ZYGOTE (ZY-goat) [Gr. yoked together]. The cell formed by the union of a sperm and an egg.

Photo Acknowledgments

Prologue Impalas, John Dominis, *Life Magazine,* © 1969 Time Inc. Vermont trees, Richard W. Brown. Bush baby, R. D. Estes. Rose quartz, E. R. Degginger, FPSA. Zebra birth, N. Meyers, Bruce Coleman, Inc. Sea anemone, Jane Burton, Bruce Coleman, Inc. Crocodile, Jonathan Blair, Woodfin Camp, Inc. Flamingoes, M. Philip Kahl, Jr., Verda International Photos. T4 virus, Lee D. Simon, Waksman Institute of Microbiology, Rutgers University, New Brunswick, N.J. Light-trapping apparatus of human eye, J. E. Dowling, *Science* 147:1965. Grana of chloroplast, Dr. L. K. Shumway, College of Eastern Utah, San Juan Center. Elk in Wyoming, Steven C. Wilson, Entheas/Communications. Corn snake, Jack Dermid. African elephant, Alpha Photo Associates. Lioness, N. Meyers, Bruce Coleman, Inc. Penguins, Eugen Schuhmacher. Starving people, Henri Cartier—Bresson, Magnum Photos, Inc. Crowded street, Nicholas DeVore, Bruce Coleman, Inc. Air pollution, E. R. Degginger, FPSA. Nursing home patients, Vidibor, Photo Researchers, Inc. Mountains, Ed Cooper Photo.

Part One opener Copyright ©, David Scharf, 1977. From *Magnifications,* by Schoken Books, Inc., N.Y.

Figure numbers precede acknowledgments.

Chapter 1 Chromosomes, p.16, J. Herbert Taylor, 43:672, 1958. 1.16(a), L. M. Beidler, Florida State University. 1.16(b), Ralph W. G. Wyckoff, University of Arizona. 1.16(c), Don W. Fawcett, M.D. 1.19, B. E. Juniper. 1.23, Courtesy of Dr. Jerome Gross, Developmental Biology Laboratory, Massachusetts General Hospital. Hairshaft, p.36, T. Fujita, J. Tokunaga, H. Inoue, Atlas of *Scanning Electron Microscopy in Medicine,* Igaku Shoin, Ltd., Tokyo, 1971.

Part Two opener Prof. Aaron Polliack, M.D., Lymphoma Unit, Hadassah University Hospital, Jerusalem, Israel.

Chapter 3 3.1, The Bettmann Archive. 3.2(a), Don W. Fawcett, M.D. 3.2(b), Courtesy of Keith R. Porter, University of Colorado. 3.2(c), Don W. Fawcett, M.D. Table 3.1, From *Biology of the Cell* by Stephen L. Wolfe. © 1972 by Wadsworth Publishing Company, Inc., Belmont, Calif. Reprinted by permission of the publisher. 3.4, Ralph W. G. Wyckoff, University of Arizona. 3.5, Courtesy of Dr. Germaine Cohen-Bazire. 3.6, J. David Robertson, Department of Anatomy, Duke University. 3.12, Peck-Sun Lin, Tufts—New England Medical Center. 3.16(a), Courtesy of Dr. James G. Hirsch. 3.16(b), B. H. Satir. 3.17(b), Omikron. 3.18, W. A. Jensen, University of California, Berkeley. 3.19(a), Photo courtesy of A. Rich. 3.20(b), C. J. Flickinger, *Journal of Cell Biology,* 47:221–226, 1971. 3.21(b), Omikron, 3.21(d), H. Fernandez-Moran, T. Oda, P. V. Blair, D. E. Green, *Journal of Cell Biology,* 22:63–100, 1964, by copyright permission of the Rockefeller University Press. 3.23(a), Porter, *Ciba Foundation, Symposium: Principles of Biomolecular Organization,* Boston, Little, Brown, 1966. 3.24(c), J. André, Université Paris. 3.25(b), Don W. Fawcett, M.D. 3.25(c), T. Tanaka, *Arch. Hist. Jap.* 39:165–175, 1976. Cilia, p.68, Ellen Roter Dirksen, Ph.D., University of California, L.A. (b), p.68, A. D. Dingle, McMaster University.

Chapter 4 4.4(d,e), L. K. Shumway. 4.5(a), S. B. Carpenter, N. D. Smith, 1975, The hidden world of leaves. *Amer. For.* 81(5):28–30. Courtesy of Dr. S. B. Carpenter, Department of Forestry, Oklahoma State University, Stillwater. 4.5(b), John H. Troughton. 4.9, FAO photo by T. Mamphuis.

Chapter 5 Mushrooms, p.92, Grant Heilman, Photography.

Chapter 6 6.5(e,f), Lee D. Simon, Waksman Institute of Microbiology, Rutgers University, New Brunswick, N.J. 6.7, R. E. Franklin and R. Gosling. Reprinted by permission from *Nature* 171:740. Copyright © 1953, MacMillan Journals Limited.

Chapter 7 7.1, Philip Sharp, MIT. 7.3(e), O. L. Miller, Jr. and B. R. Beatty. "Visualization of Nucleolar Genes." Cover from *Science,* 64:955–957, May 23, 1969. Copyright 1969 by the American Association for the Advancement of Science. Ribosomes, p.116, Alexander Rich, MIT. Nos. two, five, nine, p.121, Hewson Swift and B. J. Stevens. 7.6 Adapted from Zsolt Harsanyi: and Richard Hutton, *Genetic Prophecy: Beyond the Double Helix.* Copyright © 1981 Zsolt Harsanyi and Richard Hutton. Reprinted with permission from Rawson, Wade, Publishers, Inc.

Chapter 8 Mitosis photos, pp.131–134, Carolina Biological Supply Co. Chromatids, p.132, G. F. Bahr, *Federation Proceedings,* 34, 1975. Micrograph by W. Engler. Cytokinesis, p.135, Gary D. Wisehart. 8.2, William T. Jackson, Dartmouth College. Meiosis photos, Photomicrographs courtesy of Professor James L. Walters, University of California, Santa Barbara. Mouse, p.145, Courtesy of C. L. Markert, Yale University. 8.5(b,c) J. B. Gurdon, 8.7, Wide World Photo. a,b, p.148, Dr. M. R. Melamed and P. Saigo, Department of Pathology, Memorial Sloan-Kettering Cancer Center, NYC. 8.8, From *The Cancer Problem* by John Cairns. Copyright © 1975 by Scientific American, Inc. All rights reserved. 8.9, Photographs courtesy of The UpJohn Company.

Chapter 9 9.1, Courtesy Carl Zeiss, Inc., Thornwood, N.Y. 9.2, Courtesy of A. K. Kleinschmidt. 9.3(b), Courtesy of Dr. H. Fernandez-Moran. 9.4, Courtesy of S.T.E.M. Laboratories. 9.5, Roger M. Cole, *Journal of Bacteriology,* 132(3):950, 1977. 9.6, Reproduced with permission from Gerhardt, Pankratz, and Scherrer, Appl. *Envir. Microbiol.* 32:438–439, 1976. 9.7, Charles C. Brinton, Jr. and Judith Carnahan. 9.8, Raymond C. Valentine, University of California, Davis. 9.9, Dr. William T. Hall, Ph. D, Genetic Research Corp., Columbia, MD. 9.12(d), Courtesy of S.T.E.M. Laboratories. 9.15, Carolina Biological Supply Co. 9.16, E. B. Small, courtesy of Gregory Antipa. 9.18(a), Carolina Biological Supply Co. 9.19(c) E. E. Butler and L. J. Petersen. 9.20(a), Courtesy of R. G. Kessel and C. Y. Shih, *Scanning Electron Microscopy in Biology, A Student's Atlas on Biological Organization,* Berlin, New York, Springer Verlag, © 1974. 9.21, 01-MOE-10 Hal H. Harrison, Grant Heilman Photography. 9.24, J. M. Conrader, Photo Researchers, Inc. 9.25, Samuel Rushferth, Brigham Young University. 9.26, Jack Brotherson, Brigham Young University. 9.27(a), Jeremy Pickett-Heaps, University of Colorado. 9.27(b), 00-HYD-12 Runk/Schoenberger of Grant Heilman Photography. 9.28, Prof. W. Rauh, Institute of Systematic Botany, Heidelberg University. 9.29, Jack Dermid. 9.30(a,b), James V. Allen, Brigham Young University. 9.31(a) 04-EQA-10, Hal H. Harrison, Grant Heilman Photography. 9.31(b), Edward S. Ross. 9.32(a), USDA Forest Service Photo. 9.32(b), Jack Dermid. 9.33, Grant Heilman Photography. 9.34, USDA Forest Service Photo. 9.35(a) USDA Forest Service Photo. 9.35(b) Verna R. Johnston, Photo Researchers, Inc. 9.36, (a,b), 04-LIR-15 and 04-HIM-10, Grant Heilman Photography.

Chapter 10 10.8, 4C-OCD-10, Runk/Schoenberger of Grant Heilman Photography. 10.9, 4C-PHP-12, Runk/Schoenberger of Grant Heilman Photography. 10.11, 4D-MOE-15, Runk/Schoenberger of Grant Heilman Photography. 10.12, 4D-NEM-11, Runk/Schoenberger of Grant Heilman Photography. 10.13, 4F-TRI-13, Runk/Schoenberger of Grant Heilman Photography. 10.16, 4F-STG-12, Runk/Schoenberger of Grant Heilman Photography. 10.17, 4F-HEL-22, Runk/Schoenberger of Grant Heilman Photography. 10.18, 4F, Oct-12, Runk/Schoenberger of Grant Heilman Photography. 10.20, 4D-LUT-36, Runk/Schoenberger of Grant Heilman Photography. 10.21, 4D-HIR-11, Grant Heilman Photography. 10.22, Photography by V. J. Okulitch. 10.23, Leonard Lee Rue, III, Photo Researchers, Inc. 10.24(a), 4G4-DOL-10, Runk/Schoenberger of Grant Heilman Photography. 10.24(b), Gordon E. Smith, Photo Researchers, Inc. 10.25(a), 4G1-PAV-25, Runk/Schoenberger of Grant Heilman Photography. 10.25(b), Jack Dermid, 10.25(c), 4G1-BAB-15, Runk/Schoenberger of Grant Heilman Photography. 10.27, 4G2-SCH-10, Hal H. Harrison of Grant Heilman Photography. 10.29, Russ Kinne, Photo Researchers, Inc. 10.30(a,b), H5-116-13, H5-116-15, Hal H. Harrison of Grant Heilman Photography. 10.30(c,d), H5-116-53, H5-116-39, Grant Heilman Photography. 10.31, Sdevard Bisserot, Bruce Coleman, Inc. 10.32, Jerome Wexler, Photo Researchers, Inc. 10.33, Photomicrograph by Fritz Goro. 10.35, Stephen Dalton, Photo Researchers, Inc. 10.36, John Hendry, Photo Researchers, Inc. 10.37, 4E-ASF-18, Runk/Schoenberger of Grant Heilman Photography. 10.38, 4E-OPH-14, Runk/Schoenberger of Grant Heilman Photography. 10.40, Russ Kinne, Photo Researchers, Inc. 10.41, 4I-AMP-20, Runk/Schoenberger of Grant Heilman Photography. 10.43, R. Redden, Animals,

Animals, Inc. 10.44, Thompson, Annan Photo Features. 10.46, Peter Scoones, Colorific Photo Library. Ltd. 10.47(a) 4K-FTP-10, Hal H. Harrison of Grant Heilman Photography. 10.47(b), Karl Maslowski, Photo Researchers, Inc. 10.48(a), 4T-PAW-10, Hal H. Harrison of Grant Heilman Photography. 10.48(b), Dade Thornton, Photo Researchers, Inc. 10.50, A. Cruickshank, Photo Researchers, Inc. 10.51, 4P-PEB-29, Grant Heilman Photography. 10.52, Tom McHugh, Photo Researchers, Inc. 10.53, Gordon E. Smith, Photo Researchers, Inc. 10.54, Lynwood M. Chace. 10.55, Eric Hosking, Photo Researchers, Inc. 10.56, Leonard Lee Rue, Tom Stack & Associates. 10.57, Leonard Lee Rue, III, Photo Researchers, Inc. 10.59, Richard Ellis. 10.60, Edgar Monch, Photo Researchers, Inc. 10.63, Dr. Dorothy Warburton, Department of Human Genetics and Development, Columbia University, NY.

Part Four opener Paul B. Bell, Jr., Department of Zoology, The University of Oklahoma.

Chapter 11 11.1, The Bettmann archive.

Chapter 12 12.1, B. P. Kaufman, 12.2, The Bettmann Archive. 12.3, Peck-Sun Lin, Tufts, New England Medical Center. 12.10(a), A. M. Winchester. Barr bodies, p.259, Courtesy of M. L. Barr. 12.15, C. Steen, W. R. Centerwall, and S. S. Sakar, *Amer. J. Hum. Gen.* vol. 16, 1964, University of Chicago Press. 12.16, Culver Pictures, Inc. 12.18, President's Committee on Mental Retardation, Washington, D.C. 20201.

Chapter 13 13.1, Zig Lesczynski, Animals, Animals, Inc. 13.6, Don W. Fawcett, M.D. 13.10, William Bloom and Don W. Fawcett, *A Textbook of Histology*, W. B. Saunders, 1975. 13.16, E. S. Hafez, M.D. 13.17, 13.18(a) M. J. Tegner and D. Epel, "Sea Urchin sperm-egg interactions studied with the scanning electron microscope," *Science*. 179:685–688, 1973, Fig. 2, Fig. 1, Copyright 1973 by the American Association for the Advancement of Science. 13.18(b), Gerald Schatten, Florida State University. 13.18(c,d), Dr. Mia Tegner, Scripps Institution of Oceanography. 13.19(a-e), A. M. Winchester. Tables 13.3 and 13.4, Adapted from James E. Crouch and J. Robert McClintic, *Human Anatomy and Physiology*, 2e, New York, Wiley, 1976.

Chapter 14 14.2(b), Carolina Biological Supply Co. Triplets, p.298, Marc Anthony DeGruccio. Amniocentesis, p.300, Robert Goldstein, Photo Researchers, Inc. 14.9, Reproduced with permission from *The Birth Atlas*, published by Maternity Center Association, New York.

Part Five opener Copyright ©, David Scharf, 1977. From *Magnifications* by Schocken Books Inc., N.Y.

Chapter 15 15.1, Courtesy of F. D. Hess. 15.3, S.C. Bisserôt, Bruce Coleman, Inc. 15.7(a), Jeremy Pickett-Heaps, University of Colorado. 15.25(a,b), J. Heslop-Harrison, University of Wales, Plant Breeding Station. 15.26, Animals, Animals, Inc. 15.30(b), 04-AMA-26, Grant Heilman Photography.

Chapter 16 16.16, Triarch, Inc. 16.7(a), R. A. Popham, *Laboratory Manual for Plant Anatomy*. 16.7(b), Courtesy of F. D. Hess. 16.8, R. A. Popham, *Laboratory Manual for Plant Anatomy*. 16.10, R. A. Popham, *Laboratory Manual for Plant Anatomy*. 16.11, Triarch, Inc. 16.14, Triarch, Inc. 16.15, Triarch, Inc. 16.17, Reproduced from *Joy of Nature*, © 1977, The Reader's Digest Association, Inc. Used by permission. 16.20, Courtesy of F. D. Hess. 16.26, R. A. Popham, *Laboratory Manual for Plant Anatomy*.

Chapter 17 17.2, James E. McMurtrey, Jr., Tobacco Lab, USDA. 17.4(a,b), M. H. Zimmerman. 17.6(b) Herbert Gehr, *Life Magazine*, © 1948, Time, Inc. Somatic embryos, p.363, P. V. Ammirato, from *Plant Physiology*, 59:579–586, 1977. 17.8, Harry Borthwick, from Salisbury and Ross, *Plant Physiology*, 2 e, 1978, Belmont, Calif., Wadsworth. 17.14(a,b), Courtesy of C. H. Muller, from *Bulletin of the Torrey Botanical Club*, v. 93, 1966.

Part Six opener Scanning electron micrograph by Ch. Brucher and F. Spinelli, CIBA-GEIGY Limited, Basle, Switzerland.

Chapter 18 18.4, Syndication International Photo Trends. 18.5, from Clinical Pathological Conference, *American Journal of Medicine*, 20:133, 1956, with permission of Albert I. Mendeloff, M.D. 18.7, Camera M.D. Studios, Inc. 18.8, Centers for Disease Control, Atlanta, GA, 30333. 18.9, UPI. JFK, p.386, The John F. Kennedy Library. Napolean, p.386, The Bettman Archive. 18.11 © 1965, CIBA Pharmaceutical Company, Division of CIBA-GEIGY Corporation. Reprinted with permission from *The Ciba Collection of Medical Illustrations*, illustrated by Frank H. Netter, M.D. All rights reserved.

Chapter 19 19.1(b,c), Lennart Nilsson, from *Behold Man*, Boston, Little, Brown, 1974. 19.3(a), From *Tissues and Organs: A Text-Atlas of Scanning Electron Microscopy* by Richard G. Kessel and Randy H. Kardon. W. H. Freeman and Company. Copyright © 1979. 19.6(c), Micrograph produced by Dr. John E. Heuser of Washington University School of Medicine, St. Louis. 19.7, from E. R. Lewis, T. E. Everhart, and Y. Y. Zeevi, *Science*, 165:1140–1143, 1969. Copyright 1969 by the American Association for the Advancement of Science.

Chapter 20 20.10(a,b), Thomas Eisner, Cornell University. 20.13(a), from E. R. Lewis, Y. Y. Zeevi, and F. S. Werblin, *Brain Research 15*, 1969, Amsterdam, Elsevier. 20.12(b), From *Tissues and Organs: A Text-Atlas of Scanning Electron Microscopy* by Richard G. Kessel and Randy Kardon. W. H. Freeman and Company. Copyright © 1979. 20.16(a), Jen and Des Bartlett, Photo Researchers, Inc. 20.16(b), Arthur W. Ambler, Photo Researchers, Inc. 20.17(a), From *Tissues and Organs: A Text-Atlas of Scanning Electron Microscopy.* by Richard G. Kessel and Randy H. Kardon. W. H. Freeman and Company. Copyright © 1979. 20.20(a,b), Pamela Henry and Robert E. Preston, courtesy of Prof. Joseph E. Hawkins, D. Sc., Kresge Hearing Research Institute. 20.24(a,b), Peter N. Witt, M.D., North Carolina Foundation for Mental Health Research.

Chapter 21 21.1, From *Tissues and Organs: A Text-Atlas of Scanning Electron Microscopy* by Richard G. Kessel and Randy H. Kardon. W. H. Freeman and Company. Copyright © 1979. 21.2, Andreas Feininger, *Life Magazine*, © Time, Inc. 21.8(h), Courtesy of K. R. Porter, University of Colorado, Department of Molecular, Cellular, and Developmental Biology. 21.9, Courtesy of H. E. Huxley. *J. Molecular Biology*. 37:507–520, 1968. Copyright by Academic Press, Inc., London. 21.11, Lennart Nilsson, from *Behold Man*, Boston, Little, Brown, 1974. 21.13, Reprinted with permission from Bloom and Fawcett, *A Textbook of Histology*, 10 e, Philadelphia, W. B. Saunders. 21.14, Photograph by Carolina Biological Supply Company.

Chapter 22 22.7(a,b), Jeanne M. Riddle, Ph.D., Director, Rheumatology Research Laboratory, Henry Ford Hospital, Detroit. 22.8, From *Tissues and Organs: A Text-Atlas of Scanning Electron Microscopy* by Richard G. Kessel and Randy H. Kardon. W. H. Freeman and Company. Copyright © 1979.

Chapter 23 Table 23.2, Adapted from *Human Physiology* by Vander et al. Copyright © 1975 McGraw-Hill, Inc. Used with permission. 23.7, Animals, Animals, Inc./Len Rue, Jr. 23.8(a), Animals, Animals, Inc./Leonard Lee Rue III. 23.8(b), E. R. Degginger.

Chapter 24 24.2(b), Courtesy of Dr. H. R. Duncker, Institute of Anatomy and Cytobiology, Justus-Liebig-Universität Giessen, Fed. Rep. of Germany. 24.5 From *Tissues and Organs: A Text-Atlas of Scanning Electron Microscopy* by Richard G. Kessel and Randy H. Kardon. W. H. Freeman and Company. Copyright © 1979. 24.6, From C. Kuhn III and E. H. Finke, *J. of Ultrastructure Research*, 38:161–173, 1972. 24.7(c,d), Winchester Hospital, Winchester, Massachusetts. 24.10(a), From *Tissues and Organs: A Text-Atlas of Scanning Electron Microscopy* by Richard G. Kessel and Randy H. Kardon. W. H. Freeman and Company. Copyright © 1979. 24.10(b), Courtesy of Patricia N. Farnsworth, Ph.D. 24.11, Courtesy of N. F. Rodman of the Department of Pathology of West Virginia University. 24.17, Thomas Eisner, Cornell University. 24.20(a), Prof. Aaron Polliack, M.D., Lymphoma Unit, Hadassah University Hospital, Jerusalem, Israel. 24.20(b), Courtesy of Nanometrics, Inc., Sunnyvale, Calif.

Part Seven opener, Dr. Barbara J. Panessa-Warren, Department of Anatomical Sciences, State University of New York, Health Sciences Center, Stony Brook, N.Y.

Chapter 25 25.1, Courtesy of the Burndy Library, Inc. 25.3, The Bettmann Archive. 25.4, Arthur Ambler, Photo Researchers, Inc. 25.10, Hans Reinhard, Bruce Coleman, Inc.

Chapter 26 26.1, From W. B. Landauer and T. K. Chang, *J. Heredity*. 40:105–112, 1949. Reprinted by permission. 26.3, Dr. H. B. D. Kettlewell. 26.4(a), Courtesy of Jack Dermid. 26.4(b), Leonard Lee Rue, III, Photo Researchers, Inc. 26.4(c), John D. Cunningham. 26.4(d), Courtesy of Field Museum of Natural History, Chicago. 26.5(a), Courtesy of David Cavagnaro. 26.5(b), E. S. Ross. 26.5(c), Norman Myers, Bruce Coleman, Inc. 26.6(a,b), Courtesy of Lincoln P. Brower. 26.6(c), S. Rannels of Grant Heilman Photography. 26.6(d), Grant Heilman Photography. 26.7, John D. Cunningham. 26.8, Adapted from *Plant Systematics* by Otto Solbrig. 26.9(a), E. R. Degginger. 26.9(b), John Calwell of Grant Heilman Photography. 26.10, Richard C. Finke, Photo Researchers, Inc. 26.12(a,b), Bodger Seeds, LTD. 26.13(a,b), © Andrew Brilliant, 1978. 26.13(c,d), Grant Heilman Photography. 26.14, Merlin Tuttle, Photo Researchers, Inc. 26.15, The Bettmann Archive. 26.16(b,d), Animals, Animals, Inc. 26.16(c), Alice Kessler, Tom Stack and Associates.

Chapter 27 27.1, Abbas-Gamma-Liaison. 27.2(a), E. R. Degginger. 27.2(b), Bruce Coleman, Inc. 27.10(a), Michael Godfrey. 27.10(b), ZSH-11, Hal H. Harrison of Grant Heilman Photography. 27.11, ZSF-150, Grant Heilman Photography. 27.12(a), Michael Godfrey. 27.12(b), Klaus Paysan. 27.12(c), © Angelo Lomeo. 27.14 (left), Bruce Coleman, Inc. 27.14 (right), Jack Dermid. 27.15, Photo Researchers, Inc. 27.17, Raymond G. Valentine. 27.19, John D. Cunningham. 27.20, Bruce Coleman, Inc. 27.21, Bruce Coleman, Inc. 27.22, E. R. Degginger. 27.23, Bruce Coleman, Inc. 27.26(a,b), E. R. Degginger. 27.27(a), Wolfgang Bayer. 27.27(b), Wyman Meinzer. 27.27(c), Hans Reinhard. 27.27(d), Tom Stack and Associates. 27.29(b), Photo Researchers, Inc. 27.30, New Zealand Wildlife Service. 27.31(a,b), Courtesy of W. H. Haseler, Department of Lands, Queensland.

Chapter 28 28.1, John D. Cunningham. 28.2, E. R. Degginger. 28.3, John D. Cunningham. Stick leaf plant, p.574, Photo Researchers, Inc. Venus fly trap, p.575, Earth Scenes. 28.8, FAO photo. 28.9, Bruce Coleman, Inc. 28.10(a,b), 28.11(a,b,c), 28.12, John D. Cunningham. 28.13(a), E. R. Degginger. 28.13(b,c), Bruce Coleman, Inc. 28.14(a,b,c,d,e), 28.15, John D. Cunningham. 28.16(a), Animals, Animals, Inc. 28.16(b,c), 28.17, 28.18(a,b,c), John D. Cunningham. 28.19, USDA Forest Service. 28.20(a), Bruce Coleman, Inc. 28.20(b,c,d), 28.21, John D. Cunningham. 28.22, USDA Soil Conservation Service. 28.23, 28.24(a,b,c,d,e), John D. Cunningham. 28.25, 28.26, E. R. Degginger. 28.28, 28.29, 28.30, John D. Cunningham. 28.31, 28.34, From *Plant Communities: A Textbook of Plant Synecology* by Rexford Daubenmire. Copyright © by Rexford Daubenmire. Reprinted by permission of Harper & Row, Publishers, Inc. 28.32, John D. Cunningham. 28.35, 28.36(a), Bruce Coleman, Inc. 28.36(b,c), John D. Cunningham. 28.37, Gary Milburn/Tom Stack and Associates. 28.38(a), Bruce Coleman, Inc. 28.39(a,b), Brookhaven National Laboratory.

Chapter 29 29.2, Dr. Edwin R. Willis and Dr. Louis M. Roth. 29.4, Baron Hugo van Lawick, © National Geographic Society. 29.5, Lee Boltin. 29.7, Nina Leen, *Life Magazine*, © 1964 Time, Inc. 29.8, Lilo Hess. 29.10(a), Bruce Coleman, Inc. 29.10(b), E. S. Ross. 29.10(c), David H. Thompson, Environmental Images. 29.11, From "Annual Biological Clocks" by Eric T. Pengelley and Sally J. Asmundson. April 1971. Copyright © 1971 by *Scientific American*, Inc. All rights reserved. 29.12, R. A. Steinbrecht. 29.13, Andreas Feininger, *Life Magazine*, © 1952, Time, Inc. 29.14, F. A. Webster. 29.16, From "The Evolution of Behavior" by Konrad Z. Lorenz. Copyright © 1958 by Scientific American, Inc. All rights reserved. 29.17, Photo Researchers, Inc.

Chapter 30 30.1, National Film Board of Canada—Photothèque/Ted Grant. 30.2(a), Bruno J. Zehnder. 30.2(b), Photo Researchers, Inc. 30.3, Valerius Geist. 30.5, Tom Stack and Associates. 30.6(a), Leonard Lee Rue III, Photo Researchers, Inc. 30.6(b), Naill Rankin/Eric and David Hosking. 30.6(c), Waina Cheng Ward. 30.9, Bruce Coleman, Inc. 30.10, Bill Ogden. 30.11, Stephen Dalton, The Natural History Photographic Agency. 30.12, E. S. Ross. 30.15, Courtesy of the American Museum of Natural History. 30.16, Carl W. Rettenmeyer. 30.18, Irven DeVore Antro-Photo. 30.20, Eiji Miyazawa, Black Star.

Chapter 31 31.1, The Bettmann Archive. 31.3, Eliot Elisofon, National Museum of African Art, Eliot Elisofon Archives, Smithsonian Institution. 31.4(a), Dr. Cyril Ponnaperuma. 31.4(b), Dr. Walther Stoekenius.

Color Portfolios

I. HOW A HUMAN EMBRYO DEVELOPS 1 top, Courtesy of Dr. Landrum B. Shettles. Four sequential photos, Carnegie Institution of Washington, Department of Embryology, Davis Division. 2, top, Courtesy of Dr. Roberts Rugh. Development of hand, Carnegie Institution. 3, 4, top, Courtesy of Dr. Roberts Rugh. 4 bottom, From R. Rugh and L. Shettles, *From Conception to Birth: The Drama of Life's Beginnings,* New York, Harper and Row, 1971. 5–7, Courtesy of Roberts Rugh. 8, Drawing by Alexander Farquharson.

II. WHAT ANIMALS SEE 1, Dr. E. R. Degginger. 2–3, Marshall Henrichs.

III. HOW ANIMALS MOVE 1, Animals, Animals, © Jerry Cooke. 2–3, Alan Rokach. 4 and 5, top, © Frank Lane (A. Christiansen), Bruce Coleman, Inc.; 4 and 5, bottom, © Charles G. Summers, Jr., Bruce Coleman, Inc. 7, Animals, Animals, © Zig Leszcynski. 9, Isabelle Hunt Conant. 12, left, © Russ Kinne, Photo Researchers, Inc.; right, R. C. Snyder.

IV. HOW ANIMALS EAT 1 (cheetahs), Animals, Animals, © Fran Allan; (hyenas), © Norman Myers, Bruce Coleman, Inc. 2–3, © Norman Myers, Bruce Coleman Inc. 4 (owl), Stephen Dalton, Natural History Photo Agency; (osprey), © Laura Riley, Bruce Coleman Inc. 5 and 6 (top), © Jen and Des Bartlett, Bruce Coleman Inc.; 6 (bottom), © Tom McHugh, Photo Researchers Inc. 7 (frog), Runk/Schoenberter of Grant Heilman, Inc.; (chameleon), © Alan Blank, Bruce Coleman, Inc.; (bee), © Treat Davidson, NAS Photo, Photo Researchers, Inc.; (okapi), © K. W. Fink, Bruce Coleman, Inc. 8 (tiger), © Frank W. Lane, Bruce Coleman, Inc., (hippotamus), © John Angell Grant, Bruce Coleman, Inc.; (rodent), © Russ Kinne, Photo Researchers, Inc.; (shark), Dr. E. R. Degginger.

V. HOW ANIMALS AVOID PREDATORS 1, G. E. Schmida, F. P., Bruce Coleman, Inc. 2 (larva), © Oxford Scientific Films, Bruce Coleman, Inc.; (grasshopper), © M. P. L. Fogden, Bruce Coleman, Inc.; (hawk moths), Carl W. Rettenmyer. 3 (top), © Alan Blank, Bruce Coleman, Inc.; (middle), Richard G. Zweifel, courtesy of American Museum of Natural History; (bottom), © Hans Reinhard, Bruce Coleman, Inc. 4, Mike Logan, Tom Stack & Associates.

Index

Page numbers in *italic* type indicate illustrations. Roman numerals indicate coverage in portfolios, whose location is given in the table of contents.